COLLECTED WORKS OF

Shinya Inoué

MICROSCOPES, LIVING CELLS, AND DYNAMIC MOLECULES

COLLECTED WORKS OF

Shinya Inoué

MICROSCOPES, LIVING CELLS, AND DYNAMIC MOLECULES

Marine Biological Laboratory, Woods Hole, Massachusetts, USA

World Scientific

NEW JERSEY · LONDON · SINGAPORE · BEIJING · SHANGHAI · HONG KONG · TAIPEI · CHENNAI

Published by

World Scientific Publishing Co. Pte. Ltd.

5 Toh Tuck Link, Singapore 596224

USA office: 27 Warren Street, Suite 401-402, Hackensack, NJ 07601

UK office: 57 Shelton Street, Covent Garden, London WC2H 9HE

British Library Cataloguing-in-Publication Data
A catalogue record for this book is available from the British Library.

COLLECTED WORKS OF SHINYA INOUÉ
Microscopes, Living Cells and Dynamic Molecules
(With DVD-ROM)

ISBN-13 978-981-270-388-0
ISBN-10 981-270-388-8

Printed in Singapore by World Scientific Printers

CONTENTS

Appendixes

INTRODUCTION AND GUIDE

When I started to assemble the material for this *Collected Works*, my intent was to supplement the written text with some video sequences that illustrate the behavior of living cells and the way we study some of the underlying molecular mechanisms. However, after spending the last few months editing the video material, and wondering how the reader might best approach this volume, it occurred to me that I may have had it all backwards.

Therefore, this volume has been organized to introduce the reader first to the video, and then to the Articles and Appendices as further reading material. While this may depart from the conventional treatment of video material as a supplement, I believe that this revised sequence may be closer to how many of us learn about a subject. I recommend that you first acquire an overview of the volume by viewing the video-taped lecture that I gave in the physiology course at the MBL in the summer of 2006, along with its accompanying PowerPoint presentation, both of which can be found on the **DVD to the Collected Works of Shinya Inoué** included with this book. To activate the DVD properly, start by reading the **DVD Contents**, one copy of which is attached to the DVD while another appears as a chapter in this *Collected Works*.

This volume traces my contribution to science and technology through highlights of my published (and a few hitherto unpublished) articles.* As seen in these chronologically arranged articles, technological advances in microscopy and related fields made by me and my colleagues have often preceded new biological observations and discoveries. However, I hope that this sequence of events is not taken to mean that I had the foresight to see that particular improvisations in image improvements or experimental methods would solve riddles about certain mechanisms in life.

While some biological observations did, in fact, prompt further improvements or development of new methods or instruments, I cannot truly say that much of my interest in technological development was motivated only by my desire to answer specific biological questions. Rather, it reflects my interest in devising new instruments (Appendix I: **Development of the "Shinya Scopes"**), or means for studying living cells in action, and then in using those new devices to *explore* nature. My sense was, and remains, that because living creatures and cells are so full of surprises that defy our *logical* anticipation, it is better to improve the tools for perceiving what nature has to tell us, and then let her show us what questions we can reasonably ask.

Fortunately, what nature has revealed to us, using the instruments and methods that my colleagues and I have devised, turns out to have had significant impact on our understanding of the workings of the dividing cell and its dynamic molecular machinery. Furthermore, I have been privileged to have our efforts, both in biology and instrument development, widely recognized throughout my career. Thus, in addition to appending a **Curriculum Vita** to this volume, including the thesis topics of the

*Due to page limitations agreed to with the publisher, several interesting articles could not be included in this volume. I offer my apologies to my co-authors. I chose those articles not because they were unimportant to the relevant fields, but based on my having played a relatively minor role in the preparation of the article.

PhD students whom I have sponsored, I have taken the liberty to include some remarks made by His Majesty, Emperor Akihito, and by myself relating to the International Prize for Biology which I was honored to receive in December 2003 (Article 73).

For clarity and simplicity, the **Table of Contents** of the articles in this volume gives abbreviated titles only for each entry. Full citations for the articles can be found in the **List of Publications** at the end of this volume. Other sections of this volume, **Development of the "Shinya Scopes**," as well as the **Slides and Movies** in the PowerPoint presentation and the accompanying **Additional Material,** listed under **DVD Contents,** provide reference to article numbers. Hopefully, they will serve as a functional index for this volume.

Over the years, I owe my debt and gratitude to many persons, family, teachers, students, collaborators, and sponsors, too numerous to list here, who have made my work both productive and enjoyable. A few are listed in passing in the articles in this volume or in the presentations on the DVD. Many also appear in another volume, *Through Yet Another Eye*, which I am currently preparing.

Directly relevant to the preparation of this *Collected Works,* I wish to thank particularly members of the MBL Architectural Dynamics in Living Cells Program including Rudolf Oldenbourg for his generous support; Bob Knudson for speedily translating our conceptual plans into finished precision instruments; Michael and Elena Shribak for improving several of the drawings used; Grant Harris for extended help with the PowerPoint and DVD preparation; and Jane MacNeil who has not only assembled all of the pdf files and polished the typescripts for this volume, but who has kept the preparation of the whole volume on track. As always, my wife Sylvia has supported me with good humor and wisdom.

Shinya Inoué
Canovanas, Puerto Rico
Falmouth, Massachusetts
December, 2006

DVD CONTENTS

Appended to Collected Works of Shinya Inoué

Introduction

The attached DVD contains a video-taped lecture which can be viewed using a DVD player or by using a computer with a DVD reader and appropriate software. The disk also contains a PowerPoint presentation containing the slides and movies shown in the lecture plus additional material that can be viewed on a computer.

The lecture is titled "Early Biophysical Studies: Polarized Light Microscopy, Learning from Happy Living Cells, Symphony of Dynamic Molecules." It is a lecture that I gave at the MBL on July 24, 2006. Tim Mitchison and Ron Vale, co-directors of the MBL Physiology Course, graciously invited me to give this presentation as one of the physiology scholar lectures called "Understanding Cell Division: A Journey from Dynamics in Living Cells to the Dynamics of Molecules in a Test Tube." My lecture was video taped by Priscilla Roslansky of Woods Hole, Massachusetts, who has given us permission to use the recording in this *Collected Works*.

The slides and movies used in the lecture can be seen with greater fidelity in the PowerPoint presentation, which appears as a list on page xvii under **SLIDES AND MOVIES CONTENTS**. Although without sound, the PowerPoint presentation allows you to examine each slide and movie at your own pace and with considerably better image quality than in the video-taped lecture. Many of the frames contain references to relevant articles in this book and should serve as a dynamic index to this *Collected Works*. Some frames in the PowerPoint also contain hyperlinks to additional material as explained on page xv, and listed as **ADDITIONAL MATERIAL CONTENTS** on page xix.

Watching the Lecture on a Television with a DVD Player

The lecture with audio (but not the latter sections with PowerPoint presentation) can be viewed on a TV set using a regular DVD player. Place the DVD in the player and the video will begin automatically.

Using the DVD with a Windows Computer

On a Windows computer, simply insert the DVD. The ***AutoRun Menu (Fig. 1)*** appears and offers the following choices:

Click on ***Play Lecture Video*** to watch the lecture.[1] There may be a considerable delay while the computer loads the lecture. Once the audio starts after the visuals are advancing, you can jump back to the beginning and view the full presentation.

[1]The video is played using the open-source player VLC Media Player. The lecture video also can be played back on a PC using other DVD player software, such as "Inter/Video Win DVD." The movies are in .avi format, encoded with the high quality MPEG-4 codec called *3ivx* which is available for Windows, Macs, and other platforms. Because the movies are encoded at high quality, older or slower computers may not be capable of playing the movies smoothly, if at all.

Figure 1. The AutoRun Menu

If this is the first time that you will be viewing the slides and movies on this computer, click ***Install Codec.*** This installs the video codec required to view the movies.[2]

Click on ***View Slides and Movies*** to open and view the PowerPoint presentation. (For detail, see ***Viewing the Slides and Movies in PowerPoint*** on page xiv).

Click on ***Open Movie Folder*** to open the folder on the DVD. The folder contains a full list of the movie files and a copy of this **DVD CONTENTS** document. (For ***Viewing Movies Frame-by-Frame*** using a Windows computer, see Footnote 3).

Using the DVD with a Macintosh Computer

Using a DVD reader, you can watch the lecture by simply inserting the DVD. Mac OS X will start the video playing automatically.

In order to view the slides and movies in the PowerPoint presentation, you need to have a copy of Microsoft PowerPoint installed on your Mac. (There is no other compatible viewer that is freely available at this time.) Also, before you view the presentation for the first time, you need to install the necessary codec by clicking on *3ivx_d4_451_osx.dmg* in the folder *codec/mac os x.*[2] Once the codec is installed, re-insert the DVD and open it in the Finder. After opening the DVD in a Finder window, open the *SI_Presentation* folder and click on *SI_PPT.ppt* to open the presentation; then open the "Slide Show" menu and click "View Show."

To view the **DVD CONTENTS** document on the DVD, open the Finder and choose the *DVD CONTENTS.rtf* file.

Viewing the Slides and Movies in PowerPoint

The first section of the PowerPoint shows the slides and movies which were presented in the lecture. You can progress through all contents of the PowerPoint presentation

[2]If you experience problems playing the movies, check that the full PowerPoint application and all Microsoft Graphics Converters have been installed from the Microsoft Office disk (or reinstall these files if necessary). For additional information on playing AVI files under OS X, see http://www.thexlab.com/faqs/avidivx.html.

one frame at a time by left-clicking the mouse or by pressing the <Space bar> key. Press the <Backspace> key to go back to the previous slide. You can go to a specific slide by pressing the <g> key, typing in the slide number, and pressing the <Enter> key. (In Windows, you can right-click on the mouse to select <Go To Slide> and select by slide name as well.)

Playing Movies[3]

The movie sequences, found within the PowerPoint presentation of the lecture which are listed under **SLIDES AND MOVIES CONTENTS** (p. xvii) as well as those in **ADDITIONAL MATERIAL CONTENTS** (p. xix), are marked **MOVIE**. To play a movie, move the cursor over the image of the movie and click the (left) mouse button. (Because the *PowerPoint Viewer* insists on giving a warning when a link to a file is clicked, you will need to press [Yes] each time the "Some hyperlinks may contain viruses ..." warning appears; if Windows Media Player reports an error, simply hit [Close] and then the movie Play button — the solid triangle pointing to the right, located at the bottom of the monitor image.) PowerPoint plays movies using the default media player on your computer. Media players including *Windows Media Player* (9 or greater), *QuickTime*, and *RealPlayer 4* should be able to play the movies once the *3ivx* codec is installed. The *3ivx* code is installed automatically when you click on [Install codec] in the ***AutoRun Menu*** (Fig. 1). Notice, however, that if you are using multiple monitors, the movie may play on only one of the monitors.

Hyperlinks to Additional Material

Various slides in the PowerPoint presentation contain hyperlinks to sets of explanatory slides and to a large number of additional movies, many of which have never been published elsewhere. The links are designated by "Link to" and the name of the hyperlinked section in orange characters. Using the mouse, click on the hyperlink to view that section.

Tapping the <Space bar> or left-clicking the mouse within the hyperlinked section takes you through the sequence of the selected material. When you reach the last frame of a sequence (designated by a yellow up arrow), tapping the <Space bar> or left mouse button returns you to the slide from which you chose the hyperlink, where you can continue through the presentation. (Note that the orange color designating a hyperlink turns white when hyperlinked material has been accessed. The change in color, however, does not prevent you from accessing the same hyperlink again.

---

[3]Viewing Movies Frame-by-Frame. On a Windows computer, your media player may not allow you to examine the movies frame-by-frame. In that case, you can download and install QuickTime (download from www.apple.com/quicktime). After installing QuickTime, click on *Open Movie Folder* in the ***AutoRun menu (Fig. 1)***, then right click on the desired movie. In the menu that appears, click on "Open With" and select "QuickTime Player." Once you are in QuickTime, you can single step through the movie frames by pressing the left arrow or right arrow on the keyboard. You can also fast forward, reverse, or jump to the first or last frame of the movie by using the controls at the bottom of the display.

The white characters designating accessed hyperlinks all revert to their orange color when the PowerPoint program is restarted.)

At the end of the PowerPoint presentation is a table containing all of the links to additional material. (This can be viewed at any point by pressing the <End> key.)

To view the **DVD CONTENTS** document in Windows, click the Movie Folder in the ***AutoRun Menu*** (Fig.1) and select *DVD CONTENTS.rtf.*

Windows Media Player, *QuickTime*, and *RealPlayer* are registered trademarks of their respective owners.

SLIDES AND MOVIES CONTENTS

(File Name) follows MOVIE

Slide 31: MOVIE (GrasshopperSpematicite_Pol.avi). Spindle formation and "Northern lights flickering" (Sato H and Izutsu K, Time-lapsed ciné film, 1974)
Slide 32: Optical bench, video-enhanced Universal Pol-scope (Shinya Scope 6; MBL, ca. 1981, Appendix I)
Slide 33: Digitally processed diatom image [Articles 34, 63; see also Allen RD *et al.*, 1981, *Cell Motil* **1**: pp. 275–289]
Slide 34: MOVIE (Phalusia.avi). Fertilization and early divisions in tunicate egg [Articles 41, 49]
Slide 35: MOVIE (CPM-Chaetopterus.avi). CPM view of Ca^{2+} activation of stratified *Chaetopterus* egg [Articles 66, 67]
Slide 36: Orientation-independent DIC images of spermatocyte [Article 74]
Slide 37: Shinya's MBL Lineage in *Physiology, Cell Biology, and Microscopy*
Slide 38: Lecture Summary
Slide 39: General References [Articles 40, 71]

ADDITIONAL MATERIAL CONTENTS

(File Name) follows MOVIE

Slide 74: 4th cleavage division in egg of a sand dollar [Articles 29, 51]
Slide 75: MOVIE (Spisula.avi). Unequal cleavage in *Spisula* (clam) egg [Article 43]
Slide 76: MOVIE (CompressedLytechinusEmbryo.avi). Mitosis and cleavage in compressed sea urchin egg (Roegiers F and Inoué S, 1994, unpublished) [Articles 22, 29, 51]
Slide 77: MOVIE (Haemanthus_Pol.avi). Mitosis in endosperm cell of *Haemanthus* [Articles 19, 22]
Slide 78: Spindle anchorage and microtubule dynamics
Slides 79–82: MOVIES (ManipDimple.avi, ManipInvert.avi, ManipLast.avi, ManipReturn.avi). Micromanipulation of oocyte spindle [Articles 45, 51]
Slide 83: MOVIE (Hydrostatic.avi). Pressure-induced depolymerization of microtubules (Salmon ED, 1976, *Cold Spring Harbor Conferences* **3**: pp. 1329–1342) [Article 27]
Slide 84: Microtubule (MT) as nano-machine [Article 58]
Slide 85: Rectification
Slide 86: Back aperture of 1.25 NA oil-immersion objective [Article 14]
Slide 87: Meniscus rectifiers [Articles 14, 69]
Slide 88: Oral epithelial cell viewed with rectified optics [Article 14]
Slide 89: Pinhole and Siemens test star observed with rectified optics [Article 15]
Slides 90–91: Irradiated live sperm head observed with rectified optics [Articles 24, 40]
Slide 92: Polarized UV microbeam set up [Articles 24, 40]
Slide 93: Video Microscopy, Contrast Enhancement
Slide 94: MOVIE (ThyoneAcrosome_DIC.avi). Acrosomal reaction of *Thyone* sperm [Articles 37, 38, 39]
Slide 95: MOVIE (PyrsonymphaSaccinobacilus.avi). Axostyle (microtubule bundle) beating inside protozoa
Slide 96: MOVIE (HypermastigoteFlagella.avi). Birefringence of rootlets and beating flagella
Slide 97: MOVIE (SpirochaetWaves.avi). Birefringent waves by rotating spirochetes
Slide 98: MOVIE (AmeboidMovement.avi). Amoeboid movement by *Dictyostelium* amoeba [Article 60]
Slide 99: Video Microscopy, 3- and 4-D Microscopy
Slide 100: MOVIE (Plutius_DIC-3D.avi). Through-focus imaging and rocking stereo pair of sea urchin pluteus [Articles 29, 42]
Slide 101: MOVIE (4D-SeaUrchinGastulaDevelopment.avi). 4-D images of developing sea urchin gastrula [Articles 41, 42]
Slide 102: MOVIE (3D-GolgiStainedRatNeuron.avi). 3-D display of Golgi-stained rat neuron [Article 42]
Slide 103: Stereoscopic Apparatus, Inoué 1943 Patent [Article 2]
Slide 104: LC-Pol Scope
Slide 105: MOVIE (CraneFlyCell86_LCPol.avi). Dividing spermatocytes of crane fly (LaFountain and Oldenbourg, 2004)
Slide 106: Schematic of LC Pol-Scope (Oldenbourg, 1996, *Nature* **381**: pp. 811–812)
Slide 107: Birefringence of individual microtubules (Oldenbourg *et al.*, 1998)
Slide 108: MOVIE (CraneFlySpermatMeiosis_I.avi). Crane fly spermatocyte undergoing meiosis (LaFountain and Oldenbourg, 2000)
Slide 109: CPM–Centrifuge Polarizing Microscope

A DEVICE FOR ALTERING THE DISTANCE BETWEEN SLIDE AND COVER-SLIP AT WILL

Shinya Inoué

(Translated from the Japanese article by the author in January, 2006)

While viewing small live specimen such as sea urchin eggs or protists, one encounters situations where one wishes to exchange the medium surrounding the specimen or to hold or compress the specimen between the slide and cover-slip. For such purposes, one could, e.g., support the cover-slip by placing small fragments of cover-slips or pieces of filter paper together with the specimen under the cover-slip, then perfuse fresh media from one end of the cover-slip while removing some medium with a piece of filter paper from the other end. However, with such an approach, the precious specimen could be washed away or squashed inadvertently. Here, I wish to describe a device that I have designed in order to freely enable compressing, without squashing, various small living cells.

Figure 1 is a perspective view, and Fig. 2 shows its cross section. As shown in the figures, the cover-slip, supported above the slide by a thin glass rod (6), is gently pressed downward on side A with a thin spring (3) and on side B by plate (2). But owing to its elasticity, plate (2) is constantly attempting to open upwards, thereby pushing up against axel (4) via glass tubing (5). Since the glass tubing is cemented onto the axel eccentrically, turning the axel (4) either pushes down or releases the spring plate (2). Accordingly, side B of the cover-slip moves down or up, while the space between the slide and cover-slip opens or closes on side A.

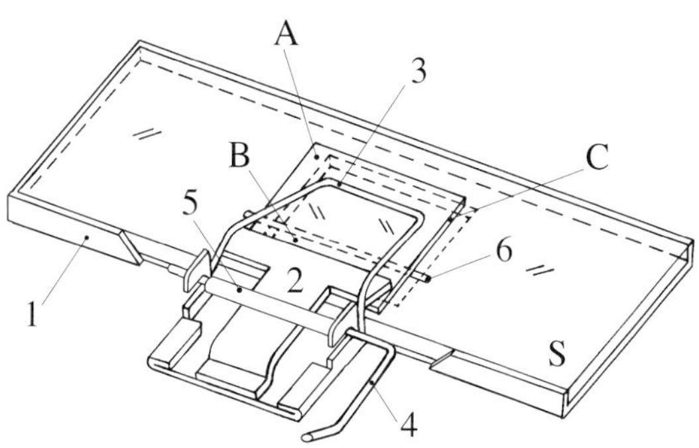

Fig. 1.

Botany and Zoology **11** (8), 669 (1943).

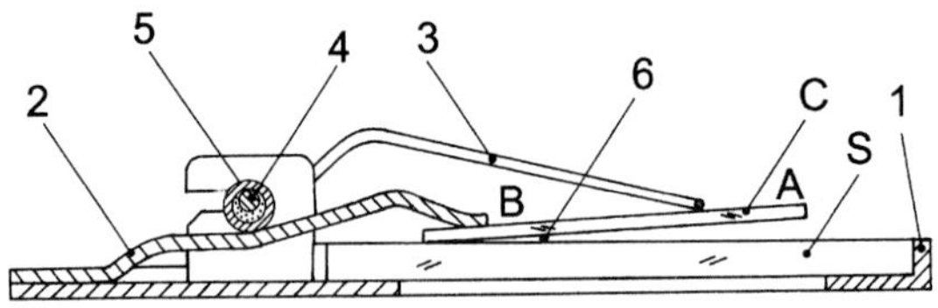

Fig. 2.

By reducing the difference in inside diameter of the glass tubing (5) and outer diameter of axel (4), B is pressed down or rises by a very small amount associated with a large rotation of axel (4). Thus, one can compress or release objects near side A. By such means, one can freely exchange the bathing medium by slightly compressing and holding onto the cell, or capture *Paramecia* and *Colpedia* during observation. Furthermore, since it is possible to use this device to compress, e.g., *Paramecia* at any desired speed to any desired degree, the device should be convenient, for example, for studying the extrusion of protoplasm from cells.

In practice, one can vary the range of application of this device by changing the diameter and location of the glass rod (6). Thereby, one can freely capture or compress various cells as large as 100 μm or as small as a few μm. However, since the cover-slip bends slightly by the mechanical forces applied, and since the slide and cover-slips are not exactly parallel to each other, the space between them varies somewhat depending on location.

Upon actual use, one may encounter considerable current flow associated with expulsion of the medium following the downward movement of the cover-slip. However, this is not a major problem since on the one hand, it is possible to avoid regions with such violent flow, and on the other hand, it is even possible to take advantage of some flow in order to displace (rotate) the specimen under observation.

Finally, with reference to the actual material that I used, support (1) was made from a thick sheet of (non-annealed) dental "Platinoid" material, the sheet spring (2) from a thin sheet of dental silver alloy, axel (4) from a stainless sewing needle, and spring (3) from a piece of violin E string.

PATENT NUMBER 166528 [JAPAN]*

Shinya Inoué

Group 3–15 Optical equipment (2, Microscopes)
Application number: 1943–11790
Application date: September 6, 1943
Patent awarded: August 18, 1944
Patent owner (Inventor): Shinya Inoué
895, Magome Higashi 2-chome, Ohta-ku, Tokyo

1. Stereoscopic Apparatus

1.1. *Summary of nature and purpose of invention*

This patent concerns stereoscopic viewing devices, featuring polarizing filters with perpendicular axes which cover two halves of an objective lens, combined with a second set of mutually orthogonal polarizing filters placed directly following the eye pieces, which transmit respectively the light transmitted through the left and right half of the objective lens, such that the left and right eye sees the specimen as transmitted through the left and right halves of the objective lens, thereby generating clear stereoscopic images which render 3-D views of the specimen using a single objective lens.

1.2. *Abbreviated description of figures*

Fig. 1 explains the principle of generating the 3-D image.

Fig. 2 shows a plane view of the polarizing filter for the objective lens used in this invention.

Fig. 3 is a schematic of an example of a microscope using this invention.

Fig. 4 is a schematic of an example of another practical application of this invention.

2. Detailed Description (Released November 20, 1944, by Japanese Patent Office)

2.1. *Detailed explanation of the invention*

This patent pertains to stereoscopic viewing devices employing polarizing filters. Firstly, to explain the principle of generating the 3-D image: As shown in Fig. 1, by producing images (A', B', C') of three point sources (A, B, C) by the objective lens (1), and by placing a focusing screen (2) in a plane between A' and (B', C'), one obtains three indistinct images (A", B", C") on the said focusing screen. At this point, when one half of the objective lens aperture (e.g., OL) is masked with an opaque screen, the images (A", B", C") become sharper, while the center of image A" shifts to the left in the diagram and those of B" and C" to the right, thereby causing the distance A"B" to become larger than A"C". Conversely, when the other half (OM) of the objective lens is masked, the distance A"C" becomes larger than A"B". Therefore, by allowing the right eye to view the image generated by masking OL, and the left eye to view the image generated by masking OM, it should be clear that one would achieve a magnified 3-D image of ABC.

*Japanese Patent (1944).

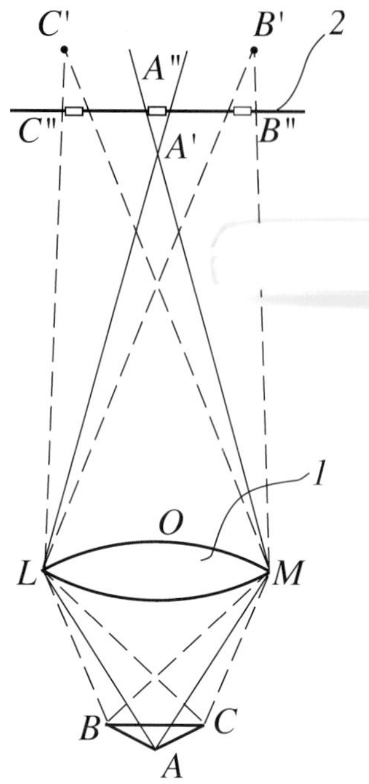

Fig 1.

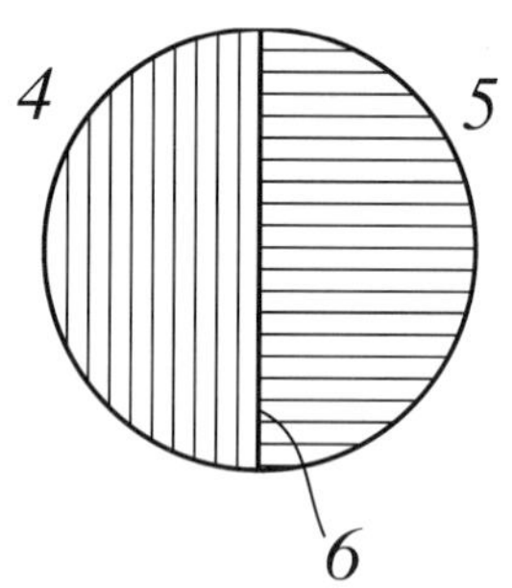

Fig 2.

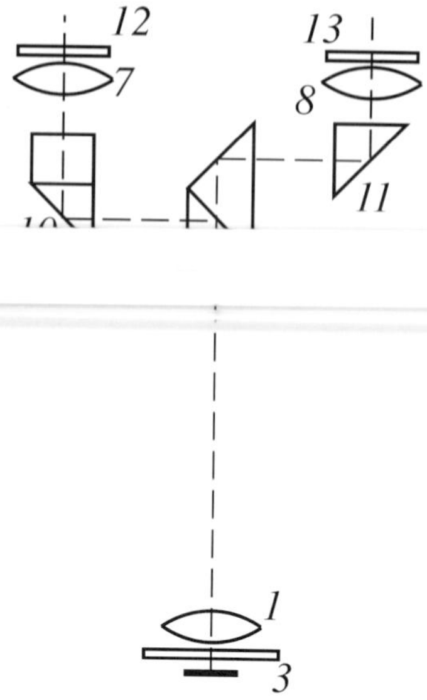

Fig 3.

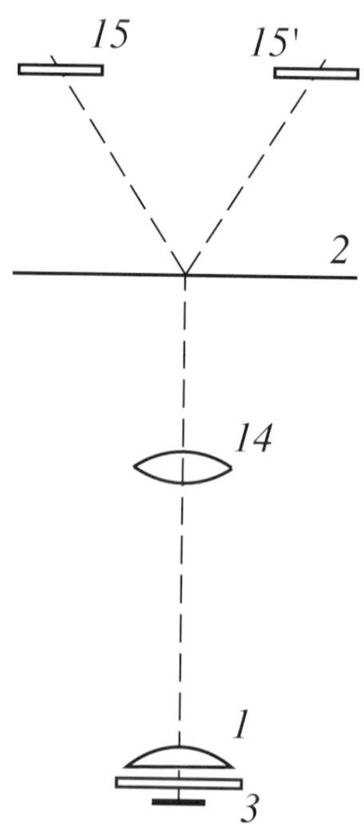

Fig 4.

The present invention employs polarizing filters in such a way as to allow the left and right eye to concurrently view one of the two images formed by masking the two halves of the objective lens according to the principle described above. As shown in Fig. 2, the polarizing filter (3) consists of two halves (4, 5) whose polarizing axes are oriented perpendicular to each other, and placed either immediately before or after the objective lens, with the line (6) joining its two halves of the filter oriented perpendicular to the line joining the viewer's left and right eyes as shown in Figs. 3 and 4.

Specifically, by placing the said polarizing filter (3) either immediately before or after the objective lens (1 in Fig. 1), and by placing polarizing filters directly above the left and right eye pieces such as to extinguish the light arriving from the left and right halves of the objective lens, one obtains the stereoscopic image. For objective lenses with multiple elements, the objective lens polarizing filter can be placed between the elements. To apply the above described principle to an epidiascope, one can place a ground glass screen at location 2 and use a standard projection device in place of the objective lens. To apply the principle to microscopes, one can place the polarizing filter

…ediately before the

… or immedi-

split the

…ring the

…ye pieces

…olarizing fil-

…

Furthermore, with microscopes equipped with only a single eye piece (14), one can view the images projected onto a screen through polarizing viewing glasses (15, 15') as used with projection devices.

Since as described, the current invention uses polarizing filters to generate stereoscopic images providing 3-D views, it can be applied effectively also to endoscopes, screw-hole examining devices, etc.

2.2. *Patent claims*

As described in detail in the purpose of this invention and as illustrated, what is claimed are stereoscopic devices characterized by a half-shade filter with two orthogonally polarizing axes, which is coupled with the objective lens, used together with two orthogonally polarizing filters placed above the left and right eye pieces such that they individually transmit light passing the left and right halves of the objective lens, with the consequence that the left and right eye views the object only as imaged through the left and right halves of the objective lens.

The following note was added by Shinya Inoué in September of 2006:

This article was translated from the Japanese patent record (three pages) by the author in January, 2006. I thank Colleen Hurter of the MBL Library for tracking down the source of this record, which can be accessed at web site <http://www.ipdl.ncipi.go.jp/homepg_e.ipdl> by selecting: Patent and Utility Model Gazette DB, Kind = C, Code = 166528, Search, then clicking on JP166258C.

When I arrived in Princeton in 1948, I came across the description of a product in a Bausch and Lomb catalog for stereoscopic attachments to binocular compound microscopes with illustrations similar to Figs. 2 and 3 in my 1944 patent. The product description disappeared soon after I pointed out this coincidence to a B&L representative. Possibly, the patent became available according to the Alien Property Custodian act, which gave US companies access to patents issued by enemy countries during WW-II.

During the early 1940s, I entertained my family with a hand-made 3-D epidiascope using a 4-inch diameter magnifier borrowed from my grandfather as the objective lens together with pieces of "Dichrome" polarizing sheets acquired from Mitsubishi Electric Company through my cousin Masao who was working as their inspector. Flowers and crawling insects projected on a non-depolarizing white wall showed up in 3-D with striking full color. Everyone was able to see their 3-D aspect except my father who had lost sight in one of his eyes after a traffic accident.

Remarks on stereoscopic acuity and various ways of generating stereoscopic images using microscopes coupled to video devices can be found in Sections 2.8, 4.8, and 10.8 in *Video Microscopy* (Inoué and Spring, 1997; also sections 4.8, 11.5, and 12.7 in Inoué, 1986). See also Articles 42 and 54 in this *Collected Works*.

BIREFRINGENCE VERSUS LENGTH OF RESITNG AND CONTRACTING SINGLE MUSCLE FIBER

Shinya Inoué*

(Article prepared by author in August 2006, based on data collected in 1947)

In 1947, two years after the end of the Second World War, I returned as a graduate student to the Zoology Department at Tokyo University and joined the laboratory of Dr. T. Kamada where his junior associates Drs. T. Yanagida and M. Tamasige were studying the excitation mechanism and electrical properties of muscle fibers.

For those studies, they meticulously isolated intact single fibers from the sartorius or semitendinsous leg muscles of the Japanese brown frog (*Rana japonica*, which has sparse connective tissues). After suspending the excised muscle, bathed in frog Ringer's solution, between two miniature hooks piercing the tendon at its two ends, physiologically intact single fibers were isolated by carefully teasing away all but one fiber with iridectomy scissors and sharpened tweezers under a dissecting microscope. Successfully isolated, intact fibers would respond for several hours to a DC (direct current) stimulus with a single twitch and completely relax without producing any contraction clot.

I felt that the isolated, intact single muscle fibers being prepared in the Kamada laboratory would be ideal for studying the fine-structural and molecular changes during physiological contraction of muscle. All previous studies used whole (smooth or striated) muscles of considerable thickness, which were therefore made up of a large number of muscle fibers, the excitable units. Thus, the measurements could be subject to complications, including partial contraction and rearrangement of the contracting muscle fibers. So I learned from my seniors and honed my skill for isolating single intact frog muscle fibers, while developing a microscope for measuring their birefringence at various fiber lengths and during contraction.

Figure 1 shows a schematic of the system that I developed, a modification of the one used by Bozler and Cottrell (1937). These authors primarily used whole smooth muscles isolated from the pharynx retractor of the edible snail *Helix pomatia*. Superimposed on the image of the muscle in the ocular of a microscope illuminated with polarized light, Bozler and Cottrell placed a wedge-shaped piece of mica with its thin edge lying perpendicular to the muscle length. The mica wedge was followed by an analyzer, with the polarizer and analyzer transmission axes lying parallel to each other and at 45° to the long axis of the muscle. Using red monochromatic illumination, they measured the area between the curved fringes on the mica [shifted according to the retardation, i.e., (birefringence per unit thickness) × (the local thickness) of the muscle region] and the straight unshifted fringes on the mica. As shown by

*Unpublished studies (1947).

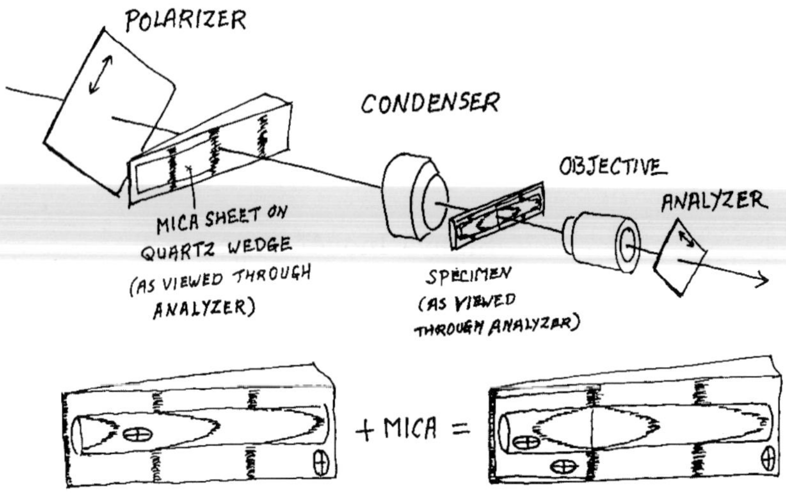

Fig. 1. Top: Schematics of polarization microscope optics used by author in 1947 (drawn from memory in 1996; see text for details). Lower left: Superimposed images of standard quartz wedge and muscle fiber with the slow axes of their birefringence indicated by the major axis of the ellipses. Lower right: With half-wave mica plate superimposed on the thinner part of the wedge, as described in the text.

these authors, that area is proportional to the (total retardation per cross section) × (the cross sectional area) of the muscle region under observation.

In my system, I used a quartz wedge placed with its long axis oriented horizontally, in front of the polarizer whose transmission axis was oriented at 45° from the horizontal and transilluminated by monochromatic green light. The microscope condenser, situated some 25 cm to this side of the quartz wedge, projected a focused image of the wedge onto the specimen plane of the microscope where the muscle fiber was placed. This way, the superimposed images of the muscle fiber and quartz wedge were observed without change in their relative magnification or position even when the magnification of the objective lens was changed, e.g., in order to measure the sarcomere lengths.

Viewed with the ocular, through an analyzer crossed with the polarizer and located behind the objective lens, the quartz wedge appears brightly on a dark background with equally-spaced extinction stripes. These dark stripes occur at the thicknesses of the quartz wedge where its retardation equals a multiple of the illuminating wave length (Fig. 1, top left).

Interacting with the birefringence of the muscle fiber, the fringes of the wedge are compensated and displaced towards the thicker side of the wedge, since the slow axis of the fiber runs along its length, while that of the wedge is oriented perpendicular to that direction (Fig. 1, lower left; see Article 40, this *Collected Works*, for further explanation). Each point of the displaced fringe thus measures the retardance of the fiber [(coefficient of birefringence) × (thickness of the fiber at that point)], so the area below the semi-elliptical fringe is proportional to the total retardance multiplied by the cross sectional area. This is the same principle used by Bozler and Cottrell even though their mica wedge was placed in the ocular slot, and hence required recalibration when the objective lens magnification was changed.

In order to improve the precision of measurement of the compensated fringe area, I added a rectangular slice of mica over the thinner part of the quartz wedge with the short edge of the mica superimposed with a dark fringe on the wedge (Fig. 1, top left). Not knowing the exact optical theory, but wishing to reverse the slow axis direction of the light coming from the thinner part of the quartz wedge, I used a piece of mica empirically, sliced

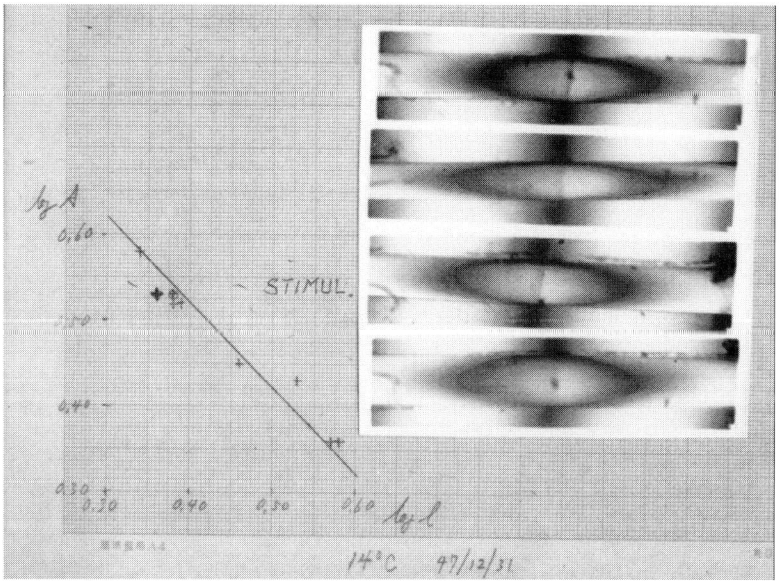

Fig. 2. Example of photographic records and plot of log (integrated birefringence per cross section) versus log (length of fiber) for one fiber. The red cross and circled black cross, together with the image of the fiber immediately to the right of the two crosses, refer to the fiber undergoing isometric contraction under AC stimulation.

into four thicknesses from a ca. 1-mm-thick clear sheet and then oriented it by trial and error. Fortunately, my hunch worked, and now I was able to measure the area surrounded by a complete "elliptical" fringe (Fig. 1, top and lower right; photographs in Fig. 2). In retrospect, what I had managed to generate was a mica half-wave plate which was oriented with its slow axis oriented parallel to the polarizer transmission axis, thus effectively turning the wedge slow axis direction by 90° (2 × 45°), basically the same principle suggested ten years later by Lem Hyde of American Optical Company for making polarization rectifiers (Article 14, this *Collected Works*).

With this microscope, I photographed the images of intact single cross-striated muscle fibers in frog Ringer's solution, stretched to different lengths (through the hooks penetrating the tendons at the two ends), and as contracted by AC (alternating current) stimulation (Fig. 2, right). The dark fringes were traced and the area surrounded by the fringe cut out from a sheet of Kent drawing paper in order to measure their weights and determine the relative areas of the "elliptical" fringes. The log of the area (A) was then plotted against the log of the relative muscle length (L), which was determined by measuring the average sarcomere length (2 μm to 4 μm) within the "ellipse." In making these measurements, it was noted that the cross-sectional shapes of the individual muscle fibers were not exact circles, but deformed towards triangles or polygons. Thus, the apparent width of the fiber, and the thickness of the region whose birefringence was compensated, was not only a function of the fiber length, but also depended on which profile of the fiber was displayed. Since the shape of the displayed profile tended to change with the length of the fiber or the tension exerted, several fibers at constant lengths were turned around their long axes at constant fiber lengths to confirm that the areas of the "ellipses" for a particular region of the fiber did not vary for any given fiber length, despite changes in the shape or the major and minor radii of the "ellipses."

As seen in Fig. 2, left, the plot of logA versus logL shows a straight line relationship with a slope of −1. In other words, logA + logL = constant, or A × L = (coefficient of birefringence) × (volume of the fiber) remains unchanged. Since the volume of the fiber can be assumed to remain more or less constant, these observations show that the coefficient of birefringence of the fiber must remain constant and be independent of fiber length.

Furthermore, Fig. 2 shows that the area of the "ellipse" of the stimulated and contracted fiber is essentially no different from (or only very slightly less than) the area of the unstimulated fiber at the same length. Interestingly, my conclusion that the birefringence of muscle under isometric contraction is unchanged for that muscle at the same length at rest agrees with what Bozler and Cottrell found with snail smooth muscle, although differing from theirs and others' earlier findings on striated muscle.

From these observations, I postulated that the coefficient of birefringence (i.e., retardation per unit thickness, or the state of folding of its polypeptide chain in a cross-striated muscle fiber) was independent of fiber length, and also whether it was isometrically contracted or not. In other words, contraction and length change of muscle could not be explained by changes in folding of its polypeptide chains.

That was a few years before A.F. Huxley and R. Niedergerke and Hugh Huxley and Jean Hanson, through observations of the A- and I-band lengths in contracting striated muscle and myofibrils, proposed the mechanism of muscle contraction by the mutual sliding of actin and myosin filaments. Prior to, and even for a few years after, their work, muscle was believed by many investigators to contract, not by sliding filaments,[1] but by folding of its polypeptide chains, much the same way that an extended rubber band shortened by folding of its isoprene chains. Associated with folding of polymer chains, one expected a substantial loss of birefringence as summarized extensively for muscle fibers by W.J. Schmidt (1937). Nevertheless, by the early 1940s, there appeared several conflicting publications regarding whether the birefringence, in fact, did or did not change during contraction of muscle.

Sadly, I never did publish my findings recorded here, since Dr. Kamada, my supervisor for the project, argued that "One cannot learn about the contraction mechanism of muscle by studying cross-striated muscle since they have such a complicated, striated structure. Instead, you must learn to isolate single smooth muscle cells and work with them!" In retrospect, it was an ironic statement indeed, in view of what the striations in skeletal muscle were to reveal in 1954. However, Dr. Kamada himself never learned of those revelations since he passed away in 1948 from terminal cancer.

References

Bozler E, Cottrell CL, *J Cell Comp Physiol* **10**, 165–182, 1937.

Huxley AF, Niedergerke R, *Nature* **173**, 971–977, 1954.

Huxley H, Hanson J, *Nature* **173**, 978–987, 1954.

Schmidt WJ, Die doppelbrechung von karyoplasma, zytoplasma und metaplasma, *Protoplasma-Monographien*, Vol. **11**. Borntraeger, Berlin, 1937.

Stephens RE, *J Cell Biol* **25**, 129–139, 1965.

[1] See Stephens (1965) for experiments using UV microbeam disruption of sarcomere regions in glycerinated myofibrils. Through these tests, he was able to convince the skeptics and rule out all but the sliding filament theory.

COVER-SLIP THICKNESS GAUGE

Shinya Inoué*

(Article prepared by author in August 2006, based on the actual gauge)

While acting as a teaching assistant in Kenneth Cooper's Cell Biology course at Princeton University in 1950, I learned that to use a microscope with high NA (numerical aperture) objective lenses, one needs to select cover slips of the proper thickness in order to get the best image. The reason was that high NA microscope objective lenses are designed to provide an image with minimum aberration only when used with the designated cover slip. In most cases, the designated thickness is 0.17 mm, coupled with the proper immersion medium and correct optical tube length, assuming the specimen is sitting close to the cover slip (see e.g., Shillaber, 1944; Inoué and Spring, 1997).

0.17-mm-thick cover slips are available commercially in boxes designated as #1.5, but those boxes usually contain cover slips whose thicknesses range from approximately 0.15 mm to 0.18 mm. In order to examine the specimen critically at high NA, one then needs to select those that do not deviate from 0.17 mm by more than +/−0.005 mm, or just a few micrometers.

To determine such thicknesses, one can use a machinist's precision caliper micrometer. However, using a caliper micrometer is somewhat cumbersome, and also one needs to scrupulously clean the contact faces of the micrometer. In order to avoid using a precision caliper micrometer altogether and to simplify the process, I designed the following cover-slip gauge that is easy to use and involves no moving parts except for the cover slip itself, which acts as the pointer for the thickness scale.

The gauge (Figs. 1 and 2) contains a narrow horizontal slit (or "gap") between two horizontally oriented edges of stainless steel razor blades. The upper blade is mounted directly onto an upright base plate, while the lower blade is mounted slightly further out by the presence of an underlying shim (Fig. 2; in the photograph, the two half-length stainless steel safety razor blades are hidden behind the brass plates which secure them in place against the thick base plate to the far left). The thick base plate is recessed behind the gap between the two blades (to the left in the figures), so that part of a cover slip which is dropped into the gap between the two blades protrudes into this narrow recess. Since the main bulk of the cover slip remains to the right of the blades, that side of the cover slip is heavier and tilts down until its tilt is constrained by the two knife edges.

The degree of tilt of the cover slip is determined by the vertical distance between the two blades and their horizontal offset (the thickness of the shim), with the lower knife edge

*Unpublished studies (1951).

Fig. 1. Photograph of cover-slip gauge.

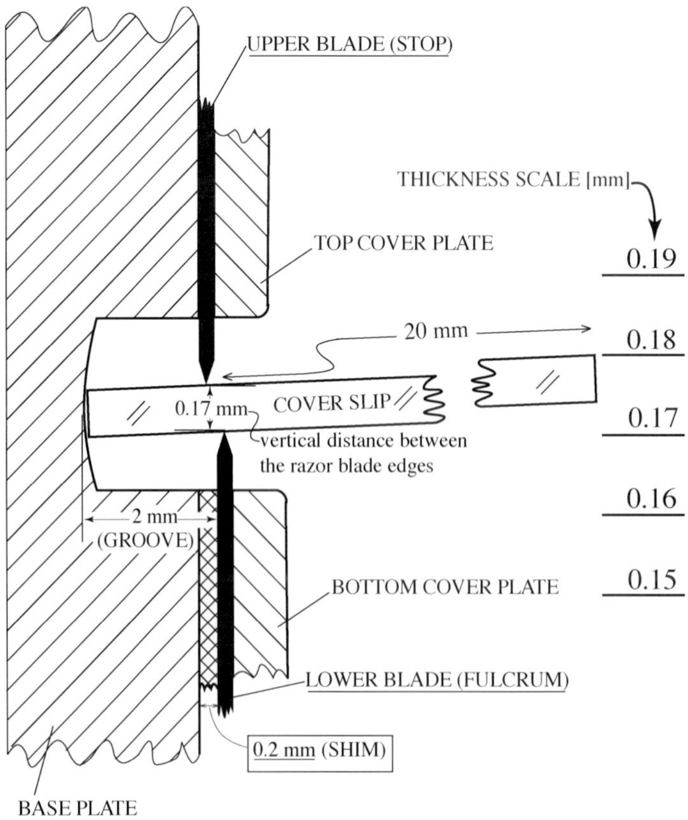

Fig. 2. Schematic of the lever system.

acting as the fulcrum, while the upper one acts as the stop that limits the tilt of the cover slip. Thus, in the gauge, the cover slips tilt differently depending on their thickness.

As shown in the figures, the actual thickness of a particular cover slip is read off from the scale inscribed on a second plate oriented at 90° to both the base plate and the slit between the razor blades.

In detail, the vertical distance between the two edges of the blades is set to 0.17 mm so that a cover slip of that thickness settles horizontally. Cover slips, whose thicknesses deviate from 0.17 mm, tilt away from the horizontal by amounts determined by their thickness and by the offset distance between the two knife edges. The shim behind the lower blade is 0.2 mm thick and defines the horizontal offset between the two blades. Since the horizontal offset between the two knife edges is 0.2 mm and the left edge of the cover slip drops into the 2 mm recess in the base plate, the right edge of a 22-mm square cover slip tips by $(22 - 2)/0.2$, or 1 mm for every 0.01 mm deviation in thickness (Fig. 2). Thus, the thickness of the cover slip, read off from the scale at the distal tip of the cover slip, is magnified 100 times.

This gauge, made in March 1951, is still used in our laboratory today after half a century. Without the fear of contaminating the surface of a carefully cleaned cover slip, its thickness is determined simply by dropping an edge of the cover slip into the gap while holding the gauge tipped counterclockwise. With the gauge brought back upright (as seen in Fig. 1), the cover slip comes to rest on the fulcrum- and stop-blades so that the scale at the distal edge of the cover slip directly indicates its thickness. This gauge is easy to use, works quickly, and with surprising accuracy.

References

Inoué S, Spring K, *Video Microscopy*, Plenum Press, New York (Sections 2.5.2 & 2.5.8), 1997.

Shillaber CP, *Photomicrography in Theory and Practice*, John Wiley and Sons, New York, 1944.

Reprinted from the *Journal of Morphology*, Vol. 89(3), pp. 423–455, November 1951.

BIREFRINGENCE OF THE DIVIDING CELL [1]

SHINYA INOUÉ [2] AND KATSUMA DAN [3]
Misaki Marine Biological Station, Kanagawa-ken and Zoology Department, Tokyo University, Tokyo

THIRTEEN FIGURES

Various structures of the living cell show weak but decided birefringence when observed with a well-adjusted polarization microscope. Double refraction of the cortex of living cells was reported by Runnström, Monné and Broman ('44) in sea-urchin eggs. This finding was confirmed through numerous tests by Runnström's associates [Monné ('45), Monné and Wicklund ('47) as well as by Monroy and Montalenti ('47)]. Double refraction of mitotic figures in the living state was found by Schmidt ('36) in sea-urchin eggs. This observation was confirmed later by Monné ('44) and by Hughes and Swann ('48). The latter authors extended the observation to the division of chick fibroblasts cultured *in vitro*.

The main object of the present paper is to find out whether birefringence data can be used for the analysis of the mechanism of cell division. Extremely important in this sense are Schmidt's findings which indicate that, in metaphase- and telophase-spindles, the traction fibers connecting the spindle

[1] The authors wish to thank Prof. J. Koana for the specially made slides and cover slips which he generously put at their disposal. They are also indebted to Miss K. Okazaki, who collected the data on the cortical birefringence of sea-urchin eggs, and to Dr. J. C. Dan for assistance in the preparation of the manuscript. Finally, the present research was made possible by a fund offered by the Nagai Women's Association, Nagai-machi, Yokosuka City, to which the authors express their gratitude.

[2] Present address, Department of Anatomy, School of Medicine, University of Washington, Seattle 5, Washington.

[3] Present address: Biology Department, Tokyo Metropolitan University, Meguro-ku, Tokyo.

PRESS OF THE WISTAR INSTITUTE OF ANATOMY AND BIOLOGY
Printed in the United States of America

pole and chromosome group (Halbspindel), as well as the astral rays (Polstrahlen), are doubly refractive.

In the first half of this study, therefore, it was attempted to repeat the observations of the previous workers by using the eggs of the medusa, *Spirocodon saltatrix,* and sea-urchins, *Clypeaster japonicus, Mespilia globulus* and *Strongylocentrotus pulcherrimus.* The results are presented with photographs. Upon assembling this material, however, the authors were struck by the unexpected fact that the positive birefringence of astral rays of early cleavage stages changes to negative birefringence during the course of the cleavage. The latter half of this paper will be devoted to elucidation of the cause of the change in the axis of the index ellipse and its connection with forces acting within dividing cells.

METHODS

All the previous works on the double refraction of living cells cited above, except that of Hughes and Swann, were performed by careful manipulation of petrographic microscopes of very high quality. The present work was carried out with a biological microscope, but with some difference in technique from Hughes' and Swann's.

Prisms. A polarizing "Nicol" prism is set below the microscope condenser and an analyzing prism is placed inside the microscope tube. For both polarizer and analyzer, prisms with perpendicularly cut ends are used in order to simplify the optical condition.

Light source. For the study of objects with weak birefringence, the use of a very brilliant light source is indispensable (see appendix). Monné ('44) used a "mercury lamp." In the present case, an air cooled super high pressure mercury lamp (0.5 amp. 200 working volts) was used together with an achromatic condenser. No indication of any ill-effect caused by the use of the intense light was observed.

In order to obtain a very sensitive condition for detecting weak birefringence, it is also necessary to insert many diaphragms in the optical system. The set-up adopted is diagrammatically shown in figure 1. Monné recommends an oil immersion lens provided with an iris diaphragm, which we also have used.

The chief difficulty in using a biological microscope lies in the fact that the objective lenses of biological microscopes have a weak birefringence intrinsic to them. This birefringence, however weak it may be, is fatal for cell studies, since the birefringence of cell components itself is so extremely weak. The cause for the birefringence of lenses is due to mechanical strains residing within the glass of the lenses. The objective lenses of petrographic microscopes marked "P" are supposed to have been freed from such strains.

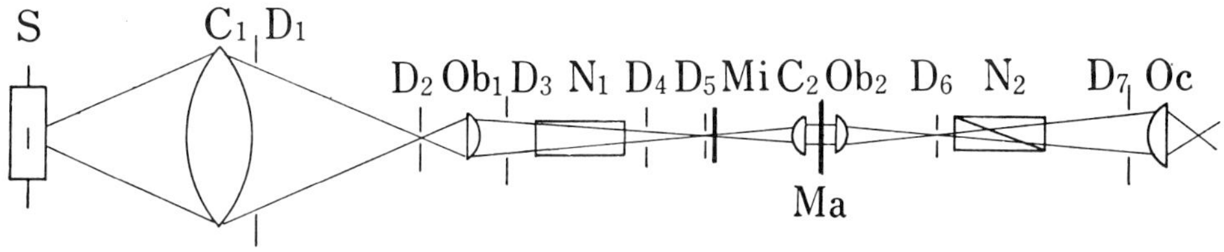

Fig. 1 Diagram of the optical system. S, light source; C_1, C_2, condenser; Ob_1, Ob_2, objective lens; N_1, N_2, "Nicol"; Oc, ocular lens; Mi, mica plate; Ma, material; D_1, D_2, ... D_7, diaphragm.

In substituting biological for petrographic microscopes, in order to minimise the disturbing effect of birefringence, objective lenses with uniformly directed birefringence are first selected, and so set that their axes of birefringence lie parallel to that of the polarizer. In the case of the microscope condenser, besides setting its axis in the proper direction, clamping it in position with a stop screw must be avoided because the clamping strains the condenser, thereby introducing a new birefringence.

Criterion for good adjustment. There is, fortunately, an objective method for testing the quality of this set-up. A short focus telescope [4] inserted in place of the ocular lens and focussed on the upper lens of the objective system, shows a

[4] A telescope provided with the Bausch and Lomb phase contrast accessories was used.

dark cross (fig. 2) in the field when proper adjustment has been attained. As the set-up departs from the ideal, the cross is correspondingly divided into two V's or missing entirely. This criterion is, therefore, useful in a two-fold sense: for the selection of good objectives and for setting them in the proper directions. The cross can be observed only when the light source is sufficiently powerful. If it is not, the field looks uniformly dark. When objective lenses of higher magnification are used, as figure 2 indicates, the central part of the cross originates from both the condenser and the objective, while the peripheral part is only from the objective.

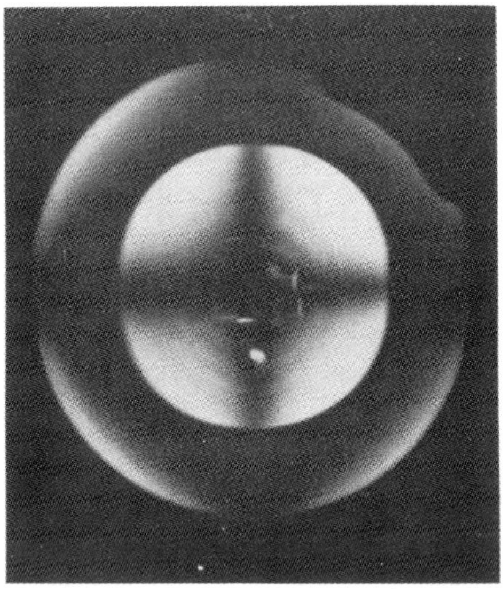

Fig. 2 Photograph of the dark cross which can be seen by a telescope inserted in place of an ocular lens under the best optical adjustment.

The authors interpret the appearance of this cross as an expression of the so-called "Landolt's fringe" phenomenon; i.e., when some very brilliant light source, such as the sun, is observed through crossed Nicols, only certain portions of the field appear dark. If the light is not strong enough, the fringe vanishes and the field becomes uniformly dark, just as in the case of the cross described above.

The cause for the Landolt's fringe is probably due to the fact that when polarized light passes through any refracting plane, the plane of polarization of the light must be altered to some extent, since polarization and double refraction ordinarily occur at interfaces. If this interpretation be correct, applied to our system, it means that only the portions of the lenses which have their planes normal or parallel to the swinging direction of the polarized light, will allow the light to pass without changing the condition of its polarization. The remaining portions will alter or disturb it in proportion to their deviation from the normal or parallel directions.

This interpretation provides an explanation for the fact that stray light coming through the bright areas around the cross is more marked with lenses of higher aperture (and therefore of stronger curvature). In order to cut down the stray light which enters through the oil immersion lens, an iris diaphragm attached to it must be closed to some extent even though at the expense of the resolving power of the lens.

Slides and cover slips. Surfaces of glass, including the lenses, slides and cover glasses, must be kept scrupulously clean, since dust particles or pieces of fiber are often strongly doubly refractive and make impossible the detection of weakly birefringent objects. Moreover, the double refraction intrinsic to the slides and cover glasses constitutes a very annoying factor. Since these glasses turn and change the brightness of the field when objects mounted on them are rotated for observation on the revolving stage of the microscope, accurate determination becomes impossible. On hearing of this difficulty, Prof. J. Koana of the Physics Department of Tokyo University generously placed at our disposal specially annealed slides and cover slips made from Jena glass, Schott BK7. Without this aid, it would have been impossible to obtain some of the results reported in this paper. The authors are happy to acknowledge their great indebtedness to Prof. Koana's act of far-reaching cooperation which has been a great inspiration to them.

Rotating mica plate. A supplementary technique which is widely adopted in polarization microscopy is the use of a rotating mica plate. If a thin paring of mica plate is inserted between the Nicols in a certain (bias) direction, the entire field of the microscope becomes brighter by virtue of the double refraction of the mica. When birefringent materials to be tested are rotated in the field of such a superimposed bias-birefringence, they no longer shine in 4 quadrants as they do ordinarily (without the mica). Instead, they shine more brightly in the two opposite quadrants where their birefringence lies in an additive direction with the bias-birefringence, and they darken in the two other quadrants where their birefringence takes a subtractive direction with respect to the bias-birefringence. Or, by a simple rotation of the mica plate, it is possible to make the objects shine or darken without moving them at all. Quite often, this is more convenient than to rotate the materials themselves. Naturally, for a complete darkening (or cancelling), the birefringence of the materials and the bias-birefringence ought to be perfectly compensatory. If the former is greater, the objects shine in 4 quadrants; but in this case, they shine more brightly in two opposite quadrants and less so in the remaining quadrants. For studies of living cells, 1/16–1/30 λ plates have been used by various workers (Schmidt, '37; Monné, '42; Monné, '47; Hughes and Swann, '48).

In the present case, many mica parings were made by hand and a piece of an appropriate thickness was attached to a filter holder, below the condenser. Since the motion of this holder was an arc, it could not produce concentric rotation of the mica plate around the optical axis of the microscope. Quite accidentally, a very convenient technique was found in this situation.

The manipulation procedure is as follows: First, the holder is set at some intermediate position most convenient for manipulation. Then by trials, a direction (relative to the holder) of the mica plate is found in which a slight back and forth movement of the holder will make the object on the stage alternately shine and darken. The mica is fixed in that direction for a rough adjustment. For the fine adjustment, the holder is brought back to the initial position and by using the telescope, the condenser is rotated and so adjusted that a clear dark cross is obtained. Under this condition, the double refraction of the condenser and that of the mica plate are almost completely cancelling each other, and a birefringent object on the stage will shine in 4 quadrants against a dark background. A slight shifting of the holder to one side immediately eliminates the shining in two of the opposite quadrants and enhances the brightness in the remaining quadrants. A shift of the holder to the other side, beyond the initial position, turns off the light in the formerly shining quadrants and turns it on in the formerly dark quadrants. In other words, by moving the holder swiftly, it is possible to make the birefringent object shine and darken in quick succession.

This is an almost ideal condition, for, by the manipulation of the holder, birefringent structures of living cells flicker and send signals while isotropic regions remain indifferent. Moreover, if the mica plate is pushed still farther, since the field becomes quite bright, ordinary microscopical observation of the structures can be combined with birefringence observation. The photographs of sea-urchin eggs accompanying this paper were so taken that both the structures visible in ordinary light and their birefringence are shown simultaneously.

For a further improvement of the technique, the authors made an analysis of the condition necessary to obtain the best visual sensitivity in the presence of stray light. The results of this study are given in the appendix.

RESULTS

Double refraction of nuclear derivatives. In 1936 and 1937, Schmidt discovered that the metaphase spindle of the eggs of *Psammechinus miliaris* has a fairly strong birefringence. The double refraction is positive with respect to the length of the spindle. As the spindle elongates, an optically isotropic gap appears in the center, separating the birefringent part in two. During the anaphase, the birefringence of the spindle gradually fades away. In his earlier papers ('36 and '37), Schmidt interpreted these doubly refractive halves of the spindle as chromosomes, but later ('39) after comparison with sectioned and stained material, he concluded that the birefringent part lies between the chromosomes and the spindle pole, thus corresponding to the region of the traction fibers ("Zugfaser").

By the aid of the technique described in the preceding section, the double refraction of the spindles in the dividing egg cells of the medusan, *Spirocodon saltatrix,* and the echinoderms, *Strongylocentrotus pulcherrimus, Mespilia globulus* and *Clypeaster japonicus* was examined. Schmidt's observations were, on the whole, confirmed when comparatively low power objectives were used. Figures 8 and 13 are the photographs showing the separating "Halbspindel" of the eggs of *Spirocodon* and of *Clypeaster.* A slight difference between the two forms lies in the fact that, in *Clypeaster,* the mitotic apparatus increases in width as well as in length.

Upon using a high dry objective and an oil immersion lens, however, two more facts were brought into view. The first is the existence of doubly refractive threads running across the "isotropic" gap between the two half-spindles. These are shown in figure 9 which was taken with a Zeiss objective D with an aperture of 0.65. For photographing, the oil immersion lens cannot be used because the spindle perceptibly elongates during the time necessary for exposure. Whether these threads correspond to the so-called interzonal fibers or are parts of the "Stemmkörper" is a point still to be decided (Schrader, '44). The axis of double refraction of these fibers is also positive with respect to their lengths.

The second fact is a weak birefringence exhibited by the spindle remnant running between the reconstructing daughter nuclei (fig. 10). The double refractive remnant of a spindle looks less fibrous than the earlier threads. In *Clypeaster* eggs, although the double refraction of the remnant is quite weak, it is retained throughout the whole process of cleavage until the furrow cuts the structure. In *Spirocodon* eggs, the spindle remnant can be followed by its double refraction until the cleavage stage in which the remnant bends (Dan and Dan, '47). After this stage, in part due to weakness of the birefringence and in part due to a close association of the cell surface with the remnant, no definite conclusion can be drawn, in spite of the fact that the remnant itself is visible with a phase contrast microscope. Sometimes, however, a birefringent streak is seen in the protoplasmic bridge between the incipient blastomeres. From all these indications, it may be thought that the spindle remnant retains its very weak double refraction also in medusan eggs. The sign of the birefringence of the spindle remnant remains positive, with respect to its length. What Monné ('44) describes as negative streaks in the two-cell stage of *Psammechinus* (his fig. 22) most closely corresponds, in shape as well as in position, to the spindle remnant here described, except that the sign of the double refraction is opposite.

A few words will be devoted to a discussion of the position of the chromosomes with reference to the half-spindles. In one isolated case, that of a *Clypeaster* egg not in optimum condition, it was possible to observe the double refraction pattern of the metaphase spindle and a group of chromosomes which showed as lines of dark dots in the middle of the spindle. When the mica plate was moved, the spindle flickered, while the chromosomes remained always dark. This apparently indicates that chromosomes are isotropic, although no definite conclusion can be drawn until the degree of light absorption and the refractive index of the chromosomes are known. At any rate, it is sufficient to show that the birefringent half-spindles correspond to the traction fibers at least in position, as proposed by Schmidt.

Double refraction of extraneous coats. The fertilization membrane and the hyaline plasma layer of the sea-urchin egg have been reported by Runnström ('28) and Runnström, Monné and Broman ('44) as negatively birefringent in the radial direction or positively birefringent in the tangential direction. The same is true of the sea-urchin eggs so far investigated by the authors.

In figure 10, the double refraction of the fertilization membranes of *Clypeaster* egg is shown, although the hyaline layer is not so evident, because of its thinness in this species. As Runnström ('48) pointed out, the strength of the double refraction of the fertilization membrane increases (from 4 mμ to

as double refraction is concerned. However this conclusion obviously contradicts numerous other cytological findings.

Beside this, there is another complicating factor which disturbs the observation of double refraction in the periphery of the aster. In one blastomere (left) of *Spirocodon* shown in figure 11 and figure 3, the center of the aster clearly indicates a positive cross. But if traced further toward the periphery, the brighter quadrants in the inner cross gradually change to dark, while the darker quadrants of the central cross change to

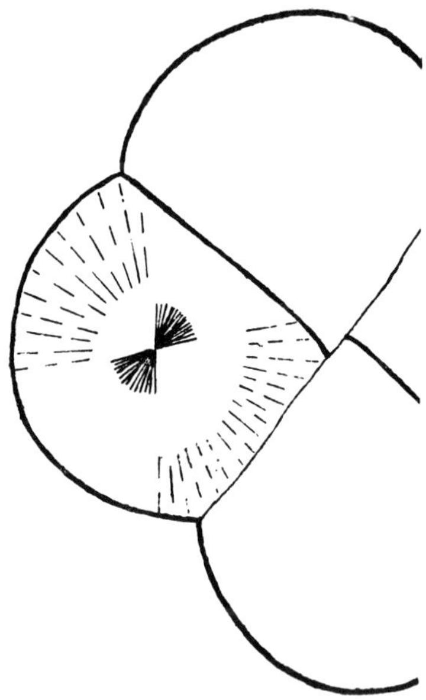

Fig. 3 Explanatory diagram for figure 11, showing a central positive cross and a peripheral negative cross.

bright. In other words, there seem to be two concentric crosses, the central one positive, and the peripheral one negative. If so, the astral rays seem to change the sign of the double refraction along their lengths. In the case of *Spirocodon* eggs, the radially negative birefringence of the distal rays is apparently overcoming the radially positive birefringence of the cortical layer. Moreover, such a reversal of the sign of double refraction is also found among sea-urchin eggs, as will be described below.

Double refraction in fertilized sea-urchin eggs during the pre-cleavage period. Monroy and Montalenti followed the change in the strength of cortical birefringence of *Psammechinus miliaris* from fertilization through early cleavages, and found cyclic changes. The same observations were carried out, using the eggs of *Strongylocentrotus pulcherrimus*, and a similar series of changes in birefringence was found. However, the stages in which the changes take place differ slightly from those observed by the above-mentioned authors. Unfertilized eggs of this species show a cortical birefringence negative in the direction of the tangent. This cortical birefringence vanishes while the fertilization membrane is being formed. However, in contrast to the finding of Monroy and Montalenti, the disappearance of the double refraction is not permanent, the cortex again shows negative birefringence immediately after the elevation of the membrane. Several minutes later, corresponding to the time of the formation of the hyaline plasma layer, the cortical birefringence becomes very weak or sometimes vanishes completely. Then it gradually regains its strength and reaches a maximum at the time of the first cleavage. Leaving aside detailed description of birefringence phenomena during cleavage, it suffices now to say that soon after the completion of cleavage, the double refraction again becomes very weak, and again regains its full strength at the second cleavage.

Concerning the double refraction of the internal structures Monné ('44a) has made a thorough study, to which nothing essential can be added, although some of our interpretations are different. Therefore only the outstanding points will be presented, together with photographs, since Monné's paper contains only diagrams.

1. When a sperm aster is formed around the sperm pronucleus, a positive cross can be seen with a polarization microscope. This means that the monaster is composed of rays which are positively birefringent in the direction of their length (Monné's fig. 14). This condition persists until the sperm and egg pronuclei meet.

2. After the encounter of the two pronuclei, one quadrant of the former positive cross of the monaster is missing on the side of the egg pronucleus. This is obviously due to the fact that as the egg pronucleus approaches the sperm pronucleus, it pushes aside the rays of the monaster.

3. Then the streak stage follows. This disk-shaped appearance of the "streak" in the stage which follows the full development of the sperm monaster, is an often-observed phenomenon which, however, has not received much attention from embryologists. According to our interpretation, some of the rays which at first extended equally in all directions from the astral center are assuming a different mutual relationship such that they become massed in a plane extending across the primary egg axis. Observed from the side, this appears as a thick disk, transparent with ordinary illumination and showing strong negative birefringence with reference to its diameter in polarized light.

Monné's description of this stage agrees essentially with ours. He writes "... This radiation figure (the monaster) is gradually flattened (fig. 16) in the direction of the egg axis ... Finally this flattened 'astrophere' of the sea-urchin egg is transformed into a clear cytoplasmic layer appearing as a streak (figs. 17, 18) when viewed from its side. This structure is known in literature as 'nucleal streak' " ('44a). He also finds this streak to possess negative birefringence, but here his explanation differs from ours. He is of the opinion that the positive rays of the early monaster have their origin in nuclear substance, while the negative rays of the streak are of cytoplasmic origin (Monné's fig. 16; our fig. 12). On the other hand, since the authors made the observation, in *Spirocodon* eggs, that the sign of the birefringence of the rays may change, they prefer to suggest the possibility that, in sea-urchin eggs also, there is only one kind of rays, and they simply reverse the sign of double refraction in different stages of development.

4. At the end of the streak stage, the rays disappear from the peripheral part of the egg. But the streak itself persists a

little longer still retaining a relatively strong negative birefringence along its length. While the negative streak is still clearly persisting, the spindle makes its appearance within the streak. The long axis of the spindle always coincides with the plane of the streak or its apparent long axis as viewed from the side. However, as was already mentioned, since the spindle has a positive birefringence along its long axis, the positivity of the spindle and the negativity of the streak stand out in a marked contrast (Monné's fig. 18; our fig. 13).

5. As the spindle grows longer, the streak becomes smaller and finally, only the last traces of the negative streak remain beyond the two poles of the spindle (fig. 13).

6. After these small negative patches disappear, two small asters are formed. But as the asters grow as positive crosses, the spindle loses its brightness and only a line of positive birefringence is left in its place, representing the spindle remnant (fig. 10).

Reversal of sign of birefringence by mechanical strain. It has been known for a long time that a rubber band, when it is stretched, acquires a positive birefringence along its length and the degree of the birefringence increases in proportion to the stretch (Spannungsdoppelbrechung). Recently, Kubo ('47) made a theoretical consideration of this phenomenon. From a theoretical basis, it can also be expected that when a rubber band is compressed, it will acquire a negative birefringence along its length. This idea was tested by direct experiment of 4 kinds of materials.

As representatives of common substance, rubber and celluloid were selected. If all strain is previously removed, these substances are isotropic. After inserting a sensitive tint plate in the optical path, a needle is stuck into a thin layer of these substances and the needle is moved in one direction. Ahead of the needle, where the material is compressed, the tint so changes as to indicate the development of a negative birefringence in the direction of the push, while behind the needle, where the material is being stretched, a color characteristic of a positive birefringence develops.

According to Schmidt, cherry gum has naturally a negative birefringence. By the above method, it is demonstrated that this substance acquires a positive birefringence when compressed and a negative one when stretched. In other words, cherry gum behaves in the opposite way from most substances.

As an example of artificial substances which most closely simulate protoplasmic gels, a soft gel of gelatin (4%) was tested. Even this soft gel decisively showed a positive birefringence in the direction of pull and a negative one in the direction of push.

As an example of a metaplasmic substance, a gel which is secreted at fertilization from the eggs of the annelid *Perinereis* was examined. A positive birefringence which this jelly naturally has, changes into negative upon compression (Inoué, '49).

As a matter of fact, Monné himself writes "Parts of the egg, exposed to the strongest pull or pressure, present the highest birefringence" ('44). Knowing that the nature of birefringence is so easily governed by imposed mechanical strains, haphazard changes in sign of birefringence of fixed chromosomes as observed by Becker ('39) and induction of birefringence in stained jelly of sea-urchin eggs as reported by Monné ('44) may become much easier to understand.

Reversal of sign of birefringence of astral rays. In the preceding section, it was shown for a number of gels that birefringence of one sign develops when these gels are stretched, while compression induces birefringence of the opposite sign. If the same situation obtains in astral rays, it can be expected that the rays will acquire a negative birefringence if they are compressed in the direction of their lengths.

The part of a dividing cell where the astral rays would be compressed most must be the place where the cell circumference decreases most during cleavage so that the aster is squeezed from all around. That such a thing is actually happening will be seen if a dividing cell is observed along the spindle axis. When viewed from this direction, the cell circumference before cleavage is obviously that of the undivided egg cell. But after the completion of cleavage, the cell circumference which is seen is that of a blastomere of the two-cell stage. Between the two, there occurs a decrease in diameter of 20%.

In actual observations with polarized light of the eggs of *Clypeaster japonicus* and *Mespilia globulus,* our expectation is satisfactorily borne out. In the beginning, we have a polar view of an aster which appears as a positive cross. As the furrow constricts and the circumference of that part decreases, the positive cross vanishes and in a moment, a negative cross appears in its place (fig. 7). In connection with a 4-cell stage of *Spirocodon saltatrix,* shown in figures 11 and 3, it was previously pointed out that in the left blastomere the double refraction of the central part of the aster is positive while the distal part is negative. That blastomere is also showing a polar view of the aster.

An experimental procedure can be resorted to in order to achieve the same result. This is to take sea-urchin eggs with well developed amphiasters and subject them to a hypertonic medium. In such a medium, the eggs shrink, and all the rays within the eggs reverse the sign of their double refraction, no matter from what direction they are observed.

Axis reversal of index ellipse of the astral rays and its relation to the mechanism of cell division. After coming this far, we are in a position to consider the reversal of sign of double refraction of the astral rays in connection with the theory of cell division. One of the authors has proposed a working hypothesis of the division mechanism (Dan, '43). According to it, the mitotic figure is a gel structure; i.e., the spindle is a rod of gel, and asters are also spheres of gelated rods radiating out in all directions. A fundamental activity involved in cell division is the elongation of the spindle, thus forcing apart the two asters. This condition is diagrammatically illustrated in figure 4a. It should be pointed out that in the majority of egg cells, there is a region around the egg where the opposed astral systems intermingle so that the distal portions of some of the rays cross each other.

Now if the spindle begins to elongate, the polar rays which extend in the direction of the spindle elongation are pushed against the cell membrane from the inside and are bent in a fountain figure (fig. 4b). The spreading of the polar rays in a fountain figure ought to make the polar surface expand. This is exactly what happens, as revealed by the movement of kaolin particles attached to the surface of this region (Dan, Yanagita and Sugiyama, '37). While the bending is taking place, however, the spindle elongation is not conveyed to the cell periphery as such but is buffered by the bending rays which are acting as a sort of spring.

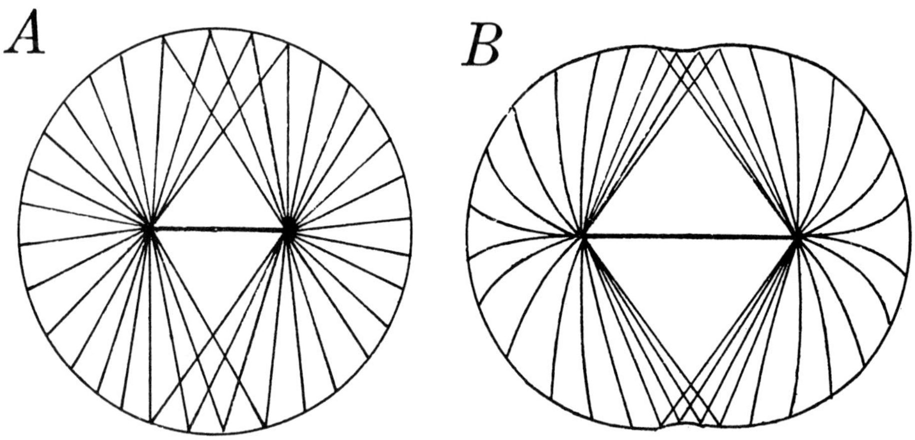

Fig. 4 Diagram showing the condition of the astral rays during early stages of cleavage. (A) All the rays are straight and the rays from the two asters are crossing each other on the median plane. (B) When the spindle begins to elongate, the crossing rays pull in the equatorial surface and make it shrink. The crossing rays remain straight during this time. The polar rays are bending, being pushed against the polar surface by the elongating spindle (fountain-figured bending).

As long as this buffering action is effective, the crossing rays at the equatorial plane pull in the equatorial surface and induce the formation of a furrow. As Dan ('43) has fully discussed, this pulling in by the crossing rays not only induces furrow formation, but also causes a shrinkage of the equatorial surface. Geometrical analyses based on the above idea can predict the depth of the furrow and the degree of the surface shrinkage for early stages, which check quantitatively with the actual readings of the koalin method.

When the fountain-figured bending reaches its maximum and the cushion effect is overcome, the equatorial surface is not pulled in by the rays any more but begins to be sucked into the gap between the two asters which are being pushed away from each other. In this stage, the equatorial surface is stretched, contrary to the behavior in the preceding stage.

From the above hypothesis, it is possible to predict that the crossing rays would exhibit a greater positive birefringence than the other rays, since they must be under tension while they are pulling on the surface. The polar rays, on the other hand, should acquire a less positive value or even a negative value.

Birefringence of astral rays in the side view of dividing cells. In order to test the validity of the above hypothesis, a careful observation was made on the double refraction of amphi-astral figures of *Clypeaster* eggs. As is shown in the photographs of figure 10 and a diagram of figure 5, it is thought to be possible to detect a stronger positive birefringence in the crossing rays.

Naturally, the detection of slightly more positive rays among other positive rays involves an extremely delicate technique. One way is to bring the crossing rays near the border of a quadrant and make the more strongly positive rays stand out over the others. Another way is to place a cell in such a direction that the crossing rays coming from one aster become bright while those from the other aster become dark, thus producing a cross figure of bright and dark rays. In these observations, it is indispensable to shift the mica plate back and forth and distinguish these particular rays by contrast. Figure 10 was taken by the latter technique. To obtain the best condition for observation, it was found advisable to remove the fertilization membrane.

Observation by photographs is convenient in the sense that as much time as needed can be spent on the analysis. But its weak point lies in the fact that it is impossible to trace a ray by changing the focus. Even when some rays are shining, since the interspaces between the rays are isotropic, what can be taken as a photograph is a complicated pattern of white and black patches. However, quite frequently, just corresponding to the place where crossing rays are found, continuous lines showing a positive birefringence can be traced from the astral

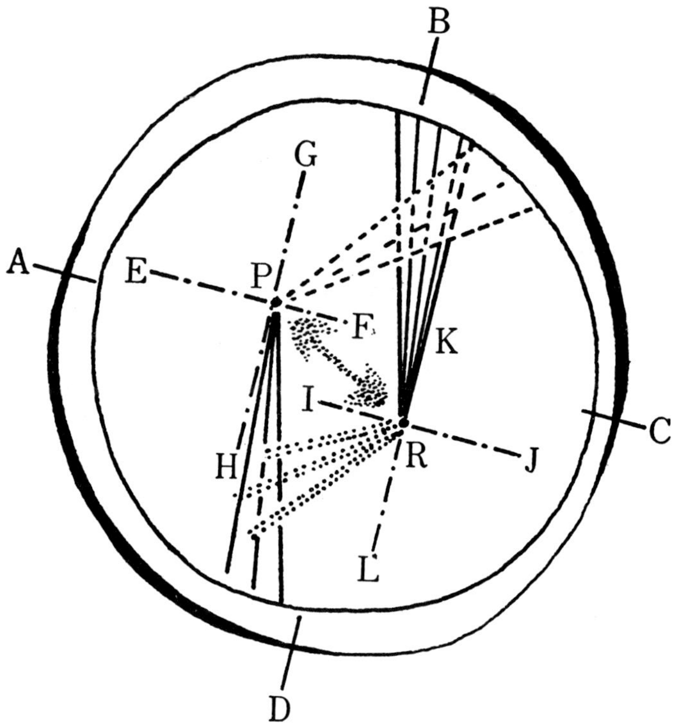

Fig. 5 Tracing of figure 10. A, B, C and D indicate the borders of the four quadrants of the fertilization membrane. Lines EPF, GPH (left side) and IRJ, KRL (right side) drawn parallel to AC and BD represent x, y axes of the four quadrants of each respective aster. The upper group of the crossing rays of the left aster, which are included within a bright quadrant, are shining more than the non-crossing rays. The upper crossing rays of the right aster are lying near the border of the dark quadrant and appear darker than the rest. On the lower half, corresponding to the position of the crossing rays, continuous dark lines can be traced between the left astral center and the equatorial surface, while continuous white lines are seen going from the right astral center to the equatorial surface.

center to the cell periphery (figs. 10 and 5 lower half). The reason for this may be either: (1) while being photographed, if a group of rays gets slightly out of focus, the weakly shining rays will soon be lost from the picture while the images of the more strongly shining rays will survive. (As a result, the chance is greater for the strongly shining rays than for the weakly shining ones to leave a photographic image.) Or (2) if the change to a negative birefringence starts from the cell periphery in *Clypeaster* as was observed in *Spirocodon,* even after the ordinary rays have begun to show a negative birefringence in the distal portion, the tensely pulled crossing rays will retain a positive birefringence along their entire lengths. In any event, analysis of photographs falls in line with the conclusion drawn from the observation of living eggs.

Concerning the double refraction of the polar rays, judgment is still more difficult. As far as our experience with *Clypeaster* eggs goes, although some rays were occasionally seen retaining a positive birefringence, in the majority of the cases, the double refraction became so weak that it was impossible to detect it definitely. Sometimes, however, some rays with a negative birefringence were also found. On the other hand, Monné described the birefringence of these polar rays as "either isotropic or weakly negatively birefringent in longitudinal direction" and further stated that "this weak birefringence may most easily be recognized in the fibrils oriented parallel to the spindle."

The second statement of Monné is particularly significant. In the preceding section, the authors advanced the argument that the polar rays must be under compression since they are bent in a fountain figure by the force from the elongating spindle. The statement itself is correct. But the fact that the rays are bending means that they are slipping away from compression in the strict sense of the term. As a result, the rays which are really compressed must be those which are in line with the spindle.

In short, it may be justifiable to conclude that the birefringence figure of the side view of dividing cells supports Dan's hypothesis.

DISCUSSION

Concerning the structure of the aster, the birefringence technique offers valuable information. Gray ('31) thinks that an aster is a solid sphere of gel within which the rays are imbedded. He attributes the chief importance to the solid gel nature and considers the rays as rather insignificant for the reason that tissue cells lack them. Dan ('43) visualizes the aster as a radiate sphere with gel rays shooting out in all directions, the interspaces between the rays being in a sol state. Under the polarization microscope, a positive birefringence in the longitudinal direction is clearly localized along the rays. This suggests that the rays are well defined entities, probably composed of a gel of protein nature. Concerning the absence of the rays among tissue cells, Hughes and Swann ('48) observed, in chick fibroblasts cultured in vitro, two small birefringent structures at the poles of the spindle, which they interpreted as asters.

Chambers ('17) proposed that astral rays are hollow tubes surrounded by gel through which liquid protoplasm is streaming. Occasionally, *Clypeaster* eggs were encountered in which individual rays could be seen very clearly by their birefringence. Careful observation of such rays under an oil immersion lens failed to offer any clues suggesting a hollow structure of the rays.

The most important line of argument presented in this paper is that the double refraction of cell components is not a static characteristic but is a dynamic phenomenon liable to change. An interpretation put forward by the authors assumes that the cause of the change of the axes of the index ellipse lies in mechanical strains imposed on these structures. In other words, the sign of the birefringence of these structures is dependent on the direction from which a force or forces act upon them.

If this be true, the polarization technique, which has, so far, been used as a means of descriptive morphology, will immediately change into a physiological weapon of attack on the analysis of forces acting within living cells. In order to pursue this line further, the fundamental assumption may require more consolidation. It may also be necessary to make the technique quantitative.

But just as a first trial, a venture was made applying the idea to the field of cell division. The fact that conclusions drawn from polarization figures of dividing cells by the application of the idea coincide well with what the hypothesis predicts, seems to be very encouraging. This statement holds good only for the astral rays. The spindle, which is supposed to play the most important part by our hypothesis, has not been studied. Nor has the cause for spontaneous reversal of the sign of the astral birefringence in the streak stage been touched upon. These points await future investigation.

SUMMARY

1. A method was developed to substitute ordinary biological microscopes for petrographic microscopes in order to make accurate studies of cellular birefringence.

2. The "traction fibers" of the mitotic spindles show a strong birefringence which persists during metaphase and telophase and fades at anaphase. In the eggs of the medusa, *Spirocodon saltatrix,* several birefringent fibers which resemble interzonal fibers in appearance are seen extending between the two half-spindles. After the anaphase, the remnant of the spindle spanning the space between the two daughter nuclei retains a weak birefringence throughout the cleavage process. The double refraction of all these components of the spindle is positive in the direction of its length.

3. In the eggs of the echinids, *Clypeaster japonicus, Mespilia globulus* and *Strongylocentrotus pulcherrimus,* the fertilization membrane and the hyaline plasma layer have birefringence which is positive in the direction of their tangents. The cortical layer in the eggs of these sea-urchins as well as that of *Spirocodon saltatrix* is negatively birefringent in the direction of the tangent.

4. In sea-urchin eggs, the rays of the monaster are positively birefringent, while those of the streak stage are negatively birefringent in the longitudinal direction of the streak. Within the negative streak, the positively birefringent spindle

appears. The rays of the diasters are positively birefringent in the longitudinal direction.

5. During cleavage, the rays crossing in the equatorial plane acquire a more positive birefringence than the non-crossing rays, while the polar rays become either isotropic or negatively birefringent. The rays visible in the end view of the spindle are positively birefringent in the beginning, become isotropic in the middle and acquire negative birefringence during the end of the cleavage process. This is true with sea-urchin as well as medusan eggs. In a hypertonic medium, all the rays within the cell change to become negatively birefringent.

6. An explanation is offered for the change in the sign of astral birefringence; viz., the astral birefringence increases in positivity when the rays are put under tension, while a negativity is produced when the rays are under compression.

7. The above idea was tested experimentally in rubber, celluloid, gelatin (4%) and the jelly of the fertilized eggs of an annelid, *Perinereis sp.,* all with confirmatory results.

8. The structure of the aster, as revealed by birefringence studies, is discussed.

APPENDIX

It was stated in the main part of this paper that when the objective lens and the condenser are properly adjusted, between a pair of Nicols, a dark cross is seen through a telescope inserted in place of the ocular lens. This fact means that it is impossible to cut off all the light; even under the best conditions stray light comes in from between the arms of the cross.

On the other hand, it was noticed that when the best adjustment is reached, the detection of extremely weak birefringence becomes more difficult than when the condenser is slightly turned away from the best adjusted position. The authors think that this is due to the increased proportion of stray light compared to the intensity of the birefringent object.

Therefore, the situation was analyzed, taking into consideration the sensitivity of the eye, and the strength of the bias-birefringence which is necessary to obtain the most sensitive detection of extremely weak birefringence was calculated.

The intensity of light, I, which passes through a birefringent object (its retardation being R) inserted in a diagonal position between a pair of crossed Nicols is

$$I = I_o \sin^2 \frac{R}{2}$$

where I_0 is the intensity of light which would pass the Nicols when they are set parallel.

When the retardation, that is, the amount of birefringence expressed in radians, of the object changes from R to $R + \Delta R$, the intensity of light changes from I to $I + \Delta I$. According to Weber-Fechner's law, in order that the eye be able to detect a change in intensity, $\frac{I + \Delta I}{I}$ or $\frac{\Delta I}{I}$ must reach a sufficient value. In other words, the ratio $\frac{\Delta I}{I}$ is the factor that decides the visual sensation, and not the absolute value I. Therefore, the change of $\frac{\Delta I}{I}$ when R is changed, or $\frac{dI}{I}/dR$ will be considered.

Considering that $\sin \Theta = \Theta$ when Θ in radians has a sufficiently small value,

$$I = I_o \sin^2 \frac{R}{2} = I_o \frac{R^2}{4}$$

$$\frac{dI}{I} \Big/ dR = \frac{dI}{dR \cdot I} = \frac{d(I_o \frac{R^2}{4})}{dR\,(I_o \frac{R^2}{4})} = \frac{\frac{I_o R}{2}}{\frac{I_o R^2}{4}} = \frac{2}{R}$$

The gain in visual sensation dS resulting from change dR would be

$$\frac{dS}{dR} = K \frac{dI}{I} \Big/ dR = \frac{2K}{R} \qquad \text{K being a constant.}$$

The above equation represents a case in which all the light entering the eye originates from the retardation R. But actually, some stray light with the intensity I' also enters the eye.

Then,

$$\frac{dS}{dR} = K \frac{dI}{(I + I')} \Big/ dR = \frac{K \cdot dI}{dR \cdot (I + I')}$$

Putting $I' = nI_o$, n being the proportion of the stray light against I_o, assuming that I' does not change with R, then

$$\frac{dS}{dR} = \frac{K \cdot I_o R/2}{\left(\frac{R^2 I_o}{4} + nI_o\right)} = \frac{2KR}{R^2 + 4n}$$

This is a fundamental equation for the present situation. The curve for $K = \text{const.}$ is represented in figure 6. When K is constant — that is, as long as Weber-Fechner's law holds, the maximum sensitivity or $\frac{dS}{dR}$ max. would be attained when $R^2 = 4n$. In other words, when a bias birefringence is inserted, the strength of which is defined as above, the maximum sensitivity is attained. Another definition is: the intensity of light resulting from the above retardation R is

$$I_R = \frac{I_o R^2}{4} = \frac{I_o 4n}{4} = nI_o$$

which is identical with I′. This means that a birefringence with a retardation R which produces the same amount of light as the intensity of the stray light, should be superimposed in order to obtain the maximum visual sensitivity.

As for the maximum value of $\frac{dS}{dR}$, or the height of the peak of the visual sensitivity curve of figure 6,

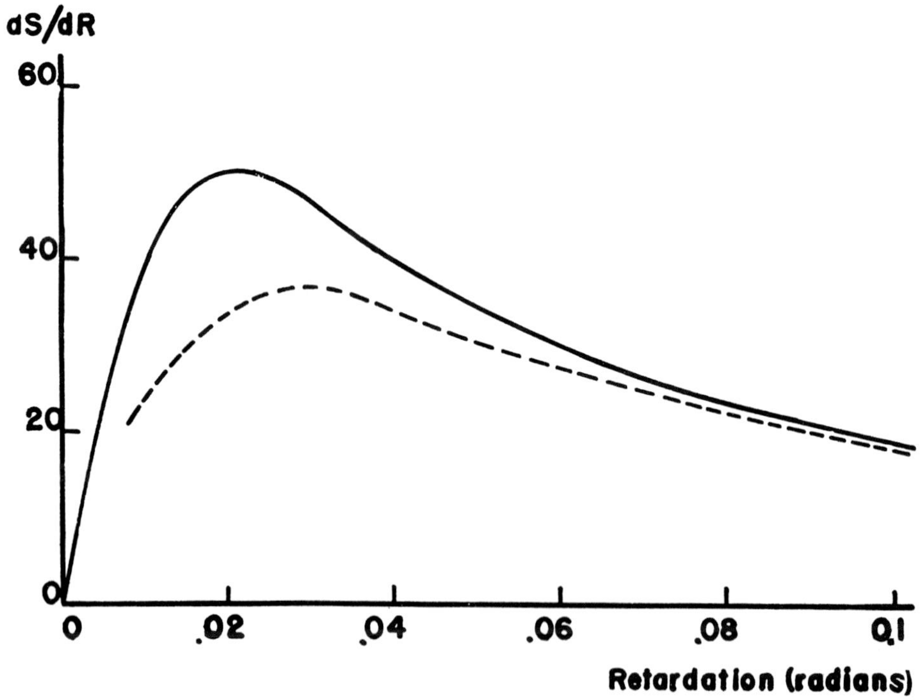

Fig. 6 Sensitivity (dS/dR) for detecting a small retardation, plotted against the retardation of the compensator. The formula for the curve was calculated from the intensity formula for a birefringent object, and Weber Fechners' Law.

$$\frac{dS}{dR}\text{ max.} = \frac{2K \cdot 2\sqrt{n}}{4n + 4n} = \frac{K}{2\sqrt{n}}$$

In order to make the peak higher, K should be made as large as possible, while n should be made as small as possible.

To make n small means to reduce the proportion of stray light. Actual methods for reducing stray light were described in the main part of this paper and it was also pointed out that, in practical cases, there is a theoretical lower limit for n.

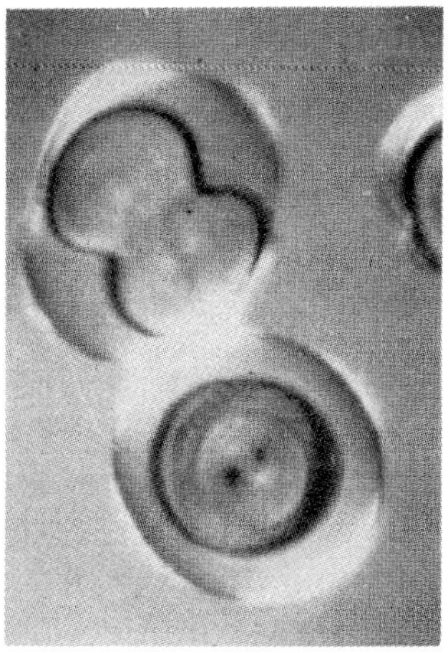

Fig. 7 Reversal of the sign of birefringence in the polar view of the aster during cleavage. Notice that the dark quadrants of the cross are lying perpendicular to the dark portions of the fertilization membrane.

K gives the coefficient of the visual acuity of the eye. The curve in figure 6 was drawn on the assumption that K does not change in the range of the curve. But it is known that this coefficient does change and becomes smaller when the size of the object diminishes and also when the intensity of light falls below a certain limit (Blackwell, '46). In the last-mentioned case, K becomes smaller as R is reduced. This, of course, would change the shape of the curve. Unfortunately, no calculation is possible beyond this point. Qualitatively speaking a decrease in the value of K would shift the curve to a lower level, as indicated by the dotted line. In some cases, this effect will be so large as to shift the curve toward the larger value of R. Nevertheless, it can be expected that the general trend of the curve will remain the same, and the curve will still have a maximum somewhere.

In short, the conclusion follows that when there is stray light a biasing retardation should be used for the detection of very weak birefringence in order to obtain maximum sensitivity. Moreover, when higher or the highest magnification is used, the fall of the sensitivity of the eye comes in as a major controlling factor. For this reason, the more powerful the light source, the better. It was for this reason that a super high pressure mercury lamp was used. It is also very important to facilitate a good adaptation of the eye by cutting down the illumination of the room.

Since the completion of this manuscript, Swann and Mitchison (J. Exp. Biol., *27:* 226–237, '50) have also published a paper describing improvements in polarization microscopy for biological purposes.

REFERENCES

BECKER, W. A. 1939 Struktur und Doppelbrechung der Chromosomen. Arch. f. exp. Zellforsch., *22:* 196.

BLACKWELL, R. H. 1946 Contrast threshold of the human eye. J. Optical Soc. Am., *36:* 624.

CHAMBERS, R. 1917 Microdissection studies. II. The cell aster: A reversible gelation phenomenon. J. Exp. Zool., *23:* 483.

DAN, K., T. YANAGITA AND M. SUGIYAMA 1937 Behavior of the cell surface during cleavage. I. Protoplasma, *28:* 66.

DAN, K. 1943 On the mechanism of cell division. J. Facult. Sci. Tokyo Imp. Univ., sec. IV. *6:* 323.

DAN, K., AND J. C. DAN 1947 Behavior of the cell surface during cleavage. VIII. On the cleavage of medusan eggs. Biol. Bull., *93:* 163.

GRAY, J. 1931 Experimental Cytology. Cambridge Univ. Press.

HUGHES, A. F., AND M. M. SWANN 1948 Anaphase movements in the living cell. A study with phase contrast and polarized light on chick tissue cultures. J. Exp. Biol., *25:* 45.

INOUÉ, S. 1949 Studies of the Nereis egg jelly with the polarization microscope. Biol. Bull., *97:* 258.

KUBO, R. 1947 Statistical theory of linear polymers. III. Double refraction. J. Phys. Soc. Japan, *2:* 84.

MONNÉ, L. 1942 Polarizationoptische Analyse des Zytoplasmas der Spermatozyten von Lithobius forficatus L. Arkiv. f. Zool., *34B,* Heft 1, No. 1.

——— 1944a Cytoplasmic structure and cleavage pattern of the sea urchin egg. Ibid., *35A,* Heft 3, No. 13.

——— 1944b The induced birefringence of the jelly coat of the sea urchin eggs. Ibid., *35B,* Heft 2, No. 3.

——— 1945 Investigations into the structure of the cytoplasm. Ibid., *36A,* Heft 4, No. 23.

——— 1947 Some observations on the polar and dorsoventral organization of the sea-urchin egg. Ibid., *38A,* Heft 3, No. 15.

MONNÉ, L., AND E. WICKLUND 1947 The influence of merthiolate on the eggs of the sea-urchin Psammechinus miliaris. Ibid., *39A,* Heft 2, No. 4.

MONROY, A., AND G. MONTALENTI 1947 Variations of the submicroscopic structure of the cortical layer of fertilized and parthenogenetic sea urchin eggs. Biol. Bull., *92:* 151.

MOTOMURA, I. 1941 Materials of the fertilization membrane in the eggs of echinoderms. Sci. Rep. Tohoku Imp. Univ. ser. IV., *16:* 345.

RUNNSTRÖM, J. 1928 Die Veränderungen der Plasmakolloide bei der Entwicklungserregung des Seeigeleies. Protoplasma, *4:* 388.

———— 1948 Further studies on the formation of the fertilization membrane in the sea-urchin egg. Ark. f. Zool., *40A*, Heft 1, No. 1.

RUNNSTRÖM, J., L. MONNÉ AND L. BROMAN 1944 On some properties of the surface layers in the sea urchin egg and their changes upon activation. Ibid., *35A*, Heft 1, No. 3.

SCHMIDT, W. J. 1936 Doppelbrechung von Chromosomen und Kernspindel und ihre Bedeutung für das Kausale Verständnis der Mitose. Arch. f. exp. Zellforsch., *19:* 352.

———— 1937 Die Doppelbrechung von Karyoplasma, Zytoplasma und Metaplasma. Protoplasma Monographien, *11.* Berlin.

———— 1939 Doppelbrechung der Kernspindel und Zugfasertheorie der Chromosomenbewegung. Chromosoma, *1:* 253.

SCHRADER, F. 1944 Mitosis. Columbia University Press.

PLATE 1

EXPLANATION OF FIGURES

8 Birefringence of the spindle and aster of the eggs of the medusa, *Spirocodon saltatrix* as observed with a low power lens. Double refraction of both structures decreases gradually and becomes hardly noticeable by the time cleavage sets in (the uppermost cell). Positive crosses of asters become visible after elimination of the brightness of the spindle by setting the structure parallel to an axis of the Nicol.

9 Double refraction figure of a dividing egg of *Spirocodon* (high dry lens). The two ends of the spindle which shine strongly represent the "traction" fibers. The shining end parts are connected by birefringent threads. Double refraction of the cell surface is particularly clear along the contact surface between the two blastomeres. Birefringence of the spindle is positive in the direction of its length while that of the surface is negative in the direction of its tangent.

BIREFRINGENCE OF DIVIDING CELL
SHINYA INOUÉ AND KATSUMA DAN

PLATE 1

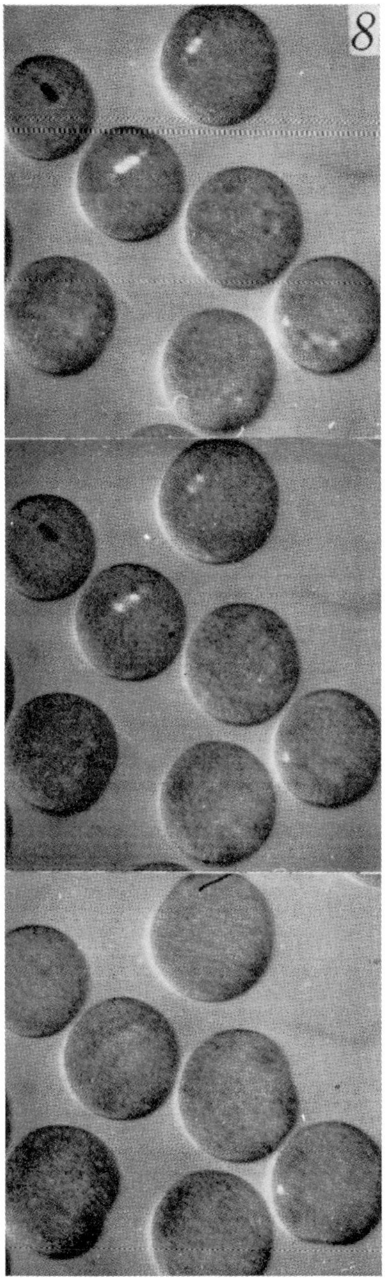

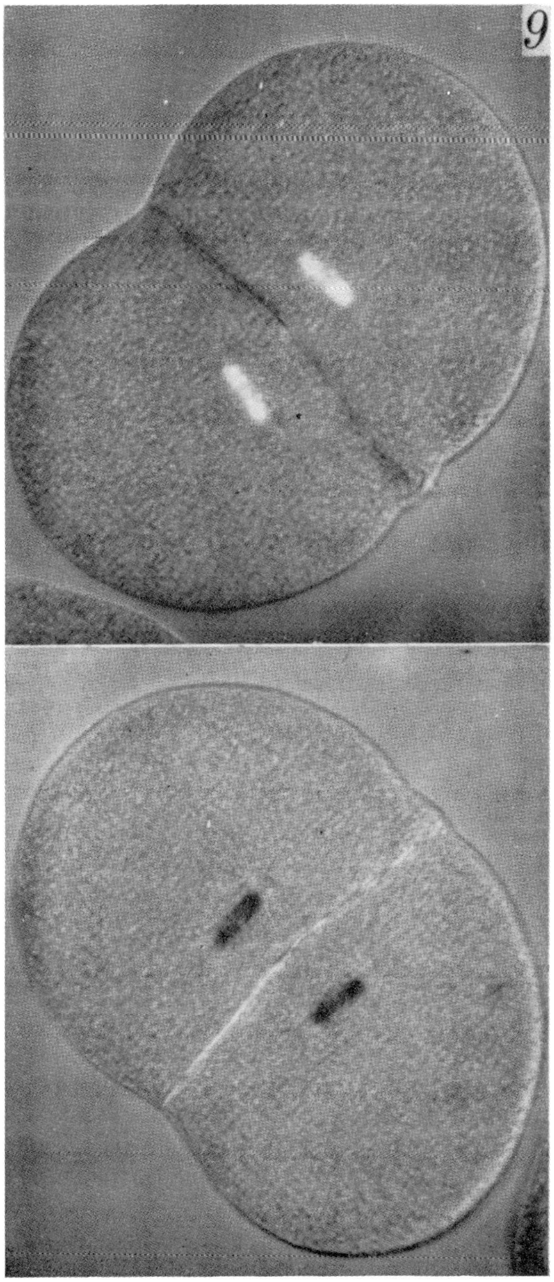

PLATE 2

EXPLANATION OF FIGURES

10 Double refraction of the spindle remnant and the crossing rays of a *Clypeaster japonicus* egg. Birefringence of the fertilization membrane is also shown. The fact that the dark spindle remnant lies parallel to the dark portion of the fertilization membrane indicates that both have the same sign of birefringence which is positive in that direction. (Cf. fig. 5, p. 20.)

11 Four-cell stage of *Spirocodon* with spindles lying in all possible directions. In the upper and the lower blastomeres the spindles are perpendicular to each other, one being bright and the other, dark. In the cell at the right the spindle is not visible because of its intermediate position; as a result, the two asters are clearly apparent. The spindle of the blastomere at the left is perpendicular to the plane of the photograph, and in consequence, a polar view of one aster shows as a positive cross in the center of the cell and a negative cross in the periphery. (See fig. 3, p. 12.)

12 Streak stage of the egg of *Clypeaster japonicus*. Notice the bending of the negatively birefringent rays and the negativity of the cortical birefringence.

13 Appearance of positively birefringent spindles within negatively birefringent streaks and later disappearance of the streaks as the mitotic apparatus increases in size.

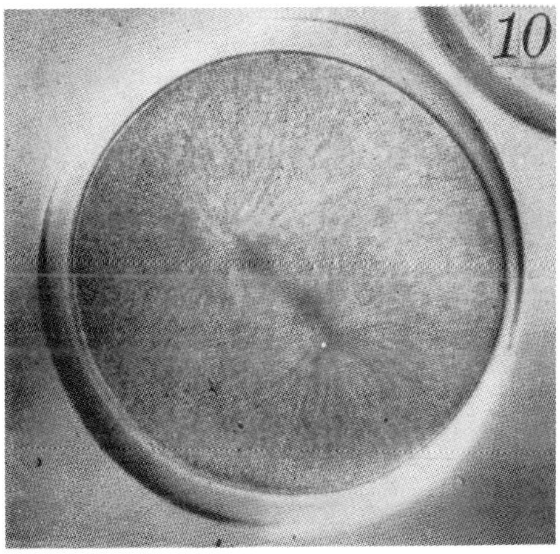
10

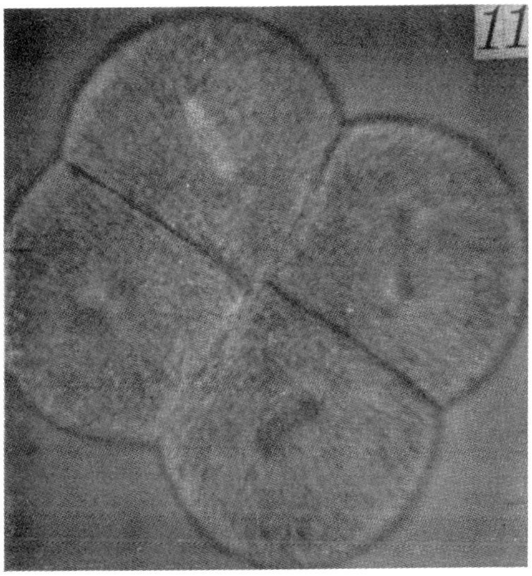
11

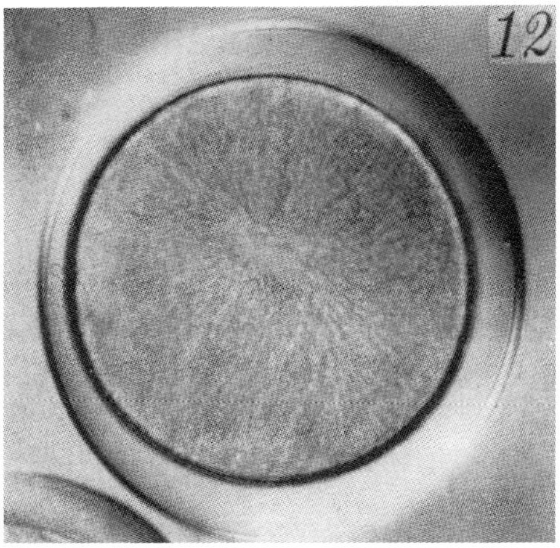
12

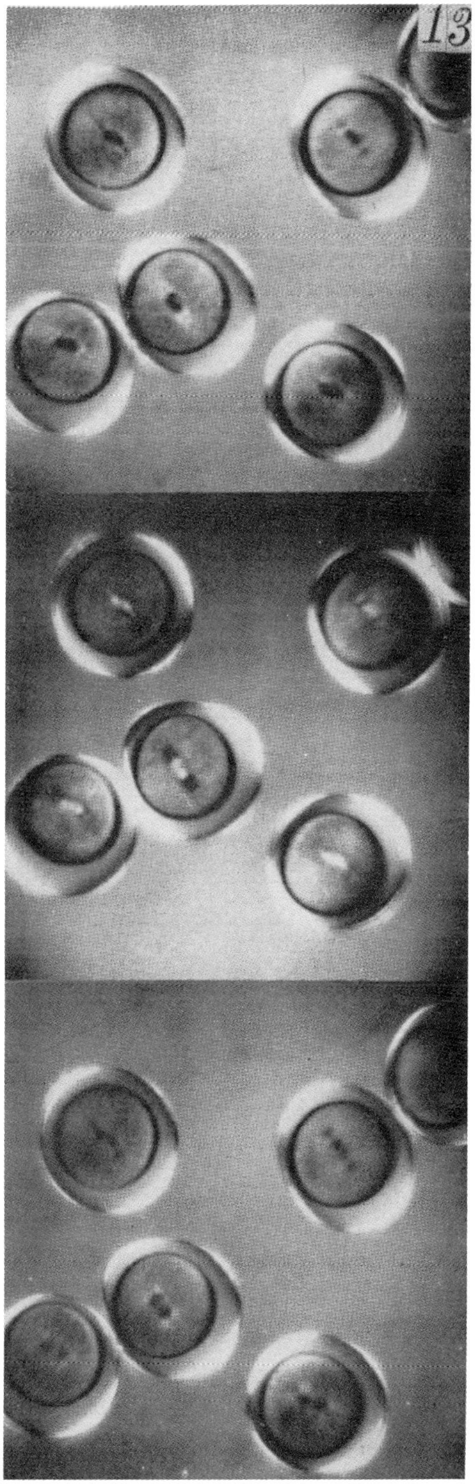
13

The following note was added by Shinya Inoué in September of 2006:

The work described in Article 5 was carried out in Professor Katsuma Dan's laboratory at the Misaki Marine Biological Station during 1947 and 1948. The station was returned by the US occupation forces to the Tokyo Imperial University in December 1945, prompted by Katy's[1] plea "The Last One to Go" (now prominently displayed in the MBL Library; see also Article 56 in this *Collected Works*).

In those days, shortly after the end of WW-II, Katy was pursuing his proposal that the cleavage furrow was induced by astral rays whose centers were pushed apart by the elongating anaphase spindle. In Katy's lab, Yoshiyuki Endo was busy analyzing the events of fertilization envelope elevation in sea urchin eggs, and Kayo Okazaki was expertly growing fragmented sea urchin blastomeres to establish their developmental capabilities including skeletal spicule formation. My collaboration with Katy was to improve on the inconclusive attempt (under the air raid black out curtains at Katy's home in 1942) to repeat W.J. Schmidt's demonstration of the mitotic spindle and aster birefringence in living sea urchin eggs.

By building the microscope shown in "Development of the 'Shinya Scopes'" (p. 970) Fig. 1, and using marine eggs available in Misaki, we were able to improve on Schmidt's images and, as described in the current article, follow the changing birefringence and morphology of the spindle and asters during the cleavage cycle and also when they are exposed to hypertonic media. The same microscope was also used extensively by Kayo for observing the biocrystalline skeletal spicules growing in sea urchin embryos, the extension of which resulted in the work described in Article 29.

Initially, I had arranged parts of the microscope on stacks of books on the bench top to align their optical axes. However, I assembled them into the more stable configuration as shown in Fig. 1, when Katy asked me to demonstrate the twinkling birefringence of the spicules in the swimming plutei to the Imperial family who came for their first post WW-II visit to the Misaki Laboratory. After my departure from Princeton in 1948, Kayo and Katy continued to use the scope, which they nicknamed the "Shinya Scope."

This article was written several years before the fine structure of cells (such as the presence of microtubules and endoplasmic reticulum) were discovered, and by assuming that the asters were gel spheres supported by astral rays. Thus, the changes in the sign (and strength) of birefringence of the astral rays, observed by exposure of the egg to hypertonic media (and with progression of the mitotic stages), were interpreted as being due to photo-elastic effects that reflected the compressive or stretching mechanical forces on those rays [cf. Inoué S, *Biol Bull* **97** (2): 258–259, 1949].

In retrospect, however, can we really assume that the astral rays, today interpreted to be radiating microtubules, would behave similarly to molecules in a gel? Or, more fundamentally, are the changes in the sign of birefringence not, as assumed in this article, a reflection of photo-elastic property of the astral rays themselves, but rather reflect some other ordered cytoplasmic component with opposite sign of birefringence, such as aligned membranes? If so, how do they become aligned, e.g., by exposure of the cell to hypertonic media, or by compression and elongation of cell parts (see Articles 65 and 67)? These points still remain unclear after all these years.

[1] "Katy" is a frivolous nickname that Katsuma Dan gave himself when asked by a female student what his initial K stood for. In the background was the ready confusion between K. Dan and Dan Mazia, Katy's graduate-school classmate and good friend at the University Pennsylvania and at the MBL, Woods Hole.

INTRODUCTION TO DOCTORAL THESIS

Shinya Inoué*

The following six articles cover my doctoral dissertation "Studies of the Structure of the Mitotic Spindle in Living Cells with an Improved Polarization Microscope" presented to the Department of Biology at Princeton University and accepted in May 1951. These reflect the work that I carried out as a graduate student at Princeton University, and summers at the Marine Biological Laboratory, Woods Hole, between September 1948 (when I arrived from Tokyo, Japan) and May 1951.

In this *Collected Works*, Articles 7 and 12 are essentially verbatim copies of Parts I and VI of my thesis, whereas 8 to 11 are copies of articles published in scientific journals based on Parts II through V of my thesis. Rather than list these articles chronologically based on dates of publication in the journals, the articles follow the sequence as they appear in my PhD thesis.

In the light of all that one has learned (and is continuing to learn) about mitosis and microtubule dynamics in the last half century, I find it amusing to re-read Part VI of my thesis which I did not publish. Dr. Franz Schrader of Columbia University, who was Kenneth Cooper's mentor, was very unhappy that I would not publish this part "because I felt that the arguments were speculative," but perhaps he had more perspective on the reasons for publication than I, who was then still quite young.

On the other hand, in re-reading Part I, Introduction, to the thesis, I am struck by how logical the whole process of the thesis project appears to be as it is presented there. However, that must have been the way I presented the aim, means and findings in the thesis in hindsight after I had finished the project. As I emphasized in the foreword to this *Collected Works*, I really did not know what I was going to find until I started looking at the living cells after improving the microscope, so that I too had presented myself as having an insight or foresight that, in fact, I never had. A hunch, yes, but not an insight or foresight! I wonder how much such misrepresentation of research dampens how people see science, or the way they perceive how scientists work?

In addition to those at Princeton whom I thank in the following acknowledgment in the thesis, I wish to record my special gratitude to Professor Katsuma Dan and Dr. Jean Clark Dan who inspired me to become a biologist and provided the opportunity for me to come to the US so soon after the second world war. As mentioned in the foreword to this *Collected Works*, the episodes surrounding those events, as well as narrations of my more general life before, during, and after my Princeton days, are being prepared for another publication, "Through Yet Another Eye."

The following are copies of the title page, table of contents, and the acknowledgment of the thesis:

*Unpublished work (2006).

STUDIES OF THE STRUCTURE OF THE MITOTIC SPINDLE IN LIVING CELLS WITH AN IMPROVED POLARIZATION MICROSCOPE

by

Shinya Inoué

A DISSERTATION

Presented to the

Faculty of Princeton University

in Candidacy for the Degree

of Doctor of Philosophy

Recommended for Acceptance by the

Department of Biology

May 1951

TABLE OF CONTENTS

ACKNOWLEDGMENTS

I wish to express my sincere thanks to the members of the faculty of the Department of Biology, Princeton University, for providing me with three years of fellowship, during which time I was able to accomplish the research described in this thesis. Especially to the very thoughtful guidance and continuous encouragement of Professor K.W. Cooper as well as his allowing me to use freely his laboratory facilities and special optical instruments, I feel deeply indebted. I am also very grateful to Professor E. Newton Harvey for the use of his super-high pressure mercury arc lamp, and to Professor A.G. Shenstone of the Physics Department, who allowed me to use his large Glan prism. Mr. Russell Mycock, who patiently and skillfully worked on the machining of the mechanical parts of the new polarization microscope, and Messrs. G.C. Crebbin and J. Benford of the Bausch and Lomb Optical Company, who provided me with specially prepared Glan-Thompson prisms and strain-free coated objectives, were people whose special services and technical assistances were indispensable. Lastly, I wish to thank my classmate Mr. J.W. Hastings for his very helpful collaboration in measuring the extinction factor of the microscope.

PART I OF DOCTORAL THESIS

Shinya Inoué*

1. Introduction

When one attempts to study a complicated physicochemical structure like a living cell, the first thing one would like to do is to know its structure. Actually, however, this is not a very easy thing to do because the structures in a living cell, which are believed to be more intimately associated with the vital processes of the cell, usually have a refractive index and absorption not very different from that of the general cytoplasm in which they are embedded. This makes the microscopical observation of the structures difficult or even impossible. It is especially true in the case of the mitotic spindle, which can be shown to possess a very important role in the division of cells and the separation of chromosomes.

Because of the importance of the function of this structure, several attempts have been made in the past to reveal the detailed structure of the spindle in a living cell while it is normally functioning and multiplying. Many such attempts failed, but in the two notable cases mentioned below, as well as in the excellent studies of Bělař, spindle structures have actually been demonstrated.

The first of these two employs special biological material in which the spindle fibers do show up in normal living cells in ordinary light. One such case described is in the blastomeres of the eggs of the grass mite, *Pediculopsis*, by Cooper[2] and another in the parasitic flagellates studied by Cleveland.[1] The structures of the spindles studied in such material, however, were considered by many biologists to be atypical, and so these observations were not given as much credit as perhaps they should have deserved.

The other case in which structure may be seen in a living spindle has been described by Schmidt[5] and by Runnström.[4] They found that the spindle of a living cell shines, although weakly, in the dark field of the polarization microscope when the Nicol prisms are crossed. Schmidt observed that an isotropic gap appeared between the two half spindles as the chromosomes separated at anaphase. Further, he observed that the birefringence of the half spindles, just as in the case of contracting muscle fibers, decreased as they shortened. He ascribed this decrease in birefringence to a folding of polypeptide chains, thus supporting the traction fiber theory of chromosome movements.[6]

Now Schmidt had undertaken the investigation of the spindle with a new method of great potential. Although the system he used had not been sufficiently sensitive to enable detection of spindle fibers if they do in fact exist, nevertheless the use of an ideal

*Unpublished work (1951).

polarization microscope should not only allow demonstration of structures which may be completely invisible in ordinary light, but should also supply information concerning the submicroscopic or micellar organization within the object. Despite the promise of Schmidt's new approach, there has actually been very little progress in this field for the past fifteen years.

The limitations of current polarization microscopes became painfully evident while studying the birefringence of spindle and asters in dividing echinoderm eggs with Professor Katsuma Dan at the Misaki Marine Biological Station. With such polarization microscopes, it is possible to determine only the existence of the spindle and astral rays, the approximate size and shape of their structures, and their signs of birefringence. It was apparent that the polarization microscope could, in principle at least, be improved to the point where details of birefringent structures could be resolved and studied in the living cell.

The main problems to be solved were: 1) the improvement of polarization optics to the point that weak birefringence (e.g., to about 0.1 millimicron) could be detected at sufficiently high apertures to guarantee good resolution; and 2) the invention of a method making possible direct measurements in living cells of very small retardations of objects whose cross sections measure only a micron or less.

The solution of these problems and subsequent analysis of the living spindle with the improved polarization microscope have comprised my doctoral research at Princeton University.

The first problem was solved following analysis of various causes of stray light in the polarization microscope, for I had previously shown that the stray light was the limiting factor for detecting small retardations.[3] Among the various sources of stray light, the only one which could not be remedied by mere use of better material and design was shown to be light which enters the microscope through rotation of the plane of polarization at the lens surfaces. The proof of this point will be presented as Part II of this thesis [Article 8]. The second problem was solved by developing a new "double half shade method" whose principles and use will be described in Part III [Article 9].

With the analysis and the new design based thereupon, and with the development of the measuring device, it has been possible to construct a greatly improved instrument. As had been anticipated, this new polarization microscope is sufficiently sensitive to make possible the precise analysis of cellular structures unresolvable or even undemonstrable by other instruments.

The photographs of the new microscope are shown in Figs. 1 and 2, and a diagram of its optical arrangement is given in Fig. 3. As shown in Fig. 4, an extinction factor as high as 10,000 at an aperture of 0.5 is attainable with this microscope. This allows the detection of retardations down to a tenth of a millimicron, with a resolution of 0.3 micron or less. Such small retardations can be directly measured by the "double half shade method" in objects as small as a micron or less in diameter [Article 9].

In Part IV, the structures of animal and plant spindles studied with this microscope are reported, and the reality of spindle fibers discussed [Article 10].

In Part V, the results of some experiments on the action of colchicine upon the metaphase spindle of the *Chaetopterus* oocyte are described [Article 11]. Lastly, in Part VI [Article 12], the submicroscopic organization of the spindle will be discussed on the basis of the observations and experiments described in Parts IV and V.

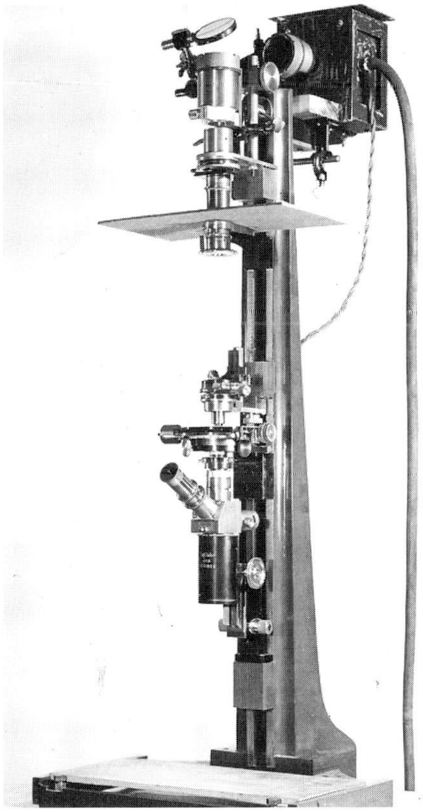

Fig. 1. The new polarization microscope. See Figs. 2 and 3 for details of the optical arrangement.

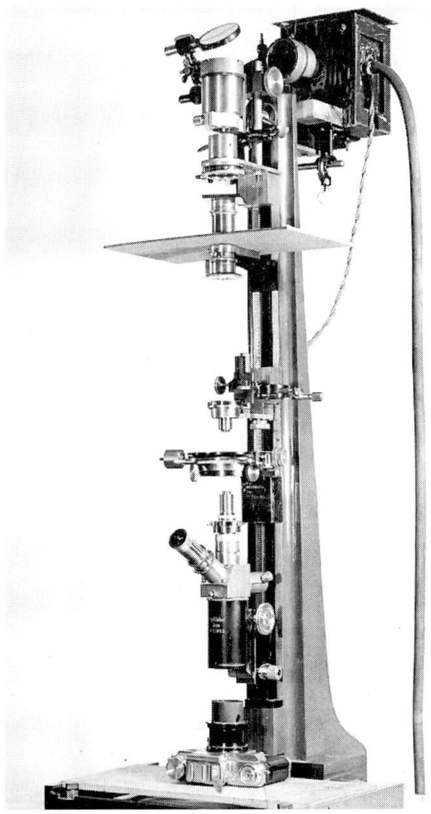

Fig. 2. Same as Fig. 1. The instrument is exploded in order to show the various parts more clearly.

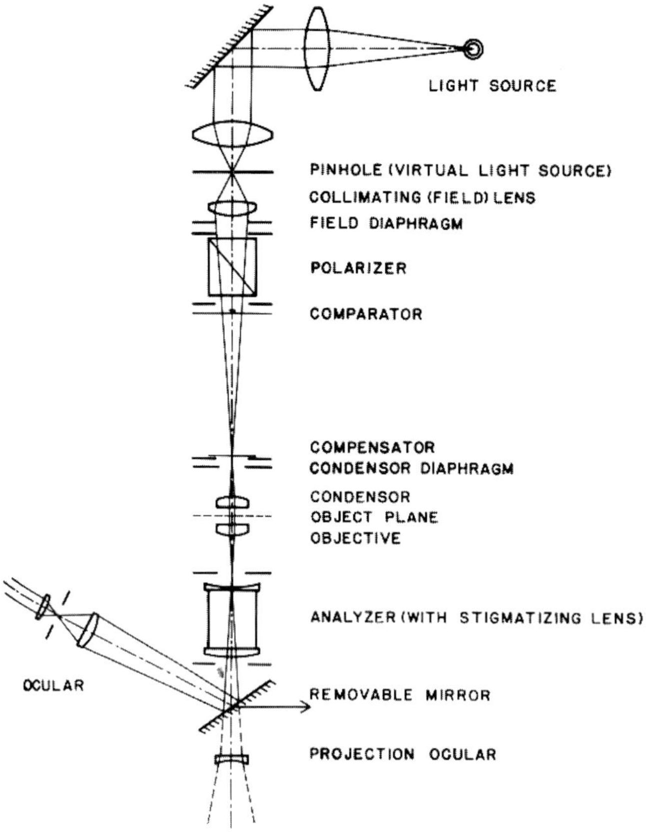

Fig. 3. Optical arrangement of the improved polarization microscope.

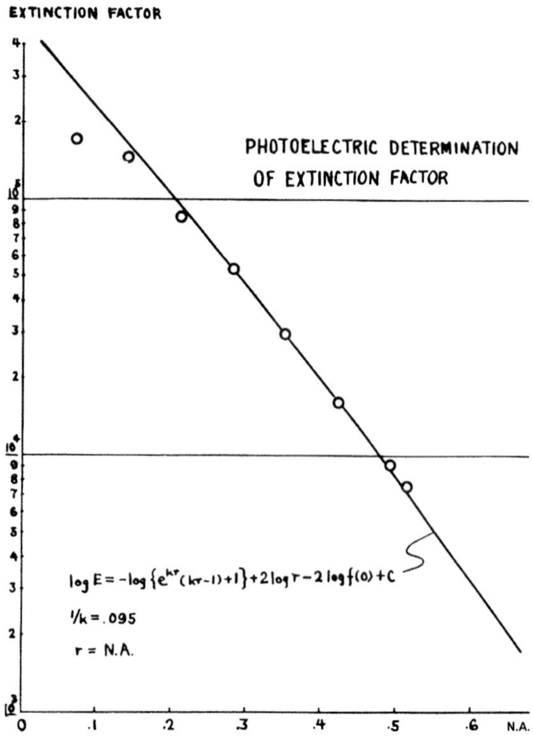

Fig. 4. Photoelectric determination of extinction factor.

References

1. Cleveland LR, *Mem Am Ac Arts Sci* **17**: 309, 1934.
2. Cooper KW, *Proc Nat Acad Sci* **27**: 480, 1941.
3. Inoué S, Dan K, *Kagaku* **19**: 111, 1949. [See also Article 5.]
4. Runnström J, *Protopl* **5**: 218, 1936.
5. Schmidt WJ, *Naturw.* **24**: 463, 1936.
6. Schmidt WJ, *Chromosoma* **1**: 253, 1939.

Reprinted from *Experimental Cell Research*, Vol. 3(1), pp. 199–208, 1952, with permission from Elsevier.

STUDIES ON DEPOLARIZATION OF LIGHT AT MICROSCOPE LENS SURFACES

I. THE ORIGIN OF STRAY LIGHT BY ROTATION AT THE LENS SURFACES[1]

SHINYA INOUÉ[2]

Department of Biology, Princeton University, Princeton, N. J.

Received June 15, 1951

It is commonly found that only low aperture lenses can be used with the polarization microscope for measuring or detecting small retardations. With low aperture, the field of the microscope can be made very dark with crossed nicols, even with a very intense light source. Under this condition a weakly birefringent object will shine brightly against a dark background.

If, on the other hand, a wide aperture objective is used and the condenser diaphragm is not stopped down, the field no longer remains dark. In this case a birefringent object does not increase its brightness proportionately with the brightness of the field, and so the contrast between the object and its background decreases. Consequently, if the retardation is not large, the birefringent image is obliterated, or nearly so, at higher apertures. For biologists who attempt to study the fine structure of minute cell organelles, this is a very distressing limitation. A high aperture is necessary for good resolution, and yet at the same time the high sensitivity required for detecting weak birefringence can only be attained at low apertures.

Inoué and Dan (2, 3) suggested that the stray light drowning the birefringent image at high apertures arises by depolarization at lens surfaces. This depolarization gives rise to the so called "polarization cross" (fig. 1), observable at the exit pupil of the objective when the ocular is removed, or with a Bertrand lens. The cross can be seen without any object on the stage, and becomes clearer the more intense the light source, and the less defective the optical system.

---

[1] Presented to the faculty of Princeton University in partial fulfilment of the requirements for the degree of Doctor of Philosophy. This research was supported in part by funds of the Eugene Higgins Trust allocated to Princeton University.

[2] Present address: Department of Anatomy, School of Medicine, University of Washington, Seattle 5, Wash., USA.

The existence of the polarization cross means that the extinction is nearly perfect only along the arms of the cross, which are parallel and perpendicular to the axes of the polarizer and the analyzer. Stray light enters into the system from the bright regions between the dark arms.

Now, the polarization cross is similar in appearance to the interference figures of monaxial crystals, but is caused in an entirely different manner. Rinne (5) first explained the cross to be caused by rotation of the plane of polarization at lens surfaces. Each of the rays which hits the lenses obliquely is considered to be separated into two vectors at every glass surface, one parallel and the other perpendicular to the plane of incidence. The two vectors are reflected at different rates, and so the two transmitted vectors no longer have the same ratio as the two entrant vectors. The combined transmitted vector, therefore, has a different direction compared to the original entrant vector (fig. 2). This results in a rotation, whose direction in each quadrant of the cross is determinate, since the vector perpendicular to the plane of incidence always suffers the greater reflection. The amount of rotation is greater the greater the angle of incidence, and the greater the angle between the plane of polarization and the plane of incidence.

Rinne's explanation was first confirmed and amplified by Césaro (1) and later by Wright (6, 7, 8). Using Fresnel's formulae both Césaro and Wright calculated the expected rotation for rays passing through tilted glass plates, as well as the amount of rotation at different points of the exit pupil of an idealized objective lens. Their calculations agree well with direct measurements of rotation, but Wright and others have suggested that rotation may not be the only cause of depolarization at high apertures. Therefore, a formula for the total amount of light introduced by rotation alone within given apertures has been developed by the present author to test this suggestion. The new formula will be shown to agree well with the amount of stray light actually measured at various apertures photo-electrically, and so the other possible factors considered by Wright to cause depolarization must be quite insignificant. The argument is as follows.

Between crossed nicols, the amount of light L introduced by a transparent plate whose area is A, and which has a rotation of Θ is

$$L = AI_0 \sin^2 \Theta \quad \ldots\ldots\ldots\ldots\ldots \quad [1]$$

where I_0 is the intensity (per unit area) of light that passes through the system when the nicols are parallel and when there is no rotation.

Knowing the amount of rotation of rays which pass through the different points of the objective exit pupil, it is possible to calculate from equation [1], the amount of stray light which would enter into the system by rotation alone within any given aperture.

Since the formulae given by Césaro and Wright are not suited for integration, approximate formulae have been developed as shown below.

CALCULATION OF THE ROTATION OF RAYS PASSING THROUGH DIFFERENT POINTS OF THE EXIT PUPIL

In the following discussions it will be assumed that "critical" (namely, Kohler's) illumination is being used. It is also assumed that the field diaphragm (Df) is made so small that all of the rays emerging from the condenser (Lc) converge into a small point at the center of the object plane (Po). Under this condition, the path of light through the microscope can be simply expressed as in fig. 3.

It is clear from fig. 3 that, if the nicols are not crossed, a solid cone of light would appear as a bright disc at the exit pupil (D_0) of the objective. The radius (r) of this disc has been shown by Mallard (4) and later by Wright (7, p. 148), to be closely proportional to the sine of the half angle Θ of the cone. Since sin Θ is the angular aperture, by studying the optical phenomena at the exit pupil of the objective, it is possible to determine what happens to each of the rays of light passing through the object plane at different angles.

Mallard's relation has been tested in the present optical system by taking two objectives with different apertures and using them alternatively in place of the condenser. The angles of the cones of light which were formed by these lenses were measured carefully and plotted against the diameter of the bright disc at the objective exit pupil. As shown in fig. 4, the agreement with Mallard's relation is very close.

The Mallard relation can be expressed graphically as in fig. 5. Here, all the beams of light pass through the center of the sphere *O*, and change their direction at its surface *O-MPA*, so that the resultant beam (*CG*) is approximately parallel to the axis of the system *OM*.

If we consider a beam *OC* which makes an angle Θ with the axis *OM* and intercepts the sphere at *C*, and draw a vertical line *CI* from *C* to plane *OPA*, calling the intercept *I*, then in triangle *COI*,

$$< CIO = < R = < MOI \quad \text{and so} \quad < OCI = < MOC = \Theta.$$

And so $OI = OC \sin \Theta$.

Therefore, the sphere *O-MPA* can be used to represent the exit pupil of the objective, while its center *O* would be the center of the object plane.

If the sphere were to be observed from an infinite distance in the direction OM, the radius of a point C on the sphere, or at the exit pupil, would be equal to the projection (OI) of MC on to plane OPA, and hence equal to $OC \cdot \sin \Theta$.

In this way, rays passing through the object plane at angles between Θ and $\Theta + d\Theta$ will emerge at the exit pupil as a ring whose radius is r to $r + dr_\theta$. The amount of rotation for rays emerging at each point along this ring shall first be considered.

In fig. 5, let POP' and AOA' include the planes of polarization of the polarizer and the analyzer respectively. The two simultaneously are at right angles to each other and to OM.

All of the optical surfaces present in a microscope are spherical and have their centers along OM, and so for a beam of light passing through O and emerging at C, the plane of incidence, refraction and reflection at every surface is always $MOBC$. The angle Φ between OP and $MOBC$ is the azimuth angle for the entrant plane polarized light. Even though the change of the plane of polarization actually occurs at several different surfaces in a microscope, it can be considered to take place in one step at point C.

In figs. 5 and 6, let CD be the vector of the plane polarized light emerging from the polarizer. This is separated into CE and CF at C; CE being perpendicular to the plane of incidence $MOBC$, and CF being in the plane.

Let the reflection for component CE be $E'E$ and for CF, $F'F$. Since the reflectivity for both components is determined only by the angle of incidence Θ and is not affected by the azimuth angle φ, for a constant angle of incidence,

$$EE' = k_n \cdot CE \quad \& \quad FF' = k_p \cdot CF$$

where $k_n \geq k_p$.

The transmitted vectors are

$$CE - E'E = CE' \quad \& \quad CF - F'F = CF'$$

and so the resulting vector CQ is

$$CQ = CE' + CF'.$$

The azimuth Φ' of this vector is no longer the same as the azimuth Φ of the entrant vector CD, for the ratio CE/CF is usually not equal to CE'/CF'; in fact this describes the rotation.

The angle of rotation $(\Phi - \Phi')$ is

$$\Phi - \Phi' = QS/CS, \quad QS = QT \cdot \cos \Phi$$

and

$$QT = QR - TR = k_n \cdot CD \cdot \sin \Phi - k_p \cdot CD \cdot \cos \Phi \cdot \tan \Phi$$

$$= CD\,(k_n - k_p) \sin \Phi$$

$$\therefore QS = CD\,(k_n - k_p) \sin \Phi \cdot \cos \Phi.$$

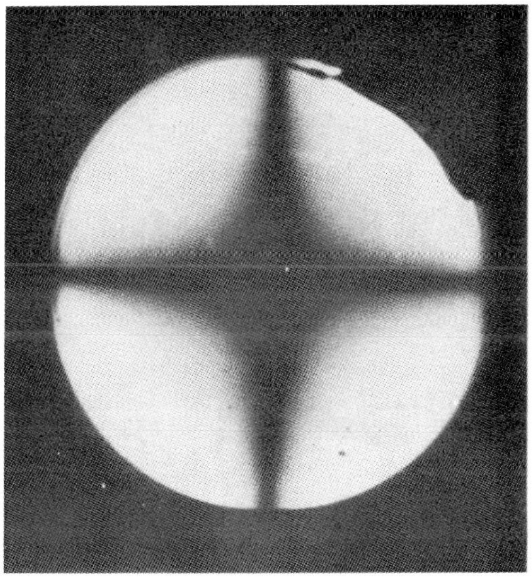

Fig. 1.

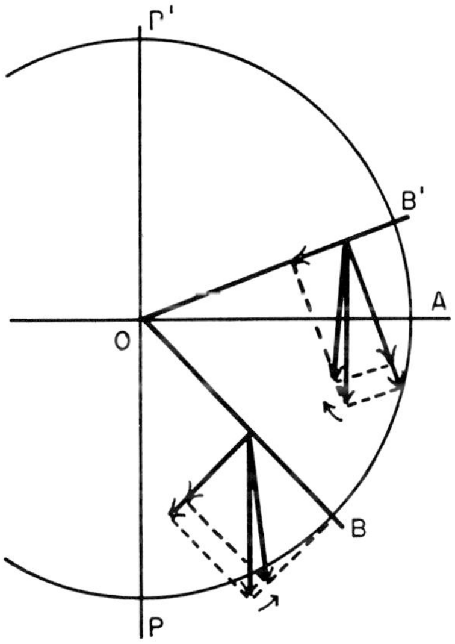

Fig. 2.

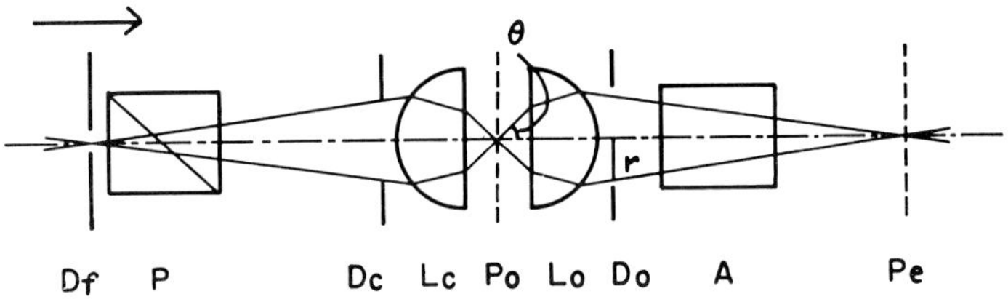

Fig. 3.

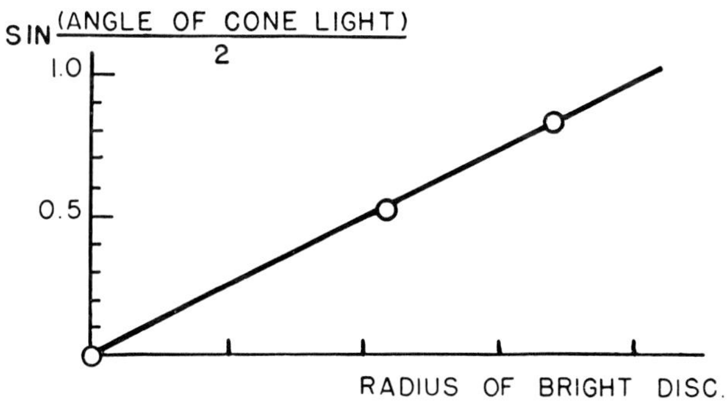

Fig. 4.

Fig. 1. Polarization cross at exit pupil of 4 mm, 0.84 N.A. strain free, coated objective; condenser also strain free lens but not coated. The irregular margin of the upper right quadrant resulted from imperfect blackening of the interior of the lens mount.

Fig. 2. Mode of rotation at lens surfaces in different quadrants. (Compare with figs. 5 and 6.) OP: plane of polarization, OB: plane of incidence.

Fig. 3. Trace of light under critical illumination with the field diaphragm (Df) closed down. The radius (r) of the bright disk at the exit pupil is proportional to sin Θ (Mallard's relation).

Fig. 4. Test for Mallard's relation.

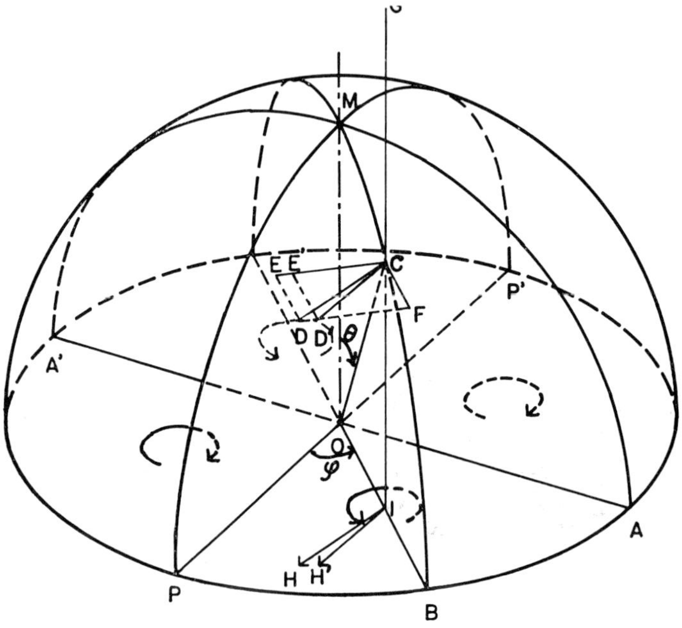

Fig. 5.

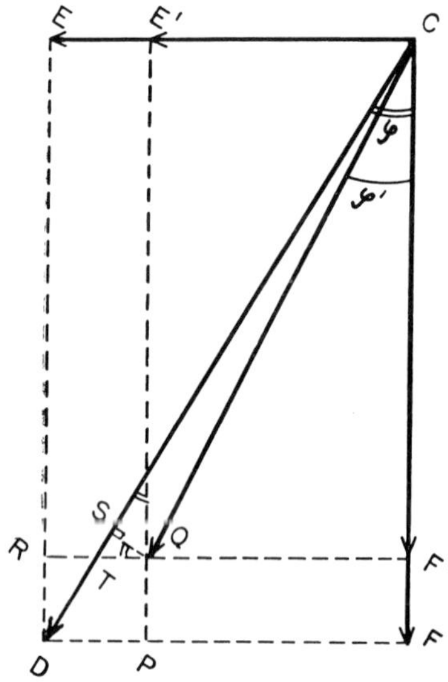

Fig. 6.

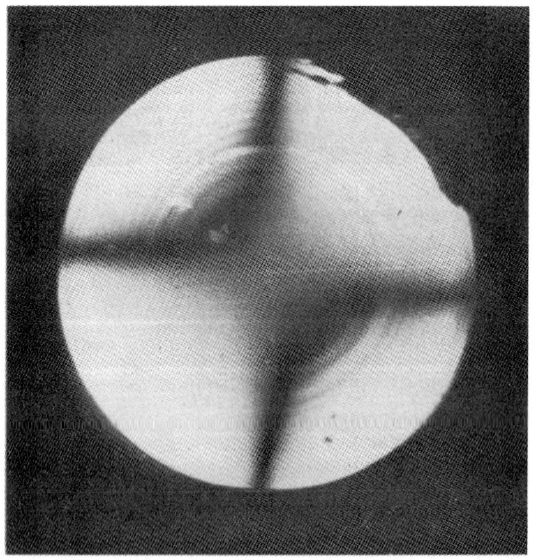

Fig. 7.

Fig. 5. Graphical expression of Mallard's relation.
Fig. 6. Vectorial analysis of rotation. (Compare with Figs. 2 and 5.)
Fig. 7. Appearance of objective exit pupil with the polarizer rotated 0.5° from its crossed position; the condition otherwise the same as in fig. 1.

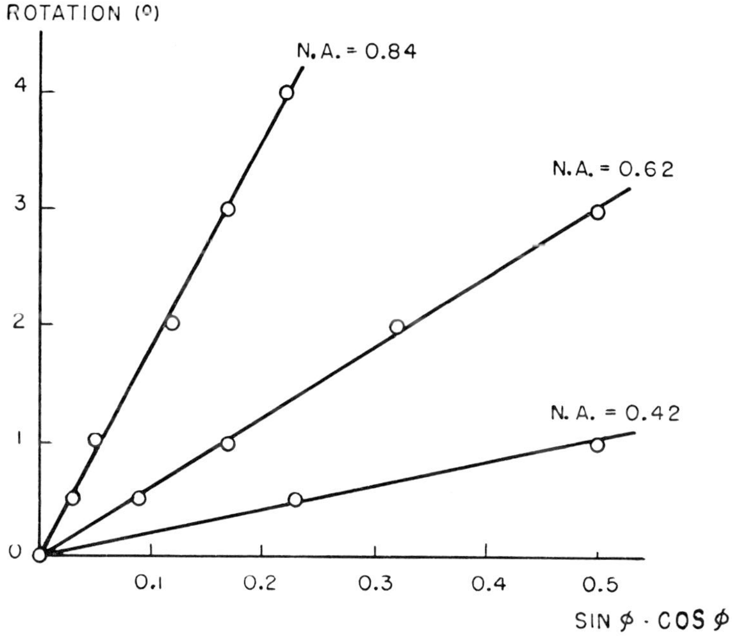

Fig. 8. Rotation measured from photographs (similar to fig. 7) plotted against sin Φ cos Φ. Slope of each curve represents twice the maximum rotation.

Actually $SD << CS$ and so

$$\Phi - \Phi' = \frac{CD\,(k_n - k_p)\cdot \sin\Phi \cdot \cos\Phi}{CD} = f(\Theta)\cdot \sin\Phi\cdot\cos\Phi \;\ldots \quad [2]$$

$(\Phi - \Phi')$ is maximum at $\Phi = \pi/4$ and $\sin\Phi\cos\Phi = 0.5$. Therefore, $f(\Theta)$ is twice the maximum rotation at any given angle of incidence.

The actual amount of rotation can be measured by rotating the polarizer or the analyzer slightly off from the crossed position. Then the polarization cross opens up into two dark hyperbolae (fig. 7) which represent the lines of equal rotation.

In fig. 8, the rotation along certain apertures measured from photographs of the dark hyperbolae are plotted against $\sin\Phi\cdot\cos\Phi$. Clearly within the range of the measurements, the approximate formula very closely describes the relation between rotation and azimuth angles at different angles of incidence.

The maximum rotation at different apertures can be determined from fig.8, for the slope of each line represents $f(\Theta)$ or twice the maximum rotation for each of the corresponding apertures. The log of $f(\Theta)$ determined in this manner has been plotted against Θ, and the relation proves to be perfectly linear (fig. 9). This corresponds to the straight line portion of a sigmoid curve described by Césaro's and Wright's formulae.

The straight line can be expressed as,

$$\log_{10} f(\Theta) - \log f(O) = \frac{\log 10}{0.44} N.A.$$

The value 0.44 was determined graphically from several sets of data and from Césaro's and Wright's formulae. Regardless of the mode of calculation, this and only this universal value was obtained.

Transferring the base of the logarithm to e,

$$0.434 \cdot ln \frac{f(\Theta)}{f(O)} = \frac{N.A.}{0.44}$$

$$f(\Theta) = f(O) e^{N.A./0.19} \quad \ldots\ldots\ldots\ldots\ldots \quad [3]$$

Equation [3] is the expression for maximum rotation at different apertures.

THE TOTAL FLUX OF LIGHT INTRODUCED BY ROTATION WITHIN A GIVEN APERTURE

From equations [2] and [3], the rotation $R(r, \Phi)$ for a beam which passes any point (r, Φ) at the exit pupil of the objective is,

$$R(r, \Phi) = f(\Theta) \sin \Phi \cdot \cos \Phi$$
$$= f(O) e^{N.A./0.19} \sin \Phi \cdot \cos \Phi.$$

Using Mallard's relation and by choosing a proper unit for the radius r of the exit pupil of the objective, namely $r = N.A.$,

$$R(r, \Phi) = f(O) e^{r/0.19} \sin \Phi \cdot \cos \Phi.$$

From equation [1], the amount of light introduced by this rotation, is,

$$L_{r,\Phi} = AI_0 \sin^2 \{R(r, \Phi)\}$$
$$= AI_0 \sin^2 \{f(O) e^{r/0.19} \sin \Phi \cdot \cos \Phi\}$$

and since the rotation is smaller than 10 degrees $\sin^2 \Theta = \Theta^2$, and

$$L_{r,\Phi} = A \cdot I_0 \cdot f^2(O) \cdot e^{2r/0.19} \sin^2 \Phi \cdot \cos^2 \Phi.$$

From fig. 10,

$$A = r\, d\Phi \cdot dr$$

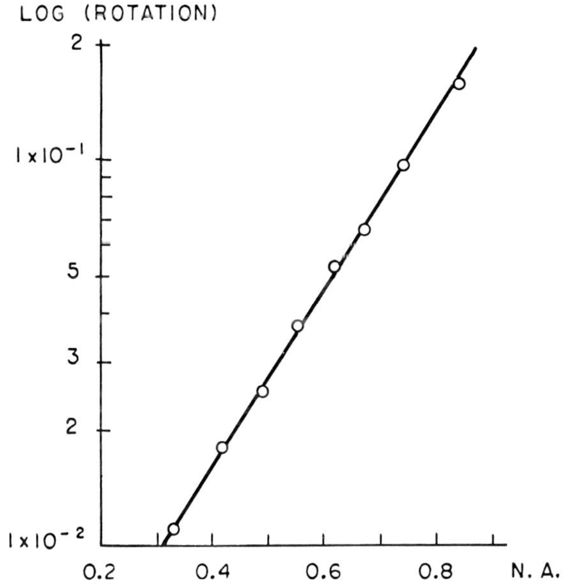

Fig. 9.

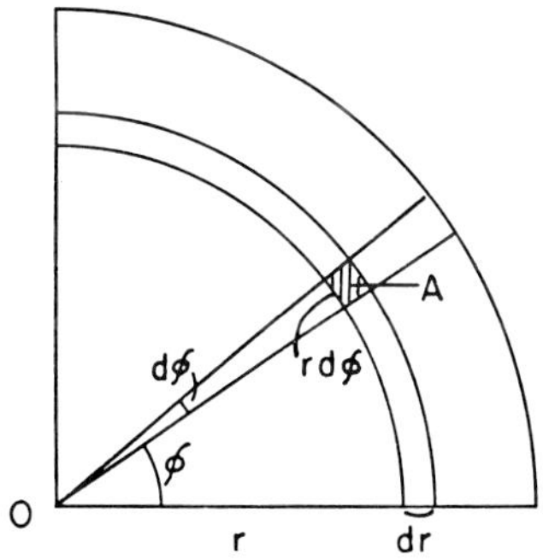

Fig. 10.

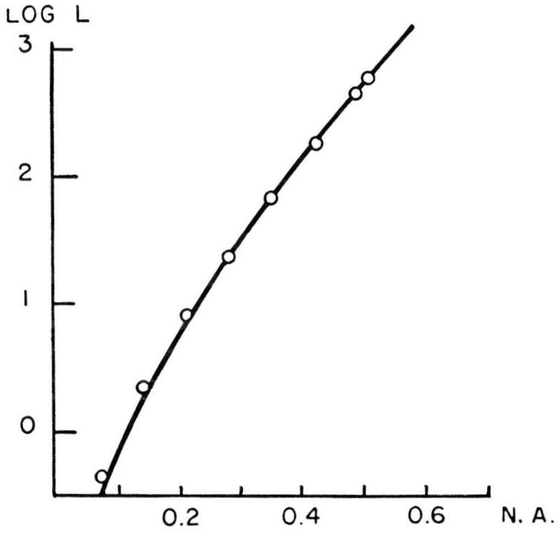

Fig. 11.

Fig. 9. Log (Maximum rotation) plotted against numerical aperture. Maximum rotation was determined graphically from figures similar to fig. 8.

Fig. 10. Construction for integration of amount of stray light introduced at different apertures.

Fig. 11. Curve giving the calculated amount of light (L) entering the system within any given aperture. Circles represent actual photoelectric measurements of stray light. Departure of measurements from calculated values at low aperture can be accounted for; it arises owing to the constant aperture of the objective diaphragms.

and so the total light L is

$$L = I_0 \cdot f^2(O) \int_0^r r \cdot e^{r/.095} \int_0^{2\pi} \sin^2 \Phi \cdot \cos^2 \Phi \cdot d\Phi \cdot dr$$

Since $$\int_0^{2\pi} \sin^2 \Phi \cos^2 \Phi \, d\Phi = \frac{1}{8} \int_0^{4\pi} \sin^2 \Phi \, d\Phi = \frac{\pi}{4}$$

$$L = \frac{I_0 f^2(O)\pi}{4} \int_0^r r \cdot e^{r/.095} \, dr$$

$$= \frac{I_0 f^2(O)\pi}{4} 0.095^2 \left[e^{r/.095}(r/.095 - 1) + 1 \right]_{r=0}^{r}$$

$$L = \frac{9 \cdot 10^{-3} I_0 \cdot f^2(O)\pi}{4} \left\{ \varepsilon^{r/.095}(r/.095 - 1) + 1 \right\} \quad \ldots\ldots \quad [4]$$

Equation [4] gives the total flux of light introduced by rotation within a given aperture r. Taking the log, we obtain

$$\log L = \log \left\{ \frac{9}{4} I_0 \cdot f^2(O)\pi 10^{-3} \right\} + \log \left\{ e^{r/.095}(r/.095 - 1) + 1 \right\} \quad \ldots\ldots \quad [5]$$

The first term on the right is a specific constant for the system, and the second term expresses the change of log L with r, the aperture.

In fig. 11, the line log L (calculated from the equation above) is plotted against r, together with the individual results of direct photo-electric measurements. The very good agreement shows quite definitely that stray light at wider apertures is caused by the rotation of the plane of polarization. The figure also shows very clearly that a small increase in the aperture causes a very great increase in the amount of stray light. Where the aperture is not very small, *a tenfold increase of stray light is brought about by an additional 0.2 numerical aperture.*

This relation then adequately explains how a great amount of stray light is introduced when the numerical aperture of the system is increased in polarization microscopy where the elements are of the usual construction and finish.

REFERENCES

1. Césaro, G., *Bull. Acad. Roy. Méd. Belg.*, 459 (1906).
2. Inoué, S., and Dan, K., *Kagaku*, **19**, 111 (1949) (in Japanese).
3. —— *J. Morphol.*, **89**, (1951).
4. Mallard, E., *Bull. Sci. Min. Fr.*, **5**, 77 (1882).
5. Rinne, F., *Zentr. Mineral. Geol.*, 88 (1900).
6. Wright, F. E., *Am. J. Sci.*, **31**, 157 (1911).
7. ——, The Methods of Petrographic Research. Carnegie Inst. of Washington, Washington, D.C., 1911.
8. —— *J. Optical Soc. Am.*, **7**, 779 (1923).

Reprinted from *Experimental Cell Research*, Vol. 2(3), pp. 513–517, 1951,
with permission from Elsevier.

A METHOD FOR MEASURING SMALL RETARDATIONS OF STRUCTURES IN LIVING CELLS[1]

SHINYA INOUÉ[2]

Department of Biology, Princeton University, Princeton, New Jersey

Received March 17, 1951

THE measurement of birefringence in living cells is accompanied by special difficulties not encountered in birefringence measurements of ordinary crystals. The retardation is very small (usually 0.1–5 mμ), and there are a great number of light scattering particles in the cell which make the birefringent image difficult or impossible to discern. Besides, the cross section of the birefringent object in the cell is usually elliptical or circular, and the area with uniform strength of birefringence is very small. Furthermore, the birefringence at the edge of the object changes gradually because of the shape of its cross section, and makes the measurement even more difficult. For these reasons, the ordinary compensation method, which can be used for measuring the retardation of thin crystals very sensitively (see Jerrard 3, for a summary of compensators), is not suitable for accurate measurement of the retardation of organelles in the cell.

Swann and Mitchison (5) have used densimetry on photographs of spindles in dividing cells to avoid the above mentioned difficulties. Their method is laborious and exacting, but will undoubtedly prove useful to some, and sufficiently accurate if precaution is taken to subtract the intensity differences caused by factors other than birefringence.

In the following is described a new direct method (which may be called the 'double half-shade method') for measuring retardation in complicated objects with small retardations.

In figure 1, 'A' is an object whose cross section is an ellipse. This is placed between crossed nicols with its axes 45° to the axes of the nicols. Another small crystal 'B' is placed beside the object with its axes also in the 45°

[1] This research was supported in part by funds of the Eugene Higgins Trust allocated to Princeton University.

[2] Charlotte Elizabeth Procter Fellow. Present address: Department of Anatomy, School of Medicine, University of Washington, Seatle 5, Washington.

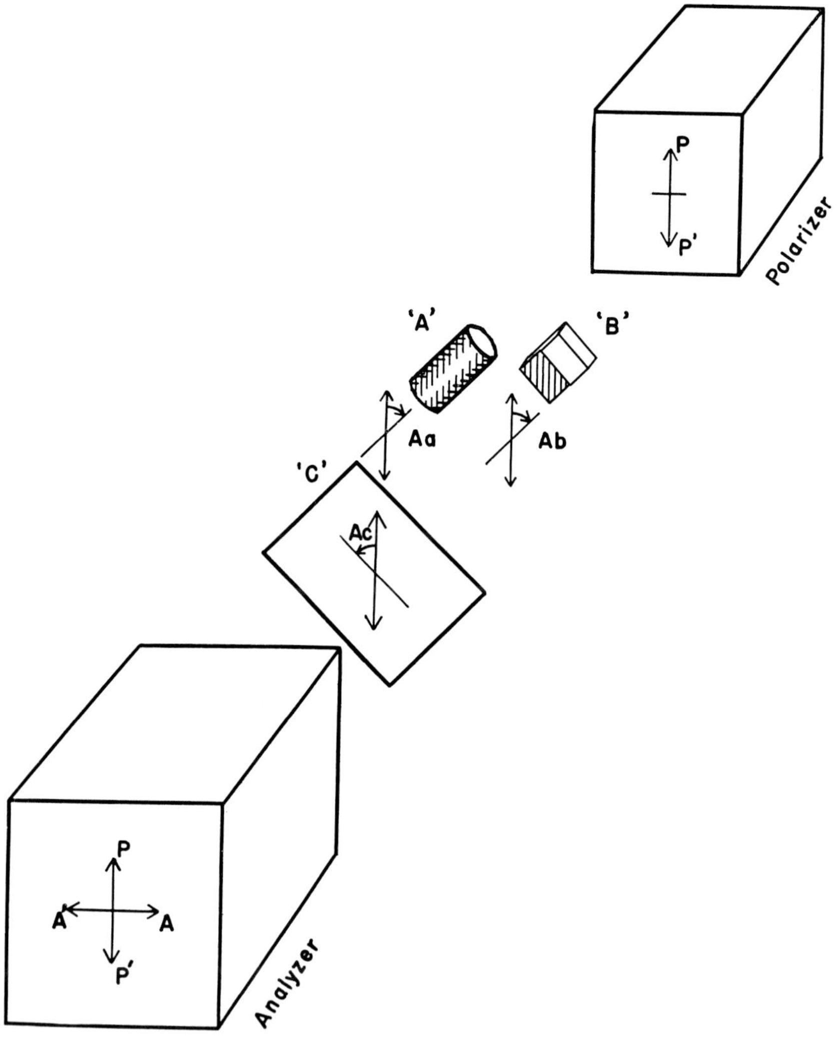

Fig. 1. Arrangement of the optical elements in the double half-shade method.

position. Part of this crystal is slightly thicker, and therefore its retardation slightly larger than the rest. In addition to the two, another large crystal 'C', the compensator, is placed as shown in the figure, so that each of the light beams which passes through the object and the small comparator crystal also passes through the compensator.

First, let us assume that the retardation of a particular part of the object and the comparator crystal are the same, and their axes parallel. If we rotate the compensator (C) under this condition, the particular part of the object and the comparator will go through extinction simultaneously. Furthermore, if the compensator is rotated a little further, a small white area will begin to appear simultaneously at the center of the object and on the thicker side of the comparator. In this way, even if light scattering objects are superimposed upon the birefringent object, *by matching the change of the intensity* of the two, it is nevertheless possible to match the birefringence of an object with the birefringence of a comparator. Thus, knowing the retardation of the comparator, one can determine the retardation of the object.

Instead of using a comparator crystal which has exactly the same retardation as the object, a crystal with a somewhat higher retardation could be used by rotating it around an axis parallel to the optical axis of the measuring device. The apparent retardation of the crystal can, in this way, be made equal to the retardation of the object as shown in the following calculation.

Let the retardation of the object, the comparator and the compensator be Ra, Rb and Rc respectively, each of their axes making an angle Aa, Ab and Ac with the polarizer axis, (fig. 1). Consider both the retardation of the object and the apparent retardation of the comparator to be equal to that of the compensator, but with an opposite sign. This brings both the object and the comparator into extinction with the compensator. The retardations R are all sufficiently small so that their sines can be equated to the retardations expressed in radians. Then from Fresnel's equations, as Köhler (4) has shown (also see Bear and Schmitt, 1),

$$\sin(2\,Aa) \times \sin(Ra/2) = -\sin(2\,Ac) \times \sin(Rc/2)$$

and also

$$\sin(2\,Ab) \times \sin(Rb/2) = -\sin(2\,Ac) \times \sin(Rc/2).$$

Therefore,

$$\sin(2\,Aa) \times \sin(Ra/2) = \sin(2\,Ab) \times \sin(Rb/2).$$

Since the object is lying in the 45° position, $\sin(2\,Aa)$ is 1, and so

$$Ra/2 = \sin(2\,Ab) \times Rb/2$$

This equates the retardation of the object with the apparent retardation of the comparator. Therefore, if the retardation (Rb) of the comparator is known, the retardation (Ra) of the object can be determined by measuring the angle Ab. At this angle, the comparator goes through extinction and becomes brighter again simultaneously with the object when the compensator is rotated around the extinction position.

The approximation used in this calculation will lead to an error of not greater than 1 percent, if the retardations of the object, comparator and compensator are each less than a thirtieth of a wave length.

In applying the described method to measurements of retardation of small objects under the microscope, it is almost impossible to prepare a comparator crystal which can be placed side by side with the object, and further to place it so that it can be rotated around its axis. The actual method is, therefore, to project an image of the comparator crystal into the object plane. The image of the comparator crystal can then be focused and compared side by side with the object.

The crystal is a thin piece of mica about 0.2–0.3 mm square, mounted at the center of a rotating drum whose angle of rotation can be measured with a vernier to a twentieth of a degree. This is placed in front of the polaizer located immediately in front of the field lens (that is, the front lens of the light source), with its image focused by the condenser, onto the plane of the object. For the present arrangement, the polarizer would have to be placed approximately 20 cm away from the condenser. Also, neither a mirror nor a prism should be used in between the polarizer and the condenser since it would disturb the condition of polarization, and so the whole microscope including the light source should be arranged on an optical bench. A microscope arranged in this fashion will be described in another paper.

The compensator can be placed at any suitable position between the polarizer and the analyzer since there are only two retardation plates in any one pathway of light. If there were three or more retardation plates, in general their positions are not inter-changeable (3, 2), and so for an accurate measurement, the lenses, slide and cover glasses etc. involved in the system should be perfectly strain free.

It has been possible to measure the retardation of spindle components to a tenth of a mμ by the method described. This has proven to be very useful in tracing the change of retardation of the spindle which is disintegrating under the influence of colchicine, or for measuring the form birefringence curve of the fixed spindle.

The method is general and accordingly of wide applicability. It has notable advantages over the photographic method, since it automatically excludes the effect of intensity differences of the image caused by factors other than birefringence, (such as absorption, refraction, scattering etc.) and because the measurements can be done very rapidly, with great accuracy and high precision.

SUMMARY

A new half shade method for measuring retardations in optically complicated structures is presented. By using a polarization microscope with an extinction of 10^5, a retardation as small as a tenth of a mμ (in objects not larger than a micron in diameter) can be measured very rapidly by this method. Especially important is the fact that such measurements can easily be made without injury, directly upon the living cells.

REFERENCES

1. Bear, R. S., and Schmitt, F. O., *J. Optical Soc. Am.*, **26**, 363 (1936).
2. Hsü, H. Y., Richartz, M., and Liang, Y. K., *ibid.*, **37**, 99 (1947).
3. Jerrard, H. G., *ibid.*, **38**, 35 (1948).
4. Köhler, A., *Z. wiss. Mikroskop.*, **38**, 29 (1921).
5. Swann, M. M., and Mitchison, J. M., *J. Exptl. Biol.*, **27**, 226 (1950).

Reprinted from *Chromosoma*, Vol. 5, pp. 487–500, 1953,
with permission from Springer Science and Business Media.

From the Department of Biology, Princeton University, Princeton, New Jersey, and The Marine Biological Laboratory, Woods Hole, Mass.

POLARIZATION OPTICAL STUDIES OF THE MITOTIC SPINDLE.
I. THE DEMONSTRATION OF SPINDLE FIBERS IN LIVING CELLS*.

By

SHINYA INOUÉ.

With 15 figures in the text.

(Eingegangen am 20. November 1952.)

Introduction.

The mechanism of chromosome movement and cell division can not be understood without an adequate knowledge of the structure of the mitotic spindle. Various optical conditions prevailing in most living cells, however, make direct microscopical observation of the spindle extremely difficult. Thus, aside from a few exceptional cases (COOPER 1941, CLEVELAND 1938), the structure of the spindle has not been satisfactorily demonstrated in most living cells. This has led many to believe that the living spindle is structureless, and that the spindle fibers and fibrils which can be demonstrated so clearly after fixation and staining do not reflect the true architecture of the cell before fixation.

The difficulty of seeing structure in the spindle of living cells with a regular microscope, however, need not necessarily be taken to mean that the spindle is structureless. In fact, as reviewed by SCHRADER (1944) and by CORNMAN (1944), there exist a number of observations favoring the view that some kind of fibrous organization must in fact exist in the living spindle during the normal course of its function. The evidence is for the most part indirect and includes: the mechanical anisotropy of the spindle that BĚLAŘ (1929a) showed following action of hypertonic media, and by his observation of non-random Brownian motion within the spindle (BĚLAŘ 1929b); the optical anisotropy in living spindles demonstrated by SCHMIDT (1939, 1941) and others;

* The main portion of the paper was presented to the faculty of Princeton University in partial fulfillment of the requirements for the degree of Doctor of Philosophy. This research was supported in part by funds of the Eugene Higgins Trust allocated to Princeton University. Thanks are also extended to the American Philosophical Society for use of equipment purchased under a grant awarded to Dr. K. W. COOPER. I am also indebted to Dr. RALPH O. ERICKSON of the University of Pennsylvania and Dr. HERSCHEL L. ROMAN of the University of Washington for supplying the plant material. Some of the observations described in this paper were done at the present address.

observations on the organization of individual chromosomal fibers in fixed material (SCHRADER 1932); the curved appearance of the spindle fibers and their presumed independent behaviour in cells fixed after centrifugation (MORGAN 1910, CONKLIN 1917, SCHRADER 1934, SHIMAMURA 1940); and finally, microdissection studies of the living cell (WADA 1935, CARLSON 1952).

Now, since the spindle and astral rays are weakly birefringent, their structure should be observable under a sensitive polarization microscope. SCHMIDT (1939), HUGHES and SWANN (1948), INOUÉ and DAN (1951), and SWANN (1951a, 1951b) have studied changes in the birefringence of the spindle and asters and related them to possible physical and chemical changes in the cell during division. Unfortunately, the instruments and technique used by these authors did not provide adequate resolution, and discrete spindle fibers could not be observed in most of these cases.

An analysis of the limiting factors in polarization microscopy (INOUÉ 1952a, 1952b) has led to the design of an improved polarization microscope for cytological purposes. With the increased resolution and sensitivity achieved by the new instrument, I have been able to observe the detailed structure of the spindle in living cells and to follow its change during mitosis. In various types of animal and plant cells, spindle fibers are clearly visible with this microscope, and their relation to the chromosomes during anaphase can be established with great certainty.

In the following, the structure of the metaphase and anaphase spindle in the oöcyte of the marine annelid, *Chaetopterus pergamentaceous*, and in the pollen mother cell of *Lilium longiflorum* will be described, and, in a subsequent paper, the submicroscopic structure of the spindle will be discussed in the light of these observations made with the improved polarization microscope.

1. The Structure of the spindle in the oöcyte of *Chaetopterus pergamentaceous.*

SCHRADER (1944, p. 7) proposes that spindle fibers be separated into the following three categories:

1. "Continuous fiber": fiber connecting the two centers or poles.
2. "Chromosomal fiber" (or half spindle component): fiber connected with the kinetochore of the chromosome. It may or may not extend to a pole or a center.
3. "Interzonal connection": connection between the separating chromosomes in anaphase and telophase.

When parapodia are cut from a female *Chaetopterus*, the eggs are shed. Thereafter the first maturation division spindle forms in about 15 minutes at 25^0 C. The spindle proceeds to metaphase and stays in this stage for several hours unless the egg is further activated (MEAD 1898, INOUÉ 1952c).

The metaphase spindle is attached at the animal pole of the egg, perpendicular to the egg's surface, and under ordinary light this region appears as an empty homogeneous space from which the granules have been expelled.

Under polarized light, the outer pole of the spindle can be seen by its birefringence when the egg is slightly compressed. However, since the granules in the *Chaetopterus* egg are highly birefringent and scatter light very strongly, the complete spindle can not be observed without greatly compressing the egg.

In order to acquire a full view of the spindle without too great a compression, the egg is centrifuged at 15,000 g for 15 minutes between sea-water and isotonic (1·1 M) sucrose. Clear quarters of the eggs which include the spindle and few yolk granules are isolated and washed in fresh sea-water. These quarters develop at least to swimming blastulae following fertilization (HARVEY 1939, INOUÉ 1952c).

Fig. 1 is a photograph of the spindle in a quarter egg prepared in this fashion. By comparison with the spindle in the whole egg, it can be demonstrated that the structure of the spindle shown in Fig. 1 is not different from the structure of the spindle in the egg before centrifugation.

In Fig. 1, a longitudinal fibrous arrangement is demonstrated in the birefringent spindle (compare with Fig. 2). Those parts which appear bright and are parallel to the spindle axis, and those parts of the astral rays which are dark and lie perpendicular to the axis, are both positively birefringent. Whether they are bright or dark merely depends on the optical quadrants in which the spindle components lie. (See INOUÉ and DAN 1951, and BENNETT 1950.)

The chromosomes also show in the figure, but this is not due to the birefringence of the chromosomes themselves. The chromosomes are much more weakly birefringent than the spindle, appearing a dull gray similar to the background of the field, and standing in strong contrast to the brightly shining spindle fibers.

By compressing the centrifuged egg it is possible to find out to which kind of spindle fiber these longitudinal structures correspond. As the egg is compressed the spindle length increases and the birefringent area becomes more and more clearly differentiated (Figs. 3 and 4) as the birefringence of the fibers increases (INOUÉ 1952c). By properly adjusting the compensator angle it is possible to demonstrate how some of the birefringent fibers run from the poles to the chromosomes and end abruptly at the chromosomes (Figs. 5 and 6). By definition, then, these are chromosomal fibers.

These chromosomal fibers are not abnormal structures caused by the compression of the eggs. This is demonstrated by the fact that just prior to the onset of, and during, anaphase, exactly the same fibers

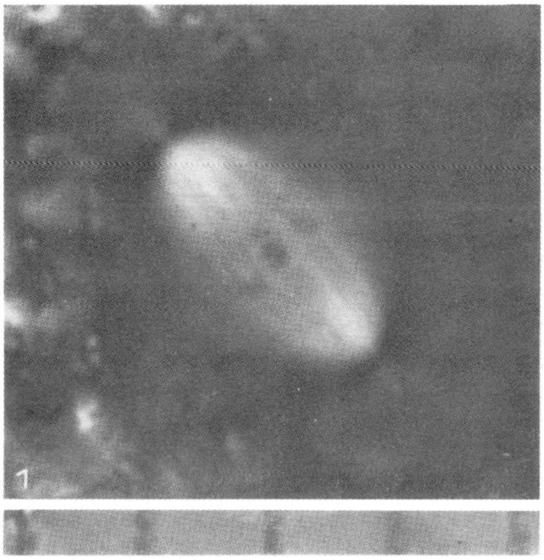

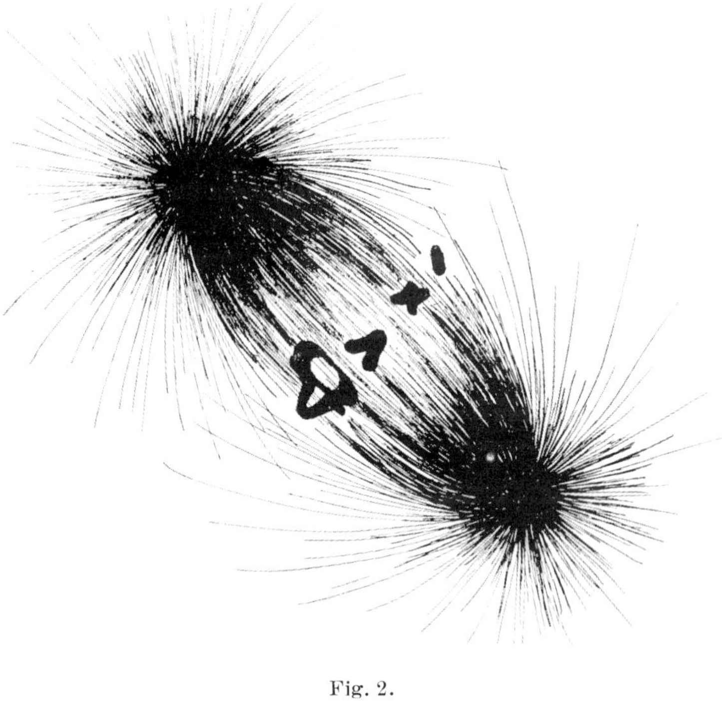

Fig. 2.

Figs. 1—3.

Fig. 1[1]. First maturation division spindle in a living oöcyte of *Chaetopterus pergamentaceous*. The bright chromosomal and continuous fibers and the dark astral rays are all positively birefringent. Notice convergence of the birefringent fibers at the poles. Chromosomes which are but weakly birefringent appear as dark grey bodies at the equatorial plate. (Compare with Figs. 2 and 9.) Scale 10 μ, for all figures except Figs. 2 and 13.

Fig. 2. Trace of Fig. 1. An image of the negative was projected and the birefringent fibers and chromosomes traced.

Fig. 3. Similar to Fig. 1, but the spindle is slightly stretched by compression of the egg. Individual chromosomal fibers and astral rays appear clearer than in Fig. 1.

[1] Photographic Data: Polarizer and analyzer (special Glan-Thompson prisms) are crossed; compensator set for ca. 3 mμ background retardation; objective, B. &. L. strain-free coated achromatic lens 43 ×, N.A. 0.84; condensor, B. &. L. strainfree coated achromat 21×, N.A. 0.50 stopped to 0.35; light source, A-H 6 mercury arc lamp with Wratten No 58 green filter; film, Kodak Panatomic-X 35 mm.

show up by their stronger birefringence in eggs which were less compressed (Fig. 7). Is should also be remarked that such chromosomal fibers may be observed in spermatocytes of various grasshoppers (*Chortophaga*, *Melanoplus*, *Dissosteira*), and of *Drosophila melanogaster*,

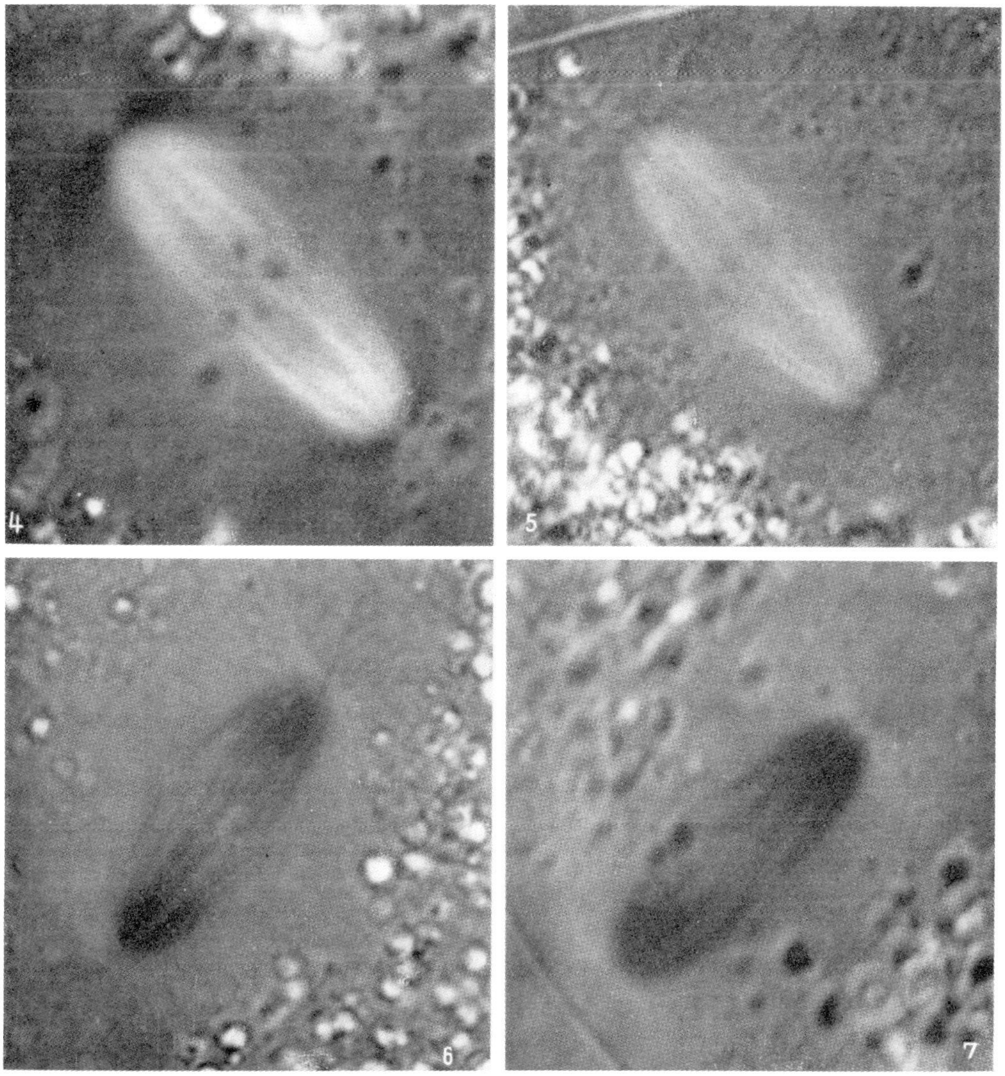

Figs. 4—7. Fig. 4. Spindle stretched even further than in Fig. 3. The large dark particle with a halo (second particle from right on the spindle equator) is not a chromosome. — Fig. 5. Compensator rotated to show the attachment of individual chromosomal fibers to chromosomes in *Chaetopterus* oöcyte. — Fig. 6. Same as Fig. 5 but compensator turned. — Fig. 7. Chromosomal fibers appearing at onset of anaphase after activation of *Chaetopterus* egg.

which have not been centrifuged or compressed, while actively dividing in Ringer's solution.

Therefore, at least part of the longitudinal birefringent structures in the metaphase spindle of an uncompressed egg (Fig. 1) represents the chromosomal fibers.

In Fig. 1 it is also clear that, aside from the birefringent element between the poles and the chromosomes, there are fibers with run by the chromosomes and seem to connect the two poles. The existence of such continuous fibers and other half spindle fibers can be demonstrated more clearly in the anaphase spindle after activation of the egg.

As the chromosomes move closer to the poles, certain birefringent fibers may be seen to lie by the chromosomes with their ends extending farther towards the equator of the spindle (Fig. 8). These fibers have

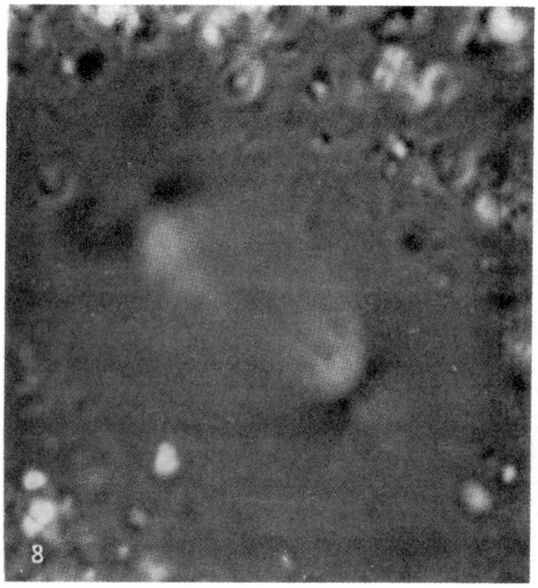

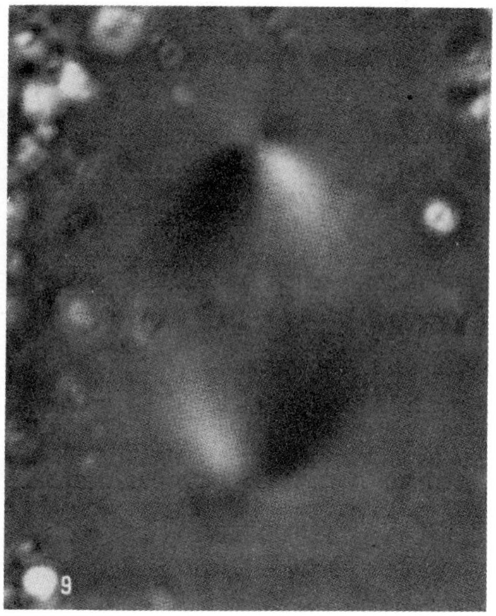

Figs. 8 and 9. Fig. 8. Anaphase of first maturation division in *Chaetopterus* oöcyte. Chromosomal fibers show between chromosomes and poles. Also continuous fibers can be seen in the central spindle region. — Fig. 9. Metaphase in *Chaetopterus* oöcyte. Spindle axis lies parallel to polarizer axis. Compare with Figs. 1 and 2. Notice fibrillar appearance of half spindle fibers and gradual transition from spindle fibrils to astral rays.

not only been seen in the anaphase spindle of *Chaetopterus* but also in all of the animal and plant cell spindles in anaphase that I have studied. Some of these fibers run from pole to pole, clearly being continuous fibers, while others seem to thin out and fade away before they reach the equator of the spindle.

The chromosomal fibers lose part of their birefringence during anaphase, but in relation to the other components of the spindle, they still maintain a stronger birefringence, and their identity is not completely lost even at a very late stage in anaphase (Fig. 8).

The spindle in the living *Chaetopterus* egg is thus composed of at least two different types of birefringent fibers. (See INOUÉ 1952c for the effect of colchicine on the various components of the spindle.) In addition, observations of the spindle in different positions (Fig. 9) indicate that the spindle fibers are actually composed of fibrils whose diameters lie close to the limit of resolution (estimated in the present system to be ca. 0.3 μ). In Fig. 9, the stage of the microscope was rotated so that the long axis of the spindle coincided with the axis of the polarizer. Hence, fibrils or fibers whose axes lie in the quadrants, including the axes of the spindles shown in Figs. 1, 3, 4 etc., appear brighter than the background; while the fibrils in the other quadrants, including the spindle axis in Figs. 6 and 7, appear darker.

The existence of fibrils as components of spindle fibers was demonstrated in fixed material by SCHRADER (1932). SWANN (1952a), from his birefringence studies on sea-urchin eggs, also suggests that the spindle and astral rays are composed of thin radiating fibrils which, however, he was unable to resolve under his microscope.

No difference in structure could be found between these fibrils and the astral rays, with which they appeared to merge without any distinct separation. Both spindle fibrils and astral rays converge to a very small area at the poles of the spindle.

When the structure of the spindle in the living cell, as studied by polarized light, is compared with some of the better figures obtained by fixation and staining, it becomes clear that some fixed preparations can preserve the configuration of structures in the living spindle quite faithfully (at least at the microscopic level).

For instance, MEAD (1898) published some figures drawn from sections of the egg of the same animal *Chaetopterus pergamentaceous* (also collected at Woods Hole), stained with Heidenhain's iron-alum haematoxylin and orange G after fixation in Boveri's picro-acetic mixture. The resemblance of these figures to the photographs of the living spindle is truly remarkable.

If one seeks for some slight difference, MEAD's figures do not actually correspond to the photographs of the spindle in the uncompressed or very slightly compressed eggs (Figs. 1, 2, 9) but resemble more closely in structure and configuration the spindles in the highly compressed eggs (Figs. 4, 5, 6). When we consider the fact that the fixatives used by MEAD almost certainly make the spindle fibers contract, it is understandable why his figures have a closer resemblance to the structure of the spindle under tension. The very fact that in both the fixed spindle and the stretched living spindle the fibers take a straighter course and the round contours at the poles of the spindles are lost argues strongly in favor of this explanation.

The demonstration of spindle fibers in living cells by means of a polarization microscope, and the very close resemblance of the structure of living spindles to figures obtained from good fixed and stained preparations, give final proof for the reality of spindle fibers in living cells. Moreover, this shows that the cytological configurations in well fixed and stained preparations can very closely resemble cellular structures in their living form. This point has already been emphasized by BĚLAŘ (1928, 1929a) for cellular organelles other than spindles.

2. The spindle in the pollen mother cell of *Lilium longiflorum*.

As an example of the structure of the spindle in a dividing plant cell, observation, under polarized light, of the dividing pollen mother cell of the Easter lily will be described. The following description of *Lilium* mitosis is based not only upon direct observations and still photographic records but also upon a time lapse motion picture of the two successive meiotic divisions taken through the new polarization microscope at a rate of 4 frames per minute.

The pollen mother cells of the lily also contain strongly birefringent granules (fat droplets) which can be removed by very mild centrifugation. All of the observations were made in a modified Ringer's solution (isotonic lactose substituted for glucose, or no sugar added at all) on the first and second meiotic mitoses in cells which had been centrifuged for 3 minutes at 1800 g.

The structure of the spindle after centrifugation can be considered to be quite normal since centrifuged cells are found to go through the two consecutive meiotic divisions without demonstrable abnormality. Supplementary observations were also made on both uncentrifuged and centrifuged pollen mother cells of *Polygonatum biflorum* and *Gasteria carinata* as well as with a species of Iris. The structure of the spindles of all these forms was similar at all stages of division, except for a difference at telophase depending on the sequence in cell plate formation.

As Fig. 10 shows, the spindle of the *Lilium* pollen mother cell under the polarization microscope shows a very striking longitudinal striation. The spindle in the figure is in early metaphase II, and the spindle poles do not show the kind of convergence found in the spindle of many animal cells. This type of apolar or multipolar spindle has been shown in plant cells by OSTERHOUT (1897) and by MOTTEIR (1897), in insect spermatocytes by HUGHES-SCHRADER (1948), and in the blastomeres of a mite by COOPER (1939).

In early metaphase (both I and II) the birefringence is strongest near the equatorial plate, and the birefringent streaks spread and become wider towards the 'poles' of the spindle. These streaks do not appear to be interrupted by the chromosomes at the equatorial plate.

By carefully focussing the spindle in side view and by observing the polar view of the spindle, it can be shown that the birefringent streaks are actually not fibers at this stage. Instead all of the plant spindles observed at early metaphase I or II are composed of sheet-like aggregations of fibrils that are oriented parallel to the spindle axis.

In cross section these sheets of fibrils appear to lie parallel to the arms of the chromosomes.

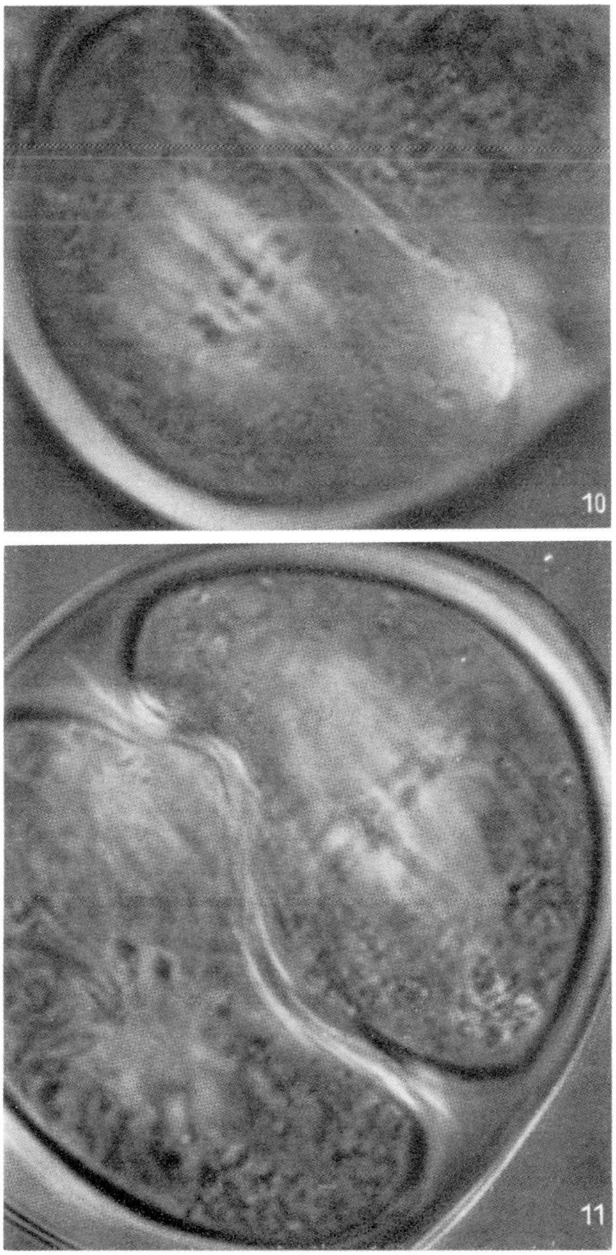

Figs. 10 and 11. Fig. 10. Second meiotic division in pollen mother cell (PMC) of *Lilium longiflorum*. Early metaphase. Sheet-like aggregate of fibrils look like spindle fibers in side view. Birefringent granules have been displaced and packed in one corner (lower right) by centrifugation. — Fig. 11. Later metaphase of *Lilium* PMC. The left cell shows oblique cross sections of birefringent chromosomal fibers. The right cell shows beginning of convergence of the spindle poles.

As anaphase approaches, the spindle poles gradually converge and the birefringence also becomes more uniformly distributed along the whole length of the spindle (Fig. 11, the right cell).

Just before the separation of the chromosomes, the regions of the spindle adjacent to the kinetochores of the chromosomes become strongly birefringent (as is also the case in *Chaetopterus* eggs and insect

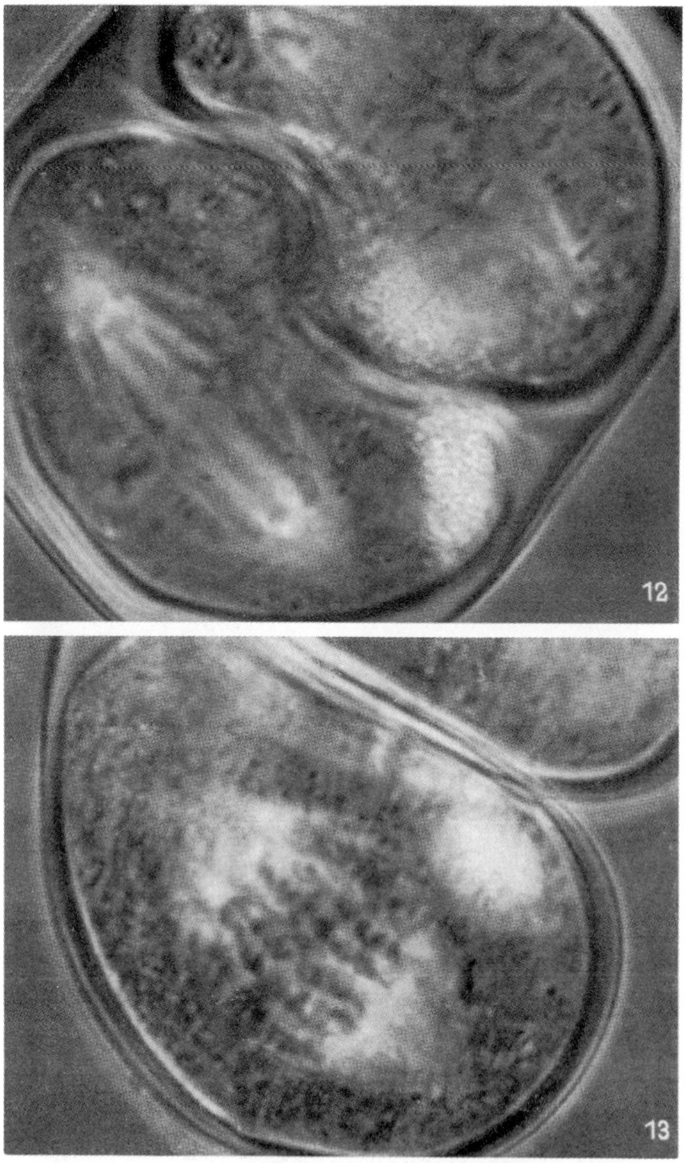

Figs. 12 and 13. Fig. 12. Anaphase of *Lilium* PMC. Birefringence is strongest in half spindle, especially adjacent to the kinetochores of the chromosomes. The central spindle region is very weakly birefringent. — Fig. 13. Anaphase of first meiotic division in *Lilium* PMC. Chromosomal fibers show very strong birefringence adjacent to kinetochores. Drawn out helices are obvious in chromosomes. Magnification lower than in other figures. Clarity of the spindle structure in this figure is due to improvement in photographic technique and does not distinguish 1st meiotic division from 2nd.

spermatocytes) and disclose the existence of strongly birefringent chromosomal fibers. As can be seen in the left cell in Fig. 11, optical sections of individual chromosomal fibers are visible when a polar view of the spindle is observed at this stage. Thus, the longitudinal striation of the late metaphase and anaphase spindles, unlike that of early metaphase, actually represents chromosomal fibers and not cross-sections of sheet-like material.

In anaphase the strongly birefringent region of the half spindles (the chromosomal fibers plus part of the continuous fibers) shifts toward the spindle poles while the poles themselves gradually move apart from each other (Fig. 12). The chromosomal fibers, especially near the kinetochores, maintain their strongest birefringence until a very late stage in anaphase, as is clearly demonstrated in Fig. 13.

At this stage helices in each of the separating chromosomes can be seen, first their gyres drawing apart as the chromosomes are stretched by the chromosomal fibers (Fig. 13), then finally recoiling, presumably after the interzonal connections between the separating chromosome pairs have broken.

During anaphase, just as in the case of the *Chaetopterus* egg, the birefringent fibers exist not only between the chromosome kinetochores and the spindle poles, but also some fibers (or fibrils) extend out farther than the tail ends of the chromosomes (Fig. 12). Some of these fibers do not appear to run much farther than the ends of the chromosomes, while others run clear through the nearly isotropic gap between the separating half spindles.

This "gap" between the separating chromosomes has been described by SCHMIDT (1936, 1939) as isotropic, and by SWANN (1951 a, b) as weakly birefringent. INOUÉ and DAN (1951) found weakly birefringent strands here but were unable to determine to which component of the spindle these birefringent strands belong. It now seems clear that continuous fibers mainly contribute to the birefringence on this region.

In telophase the continuous fibers regain their strong positive birefringence, and numerous fibers connecting the two daughter nuclei can be observed (Fig. 14, the left cell). The birefringence of these fibers continues to increase, especially around the old equatorial plate, and finally the fibers form the phragmoplast. Very shortly after the phragmoplast is completed, the cell plate forms in the middle of the phragmoplast by a rapid fusion of small granules which collect in that region (Fig. 14, the right cell).

The new cell plate has a birefringence positive with respect to the plane in which it lies, and consequently appears darker than the birefringent phragmoplast whose sign is positive relative to the long axis of the spindle. As the new cell wall develops the daughter nuclei move closer together, perhaps because of a pulling action of the remaining continuous fibers. Fig. 15 shows the completion of the division of the same cells shown in Figs. 12 and 14.

Therefore, in the *Lilium* pollen mother cell, the existence of various fibrous components has been demonstrated in the spindle of living and normally functioning cells, just as in the animal cells described

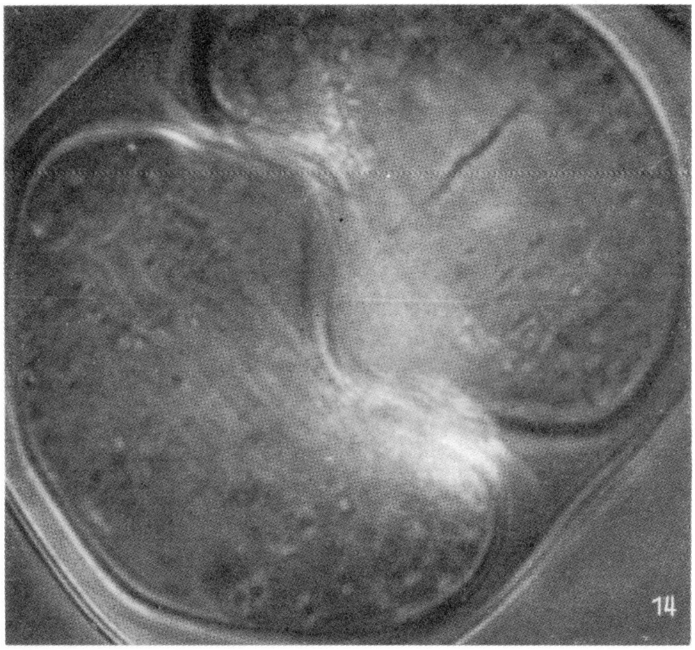

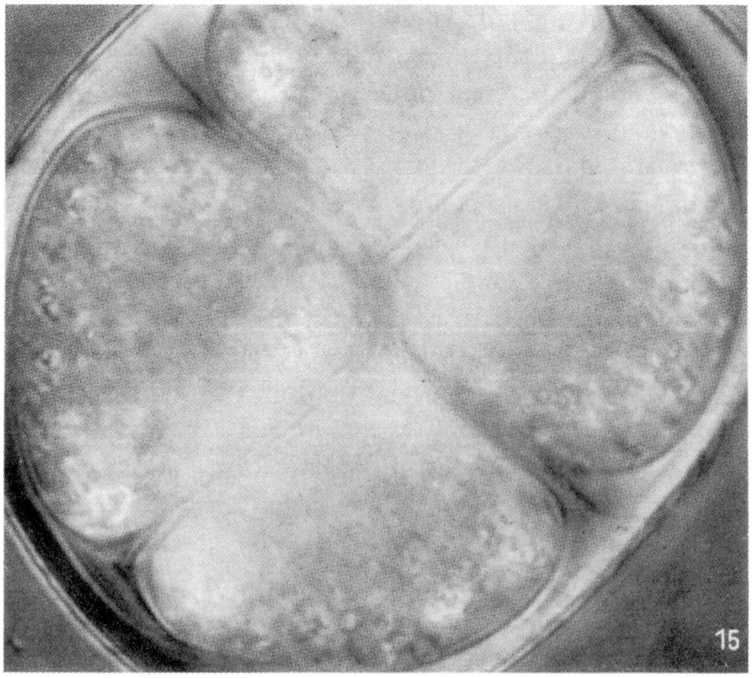

Figs. 14 and 15. Fig. 14. Telophase of *Lilium* PMC. Continuous fibers between the daughter nuclei become strongly birefringent again (the left cell) and form the phragmoplast (the right cell). The new cell plate is positively birefringent with respect to a plane perpendicular to the spindle axis and appears dark. The birefringent granules have started to disperse. Fig. 15. The same cells shown in Figs. 12 and 14 after completion of the 2nd meiotic division. Birefringent granules have filled the cells again.

above. Furthermore, the resemblance of the fixed and stained figure with the structure of the spindle observed under polarized light is again quite remarkable (for instance see MOTTIER 1897), although the similarity is not quite so good as in the case of the *Chaetopterus* egg.

Summary.

1. With an improved polarization microscope, spindle fibers and fibrils can be seen in living, normally dividing cells.

2. Chromosomal and continuous fibers are demonstrated in photographs of oöcytes of *Chaetopterus pergamentaceous*, as well as in pollen mother cells of *Lilium longiflorum*.

3. Supplementary observations on spermatocytes of several species of grasshopper and of *Drosophila melanogaster*, as well as on pollen mother cells of *Polygonatum biflorum*, *Gasteria carinata* and a species of *Iris*, confirm the findings mentioned above.

4. Some of the structural changes observed in the spindle and chromosomes during division are described.

5. The appearance of spindle structures in certain fixed preparations closely resembles structures observable in living cells under polarized light.

Literature.

BĚLAŘ, K.: Über die Naturtreue des fixierten Präparats. Z. Vererbungslehre Suppl. **1**, 402—406 (1928). — Beiträge zur Kausalanalyse der Mitose. II. Untersuchungen an den Spermatocyten von *Chorthippus (Stenobothrus) lineatus* PANZ. Roux' Arch. **118**, 359—484 (1929a). — Beiträge zur Kausalanalyse der Mitose. III. Untersuchungen an den Staubfadenhaarzellen und Blattmeristemzellen von *Tradescantia virginica*. Z. Zellforsch. **10**, 73—134 (1929b). — BENNETT, H. S.: The microscopical investigation of biological material with polarized light. In McCLUNG's Handbook of Microscopical Technique, 3. ed., p. 591—677. 1950. — CARLSON, J. G.: Microdissection studies of the dividing neuroblast of the grasshopper *Chortophaga viridifasciata* (DE GEER). Chromosoma **5**, 199—220 (1952). — CLEVELAND, L. R.: Origin and development of the achromatic figure. Biol. Bull. **74**, 41—55 (1938). — CONKLIN, E. G.: Effects of centrifugal force on the structure and development of the eggs of *Crepidula*. J. of Exper. Zool. **22**, 311—419 (1917). — COOPER, K. W.: The nuclear cytology of the grass mite, *Pediculopsis graminum* (REUT.), with special reference to karyomerokinesis. Chromosoma **1**, 54—103 (1939). — Visibility of the primary spindle fibers and the course of mitosis in the living blastomeres of the mite *Pediculopsis graminum* REUT. Proc. Nat. Acad. Sci. U.S.A. **27**, 480—484 (1941). — CORNMAN, I.: A summary of evidence in favour of the traction fiber in mitosis. Amer. Naturalist **78**, 410—422 (1944). — HARVEY, E. B.: Development of half eggs of *Chaetopterus pergamentaceous* with special reference to parthenogenetic merogony. Biol. Bull. **76**, 384—404 (1939). — HUGHES, A. F., and M. M. SWANN: Anaphase movements in the living cell. A study with phase contrast and polarized light on chick tissue cultures. J. of Exper. Biol. **25**, 45—70 (1948). — HUGHES-SCHRADER, SALLY: Cytology of coccids *(Coccoidea-Homoptera)*. Adv. Genet. **2**, 127—203 (1948). — INOUÉ, S.: Studies on depolarization of light at microscope lens surfaces. I. The origin of stray light by rotation at the lens surfaces. Exper. Cell Res. **3**, 199—208 (1952a). — Improvements of the Polarization Microscope for Biological Purposes. Trans. Amer. Microsc. Soc. **71**, 311—312 (1952b). — The effect of colchicine on the microscopic and submicroscopic structure of the mitotic spindle. Exper. Cell Res. Suppl. **2**, 305—318 (1952c). — Inoué, S., and K. Dan: Birefringence of the dividing cell. J. of Morph. **89**, 423—455 (1951). — Mead, A. D.: The origin and behavior of the centrosomes in the annelid egg. J. of Morph. **14**, 181—218 (1898). — Morgan, T. H.: Cytological studies of centrifuged eggs. J. of exper. Zool. **9**, 593—656 (1910). — Mottier, D. M.: Beiträge zur Kenntnis der Kernteilung in den Pollenmutterzellen einiger Dikotylen und Monokotylen. Jb. wiss. Bot. **30**, 169—204 (1897). — Osterhout, W. J. V.: Über Entstehung der karyokinetischen Spindel bei *Equisetum*. Jb. wiss. Bot. **30**, 159—168 (1897). — Schmidt, W. J.: Kernspindel und Chromosomen im lebenden, sich furchenden Ei von *Psammechinus miliaris* (Müll.). Ber. oberhess. Ges. Naturheilk. **17**, 140 (1936). — Doppelbrechung der Kernspindel und Zugfasertheorie der Chromosomenbewegung. Chromosoma **1**, 253—264 (1939). — Die Doppelbrechung des Protoplasmas und ihre Bedeutung für die Erforschung seines submikroskopischen Baues. Erg. Physiol. **44**, 27—95 (1941). — Schrader, F.: Recent hypotheses on the structure of spindles in the light of certain observations in *Hemiptera*. Z. Zool. **142**, 520—539 (1932). — On the reality of spindle fibers. Biol. Bull. **67**, 519—533 (1934). — Mitosis. The movements of chromosomes in cell division. Columbia University Press 1944. — Shimamura, T.: Studies on the effect of the centrifugal force upon nuclear division. Cytologia **11**, 186—216 (1940). — Swann, M. M.: Protoplasmic structure and mitosis. I. The birefringence of the metaphase spindle and asters of the living sea-urchin egg. J. of Exper. Biol. **28**, 417—433 (1951a). — Protoplasmic structure and mitosis. II. The nature and cause of birefringence changes in the sea-urchin egg at anaphase. J. of Exper. Biol. **28**, 434—444 (1951b). — Wada, B.: Mikrurgische Untersuchungen lebender Zellen in der Teilung. II. Das Verhalten der Spindelfigur und einige ihrer physikalischen Eigenschaften in den somatischen Zellen. Cytologia **6**, 381—406 (1935). — The mechanism of mitosis based on studies of the submicroscopic structure and of the living state of the *Tradescantia* cell. Cytologia **16**, 1—26 (1950).

Dr. Shinya Inoué, Department of Anatomy,
School of Medicine, University of Washington,
Seattle 5, Washington, USA.

Druck der Universitätsdruckerei H. Stürtz AG., Würzburg

Reprinted from *Experimental Cell Research*, Vol. 2, pp. 305–318, 1952, with permission from Elsevier.

THE EFFECT OF COLCHICINE ON THE MICROSCOPIC AND SUBMICROSCOPIC STRUCTURE OF THE MITOTIC SPINDLE*

SHINYA INOUÉ**

Department of Biology, Princeton University, Princeton, New Jersey, and the Marine Biological Laboratory, Woods Hole, Massachusetts

The plant alkaloid, colchicine, suppresses mitosis and hence cell division in both higher animals and plants. On the other hand, the multiplication of the chromosomes, although delayed, may not be interrupted, and so colchicine treatment can result in cells with abnormally large numbers of chromosomes (10, 2, 12).

Mainly from observations of the behavior of chromosomes, it has been suggested that colchicine quite specifically affects the submicroscopic organization of the mitotic spindle (1, 13, 14, 17). Since, however, the submicroscopic structure of the mitotic spindle has not been adequately understood, and since the structure of the spindle in most cells could not be seen under the microscope without fixing and staining the cells, the proposed effects of colchicine on the fine structure and mechanism of the spindle have been somewhat speculative.

With recent improvements in the polarization microscope, which will be described elsewhere, it has become possible in living and functioning cells of various organisms to see the spindle fibers, owing to their weak but definite birefringence. Furthermore, a new technique enables rapid and precise measurement of the amount of double refraction characterizing these spindle fibers (5). These improvements have made possible qualitative and quantitative study of structural changes in the mitotic spindle under the influence of various agents.

The present paper reports some observations of the effect of colchicine on the microscopic and submicroscopic structure of the first maturation division spindle of the egg (oocyte) of a marine annelid worm, *Chaetopterus pergamentaceous.*

MATERIALS AND METHODS

The first maturation division spindle of *Chaetopterus pergamentaceous* forms in about 15 min at 25°C after the egg is shed in sea water. The metaphase spindle then attaches itself by one of its poles to the cell surface at the animal pole of

*Presented to the faculty of Princeton University in partial fulfillment of the requirements for the degree of Doctor of Philosophy. This research was supported in part by funds of the Eugene Higgins Trust allocated to Princeton University. Thanks are also extended to the American Philosophical Society for use of equipment purchased under a grant awarded to Dr. K. W. Cooper.

**Charlotte Elizabeth Proctor Fellow. Present address: Department of Anatomy, School of Medicine, University of Washington, Seattle 5, Washington.

the egg (11). After this, the spindle stays at metaphase for several hours without proceeding into anaphase unless the egg is further activated. The metaphase spindle is thus maintained over a period of several hours, and provides a very convenient material for studying the effect of various agents upon the fine structure of the mitotic spindle.

The *Chaetopterus* used in the present studies were collected during July and August of 1950 and 1951 by the Marine Biological Laboratory, Woods Hole, Massachusetts. The females were kept separately from the males in large finger bowls aerated with running sea water. The oocytes were collected from amputated parapodia of female worms. After removing tissue fragments from among the eggs with a medicine dropper, the eggs were washed in three changes of fresh, filtered sea water.

Approximately 1 hr after the eggs were shed, they were centrifuged in an air turbine centrifuge between a layer of isotonic sucrose solution (ten parts 1.1 *M* sucrose-distilled water mixed with one part filtered sea water) and sea water. The centrifugation threw the birefringent and light-scattering yolk granules and oil drops to the centrifugal and centripetal poles, respectively, and eventually pinched the eggs into four fragments. One of these quarters contained the intact spindle and some yolk granules (4). Such quarter eggs, which formed a middle layer in the centrifuge tube, were collected and immediately washed twice with fresh sea water.

After the centrifugation, the quarters still had to be compressed to about 25 μ thickness, so that the strongly birefringent yolk granules were sufficiently displaced to show the detailed structure of the spindle under the polarization microscope. The compression, however, did not seem to interfere with the essential structure or function of the spindle, since inseminated quarters would still go through the subsequent maturation divisions in the compressed state. Furthermore, by subjecting the eggs to low temperature, the spindles in the compressed quarter eggs can be made to disappear completely; thereafter they reappear as perfectly normal spindles once again in the compressed eggs when returned to room temperature. The structure of the spindle in the compressed quarters, therefore, can be considered to be quite normal.

Colchicine solutions were applied to the eggs in the following manner (16). With a braking pipette, a known quantity of colchicine solution was placed on the slide, and similarly, a known quantity of egg suspension on the cover glass. The volumes were so chosen that when the two were mixed upon contact with each other and the cover glass laid flat, the eggs were compressed to the desired thickness. The lapse of time from the application of the colchicine to the observation was thus minimized.

Slides, cover glasses, and all glassware were washed thoroughly in calgonite, rinsed with running tap water, immersed in hydrochloric acid-alcohol, and finally dried from double distilled water after thorough rinsing in tap water and distilled water.

The improved polarization microscope, whose design and characteristics will be described in detail elsewhere, was used under the following conditions. The light source was an A-H6 water-cooled high pressure mercury arc lamp with a Wratten No. 58 green filter. A 45-mm, 0.85 NA coated objective, especially selected free from strain, was used with a Homal ocular for photography, and an 18x compensating ocular for visual observation. The condenser was an 8-mm 0.50 NA coated strain-free objective lens.*

A thin rotating mica plate compensator was inserted in the optical system between the polarizer and condenser in order to improve the sensitivity of the instrument for detecting small retardations (6).

The retardation of the spindle and its components was measured by a newly developed method which employs a small comparison crystal (5). The sensitivity of this method in the present studies was 1 mμ retardation, and each measurement was completed within about 20 sec.

The polarizers were specially prepared Glan-Thompson prisms, and the extinction factor of the whole system with the above mentioned lenses was approximately 40,000 at a working aperture of 0.5.

OBSERVATIONS ON THE EFFECT OF COLCHICINE

Plate I shows the first maturation division spindle in a living *Chaetopterus* oocyte as it appeared under the improved polarization microscope. In the photograph, the main axis of the positively birefringent spindle lies parallel to the slow axis of the compensator (additive position), and so the chromosomal and continuous fibers shine brightly (see reference (15) for definition of the spindle components).

Those astral rays lying perpendicular to the spindle axis appear darker than the background, even though they are also positively birefringent. The axes of these astral rays are perpendicular to the slow axis of the compensator (subtractive position), and so the effects of their retardations cancel each other out (6).

The *Chaetopterus* chromosomes display a much weaker birefringence than the spindle fibers, but by contrast, show up as dark grey bodies against the brilliant spindle fibers.

In fresh sea water, the spindle was observed to stay in this state for several hours at room temperature, provided that the slide and cover glass were properly cleaned and the preparation sealed with vaseline in order to prevent evaporation of the sea water and further compression of the egg.

When such an egg (centrifuged or uncentrifuged) was immersed in colchicine sea water, the spindle disappeared in a few minutes or in the course of an hour or so, depending on the concentration of colchicine. Figure 1 shows the relation between

*The author wishes to express his gratitude to the Bausch and Lomb Optical Company for supplying him with the specially selected and prepared objectives and Glan-Thompson prisms.

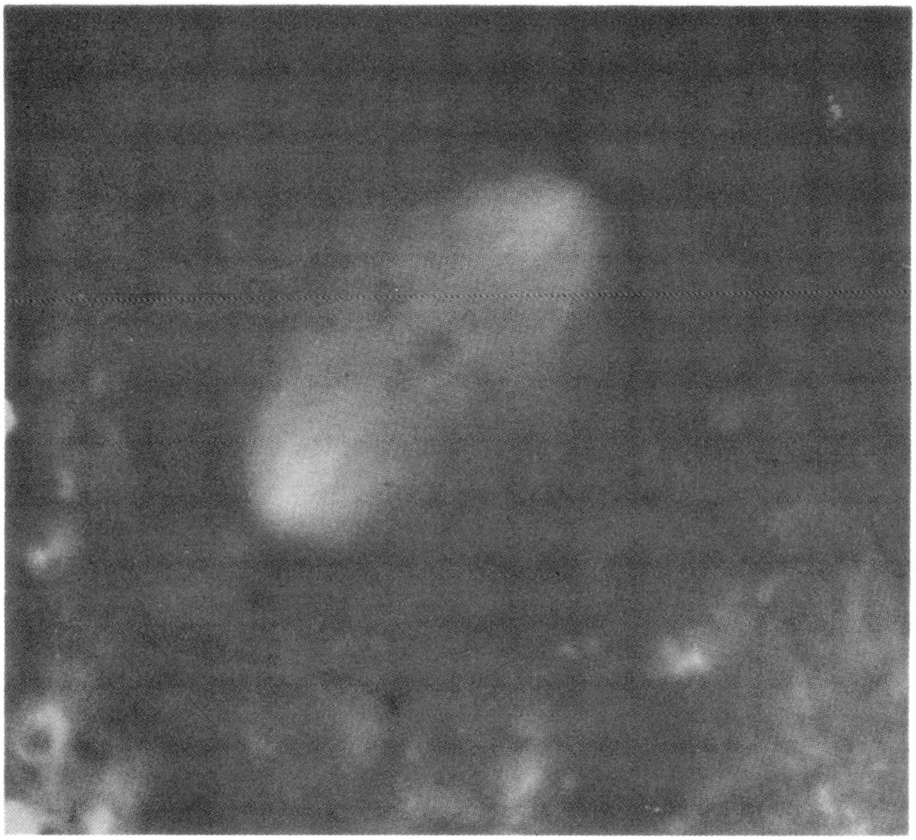

Plate I. First maturation division spindle in a living *Chaetopterus* oöcyte.

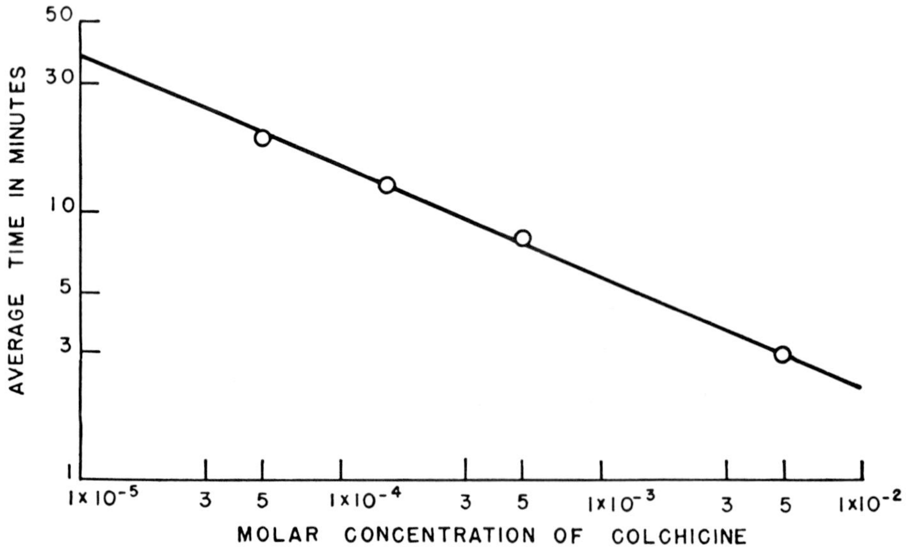

Fig. 1. Average time for disappearance of the *Chaetopterus* metaphase spindle in various concentrations of colchicine sea water (25° C).

colchicine concentration and the approximate time for the complete disappearance of the spindle at 25°C.

Soon after the *Chaetopterus* egg was immersed in colchicine sea water, the birefringence of the astral rays and the continuous fibers began to decrease, while the spindle length also commenced to shorten (Plate II, a, b, e, f). As the spindle shortened, the birefringence of the astral rays and continuous fibers gradually disappeared (Plate II, c, g), and the thick chromosomal fibers also lost their birefringence, though less rapidly than the other components of the spindle. Finally the birefringence of the shortening chromosomal fibers, which by that time appeared collectively to resemble the shape of an empty cage (Plate II, h), also disappeared and so the whole spindle vanished from sight.

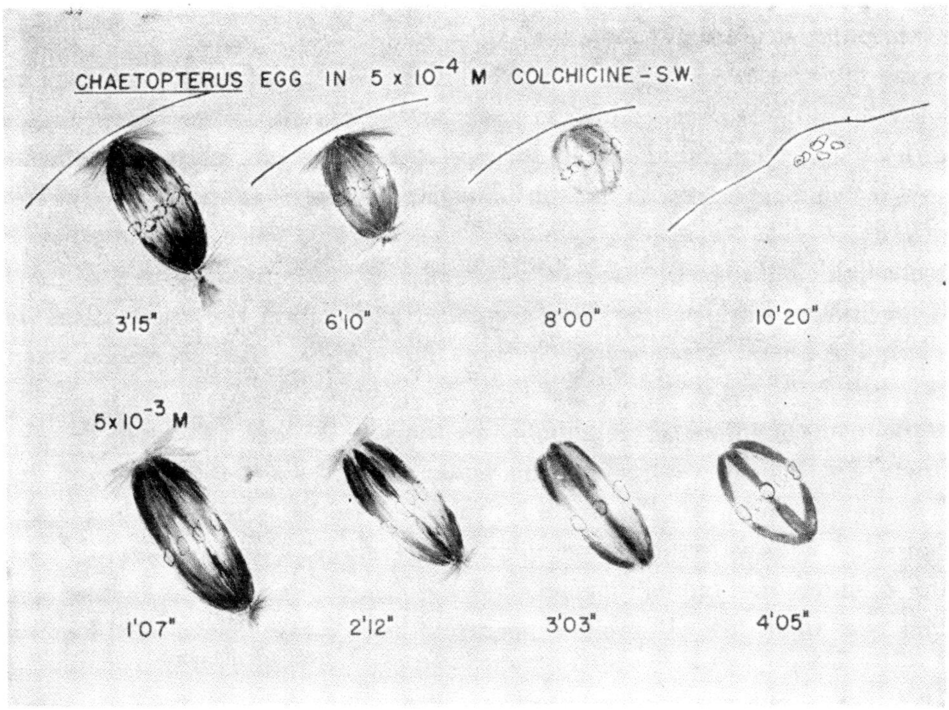

Plate II. a-d. (Left to right above), e-h (left to right below). Change of the *Chaetopterus* spindle under the influence of colchicine, traced from photographs. Time in minutes and seconds after immersion of the eggs in colchicine solutions.

The length of the spindle at which the birefringence of the spindle fibers disappeared depended on the concentration of colchicine, but in all cases the chromosomes stayed regularly arranged on the equatorial plate until the birefringence of the spindle had become undetectable. But then, as soon as the birefringent spindle had disappeared, the chromosomes began to scatter, regardless of the length of the spindle at the time. This indicates that there is a close relationship between the birefringence and mechanical integrity of the spindle, as well as between the integrity of the spindle and the maintenance on the metaphase plate of chromosomes.

The maturation division spindle of the *Chaetopterus* egg is normally attached by one of its poles to the peripheral layer of the egg (Plate II, a) (11). As the spindle shortened under the influence of colchicine, the other pole of the spindle and the chromosomes on the equatorial plate were seen to move towards the cell surface to which the spindle was anchored (Plate II, c, d). During this movement the equatorial plate always stayed half way in between the spindle poles as shown in Plate II. Such movements of the spindle pole and equatorial plate indicate actual contraction of the spindle fibers.

With a colchicine concentration of $\sim 5 \times 10^{-4}$ *M*, the birefringence of the spindle did not disappear until the spindle length vanished, and the chromosomes on the equator were carried along until they came very close to the surface of the cell (Plate II, d). There, the birefringence of the spindle disappeared completely and the chromosomes scattered.

After the chromosomes began to scatter, they were frequently observed to move inward for a little distance. Also, as soon as the chromosomes began to scatter, a characteristic bulge appeared at the surface of the egg just above the chromosomes and showed what appears to be an abortive attempt at polar body formation (Plate II, d).

With a colchicine concentration of 5×10^{-3} *M*, the chromosomes did not reach the cell periphery because the birefringence, and hence the mechanical integrity of the spindle fibers, disappeared before the spindle had shortened very much (Plate II, h).

With the concentration of colchicine as low as 10^{-5} *M*, the spindle contracted very slowly. In such concentrations the polar regions were the first to disintegrate, and as many as seven small parallel "spindles" were observed in a single cell, each "spindle" with a chromosome pair at its center.

Figure 2 shows the typical changes of the length and width of the spindle in colchicine sea water of different concentrations. While the width hardly shows

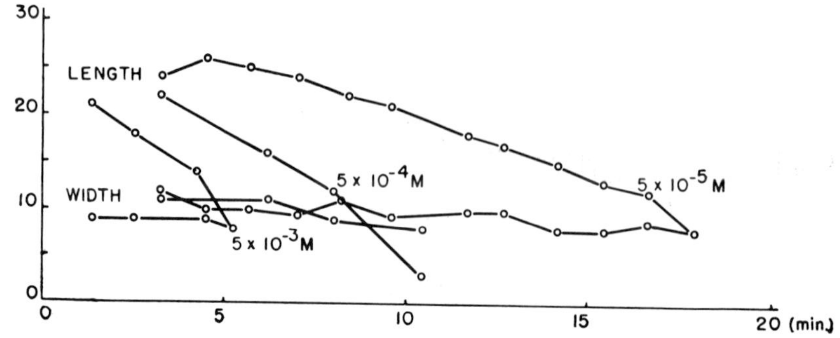

Fig. 2. Shortening of the *Chaetopterus* spindle in colchicine sea water, against time in colchicine solution.

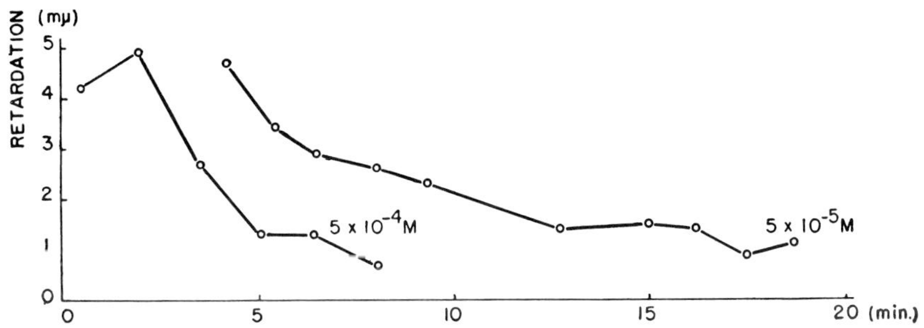

Fig. 3. Retardation change of the *Chaetopterus* spindle in colchicine sea water, against time in colchicine solution.

any change, the length of the spindle decreases gradually at first and then more rapidly as it contracts.

The simultaneous change of retardation of the spindle is graphically represented in Fig. 3. The decrease of retardation is first rapid and then slows down as the remaining retardation becomes weaker. It actually seems to follow an exponential decay curve.

The effect of colchicine on the *Chaetopterus* oocyte is reversible. Eggs treated for 5 and 10 min in 10^{-4} *M* colchicine at 23°C were washed in three changes of fresh sea water. The spindle had then completely disappeared, and the characteristic bulge at the animal pole was present in every egg observed. This condition persisted for over an hour until gradually the bulge receded and a small spindle and asters began to appear. In approximately 3½ hr the spindle had grown to approximately the original size, and many cells showed a single, double or triple spindle depending on whether the centers had separated during the treatment or not. Three hours after the treatment, the eggs could be inseminated and formed normal looking polar bodies.

DISCUSSION

Various authors have suggested that the primary effect of colchicine on the mitotic figure is to cause a sphering of the spindle micelles which results in a dissolution of the spindle (1, 13, 17, 3). The bases for this suggestion are: 1. The existence of mechanical and optical anistropy in the untreated spindle; 2. the behavior of the chromosomes in treated cells (c-mitosis); 3. the disappearance of the achromatic figure observed in treated and fixed cells; 4. the diminution of the spindle "area" in treated, unfixed and fixed cells.

Although such observations do favor the above-mentioned hypothesis, it is nevertheless essential in evaluating the hypothesis critically to study birefringence data obtained from living cells. In the following, therefore, the changes in the submicroscopic structure of the treated spindle will be discussed in light of the polarization optical studies described in the previous section.

When the *Chaetopterus* egg, which has not been treated with colchicine, is compressed between slide and cover glass, the length of the spindle will be found to increase proportionately with the diameter of the egg. If the egg is compressed very gradually, the spindle can be greatly elongated, but if the compression is rapid, the spindle apparently loses its mechanical connection with the more peripheral part of the egg and begins to contract.

The width of the extended spindle in a compressed egg does not change very much at first, but as the compression of the egg and consequently the extension of the spindle becomes very great, the width decreases. The change in the shape of the spindle during compression of the egg leads one to believe that the extension of the spindle is not caused by an over-all flattening of a gelated area in which the spindle lies, but that the two poles of the spindle are actually being pulled apart by some mechanical connection of the poles (the astral rays) with the cell periphery. This idea is further supported by the strict proportionality between the length of the spindle and the diameter of the compressed egg (Fig. 4).

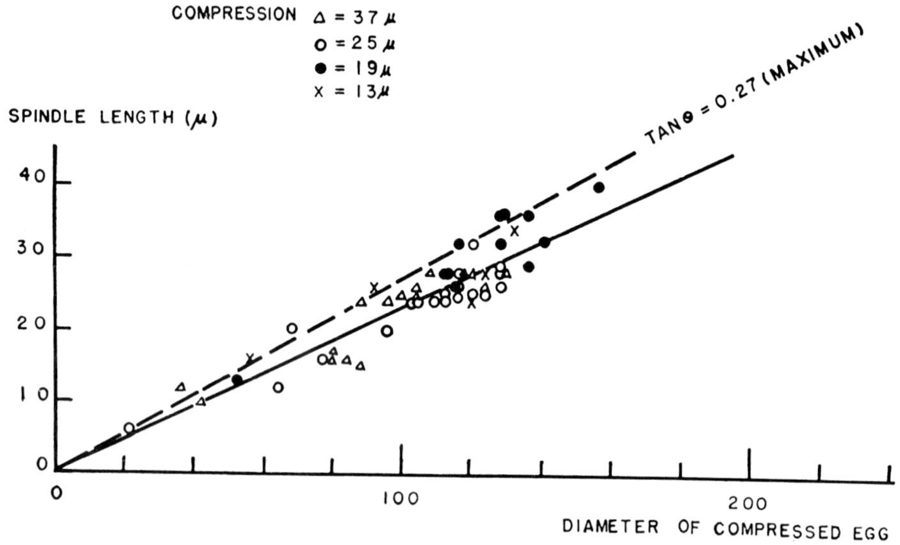

Fig. 4. Spindle length in compressed egg fragments.

When the spindle is extended by compressing the egg, the retardation of the spindle increases. Conversely, when the spindle begins to contract after detaching from the cell periphery, the retardation decreases. The width of the spindle decreases in the latter case together with the length. By plotting the retardation against the length of the spindle, it can be shown that the retardation increases very rapidly as the spindle length is increased (Fig. 5).

At equivalent lengths the retardation of the spindle contracting under the influence of colchicine is much smaller than the retardation of a non-treated egg. Furthermore, the width of the spindle in a treated egg remains constant, and so the

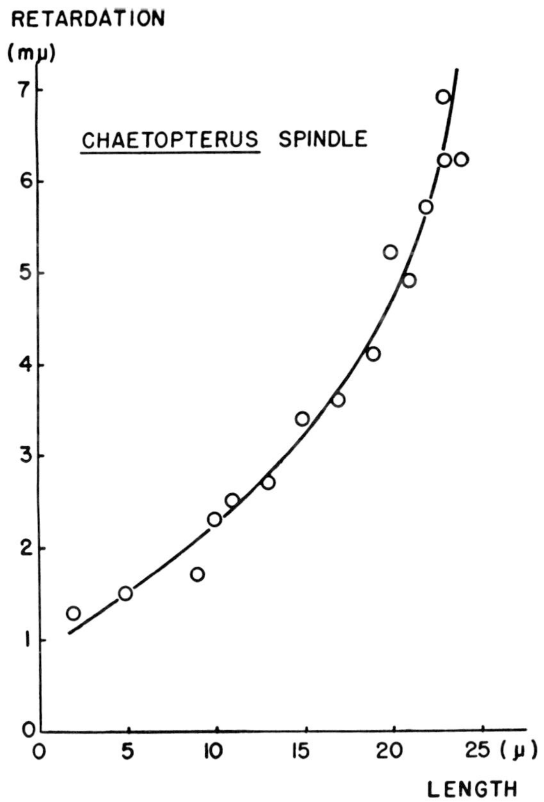

Fig. 5. Retardation of a spindle at various lengths. The initial length of the spindle was 20 μ. See text for the method used in changing the spindle length.

birefringence per unit thickness of the spindle is greatly reduced. This must mean a very high degree of micellar disorientation in the colchicine treated spindle.

At the same time, the spindle affected by colchicine is able to perform mechanical work and actually pull the poles and chromosomes together. The disorientation of the spindle micelles, therefore, is accompanied by active contraction of some of the spindle elements.

Furthermore, since the volume of the spindle in the colchicine treated egg may be reduced almost to zero, it should be assumed that many of the disoriented micelles have been excluded from the main bulk of the spindle where lie the micelles that have not yet been disoriented. This strongly suggests a breakdown of linkages either within or between the spindle micelles.

In this connection, it is interesting to note that Gaulden and Carlson (3) have described the appearance of a hyaline globule in grasshopper neuroblasts after treatment with colchicine. They interpret this globule as having arisen from the spindle substance whose molecular configuration had changed and which had escaped from its original position in the cell.

In summary then, the action of colchicine on the submicroscopic organization of the spindle of the *Chaetopterus* egg is to disorganize the orientation of the micelles in the astral rays and spindle fibers, most likely by breaking down some chemical bond in or in between the spindle micelles and simultaneously causing some of the remaining micelles to contract as well as further breaking down the linkage between them. It is very likely that the action of colchicine on the spindle is to antagonize the action of a component in the cell which keeps the spindle substance polymerizing and which keeps the spindle micelles in their extended form, in effect preventing the micelles from dissociating from each other and returning to their more stable spherical form. On this score Lettré (8, 9) discusses possible chemical groups involved in such a reaction.

References

1. Barber, H. N. and Callan, H. G., *Proc. Roy. Soc. London* **B131**, 258 (1943)
2. Eigsti, O. J. and Dustin, P., Jr., *Lloydia* **12**, 185 (1949)
3. Gaulden, M. E. and Carlson, J. G., *Exptl. Cell Research* **2**, 416 (1951)
4. Harvey, E. B., *Biol. Bull.* **76**, 384 (1939)
5. Inoué, S., *Exptl. Cell Research* **2**, 513 (1951)
6. Inoué, S. and Dan, K., *J. Morphol.* **89**, 423 (1951)
7. Lehmann, F. E. and Hardon, H., *Helv. Physiol. et Pharmacol. Acta.* **4**, 11 (1946)
8. Lettré, H., *Naturwissenschaften* **30**, 34 (1942)
9. Lettré, H., *Naturwissenschaften* **34**, 127 (1947)
10. Levan, A., *Hereditas* **24**, 471 (1938)
11. Mead, A. D., *J. Morphol.* **14**, 181 (1898)
12. Melander, Y., *Hereditas* **37**, 288 (1951)
13. Östergren, G., *Hereditas* **30**, 429 (1944)
14. Östergren, G., *Hereditas* **36**, 371 (1950)
15. Schräder, F., Mitosis. Columbia University Press, New York, 1944
16. Tyler, A., *Collecting Net* **19** (1946)
17. Wada, B., *Cytologia* **15**, 88 (1949)

Discussion

(Editor's note: The following notes were prepared by Dr. Inoué to describe his motion picture which preceded the discussion.)

Several lantern slides and a motion picture of normally dividing cells as seen through the improved polarization microscope were demonstrated. The description of most of these will be reported in another paper. Since, however, much of the following discussion centers around the motion picture, a brief description of the film is given below.

The material is a pollen mother cell of *Lilium longiflorum*. The picture covers the period from the first maturation division metaphase to the second division telophase. The cell was centrifuged mildly (3 min at 1800 *g*) while it was still in the anther, in order to displace the birefringent granules which would otherwise obscure the appearance of the spindle. The cell is immersed in ordinary Ringer's solution (Belar's solution) with isotonic lactose substituted for glucose.

Four exposures were made per minute with approximately 3-sec intervals; the exposures are therefore approximately 12 sec each.

The picture was taken through an improved polarization microscope, using a Wratten 58 green filter, and an AH-6 mercury arc lamp as the light source. The field is compensated with a very weak retardation plate (2.8 mμ) so that the positively birefringent spindle appears bright, while the new cell wall lying perpendicular to the spindle appears dark.

The main points that the picture shows are:

1. An unequivocal demonstration of longitudinal arrays of oriented material (spindle fibers) in living and dividing cells.
2. The shortening of the chromosomal fibers.
3. The over-all elongation of the spindle during anaphase.
4. The overshooting of the chromosome groups beyond the spindle poles in telophase.
5. The existence of numerous continuous fibers which regain their birefringence after the daughter nuclei are formed.
6. The formation of the phragmoplast from these continuous fibers.
7. The sequence of the cell plate formation in the phragmoplast.

Novikoff: I would like to inquire about the "pulling" of the chromosomes by the spindle fibers. Might not the fact that the chromosomes move beyond the fibers suggest that the movement of the chromosomes does not require "pulling" by the spindle?

Inoué: Although your proposition is not totally improbable, I do not believe that to be the case. Experiments in which the spindle is elongated by compressing the egg, or by changing the temperature, indicate a close mechanical connection of the chromosomes to the spindle fibers at metaphase and anaphase. I believe that the overshooting of the chromosomes in telophase may be caused by a different or separate mechanism from anaphase movement. The appearance of the chromosomes changes by the time the overshooting takes place and shows a swelling of the chromosome vesicles. Perhaps it is this swelling and aggregation process which pushes the chromosomes beyond the pole.

Sparrow: There is another possible explanation for this so-called overshooting that most cytologists would think of. That is, the great attraction the chromosomes have for each other as they go into telophase. At anaphase chromosomes tend to repulse each other but at telophase they seem to have a strong mutual attraction. Now, if they were near the end of a spindle-shaped body and developed this great attraction for each other, I think they would as a result slip off the end of the spindle and thus give rise to the so-called overshoot.

Stern: I should like to make some physical-chemical comments at this point. One concerns the subject of the reality of the spindle apparatus. From the work of Schmidt it seems that the spindle fibers become birefringent in polarized light only when the cell has been fixed. This might be taken to indicate that the spindle fibers are a fixation artifact. Another point on which I would like you to comment is, what interpretation do we have to place on the phenomenon itself? There are three types of birefringence which you can observe under these conditions. The first one is intrinsic birefringence. The particle itself is anisotropic, like the crystal which you put into the field of the microscope. The second type is form birefringence, or rodlet birefringence which would only occur if individual particles which are asymmetric, although possibly isotropic, are oriented along flow gradients. It has been suggested that spindle "fibers" are produced by such an alignment of asymmetric particles. The third type is one which forms the basis of Pfeiffer's observations on the birefringence of protoplasmic fibers when they are put under mechanical tension. This is the so-called strain birefringence. You have mentioned in your talk that many of these pictures were taken while the cells are under mechanical pressure and I wonder whether you would comment on which of these different types of birefringence you believe are observed in your pictures.

Inoué: As to the first point, I believe that Schrader and others have shown quite adequately that spindle fibers are not mere artifacts. However, since many people were still not satisfied without a direct visualization of the fibers in living cells, I would like to emphasize that all of the pictures I have shown you here were taken with living material. Incidentally, W. J. Schmidt's description of the birefringence of the spindle was made in living as well as fixed material.

The second point, about what kind of birefringence we have here, raises a very interesting question. I have taken form birefringence curves of the spindle, but they have been on Bouin fixed material and I didn't want to present the curves here because the fixation itself may have caused all kinds of modifications on the sub-microscopic structure. In fact, I believe that we ought to study with the polarization microscope the changes which occur during fixation so that we can tell how good an image we really get in microscopy and electron micrography. Anyhow, with the Bouin fixed material, most of the birefringence was rodlet birefringence and very little was intrinsic birefringence. The refractive index of the micelles themselves was 1.54 as you find for most proteins.

Also, there is the factor of tension. Whether we can call this strain birefringence or not is a problem about which I have been thinking. When we pull glass we get strain birefringence as also in some organic substances, but the signs of birefringence may be different. For instance, W. J. Schmidt has shown that in cherry gum, negative birefringence is developed in the direction of the pull and a positive birefringence in the direction of compression. This is opposite to the behavior of glass and of many other substances. If, however, you pull cherry gum very rapidly, you can get a positive birefringence which quickly changes to the negative. The first, positive part, is supposed to be a true strain birefringence and the second negative due to reorientation. In a very soft gel, I think, it is difficult to conceive of true strain birefringence and what we call strain birefringence there is actually an orientation birefringence. Kunitz, in fact, has shown with gelatin models, about 1938, that by making them swell anisotropically, you can get so-called strain birefringence which is actually a non-isotropic orientation birefringence. You can calculate the amount of swelling in each direction and from the concentration determine exactly what the amount of this birefringence ought to be. So that, I think, in gel systems such as the spindle, we can assume that the birefringence is not pure strain birefringence but orientation birefringence. But this also indicates how much strain there is on the spindle since strain is one of the factors that orients the micelles.

Tahmisian: This really intrigues me very much because I have been working on a very similar thing using instead of polarized light, a standard phase contrast system. I have some pictures showing that the spindle is actually a real thing. In cells which have been treated with colchicine 24 hr before examination, there is no spindle formed. I would like to answer Dr. Novikoff's question as to whether you can have cell division in the absence of a spindle. We have been able to knock out spindles by irradiation and the cell will go ahead and divide in an anastrous division, but if no spindle is present the morphological identity of the chromosomes is lost; the chromatin mass is pinched in two. However, in order to have a normal cell with normal chromatin arrangement, it appears that a spindle is necessary.

Inoué: I do not wish to give the impression that the chromosomal fibers are the only active participants in chromosome movement. Actually in order to separate the chromosomes there has to be a point of attachment for the chromosomal fibers, and I myself believe that there is a mass of structure in the center of the spindle, probably swelling, which supports the spindle poles and holds or even pushes them apart, while the chromosomal fibers which are attached to the poles are being pushed away and also contracting at the same time.

Lessler: These excellent motion pictures of cell division show a phenomenon which I think has been overlooked. After the chromosomes have arranged themselves in a metaphase plate, and just prior to their anaphase movement, there is a considerable "swelling" of the spindle in the region of the metaphase plate. This swelling is parallel to the metaphase plate and at right angles to the motion of the separating chromosomes. The primary spindle "elongation" is thus not poleward but in a direction at right angles to a line drawn between the poles.

The first time I saw this phenomenon was in the phase microscope motion pictures of isolated sperm cells of the grasshopper *Psophus stridulus* (L.). These were taken by Dr. K. Michel of the Zeiss Company during the last war (now War Department Official Film Misc. 1298). These phase microscope pictures, as in Dr. Inoué's birefringence film, show a primary spindle "elongation" at right angles to the poles in the region of the metaphase plate. This probably is a gelated phase, occurring at the spindle center, induced by the orientation of the chromosomes. Such a wave of gelation, if progressive, could push the chromosomes poleward. Thus the primary anaphase motion of the chromosomes may be due to the oriented kinetochores inducing between the chromosomes a region of gelation which pushes them apart. Such a pushing could be, especially in fixed preparation, misinterpreted as a pulling. This explanation of the poleward motion of the chromosomes is consistent with the motion picture studies of cell division and provides a reasonable mechanism for the chromosomes overshooting the ends of the spindle in late anaphase.

Stern: I would like to mention in connection with this old controversy the *Sciara* chromosomes which Metz has studied in such detail. He has shown that there are between the chromosomes, as they go away from the metaphase plate, fibers, and he is firmly convinced that these fibers push the chromosomes whereas others from the pole pull them. That might perhaps be a happy compromise.

Levine: With the interesting pictures of the oocytes of *Chaetopterus* as a background, it seems of interest to call attention to the behavior of the somatic cells of the higher plants under the influence of colchicine. In *Allium cepa* root-tips, colchicine arrests nuclear divisions in metaphase. The chromosomes divide and multiploidy results. Observations on the fate of the spindle fibers have been viewed differently. While most workers state that the spindle fibers under the influence of colchicine disappear as shown by Dr. Inoué, Shimamura contends that the spindle substance takes the form of an achromatic sphere; this mass divides after forming a dumb-bell shaped body. Does the polarization microscope offer any evidence on this phase of the spindle behavior? Do the chromosomes in the egg increase in number?

Inoué: In animal cells also there are some people who have obtained polyploidy by colchicine treatment, as for instance by treating pig sperm. I think the general effects are the same. I didn't mention anything about the chromosomes, since I was mainly discussing the effect of colchicine on the spindle fibers. In plant cells nobody has made polarization optical studies of the effects of colchicine on spindle fibers in living cells. I have a feeling, however, that it is going to turn out the same way.

About this pushing and pulling action that we have been talking so much about, I hope in the future to be able to show you actually elongating fibers and pushing of the chromosomes.

In the *Chaetopterus* cells when you lower the temperature the spindle disappears but when it comes back at room temperature it actually elongates and pushes the chromosomes. Thus, it is not at all surprising that the spindle can both push and pull even in cells in which ordinarily pulling is the main contribution to chromosome movement. The spindle is probably some kind of gel whose micelles line up and elongate the spindle, while disorientation of the micelles causes contraction of the fibers.

PART VI OF DOCTORAL THESIS

Shinya Inoué*

1. The Submicroscopic Structure of the Spindle in Living Cells

The reality of spindle fibers in living animal and plant cells, and the distinct differentiation of the chromosomal fibers and other birefringent components have been demonstrated in Part IV of this thesis [Article 10].

Now that the spindle fibers (and fibrils) can no longer be considered to be fixation artifacts, several of the proposed hypotheses on the submicroscopic structure of the spindle and the mechanism of chromosome movements are automatically eliminated (see Schrader[27,28] for an extensive catalog of the various hypotheses which have been proposed).

Here, I propose to discuss three hypotheses concerning the spindle (and astral) structure, which have been, at least in part, based on the existence of optical anisotropy.

Soon after the discoveries of Schmidt[22] and Runnström[20] that the spindle is birefringent, Freundlich[8] pointed out the resemblance of the shape and birefringence of the spindle to a tactoid. A tactoid is a kind of liquid crystal which is composed of needle-shaped particles like vanadium pentoxide crystals or elongated protein micelles like the tobacco mosaic virus. Thc lateral distance between the micelles may reach up to several hundred Ångstroms, and the micelles are believed to be held together by a force similar to Van der Waals' force or the forces as described by London.[14] This lateral force must be very small, for Freundlich himself has observed the (lateral) Brownian motion of crystals in a vanadium pentoxide tactoid. Although he had suggested that the spindle may be a tactoid, Freundlich, however, does not describe how a tactoid could perform mechanical functions such as the separation of bodies embedded within or attached to it, or the initiation of a cleavage of the medium in which a tactoid lies.

Later, Bernal[4] proposed a new tactoid hypothesis of the spindle, in which he assumes the existence of negative tactoids. A negative tactoid is a spindle-shaped region in an ordinary (positive) tactoid lacking the micellar element which composes the positive tactoid. He thinks that the chromosome movements could be explained if the negative tactoids, in which he considers the chromosomes to lie, carried the chromosomes along to the spindle pole as they themselves were shifted to the poles by certain surface forces.

Barber and Callan[2] studied the chromosomal configuration in newt cells after treatment with cold and with colchicine and gave an interpretation to the micellar disorganization of the spindle which was entirely based on Bernal's tactoid hypothesis. A strongly damaging weakness of their argument, as in most of the other cases, is the lack of any information

*Unpublished work (1951).

concerning the micellar orientation of the spindle they had studied.

Östergren[15] also explains the effect of colchicine to be the rounding up and disorientation of the spindle tactoid micelles. His adherence to the tactoid hypothesis stems from his own observations on mitosis in *Luzula*, which seemed to show that chromosomal fibers could be traversed by chromosomes at metaphase without ultimate destruction of their mechanical integrity.[16]

Similar reversible rifts of spindle fibers or astral rays can probably be made to occur very easily, for instance as in the microdissection studies of Chambers[5,6] or when the spindle is treated by cold.[9]

Although this reversible growth process of the spindle fibrils and astral rays does offer the greatest interpretive challenge to the micellar structure of the spindle, I do not see how the capacity of the fibers or fibrils that go through such reversible changes would especially favor the tactoid hypothesis. In any event the following discussions, based on observations of the birefringence of spindles in living cells, would seem to rule out the possibility of a tactoid hypothesis.

The first of the possibilities, namely, that the spindle as a whole is a single tactoid, is eliminated simply by the complicated distribution of retardation in a spindle and its differentiation into fibers and fibrils. The second alternative, for example, Bernal's scheme, may not be as simply dismissed as the first, for the weaker retardation of the equatorial region of the metaphase *Chaetopterus* spindle (Fig. 2 in Ref. 12) would, at first, be thought to be attributed to the lack of birefringent material in the negative tactoids. However, the demonstration of strongly birefringent chromosomal fibers is contradictory to the expectation of Bernal's hypothesis. Also, the divergence of birefringent material towards the polar region of the early metaphase spindle in *Lilium* (and other plant) pollen mother cells could not possibly be explained by Bernal's scheme, for such a configuration of the tactoid would be extremely unstable.

The last alternative is, then, that each of the spindle fibers is a tactoid. One of the main difficulties here would be the high mechanical integrity that the spindle fibers can show when they are laterally deformed (for instance, by centrifugation, Schrader,[26] Shimamura,[21] and others) or when they are stretched gently by compression of the egg. Even if we allow certain unknown forces to account for this, the divergence of each of the fibers at the polar side of the *Lilium* spindle in early metaphase is sufficient evidence to discredit the last of these alternatives.

All in all, then, the tactoid hypothesis can no longer be considered to be a general hypothesis for explaining the spindle structure and, consequently, for explaining chromosome movements.

From his findings in microdissection studies, Chambers[5,6] maintains that the astral rays are watery channels which lie in a gelated astrosphere. He judged by eye that the spherical empty spaces at the poles of the spindle in several marine eggs offered no resistance to the microneedle he inserted and moved around. He also saw centripetal movements of granules in the eggs along the length of astral rays, and therefore concluded that the polar regions of the spindle were liquid pools, where liquid material flowing in through the astral rays accumulates.

The staining properties of the astral rays and spindle fibers, especially at the poles of the spindle, offer great difficulty to the acceptance of Chambers' idea.

Pollister,[19] in connection with his studies of cytoplasmic diffusion in secretory cells, slightly modified, but fundamentally followed, Chambers' idea of the astral structure. He proposed that instead of the astral rays being "watery" channels, they were lines of flow of rather dense cytoplasmic material which contained elongated particles or micelles. Pollister thought that the staining property of the astral rays could be accounted for in this way, and also that the birefringence exhibited by the astral rays was due to the birefringence of flow of the elongated particles.

However, the very strong convergence of the birefringent fibers and fibrils at the spindle poles in the *Chaetopterus* egg (Figs. 1 through 10 in Ref. 12) and in all other animal spindles studied, where asters existed, seems to contradict the notion of a liquid pool at the poles of the spindle. Furthermore, this very strong polar convergence in *Chaetopterus* eggs is maintained for several hours and there is no indication of an accumulating pool (which would have to be isotropic) during this period. The evidence is, therefore, quite strongly against Pollister-Chambers "line of flow" concept of the astral rays.

Schmidt,[23] in his pioneer studies, had already discovered that an "isotropic gap" appears between the two half spindles as the chromosomes separated. He also found that the retardation of the contracting half spindle decreased qualitatively as the chromosomes moved towards the poles, and pointed out the similarity of the change of birefringence of the half spindle component during anaphase to a muscle fiber contracting isotonically. From this, he considered that polypeptide chain folding may occur in the spindle during anaphase.[24]

More recently, Hughes and Swann[10] and Swann and Mitchison,[30] using a photodensimetric method, determined the change of the retardation of a spindle along its long axis during anaphase. Since, however, they did not consider the possible existence of various differentiated structures in the spindle, as I have shown to exist in this thesis, their measurements can only be taken as perhaps average values of the local changes. The hypothesis of chromosome movement proposed by Swann[29] which is based upon these measurements cannot, therefore, be taken too seriously. However, Hughes and Swann have more than sufficiently verified the validity of Schmidt's classical observation as far as the overall birefringence of the half spindle is concerned.

Wada,[32] through his continued studies on the spindle in the living stamen hair cells of *Tradescantia*, also proposes that the spindle (or in general, his atractoplasm) is a protein gel structure. He also gave an explanation for the (expected) change in micellar organization of colchicine-treated spindles, from the standpoint of a protein chain hypothesis.[31]

The protein chain hypothesis does not meet with the morphological difficulties which greatly discredit the tactoid hypothesis. For, in a gel, there could be sufficient (lateral as well as longitudinal) forces holding the micelles in position together, which then accounts for the existence of localized regions in the spindle with higher degrees of orientation and perhaps a denser packing of the micelles. Both the differentiation of strongly birefringent chromosomal fibers and the polar divergence of plant spindle fibers in early metaphase, therefore, cause no great difficulties for the protein chain hypothesis, as long as we can find an adequate explanation for the forces which make the spindle or spindle components take the particular form that they do.

The local increase in birefringence which reveals the chromosomal fibers at anaphase, or when the spindle is stretched, is a phenomenon which would be definitely expected if the spindle or spindle fibers are composed of protein gels. A similar increase in the retardation of a spindle stretched by centrifugation was observed through a centrifuge microscope by Pfeiffer,[18] although he does not commit himself to a definite explanation for the increase in retardation.

The part of the spindle in anaphase between the separating chromosomes which Bělăr[3] called the "Stemmkörper" has been shown by many to be much less resistant to deformation than the half spindle region.[21,26] The low mechanical integrity of the Stemmkörper is also associated with a weaker staining property in well fixed preparations, as in Bělăr's grasshopper spermatocytes.[3] It is very likely that the Stemmkörper is composed of continuous fibers which have become highly hydrated and swollen. The appearance of very discrete wavy and sparse continuous fibers in many fixed preparations strongly supports this idea. It is therefore not surprising at all that the phragmoplast forms a continuous structure from the continuous fibers in the Stemmkörper region at telophase, as the motion picture of the *Lilium* pollen mother cell division clearly shows. The very weak birefringence of the Stemmkörper at anaphase is, therefore, completely in agreement with the expectations from other cytological evidences, since, as Kunitz[13] and others showed, a high degree of hydration and swelling would eventually result in very low birefringence.

The experiments with colchicine on the *Chaetopterus* spindle, where an active contraction of the decomposing spindle was demonstrated, could also be quite adequately explained from the protein chain hypothesis where side linkages of the micelles are naturally expected.[11]

As mentioned before, the most difficult question in the polypeptide chain hypothesis is how the spindle fibrils or astral rays grow.[1,7,25] Soon after fertilization, when the monaster grows, and at each successive division of cells, how do these fibers form over and over again very rapidly and as very fine straight threads? The problem can probably be solved in the future by studying under the polarization microscope the effects of various agents, such as cold[9] colchicine,[11] or high hydrostatic pressure,[17] which make the spindle fibers and astral rays disappear reversibly.

Once the mechanism of these fibril formations is found, the reversible disappearance of the astral rays in Chambers' microdissection studies and of the spindle fibers in Östergren's *Luzula* case will perhaps find its explanation as a rapid process of new formation. As indicated above, microdissection of living spindles and astral rays under the polarization microscope is very greatly desired.

The protein chain hypothesis, then, seems to provide the most natural explanation for the submicroscopic organization of the spindle, without contradicting any of the findings of spindle birefringence. This, however, by no means shows that the problem of the submicroscopic organization of the spindle and astral rays is anywhere near settled. Rather, it should be considered that we have just reached the stage where we are at last able to study the structure and function of the spindle and several other cell organelles in their living and active form. Several experiments, simultaneously combined with polarization optical observations, will have to be made before we can truly formulate what the molecular structure and their changes are in the spindle of living cells.

References

1. Astbury WT, *Nature* **135**: 765, 1935.
2. Barber HN, Callan HG, *Proc Roy Soc London* **131**: 258, 1943.
3. Bělăr K, *Arch Entwmk* **118**: 359, 1929.
4. Bernal JD, *Publ Am Ass Adv Sci* **14**: 199, 1940.
5. Chambers R, *General Cytology* p. 235, Univ of Chicago Press, 1924.
6. Chambers R, *J Exp Zool* **23**: 483, 1971.
7. Elliot A, Ambrose EJ, Robinson C, *Nature* **166**: 194, 1950.
8. Freundlich H, *J Phys Chem* **41**: 1151, 1937.
9. Hertwig O, *The Cell*, Chapter 7, MacMillan, New York, 1895.
10. Hughes AF, Swann MM, *J Exp Biol* **25**: 45, 1948.
11. Inoué S, *Exp Cell Res Suppl* **2**: 305–318, 1952 [Article 11].
12. Inoué S, *Chromosoma* **5**: 487–500, 1953 [Article 10].
13. Kunitz M, *J Gen Physiol* **13**: 565, 1930.
14. Jury SH, Ernst RC, *J Phys Coll Chem* **53**: 609, 1948.
15. Östergren G, *Hereditas* **30**: 429, 1944.
16. Östergren G, *Hereditas* **35**: 445, 1947.
17. Pease DC, *J Morph* **69**: 405, 1941.
18. Pfeiffer HH, *Biodynamica* **2**: 35, 1938.
19. Pollister AW, *Physiol Zool* **14**: 268, 1941.
20. Runnström J, *Protopl* **5**: 218, 1936.

21. Shimamura T, *Cytologia* **11**: 186, 1940.
22. Schmidt WJ, *Naturw* **24**: 463, 1936.
23. Schmidt WJ, *Chromosoma* **1**: 253, 1939.
24. Schmidt WJ, *Protopl* **35**: 1, 1940.
25. Schmitt FO, *The Harvey Lectures* **40**: 249, 1945.
26. Schrader F, *Biol Bull* **67**: 519, 1934.
27. Schrader F, *Mitosis*, Columbia Univ Press, 1944.
28. Schrader F, *Symp on Cytology*, p. 37, Michigan State College Press, 1951.
29. Swann MM, *VIIth Internatl Conf Cell Biol*, 1950.
30. Swann MM, Mitchison JM, *Nature* **164**: 131, 1949.
31. Wada B, *Cytologia* **15**: 88, 1949.
32. Wada B, *Cytologia* **16**: 1, 1950.

Reprinted from *The Biological Bulletin*, Vol. 103(2), pp. 316–317, 1952.

EFFECT OF TEMPERATURE ON THE BIREFRINGENCE OF THE MITOTIC SPINDLE

Shinya Inoué

The birefringence of the metaphase spindle of the oocyte of *Chaetopterus pergamentaceus* can be abolished when the temperature is either lowered below 4°C–5°C or raised above 30°C–35°C. Upon returning the preparation to intermediate temperatures, a smaller spindle reappears, almost immediately, at the position where the original spindle had disappeared. At 25°C, the new spindle regains the full size in approximately 15 minutes. In eggs treated with cold, the time required for the re-establishment of the spindle is not affected by the duration of the cold treatment, even up to several hours.

During the reconstitution, the spindle goes through a regular series of morphological changes. Firstly, the small spindle migrates rapidly to the cell periphery, then moves slowly away from the surface again as the retardation and length of the spindle increase. At this stage, continuous fibers show strong birefringence, but after the retardation has reached a maximum, chromosomal fibers replace the continuous fibers. The retardation of the individual chromosomal fibers fluctuates for some time until finally the chromosomes are regularly arranged on the metaphase plate. Once the spindle is completed, the birefringence of the spindle is fairly stable for a given temperature.

The equilibrium retardation is a function of temperature and is greater the higher the temperature up to a maximum at 28°C–30°C. At the same time, the retardation of the spindle increases proportionately with spindle length. Since the spindle tends to be longer at higher and shorter at lower temperatures, it would appear as if the temperature dependence of the retardation were correlated with the change in length. However, it can be shown that even at equivalent lengths, the retardation of the same spindle is greater at a higher temperature.

The following note was added by Shinya Inoué in September of 2006:

This abstract records the first observations on the effect of low and high temperatures on the birefringence of spindle fibers in living cells (i.e., the assembly states of the dynamic submicroscopic fibrils, later identified as microtubules). The observations are further documented and analyzed in Articles 16, 22, 25, and 27.

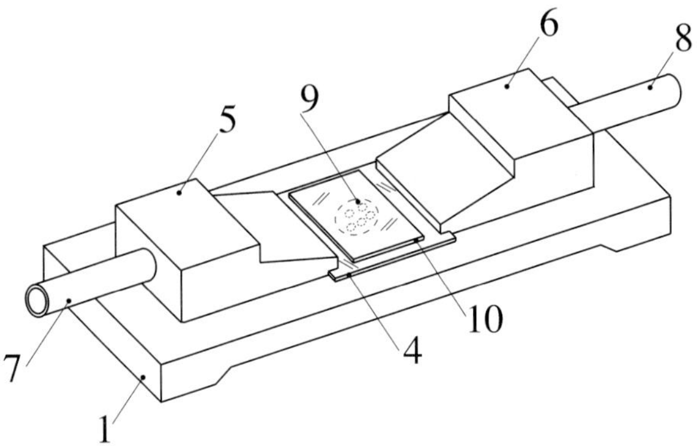

Fig. 1. Overview of temperature control microscope slide.

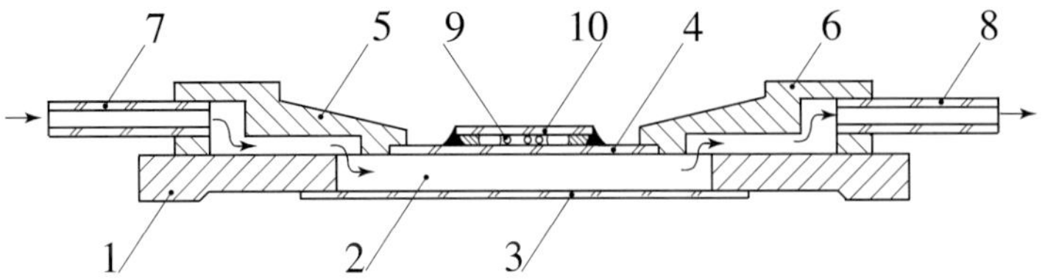

Fig. 2. Median cross section. A 1.6-mm-thick, 30-mm-wide, by 75-mm-long stainless steel base (1), with bottom recessed to accommodate the lower cover slip (3), is provided with an 8-mm-wide, 36-mm-long slot (2) through which the temperature control water flows. The slot is sealed on the bottom by a 40-mm-long cover slip (3) and on top by a 22-mm-by-22-mm cover slip (4) together with plastic manifolds (5 and 6) to which the short inlet- and output-glass tubes (7 and 8) are attached. The specimen (9) in a micro-drop surrounded by air space and moistened filter paper are sandwiched between the cover slips (4) and (10) and sealed with Valap to prevent evaporation (see Article 41 for further details on micro-drop preparations). The specimen is separated from the water flowing through 7, 5, 2, 6, and 8 by the single cover slip (4) so that its temperature can be kept constant or changed quite rapidly by changing the temperature of the flowing water. (For a chamber allowing more rapid change of the specimen temperature, see the advanced design described in Article 27.) To prevent condensation of the slide cooled below room temperature, the whole slide is placed in a plastic "sandwich box" through which dried air is gently blown. The air escapes through the thin space beneath plastic washers attached to the objective and condenser lenses, which cover the over-sized holes in the top and bottom of the sandwich box.

Figure 1 shows an overview, and Fig. 2 shows the median cross section, of the temperature control slide used in these studies. A video sequence showing the birefringence change of the spindle in a developing sea urchin egg exposed to repeated chilling is shown in the DVD lecture record of July 24, 2006, appended to this *Collected Works*.

Reprinted from *Journal of Biophysical and Biochemical Cytology*, Vol. 3(6), pp. 831–838, 1957.

STUDIES ON DEPOLARIZATION OF LIGHT AT MICROSCOPE LENS SURFACES

II. The Simultaneous Realization of High Resolution and High Sensitivity With the Polarizing Microscope*

By SHINYA INOUÉ‡, Ph.D., and W. LEWIS HYDE, Ph.D.

(From the Department of Biology and Institute of Optics, University of Rochester, Rochester, New York; and the Research Center, American Optical Company, Southbridge, Massachusetts)

Plates 262 to 264

(Received for publication, June 28, 1957)

I. The Value and Limiiation of Polarization Microscopy in Biology:

The uniqueness of polarization microscopy lies in the use of light beams with controlled states of polarization. With such beams, optical properties such as birefringence, dichroism, and optical activity are detectable, and optical constants are determinable in an exceedingly small sample volume. Since birefringence and dichroism generally reflect the arrangement or properties of material the dimensions of which lie below the limit of resolution of the light microscope, the polarizing microscope can in principle be used for studying the optically unresolvable "submicroscopic structures" of cells and tissues. Such studies can be made directly on living cells and the submicroscopic events followed as the cells undergo physiological activities. In contrast, the electron microscope gives a vastly improved resolving power and enables a direct visualization of the submicroscopic structures, but only after their fixation and desiccation. The two instruments, therefore, are complementary to one another, and together they make up for each other's shortcomings. Indeed Sjöstrand (9, p. 511) has commented on these aspects of electron microscopy, concluding that "... the polarized light microscope should be the standard equipment of the electron microscopists. But it has to be used on living cells and it has to be equipped for analyzing very weak birefringence. ..."

Although potentially endowed with such desirable and unique attributes, the polarizing microscope has not yet been used very extensively for studying living cells. This is not due to the absence of a submicroscopic organization in cells, for such has been amply demonstrated by recent electron microscope

* This investigation was supported in part by a research grant C-3002 (C) from the National Institutes of Health, United States Public Health Service.

‡ Scholar in Cancer Research of the American Cancer Society, Inc.

studies, nor to the lack of birefringence and dichroism in the cytoplasmic and nuclear elements, for W. J. Schmidt in his celebrated monograph (8) has long ago demonstrated their wide distribution.

The optical problems that confront the biologist who wishes to apply a polarizing microscope to the study of live cellular structures are twofold:

1. The method must be sensitive enough to produce perceptible contrast in objects whose maximum retardation (strength of birefringence × thickness) may lie well below a hundredth of a wave length. In general, a sensitivity of an angstrom unit or less is required.

2. The retardation must be detectable and rapidly measurable in regions or in structures whose dimensions may lie close to the limit of resolution of the light microscope.

Methods for obtaining maximum sensitivity for detecting weak birefringence have been described by Schmidt (7), Swann and Mitchison (10), and Inoué and Dan (5) (also see Wright 11). Methods for measurement of low retardation of small objects have been described by Swann and Mitchison (10), and by Inoué (3). In all of these references, strain birefringence and other sources of stray light are removed to gain maximum extinction; a very bright light source is used to aid the sensitivity of the eye and to reduce the exposure time for photographing the object which may constantly be changing; and a mica compensator is used to improve contrast and field brightness.

When these precautions are taken, the sensitivity and resolution of the polarizing microscope ultimately become limited by depolarization of light at the lens and slide surfaces. The depolarization, whose nature is explained below, results in stray light whose intensity increases approximately exponentially as the numerical aperture is increased. The stray light reduces contrast and drowns the image of weakly birefringent objects. This necessitates the use of low numerical aperture objectives and/or considerable stopping down of the condenser diaphragm. In either case the resolving power is lowered. Indeed *high sensitivity and high resolution have appeared to be mutually exclusive qualities of polarizing microscopes* (1).

II. The Polarization Rectifier:

In a conventional polarizing microscope equipped with high extinction polarizers and in excellent adjustment, the extinction diminishes rapidly as the numerical aperture is increased. Inoué (4) has earlier shown that the increased amount of stray light at higher numerical apertures measured photoelectrically agrees closely with the amount of light calculated to enter the system by depolarization, assumed to occur solely by the rotation of the plane of polarization at lens and slide surfaces. At these surfaces, the component polarized in the plane of incidence suffers less reflection loss than the component polarized perpendicular to it, and the plane of polarization of the transmitted light is rotated towards the plane of incidence. Only when the plane of incidence lies parallel or perpendicular to the plane of polarization is the beam transmitted without rotation. When the polarizer and analyzer are crossed, a dark polarization cross therefore results, which can be seen at the back aperture of the objective. The rotation can be reduced somewhat by low reflection coating of the glass surfaces and by the use of oil immersion lenses, but even then it continues seriously to limit the extinction and the polarization cross persists.

When the condenser and objective lenses and the specimen slide are free from (lateral) strain birefringence the polarization cross is symmetrical and complete (Fig. 1 *a*). As the polarizer is rotated, the cross opens up into two dark V's which move out symmetrically to the edge of the aperture (Fig. 1 *b*). Loss of contrast in the V's is generally an indication of (radial) strain birefringence or elliptical polarization due to the coating on the lenses. Rotation of the polarizer in the opposite direction results in V's in the two other quadrants. These V's represent regions of the objective aperture through which rays suffering equal degrees of rotation have traveled. The degree of rotation thus varies with aperture and azimuth and the sense of rotation is reversed in adjacent quadrants.

For any particular ray the sense of rotation is the same in the condenser, objective, and specimen slide, and cannot be reduced by the simple expedient of adding lenses to either. The rotation can be reduced or even removed, however, by a second spherical lens system if the rotatory sense of the error introduced by the two systems can be made opposite. This can be accomplished by inserting a half-wave plate[1] between the two optical systems with its axis parallel to the polarizer (Text-fig. 1). Those rays for which the plane of polarization is rotated before entering the half-wave plate will emerge from it with an error of the same magnitude but opposite sense. This effect is found irrespective of the magnitude and sense of the original error.

In principle, such a half-wave plate might be placed between the condenser and the objective, letting the rotation of the one compensate for that of the other. This is not feasible because no available half-wave plate has uniform properties at all the angles of incidence found in this region. The half-wave plate must lie in a path where the beams are nearly parallel. It could, for example, be inserted between two entirely separate but duplicate optical systems, but this would be awkward and expensive. A simpler and more practical system we have devised is based on the following reasoning.

It is found that the rotation imposed on each and all of the rays by the regular condenser-slide-objective system increases proportionately when coverglasses are added (without oil immersion) between the condenser and objective lenses (Inoué, unpublished data). By choosing a proper number of coverglasses

[1] For the explanation of rotation of the plane of polarization by a half-wave plate, see *e.g.* references 2, 6.

or by altering their refractive indices, the rotation of each ray can be exactly doubled. If now the coverglasses are transferred below the condenser and given a suitable curvature, the resulting meniscus would continue to have the same effect on rotation, since the radius of each zone at the front focal plane of the condenser is proportional to the sine of the angle of the corresponding ray in the object plane. If a half-wave plate is now inserted between the meniscus and the condenser, it will reverse the rotation of each ray, and rotation at the meniscus will cancel out the rotation introduced by the condenser, objective lens, and

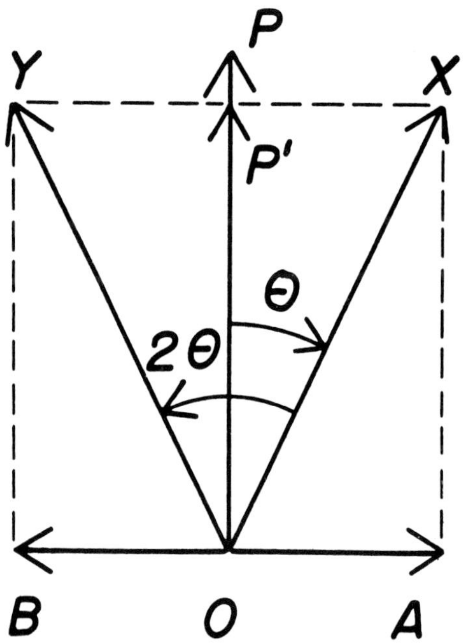

TEXT-FIG. 1. Effect of a half-wave plate on a ray whose plane of polarization OX was deviated by an angle θ away from the polarizer axis OP. The half-wave plate, whose axis is parallel to OP separates $\overrightarrow{OX}$ into $\overrightarrow{OA}$ and $\overrightarrow{OP'}$. Then after a half-wave retardation, $\overrightarrow{OB}$ and $\overrightarrow{OP'}$ are combined to give $\overrightarrow{OY}$. The plane of polarization OX is thus rotated -2θ to OY and the deviation is effectively reversed.

the specimen slide. Such a combination of a half-wave plate and a meniscus we have named the *polarization rectifier.*

The rectifier can be adjusted by moving it along the optical axis of the microscope until the rotation at each point of the meniscus is precisely matched with the rotation at each point of the lens system. This adjustment is possible because the gradually narrowing cone of rays approaching the center of the field stop (at the light source) passes a more or less steeply curved region of the meniscus.

The rectifier can be placed under the condenser (Text-fig. 2 B), above the objective (Text-fig. 2 A), or at both places. The first method is adequate, when only high extinction is required, or when there is no marked deviation of the beam by the object. Whenever image quality is emphasized, however, the last method is preferred, since otherwise the beams diffracted and deviated by the object will be incorrectly rectified in the objective. The result is that some objects will become visible merely because of their small size, whether birefringent or not.

The rectifying rotation can be provided either by a deeply curved zero-power lass shell (Text-fig. 2 B) or a thin air bell produced between a pair of convex

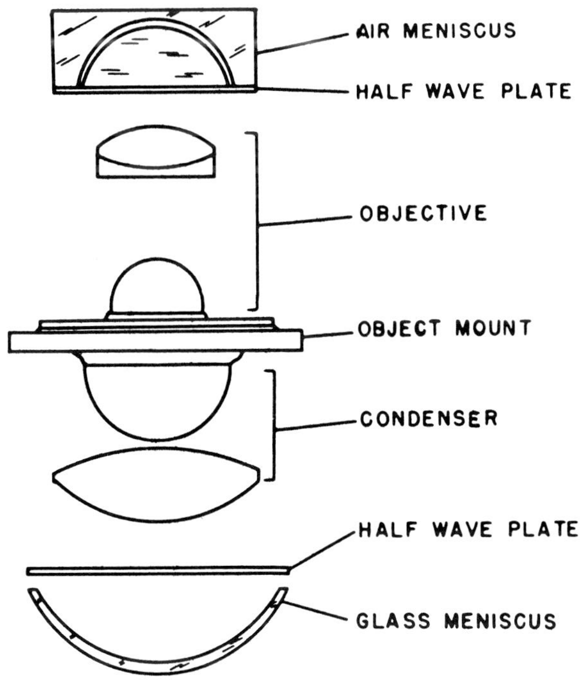

TEXT-FIG. 2. Arrangement of polarization rectifiers. Rectifiers are placed above the objective (*A*), below the condenser (*B*), or at both *A* and *B*. The glass or air meniscus introduces additional rotation to each beam of light which equals the amount introduced at the lenses and the specimen slide. The sense of rotation is, however, reversed by the half-wave plate so rectification is achieved over the whole aperture. All elements must be free from strain.

and concave glass surfaces (Text-fig. 2 A). In practice, it may also be possible to place the half-wave plate within the objective or the condenser, adjusting the amount of rotation of some or all of the surfaces by low or high reflection coating and letting part of the elements provide rectification for the other elements.

It should be remarked here that the half-wave plate of the rectifier oriented properly changes the sign, but not the magnitude of the ellipticity of the light beam. The rectifier, therefore, does not affect the measurement of the amount of retardation but, since it effectively changes the two axes of the ellipse, a compensator placed beyond a single half-wave plate acts as though it were oriented in the wrong quadrants.

Although the rectifier is best used with monochromatic light, it has also been

PLATE 264

FIG. 4. A high magnification image (× *ca.* 3600) taken with the rectified optics showing birefringent surface structures (dots and ridges) of a portion of a cell similar to the one shown in Figs. 2 and 3. 97× 1.25 N.A. strain-free coated objective and 97× 1.25 N.A. strain-free coated condenser are both used at full aperture and separately rectified. Polarizers crossed, background retardation *ca.* 5 mμ.

FIG. 5. Phase contrast image (× *ca.* 3000) showing the surface structure of an adjacent portion of the same cell shown in Fig. 4. 97× 1.25 N.A. phase contrast objective with annulus diaphragm in the condenser. The resolution is comparable to the polarization optical image in Fig. 4, but contrast is produced by entirely different principles.

THE JOURNAL OF
BIOPHYSICAL AND BIOCHEMICAL
CYTOLOGY

PLATE 264
VOL. 3

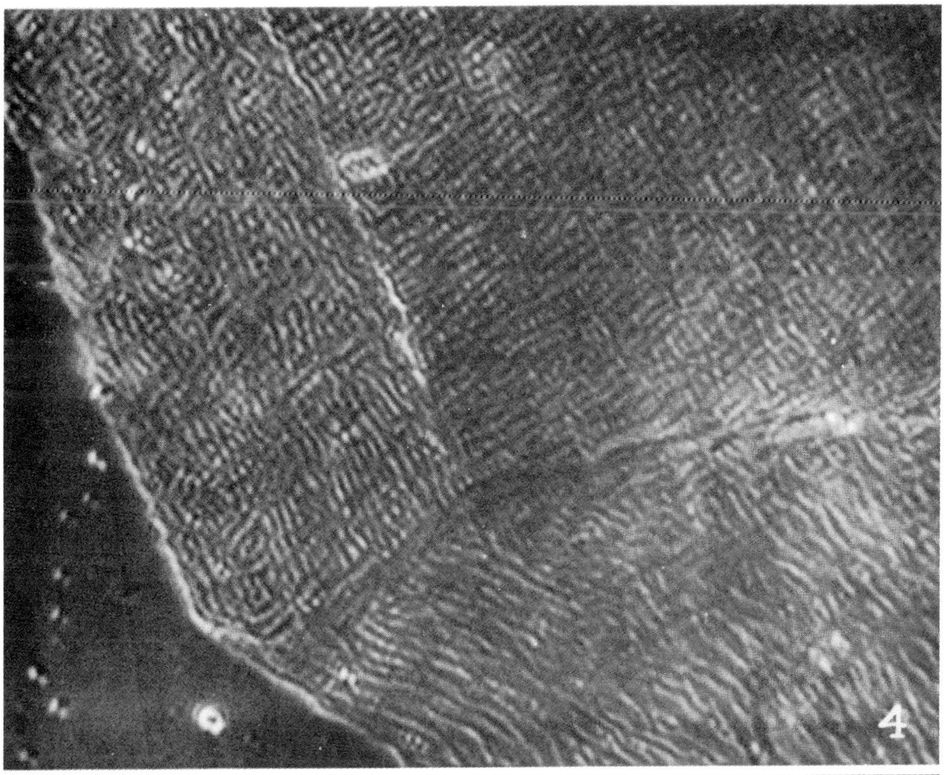

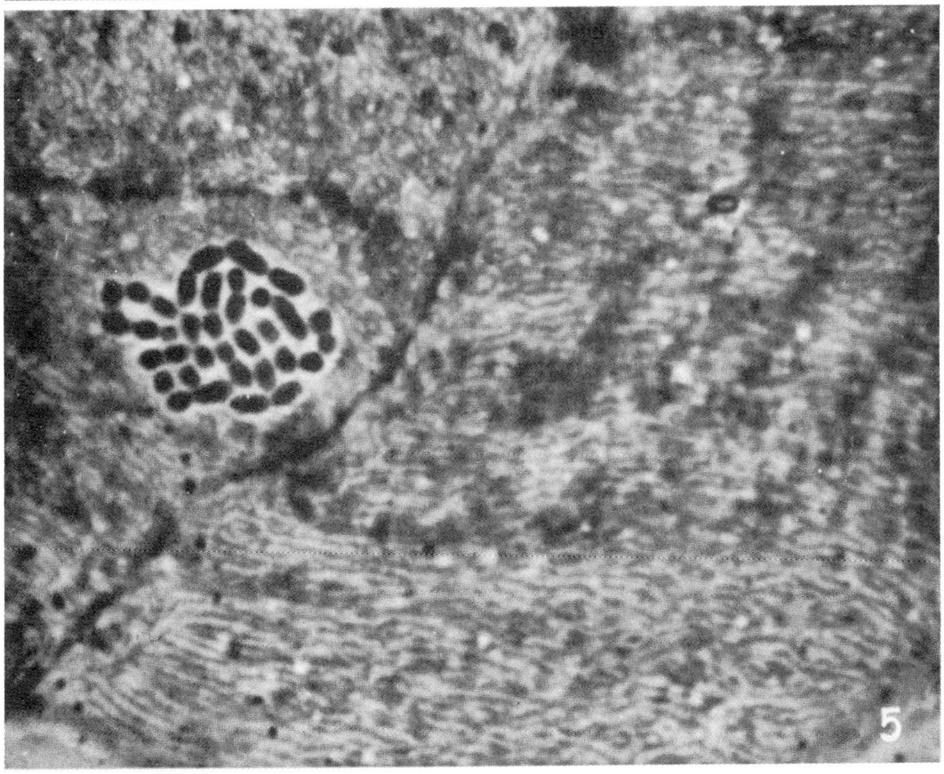

(Inoué and Hyde: Studies on the polarizing microscope. II)

Reprinted from *Nature*, Vol. 182, pp. 1725–1726, 1958.

Diffraction Anomaly in Polarizing Microscopes

THE image quality of the polarizing microscope has never been critically examined, perhaps because the instrument was used primarily for determining optical constants of crystals and fibres, rather than for studying fine details of the object. We have recently re-examined its reliability and find that unless the rotation at the lens surfaces is rectified, the unit diffraction pattern is not an Airy disk but a four-leaf-clover pattern, and that the resolution and contrast may be spurious.

In a conventional polarizing microscope the field between crossed polarizers illuminated by a bright source is never completely dark. If the microscope is equipped with lenses completely free from strain birefringence a polarization cross is observed at the back aperture of the objective lens. The light in this cross is introduced almost exclusively by rotation of the plane of polarization at the oblique interfaces between the polarizer and the analyser[1,2]. The rotation is a consequence of differential transmittance of the vectors polarized parallel and perpendicular to the planes of incidence and may reach 6 deg. or more at the edge of the aperture. This not only lowers the extinction of the polarizing microscope at high numerical apertures, but also modifies the aperture function of the system even when the condenser diaphragm is closed. As a result, the diffraction image of a pin-hole between crossed polarizers is no longer a regular Airy disk (Fig. 1) but a clover-leaf pattern possessing a dark centre and a dark cross (Fig. 2). Its intensity is expressed by :

$$I \simeq \sin 2\theta \, J_3(r)/r \qquad (1)$$

where θ and r are polar co-ordinates in the image plane (θ is measured from the plane of polarization of the polarizer and r is measured in diffraction units) and $J_3(r)$ is a Bessel function of the third order.

Densitometry of photographs of sub-resolution pinholes taken through a special strain-free polarizing microscope agrees well with the distribution of intensity calculated from this formula. The special polarizing microscope used for this work was designed jointly by one of us (S. I.) and the American Optical Co. and was constructed at the University of Rochester with generous support from the former. The distribution of intensity along $\theta = \pm\pi/2$

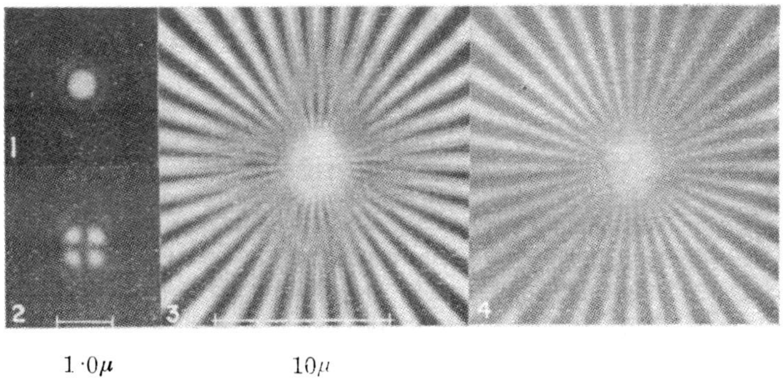

Fig. 1. Airy disk in the absence of a polarizer. The diffraction pattern was obtained by optically reducing a pinhole 20μ in diameter located in front of the lamp to 0·2μ and then re-magnifying. A pair of identical strain-free 97 × oil-immersion objectives with 1·25 numerical aperture was used. The image produced by this system is identical with the image of a 0·2-μ pinhole located at the object space, except that the lens errors are doubled

Fig. 2. As for Fig. 1, but between crossed polarizers. Scale for Figs. 1 and 2, 1·0μ

Fig. 3. Image of a Siemens's test chart between crossed polarizers. Except for the Siemens's star which replaced the pinhole, system identical with that of Fig. 2. Notice reversal of contrast and spurious resolution. Scale for Figs. 3 and 4, 10μ

Fig. 4. Image of same test-chart as in F g. 3, between crossed polarizers but viewed through the same objectives after they were rectified. The diffraction anoma y has disappeared

(Fig. 5) shows that the radius of the central maximum in the clover-leaf pattern coincides approximately with the first minimum of the Airy disk. The ratio of the intensity of the central to that

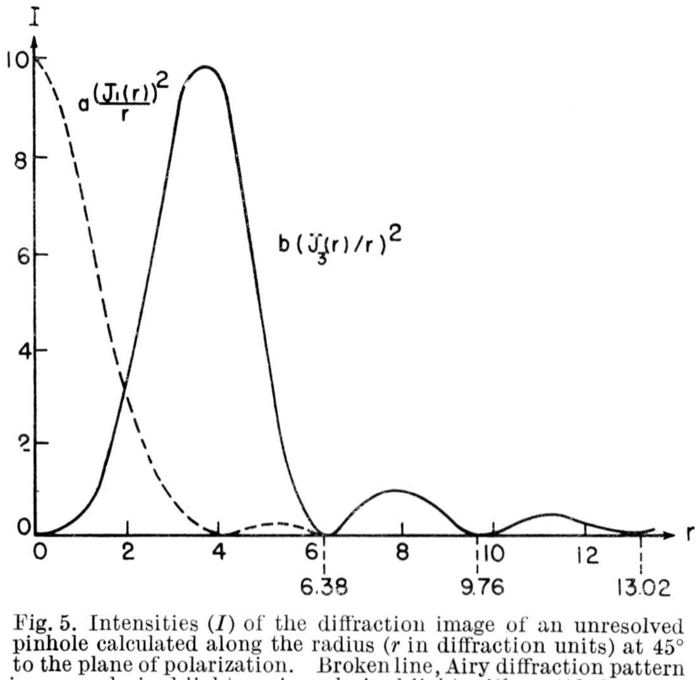

Fig. 5. Intensities (I) of the diffraction image of an unresolved pinhole calculated along the radius (r in diffraction units) at 45° to the plane of polarization. Broken line, Airy diffraction pattern in non-polarized light, or in polarized light with rectified optics. Solid line, with non-rectified optics between crossed polarizers. The curves have been normalized to make the central maxima equal. Microdensitometer tracings of Figs. 1 and 2 agree with the calculated curve to within a few per cent

of the first diffraction maxima is 10 : 1 in the clover leaf as opposed to 50 : 1 in the Airy disk.

As the polarizer is rotated away from the crossed orientation the clover leaf changes to a diagonal figure of eight, an ellipse, and finally to the regular Airy disk when the polarizers are parallel. Similar changes are observed when compensators are introduced between crossed polarizers.

When, on the pinhole, a uniaxial crystal is superimposed with its optic axis parallel to the axis of the microscope (Z-cut), the diffraction image of the pinhole between crossed polarizers is again expressed by equation (1) even in the absence of rotation. With parallel polarizers a diffraction image similar to that in an astigmatic system is formed by the extraordinary rays, superimposed on the Airy disk formed by the ordinary rays. These are also the diffraction images of pinholes formed by catadioptric (reflecting) polarizing microscopes without any crystals.

When a crystal is introduced with its axis normal to the microscope axis (X-cut), a highly astigmatic diffraction pattern identical with those reported by Nienhuis for conventional astigmatism is superimposed on the Airy disk. Theoretical formulæ for all these diffraction patterns, except for the X-cut crystal, were calculated by one of us (H. K.) ; they are reported in ref. 3 together with their experimental verifications.

The effect of anomalous diffraction on a periodic absorbing object was determined by observing a black-and-white test chart through a pair of oil-immersion lenses showing a medium degree of rotation. As shown in Fig. 3, contrast reversal and spurious resolution are observed to varying degrees at different orientation of the periodic object.

These diffraction anomalies clearly indicate the danger of using polarizing microscopes or other optical instruments employing polarized light at high angles of incidence[2] without due consideration of the image quality. On the other hand, some of the properties described may be used to advantage for modifying the aperture function and even for obtaining higher 'resolution' in special cases. In general, the Rayleigh criterion of resolving power is not applicable and, instead, it is suggested that the transfer (or response) function of the system be used to evaluate the image.

The diffraction anomalies and the resulting spurious image are practically eliminated with the polarization rectifier[2]. The rectifier corrects the rotation so that the state of polarization is no longer irregular, and a uniformly dark aperture is obtained. Fig. 4 shows the correct image of the same test-chart viewed through a pair of rectified oil-immersion objectives. With rectification, the maximum resolving power of the light microscope can be realized, and the system has been utilized for exact studies of biological objects possessing birefringence down to a fraction of an angstrom unit.

This research was supported in part by grant (*C*-3002-*C*) from the National Institutes of Health, U.S. Public Health Service, awarded to Shinya Inoué, a National Science Foundation grant awarded to Dr. M. Parker Givens, Institute of Optics, University of Rochester, and by the American Optical Company.

SHINYA INOUÉ*

Department of Biology and
Institute of Optics,
University of Rochester.

HIROSHI KUBOTA†

Institute of Optics,
University of Rochester.

* Scholar in Cancer Research of the American Cancer Society.

† Fulbright Visiting Professor on leave from the Institute of Industrial Science, University of Tokyo (permanent address).

[1] Inoué, S., *Exp. Cell Res.*, **3**, 199 (1952).

[2] Inoué, S., and Hyde, W. L., *J. Biophys. and Biochem. Cytol.*, **3**, 831 (1957).

[3] Kubota, H., and Inoué, S. (in the press).

Printed in Great Britain by Fisher, Knight & Co., Ltd., St. Albans.

Reprinted from *Reviews of Modern Physics*, Vol. 31(2), pp. 402–408, 1959, with permission from American Physical Society.

44
Motility of Cilia and the Mechanism of Mitosis*

SHINYA INOUÉ
Department of Biology, The University of Rochester, Rochester 20, New York

THE movement of cilia and chromosomes are interpreted in terms of the microscopic and fine structures in cells. Section 1 discusses structure and motility of cilia; Sec. 2, microscopic structure of the mitotic spindle; Sec. 3, centers of organization in spindle and cilia; and Sec. 4, on the physicochemical nature of the spindle; and Sec. 5, relation of cilia and chromosome movement to muscle contraction.

1. STRUCTURE AND MOTILITY OF CILIA

Many small organisms—bacteria and protozoa, sperm and embryos of larger organisms—are propelled by beating their thin whip-like cilia or flagella (used interchangeably here). Also, ctenophores even a foot long and several inches wide can swim about or adjust their gravitational orientation by coordinated beating of their ciliary bundles. When the cell or organism is fixed, as in the gills of mussels and clams and in the human trachea, cilia create a current capable of pumping a considerable quantity of liquid or mucous material.[1–3]

The length of cilia may range from a few microns to several millimeters, and there may be one to several thousand cilia per cell but their diameter is quite constant, usually between a tenth and a half micron. Where a number of cilia occur in a row, adjacent cilia beat slightly out of phase with each other and a regular propagating wave is observed.

At the base of each cilium is found a characteristic bulbous enlargement, the basal granule. Further, proximal to the basal granule, rootlets sometimes are found which may function as anchorage or serve to conduct impulses. The membrane of the cilium appears continuous with that of the cell body.

Examined with an electron microscope (Fawcett and Porter,[4] and also dicussed in the following), cilia reveal a characteristic inner fibrillar pattern of amazing uniformity. Near the periphery of the cilium, there are usually nine (or occasionally a multiple thereof) fibrils, each composed of two filaments (or tubules?) some 100 to 200 A in diameter. Surrounded by the outer nine are two additional central fibrils of somewhat smaller diameters (Figs. 1 and 2). The nine outer fibrils reach the base of the cilium and appear to merge laterally into a hollow tube to make up the basal granule. The two central filaments also extend the whole free length of the cilium but apparently do not reach the basal granule. At the tip of the cilium, the outer nine and inner two fibrils are said to merge. In cilia with distinct directional beating, the line intersecting the two central filaments lies at right angles to the direction of beat.

The pattern of nine plus two has been found in cilia from a wide variety of cells, in fact, in practically all cilia observed with adequate resolution. The eleven fibrils are unlikely to be artifacts formed during preparation for electron microscopy, as sperm-tail flagella macerated in distilled water also show the frayed eleven fibrils in dark field illumination.

Given this structure, how does one explain the mechanism of ciliary beat? The pattern of beat may be relatively simple, as shown in Fig. 3(a). There is a recovery stroke in which the limp cilium stiffens from the base up, and an effective stroke where bending is mostly at the base and the rest of the cilium acts as if it were stiff. The same flagellate organism which swims forward by this

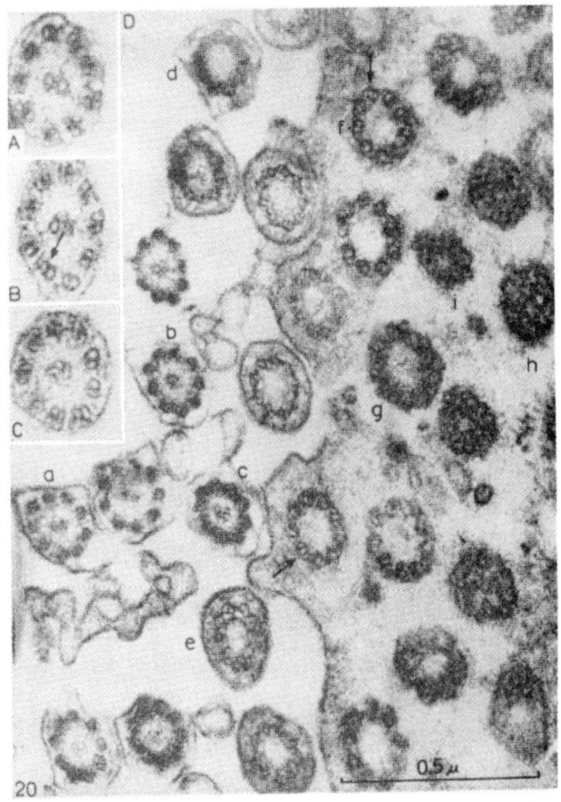

FIG. 1. Electron micrograph of cross sections of rat-trachea cilia. Compare with interpretive diagram, Fig. 2 [from J. Rhodin and T. Dalhamn, Z. Zellforsch. u. mikroskop. Anat. **44**, 345 (1956)].

* Original work appearing in this paper was supported in part by the Public Health Service, U. S. Department of Health, Education, and Welfare (G-3002-C), and by the American Cancer Society.

beat may swim backward, also sidewise or circularly, as shown in Fig. 3(b).[2,5] Bradfield[6] postulates that waves of contraction proceed along the length of the outer nine fibrils with a message perhaps traveling in advance along the two inner ones. The waves of contraction may be started at the base by a commutator-like device which would result not in a synchronous contraction but in waves with various phase lags. It is more likely, however, that the outer fibrils may be the conductive elements, the inner two at least partaking a more active function in beating. For, in certain sensory cells (see following) and at the embedded base of each cilium[4,7] where one would expect conduction but not contraction, one in fact finds the outer nine fibrils and not the inner two.

It was tacitly assumed in the foregoing that contraction of the fibrils was the basis for cilia beat. Actually, the only evidence for contraction of cilia components (at the molecular level) appears to lie in the x-ray diffraction studies of Astbury *et al.*[8] There they find in flagella collected from bacteria, in addition to an α-protein pattern (which is characteristic of many fibrous proteins, such as keratin, myosin, elastin, etc., and is believed to reflect the fundamental spacing of the polypeptide backbone), a folded β pattern. This supercontracted pattern they believe reflects the folding of a fraction of the polypeptide chains responsible for contraction. The protein isolated from this bacterial flagella preparation lacks sulfur-containing amino acids and appears dissimilar to any of the muscle components so far known.

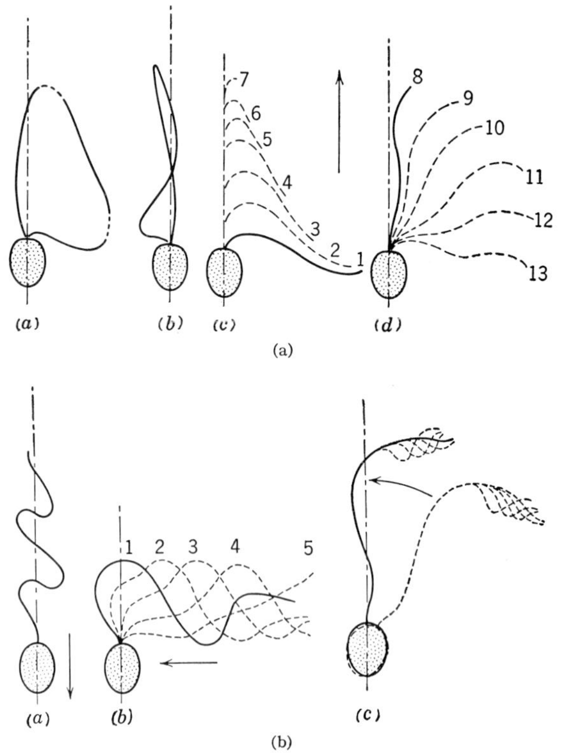

FIG. 3. Various patterns of beat of a *Monas* flagellum. Arrows indicate directions of swimming [from M. Hartmann, *Allgemeine Biologie* (Gustav Fischer Verlag, Stuttgart, 1953), fourth edition; after Krijgsman[5]].

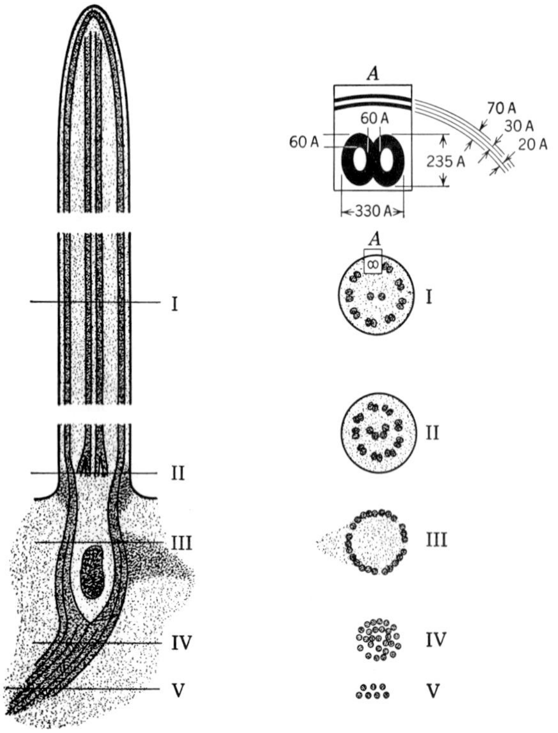

FIG. 2. Interpretation of electron micrograph (Fig. 1) showing fine structure of cilia [from J. Rhodin and T. Dalhamn, Z. Zellforsch. u. mikroskop. Anat. 44, 345 (1956)].

Aside from localized contraction, the mechanism of ciliary beat has been interpreted by other schemes also, such as local swelling, reciprocal pumping of a liquid in and out of the cilium, or discontinuous flow of material through the cilium.[1,3,9] The exact site of the motor function in the cilium is also disputed, but nevertheless there exist certain observations (Secs. 3 and 5) which link the structure and function of these minute structures to other specialized motile structures of the cell.

DeRobertis, Sjöstrand, and others[10,11] made an interesting discovery related to the structure of cilia in the filaments of the retinal-rod cells and of the sensory-hair cells of the inner ear. These fibrous elements, believed for some time to be derived from embryonic cilia, also show a fine structure similar to that of cilia described in the foregoing. In these apparently nonmotile fibers, the same outer nine fibrils are found, but the inner two are missing.

2. MICROSCOPIC STRUCTURE OF THE MITOTIC SPINDLE

Cilia may beat as frequently as a hundred cycles per second. Chromosomes, on the other hand, move extremely slowly, the maximum velocity being a few

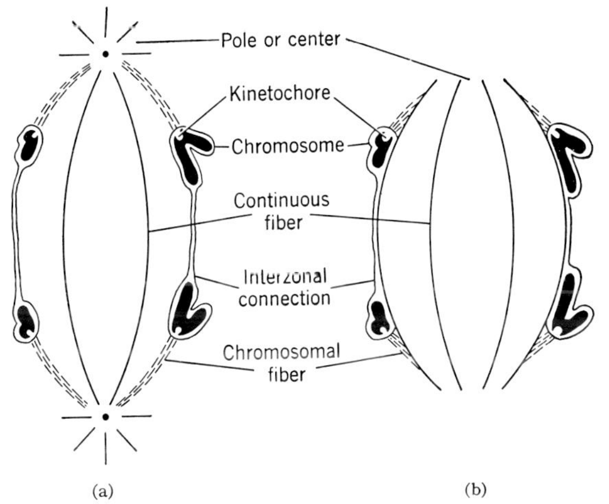

FIG. 4. Schematic diagram of mitotic spindle, (a) with centrioles and (b) without (modified from Schrader[19]).

microns per minute at anaphase (for reviews of mitosis see references 12 to 21). However, certain elements of the mitotic apparatus responsible for chromosome movement may be identical to, or at least have properties common to, portions of cilia. Before discussing this problem in Sec. 3, the structure of representative mitotic apparatuses are described.

Using Schrader's description,[19] the following terms are used. A dot or rod-like structure, the "centriole," is found at the poles of the mitotic spindle of many animal cells and occasionally in plant cells. The centrioles are morphologically the focal points for the spindle fibers and the astral rays [Fig. 4(a)]. In an average plant cell [Fig. 4(b)], both the centrioles and the astral rays are missing and the spindle fibers show less tendency to converge at the poles. Between the two spindle poles lie "continuous fibers." From the "kinetochores," a specific region of the chromosomes "chromosomal fibers" extend to or toward the spindle poles.

For half a century, the reality of these fibers in *living* cells has been disputed, for, with very few exceptions, they could be seen only in cells after fixation and staining. Recently, however, with improvements of the polarizing microscope, the author has been able to show these fibers clearly in living cells of many animals and plants by virtue of their positive birefringence (strength of birefringence $10^{-3}-10^{-4}$).[22,23] In animal cells, the fibers converge and their birefringence is stronger adjacent to the chromosome kinetochores and near the centrioles [Fig. 5(a)]. In plant cells, the situation is similar at the kinetochore region, but toward the "poles" the birefringence is weaker and the fibers are more diffuse [Fig. 5(b)]. During anaphase movement, the birefringence of the chromosome fibers persists and always is strongest adjacent to the kinetochores and the centrioles. The birefringence of the continuous fibers falls once and then rises again after complete separation of the chromosomes. The midregion containing fibers with strong secondary birefringence becomes the phragmoplast of plant cells [Fig. 5(c)]. Within the phragmoplast, small granules align, fuse, and become the cell plate, which divides the original cell into two. In animal cells, the cytoplasm generally cleaves inward at right angles to the spindle remnant forming two new cells with one nucleus each. Time-lapse motion pictures of these processes have been made by the author using a special polarizing microscope and were shown at the meeting.

The material of the mitotic apparatus has been isolated from sea urchin and other eggs by Mazia, Dan, and their collaborators in quantities sufficient for chemical analyses.[24–26] Amino-acid composition of their protein fraction (molecular weight *ca* 45 000) shows, unlike the bacterial flagella protein, a fair content of sulfhydryl-containing amino acids.

Although the spindle isolated after alcohol treatment by Mazia and Dan is stable, the fibers of the mitotic spindle in living cells are apparently extremely labile. The spindle fibers may disappear by slight mechanical agitation or by low-temperature treatment of the cell, only to reform in the course of a few minutes (see Sec. 5, also Carlson[13,27], Chambers,[28] Inoué,[29] and Östergren[30]).

3. CENTERS OF ORGANIZATION IN SPINDLE AND CILIA

Although the apparent velocity and very probably the stability of the spindle fibers differ by orders of magnitude from the cilia, the fibrous elements of the two structures may be formed or organized in a very similar fashion. The argument follows.

(A) The birefringence of the spindle fibers and astral rays is strongest adjacent to the kinetochores and centrioles throughout metaphase and anaphase (Inoué[23] and Schmidt[31]). The fibers are arranged radially (within restricted cones, in the case of the kinetochores) from these centers, and the growth of the spindle (at least in animal cells) takes place by lengthening of the fibers joining the centers. This strongly suggests that both the centrioles and the kinetochores are centers of fiber orientation.

(B) Growth of the axial filament of the sperm-tail flagellum starts from the basal granule, which same structure during the last mitosis acted as a centriole of the spindle.[21,32] In certain protozoa, flagella and the mitotic spindle both grow simultaneously from common giant centrioles.[14,32]

(C) In abnormal divisions of snail spermatocytes, some chromosomes lose their kinetochores and cannot partake in mitosis. The Pollisters[33] have shown that a clear correlation exists between the number of such chromosomes and the number of supernumerary basal granules, which migrate to the cell periphery and form the same number of extra sperm tails. Also, those kinetochores earlier dissociated from chromosomes form small extra astral rays while the spindle for the next division is formed.

spindle fibers are undoubtedly almost fluid in nature and are not stably crosslinked gels as the term "fibers" may imply. With dehydrating agents (e.g., alcohol) and in an acidic environment the crosslinking is probably enhanced until the spindle is finally "fixed."

On this basis and from observations described in Sec. 2, anaphase movement of chromosomes may be explained by local reduction in the quantity of oriented material and the consequent shortening of chromosomal fibers. The orienting forces of kinetochore and centriole must be just as actively at work throughout this process. The continuous fibers, either actively elongating or at constant length, could function as supports to counteract the pulling action of the chromosomal fibers. This mechanism of contraction of the chromosomal fibers is similar to that postulated by the author for the action of low concentrations ($<10^{-3}M$) of colchicine.[22] It is, however, in distinct contrast to the mechanism suggested by Swann[47,48] whose microscope lacked the power of resolution for detecting individual spindle fibers.[49]

5. RELATION OF CILIA BEAT AND MITOSIS TO MUSCLE CONTRACTION

The sole evidence for molecular contraction in cilia appears to lie in the x-ray diffraction data on isolated bacterial flagella (Sec. 1). Hypotheses involving mechanisms other than contraction (e.g., differential swelling) also have been postulated, but discriminating experiments are lacking.

Anaphase movement of chromosomes was explained by an orientation equilibrium of spindle fibers (Sec. 4). The action of centrioles and kinetochores—the center of foci of orientation in the spindle—appears similar to that of basal granules of cilia during fibrogenesis (Sec. 3).

Occasionally, cilia are resorbed or re-formed, but the process is much slower than the formation and disappearance of the mitotic spindle at each cell division. It appears that the cilia fibrils are quite stable while the molecules in spindle fibers probably are barely crosslinked (Sec. 4). In comparison, the contractile material in muscle may have a stability lying in between that of cilia and spindle fibers. The primary function of muscle and cilia is repeated rapid contraction, while with the spindle it is a single successful partition of the chromosomes into two new cells.

Mechanisms of muscle contraction are discussed in other papers of this symposium. It is interesting that one of the most widely discussed (and rather widely accepted) current hypotheses is that involving the creeping of two sets of filaments past each other.[50,51]

Regardless of the exact mechanism of choice, one may not overlook the muscle model systems which can contract and produce the same force per cross section as live muscles. This is true in muscle fibers extracted with 50% chilled glycerol, and in an oriented gel fiber formed by mixing two purified muscle proteins, actin and myosin. In either case, contraction is induced specifically by the addition of ATP (adenosine triphosphate) in the presence of magnesium and potassium ions.[52–54]

Hoffmann-Berling has shown further that motility can be induced in glycerol-extracted cells other than muscle, again by the addition of ATP. Thus, he was able to induce glycerinated sperm-tail flagella to undergo prolonged beating, chromosomes to separate, and extracted dividing cells to complete formation of their cleavage furrow (motion picture commercially available[54,55]). These models respond to approximately the same concentration of ATP as muscle models.

To what extent the movements induced by ATP in various cells reflect the same molecular mechanisms still is not clear. For example, the elongation of the central spindle, apparently responsible for the separation of chromosomes in the cell model, is not prevented by the same poisons to which the muscle and cilia model are very sensitive. Furthermore, although the organic triphosphate ATP (and ITP) specifically induces movements in extracted cells, the responding proteins show significant difference in their amino-acid compositions (see Secs. 1 and 2). It is nevertheless encouraging that movements closely resembling those found in living cells can be induced by the same reagent in cells from which much of the complex structures and materials have been removed.

In conclusion, evidence for the long-sought molecular folding is still weak, and no unifying molecular mechanism has been found for cilia beat, anaphase chromosome movement, and muscle contraction. Developments in recent structural and physicochemical analyses are, however, encouraging, and with a concerted intelligent approach, one may acquire before too long a much clearer understanding of the mechanisms underlying these cellular movements.

BIBLIOGRAPHY

1 J. Gray, *Ciliary Movement* (The Macmillan Company, New York, 1928).

2 M. Hartmann, *Allgemeine Biologie* (Gustav Fischer Verlag, Stuttgart, 1953), fourth edition.

3 L. V. Heilbrunn, *An Outline of General Physiology* (W. B. Saunders Company, Philadelphia, 1952), third edition.

4 D. W. Fawcett and K. R. Porter, J. Morphol. **94**, 221 (1954).

5 B. J. Krijgsman, Arch. Protistenk. **52**, 478 (1925).

6 J. R. G. Bradfield, Symposia Soc. Exptl. Biol. **9**, 306 (1955).

7 D. R. Pitelka and C. N. Schooley, J. Morphol. **102**, 199 (1958).

8 W. T. Astbury, E. Beighton, and C. Weibull, Symposia Soc. Exptl. Biol. **9**, 282 (1955).

9 W. J. Schmidt, *Protoplasma-Monographien* (Gebrüder Borntraeger, Berlin, 1937), Vol. II.

10 E. De Robertis, J. Biophys. Biochem. Cytol. **2**, 319 (1956).

11 F. S. Sjöstrand, J. Cellular Comp. Physiol. **42**, 45 (1953).

12 N. G. Anderson, Quart. Rev. Biol. **31**, 243 (1956).

13 J. G. Carlson, Science **124**, 203 (1956).

14 L. R. Cleveland, J. Protozool. **4**, 230 (1957).

15 A. Hughes, *The Mitotic Cycle, the Cytoplasm and Nucleus during Interphase and Mitosis* (Academic Press, Inc., New York, 1952).

[16] H. Ris in *Analysis of Development*, B. H. Willier, P. A. Weiss, and V. Hambürger, editors (W. B. Saunders Company, Philadelphia, 1955), p. 91.
[17] F. Schrader, Biol. Bull. **67**, 519 (1934).
[18] F. Schrader, "A critique of recent hypotheses of mitosis," *Symposium on Cytology* (Michigan State College Press, East Lansing, Michigan, 1951).
[19] F. Schrader, *Mitosis* (Columbia University Press, New York, 1953), second edition.
[20] C. P. Swanson, *Cytology and Cytogenetics* (Prentice-Hall, Inc., Englewood Cliffs, New Jersey, 1957).
[21] E. B. Wilson, *The Cell in Development and Heredity* (The Macmillan Company, New York, 1928), third edition.
[22] S. Inoué, Exptl. Cell Research. Suppl. 2, 305 (1952).
[23] S. Inoué, Chromosoma **5**, 487 (1953).
[24] D. Mazia, Symposia Soc. Exptl. Biol. **9**, 335 (1955).
[25] D. Mazia, Advances in Biol. and Med. Phys. **4**, 70 (1956).
[26] D. Mazia and K. Dan, Proc. Natl. Acad. Sci. U. S. **38**, 826 (1952).
[27] J. G. Carlson, Chromosoma **5**, 199 (1952).
[28] R. Chambers in *General Cytology*, E. V. Cowdry, editor (The University of Chicago Press, Chicago, 1924), p. 235.
[29] S. Inoué, Biol. Bull. **103**, 316 (1952).
[30] G. Östergren, Hereditas **35**, 445 (1949).
[31] W. J. Schmidt, Chromosoma **1**, 253 (1939).
[32] K. Bělař, Ergeb. Fortschrift. Zool. **6**, 235 (1926).
[33] A. W. Pollister and P. F. Pollister, Ann. N. Y. Acad. Sci. **45**, 1 (1943).
[34] M. Bessis and J. Breton-Gorius, Bull. microscopic appl. **7**, 54 (1957).
[35] E. de Harven and W. Bernhard, Z. Zellforsch. u. mikroskop. Anat. **45**, 378 (1956).
[36] J. Rhodin and T. Dalhamn, Z. Zellforsch. u. mikroskop. Anat. **44**, 345 (1956).
[37] Ch. Rouiller and E. Fauré-Fremiet, J. Ultrastructure Research **1**, 289 (1958).
[38] M. H. Burgos and D. W. Fawcett, J. Biophys. Biochem. Cytol. **2**, 223 (1956).
[39] D. W. Fawcett, Intern. Rev. Cytol. **7**, 195 (1958).
[40] F. Meves, Verhandal. Anat. Ges. (Jena) Halle, 152 (1902).
[41] P. R. Gross, Trans. N. Y. Acad. Sci. **20**, 154 (1957).
[42] K. R. Porter in *Harvey Lectures* (Academic Press, Inc., New York, 1957), Vol. LI, p. 175.
[43] J. C. Dan, Physiol. Zool. **21**, 191 (1948).
[44] S. Inoué and K. Dan, J. Morphol. **89**, 423 (1951).
[45] M. M. Swann, Quart. J. Microscop. Sci. **94**, 369 (1953).
[46] Thermodynamical analysis of the data was carried out in 1958 by M. Morales of Dartmouth College Medical School. The author is deeply indebted for his contribution.
[47] M. M. Swann, J. Exptl. Biol. **28**, 417 (1951).
[48] M. M. Swann, J. Exptl. Biol. **28**, 434 (1951).
[49] A. F. Hughes and M. M. Swann, J. Exptl. Biol. **25**, 45 (1948).
[50] J. Hanson and H. E. Huxley, Symposia Soc. Exptl Biol. **9**, 228 (1955).
[51] A. F. Huxley, Progr. in Biophys. and Biophys. Chem. **7**, 255 (1957).
[52] T. Hayashi and R. Rosenbluth, J. Cellular Comp. Physiol. **40**, 495 (1952).
[53] A. Szent-Györgyi, *Chemistry of Muscular Contraction* (Academic Press, Inc., New York, 1951), second edition.
[54] H. H. Weber, *The Motility of Muscle and Cells* (Harvard University Press, Cambridge, 1958).
[55] H. Hoffmann-Berling, Fortschr. Zool. **11**, 142 (1958).

Reprinted from *Journal of the Optical Society of America*, Vol. 49(2), pp. 191–198, 1959, with permission from The Optical Society of America.

Diffraction Images in the Polarizing Microscope*

HIROSHI KUBOTA,† *Institute of Optics, University of Rochester, Rochester, New York*

AND

SHINYA INOUÉ‡ *Department of Biology and Institute of Optics, University of Rochester, Rochester New York*

(Received February 12, 1958)

In the polarizing microscope set for extinction, only that image whose polarization has been altered is available to form an image. The lenses themselves introduce such an alternation by rotation of the plane of polarization of rays having oblique incidence. This paper shows that the diffraction image of a pinhole has the form $\sin 2\theta \cdot J_3(r)/r$, where θ and r are polar coordinates in the image plane. The image has four bright zones separated by a dark cross and the central pattern becomes a four-leaf clover form. The diffraction image of a point source through a plate of uniaxial crystal cut perpendicular to its optic axis (z-cut) is also a four-leaf clover when the polarizers are crossed. When the polarizers are parallel, the diffraction image becomes similar to that of an astigmatic system.

1. INTRODUCTION

IN an optical system, amplitude of the image $A(r,\theta)$ in a plane conjugate to the object plane is given as the Fourier transform of the amplitude of light in the exit pupil, $P(\rho,\varphi)$, when the aperture is not so large. Using a proportional factor of 1 for simplicity, this is given as

$$A(r,\theta)=\int_0^a\int_0^{2\pi} P(\rho,\varphi)\exp\{iz\rho\cos(\theta-\varphi)\}\rho d\rho d\varphi, \quad (1)$$

where (ρ,φ) and (r,θ) are the polar coordinates of the exit pupil and in the image plane, a is the radius of the exit pupil and

$$z=2\pi r/\lambda l.$$

Here l is the distance between the exit pupil and the image plane which is large as compared with a, and λ the wavelength. $P(\rho,\varphi)$ is called the pupil function, and in an ideal optical system, it is constant $(=1)$ over the whole pupil for a point source at the center of the object plane. The amplitude in this case is given from (1) as

$$A(r,\theta)=2\pi\int_0^a J_0(z\rho)\rho d\rho=2\pi a^2 J_1(az)/az. \quad (2)$$

This describes the Airy disk referred to hereafter. $P(\rho,\varphi)$ is usually written in the form

$$P(\rho,\varphi)=E(\rho,\varphi)\exp\{iV(\rho,\varphi)\}.$$

The effect of the aberration function $V(\rho,\varphi)$ on the diffraction image has been studied thoroughly by Nijboer,[1] Mahan,[2] etc. When polarized light is used, an inherent modification of the pupil function takes place and the diffraction pattern becomes different from that of the Airy disk. This effect is especially emphasized in the polarizing microscope. In the following, we shall calculate these anomalous diffraction patterns and compare them with observed patterns.

2. DIFFRACTION IMAGE

Let Fig. 1 be the diagram of an optical system having a polarizer and an analyzer. UZ is the plane of polarization of the polarizer and VZ is the plane perpendicular to it. PZ is the plane of incidence at a typical point of the refracting (reflecting) surfaces which makes an angle φ with UZ. Then the components of amplitude of light E_p and E_n parallel and perpendicular to the incident plane are

$$E_p=\cos\varphi, \quad E_n=\sin\varphi. \quad (3)$$

The amplitude of light coming from the polarizer is assumed to be 1 over the pupil. If the amplitude loss factor of the optical system for the components of light polarized parallel and perpendicular to the plane of incidence are K_p and K_n, then the amplitude of light passing through the optical system is given by multiplying them by K_p and K_n, respectively. They can be complex numbers and in an uncoated optical system of no absorption near normal incidence, $K_p=K_n \doteqdot (0.96)^N$, where N is the number of surfaces.

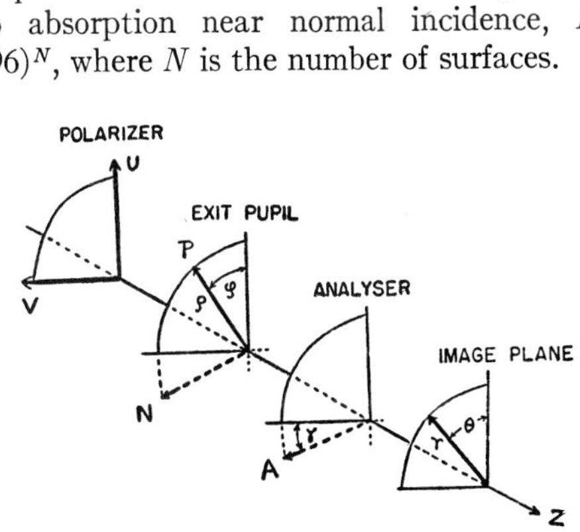

FIG. 1. Optical system.

* Supported in part by the National Science Foundation and by the National Institute of Health (C-3002) United States Public Health Service.

† On leave as Fulbright visiting professor from the Institute of Industrial Science, University of Tokyo, Chiba-City, Japan (present address).

‡ Scholar in Cancer Research of the American Cancer Society.

[1] B. R. A. Nijboer, Thesis, Groningen (1942).

[2] A. I. Mahan, J. Opt. Soc. Am. 40, 664 (1950), etc.; F. Zernike, J. Opt. Soc. Am. 47, 466 (1957). Dr. R. E. Hopkins kindly called our attention to this literature.

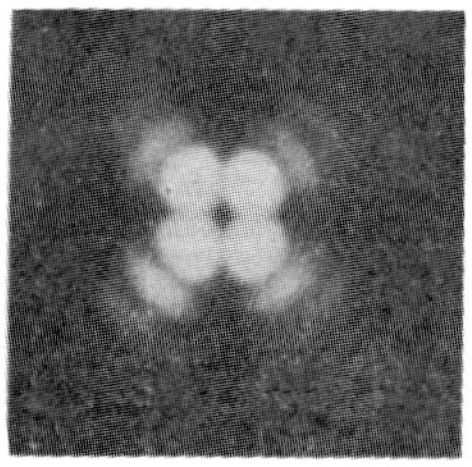

FIG. 2. Diffraction image (crossed polarizers).

Finally the plane of polarization of the analyzer AZ is assumed to make an angle γ with VZ (Fig. 1). Then the component of light E_A in the plane AZ, that is, the amplitude of light coming out from the analyzer is

$$E_A = K_p \cos\varphi \cdot \sin(\varphi-\gamma) - K_n \sin\varphi \cdot \cos(\varphi-\gamma)$$
$$= \tfrac{1}{2}[\sin(2\varphi-\gamma)(K_p-K_n) - \sin\gamma \cdot (K_p+K_n). \quad (4)$$

If K_p and K_n were expressed as functions of ρ and φ, this is the pupil function (of a point source at the center of the object plane) of the optical system, when the off-cross angle between the polarizer and analyzer is γ. The amplitude of the diffraction image is given by putting E_A into (1) in place of $P(\rho,\varphi)$. When K_p and K_n are functions of ρ only, we first perform the integration with respect to the angular variable and obtain

$$A(r,\theta) = \pi \sin(\gamma-2\theta) \int_0^a (K_p-K_n) J_2(z\rho)\rho d\rho$$
$$-\pi \sin\gamma \int_0^a (K_p+K_n) J_0(z\rho)\rho d\rho. \quad (5)$$

To calculate the radial intensity distribution from (5), we have expanded $(K_p \pm K_n)$ into a series of circle polynomials as[3]

$$(K_p-K_n) = \sum_{n=1} \alpha_n R_{2n}^{2}(\rho/a), \quad (6)$$
$$(K_p+K_n) = \sum_{n=0} \beta_n R_{2n}^{0}(\rho/a). \quad (7)$$

Then we can utilize an excellent property of the circle polynomials[4]:

$$\int_0^a R_{2n}^{m}(\rho/a) J_m(z\rho)\rho d\rho = (-1)^{n-m/2} a^2 J_{2n+1}(az)/az,$$

and we have, as the amplitude of the diffraction image,

$$A(r,\theta) = \pi a^2 \{\sin(\gamma-2\theta)F_1(z) + \sin\gamma F_2(z)\}, \quad (8)$$

where

$$F_1(z) = \{\alpha_1 J_3(az) - \alpha_2 J_5(az) + \alpha_3 J_7(az) - \cdots\}/az,$$
$$F_2(z) = -\{\beta_0 J_1(az) - \beta_2 J_3(az) + \beta_3 J_5(az) - \cdots\}/az. \quad (9)$$

When z is very large, the fomula for asymptotic expansion of the Bessel function shows that

$$J_{2n+1}(az) \sim (-1)^n J_1(az);$$

accordingly,

$$A(r,\theta) = \pi a^2 \{\alpha \sin(\gamma-2\theta) + \beta \sin\gamma\} J_1(az)/az, \quad (10)$$

where

$$\alpha = -\sum_{n=1} \alpha_n, \quad \beta = -\sum_{n=0} \beta_n.$$

That is, the intensity distribution for any fixed value of θ becomes that of an Airy disk.§

3. REFRACTING OBJECTIVE

We shall apply these formulas to the polarizing microscope with refracting objective. If we assume the objective has no aberration, then K_p and K_n are real and we shall denote them with k_p and k_n. They are given by Fresnel's formula and different for the components parallel and perpendicular to the plane of incidence.

(a) Crossed Polarizers

When the polarizers are crossed ($\gamma=0$), the amplitude of the diffraction image is given, from (8) and (9) as

$$A(r,\theta) = -\pi \sin 2\theta \int_0^a (k_p-k_n) J_2(z\rho)\rho d\rho$$
$$= -\pi a^2 \sin 2\theta \cdot F_1(z). \quad (11)$$

So, the intensity along $\theta=0$ and $\pi/2$ is zero. This is due to the fact that the pupil function here is nonuniform, and the light comes from four zones of the aperture and the alternate zones are shifted π in phase in such a way as to interfere destructively on the axes and give the dark cross.

A photograph of the diffraction image taken with nearly aberrationless, noncoated objective (that is, with an objective which gives a nearly perfect Airy disk when polarizers are removed) is shown in Fig. 2.

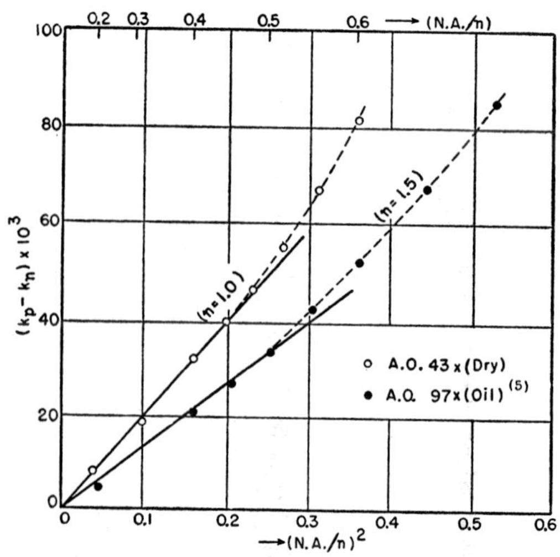

FIG. 3. Examples of the value of $(k_p - k_n)$.

[3] H. Gamoh, Ôyō Butsuri 26, 102 (1957) (in Japanese); H. Slevogt, Optik 14, 383 (1957).
[4] F. Zernike, Physica 1, 689 (1934).

§ This fact was also noticed by Nijiboer reference 1, p. 57, in the case of the pattern of astigmatism.

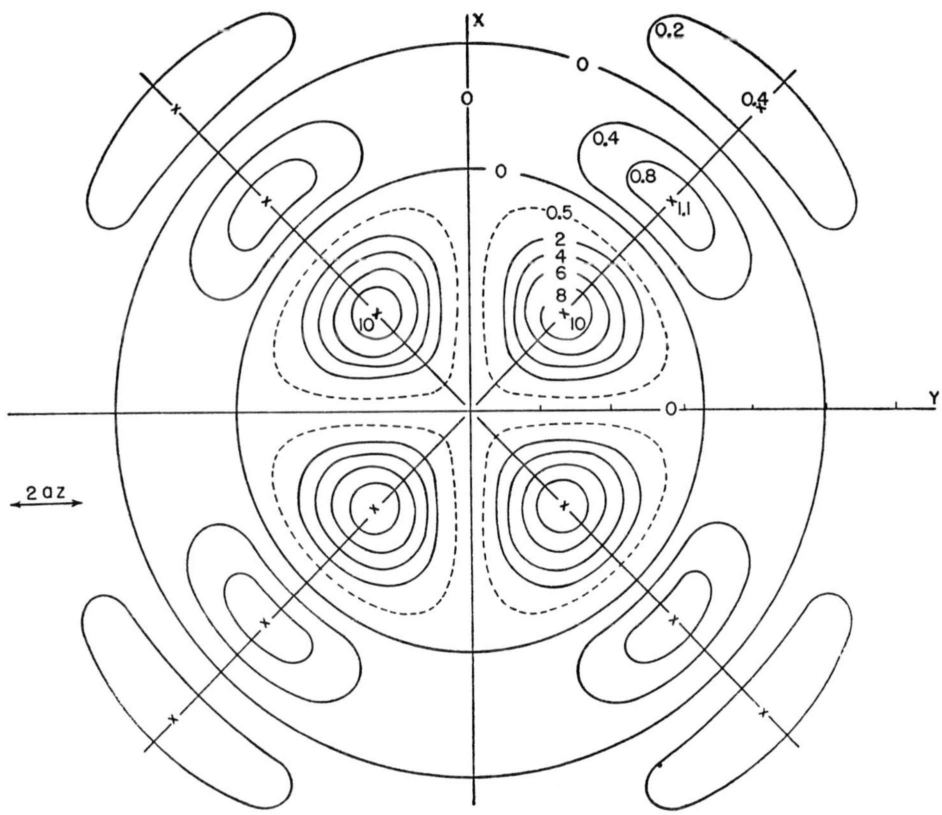

FIG. 4. Contours of equal intensity of the diffraction image.

A dark cross is clearly seen and the central pattern is four-leaf clover, quite different from the Airy disk. To calculate the radial intensity distribution when z is not so large, the coefficients α_n of (9) must be known. These coefficients are different for different objective and we have to know them for each case. But, when the aperture is not so large (about $(N.A./n)<0.5$, n is the refractive index of the ambient medium), as is seen from the examples shown in Fig. 3,[5] we may put as

$$(k_p-k_n)=\alpha\,\rho^2,$$

and the general expression of the intensity for any objectives is, from (11),

$$I=\text{const}\times\{\sin 2\theta\cdot J_3(az)/az\}^2. \qquad (12)$$

The contours of equal intensity calculated are shown in Fig. 4, the intensity being so normalized that the first maxima become 10. The intensity distribution along the diagonal is shown in Fig. 5, together with the roots of $J_3(az)=0$. The radius of the first dark ring is $az=6.38$, about 1.7 times larger than that of the Airy disk, so the resolving power must be lowered considerably. However, as the form of the diffraction image, differing from the Airy disk, has a remarkable periodic structure, and Rayleigh's definition of resolution cannot be applied as it is. For estimating the resolving ability of the objective, it may be best to calculate the response (frequency transfer) function. This will be shown in another paper.

[5] C. J. Koester (private communication).

The maxima of the intensity are given as the roots of

$$\frac{\partial}{\partial z}\{J_3(az)/az\}=0, \quad \text{e.g.,} \quad J_2(az)=2J_4(az).$$

The calculated maxima are shown in Table I, they agree well with the observed values. The intensity ratio between the first and the higher order maxima is also very large as compared with that of the Airy disk. So the halo rings are very bright.

(b) Parallel Polarizers

In the case of parallel polarizers ($\gamma=-\pi/2$), (k_p-k_n) is very small compared to $(k_p+k_n)\doteqdot 2$, and so, neglect-

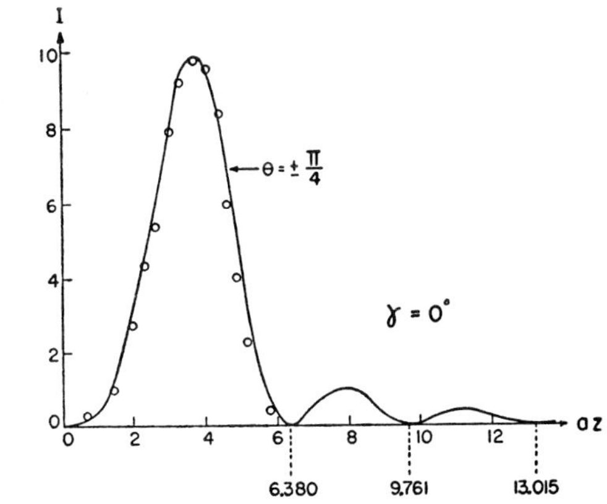

FIG. 5. Intensity distribution along $\theta=\pi/4$.

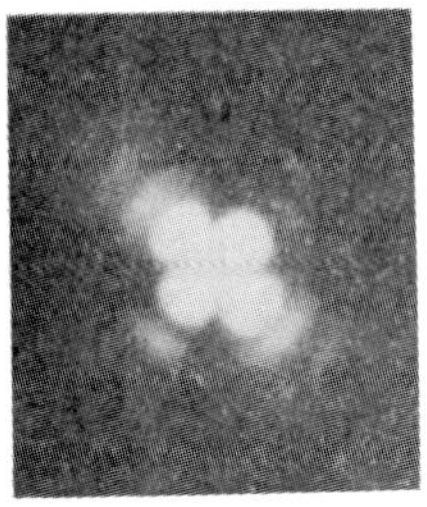

(6a)

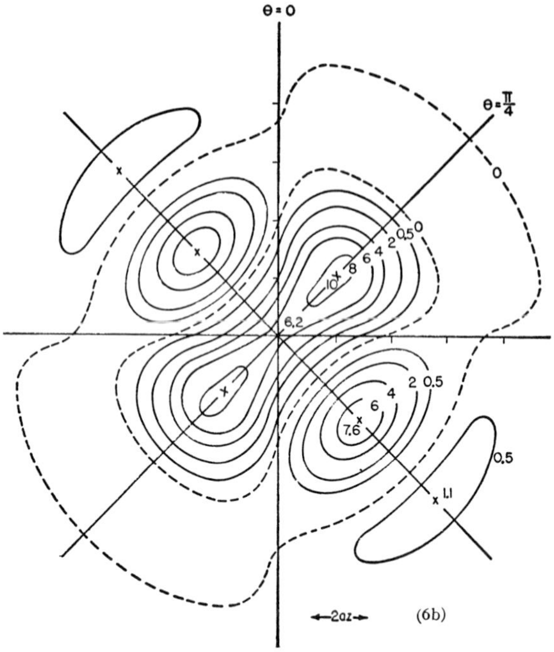

(6b)

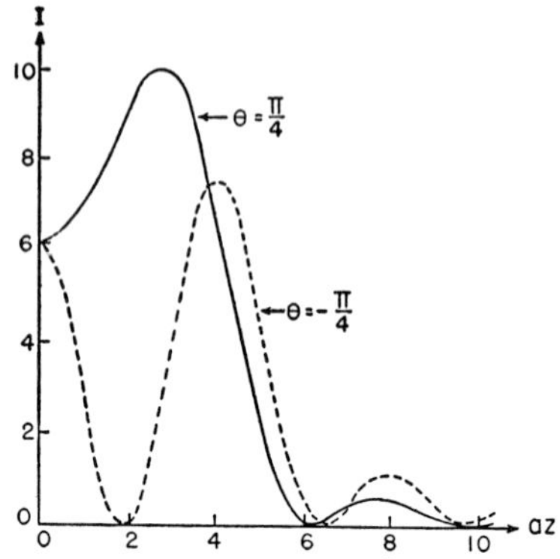

(6c)

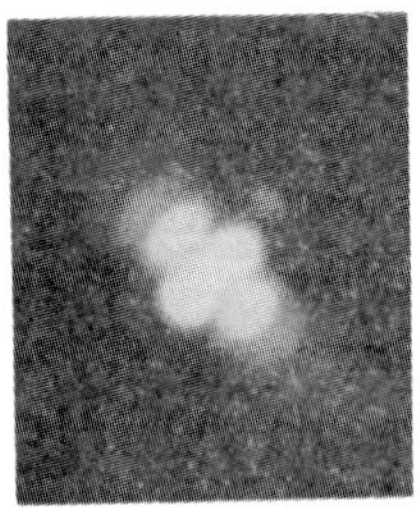

(7a)

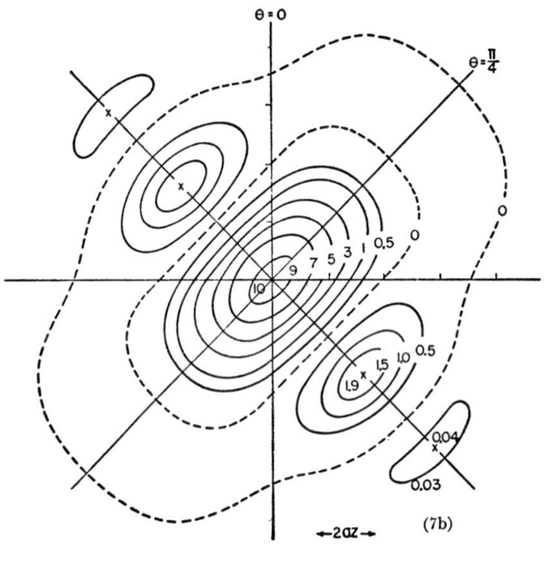

(7b)

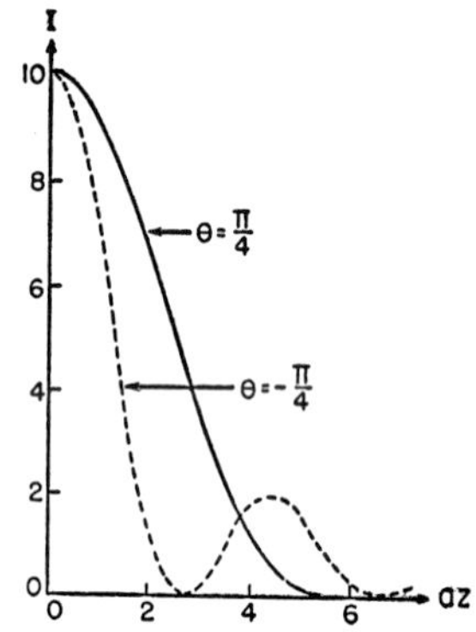

(7c)

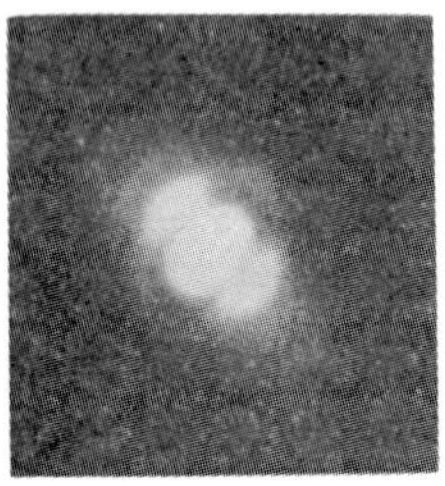

(8a)

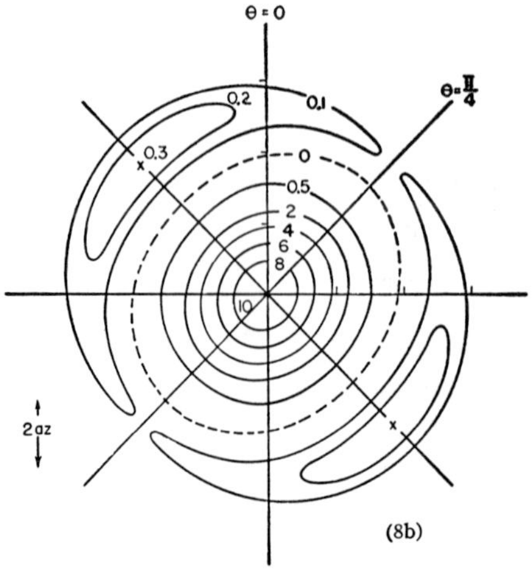

(8b)

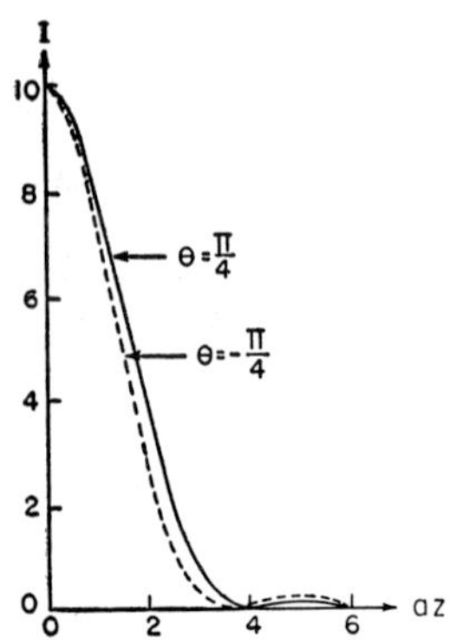

(8c)

FIGS. 6, 7, and 8. Diffraction images (a); contours of equal intensity (b); and intensity distribution along $\theta=\pm\pi/4$ (c) ($g=\alpha_1/2\gamma$, γ is the off-cross angle). Fig. 6. $g=5$. Fig. 7. $g=2.5$. Fig. 8. $g=1.75$.

ing the former and putting the latter equal to 2, we have as the amplitude

$$A(r,\theta)=2\pi\int_0^a J_0(z\rho)\rho d\rho.$$

This is just the amplitude of the Airy disk given in (2).||

(c) When the Off-Cross Angles is γ

When the off-cross angle is γ, we have from (8), taking the first terms of (9) and putting $\beta_0=2$ and assuming γ is small;

$$A(r,\theta)=-2\pi a^2\gamma\{J_1(az)+g\sin 2\theta\cdot J_3(az)\}/az,$$

where, $g=\alpha_1/2\gamma$. The intensity along $\theta=0$ and $\theta=\pi/2$ is the same as that of the Airy disk and the intensity distribution is symmetrical with respect to the $\theta=\pm\pi/4$ axes. The photographs of the diffraction patterns when $g=5$, 2.5, and 1.75 (these correspond to the cases of $\gamma=0.5°$, 1.5°, and 10° in the present experiment) are shown in Figs. 6, 7, and 8, respectively. Contours of equal intensity and the intensity along the $\theta=\pm\pi/4$ axes, calculated by the above formula are also shown

TABLE I. Maxima of $\{J_3(az)/az\}^2$.

az	I (ratio)
3.611	10.00
7.869	1.09
11.251	0.37
14.514	0.17
17.731	0.09

in the same figures. Zeros and maxima of the intensity along the $\theta=-\pi/4$ axis are obtained as the roots of

$$J_1(az)=gJ_3(az) \quad \text{and} \quad (g+3)J_2(az)=2gJ_4(az).$$

The solid lines in Fig. 9 show the calculated values of maxima and zeros on the $\theta=-\pi/4$ axis. Superimposed are actual measurements made from the photographs. The agreement can be said to be good if we consider the difficulty of removing the residual strain-birefringence completely from the optical system.

4. CATADIOPTRIC SYSTEM

In the case of the polarizing microscope with a catadioptric objective,

$$K_p=k_p\exp(i\delta_p) \quad \text{and} \quad K_n=k_n\exp(i\delta_n),$$

where δ_p and δ_n are the phase change of light when it is reflected at the metallic surfaces of the mirrors. If the objective is of a pure reflecting type, we can put $k_p=k_n\doteq 1$. However, in these objectives, the central part of the aperture is obstructed by the auxiliary mirror. For the case of crossed polarizers, the amplitude

|| When $k_p=k_n$ and $k_p+k_n=2-b(\rho/a)^2$, where b is a constant, was discussed by H. H. Hopkins [Proc. Phys. Soc. (London) **B62**, 22 (1949)] and $A(r,\theta)=\{(4-b)J_1(az)+bJ_3(az)\}/az$.

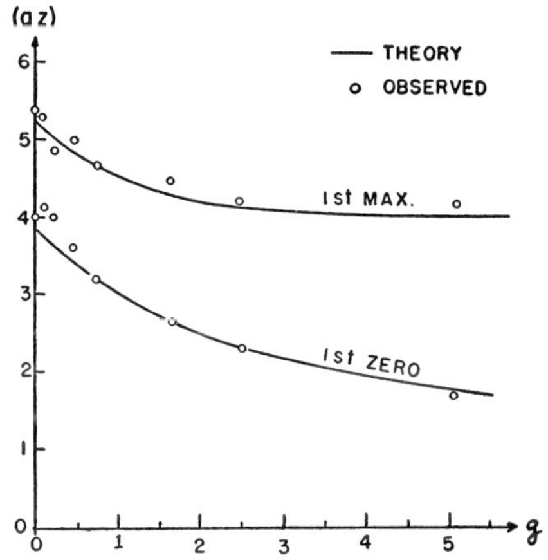

FIG. 9. Maxima and zeros of the diffraction images.

of the diffraction image is given, in place of (11), by

$$A(r,\theta)=-\pi\sin 2\theta\int_{a_0}^a\{\exp(i\delta_p)-\exp(i\delta_n)\}J_2(z\rho)\rho d\rho,$$

where a_0 is the projected radius of the auxiliary mirror at the exit pupil. When the numerical aperture is so small that the terms higher than ρ^2 can be neglected, $[\exp(i\delta_p)-\exp(i\delta_n)]$ is approximately $i\alpha_1(\rho/a)^2$. Then the amplitude is

$$A(r,\theta)=-i\pi a^2\alpha_1\sin 2\theta\left\{\frac{J_3(az)}{az}-s^4\frac{J_3(saz)}{saz}\right\}, \quad (13)$$

where $s=a_0/a$. The term for the central obstruction [the second term in (13)] is multiplied by s^4, different from the usual case of the Airy disk,[6] in which it is multiplied by s^2. So the effect of obstruction in this case is very small so long as s is not large (Fig. 10).¶ Accordingly we may say that in practice, the auxiliary

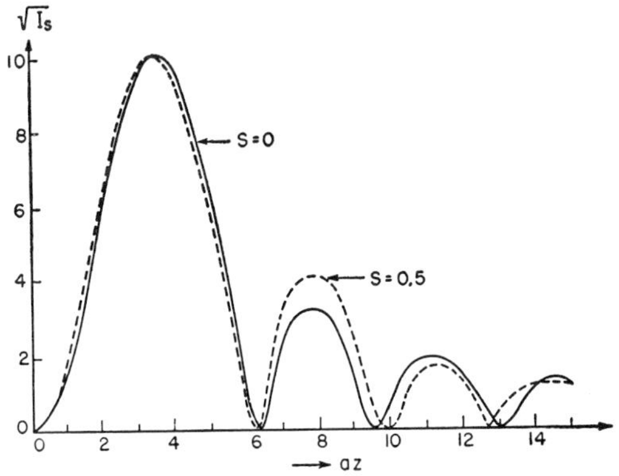

FIG. 10. Intensity distribution of the diffraction image along $\theta=\pi/4$ (annular aperture, $s=a_0/a$).

[6] E.g., M. Françon, *Handbuch der Physik* (Berlin, 1956), Vol. 24, p. 280.

¶ The radius of the first dark ring is reduced only 2% when $s=0.5$ in the present case, while in the usual case, it is 18% when $s=0.5$.

TABLE II. Values of az.

s	0	0.2	0.3	0.4	0.5
1st max.	3.611	3.611	3.609	3.600	3.576
1st zero	6.380	6.379	6.370	6.333	6.243
2nd max.	7.869	7.870	7.872	7.875	7.856
2nd zero	9.761	9.765	9.793	9.872	9.761
I_2/I_1[a]	0.110	0.109	0.106	0.096	0.078

[a] Intensity ratio between the 1st and the 2nd maxima.

mirror has little effect on the diffraction image. This means that the light coming from the central part of the aperture does not contribute much to the image in the case of crossed polarizers.

Values of az which give zeros and maxima of the intensity distribution, calculated from (13), are given in Table II.

5. z-CUT UNIAXIAL CRYSTAL

(a) Crossed Polarizers

Now, we shall examine the diffraction image of a pinhole through a crystal plate (Fig. 11). If the crystal is uniaxial and cut perpendicular to its optic axis (z-cut), then the plane of polarization of the ordinary ray is the plane of incidence and that of the extraordinary ray is perpendicular to it. Accordingly, if we assume $k_p = k_n = 1$, which means that the aperture is small enough to ignore the effect discussed in this paper, we have

$$K_p = \exp(i\Delta_0) \quad \text{and} \quad K_n = \exp(i\Delta_e), \tag{13}$$

where Δ_0 and Δ_e are advance in phase due to the insertion of crystal for the ordinary and the extraordinary rays, respectively. They are given by, putting the thickness of crystal as d:

$$\Delta_0 = (2\pi d/\lambda)(n_0 \cos r_0 - n \cos i),$$
$$\Delta_e = (2\pi d/\lambda)(n_e \cos r_e - n \cos i),$$

where n_0, n_e and r_0, r_e are the refractive indexes and the angle of refractions of the ordinary and the extraordinary rays in the crystal and n and i are the refractive index of the ambient medium and the angle of incidence. If we put the principal refractive indexes of the crystal as n_1 $(=n_0)$ and n_2, we have[7]

$$n_0 \cos r_0 = n_1\{1 - (n \sin i/n_1)^2\}^{\frac{1}{2}},$$
$$n_e \cos r_e = n_1\{1 - (n \sin i/n_2)^2\}^{\frac{1}{2}}.$$

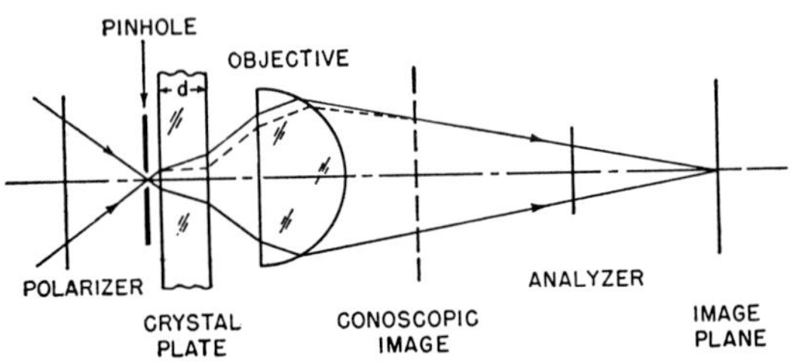

FIG. 11. Diffraction image of a pinhole through a crystal plate.

[7] E.g., reference 6, p. 443.

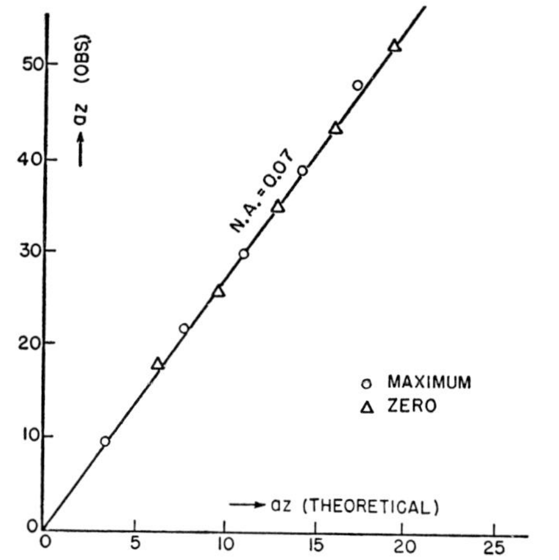

FIG. 12. Observed values of the maxima and zeros (z-cut uniaxial crystal) in arbitrary scale.

Putting them into Δ_0 and Δ_e, and assuming $\sin i = (\rho/f)$ approximately, we have as the series expansion of the retardations as

$$\Delta_0 = \Delta + B_0\rho^2 + \cdots, \quad \Delta_e = \Delta + B_e\rho^2 + \cdots,$$

where

$$\Delta = (2\pi d/\lambda)(n_1 - n), \quad B_0 = (\pi d/\lambda)\left(\frac{1}{n_1} - \frac{1}{n}\right)\left(\frac{n}{f}\right)^2,$$

$$B_e = (\pi d/\lambda)\left\{\frac{1}{n_1}\left(\frac{n_1}{n_2}\right)^2 - \frac{1}{n}\right\}\left(\frac{n}{f}\right)^2, \tag{15}$$

where f is the focal length of the objective.

The amplitude of the diffraction image is given by (8) for the general case and by (11) for crossed polarizers, with the coefficients given from (15). If the aperture is so small that all the terms higher than ρ^2 in (15) can be neglected, the coefficients α_n and β_n in (8) are

$$\alpha_1 = i(B_0 - B_e), \quad \beta_0 = 2, \quad \beta_1 = i(B_0 + B_e). \tag{16}$$

The intensity in this case is given by a formula similar to (12), that is

$$I = \text{const} \times \{\sin 2\theta \cdot J_3(az)/az\}^2, \tag{17}$$

where

$$\text{const} = \pi^2 a^4 (B_0 - B_e)^2.$$

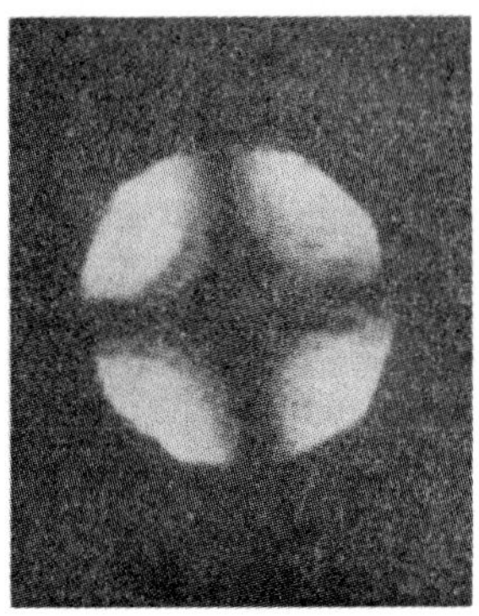

FIG. 13. z-cut uniaxial crystal (crossed polarizers), conoscopic image.

Accordingly, the diffraction image in this case is the same as Fig. 2. Radii of the dark and bright rings of the image measured in the photograph (with z-cut calcite) are plotted in Fig. 12. The abscissa of the figure is the maxima and zeroes of $(J_3(az)/az)^2$, given in Table I and in Fig. 5. The aperture is $N.A.=0.07$, so small that the amplitude of the image is expressed by (17) with good approximation; agreement between the calculated and observed values is very good. Numerical aperture of 0.07 is so small that only the first interference fringe of the conoscopic image is in the view (Fig. 13).

(b) Parallel Polarizers

Figures 14 and 15(a) show the diffraction images of a point source through a z-cut uniaxial crystal at different image planes, when the polarizers are parallel ($N.A.=0.07$). In this case, we cannot neglect (K_p-K_n) as being small as we did in Sec. 3(b). Here we shall, in order to discuss the image in different planes, multiply the terms of the displacement of the image plane $\exp(ic\rho^2)$ by the pupil function, where c is a constant proportional to the displacement of the image plane.[8] The amplitude of the diffraction image in case of parallel polarizers is given, in place of (4), by

$$A(r,\theta)=\pi\cos 2\theta\int_0^a \exp(ic\rho^2)(K_p-K_n)J_2(z\rho)\rho d\rho$$
$$-\pi\int_0^a \exp(ic\rho^2)(K_p+K_n)J_0(z\rho)\rho d\rho.$$

The general expression for the amplitude is obtained by expanding the integrands into series of circle polynomials. If the aperture is so small that the terms higher than ρ^2 in (15) can be neglected, we have from (16):

$$A(r,\theta)=\pi a^2\exp(i\Delta)\{J_1(az)+i[(B_0-B_e)\cos 2\theta+(B_0+B_e+2c)]J_3(az)\}/az.$$

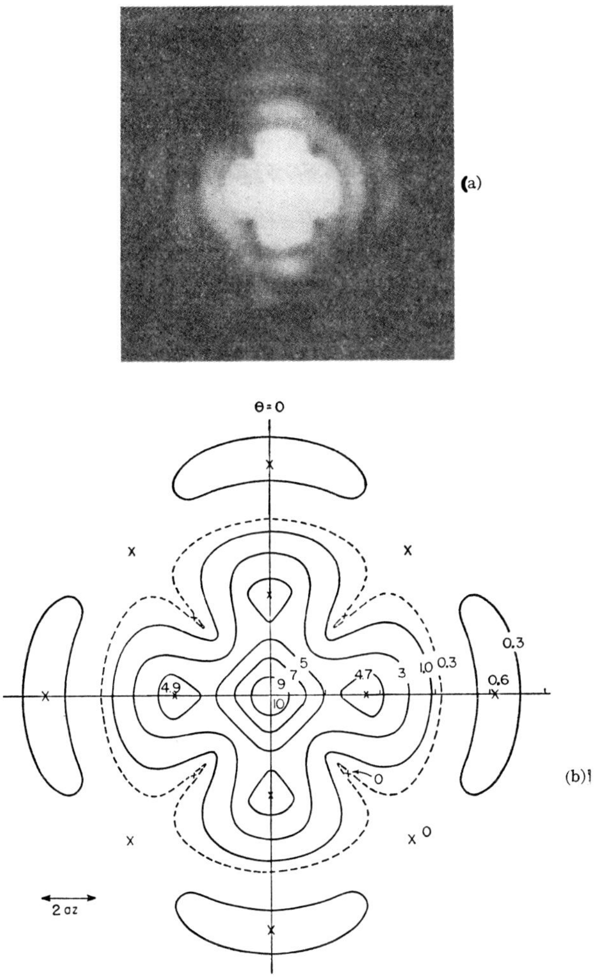

FIG. 14. Midway image plane.

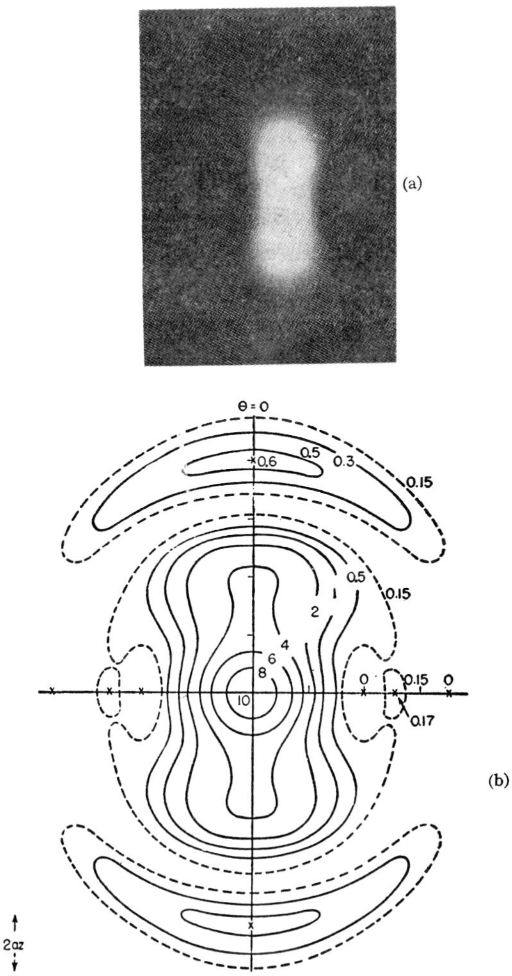

FIG. 15. Meridional image plane. (To get the one in the sagittal image plane, rotate this figure 90°.)

FIGS. 14 and 15. z-cut uniaxial crystal (parallel polarizers). (a) Diffraction image. (b) Contours of equal intensity.

[8] E.g., reference 1, p. 44.

28, 32, 33; see especially 5 for general reference and 33 for a thorough theoretical treatment of the polarizing microscope).

This article will give special attention to methods and devices for obtaining maximum sensitivity with the polarizing microscope. As described later, the combined sensitivity, resolution, and image quality of the polarizing microscope has been vastly improved in the past few years, and retardation (coefficient of birefringence × thickness) of 0.1 Å unit can be detected on objects 0.2μ wide. These advances now permit application of polarization microscopy to new realms such as analyses of fine structure in living cells, or experimental studies of rigorous diffraction theory of the kind which could not be performed with equipment available in the past.

Specification of the System for Obtaining Maximum Sensitivity

For the sensitive detection of birefringence (BR) and dichroism, precise measurement of retardation, optical activity (rotation of plane-polarized light), extinction angle, etc., detectable contrast must be introduced with only minimal differences in a parameter. Contrast is maximized for a given object by reducing stray light in the system (11, 30), and by using appropriate compensators or half-shade plates (3, 7, 11, 13, 15, 18, 27, 28, 30, 31). The degree of success in reduction of stray light, or increase in extinction factor (EF), can be expressed numerically as EF = intensity of light with parallel polars, divided by intensity of light with crossed polars (30). The effect of various components on the EF is discussed in the next section.

The ability of the eye to perceive contrast is drastically lowered at low levels of image brightness, a condition which prevails at high EF's. A maximally bright light source is therefore required. Contrast discrimination can also be improved by darkening the room and by masking off unwanted sources of bright light in the image (e.g., retardation can be measured to λ/1000 with a quartz wedge when the regions outside of the dark fringes are masked).

Specification of the Components

Polars. Modern sheet polaroids (general properties outlined in 21) can, but do not always, have EF as high as 3×10^5 for green light at normal incidence. For most work EF of this magnitude is adequate, but two additional factors should be borne in mind for critical application. The parallel transmission of a pair of high extinction polaroids is of the order of 10 ~ 15%, hence a considerable light loss takes place. Most sheet polaroids have microcrystalline textures or inclusions which can cause "hot spots" affecting the diffraction image.

Calcite polars coated with a low reflection film may show virtually no light loss (parallel transmission ≒50%) and an extremely high EF (~3×10^6). Of the various types of calcite prisms available (17, Lambrecht 20 has supplied excellent quality prisms), the square-ended Glan-Thompson prisms appear to be the most effective. The optical quality and EF of calcite prisms depend to a great extent on the skill of manufacture and subsequent care of handling since the soft calcite surfaces are prone to scratching and pitting. The beam passing through the analyzing calcite prism should be made parallel by the stigmatizing lenses to minimize the astigmatism resulting from the double refraction of calcite.

Condenser and Objective Lenses. All optical components lying between the polarizer and the analyzer must be scrupulously clean. Not only may any dust particle, greasy surface, etc. scatter enough light to lower the EF and seriously reduce object contrast, but by acting as loci having high transmittance, such entities would give the effect of oblique illumination and distort the diffraction image.

The deleterious effect of large amounts of strain BR in the condenser and objective

POLARIZING MICROSCOPE

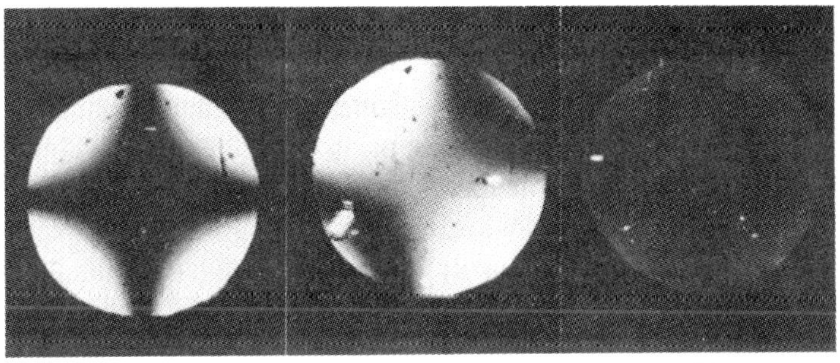

(a) (b) (c)

FIG. 1. Appearance of the back aperture of a 97 × 1.25 NA strain-free coated objective with and without rectifiers. The condenser, which is identical with the objective, is used at full aperture. Collimated mercury green light (546 mμ) used for illumination. (a) Crossed polarizers, no rectifier. (b) Polarizer turned 2°, no rectifier. (c) Crossed polarizers, with rectifiers in both condenser and objective. Photographs a, b, and c were given identical exposures. From Inoué, S. and W. L. Hyde (12).

lenses has long been recognized and polarizing microscopes are generally furnished with "strain-free" lenses. The magnitude of strain BR in these lenses is frequently still considerable and for realizing the utmost sensitivity, lenses with exceptionally low strain BR must be selected. The procedure for selecting these follows.

Preparation for the Test. Provide a polarizing microscope with a very high intensity light source, such as a high pressure mercury arc lamp (General Electric A-H6 or Osram HBO 200) or other lamp of equivalent brightness. The components of the microscope should be adjusted or repaired so that the EF of the system without the lenses is greater than 10^5. Close down the field stop and use Köhler illumination, employing a nominally identical objective as the condenser. It is convenient to screw the test lens into an objective holder which in turn is placed inverted on the rotating stage and centered to the optical axis. This acts as the condenser (or objective in an inverted system as in Fig. 2) and the selected mate remains fixed. In addition prepare a removable rotating mica compensator of the order of $\lambda/20$ to be placed in the slot above the objective or between the condenser and the polarizer.

Test for Freedom from Strain BR. When the pair of lenses is aligned, the polars crossed and the image of the field stop properly focused a dark cross (Fig. 1a) should appear at the back aperture of the objective. In the absence of strain BR the cross is observable with objectives of NA as low as 0.1 with no object lying between the condenser and objective lenses. The cross must be symmetrical and very dark between crossed polars and should open symmetrically into two hyperbola-like fringes the "V's" (Fig. 1b) when the polarizer or analyzer is rotated. The arms of the V's should remain dark until they disappear beyond the edge of the aperture. If the lenses are free from local and lateral strain these dark fringes remain undistorted when the stage is rotated and the test lens is examined at various orientations.

Some lenses seemingly passing the above test may still suffer from radially symmetric (strain) BR. This is checked by inserting the mica compensator. When the lenses are free from radial as well as lateral BR the cross loses contrast and fades away with little change in its shape or position when the compensator is rotated. If lenses possess radial BR the cross will open into two V's as though with a strain-free lens the analyzer had been turned.

Rectifiers. Lenses selected for freedom from strain by the methods described can give extremely high EF at low NA's (e.g., 3×10^5 at condenser NA = 0.1). However,

17. JOHANSEN, A., "Manual of Petrographic Methods," McGraw-Hill, New York, 1918, second edition.
18. KÖHLER, A., "Ein Glimmerplättchen Grau I. Ordnung zur Untersuchung sehr schwach doppelbrechender Präparate," *Z. wiss. Mikroskop.*, **38**, 29–42 (1921).
19. KUBOTA, H. AND INOUÉ, S. "Diffraction images in the polarizing microscopes," *J. Opt. Soc. Am.*, **49**, 191–198 (1959).
20. LAMBRECHT, K., See Catalogue "Polarizing Optics." (4318 N. Lincoln Ave., Chicago 18, Ill.)
21. LAND, E. H., "Some aspects of development of sheet polarizers," *J. Opt. Soc. Am.*, **41**, 957–963 (1951).
22. OSTER, G., "Birefringence and Dichroism," in Oster and Pollister: "Physical Techniques in Biological Research," Vol. I, Chapter 8, Academic Press, New York, 1955.
23. PFEIFFER, HANS H., "Das Polarisationsmikroskop als Messinstrument in Biologie und Medizin," Friedr. Vieweg and Sohn, Braunschweig, 1949.
24. RINNE, F. W. B. AND BEREK, M., "Anleitung zu optischen Untersuchungen mit dem Polarizationsmikroscop," Leipzig, 1953.
25. RUCH, FRITZ, "Birefringence and Dichroism of cells and tissues," in Oster and Pollister: Physical Techniques in Biological Research, Vol. III, 149–176 Academic Press, New York, 1956.
26. SCHMIDT, W. J., "Die Bausteine des Tierkörpers," F. Cohen, Bonn, 1924.
27. SCHMIDT, W. J., "Die Doppelbrechung von Karyoplasma, Zytoplasma und Metaplasma," Bd. II, Protoplasma-Monographien, Berlin, 1937.
28. SCHMIDT, W. J., "Die Doppelbrechung des Protoplasmas und ihrer Bedeutung für die Erforschung seines submikroskopischen Baues," *Ergeb. Physiol.*, **44**, 27 (1944).
29. SKINNER, C. A., "A universal polarimeter," *J. Opt. Soc. Am.*, **10**, 491–520 (1925).
30. SWANN, M. M. AND MITCHISON, J. M., "Refinements in polarized light microscopy," *J. Exp. Biol.*, **27**, 226 (1950).
31. TUCKERMAN, L. B., "Doubly refracting plates and elliptic analyzers," *Univ. Nebraska Studies*, pp. 173–219 (1909).
32. WAHLSTROM, E. E., "Optical Crystalography," John Wiley and Sons, New York, 3rd. Ed., 1960.
33. WRIGHT, F. E., "The Methods of Petrographic Microscopic Research," (Carnegie Institution of Washington, Washington, D.C., 1911).

SHINYA INOUÉ

FIBERS (TEXTILES). *See* GENERAL MICROSCOPY, p. 343.

INDUSTRIAL RESEARCH, APPLICATION TO. *See* GENERAL MICROSCOPY, p. 363.

PLASTICS, *See* GENERAL MICROSCOPY, p. 390.

PULP AND PAPER. *See* GENERAL MICROSCOPY, p. 394.

Reprinted from *Chromosoma*, Vol. 12, pp. 48–63, 1961,
with permission from Springer Science and Business Media.

From the University of Rochester, Rochester, New York

BIREFRINGENCE IN ENDOSPERM MITOSIS* **

By

SHINYA INOUÉ and ANDREW BAJER

With 25 Figures in the Text

(Received February 11, 1961)

Introduction

With a sensitive polarizing microscope the microscopic structure and the submicroscopic organization of the mitotic spindle can be followed in living cells by virtue of the birefringence exhibited by the spindle fibers (INOUÉ 1953). To date INOUÉ has observed chromosomal and continuous spindle fibers (following SCHRADER's definition, see SCHRADER 1953) in the following *living* cells in division—pollen mother cells of *Lilium longiflorum* (centrifuged), *Gasteria carinata*, *Polygonatum biflorum* and an iris; oöcytes of the marine worms *Chaetopterus pergamentaceous* (centrifuged), *Mesochaetopterus taylori* (centrifuged), and a mussel *Mytilus californianus* (centrifuged or flattened); developing eggs of *Halistaura cellularia*, *Aglantha digitale* and other medusae, *Bolinopsis sp.* (Friday Harbor, Wash.), *Pleurobrachia sp.* (Friday Harbor, Wash.) and other ctenophores, *Asterias sp.* (Woods Hole, Mass.) and other echinoderms; and spermatocytes of *Drosophila melanogaster* as well as of *Dissosteira carolina*, *Chloealtis conspersa*, *Chortophaga viridifasciata* and other Orthoptera. Except where otherwise specified the spindle fibers were observed without flattening or centrifuging the cells. In meiotic cells two consecutive divisions were often observed and in developing eggs several successive divisions in early cleavage were followed. In all cases observations were continued several hours after division to make sure the cells were developing normally. Time-lapse motion pictures showing birefringence of the spindle fibers in *Lilium longiflorum*, *Halistaura cellularia*, *Bolinopsis sp.* (Friday Harbor, Wash.), *Aglantha digitale*, *Dissosteira carolina* and *Haemanthus katharinae* have also been made[1]. Every observation confirmed the original findings (INOUÉ 1953)—namely, the presence, in actively dividing living cells, of chromosomal fibers whose birefringence[2] throughout anaphase remains strongest

---

* This paper is dedicated to Professor FRANZ SCHRADER on the occasion of his seventieth birthday.

** This research was supported in part by USPH grant C-3002 awarded to SHINYA INOUÉ and by the Rockefeller Foundation travel grant GA BMR 5865 awarded to ANDREW BAJER.

[1] Copies available by contacting senior author.

[2] In this paper the term birefringence is used in the same sense as retardation and not as coefficient of birefringence.

adjacent to the kinetochores (and to the centrioles when present), continuous fibers whose birefringence diminishes during anaphase but recovers during telophase, and in plant cells transformation of the continuous fibers into the more strongly birefringent phragmoplast fibers.

BAJER earlier found that details of individual chromosome behavior could be followed during mitosis by phase contrast microscopy in living endosperm cells of a variety of monocotyledonous plants (BAJER and MOLÈ-BAJER 1954—1958). Many sequences of time-lapse motion pictures were taken especially on endosperm cells of *Haemanthus katharinae* (*Amaryllidaceae*)—an exceptionally favorable material—and the movements of chromosomes and their arms were analyzed frame by frame. However, although chromosomes were clearly visible, spindle fibers could not be seen in healthy cells with phase contrast microscopy.

We have recently found that the same *Haemanthus* endosperm cells show qualities uniquely favorable for observing mitosis in polarized light. Spindle fibers can be seen clearly in living cells without flattening or centrifuging since the cells lack a cell wall, starch grains and other highly birefringent or light scattering elements. In addition, when the cells are flattened until the chromosomes spread into one plane the relation of the spindle fibers to regions of individual chromosomes can be clearly established. Both flattened and non-flattened cells prepared in prophase successfully enter mitosis, divide and complete the formation of their cell plates. With this material we have not only further confirmed the reality of spindle fibers and followed their behavior in greater detail, but were also able to study aspects of mitosis which could not be seen in the other cells previously employed. The following is a report on the observations we have made.

Technique

The endosperm cells were collected from translucent young seeds (ca. 6—8 mm long) of *H. katharinae*[1]. After excess liquid was removed from the seed the cells were extruded onto a cover glass coated with a thin layer of 0.5% agar dissolved in 3.5% glucose. Special precautions were taken to prevent any drying of the agar and the cells. This same cover glass which was ringed with vaseline-paraffin beforehand was placed on another glucose-agar coated cover glass and sealed. When it was desirable to flatten the cells, the preparation was tilted to allow excess liquid to drain onto the lower cover glass or onto a piece of filter paper inserted between the covers. With proper adjustment the isolated endosperm cells were slowly flattened until details of individual chromosomes were clearly visible in one focal plane. The liquid contact between top and bottom cover glasses was then broken and the preparation stabilized (see BAJER 1954a, b for more detail).

[1] We are deeply indebted to Professor R. D. GIBBS of the Department of Botany, McGill University, Montreal, Canada for providing us with the single plant in the proper flowering stage which could be located in continental North America. Many thanks are also due to Mr. RAY J. MAAS of the University of Rochester for locating this plant. The plant bore over 30 fertile fruits.

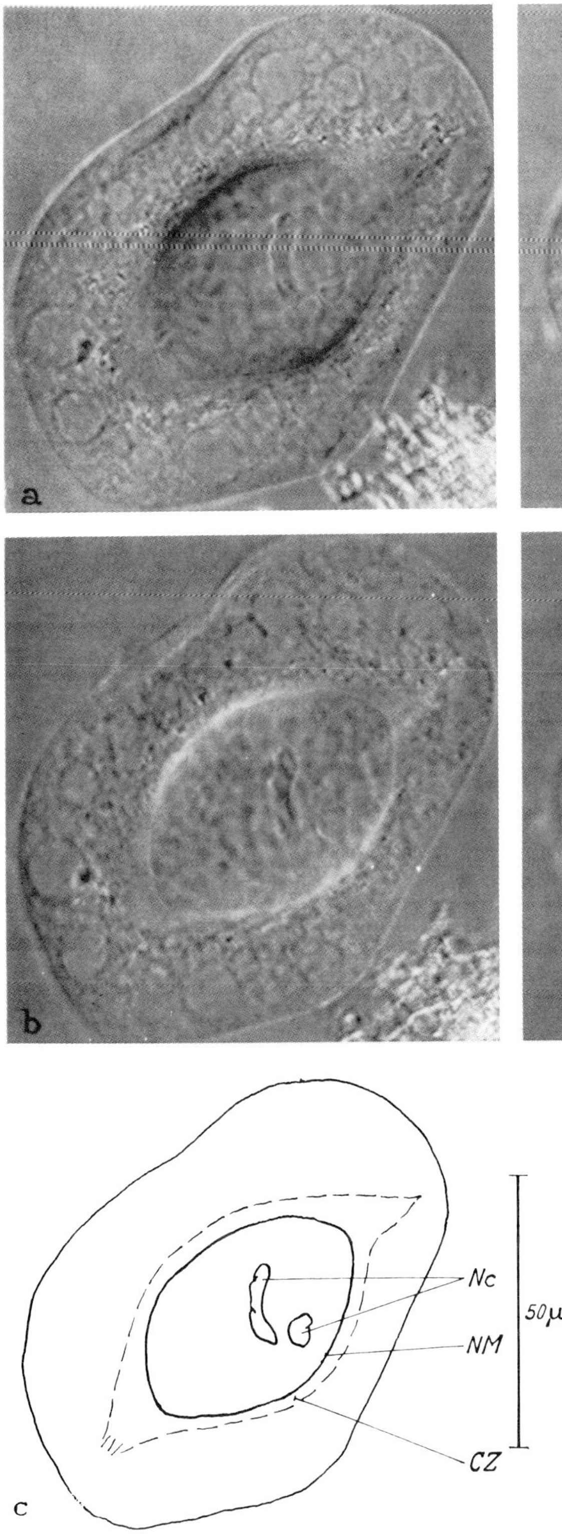

Fig. 1 a—c

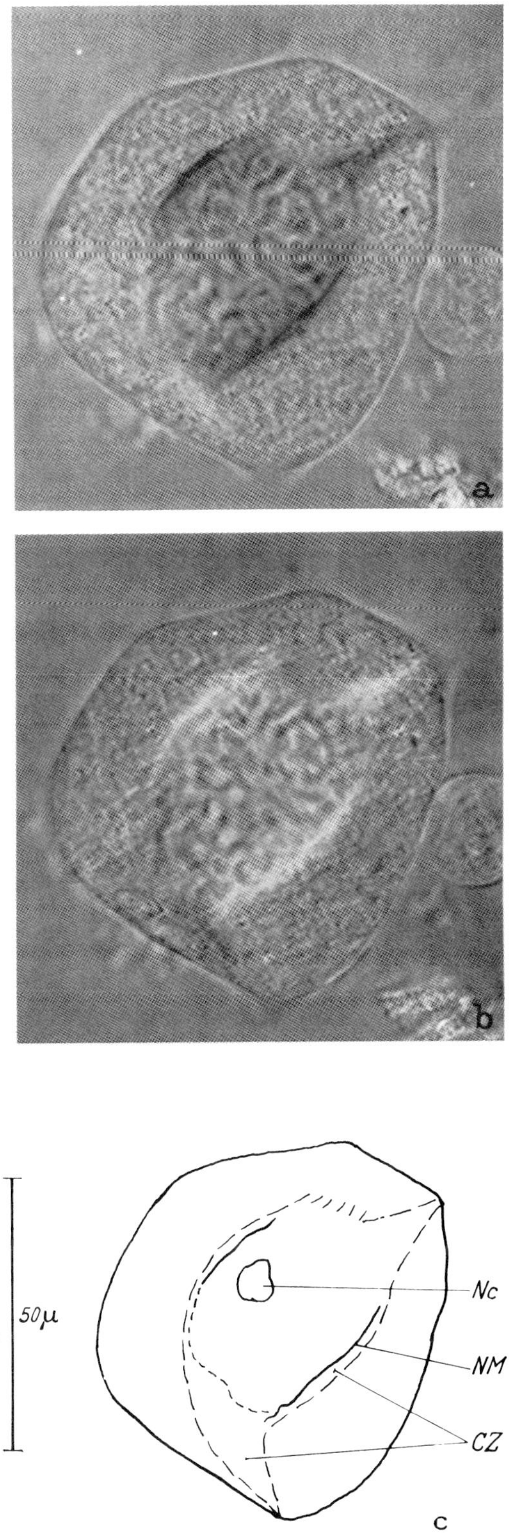

Fig. 2 a—c

Birefringence of the cells was observed and recorded by still and motion pictures with a special polarizing microscope[1]. For these studies a non-rectified coated stain-free 25× 0.65 N.A., Leitz oil-immersion objective was used in conjunction with a 10×0.25 N.A., American Optical coated stain-free objective as the condenser. Pure green light (546 mμ) from a high pressure mercury arc lamp (A-H 6, General Electric) was isolated with a multi-layer high transmission interference filter (Baird Associates, nominal half band widths 70 Å) and its accessory high and low cut-off filters.

Observations

The salient features of our observations are:

1. In prophase a few hours before nuclear membrane breakdown, a narrow birefringent zone appears outside of the nuclear membrane (Figs. 1—3[2]). In position and shape this corresponds to the "clear zone" observable by phase contrast and described earlier by Bajer (1957) as a structureless semi-liquid region. At an early stage of "clear zone" formation two or more pointed arms of the zone are frequently found extending in opposite directions (Fig. 1). The contents of these arms (the classical "polar caps") are also birefringent and show a fibrous texture running in the direction of the protrusions. The slow axis of the birefringence lies parallel to the arms and continues tangentially to the nuclear membrane, i.e., the fibrils show tangential positive birefringence. No distinction or separation is seen between the "polar caps" and the rest of the "clear zone" surrounding the nucleus.

[1] Designed jointly by Shinya Inoué and the research staff of the American Optical Co., and built at the University of Rochester and the A. O. Research Center with generous support from the latter. We are especially grateful to Dr. Stephen M. McNeille of the American Optical Co., and Profs. Robert E. Hopkins and Kenneth W. Cooper of the University of Rochester for making possible the building of this instrument. For a brief description of this instrument see Inoué (1961 b).

[2] Magnification of all prints is identical. No photographs are retouched. The scale appears by Fig. 2c and below Fig. 24. Scale 10 μ interval for all photographs in this paper. In figures marked **a**, the slow axis of the revolving mica compensator (approximate value λ/30) is oriented in the quadrangle opposite to that of the clear zone and the spindle. Thus the material of the clear zone and the spindle fibers appear dark. Regions of the nuclear membrane oriented in this direction are also dark while those at right angles are bright. The chromosomes and nucleoli exhibit a complex pattern of light scattering and birefringence. In figures marked **b**, the orientation of the compensator and the contrast of the birefringent areas is reversed.

Fig. 1 a—c. Mid-prophase showing clear zone, nuclear membrane and nucleoli. The fusiform clear zone (*CZ*, the classical "clear zone" and "polar caps"), exhibits a positive birefringence with reference to its long axis. The intact nuclear membrane (*NM*) also shows a tangential positive birefringence. The nucleoli (*Nc*) are still intact and highly refractile. The cell which is slightly compressed is in a stage just before the cell in Fig. 3

Fig. 2 a—c. Late prophase during nuclear membrane breakdown. Chromosomes have become more differentiated. The nucleolus generally disappears at or shortly after this stage. The nuclear membrane although deformed is still evident near the equatorial region of the fusiform clear zone. Cell slightly compressed.

Figs. 1c and 2c. Tracings of Figs. 1a, b and 2a, b

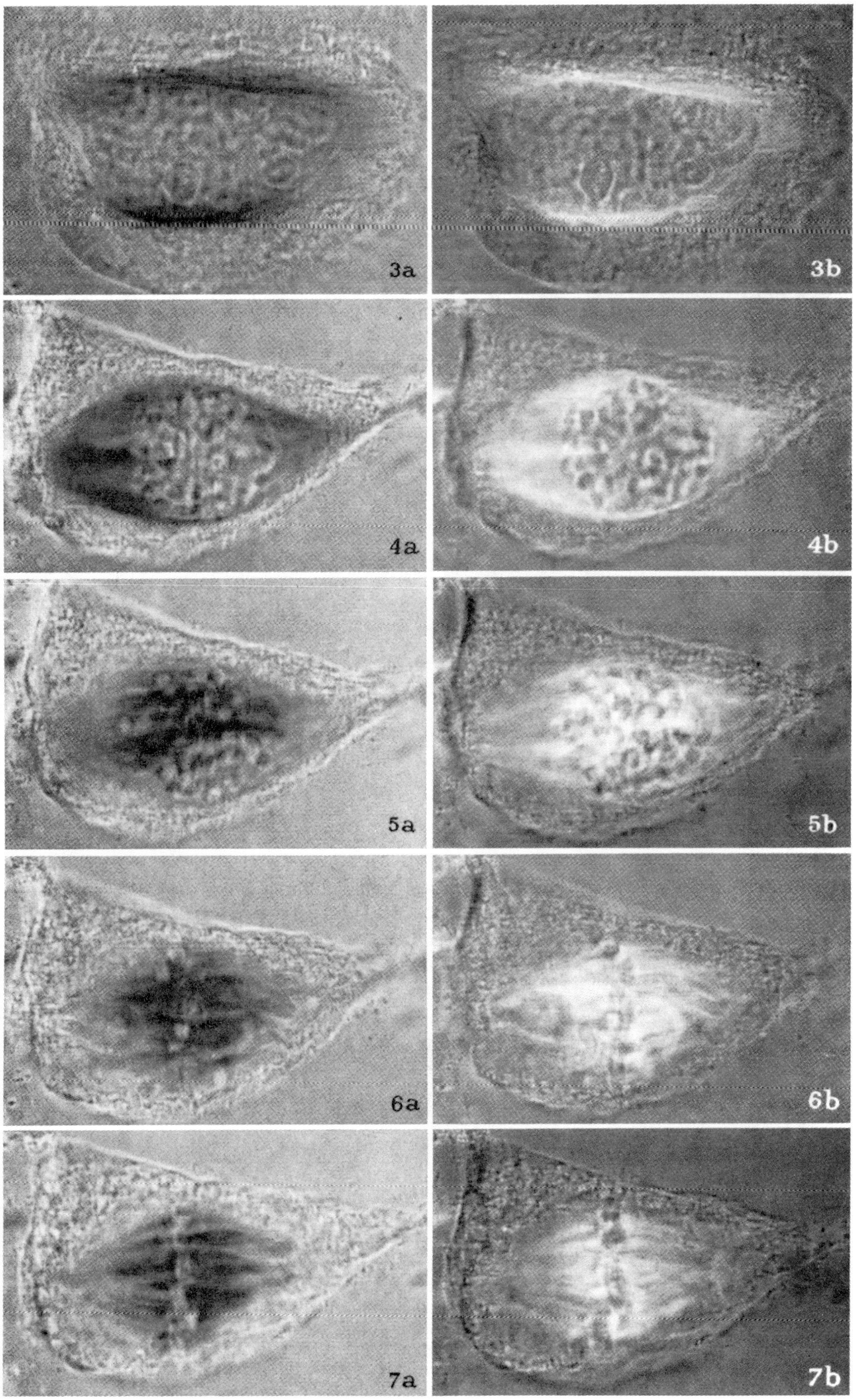

Figs. 3—7

16), as was observed in *Lilium* pollen mother cells and many other cells (INOUÉ 1953 and Introduction).

7. During early anaphase the positive birefringence of the continuous fibers is frequently inconspicuous (Figs. 8—10, 15, 16) but at late anaphase or telophase it again becomes apparent (Figs. 11, 12, 17, 24). Then the continuous fibers condense towards the original metaphase plate and become the phragmoplast fibers (Fig. 18). Time-lapse motion pictures of this stage show a successive transformation of the continuous fibers into phragmoplast fibers. Within the fibrous phragmoplast small granules or vacuoles accumulate to the mid-plane, fuse and become the cell plate (Figs. 19, 20). (BAJER believes these "granules" may actually represent localized droplets of phragmoplast fibers.) The behavior of the continuous fibers, and the formation of the cell plate from the birefringent phragmoplast is similar to that earlier described for *Lilium* and other pollen mother cells (INOUÉ 1953).

8. While some cells were greatly flattened (Figs. 13—20) others were observed without flattening (Figs. 3—12, 21—25). In extremely flattened cells the continuous fibers were less conspicuous and the divisions somewhat slower. The chromosomal spindle fibers may have been somewhat shorter but the pattern of distribution of birefringence in the fibers was not affected by flattening.

9. Many cells were continuously exposed to the very intense monochromatic mercury green light from prophase before nuclear membrane breakdown through completion of cell division. Contrary to earlier observations where endosperm cells were found highly sensitive to illumination (BAJER and MOLÈ-BAJER 1956, MOLÈ-BAJER 1958) no anomaly due to the illumination could be detected.

Discussion and conclusion

Our observations of a tangential positive birefringence in the "clear zone" and its "polar caps" and their subsequent behavior in prophase adds further strength to BAJER's contention that the "clear zone" and polar cap substance is in fact a spindle precursor material (BAJER 1957 as contrasted to WADA 1950, also see OSTERHOUT 1902 and WILSON 1928, pp. 150—156). However, the pattern of birefringence we observed makes it unlikely that the "clear zone" is an "unoriented semi-liquid" (BAJER 1957), although it could very well be, as may also the spindle itself, a labile thixotropic gel.

Figs. 8—12. Fig. 8. 14^{00} *h*, Fig. 9. 14^{06} *h*, Fig. 10. 14^{13} *h*, Fig. 11. 14^{24} *h*, Fig. 12. 14^{32} *h*. Anaphase. Strong birefringence of chromosomal spindle fibers is seen adjacent to chromosome kinetochores throughout anaphase. Chromosomal fibers gradually shorten and their overall birefringence drops towards telophase but local birefringence remains strongest adjacent to kinetochores. The weak birefringence of the continuous fibers, also longitudinally positive, is more apparent in the later stages of anaphase (Figs. 11, 12)

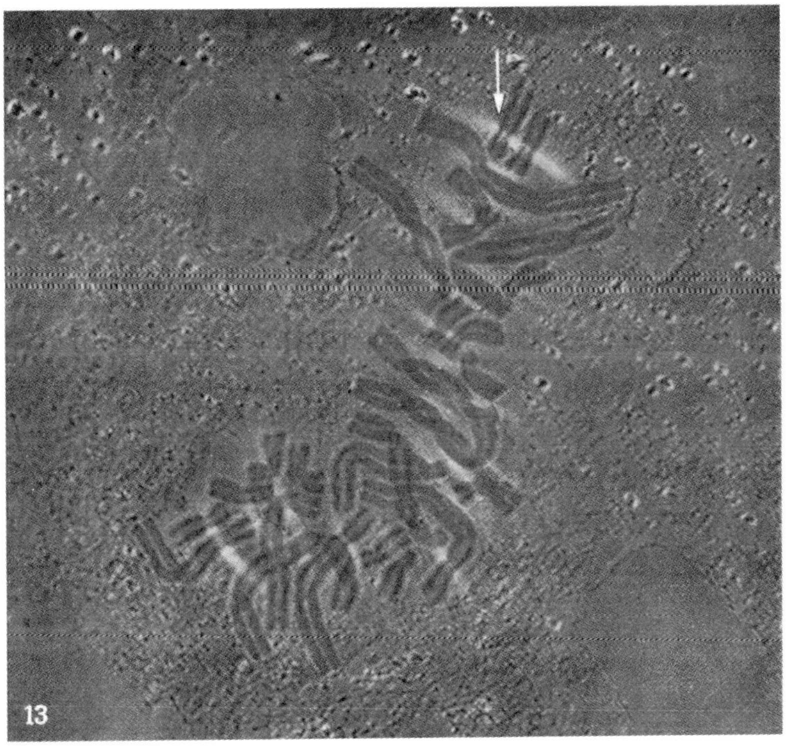

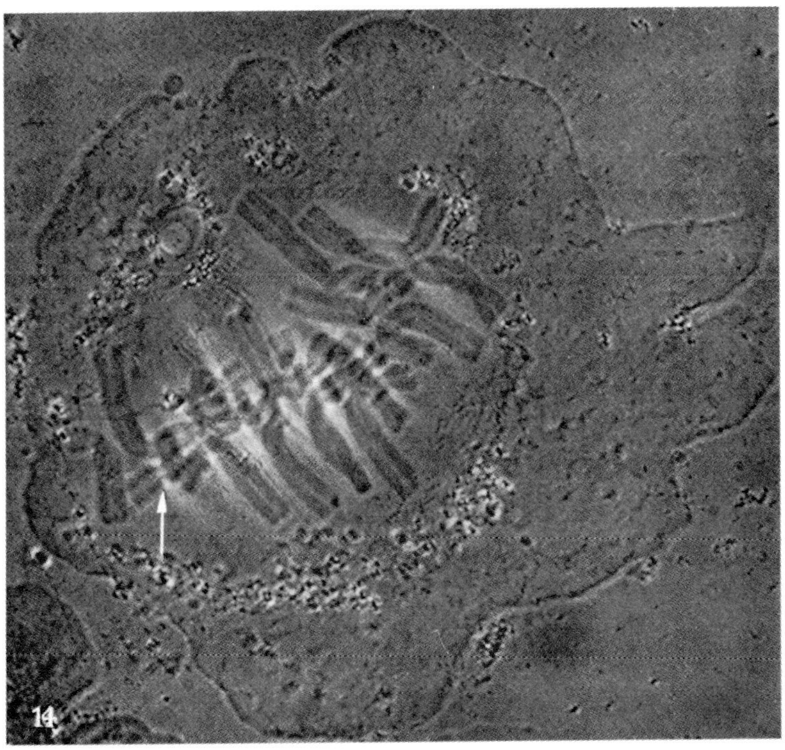

Fig. 13. Late pro-metaphase. Highly flattened cell showing relation of chromosomal spindle fibers to individual chromosomes. To the primary constriction or kinetochore region (indicated by arrow) of each chromosome is attached the bright, strongly birefringent region of the chromosomal spindle fiber. The fiber birefringence dwindles towards the poles. Continuous fiber birefringence at this stage is too weak to be seen in these photos

Figs. 14—20. A highly compressed cell from metaphase to formation of cell plate. Magnification same as other pictures. (Data series 58 j 2.)

Fig. 14. *17*^{33} *h*, Metaphase. See explanation for Fig. 13

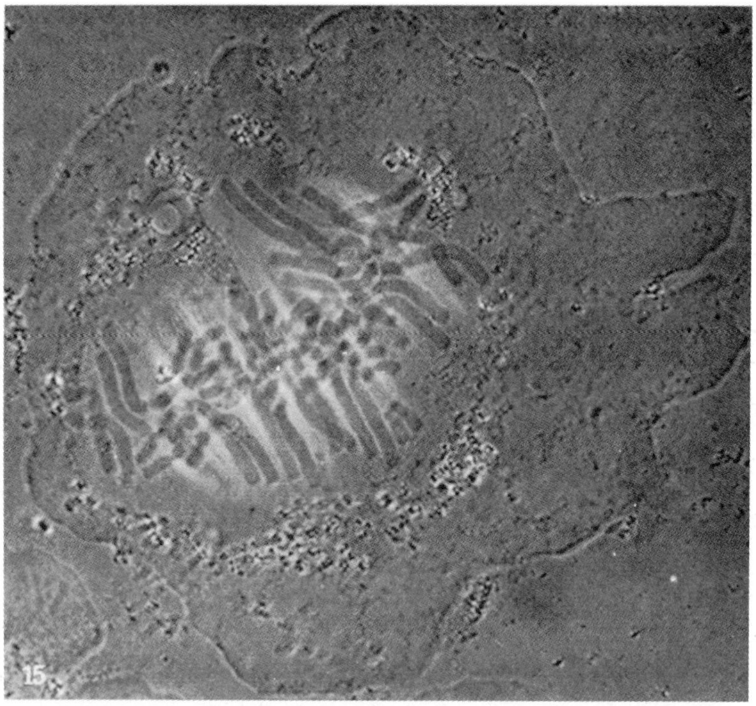

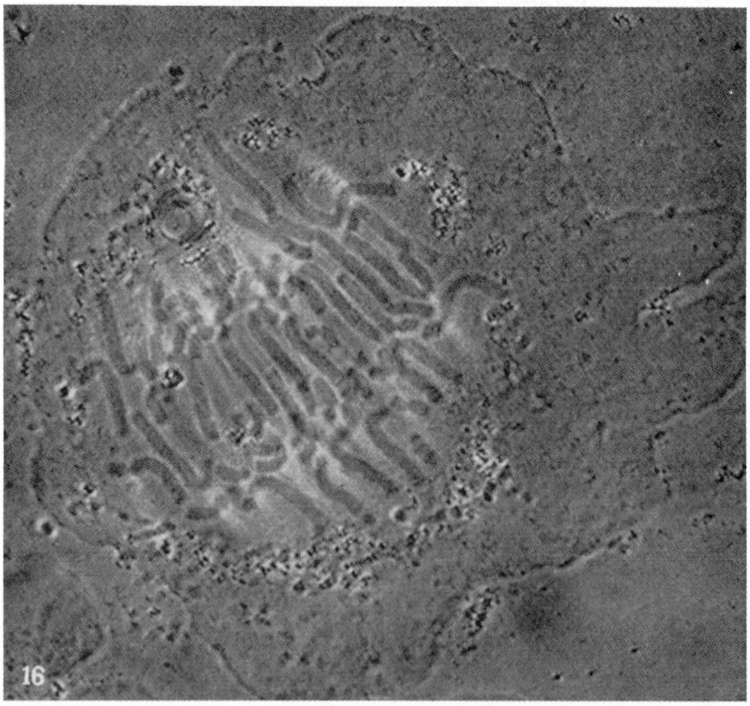

Fig. 15. *17*^{58} *h*, Early anaphase. Kinetochores and chromosome arms separating. Chromosomes are led towards the diffuse spindle poles by the bright chromosomal spindle fibers

Fig. 16. *18*^{14} *h*, Anaphase. Chromosomes mostly separated. Chromosomal spindle fibers leading the kinetochore region of chromosomes still show strongest birefringence adjacent to latter

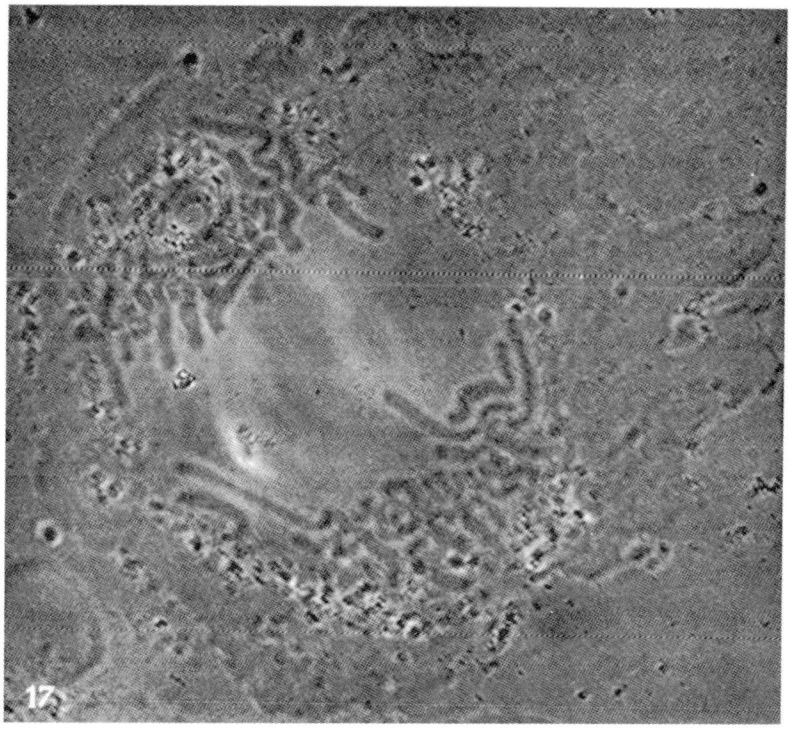

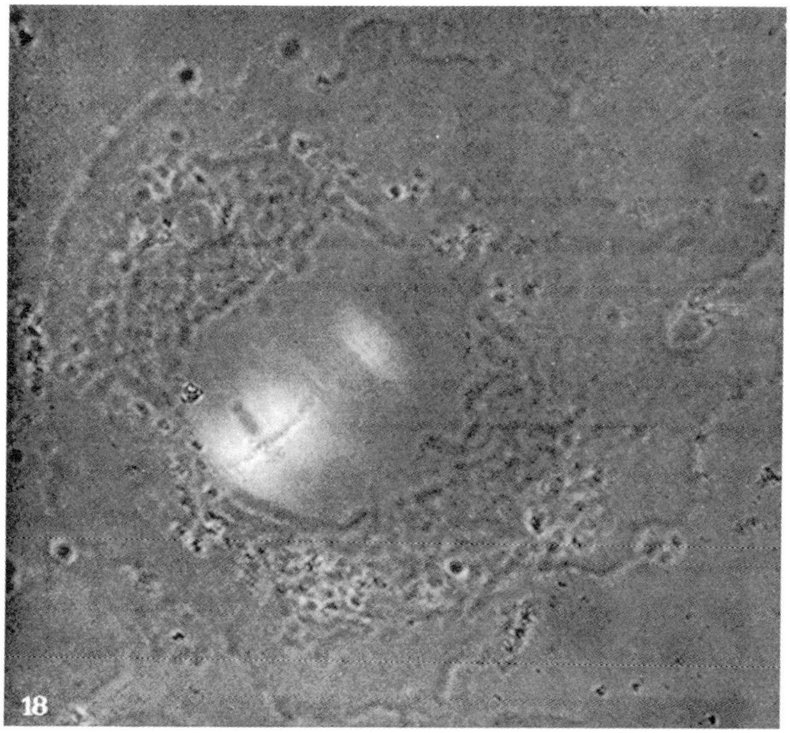

Fig. 17. *19^{07} h*, Early telophase. Chromosomal spindle fibers are no longer visible. The continuous fibers become more strongly birefringent throughout the gap between the separated chromosome groups. Also see Fig. 24

Fig. 18. *19^{47} h*, Telophase. Phragmoplast formation. The birefringence of the continuous fibers becomes weaker near the chromosomes and strong in the equatorial region of the spindle. In the middle of the phragmoplast thus formed small granules (vacuoles) are aligned

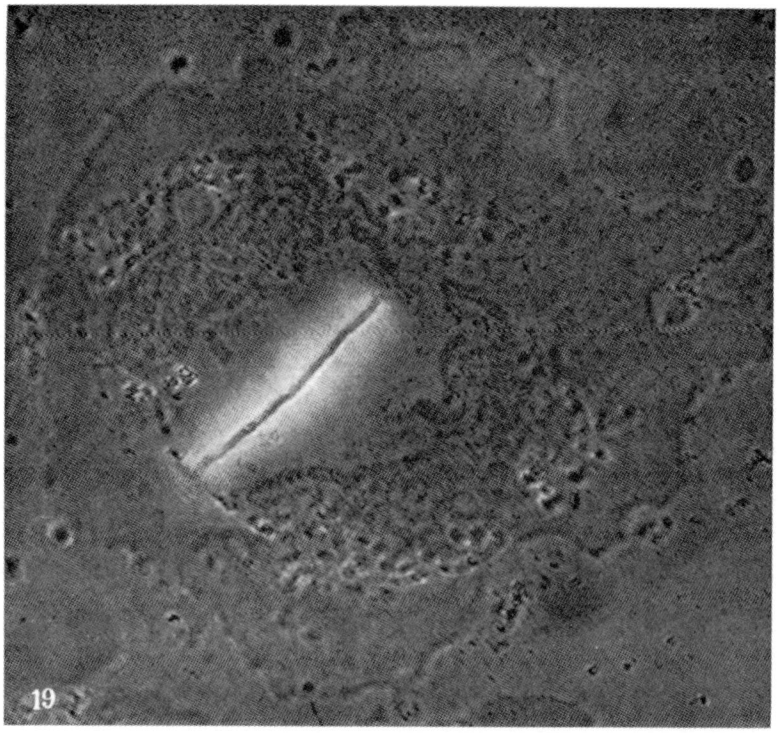

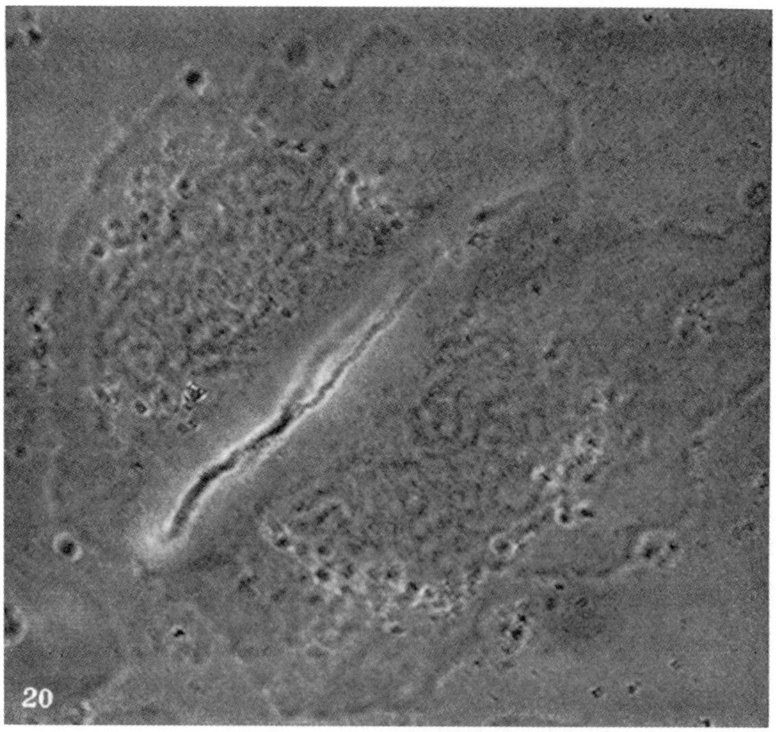

Fig. 19. 20^{11} *h*, Phragmoplast stage. The continuous fibers have been converted completely into phragmoplast fibers and the equatorial granules have merged to form the cell plate. The longitudinally positive birefringence of the phragmoplast fibers is strongest adjacent to the cell plate and gradually diminishes towards the chromosome groups

Fig. 20. 21^{25} *h*, Late stage of cell plate formation. The phragmoplast birefringence has diminished in the mid region of the cell where cell plate formation is nearly complete. Birefringence of the phragmoplast fibers is stronger near the cell periphery where cell plate formation is still proceeding. Also see Fig. 25

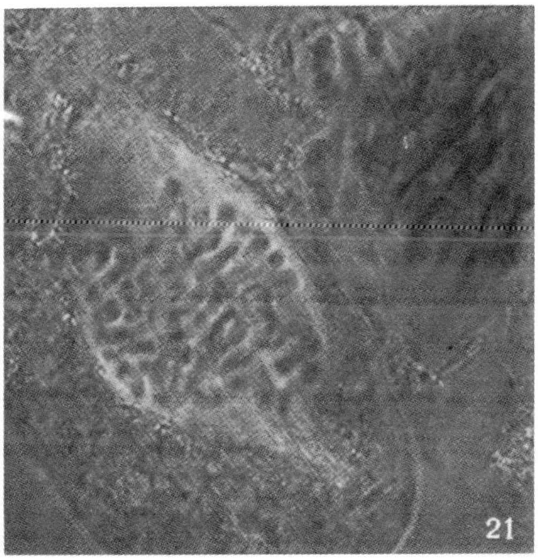

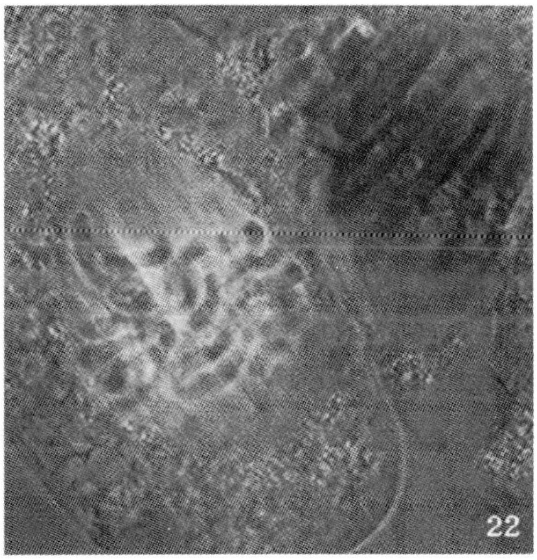

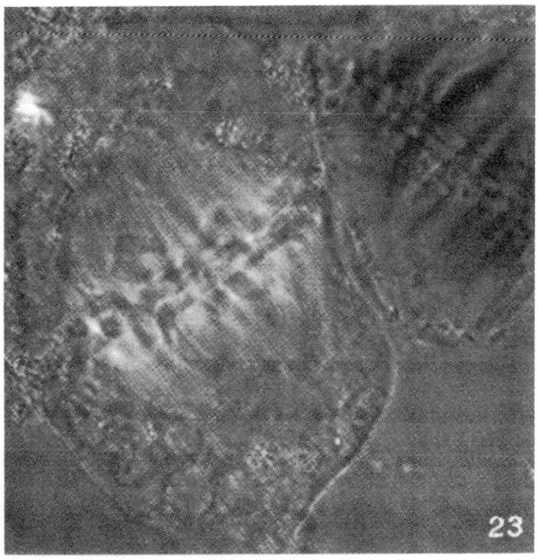

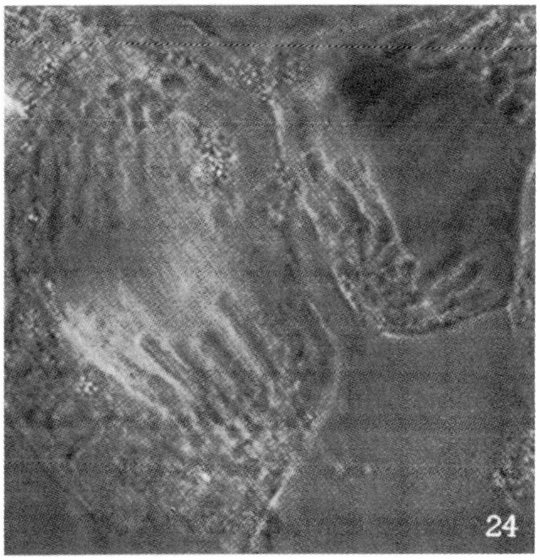

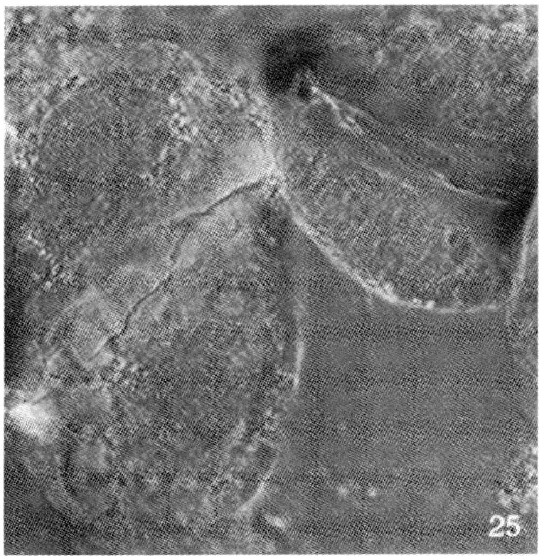

Figs. 21—25. Two unflattened cells from early contraction stage through cell plate formation. (From series 58 j 2.) Fig. 21. 10^{53} *h*, Left cell is at a stage between cells shown in Figs. 2 and 4, immediately following nuclear membrane breakdown. Right cell is in a slightly advanced stage. Fig. 22. 11^{10} *h*, Left cell in stage between Figs. 5 and 6. Fig. 23, 12^{24} *h* Left cell in stage between Figs. 6 and 7. Right cell in metaphase. Fig. 24. 13^{26} *h*, Anaphase. See legend for Fig. 17. Fig. 25. 15^{00} *h*, Cell plate formation. See legend for Fig. 20. Scale 10 μ interval for all photographs in this paper

Very shortly after breakdown of the nuclear membrane, chromosomal spindle fibers are established at the kinetochores. These fibers appear to be regions of the general spindle (continuous fibers and fibrils) converted and organized into fibers with a higher degree of orientation. The kinetochore in *Haemanthus katherinae* then possesses an orienting or organizing faculty from early stages of pro-metaphase. The distribution of birefringence at this stage of spindle development has not been clearly observable in other cells previously studied.

The occurrence of birefringent chromosomal and continuous fibers in a wide variety of dividing cells establishes the strong likelihood that they are general features of mitosis wherever spindles occur. Whenever the optical conditions in the cells are favorable, the cells healthy, and the instrument adequately sensitive and with appropriate resolving power, these fibers have been observed in living cells in polarized light (see introduction). The objections of HEILBRUNN (1956) and others concerning the non-existence of these fibers in normal cells are therefore no longer admissible.

Outside of a very few exceptional cases (CLEVELAND 1953, COOPER 1941, see SCHRADER 1953), spindle fibers cannot be observed by ordinary light or phase contrast microscopy in living cells unless the cells are somewhat damaged or when they appear to be in an abnormal state, such as exposure to hypertonic or acidic media (e.g. LEWIS 1923). Spindle fibers in many, if not most, healthy living cells therefore represent fibrous regions of the spindle with specific orientation (therefore visible in polarized light) but whose refractive index varies little from its suroundings. This optical condition would prevail if the fibers lay within a milieu of the same material with approximately the same concentration but which has rather less orientation than the fibers themselves.

In all of the cell types so far studied including *Haemanthus*, the birefringence of chromosomal fibers is strongest adjacent to the kinetochore *throughout* anaphase. This observation disproves the postulate of SWANN (1951) that during anaphase a disorienting substance is secreted from the kinetochore, thus inducing disorientation and contraction of the spindle fibers. His whole argument was based on birefringence measurement of half spindle areas in developing sea urchin eggs, but unfortunately with the material and instrument employed he was neither able to resolve the chromosomal spindle fibers nor to determine the exact location of the chromosomes in the living cells.

INOUÉ (1959) has postulated that the micelles within the spindle fiber are barely cross-linked and tend to seek an equilibrium with the unoriented "pool material" in their surrounding. In this equilibrium, which is temperature sensitive, it was explained that alignment of the micelles is governed in part by orienting forces thought to be present at the kinetochores and centrioles. Were this interpretation correct, one may conclude from the observed persistence of birefringence near the kinetochore that an orienting force is continuously at work there throughout anaphase. The chromosomal fibers may then contract or elongate depending on whether more oriented material leaves or enters the fibers, as compared to the equilibrium state. A similar argument would apply to continuous fibers at various stages of mitosis and to spindle fibers treated with colchicine (INOUÉ 1952, 1961a).

From the several observations described in this paper and earlier (INOUÉ 1952—1959, BAJER 1954—1958), we would not at all be surprised if the clear zone material, the continuous fibers, the chromosomal spindle fibers and the phragmoplast fibers all turned out to be transformations of one and the same material. The difference would primarily depend on the mode and degree of orientation and the active centers organizing the particular fiber species. In fact such notions have been proposed long ago by OSTERHOUT (1902) and others and have later been disfavored by workers who were unable to demonstrate the fibers in carefully fixed cells (see WILSON, pp. 155, 161, SCHRADER pp. 42, 47). Since birefringence of the fibers can now actually be seen in living healthy cells, the latter objections could hardly hold any longer, although it is no doubt true that the very sharp fibers and fibrils depicted by the earlier workers are somewhat artifactual.

Intense green light is apparently harmless to *Haemanthus* cells as was found true for most of the cells observed previously by INOUÉ. On the other hand he has observed a number of cases in which the blue spectrum of visible light rapidly injures or kills the cells or organelles illuminated. Therefore, the injurious effect of "green" light observed by BAJER on *Haemanthus* division is very likely due to inadequate filtration of the blue end of the spectrum (the far red spectrum was removed by a thick layer of Mohr's solution).

Endosperm cells of *Haemanthus katherinae* offer excellent opportunities for further experimental analysis of mitosis in polarized light. Observation is possible on individual cells from early prophase through completion of division. The primary constriction or the kinetochore region of the large chromosomes are clearly seen and the attachment of the spindle fibers can be followed in living cells. Correlation of chromosome movement with changes in chromosomal fiber birefringence may yield considerable insight on the mechanisms at work during mitosis.

References

BAJER, A.: Ciné-micrographic studies on mitosis in endosperm. I. Acta Soc. Bot. Polon. **23**, 382—412 (1954). — Living smears from endosperm. Experientia (Basel) **11**, 221—222 (1955). — Ciné-micrographic studies on mitosis in endosperm. III. The origin of the mitotic spindle. Exp. Cell Res. **13**, 493—502

(1957). — Ciné-micrographic studies on mitosis in endosperm. IV. The mitotic contraction stage. Exp. Cell Res. **14**, 245—256 (1958a). — Ciné-micrographic studies on chromosome movements in β-irradiated cells. Chromosoma (Berl.) **9**, 319—331 (1958b). — Ciné-micrographic studies on mitosis in endosperm. V. Formation of the metaphase plate. Exp. Cell Res. **15**, 370—383 (1958c).

BAJER, A., and J. MOLÈ-BAJER: Endosperm, material for study on the physiology of cell division. Acta Soc. Bot. Polon. **23**, 69—98 (1954). — Ciné-micrographic studies on mitosis in endosperm. II. Chromosome, cytoplasm and Brownian movements. Chromosoma (Berl.) **7**, 558—607 (1956).

CLEVELAND, L. R.: Studies on chromosomes and nuclear division. IV. Photomicrographs of living cells during meiotic divisions. Trans. Amer. Philos. Soc., N.S. 43 (1953).

COOPER, K. W.: Visibility of the primary spindle fibers and the course of mitosis in the living blastomeres of the mite *Pediculopsis graminum* REUT. Proc. Nat. Acad. Sci. (Wash.) **27**, 480—483 (1941).

HEILBRUNN, L. V.: The dynamics of living protoplasm. New York: Academic Press 1956.

INOUÉ, S.: The effect of colchicine on the microscopic and submicroscopic structure of the mitotic spindle. Exp. Cell Res., Suppl. **2**, 305—318 (1952). — Polarization optical studies of the mitotic spindle. I. The demonstration of spindle fibers in living cells. Chromosoma (Berl.) **5**, 487—500 (1953). — Motility of cilia and the mechanism of mitosis. Rev. Mod. Phys. **31**, 402—408 (1959). — On the physical properties of the mitotic spindle. Ann. N.Y. Acad. Sci. **90**, 529—530 (1961a). — In: Encyclopedia of Microscopy, GR. CLARK edit., pp. 480—485. New York: Rheinhold 1961.

LEWIS, M. R.: Reversible gelatin in living cells. Bull. Johns Hopkins Hosp. **34**, 373—379 (1923).

MOLÈ-BAJER, J.: Ciné-micrographic analysis of C-mitosis in endosperm. Chromosoma (Berl.) **9**, 332—358 (1958).

OSTERHOUT, W. J. V.: Cell studies I. Spindle formation in *Agave*. Proc. Calif. Acad. Sci., Ser. III **2**, 255—285 (1902).

SCHRADER, F.: Mitosis. The movements of chromosomes in cell division. 2nd ed. New York: Columbia University Press 1953.

SWANN, M. M.: Protoplasmic structure and mitosis II. The nature and cause of birefringence changes in the sea-urchin egg at anaphase. J. exp. Biol. **28**, 434—444 (1951).

WADA, B.: The mechanism of mitosis based on studies of the submicroscopic structure and of living state of the *Tradescantia* cell. Cytologia (Tokyo) **16**, 1—26 (1950).

WILSON, E. B.: The cell in development and heredity. 3d ed. New York: MacMillan 1928.

Professor Dr. SHINYA INOUÉ,
Department of Cytology,
Dartmouth Medical School,
Hanover, New Hampshire, USA

Docent Dr. ANDREW BAJER,
Laboratory of Plant Physiology,
Jagellonian University,
53 Grodzka, Cracow, Poland

Druck der Universitätsdruckerei H. Stürtz AG., Würzburg

Reprinted with permission from *The Biological Bulletin*, Vol. 125(2), pp. 380–381, 1963.

HEAVY WATER ENHANCEMENT OF MITOTIC SPINDLE BIREFRINGENCE

Shinya Inoué, Hidemi Sato, and Robert W Tucker

When developing eggs of the sea urchin *Lytechinus variegatus* were placed in sea water made up in heavy water (D_2O) and H_2O, the cells in mitosis but not cleavage were reversibly arrested at D_2O concentrations of 45% and higher. In sea water containing 30% to 40% D_2O, the birefringence of the spindle fibers rose above normal but mitosis proceeded. When sea water with 45% to 60% D_2O was applied during mitosis, the spindle birefringence rose higher and then fell; mitosis was arrested unless the cell was exposed to D_2O during or after anaphase. At these concentrations, the effect of raising the spindle birefringence is a sensitive function of the exact stage of mitosis at which the D_2O is applied. At 50% D_2O, the retardation can be doubled and may reach a maximum of 7 mμ only if D_2O reaches the spindle at a particular 10-second interval in anaphase. This differential sensitivity of the spindle fiber birefringence to D_2O at different stages of mitosis closely parallels its sensitivity to temperature. The birefringence increases at elevated temperatures and vanishes reversibly at lower temperatures. The sensitivity rises once during spindle formation, becoming minimum at metaphase, then rises to a higher maximum again in anaphase and gradually decreases towards telophase. The rise of spindle fiber birefringence in D_2O adds to the list of agents which similarly affect (1) the spindle fiber birefringence, (2) the "gel strength" of cleaving sea urchin eggs, and (3) the polymerization of tobacco mosaic virus S proteins. This fact increases the likelihood that a common mechanism underlies these various processes.

Reprinted with permission from *The Biological Bulletin*, Vol. 125(2), p. 395, 1963.

RAPID EXCHANGE OF D_2O AND H_2O IN SEA URCHIN EGGS

Robert W Tucker and Shinya Inoué

When *Arbacia punctulata* and *Lytechinus variegatus* eggs from normal (H_2O) sea water are layered on top of a heavy (D_2O) sea water solution (concentration of D_2O greater than 98%) and then centrifuged at 200 g for two seconds, the eggs fall even though the density of the D_2O sea water (1.127 gm/ml) is greater than that of the eggs in H_2O sea water (1.084 gm/ml for *Arbacia punctulata*). The converse experiment of eggs which have been equilibrated in D_2O sea water falling under gravity in H_2O sea water (1.02 gm/ml) shows that the eggs fall at a large initial rate, gradually slowing down to the terminal velocity determined by Stokes' law. Eggs in H_2O sea water when forced under D_2O sea water rise for a maximum of 3.5 seconds and then fall under gravity. All three experiments can be explained by the fact that at least 30% of the heavy D_2O exchanges reversibly for the light H_2O in less than two to three seconds.

Reprinted from *Primitive Motile Systems in Cell Biology*, pp. 549–598, 1964,
with permission from Academic Press.

Organization and Function of the Mitotic Spindle[1]

SHINYA INOUÉ

Department of Cytology, Dartmouth Medical School, Hanover, New Hampshire

Introduction

The occurrence of birefringence in contractile elements of living cells may have a deeper significance than the mere localization of oriented molecules or micelles. This assertion is supported by two generalizations, the first made by Engelmann in 1875, stating that contractility in living systems depended upon the presence of birefringence (also Engelmann, 1906; but see Schmidt, 1937, p. 251). Whether strictly true or not, the statement takes on added significance in the light of the second generalization made by Dr. Aaron Katchalaski at the February, 1963 Meeting of the Biophysical Society. On thermodynamic grounds, he argued that in order to convert chemical energy (a scalar quantity) *directly* into contractile mechanical force (a vectorial quantity), structural anistotropy is required. In other words, the molecules giving rise to contractile force cannot be randomly arranged, but must form an organized structure whose properties vary in different directions. In general, structural anisotropy is associated with optical anisotropy, i.e., a different response of the material to light vibrating or traveling in different directions. When the velocity of light propagation in a material varies in different directions, the material is said to be birefringent or to exhibit double refraction.

The birefringence of muscle fibers, for example, is strong enough to be seen with conventional polarizing microscopes. The mitotic spindle, on the other hand, exhibits a considerably lower retardation than do muscle fibers and can be detected only by the use of refined polarization microscopic techniques. The contrast due to weakly birefringent specimens must be maximized by reducing all possible sources of stray light and by judicious use of a compensator (Schmidt, 1934; Swann and Mitchison, 1950; Inoué and Dan, 1951; Mitchison, 1953; Inoué, 1961). Conventional polarizing microscopes are not satisfactory for the study of

[1] Supported in part by grants from the National Science Foundation (G 19487) and National Cancer Institute, U.S. Public Health Service (CA 04552).

detailed structures of the mitotic spindle, not only because of excessive stray light, but also because of the presence of a troublesome diffraction anomaly which limits effective lateral resolution (Inoué and Kubota, 1958; Kubota and Inoué, 1958). Some years ago we developed the polarization rectifier which corrects both of these defects (Inoué and Hyde,

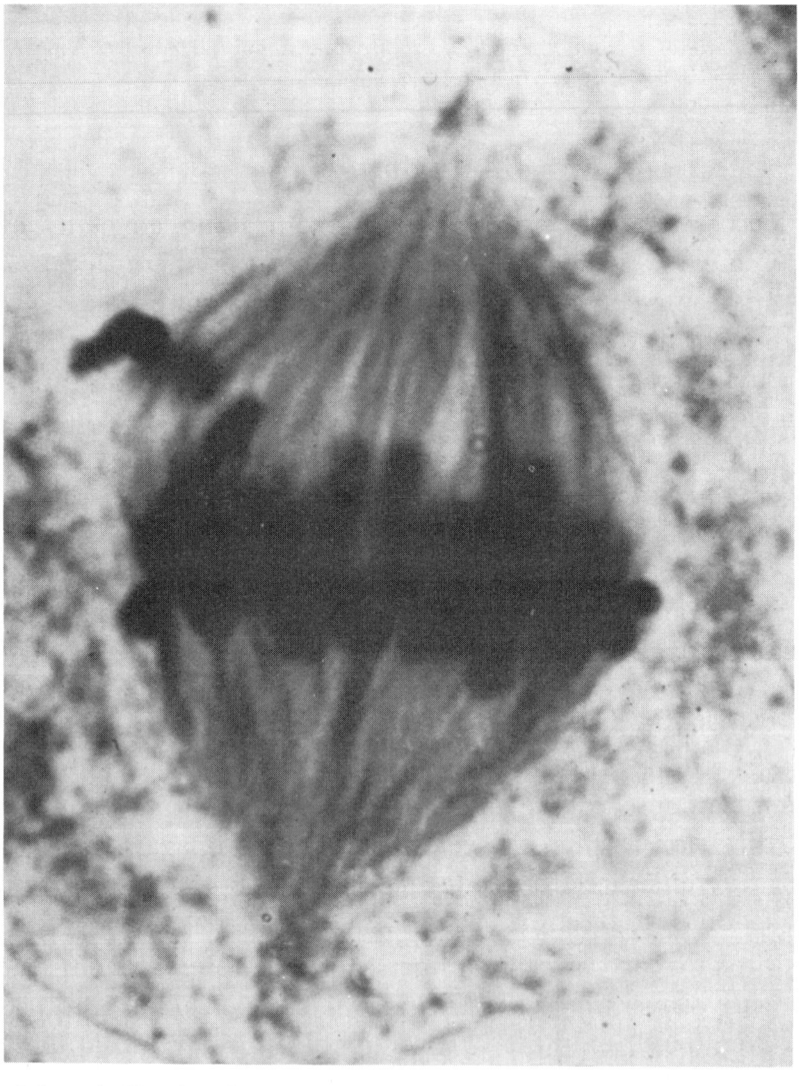

FIG. 1. The spindle of meiosis I as seen in fixed and stained preparation of a hybrid wheat pollen mother cell. (Courtesy, Dr. G. Östergren, University of Lund, Sweden.)

1957). The polarizing microscope used in these studies (Figs. 57, 58) has been described elsewhere (Inoué, 1961).

In this paper the term "mitotic apparatus" will be used in a broad sense, as Mazia and Dan (1952) defined it, to include astral rays, centrioles, chromosomal and continuous spindle fibers, and so on, i.e., all that which makes up the assembly that comes out as one piece when the

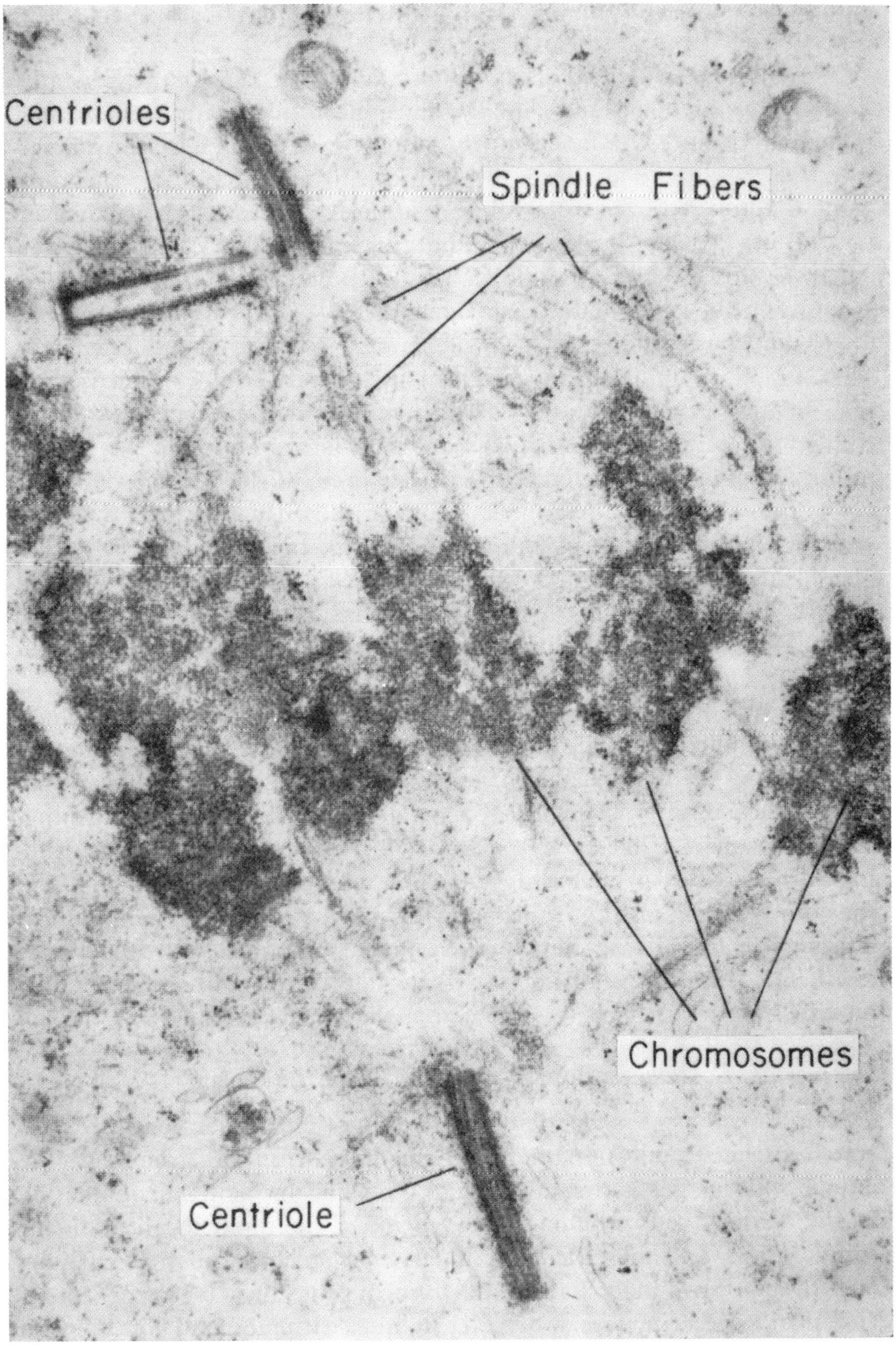

FIG. 2. Electron micrograph of a mitotic spindle from the spermatocyte of the domestic fowl. (From Bloom and Fawcett, 1962, p. 36.)

"spindle" is isolated from the cell in division (Mazia and Dan, 1952; Kane, 1962).

The fibrous structures of the mitotic apparatus (Fig. 1) as generally depicted in textbooks of cytology have been observed in fixed and stained slides since the last century, as also recently by electron microscopy (Fig. 2; also see article by Roth in this Symposium). The functions of these fibrous structures have been variously discussed (see summaries by Schrader, 1953; Mazia, 1961), but there have also been numerous debates as to whether the fibers really existed in living cells. The grounds for the debate have been chiefly that (1) with careful fixation the fibers sometimes could not be demonstrated (see Porter and Machado, 1960, as an example at the electron-microscopic level), whereas with poor fixation they appeared coarser and (2) chromosomes exhibited strange mitotic movements which could not be explained by the simple pulling or pushing by spindle fibers (Östergren, 1949; Schrader, 1953).

As to the reality of these fibers in living cells, I have proved over the years by the use of sensitive polarizing microscopes (Inoué, 1961) that fibers as depicted by the best classic work (e.g., Bělař, 1929a,b) in fact do exist in a wide variety of healthy dividing cells (Inoué, 1953; Inoué movie, 1960b; Inoué and Bajer, 1961). Additional evidence along this line is also presented in this paper.

The main features which are stressed in this paper are the following: (1) The spindle fibers are not static structures but exist in a dynamic state of flux; (2) the fibers are oriented and organized by "centers" (Boveri, 1888; Wilson, 1928) such as centrioles and kinetochores and other organelles; these will be referred to as "orienting centers"; (3) depending on the activity of such centers and the physiological state of the cell, the spindle fibers can be readily built up, broken down, or reorganized. The same material can, therefore, be made into one kind of fiber or another or transformed from one type of fiber to another, depending on which center happens to be active at that time.

Description of Spindle Birefringence

Very broadly speaking, there are two types of spindles, one with discrete centrioles and the other without (Fig. 3). Most commonly in animal cells one encounters the former type and in plant cells the latter. In the animal cell spindle with centrioles the birefringence of the spindle fibers is stronger, and the fibers converge, toward the centrioles and toward the kinetochores. The birefringent fibers between kinetochores and the centrioles we shall call the "chromosomal fibers." The birefringent fibers running from pole to pole we shall call "continuous fibers."

In spindles of plant cells, lacking centrioles, asters are generally missing, and there may or may not be convergence toward the poles. Whether or not there is convergence, the birefringence of the chromosomal fibers is invariably weak at the pole. The birefringence is strongest adjacent to the kinetochores and gradually becomes weaker toward the poles. Continuous fibers are generally present.

If a living cell, for example, a grasshopper spermatocyte (Fig. 4), is observed under a phase-contrast microscope, one may see the centrioles in an appropriate focus. The spindle region is outlined by mitochondria

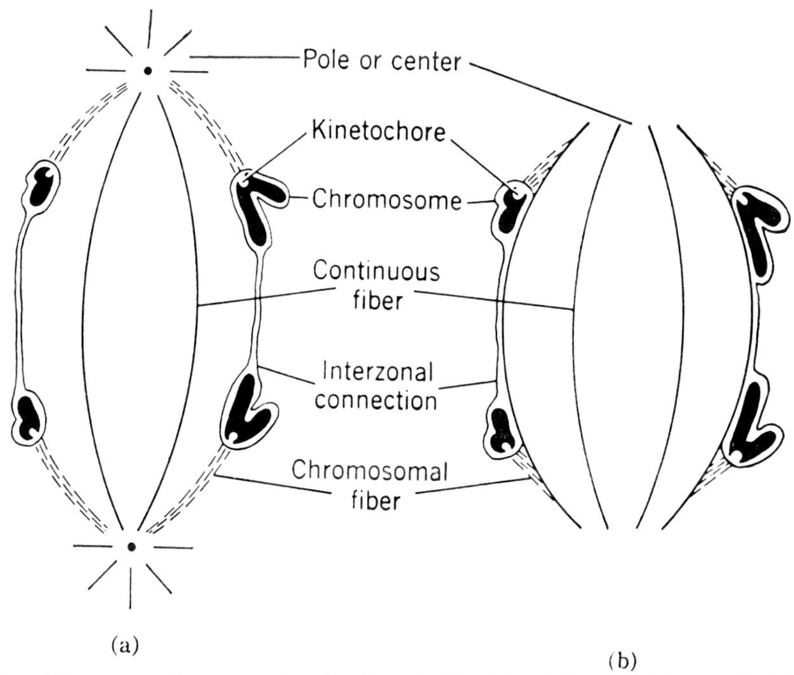

FIG. 3. Schematic diagram of mitotic spindle (a) with centrioles and (b) without. (Modified from Schrader, 1953.)

which are often elongated and surround the spindle as a sheath. The region which should show spindle fibers, however, generally appears structureless under the phase-contrast microscope. "Structureless" here means that there is not an adequate optical path difference (owing in this case to difference of refractive index between the fibers and their surroundings) to produce detectable contrast. It is significant that in living cells in general the spindle fibers do not exhibit such refractive index difference unless the cell is fixed, exposed to acid (Lewis, 1923), or otherwise maltreated. Exceptions to this generalization are found in the flagellate spindle (Cleveland, 1953, 1963), in mite eggs (Cooper, 1941), and occasionally in various other healthy cells just at the onset of anaphase.

Observing a living cell with a sensitive polarizing microscope, one can clearly see these fibers in regions which appear empty under the phase-contrast microscope. In Fig. 5 we see a plant cell, a pollen mother cell of the Easter lily, *Lilium longiflorum,* in anaphase. The spindle converges toward the poles somewhat, but the birefringence of the chromosomal fibers is strongest adjacent to the kinetochores and weaker

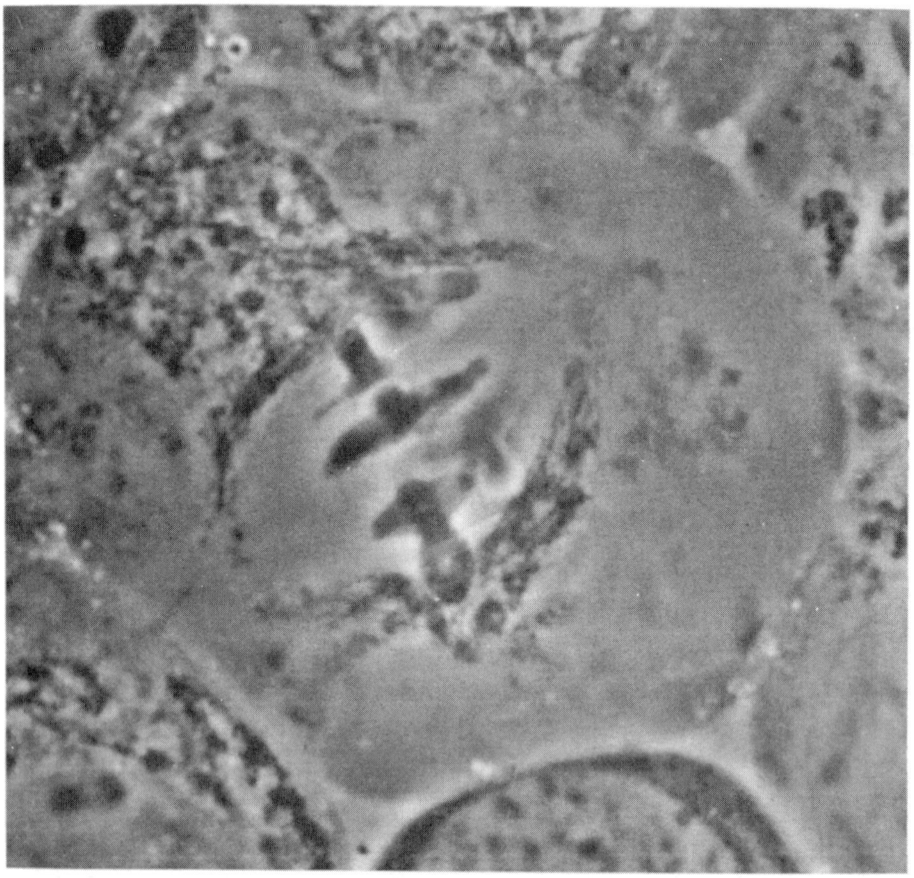

FIG. 4. Living grasshopper spermatocyte (*Chloealtis genicularibus*) as seen with a phase-contrast microscope. Notice early anaphase chromosomes still at equator and mitochondria outlining the spindle. Half-spindles appear empty, and continuous and chromosomal fibers are not visible. (Courtesy, Professor K. Shimakura, Faculty of Agriculture, Hokkaido University, Sapporo, Japan.)

toward the poles. The continuous fibers of this stage are less birefringent and not readily detectable. During anaphase movement, the birefringence of the chromosomal fibers adjacent to the kinetochore remains strong and the fibers become shorter but not detectably thicker. In late anaphase, continuous fibers can be seen again by their increased birefringence.

As an example of an animal cell, Fig. 6 shows the first maturation division spindle in a oocyte of a marine worm, *Chaetopterus pergamen-*

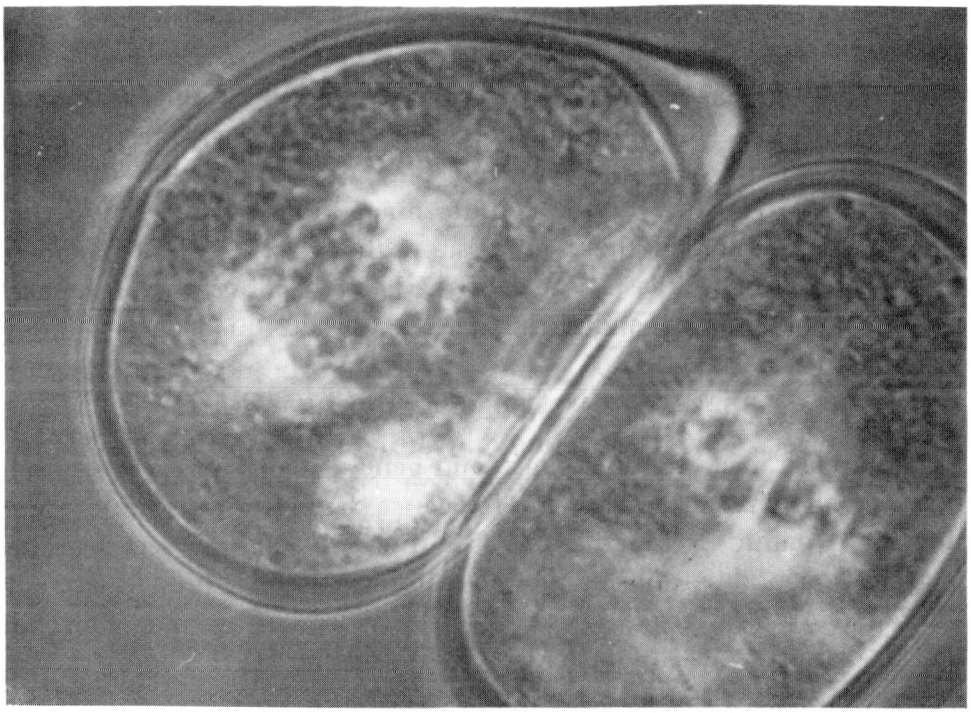

FIG. 5. Living pollen mother cell of the Easter lily, *Lilium longiflorum,* as seen with a sensitive polarizing microscope. The chromosomal fibers show strong birefringence adjacent to the helical chromosomes. (Original photo.)

FIG. 6. Birefringent spindle fibers in a living oocyte of a marine worm, *Chaetopterus pergamentaceus*. Egg centrifuged to move away the birefringent yolk granules. (From Inoué, 1953.)

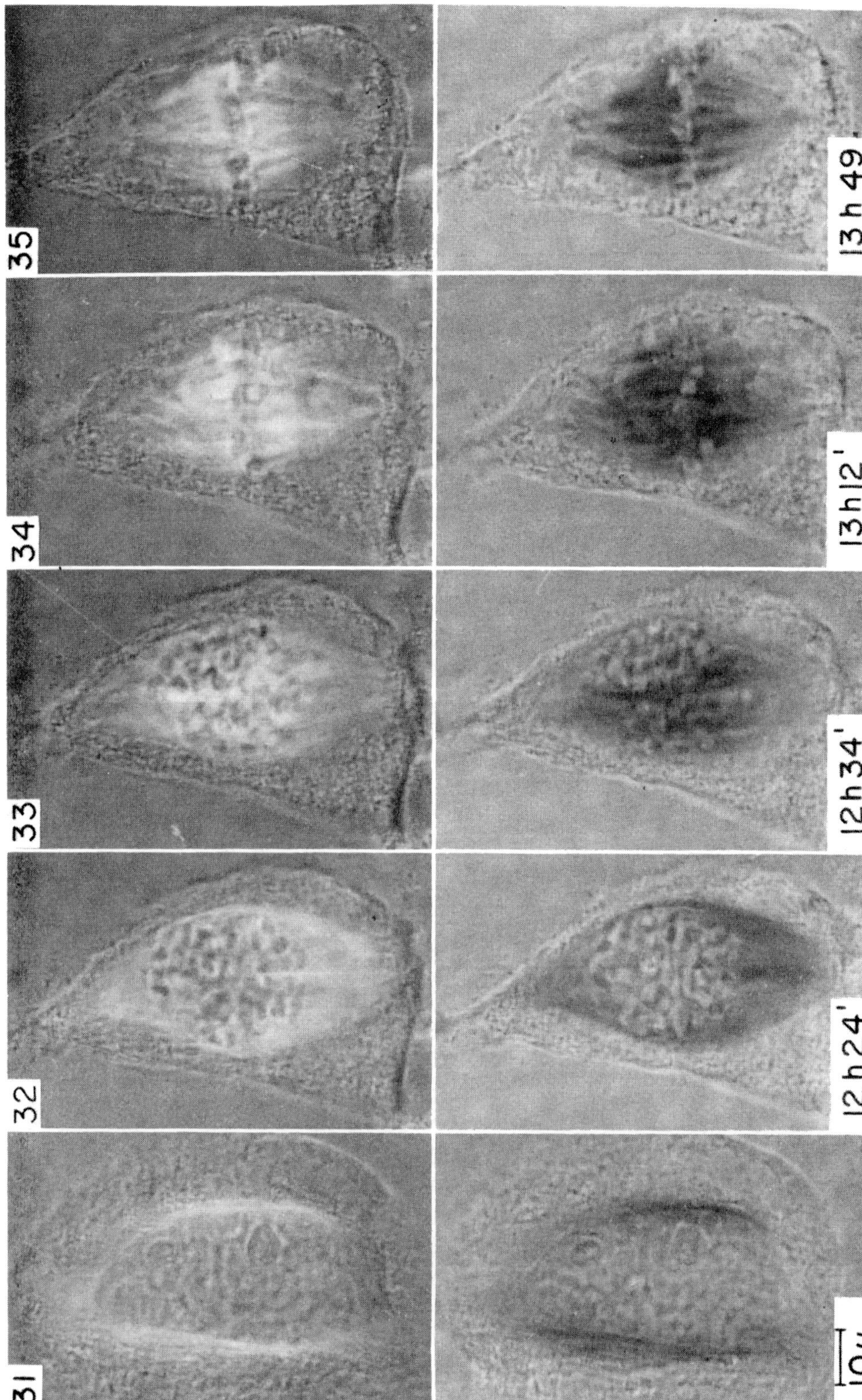
35
34
33
32
31
13 h 49
13h12'
12h34'
12h24'
10μ

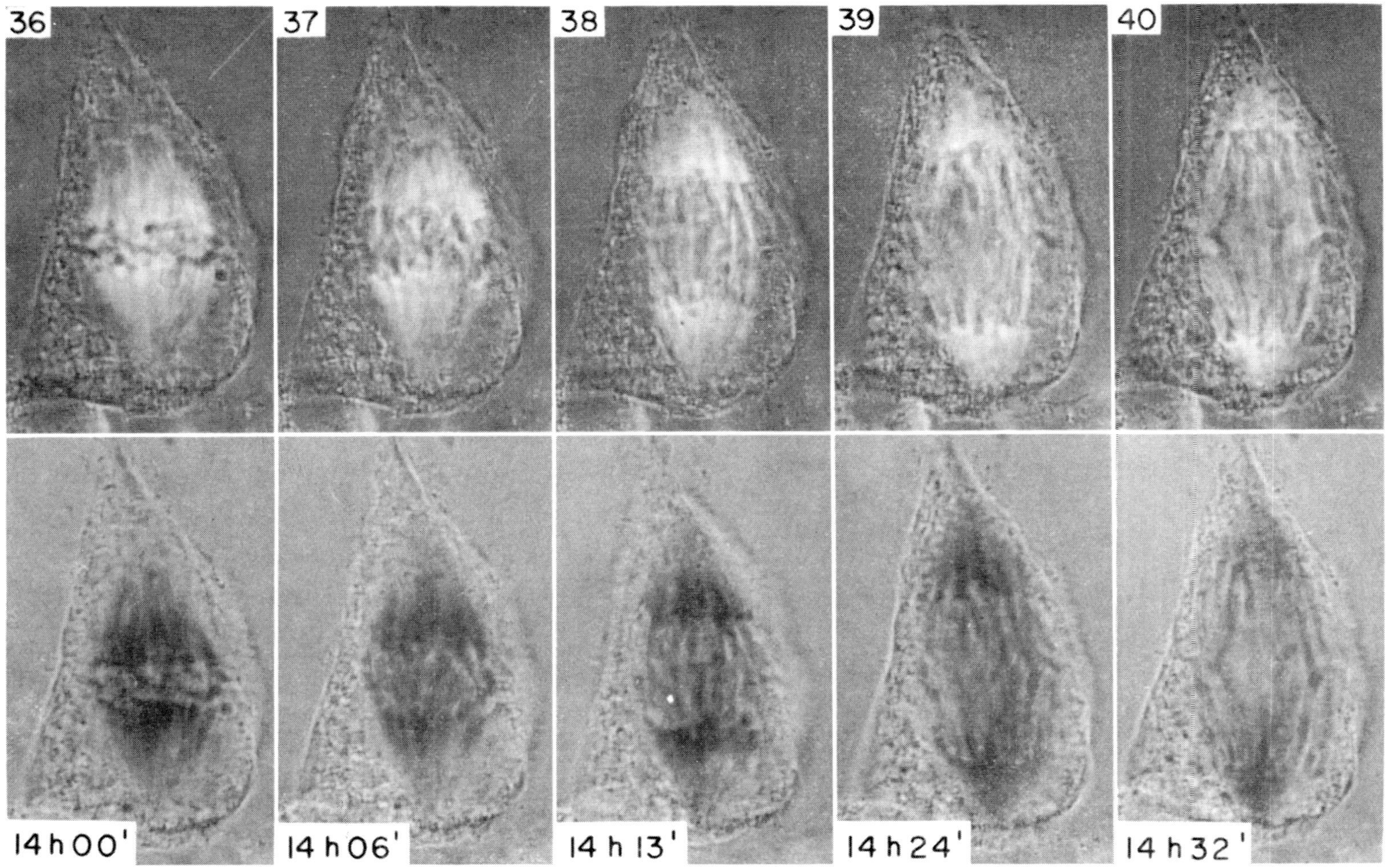

FIGS. 31 to 40. Development of the birefringent spindle fibers and anaphase movement in the endosperm cell of the African blood lily, *Haemanthus katherinae*. Pairs of photos are taken with opposite compensation. Same magnification as Figs. 41 to 48. See Fig. 31 for 10 μ scale. (From Inoué and Bajer, 1961.)

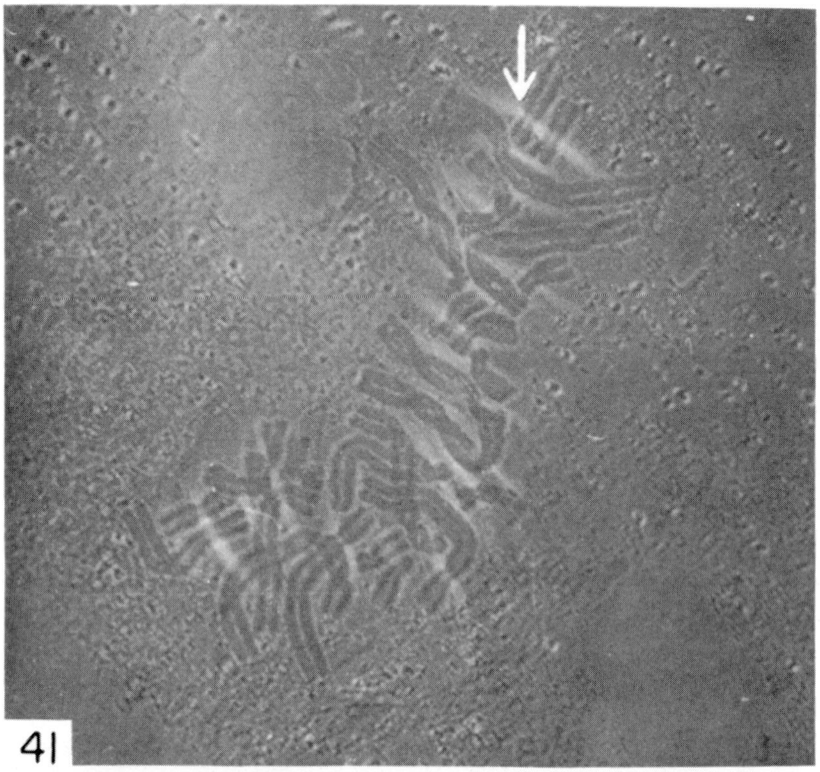

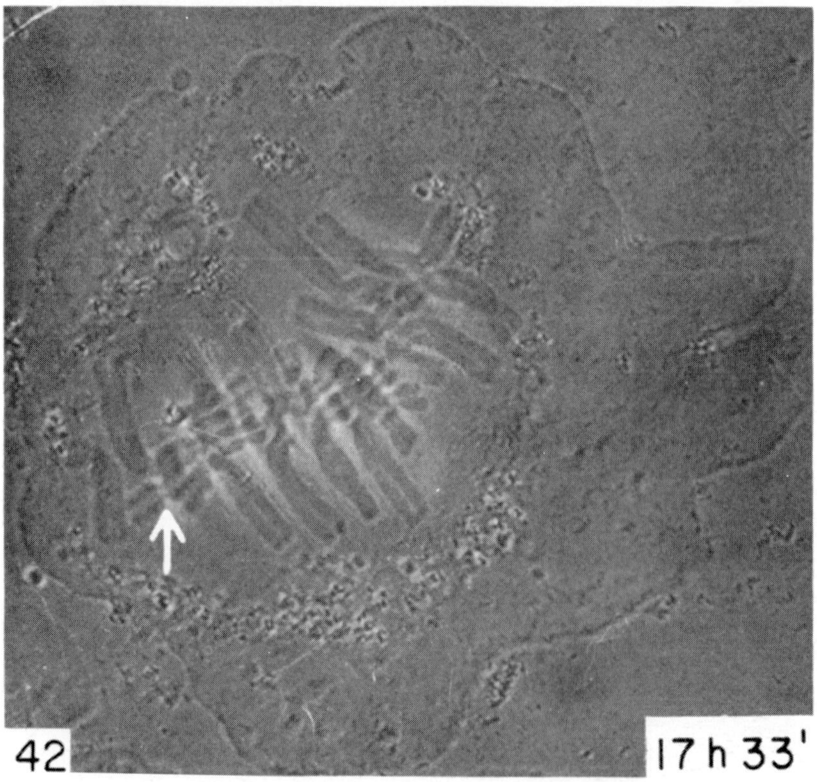

FIGS. 41 to 48. Chromosomes and birefringent spindle fibers in flattened cells of *Haemanthus katherinae*. The strong birefringence of the chromosomal fibers adjacent to the kinetochore is indicated by arrows in Figs. 41 and 42. Continuous fibers visible

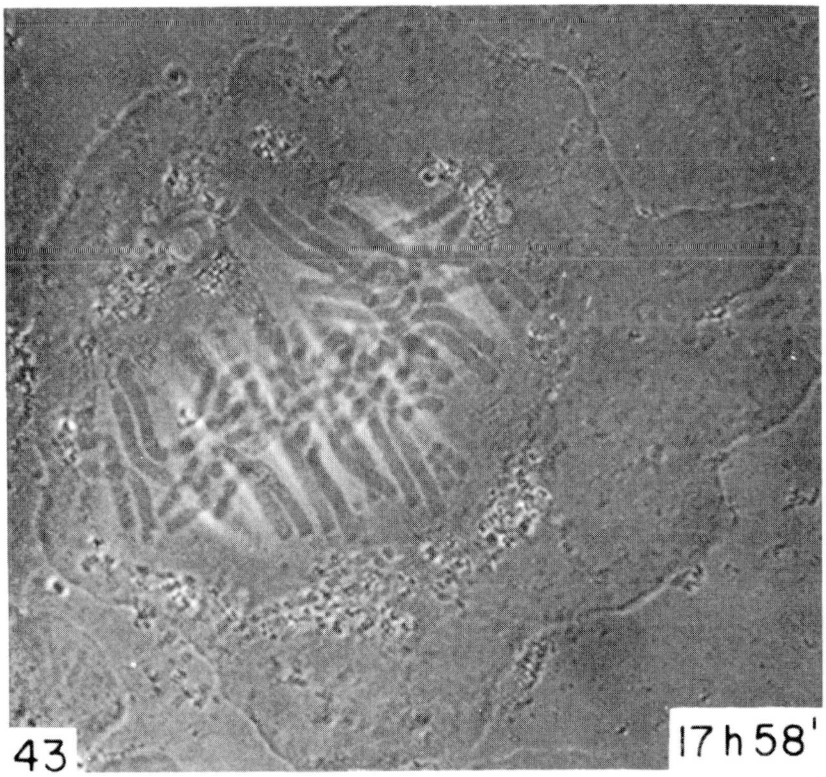

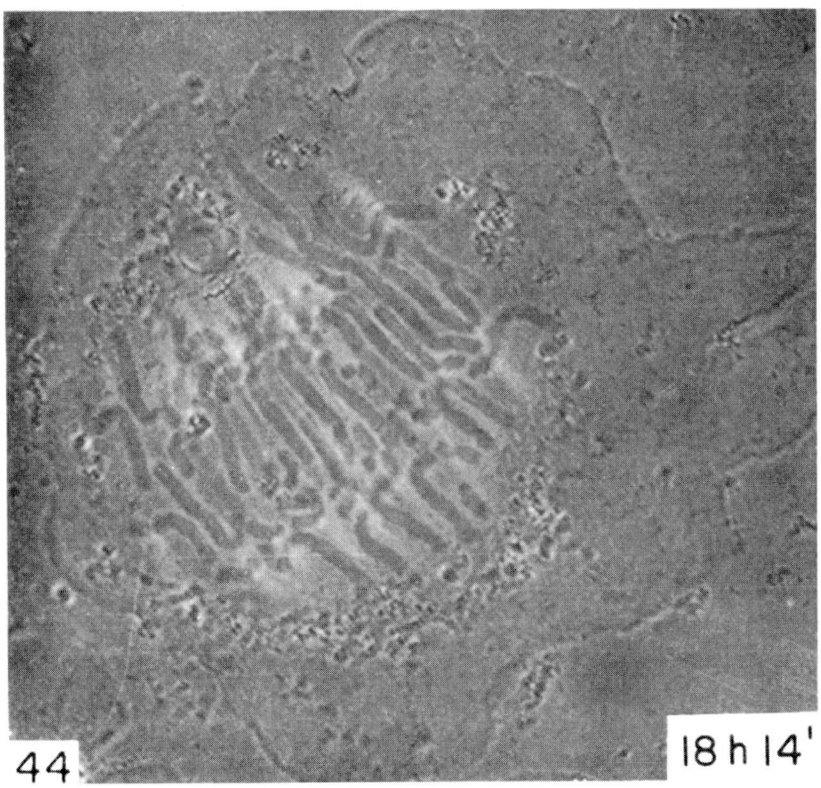

in Figs. 45 and 46. Phragmoplast fibers visible in Figs. 47 and 48. Same magnification as Figs. 31 to 40. A 10 μ scale is shown on Fig. 31. (From Inoué and Bajer, 1961.)

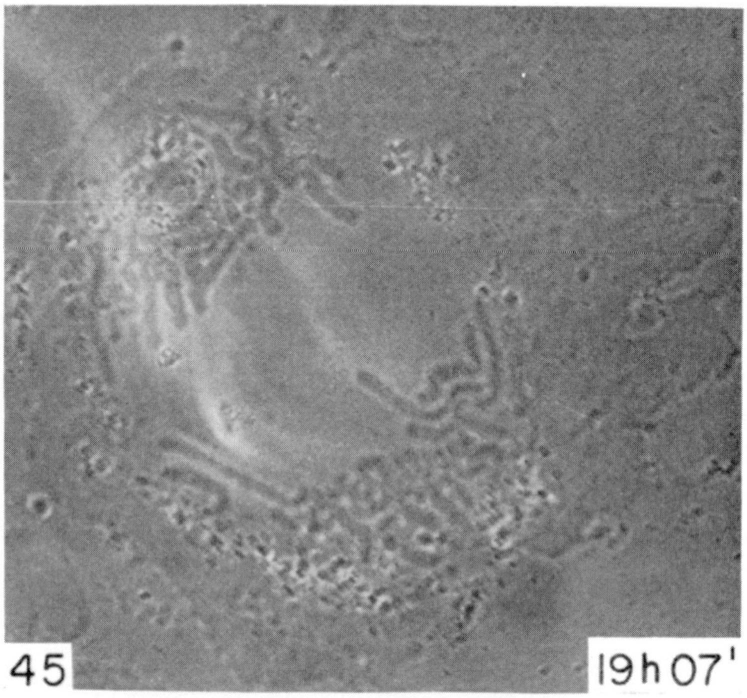

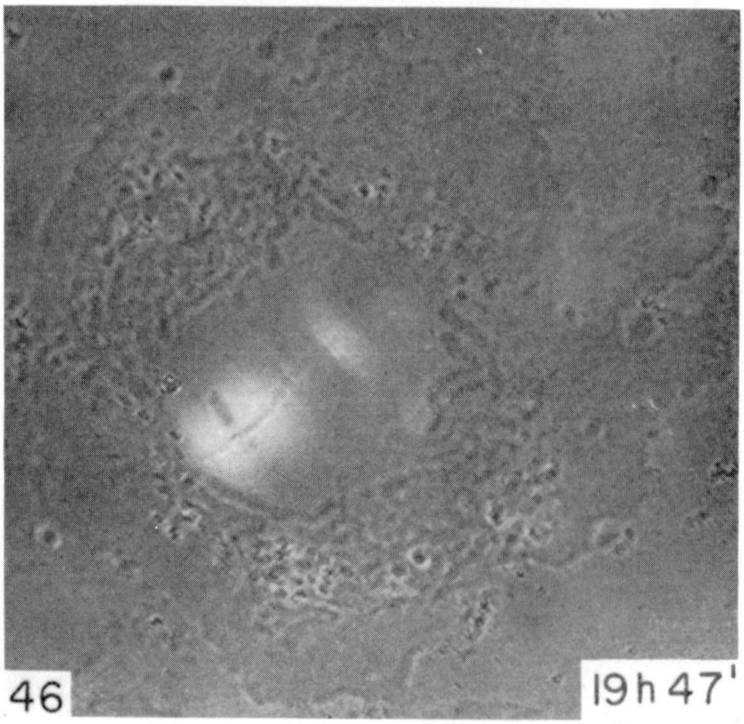

For legend see pages 566 and 567.

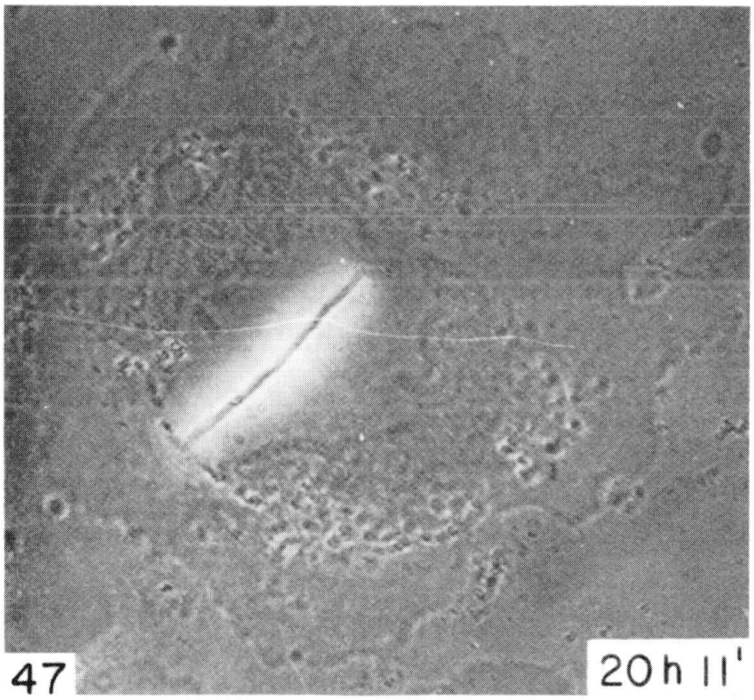

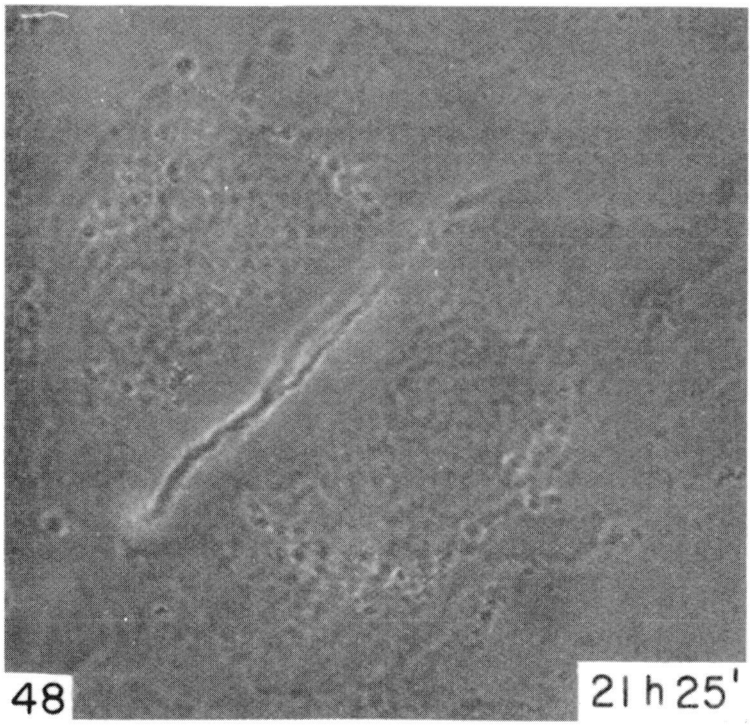

For legend see pages 566 and 567.

of positive birefringent material in agreement with the postulate of Wada (1950). Continuous fibers thus appear, then chromosomal fibers, then gradually the chromosomes are moved up and down until the metaphase plate is formed (Figs. 33 to 35). Here is another example of transformation of the clear-zone fiber to continuous spindle fibers and to chromosomal fibers (Inoué and Bajer, 1961).

Continuing the series (Figs. 36 to 40) throughout anaphase, the birefringence of the chromosomal fibers remains strongest adjacent to the kinetochore region. In these unflattened cells, the individual spindle fibers, although visible, are difficult to distinguish, but if we flatten the endosperm cells (Figs. 41 to 48) we clearly see the individual continuous fibers, the pairs of chromosomes, and their chromosomal fibers. The flattened endosperm cells continue to divide, and in the movie, the unwinding of the chromosome pairs and their pulling apart by the chromosomal fibers can be clearly seen.

The continuous fibers in late anaphase (Fig. 45) become more strongly birefringent in the mid-region and turn into phragmoplast fibers (Figs. 46 to 48); small vacuoles accumulate at the equator amidst the phragmoplast fibers and form the cell plate. As the cell continues to divide, the phragmoplast birefringence extends toward the cell periphery where the new cell plate is laid down.

Motion Picture of Living Cells in Division

A time-lapse motion picture film[3] showing spindle fiber birefringence during division of various animal and plant cells was shown at this Symposium. The pictures taken with the sensitive polarizing microscope developed by the author demonstrate the following points.

1. In the eggs of the jellyfish, *Halistaura cellularia,* flattening the cell places the spindle in a single plane of focus and reveals the synchrony of cell division. The nucleus normally lies at the periphery of the cell; flattening the cell may displace the nucleus and, therefore, the final position of the spindle. Regardless of which direction or where the spindle comes to lie, the cell divides at right angles to the axis of the

[3] The film required exposure times of from 5 to 20 sec/frame because of the low brightness of the specimen. The interval between frames was identical with the exposure. This film is now available at cost (Inoué, 1960b).

FIGS. 49 and 50. First and second cleavage in the egg of a jellyfish, *Aglantha digitale.* The positive birefringence of the spindle fibers and the negative birefringence of the vacuolar walls provide contrast in these illustrations. The vacuolar walls are deformed and point to the astral centers at these stages. (Original photos.)

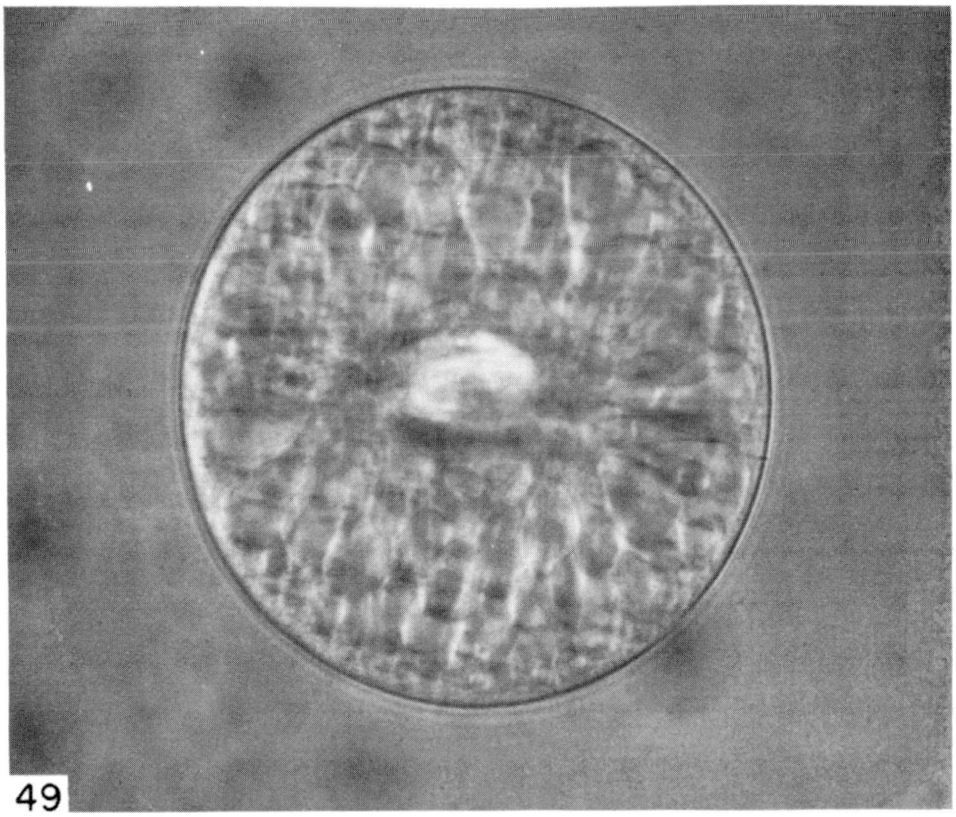

49

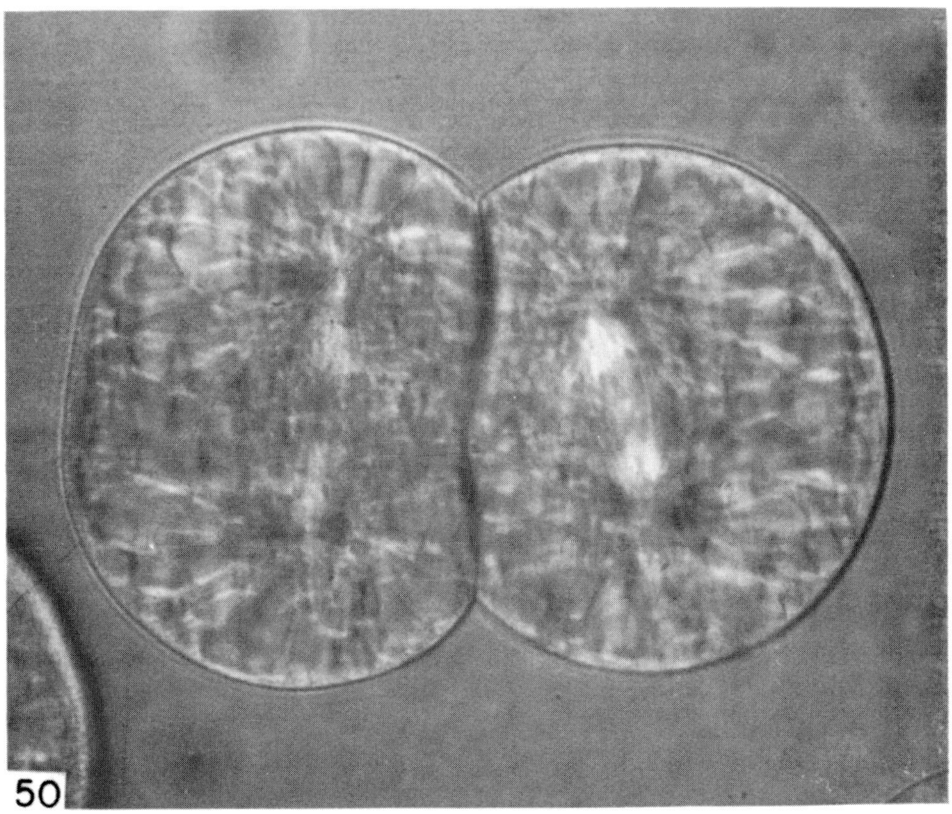

50

spindle, and the furrow begins to form at the surface nearest the spindle. Comparison with the sequence of cleavage in a noncompressed egg (Figs. 7 to 14) demonstrates that the birefringence of the spindle and its relation to the cleavage furrow is not altered by compression.

2. In the eggs of another jellyfish, *Aglantha digitale* (Figs. 49 and 50), which are even more transparent than those of the previous species, as the spindle and asters develop, the cytoplasmic vacuoles become tear-shaped, their apices pointing toward the astral centers. Prior to the separation of the chromosomes, a striking oscillation (or "rocking") of the whole spindle of the kind reported in nematode eggs (Ziegler, 1895) and in chick fibroblasts (Hughes and Swann, 1948) is observed. In the *Aglantha* eggs, the vacuoles change back and forth between tear-drop and round shape as the spindle rocks, as if the astral rays were attached to them and exerting a pull. After mitosis, the vacuoles regain their spherical shape.

3. In spermatocytes of the grasshopper, *Dissosteira carolina* (Figs. 15 to 26), the "northern lights" flickering of individual chromosomal fibers is clearly demonstrated in prometaphase (Figs. 15 to 19). Just prior to anaphase, the birefringence of the chromosomal fibers becomes stronger. While the chromosomes are pulled apart and the chromosomal fibers shorten, their birefringence remains strong until late anaphase (Figs. 18 to 21). The birefringence of the astral rays (Figs. 15 to 22) and of the continuous fibers in the interzonal region (Figs. 19 to 21) is much weaker than that of the chromosomal fibers. The mitochondrial sheath acquires a very strong positive birefringence in telophase (Figs. 19 to 23 and 26) and often becomes twisted in late telophase (Fig. 24).

4. In pollen mother cells of *Lilium longiflorum* (Figs. 27 to 30), chromosomal fiber birefringence becomes strongest just at the onset of anaphase, the birefringence being strongest adjacent to the kinetochores (Fig. 27). The chromosomal fibers retain their strong birefringence as they lead the chromosomes to the poles (Fig. 28). Continuous fiber birefringence is initially weak but becomes stronger during the later stages of anaphase, especially near the equatorial region (Fig. 29) where it finally transforms into the phragmoplast (Fig. 30). At this time, vacuoles come together in the equator and merge to form the cell plate—a structure which is also birefringent but with a different character.

5. Endosperm cells of *Haemanthus katherinae*, unlike most other types of plant cells, are not surrounded by conventional cell-wall material and may thus be flattened like an animal cell in tissue culture (see footnote 2 on p. 557). In the flattened cells, the relatively large chromosomes are sufficiently dispersed (Figs. 41 to 48) so that the individual chromosomes, their kinetochores, the chromosomal and continuous and the

phragmoplast fibers are all clearly visible in polarized light. The chromosomes are untwisted and pulled at their kinetochores by the chromosomal fibers.

As stated previously, the fibers in all these cells are not visible with phase-contrast microscopy in untreated dividing cells.

The motion picture of living cells in division shows not only the reality of the fibers but how the spindle fibers appear to transform from one type of fiber to another, depending on the sequence of events. The various spindle movements and the fluctuation in the fiber birefringence which also reflect the dynamic state of these fibers in the dividing cells are demonstrated.

Effect of Low Temperature[4]

Low-temperature experiments further demonstrate the dynamic state of the spindle fibers.

A pollen mother cell of *L. longiflorum* in early anaphase was chilled to 3°C (Fig. 51) resulting in complete disappearance of spindle birefringence. On raising the temperature to 27°C, the birefringence reappeared, at first as continuous fibers; after about 8 min, both the continuous and chromosomal fibers were fully reorganized. This process of spindle re-formation after chilling is faster than, but otherwise similar to, spindle formation at prometaphase in normal mitosis.

After re-establishment of the spindle birefringence and structural organization, the chromosomes began to move again and completed mitosis.

If we cool a cell at a later stage in anaphase (Fig. 52), by which time the chromosomes have become stretched but not yet separated, the chromosomes are found to "recoil" as the spindle birefringence disappears. In this case, when the temperature is raised, birefringent chromosomal fibers reappear first. Then the chromosomes become stretched, the kinetochore distance increases, and the cell completes mitosis, but only after the size and birefringence of the chromosomal fibers have fully recovered.

There thus exists a direct correlation between spindle birefringence and deformation and movement of the chromosomes (also see Inoué, 1952a on the effect of colchicine).

This temperature treatment can be performed on many types of cells over and over again during a single division. Thus one can interrupt the mitotic process experimentally without impeding the ability of the spindle fibers to become reorganized again.

---

[4] Original report.

been calculated by Dr. Manuel Morales (Inoué, 1959, see correction in 1960a).

The result of this analysis showing a great increase of entropy at higher temperatures is similar to that for globular to filamentous actin transformation (Asakura *et al.,* 1960) and to that for tobacco mosaic virus A protein association (Lauffer, 1958; Ansevin and Lauffer, 1963) and may suggest a common mechanism (see Inoué, 1959; also Kauzmann, 1959, Singer, 1962, for pertinent reviews).

The orientation equilibrium of the spindle elements is schematically shown in Fig. 56. On the chromosomes, the kinetochores act as orienting centers which determine the localization of the oriented material. There is a pool of unoriented or perhaps slightly oriented material around the chromosomal fibers, depending on the stage, which can be transformed into oriented chromosomal fibers. Conversely, the chromosomal fiber material may become disoriented. This equilibrium is temperature-sensitive.

Orienting Centers—Ultraviolet Microbeam Experiments[5]

Experiments were done to test further the notion of orienting centers and orientation equilibrium. If we somehow stop the action of the orienting center, then the oriented material should go into the disoriented state. If we could locally disorient the material distal to the orienting center, then the material proximal to it should not be affected because the orienting center remains, but distally the orientation and the birefringence should disappear.

If we remove the disorienting influence, we should again recover the orientation by the incorporation of the nonoriented material into the oriented state. These predictions have been fulfilled by an experiment in which the spindle fibers were irradiated with a small spot of ultraviolet light.

Figure 57 shows a part of the instrumentation, the high extinction polarizing microscope built onto a stable optical bench. As shown in the schematic diagram (Fig. 58), the visible light source is at the top of the instrument. The light coming through the polarizer, a compensator, and condenser, illuminates the object. Through the objective, analyzer, and eyepiece, we see the birefringence of the specimen. A mercury arc lamp (Osram, HBO-200) shines the ultraviolet light onto a small first surface mirror placed in front of the visible source diaphragm. The

[5] This section reports original work carried out in collaboration with Dr. Hidemi Sato of the Department of Cytology, Dartmouth Medical School.

Fig. 57. Rectified polarizing microscope designed by author with help of the staff of the Research Center, American Optical Co. and the Institute of Optics, University of Rochester. Compare with schematic diagram, Fig. 58. (Original photo.) See Inoué, 1961, for further details.

image of the mirror is projected into the specimen plane and delineates the area of the specimen receiving the ultraviolet exposure.

Figure 59 shows a prophase *Haemanthus* endosperm cell in which a small spindlelike body spontaneously developed. This body was irradiated in its mid-region (Figs. 59 and 60), where the birefringence instantly disappeared. After irradiation, the two halves that remained birefringent quickly came together (Fig. 61) and then merged to form another single spindle-shaped body (Fig. 62).

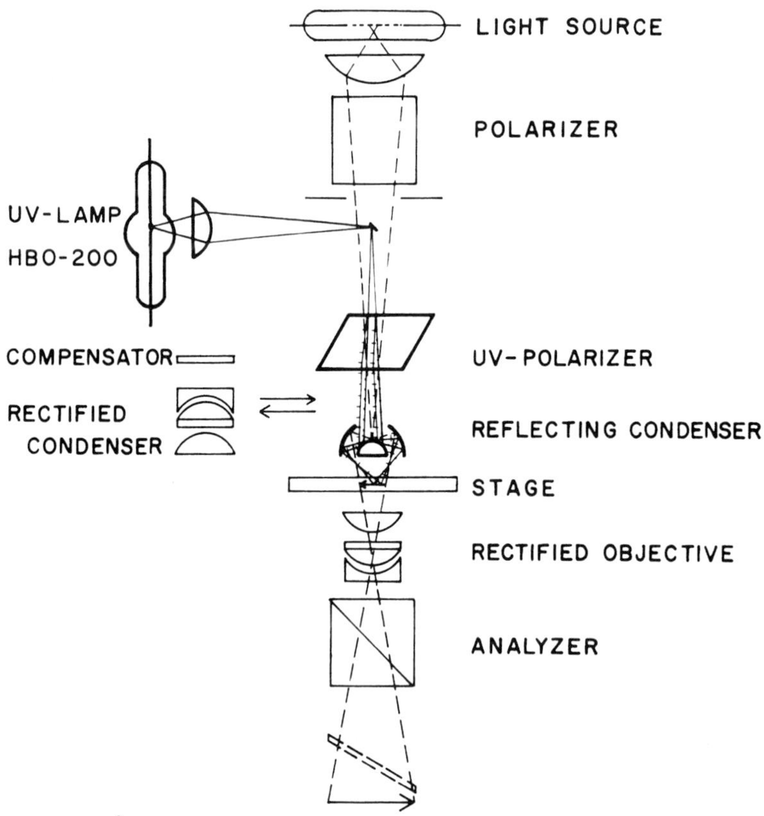

FIG. 58. Schematic diagram of (polarized) ultraviolet-microbeam irradiating system which was incorporated into the rectified polarizing microscope shown in Fig. 57. (Original.)

When a cell in early anaphase is irradiated so that the ultraviolet beam covers the basal regions of some chromosomal fibers and their kinetochores, the whole length of the chromosomal fibers including the distal unirradiated part is lost and does not reappear for a long time (Figs. 63 and 64).

If instead we irradiate only the chromosomal fibers and avoid the kinetochores (Figs. 65 and 66), we find the birefringence has disappeared from the irradiated region as well as from the distal part, but is still intact between the kinetochores and the irradiated region (Fig. 67). In the course of a few minutes the birefringence of the irradiated and distal regions returns (Fig. 68).

These experiments not only support the notion of the dynamic equilibrium and the activity of the organizing centers, but provide us with a means for establishing the locations of organizing centers in general.

For example, in the case of the phragmoplast, the birefringence is stronger next to the cell plate, but the presence of an organizing center in that region of the phragmoplast had never been suspected.

However, when we irradiated a narrow diagonal area of the phragmoplast, the birefringence was lost from the full length of the phragmoplast fibers where the cell plate itself was irradiated (Figs. 69 to 71). If only the distal parts of the fibers, and not the cell plate, were irradiated, the portions of the fibers nearer the cell plate remained birefringent. The birefringence of the distal parts eventually returned.

Figure 72 shows a late phragmoplast in which the birefringence is now confined to the peripheral regions; in the middle, the cell plate has already been laid down. When the center of one phragmoplast is irradiated, as shown in Fig. 73, we find that all birefringence is gone from the phragmoplast fibers on that side (Fig. 74). Here then exists an unsuspected orienting center, the activity of which moves progressively with the phragmoplast as it moves toward the cell periphery.

As shown by these experiments, the phragmoplast fibers behave very much like chromosomal and continuous spindle fibers, not only in their birefringence but in terms of their response to ultraviolet irradiation. This provides further support for the notion that the composition and molecular organization of the phragmoplast fibers is similar or identical to those of the chromosomal fibers, continuous fibers, and the clear-zone fibers.

Conclusion

In summary, then, we have seen in a wide variety of living, dividing cells "fibers" which could explain, at least topologically, the movement of chromosomes. This does not explain the molecular mechanism of chromosome movement yet, but at least we know there is an anisotropic distribution of material which could one way or another account for the development of mechanical forces required for pulling (or pushing) the chromosomes, most likely dependent on the shift of orientation equilibrium.

It may well be that as material is removed from the chromosomal fibers, while the organizing centers are still active, the fibers shorten in length simply to reach a new orientation equilibrium. The continuous

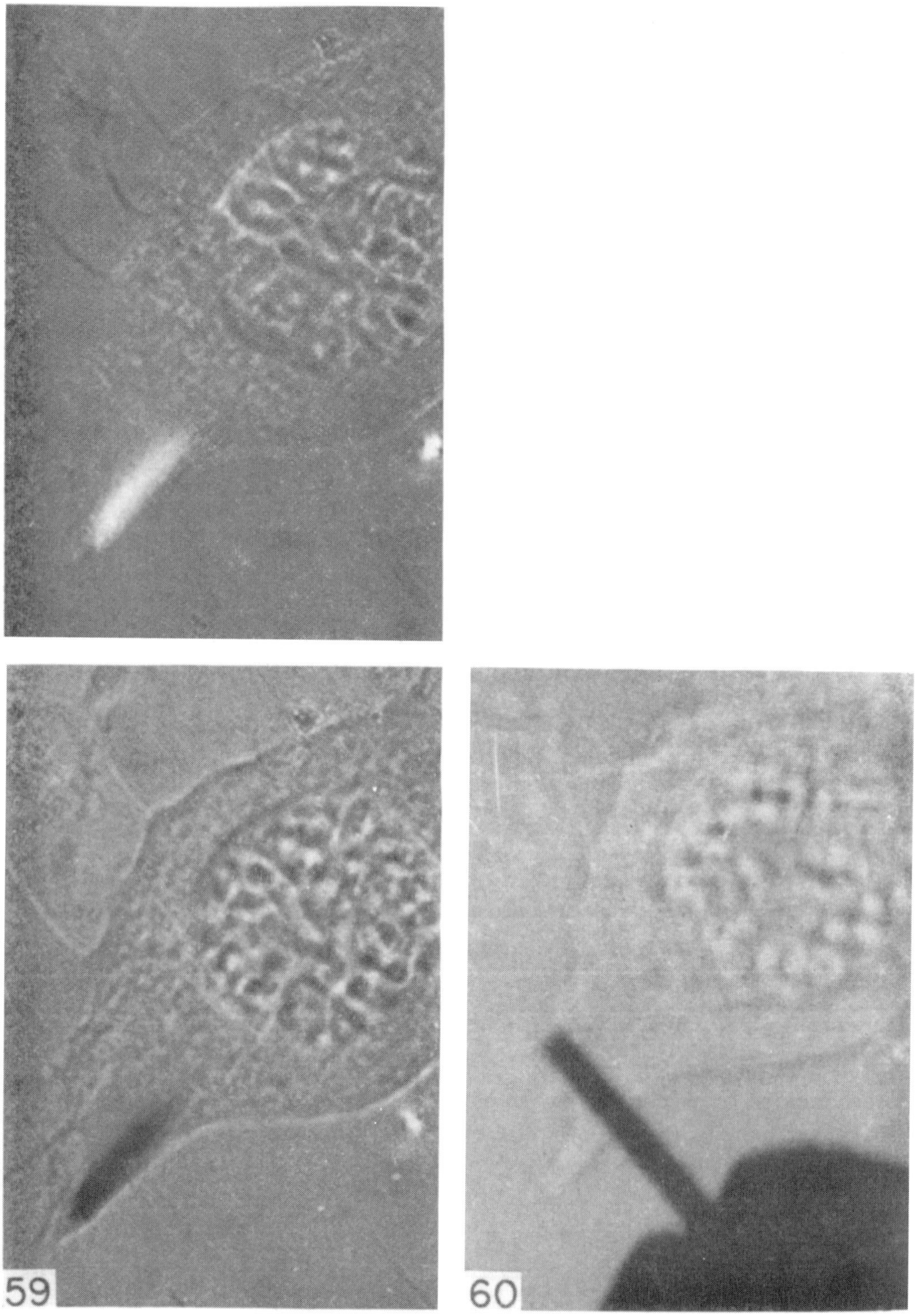

FIGS. 59 to 62. Irradiation of cytoplasmic spindle spontaneously formed in prophase of a *Haemanthus* endosperm cell. Figure 59, before irradiation; Fig. 60, the dark bar indicates the area irradiated with ultraviolet; Fig. 61, 1 min after irradiation

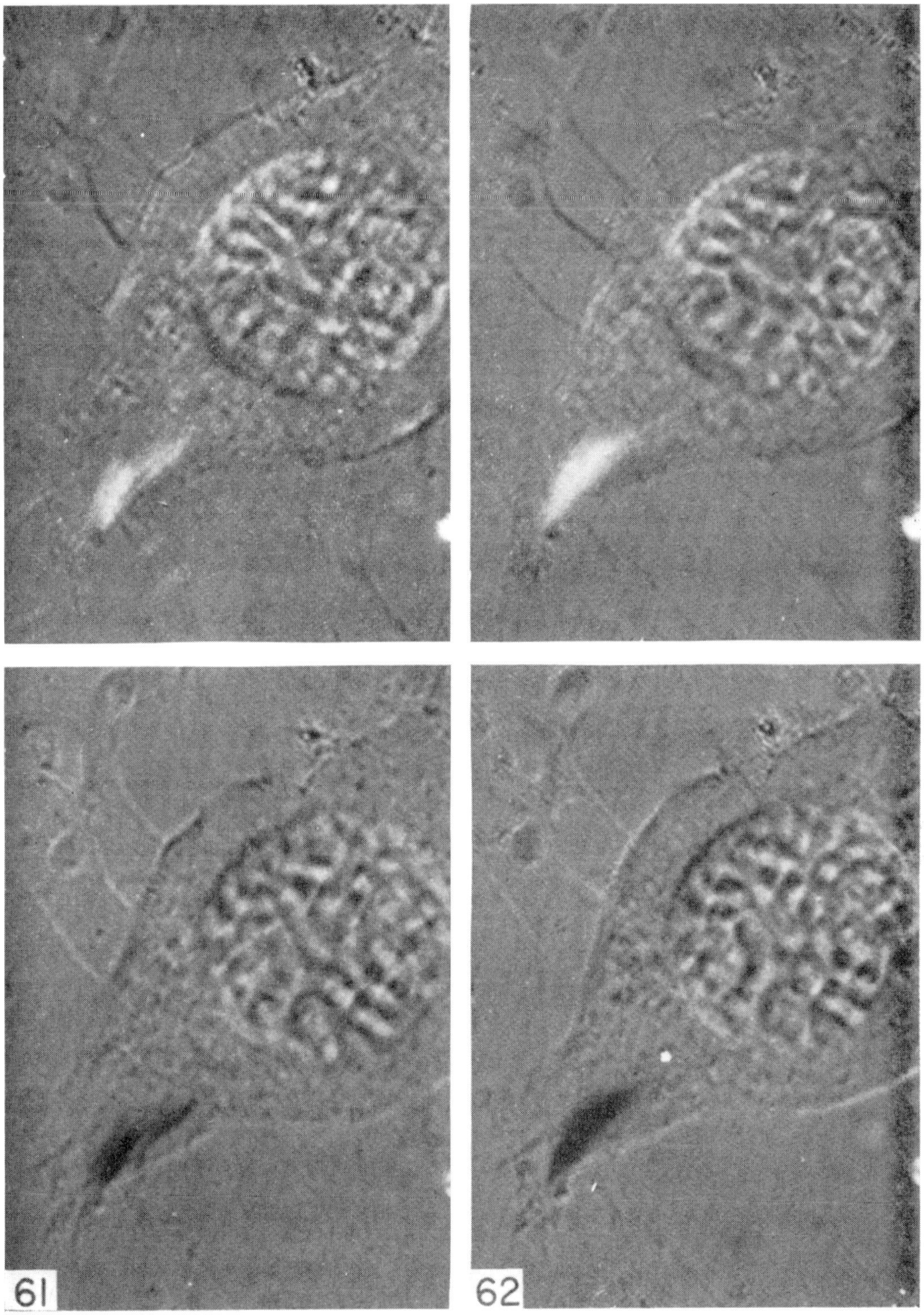

—the irradiated area lost birefringence, the nonirradiated parts have moved together to close the gap; Fig. 62, ca. 5 min after irradiation, the spindle has become reorganized into a single body again. (Original photos.)

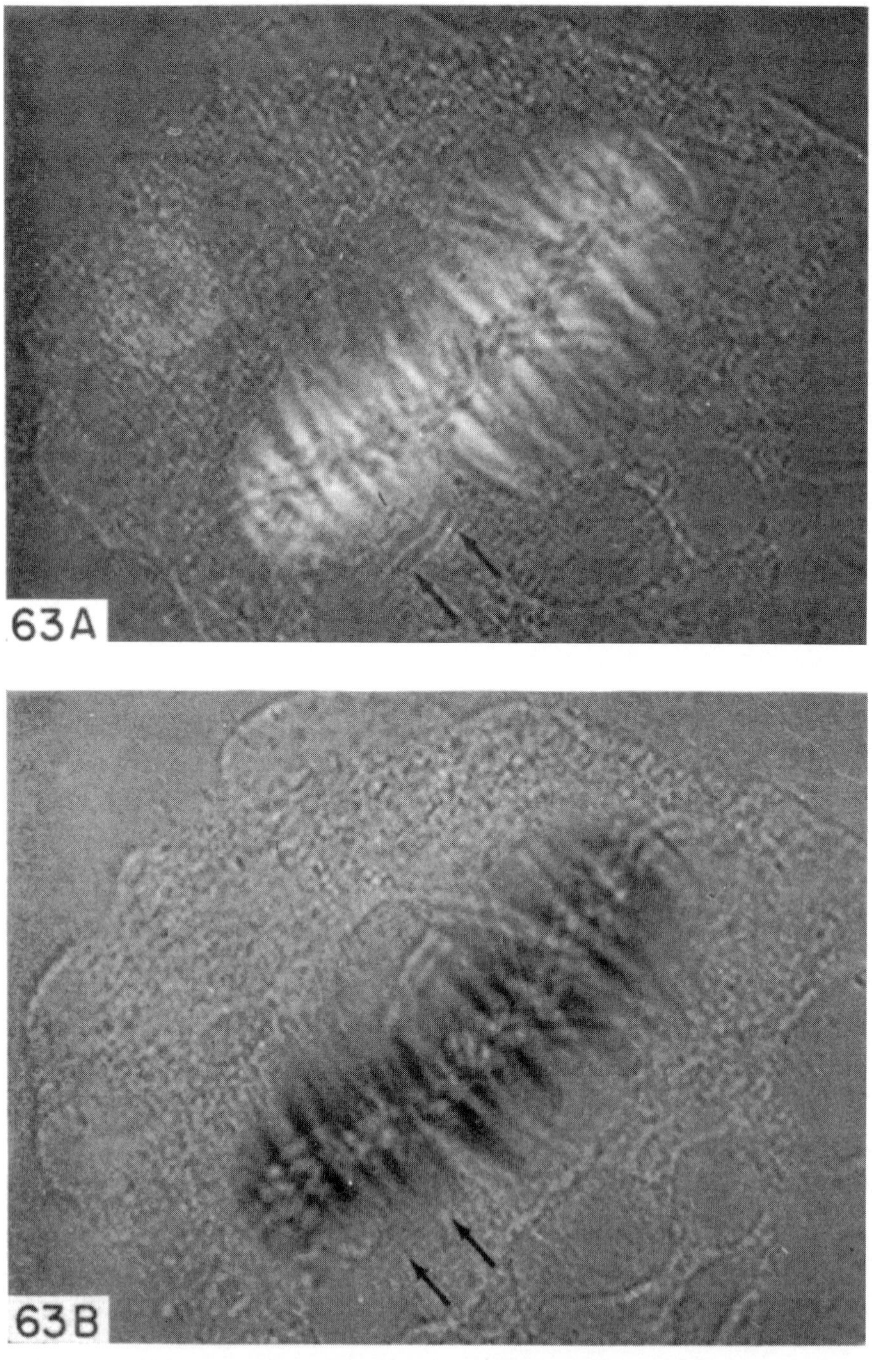

FIGS. 63 and 64. Ultraviolet-microbeam irradiation of chromosomal fibers and kinetochores in one-half spindle of *Haemanthus* endosperm cells. The irradiating ultraviolet beam covered a square area between the tips of the arrows and the

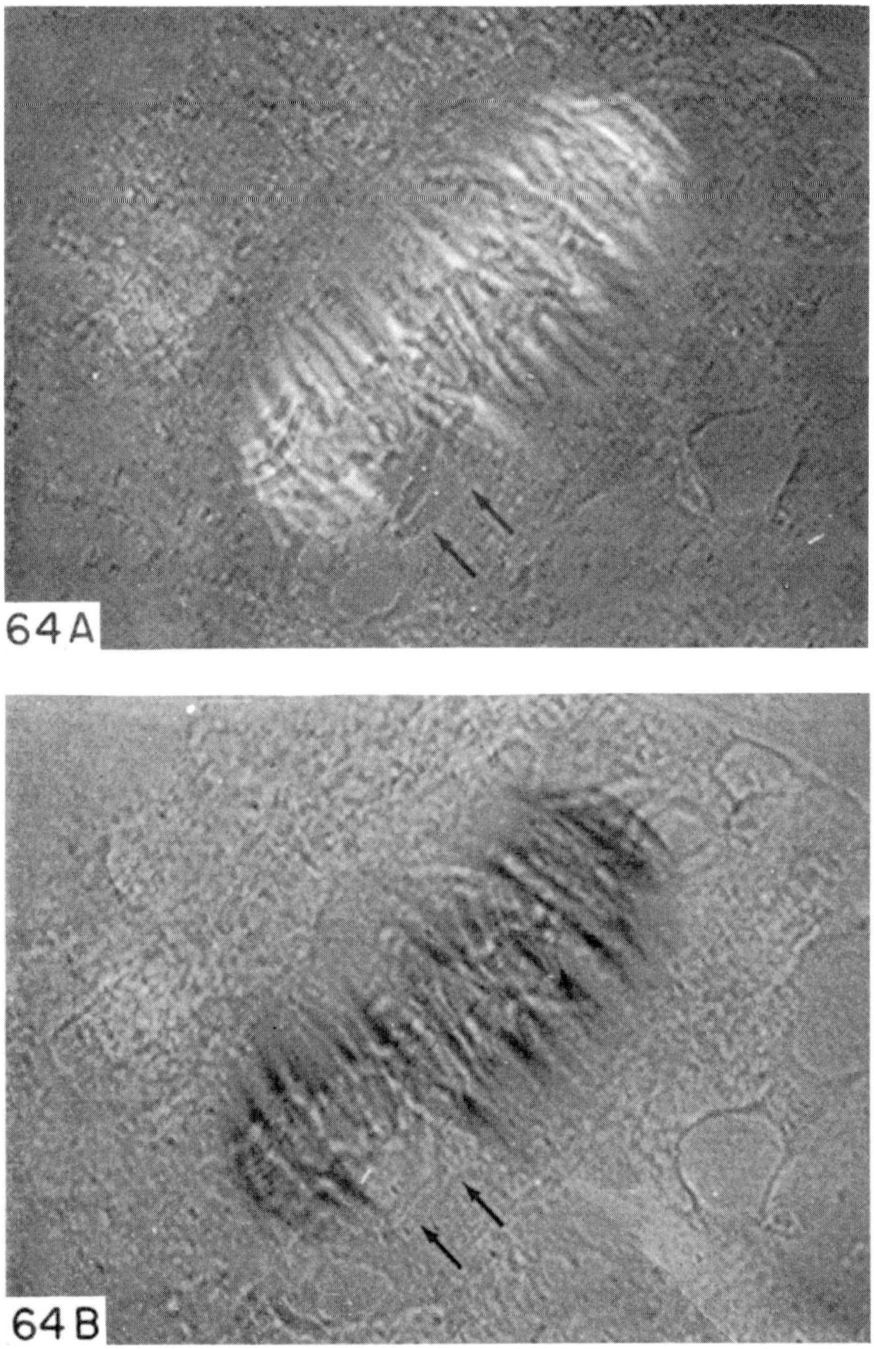

nearby kinetochores. Figure 63, before, Fig. 64, after irradiation. No recovery of the chromosomal fibers can be seen. (Original photos.)

DR. INOUÉ: I can only answer this question operationally. If one fixes a cell with one of the classic fixatives, in which the fiber birefringence appears to remain, let's say in Bouin's solution, and then changes the refractive index of the medium with organic solvents, the birefringence then disappears at the matching refractive index and returns again at a higher index. Therefore, according to this operational definition there is form birefringence, but since fixation involves very drastic changes, that is the extent of the answer I can give.

CHAIRMAN BISHOP: I can understand your reticence in naming molecules, but could you indicate which part of the spindle structure might contain a myosin-like structure?

DR. INOUÉ: May I ask you whether anybody has shown a myosin-like material?

CHAIRMAN BISHOP: I am not at all sure that what Dr. Mazia has found is myosin, but can you identify what he claims is myosin-like, or others have discussed as being myosin-like, with respect to the position of the fibrils here?

DR. INOUÉ: Dr. Mazia has isolated spindles by various means and put the material into solution, in general using rather strong reagents. Then he finds a material, which is predominantly protein mixed with some ribonucleic acid. Presumably this is the material which is making up the fibers; but I have a feeling that what we are seeing in this kind of preparation is material which has already become rather insoluble, or denatured and vulcanized. I doubt that this final step would be reversible.

But if we take the intermediate state, where the reaction is still reversible, for example, the spindle isolated by Dr. Kane, in distilled water at pH of about 6.5, if the pH is raised by 0.2 of a unit the spindle will instantly dissolve and go into solution. If one observes the spindle with a metaphase plate, the chromosomes will be seen to fall as a flake onto the slide. In such a preparation I think one can start worrying about what the state of aggregation is and whether this is myosin-like or not in terms of the over-all amino acid analysis. I believe Dr. Andrew G. Szent-Györgyi pointed out there is no resemblance to myosin. There was some similarity to actin, but I gather the data are not critical enough to say yes or no. Is that right?

DR. ANDREW G. SZENT-GYÖRGYI: Yes, you are right.

DR. GOLDACRE: If you can disorient the spindle by lowering the temperature, how do you account for formation of tubes, as mentioned by Dr. Evans Roth?

DR. INOUÉ: I do not quite follow your suggestion or question, Dr. Goldacre.

DR. GOLDACRE: I gather the electron microscope shows tubular structure in the spindles.

DR. INOUÉ: In cross section there is an outer dense region, and in longitudinal section, two lines. If we call it a tube, this is one of the ways the spindle filaments can appear under the electron microscope. Conversely, Dr. Kane has changed the conditions (the salt concentrations, etc.) of the isolated spindle before fixation and then he sees solid structures. I am at a loss to say just what temperature and disorientation may have to do with the tubular aspect.

DR. GOLDACRE: Apparently you may be building up tubes by allowing temperature to rise after the spindle material has been disoriented and disaggregated by lowering the temperature.

DR. INOUÉ: The over-all morphology and the volume of the spindle determine birefringence changes to such an extent that I don't think we can explain the temperature effect simply by a transformation of the over-all birefringent material into tubes. I gather you are suggesting this would be less birefringent.

DR. GOLDACRE: I just cannot understand how a tubular structure could be built up by raising the temperature of disoriented material.

DR. INOUÉ: Shouldn't we be more sure we are dealing with a tube before we worry about it?

DR. GOLDACRE: I suppose so.

DR. WOLPERT: Do you have any idea whether a reversible process of random disorientation leading to orientation could give rise to small movements in cytoplasm? In other words, could it exert a force for such a process as saltatory motion of particles?

DR. INOUÉ: Well, I have only limited experience on this matter. If you form liquid crystals of detergent, let's say, and watch the front of the crystals forming, this does not appear to displace granules which seem to be floating around freely in the medium.

DR. HOFFMANN-BERLING: May I comment on the energy characteristics? Couldn't they be those of another process which regulates spindle production?

DR. INOUÉ: Dr. Hoffmann-Berling would like to caution us, I believe, that what we observe as entropy change and free-energy change may not be the primary change in the spindle material itself. I agree with you whole-heartedly, and, in fact, I had this data lying around for 6 or 7 years before we actually put this interpretation on because of this danger. However, it appears that there are many encouraging results in isolated systems such as I mentioned, the G-F actin transformation and the tobacco mosaic virus A protein polymerization.

DR. HOFFMANN-BERLING: A low entropy and high ΔH.

DR. INOUÉ: High ΔH and huge ΔS. So I am somewhat encouraged that perhaps we may be dealing with something rather close to the spindle itself.

DR. EDWIN TAYLOR: While you are on the subject, I wish you would clarify something that has always bothered me. That is in making this calculation, Dr. Morales either knew the molecular weight of the spindle protein or I do not know what these numbers could refer to.

DR. INOUÉ: Dr. Kauzmann, would you answer this?

DR. KAUZMANN: If the model used by Dr. Inoué is correct, then he does not need to know the molecular weight of the spindle protein. If this protein can exist in only two states—one of which is birefringent and the other of which is not—and if the relative amounts of the two states is proportional to the amount of birefringence, then one can deduce an equilibrium constant for the two states. From this equilibrium constant one can find the difference in free energy (ΔF) of the two states (equilibrium constant $= K = \exp(-\Delta F/RT)$) without knowing the molecular weights. This is one of the wonderful consequences of thermodynamics. Remember, however, that the reality of this ΔF depends upon the adequacy of the assumed model. If there were several kinds of protein in the spindle, each having a different equilibrium constant, or if there were a series of intermediate states, then the ΔF value that Dr. Inoué has calculated would not have much meaning.

DR. TAYLOR: What reaction does this ΔF refer to?

DR. KAUZMANN: It could refer to the direct aggregate reaction or it could refer to some controlling equilibrium that indirectly determines the birefringence.

Concerning the point that has been raised about the pH, could one not look for pH changes by introducing an inert, indicator dye and look for color changes that would result from changes in the pH?

DR. INOUÉ: Yes. Intracellular pH has been measured by colors of indicators, and I think such an experiment would be worth trying. There is also the possibility, for instance, of calcium ion release which I gather affects active polymerization a great deal; there are all kinds of other things that could go on.

DR. ALLEN: I just wanted to ask you how well you can separate the birefringence of the spindle from that of the mitochondrial sheath that is above and below it in some cases. Doesn't the sheath act as a compensator sometimes?

DR. INOUÉ: Yes, sometimes. The mitochondrial sheath in telophase has a much stronger birefringence than the continuous fibers in grasshopper spermatocytes. However, this is not always true with other types of cells. I can generally distinguish the birefringence of spindle structures both morphologically and by optical sectioning from the birefringence of the surrounding material.

Reprinted from *The Biological Bulletin*, Vol. 129(2), pp. 409–410, 1965.

Counteraction of Colcemid and heavy water on the organization of the mitotic spindle. SHINYA INOUÉ, HIDEMI SATO AND MICHAEL ASCHER.

(1) Treatment of metaphase-arrested spindles of *Pectinaria* oocytes with Colcemid (N-desacetyl-N-methylcolchicine, a colchicine derivative with greater effectiveness and less toxicity), reduces spindle length and birefringence reversibly, confirming the action of colchicine reported on *Chaetopterus*. (2) D_2O increases spindle length and birefringence, as shown earlier in *Lytechinus* eggs. The effect is maximal at 45% concentration, is rapidly reversible, and can be repeated many times on the same egg. (3) D_2O delays Colcemid-induced reductions of spindle length and birefringence without altering actual rates of reduction. (4) The spindle length and birefringence begin to recover immediately when cells treated with Colcemid are washed with 45% D_2O or puromycin (10^{-5} *M*), provided spindle disappearance has not proceeded too far. (5) Cells which have completely lost their spindles by Colcemid treatment begin to recover their spindles after a lag period of one-half to one hour if washed with sea water or D_2O sea water. (6) The lag period could not represent the need for protein synthesis since it is not affected by chloramphenicol (10^{-5} *M*) and is even reduced to one-third or one-quarter when cells are washed with puromycin (10^{-5} *M*) or actinomycin D (10^{-5} *M*). (7) These observations further substantiate Inoué's dynamic equilibrium model of the spindle.

Aided by grants from the National Science Foundation, GB-2060, and the U.S.P.H.S., CA 04552.

Reprinted from *Molecular Architecture in Cell Physiology*, pp. 209–248, 1966.

Deoxyribonucleic Acid Arrangement in Living Sperm[1]

Shinya Inoué and **Hidemi Sato**
Department of Cytology
Dartmouth Medical School
Hanover, New Hampshire

Birefringence of the Sperm Head

During the last decade much has been learned about the structure of deoxyribonucleic acid (DNA) molecules (Watson and Crick, 1953; Crick and Watson, 1954; Wilkins, 1963) and their genetic role in protein synthesis. In spite of the remarkable advances at the chemical level, we still know very little about the precise arrangement of DNA molecules within the chromosomes (Kaufman *et al.*, 1960; Ris, 1963). The problem was therefore to devise means to study and measure the arrangement of DNA in chromosomes in a living cell.

In order to undertake such a study, we have selected certain sperm cells in which the molecular arrangement of DNA is reflected in a measurable physical parameter (Inoué and Sato, 1962).

Figure 1 (figures follow text) depicts several stages in the process of sperm maturation in a living squid testis smear as observed through the phase contrast microscope. During spermiogenesis the chromosomes gradually become condensed and aligned as the nucleus decreases in volume. In the mature sperm the chromosomes are very dense and packed together so tightly that they can no longer be resolved as separate strands (Sato and Inoué, 1964).

[1] Supported in part by grants from the National Science Foundation (G 19487) and National Cancer Institute, United States Public Health Service (CA 04552).

When the same process is studied through the polarizing microscope, one sees a gradual increase in contrast of the nucleus during sperm maturation (Fig. 2). The contrast is caused by the negative birefringence of DNA molecules in the chromosomes, and the increase in contrast reflects the increased alignment and packing of the DNA molecules as the sperm matures (Seeds and Wilkins, 1950; Wilkins and Bataglia, 1953; Wilkins and Randall, 1953; Sato and Inoué, 1964).

The alignment of DNA in the mature squid sperm is nearly perfect or virtually crystalline, for an X-ray diffraction pattern of DNA in the B form (Fig. 3) was demonstrated in a bundle of fresh squid sperm by Wilkins and Randall (1953) and Wilkins (1963).

Figure 4 shows the needle-shaped heads and portions of the tails of sperm of the cave cricket *Ceuthophilus nigricans* Scudder as observed under the polarizing microscope. The head appears bright in one orientation and dark in the other, while the tail shows the reverse contrast. The contrast and the orientation of the compensator axes show that the head exhibits a negative birefringence while the tail and the acrosome exhibit a positive birefringence with respect to the length of the individual sperm (Pattri, 1932; Schmidt, 1937). The negative birefringence is intrinsic. It is caused, not by the texture of submicroscopic structures, but by the optical character of the constituent DNA molecules (Wilkins, 1951; Inoué and Sato, 1962). The coefficient of birefringence of the head measures -2×10^{-2}, the same order of magnitude as that found for stretched gels of pure DNA (Wilkins, 1951; Seeds, 1953; Beaven *et al.*, 1955; Ruch, 1956). Protein fibers in general exhibit a positive birefringence (Schmidt, 1937).

Effect of Polarized Ultraviolet Irradiation

We have found that irradiation of a small part of a living sperm with a beam of ultraviolet light (UV) results in the disappearance, in visible light, of the birefringence from that part (Inoué and Sato, 1962) (Fig. 5). This signifies that the molecules responsible for the birefringence could be changed structurally, destroyed, or lost. However, the last possibility can be excluded, for the dry mass of the irradiated area, as observed with the phase contrast microscope, does not change (Fig. 6).

As we have earlier stated, the oriented DNA molecules are mainly responsible for the sperm birefringence (Schmidt, 1937; Wilkins, 1951; Wilkins and Randall, 1953; Inoué and Sato, 1962; Sato and Inoué, 1964). Disappearance of sperm-head birefringence induced by irradiation in the 2,500–3,000-Å range suggests that DNA is directly absorbing the UV and losing its birefringence.

As seen in Fig. 7 (Crick and Watson, 1954; Feughelman *et al.*, 1955), the purine and pyrimidine bases in the B-form DNA lie in planes at 90° to the backbone of the molecule and form the steps of a spiral staircase. This is the form of DNA that Wilkins and his coworkers found in the fresh squid sperm by X-ray diffraction analysis (Wilkins and Bataglia, 1953; Wilkins and Randall, 1953). These bases are responsible for the greater refractive index measured at right angles to the backbone of the molecule compared with that measured along the long axis (Seeds and Wilkins, 1950; Thorell and Ruch, 1951; Seeds, 1953; Beaven *et al.*, 1955). In other words, the arrangement of the bases is responsible for the negative birefringence of DNA. The bases also absorb more light energy when UV vibrates parallel to their planes than when it vibrates perpendicular to them (Kasha, 1961). The arrangement of the bases, therefore, gives rise to a negative dichroism of DNA in UV in addition to the negative birefringence that is seen with visible light. Figure 8 shows the UV dichroism of stretched DNA films measured by Seeds and Wilkins (1950) and Seeds (1953). At 90% relative humidity, the condition giving the B-form X-ray pattern for sodium DNA, the dichroic ratio is 3.5 to 4 throughout the wavelength range 2,400 Å to 3,000 Å.

Thus, irradiation of cave-cricket sperm with the UV ray polarized with its electric vector (E-vector: indicated in figures by double arrow) vibrating parallel to the base planes should remove the birefringence more effectively than should irradiation with the UV ray polarized with its E-vector perpendicular to the base planes. This is found to be true, as shown in Fig. 9. (See Fig. 10 for a schematic diagram of the polarized-UV microbeam system.) Induction of the same degree of birefringence loss takes three to four times longer with the UV E-vector parallel to the sperm axis than with it perpendicular. This is in agreement with the UV dichroism of a pure DNA fiber that loses its birefringence two to three times faster with the UV E-vector perpendicular to the fiber axis than with it parallel (Fig. 11). Again, little loss of dry mass is observed with the phase contrast microscope. Further, Seeds (1953) has shown that the parallel absorption of UV in an irradiated DNA fiber *rises* to offset the drop of perpendicular absorption at 260 mμ, which indicates, not that the bases are being destroyed, but rather that their orientations are becoming randomized. The evidence is therefore strong that the UV-induced birefringence change directly reflects disorientation of DNA bases in the sperm head.

The UV-induced change in birefringence can be used to study the detailed arrangement of DNA molecules in the living sperm head. Figure 9 shows the internal structure of the cave-cricket sperm that can be discerned under polarized light. The pictures in this figure were taken under the rectified polarizing microscope, which was designed to give a highly corrected image of weakly birefringent specimens at the maximum resolution obtainable with a light microscope (Inoué and Hyde, 1957; Inoué and Kubota, 1958; Inoué, 1961). This structure that is seen with polarized light, but which is not visible with the phase contrast microscope, gives the impression of a helical configuration of birefringent regions. Rotation of the specimen on the microscope stage shows that the alternate bright and dark regions reflect minute domains where the crystalline axes are not exactly parallel to the long axis of the sperm but slightly tilted to the left and to the right. In other words, although on the average the sperm head shows a negative birefringence as if the backbone of the DNA molecules were precisely parallel to the sperm long axis, in detail the molecules are tilted slightly and form zigzag microdomains.

Response of Sperm to Irradiation by UV Ray Polarized at 45°

When the irradiating UV ray is polarized perpendicular to the sperm axis, the UV E-vector is nearly parallel to the DNA bases in all of the microdomains, and absorption of UV and loss of birefringence are uniform and maximum. Contrastingly, with the UV E-vector parallel to the sperm axis, the DNA bases are all nearly perpendicular to the E-vector, and absorption and birefringence loss are low. These effects are shown in Fig. 9.

When the UV is polarized with its E-vector tilted 45° to the sperm axis, microdomains with bases tilted in the same direction lose their birefringence faster than those with base planes tilted in the opposite direction. After appropriate exposure, the former microdomains lose their birefringence but the latter microdomains retain some of their birefringence. Therefore the *average* axis of birefringence, which before irradiation was parallel to the sperm axis, is now rotated toward the E-vector of the irradiating UV, as shown schematically in Fig. 12 and in the composite photograph in Fig. 13. Each pair of pictures in Fig. 13 was taken with the compensator set for reversed contrast (Köhler, 1921; Swann and Mitchison, 1950; Hartshorne and Stuart, 1960). As shown in the middle set, where the sperm axes lie parallel to the polarizer axis, the unirradiated regions show alternate black-white stripes representing the zigzag tilt of the intact microdomain axes. After irradiation with a UV beam polarized at 45° with respect to the sperm axis, as indicated by the double arrow, all microdomains in this region appear either white and gray or black and gray, showing the elimination of birefringence in individual microdomains with one direction of tilt.

The tilt angle or azimuth of the microdomains can be determined from the extinction angle of the irradiated region. As shown in Fig. 13, the contrast of the irradiated region is minimum when the sperm axis is turned 15° *counter-clockwise.* Thus the rotation *clockwise* of the optical axis due to irradiation must be approximately 15°, for in this position the altered optic axis of the irradiated region is nearly parallel with the polarizer axis. Alternatively, if the sperm axis is turned 15° *clockwise*, the average axis of the irradiated region will now be 15° + 15°, or 30°, off from the polarizer axis, that is, near the angle providing maximum contrast. This experiment again supports the view that the individual DNA bases directly absorb polarized UV and lose their birefringence according to their submolecular orientation relative to the UV E-vector.

Some of the microdomains are as small as 0.2 μ in width. The arrangement of DNA bases can be shown to exist in zigzag array even when the microdomains are not resolved, for their arrangement manifests itself as a rotation of the average birefringence axis following irradiation with UV polarized at 45°. For example, the sperm head of the fruit fly (*Drosophila busckii*) is only 8 μ long and too small to reveal microdomains even under the highest resolution of the rectified polarizing microscope. However, irradiation of this sperm head with UV at 45° from the sperm axis brings about a measurable rotation (9°) of the average birefringence axis, indicating a similar zigzag arrangement of the DNA. Figure 14 shows the loss of birefringence in such an irradiated sperm head.

We have thus far shown the presence of zigzag microdomains that suggests a helical structure. However, a packing problem arises regarding the arrangement of the DNA molecules within such a structure, because it is known that the length of such molecules is of the order of at least a few hundred micra (Cairns, 1961, 1963; Kleinschmidt *et al.*, 1962; Hershey *et al.*, 1963).

Photographic Microdensitometry

To ascertain the arrangement of DNA molecules within each microdomain, it is necessary to measure the birefringence and tilt angle of individual microdomains, and for this purpose we have devised the following method of photographic microdensitometry.

A set of photographic negatives of a cave-cricket sperm is taken at a particular orientation with respect to the axis of the polarizer and with the compensator (Köhler, 1921; Swann and Mitchison, 1950) set to various angles (Fig. 15). As the compensator is turned, the background intensity changes from bright to dark and back again to bright. The oriented microdomains also change contrast but reach darkness at different settings of the compensator (e.g., see point *V*, Fig. 15). It is important that such sets of photographic negatives be processed simultaneously.

Each image and its background are then scanned with a Joyce-Loebel microdensitometer (Fig. 16). The densities of the negatives are translated into densitometer readings, and the heights of the readings for each microdomain and its background are recorded (Fig. 17). When these readings are plotted against compensator settings (Fig. 18), two similar "parabolic" curves are obtained, one for the microdomain and one for the background. The lateral displacement of these two curves gives the orientation of the compensator so that the cross-section is elliptical rather than circular (as in Fig. 27, *a*), more of the DNA backbones would fit the average tilt angles required, and the tilt of all of the microdomains would fit the measured azimuth angles without distorting the helices locally.

Taking into account the azimuth distribution across the width of the sperm, the model can be fitted to the data without locally distorting the gyres only if the double helices were interrupted at specific loci where the ends of the adjoining breaks would then lie parallel to each other. This yielded a resynthesized image of the DNA-containing structure as shown in Fig. 28. The notion that the "breaks" may represent chromosome ends and that the chromosomes in certain insect sperm may be arranged in tandem (also see Cooper, 1952; Nakanishi, 1957; Reitberger, 1964) as illustrated is strengthened by the tritiated-thymidine labeling experiment by Taylor (1964). He finds that grasshopper sperm labeled during meiosis often show a section that lacks labeling yet shows a Feulgen-positive reaction. He interprets this as the locus of the X-chromosome, which has been demonstrated to incorporate thymidine into DNA later than do the other chromosomes in grasshopper cells.

We find that in the three sperm heads thoroughly analyzed, the breaks all occur in the same position, shorter segments and long segments always appearing respectively in the same positions. This suggests that the chromosomes are arranged not only in tandem but with a sequence that is nonrandom and unique. The average length between breaks is 3.4 μ. This is in reasonably good agreement with the length of the sperm head (41μ) divided by the chromosome numbers 18 and 19.

The arrangement of DNA proposed in the model apparently does not conform to published electron micrographs of other sperm (Ris, 1959; Sotelo and Cenóz, 1960). This is not too disturbing, for we found that immersion of cave-cricket sperm in dimethylsulfoxide (DMSO) does not change the magnitude or distribution of birefringence (Fig. 29), and that the DMSO-treated sperm can be embedded in plastic and sectioned for electron microscopy without further dehydration or alteration of birefringence. Dr. R. E. Kane, Department of Cytology, Dartmouth Medical School, has taken electron micrographs of such sections, shown in Fig. 30. A structural organization of DNA not inconsistent with our model is then demonstrable in the sperm nucleus with the electron microscope.

Concluding Remarks

We have developed a new method for fine structure analysis that can reveal with high precision the distribution and orientation of dichroic molecules in living cells. Several new techniques, which may find application in related fields of study, have been introduced in this paper. The biological findings may be significant in understanding UV-induced mutations as well as the arrangement of DNA in chromosomes generally.

The probable nonrandom sequence found in the living needle-shaped sperm head should provide exceptional opportunities for controlled modifications and analyses of chromosomes and genes.

ADDENDUM

Recently, Maestre and Kilkson (1965) computed the intrinsic birefringence of multiple-coiled DNA and proposed a valuable equation. They applied their theoretical formula to our model and measurements of cave-cricket sperm and calculated that the azimuth oscillation of each microdomain should be $\pm 9°$ and the angle of coiling of the secondary helix 15.3°, agreeing quantitatively with our experimental results.

REFERENCES

Beaven, G. H., R. R. Holiday, and E. A. Johnson. 1955. Optical properties of nucleic acids and their components. In *The Nucleic Acids, Chemistry and Biology*, Vol. 1, ed. E. Chargaff and J. N. Davidson. New York: Academic Press, Inc. Pp. 493–553.

Cairns, J. 1961. An estimate of the length of the DNA molecule of T_2 bacteriophage by autoradiography. *J. Mol. Biol.* 3: 756–761.

Cairns, J. 1963. The bacterial chromosome and its manner of replication as seen by autoradiography. *J. Mol. Biol.* 6: 208–213.

Cooper, K. W. 1952. Studies on spermatogenesis in *Drosophila. Trans. Am. Phil. Soc.* 1951: 146–147.

Crick, F. H. C., and J. D. Watson. 1954. The complementary structure of deoxyribose nucleic acid. *Proc. Roy. Soc.* (*London*) *Ser. B* 223: 80–96.

Feughelman, M., R. Langridge, W. E. Seeds, A. R. Stokes, H. R. Wilson, C. W. Hooper, M. H. F. Wilkins, R. K. Barclay, and L. D. Hamilton. 1955. Molecular structure of deoxyribose nucleic acid and nucleoprotein. *Nature* 175: 834–838.

Hartshorne, N. H., and A. Stuart. 1960. *Crystals and the Polarizing Microscope*, 3rd ed. London: Edward Arnold Ltd.

Hershey, A. D., E. Burge, and L. Ingraham. 1963. Cohesion of DNA molecules isolated from phage lambda. *Proc. Natl. Acad. Sci. U.S.* 49: 748–755.

Inoué, S. 1961. Polarizing microscope: design for maximum sensitivity. In *The Encyclopedia of Microscopy*, ed. G. L. Clark. New York: Reinhold Publishing Corp. Pp. 480–485.

Inoué, S., and W. L. Hyde. 1957. Studies on depolarization of light at microscope lens surfaces. II. The simultaneous realization of high resolution and high sensitivity with the polarizing microscope. *J. Biophys. Biochem. Cytol.* 3: 831–838.

Inoué, S., and H. Kubota. 1958. Diffraction anomaly in polarizing microscopes. *Nature* 182: 1725–1726.

Inoué, S., and H. Sato. 1962. Arrangement of DNA in living sperm: a biophysical analysis. *Science* 136: 1122–1124.

Kasha, M. 1961. The nature and significance of $n \rightarrow \pi^*$ transition. In *A Symposium on Light and Life*, ed. W. D. McElroy and B. Glass. Baltimore: The Johns Hopkins Press. Pp. 31–38.

Kaufmann, B. P., H. Gay, and M. McDonald. 1960. Organizational patterns within chromosomes. *Intern. Rev. Cytol.* 9: 77–127.

Kleinschmidt, A. K., D. Lang, D. Jacherts, and R. K. Zahn. 1962. Darstellung und Längenmessungen des gesaten Desoxyribonucleinsäure-Inhaltes von T_2-Bacteriophagen. *Biochim. Biophys. Acta* 61: 857–864.

Köhler, A. 1921. Ein G. Glimmerplätchen Grau. I. Ordnung zur Untersuchung sehr Schwach doppelbrechender Präparate. *Z. Wiss. Mikroskopie* 38: 29–42.

Maestre, M. F., and R. Kilkson. 1965. Intrinsic birefringence of multiple-coiled DNA, theory and applications. *Biophys. J.* 5: 275–287.

Nakanishi, Y. H. 1957. Some observations on the internal structure of the sperm head of the grasshopper, *Oxya yezoensis*, after dehydration and hydration treatments. *J. Fac. Sci. Hokkaido Univ.* Ser. VI. 13: 276–280.

Pattri, H. O. E. 1932. Über die Doppelbrechung der Spermien. *Z. Zellforsch. Mikroskop. Anat.* 16: 723–744.

Reitberger, A. 1964. Lineare Anordnung der Chromosomen im Kern des Spermatozoids des Lebermooses *Sphaerocarpus donnellii. Naturwiss.* 51: 395–396.

Ris, H. 1959. Die Feinstruktur des Kerns während der Spermiogenese. In *Chemie der Genetik*. Berlin: Springer-Verlag. Pp. 1–30.

Ris, H. 1963. Ultrastructure of the cell nucleus. In *Funktionelle und Morphologische Organization der Zelle*. Berlin: Springer-Verlag. Pp. 1–14.

Ruch, F. 1956. Birefringence and dichroism of cells and tissues. In *Physical Techniques in Biological Research*, Vol. 3, ed. G. Oster and A. W. Pollister. New York: Academic Press, Inc. Pp. 149–176.

Sato, H., and S. Inoué. 1964. Condensation of the sperm nucleus and alignment of DNA molecules during spermiogenesis in *Loligo pealii*. *Biol. Bull.* 127: 357.

Schmidt, W. J. 1937. Die Doppelbrechung von Karyoplasma, Zytoplasma und Metaplasma. *Protoplasma Monogr.* 11. Berlin: Gebrüder Borntraeger Verlag.

Seeds, W. E. 1953. Polarized ultraviolet microspectrography and molecular structure. *Progr. Biophys. Biophys. Chem.* 3: 27–46.

Seeds, W. E., and M. H. F. Wilkins. 1950. Ultraviolet microspectrographic studies of nucleoproteins and crystals of biological interest. *Discussions Faraday Soc.* 9: 417–423.

Sotelo, J. R., and O. Trujillo Cenóz. 1960. Electron microscope study on spermatogenesis. Chromosome morphogenesis at the onset of meiosis (cyte 1) and nuclear structure of early and late spermatids. *Z. Zellforsch. Mikroskop. Anat.* 51: 243–277.

Swann, M. M., and J. M. Mitchison. 1950. Refinements in polarized light microscopy. *J. Exptl. Biol.* 27: 226–237.

Taylor, J. H. 1964. The arrangement of chromosomes in the mature sperm of the grasshopper. *J. Cell Biol.* 21: 286–289.

Thorell, B., and F. Ruch. 1951. Molecular orientation and light absorption. *Nature* 167: 815.

Watson, J. D., and F. H. C. Crick. 1953. A structure for deoxyribose nucleic acids. *Nature* 171: 737–738.

Wilkins, M. H. F. 1951. I. Ultraviolet dichroism and molecular structure in living cells. II. Electron microscopy of nuclear membrane. *Pubbl. Staz. Zool. Napoli* 23: Suppl. 104–114.

Wilkins, M. H. F. 1963. Molecular configuration of nucleic acids. *Science* 140: 941–950.

Wilkins, M. H. F., and B. Bataglia. 1953. Note on the preparation of specimens of oriented sperm heads for X-ray diffraction and infra-red absorption studies and on some pseudo-molecular behavior of sperm. *Biochim. Biophys. Acta* 11: 412–415.

Wilkins, M. H. F., and J. T. Randall. 1953. Crystallinity in sperm heads, molecular structure of nucleoprotein *in vivo*. *Biochim. Biophys. Acta* 10: 192–194.

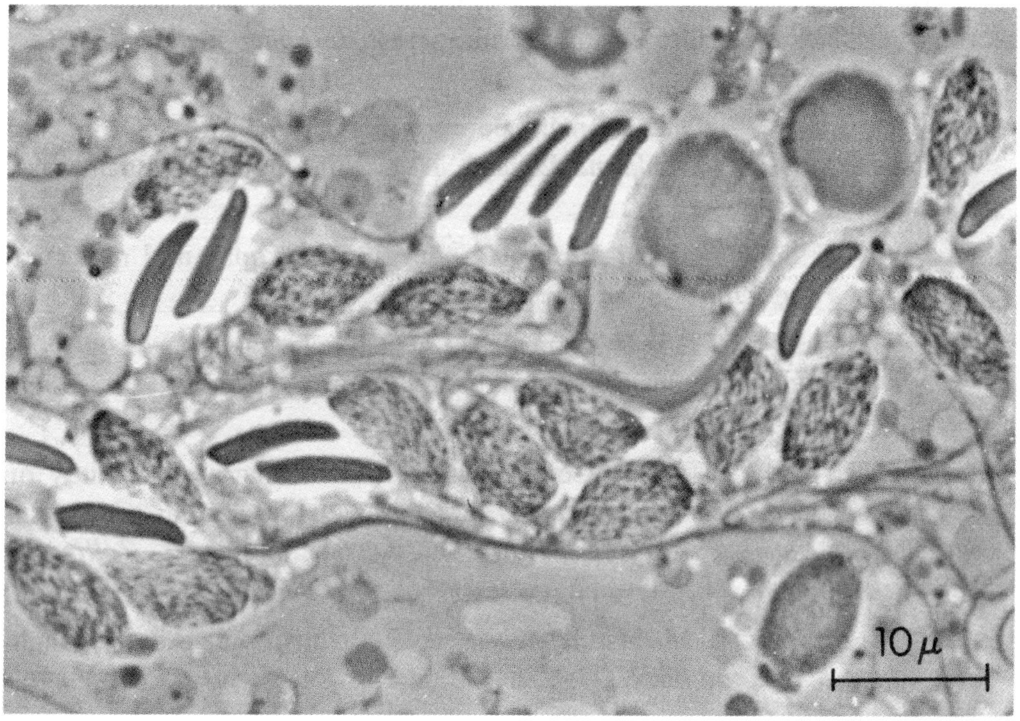

Fig. 1. Photograph through phase contrast microscope of spermatid nuclei of the squid *L. pealii,* at various stages of sperm maturation. Teased fresh testis imbibed in nontoxic Kel F-10 oil to reduce Brownian movement and improve image quality.

During spermiogenesis, nucleus of young spermatid goes directly from "telophase" to "prophase." In this stage, many coils of chromonemata are arranged more or less longitudinally in the sperm head. The oriented chromonemata shorten and become more compact and more densely packed as the sperm approaches maturity. No structure in the mature sperm head is resolvable. Nucleus gradually condenses laterally but does not change in length (*ca.* 7.8 μ) during maturation process.

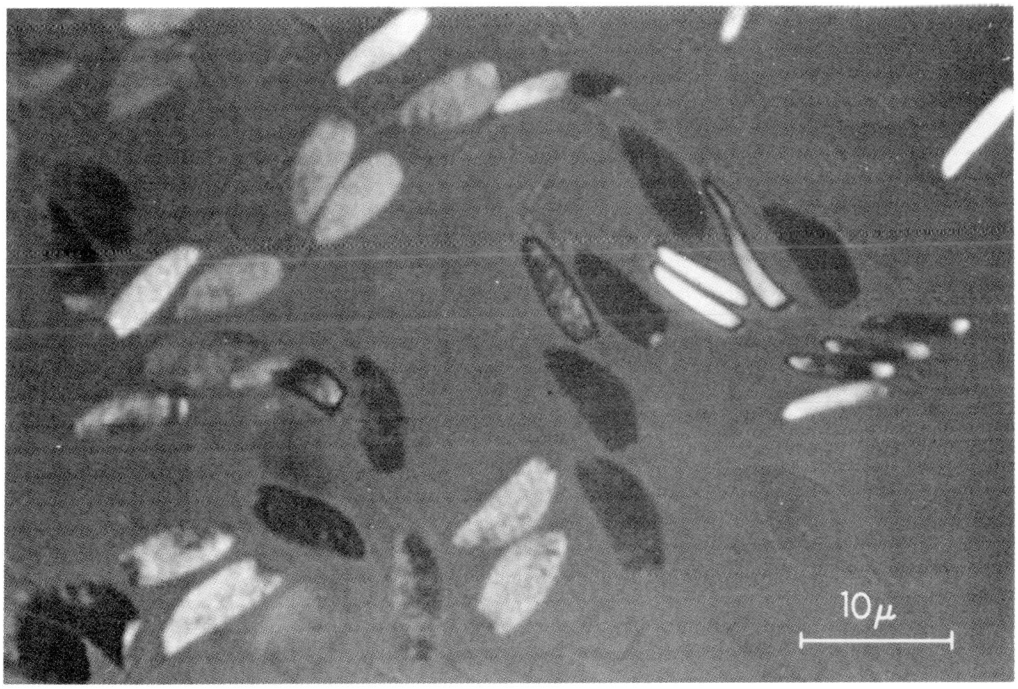

Fig. 2. Photograph through polarizing microscope of fresh smear of squid testis imbibed in Kel F-10 oil, as for Fig. 1. Compare weak birefringence (almost zero) in young spermatid nuclei, increasing birefringence of nuclei in developing sperm, and strong (-30 mμ) birefringence in mature sperm. Since the width of the mature sperm is 1.4 μ, the coefficient of birefringence reaches -2×10^{-2}. A negative coefficient of birefringence is characteristic of well-oriented DNA gels.

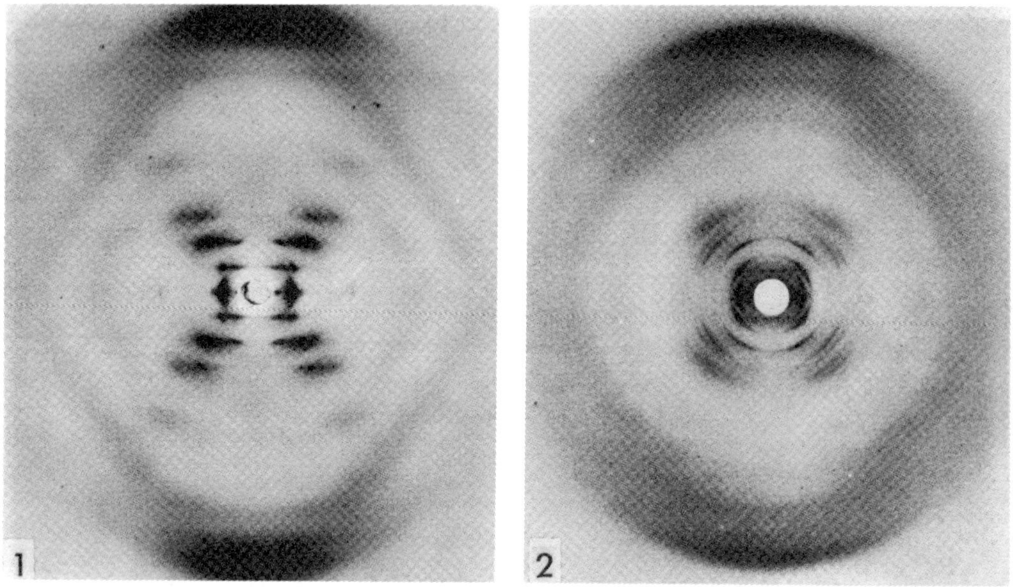

Fig. 3. 1. Early X-ray diffraction photo of fibers of sodium DNA in B configuration at high humidity. Fibers oriented vertically. The 3.4-Å reflection is at top and bottom. Angle in pronounced X shape, made by reflections in central region, corresponds to constant angle of ascent of polynucleotide chains in helical molecule.

2. X-ray diffraction photograph of fibers of oriented *Sepia* sperm. Axes of DNA molecules in sperm heads are vertical. The 3.4-Å internucleotide spacing corresponds to strong diffraction at top and bottom of pattern. X-shape reflections in central region show molecules are in crystalline array. (From Wilkins, 1963, *Science* 140: 941–950 copyright 1963 by the American Association for the Advancement of Science.)

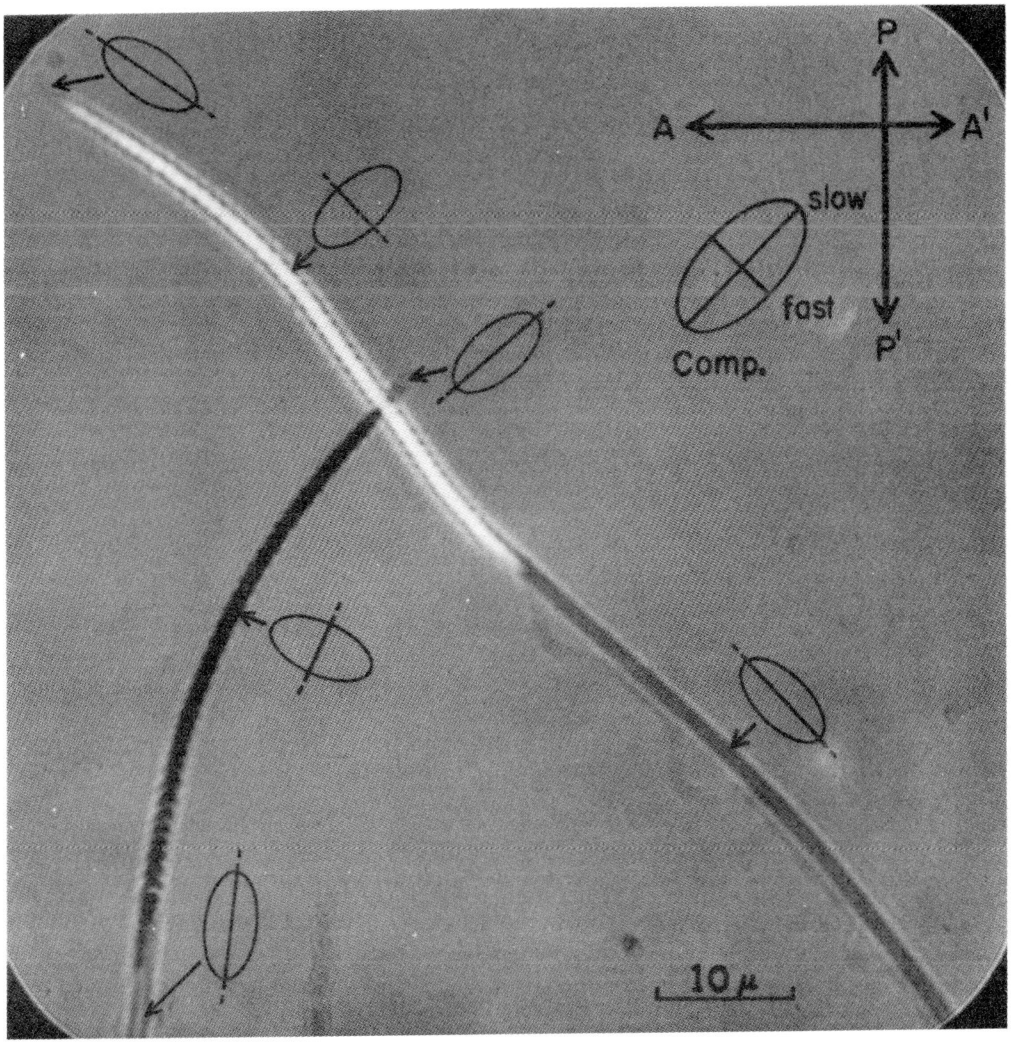

Fig. 4. Two sperm heads and proximal parts of tails.[3] Contrast shows birefringence of head to be negative. Coefficient of birefringence, *ca.* -2×10^{-2}. Sperm tail and acrosome show weak positive birefringence. Picture taken through polarizing microscope at relatively low power between crossed polars, using a $\lambda/24$ revolving mica compensator. Fast and slow axes of compensator (*Comp.*) indicated. $P \longleftrightarrow P'$: transmission direction of polarizer; $A \longleftrightarrow A'$: transmission direction of analyzer.

[3] Unless otherwise specified, sperm are from the cave cricket, *Ceuthophilus nigricans* Scudder.

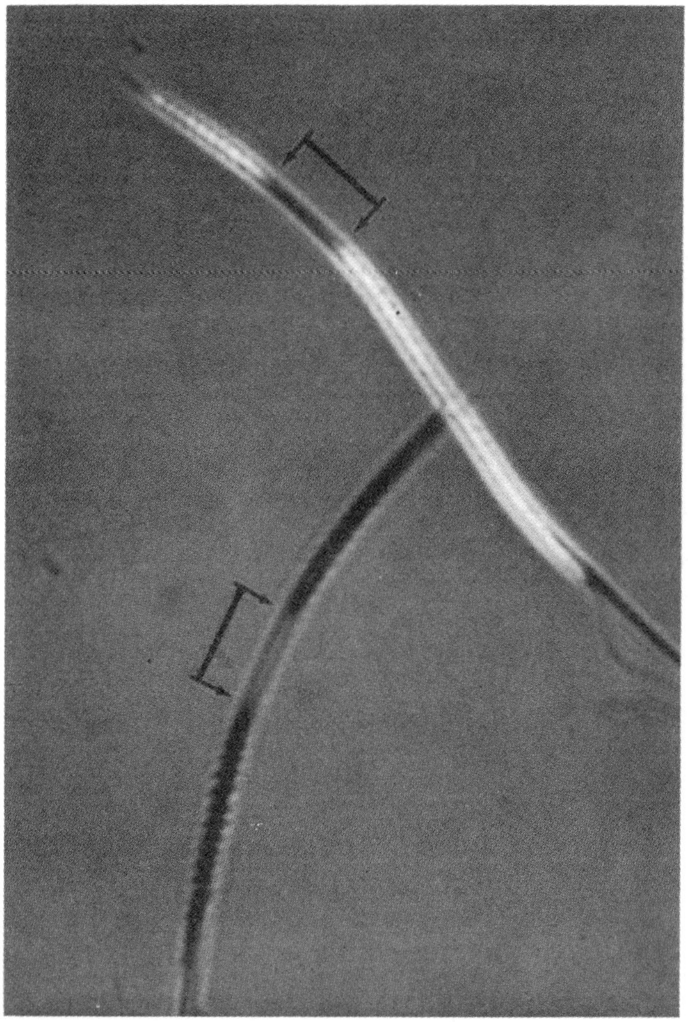

Fig. 5. Effect of UV-microbeam irradiation on sperm heads. Negative birefringence observed in visible light is lost irreversibly from indicated irradiated areas. Same sperm as in Fig. 4.

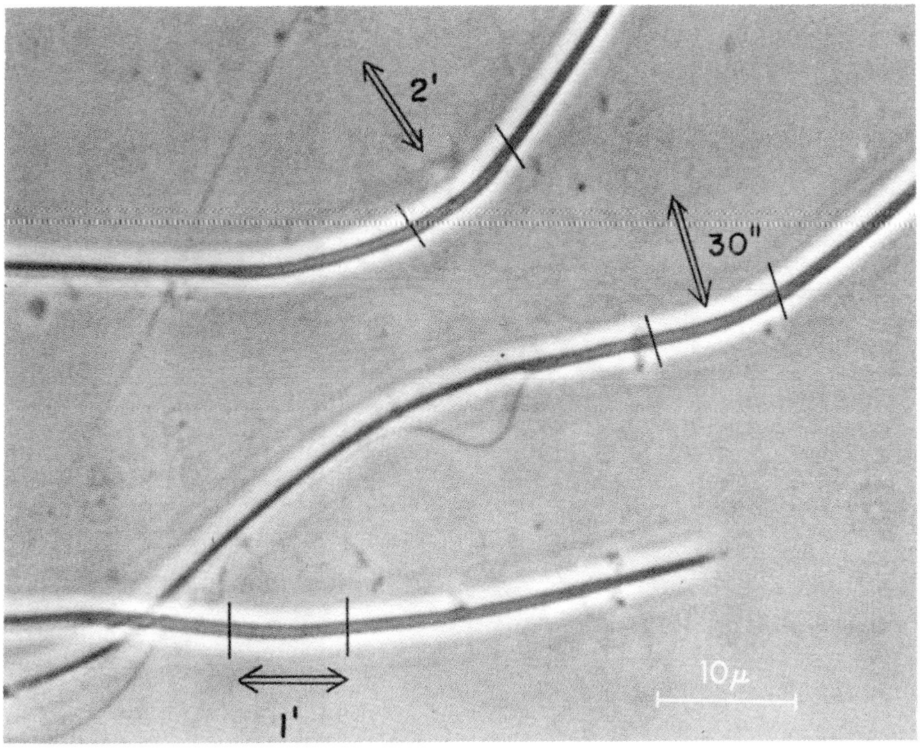

Fig. 6. Photograph through phase contrast microscope of sperm heads previously irradiated by polarized-UV microbeam. No change in contrast or structure is detectable. Exposure times indicated. Double arrows here and in other figures show direction of E-vector of polarized-UV ray.

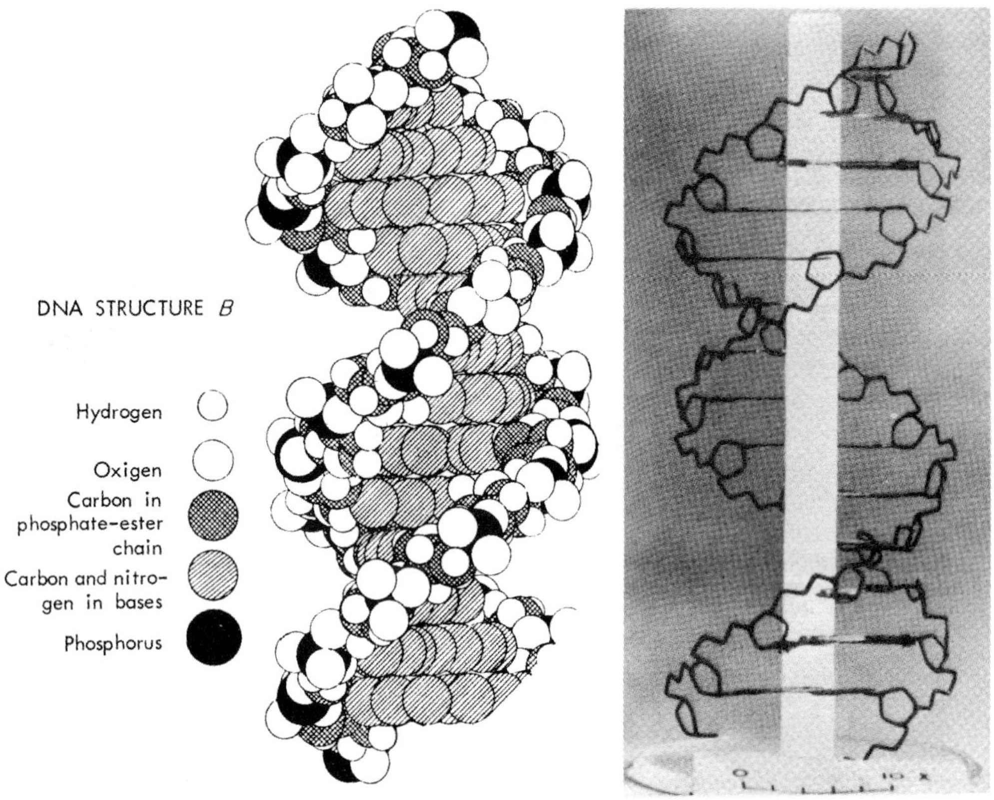

Fig. 7. Models of molecular structure of DNA in the B form. (Left, from Feughelman *et al.*, 1955, *Nature* 175: 834–838; right, from Crick and Watson, 1954, *Proc. Roy. Soc. (London) Ser. B* 223: 80–96.)

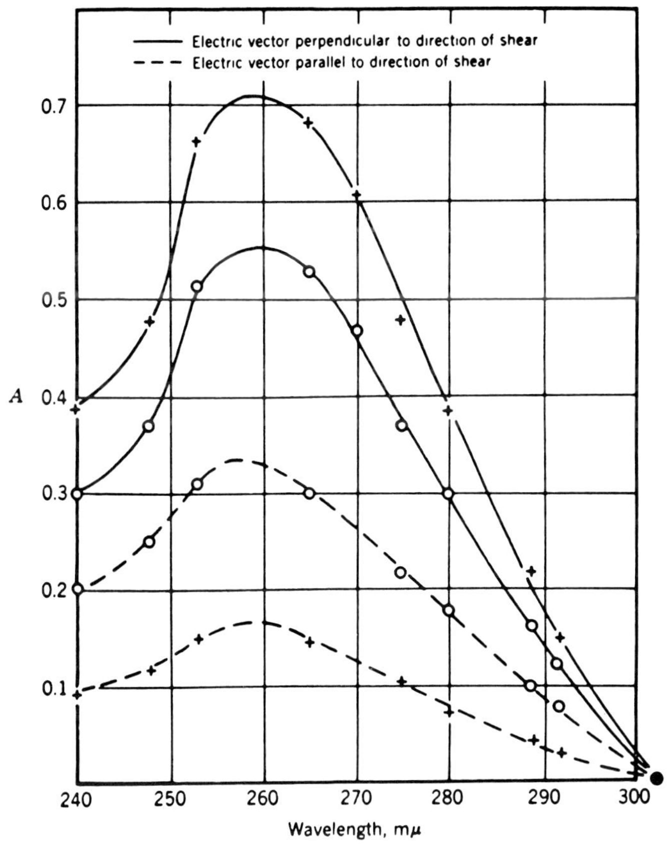

Fig. 8. UV-absorption spectra of oriented films of thymus sodium DNA. +: at 90% relative humidity; ○: at 60% relative humidity. (From Seeds, 1953, *Progr. Biophys. Biophys. Chem.* 3: 27–46.)

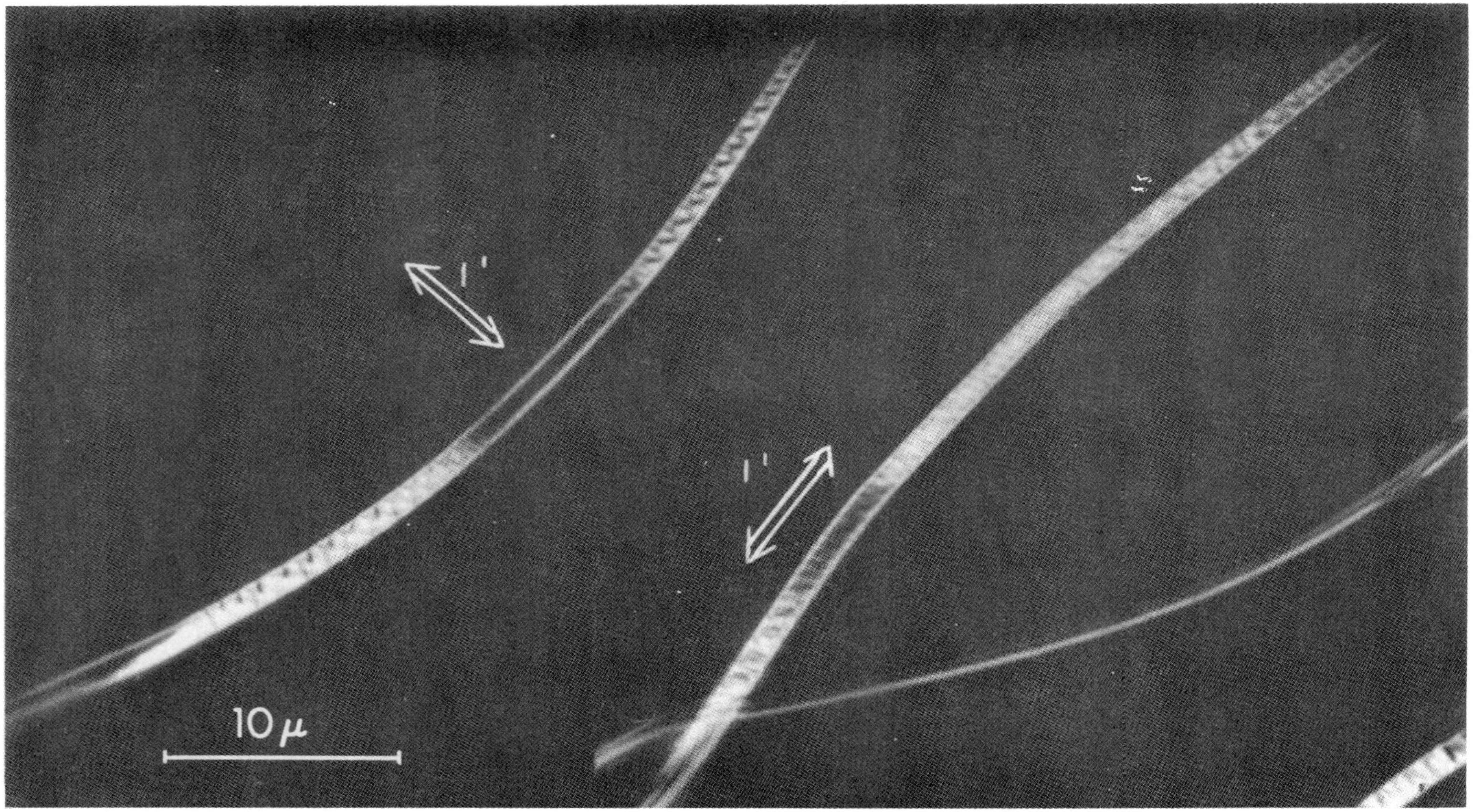

Fig. 9. Photograph of sperm, in visible light through rectified polarizing microscope, after irradiation with polarized UV. Irradiation for 1 min with E-vector perpendicular to axis of sperm virtually eliminated birefringence; with E-vector parallel, appreciable birefringence remained. To achieve equivalent birefringence reduction with E-vector parallel, exposure three or four times longer was required.

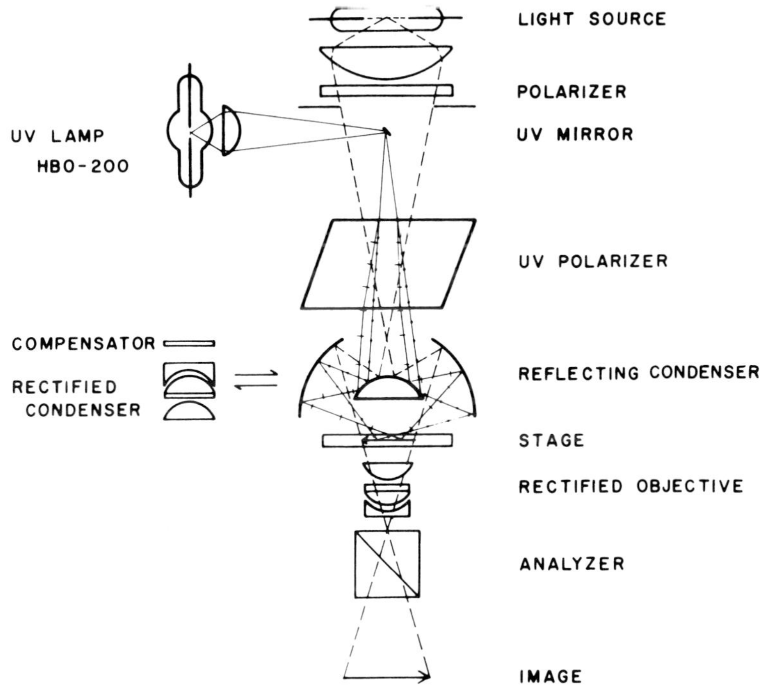

Fig. 10. Schematic diagram of polarized-UV microbeam apparatus.

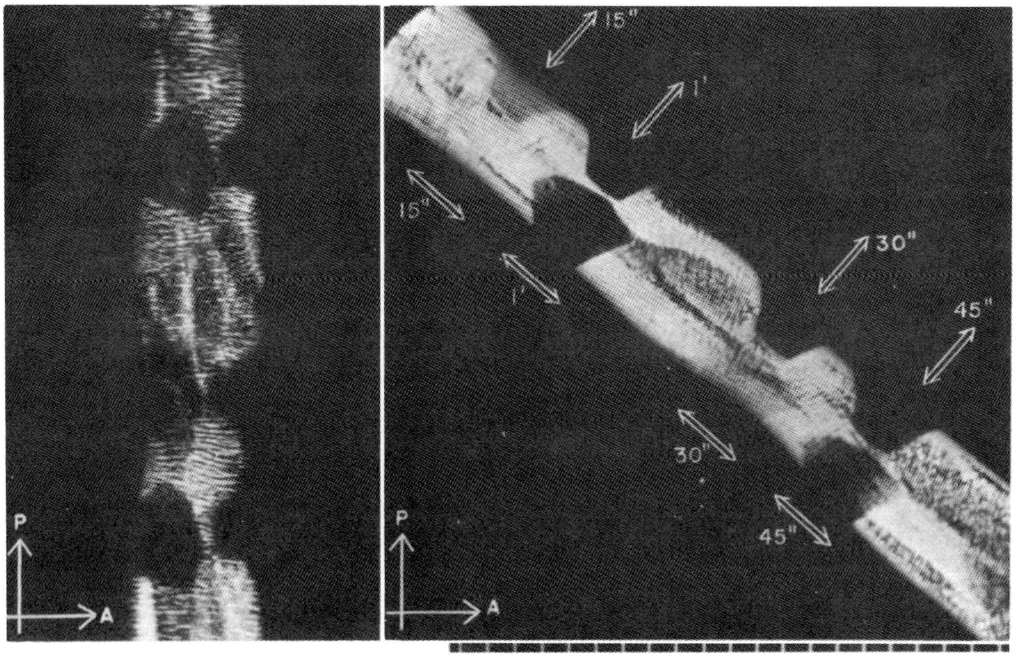

Fig. 11. Stretched gel strand, 50 μ wide, of sodium DNA at 92% relative humidity in two orientations as seen between crossed polars. Large dark areas or "shadows" are irradiated areas that have lost their birefringence to varying degrees. These paired shadows are the areas that were covered by the upper and lower images of the irradiating UV source, split and polarized perpendicular to each other by the calcite crystal (see Fig. 10). A 15-sec irradiation with UV E-vector polarized perpendicular to the fiber axis reduces DNA birefringence to the same degree as does a 35- or 40-sec irradiation with the UV E-vector parallel. Crossed arrows (P, A): transmission directions of polarizer and analyzer, respectively. Scale interval: 10 μ

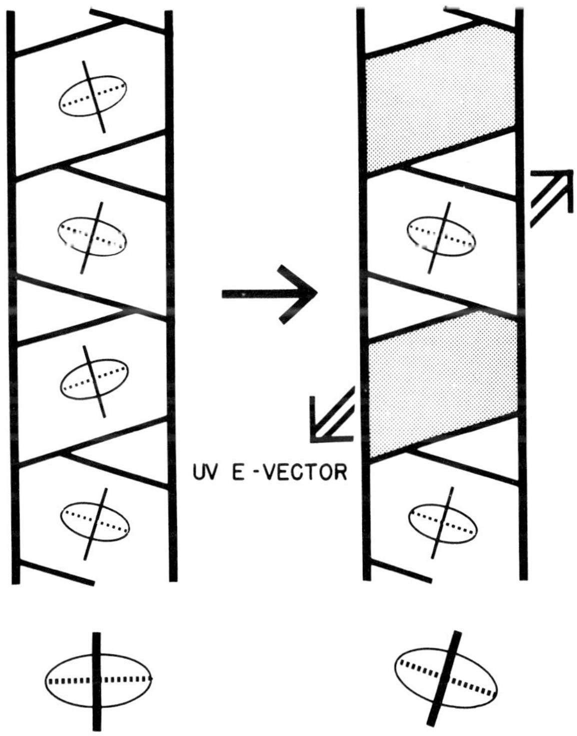

Fig. 12. Schematic diagram of rotation of average axis of birefringence toward E-vector after irradiation of sperm by UV polarized at 45° to sperm axis.

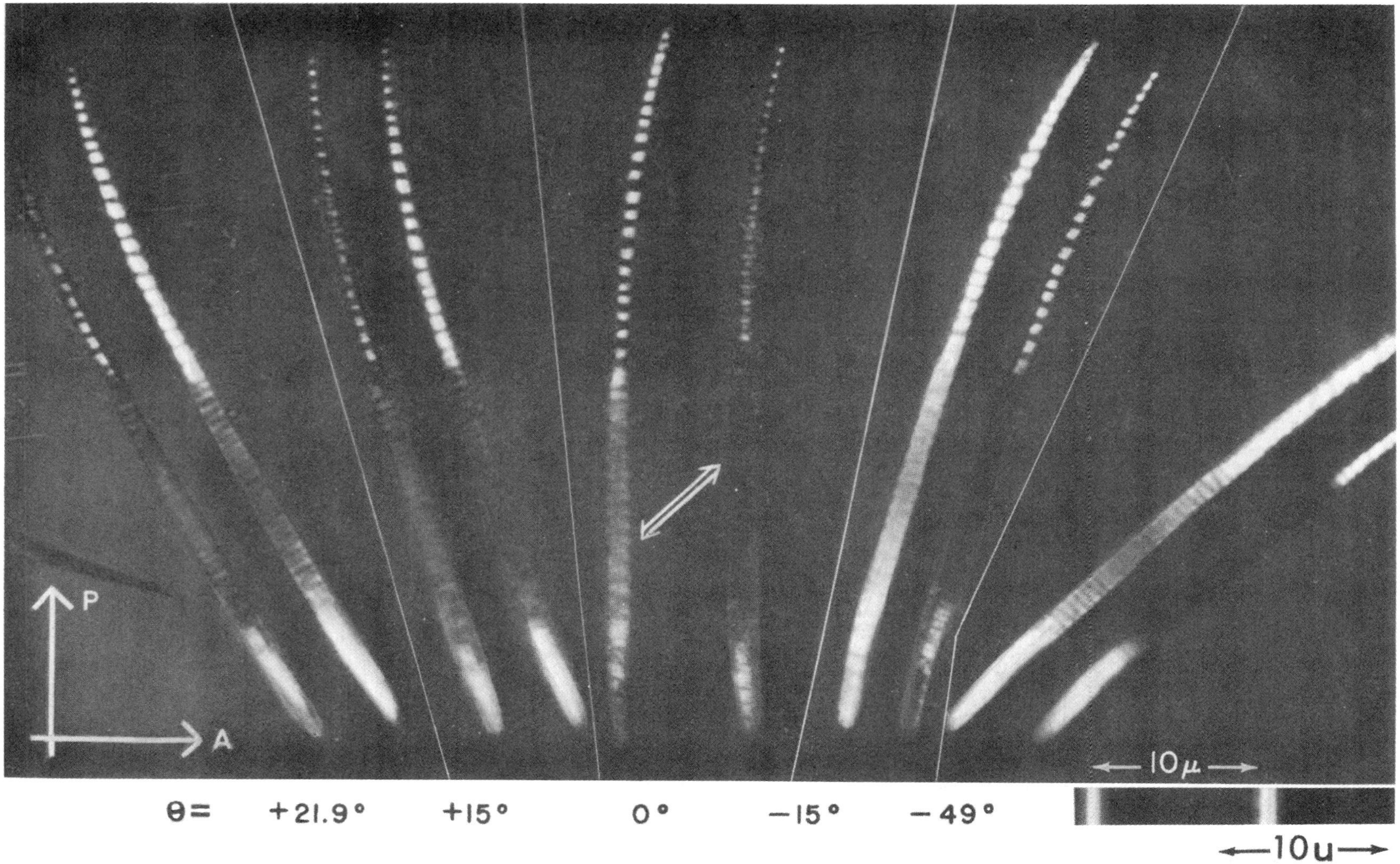

Fig. 13. Paired photographs of a single sperm head taken between crossed polars at various stage angles. In each pair, compensator slow axis was turned counter-clockwise from polarizer axis in the left image, and clockwise in the right. Anterior (narrower) portion of head shows normal microdomains. Posterior was irradiated by UV polarized at 45° to sperm axis as indicated in the middle pair. See text for further explanation. Sperm head immersed in quinoline-dioxane mixture with matching refractive index to reduce form birefringence and light scattering.

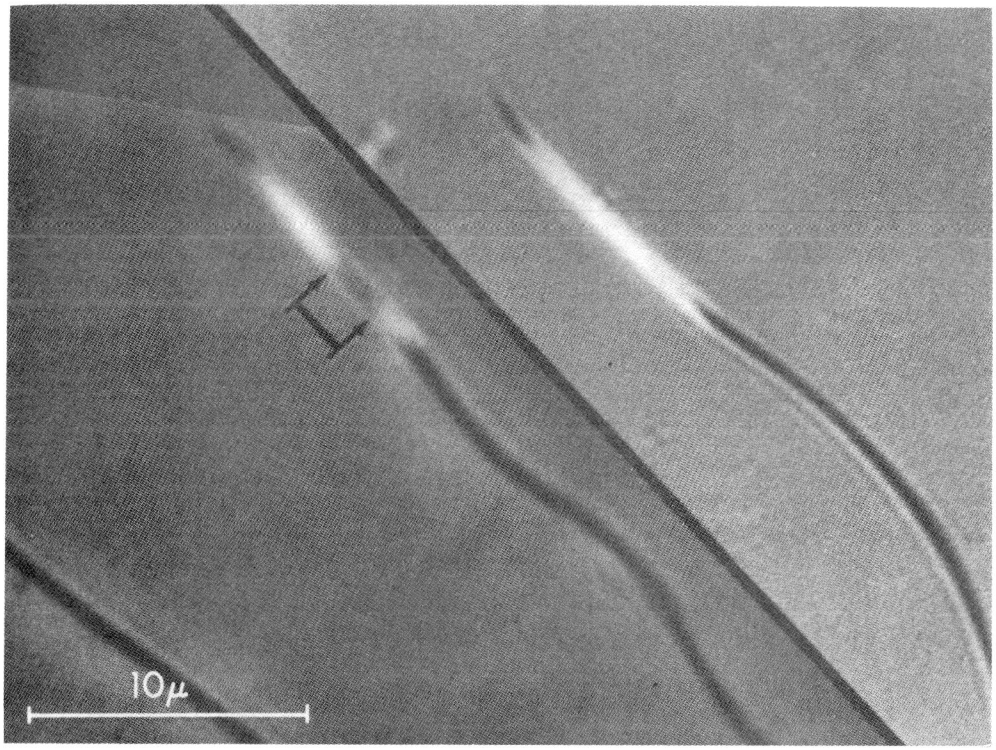

Fig. 14. Fruit-fly (*D. busckii*) sperm with heads oriented in white compensation, tails and acrosomes in black compensation. Negative birefringence of head is reduced after irradiation with UV (between arrows) (cf Figs. 4, 5). Although no structures are resolvable with highest resolution of rectified polarizing microscope, their presence is deduced from behavior of area irradiated with UV polarized at 45° to sperm axis. See text.

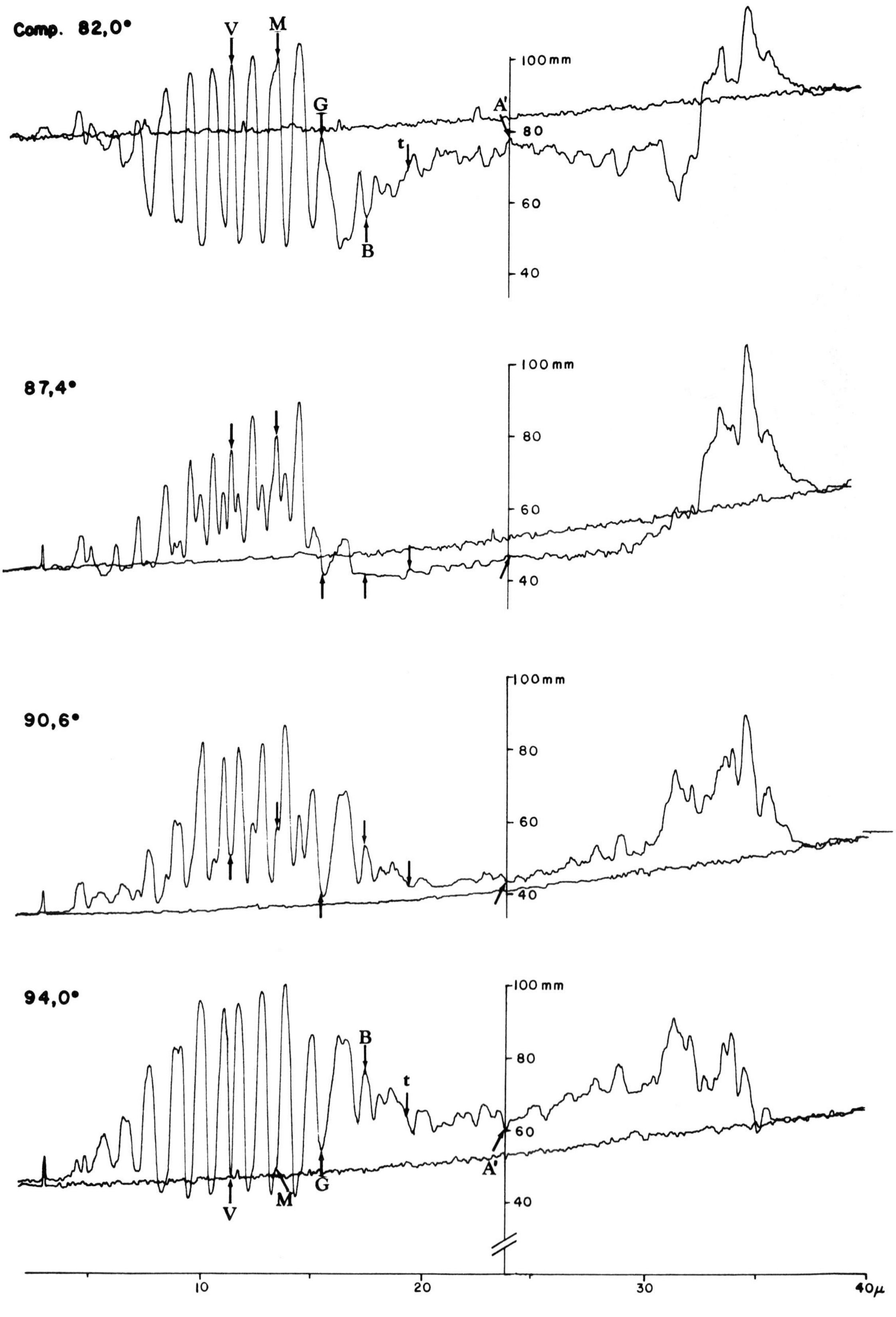

Fig. 17. Microdensitometer traces at various compensator settings, sperm axis oriented parallel to polarizer axis.

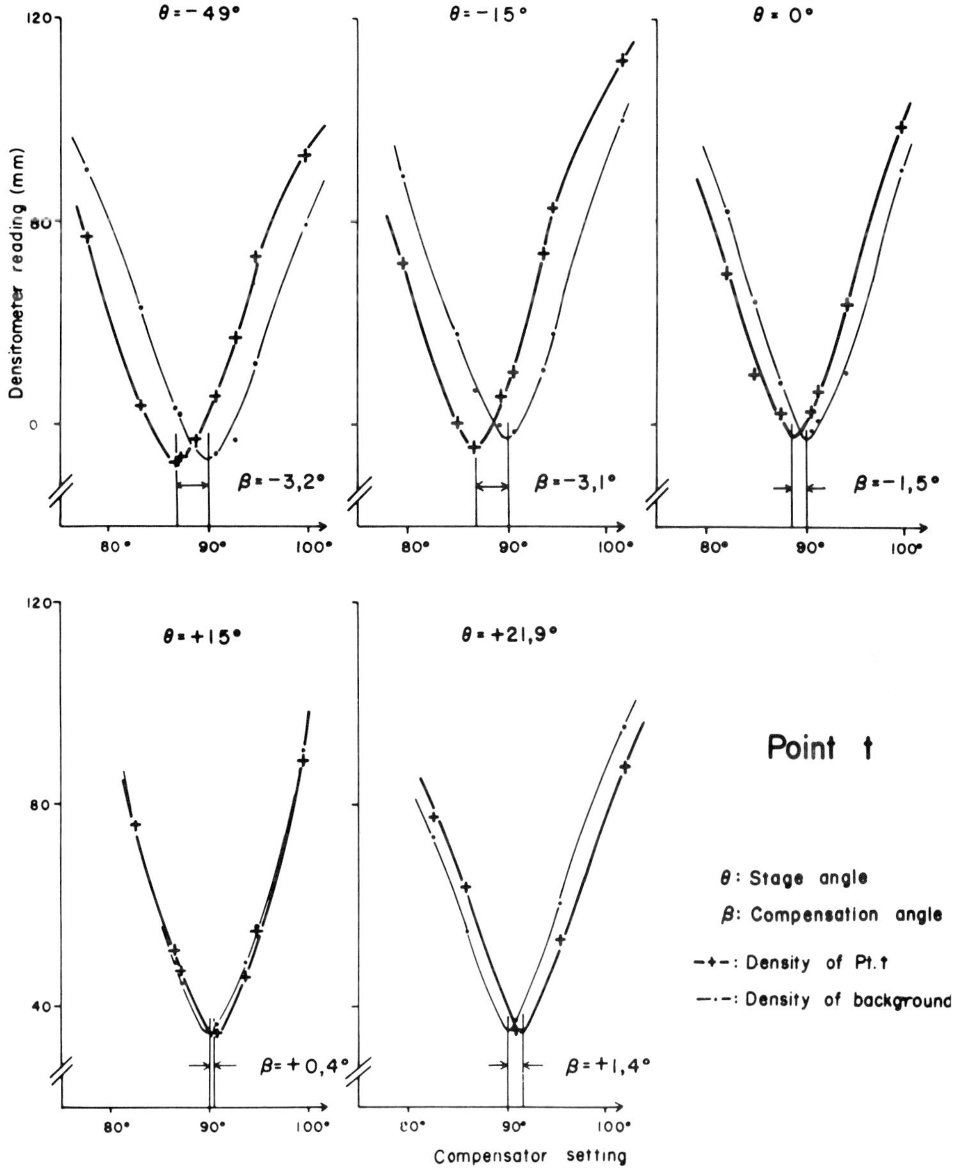

Fig. 18. Determination of compensation angles for sperm axis oriented at various angles to polarizer axis. Point *t* chosen as an example.

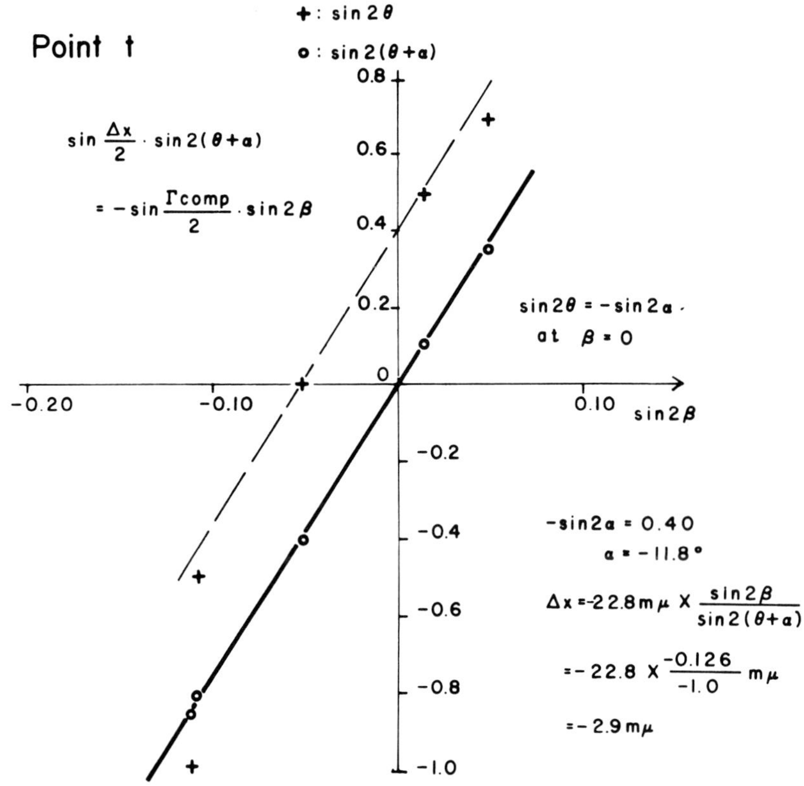

Fig. 19. Graphic computation of retardation and azimuth angle of point *t*. See text.

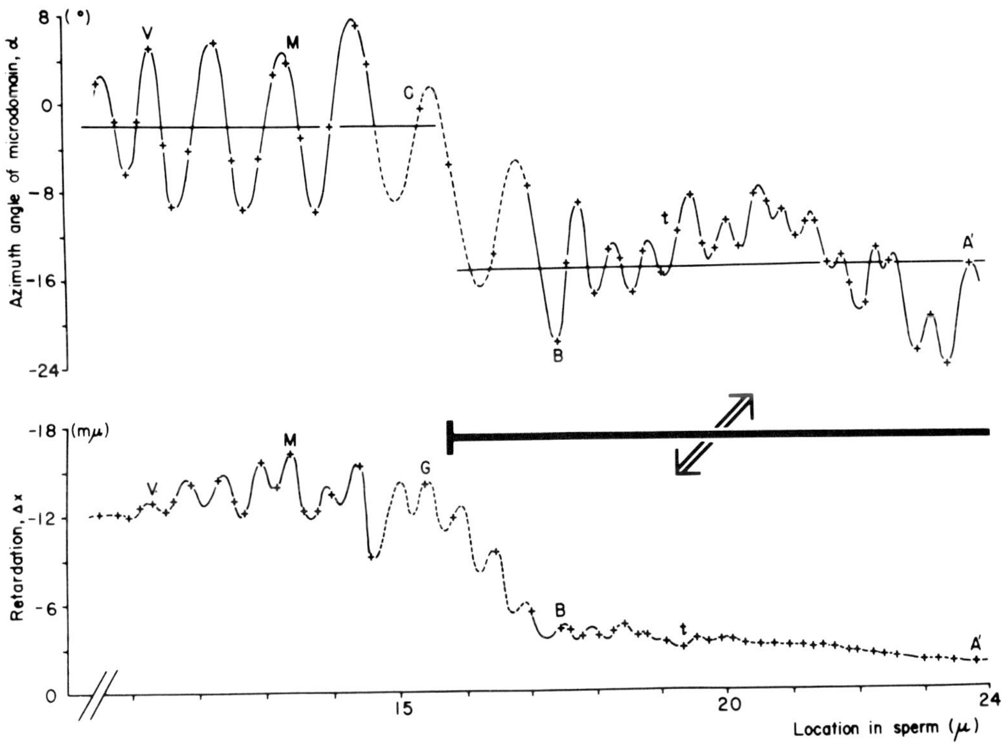

Fig. 20. Distribution of azimuth angles and retardations in microdomains measured along a 14-μ length of sperm head. Dark bar: region irradiated by UV polarized at 45°. Frequency of fluctuation of retardation is twice the frequency of oscillation of angles. (From Inoué and Sato, 1962, *Science* 136: 1122–1124; copyright 1962 by the American Association for the Advancement of Science.)

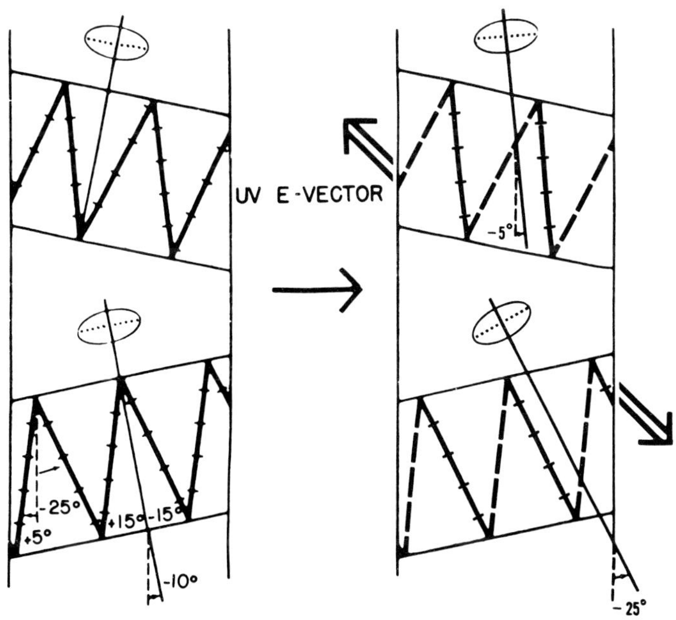

Fig. 21. Schematic figure explaining effect of polarized UV with E-vector vibrating 45° to sperm axis. See text.

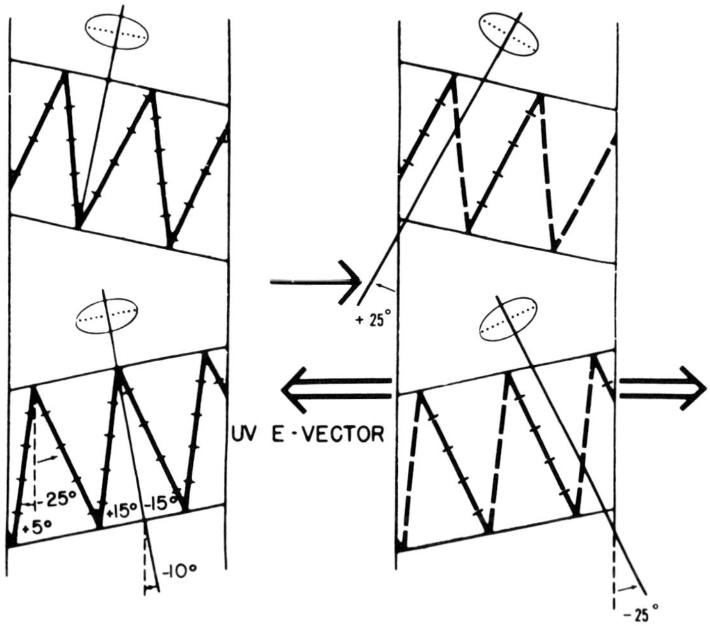

Fig. 22. Schematic figure predicting alteration of azimuth angles of microdomains following irradiation by UV polarized at 90° to sperm axis. Azimuth angles of microdomains should become ±25°.

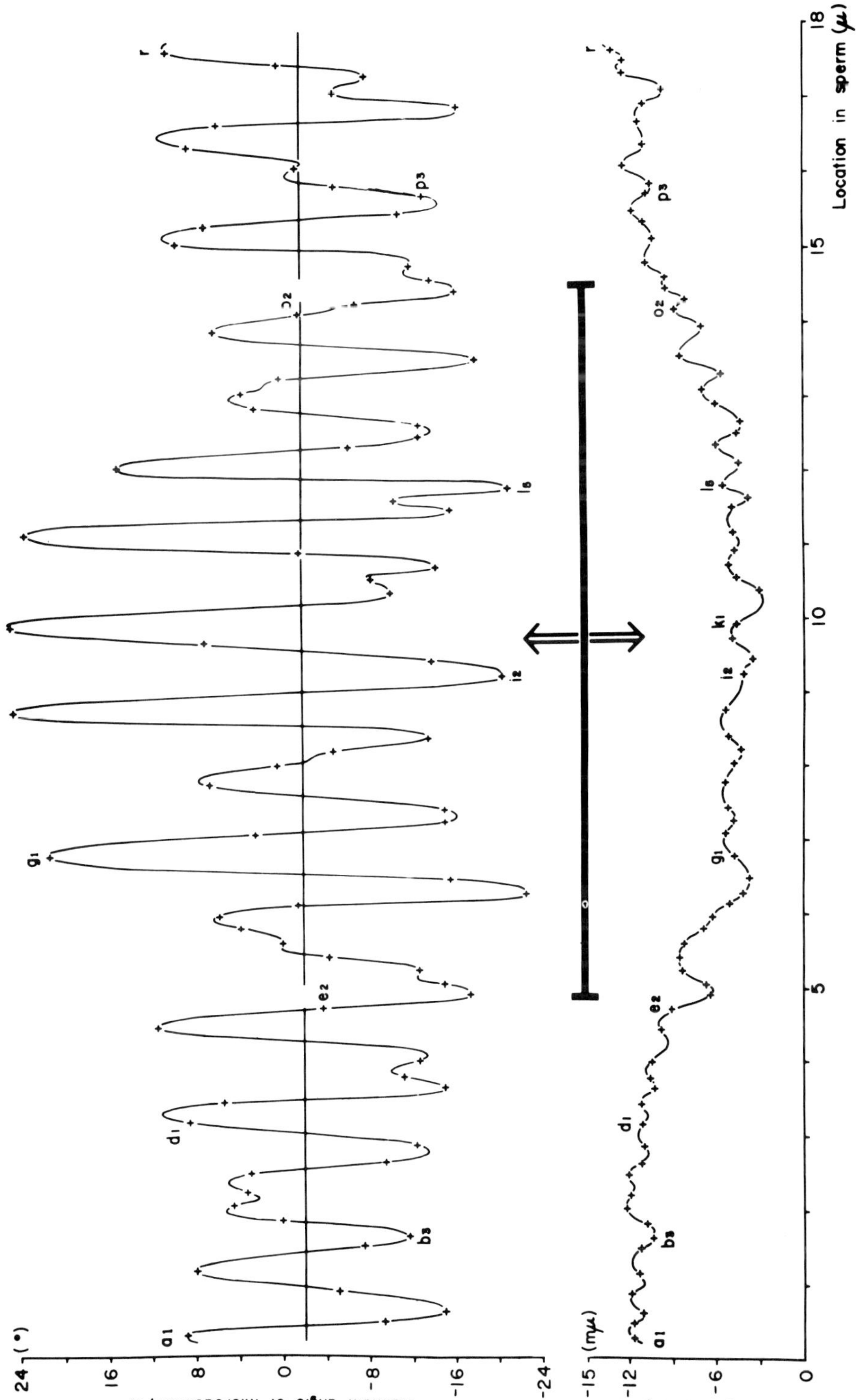

Fig. 23. Distribution of azimuth angles and retardations in microdomains (some labeled) measured along an 18-μ length of living sperm head. Dark bar: region irradiated by UV polarized with E-vector vibrating at 90° to sperm axis.

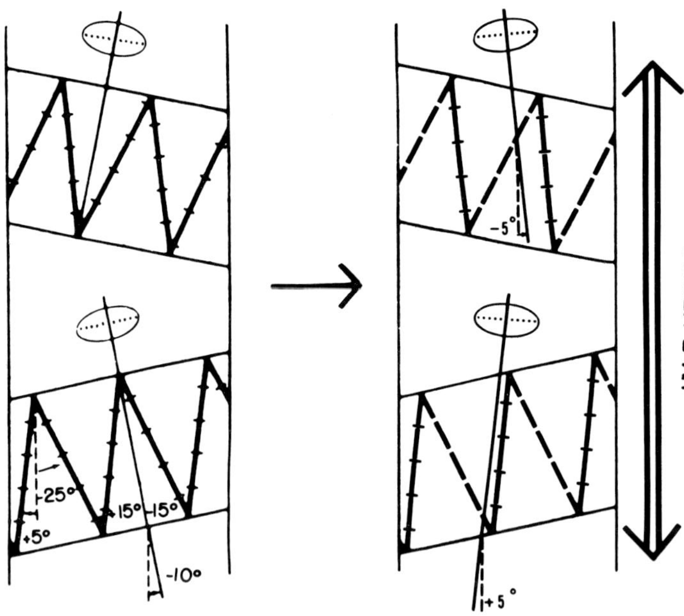

Fig. 24. Schematic figure predicting effect of UV with E-vector vibrating parallel to sperm axis; azimuth oscillations or the average axis of microdomains should become $\mp$ 5°.

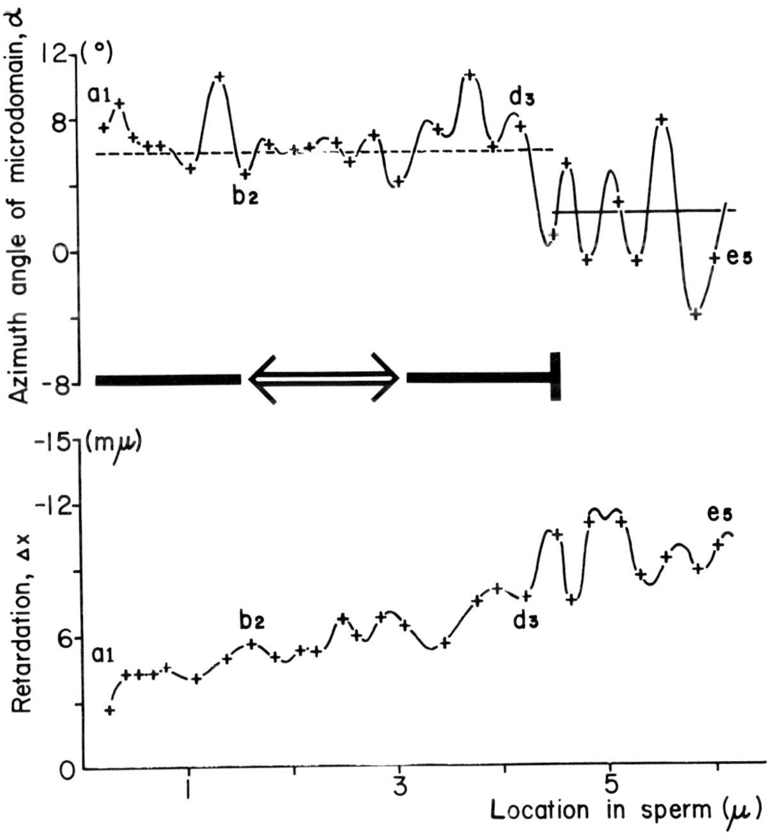

Fig. 25. Measured distribution of azimuth angles and retardations of microdomains irradiated with UV polarized parallel to sperm axis. Oscillation of microdomain azimuth angles in irradiated region (dark bar) considerably reduced as compared to unirradiated region (cf Figs. 20, 23).

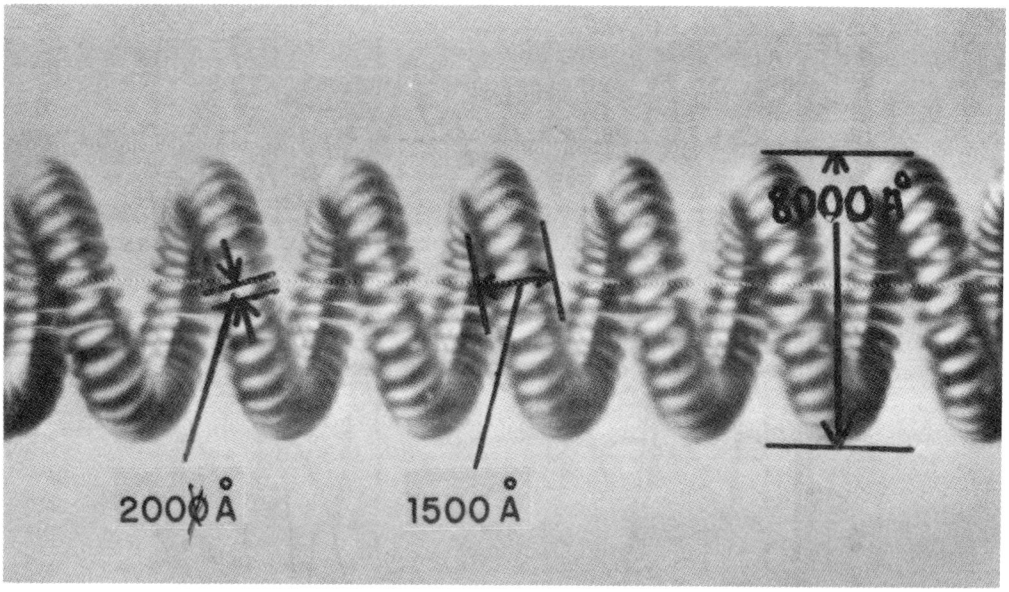

Fig. 26. Preliminary model of DNA arrangement in sperm head. Considering the great length of DNA molecules and examples of cytologically demonstrated helical structures of chromosomes, one can reasonably expect that the DNA molecules in the sperm nucleus are wound into a coiled coil. Our data support such a model and provide the lead angle of the different orders of helices: the B-form DNA, presumably in parallel bundles of Watson–Crick type helices *ca.* 200 Å thick, wind into a coil *ca.* 1,500 Å in diameter with a lead angle of 15°; this coil, in turn, is wound into a larger coil *ca.* 8,000 Å in diameter with a lead angle of 10°.

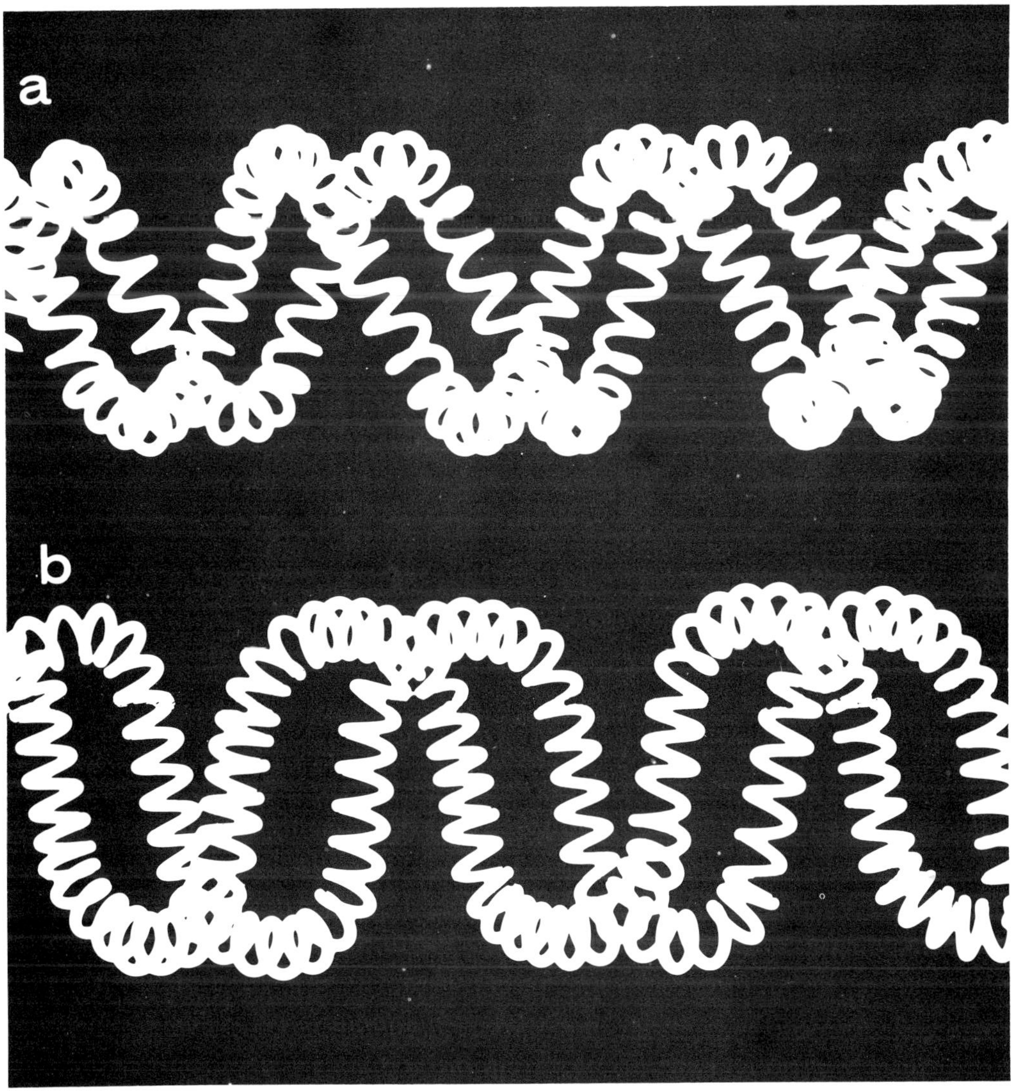

Fig. 27. Two models, constructed of copper wire, of DNA arrangement in the sperm head. Model *a* has a circular cross-section. Model *b* is formed by compressing model *a* so that the cross-section becomes elliptical. Only model *b* satisfies all our measurements. If, as in model *b*, DNA were arranged in a coiled coil with the major gyres of the intertwined chromonemata alternately closer together to one side than to the other, the DNA would fill the sperm head and the microdomains would represent the central portions of the barely resolved chromonemal coil. The double-peaked distribution of retardations (Figs. 20, 23, 25) reflects the two major gyres of the intertwined chromonemata.

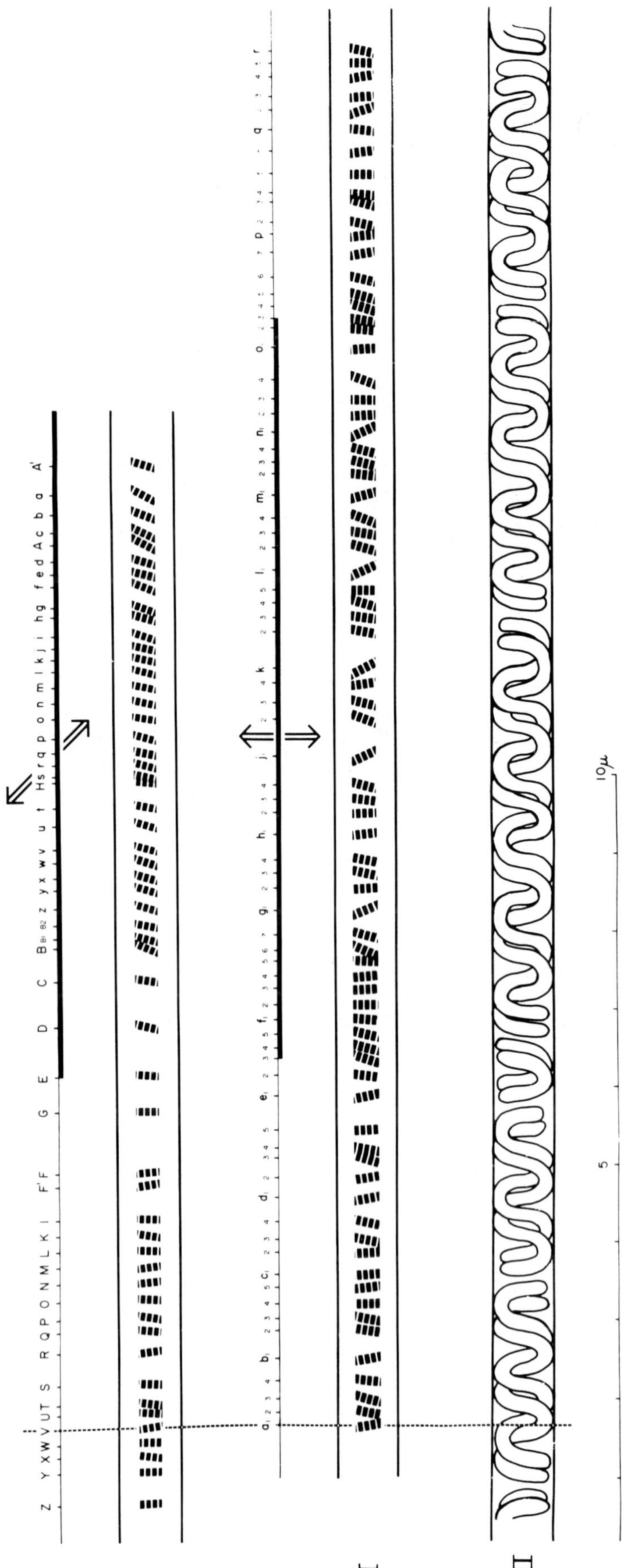

Fig. 28. Resynthesized image (*I*) of chromosomes with double chromonemata in the sperm head. Gyres and breaks were located from the measured distributions of azimuth angles and retardations (*II*, *III*; and Figs. 20, 23). The two other lines are indices to microdomains.

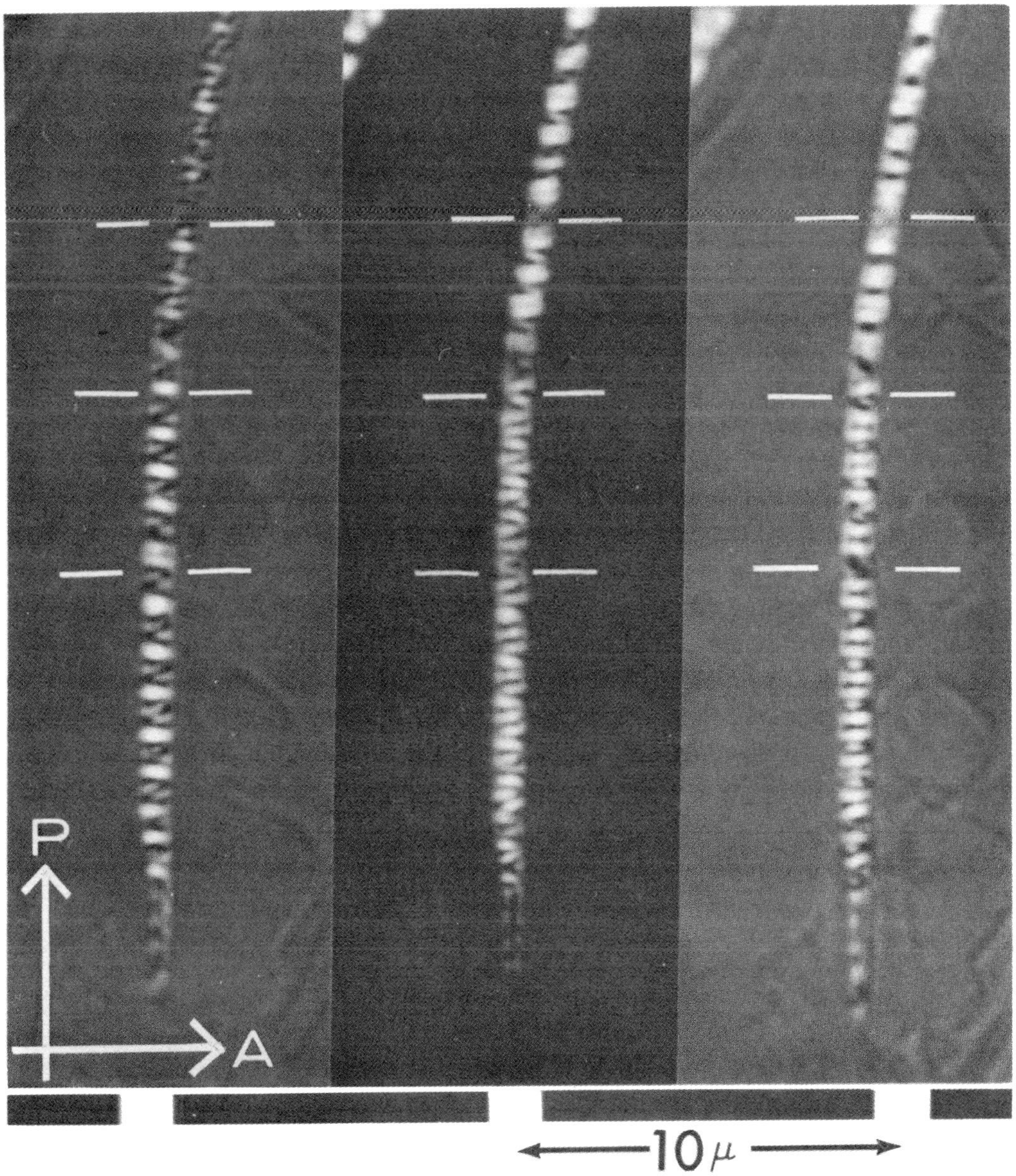

Fig. 29. Sperm head observed with the rectified polarizing microscope at three different settings of mica compensator. Detailed distribution of birefringence in these chromosomes is shown with great clarity by immersion in dimethyl sulfoxide (N_D^{20} 1.475). White bars: positions of chromosomal "breaks"; probably correspond to the end of each chromosome.

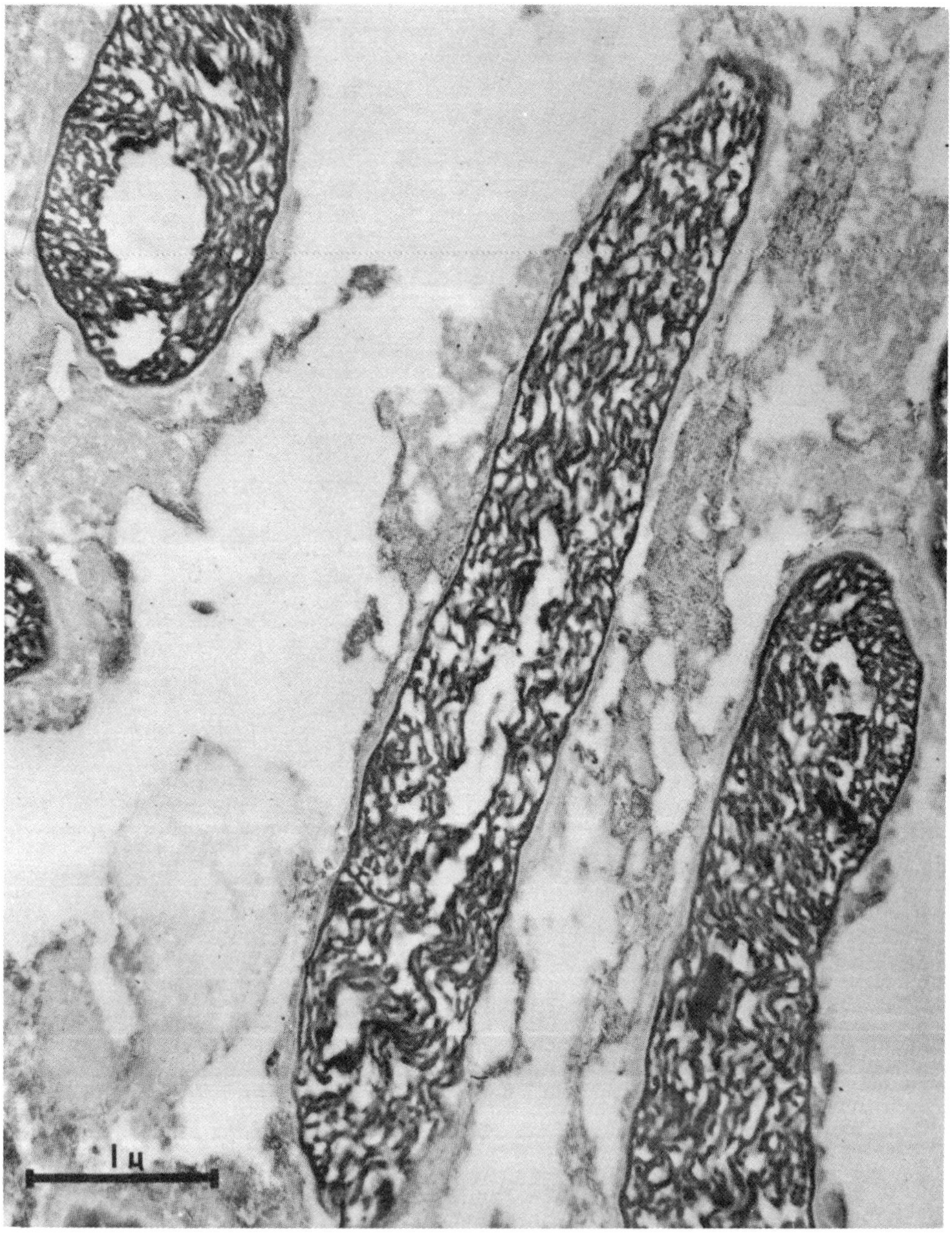

Fig. 30. Electron micrograph of sections of mature sperm head. We believe the bundle of wavy filaments represents the substructure of the wound sperm chromosomes, and the clear area in the central region of the sperm in cross-section is probably the center of the helix. Specimen was gradually dehydrated with a series of graded mixtures of dimethylsulfoxide and Belar solution for Orthoptera cells at 4°C, then imbibed in methacrylate, sliced, and stained with uranyl acetate. No fixative was employed during the procedure.

Reprinted from *The Journal of General Physiology*, Vol. 50(6), pp. 259–292, 1967.

Cell Motility by Labile Association of Molecules

The nature of mitotic spindle fibers and their role in chromosome movement

SHINYA INOUÉ and HIDEMI SATO

From the Laboratories of Biophysical Cytology, Department of Biology, University of Pennsylvania, Philadelphia

ABSTRACT This article summarizes our current views on the dynamic structure of the mitotic spindle and its relation to mitotic chromosome movements. The following statements are based on measurements of birefringence of spindle fibers in living cells, normally developing or experimentally modified by various physical and chemical agents, including high and low temperatures, antimitotic drugs, heavy water, and ultraviolet microbeam irradiation. Data were also obtained concomitantly with electron microscopy employing a new fixative and through measurements of isolated spindle protein. Spindle fibers in living cells are labile dynamic structures whose constituent filaments (microtubules) undergo cyclic breakdown and reformation. The dynamic state is maintained by an equilibrium between a pool of protein molecules and their linearly aggregated polymers, which constitute the microtubules or filaments. In living cells under physiological conditions, the association of the molecules into polymers is very weak (absolute value of $\Delta F_{25°C} < 1$ kcal), and the equilibrium is readily shifted to dissociation by low temperature or by high hydrostatic pressure. The equilibrium is shifted toward formation of polymer by increase in temperature (with a large increase in entropy: $\Delta S_{25°C} \simeq 100$ eu) or by the addition of heavy water. The spindle proteins tend to polymerize with orienting centers as their geometrical foci. The centrioles, kinetochores, and cell plate act as orienting centers successively during mitosis. Filaments are more concentrated adjacent to an orienting center and yield higher birefringence. Astral rays, continuous fibers, chromosomal fibers, and phragmoplast fibers are thus formed by successive reorganization of the same protein molecules. During late prophase and metaphase, polymerization takes place predominantly at the kinetochores; in metaphase and anaphase, depolymerization is prevalent near the spindle poles. When the concentration of spindle protein is high, fusiform bundles of polymer are precipitated out even in the absence of obvious orienting centers. The shift of equilibrium from free protein molecules to polymer increases the length and number of the spindle microtubules or filaments. Slow depolymerization of the polymers, which can be brought about by low concentrations of colchicine or by gradual cooling, allows the filaments to shorten and perform work. The dynamic equilibrium controlled by orienting centers and other factors provides a plasusible mechanism by which chromosomes and other organelles, as well as the cell surface, are deformed or moved by temporarily organized arrays of microtubules or filaments.

In this paper, we shall attempt to relate the mitotic movement of chromosomes to the structure, physiology, and function of the mitotic spindle.

At each division of a eukaryotic cell a spindle is formed, and the chromosomes are oriented, aligned, and then separated regularly into two daughter cells. The spindle is disassembled when the task is done.

In a living cell, the fibrous elements of the mitotic spindle and its astral rays are weakly birefringent. The appearance and growth, and contraction and disappearance, of the fibers can therefore be observed in living cells with a sensitive polarizing microscope. The change in birefringence can also be measured with the polarizing microscope. Correlations can then be established between spindle fiber birefringence and alteration of fiber fine structure in cells altered physiologically or experimentally.

Many conditions and agents have been found in which spindle morphology and fiber birefringence are systematically and reversibly altered. From the various observations described or reviewed in this paper, we conclude that spindle fibers are composed of parallel arrays of thin filaments which are formed by a reversible association of globular protein molecules. The molecules in functioning spindle fibers are not stably aligned and cross-linked as in more stable fibers. Instead, spindle fibers are labile structures existing in a dynamic equilibrium with a large pool of unassociated molecules. The association of the molecules and formation of filaments are controlled by the activities of orienting centers and concentrations of active pool material.

We hypothesize that the contraction and elongation of the spindle fibers are responsible for regular mitotic movement of chromosomes. The spindle fibers, however, do not contract and elongate by folding and unfolding of polypeptide chains in the protein molecules. For example, the fiber birefringence remains unchanged during anaphase movement. Instead, the fibers are believed to elongate by increased addition and alignment of the molecules, which contribute to a pushing action. Conversely, they shorten and pull chromosomes as molecules are slowly removed from the filaments.

We view this hypothesis as providing a plausible mechanism, not only for mitotic movement of chromosomes, but also for movements of various other organelles and mechanical modulations of the cell surface.

In the following, we shall:

1. Review the visible changes of spindle fiber birefringence during normal mitosis
2. Discuss the mechanical integrity of the spindle fibers
3. Describe the dynamic nature of the spindle fibers
4. Discuss the dynamic equilibrium model
5. Describe the effect of heavy water on spindle fiber birefringence, protein content, and electron microscopic appearance
6. Provide evidence that the bulk of the spindle fiber protein comes from a preformed pool
7. Describe factors which control the formation and orientation of the fibers
8. Discuss the extension and contraction of spindle fibers
9. Discuss the nature of the spindle protein molecules
10. Relate spindle fiber birefringence to microtubules and filaments observed in fixed cells with the electron microscope
11. Discuss cell movements which appear to be brought about by labile association of molecules

1. VISIBLE CHANGES OF SPINDLE FIBER BIREFRINGENCE DURING MITOSIS

Spindle fibers and astral rays generally possess a positive birefringence whose magnitude is of the order of 1 mμ, or $2/1000$ of a wavelength of green light. With a sensitive polarizing microscope, they can be clearly visualized (Fig. 1) and their fine structural changes can be followed by interpreting the measured change in birefringence[1] (Bajer, 1961; Dietz, 1963; Forer, 1965, 1966; Inoué, 1953, 1964; Inoué and Bajer, 1961; Schmidt, 1937, 1939, 1941; Taylor, 1959; for a thorough discussion and general bibliography of mitosis, and for definition of terms, also see Schrader, 1953, and Mazia, 1961).

The change in spindle fiber birefringence during mitosis in a wide variety of animal and plant cells has been documented photographically (Inoué, 1964; also 1953). The same paper (Inoué, 1964) also describes a sequence of time-lapse motion pictures depicting the birefringence changes during mitosis and cytokinesis in several types of cells.

The major patterns of birefringence changes can be summarized as follows. In prophase before nuclear membrane breakdown, birefringent fibers are formed as astral rays in animal cells (Fig. 2 *A*) (Inoué and Dan, 1951; Forer, 1965; also see Cleveland, 1938, 1953, 1963), or as clear zone fibers in some plant cells (Fig. 3 *A*) (Inoué and Bajer, 1961; also see Bajer, 1957). In newt fibroblasts, linear growth of the birefringent central spindle has been measured

[1] In this paper, the term "birefringence" is used synonymously with retardation [i.e. $d(n_1 - n_2)$] and should not be interpreted to mean coefficient of birefringence $(n_1 - n_2)$, or retardation per unit specimen thickness.

by Taylor (1959). In the central spindle formed between the separating asters, and in the clear zone material, the birefringence fiber axis (slow axis) corresponds to the direction in which more continuous fibers, sheath fibers, and chromosomal fibers appear. The central (continuous) spindle fibers and clear zone fibers thus establish a fine structural foundation for subsequent orientation of other fibers (also see Wada, 1965; however, see Inoué and Dan, 1951, for anomalous situations regarding axis of streak material vs. aster and spindle, and for changes to negative birefringence under presumed compression).

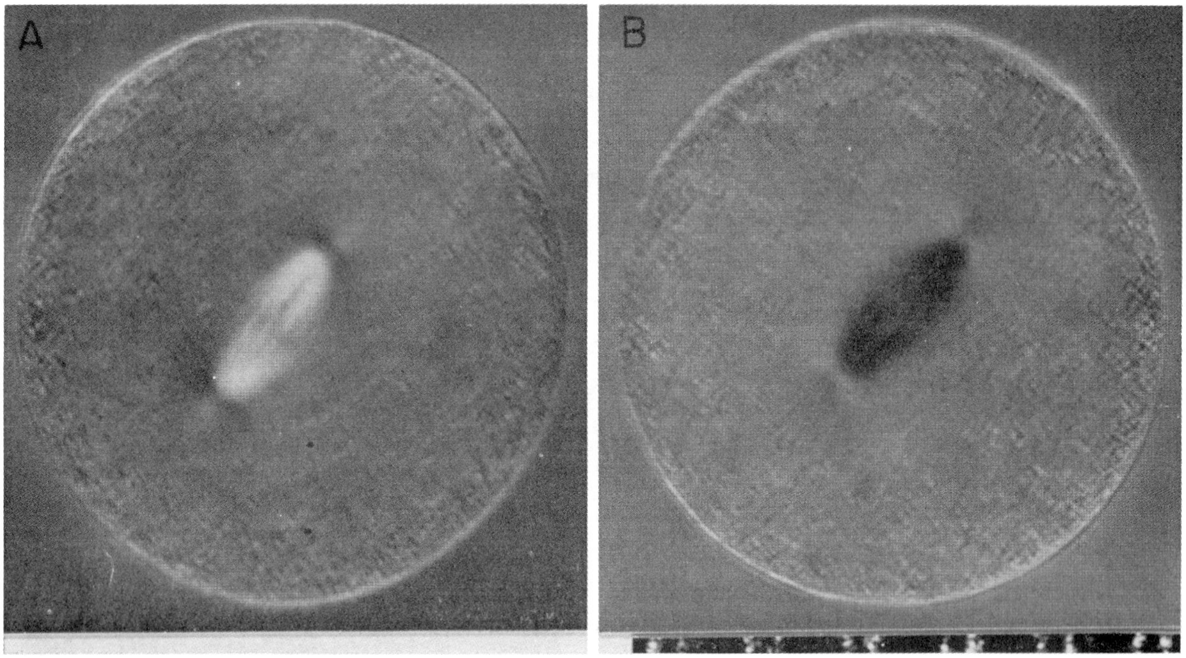

FIGURE 1. Mitotic spindle in arrested metaphase oocyte of *Pectinaria gouldi*. The over-all birefringence[1] of the spindle has been enhanced with 40% heavy water (D_2O)–sea water to approximately 3 mμ. Polarization microscopy. *A*, chromosomal fibers appear bright; *B*, compensator axis reversed, chromosomal fibers are dark. Scale interval, 10 μ.

In prometaphase following nuclear membrane breakdown (Fig. 3 *B*), more spindle material becomes oriented around the chromosomes to form sheath fibers in a crane fly, *Pales crocata* (Dietz, 1963; Went, 1966; also see Dietz, 1959, for special attributes of crane fly spindle poles), and in pollen mother cells of the Easter lily (Inoué, 1953, 1964). The birefringent sheath fibers represent a distinct accumulation of oriented material. In many cases, sheath fibers are not observed, and chromosomal fibers arise parallel to continuous fibers directly as in Fig. 3 *C*. In either case, the continuous or sheath fiber material appears to transform into chromosomal fibers. The birefringence of the former decreases as the birefringence of the latter increases (Fig. 3 *B* and *C*).

The birefringent chromosomal fibers are established parallel to the continuous fibers or sheath fibers of prometaphase. Their orientation in turn forecasts the direction of movement of the chromosomes (Figs. 2 *B–E* and 3 *B–E*). The chromosomal fibers are attached to the kinetochore of the chromosomes. In plant cells devoid of centrioles, the fiber birefringence is strongest adjacent to the kinetochore and weaker toward the poles (Fig. 3 *B–D*). Chromosomal spindle fibers with strong birefringence from kinetochore to poles are found in animal cells with active centrioles (Fig. 2 *B–D*).

In prometaphase, chromosomal spindle fibers also appear to participate in the orientation of the chromosomes and their alignment on the metaphase plate. (See Östergren, 1951, for a proposed mechanism of metaphase equilibrium and orientation.)

During prometaphase, fluctuation of birefringence is observed in the chromosomal spindle fibers of a grasshopper spermatocyte, *Dissosteira carolina*. In a time-lapse motion picture, the fluctuation of birefringence resembles that of northern lights (Inoué, 1964). It is interpreted to reflect the fluctuation in the amount of molecules oriented in the chromosomal spindle fibers at this stage. Such changes might contribute to the achievement of exact alignment of chromosomes on to the metaphase plate as their equilibrium position (Östergren, 1951).

In anaphase, the chromosomes are led by the strongly birefringent chromosomal spindle fibers to the poles of the spindle. The birefringence adjacent to the kinetochore and in much of the chromosomal spindle fiber remains unchanged for the major part of anaphase movement (Figs. 2 *C–D* and 3 *C–E*). The chromosomal spindle fiber may or may not shorten during most of the anaphase, depending on the exact mode of chromosome separation relative to spindle elongation.

Continuous fibers whose birefringence is extremely low in anaphase regain birefringence during late anaphase and telophase and establish a phragmoplast in the case of a plant cell (Fig. 3 *E*). The phragmoplast fibers guide the accumulation of vesicles onto its midplane (Inoué, 1953; Bajer, 1965). The vesicles fuse and form the primary cell wall (Fig. 3 *F*). (For electron microscopy of this process, see Frey-Wyssling et al., 1964, and Porter and Machado, 1960). In animal cells, continuous spindle fibers also become more strongly birefringent, and presumably more stable in telophase, forming the core of the stem (spindle rest, stem *Körper*) connecting the separating cells (Fig. 2 *F*). In this region electron micrographs show an especially large number of microtubules (e.g. Robinson and Gonatas, 1964).

Thus, what appears to be a transition of birefringent material from one component of the spindle to another is seen throughout division. Much of this is presumably brought about by successive activation and inactivation of orienting centers (Inoué, 1964). At the same time, the orientation of the birefringent fibers of an earlier stage appears to contribute to the orientation

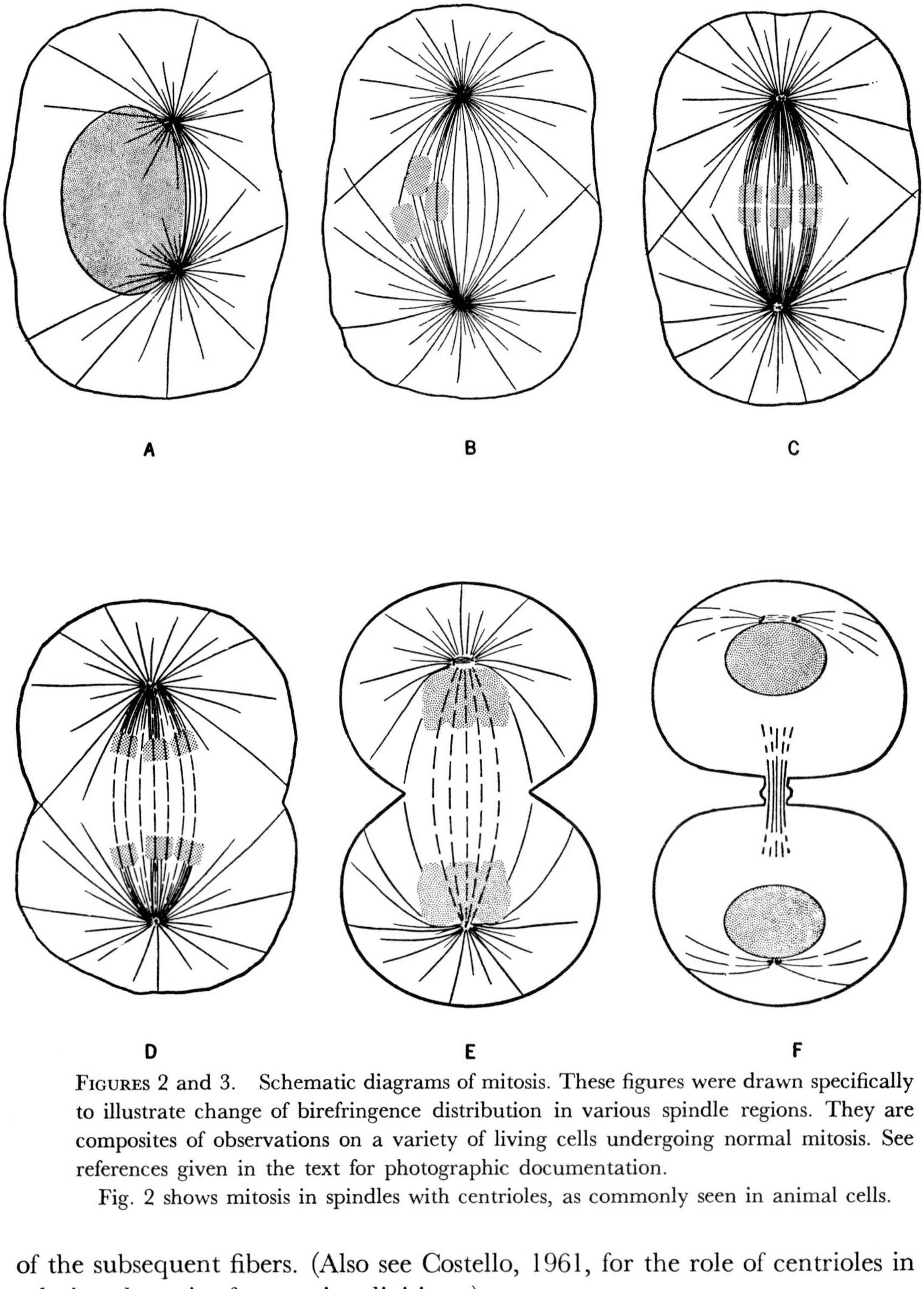

FIGURES 2 and 3. Schematic diagrams of mitosis. These figures were drawn specifically to illustrate change of birefringence distribution in various spindle regions. They are composites of observations on a variety of living cells undergoing normal mitosis. See references given in the text for photographic documentation.

Fig. 2 shows mitosis in spindles with centrioles, as commonly seen in animal cells.

of the subsequent fibers. (Also see Costello, 1961, for the role of centrioles in relating the axis of successive divisions.)

The successive fibers formed appear to co-orient and align chromosomes to the metaphase plate, pull them apart to the poles, and correlate cytokinesis with karyokinesis.

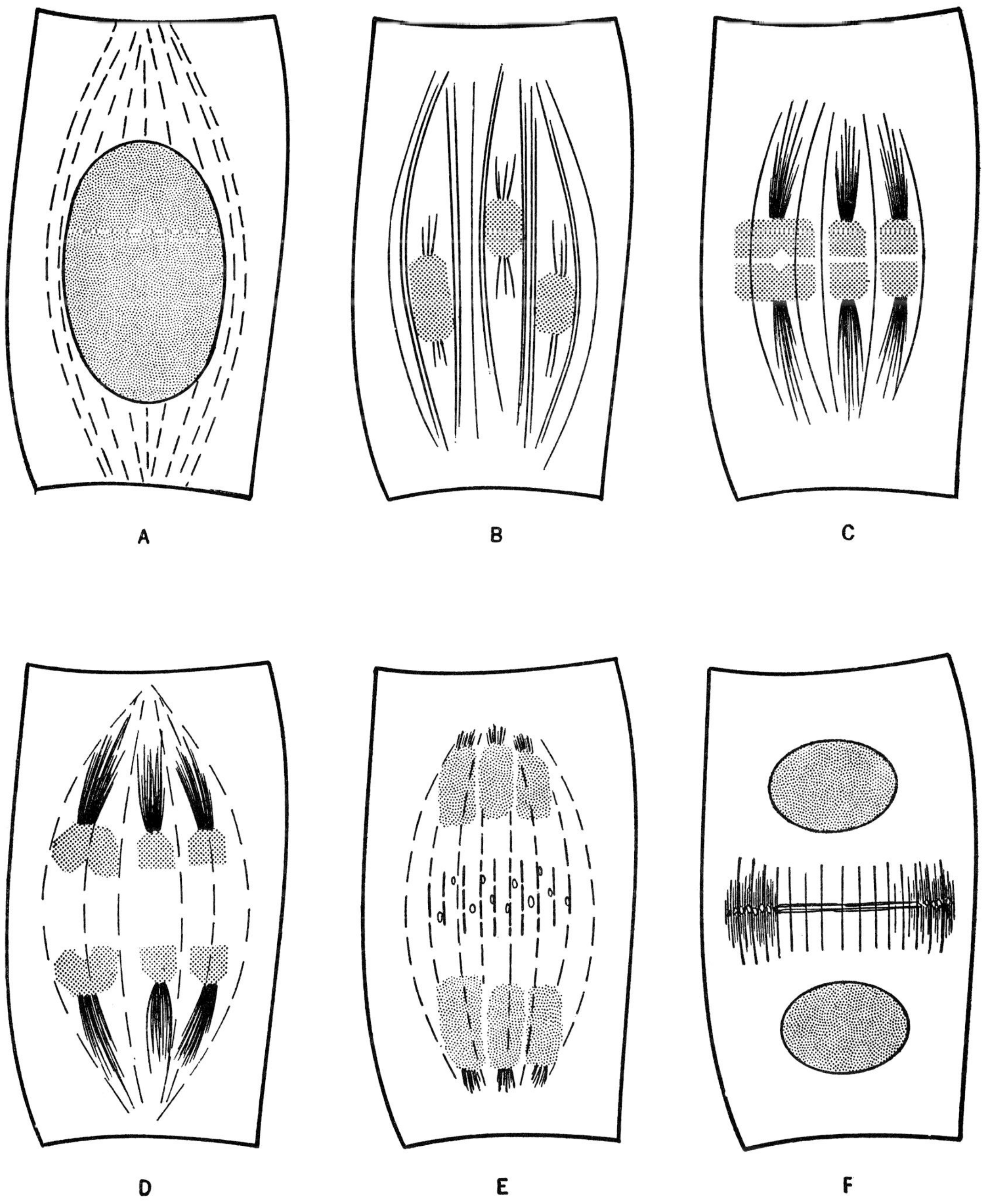

Fig. 3 shows mitosis in spindles without centrioles. Certain features shown in this illustration may appear in cells with the type of spindle shown in Fig. 2, as explained in the text.

2. MECHANICAL INTEGRITY OF SPINDLE FIBERS

Recently, the mechanical integrity of spindle fibers in prometaphase and anaphase movement was demonstrated directly by elegant micromanipulation experiments (Nicklas, 1965, 1967; also see earlier experiments of Carlson, 1952). Taking advantage of the Ellis micromanipulator (Ellis, 1962), Nicklas managed to pull and stretch chromosomes which were attached by their kinetochores to chromosomal spindle fibers. He was also able to sever the chromosomal spindle fiber, turn the chromosome, and point its kinetochore to the opposite spindle pole. The chromosome then moved toward the new pole, presumably acquiring a new spindle fiber attachment to that pole. The chromosomal spindle fibers are thus capable of resisting extension when the same force deforms the chromosome considerably. The fiber maintains a mechanical integrity within limits, but a new fiber can be reformed rapidly if the old one is broken.

The mechanical integrity of the chromosomal spindle fibers and the resistance of the central spindle to compression were also seen in centrifugal experiments performed by Shimamura (1940) and by Conklin (1917).

3. DYNAMIC NATURE OF SPINDLE FIBERS

The reformation of the spindle fibers following direct mechanical disruption by micromanipulation (Nicklas, 1965, 1967; also see Chambers, 1924) was described above. The spindle fibers have also been shown to reform rapidly after other means of disruption. Their birefringence is abolished in a matter of seconds by treatment with low temperature in *Chaetopterus* oocytes (Inoué, 1952 *a*) and in *Lilium* pollen mother cells (Inoué, 1964). Upon return to normal temperature, they recover in the course of a few minutes, after which chromosome movement can continue (Inoué, 1964). Low temperature disintegration of the spindle can be repeated for as many as 10 times in the same cell. In *Chaetopterus* oocytes, *Lilium* pollen mother cells, *Halistaura* developing eggs, and *Dissosteira* spermatocytes, recovery from low temperature treatment is not affected by the duration of chilling, even up to several hours or longer (Inoué, 1952 *b*, and unpublished data).

Colchicine, Colcemid, and many other drugs eliminate the spindle birefringence (Inoué, 1952 *a*; Inoué, Sato, and Ascher, 1965; also see Gaulden and Carlson, 1951, and Molé-Bajer, 1958). The effects are reversible, and *Pectinaria* oocytes treated with 10^{-5} M griseofulvin may recover their birefringent spindles in as little as 5.5–11 min when the cells are washed with normal sea water (Malawista and Sato, 1966).

High hydrostatic pressure is known to destroy the spindle organization reversibly (Pease, 1946; Zimmerman and Marsland, 1964; also see Marsland, 1966, on synergic action of colchicine and pressure). UV microbeam irradiation of portions of spindle fibers introduces temporary lesions which recover rapidly (Campbell and Inoué, 1965; Forer, 1965, 1966; Inoué and Sato, 1964; Izutsu, 1961 *a*, *b*).

The natural fluctuation of birefringence in prometaphase has already been described. The apparent assembly of the spindle material, first into the astral ray or clear zone material, then into the continuous fiber and sheath fiber material, from there to the chromosomal fiber, from chromosomal fiber to the continuous fiber, and, finally, to the phragmoplast fiber material, has also been described. We interpreted these changes to reflect the orderly assembly and disassembly of the same material, sequentially into different fibrous structures to perform different functions.

The various fibers of the spindle are thus extremely labile and their birefringence can be readily abolished. The spindle is also a dynamic structure, and, in most cases, the birefringence recovers rapidly when the external agent is removed. The birefringence can also be reduced or increased to another level by many agents and maintained at an equilibrium value.

4. THE DYNAMIC EQUILIBRIUM MODEL

In order to explain the dynamic nature of the spindle fiber and the response of the spindle birefringence and structure to various experimental alterations, Inoué (1959, 1960, 1964) earlier proposed a dynamic equilibrium model. In the model, spindle fibers are made up of oriented polymers which are in an equilibrium with dissociated molecules. The equilibrium is temperature-sensitive, the polymers dissociating at lower temperatures and the molecules associating to form oriented polymers at higher temperatures, up to a maximum.

Assuming that the birefringence measures the amount of oriented molecules, that the total amount of material which can be oriented is constant, and that the equilibrium between oriented and nonoriented material respresents a simple thermodynamic system, a van't Hoff plot was made of the equilibrium constant, log [(birefringence)/(maximum birefringence minus birefringence)] vs. $1/T$ (where T is absolute temperature). A straight line relationship was obtained which showed a very large *increase* in entropy (100 eu at 25°C) and a large heat of activation (29 kcal/mole) as the molecules polymerized. The standard free energy at 25°C was less than -1 kcal. Similar thermodynamic data have now been obtained, using *Pectinaria* oocytes with metaphase equilibrium spindles, in ordinary sea water and in sea water substituted with approximately 40% heavy water (Carolan et al., 1965, 1966).[2]

The high heat and the very high entropy increase associated with the formation of the polymers could be explained primarily by the loss in regularity of the water molecules, which is associated with the free protein molecules (see Kauzmann, 1957; Klotz, 1958, 1960; Robinson and Jencks, 1965; Scott and Berns, 1965; and Scheraga, 1967, for relevant discussions).

Higher temperatures are thought to dissociate bound water from the protein "subunit" molecules and allow them to interact closely (very likely by

[2] As of now, the thermodynamic data obtained comparing behavior of spindle birefringence in deuterated and normal sea water at different temperatures can be explained only by assuming an increased active pool size in D_2O.

hydrophobic interaction) and to associate. The low free energy reflects the weakness of the forces holding the "polymer" together.

The higher degree of alignment of material at higher temperature, as well as the thermodynamic parameters determined, resemble closely those obtained in the polymerization of tobacco mosaic virus (Lauffer et al., 1958) and in G- to F-actin transformation (Asakura et al., 1960; Grant, 1965; Oosawa et al., 1965).

Stevens and Lauffer (1965) have recently measured the buoyancy change of tobacco mosaic virus A-protein as it associated into virus-shaped rods. They concluded that in fact some 150 moles of water are dissociated from each 10^5 g of protein during their polymerization.

These simple model systems also respond to high hydrostatic pressure and to heavy water in a manner similar to the spindle in living cells (Ikkai and Ooi, 1966; Grant, 1965; Khalil et al., 1964).

5. EFFECT OF HEAVY WATER ON SPINDLE BIREFRINGENCE

In the dynamic equilibrium model, structured water is believed to dissociate from the protein molecules upon polymerization. Substitution of heavy water (D_2O) for normal water (H_2O) might then be expected to alter the equilibrium. Gross and Spindel (1960 *a*, *b*) and Marsland and Zimmerman (1965) have shown high concentrations of D_2O to "freeze" mitosis, or to overstabilize the gel structure of the spindle. Marsland and Hiramoto (1966) have shown the stiffness of the sea urchin egg to rise with D_2O.

According to Sidgwick (1950), heavy water molecules are held together more tightly than ordinary water molecules. Heavy water then may be expected to strip the protein molecules of ordered water and shift the equilibrium toward greater association of the molecules. This would enhance polymerization and birefringence of the spindle fibers (see Tomita et al., 1962, and Némethy and Scheraga, 1964, for other possible effects of heavy water on protein structure.)

This was, in fact, found to be so, as shown in Figs. 4–7, 9, and Table I, both in *Pectinaria* metaphase equilibrium oocyte and in dividing eggs of a sea urchin, *Lytechinus variegatus* (Inoué et al., 1963; Sato et al., 1966). When 45% D_2O was applied at the appropriate stage of mitosis, the spindle birefringence was found to increase at least 2-fold, and the volume occupied by the birefringent spindle, to increase as much as 10-fold at times. Fig. 7 illustrates the rapid change in the responsiveness of the spindle to D_2O (and temperature treatment) during mitosis in developing sea urchin eggs.

The action of heavy water is rapid, being 80% complete within 40 sec (see Tucker and Inoué, 1963, for determination of the rapid penetration rate of D_2O into sea urchin eggs). It is completely reversible, and the experiment can

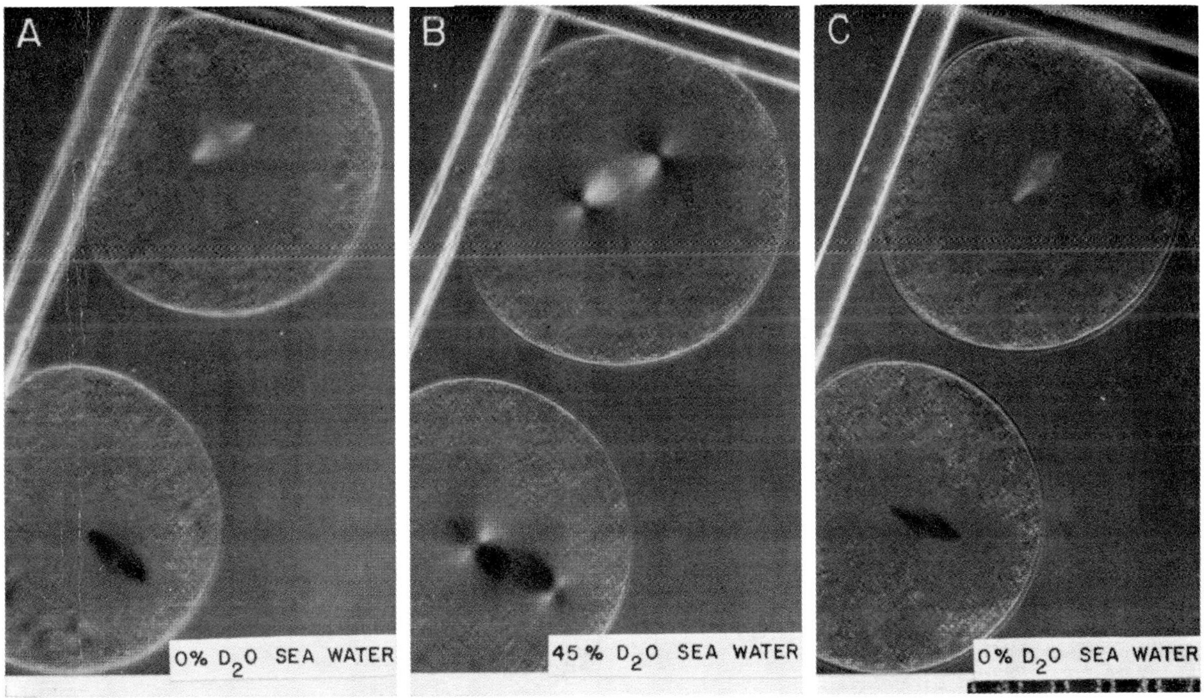

FIGURE 4. Reversible enhancement of spindle volume and birefringence by heavy water in arrested metaphase oocyte of *Pectinaria gouldi*. Cells supported by Butvar film and slightly compressed. Polarization microscopy. Scale interval, 10 μ. *A*, before D_2O–sea water perfusion; *B*, perfused for 2 min in 45% D_2O–sea water; *C*, approximately 3 min after return to normal sea water perfusion.

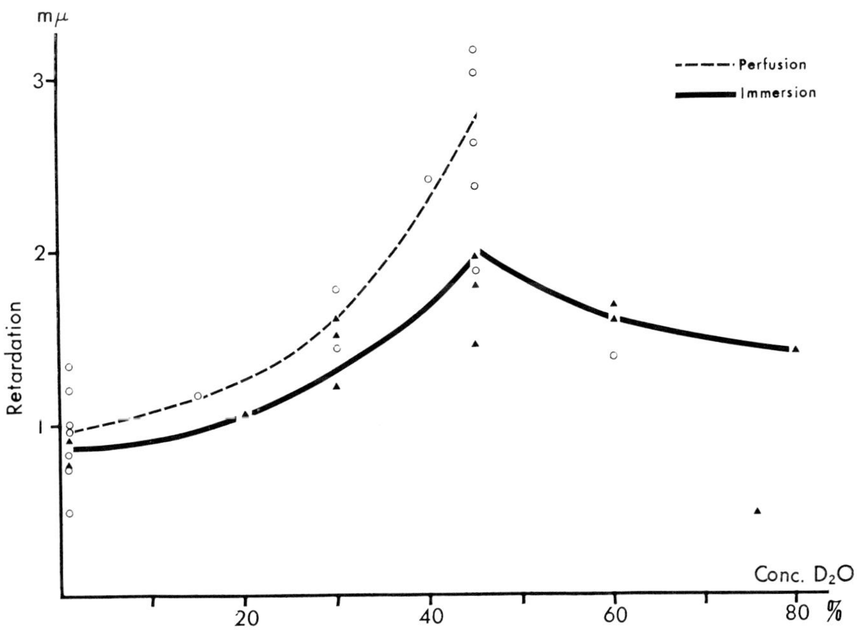

FIGURE 5. Effect of various concentrations of D_2O on the measured birefringence (retardation) of the mitotic spindle.

be repeated many times over on the same metaphase equilibrium cell. In many respects, it is similar to the effects of higher temperature and opposite to that of chilling.

Electron microscopy of spindles fixed with a new fixative, which prevents alteration of spindle birefringence during fixation (see section 10), showed the heavy water–treated spindle to have a larger bulk and more spindle filaments (Fig. 9 *B*) than the control (Fig. 9 *A*). The density of filaments was approximately equal to the control (Inoué, Kane, and Sato, unpublished data). Thus there is a parallel between birefringence observed in living cells under

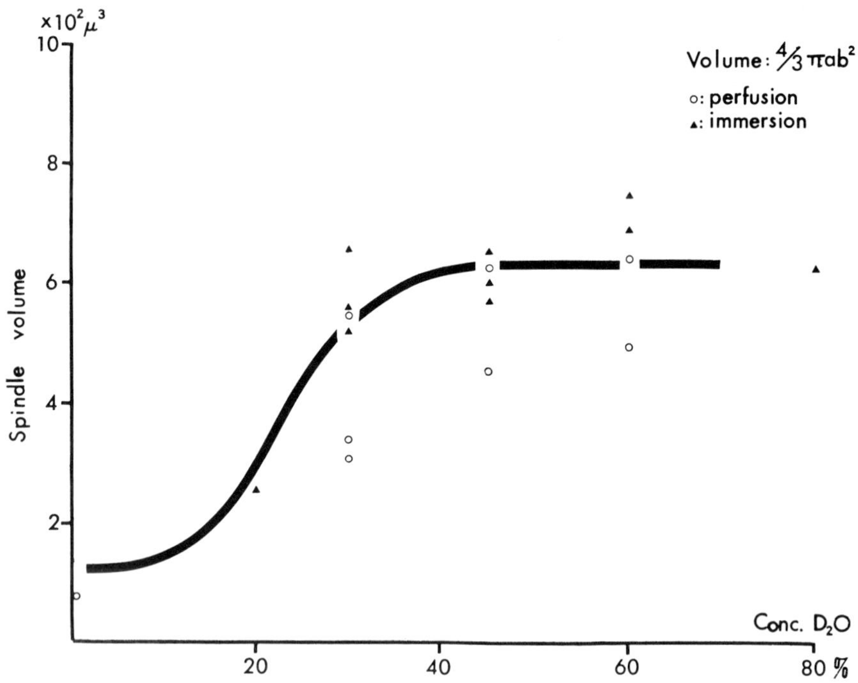

FIGURE 6. Spindle volume vs. D_2O concentration of *Pectinaria* oocyte. Temperature, 22° ± 1°C.

polarized light and the number of filaments (microtubules) observed with the electron microscope.

Similarly, a parallel increase is found in the quantity of the major (22S) protein extractable from an isolated spindle with and without heavy water treatment (Table I). In this connection, it should be pointed out that the total 22S protein extracted from each whole unfertilized egg treated with heavy water was no greater than that found in the untreated control eggs.

6. ASSEMBLY FROM POOL, NOT DE NOVO SYNTHESIS

Given the lability of the spindle fiber and its ability to recover rapidly from disruption and even to rapidly add more birefringent material, are the spindle fiber molecules synthesized rapidly each time they are required, or are they

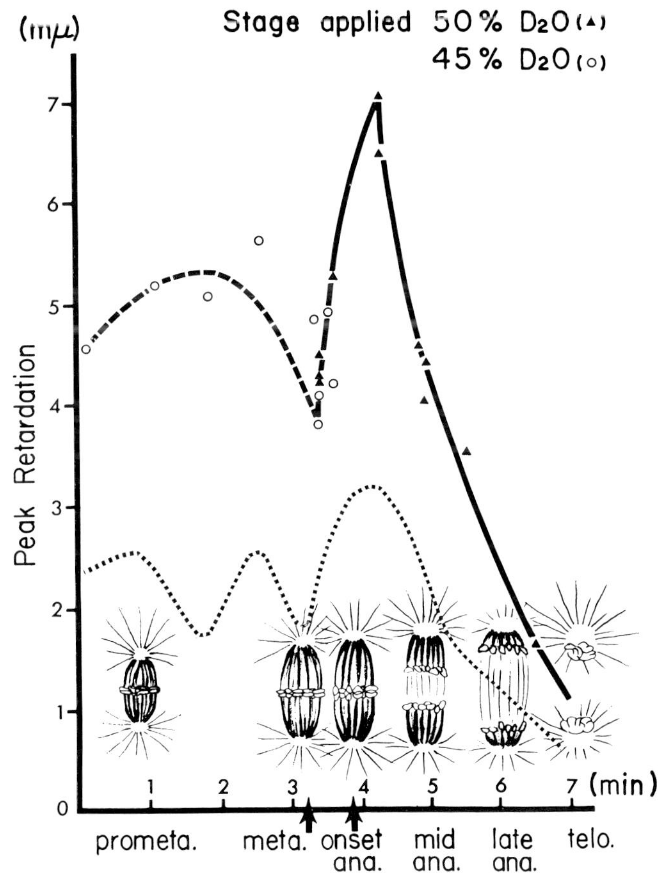

FIGURE 7. In developing eggs of sea urchin *Lytechinus variegatus* placed in D_2O–sea water, the spindle volume and birefringence increase, but with different sensitivities at different stages of mitosis. In 30–40% D_2O–sea water, mitosis proceeds. In 45–60% D_2O–sea water, mitosis is arrested unless the cell is exposed to D_2O during or after anaphase. At 50% D_2O, the retardation can be doubled and may reach a maximum of 7 mμ only if D_2O reaches the spindle at a particular 10 sec interval in early anaphase. (See Tucker and Inoué, 1963, for D_2O penetration rate.) The differential sensitivity of the spindle fiber birefringence to D_2O at different stages of mitosis closely parallels its sensitivity to chilling (Inoué, unpublished data). Heavy broken line and heavy black line in this figure indicate measured peak retardation achieved by spindle fiber with 45 and 50% D_2O–sea water at different stages of mitosis. Thin broken line indicates control.

TABLE I

ANALYTICAL ULTRACENTRIFUGATION DATA

Material as in Fig. 11 (Sato et al., 1966).

Concentration of D_2O in sea water	22S protein per spindle	Ratio*
%	*mg* × 10^{-7}	
0	2.4–5.1‡	2.6–10.0
40	9.0–35.0	

* Ratios of 22S protein per spindle in D_2O vs. H_2O sea water eggs, calculated for each experi ment.

‡ Average = 3.5.

9. CHEMISTRY OF THE SPINDLE PROTEIN MOLECULES

Kane (1967) has prepared a pure 22S protein from isolated mitotic apparatuses and from whole cells. Stephens (1965, 1967) has characterized this protein. It is monodisperse, has a particle weight of 880,000, and can be broken down into units of 110,000 particle weight. The 22S material makes up more than 90% of the KCl-soluble proteins of the spindle after isolation with hexylene glycol. Its quantity in the spindle varies parallel with spindle birefringence and with the number of microtubules during D_2O treatment. It is, therefore, a likely candidate for the spindle fiber material. The 22S proteins extracted from the unfertilized whole egg and from the spindle show identical amino acid compositions (Stephens, 1965, 1967).

On the other hand, Sakai (1966) and Kiefer et al. (1966) isolated a 2.5S protein from the mitotic apparatus with a particle weight of approximately 34,000. This particle would appear to have a dimension better fitting the electron microscope periodicity seen in microtubules of spindle and other cellular structures. At this writing, it is not clear whether these are two distinctly different proteins or are the same material polymerized to different degrees (see Introduction by Mazia).

10. RELATION OF SPINDLE FIBER BIREFRINGENCE TO MICROTUBULES AND TO FILAMENTS OBSERVED BY ELECTRON MICROSCOPY

Earlier electron micrographs of dividing cells often failed to show any fibrous elements in the spindle region. The number of filaments or microtubules that were observed in the spindle fiber region was often very small. With the recognition that low temperature can rapidly destroy spindle birefringence and, hence, the oriented spindle material, and following introduction of improved fixatives, increasing numbers of microtubules and filaments have been observed in spindle regions which are birefringent in life. There now exists a good correlation with the distribution of spindle fiber birefringence and density of intact microtubules or filaments (compare Inoué, 1964, with de-Thé, 1964; Harris and Bajer, 1965; Porter, 1966; and Robbins and Gonatas, 1964). Also, when birefringence declines in isolated spindles, a parallel disruption of filaments in the electron micrographs is found (Kane and Forer, 1965).

Even with glutaraldehyde fixation, which generally shows large numbers of microtubules in the electron micrographs, we find that the *Pectinaria* oocyte spindle birefringence had declined by 50% in the fixative. Further reduction was found with neutral osmium fixation. However, mixture of these fixatives with the spindle-isolating medium, hexylene glycol, prevented the decline of birefringence. Furthermore, the elevated birefringence in cells treated with D_2O and the reduced birefringence of cells treated for a short period with Colcemid retained their values exactly when fixed in this new fixative (Inoué, Kane, and Sato, unpublished data). The number of spindle filaments thus observed with the electron microscope and the birefringence of the spindle fibers observed in the polarizing microscope showed a close

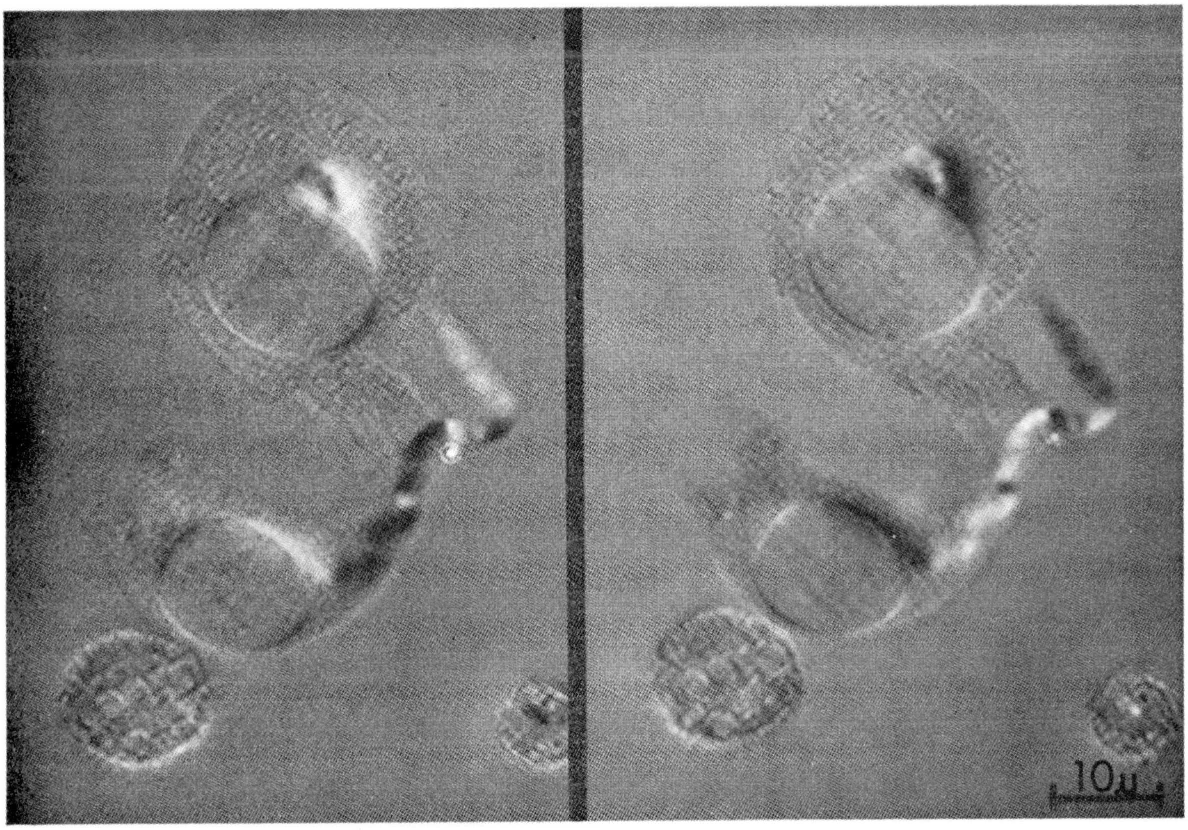

FIGURE 12. Polarization microscopy of Balb/C+ renal tumor cell from the mesenteric vein of a mouse. Live cells were embedded in Kel-F 10 oil. The fibrous structures seen near the nucleus and cytoplasmic process have a positive sign of birefringence. They correspond to the aggregated bundle of microfilaments demonstrated in electron micrographs in Fig. 13.

parallel for D_2O-treated eggs (Fig. 9 *B*) and their untreated controls (Fig. 9 *A*).

Microtubules and filaments have been observed by electron microscopy following glutaraldehyde fixation, not only in the mitotic spindle, but also in various other regions of the cell (see review by Porter, 1966). Recently, we had the good fortune of observing renal tumor cells with Professor A. Claude. Examination of living cells under polarized light revealed positively birefringent bundles of material (Fig. 12) running parallel to the bundle of microfilaments, which Professor Claude had shown in his electron micrograph (Fig. 13) with a special fixative in 1960 (Claude, 1961 *a,b*). Wherever we have tested, we have found that the long axes of such microtubules and filaments lie parallel to the fiber axes of positively birefringent structures in living cells.

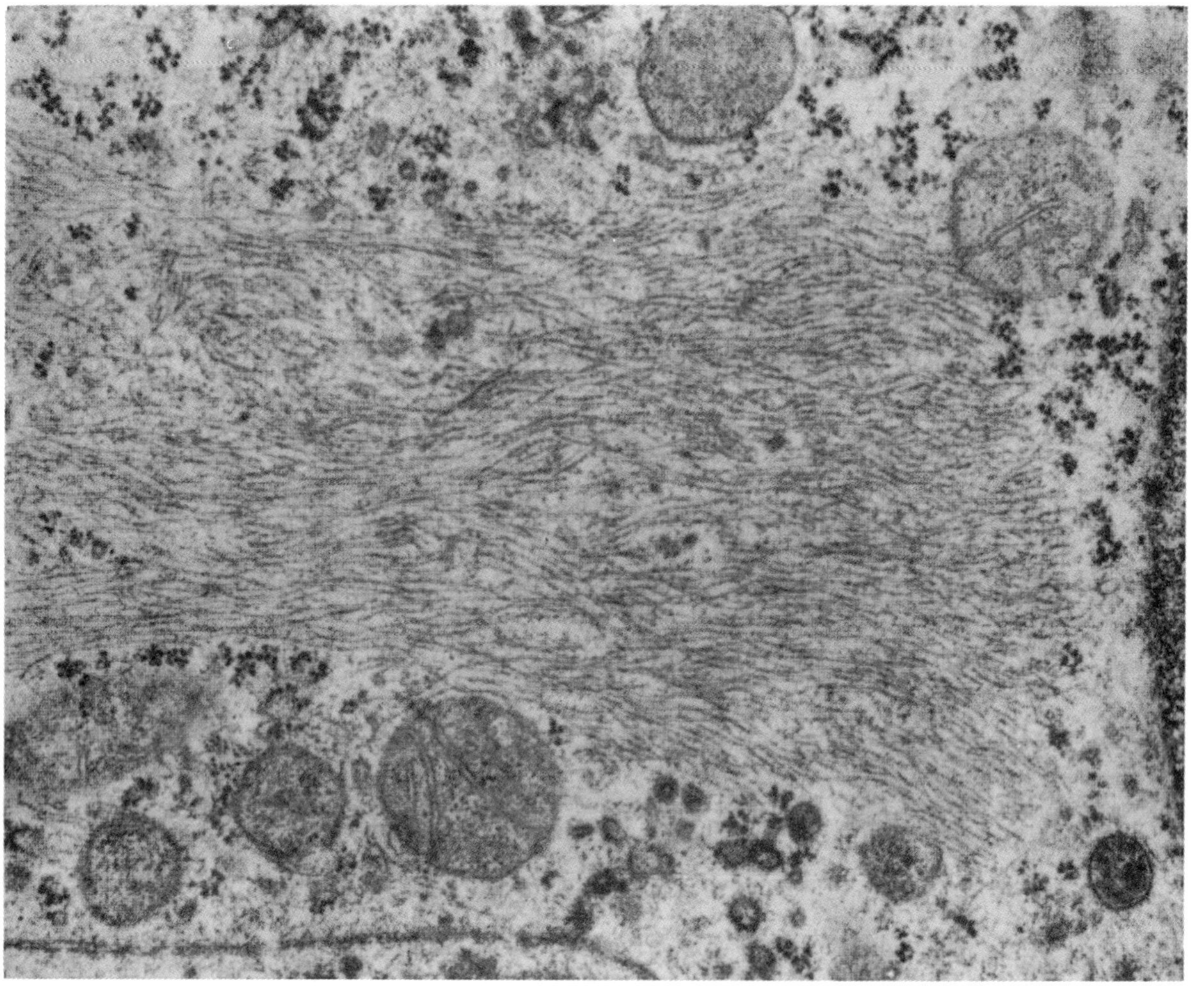

FIGURE 13. Electron micrograph of Balb/C+ renal tumor cell of a mouse (Claude, 1961 *a,b*, 1965). × 33,400. *Figure reprinted by permission from La Biologie, Acquisitions Récentes, Centre International de Synthèse, Editions Aubier-Montaigne, Paris, 1965, 13.*

In general, conditions which give rise to decreased birefringence also give rise to the reduction of the microtubules and filaments. Conditions which give rise to increase in birefringence give rise to increased microtubules or filaments and, with D_2O, also to the 22S protein in the spindle. This correlation is found in spindle fibers and in axopods of *Actinosphaerium* and in other systems also (see sections 4 and 11).

It now seems reasonable, then, to assume that the measured birefringence of spindle fibers under a variety of conditions closely follows the number of microtubules or filaments making up the fibers (see Engelmann, 1875, 1906; Schmidt, 1937, 1941; and Picken, 1950, for general discussions relating birefringence or anisotropic fine structure to contraction).

11. CELL MOTILITY AND LABILE ASSOCIATION OF MOLECULES

In the foregoing discussions, the dynamic orientation equilibrium hypothesis was used to explain the organization and function of the spindle. A species of ubiquitous protein molecule available in a pool was reversibly assembled into filaments and fibers upon demand. The material could perform various mechanical functions in the mitotic apparatus during division of the cell, and, when not in division, the same material could be organized in different parts of the cell to perform other functions. It could maintain or alter cell shape. It could partake in pinocytosis, phagocytosis, ameboid movement, streaming, etc. In this connection it is interesting to recall that cells generally stop wandering, and round up, prior to mitosis as though localized mechanical modulations of the cell had disappeared. Cells also generally resume their various mechanical activities shortly following mitosis. The presence of microtubules and filaments in association with these activities has been described, among others: for streaming, by Wolfarth-Bottermann (1964), Nakajima (1964), and Nagai and Rebhun (1966); for melanophore pigment migration, by Bikle et al. (1966); for pinocytosis and ameboid movement, by Marshall and Nachmias (1964) and Nachmias (1964); and in differentiating cells associated with their shape changes, by Arnold (1966), Byers and Porter (1964), Taylor (1966), Tilney and Gibbins (1966), and Tilney et al. (1965, 1966) (also see review by Porter, 1966). In those cases tested, the labile microtubules responded to temperature, hydrostatic pressure, heavy water, and colchicine or their combinations in a manner virtually identical with the response of the spindle material, all being reversible as in the spindle (e.g. Tilney et al., 1965, 1966; Malawista, 1965; Marsland, 1965, 1966; Marsland and Hiramoto, 1966).

We are considering here primarily slow movements such as movements of chromosomes and changes in shape of developing cells. These generally involve velocities of 1 to a few micra per second. In general, the growth and dissolution of the filaments or microtubules are believed to impose an anisotropic modulation of the mechanical properties that could cause slow extension or contraction of local regions of cells. Faster movements may also arise secondarily. For example, cytoplasmic streaming in the slime mold *Physarum* could result from formation and dissolution of the birefringent fibers observed by Nakajima (1964) in their ectoplasm. Local changes in the intracellular pressure, i.e. an interplay between the growth and contraction of the filaments and the elastic properties of the slime mold surface, could indirectly drive the streaming.

We have outlined our current hypothesis regarding the relation of the birefringent structures to slow movement of cell parts by reversible association of molecules. An alternative interpretation would describe the observed changes in the birefringent structures as simply the formation and dissolution of the fibrous fine structure necessary to make the movement possible. The motor force itself could result from a separate mechanism, as in the sliding filament model of muscle contraction, or from a mechanism such as that proposed by Thornburg (1967), in which fine pulses travel along the length of the fine anisotropic filaments. The dynamic equilibrium hypothesis would then account for motile structure organization rather than for production of force per se. The distinction may, however, turn out to be more subtle than is apparent and awaits analysis of the various systems in further detail (for example, see Szent-Györgyi and Prior, 1966, for the behavior of actin filaments during muscle contraction.)

POSTSCRIPT

This article portrays the authors' best current picture of the molecular mechanism of reversible fibrogenesis and mitotic chromosome movements. Clearly, much information is still wanting, and the arguments are often less than tight. Nevertheless, we hope that the article may convey enough useful information to stimulate further experimentation and to bring to focus some of the critical questions about mitosis and spindle formation which now may be asked.

Supported in part by NIH Grant CA 10171 (formerly 04552) from the National Cancer Institute and by Grant GB 5120 (formerly 2060) from the National Science Foundation.
Much of our work reported here was done at the Department of Cytology, Dartmouth Medical School, and at the Marine Biological Laboratories, Woods Hole, Massachusetts. The authors gratefully acknowledge the efficient assistance provided by the devoted staff, as well as the encouragement and critical discussions offered by their colleagues and students.

REFERENCES

Anderson, N. G. 1956. Cell division. II. A theoretical approach to chromosomal movements and the division of the cell. *Quart. Rev. Biol.* **31**:243.

Arnold, J. M. 1966. Squid lens development in compounds that affect microtubules. *Biol. Bull.* **131**:383.

Asakura, S., M. Kasai, and F. Oosawa. 1960. The effect of temperature on the equilibrium state of actin solutions. *J. Polymer Sci.* **44**:35.

Bajer, A. 1957. Cine-micrographic studies on mitosis in endosperm. III. The origin of the mitotic spindle. *Exptl. Cell Res.* **13**:493.

Bajer, A. 1961. A note on the behavior of spindle fibres at mitosis. *Chromosoma.* **12**:64.

Bajer, A. 1965. Cine-micrographic analysis of cell plate formation in endosperm. *Exptl. Cell Res.* **37**:376.

BIKLE, D., L. G. TILNEY, and K. R. PORTER. 1966. Microtubules and pigment migration in the melanophores of *Fundulus heteroclitus* L. *Protoplasma*. **61**:322.

BYERS, B., and K. R. PORTER. 1964. Oriented microtubules in elongating cells of the developing lens rudiment after induction. *Proc. Natl. Acad. Sci. U.S.* **52**:1091.

CAMPBELL, R. D., and S. INOUÉ. 1965. Reorganization of spindle components following UV micro irradiation. *Biol. Bull.* **129**:401.

CARLSON, J. G. 1952. Microdissection studies of the dividing neuroblast of the grasshopper, *Chortophaga viridifasciata* (de Geer). *Chromosoma*. **5**:199.

CAROLAN, R. M., H. SATO, and S. INOUÉ. 1965. A thermodynamic analysis of the effect of D_2O and H_2O on the mitotic spindle. *Biol. Bull.* **129**:402.

CAROLAN, M., H. SATO, and S. INOUÉ. 1966. Further observations on the thermodynamics of the living mitotic spindle. *Biol. Bull.* **131**:385.

CHAMBERS, R. 1924. The physical structure of protoplasm as determined by microdissection and injection. *In* General Cytology. E. V. Cowdry, editor. University of Chicago Press, Chicago. 237.

CLAUDE, A. 1961 *a*. Mise en évidence par microscopie électronique d'un appareil fibrillaire dans le cytoplasme et le noyau de certaines cellules. *Compt. Rend.* **253**: 2251.

CLAUDE, A. 1961 *b*. Problems of fixation for electron microscopy. Results of fixation with osmium tetroxide in acid and alkaline media. *Ext. Pathol. Biol.* **9**:933.

CLAUDE, A. 1965. Naissance de la biologie moléculaire. Vingt années d'invention technique et de progrès dans l'exploration de la cellule. *In* La Biologie, Acquisitions Récentes. Centre International de Synthèse. Éditions Aubier-Montaigne, Paris. 13.

CLEVELAND, L. R. 1938. Origin and development of the achromatic figure. *Biol. Bull.* **74**:41.

CLEVELAND, L. R. 1953. Studies on chromosomes and nuclear division. *Trans. Am. Phil. Soc.* **43**:809.

CLEVELAND, L. R. 1963. Functions of flagellate and other centrioles in cell reproduction. *In* The Cell in Mitosis. L. Levine, editor. Academic Press, Inc., New York. 3.

CONKLIN, E. G. 1917. Effects of centrifugal force on the structure and development of the eggs of *Crepidula*. *J. Exptl. Zool.* **22**:311.

COSTELLO, D. P. 1961. On the orientation of centrioles in dividing cells, and its significance: a new contribution to spindle mechanics. *Biol. Bull.* **120**:285.

DE-THÉ, G. 1964. Cytoplasmic microtubules in different animal cells. *J. Cell Biol.* **23**:265.

DIETZ, R. 1959. Centrosomenfreie Spindelpole in Tipuliden-Spermatocyten. *Z. Naturforsch.* **146**:749.

DIETZ, R. 1963. Polarisationsmikroskipische Befunde zur chromosomeninduzierten Spindelbildung bei der Tipulide *Pales crocata* (Nematocera). *Zool. Anz.* **26** (Suppl.), 131.

ELLIS, G. W. 1962. Piezoelectric micromanipulators. Electrically operated micromanipulators add automatic high-speed movement to normal manual control. *Science*. **138**:84.

ENGELMANN, R. W. 1875. Contractilität und Doppelbrechung. *Arch. Ges. Physiol.* **11**:432.

ENGELMANN, R. W. 1906. Zur Theorie der Contractilität. *Sitzber. Kgl. Preuss. Akad. Wiss.* 694.

FORER, A. 1965. Local reduction of spindle fiber birefringence in living *Nephrotoma suturalis* (Loew) spermatocytes induced by ultraviolet microbeam irradiation. *J. Cell. Biol.* **25**:95.

FORER, A. 1966. Characterization of the mitotic traction system, and evidence that birefringent spindle fibers neither produce nor transmit force for chromosome movement. *Chromosoma.* **19**:44.

FREY-WYSSLING, A., J. F. LOPEZ-SAEZ, and K. MÜHLETHALER. 1964. Formation and development of the cell plate. *J. Ultrastruct. Res.* **10**:422.

GAULDEN, M. E., and J. G. CARLSON. 1951. Cytological effects of colchicine on the grasshopper neuroblast in vitro with special reference to the origin of the spindle. *Exptl. Cell Res.* **2**:416.

GRANT, R. J. 1965. Doctorate Thesis. Columbia University, New York.

GROSS, P. R., and W. SPINDEL. 1960 *a*. Mitotic arrest by deuterium oxide. *Science.* **131**:37.

GROSS, P. R., and W. SPINDEL. 1960 *b*. The inhibition of mitosis by deuterium. *Ann. N.Y. Acad. Sci.* **84**:745.

HARRIS, P. 1962. Some structural and functional aspects of the mitotic apparatus in sea urchin embryos. *J. Cell Biol.* **14**:475.

HARRIS, P., and A. BAJER. 1965. Fine structure studies on mitosis in endosperm metaphase of *Haemanthus katherinae* bak. *Chromosoma.* **16**:624.

HIRAMOTO, Y. 1964. Mechanical properties of the starfish oöcyte during maturation divisions. *Biol. Bull.* **127**:373.

IKKAI, T., and T. OOI. 1966. The effects of pressure on F-G transformation of actin. *Biochemistry.* **5**:1551.

INOUÉ, S. 1952 *a*. The effect of colchicine on the microscopic and submicroscopic structure of the mitotic spindle. *Exptl. Cell Res.* Suppl. 2:305.

INOUÉ, S. 1952 *b*. Effect of temperature on the birefringence of the mitotic spindle. *Biol. Bull.* **103**:316.

INOUÉ, S. 1953. Polarization optical studies of the mitotic spindle. I. The demonstration of spindle fibers in living cells. *Chromosoma.* **5**:487.

INOUÉ, S. 1959. Motility of cilia and the mechanism of mitosis. *Rev. Mod. Phys.* **31**:402.

INOUÉ, S. 1960. On the physical properties of the mitotic spindle. *Ann. N.Y. Acad. Sci.* **90**:529.

INOUÉ, S. 1964. Organization and function of the mitotic spindle. *In* Primitive Motile Systems in Cell Biology. R. Allen and N. Kamiya, editors. Academic Press, Inc., New York. 549.

INOUÉ, S., and A. BAJER. 1961. Birefringence in endosperm mitosis. Chromosoma. **12**:48.

INOUÉ, S., and K. DAN. 1951. Birefringence of the dividing cell. *J. Morphol.* **89**:423.

INOUÉ, S., and H. SATO. 1964. See Inoué (1964). 580.

INOUÉ, S., H. SATO, and M. ASCHER. 1965. Counteraction of Colcemid and heavy water on the organization of the mitotic spindle. *Biol. Bull.* **129**:409.

INOUÉ, S., H. SATO, and R. W. TUCKER. 1963. Heavy water enhancement of mitotic spindle birefringence. *Biol. Bull.* **125**:380.

IZUTSU, K. 1961 *a*. Effects of ultraviolet microbeam irradiation upon division in grasshopper spermatocytes. I. Results of irradiation during prophase and prometaphase I. *Mie Med. J.* **11**:199.

IZUTSU, K. 1961 *b*. Effects of ultraviolet microbeam irradiation upon division in grasshopper spermatocytes. II. Results of irradiation during metaphase and anaphase I. *Mie Med. J.* **11**:213.

KANE, R. E. 1962. The mitotic apparatus: isolation by controlled pH. *J. Cell Biol.* **12**:47.

KANE, R. E. 1965. The mitotic apparatus: physical-chemical factors controlling stability. *J. Cell Biol.* **25**:137.

KANE, R. E. 1967. The mitotic apparatus: identification of the major soluble component of the glycol-isolated mitotic apparatus. *J. Cell Biol.* **32**:243.

KANE, R. E., and A. FORER. 1965. The mitotic apparatus: structural changes after isolation. *J. Cell Biol.* **25**:31.

KAUZMANN, W. 1957. The physical chemistry of proteins. *Ann. Rev. Phys. Chem.* **8**:413.

KHALIL, M. T., R. A. SHALABY, and M. A. LAUFFER. 1964. Reversible polymerization of TMV protein in D_2O and versene solutions. Abstracts of the Biophysical Society 8th Annual Meeting. Chicago. TC3.

KIEFER, B., H. SAKAI, A. J. SOLARI, and D. MAZIA. 1966. The molecular unit of the microtubules of the mitotic apparatus. *J. Mol. Biol.* **20**:75.

KLOTZ, I. M. 1958. Protein hydration and behavior. *Science.* **128**:815.

KLOTZ, I. M. 1960. Non-covalent bonds in protein structure. *In Brookhaven Symp. Biol.* **13**:25.

LAUFFER, M. A., A. T. ANSEVIN, T. E. CARTWRIGHT, and C. C. BRINTON, JR. 1958. Polymerization-depolymerization of tobacco mosaic virus protein. *Nature.* **181**: 1338.

LEDBETTER, M. C., and K. R. PORTER. 1963. A "microtubule" in plant cell fine structure. *J. Cell Biol.* **19**:239.

LEWIS, M. R. 1923. Reversible gelation in living cells. *Bull. Johns Hopkins Hosp.* **34**:373.

MALAWISTA, S. E. 1965. On the action of colchicine: the melanocyte model. *J. Exptl. Med.* **122**:361.

MALAWISTA, S. E., and H. SATO. 1966. Vinblastine and griseofulvin reversibly disrupt the living mitotic spindle. *Biol. Bull.* **131**:397.

MARSHALL, J. M., and V. T. NACHMIAS. 1964. Cell surface and pinocytosis. *J. Cell Biol.* **13**:92.

MARSLAND, D. 1965. Partial reversal of the anti-mitotic effects of heavy water by high hydrostatic pressure: an analysis of the first cleavage division in the eggs of *Strongylocentrotus purpuratus*. *Exptl. Cell Res.* **38**:592.

MARSLAND, D. 1966. Anti-mitotic effects of colchicine and hydrostatic pressure; synergistic action on the cleaving eggs of *Lytechinus variegatus*. *J. Cell Physiol.* **67**:333.

MARSLAND, D., and Y. HIRAMOTO. 1966. Cell division: pressure-induced reversal of the antimeiotic effects of heavy water in the oocytes of the starfish, *Asterias forbesi* *J. Cell Physiol.* **67**:13.

MARSLAND, D., and A. M. ZIMMERMAN. 1965. Structural stabilization of the mitotic apparatus by heavy water, in the cleaving eggs of *Arbacia punctulata*. *Exptl. Cell Res.* **38**:306.

MAZIA, D. 1961. Mitosis and the physiology of cell division. *In* The Cell. J. Brachet and A. E. Mirsky, editors. Academic Press, Inc., New York. **3**:77.

MERCER, E. H. 1952. The biosynthesis of fibers. *Sci. Monthly.* **75**:280.

MOLÉ-BAJER, J. 1953 *a*. Influence of hydration and dehydration on mitosis. II. *Acta Soc. Botan. Polon.* **22**:33.

MOLÉ-BAJER, J. 1953 *b*. Experimental studies on protein spindles. *Acta Soc. Botan. Polon.* **22**:811.

MOLÉ-BAJER, J. 1958. Cine-micrographic analysis of C-mitosis in endosperm. *Chromosoma.* **9**:332.

NACHMIAS, V. T. 1964. Fibrillar structures in the cytoplasm of *Chaos chaos*. *J. Cell Biol.* **23**:183.

NAGAI, R., and L. I. REBHUN. 1966. Cytoplasmic microfilaments in streaming *Nitella* cells. *J. Ultrastruct. Res.* **14**:571.

NAKAJIMA, H. 1964. The mechanochemical system behind streaming in *Physarum*. *In* Primitive Motile Systems in Cell Biology. R. Allen and N. Kamiya, editors. Academic Press, Inc., New York. 111.

NÉMETHY, G., and H. A. SCHERAGA. 1964. Structure of water and hydrophobic bonding in proteins. IV. The thermodynamic properties of liquid deuterium oxide. *J. Chem. Phys.* **41**:680.

NICKLAS, R. B. 1965. Experimental control of chromosome segregation in meiosis. *J. Cell Biol.* **27**:117A.

NICKLAS, R. B. 1967. Chromosome micromanipulation. I. The mechanics of chromosome attachment to the spindle. II. Induced reorientation and the experimental control of segregation in meiosis. *Chromosoma.* **21**:1, 17.

OOSAWA, F., S. ASAKURA, S. HIGASHI, M. KASAI, S. KOBAYASHI, E. NAKANO, T. OHNISHI, and M. TANIGUCHI. 1965. Morphogenesis and motility of actin polymers. *In* Molecular Biology of Muscular Contraction. Igakushoin, Tokyo. 56.

ÖSTERGREN, G. 1951. The mechanism of co-orientation in bivalents and multivalents. *Hereditas.* **37**:85.

PEASE, D. C. 1946. High hydrostatic pressure effects upon the spindle figure and chromosome movement. II. Experiments on the meiotic divisions of *Tradescantia* pollen mother cells. *Biol. Bull.* **91**:145.

PICKEN, L. E. R. 1950. Discussion on morphology and fine structure. Fine structure and the shape of cells and cell-components. *Proc. Linnean Soc. London.* **162**:72.

PORTER, K. R. 1966. Cytoplasmic microtubules and their functions. *Ciba Found. Symp., Principles Biomol. Organ.* 308.

PORTER, K. R., and R. D. MACHADO. 1960. Studies on the endoplasmic reticulum. IV. Its form and distribution during mitosis in cells of onion root tip. *J. Biophys. Biochem. Cytol.* **7**:167.

REES, A. L. G. 1951. Directed aggregation in colloidal systems and the formation of protein fibers. *J. Phys. Chem.* **55**:1340.

ROBBINS, E., and N. K. GONATAS. 1964. The ultrastructure of a mammalian cell during the mitotic cycle. *J. Cell Biol.* **21**:429.

ROBINSON, D. R., and W. P. JENCKS. 1965. The effect of compounds of the urea-guanidinium class on the activity coefficient of acetyltetraglycine ethyl ester and related compounds. *J. Am. Chem. Soc.* **87**:2462, 2470, 2480.

SAKAI, H. 1966. Studies on sulfhydryl groups during cell division of sea-urchin eggs. VIII. Some properties of mitotic apparatus proteins. *Biochim. Biophys. Acta.* **112**:132.

SATO, H., S. INOUÉ, J. BRYAN, N. E. BARCLAY, and C. PLATT. 1966. The effect of D_2O on the mitotic spindle. *Biol. Bull.* **131**:405.

SCHERAGA, H. A. 1967. Contractility and conformation. *J. Gen. Physiol.* **50**(6, Pt. 2):5.

SCHMIDT, W. J. 1937. Die Doppelbrechung von Chromosomen und Kernspindel und ihre Bedeutung für das kausale Verständniss der Mitose. *Arch. Exptl. Zellforsch.* **19**:352.

SCHMIDT, W. J. 1939. Doppelbrechung der Kernspindel und Zugfasertheorie der Chromosomenbewegung. *Chromosoma.* **1**:253.

SCHMIDT, W. J. 1941. Die Doppelbrechung des Protoplasmas und ihre Bedeutung für die Erforschung seines submikroskopischen Baues. *Ergeb. Physiol. Biol. Chem. Exptl. Pharmakol.* **44**:27.

SCHRADER, F. 1953. Mitosis. The Movements of Chromosomes in Cell Division. Columbia University Press, New York. 2nd edition.

SCOTT, E., and D. S. BERNS. 1965. Protein-protein interaction. The phycocyanin system. *Biochemistry* **4**:2597.

SHIMAMURA T. 1940. Studies on the effect of the centrifugal force upon nuclear division. *Cytologia.* **11**:186.

SIDGWICK, N. V. 1950. Chemical Elements and Their Compounds. Clarendon Press, Oxford. **1**:44.

STEPHENS, R. E. 1965. Characterization of the major mitotic apparatus protein and its subunits. Doctorate Thesis. Dartmouth College, Hanover, N.H.

STEPHENS, R. E. 1967. The mitotic apparatus. Physical chemical characterization of the 22S protein component and its subunits. *J. Cell Biol.* **32**:255.

STEVENS, C. L., and M. A. LAUFFER. 1965. Polymerization-depolymerization of tobacco mosaic virus protein. IV. The role of water. *Biochemistry.* **4**:31.

SZENT-GYÖRGYI, A. G., and G. PRIOR. 1966. Exchange of adenosine diphosphate bound to actin in superprecipitated actomyosin and contracted myofibrils. *J. Mol. Biol.* **15**:515.

TAYLOR, A. C. 1966. Microtubules in the microspikes and cortical cytoplasm of isolated cells. *J. Cell Biol.* **28**:155.

TAYLOR, E. W. 1959. Dynamics of spindle formation and its inhibition by chemicals. *J. Biophys. Biochem. Cytol.* **6**:193.

THORNBURG, W. 1967. Mechanisms of biological motility. *In* Theoretical and Experimental Biophysics. A. Cole, editor. Marcel Dekker, New York **1**:77.

TILNEY, L. G., and J. R. GIBBINS. 1966. Microtubules and morphogenesis. The role of microtubules in the development of the primary mesenchyme in the sea urchin embryo. *Biol. Bull.* **131**:378.

TILNEY, L. G., Y. HIRAMOTO, and D. MARSLAND. 1966. Studies on the microtubules in heliozoa. III. A pressure analysis of the role of these structures in the formation and maintenance of the axopodia of *Actinosphaerium nucleofilum* (Barrett). *J. Cell Biol.* **29**:77.

TILNEY, L. G., and K. PORTER. 1965. Studies on microtubules in heliozoa. I. The fine structure of *Actinosphaerium nucleofilum* (Barrett), with particular reference to the axial rod structure. *Protoplasma.* **60**:317.

TOMITA, K., A. RICH, C. DE LOZÉ, and E. R. BLOUT. 1962. The effect of deuteration on the geometry of the α-helix. *J. Mol. Biol.* **4**:83.

TUCKER, R. W., and S. INOUÉ. 1963. Rapid exchange of D_2O and H_2O in sea urchin eggs. *Biol. Bull.* **125**:395.

WADA, B. 1965. Analysis of mitosis. *Cytologia.* **30** (Suppl.):1.

WENT, M. A. 1966. The behavior of centrioles and the structure and formation of the achromatic figure. *Protoplasmatologia.* **6** (G 1):1.

WOHLFARTH-BOTTERMANN, K. E. 1964. Differentiations of the ground cytoplasm and their significance for the generation of the motive force of ameboid movement. *In* Primitive Motile Systems in Cell Biology. R. Allen and N. Kamiya, editors. Academic Press, Inc., New York. 79.

ZIMMERMAN, A. M., and D. MARSLAND. 1964. Cell division: effects of pressure on the mitotic mechanisms of marine eggs (*Arbacia punctulata*). *Exptl. Cell. Res.* **35**:293.

Discussion

Dr. Bárány: I would like to make two brief comments on the very nice work just presented by Dr. Inoué. You suggested that hydrophobic bonds are involved in the aggregation of the protein isolated from the spindles. The amino acid composition of this protein exhibited a rather polar character. There was no apparent excess in apolar amino acids, like alanine, valine, isoleucine, and leucine. Interestingly, the proline content of this material was extremely high. Do you think this amino acid composition supports the concept of the hydrophobic interaction? Second, you showed that the aggregation phenomenon has a positive temperature coefficient. As you know, hydrophobic interactions have a negative temperature coefficient. Would you care to comment on this?

Dr. Inoué: Let me try and answer this question. It was pointed out by Dr. Bárány that the amino acid composition did not fit in with what one would expect from molecules assembling with hydrophobic interaction. I think we are all right on this point. Van Holde (1966. The molecular architecture of multichain proteins. *In* Molecular Architecture in Cell Physiology. T. Hayashi and A. G. Szent-Györgyi, editors. Prentice Hall, Inc., Englewood Cliffs, N.J. 81) has examined a large number of globular proteins which are known either to form or not to form aggregates naturally, and looked at their amino acid compositions. He showed that all the proteins with more than 31% (and some above 29%) form aggregates without exception. All those with less than 28% (and some with less than 31%) hydrophobic amino acid did not. The Kane-Stephens spindle protein has slightly above 30% hydrophobic amino acids so that it does fit in the right range. The second point had to do with the temperature effect. Temperature response certainly cannot be an unequivocal clue as to what kind of interaction is taking place. And, furthermore, the situation is much worse because we are not dealing with an obviously isolated closed system, which is the necessary condition for doing thermodynamic analysis. We've taken whole cells and seen if we could make any sense. It is actually surprising that we should get as good a fit to the van't Hoff relation. Again, however, the signs and magnitudes of the thermodynamic parameters are consistent with hydrophobic interaction (see Scheraga, this Symposium). Now, we must obviously do more experiments. One important point I had not gotten around to mentioning is that no person up to now can take isolated spindles and make them respond in ways similar to the spindle in living cells. This is a huge gap and I think we should be very conscious of this limit of our knowledge now. The moment eggs are treated, for example, with hexylene glycol for isolating the spindle, all of the characteristics disappear; that is, the colchicine response, temperature response, heavy water response, and so on. We have no idea as to what factor is missing here, but we're clearly dealing with incomplete systems.

On the other hand, there is some encouragement to be gained from model systems. If one takes the association of tobacco mosaic virus A-protein or G–F transformation of actin and other types of protein subunit association, these respond to temperature, to heavy water, and, I'm told, to colchicine in just the same way.

This is a nice correlation. I really don't know yet what is happening during spindle isolation, but I have a feeling that if we get the right factors we might be able to get the isolates to respond properly. But you are perfectly right in saying that the mechanism we postulate has not yet been proven.

Question from the Floor: I have a question for the Chairman. I wasn't able in the micrograph that you showed of the 40 A globules to see whether or not there were 13. But I was aware of your counting them in that recent paper. Do you still see 13 distinctly at all times?

Dr. Mazia: Not at all times. They bend over on themselves. If one takes unspread ones, one sees 6 or 7; in the best spread ones, one often sees 12 and sometimes 13.

Dr. Jerome J. Freed: I'd like to add a few comments to the discussion on saltatory movements. Last year (1965) we reported at the meetings of the American Society for Cell Biology an experimental approach to the problem, in which we examined the movements of various particles in cultured cells and found that these did have the to-and-fro characteristics so beautifully shown in Dr. Rebhun's pictures taken with the Nomarski microscope. These movements were sensitive to the action of colchicine and colchicine-like drugs. We consider this as evidence that saltatory movements may be associated with the microtubules. We have since looked at the distribution of microtubules in these cells using rather thick methacrylate sections and the stereoreconstruction technique of Francis Ashton and Jack Schultz. The predominant directions of the saltatory movements and of the microtubules seem to coincide to the extent that we can associate them. The interesting part of this, as one might also expect, is that the array of microtubules seems to be centered upon the juxtanuclear region, or "nuclear *hof*," where the centrioles are located.

Dr. Rebhun: I'm sorry I didn't get a chance to mention Dr. Freed's work in the talk. It is discussed in the written paper. Basically, we found that saltatory motion of echinochrome granules in *Arbacia* is colchicine-independent although colchicine gets into the egg.

Dr. Freed: Were you able to confirm that the asters, or microtubular arrangement, had been destroyed by the colchicine treatment?

Dr. Rebhun: No. We have not been able to; I think that practically nobody has seen microtubules as such in unfertilized eggs. We have under certain special conditions of rapid freezing in electron microscopy, but not in other ways. So I don't know quite what to make of this except here is a case in which colchicine does not affect saltatory movements.

Dr. Edwin W. Taylor: I think that at this stage in our understanding of the chemistry of mitosis and the mitotic spindle we have to be prepared to go on disagreeing with each other, at least for a little while to come. And I find it difficult to accept the idea that the 22S protein is in fact the subunit of the spindle microtubule. If this is the subunit, you would expect to get this protein if you started with some other source of tubules and, in particular, if you use cilia.

If you remove essentially only the central pair of microtubules from cilia, under very mild conditions, that is, just by dialyzing at low ionic strength and pH 7, you get essentially one protein and it has a sedimentation constant of 6S. The outer nine go on into solution to give a protein with a sedimentation constant of 4S. The properties of the 4S protein seem to agree fairly well with the protein isolated by Sakai and Professor Mazia from the mitotic apparatus. We do not find the 22S protein in cilia.

Our other evidence is, of course, that if colchicine has the effect of breaking down the spindle, you might expect to demonstrate some interaction between colchicine and this protein subunit. A long series of experiments along these lines done in my laboratory by Dr. Borisy, Dr. Shelanski, and Mr. Weisenberg show that colchicine does in fact form a fairly strong complex with an equilibrium constant of around 10^6 with the 6S protein that we get from cilia, and also from the mitotic apparatus and nervous tissue; that is, other systems which are rich in microtubules. Colchicine shows no affinity whatsoever for the 22S protein.

Dr. Inoué: I should like to report what Stephens has done (1965). He has been able to get the 22S protein from cilia as well as from the spindle of the same species of sea urchins and has looked at their amino acid compositions. The compositions are virtually identical. With different species there is more variation.

Now, I think we all are aware that there are other 22S protein–containing particles that come out of cells, and I think we should distinguish which particular one we're talking about. I have a feeling that the Sakai protein and the 22S are not different, that these are different states of aggregation of the same protein. I believe Dr. Mazia has more or less the same view. The reason why one gets the 22S with Kane's preparation and 3.5 or 6S with Sakai's isolation we don't know, but it might turn out to be a methodological problem rather than a basic difference in materials that we're dealing with. I do think one should define the amino acid composition of material clearly to make the argument meaningful.

Dr. Mazia: I don't know if the audience is interested in this small point. In fact, the difference has to be methodological. Sakai also obtained a lot of 22S when he used the same method as did Stephens and Kane. The 22S can't be a subunit. One need only look at the picture to see that subunit is just too small to be the 22S particle.

Dr. Hayashi: In line with this discussion I would like to comment that in our laboratory Mrs. Forsheit and I have been examining the effect of colchicine on contractile proteins and we can report now that colchicine has no effect on myosin ATPase and on actomyosin ATPase. However, with regard to the temperature-reversible polymerization of G-ADP actin, colchicine has an inhibitory effect but it does not depolymerize already polymerized actin. The monomer-to-polymer change of the actin system does not seem to be on as fine a hairline as the mitotic spindles as described by Dr. Inoué. Making exactly the same assumption as was done for the mitotic spindle, Grant has obtained values for the thermodynamic parameters in which the ΔS and the ΔH for the actin are somewhat larger. This may explain the fact that the colchicine does not seem to bring about a depolymerization, but it does inhibit the polymerization and this is essentially in the same direction as the colchicine effect as presented by Dr. Inoué.

Dr. John J. White: Dr. Mazia, I believe that some years ago you reported the association of adenosinetriphosphatase activity with the spindle tubules. Is this information still valid?

Dr. Mazia: Yes, we did find ATPase in the association with the spindle, but the work was done at a time when we couldn't distinguish tubules from the whole apparatus. I really meant to imply in my introduction that we don't know whether ATPase is associated with the tubule or whether the ATPase could lead us to a second protein which has not been found in a search for a protein having the physical-chemical properties of myosin.

Dr. White: Has there been any evidence that the resolved microtubule protein does have ATPase activity?

Dr. Mazia: The best evidence, speaking of microtubules in general, is the evidence given by Dr. Satir this afternoon; namely, that the ATPase in the form of dynein is associated with the "arms" but not with the structures we would call microtubules proper.

Dr. Rebhun: I would just like to make one comment on the effects of some of the agents which have been used for isolating spindles on the spindle itself. We became interested for a variety of reasons in trying things like hexylene glycol, dimethylsulfoxide, ethylene glycol, dithiodipropanol, ethanol, or acetone, or what-have-you, all of which can be used for isolating spindles under the same conditions on the in vivo mitotic apparatus. If one is going to isolate spindles then one would like to have a reagent which doesn't affect the in vivo spindle, at least to a great extent. It turns out that all of these agents have rather phenomenal effects on the spindles. You can divide the agents in two groups; half of them have the same kinds of effects in general as D_2O, although the concentration of the agent varies. Thus, 2 or 3% hexylene glycol in sea water rather than in a buffer will increase both size and retardation of the in vivo spindle reversibly. The effects on sea urchin spindles, on *Pectinaria*, and on clam spindles are similar. Dimethylsulfoxide and ethylene glycol at higher concentrations will also have these effects.

On the other hand, acetone, ethanol, dithiodipropanol, and *t*-butyl alcohol have diametrically opposite effects. That is, they disperse the spindle, generally reversibly, the effect looking somewhat like a colchicine type of effect. However, I don't know if it is depolymerization or separation of the mitotic filaments on the ultrastructural level. The time constants of both these effects are rather rapid. That is, the effects begin within about 0.5–1 min of the application of reagent, which is certainly the time during which swelling takes place when these agents are used as isolating media. So I think one has to be a little bit careful, since no matter which of these agents you use, you get very nice birefringent mitotic apparatuses with microtubules if you use it as an isolating medium. Thus, these agents which have been used in spindle isolation are not innocuous with respect to spindle structure. That should be kept in mind and should be a precaution used in interpreting the results of chemical analysis of isolated mitotic apparatuses.

Dr. Arthur Zimmerman: Dr. Inoué, would you care to speculate as to perhaps why there's an accelerated recovery following the Colcemid treatment when you have used actinomycin D? and puromycin?

Dr. Inoué: Why is recovery faster in actinomycin D and puromycin? Well, we have no clear answer. But it is a striking phenomenon. It might be—a real might—that these agents have a high enough binding constant with Colcemid to pull out the Colcemid faster from cells. However, we have no proof of this explanation.

The following note was added by Shinya Inoué in September of 2006:

This article, published in 1967, together with the Editors' comments, was selected for publication in Gall JG, McIntosh JR eds., *Landmark Papers in Cell Biology*, Cold Spring Harbor Laboratory Press, Cold Spring Harbor, New York, pp. 177–207 (2001).

Reprinted from *Journal of Cell Biology*, Vol. 45(2), pp. 470–477, 1970.

REVERSAL BY LIGHT OF THE ACTION OF *N*-METHYL *N*-DESACETYL COLCHICINE ON MITOSIS

JOHN ARONSON and SHINYA INOUÉ. From the Department of Biology, University of Pennsylvania, Philadelphia, Pennsylvania 19104

INTRODUCTION

Colchicine, in low concentrations, inhibits cell division in a wide variety of cell types through an effect which results in dissolution of the spindle (1, p. 62 *et seq.*) This is especially clearly shown in the work of Inoué (2).

When exposed to sunlight, colchicine undergoes a photochemical rearrangement to give α, β, and γ lumicolchicines (3, 4) (Fig. 1). There is also evidence that exposure to sunlight and air reduces the biological activity of colchicine (1, p. 374). This suggests the possibility of photochemically reversing colchicine effects in living cells.

The *N*-methyl derivative of *N*-desacetyl colchicine (Ciba's trade name is Colcemid) was used in this study in preference to colchicine because of its effectiveness at lower external concentrations and because of fewer irreversible effects.

MATERIALS AND METHODS

N-methyl *N*-desacetyl colchicine[1] (Ciba Colcemid lot No. M 1168) had a molar extinction coefficient of 1.7×10^4 at 355 nm in water (assuming a molecular weight of 371) and an $\epsilon_{290}/\epsilon_{355}$ ratio of 0.25. For irradiation, either a G.E. AH-4 or an Osram HBO-200 mercury arc was used and the 366 nm complex was isolated by a 2 mm Jena glass UG-1 filter or a Kodak 18 A filter. These were used in conjunction with a 3 mm KG-1 heat filter and with a 2 mm BG 38 filter.

Quantum efficiencies were determined by ferric oxalate actinometry (5) with a Zeiss Model PMQ II spectrophotometer. The absorption curves shown in Fig. 2 were obtained with a Beckman Model DB-G spectrophotometer.

Several species of sea urchin, *Strongylocentrotus drobachiensis*, *Strongylocentrotus purpuratus*, and *Lytechinus variagatus*; one species of starfish, *Pisaster ochraceous*, and the annelid *Pectinaria gouldii* were used. These specimens were obtained from commercial sources and shipped to Philadelphia for use during the spring of 1969. Maturation in the starfish was induced by motor nerve extract (6).

Artificial seawater prepared according to the Woods Hole Marine Biological Laboratories formulation was used throughout.

Microscope observations on *L. variagatus* and *P. gouldii* were made at room temperature (22°C) and on *P. ochraceous* below room temperature (13–17°C) with a Leitz polarizing microscope and objective (UMK 32. NA about 0.4) in conjunction with an American Optical Company (Southbridge, Mass.) "Super Bio" rectified condenser.

RESULTS

Chemical

Irradiation of Colcemid at 366 nm caused a large decrease in absorbance at 355 nm which was comparable to that seen on irradiating colchicine (Fig. 2).

The quantum efficiency of this reaction is about 0.003 in water and is 12 times greater in *n*-butanol containing ca. 3% of water. It may be possible to estimate the quantum efficiency for Colcemid bound to the cell in vivo which would give information about the binding site. An order of magnitude value is 0.003 based on an OD of 0.07 for the egg in 10^{-6} M Colcemid, an incident intensity at 366 nm of 4×10^8 quanta/μ^2 per sec and assuming, on the basis of birefringence recovery, that 1 min of irradiation inactivates half the bound Colcemid.

Cell Irradiation

Cleavage of sea urchin eggs was inhibited, when they were placed, soon after fertilization, in Colcemid at a concentration of 3×10^{-7} M or greater. Continuous exposure to 366 nm radiation from a 100 watt mercury arc[2] reversed this inhibition with

[1] *N*-methyl *N*-desacetyl colchicine is listed as demecolcine in the Merck Handbook, 7th edition, and is sold by Ciba Pharmaceutical Company, Fairlawn, N.J., under the trade name of Colcemid[R]. Since the name Colcemid is in fairly common usage, we have used it throughout this paper.

[2] We used a 100 watt G.E. AH-4 mercury arc lamp and a Kodak 18A filter mounted in a Spencer 370-A microscope lamp housing and imaged about 15 in. away.

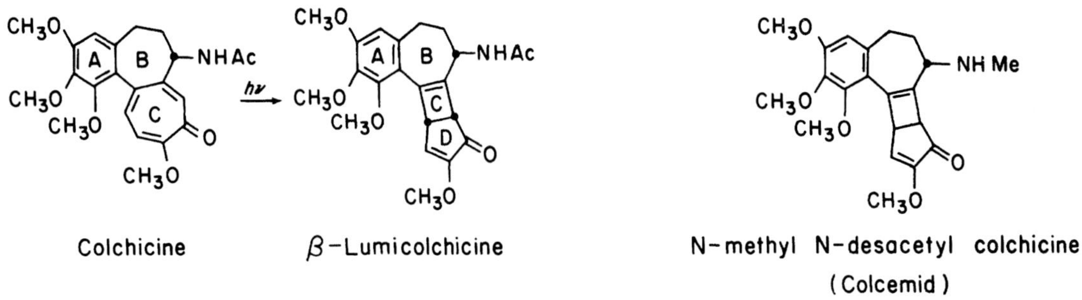

FIGURE 1 Formulae for photoconversion of colchicine to β-lumicolchicine and formula of Colcemid. α- and γ-lumicolchicine, dimer and isomer of β-lumicolchicine, are also found on irradiation.

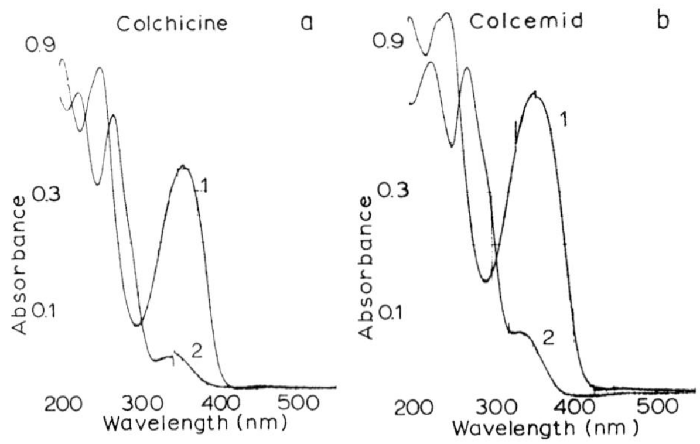

FIGURE 2 Changes in absorbance seen when solutions of colchicine and of *N*-methyl *N*-desacetyl colchicine (Colcemid) were irradiated at 366 nm; (*1*) before irradiation; (*2*) after irradiation. The spikes at about 330 nm and the discontinuities at 355 nm are instrument artefacts. (*a*) 2×10^{-5} M colchicine in water. (*b*) 3×10^{-5} M *N*-methyl *N*-desacetyl colchicine in water.

no delay for eggs in 1×10^{-6} M Colcemid, with a slight delay for eggs in 3×10^{-6} M Colcemid, and had no obvious effect on eggs in 3×10^{-5} M Colcemid. Irradiation of eggs in Colcemid-free seawater under the same conditions did not cause a decrease in the number of cleaving eggs or a definite change in the time of cleavage.

Most of the cytological observations were made on the second cleavage division of *L. variagatus* eggs. We routinely used artificial seawater containing 3×10^{-5} M Colcemid for 5 min to destroy the spindle birefringence, followed by 1×10^{-6} M Colcemid in artificial seawater to provide a defined background concentration. Healthy cells already in anaphase continued through cleavage, while cells in other stages were blocked from cleavage or spindle formation. *L. variagatus* spindles showed a major decrease in length and in birefringence within 2 min in 3×10^{-5} M Colcemid which gives an indication of the time to affect the spindle. In the absence of Colcemid, the time from nuclear membrane breakdown to anaphase was about 7 min.

A specific cell was followed under the polarizing microscope. As the spindle in this cell reached the desired stage (usually midmetaphase), Colcemid was applied by perfusion, and the resulting decrease in birefringence was observed. When the birefringence loss was complete, the cell was irradiated at 366 nm. The general approach was to irradiate for a short time until a localized increase in birefringence was seen; to wait a few minutes for obvious time-dependent effects; to irradiate so as to maximize the birefringence increase; then to irradiate just because it might help the cell pass through anaphase and cleavage. The usual radiation intensity[3] on the cell was about 4×10^{8} quanta/μ^2 per sec.

[3] This value was obtained with an HBO-200 lamp using a Leitz lamp housing, microscope base, and Ortholux stand, and with the condenser unimmersed but the diaphragm fully open.

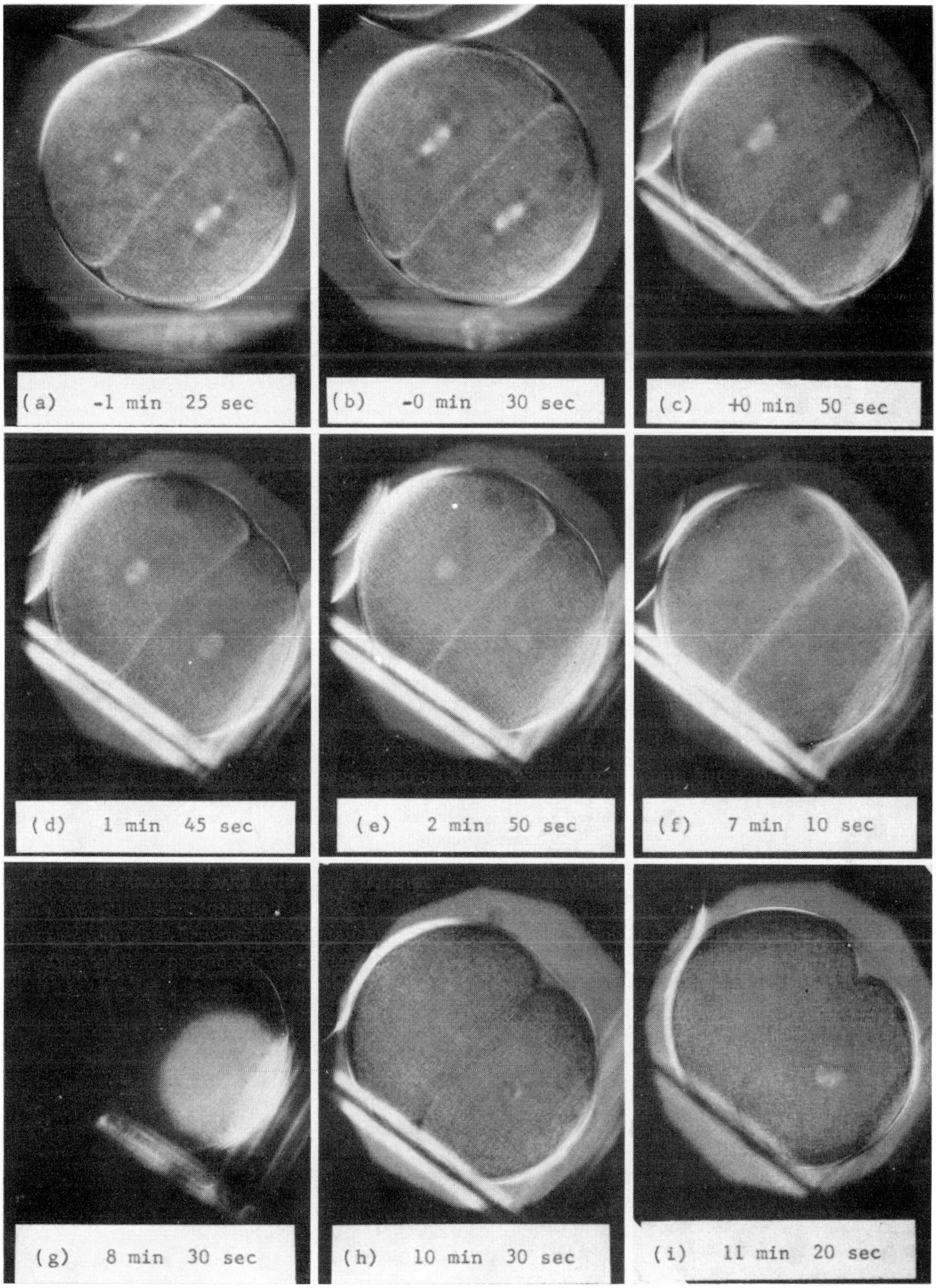

FIGURE 3 Initial development of spindle birefringence centers on the asters (*a*) and develops toward the metaphase plate to give a definitive spindle (*b*). At time *0*, 3 x 10^{-5} M Colcemid was added and the changes in the spindle seen in *c–f* were observed. 1 x 10^{-6} M Colcemid was added at time *5′20″*, and this level of Colcemid was maintained for the rest of the experiment. *g* shows the area in the lower cell to be irradiated for 30 sec at time *9′15″*, and *h–j* show some of the change observed. Further irradiation

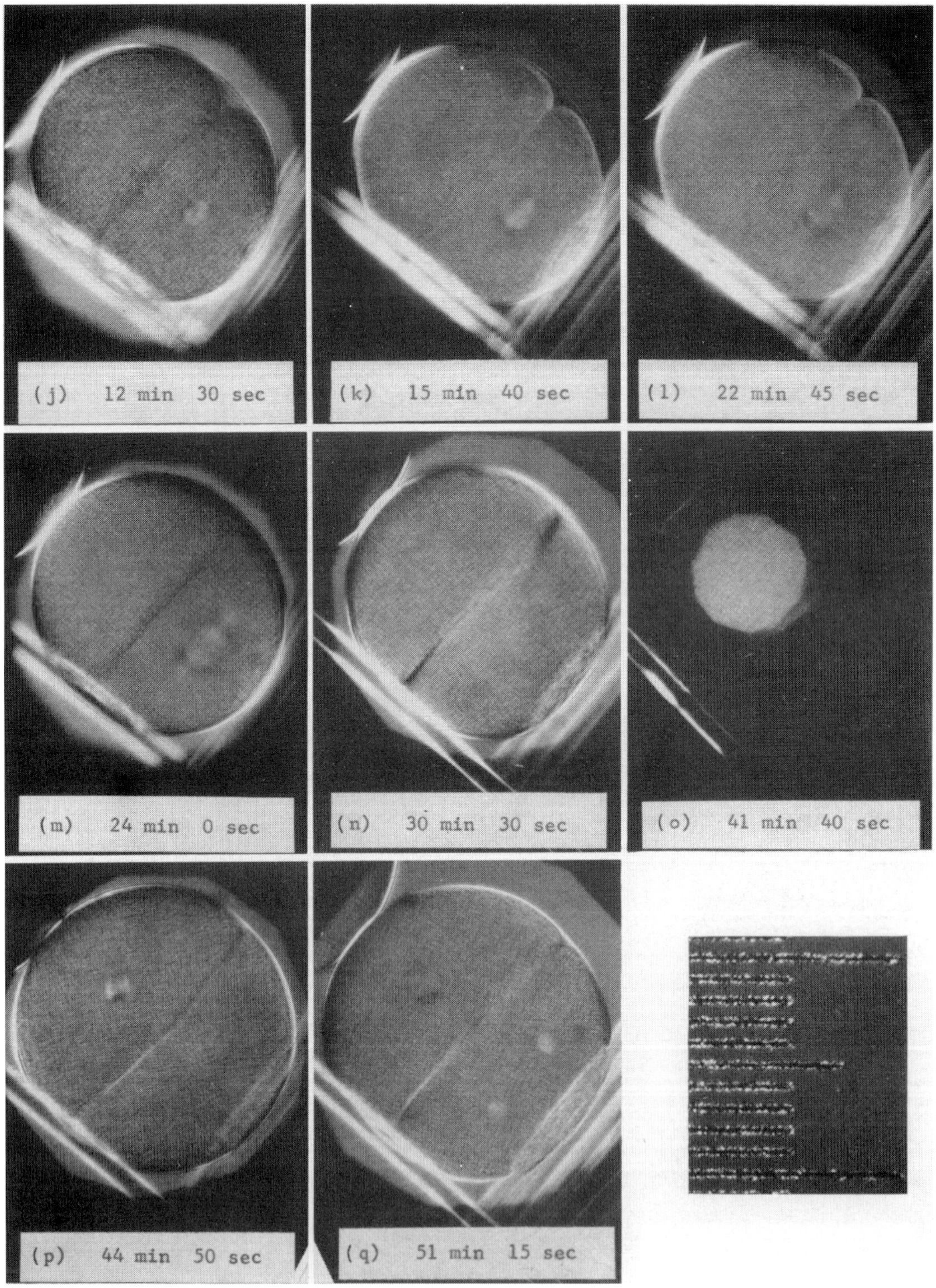

in the region shown in *g*, given at reduced intensity over a period of minutes, carried the irradiated blastomeres through anaphase as shown in *k–n*. There was a suggestion of cleavage only.

The previously unirradiated blastomere was irradiated over the area shown in *o* for 30 sec at time 42′40″. Further low-intensity irradiation was followed by the appearance of the next set of spindles (*q*).

The scale has a spacing of 10 nm. Figs. 3 and 4 show fertilized eggs of *Lytechinus variagatus*.

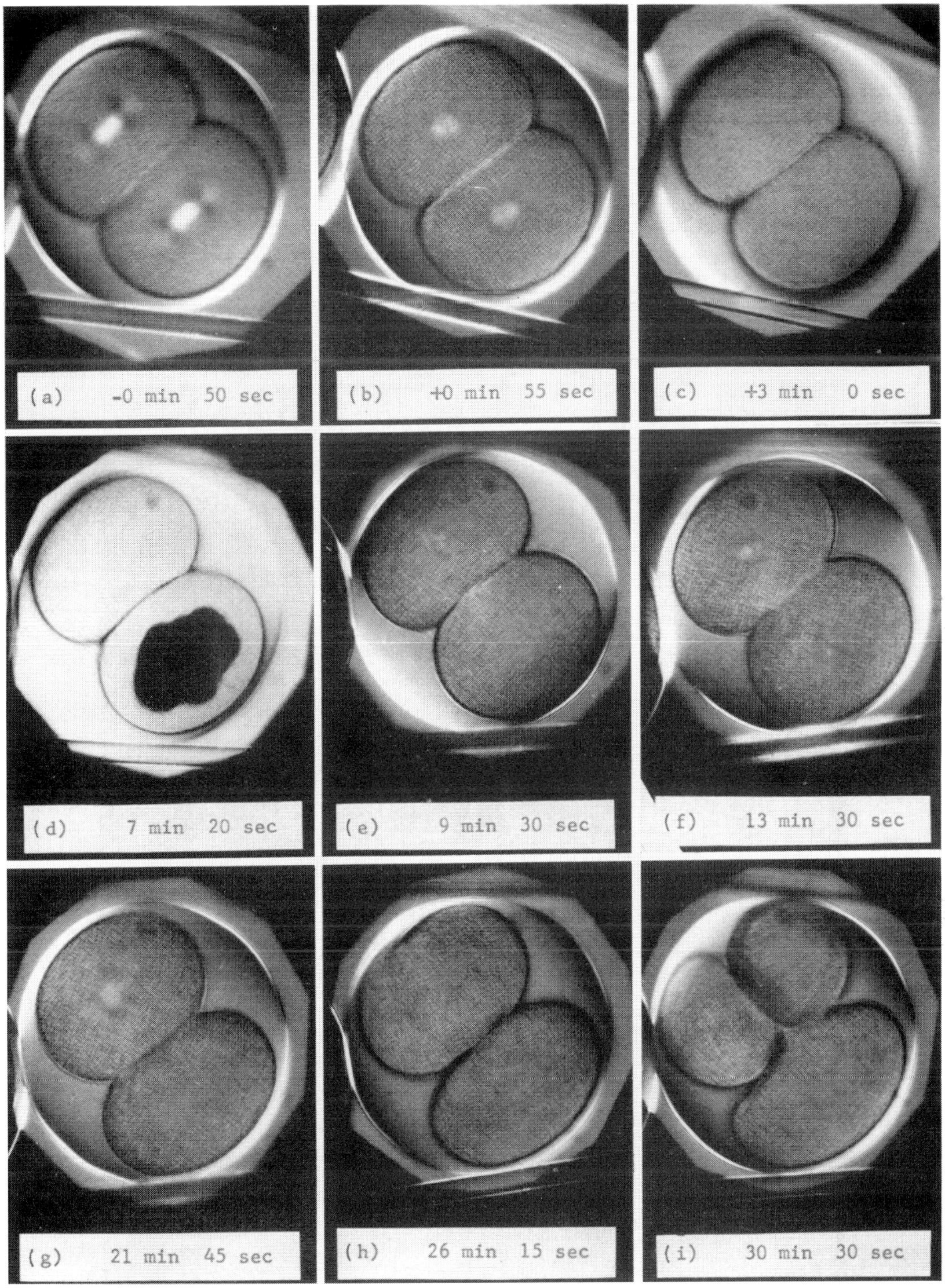

FIGURE 4 An egg at the two-cell stage with well-developed metaphase spindles (*a*) was treated with 3×10^{-5} M Colcemid at time *0* which was followed by a rapid loss of spindle birefringence (*b* and *c*). The level of Colcemid was reduced to 1×10^{-6} M at time *5′10″* and kept at this level for the rest of the experiment. For irradiation, an opaque spot in the plane of the field diaphragm was imaged in green light on the former spindle region of one cell (*d*). After 15 sec of 366 nm irradiation at time *7′30″*, there was recovery of birefringence in the unshielded cell and no detectable recovery in the shielded cell (*e*). Further localized irradiation increased the amount of birefringent material in the unshielded cell and caused some birefringence to appear in the shielded region (*f*, *g*). The fully irradiated cell cleaved after being irradiated at a reduced intensity level for about 3 more min over the next 10 min (*h*, *i*). The magnification is the same as in Fig. 3.

For *L. variagatus* eggs, 15 sec of irradiation of the whole cell at 366 nm often brought back detectable spindle fibers within 30 sec. These fibers were usually short with weak retardation and showed some further increase in birefringence during the next minute without further irradiation. This increase is opposed by the finite background level of Colcemid, and the birefringence can be seen to again decrease with time. Further irradiation in many instances increased the number of recognizable spindle fibers, made the existing ones longer, wider, and more strongly birefringent, and improved spindle organization.

The birefringence recovered was usually less and the spindles were smaller than those seen in normal divisions, although, in several instances, irradiation for 3 min gave recovery approaching that expected for a normal spindle. The degree of organization of the radiation-induced spindle was variable; well-formed bipolar spindles were seen, but most were more complex.

A stage comparable to anaphase was recognizable when the spindle was fairly well organized. It was characterized by spindle fibers with a less birefringent middle region whose occurrence was correlated with an increase in spindle length and

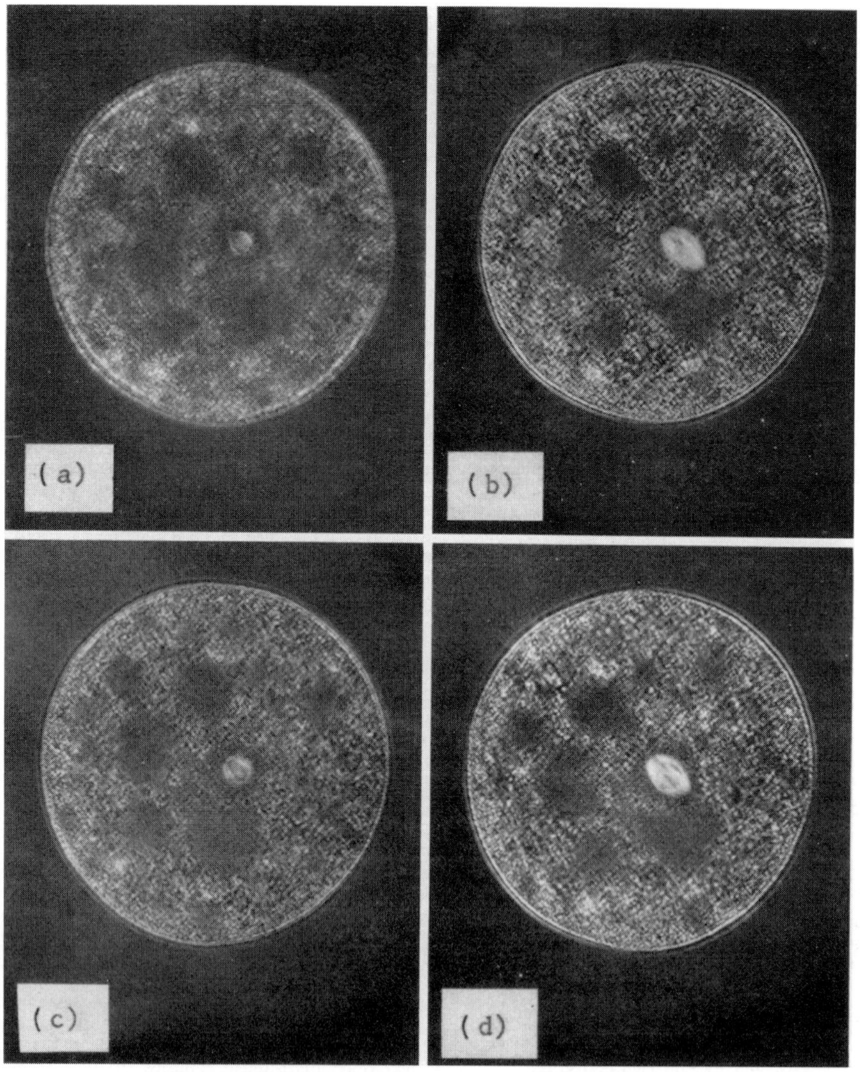

FIGURE 5 This figure shows an egg from *P. gouldii* in 1 x 10^{-6} M Colcemid seawater: (*a*) before irradiation, (*b*) then after irradiation at 366 nm, (*c*) then after 25 min in 1 x 10^{-6} M Colcemid, (*d*) then 1 min after another irradiation at 366 nm for 10 sec. × 650.

was usually followed by a rapid loss of birefringence and frequently by cleavage. Some of these observations are apparent in Figs. 3 and 4.

Comparable experiments were done with *Lytechinus pictus* eggs in which the spindles were inhibited before forming by putting cells in 3×10^{-6} M Colcemid either at 30 min after fertilization or just after first cleavage. Irradiation of these cells over the next 20 min or so routinely gave normal spindles and regular cleavage.

The first meiotic spindles of *P. ochraceous* eggs (starfish) and of *P. gouldii* eggs (annelid) have a prolonged metaphase which makes it possible to demonstrate clearly that the spindle birefringence regained on irradiation is still sensitive to Colcemid. In some experiments of this sort with *Pectinaria* eggs (such as the one shown in Fig. 5), Colcemid was applied before germinal vesicle breakdown so that a normal-size spindle had never been formed. This early application of Colcemid did not decrease the speed of recovery, which, if anything, was faster than in eggs in which the spindle was removed with 3×10^{-5} M Colcemid, nor did it obviously affect the final size of the spindle. The recovery of birefringence, as with *L. variagatus* eggs, was rapid with most of the change occurring within 30 sec.

Spindle birefringence did not reappear when Colcemid-blocked cells were exposed to large doses of radiation peaking at 436 nm given at incident intensities greater than those used at 366 nm.

DISCUSSION

The present work shows that it is possible to reverse the effects of Colcemid on the spindle and on cell division by irradiation. It is likely that this result is due to photo-inactivation of the Colcemid by light absorbed in the 355 nm peak, although we have not yet obtained direct evidence for this, such as an action spectrum and a decrease in absorbance at 355 nm on irradiation.

The cytological results require more work. However, the state of the spindle at the time Colcemid was applied and the time at which irradiation occurred were clearly important parameters. Partial cell irradiation experiments, such as shown in Fig. 4, often resulted in the recovery of birefringence behind the image of the opaque spot, but we are hesitant to interpret this until scattering and focal effects are more carefully ruled out.

If colchicine and Colcemid act on the association-dissociation reaction (7) of the microtubule protein, as seems reasonable from colchicine-binding studies of Borisy and Taylor (8), then the photochemical inactivation of Colcemid should be a means of varying the pool size or the local concentration of functionally effective microtubule protein. On this basis, one might liken the effect of a short exposure to light as a "concentration jump" which is followed by changes whose rate and extent may give information about the in vivo development of microtubules. The rate at which spindle length increased in some of our experiments could be placed at better than 10 μ/min, and this is a low estimate for the initial rate since it is certainly affected by the speed with which Colcemid diffused back in, and probably by diffusion of the microtubule protein by limited monomer pool size and by a possible requirement for nucleating centers.

The delay in or lack of cell cleavage, which often occurred when Colcemid-treated cells were irradiated to give recovery of spindle birefringence, may have resulted from experimental factors related to irradiation or the application of Colcemid or to characteristics of the mitotic cycle. Whatever the explanation, the use of controlled activation of the spindle by light for studying the relation of the metaphase spindle to anaphase and to cell cleavage seems promising.

Other microtubule-associated colchicine-sensitive processes have been described (1, 7), and the approach used on the spindle may be experimentally useful in their study.

SUMMARY

The colchicine derivative Colcemid[R] was used for causing dissolution of the spindle and inhibiting cleavage in early divisions of *Lytechinus variagatus* eggs. Irradiation of such Colcemid-blocked cells at 366 nm reversed these effects as determined by the recovery of spindle birefringence and by cleavage. We tentatively conclude that 366 nm radiation causes a photochemical rearrangement in the Colcemid which reduces or destroys its effect on the spindle.

It is a pleasure to thank Dr. Hidemi Sato for the biological material used in this study and Mr. Richard Markley for the photography.

This work was supported by National Science

Foundation Grant No. 1520 and by National Institutes of Health Grant No. 10171-04.

Received for publication 15 October 1969, and in revised form 14 November 1969.

BIBLIOGRAPHY

1. Eigsti, O. J., and P. Dustin. 1955. Colchicine in Agriculture Medicine, Biology and Chemistry. Iowa State College Press, Ames, Iowa.
2. Inoué, S. 1952. *Exp. Cell Res. Suppl.* **2**:305.
3. Forbes, E. J. 1955. *J. Chem. Soc.* 3864.
4. Chapman, O. L., H. G. Smith, and R. W. King. 1963. *J. Amer. Chem. Soc.* **85**:803, 806.
5. Hatchard, C. G., and C. A. Parker. 1956. *Proc. Roy. Soc. Ser. A.* **235**:518.
6. Chaet, A. B. 1964. *Biol. Bull. (Woods Hole).* **126**:8.
7. Inoué, S., and H. Sato. 1967. *J. Gen. Physiol.* **50** (pt. 2):259.
8. Borisy, G., and E. Taylor. 1967. *J. Cell Biol.* **34**: 535.

Note Added in Proof: Wilson and Frieden (*Biochemistry.* 1967. **6**:126) have shown that 2×10^{-6} M lumicolchicine does not inhibit cell division in grasshopper embryos nor does it inhibit the binding of colchicine in homogenates of such cells.

Reprinted from *Biophysical Journal*, Vol. 15, pp. 725–744, 1975.

FUNCTIONAL ORGANIZATION OF MITOTIC MICROTUBULES

PHYSICAL CHEMISTRY OF THE IN VIVO EQUILIBRIUM SYSTEM

SHINYA INOUÉ, JOHN FUSELER, EDWARD D. SALMON, *and* GORDON W. ELLIS

From the Program in Biophysical Cytology, Department of Biology, University of Pennsylvania, Philadelphia, Pennsylvania 19174, and Marine Biological Laboratories, Woods Hole, Massachusetts 02543

ABSTRACT Equilibrium between mitotic microtublues and tubulin is analyzed, using birefringence of mitotic spindle to measure microtubule concentration in vivo. A newly designed temperature-controlled slide and miniature, thermostated hydrostatic pressure chamber permit rapid alteration of temperature and of pressure. Stress birefringence of the windows is minimized, and a system for rapid recording of compensation is incorporated, so that birefringence can be measured to 0.1 nm retardation every few seconds. Both temperature and pressure data yield thermodynamic values ($\Delta H \simeq 35$ kcal/mol, $\Delta S \simeq 120$ entropy units [eu], $\Delta \bar{V} \simeq 400$ ml/mol of subunit polymerized) consistent with the explanation that polymerization of tubulin is entropy driven and mediated by hydrophobic interactions. Kinetic data suggest pseudo-zero-order polymerization and depolymerization following rapid temperature shifts, and a pseudo-first-order depolymerization during anaphase at constant temperature. The equilibrium properties of the in vivo mitotic microtubules are compared with properties of isolated brain tubules.

INTRODUCTION

In contrast to the relatively stable and repeatedly operable force-generating structures found in muscle and flagella, the mitotic spindle is a transient structure. It is newly assembled each time a cell prepares to divide and is disassembled with completion of mitosis. The fibrous elements of the mitotic spindle provide the transient, structural framework and force-producing system which position and distribute chromosomes and other organelles.

The spindle fibers are birefringent and can be visualized with a sensitive polarizing microscope. At each division of a living cell, birefringent fibers can be seen to arise, dynamically organize into the mitotic spindle, and distribute chromosomes and organelles to the daughter cells. These events are vividly displayed in time-lapse motion pictures of developing eggs, spermatocytes, pollen mother cells, and endosperm cells, for example, as in film sequence 1, 3–5 shown at the Minneapolis meetings. (See Inoué,

1964; Inoué and Sato, 1967; Sato and Izutsu, 1974; and Fuseler, 1975, for photographs and description of birefringence changes.)

In electron micrographs, spindle fibers appear as a bundle of ca. 24 nm diameter tubular protein filaments, or microtubules. The oriented array of microtubules accounts for the positive birefringence of spindle fibers[1] (Sato et al., 1971; reviewed by Inoué and Ritter, 1975). In living dividing cells, the change of spindle birefringence reflects fine-structural reorganization of spindle fibers, and in many cases the assembly and disassembly of mitotic microtubules (reviewed in Inoué and Ritter, 1975).

In addition to the organizational changes of spindle fibers which occur naturally during mitosis, one can artificially induce rapid disassembly or assembly of mitotic microtubules. Cold, hydrostatic pressure, colchicine, and Colcemid (Ciba Pharmaceutical Company, Summit, N.J.) can abolish birefringence, depolymerize mitotic microtubules, and halt mitosis in 1–2 min. Removal of these depolymerizing agents, even after prolonged application, allows rapid microtubule reassembly, spindle reorganization, and resumption of mitosis. Protein synthesis is not required for reassembly, and the normal amount of mitotic microtubules can be reversibly doubled or even further increased by temperature elevation or by addition of D_2O (Inoué and Sato, 1967) or glycols (Rebhun et al., 1974). In living cells therefore, mitotic microtubules exist in an equilibrium with a pool of polymerizable subunits (Inoué, 1964; Inoué and Sato, 1967), presumably the 110,000 mol wt heterodimer of tubulin (see e.g., Bryan, 1974).

In 1972 Weisenberg successfully polymerized microtubules in vitro in solutions of tubulin extracted from brain. The microtubules were reversibly disassembled by cold, their polymerization was inhibited by colchicine and the repolymerized microtubules were microscopically undistinguishable from native tubules (reviewed by Olmsted and Borisy, 1973 *a*).

In the present article we report our contribution to the in vivo quantitation of the equilibrium system and discuss its thermodynamic significance. The general features of the in vivo equilibrium system, the justification for using birefringence measurements for study of the equilibrium, and the significance of the equilibrium system to chromosome movements are discussed in a companion article (Inoué and Ritter, 1975).

TEMPERATURE CONTROL AND RETARDATION RECORDING SYSTEM

Time-lapse film sequence 2 shown at the Minneapolis meetings demonstrates the rapid disappearance of spindle birefringence in sea urchin (*Lytechinus variegatus*) eggs exposed to cold, and upon warming, the quick reappearance of birefringence followed by normal cell division (Fig. 1 and 19). Photographic recording and measurement of the low values of spindle birefringence (retardation $< \lambda/1{,}000$) during these transient changes necessitated the development of a new, stress birefringence-free temperature

[1] Sato, H., G. W. Ellis, and S. Inoué. 1975. Microtubular origin of mitotic spindle form birefringence. Submitted for publication.

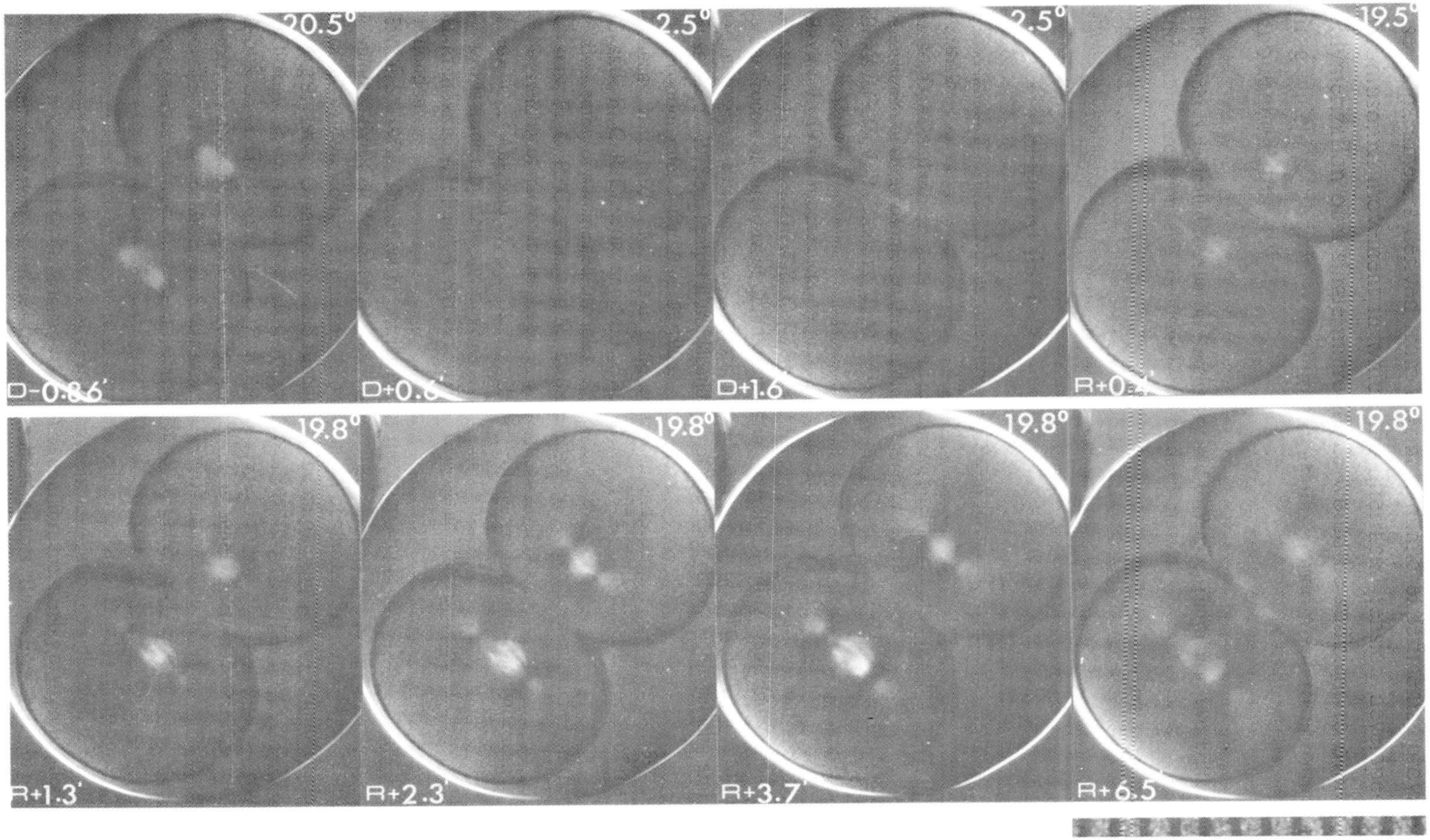

FIGURE 1 Two-cell stage sea urchin *Lytechinus variegatus* eggs exposed to cold during prometaphase (upper cell) and metaphase (lower cell). At time *D*, temperature is dropped from 20.5°C to 2.5°C. Spindle birefringence disappears in ca. 1 min. After 6.2 min in the cold the cell is rewarmed to 19.8°C at *R*. Birefringence recovers rapidly, the spindle is reformed, and anaphase progresses (at $R + 6.5$ min in the lower cell). Time in minutes. Scale interval 10 μm.

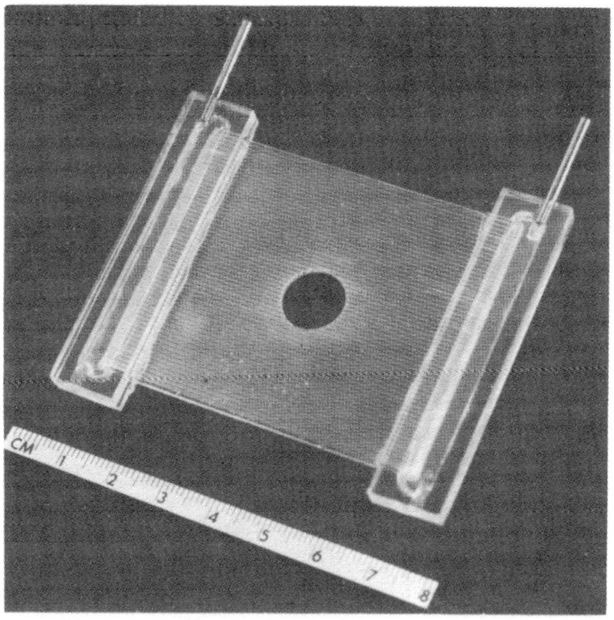

FIGURE 2 Temperature control slide before mounting the two 22 × 22 mm stress-free glass slips which cover the viewing port in the middle of the slide.

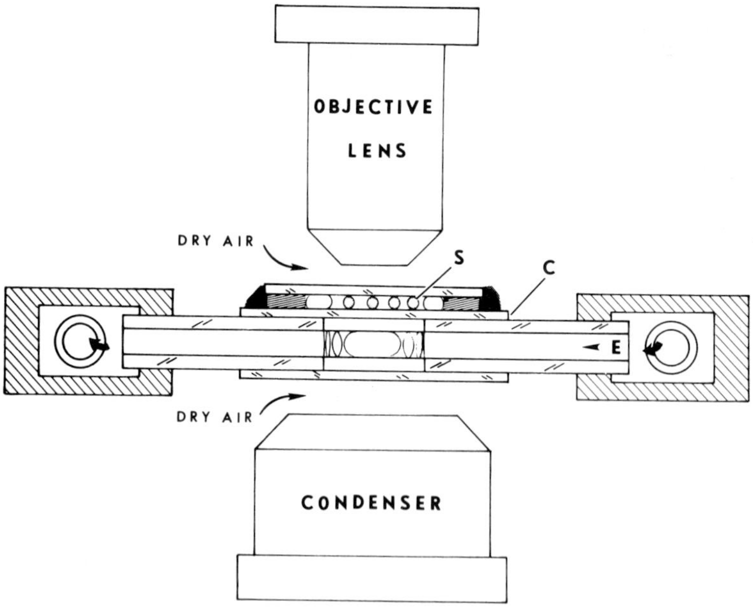

FIGURE 3 Schematic cross section of temperature control slide. Temperature controlled 50% ethanol (*E*) flows through the slide. Specimen (*S*), sandwiched between two cover slips in a column of culture medium surrounded by humidity-equilibrated gas phase, is separated from the alcohol solution by a single 0.17 mm thick cover glass (*C*) and closely follows the alcohol temperature as shown in Fig. 7.

control slide and a system for measuring retardation every few seconds (Inoué, et al., 1970).

A temperature control slide of the following construction permitted rapid shift of temperature upwards and downwards, free from stress birefringence due to differential or anisotropic expansion. A single row of about 40, 0.8 mm OD glass capillaries laid side by side was bonded with silicone or epoxy cement between two, 75 × 35 mm cover slips, previously etched on their inner faces for improved adhesion. The viewing port, a 10 mm diameter hole cut with an abrasive jet through the center of the glass laminate, was covered on both sides with 22 × 22 mm birefringence-free coverslips. The two ends of the laminate where the capillaries open were cemented into two plastic manifolds containing, respectively, the inflow and exit pipes for the temperature control fluid (Fig. 2). The specimen were mounted externally against the 22 × 22 mm glass slip covering the 10 mm port as shown in Fig. 3.

Through this slide, temperature-regulated fluid (50% ethanol) is perfused as shown in Figs. 4 and 5. Fluid from the thermostatically controlled hot and cold reservoirs flows by gravity at a constant rate (100–200 ml/min) to a specially designed mixing

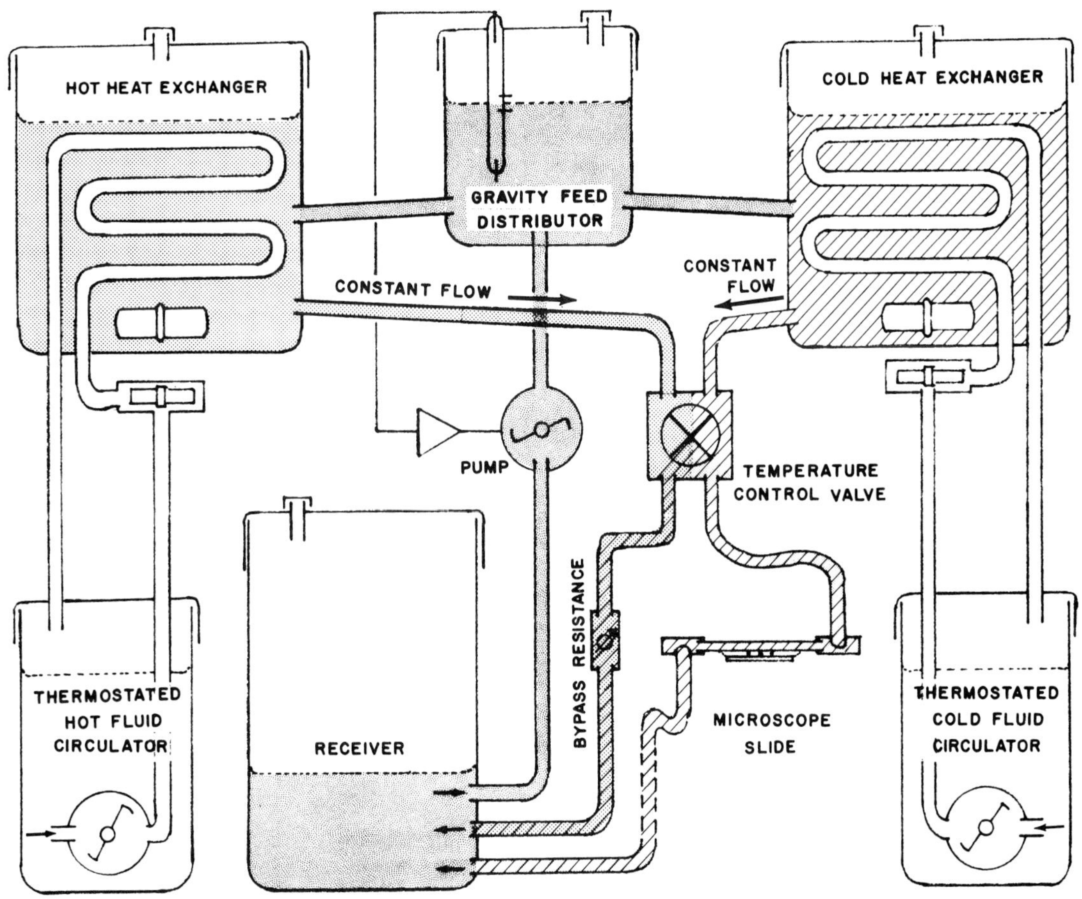

FIGURE 4 Schematic diagram of temperature control system.

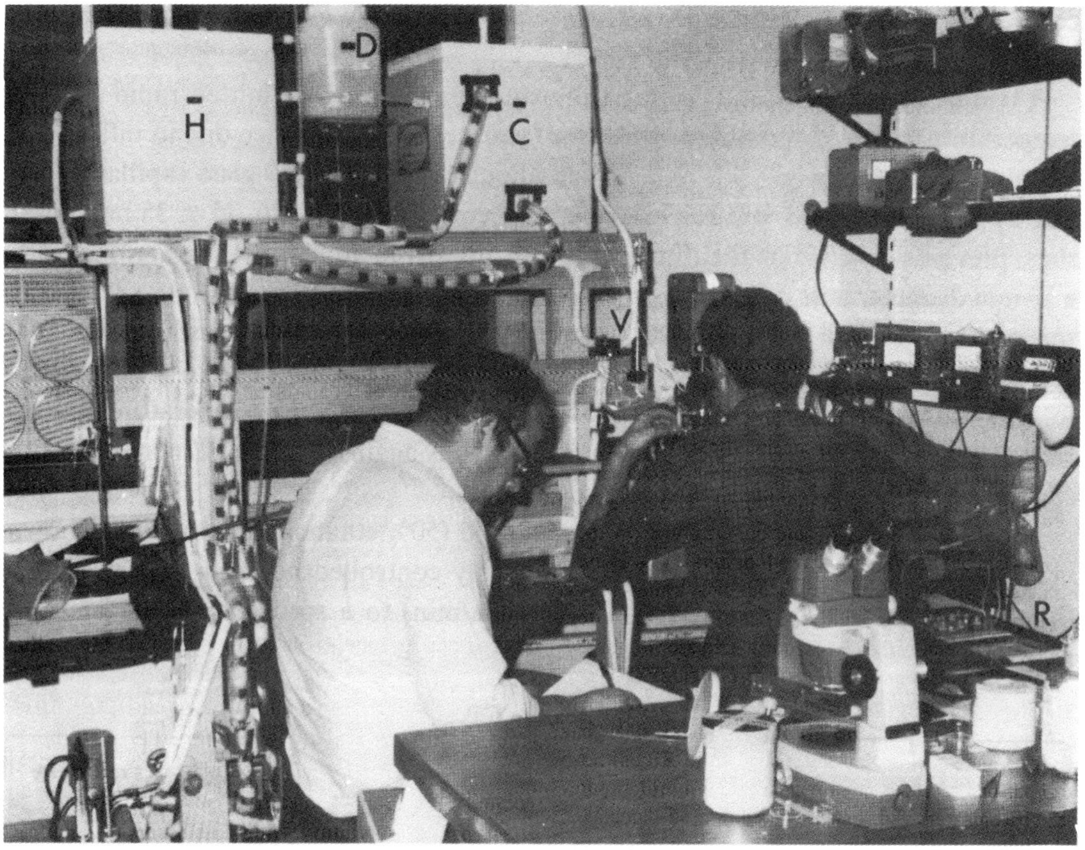

FIGURE 5 Temperature control system and recording compensator in operation. Hot (*H*) and cold (*C*) storage tanks, gravity-feed distributor (*D*) at top left. Temperature controlling mixing valve (*V*) above operator's left hand which is shown controlling the Brace-Köhler compensator. The microswitches, which signal for recording of compensator position, are placed near the operator's right hand which also focuses the microscope. Chart recorder (*R*), right of the operator.

valve (Fig. 6) which diverts the required proportions of hot and cold fluid to the slide and delivers the remainder to the waste receiver. The low heat capacity of the microscope slide and mixing valve and the short tube connecting the two allows the temperature of the specimen to be changed with a time constant of less than 3 s (Fig. 7). The waste cooling or heating fluid is pumped back up to a de-gassing distributor from which the hot and cold tanks are replenished to a constant level. Accidental spillage of temperature control fluid onto the microscope is minimized by keeping the pressure head at the slide slightly below atmospheric pressure. The constant-head gravity feed design facilitates the maintenance of a pressure differential and provides a constant flow free from pulsation. Except for vent holes the circulating fluid is enclosed in order to keep out birefringent and light-scattering contaminants.

Birefringence of the specimen is measured as retardation with a Brace-Köhler compensator (see e.g., Hartshorne and Stuart, 1960) on whose control axis we mounted a miniature precision potentiometer. Supplied with a constant DC voltage, the linear potentiometer provides a voltage output proportional to the compensator orientation.

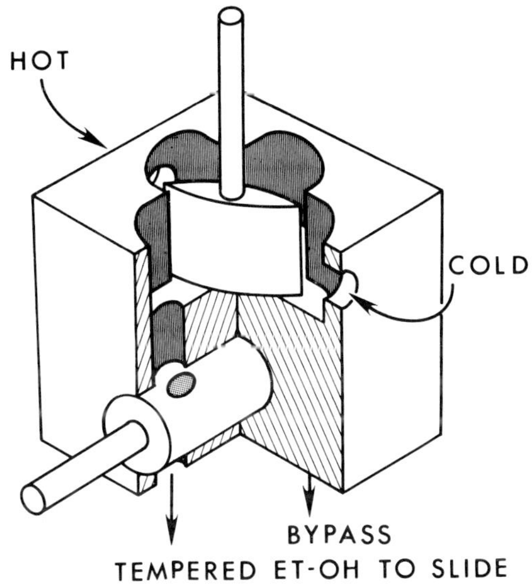

FIGURE 6 Schematic cut-away view of temperature-control mixing valve. Top, front and back covers and O-ring gaskets are not shown. The body of the valve is machined out of polycarbonate resin for alcohol resistance. Temperature of the 50% ethanol (ET-OH) to be perfused is regulated by the top spindle; boat shaped vane proportions hot and cold solutions so that the flow rate through the slide and the bypass circuit each remain constant. The lower valve controls the amount of flow and the pressure head at the slide.

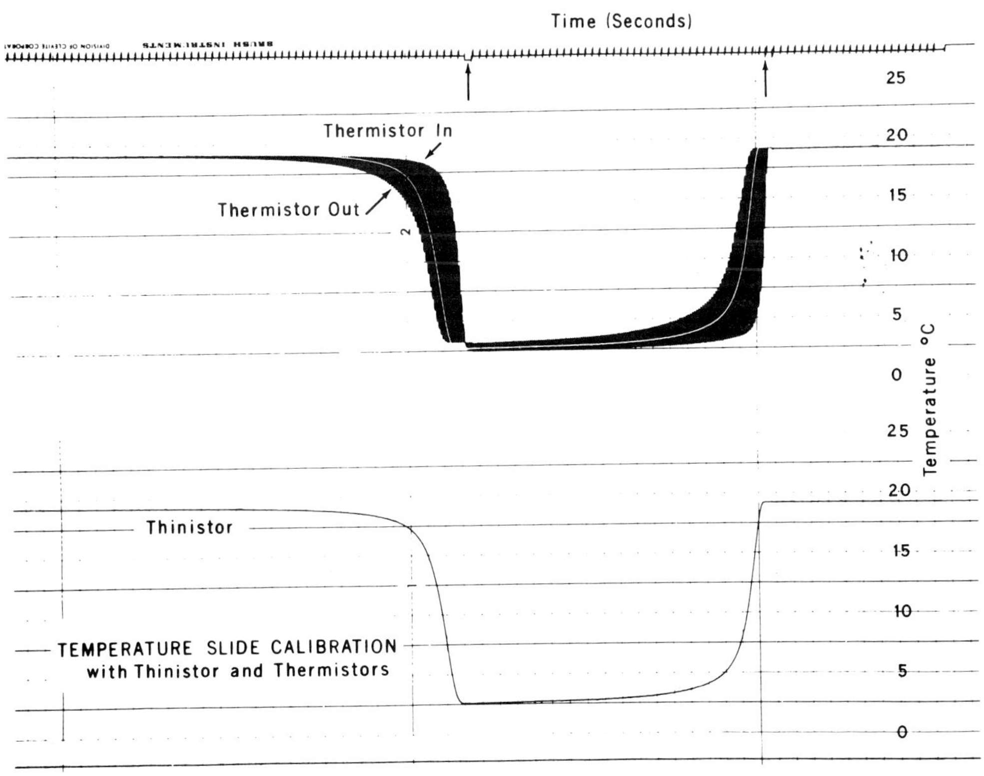

FIGURE 7 Chart record showing response time of temperature control slide. The temperatures of the alcohol solution, measured by thermistor beads placed in the stream near the in-flow out-flow ports of the slide, are alternately monitored four times a second and displayed on the upper analogue channel. The output of a thin, flat thermistor (Thinistor, Victory Engineering Corp., Springfield, N.J.) mounted in place of the specimen is recorded on the lower channel. The time constant for shifting temperature up or down as detected by the Thinistor is about 3 s. At equilibrium the specimen temperature is within 0.1°C of averaged in-flow out-flow solution temperatures.

Once the compensator is adjusted to extinguish the specimen, the observer actuates a microswitch which triggers a one-shot timed pulse thus recording the compensator orientation onto a strip chart recorder (Fig. 8). A time of day signal is recorded by an event marker and the inflow and exit control fluid temperature, measured by small thermistors, are alternately recorded on another analog channel. The recorder also registers the setting of the compensator with each exposure of the still or movie camera.

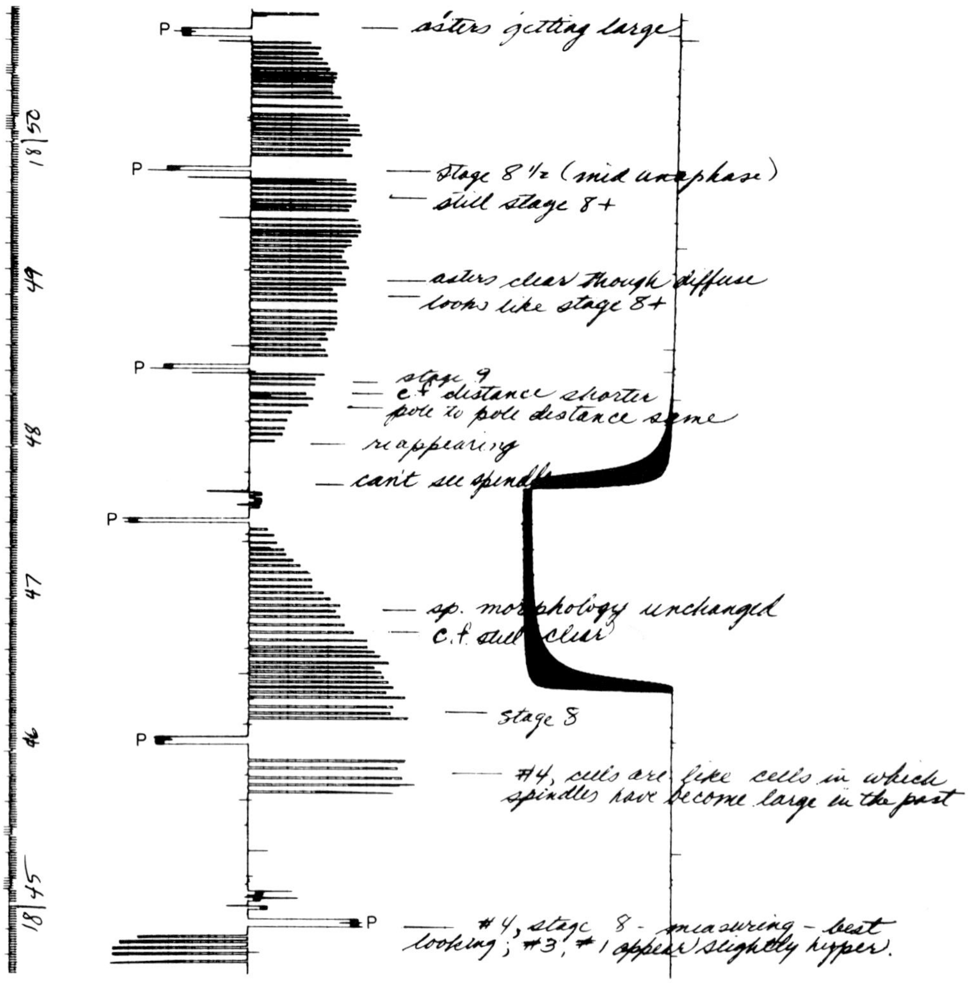

FIGURE 8 Chart record showing: time of day, photodocumentation, retardation measurements, and slide temperature. Far left channel, interval marks for 1 s (right toggle), 10 s (left and right toggle), and 1 min (left toggle only). Every 5 min, time of day signal is given by the following code: during the first 9 s, the number of toggles to the left indicates the number of tens of minutes—if no toggling (to the left) occurred previously at the minute mark then 5 min is added to the minutes reading. During the 21–29 s interval, the intergral number of times the hours is divisible by six is indicated, and during the 11–19 s interval the remaining hours are indicated. The toggling of the pen is generated by a series of sequential CLARE no. 11 and no. 26 stepping switches driven by a 1 s synchronous motor timer. Left analogue channel, compenstor angle measuring spindle extinction (1 s bars), background extinction (short 2 s bars), and compensator setting during photographic exposure (*P*). Although not shown here, photoexposure also toggles far right event marker. Width of bar indicates duration of exposure. Right analogue channel, slide temperature as described for Fig. 7.

THERMODYNAMICS OF THE EQUILIBRIUM SYSTEM

When a cell in mitosis is chilled to an intermediate temperature rather than to a temperature so low as to completely depolymerize the spindle microtubules, the spindle birefringence reaches an intermediate value as shown in Fig. 9. The same equilibrium value is reached whether the final temperature is approached from above or below. Johnson and Borisy (1975) recently showed, for a purified solution of tubulin extracted from brain, that microtubules in vitro can also be in a true equilibrium with the 110,000 mol wt (hetero-) dimer of tubulin.

As shown earlier (Inoué, 1959), the equilibrium birefringence of a metaphase arrested spindle in the oocyte of *Chaetopterus* approaches a plateau value (A_o) at the higher range of physiological temperatures. These data, relating equilibrium retardation to temperature, are plotted in Fig. 10 together with recent data on *Chaetopterus* obtained by Salmon (1975). The solid, sigmoid curve shown in Fig. 10 was theoretically specified by the straight line in Fig. 11 as discussed below.

The birefringence (B) measures the concentration of tubulin in the oriented microtubules (Inoué and Ritter, 1975; Sato et al., 1971, 1975). Concentration of the free subunits is given by ($A_o - B$) since A_o represents the tubulin concentration when the equilibrium is pushed all the way towards the polymer.

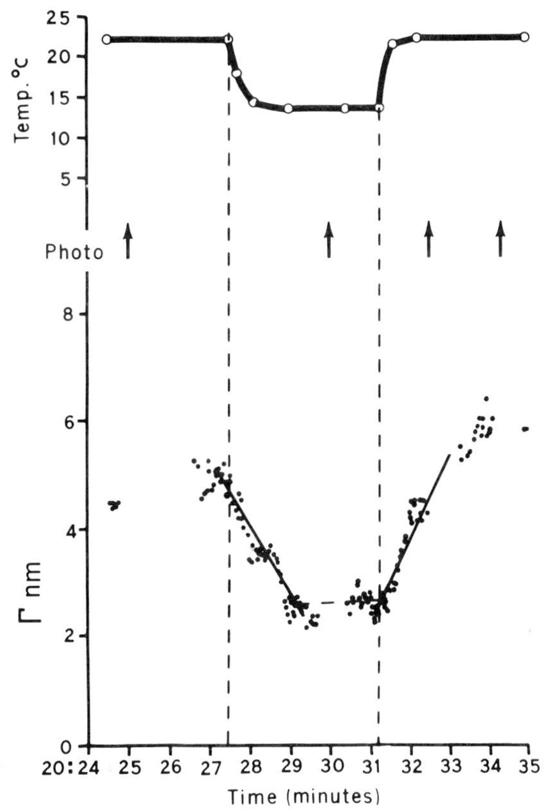

FIGURE 9 Birefringence measured as retardation (Γ) of a metaphase-blocked spindle in a *Chaetopterus* oocyte treated with 45% D_2O sea water and then chilled from 22.5° to 13.5°C. Equilibrium retardation is reached within 2 min.

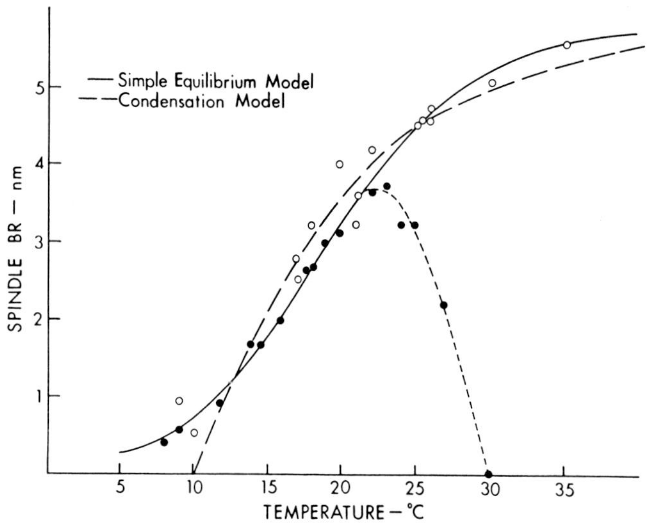

FIGURE 10 Equilibrium values of *Chaetopterus* oocyte spindle retardation vs. temperature. Open circles from Inoué, (1959), solid circles from Salmon (1973). The solid curve is defined by Eq. 4 and plotted for constant ΔH and ΔS values obtained by least squares fitting of data as shown in Fig. 11. For a discussion on the fit of the broken curve "condensation model" see Salmon (1975). (From Salmon, 1975.)

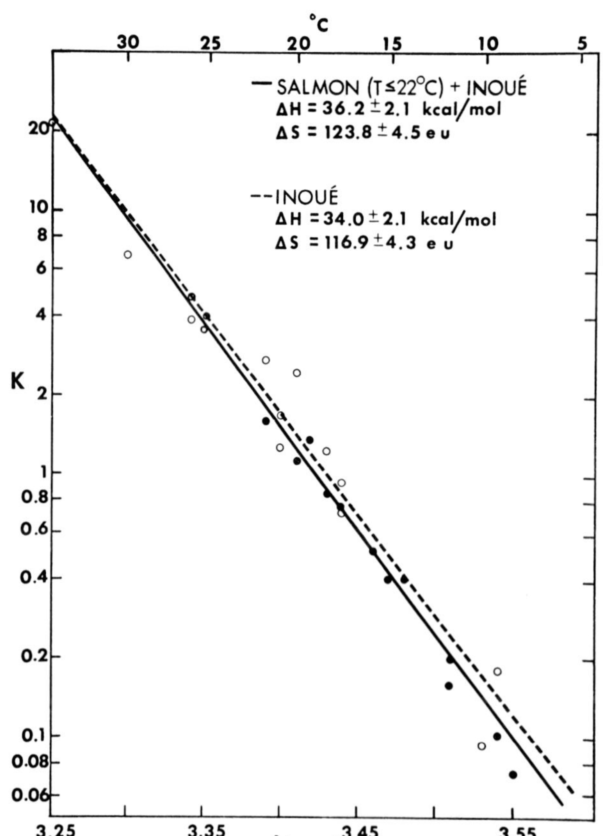

FIGURE 11 Data in Fig. 10 shown as van't Hoff plot: $K = \ln\,[B/A_o - B)]$ vs. inverse absolute temperature. A_o is high temperature asymtote in Fig. 10. (From Salmon, 1975.)

Thus

$$A_o - B \overset{K(T,P)}{\rightleftarrows} B. \quad (1)$$

The equilibrium constant $K(T,P)$ is given by

$$K(T,P) = [B]/[A_o - B]. \quad (2)$$

Since the free energy change (ΔG^o) of an equilibrium chemical system is given by:

$$K(T,P) = e^{-\Delta G^o/RT},$$
$$-\Delta G^o = RT \ln K(T,P) = RT \ln ([B]/[A_o - B]). \quad (3)$$

At 1 atm, the Gibbs free energy ΔG_1^o is:

$$\Delta G_1^o = \Delta H - T\Delta S,$$

where ΔH is the enthalpy increment, T is the absolute temperature, and ΔS is the entropy change per mole of reactant. Combined with Eq. 3,

$$-\ln ([B]/[A_o - B]) = (\Delta H/RT) - (\Delta S/R). \quad (4)$$

As shown in Fig. 11, the van't Hoff plot, or ln $[B/(A_o - B)]$ plotted against $1/T$, yields a straight line whose slope provides a ΔH value of 35 kcal/mol of subunit associated, and its intercept gives a positive ΔS of 120 entropy units (eu).

These measurements were made on metaphase arrested spindles in *Chaetopterus* oocytes. Similar thermodynamic behavior has been observed in active metaphase spindles of plant and animal cells, both in meiosis and in mitosis (Stephens, 1973; Fuseler, 1973*a*). In all cases, polymerization is accompanied by a high positive entropy (ΔS = 100–300 eu) and positive enthalpy (ΔH = 25–40 kcal). At physiological temperatures the free energy (ΔG^o) released upon polymerization is less than 1 kcal/mol.

This endothermic, entropy driven reaction appears to reflect destructuring of water from the tubulin subunits upon polymerization. In other words the assembly of microtubules appears to be driven by hydrophobic interactions. Similarly high values of positive ΔH and ΔS are characteristically observed in hydrophobic associations of proteins generally.

As with several other hydrophobically mediated protein polymerizing systems, equilibrium between tubulin and microtubules is shifted towards polymerization by heavy water, with peak effect at ca. 45% D_2O volume concentration (Inoué and Sato, 1967; Olmsted and Borisy, 1973*b*). Compared with H_2O, D_2O forms a tighter association with itself (e.g., Sidgwick, 1950) and forms micelles at lower concentration of detergent than in H_2O (Kresheck et al., 1965).

FIGURE 12

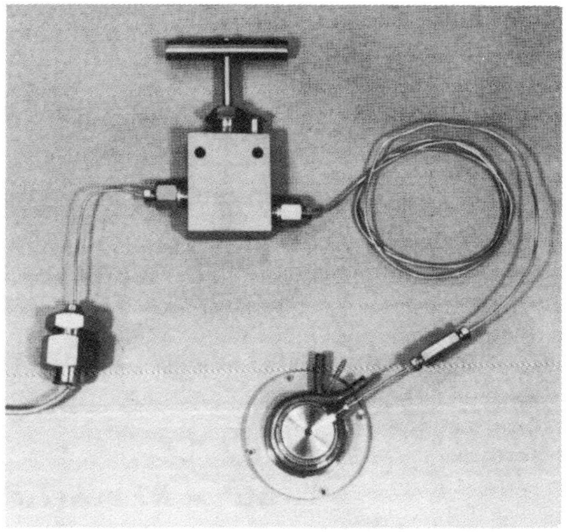

FIGURE 13

FIGURE 12 Miniature pressure chamber designed for use with polarization and phase contrast microscopes shown inverted with bottom (left) removed. At 5,500 lb/in^2, window stress birefringence is still sufficiently low to permit polarized light observations of spindle birefringence in living cells. Total retardation of the spindle at 1 atm is less than one hundredth of a wave length of light. (From Salmon and Ellis, 1975.)

FIGURE 13 Microscope pressure chamber with "pump" and pressure gauge attached. Because of the very small displacement volumes involved, several thousand pounds per square inch of pressure can be generated by a single turn of a needle valve (center of photo) which replaces a conventional pump. The double hexagonal unit to the left, houses the strain-gauge pressure transducer whose electrical resistance provides the pressure reading. (From Salmon and Ellis, 1975.)

PRESSURE INDUCED CHANGE OF EQUILIBRIUM

Thermodynamic analysis of the polymerization equilibrium in *Chaetopterus* spindle was extended, using pressure as a variable. A new microscope chamber (Figs. 12–14) permits rapid application of hydrostatic pressures up to 15,000 lb/in^2, with sufficiently low window stress birefringence to permit polarized light microscopy of spindles up to ca. 5,500 lb/in^2 and phase contrast observation of chromosomes up to ca. 10,000 lb/in^2 (Salmon and Ellis, 1975).

In living cells, application of hydrostatic pressure reduces spindle fiber birefringence signaling the depolymerization of spindle microtubules. As with temperature, birefringence change is totally reversible and a *Chaetopterus* egg subsequently fertilized can develop as a normal embryo.

At constant temperature, an equilibrium birefringence is reached for each given pressure (Fig. 15). The equilibrium retardation measured at three temperatures is shown in Fig. 16. As the figure shows, the data closely fit theoretical curves derived from Eqs. 2 and 5 for a constant molar volume increment of association ($\Delta \overline{V}$) of 400 ml/mol of polymerizing subunit. $\Delta \overline{V}$ was derived from the van't Hoff plots as described below.

When pressure is altered, the thermodynamic Eq. 3 takes the general form (Salmon, 1975),

$$-\ln K(T,P) = \Delta G_1^o/RT - (P - 14.7)\,\Delta \overline{V}/RT, \qquad (5)$$

FIGURE 14 Pressure chamber mounted on the stage of a polarizing microscope. The chamber fits in a water jacketed sleeve. (From Salmon and Ellis, 1975.)

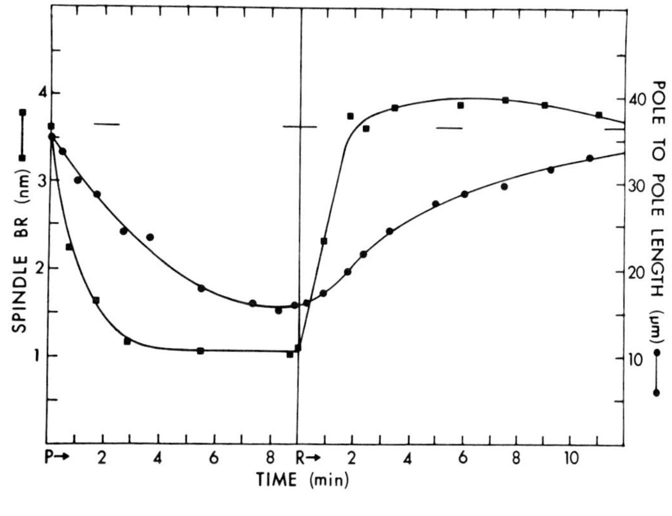

FIGURE 15 Birefringence retardation and length of *Chaetopterus* spindle exposed at *P* to 2,000 lb/in^2. At *R* pressure is returned to 1 atm (=14.7 lb/in^2). (After Salmon, 1975.)

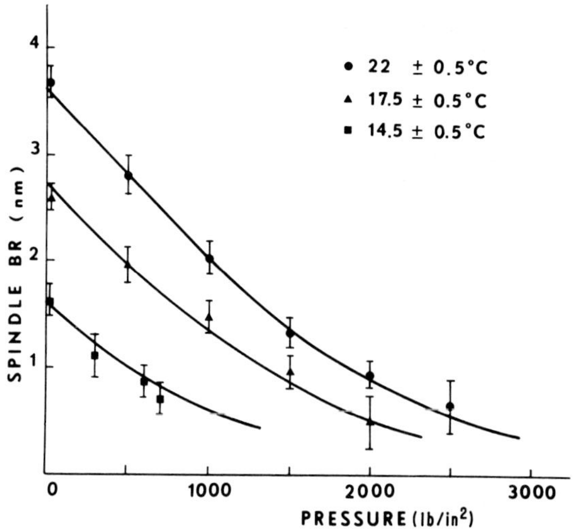

FIGURE 16

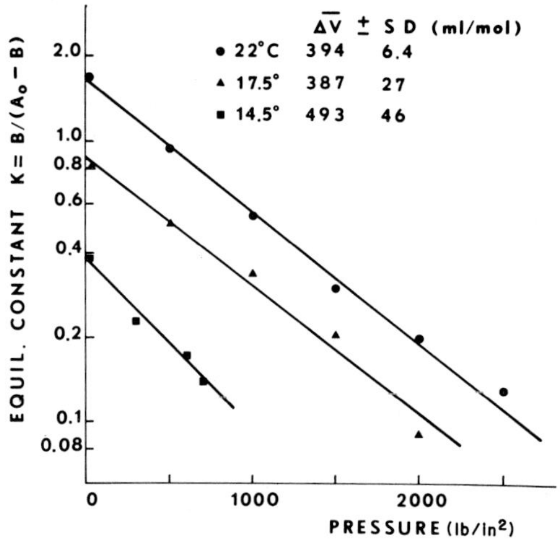

FIGURE 17

FIGURE 16 Equilibrium retardation measured as function of pressure for *Chaetopterus* spindle at three temperatures. The three curves are theoretical curves defined by Eqs. 2 and 5, and are drawn for constant molar volume increments of 400 ml/mol of subunit polymerized. ΔG_1^o for each of the three curves is obtained from Eq. 3 and Fig. 11.

FIGURE 17 van't Hoff plot of data shown in Fig. 16; plotted as ln $[B/(A_o - B)]$ vs. P for three temperatures. Each straight line which is least squares fitted to the data, yields the $\Delta \bar{V}$ value indicated. (After Salmon, 1975.)

where ΔG_1^o is the Gibbs free energy at 1 atm (=14.7 lb/in^2, and pressure P is expressed in pounds per square inch. Consistent with this equation for constant ΔV and with ln $K(T,P)$ as defined by Eq. 2 ln $[B/(A_o - B)]$ plots against P as a straight line for each temperature (Fig. 17). Measurements made at three temperatures yield straight lines displaced as predicted from Eqs. 2 and 5 and the values of ΔG_1^o obtained by varying temperature at unit atmosphere.

The molar volume change ($\Delta \bar{V}$) is given by the slope of the lines as ca. 400 ml/mol (Fig. 17). This value, as with the ΔH and ΔS values, is comparable with the molar volume changes of association seen in other hydrophobic, entropy-driven polymerization systems such as tobacco mosaic virus, flagellin, and actin. Depending on the system involved, however, different equations have been used to describe the polymerization-depolymerization equilibria (see Stephens, 1973; Fuseler, 1973*a*; Salmon, 1975; and Johnson and Borisy, 1975, for discussions of this point and for tables comparing thermodynamic parameters for the various entropy-driven, polymerization systems).

KINETICS OF SPINDLE RETARDATION CHANGE

Figs. 18 and 19 illustrate the kinetic changes of spindle retardation upon sudden chilling, sudden rewarming, and as seen in natural anaphase at constant temperature. In this sea star (*Asterias forbesi*) egg cooled at −1°C, the retardation of the spindle fibers, measured halfway between the pole and chromosomes, decays lineraly to zero at approximately 3.3 nm/min; upon rewarming to 17°C, retardation recovers at the same

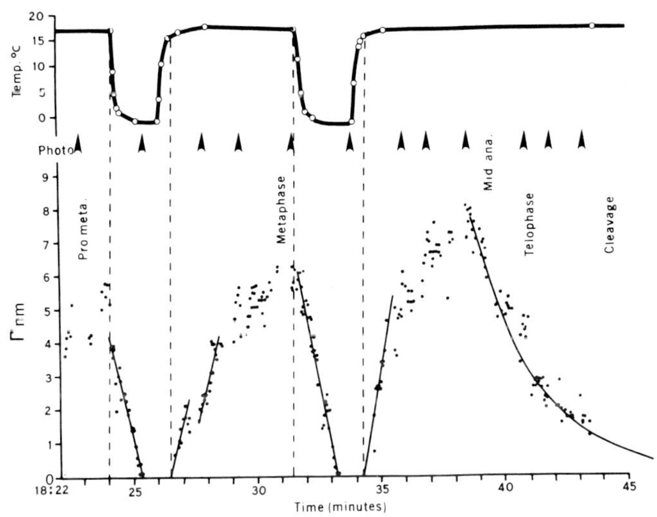

FIGURE 18 Birefringence changes, measured as retardation (Γ), of mitotic spindle in the egg of a sea star *Asterias forbesi*. The first two drops in retardation measure the rapid depolymerization of mitotic microtubules upon chilling. Upon rewarming, spindle birefringence recovers just as fast as it disappeared, reflecting the rapid reassembly of microtubules. After initial reassembly, retardation fluctuates as spindle is reorganized (see Fig. 1). The third drop in retardation takes place without chilling. This reflects the natural depolymerization of microtubules in anaphase. The anaphase retardation decay is logarithmic in contrast to the linear decay induced by sudden chilling.

rate linearly, until fiber reorganization gives rise to secondary changes. Similar "pseudo-zero-order" kinetics have been observed on spindle isolated in tubulin-containing solutions and depolymerized in vitro by sudden chilling (Inoué et al., 1974).

Taken at face value this pseudo-zero-order decay would seem to suggest: (*a*) that the monomer-polymer equilibrium between tubulin and microtubules is shifted strongly in favor of the monomer so that little impetus toward repolymerization is felt by the depolymerizing tubules; and (*b*) that depolymerization is confined to tubule ends (so that depolymerization proceeds from a fixed number of sites) rather than elsewhere along the tubule. We view this interpretation with some reservation, however, since Gaskin et al. (1974) have shown that the number of microtubule ends increased by sonic fragmentation does not alter the rate of cold induced depolymerization in vitro. Johnson and Borisy (1975) on the other hand find enhanced depolymerization by cold in more mildly sheared microtubule solutions. In vivo, the retardation appears to drop uniformly along the length of fibers when cells are chilled sufficiently rapidly so that the fibers do not shorten during chilling (see Fig. 1 and Inoué and Ritter, 1975).

The rates of retardation change vary within a limited range in cells from different species but are relatively unaffected by the medium in which the cell is immersed (Table I). Mitotic stage-dependent variation is consistently observed for some cells but a fixed pattern has not yet emerged for different cell types.

In contrast to the linear retardation decay encountered in cold-induced depolymerization, the spontaneous retardation decay seen in anaphase follows a logarithmic course (Fig. 18). The retardation decay in colchicine (and Ca^{++}) induced depolymeri-

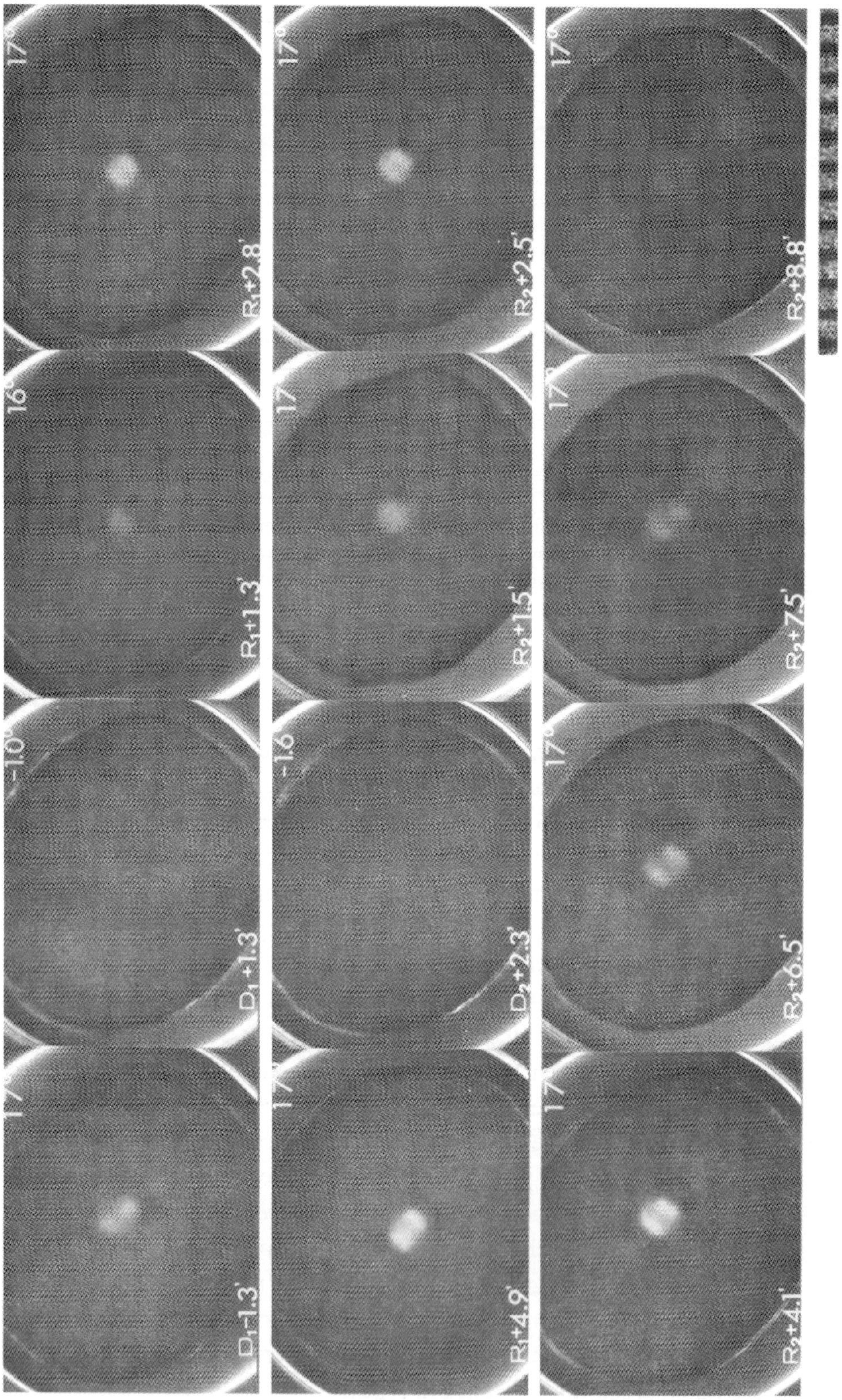

FIGURE 19 Photographs of *Asterias* egg taken at the times indicated by arrows in Fig. 18.

TABLE I

RATE OF RETARDATION CHANGE OF MITOTIC SPINDLE IN VIVO UPON SUDDEN DROP OR RISE OF TEMPERATURE

Cells	Temperature		Retardation change rate for step change in temperature		No. of cells	Observed in
	Upper	Lower	Drop	Rise		
PLANT ENDOSPERM	°*C*		*nm/min*			
Haemanthus katherinae						
Meta.	20	4	1.4 ± 0.3	1.3 ± 0.4	6	Endosperm fluid
Phragmopl.	20	4	0.5 ± 0.1	0.4 ± 0.1	3	Endosperm fluid
Tilia americana						
Meta.	22	3.5	1.9 ± 0.5	2.2 ± 0.8	5	Endosperm fluid
Phragmopl.	22	2	1.4 ± 0.6	2.4 ± 1.1	2	Endosperm fluid
ANIMAL CELLS						
Meiotic metaphase oocytes						
Asterias forbesi	18	0.5	2.5 ± 0.9	1.8 ± 0.6	8	A.S.W.
	18	0.5	2.4 ± 0.8	1.6 ± 0.5	10	45% D_2O-A.S.W.
	18	0.5	2.5 ± 0.5	1.4 ± 0.4	9	Low Ca, D_2O-S.W.
Meiotic meta. arrested oocytes						
Pectinaria gouldi	24	0.5	3.3 ± 0.8	2.3 ± 0.3	8	A.S.W.
Chaetopterus pergamentaceous	24	0.4	5.1 ± 0.8	2.5 ± 0.8	4	A.S.W.
Spermatocytes, meiosis I						
Nephrotoma sp. Meta.	20	11	0.4 ± 0.2	0.4 ± 0.1	4	Testis fluid in Kel-F10 oil
" Meta.	20	1.8	0.8 ± 0.2	1.9 ± 0.5	9	Testis fluid in Kel-F10 oil
" Ana.	20	1.8	1.0 ± 0.1	1.1 ± 0.1	5	Testis fluid in Kel-F10 oil
Cleavage mitosis metaphase						
Lytechinus variegatus	20	1	4.9 ± 1.4	4.2 ± 0.8	9	A.S.W.
	17.5	1.7	4.9 ± 0.9	3.8 ± 0.8	13	30% D_2O-A.S.W.
	20	2.2	6.0 ± 1.7	3.5 ± 1.0	10	Ca-free-A.S.W.
	17.7	1	4.7 ± 0.1	3.0 ± 0.5	4	1 mM Caffeine-A.S.W.
Asterias forbesi	18	1	4.3 ± 1.6	2.5 ± 1.2	17	A.S.W.
	18	1	3.6 ± 1.5	2.6 ± 0.9	5	45% D_2O-A.S.W.
	18	1	3.3 ± 1.2	2.7 ± 1.1	6	55% Ca-A.S.W.
	18	1	3.3 ± 1.0	1.7 ± 0.4	7	Low Ca, D_2O-S.W.

Meta., metaphase; Ana., anaphase; Phragmopl., phragmoplast; A.S.W., artificial sea water according to Kavanaugh (1956), "Formulae and Methods," Marine Biological Laboratories, Woods Hole, Mass.; Low Ca, D_2O-S.W., 55 vol of A.S.W. mixed with 45 vol of Ca-free D_2O sea water; D_2O, deuterium oxide 99.84% from BioRad Lab, Richmond, Calif.; Kel-F 10, fluorocarbon oil supplied by Minnesota Mining and Manufacturing Co., but now discontinued. Halocarbon HC 10-25, from Halocarbon Products Corp., Hackensack, N.J., has appeared to be an adequate substitute.

zation likewise follows a logarithmic time course in vivo (see Inoué, 1952, for colchicine-induced decay) as well as in vitro (Inoué et al., 1974; Borisy et al., 1974; Haga et al., 1974; Gaskin et al., 1975).

In these cases the reaction is proceeding as though: either (*1*) significant back pressure from subunits rate limits depolymerization, or (*2*) the equilibrium is otherwise limited by a first-order reaction. D_2O and temperature step-up experiments which we have performed favor the second interpretation for anaphase birefringence decay.

The time constant for the logarithmic retardation decay in anaphase is a sensitive, nonlinear function of temperature. The same function governs the velocity of anaphase chromosome movement. This observation plays an important role in considerations of chromosome movement mechanisms (Fuseler, 1973 *a* and *b*; Inoué and Ritter, 1975).

DISCUSSION

We are today not yet in a position to compare directly the kinetics of the in vitro assembly system with mitotic microtubule assembly in vivo. The in vivo system is in a dynamic rather than static equilibrium (Inoué and Ritter, 1975). In the in vitro system nucleation steps affect the kinetics of polymerization. In vivo, the need for spontaneous nucleation may be circumvented by the presence of specialized nucleating sites such as centrioles and kinetochores or by other cellular control mechanisms (see Inoué, 1964; Inoué and Sato, 1967; Pickett-Heaps, 1969; and Weisenberg, 1973, 1975).

These reservations notwithstanding, equilibrium behavior of the in vitro tubulin-microtubule assembly system is found to parallel the in vivo system to a remarkable degree (reviewed in Olmsted and Borisy, 1973 *a*; Inoué and Ritter, 1975; and Johnson and Borisy, 1975).

The dynamic equilibrium concept offers valuable insight into the properties of the mitotic spindle and related labile microtubular systems in living cells. We have shown that it describes the behavior of the spindle in response to changes in temperature and pressure, and a variety of chemical agents that affect the equilibrium. To lengthen, microtubules must polymerize more tubulin; to shorten, the tubulin polymer must be depolymerized. Since the mechanical connection of chromosomes to the spindle has been shown to be coextensive in time and space with the birefringent (microtubular) chromosomal fiber (Begg and Ellis, 1974), chromosome movement involving chromosomal fiber shortening without thickening must at least be rate limited by microtubule depolymerization. We have long suggested that microtubule depolymerization may provide the motive force for this type of chromosome movement.[2] We now have found that by regarding the microtubule as a cylindrical micelle, we can, through the

[2] At the Biophysics-Biochemistry Society Meeting, time-lapse motion pictures demonstrating the following were shown in addition to sequences 1–5 already described in the text. Sequence 6: 3 mM colchicine induces depolymerization and shortening of *Chaetopterus pergamentaceous* spindle fibers, resulting in transport of chromosomes to the cell cortex to which the outer spindle pole is anchored by its astral rays. 5 mM Colcemid dissolves the spindle *in situ* without appreciable fiber shortening or chromosome transport. Spindle fibers shown with polarized light, chromosomes with phase contrast microscopy. Sequence 7: Colchicine-

use of parameters derived from thermodynamic experiments, provide a model for chromosomal movement in which the interfacial tension developed during the depolymerization of a single microtubule would be sufficient to overcome the viscous drag which is the primary nonmicrotubular impediment to chromosome movement (Inoué and Ritter, 1975).

This paper is dedicated to Aharon Katchalsky and to Wayne Thornburg, two brilliant individuals whose stimulating discussions and friendships shall be dearly missed.

This work was supported in part by National Institutes of Health grant CA-10171 and National Science Foundation grant GB-31739.

REFERENCES

ARONSON, J. F. 1973. Nuclear membrane fusion in fertilized *Lytechinus variegatus* eggs. *J. Cell Biol.* **58**:126.

BEGG, D. A., and G. W. ELLIS. 1974. The role of birefringent chromosomal fibers in the mechanical attachment of chromosomes to the spindle. *J. Cell Biol.* **63**:18*a*.

BORISY, G. G., J. B. OLMSTED, J. M. MARCUM, and C. ALLEN. 1974. Microtubule assembly *in vitro*. *Fed. Proc.* **33**:167

BRYAN, J. 1974. Biochemical properties of microtubules. *Fed. Proc.* **33**:152.

FUSELER, J. 1973 *a.* The Effect of Temperature on Chromosome Movement and the Assembly-Disassembly Process of Birefringent Spindle Fibers in Actively Dividing Plant and Animal Cells. Ph.D. Thesis, University of Pennsylvania. University of Michigan Film Library, Ann Arbor, Mich.

FUSELER, J. 1973 *b.* Temperature dependence of anaphase chromosome velocity and microtubule depolymerization rate. *J. Cell Biol.* **59**:106*a*.

FUSELER, J. W. 1975. Mitosis in *Tilia americana* endosperm. *J. Cell Biol.* **64**:159.

GASKIN, F., C. R. CANTOR, and M. L. SHELANSKI. 1974. Turbidimetric studies of the *in vitro* assembly and disassembly of porcine neurotubules. *J. Mol. Biol.* **89**:737.

GASKIN, F., C. R. CANTOR, and M. L. SHELANSKI. 1975. Biochemical studies on the *in vitro* assembly and disassembly of microtubules. *Ann. N.Y. Acad. Sci.* In press.

HAGA, T., T. ABE, and M. KUROKAWA. 1974. Polymerization and depolymerization of microtubules *in vitro* as studied by flow birefringence. *FEBS Lett.* **39**:291.

HARTSHORNE, N. H., and A. STUART. 1960. Crystals and the Polarizing Microscope. Edward Arnold Ltd., London.

INOUÉ, S. 1952. The effect of colchicine on the microscopic and submicroscopic structure of the mitotic spindle. *Exp. Cell Res. Suppl.* **2**:305.

INOUÉ, S. 1959. Motility of cilia and the mechanism of mitosis. *Rev. Mod. Phys.* **31**:402.

INOUÉ, S. 1964. Organization and function of the mitotic spindle. *In* Primitive Motile Systems in Cell Biology. R. D. Allen and N. Kamiya, editors. Academic Press Inc., New York. 549–598.

INOUÉ, S., G. G. BORISY, and D. P. KIEHART. 1974. Growth and lability of *Chaetopterus* oocyte mitotic spindles isolated in the presence of porcine brain tubuline. *J. Cell Biol.* **62**:175.

INOUÉ, S., G. W. ELLIS, E. D. SALMON, and J. W. FUSELER. 1970. Rapid measurement of spindle birefringence during controlled temperature shifts. *J. Cell Biol.* **47**:95*a*.

INOUÉ, S., and H. RITTER. 1975. Dynamics of mitotic spindle organization and function. *In* Molecules and Cell Movement. S. Inoué and R. E. Stephens, editors. Raven Press, N.Y. In press.

INOUÉ, S., and H. SATO. 1967. Cell motility by labile association of molecules. *J. Gen. Physiol.* **50**:259.

JOHNSON, K. A., and G. G. BORISY. 1975. The equilibrium assembly of microtubules *in vitro*. *In* Molecules and Cell Movement. S. Inoué and R. E. Stephens, editors. Raven Press, N.Y. In press.

inhibited migration of sea urchin (*Lytechinus variegatus*) pronuclei, was reinitiated by photoconversion of colchicine to the inactive lumicolchicine. The female pronucleus, towards which the male pronucleus migrates in the absence of this treatment, rapidly approached the male pronucleus and syngamy is completed. Differential interference contrast microscopy. (Courtesy Dr. John F. Aronson. See Aronson, 1973).

KRESHECK, G. C., H. SCHNEIDER, and H. A. SHERAGA. 1965. The effect of D_2O on the thermal stability of proteins. Thermodynamic parameters for the transfer of model compounds from H_2O to D_2O. *J. Phys. Chem.* **69**:3132.

OLMSTED, J. B., and G. G. BORISY. 1973 *a*. Microtubules. *Ann. Rev. Biochem.* **42**:507.

OLMSTED, J. B., and G. G. BORISY. 1973 *b*. Characterization of microtubule assembly in porcine brain extracts by viscometry. *Biochemistry.* **12**:4282.

PICKETT-HEAPS, J. D. 1969. The evolution of the mitotic apparatus: an attempt at comparative ultrastructural cytology in dividing plant cells. *Cytobios.* **3**:257.

REBHUN, L. I., M. MELLON, D. JEMIOLO, J. NATH, and N. IVY. 1974. Regulation of size and birefringence of the *in vivo* mitotic apparatus. *J. Supramol. Struct.* **2**:466.

SALMON, E. D. 1973. The Effects of Hydrostatic Pressure on the Structure and Function of the Mitotic Spindle: an in vivo Analysis with a Newly Developed Microscope Pressure Chamber. Ph.D. Thesis, University of Pennsylvania. University of Michigan Film Library, Ann Arbor, Mich.

SALMON, E. D. 1975. Pressure-induced depolymerization of spindle microtubules. II. Thermodynamics of in vivo spindle assembly. *J. Cell Biol.* In press.

SALMON, E. D., and G. W. ELLIS. 1975. A new miniature hydrostatic pressure chamber for microscopy: strain-free optical glass windows facilitate phase contrast and polarized light microscopy of living cells. Optional fixture permits simultaneous control of pressure and temperature. *J. Cell Biol.* **65**:587.

SATO, H., S. INOUÉ, and G. W. ELLIS. 1971. The microtubular origin of spindle birefringence: experimental verification of Wiener's equation. *Proc. 11th Annu. Meeting Am. Soc. Cell Biol.* 261*a* (Abstr.).

SATO, H., and K. IZUTSU. 1974. Birefringence in Mitosis of the Spermatocyte of Grasshopper *Chrysochraon japonicus*. Time-lapse motion picture. Available from George W. Colburn Laboratory, Inc., Chicago, Ill.

SIDGWICK, N. V. 1950. Chemical Elements and Their Compounds. Vol. 1. Clarendon Press, Oxford.

STEPHENS, R. E. 1973. A thermodynamic analysis of mitotic spindle equilibrium at active metaphase. *J. Cell Biol.* **57**:133.

WEISENBERG, R. C. 1972. Microtubule formation *in vitro* in solutions containing low calcium concentrations. *Science (Wash. D.C.).* **177**:1104.

WEISENBERG, R. C. 1973. Regulation of tubulin organization during meiosis. *Am. Zool.* **13**:981

WEISENBERG, R. C., and A. C. ROSENFELD. 1975. In vitro polymerization of microtubules into asters and spindles in homogenates of surf clam eggs. *J. Cell Biol.* **64**:146.

Reprinted from *Journal of Cell Biology*, Vol. 67(3), pp. 501–517, 1975.

MICROTUBULAR ORIGIN OF MITOTIC SPINDLE FORM BIREFRINGENCE

Demonstration of the Applicability of Wiener's Equation

HIDEMI SATO, GORDON W. ELLIS, and SHINYA INOUÉ

From the Program in Biophysical Cytology, Department of Biology, University of Pennsylvania, Philadelphia, Pennsylvania 19174

ABSTRACT

Meiosis I metaphase spindles were isolated from oocytes of the sea-star *Pisaster ochraceus* by a method that produced no detectable net loss in spindle birefringence. Some of the spindles were fixed immediately and embedded and sectioned for electron microscopy. Others were laminated between gelatine pellicles in a perfusion chamber, then fixed and sequentially and reversibly imbibed with a series of media of increasing refractive indices.

Electron microscopy showed little else besides microtubules in the isolates, and no other component present could account for the observed form birefringence. An Ambronn plot of the birefringent retardation measured during imbibition was a good least squares fit to a computer generated theoretical curve based on the Bragg-Pippard rederivation of the Wiener curve for form birefringence. The data were best fit by the curve for rodlet index (n_1) = 1.512, rodlet volume fraction (f) = 0.0206, and coefficient of intrinsic birefringence = 4.7×10^{-5}. The value obtained for n_1 is unequivocal and is virtually as good as the refractometer determinations of imbibing medium index on which it is based. The optically interactive volume of the microtubule subunit, calculated from our electron microscope determination of spindle microtubule distribution ($106/\mu m^2$), 13 protofilaments per microtubule, an 8 nm repeat distance and our best value for f, is compatible with known subunit dimensions as determined by other means.

We also report curves fitted to the results of Ambronn imbibition of Bouin's-fixed *Lytechinus* spindles and to the Noll and Weber muscle imbibition data.

Fine structure and molecular orientation in cell organelles can be analyzed by polarized light microscopy. By nondestructive means, and in a continuous temporal framework, one can use birefringence (BR) analysis to pinpoint the occurrence and alteration of anisotropic molecular organization in volumes under 10^{-9} μl.

The physiology of spindle fibers has been extensively studied in living cells through measurement of BR changes (Inoué and Sato, 1967; Sato, 1975; Inoué and Ritter, 1975). Reversibly depolymerizable labile fibrils were shown to make up the birefringent fibers, and the nature of the molecular equilibrium was determined.

The positive BR of spindle fibers parallels the distribution and behavior of microtubules seen with the electron microscope. Rise and fall of fiber BR is accompanied by increase and decrease of tubule concentration. We therefore postulated that microtubules are responsible for spindle fiber BR (Inoué and Sato, 1967). In isolated spindles, BR shows exclusive proportionality to the quantity of tubulin, the subunit protein of microtubules (Stephens, 1972).

Recently pure tubulin has been isolated and polymerized to form labile microtubules in vitro (Weisenberg, 1972; Olmsted and Borisy, 1973). As reviewed elsewhere (Inoué and Ritter, 1975; Inoué et al., 1975), the associative properties of this protein follow closely the polymerization-deploymerization properties of microtubules deduced from spindle fiber BR in living cells.

We now report our analysis of the microtubular origin of spindle fiber BR by demonstrating: first, the fit of the Wiener form BR curve to the measured BR of isolated carefully imbibed spindles; and secondly, the concordance of the rodlet volume fraction defining the best fit Wiener curve with the volume fraction of microtubules calculated from electron microscopy, X-ray diffraction, and hydrodynamic data.

The BR of mixed bodies composed of oriented rodlets or platelets whose thicknesses are well below the wavelength of light is believed to arise from two sources: (a) an intrinsic BR due to the intramolecular anisotropy of the rodlets or platelets; and (b) a textural or form BR due to the anisotropic arrangement, or texture, of the fine structure. Wiener (1912) and Bragg and Pippard (1953) have derived equations which relate the form BR in terms of dielectric anisotropy to: the volume fraction occupied by the rodlets or platelets; their refractive index (n_1) or dielectric constant ($\epsilon_\nu = n_\nu{}^2$, for nonabsorbing material measured at frequency ν, see e.g. Born and Wolf, 1959, p. 91); and the refractive index (n_2) of the second phase in which the rodlets or platelets are immersed.

Experimentally, Ambronn and Frey (1926) have devised an imbibition method which allows one to determine the form and intrinsic BR contributions in mixed bodies. The intrinsic BR is taken as constant while the form BR varies, as described by the Wiener formula, with the refractive index of the imbibing medium (n_2). The measured BR is the sum of the two and therefore equals the intrinsic BR when the form contribution becomes zero, namely when $n_1 = n_2$.

MATERIAL AND METHODS

Choice of Experimental Material

For analyzing the form BR contributed to spindles by microtubules, we have chosen to use isolated meiotic spindles from oocytes of the sea-star *Pisaster ochraceus*. The choice of isolated spindles rather than of spindles in fixed whole cells was predicated upon the following considerations. (a) Thorough imbibition of intertubular space with media of defined refractive index is essential for exact form BR analysis. (*b*) The dimensions of the specimen must be readily measurable. The dimensions moreover must not change by fixation, imbibition and dehydration. (*c*) Light scattering and BR from extraneous sources must be held to a minimum. Isolated spindles can be sufficiently clean that optical and chemical influences of surrounding cytoplasm are minimum.

The meiosis I spindle can be isolated in large quantity from *P. ochraceus* oocytes. These isolates are reliably clean (Fig. 1), and, in the regions between the metaphase chromosomes and the poles, show little else than microtubules (Figs. 2 and 3). Unlike the spindle in mitotic division, whose dimensions, overall BR, and BR distribution vary rapidly with time and progression of mitosis, the *Pisaster* meiosis I spindle is arrested in metaphase and yields isolates of uniform BR and dimensions. Moreover, the BR of the isolates, which remained close to the value in life, was unchanged upon fixation; and the isolates could be imbibed in a way that the same BR returned when the imbibing medium index was brought back to the original value.

Isolation of Pisaster Spindle

Mature oocytes were obtained by treating isolated sea-star ovaries with 10^{-6} M 1-methyl adenine (Kanatani et al., 1969) in artificial sea water (Cavanaugh, 1964). 1 mg/ml Pronase (ICN K & K Laboratories, Inc., Plainview, N. Y.) and 0.1% mercaptoethylgluconamide (MEGA; Cyclo Chemical, Division of Travenol Laboratories Inc., Los Angeles, Calif.) have been used to activate and remove the fertilization membrane. Spindles are isolated in quantity from these demembranated oocytes at 13°C in 12% hexylene glycol, pH 6.3 following the method described by Bryan and Sato, 1970. The isolation medium is essentially the same as that described for sea urchins by Kane (1962 and 1965). Isolated spindles are stored in isolation medium at 4°C. All imbibition experiments were carried out on spindles not more than 3 h (typically within 40 min) from isolation.

Imbibition Procedure

We took freshly isolated *Pisaster* spindles, transferred them in the isolation medium (12% hexylene glycol) onto a stress-free microscope cover slip flamed and coated with a thin pellicle of gelatin, covered the spindles with another pellicle of gelatin, and assembled the whole

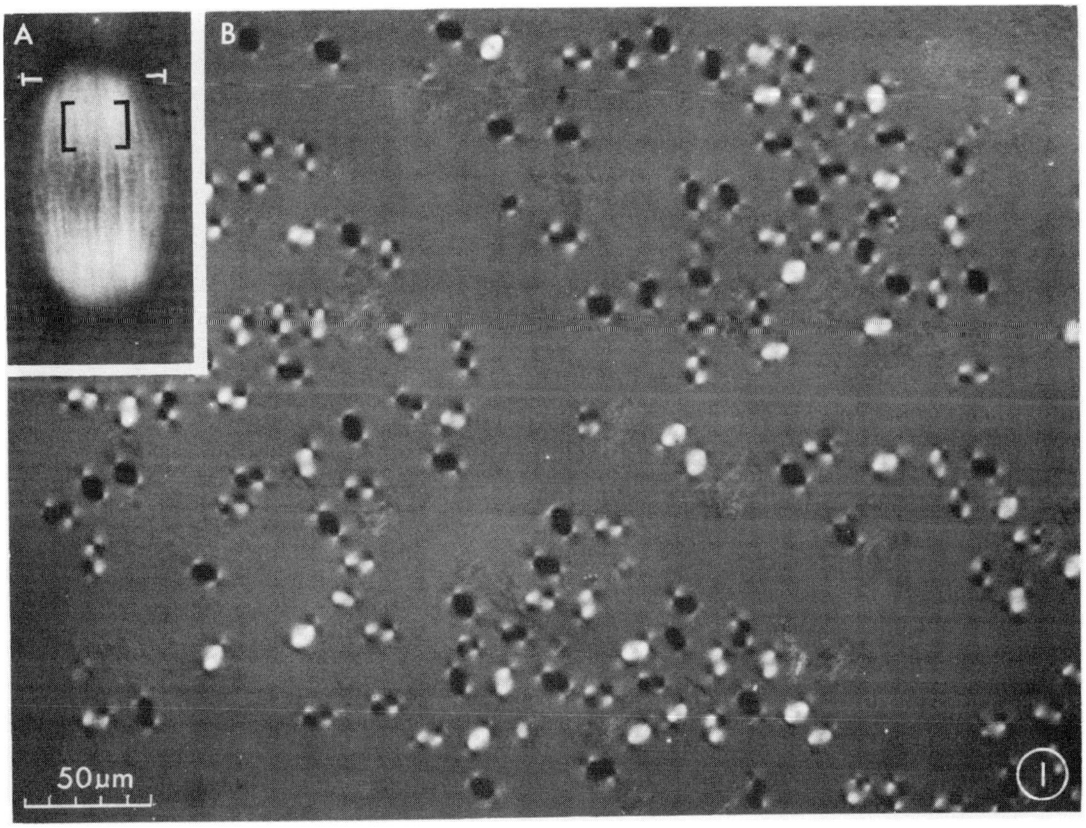

FIGURE 1 Meiosis I spindles from mature oocytes of *P. ochraceus* isolated in hexylene glycol after pronase-MEGA treatment. (A) Spindle birefringence is determined by measuring the retardation (see Fig. 4) in the area surrounded by black brackets. White bars specify the position corresponding to the electron micrograph in Fig. 3. × 2,250 (B) Clean mass isolation of *Pisaster* spindles is shown at lower magnification. Polarization microscopy, × 330.

sandwich as the upper element in a modified "Rose" type perfusion chamber. The fixative and imbibition media were perfused through this chamber with constant velocity so that they penetrated the spindles by diffusion through the gelatin sandwich which also held the spindles securely in place. The gelatin pellicles were prepared at the time of use by dipping a 3 cm diameter platinum loop into a warm (32–38°C) solution of 1%–2% gelatin-isosmolar salt solution (19:1 mixture of 0.56 M NaCl and 0.56 M KCl). Application of this method of cell support and perfusion (developed by Molè-Bajer and Bajer, 1968) was essential for successful perfusion without distortion or loss of the spindles.

After fixation in 3% glutaraldehyde-12% hexylene glycol (pH 6.3 with 10 mM phosphate buffer), the spindle was imbibed with the sequence of solutions listed in Table I. Each successive medium was serially diluted and exchanged gradually to avoid spindle collapse or irreversible BR change. Also, slow exchange is needed to dissipate the heat evolved with each successive mixing of dimethyl sulfoxide (DMSO) with aqueous solution. The transition from DMSO to benzene and the reverse are especially critical points since at these transitions the spindle structure and BR may change irreversibly. This transition into the nonpolar medium appears to be critical for successful imbibition of many cellular structures. The use of benzyl alcohol between DMSO and benzene helps to alleviate this difficulty.

Even with these precautions, some reagents appear to introduce inherently irreversible changes. For this reason, we have avoided the use of glycerol above 30% concentration or high concentrations of hexylene glycol.

We stress the need to qualify all data by the success of reversibility. From the many measurements we have made, we have used data points only from those specimens which showed successful and complete reversal of BR and which showed no change in their microscopic dimensions. Once the data were collected according to these criteria, no data were discarded for the form BR analysis.

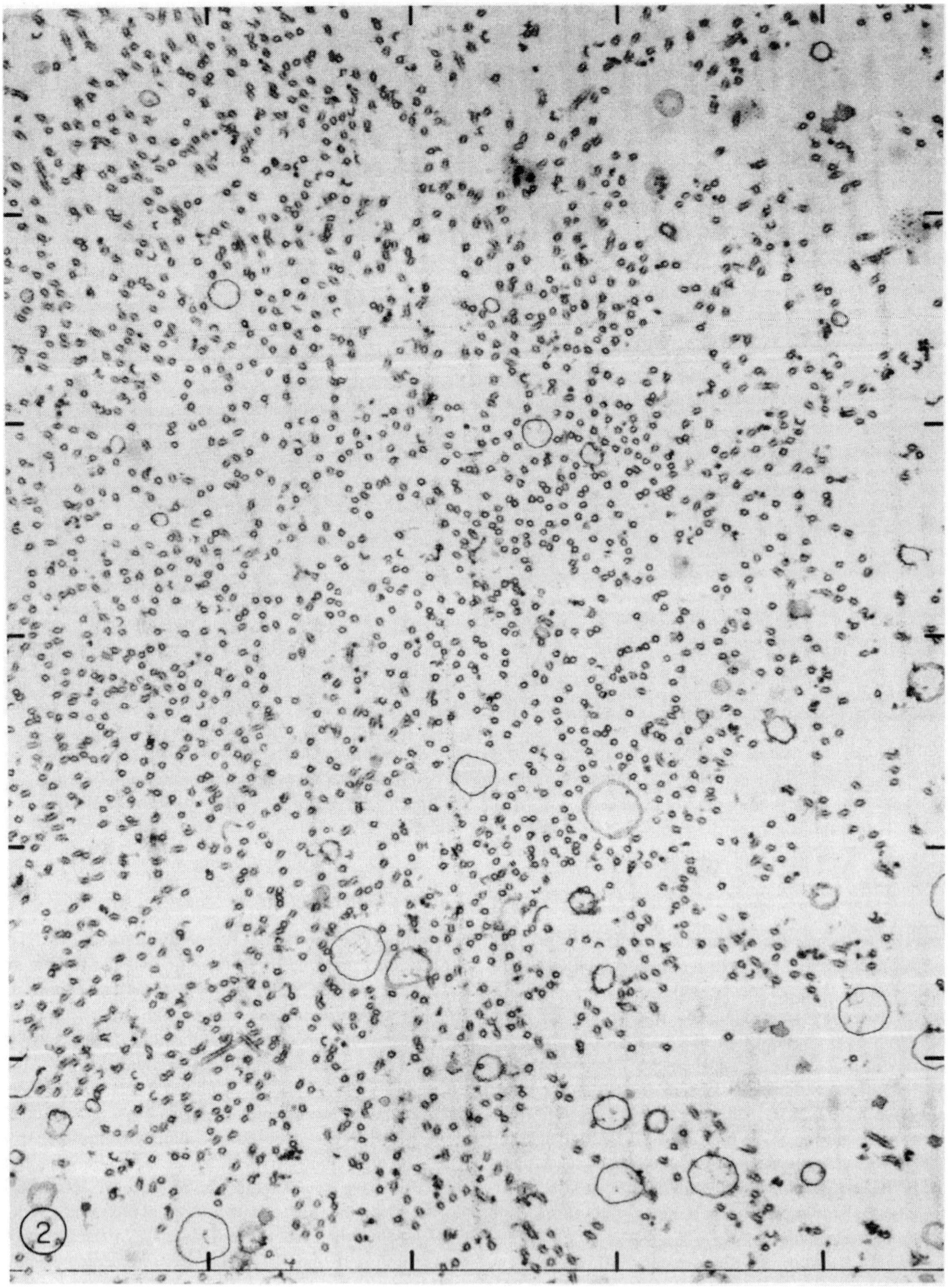

FIGURE 2 Thin cross-section electron micrograph of isolated spindle of *P. ochraceus*. The spindle was sectioned in the mid-region indicated by the bracket in Fig. 1 A. The microtubules in the mid-spindle region are mostly aligned parallel to the spindle axis. This is one of the micrographs used for counting the number of microtubular cross sections per unit area. Grid coordinates 1 μm interval. × 30,500.

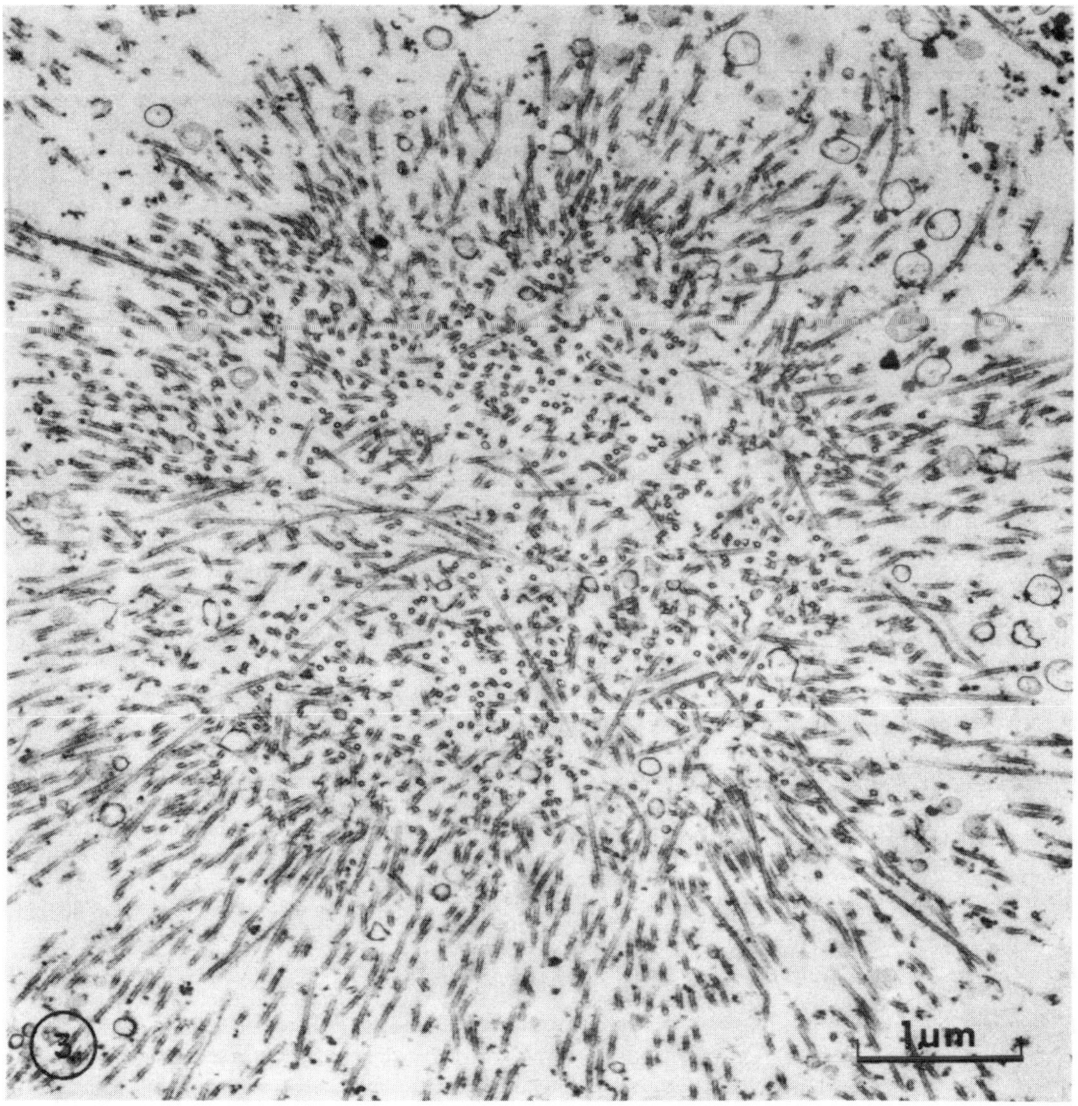

FIGURE 3 Cross section of the polar region (Fig. 1 A white bars) of an isolated metaphase spindle of *P. ochraceus* is mainly occupied by microtubules and little else. Some of the vesicles in the photograph are degenerated mitochondria resulting from the isolation and fixation procedures. × 22,000.

Retardation Measurements

A Leitz Ortholux-Pol (E. Leitz, Inc., Rockleigh, N. J.) microscope was equipped with rectified (Inoué and Hyde, 1957) strain-free optics (American Optical Corp., Scientific Instrument Div., Buffalo, N. Y. and Nippon Kogaku K. K., Toyko, Japan), and with a Polaroid HN-22 nonlaminated sheet polarizer (Polaroid Corp., Cambridge, Mass.) mounted beneath the condenser. All measurements were made with the 40× numerical aperture (NA) 0.65 objective. With the built-in Leitz analyzer, the extinction factor ($I_{parallel}/I_{crossed}$) of the system reached 5×10^4. The output of an Osram HBO-200W L-2 mercury are lamp (Osram GmbH, Berlin-München, West Germany) was filtered with at least one Corning no. 4602 (Corning Glass Works, Optical and Filter Division, Corning, N. Y.) heat-absorbing filter followed by a high transmission type B-2 Baird Atomic Interference filter (Baird Atomic, Inc., System Components Division, Bedford, Mass.) to provide intense illumination by mercury green light at 546 nm. A Zeiss Brace-Köhler compensator (Carl Zeiss, Inc., New York, N. Y.) with a retardance of $\lambda/25$ was used to measure spindle retardation. The isolated spindle

TABLE I
Sequence of Imbibition Media

	Refractive index of imbibition medium (n_2)
1. 12% Hexylene glycol, pH 6.3*	1.348
2. 12% Hexylene glycol-3% glutaraldehyde, pH 6.3*	1.348
3. 12% Hexylene glycol, pH 6.3*	
4. 2% DMSO-12% hexylene glycol	
5. 4% DMSO-12% hexylene glycol	
6. 8% DMSO-12% hexylene glycol	
7. 14% DMSO-12% hexylene glycol	1.367
8. 20% DMSO-12% hexylene glycol	
9. 30% DMSO-12% hexylene glycol	1.390
10. 40% DMSO-12% hexylene glycol	1.404
11. 50% DMSO-12% hexylene glycol	
12. 60% DMSO-12% hexylene glycol	
13. 70% DMSO-12% hexylene glycol	
14. 80% DMSO-12% hexylene glycol	1.455
15. 90% DMSO-12% hexylene glycol	
16. DMSO	1.473
17. DMSO-benzyl alcohol (50:50)	
18. Benzyl alcohol	
19. Benzyl alcohol-benzene (50:50)	
20. Benzene	1.501
21. Benzene-nitrobenzene (75:25)	
22. Benzene-nitrobenzene (50:50)	
23. Benzene-nitrobenzene (25:75)	
24. Nitrobenzene	1.556
25. Nitrobenzene-iodobenzene (75:25)	
26. Nitrobenzene-iodobenzene (50:50)	
27. Nitrobenzene-iodobenzene (25:75)	
28. Iodobenzene	1.622
29–55 Reversal of above sequence to step 3	
56. Artificial sea water (MBL-formula)	1.339

The refractive indices of solutions were determined by a Leitz-Jelly Micro-Refractometer (Leitz Wetzlar, West Germany) using 546 nm light at 20°C. Hexylene glycol lot number 916146, obtained from Union Carbide, Morristown, N. J. Dimethyl sulfoxide (DMSO) obtained from Crown Zellerbach Corp., Camas, Wash. Benzyl alcohol, benzene, nitrobenzene and iodobenzene obtained from Eastman Organic Chemicals Div., Eastman Kodak Co., Rochester, N. Y.

* pH adjusted with ca. 10 mM phosphate buffer.

was oriented at 45° to the polarizer axis, and the region indicated in Fig. 1 A was compensated (Fig. 4) to measure the retardation arising from the BR of parallel microtubules in that region.

Spindle retardation was determined according to the formula:

$$\sin 2\theta \cdot \sin(\Delta/2) = -\sin 2\theta_c \cdot \sin(\Delta_c/2) \quad (1)$$

where θ is the azimuth orientation of the specimen (here maintained at 45° ± 5°); θ_c is the azimuth angle at which the Brace-Köhler compensator compensates (Fig. 4) the specimen; Δ is the specimen retardation and Δ_c the retardance of the compensator, both expressed in degrees (e.g. see Hartshorne and Stuart, 1960).

Effect of Numerical Aperture on Retardation Measurement

Some experimenters have expressed concern regarding the accuracy of birefringent retardation measurements made at the high objective and condenser apertures required for good resolution. The common thought is that the oblique rays traveling through the specimen follow a longer path through the birefringent material and therefore may suffer a greater retardation than for rays nearer the microscope axis. In the present case the shape of the spindle is such that all rays traveling nearly perpendicular to the spindle axis will have a similar optical path, while the oblique rays traveling in a plane parallel to the spindle axis have a somewhat longer distance to travel, but will experience a decreasing BR. This and the convergent geometry of the microtubules at the spindle ends precludes an easy prediction of the optical path for such rays. Consequently we have measured the birefringent retardation of a number of freshly isolated spindles from *Lytechinus pictus* eggs at three different values of objective and condenser numerical aperture (Nikon rectified optics: 10×, 0.25 NA, objective and condenser matched; 20×, 0.4 NA, objective and condenser matched; and 40×, 0.65 NA, condenser NA = 0.52). For 45 spindles, we obtained the following numerical averages of the retardation: 10×, 3.242 ± 0.095 nm; 20×, 3.207 ± 0.113 nm; 40×, 3.193 ± 0.103 nm. Therefore, we conclude that for measurement of spindles with rectified-optics, no difference in retardation will be found with any of the dry objectives.

Electron Microscopy

Metaphase spindles isolated at 13°C, the optimum physiological temperature for this specimen (Sato and Bryan, 1968), were immediately cooled to 4°C after isolation and washed twice in cold isolation medium by centrifugation at 1,250 *g* for 5 min.

All of the following steps for fixation and dehydration were carried out at 4°C. The use of hexylene glycol in the fixative and dehydration solutions prevents the alteration of spindle BR during these steps (Inoué and Sato, 1967).

Small pellets of isolated spindles were fixed for 30 min in 3% glutaraldehyde-12% hexylene glycol solution at

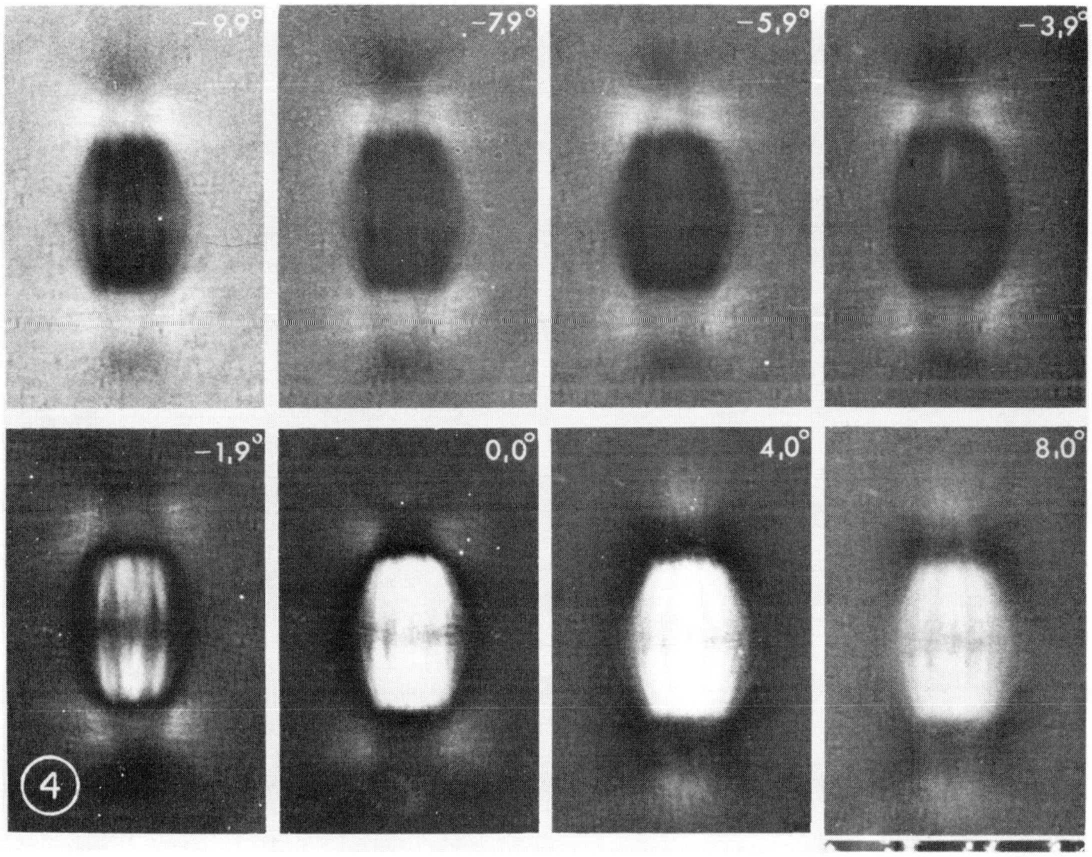

FIGURE 4 Appearance of a single spindle at various compensation angles (white letters on the right shoulder of each photograph). Between crossed polarizers, the spindle is oriented with its slow axis at +45°. As the λ/25 Brace-Köhler compensator is turned, the areas with lower retardation are first compensated (−1.9°) leaving the more highly retarding spindle fibers shining above the dark, compensated areas. With further rotation of the compensator (−3.9°, −5.9°) the chromosomal fibers are still not yet compensated. At −7.9° they are exactly compensated, while at −9.9° the more highly retarding fibers are clearly darker than their surrounding. At 0.0° the background is at extinction position. First mitotic spindle of the fertilized egg of *L. pictus* immediately after stabilization with 12% hexylene glycol, pH 6.3, at 22°C. Polarization microscopy, × 1,100. Scale division, 10 μm.

pH 6.3, washed twice with the isolation medium, postfixed for 30 min in 1% osmium tetroxide-12% hexylene glycol solution at pH 6.3, and washed twice for 5 min each with the isolation medium. Dehydration started with 30% ethanol-12% hexylene glycol solution, then 50% ethanol-12% hexylene glycol. Hexylene glycol was omitted from the graded concentrations of ethanol above 70%. Fixed isolated spindles can be stored in 100% ethanol overnight without altering their size and BR. After dehydration, spindles were placed in propylene oxide, followed with a mixture of equal proportions of propylene oxide and Epon (Epon 812, Shell Chemical Corp., San Francisco, Calif.). Embedding was accomplished in hard Epon (Luft, 1961) rather than Araldite to provide good preservation of fine structure without shrinkage or compression upon sectioning (Page and Huxley, 1963). Thin sections were cut with a diamond knife on a Porter-Blum Ultramicrotome MK-II (DuPont Instruments, Sorvall Operations, Newtown, Conn.), stained with uranyl acetate and lead citrate, and examined with a Philips model 200 electron microscope.

Although not commonly noted, rapid alteration of pH from 6.3 to 6.8 during spindle isolation produces dissociation and lateral splits of microtubules. Likewise, in the early steps of dehydration, change in pH or osmolarity can induce "C tubules" within isolated spindles even during or after glutaraldehyde or osmium tetroxide fixation (Sato, 1975).

Calculations

In Wiener's equation the form BR of rodlets is given as $(n_e^2 - n_o^2)$, the difference of dielectric constants or of the squares of the extraordinary and the ordinary refractive indices. This value is difficult to compare with specimen retardation which is the parameter measured. Fortunately, Bragg and Pippard (1953) have rederived the Wiener equation in a more general form which permits expression of form BR as $(n_e - n_o)$ which is the coefficient of BR (BR_{coeff}).

Bragg and Pippard give the equation for the dielectric constant experienced by an electric field in a particular direction relative to the orientation direction of a group of ellipsoids embedded in a medium of different refractive index as:

$$\epsilon = \epsilon_2 + \frac{f(\epsilon_1 - \epsilon_2)}{1 + (1 - f)\{(\epsilon_1 - \epsilon_2)/\epsilon_2\}L} \quad (2)$$

where ϵ is the average dielectric constant in the direction specified, ϵ_1 is the dielectric constant of the ellipsoids[1], ϵ_2 is the dielectric constant of the medium, f is the fraction of the total volume occupied by the ellipsoids, and L is the depolarizing coefficient for the given direction. Values of L, a geometric constant dependent on the ratios of the axes of the ellipsoids, have been tabulated by Stoner and cited in Bragg and Pippard. For ellipsoids of revolution whose polar axes are much greater than their equatorial axes (rodlets), L becomes zero for the field parallel to the polar axis and 0.5 for the field perpendicular to the polar axis. Since the index of refraction for nonabsorbing material is given by the square root of the dielectric constant experienced by its electric vector, the coefficient of BR for an assembly of rodlets is:

$$(n_e - n_o)_F = (\epsilon^{1/2} - \epsilon_\perp^{1/2})$$

$$= [\epsilon_2 + f(\epsilon_1 - \epsilon_2)]^{1/2} - [\epsilon_2 + \frac{f(\epsilon_1 - \epsilon_2)}{1 + (1 - f)\{(\epsilon_1 - \epsilon_2)/\epsilon_2\}0.5}]^{1/2} \quad (3)$$

where ϵ is the dielectric constant (at light frequency) in the direction of the long axes of the rodlets (optic axis direction), and $\epsilon_\perp$ is the dielectric constant when the electric vector is perpendicular to the optic axis.

Since specimen BR was determined from the retardation it produced, we have chosen to analyze our data as BR retardation, $\Gamma = T(n_e - n_o)$, rather than as coefficient of BR, $(n_e - n_o)$. Additionally, the actual retardation may include some level of intrinsic BR. Accordingly we have included specimen thickness (T) and coefficient of intrinsic BR (I) as variables in the equation we use to plot our curves. Hence:

$$\Gamma = T\{I + (n_e - n_o)_F\}$$

$$= T\{I + [n_2^2 + f(n_1^2 - n_2^2)]^{1/2} - [n_2^2 + \frac{f(n_1^2 - n_2^2)}{1 + (1 - f)\{(n_1^2 - n_2^2)/n_2^2\}0.5}]^{1/2} \quad (4)$$

where n_1 is the index of refraction of the rodlets, n_2 is the index of refraction of the medium and f is fraction of the volume occupied by the ellipsoids, which are now rodlets.

Calculations were performed on a programmable calculator (Hewlett Packard 9830A, Hewlett Packard Co., Loveland Division, Loveland, Colo.) with its associated printer and plotter. The calculator was programmed to calculate Γ according to the equation above for specified values of n_1, T, I, and f at each value of n_2 for which we had an observation. This calculated value was subtracted from each observed value and the result squared. The data points were then plotted on the chart together with the calculated curve. At the end of the curve the plotter printed the sum of the squared error.

Of the variables (n_1, f, I, and T), T can be determined at the time of retardation measurement to within about 1 μm, but the others are not directly measurable. Given a firm value for T and a set of data points, from a specific Wiener curve, the values of n_1, I, and f belonging to that curve may be uniquely determined by curve fitting. The calculator program facilitates this process by providing both a visual comparison of the curve with the data and a quantitative rating of the fit for each curve plotted.

The deviation of the calculated curve from the plotted data points reveals which variable is in need of adjustment. If the chosen n_1 is too great, the curve shifts toward a larger index relative to the data points; if n_1 is too small, the curve shifts toward lower index (Fig. 5). If the value used for f is too large, the extremes but not the trough of the curve are high relative to the data points and low if f is too small. Changes in I simply translate the curve *in toto* to higher or lower values with increase or decrease of I. An increase in T has the same effect as a simultaneous increase in I and f.

By this means, a good visual fit of the curve to the data points is quickly achieved. The values of n_1, f, and I obtained are good approximations of the final values and serve as excellent starting values to achieve rapid convergence in the optimization process to follow.

To improve the fit beyond that which could be achieved visually, we devised a program which employed the method of least squares (Frazer and Suzuki, 1973) to optimize the values of n_1, f, and I to fit the observed data and a given value of T. In this program the unweighted standard deviation (σ = square root of the mean squared error) was used to quantitate the fit of the curve to the data. The curve which best fit the data would be that for which σ was minimum. To find this curve, the values of

[1] Our use of subscript 1 for the ellipsoids and subscript 2 for the medium conforms with Wiener's original usage rather than that of Bragg and Pippard which is reversed.

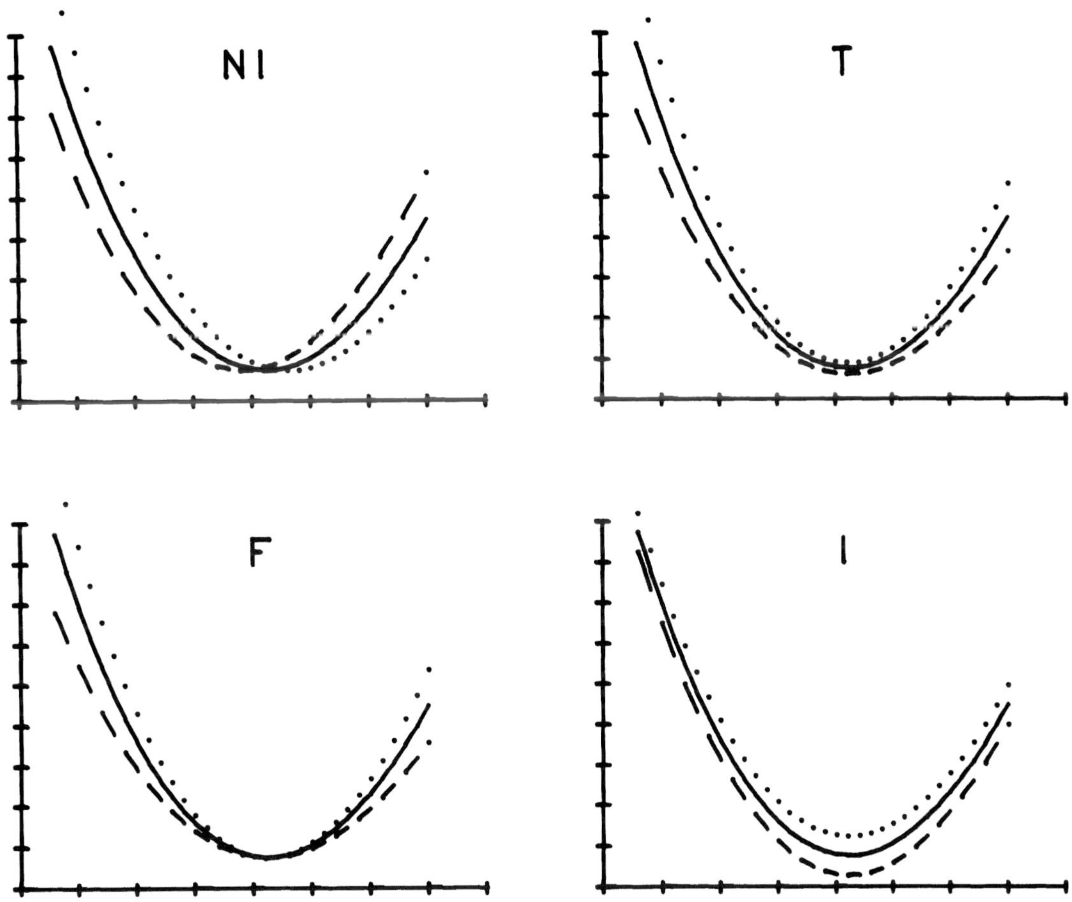

FIGURE 5 Plots of calculated Wiener curves showing effect of variation in the parameter with which each is labeled. For each parameter shown, the calculations have been made with the other parameters held constant at the values used for their solid curve. *N*1: solid curve, n_1 = 1.512; dotted curve, n_1 = 1.532; dashed curve, n_1 = 1.492. *F*: solid curve, f = 0.0206; dotted curve, f = 0.0256; dashed curve, f = 0.0156. *T*: solid curve, T = 8.0 μm; dotted curve, T = 9.5 μm; dashed curve, T = 6.5 μm. *I*: solid curve, $I = 4.7 \times 10^{-5}$; dotted curve, $I = 7.7 \times 10^{-5}$; dashed curve, $I = 1.7 \times 10^{-5}$.

n_1, f, and I were varied systematically until σ was at a minimum with respect to alteration of the sixth significant digit of each of the variables. The resulting values of n_1, f, I, and σ were then accepted as optimum for fit for the curve to the data.

These values were then used in a second version of the first program, which plotted the data points and the best fit curve together with dotted curves marking the unweighted standard deviation (Figs. 6, 7, and 8).

FORM BR OF ISOLATED *PISASTER* SPINDLE

Results of Imbibition

Spindles isolated in 12% hexylene glycol and held at room temperature slowly lose their BR unless fixed (Kane and Forer, 1965). Consequently, during the period required for mounting the isolated spindles between gelatin pellicles in the Rose chamber before fixing, some spindle BR is lost. We estimate that the loss is no more than 10% on the basis of comparison with retardation measurements made on 340 fresh, unfixed isolates in 12% hexylene glycol in which we found a retardation of 3.82 nm ± 0.54 nm. This measurement was made on spindles unselected for size and cannot rigorously be compared with those used in the imbibition series since the latter were selected for similar spindle width (8 μm ± 0.5 μm). The mounted spindles had the same retardation, ~3.5 nm, in 12% hexylene glycol before and after fixation.

The imbibition media and their refractive in-

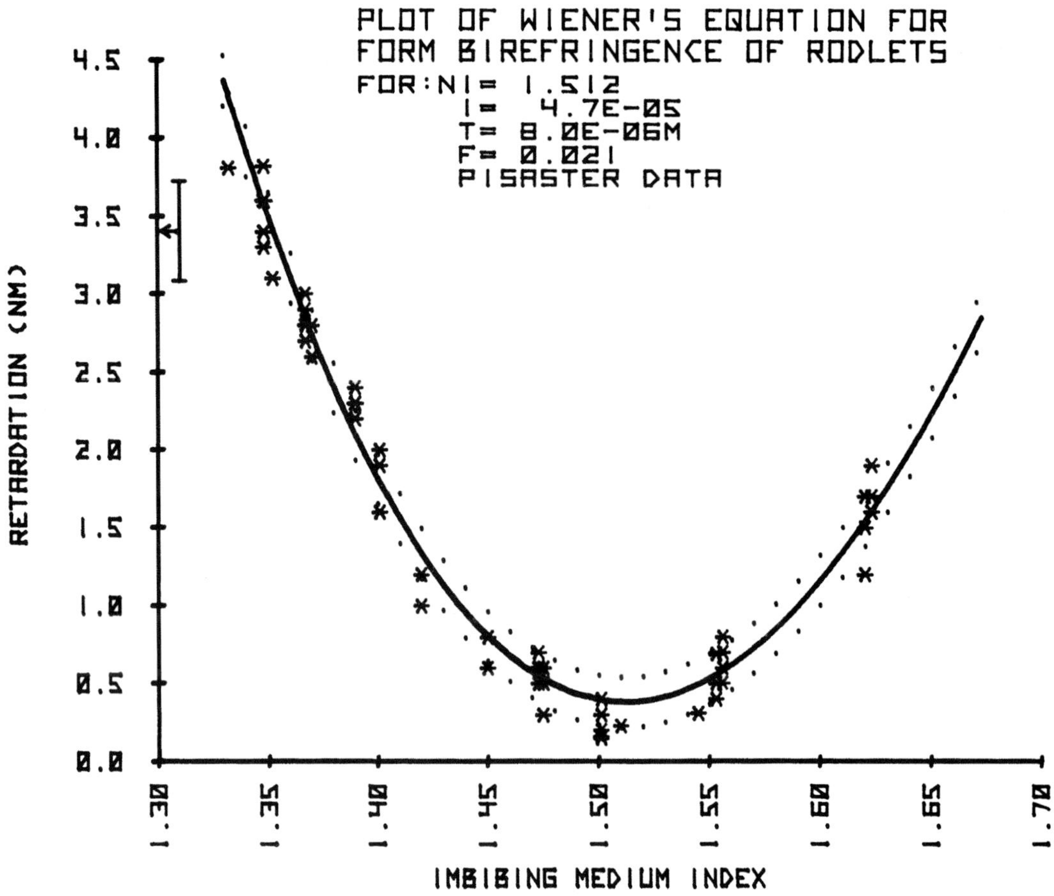

FIGURE 6 Machine plot of the best-fit Wiener curve (solid curve) with the *Pisaster* imbibition data points to which the curve was fitted. The total number of plotted points is 89, of which nearly half are coincident with others. The dotted curves represent the theoretical retardation at each point plus or minus the unweighted standard deviation of the theoretical curve as compared with the data as a whole. The error bar with arrow shown at the ordinate represents the mean and standard deviation of the retardation measured for 30 spindles in intact living *Pisaster* oocytes.

dices (n_2) are summarized in Table I. The points in Fig. 6 represent all the measurements on the 14 isolated spindles which showed total reversibility of retardation when returned to the initial medium of n_2 = 1.348. Each point represents the mean of triplicate measurements with an estimated error of not greater than 0.1 nm. The solid curve in Fig. 6 is the calculated Wiener curve optimized for a spindle thickness (T) of 8 μm. The dotted lines mark the unweighted standard deviation ($\pm$ 0.16 nm) for the fit of the calculated curve to the aggregate of all the data points. The agreement between the values measured for the isolated spindles and the calculated curve is excellent.

For the given spindle thickness the value obtained for f is 0.0206, for I is 4.7×10^{-5}, and the value for n_1 is 1.512. Since the values obtained for f and I are dependent on T, the optimization routine was performed for spindle thicknesses of 7.5 μm and 8.5 μm to determine the range of variation to be expected for f and I. At T = 7.5 μm, f = 0.0219 and I is 5.0×10^{-5}. At T = 8.5 μm, f = 0.0193 and I becomes 4.4×10^{-5}.

We have also indicated at the ordinate axis the range of retardations we have observed at 13°C in living *Pisaster* spindles. This is found to be 3.4 nm $\pm$ 0.32 nm. If compared with our calculated Wiener curve, this would suggest that the living spindle is in a medium (cytoplasm) having a refractive index of about 1.352 $\pm$ 0.008.

Effective Volume of Microtubule Subunit

For a variety of microtubule types, Tilney et al. (1973) have verified that each turn of the microtubule has 13 subunits or protofilaments. X-ray diffraction of wet microtubules (Cohen et al., 1971), in agreement with negative staining electron microscope measurements (Grimstone and Klug, 1966), shows both 4-nm and 8-nm longitudinal periods. The 8-nm repeats are stronger and may be considered to correspond to the length of the tubulin dimer. Both methods arrive at a protofilament width of about 5 nm. The surface lattice derived from X-ray diffraction (Cohen et al., 1971) was first thought to differ from that of Grimstone and Klug (1966), but recent findings have reconciled the difference in favor of the earlier lattice (Cohen et al., 1975). Erickson (1974), using optical reconstruction, reports a relatively complex shape for the subunits. Neither this method nor X-ray diffraction has yet yielded exact measurement of subunit volume.

Since we have determined the volume fraction of microtubules in the spindle, and since the microtubules approximate infinitely long rodlets, we can independently calculate the subunit volume, or the volume defended by the protofilament subunit as follows.

Electron micrographs of our isolated *Pisaster* spindles (Figs. 2 and 3) showed microtubules, almost exclusively, to be the predominant structural component. Measurements from cross sections (Fig. 2) taken from the same region that we used for retardation measurements (Fig. 1 A) showed the average density of microtubules across the 8 μm diameter to be 106 microtubules/μm^2.

Taking our values for f obtained from curve fitting and for the number of microtubules/μm^2 (M), we find as follows:

$$f = 0.0206$$

$$= \frac{\text{volume defended by rodlets}}{\text{unit volume of structure}}$$

$$= V_s \times S \times P \times M/\mu\text{m}^3$$

$$= V_s \times 125 \times 13 \times 106 \div 10^9\ \text{nm}^3/\mu\text{m}^3,$$
$$\text{and } V_s = 119.6\ \text{nm}^3 \quad (5),$$

where V_s is the protofilament subunit (dimer) volume, S is the number of subunits per μm of protofilament, and P is the number of protofilaments per microtubule.

If the dimeric subunits were right circular cylinders, their diameter would be about 4.4 nm. The presence of the 4-nm longitudinal repeat period in the X-ray data suggests that the dimers are not simple cylinders. If they were made up of two spheres merged at the center for a total length of 8 nm, the sphere diameter would be about 5.0 nm. We have not attempted to use the shape suggested by Erickson (1974) since it is undefined in the third dimension. Obviously an unlimited variety of shapes could be imagined which would equally fit the present constraints. However, if the shape deviates too much from being essentially cylindrical as a protofilament, this must be considered in the choice of the depolarizing coefficient L in Eq. 2.

Although the detailed shape and exact dimensions of the tubulin molecule *in situ* in the microtubule are yet to be determined, the preceding calculations show a good fit between the structural parameters obtained from X-ray diffraction, electron microscopy, and hydrodynamic considerations and the volume fraction value we obtained from the optimized Wiener curve.

FORM BR IN OTHER MATERIAL

Spindle in Bouin's-Fixed Cells

Fig. 7 shows imbibition data for a fixed spindle at anaphase onset in a 7-μm thick section of a sea urchin egg *Lytechinus* (*Toxopneustes*) *variegatus*. The slide with sections fixed in Bouin's solution and unstained was a gift of the late Dr. E. B. Harvey. Retardations were measured with a 29.4-nm Brace-Köhler compensator with the mercury green line, in Schillaber's oil (R. P. Cargille Labs Inc., Cedar Grove, N. J.), in the following sequence of imbibing medium refractive index: 1.51 → 1.80 → 1.70 → 1.60 → 1.50 → 1.40 → 1.45 → 1.55 → 1.65 → 1.75. The best fit Wiener rodlet BR curve was drawn by computer as described for the isolated spindle.

The BR of the fixed spindle at low imbibing medium index is severalfold higher (BR_{coeff} = $4.9 \cdot 10^{-3}$ for the form curve extrapolated to n_2 = 1.33) than the BR of *L. variegatus* spindles in living cells ($2\text{–}3 \cdot 10^{-4}$ depending on exact stage of division and on temperature) or of the isolated *P. ochraceus* spindle. We have noticed sharp rises in the BR of spindles in a variety of cells fixed with histological fixatives in which the spindles appear coarsely fibrillar (also see Satô, 1958).

The matching refractive index of the Bouin's-fixed *Lytechinus* spindle (1.552) is higher, and at match point the BR ($BR_{coeff} = 3.0 \cdot 10^{-4}$) is larger, than the value for the isolated *Pisaster* spindle (Fig. 6). The volume fraction, f = 0.135, is

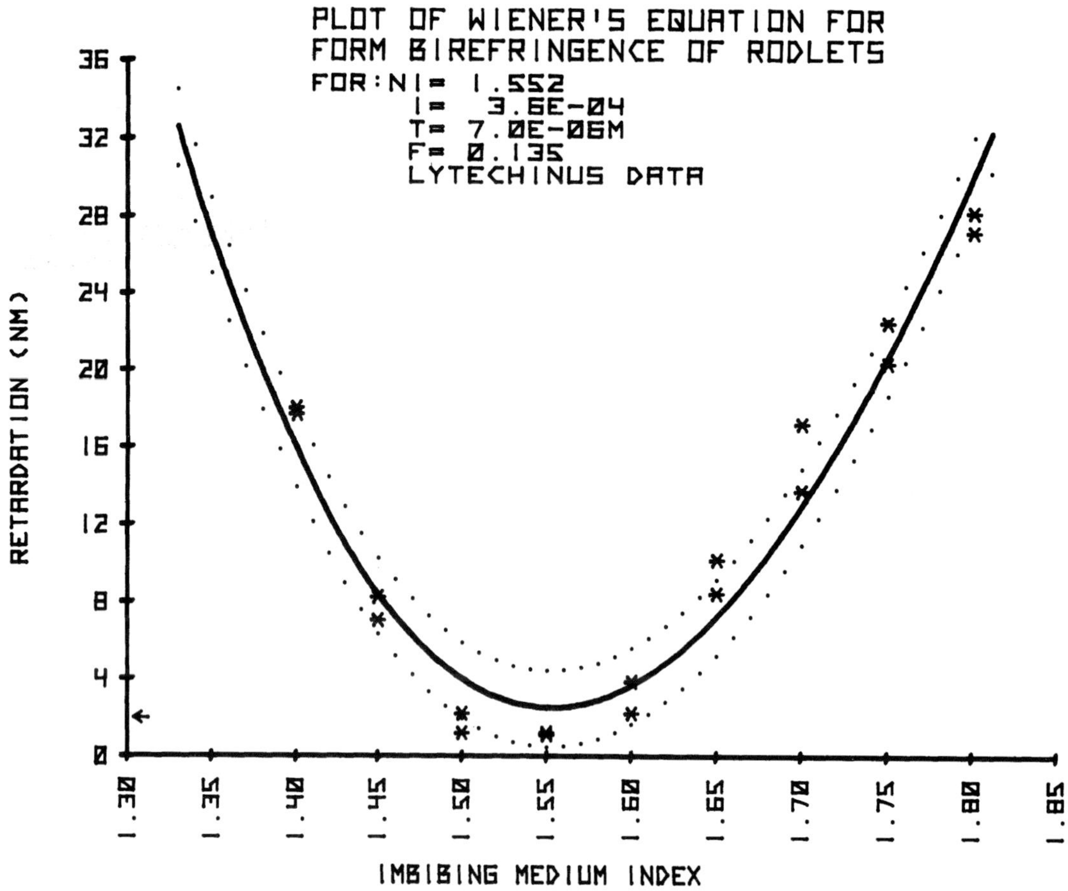

FIGURE 7 The data shown on this plot were obtained by imbibition of Bouin's-fixed spindle of *L. variegatus* zygote. It shows a large increase in rodlet volume fraction (f) and a slight increase in rodlet refractive index (n_1), both of which suggest that the fixative has produced a condensation of cytoplasmic materials onto the spindle microtubules. The arrow at the ordinate axis shows the average birefringent retardation measured for metaphase spindles in living *L. variegatus* zygotes.

severalfold larger than that obtained with isolated *Pisaster* spindles.

We interpret these observations to indicate the dense deposition of cytoplasmic components onto the microtubules which in turn may have formed coarse fibrillar aggregates.

Form BR of Skeletal Muscle

We have also examined the Wiener rodlet form BR in skeletal muscle. Fig. 8 shows imbibition data for the Q (= A) band obtained by Noll and Weber (1934) on frog cutaneous pectoris muscle fixed at 0–4°C with 5–10% formalin for 24 h (imbibition sequence shown in their Table IV), together with our computer derived best fit form BR curve. We obtain $n_1 = 1.526$, $I = 8.6 \times 10^{-4}$, and $f = 0.064$.

The form BR values given by our curve differ from those calculated by Noll and Weber who assumed the rodlet volume fraction to equal the fractional dry matter volume. The latter was calculated from the diameters of the muscle fiber measured in each imbibing medium compared to their diameter upon drying. The form BR calculated on this basis does not agree with their measured BR, varying with imbibing medium and showing discrepancies as high as fivefold (see Noll and Weber, 1934, their Table IV a).

Instead of total dry matter, we calculated the rodlet partial volume from cross-sectional dimensions of the overlapping A and I filaments seen in electron micrographs. The thick filaments measure 10–11 nm, and the thin filaments 5–6 nm in diameter (Huxley, 1960). X-ray diffraction studies of frog living sartorius muscle at rest length show the thick filaments to be hexagonally arranged with a center-to-center distance of 45 nm (Huxley,

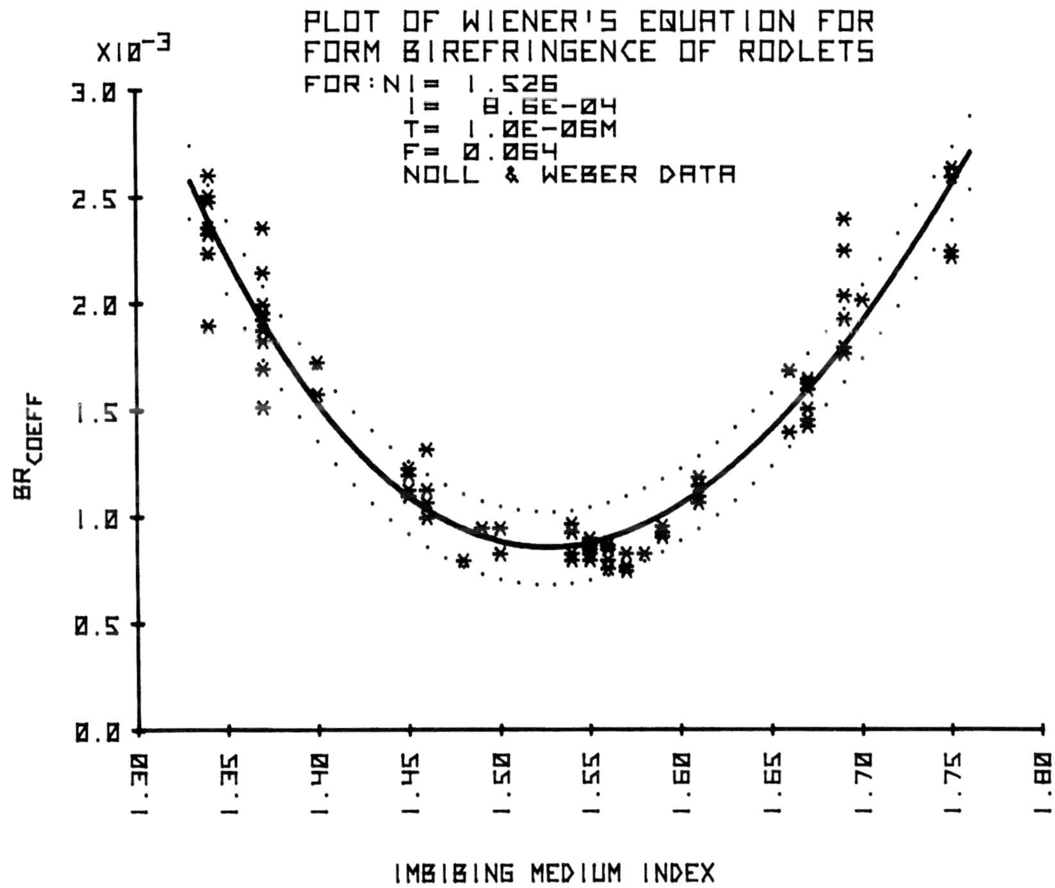

FIGURE 8 The data plotted are from Noll and Weber (1934) for the *A* band of frog sartorius muscle. In this case the ordinate represents coefficient of birefringence. The computer optimized Wiener curve shows a reasonably good fit to their data and yields a rodlet volume fraction (f = 0.064) close to the value we calculated from electron micrographs (see text).

1960). One thin filament is found at each trigonal point. The cross-sectional area which includes one thin and three one-sixths thick filaments is then 877 nm^2. The volume fraction of the thick and thin filaments together thus becomes 0.067–0.086. These values are in reasonable agreement with the partial volume 0.064 for the Wiener best fit curve shown in Fig. 8.

From myosin and actin contents of muscle, Cassim et al., (1968) calculate the *A*-band rodlet partial volume as 0.1–0.15. If, instead of total myosin and actin, one uses the lengths, particle weights, and partial specific volumes of the molecules making up the shaft portion only of the thick filament (12 1.6 nm diameter light meromyosin C rods per cross-section: Young et al., 1972) and the thin filament (two 70,000 dalton tropomyosin and 14 46,500 dalton actin monomers per 385 Å length: Pepe, 1972), one obtains a volume fraction of 0.041. Considering the variable contribution of myosin arms etc, one would expect the effective volume fraction to lie in between these values.

DISCUSSION

The validity of the Wiener formulation has been questioned on the basis of flow BR measurements of several "fibrous" proteins and rod-shaped viruses (Taylor and Cramer, 1963; Cassim and Taylor, 1965; Cassim et al., 1968). In its place, Cassim et al. propose the use of an empirical formula which equates BR_{coeff} to partial volume of the rodlets times specific BR in water where BR = $(1.7 \pm 1.3)\cdot 10^{-2}$. Using this formula, they calculate that for vertebrate skeletal muscle the total BR_{coeff} should be $(1.7–2.5)\cdot 10^{-3}$ and the intrinsic BR_{coeff} $(1–1.5)\cdot 10^{-3}$. Applied to the isolated *Pisaster* spindle, the BR_{coeff} should be $(1.4–2.0)\cdot 10^{-2}\cdot 0.021 = (2.9–4.2)\cdot 10^{-4}$ in water. While some of these values match the observed ones, the results of their formulation can depart as much as severalfold (see Cassim et al. 1968, Fig. 6) from those obtained by using Wiener's equation which we have demonstrated to fit the measured BR over a wide range of imbibing medium refractive index.

In the application of the Ambronn imbibition method, alteration of the volume fraction f, or the index of refraction n_1, or of any factor necessitating the use of an alternate depolarizing factor L, for example, by interaction of the particles with the media, could lead to a departure from the Wiener relationship. This would be particularly important when the rodlets are not supported by an organized structure. Large changes in rodlet shape (for example by bending or coiling or conversely by stiffening) could have a drastic effect on the predictive value of the equation unless the changes are accounted for by selection of appropriate values of L.

The imbibition data of striated muscle *A* band obtained by Noll and Weber (1934) can be fitted to a Wiener curve yielding a reasonable volume fraction value. However, Noll and Weber found considerable lateral and some longitudinal shrinkage of the muscle exposed to the dehydrating and imbibition media. The index "match point" given for their dehydrated muscle is higher than the minimum of the best fit Wiener curve.

Further imbibition studies on myofibrils free from shrinkage and irreversible BR changes would seem to be in order since the observed reduction of the BR by overlap of the *A*- and *I*-filaments (Colby, 1971, and D. L. Taylor, personal communication) is likely to provide significant insight into the dynamics of actomyosin interaction in muscle contraction.

In the mitotic spindle, careful perfusion of the glutaraldehyde-fixed isolated spindles, sandwiched with pellicles of gelatin, introduced little microscopically visible dimensional change, and the BR was totally reversible when returned to the original low index medium. The fact that those preparations which maintained constant width gave reproducible BR values fitting the Wiener formula gives us confidence that Ambronn's imbibition method is indeed applicable to form BR analysis, and that the Wiener formulation is itself accurate. Furthermore, we note that there do exist many reports of imbibition results which display typical Wiener curves (see e.g. Ambronn and Frey, 1926; Chinn and Schmitt, 1937; Schmidt, 1951; and Bendet and Beardon, 1972).

The various parameters for Wiener rodlet form BR curves for the isolated, clean *Pisaster* meiotic spindles are summarized in Table II. The imbibed isolated spindles follow a theoretical Wiener curve for $f = 0.0206$. The retardation of the same spindle measured in intact oocytes at 13°C corresponds to the value of the Wiener curve at $n_2 = 1.352$.

On the basis of a disparity between the BR observed in living cells and the (very low) number and unexpected distribution of microtubules observed in their electron microscope sections, Behnke and Forer (1967) argued that microtubules are not responsible for spindle fiber BR in crane fly spermatocytes. Their argument has been disputed by Rebhun and Sander (1967) and Goldman and Rebhun (1969), and LaFountain (1974) has obtained electron micrographs of crane fly spermatocytes showing approximately 100 microtubules per μm^2 spindle cross section, clustered in regions corresponding to chromosomal fibers with higher BR. While concluding, as do Rebhun and his co-workers, that microtubules must be responsible for the major part of the form BR of the living spindle observed, LaFountain questions whether form BR of microtubules accounts for all of the spindle BR.

An exact accounting of all of the BR in a living spindle, as contrasted to a clean isolated spindle, demands that we at least know: (*a*) the refractive index of the intertubular material, (*b*) the distribution of angular divergence and lengths of the microtubules at each measured volume in the spindle, and (*c*) the fraction of microtubules and other components preserved for electron microscopy. Exact data are still missing on these points, and in fact the second and third points are known to be sensitive to many physiological and methodo-

TABLE II

Summary of Data from Isolated Pisaster Spindles

Rodlet refractive index	1.512
Rodlet volume fraction	0.0206
Average number of microtubules per μm^2	106
Average number microtubules per μm^2 for 6 μm diameter spindle core	130
Optically effective volume per tubulin dimer	120 nm^3
Coefficient of BR	5×10^{-4}
Intrinsic BR at match point (= 1.512)	4.7×10^{-5}
Estimated refractive index of cytoplasm	1.352

logical variables. Spindle BR in intact cells varies with temperature, pressure, osmotic pressure, pH, calcium ion concentration, tubulin availability, and exact stage of division. Parallel changes in microtubule behavior are found in vitro. Thus, these parameters would have to be carefully controlled at fixation and dehydration, and the change of BR, taking into account the refractive index of the imbibing media, would have to be closely monitored.

The data on Bouin's-fixed *Lytechinus* spindle presented above, and in Pfeiffer's (1951) report on the extraordinarily high form ($1 \cdot 10^{-3}$) and intrinsic ($4 \cdot 10^{-3}$) BR of spindles in a copepod egg and in the pollen tube of *Clivia* sp., clearly reflect examples of drastic alteration of spindle fine structure during fixation or dehydration, very probably by dense deposition of cytoplasmic components onto microtubules which in turn may have formed coarse aggregates. In fact Rebhun and Sander (1967), Goldman and Rebhun (1969) and Forer and Goldman (1972) have noted the presence of large amounts of ribosomes and other material paralleling the microtubules of isolated sea urchin spindles and have discussed their possible influence on BR. The intertubular materials adhere to the tubules to varying degrees depending on the pH and divalent cations present at the time of spindle isolation. Using interference microscopy, Forer and Goldman (1972) found the dry mass of washed isolated spindles to vary as a function of decreasing pH, from not more than 20% (pH 7.3) to 60% (pH 5.3) of their dry mass in intact eggs. Even in isolated spindles, Forer and Goldman (1972) have reported, especially at low pH, a rise of dry mass concentration when sea urchin spindles were preserved in a test tube. These effects need to be distinguished from genuine growth of microtubules in spindles and asters isolated in microtubule polymerizing solutions (Weisenberg, 1973; Weisenberg and Rosenfeld, 1975; Inoué et al., 1974; Rebhun et al., 1974).

Our estimate of the refractive index of the intertubular medium in living *Pisaster* oocytes (Table II) along with the data used to calculate microtubular subunit volume provides a means to calculate the fraction of the dry mass due to microtubules in intact cells. Where C is the concentration of dry mass in g/100 ml, and n is the refractive index (1.352), of the intertubular cytoplasm; n_s is the refractive index of the solvent (1.334); and α, the specific refractive increment is 0.0018, then: $n = n_s + \beta C$ (Barer and Joseph, 1954)

$$\text{or} \quad C = \frac{n - n_s}{\alpha} = \frac{1.352 - 1.334}{0.0018} = 10\ \text{g}/100\ \text{ml}.$$

Using the notation of Eq. 5 and MW and N as subunit molecular weight and Avogadro's number, respectively, then the mass of microtubule per unit volume is:

$$\begin{aligned} S \times P \times M \times MW \div N \\ &= 125 \times 13 \times 106 \times 110{,}000 \div 6.022 \times 10^{23}\ \text{g}/\mu\text{m}^3 \\ &= 3.15 \times 10^{-14}\ \text{g}/\mu\text{m}^3 \\ &= 3.15\ \text{g}/100\ \text{ml}^{2} \end{aligned}$$

If the total dry mass concentration in the intact living spindle is given by the sum of these, then microtubular dry mass contribution to the total is given by:

$$\frac{3.15}{10 + 3.15} = 0.24$$

Since some microtubules have probably been lost in spindle isolation (about 10%), this suggests that in the living *Pisaster* spindle the microtubular contribution to the dry mass is a minimum of 24%.

In the absence of extraneous material the microtubules yield an apparent refractive index of 1.512. This value is appreciably lower than the range of refractive index of dried protein cited in Taylor and Cramer (1963) (1.57–1.60). As Bragg and Pippard (1953) noted, the match point index does occur at a value considerably lower than that for dried protein, and they attributed the difference to a hydration shell. Barer and Joseph (1954) have called attention to a similar problem in cell refractometry and have cautioned that one cannot reliably go from the refractive index of a dried protein to its specific refractive increment in solution, or the reverse. Our finding reinforces these and points out the hazard inherent in attempting the prescriptive use of refractive index values from dried protein as starting points in Wiener calculations.

A number of authors have asked whether strain-induced BR may account for some of the spindle BR in living cells (Inoué and Dan, 1951; Allen,

[2] This is equivalent to a tubulin "concentration" of 31.5 mg/ml.

1972; LaFountain, 1972). Indeed, mechanically deformed gels do exhibit BR as do gels anisotropically swollen or dehydrated (Kunitz, 1930; Inoué, 1949). Whether similar strain BR does contribute to spindle BR can only be answered with further experimentation, but it would appear unlikely that the well oriented microtubules could appreciably contribute to strain BR. Strain BR would be expected by deformation of less well oriented gels such as those seen in isolated amoeba cytoplasm (Taylor et al., 1973), while increase in microtubular BR would be expected to reflect greater ordering and/or increase in microtubular concentration. We have found that change in orientation of even small clusters of (several to about ten parallel) microtubules can be detected with sensitive polarizing microscopes and, we believe, is distinguishable from changes in the quantity of regularly oriented microtubules.

This work was supported in parts by grant CA-10171 from the National Institutes of Health and grant GB-31739X from the National Science Foundation.

Received for publication 7 April 1975, and in revised form 1 July 1975.

REFERENCES

ALLEN, R. D. 1972. Pattern of birefringence in the giant amoeba, *Chaos carolinensis*. *Exp. Cell Res.* **72**:34–45.

AMBRONN, H., and A. FREY. 1926. Das Polarisationsmikroskop. Akademische Verlagsgesellschaft mbH. Leipzig, Germany.

BARER, R., and S. JOSEPH. 1954. Refractometry of living cells. I. Basic principles. *Q. J. Microsc. Sci.* **95**:399–423.

BEHNKE, O., and A. FORER. 1967. Some aspects of microtubules in spermatocyte meiosis in a crane fly (*Nephrotoma suturalis* Loew)—Intranuclear and intrachromosomal microtubules. *C. R. Trav. Lab. Carlsberg.* **35**:437–455.

BENDET, I. J., and J. BEARDEN, JR. 1972. Birefringence of spermatozoa. II. Form birefringence of bull sperm. *J. Cell Biol.* **55**:501–510.

BORN, M., and E. WOLF. 1959. Principles of Optics. Pergamon Press, Inc., New York.

BRAGG, W. L., and A. B. PIPPARD. 1953. The form birefringence of macromolecules. *Acta Crystallogr. Sect. B. Struct. Crystallogr. Chryst. Chem.* **6**:865–867.

BRYAN, J., and H. SATO. 1970. The isolation of the meiosis I spindle from the mature oocyte of *Pisaster ochraceous*. *Exp. Cell Res.* **59**:371–378.

CASSIM, J. Y., and E. W. TAYLOR. 1965. Intrinsic birefringence of poly-γ-benzyl-L-glutamate, a helical polypeptide, and the theory of birefringence. *Biophys. J.* **5**:531–551.

CASSIM, J. Y., P. S. TOBIAS, and E. W. TAYLOR. 1968. Birefringence of muscle proteins and the problem of structural birefringence. *Biochim. Biophys. Acta.* **168**:463–471.

CAVANAUGH, G. M. 1964. Formulae and Methods V. of the Marine Biological Laboratory Chemical Room. Marine Biological Laboratories, Woods Hole, Massachusetts.

CHINN, P., and F. O. SCHMITT. 1937. On the birefringence of nerve sheaths as studied in cross section. *J. Cell. Comp. Physiol.* **9**:289–296.

COHEN, C., D. DEROSIER, S. C. HARRISON, R. E. STEPHENS, and J. THOMAS. 1975. X-ray patterns from microtubules. *Ann. N. Y. Acad. Sci.* **253**: 53–59.

COHEN, C., S. C. HARRISON, and R. E. STEPHENS. 1971. X-ray diffraction from microtubules. *J. Mol. Biol.* **59**:375–380.

COLBY, R. H. 1971. Intrinsic birefringence of glycerinated myofibrils. *J. Cell Biol.* **51**:763–771.

ERICKSON, H. P. 1974. Microtubule surface lattice and subunit structure and observations on reassembly. *J. Cell Biol.* **60**:153–167.

FORER, A., and R. D. GOLDMAN. 1972. The concentrations of dry matter in mitotic apparatuses *in vivo* and after isolation from sea urchin zygotes. *J. Cell Sci.* **10**:387–418.

FRAZER, R. D. B., and E. SUZUKI. 1973. The use of least squares in data analysis. *In* Physical Principles and Techniques of Protein Chemistry. Part C. Syndney J. Leach, editor. Academic Press, Inc., New York.

GOLDMAN, R. D., and L. I. REBHUN. 1969. The structure and some properties of the isolated mitotic apparatus. *J. Cell Sci.* **4**:179–209.

GRIMSTONE, A. V., and A. KLUG. 1966. Observations on the substructure of flagellar fibres. *J. Cell Sci.* **1**:351–362.

HARTSHORNE, N. H., and A. STUART. 1960. Crystals and the Polarizing Microscope—A Handbook for Chemists and Others. 3rd. ed. Edward Arnold Publishers Ltd. London.

HUXLEY, H. E. 1960. Muscle cells. *In* The Cell. J. Brachet and A. E. Mirsky, editors. **4**:365–481. Academic Press, Inc., New York.

INOUÉ, S. 1949. Studies of the *Nereis* egg jelly with the polarization microscope. *Biol. Bull.* (*Woods Hole*). **97**:258–259.

INOUÉ, S., G. G. BORISY, and D. P. KIEHART. 1974. Growth and lability of *Chaetopterus* oocyte mitotic spindles isolated in the presence of porcine brain tubulin. *J. Cell Biol.* **62**:175–184.

INOUÉ, S., and K. DAN. 1951. Birefringence of the dividing cell. *J. Morphol.* **89**:423–455.

INOUÉ, S., J. FUSELER, E. SALMON, and G. W. ELLIS. 1975. Functional organization of mitotic microtubules: physical chemistry of the *in vivo* equilibrium system. *Biophys. J.* **15**:725–744.

INOUÉ, S., and W. L. HYDE. 1957. Studies on depolarization of light at microscope lens surfaces. II. The simultaneous realization of high resolution and high sensitivity with the polarizing microscope. *J. Biophys. Biochem. Cytol.* **3**:831–838.

INOUÉ, S., and H. RITTER, JR. 1975. Dynamics of mitotic spindle organization and function. *In* Molecules and Cell Movement. R. E. Stephens and S. Inoué, editors. Raven Press, New York. 3–31.

INOUÉ, S., and H. SATO. 1967. Cell motility by labile association of molecules. The nature of mitotic spindle fibers and their role in chromosome movement. *J. Gen. Physiol.* **50**:259–292.

KANATANI, H., H. SHIRAI, K. NAKANISHI, and T. KUROKAWA. 1969. Isolation and identification of meiosis inducing substance in starfish *Asterias amurensis*. *Nature* (*Lond.*). **221**:273–274.

KANE, R. E. 1962. The mitotic apparatus: isolation by controlled pH. *J. Cell Biol.* **12**:47–55.

KANE, R. E. 1965. The mitotic apparatus: physical-chemical factors controlling stability. *J. Cell Biol.* **25**:137–144.

KANE, R. E., and A. FORER. 1965. The mitotic apparatus. Structural changes after isolation. *J. Cell Biol.* **25**:31–39.

KUNITZ, M. 1930. Elasticity, double refraction and swelling of isoelectric gelatin. *J. Gen. Physiol.* **13**:565–606.

LAFOUNTAIN, J. R. 1972. Changes in the patterns of birefringence and filament deployment in the meiotic spindle of *Nephrotoma suturalis* during the first meiotic division. *Protoplasma.* **75**:1–17.

LAFOUNTAIN, J. R. 1974. Birefringence and fine structure of spindles in spermatocytes of *Nephrotoma suturalis* at metaphase of first meiotic division. *J. Ultrastruct. Res.* **46**:268–278.

LUFT, J. H. 1961. Improvement in epoxy resin embedding methods. *J. Cell Biol.*, **9**:409–414.

MOLÈ-BAJER, J., and A. BAJER. 1968. Studies of selected endosperm cells with the light and electron microscope. The technique. *LaCellule.* **67**:257–265.

NOLL, D., and H. H. WEBER. 1934. Polarisationsoptik und molekularer Feinbau der Q-Abschnitte des Froschmuskels. *Pflügers Arch.* **235**:234–246.

OLMSTED, J. B., and G. G. BORISY. 1973. Microtubules. *Ann. Rev. Biochem.* **42**:507 540.

PAGE, S. G., and H. E. HUXLEY. 1963. Filament lengths in striated muscle. *J. Cell Biol.* **19**:369–390.

PEPE, F. A. 1972. The myosin filament: immunochemical and ultrastructural approaches to molecular organization. *Cold Spring Harbor Symp. Quant. Biol.* **37**:97–108.

PFEIFFER, H. H. 1951. Polarisationsoptische Untersuchungen am Spindelapparat mitotischer Zellen. *Cytologia* (*Tokyo*) **16**:194–200.

REBHUN, L. I., J. ROSENBAUM, P. LEFEBVRE, and G. SMITH. 1974. Reversible restoration of the birefringence of cold-treated isolated mitotic apparatuses of surf clam eggs with chick brain tubulin. *Nature* (*Lond.*). **249**:113–115.

REBHUN, L. I., and G. SANDER. 1967. Ultrastructure and birefringence of the isolated mitotic apparatus of marine eggs. *J. Cell Biol.* **34**:859–883.

SATO, H. 1975. The mitotic spindle. *In* Aging Gametes. R. Blandau, editor. S. Karger AG, Basel. 19–49.

SATO, H., and J. Bryan. 1968. Kinetic analysis of association-dissociation reaction in the mitotic spindle. *J. Cell Biol.* **39**(2, Pt. 2):118 *a* (Abstr.).

SATÔ, S. 1958. Electron microscope studies on the mitotic figure. I. Fine structure of the metaphase spindle. *Cytologia* (*Tokyo*) **23**:383–394.

SCHMIDT, W. J. 1951. Polarisationsoptische Analyse der Verknupfung von Protein und Lipoidmolekeln, erlautert am Aussenglied der Sehzellen der Wirbeltiere. *Publ. Stn. Zool. Napoli.* **23**(Suppl.):158–183.

STEPHENS, R. E. 1972. Studies on the development of the sea urchin *Strongylocentrotus droebachiensis*. II. Regulation of mitotic spindle equilibrium by environmental temperature. *Biol. Bull.* (*Woods Hole*) **142**:145–159.

TAYLOR, D. L., J. S. CONDEELIS, P. L. MOORE, and R. D. ALLEN. 1973. The contractile basis of amoeboid movement. I. The chemical control of motility in isolated cytoplasm. *J. Cell Biol.* **59**:378–394.

TAYLOR, E. W., and W. CRAMER. 1963. Birefringence of protein solutions and biological systems. I. *Biophys. J.* **3**:127–141.

TILNEY, L. G., J. BRYAN, D. J. BUSH, K. FUJIWARA, M. S. MOOSEKER, D. B. MURPHY, and D. H. SNYDER. 1973. Microtubules: evidence for 13 protofilaments. *J. Cell Biol.* **59**:267–275.

WEISENBERG, R. C. 1972. Microtubule formation *in vitro* in solutions containing low calcium concentrations. *Science* (*Wash. D. C.*). **177**:1104–1105.

WEISENBERG, R. C. 1973. Regulation of tubulin organization during meiosis. *Am. Zool.* **13**:981–987.

WEISENBERG, R. C., and A. C. ROSENFELD. 1975. In vitro polymerization of microtubules into asters and spindles in homogenates of surf clam eggs. *J. Cell Biol.* **64**:146–158.

WIENER, O. 1912. Die Theorie des Mischkörpers für das Feld der stationären Strömung. *Abh. Sächs. Ges. Akad. Wiss., Math.-Phys. Kl.* No. 6, **32**:509–604.

YOUNG, M., M. V. KING, D. S. O'HARA, and P. J. MOLBERG. 1972. Studies on the structure and assembly pattern of the light meromyosin section of the myosin rod. *Cold Spring Harbor Symp. Quant. Biol.* **37**:65–76.

Reprinted from *Development, Growth and Differentiation*, Vol. 18(4), pp. 413–434, 1976, with permission from Blackwell Publishing.

CRYSTAL PROPERTY OF THE LARVAL SEA URCHIN SPICULE*

KAYO OKAZAKI** AND SHINYA INOUÉ***

The Marine Biological Laboratory,
Woods Hole, Mass. 02543, U.S.A.

A pair of pluteus skeletal spicules arises from a pair of calcareous granules via the triradiate form. In polarized light, each spicule behaves as though carved out of a single crystal of magnesian calcite. The optic axis lies perpendicular to the plane of the triradiate and parallel to the body rod of the pluteus. However, in the scanning electron microscope, the spicule surface appeared smooth or somewhat spongy and manifested no crystal faces. Neither etching nor fracturing revealed underlying crystalline texture. Nevertheless, rhombohedral calcite crystals could be grown epitaxially onto isolated spicules immersed in a medium containing $CaCl_2$ and $NaHCO_3$. The optic axes of all crystals coincided with the optic axis of the spicule on which they were grown. Corresponding faces of the crystals were all aligned parallel to each other despite the complex shape of each spicule. Where the left and right spicules joined, two mutually tilted sets of crystals were observed but not crystals of intermediate orientation. Thus, the sea urchin larval spicule is built from a stack of molecularly contiguous microcrystals but its overall shape is generated by the mesenchyme cells independent of the magnesian calcite crystal habit.

A bilaterally symmetric, mutually inclined pair of skeletal spicules develop in the pluteus larva of sea urchins from a pair of calcareous granules formed in the gastrular mesoderm. The rudimental granule soon develops into a triradiate spicule which in turn develops into the pluteus spicule. In polarized light, each triradiate and left and right pluteus spicule exhibits a single optic axis despite its complex, species-specific shape. Each spicule is therefore a "biocrystal" (30, 35), or acts as though carved out of a single large crystal. The optic axis (c-axis) of this biocrystal lies normal to the plane defined by the triradiate, three radii of which are included in a single plane. The c-axis lies along the major axis of the pluteus spicule (29).

The spicule grows in the pseudopodial cable formed by the primary mesenchyme cells (6, 19, 22, 36, 38, 39), the shape of the spicule being foreshadowed by the pseudopodial cable (21). As pointed out by several workers, the growth of the spicule seems to be controlled by a combination of two mechanisms, biological and crystallographic, the latter depending on the crystal properties of the mineral component of the spicule, the former on the pseudopodial activity of the primary mesenchyme cells.

There have been numerous reports concerning the crystallography of plates and spines of adult sea urchins. Under polarized light and by x-ray diffraction, the plate or spine behaves as a single crystal of magnesian calcite (5, 18, 28). On the basis of electron microscopic studies

* This paper is dedicated to Prof. Emer. Katsuma Dan for the celebration of his 70th birthday. Under his tutelage and inspiration it was the privilege of the authors to start this work at the Misaki Marine Biological Station in 1947.

** Present address: Department of Biology, Tokyo Metropolitan University, 2-1-1 Fukazawa, Setagaya-ku, Tokyo 158, Japan.

*** Present address: Department of Biology, University of Pennsylvania G7, Philadelphia, Pa. 19174, U.S.A.

of thin sections, however, TRAVIS (33) emphasizes that the mineralized plates and spines of echinoderms are not single crystals but are composed of polycrystalline aggregates which are structurally organized into sheet-like layers. The constituent crystallites are roughly rectangular in shape and lie in well ordered, closely packed, parallel rows with inter-crystalline separations. The long axes of the crystallites parallel the axes of collagen fibers among which they have developed.

In contrast to the exhaustive studies on adult endoskeleton, only fragmentary data are available on the larval spicule. The mineral component of the larval spicule is magnesian calcite, the ratio of $Mg^{2+}/(Ca^{2+} + Mg^{2+})$ being ca. 5/100 (26). The larval spicule also contains organic substances of ca. 1% by weight (hexose, 0.023%; amino-sugar, 0.024%; reducing sugar, 0.036%; protein, 0.36%; 34). Labelled proline is incorporated as hydroxyproline in the spicule-bound protein fraction which shows solubility properties characteristic of collagen (27). Electron microscopically, however, neither collagen fibers nor the texture of the mineral component have been observed. Spicules are missing in thin sections leaving a hole in the pseudopodial cable, owing to dissolution either during or following the cutting process (6, 16). Consequently, it is still unclear whether magnesian calcite of the larval spicule is a polycrystalline aggregate or it is a molecularly contiguous single crystal.

For some time the authors wondered whether mesenchyme cells first formed small crystallites which were deposited like regular stacks of brick onto the growing spicule, or whether deposition was via a material more molecularly dispersed. Occasionally, we observed that minute birefringent crystals travelled along mesenchymal pseudopodia and, upon approaching the developing spicule, flickered and suddenly disappeared as though becoming incorporated into the spicule. Moreover, in the polarizing microscope, the junction of the left and right ventral transverse rods, which meet and fuse at the center of the ventral side of the larva at the pluteus stage, exhibited a complex extinction pattern (see Fig.1). This pattern may well have arisen from crystallites randomly stacked at the junction (3, 23).

These circumstances prompted the authors to study the surface of isolated spicules with the scanning electron microscope. As described later, initial attempts were disappointing owing to a singular lack of surface texture suggestive of the spicule's crystalline composition. Therefore, the spicules were "decorated" with calcite, whose crystalline arrangement indicated a molecularly contiguous stack of microcrystals in the spicule.

In advance of the studies with the scanning electron microscope, we confirmed, on several species of sea urchins, RUNNSTRÖM's previously mentioned observation on *Paracentrotus lividus* (29). In the description hereafter, the following three terms will be used to refer to the larval spicule: "triradiate" for the triradiate spicules formed at the gastrula stage; "skeletal spicule" for the spicules of the pluteus stage; "spicule" as a general term for the spicules of various developmental stages.

MATERIALS AND METHODS

Materials *Arbacia punctulata* was principally used for our studies, supplemented by *Lytechinus pictus*, *Hemicentrotus pulcherrimus*, *Pseudocentrotus depressus* and *Anthocidaris crassispina*.

For observation under the polarizing microscope, larvae were followed in the living state or viewed as total mounts in Canada balsam after fixation by methanol. Where necessary, spicules were isolated from the larvae. For observation by the scanning electron micrsocope, only isolated spicules were used.

Isolation of spicules Cells were lysed by exposing the larvae to distilled water, previously saturated with magnesian calcite (by overnight immersion of fragments of adult calcareous test) and adjusted to pH 8.0 with NaOH. After 2 washes in a hand centrifuge, the spicules were cleaned by immersion in N/5 NaOH or Clorox (5% sodium hypochlorite) for 2 hr with intermittent agitation with a pipette. After 5 washes with the magnesian calcite saturated distilled water, the white spicule pellet was dehydrated sequentially in 70%, 90% and 100% ethanol. Spicules were either kept in 100% ethanol or as a dry powder which was resuspended in 100% ethanol prior to use.

Scanning electron microscopy Spicules suspended in 100% ethanol was affixed to a ca. 10 mm square microscope cover slip by evaporation of the alcohol. The cover slip was attached onto the scanning electron microscope specimen holder by double-stick Scotch tape. The surface of the cover slip was coated with gold-palladium in a Tousimis Samsputter™.

For decoration with calcite crystals, spicules were immersed for a few minutes in M/10 $CaCl_2$ (or M/10 $NaHCO_3$), then M/10 $NaHCO_3$ (or M/10 $CaCl_2$) was added. When crystals grew to the desired extent, the spicules were washed in distilled water, serially dehydrated in ethanol and affixed to the cover slip. Alternately, reagents were added onto spicules already dried down on the cover slip; after examination of crystal growth, the cover slip with spicules was washed with distilled water and dehydrated.

A JEOL M-35 scanning electron microscope was used for observation and photographic recording. Anode voltage ranged from 10 KV for lower power to 30 KV for higher power views.

Results

(1) Observation with the polarizing microscope

Viewed in a polarizing microscope between crossed polars, the triradiate as well as the skeletal spicule behaved optically as though chiselled out from a single crystal of calcite. This was true in all species of sea urchins we used (Figs.1–3). In other words, a spicule alternately brightens and darkens completely four times in a 360° rotation of the stage of the

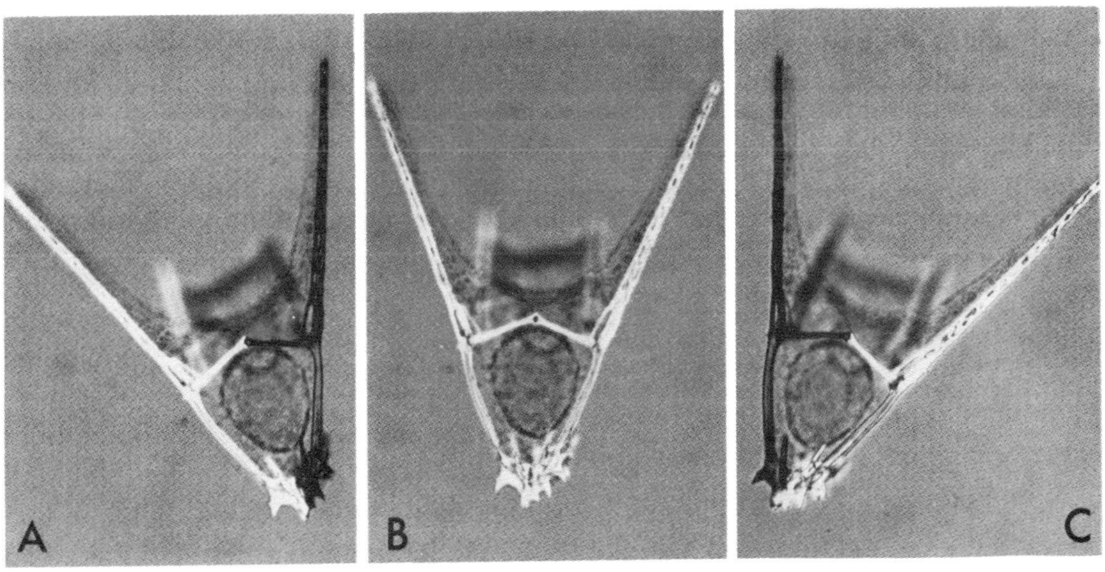

Fig.1. Pluteus of *Arbacia punctulata* fixed by methanol, mounted in Canada balsam and observed with a polarizing microscope.

A pair of skeletal spicules shows high negative birefringence. Despite its complex morphology, the whole left (C) or right (A) spicule is extinguished when the body rod becomes oriented parallel to the polarizer or analyzer axis. The optic axis of the spicule lies along the body rod. Optically, the left and right spicule thus each behaves as though carved out of a single crystal of calcite (cf. Fig.4).

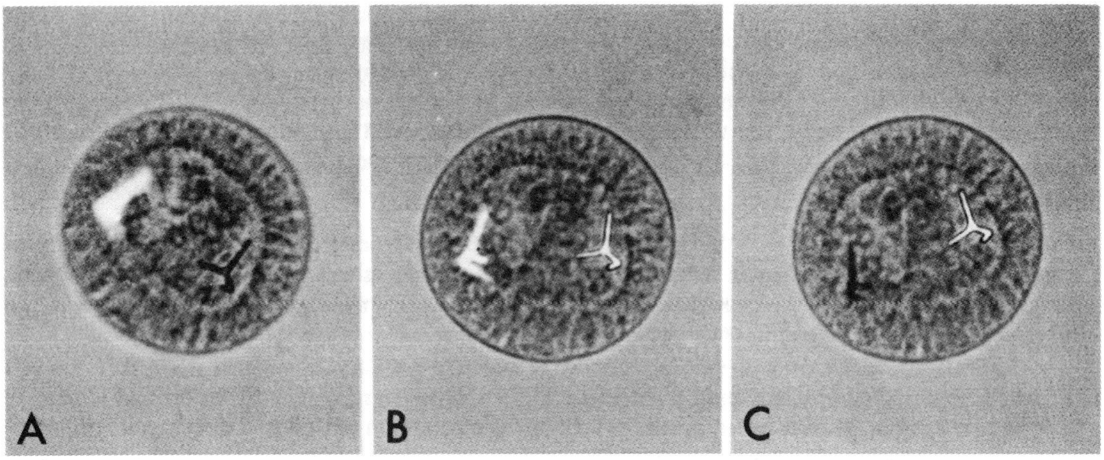

Fig. 2.

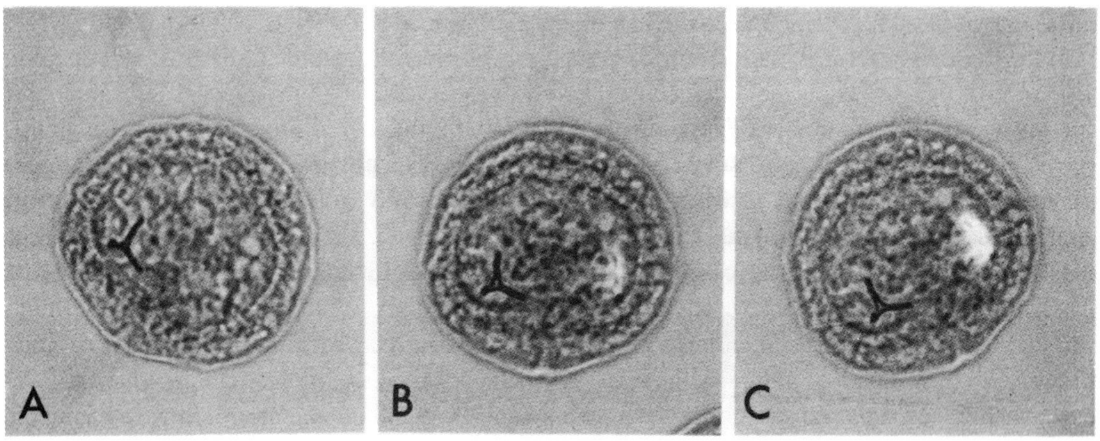

Fig. 3.

Figs.2,3. Gastrula of *Arbacia punctulata* fixed by methanol, mounted in Canada balsam and observed with a polarizing microscope. One radius of the triradiate spicule has already bent to form the body rod.

Fig.2. When the gastrula is rotated between crossed polars, the two triradiates alternately brighten and darken, each independently.

Fig.3. One triradiate seen from the direction of the optic axis does not show birefringence, while another spicule (out of focus) glows and darkens when the gastrula is rotated; in the former, the primordial body rod is barely seen on the shortest arm since it elongates along the microscope axis (cf. right spicule in Fig.2).

microscope, provided the optic axis of the spicule does not coincide with the microscope axis. The darkening occurs when the optic axis lies in the plane of vibration of the polarized light or in the plane at right angles to this.

When the triradiate is observed from a direction such that two of the radii lie in a single straight line, the remaining radius is barely observable as a point at the center of the two radii. The fact indicates that the three radii of the triradiate are included in a single plane. If the triradiate is observed along a direction perpendicular to this plane, the spicule ceases to shine at any orientation of the microscope stage (Fig.3). This condition is easily observed in isolated spicules since the young triradiate spicules tend to lie flat on the microscope slide. Therefore, it is clear that the direction of the optic axis of the triradiate is perpendicular to the plane of the spicule.

At the late gastrula stage, one radius of the triradiate elongating towards the dorsal side of the larva bends at right angles to the plane of the spicule to form the body rod of the pluteus (Figs.2, 3 and 4). The evidence for the angle of bending is as follows: 1) if the spicule is seen from the direction perpendicular to the plane containing the initial triradiate spicule, the spicule shows no birefringence and the primordial body rod is barely observable because it lies parallel to the microscope axis (Fig.3); 2) if the spicule is seen from a direction such that the two unbent radii lie in a straight line, the spicule appears T shaped. Although the point of bending varies from species to species, a general statement can be made that it is only several μm from the center of the triradiate in sea urchin larvae with small blastocoels, e.g. *Arbacia punctulata,* while it is over 10 μm in larvae with large blastocoels, e.g. *Lytechinus pictus.*

From the facts so far described, we should expect that the direction of the optic axis of the skeletal spicule coincides with the morphological long axis of the body rod. This was indeed the case with all the species of sea urchins examined. Because of the shape of the pluteus skeletal spicule, it is difficult to see the spicule from a direction perpendicular to the plane of the initial triradiate spicule. Therefore, the direction of the optic axis was determined by one of the following two criteria based on the fact that the crystal form of the spicule is that of calcite (4, 26). 1) Orient the pluteus or isolated skeletal spicule mounted in Canada balsam so that one of the pair of skeletal spicules extinguishes, then remove the analyzer: when the optic axis lies in the plane of vibration of the polarized light the spicule refractive index approximates the balsam and the spicule appears faint; at right angles to that plane, the spicule appears distinct owing to the greater difference between the refractive indices of calcite and balsam. 2) Using a quartz wedge, the direction of the slow axis of the spicule is determined: the direction of the optic axis is perpendicular to the slow axis in the case of calcite.

The major component rods of the skeletal spicule are the body, post-oral, antero-lateral and ventral-transverse rods. The developmental courses of these rods other than the body rod will be mentioned here. In *Hemicentrotus, Pseudocentrotus* and *Lytechinus,* in which the post-oral rod is a simple slender rod, a new branch formed near the bending point for the body rod extends in the opposite direction from the body rod and develops into the post-oral rod (see Fig.4). In *Anthocidaris* and *Arbacia,* in which the post-oral rod is fenestrated, three parallel branches are formed toward the opposite direction from the body rod, one on each radius of the triradiate. These three branches elongate, twist and further elongate while forming cross bars to unite the three branches into a fenestrated rod (see Fig.6). The antero-lateral rod is formed by bending of the radius extending toward the animal pole of the larva, at an obtuse angle toward the oral lobe. The last radius extending toward the ventral side of the larva continues its elongation without bending to form the ventral-transverse rod which meets and joins to the corresponding rod of the opposite side of the larva. As has repeatedly been pointed out, the optic axis of the spicule is not changed at all through the course of its development irrespective of the angle of bending or budding of new branches. If an entire spicule had been carved out of a single crystal of calcite, the spicule would then be oriented in the crystal as shown in Fig.4. The body and post-oral rods run parallel to the crystal's optic axis (C-C' in Fig.4) and the plane of the triradiate lies perpendicular to this direction, i.e., parallel to the plane including the trigonal, or three equivalent crystal axes (the three a-axes). The directions of elongation of the three radii of the triradiate coincide with the negative a-axes of the calcite

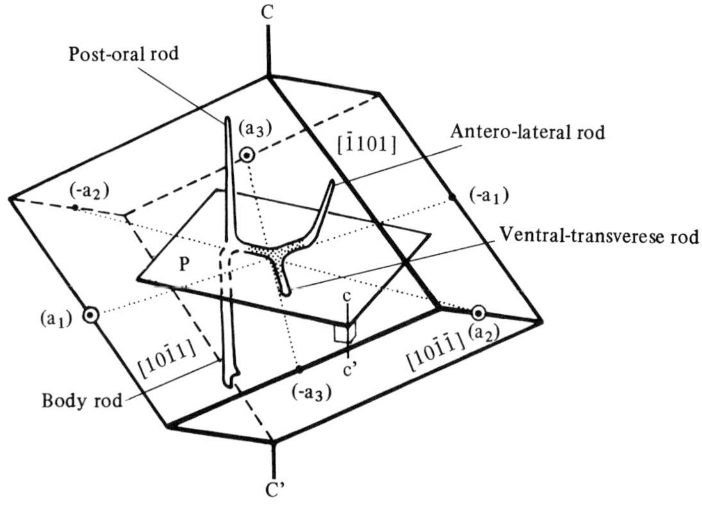
C
Post-oral rod
(a3)
[1̄101]
Antero-lateral rod
(-a2)
(-a1)
Ventral-transveresе rod
P
c
(a1)
[101̄1]
c'
[101̄1̄]
(a2)
(-a3)
Body rod
C'

Fig. 4.

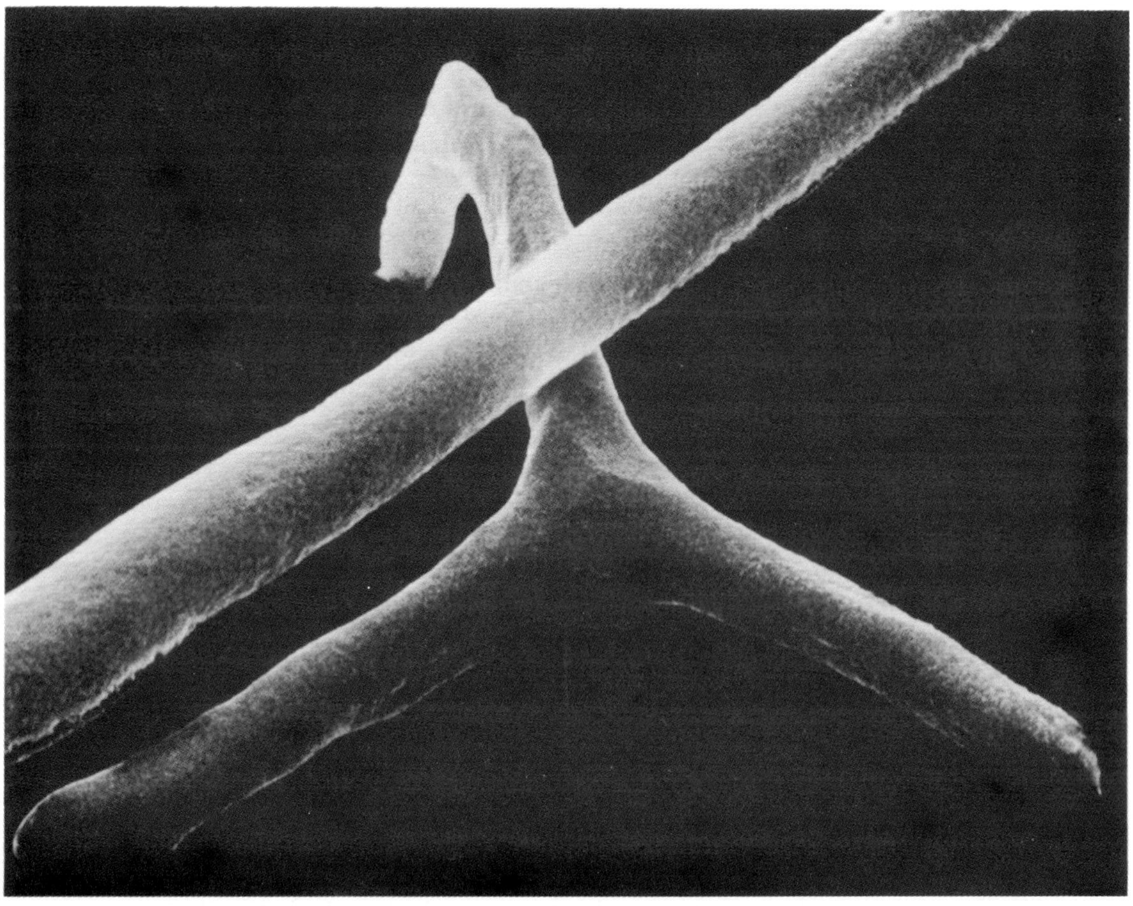

Fig. 5.

crystal, evidence of which will be described later.

(2) Observation of spicule surface with the scanning electron microscope

(a) Isolated, untreated spicule

Neither triradiate nor skeletal spicule isolated from larvae of *Arbacia, Lytechinus* or *Hemicentrotus* exhibited any texture suggestive of crystal faces. Mostly, surfaces of the spicules appeared as though encrusted by a thin spongy skin (Figs.5–8 and 11–14), but other surfaces appeared relatively smooth even at the highest resolution attained (estimated at ca. 100 Å), in agreement with MILLONIG's observation on *Paracentrotus lividus* (16). Occasionally, the surface roughness varied from portion to portion even on a single spicule. Isolation by NaOH and Clorox gave no essential differences on the texture of the surfaces of the spicules.

(b) Etched spicule

The surface of isolated spicules treated with 10^{-2} N HCl or with 10^{-5} M EDTA occasionally showed jagged edges (Fig.9). However, compared with similarly treated crystals of pure calcite, the treated spicules could not definitely be said to exhibit etch patterns characteristic of calcite.

(c) Fracture planes

Since calcite crystals exhibit striking cleavage patterns when fractured, attempts were made to look for cleavage patterns on the fracture plane of isolated skeletal spicule. Spicules were fractured by the following means: affix the isolated skeletal spicule on the cover glass by the method previously mentioned; under the disecting microscope, select a spicule attached by its posterior end to the cover glass and touch its anterior part with a fine glass needle so that the spicule breaks with a snap. When the fractured plane was flat, the fractured surface appeared to be completely smooth (Fig.6), while ridges were observed along the edge when the fractured plane was concave (Fig.8). The ridges were seen at intervals of several hundred Å. These may be related to the granulated texture of the spicule surface, but no fracture face was clearly identifiable as calcite crystalline cleavage patterns.

(d) Triradiate spicules in low Ca^{2+}-Mg^{2+} reared larvae

If normal living gastrulae with triradiate spicules are transferred into artificial sea water containing reduced concentrations of both calcium and magnesium ions (1/10 of natural sea water), the spicules thicken without elongating and the thickened spicules exhibit calcite-like

---

Fig.4. Schematic diagram showing the crystallographic orientation of a larval sea urchin spicule.

If the spicule had been carved out of a single crystal of calcite, it would be oriented as shown here. The pluteus skeletal spicule arises from the triradiate spicule (stippled in the figure). Three radii of the triradiate are included in a single plane (P) which is perpendicular to the c-axis of the calcite rhomb, namely parallel to the plane including the three a-axes. The three radii of the triradiate elongate parallel to the directions of the three a-axes of the crystal in negative directions. Since the body rod extends posteriorly, while the post-oral rod extends anteriorly, at right angles to the plane of the triradiate, these spicular rods run parallel to the c-axis of the crystal. See text for further explanation. C-C': optic-axis of the calcite rhomb. c-c': optic axis of the spicule. $(a_1)-(-a_1)$, $(a_2)-(-a_2)$ and $(a_3)-(-a_3)$: three a-axes of the calcite rhomb (dotted lines). $[10\bar{1}1]$, $[\bar{1}101]$ and $[01\bar{1}\bar{1}]$: Miller indices of the three crystal faces outlined by the continuous lines.

Fig.5. Scanning electron micrograph of triradiate spicule and a part of the young body rod isolated with Clorox from gastrula and prism larvae of *Arbacia punctulata*.

A radius of the triradiate has already bent to form the body rod. The surfaces of the spicules look spongy and no obvious crystal faces are seen. The triradiate ridges seen at the center of the spicule run ca. 30° counter-clockwise from the extension directions of the spicule arms. The central ridges thus display the three rhombohedral vertices surrounding the c-axis (cf. Fig.4). X 10,400.

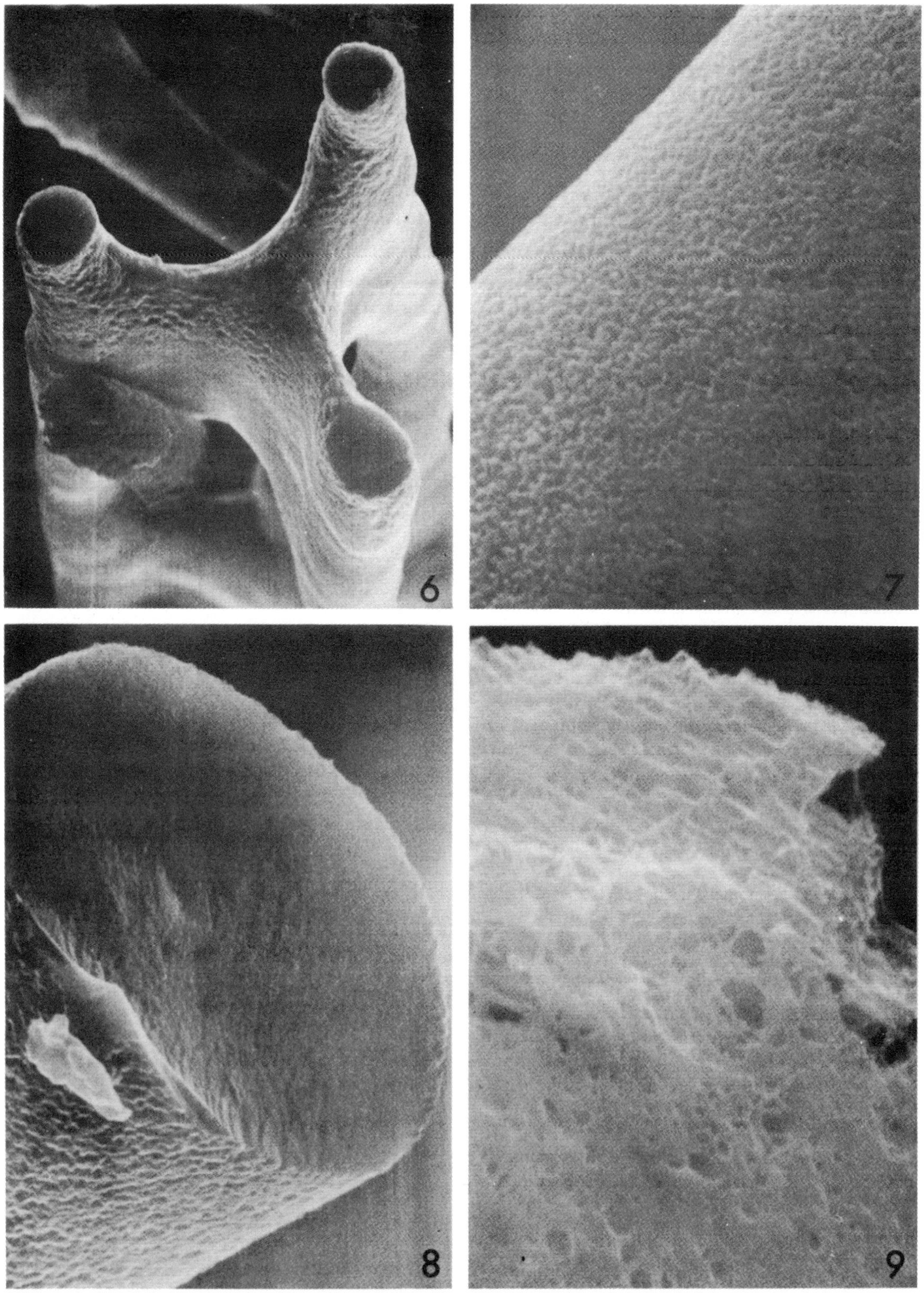

Figs. 6–9.

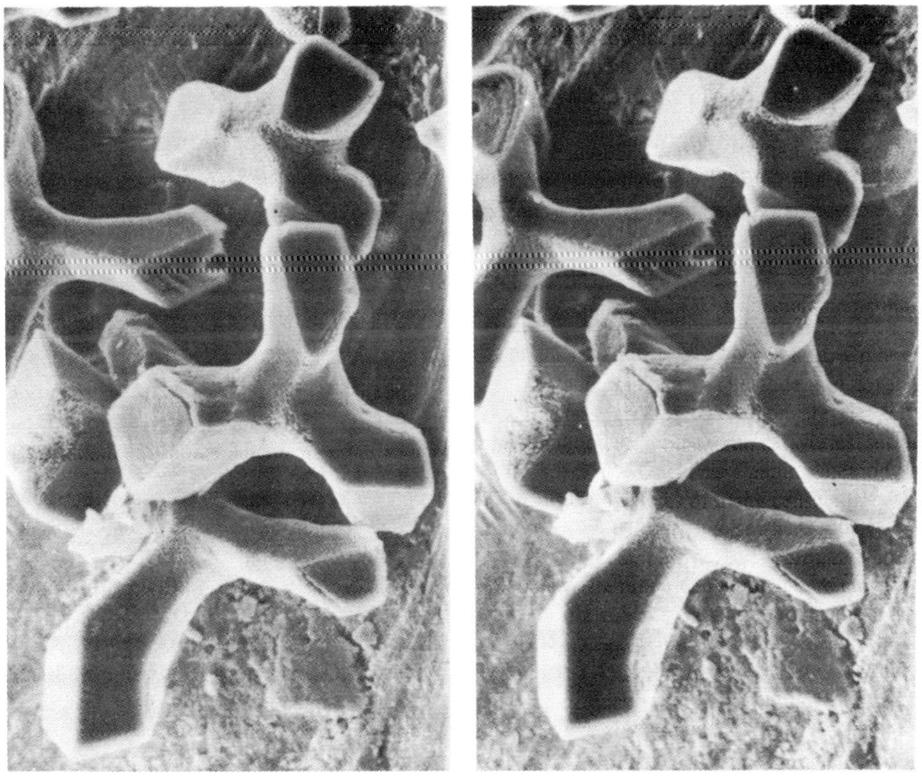

Fig.10. Stereo-pair scanning electron micrograph of thickened triradiate spicules from *Arbacia* larvae grown in 1/10 Ca^{2+}-Mg^{2+}-sea water.

When gastrulae with triradiate spicules are transferred to sea water containing 1/10 normal amount of Ca^{2+} and Mg^{2+}, the spicules cease to grow slender extensions. Instead, the spicules thicken and each tip of the triradiate acquires a "college cap". The cylindrical arms change into rods with rhomboid cross sections. When these triradiates are isolated and viewed as a scanning electron micrograph stereo-pair, side by side with a half-inch rhomb of calcite crystal, the faces on the spicule can clearly be identified as those of calcite. The calcite rhomb should be oriented so that a corner where three obtuse angles meet is pointing up, and with the line symmetrically transsecting the three faces at that corner likewise transsecting the triradiate arms. The flat faces on the spicule can then be exactly superimposed with the faces on the calcite rhomb. X 2,700.

faces (20, 21). A scanning electron microscope stereo-pair of such spicules is shown in Fig.10. Observing the pair with a stereo-viewer together with a half-inch rhomb of calcite (Iceland spar), the rhomb can be oriented so that each of the faces on the triradiate lies parallel to a face of the calcite rhomb. By this means, the following information is obtained.

Figs.6–9. Scanning electron micrographs of spicules isolated from larvae of *A. punctulata*.
Spicules were isolated with Clorox (Figs.6, 8, 9) or with N/5 NaOH (Fig.7).

Fig.6. Front view of fractured surface of the fenestrated post-oral rod. Three parallel columns and cross bridges are clearly seen. In contrast to the spongy appearance of the spicule surface, fractured faces of the columns are completely smooth. X 12,000.

Fig.7. A part of the body rod of a 4-armed pluteus. The spicule surface shows spongy texture. X 25,000.

Fig.8. Surface texture and fracture of the body rod of 4-armed pluteus. By chance, the fracture is concave and small ridges are seen especially along the edge. The ridges are not identifiable as crystalline cleavage patterns. The spicule surface again looks spongy but the texture looks different from that of Fig.7. X 20,000.

Fig.9. A part of the post-oral rod treated with 10^{-5} M EDTA. Although jagged edges and rectangular compartments are observed, these can not be identified as etch patterns characteristic of calcite. X 18,000.

Each spicule tip is enclosed by four planes corresponding to four cleavage faces of the calcite rhomb. The spicule apex displays the vertex between two faces which meet at an acute angle, i.e., the faces $[10\bar{1}1]$ and $[01\bar{1}\bar{1}]$, $[\bar{1}101]$ and $[\bar{1}01\bar{1}]$, $[0\bar{1}11]$ and $[1\bar{1}0\bar{1}]$. The three arms of the triradiate elongate in the $-a_1$, $-a_2$ and $-a_3$ directions, i.e., along the three trigonal axes of calcite but in the negative directions. On each arm of the triradiate, one pair of parallel faces of the calcite rhomb appear smooth (only the upper faces are seen in the photographs). These faces lie parallel to the crystal axes along which the arms elongate.

Looking again at normal triradiate spicules (Fig.5), hereupon it is noticed that the triradiate ridges at the center of the spicule run ca. 30° counterclockwise from the extension direction of the spicule arms. The central ridges thus display the three rhombohedral vertices surrounding the c-axis (cf. Fig.4).

The surfaces of the three arms of such thickened triradiates exhibited spongy appearances similar to those of untreated normal spicules (Figs.10–12). On the other hand, the surfaces at the spicule tips varied from face to face: sometimes the appearance was exactly the same as the surface of the arm; sometimes they appeared smooth; and occasionally, only a part of a face was smooth as if the granulated skin were ablated (Fig.12).

By chance, concentric lamination with a central hole was found at the tip of some spicules, each layer being composed of small structures (crystals?) of various sizes within the range of roughly 100–1,000 Å in width (Figs.11, 13 and 14). The photographs shown in Figs.12–14 suggest that the surface of the spicule may be encrusted by a granulated skin, whose granules reflect the arrangement of microcrystals inside the spicule. Thereupon, we made the attempt which is described in the next section.

(e) Calcite crystals epitaxially grown on isolated spicules

When isolated spicules are immersed in a solution containing M/10 $NaHCO_3$ and M/10 $CaCl_2$, the spicules become fatter and acquire jagged edges within several minutes. The new material deposited on the spicule is strongly birefringent, yet each spicule still extinguishes at once between crossed polars, the initial direction of the optix axis remaining unchanged. When such spicules are viewed with the scanning electron microscope, the material on the spicule turns out

---

Figs.11–14. Scanning electron micrographs of spicules formed in 1/10 Ca^{2+}-Mg^{2+}-sea water.

Fig.11. Triradiate spicules displaying crystal faces. In the three spicules seen at the periphery of the photograph (two, upper; one, lower), the spicule tips are covered with crystalline flat faces, surfaces of which are smooth in comparison with the granulated surfaces of the arms. Clefts, perhaps produced by desiccation, are seen at the vertices between two faces meeting at the spicule tips or at the borders of crystalline faces and arms. In the spicule at the center of the photograph, concentric lamination of small rectangular structures and a central hole are seen at each spicule tip as if the cover concealing the inside texture of the spicule were removed. A triradiate cleft is seen on the spicule; probably produced by desiccation, corresponding to the borders between the flat sides of the arms exhibiting crystal faces and the other sides not displaying crystal faces (compare with Fig.16). × 4,000.

Figs.12 and 13. Higher magnification of portions of Fig.11: the middle of the right side (Fig.12) and a part of the tip of the spicule at the center (Fig.13).

Fig.12. At the top of the photograph, a face at the spicule tip is shown; note that only a part of the surface appears smooth as if a granulated lid had been removed. Down below, a spicule tip is seen, a corner of which is accidentally broken; note that the inside of the spicule looks spongy. × 12,000.

Fig.13. Note concentric laminated structure and a central hole. This figure is very similar to that of a sponge spicule treated for 24 hours with 10% potash [Taf. III, Fig.45 in VON EBNER (35)]. × 30,000.

Fig.14. A rhombohedral spicule formed in a larva transferred into 1/10 Ca^{2+}-Mg^{2+}-sea water. The larva was transferred at the early gastrula stage when it had developed a pair of calcareous granules. The spicule appears to be somewhat corroded, since a part of the surface has disappeared and the inside layered texture is exposed. × 7,800.

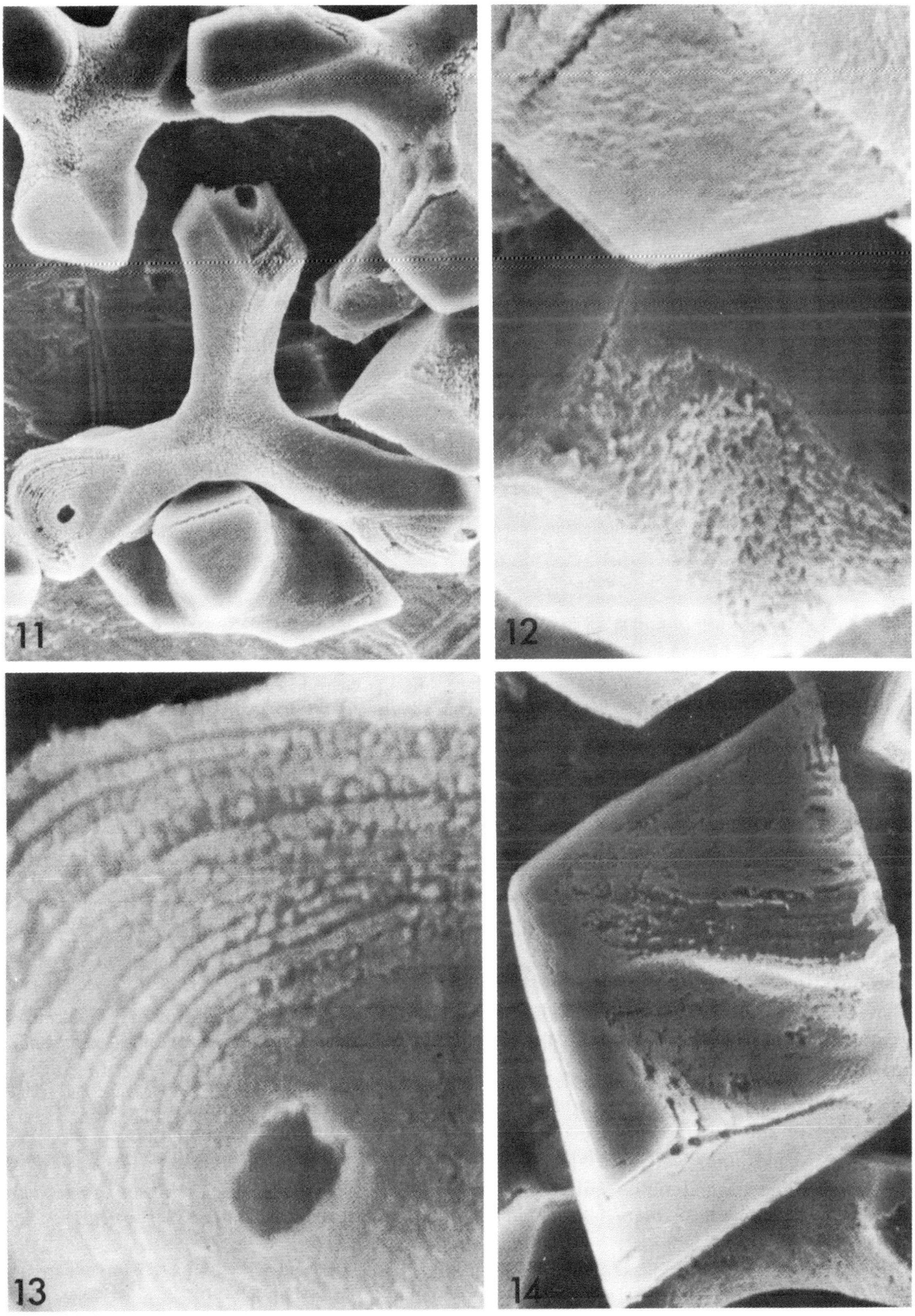

Figs. 11–14.

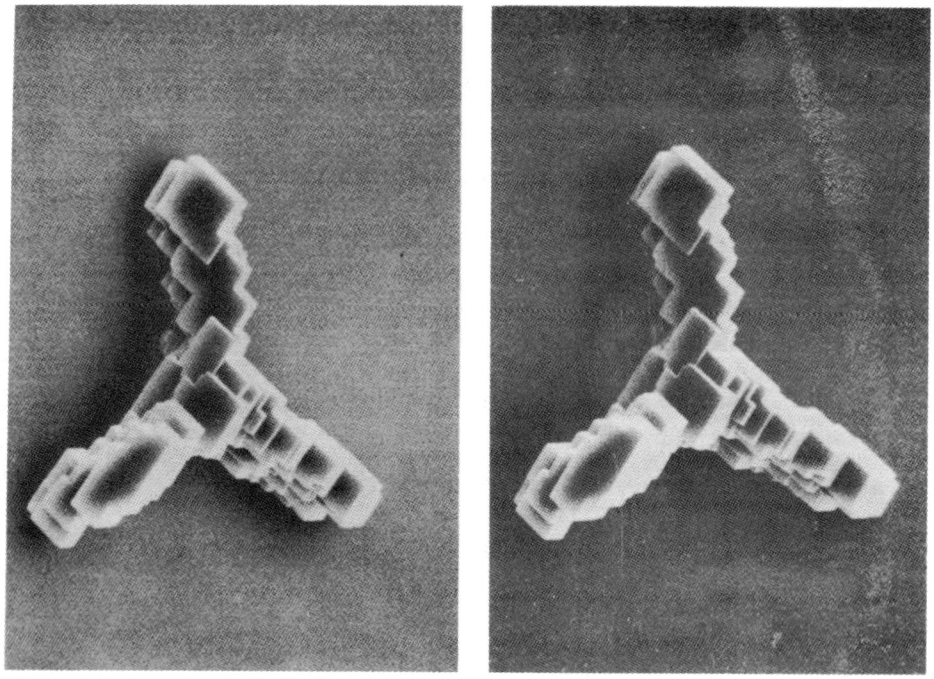

Fig.15. Stereo-pair scanning electron micrograph of a triradiate spicule isolated from *Arbacia* gastrula and decorated with calcite crystals.

Viewed with a stereo-viewer, the three dimensional shape and the directions of the c- and a-axes of the crystals are clearly recognizable. Note the coincidence of the directions of arm elongation with those of the minus a-axes of the crystals. Each spicule apex displays the vertex between two faces which meet at an acute angle as in the spicule in Fig.10. X 3,900.

to be stacks of crystals, whose form is identical to a cleavage rhombohedron of calcite. Whether a few crystals are sparsely grown without contacting each other (Fig.26) or many crystals are staggered on top of each other (Figs.15–25), the corresponding faces of the crystals all lay precisely parallel to each other over a whole spicule. The calcite crystals must therefore have grown epitaxially onto a regular magnesian calcite lattice existing in the normal spicule.

Fig.15 is an example of a stereo-pair of a young triradiate of *Arbacia* epitaxially decorated with stacked crystals of calcite. When viewed with a stereo-viewer together with a half-inch rhomb of calcite (Iceland spar) set in the same orientation as the crystals on the spicule, the following can be concluded: the shape of the crystals on the spicule is identical to the cleavage rhombohedron of calcite; corresponding faces of the crystals all lie in parallel planes; the optic axes of the crystals all lie parallel to each other and are perpendicular to the plane of the triradiate; the three arms of the triradiate run parallel to and grow in the negative directions of the three crystal axes of the calcite. These conclusions held true for all of the several examples on which stereo-pair photographs were made. Furthermore, the full set of crystallographic characteristics noted for spicules grown in low Ca^{2+}-Mg^{2+}-sea water (section d above) held true also for the calcite crystals epitaxially grown onto the smooth surface of the isolated triradiate.

Fig.16 is a similarly decorated *Lytechinus* triradiate. Although the arms of the triradiate are not straight and more or less wavy, which is characteristic of the *Lytechinus* spicule, the corresponding faces of the crystals all lie in parallel planes as in the case of *Arbacia.* The spicule

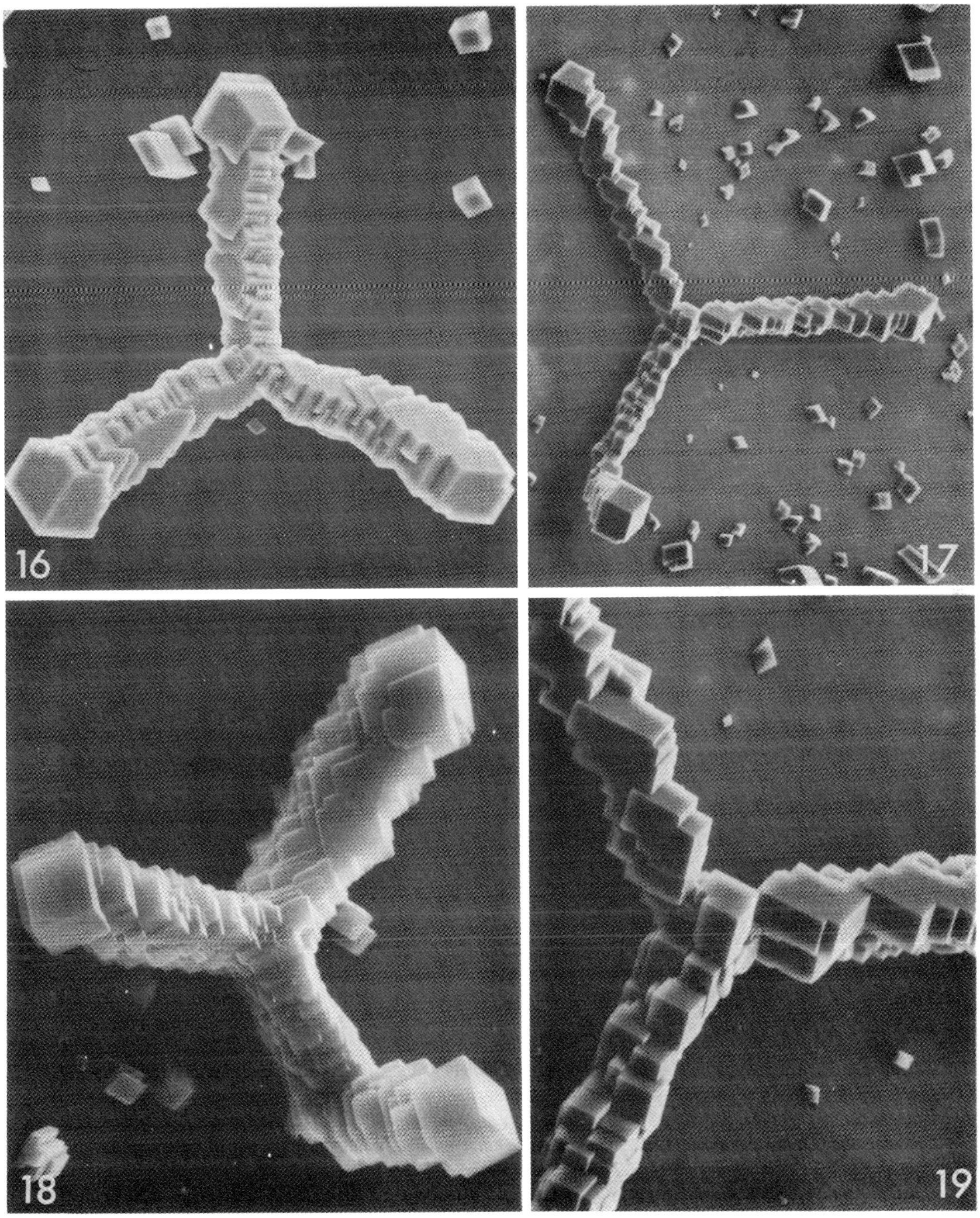

Figs.16–19. Scanning electron micrographs of triradiate spicules isolated from *Arbacia* (Fig.18) or *Lytechinus* (Figs.16, 17 and 19) gastrulae and decorated with calcite crystals.

Fig.16. The decorated spicule is photographed from much the same angle as the spicule in the middle of Fig.10. Crystallographic characteristics are all the same as the spicule in Fig.10. X 2,000.

Figs.17 and 18. The triradiate, whose one radius had already bent to form the body rod, was decorated with calcite. Note the same orientation of the calcite crystals on the young body rod as those on the triradiate arms. X 1,200 (Fig.17), X 3,600 (Fig.18).

Fig.19. Higher magnification of central portion of the spicule shown in Fig.17. X 3,000.

Throughout Figs.16–19, note that 1) corresponding faces of the calcite crystals on each spicule all lie precisely parallel to each other; 2) on each arm of the triradiate, crystal faces which are parallel to the elongation direction of the arm are contiguous and form a relatively flat plane, while the other crystal faces arrange themselves forming tiers; 3) the orientations of crystals not growing on the spicule are random and completely independent from those on the spicule even when the crystals lie in very close proximity to the spicule.

in Fig.16 is photographed from much the same angle as the spicule shown in the middle of Fig.10. Comparing these photographs, the shape of the crystal faces, the orientation of the crystals and the mode of appearance of crystal faces in relation to the three arms of the triradiate are all found to be the same for the spicule formed *in vivo* in low Ca^{2+}-Mg^{2+}-sea water (Fig.10) and the spicule artificially decorated with calcite crystals *in vitro* (Fig.16).

In the cases shown in Figs.17 (*Lytechinus*) and 18 (*Arbacia*), crystals were grown on relatively old triradiate spicules, one radius of which had already bent to form the body rod. Although the orientations of the crystals on the young body rod are the same as those on the initial radius, the direction in which the crystals stack themselves has changed at the bending point. This fact indicates that bending of the spicular rod is caused by alteration of the direction in which building material of the spicule is piled up, without changing the orientation of the individual construction units.

An example of *Arbacia* pluteus skeletal spicule decorated with calcite crystals is shown in Figs.20–25. Clearly, corresponding faces of the crystals lie parallel to each other over the entire left or right skeletal spicule. Complete calcite rhombohedra can be seen at the tip of the post-oral and antero-lateral rods (Figs.20, 21). In Fig.22, the bases of the post-oral (upper part), body (lower part) and ventral-transverse (left side) rods are displayed. Again, the orientation of the crystals are all the same in the initial triradiate (ventral transverse rod), in the rod after bending (body rod) and on new branches for the post-oral rod. Examined with the polarizing microscope, the optic axes of the crystals all lie parallel to each other and to the long axis of the body rod, i.e., perpendicular to the plane of the triradiate.

Since the optic axes of the left and right spicules are tilted about 45° to each other, the crystal faces are likewise tilted between the two spicules. Such relations are plainly seen at the junctions of the left and right ventral-transverse rods (Figs.24 and 26) and of the body rods (Fig.25). The two tilted sets of crystals interdigitate and form a jagged but compact junction. No crystals of intermediate orientation are found at the junctions.

Discussion

Uniform crystal lattice in the sea urchin larval spicule

As already mentioned, endoskeletal plates and spines of adult sea urchins have long been known to behave as single crystals. Although polarized light study and x-ray diffraction suggest

Figs.20–25. Scanning electron micrographs of skeletal spicule isolated from a 4-armed pluteus of *Arbacia* and decorated with calcite crystals.

Fig.20. Full portrait in low magnification. Small spots on the lateral sides of the post-oral and antero-lateral rods are the crystals formed on the lateral spines of the rods (cf. Fig.1). Under the polarizing microscope, the whole left and right decorated spicule each ceases to shine and extinguishes uniformly when the body rod is oriented parallel to the polarizer or analyzer axis just as in the spicule shown in Fig.1. X 360.

Figs.21–25. Higher magnifications of portions of the right spicule illustrated in Fig.20.

Fig.21. Tip of the fenestrated post-oral rod composed of three parallel rows (cf. Fig.6). On the crystals grown at the tip of the three rows, three crystal faces are seen in the photograph: assuming that the body rod had developed from radius along $-a_2$ axis as in Fig.4, the faces are $[0\bar{1}11]$, $[\bar{1}101]$ and $[\bar{1}011]$. X 6,000.

Fig.22. Bases of the post-oral (upper), body (lower) and ventral-transverse (left) rods. Despite the bends in the spicule, the faces of the calcite crystals decorating the spicule all lie parallel to each other. On the lateral spines (protruding at the right margin of the spicule), crystal orientation is also the same as the main part of the spicule. X 6,000.

Fig.23. A part of the body rod. Although the crystal faces are not fully displayed because of the slate-like stack of crystals, the vertices, on the same assumption as Fig.21, between $[0\bar{1}11]$ and $[\bar{1}01\bar{1}]$, $[0\bar{1}11]$ and $[1\bar{1}0\bar{1}]$, $[\bar{1}101]$ and $[\bar{1}01\bar{1}]$ are clearly seen in the photograph. X 6,000.

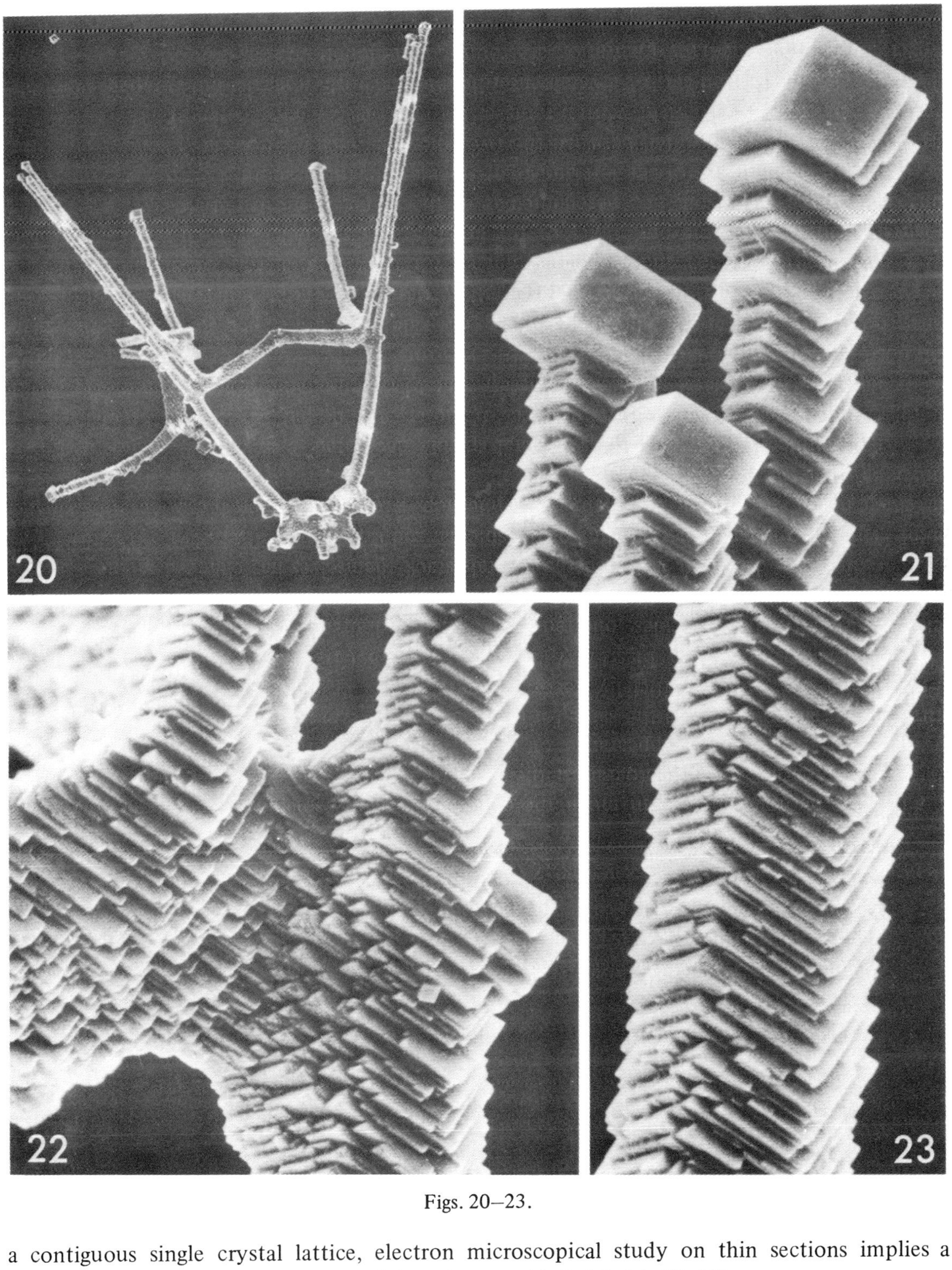

Figs. 20–23.

a contiguous single crystal lattice, electron microscopical study on thin sections implies a polycrystalline array with near parallel optic axes (4, 5, 17, 18, 28, 33).

Larval spicules of sea urchins also show single optic axes. When microcrystals were grown onto the isolated spicule, not only did the optic axes of the decorating crystals parallel the optic axis of the spicule, but with scanning electron microscopy it was shown unequivocally that corresponding faces of all of the crystals on a single spicule lie parallel to each other. Furthermore, these faces lie parallel to the faces which develop on the triradiate spicules in larvae grown in 1/10 Ca^{2+}-Mg^{2+}-sea water. These facts indicate crystalline continuity from decorating crystals to spicule calcite. Judging from the arrangement of these crystals, the larval sea urchin spicule is not made up of an array of microcrystals with parallel optic axis whose other crystallographic axes are oriented helter skelter or inverted as in twinned crystals. Rather, the crystal lattice is uniform throughout a whole single spicule no matter how complex the external morphology.

When paired skeletal spicules were decorated with crystals, only two sets of crystal axes were found, one for each of the left and right spicules. No microcrystals with other orientation were observed even at the junction of the two spicules. Therefore, the complex extinction pattern seen at these junctions between crossed polars could not be due to randomly stacked crystallites. Instead it must be explained by the complex interdigitation of the left and right spicules which, because of overlap, provide minute regions whose apparent optic axes do not parallel the axes of the rest of the spicules. There is then no sign that preformed crystallites were jammed into the final junction of the spicules as we earlier believed.

Morphologic-crystallographic relation of the spicule

One of the important questions regarding spicule formation is the relationship between spicule morphology and symmetry axes of the constituent crystal. Morphologic-crystallographic relation in sea urchins, however, is poorly known except in terms of c-axis orientation. In the adult endoskeleton, the c-axis is always parallel to the long direction of the spine but either tangential or perpendicular to the surface of a plate depending on the species of sea urchin (28). As has been repeatedly mentioned, the c-axis of the larval spicule is perpendicular to the plane of the triradiate and parallel to the long direction of the body rod.

In contrast to the striking parallelism between the c-axis and the morphology of adult skeleton, the directions of the a-axes are reported to be various, although one of the a-axes generally points approximately toward the anus (5, 18). Since only a few species have been used for the study of a-axes, more investigations seem to be desired. Furthermore, to the best of our knowledge no report has been published on the a-axes of the larval spicule.

The directions of the a-axes can be determined if crystal faces or etched figures were observable relative to the direction of the c-axis. Unfortunately, crystal faces and etched figures are rarely displayed in intact, fractured or etched surfaces both in adult and larval spicules. In adult skeletons, intact surfaces appear noncrystalline, consisting of smooth or spongy textures. Cleavage planes and etched figures are difficult to obtain (1, 5, 16, 18, 37). In larval

Fig.24. Junction of the left and right ventral-transverse rods. The two sets of crystals intermesh tightly at the junction and there appear to be no randomly oriented crystals which join the two sets. Since the body rods of the left and right spicules make an angle of about 45° to each other (see Fig.20), their optic axes are likewise inclined (see also Fig.1). The calcite crystals which grew on the left spicule are therefore inclined by that angle relative to those on the right spicule. X 7,800.

Fig.25. Junction of the left and right body rods. The left and right body rods join at the posterior end of the pluteus and form, depending on the species, simple junctions or complex basket shaped structures. In *Arbacia*, this region of the body rod takes on a shape similar to a moose antler. Despite the complex morphology, the calcite crystals on the left and right spicule each maintain their orientation. X 3,000.

Fig.26. Another example showing crystal arrangement on the left and right ventral-transverse rods and their junction. Although small crystals are sparsely distributed on the mid portion of the rods, crystallographic characteristics are all the same as the case shown in Fig.24. In Figs.24 and 26, the axis of the pluteus has been turned ca 90° clockwise compared to Fig.20. X 4,000.

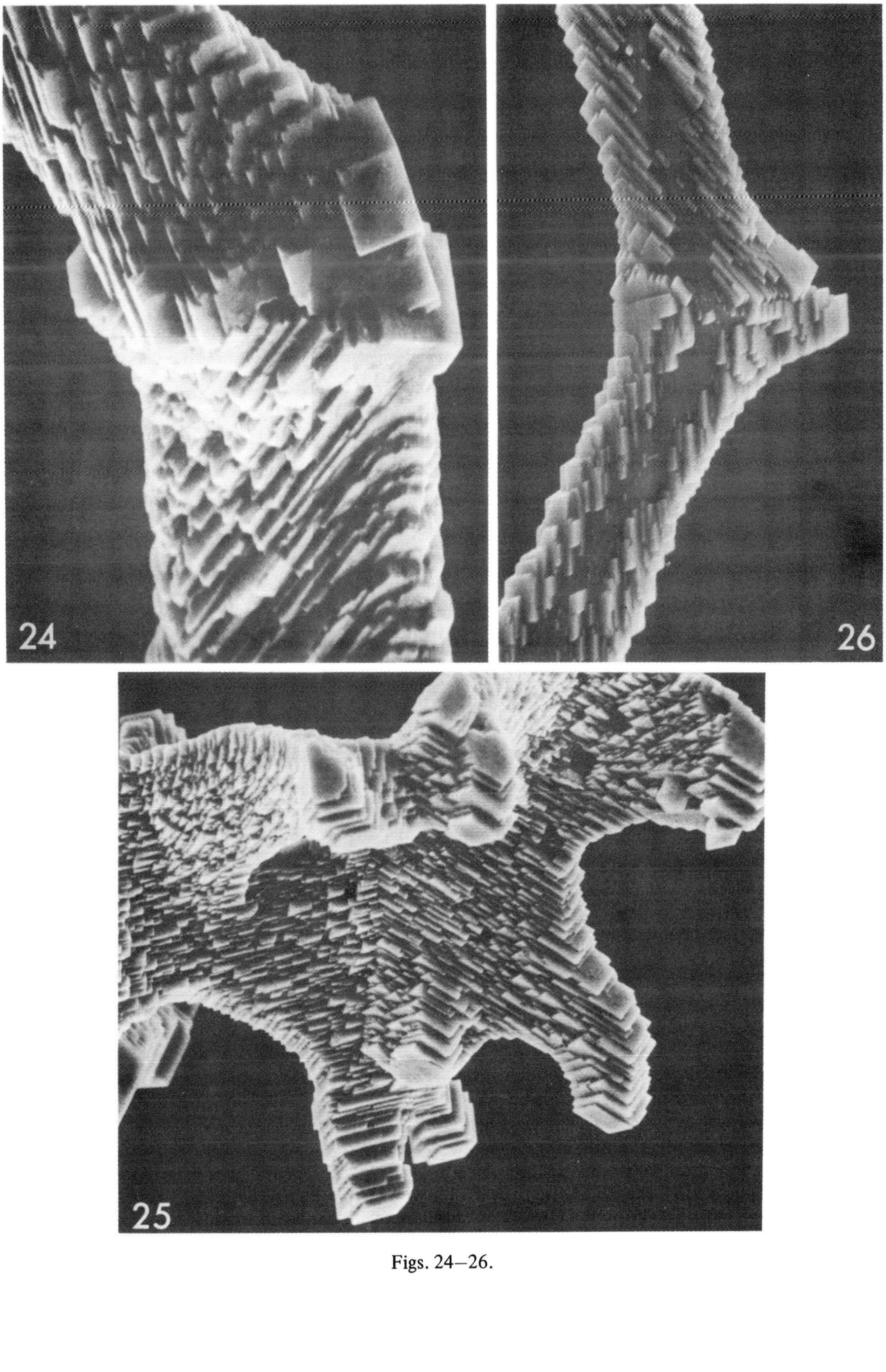

Figs. 24–26.

sea urchin spicule, we were also unable to demonstrate convincing etch patterns or cleavage faces. However, we were able to epitaxially grow calcite microcrystals onto the isolated triradiate and pluteus skeletal spicules. Since the crystals grown on the spicule take the form of cleavage rhombohedra, one can determine the relation between the shape of the spicule and symmetry axes of the crystal. On such basis, the course of spicule development relative to the crystal axes was deduced as follows.

A granule of magnesian calcite, the spicule rudiment, extends three arms in the negative directions of the three a-axes to form a triradiate spicule. As a matter of fact, the three radii of the triradiate form an angle of almost 120° between each other in a plane perpendicular to the c-axis. When the triradiate has grown to a certain size, one radius pointing to the dorsal side of the larva bends at right angles and elongates straight in this direction to form the body rod so that the morphologic long axis of the body rod lies parallel to the c-axis as a natural consequence. A new branch or branches for the post-oral rod elongates in a direction opposite to the body rod; therefore its long direction also parallels the c-axis. Another radius pointing to the animal pole of the larva bends at an obtuse angle and elongates toward the oral lobe to form the antero-lateral rod. The elongation direction of this rod, therefore, intersects both the a- and c-axes. The remaining third radius continues its elongation without bending to form the ventral-transverse rod; i.e., the rod lies parallel to one of the a-axes in the initial plane of the triradiate. By what mechanism do the three radii of the triradiate each follow individual elongation directions relative to the symmetry axes of calcite?

Descendants of micromeres isolated at the 16-cell stage and cultured *in vitro* sometimes form a triradiate spicule, from which a three dimensional skeletal spicule similar to the *in vivo* spicule develops (24, 25). Under such a condition, where the spicule forming cells are completely free from the influence of the ectodermal wall, the three radii of the triradiate equally bend at right angles, forming three rods parallel to each other and to the c-axis. It seems noteworthy that no radius of the triradiate continues to elongate without bending as does the ventral-transverse rod *in vivo*. The fact suggests that spicule elongation away from the c-axis as in the ventral-transverse and antero-lateral rods in normal larvae require complex influence by biological factors. When spicules grow under conditions completely governed by crystalline characteristics, because of debility of the biological factors, spicules may increase their size as in simple mineral crystals, e.g. as seen in the larvae reared in 1/10 Ca^{2+}-Mg^{2+}-sea water (21).

Why do spicular rudiments first extend three arms along the a-axes? As will be discussed later, unique biological conditions which enhance the crystal growth along the a-axes rather than the c-axis may develop in the pseudopodial cable when the spicule develops into the triradiate form. It must be mentioned here that a rudimental granule of the spicule frequently elongates along the c-axis from the beginning and forms a long straight spicular rod in culture of isolated micromeres.

Calcareous and silicious spicules from sponges have earlier been characterized and shown to behave like a single crystal between crossed polars. In spite of the fact that calcareous sponge spicules are also composed of magnesian calcite (13) and are similar in shape to larval sea urchin spicule, the former are different from the latter with respect to the morphologic-crystallographic relation. In an elegant study, SOLAS (31) decided from the ordinary- and extra-ordinary refractive indices that calcareous spicules of sponge belonged to the calcite rather than aragonite type. Furthermore, he observed striated etch figures and cleavage faces typical of calcite, and even discovered regularly oriented minute crystals of calcite growing on acerate and triradiate spicules which were left to stand for several days in water containing an excess of calcium carbonate (32). Shortly thereafter, VON EBNER (35) followed with an extensive study of calcareous sponge spicules. He used a combination of etch figures obtained by treatment with formic acid and the direction of c-axis determined by optical methods and found a remarkable correspondence between the crystal axes and spicule morphology in the large regular tetraxon. In this spicule, three rays lie in one plane and the fourth is normal to this plane; the fourth ray parallels the c-axis and the other three rays each bisect the angles formed by the three a-axes. In the spicules of another type, however, such clear correspondence was not observed, even the direction of the c-axis being not necessarily parallel or perpendicular to one of the rays.

More recently, JONES (8, 9, 11) confirmed and extended these earlier observations on spicules of *Leucosolenia complicata*. In *L. complicata*, the triacts and tetracts are embedded in the mesohyl of the wall of the oscular tube, while the monacts project from the outer surface of the tube. The slender monacts have their long axes parallel to the c-axis, while the curved monacts start approximately at right angles to the c-axis but then curve in a principal plane as they grow. The c-axis of the triacts and tetracts, however, lies in the plane containing the basal ray and bisecting the angle between the oscular rays. The optic angle (the angle between the c-axis and the basal ray) varies with the distance of the spicule from the oscular edge. The shape of the spicule likewise varies with the position in the tube, but is not closely correlated with the orientation of the c-axis: two spicules of the same shape can have different optic angles, or the same optic angle and yet different shapes. He takes a view that the c-axis of the first formed spicular rudiment lies in the direction of cell division just prior to spicule formation.

Furthermore, using crystallization technique, with which rhombohedral calcite crystals can be grown on various kinds of spicules of a few species, JONES (9, 11) examined the direction of spicule rays in relation to the faces of the cleavage rhombohedra of calcite. As a general rule, whereas the spicule rays tend to grow in planes containing the c-axis of calcite, the directions taken within the planes appear to be guided by the bounding surface of the mesohyl. Some of the triacts have rays of equal length and the rays lie in the three principal planes, each ray making the same angle with the c-axis of calcite. The c-axis is thus oriented perpendicular to the facial plane (the plane upon which the ray tips lie) of the spicule. These spicules appear to be similar to sea urchin larval spicules as regards to spicule form and crystallographic orientation of calcite, but not exactly the same, the shape of the latter relating more strictly to the symmetry axes of calcite.

Biological factors controlling the spicule shape

In normal sea urchin larvae, the spicules develop within the pseudopodial cable. The cable is formed by fusion of pseudopodia of the primary mesenchyme cells and lies between the ectodermal wall and the mesenchyme cell bodies. The shape of the growing spicule is always foreshadowed by the pseudopodial cable. The cell bodies containing the nucleus and much of cytoplasm are connected to the cable by stalks about 0.5 μm in diameter and 1–2 μm in length. Fine processes extend from the cable and make contact with the ectodermal wall, sometimes appearing to create tension in the cable. Although such situations are observable with the light microscope (19, 21, 22, 38), further confirmation has been made with the electron microscope (6, 15, 16, 34).

The cable cytoplasm contains a large number of mitochondria, microtubules and vesicles, but not nuclei, yolk granules or the Golgi complexes. The Golgi complexes are found in the cell body, adjacent to the nucleus on the side facing to the cable. Microtubules in the stalks run parallel to the long axis of the stalk and extend into the cable. Within the cable, microtubules run parallel to the cable long axis except at the junctions with the stalks where they are oblique. Although the spicules dissolve out of thin sections, the position of the spicule can be recognized by the hole left behind. The hole is found within a membrane-limited vacuole which has a lining of some electron dense material. Small vesicles (coated vesicles in 6) containing material of the same electron density as the lining of the vacuole are found in the cable especially near the vacuole. Furthermore, micrographs have been obtained showing the coated vesicle in the process of fusion with the vacuole (34). KOIZUMI (15) found the coated vesicles not only in the cable but also in the stalks and cell bodies near the Golgi region. It seems interesting to note the report that magnesian clacite can be precipitated from calcium bicarbonate solution containing magnesium ions even at normal room temperature and pressure, if certain organic substances such as organic acids are added to the solution (14).

Putting these facts together, the following speculation may be allowed. 1) The coated vesicles contain organic substances in which magnesian calcite are ready to be precipitated. 2) The coated vesicles are formed at the Golgi region of the cell bodies and transported along microtubules into the cable through the stalks. 3) At the initial stage of spicule formation, the coated vesicles fuse and form a vacuole in which the spicule develops. As a matter of fact, a small calcareous granule is frequently found in a vacuole in *Echinarachnius parma* (2) and *Mespilia globulus* (19). Furthermore, at the very beginning of spicule formation, the rudimental granule invariably takes a shape of a rhombohedron. It is therefore suspected that the rudimental granule lies free in the vacuole even if the vacuole is not always visible at the level of the light microscope, so that the rudiment may grow as a crystal in a glass container. 4) The vacuole including the spicular rudiment may change its shape to a flattened equilateral triangle owing to stretching from three apices of the triangular mesenchymal aggregate. Such a situation may naturally induce the spicular rudiment calcite crystal to orient so that its a-axes become coincident with the three median lines of the triangular vacuole. Simultaneously, the three arms of the spicule could start to elongate along negative directions of the a-axes since the coated vesicles may be concentrated on the three apices of the triangular vacuole assisted by microtubules in the cable. If the vacuole is stretched from two opposing directions, the vacuole may take on a cylindrical shape and the calcite crystal may orient with its optic axis parallel to the long direction of the vacuole. 5) Following rapid growth of the spicule, the vacuole itself, being in contact with the surface of the spicule, may change its shape to a triradiate. 6) The vacuole elongates three arms and increases the amount of organic substances by successive fusion of the coated vesicles. 7) After a triradiate spicule has grown to a certain size, each radius follows its individual elongation direction as has been described. In this case, the vacuole may change its shape following the change of elongation direction of the cable which is guided by the pattern of arrangement of the primary mesenchyme cells. When the spicule changes its elongation direction toward the optic axis, the crystalline characteristic of the spicule seems to predominate, although the bending vacuole may also provide the opportunity to change the course of calcium deposition. In short, the vacuole seems to provide an ideal container, within which a crystal of magnesian calcite is able to grow as a spicule. The shape of the vacuole is easily changed concomitantly with form change of the cable, more or less being affected by the shape of the already formed spicule. It is probable that the organic component of the spicule is laid down first in the vacuole and the deposition of the inorganic component follows thereafter. It is not known, however, what organic substances are contained in the coated vesicles and how the inorganic components are transported into the vacuole.

In gastrulae transferred from normal sea water to 1/10 Ca^{2+}-Mg^{2+}-sea water, crystal faces appear at the growing tips and on two lateral sides of each arm of the triradiate spicule. These faces parallel the faces of calcite crystals epitaxially grown onto the isolated spicules. Furthermore, it has been observed with the light microscope that a vacuole surrounding the spicule appears when larvae are transferred into 1/10 Ca^{2+}-Mg^{2+}-sea water, and conversely, the vacuole disappears and slender spicular rods elongate from the corners of the rhombohedron when larvae are returned to normal sea water (21). It therefore appears that the conditions produced by healthy mesenchyme cells suppress the natural tendency of crystal faces to grow.

Micrographs shown in Figs.11, 13 and 14 give us the impression that the spicule is composed of an accumulation of crystallites and that each rectangular structure indicates a unit of inorganic component. Since such texture was frequently found when spicules were isolated from larvae debilitated in 1/10 Ca^{2+}-Mg^{2+}-sea water, it is possible that organic substances filling up the gaps of inorganic components have been dissolved. Even if this is the case, however, questions still remain what is the central hole and why only two faces at the spicule tip are so easily corroded. Similar concentric lamination is reported to appear when giant rays of sponge are treated for 24 hr with 10% potash solution (35). In this case, it is possible that the concentric lamination and the central hole are caused by difference of magnesium content owing to alternate deposition of less magnesium and more magnesium. Examination by means of electron probe micro-analysis, however, showed no variation in composition corresponding to the lamination (12). Although the central hole reminds us of the so-called axial filament in sponge spicules, it has been demonstrated that no organic filament exists in the sponge spicule (10). At any rate, examination of spicules isolated from sea urchin larvae once normally developed and then debilitated in normal sea water seems to be necessary. Nevertheless, it should not be overlooked that UEMURA (34) has succeeded in making thin sections of the spicules isolated from normally developed larvae and observed regular arrays of slender rectangular structures of roughly 100–1,000 Å in width and of various lengths. His photographs are similar to those shown by TRAVIS (33) on adult sea urchin plates, although the dimension of the structural unit is somewhat different.

The intricate morphology of the pluteus skeletal spicule is genetically determined (7). Spicules with species-specific morphology can now be produced in culture by a cluster of as few as 20 mesenchyme cells (24, 25). Further elucidation of spicule growth mechanism is expected to provide important insight into the basic mechanism of morphogenesis.

This work was performed in the summer of 1975 while K. Okazaki was Lillie Fellow of the Marine Biological Laboratory, Woods Hole, Mass., U.S.A. Support from the Lillie Fellowship Foundation, grant-in-aid from the Ministry of Education of Japan (Nos. 001012 and 001030), and grants from the National Institutes of Health grant no. CA 10171 and the U.S. National Science Foundation grant no. BMS-00473 are greatefully acknowledged. We are especially indebted to Mr. Kent McDonald of the University of Colorado for his cooperation with the scanning electron microscopy.

REFERENCES

1. ALEXANDERSSON, E. T., 1975. Science, **189**, 47–48.
2. BEVELANDER, C. and H. NAKAHARA, 1960. In "Calcification in Biological Systems." (R. F. Sognnaes,

ed.), pp. 41–56. Amer. Associ. Adv. Sci., Washington, D.C.
3. DAN, K., 1960. Inter. Rev. Cytol., **9**, 321–367.
4. DONNAY, G., 1956. Carnegie Inst. Wash. Yearb., **55**, 205–206.
5. ——— and D. L. PAWSON, 1969. Science, **166**, 1147–1152.
6. GIBBINS, J. R., L. G. TILNEY and K. R. PORTER, 1969. J. Cell Biol., **41**, 201–226.
7. HÖRSTADIUS, S., 1973. Experimental Embryology of Echinoderms. Clarendon Press., Oxford.
8. JONES, W. C., 1954. Quart. J. Microscop. Sci., **95**, 33–48.
9. ———, 1955. Quart. J. Microscop. Sci., **96**, 129–149.
10. ———, 1967. Nature, **214**, 365–368.
11. ———, 1970. Symp. Zool. Soc. Lond., **25**, 91–123.
12. ——— and D. W. F. JAMES, 1969. Micron, **1**, 34–39.
13. ——— and D. A. JENKINS, 1970. Calcif. Tiss. Res., **4**, 314–329.
14. KITANO, Y. and N. KANAMORI, 1966. Geochemical J., **1**, 1–10.
15. KOIZUMI, Y., Unpublished.
16. MILLONIG, G., 1970. J. Submicro. Cytol., **2**, 157–165.
17. NISSEN, H. U., 1963. Neues Jb. Geol. Abh., **117**, 230–234.
18. ———, 1969. Science, **166**, 1150–1152.
19. OKAZAKI, K., 1960. Embryologia, **5**, 283–320.
20. ———, 1961. Biol. Bull., **120**, 177–182.
21. ———, 1962. Embryologia, **7**, 21–38.
22. ———, 1965. Exp. Cell Res., **40**, 585–596.
23. ———, 1970. Collagen Symp. (in Japanese), 8, 113–129.
24. ———, 1975a. Amer. Zool., **15**, 79–93.
25. ———, 1975b. Biol. Bull., **149**, 439–440.
26. ——— and T. HAYASHI, 1967. Zool. Mag. (in Japanese), **76**, 413.
27. PUCCI-MINAFRA, I., C. CASANO and C. LA ROSA, 1972. Cell Differentiation, **1**, 157–165.
28. RAUP, D. M., 1966. In "Physiology of Echinodermata." (R.A. Boolootian ed.), pp. 359–395. John Wiley & Sons, N. Y.
29. RUNNSTRÖM, J., 1931. Roux' Arch., **124**, 273–297.
30. SCHMIDT, W. J., 1930. Zool. Jb. (Physiol.), **47**, 357–509.
31. SOLAS, W. J., 1885. Sci. Proc. Roy. Dublin Soc., **4**, 374–392.
32. ———, 1886. Sci. Proc. Roy. Dublin Soc., **5**, 73.
33. TRAVIS, D. F., 1970. In "Biological Calcification: Cellular and Molecular Aspects." (H. Schraer ed.), pp. 203–311. Appleton Century Crofts, N. Y.
34. UEMURA, I. Unpublished.
35. VON EBNER, V., 1887. Sitz. Akad. Wiss. Wien, **95**, 55–149.
36. VON UBISCH, L., 1937. Z. Wiss. Zool., **149**, 402–476.
37. WEBER, J., R. GREER, B. VOIGHT, E. WHITE and R. ROY., 1969. J. Ultrastruct. Res., **26**, 355–366.
38. WOLPERT, L. and T. GUSTAFSON, 1961. Exp. Cell Res., **25** 311–325.
39. WOODLAND, W., 1906. Quart. J. Microscop. Sci., **49**, 305–325.

(Received: 2 August, 1976)

Reprinted from *Journal of Cell Biology*, Vol. 77(3), pp. 638–654, 1978.

MITOSIS IN *BARBULANYMPHA*

I. Spindle Structure, Formation, and Kinetochore Engagement

HOPE RITTER, JR., SHINYA INOUÉ, and DONNA KUBAI

From the Department of Zoology, University of Georgia, Athens, Georgia 30602, Program in Biophysical Cytology, Department of Biology, University of Pennsylvania G7, Philadelphia, Pennsylvania 19104, and the Department of Zoology, Duke University, Durham, North Carolina 27708

ABSTRACT

Successful culture of the obligatorily anaerobic symbionts residing in the hindgut of the wood-eating cockroach *Cryptocercus punctulatus* now permits continuous observation of mitosis in individual *Barbulanympha* cells. In Part I of this two-part paper, we report methods for culture of the protozoa, preparation of microscope slide cultures in which *Barbulanympha* survived and divided for up to 3 days, and an optical arrangement which permits observation and through-focus photographic recording of dividing cells, sequentially in differential interference contrast and rectified polarized light microscopy. We describe the following prophase events and structures: development of the astral rays and large extranuclear central spindle from the tips of the elongate-centrioles; the fine structure of spindle fibers and astral rays which were deduced in vivo from polarized light microscopy and seen as a particular array of microtubules in thin-section electron micrographs; formation of chromosomal spindle fibers by dynamic engagement of astral rays to the kinetochores embedded in the persistent nuclear envelope; and repetitive shortening of chromosomal spindle fibers which appear to hoist the nucleus to the spindle surface, cyclically jostle the kinetochores within the nuclear envelope, and churn the prophase chromosomes. The observations described here and in Part II have implications both for the evolution of mitosis and for understanding the mitotic process generally.

KEY WORDS mitosis · protozoa · birefringence · anaerobe · culture · microtubule

In studying vital activities of cells, one occasionally encounters bizarre forms which nevertheless uniquely illuminate or illustrate general principles of life. Through his pioneering, now classical studies, L. R. Cleveland pointed out that the life cycles of hypermastigote protozoa, residing in the hindgut of the wood-feeding cockroach *Cryptocercus punctulatus*, exemplify just such a unique gift of nature. The coiling and segregation of chromosomes and the structure of the achromatic mitotic apparatus are exquisitely displayed in these extraordinary anaerobes (e.g. Cleveland, 4, 7).

The hindgut protozoa of *Cryptocercus*, symbionts of the mutualist type, exhibit an extraordinary diversity. Individuals of 30 species fill an anaerobic niche along with equally diverse prokaryotes and particles of ingested cellulose. The protozoa and prokaryotes digest the cellulose ingested by the host, making the metabolic products available to it. These microorganisms are unique with respect to their structure, physiology, and reproduction to the extent that one is tempted to regard them as archaic forms resident in a host that itself is a primitive insect (34). *Barbulanympha* is one of 14 protozoan genera in *Cryptocercus*. Four species of *Barbulanympha* dominate the hindgut mass of cells in terms of size, though not in numbers.

Cleveland's remarkable accounts of *Barbulanympha* reveal the potentialities of these species for gaining further insight into the dynamics of mitosis (4–17; see also Hollande and Valentin, 21–25). Specifically, *Barbulanympha* provides (*a*) a relatively large nucleus with a nuclear envelope that remains intact throughout mitosis; (*b*) an extranuclear achromatic figure consisting of large, well-separated parts that afford a clear basis for functional interpretation; and (*c*) synchronous mitosis among individuals, the onset of which is predictable in cells obtained from a host that is close to the time of its molt (4, 13, 16).

Despite Cleveland's extensive and precise descriptions, his studies were limited to light microscope observations of fixed and stained samples, or 15–30-min observations of supra vital cells dying from the presence of toxic oxygen in the atmosphere.

After 15 years of search, Ritter succeeded in developing a synthetic medium permitting long-term culture of the symbiont fauna in an oxygen-scrubbed atmosphere (36, 37). We have since developed a method for microscope slide preparation which permits observation and recording of mitotic events of cultured individuals living for over 3 days.

We now report on several dynamic features of mitosis and fine-structural organization of the achromatic apparatus of *Barbulanympha*. Studies on individual living cells were made by through-focusing with two alternative modes of light microscopy that take advantage of the high contrast, resolution, and image quality now available with rectified polarized light (30, 31) and Smith T system differential interference contrast optics (E. Leitz, Inc., Rockleigh, N.J.). Fine structure was analyzed by polarization optical studies of the birefringent spindles in living cells combined with thin-section electron microscopy of cultured cells preserved with diluted Karnovsky's solution (33).

We confirm Cleveland's descriptions of chromosomal engagement with astral rays by way of kinetochores embedded in the intact nuclear envelope, and of mitosis in *Barbulanympha* characterized by the absence of a metaphase configuration. We find, in addition, that repetitive shortening of the (chromosomal) astral rays appears to participate in nuclear morphogenesis as well as in chromosome movement to the spindle poles. We also find that mitosis in *Barbulanympha* progresses through two discrete, successive stages, (*a*) chromosomal spindle fiber shortening to the point at which the nucleus becomes wrapped around the central spindle and the chromosome sets have reached their respective poles without change in central spindle length, and (*b*) central spindle elongation which induces further chromosome separation accompanied by an intricate topological maneuver of the nuclear envelope, resulting in karyokinesis.

These anaphase events and the dramatic events of nuclear morphogenesis by which *Barbulanympha* completes karyokinesis are described in Part II. A brief synopsis of some of our observations has been published elsewhere (32).

MATERIALS AND METHODS

Culture of the Protozoa

Nymphs of *Cryptocercus* collected in May from the mountains of Georgia and North Carolina were maintained in the laboratory at 13°C. Small blocks of wood, chopped from the water-saturated logs that supplied the collection, provide the environment required for a stock culture. Necessary maintenance is limited to a weekly transfer of wood and cockroaches to a clean, dry container with a tightly fitting lid.

During July and August, individual cockroach nymphs close to molting, molting, or a few hours past molt were selected to provide the inoculum for cell cultivation. These *Barbulanympha* have already been induced by hormone of the host to undergo reproduction in synchrony. At this stage, the hindgut is free of highly birefringent cellulose particles, as the host preparing to molt ceases to feed. Mitotic events of all four species of *Barbulanympha* are uniform (4, 13, 16). Our observations focused primarily on the largest of these species, *Barbulanympha ufalula* ($\sim 285 \times 205\ \mu m^2$).

Because of the extreme anaerobic requirement that *Cryptocercus* hindgut protozoa exhibit in their natural environment (12, 36), all precautions are observed for anaerobic maintenance of the synthetic environment. The liquid medium prepared by Ritter contains reduced glutathione (Table I). Inoculation, cultivation, and slide preparation procedures are conducted in a gas atmosphere of 95% N_2 and 5% CO_2, free of O_2 within the range 0.8–20 μM. The mechanical system utilized to maintain this atmosphere is described in detail elsewhere

TABLE I

Cellulose-Free Synthetic Medium for In Vitro Cultivation of Cryptocercus Hind-Gut Microorganisms*

	(g/liter)
$NaH_2PO_4 \cdot H_2O$	0.290
$KHCO_3$	3.330
KCl	2.494
$CaCl_2$	0.820
$MgSO_4 \cdot 7H_2O$	0.752
d-Glucose‡	0.100
d (+) Trehalose‡	1.000
Glycogen‡	1.000
Casein hydrolysate (enzymatic)§	0.300
Yeast extract‖	0.100
Loeffler blood serum‖	0.100
Glutathione (reduced)‡	0.219
Water, glass distilled, to prepare 1 liter	

* pH adjusted to 6.8 with KOH, if necessary.
‡ Nutritional Biochemicals, Corp., Cleveland, Ohio.
§ General Biochemicals, Inc., Chagrin Falls, Ohio.
‖ Difco Laboratories, Detroit, Mich.

(37). In brief, this assembly consists of three units connected in tandem; (*a*) a gas supply and O_2-scrubbing facility, (*b*) a sealed, glove-box culture chamber, and (*c*) a gas collecting reservoir that functions as a surge tank.

The medium is prepared, deoxygenated, and introduced into the chamber at a room temperature of 20°C. Working within the chamber, 15 ml of medium is delivered to a 25-ml Erlenmeyer flask, the culture vessel. This provides optimum surface to volume ratio for gas equilibrium. In the anaerobic chamber, the hindgut from the donor is then removed with two stainless steel forceps, one applied to the thorax, the other clamped to the caudal abdominal segment. A quick pull ruptures the intersegmental membranes and allows withdrawal of the entire alimentary canal. Once removed, adhering fat body is teased away from the surface of the relatively large hindgut; the anterior portion is pinched off with forceps at its juncture with the hindgut, as is the segmental remnant posterior to the hindgut (see Cleveland, reference 4 for anatomy). In this condition, the hindgut is dropped into the medium. The hindgut sinks to the bottom of the flask and peristalsis ejects the contents (0.4–0.8 ml, depending on the size of the cockroach) in a manner resembling discharge of paste from a tube. The mass of symbionts gradually scatters over the vessel's bottom, where all but the smallest flagellate species remain. Sterile technique is not observed, and culture vessels remain open to the anaerobic atmosphere. The protozoa continue to grow and reproduce actively under these conditions for an indefinite period, generally exceeding many weeks.

Slide Preparation

Slide preparation for light microscope examination of *Barbulanympha* is begun when cultivated cells exhibit the first signs of an achromatic figure, usually 7–10 h after molt at 20°C and 13–18 h at 13°C.

Important considerations for slide preparation include (*a*) procurement of an optimum volume of liquid containing synchronously dividing cells, (*b*) adjustment of cover glass height above the slide surface just sufficient for cell immobilization and flattening to achieve optical clarity without inducing trauma, and (*c*) application of sealant around the sample making possible observation outside the anaerobic chamber for extended periods.

Strain-free slides and cover glasses are utilized in these procedures (M6145 slides from Scientific Products Div., American Hospital Supply Corp., McGraw Park, Ill.; Selex 18 × 18 mm cover glasses from Dolbey Scientific Co., Philadelphia, Pa.). The slide and cover glasses as well as glassware for cultivation are cleaned to avoid contaminating materials of all categories, including even minor specks of lint. This cleaning method, which is a prerequisite to studies of living cells with rectified polarization optics, is reported elsewhere (20).

Our slide chambers contain *Barbulanympha* cells appropriately flattened and uniformly distributed. This is managed by collecting a 50-μl vol of medium and cells from the bottom of the culture flask and transferring this volume to a 450-μl plastic microcentrifuge tube. *Barbulanympha* quickly settle by gravity into the conical bottom. A 5-μl volume of bottom solution is drawn into a 3-cm length of 0.58 mm inside diameter polyethylene tubing fitted to the tip of a calibrated Hamilton microsyringe and transferred to an 18-mm square cover glass.

The 18-mm square cover glass is previously prepared to receive the sample. A 1.7-mil (43 μm) polyethylene membrane, with a square hole 14 mm to the side, is cut to conform to the outside dimensions of the cover glass. The frame remaining is 2 mm wide on each of its four sides. This spacer is aligned with the edges of the cover glass and applied to the glass surface previously rimmed with a small amount of silicone grease (high vacuum grease from Dow Corning Corp., Midland, Mich.). Uniform pressure from a flat tool moved around the surface of the spacer squeezes out excess grease and gas bubbles. The exposed face of the spacer is then rimmed with more silicone grease.

With the 5-μl sample delivered to the center of the well formed by this assembly, a glass slide is carefully centered and lowered to contact the sample. For optimum compression (to ~100-μm thickness) of the large *Barbulanympha* cells, the sample should spread to a drop approximately 8 mm in diameter as the slide is gently pressed down to complete the silicone seal. Finally, the slide culture prepared in the anaerobic chamber is removed for subsequent examination in room atmosphere.

In practice, a number of slide cultures are prepared to provide a large number of cells from which to select those in suitable orientation for study. Except for occasional preparations in which cells fail to complete their cycle, most acceptable slide cultures contain viable cells up to 3 days. This time greatly exceeds the time required for *Barbulanympha* to complete its cell division.

Microscopy

Observations were made on individual dividing cells alternately with polarized light and differential interference contrast microscopy. Rectified 20× and 40× Nikon objective lenses (Nikon Inc. Instrument Div., Garden City, N. Y.) were used for polarized light, and 16× and 40× Leitz Smith T system objectives were mounted on the same quadruple revolving nosepiece[1] for differential interference contrast microscopy. A strain-free condenser for the Leitz differential interference contrast system, with turret mounted Wollaston prisms and blank space, served for both types of microscopy. The optical components of the inverted microscope were mounted on a rigid cast steel vertical optical bench designed by us (29). Collimated light from a 100-watt concentrated mercury arc lamp (#110 from Illumination Industries, Inc., Sunnyvale, Calif.), filtered through two heat cut filters (#4602 from Corning Glass Works, Science Products Div., Corning, N.Y., or #4010 from Edmund Scientific Co., Barrington, N.J.) and a 546-nm narrow band pass interference filter (#10-98-2 from Baird Atomic, Inc., Bedford, Mass.) was linearly polarized by a Glan-Thompson calcite prism (20 mm square from Karl Lambrecht Corp., Chicago, Ill.). A rotary sheet of mica λ/25 in retardation and placed before the condenser, and a second Glan-Thompson prism with directly cemented stigmatizing lenses mounted in the body tube, served as compensator and analyzer for both polarized light and differential interference contrast.

With the optical system described, one can quickly switch between rectified polarized light and differential interference contrast observations by rotating the quadruple nosepiece and condenser turret. Also, by adjustment of the λ/25 Brace-Köhler compensator, some spindle birefringence can be detected, superimposed on the differential interference image. For differential interference contrast not used at peak sensitivity, or near extinction, a third swing-out polarizer built into the Leitz condenser was introduced to attenuate image brightness and reduce the intensity of illumination reaching the specimen. For observations or time-lapse photography over several hours, we found that cells proceed through division better and especially enter anaphase-B more readily when illumination is further reduced by 2–3 sheets of an amber filter (#5 from Kliegl Bros., Universal Electric Stage Lighting Co., Inc., Long Island City, N.Y.). These filters may have reduced any stray short wavelength visible light or long wavelength ultraviolet, or may have simply acted as an attenuator for the 546-nm mercury green light. Eventually, our best records were obtained by mounting one to two such filters on a solenoid shutter synchronized to move out of the light path with each exposure of the still and movie cameras, another amber filter being fixed permanently in the light path.

Serial, through-focus 35 mm still pictures were taken on Kodak Plus-X film with a Nikon Automatic Microflex motor-driven camera with the beam splitter oriented at 45° to the analyzer transmission direction. The film was developed at 18°C in 1:3 dilution Kodak Microdol X. Time-lapse 16 mm films were taken with an Arriflex camera on Kodak Plus-X negative film and processed commercially. All observations were made in a room thermostated to 18° ± 1°C.

For electron microscopy, organisms were prepared as described by Kubai (33), except that the more dilute first fixative contained 0.036% paraformaldehyde, 0.23% glutaraldehyde, and 0.02m phosphate buffer, pH 7.0.

RESULTS

Development of Astral Rays and Central Spindle

Approximately 18 h after inoculation of *Barbulanympha* into culture, two prominent, elongated structures are evident at the anterior end of the organism beneath the flagellar tuft (Fig. 1). These elongate-centrioles exhibit positive birefringence and extend in a postero-lateral direction from their origin near the medial end of the corresponding parabasal-axostylar lamella (Figs. 1, 2*a*, *b*, 3*a*, *b*). A spherical centrosome ~8.5 μm in diameter surrounds the posterior tip of each elongate-centriole (Figs. 1, 2*a*, *b*, 3*a*, *b*). Although not easily detectable at this early stage, astral rays radiate from the centrosomal surface. These can be seen by phase contrast (Fig. 1), differential interference contrast (Fig. 3*a*), and polarized light (Fig. 3*b*) microscopy. Furthermore, in polarized light the positive birefringence of the astral microtubules can be traced through the centrosome to their origin at or near the tip of the elongate-centriole (Figs. 3*b*, *c*; see also text figure 20 in reference 4). The astral microtubules arise from the tip of the elongate-centriole, extend through the centrosome, and coalesce to form refractile fibers visible with phase and interference contrast optics beyond the centrosomal surface.

As described in fixed and stained preparations and in living cells by Cleveland (4, 5, 14, 17), some of the astral rays from the two elongate-

[1] The Leitz Ortholux quadruple nosepiece was modified by Mr. Edward Horn of the University of Pennsylvania, Biology Instrument Shop, according to his unique design, so that each of the four objective lenses can be independently rotated in its place without decentration.

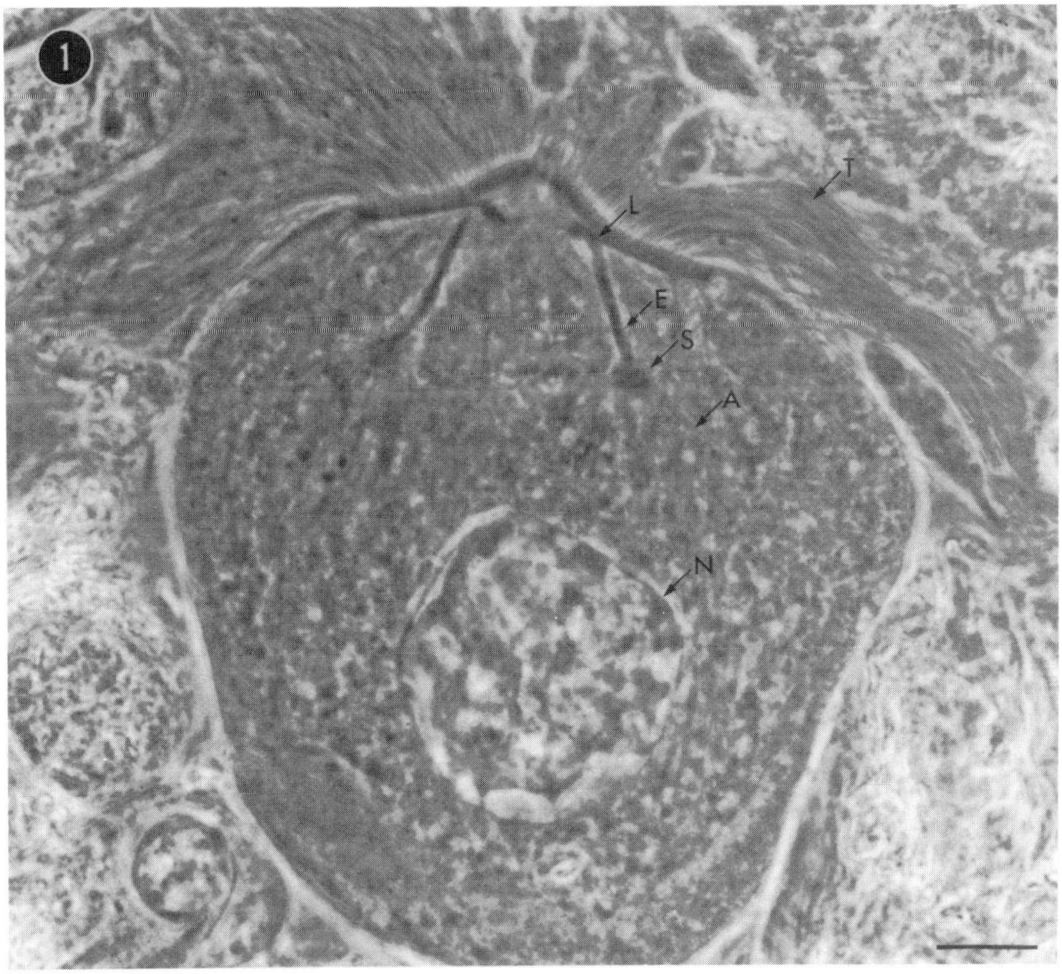

FIGURE 1 Phase-contrast micrograph of *Barbulanympha* very early in division. The nucleus (*N*) still lies some distance away from the two bulbous centrosomes (*S*). The centrosomes cap the distal tips of the elongate-centrioles (*E*) which are seen as prominent twisted rods at the anterior end of the organism. Asters (*A*) radiate from the centrosomes. The converging proximal ends of the two elongate-centrioles bend medially and merge with the parabasal axostylar lamellae (*L*), seen in this optical section as thin dark lines running at the base of the flagellar kinetosomes. The left and right flagellar tufts (*T*) have not yet started to separate. Many species of protozoa cohabiting the wood cockroach intestine crowd around the *Barbulanympha*. Bar, 10 μm. × 1,370.

centrioles become aligned parallel to each other and coalesce to form the early central spindle fibers. As seen in our Fig. 3*a* and *b*, some of these fibers originate at the tip of the elongate-centrioles and others arise at additional loci some distance proximal (anterior) to the centrosomes. With time, the number and length of astral rays, as well as number of fiber bundles which make up the central spindle, progressively increase as manifested in the striking rise in their birefringence (Figs. 3*c*, *d*, 4*a* to *d*).

The developing central spindle often bows anteriorly while further fibers are added between this part and the nucleus located somewhat posteriorly (Fig. 3*c*, *d*). This pattern of spindle fiber formation may be responsible for the bipartite organization and helical arrangement of the fibers found in the mature central spindle. During these earlier stages of spindle development, the birefringence of astral and central spindle fibers remains converged to the elongate-centriolar tip.

Progressively, more fibers are added until eventually the central spindle reaches its full length, diameter, and birefringence (Fig. 4*a* to *d*). Once the spindle has thus matured, these parameters remain unchanged until the end of anaphase-A,

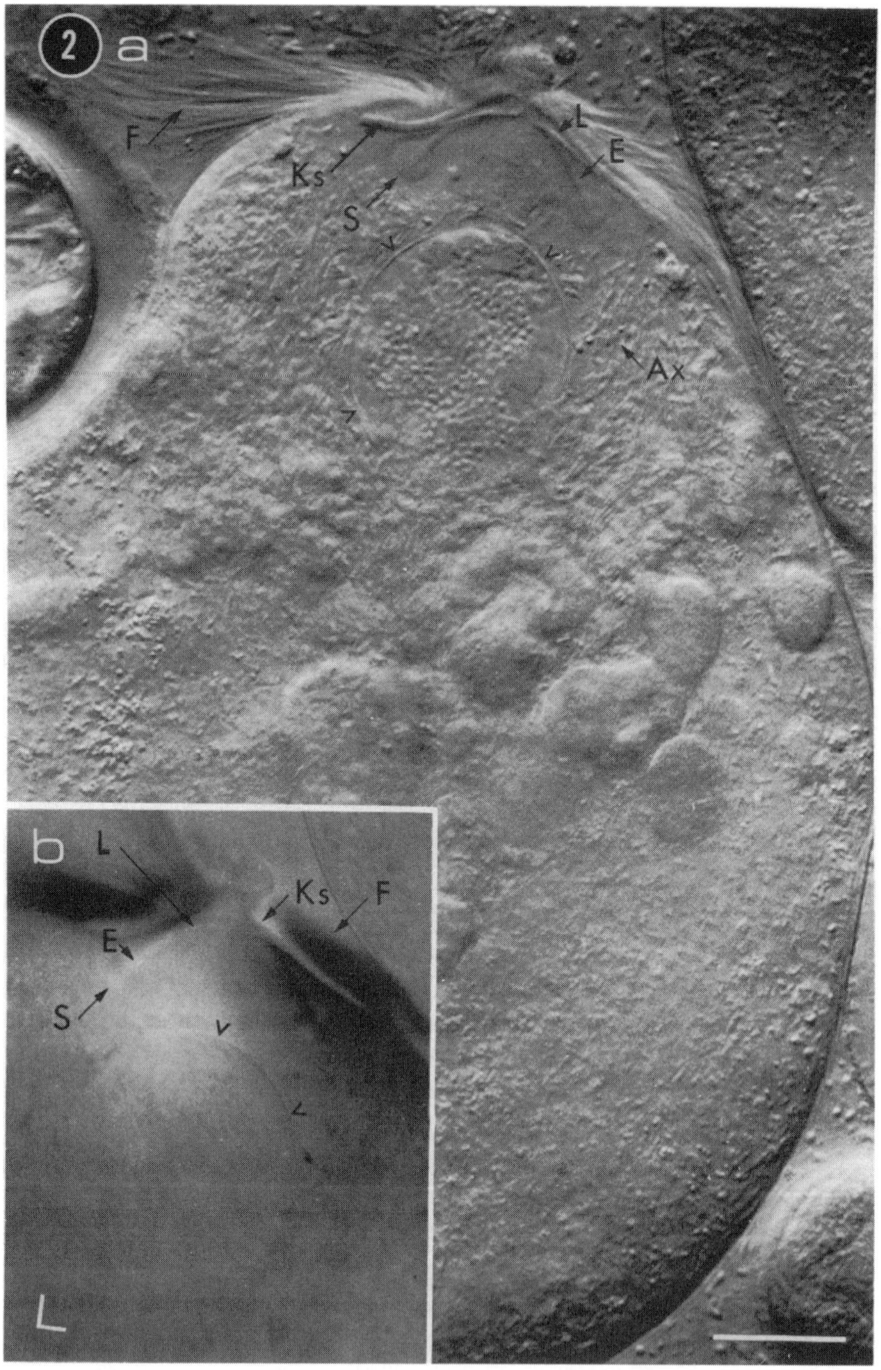

FIGURE 2 (*a*) Differential interference contrast image of ***Barbulanympha*** early in prophase. The organism was flattened for observation. Chromosomes are condensing within the nuclear envelope (ν's). One spherical centrosome (*S*) each caps the distal tips of the two elongate-centrioles (*E*). The base of the elongate-centrioles merge with the parabasal axostylar lamellae beneath the rows of flagellar kinetosomes (*Ks*). Asters are not obvious. Instead, long thin parabasals and axostyles (*Ax*) which trail posteriorly from the parabasal axostylar lamellae surround the nucleus. (*b*) inset, polarized light image of anterior portion of the same individual. Most of the flagella (*F*) which run in the NW and SE quadrants appear in dark compensation as do the right hand elongate-centriole and parabasal axostylar lamella. In contrast, the right hand row of kinetosomes (each kinetosome is oriented perpendicular to the row), the left hand elongate-centriole, and the adjoining region of the parabasal axostylar lamella appear in bright compensation. Each flagellum, kinetosome, elongate-centriole, and parabasal axostylar lamella, as well as the nuclear envelope (cf. esp. Fig. 3*b, c, d*), all display tangential positive birefringence. At the compensator setting used to take this photograph, sectors of the centrosomes in the NW and SE quadrants appear dark, and in the NE and SW quadrants they appear bright. This pattern reflects the positively birefringent material which is oriented radially in the centrosomes. *L*: Axes of crossed polarizer and analyzer and quadrant containing the compensator slow axis. Bar, 30 μm. × 570.

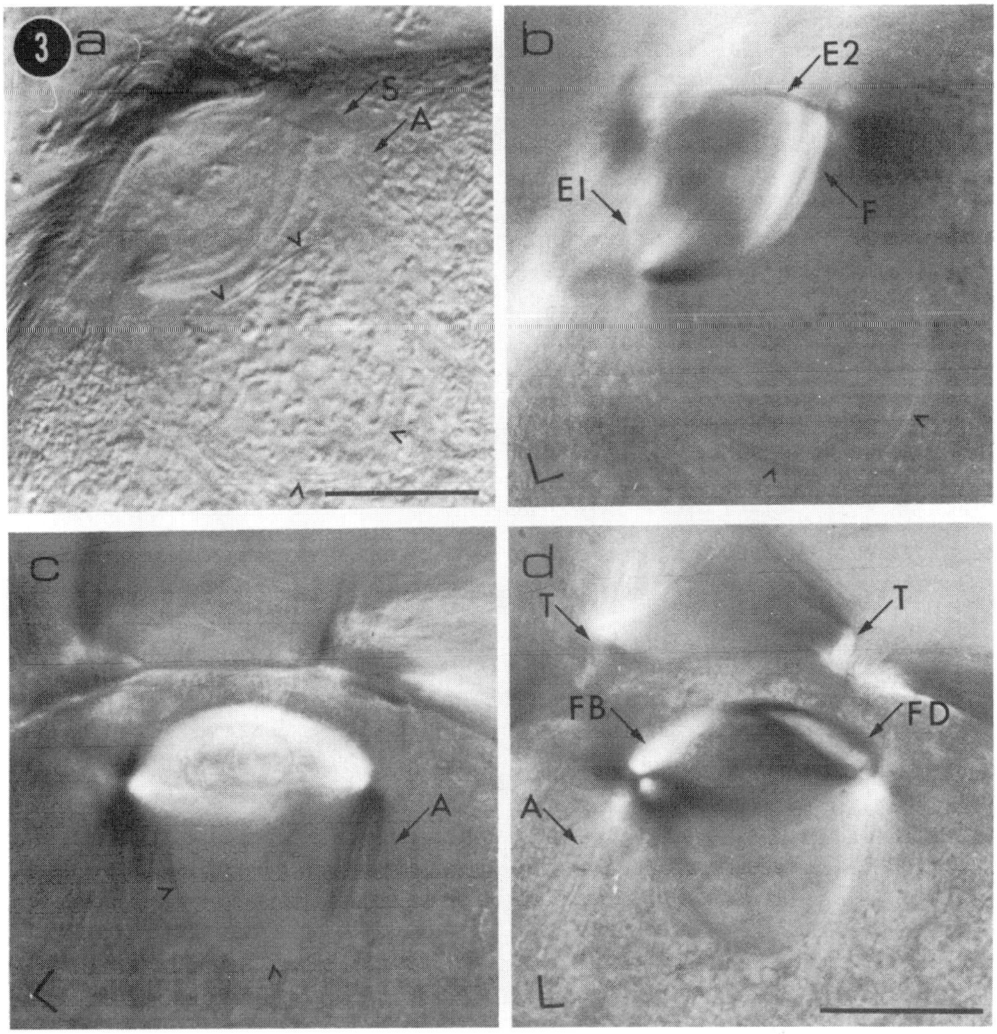

Figure 3 Early spindle formation. (*a*) Differential interference contrast. (*b*) Polarized light image of the same cell. (*c, d*) Another cell at a somewhat later stage both in polarized light but with polarizer and compensator axes oriented differently. *L*, axes of crossed polarizer and analyzer and quadrant containing slow axis of compensator. In (*a*) and (*b*) bundles of spindle fibers (*F*), which were formed by merging astral rays growing from opposing elongate-centrioles, show as segments of arcs running between the elongate-centrioles. In (*b*), the right elongate-centriole (*E2*) appears threadlike in dark compensation, the left one in bright compensation (*E1*). The flagellar tufts have not yet separated so that the two elongate-centrioles converge anteriorly. Some spindle fibers originate at the distal tips of the elongate-centrioles, others arise more proximally. In the differential interference contrast image (*a*), astral rays (*A*) which are not incorporated in the central spindle can be seen as if radiating from the (right hand) centrosomal surface (*S*). However, the polarized light images show that the birefringent microtubules of these newly formed astral rays in fact run through the centrosome and reach the elongate-centriole from whose surface they originate (*b*). In (*a*), the optical section of the nuclear envelope (*v*'s) appears shield-shaped. The nucleus has been drawn closer to the extranuclear spindle, and a segment of the envelope appears as a straight line where it is apposed to the spindle. The parabasals and axostyles have not yet fully contracted and appear as mottled birefringent structures to the left and right of the nucleus in (*b*). In (*c, d*), the flagellar tufts (*T*) have separated. The spindle, though still not yet fully mature, exhibits a higher birefringence now being made up of many more fibers. Spindle fiber bundles appear bright (*FB*) where they run parallel to the slow axis of the compensator and generally dark (*FD*) where they run in the opposite quadrants. In (*d*), however, white "eyes" appear in the middle of the dark fiber bundles (lying in the NW and SE quadrants) where the fiber concentration is so high that the spindle retardation far exceeds the effective retardation introduced by the compensator. (See 40, for how the compensator setting affects the appearance of the image of mitotic spindles). The spindle appears bipartite possibly because fiber formation is obstructed in the middle by the nuclear sleeve (structure seen in the middle of the spindle in (*c*) and (*d*)?). The astral rays (*A*), especially those running posteriorly and parallel to the nuclear envelope, have grown in number, length, and birefringence. They appear in dark compensation in (*c*), and are bright in (*d*). Bar, 30 μm. × 690.

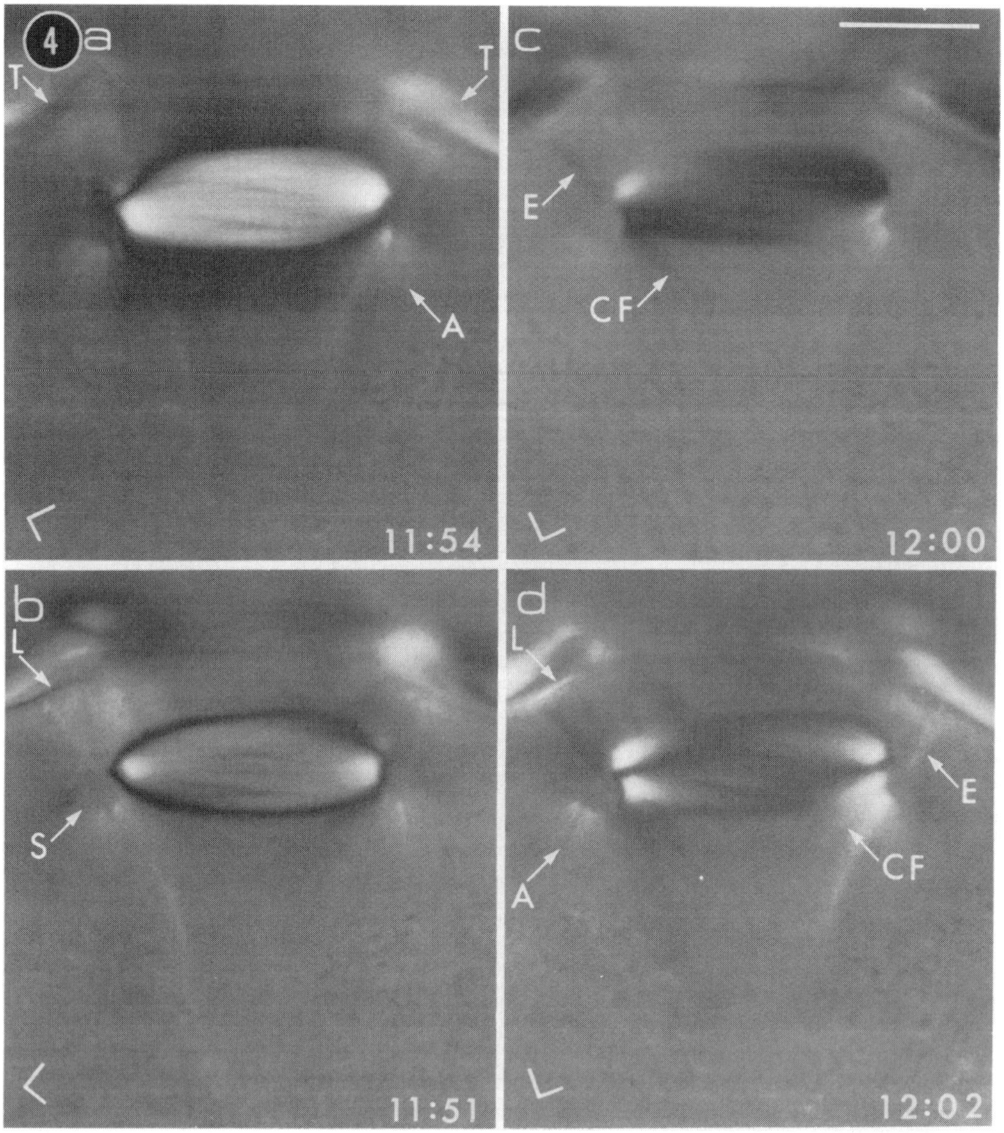

FIGURE 4 Birefringence of mature, central spindle in prophase. Flagellar tufts (*T*) and their basal structures are now sufficiently separated so that the elongate-centrioles (*E*) diverge anteriorly. All four pictures were taken with Nikon 40x rectified polarized light objectives, with crossed polars and compensator axes oriented variously relative to the spindle axis. *L*, Axes of crossed polarizer and analyzer and quadrant in which compensator slow axis is oriented. These photographs show how the central spindle is clearly composed of birefringent fiber bundles which are somewhat twisted relative to the spindle axis. Compare with electron micrograph Fig. 5. Where the concentration of fibers is greatest, adjacent to the centrosomal surface, individual fibers are not resolved and the birefringence is highest. Both spindle and aster birefringence appear to terminate at the surface (*S*) of the spherical centrosome, but in fact a weak, radially positive birefringence exists within the centrosomes (cf. Fig. 2*b*). The dark conical tip seen at the left spindle pole in (*b*) is in fact made up of birefringent fibers which continue from the central spindle through the centrosome and terminate at the tip of the elongate-centriole. (Also, see in Part II, Figs. 2*a*, *b*, *f* and 14*a*, *c*, *e*, and *f*). In (*b*), the fibers of the central spindle outside of the centrosome are more birefringent; they possess the same sign of birefringence, and also lie in the same orientation as the fibers within the centrosome, but they overcompensate and appear bright. In (*b*), the distal portion of the left hand elongate-centriole appears in bright compensation. In (*a*), (*c*), and (*d*), it appears in dark compensation. The distal portion of the right hand elongate-centriole which is mostly out of focus appears in bright compensation in (*a*), (*c*), and (*d*). Long astral rays (*A*), some running tangential to the teardrop-shaped nucleus, are seen in bright compensation in (*a*), (*b*), and (*d*). In (*c*) and (*d*), what appear to be short astral rays which converge postero-medially towards the spindle axis are in fact chromosomal spindle fibers (*CF*). They terminate abruptly on kinetochores which are permanently embedded in the nuclear envelope. Although difficult to photograph, a miniature, radially positive spherulite is visible in the rectified polarizing microscope at the tip of each of these astral rays. These spherulites are in fact the kinetochores. Time in h:min of day on 74g26. Bar, 30 μm. × 630.

when all of the chromosomes have arrived at the spindle poles (see Part II).

Structure of the Mature Spindle

A fully mature central spindle takes on the configuration illustrated in Figs. 4*a* to *d* and 7*a* to *c*. These polarized light images show varied specimen and compensator orientations (relative to the crossed polarizers) to emphasize different structural features. Birefringent fibers that correspond to bundles of microtubules in the central spindle are especially clear in Fig. 4*b*. They diverge near the spindle axis and give rise to a helical appearance in polarized light. The more intense pericentrosomal birefringence reflects the higher packing density of these microtubules toward the spindle poles.

The grouping of microtubules into thin bundles, their helical course, and their higher packing density at the centrosomal surface are also apparent in electron micrographs (Figs. 5 and 6).

The long astral rays, some of which run posteromedially tangential to the nuclear envelope, can be seen in varying contrast in Figs. 4*a* to *d* and 7*a* to *c*. Some of the astral rays which terminate on the chromosomal kinetochores, embedded in the persistent nuclear envelope, are seen as medially pointing, discrete strands in Figs. 4*c*, *d* and 7*a* to *c*.

The centrosomes are now sharply outlined by the astral rays and spindle fibers (Figs. 4*a* to *d* and 7*a* to *c*) in contrast to their earlier appearances seen in Fig. 3*b* to *d*. The birefringence of the microtubular bundles now appears to end abruptly at the centrosomal surfaces. Although much of the birefringence no longer penetrates the centrosome, closer inspection with appropriate compensator adjustment reveals that in fact some positively birefringent strands do reach the tip of the elongate-centriole. Some of the central spindle fibers and astral and chromosomal microtubules are still connected to the centriolar tip embedded in the centrosome (Figs. 4*b*; and in Part II, 2*a*, *b*, *f*, 14*a*, *c*, *e*, *f*, 15*f*, *h*). Some microtubules can be seen likewise within the centrosome in the electron micrograph (Fig. 5).

With time, the two tufts of flagella and the parabasal-axostylar lamellae move apart and the anterior bases of the elongate-centrioles diverge (Figs. 3*c*, *d*, 4*a* to *d*, 7, 8*a* to *c*). In Figs. 4*c*, *d*, and 7*a*, the two elongate-centrioles are oriented at 90° to each other in opposite quadrants, hence appearing in dark (left) and bright (right) contrast in the compensated polarized light image.

Chromosome and Nuclear Hoisting

During or shortly after central spindle maturation, some astral rays are seen terminating on the chromosome kinetochores which are embedded in the persistent nuclear envelope. These rays, now designated chromosomal spindle fibers (Figs. 4*c*, *d*, 7*a* to *c*, 8*a*, *b*), appear to form and contract repeatedly. In the posterior tapered region of the nucleus that otherwise appears empty, chromosomes are occasionally observed with their kinetochores displayed in the lateral margins of the nucleus (Fig. 1*a* to *d* in Part II). Such a kinetochore with its chromosome travels along the nuclear envelope towards the adjacent spindle pole for several tens of micrometers until they merge with the majority of chromosomes. In a few minutes, another chromosome, not earlier evident, again appears in the "empty" region, displaying its kinetochore in the lateral nuclear envelope, and repeats that poleward migration. Reflecting these movements, time-lapse motion pictures show the chromosomes in *Barbulanympha* prophase to churn violently within the nucleus.

On the whole, the chromosomes move kinetochore foremost, first toward the side of the spindle and then repetitively toward the spindle poles. Chromosomes at this prophase stage are shown in differential interference contrast in Fig. 7*d* to *f*.

As the kinetochores are drawn toward the spindle and to the spindle poles, so also is the nuclear envelope hoisted forward. In Fig. 7*d* to *f*, the anterior margin of the nuclear envelope has wrapped about halfway around the spindle. By the stage shown in Fig. 8, the nucleus has completely wrapped around the central spindle. The posterior margin of the nucleus at this stage often takes on an inverted bell shape, or the shape of a hanging bubble.

In detail, the nuclear envelope is dynamically deformed at each kinetochore insertion. In Part II, we document these deformations and analyze the force operating on the kinetochores and the nuclear envelope after we describe the events of the two-stage anaphase, nuclear morphogenesis, spindle, and aster dynamics and cytokinesis.

DISCUSSION

Successful cultivation of the anaerobic, hindgut protozoa of *Cryptocercus* now enables us to follow

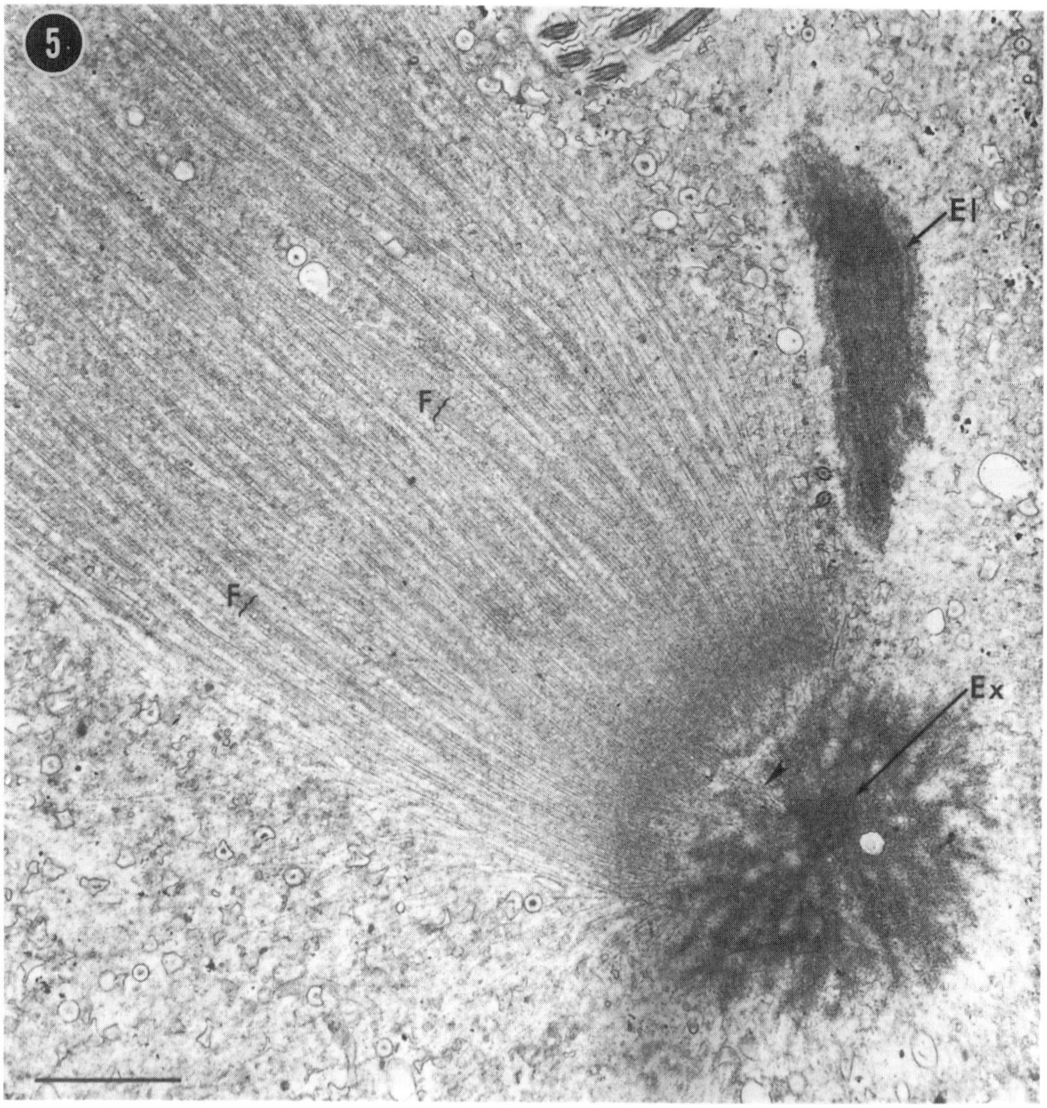

FIGURE 5 Low-power electron micrograph of relatively young central spindle, centrosome, and portions of elongate-centriole. The fibrous elongate-centriole seen in longitudinal section in the upper right (*E*) bends (medially) towards, rather than away from, the spindle equator. Thus, the flagellar tufts have apparently not yet separated. The amorphous centrosome (lower right) shows an irregular radiating pattern and no limiting membrane. The dense core seen in the centrosome is a cross section of the elongate-centriole (*Ex*) near its tip. The elongate-centriole may contain microtubules, but the cartwheel-like arrangement characteristic of centrioles in other organisms has not been observed. The disposition of microtubules in the central spindle correspond exactly to that deduced from polarized light observations in vivo. The microtubules are clustered into longitudinal bundles, namely the continuous spindle fibers (*F*). They in turn form upper and lower spindle portions which are somewhat twisted relative to each other and to the spindle axis. Near the centrosomal surface, the microtubules of the central spindle are highly concentrated and microtubules no longer appear clustered into bundles (i.e., into discrete spindle fibers). This accounts for the higher birefringence, and also the radial disposition of the birefringence axis in the pericentrosomal central spindle. From this region, a few bundles of microtubules can be seen penetrating the centrosome and reaching the elongate-centriole (arrowhead). These microtubules apparently account for the weak radially positive birefringence in the centrosomes. Astral microtubules are visible above the central spindle, below the longitudinal section of the elongate-centriole. Other microtubules appearing below the spindle are presumably reaching out towards the nucleus. Bar, 2 μm. × 9,450.

FIGURE 6 Higher magnification electron micrograph of a portion of the central spindle shown in Fig. 5. To the right, the microtubules converge towards the centrosomal surface. From the middle left to the lower right margins of the picture, one sees the junction of the intertwined bundles of microtubules. Lateral arms of microtubules are not obvious in this picture but are present. They appear prominently when two transparencies of this same photograph are superimposed and displaced along the microtubule axis by 1.5 to 3 times the microtubule diameter (See Fig. 16 in Inoué and Ritter, 32). Fixed in diluted Karnovsky's solution (33). Bar, 500 nm. × 29,200.

dynamic events of division in individual *Barbulanympha* cells.

The anaerobic cultivation system, the nutrient medium, and techniques utilized in maintaining *Barbulanympha* in vitro are the result of many years of effort. Details of the chemical medium that supports growth and reproduction of *Cryptocercus* hindgut protozoa will be described in a separate publication along with cultivation techniques. Table I lists the composition of the cellulose-free medium specifically applied to these studies. The culture chamber and associated equipment, assembled for regulation and maintenance of an O_2-free gas atmosphere, is now an adequate substitute for the physical environment provided by the hindgut. This system facilitates limited microscope observations and experimental procedures requiring an atmosphere containing as little as $0.8-20$ μM O_2 (37).

In contemplating the application of highly refined optics to *Barbulanympha* studies in vitro, it became obvious that this cultivation chamber would not be readily adaptable. As a first alternative, a perfusion chamber was mounted on the stage of the inverted optical-bench microscope designed by Inoué and Ellis. Although perfusion chambers designed by Dvorak (19) and Ellis[2] were tried, neither was satisfactory for this application. However, in these trials, several requirements for prolonged microscope observation of *Barbulanympha* cells became evident. (*a*) A seal or chamber wall made of silicone grease was harmless and sustained activities of crowded *Cryptocercus* protozoa and prokaryotes for several days. In spite of the high oxygen solubility of silicone grease, and perhaps because of an even higher solubility of carbon dioxide, the collection of microorganisms appeared able to maintain the strict anaerobic environment within the microchamber. (*b*) Compression of the large *Barbulanympha* cells to ~100-μm thickness provided optimal combination of optical clarity and suppression of movement while affording good viability. As shown in Fig. 8*d*, *Barbulanympha* are covered with bacterial symbionts (see also 25, 38). The microscope image becomes fuzzy unless growth of a thick lawn of bacteria on the slide and cover glass surfaces is prevented by compression of the *Barbulanympha* held against these surfaces.

These findings suggested that the conditions required for prolonged observation of individual dividing cells could be met most simply with a silicone grease-sealed slide chamber in the absence of perfusion. This led to the design of the slide culture chamber employing a 1.7-mil thick polyethylene spacer-gasket as described.

For observations with the light microscope, we chose to combine rectified polarized optics with the Leitz Smith T system. With polarized light optics, the birefringent achromatic regions of the mitotic apparatus are vividly displayed. Furthermore, the axes and magnitude of birefringence permit interpretation of the fine structure, changes in which can be followed with time (review, 29, 32). The T system differential interference contrast optics provides a crisp, shallow depth-of-field image displaying refractive index boundaries in relief with little interference from out-of-focus objects.

As results indicate, the combination of these two systems mounted on a common microscope base enabled recording both "achromatic" and "chromatic" structures in considerable detail. Through-focus observations (e.g. Fig. 12 in Part II) with these optical methods provide both visibility of image detail and the opportunity for uninterrupted observation lasting many hours. Hughes and Swann (27) first used phase-contrast and polarized light alternately to study mitosis in chick fibroblast cells. Improvements in microscope optical systems through the intervening years, as well as the larger size and unique features of mitosis in the *Barbulanympha* cell, have provided us with much more detailed information than that contributed by earlier workers.

We have noted with particular interest the similarity of the electron microscope appearance of the central spindle compared to its polarized light image and the fine structure deduced therefrom. This consistency should be expected in view of the contribution of microtubules to the formation of spindle fibers and astral rays (review, 29, 32), and to the accounting of their form birefringence (39, 40). Nevertheless, in view of the skepticism expressed by some regarding the interpretation of spindle birefringence, it is gratifying to find that the exact disposition of microtubular bundles deduced from our rectified polarized light images in vivo is, in fact, precisely depicted in Kubai's thin-section electron micrographs of the *Barbulanympha* spindle (Figs. 5 and 6). The very high pericentrosomal birefringence of central spindle fibers and astral rays, and the low birefringence within the centrosome of the mature anaphase-A spindle are likewise revealed in the microtubule distribution seen in the electron micro-

[2] An unpublished, compact design perfusion chamber.

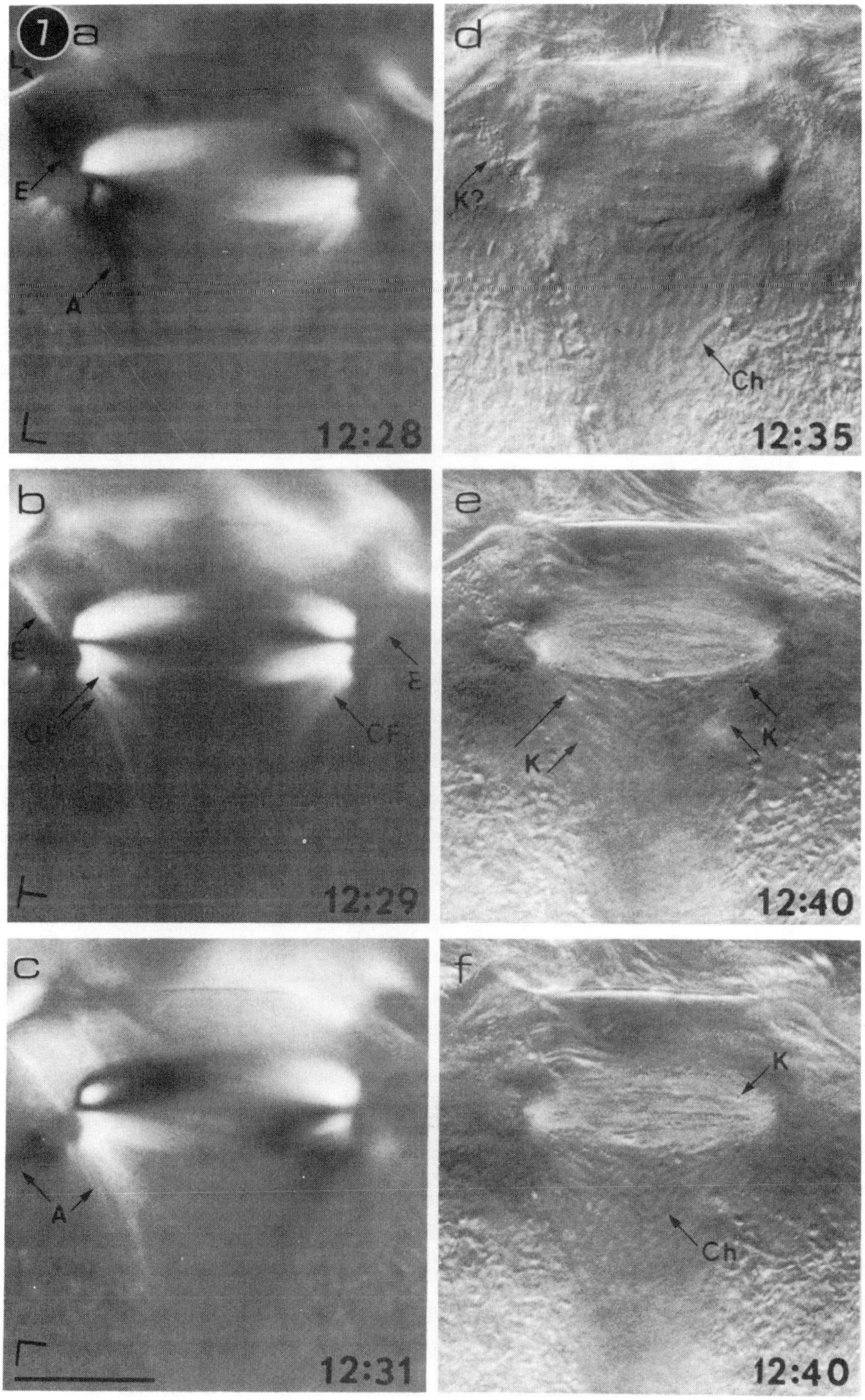
7
a
L
E
A
12:28
b
E
CF
E
CF
12:29
c
A
12:31
d
K?
Ch
12:35
e
K
K
12:40
f
K
Ch
12:40

graphs.

For the earlier stages of spindle formation, we described the convergence of birefringent fibers to a point at the posterior tip of the elongate-centriole. We believe this to be a reflection of the microtubule organizing capability of the elongate-centriole. Later in division, centrosomal birefringence decreases (e.g., compare Fig. 3*b*, *d* with Figs. 4 and 7*a* to *c*). This seems to reflect the reduction of microtubules which reach the elongate-centriole as the spindle fiber and astral ray organizing centers presumably migrate to the surfaces of the centrosomes. This may well represent a rather general phenomenon. In the early stages of division in many marine eggs and insect spermatocytes, spindle fibers and astral rays converge to a point so small as not to be resolvable in the polarizing microscope. (It should be noted that, even limited by the wavelength of light, rectified polarization optics affords a resolution of ~200 nm). Shortly after spindle formation, the birefringent fibers can be seen to terminate farther away from the spindle poles (28).

In careful studies of electron micrographs, many have questioned whether mitotic microtubules, in fact, reach or grow from the centriolar surfaces. Much has been made of the fact that microtubules are commonly found to grow from pericentriolar satellites rather than directly from the surface of the centrioles (43). However, Brinkley and Stubblefield (3) have found that, in the early stages after recovery from colchicine or cold depolymerization of microtubules, the newly formed microtubules, in fact, do grow attached directly to the centriole as far as electron microscope resolution permits one to decide. These electron microscope findings are consistent with our interpretation that microtubules initially grow directly from centrioles and that later the organizing centers migrate some distance away from the centriole surface.

At times in the very early process of central spindle formation, we have observed the emanation of microtubule bundles from locations on the elongate-centriole at points somewhat proximal to the tip (Fig. 3*a*, *b*). This pattern is also reported by Cleveland (e.g. Fig. 76 in reference 4) and clearly illustrates an extensive microtubule organizing capability of the elongate-centriole. In addition to this activity at the distal, posterior end, the anterior end of the organelle organizes both the daughter elongate-centriole and the parabasal-axostylar lamella (4, 17, 25). For these reasons, we adhere to usage of the term elongate-centriole as originally designated by Cleveland even though the electron microscope has not revealed a "cartwheel" organization typical in centrioles of many cell types.[3] It should be noted that one of us

[3] We have not adopted Hollande and Valentin's term "pseudocentriole" (21) nor their term "atractophore" (fr. Gr. atraktos = spindle, arrow) proposed more recently (22, 23, 25). "Mimocentriole" (*mimo* = to mimic; kindly suggested by Drs. William Coleman and Camille Limoges of the History of Science Program, Marine Biological Laboratory, Woods Hole, Mass.) appeared better-suited until reviewing Boveri's definition of centriole. Apart from having priority, centriole is a workable term that will avoid needless confusion and plethoric terminology otherwise designed to handle structural variations. Application of the term in this classical sense appears extensively in the literature of Boveri (2), who defined centriole, and of E. B. Wilson (44), Schrader (42), Cleveland, and countless others. Furthermore, Cleveland (4, 17) provided the first clear evidence for the dual functional nature of the centriole based in part on studies of *Barbulanympha*. However, in order not to imply the tubular cartwheel arrangement seen by electron microscopy in conventional centrioles, we have hyphenated elongate-centriole to make it a single term.

FIGURE 7 Mature central spindle in late prophase. Same spindle continued from Fig. 4, approximately one-half hour to 50 min later. (*a*), (*b*), (*c*), In polarized light with spindle axis oriented parallel to the polarizer. (*d*), (*e*), (*f*), Differential interference contrast displaying overlapping prophase chromosomes (*Ch*) whose dotlike kinetochores (*K*, see Part II Figs. 1, 9, 10) point to the spindle poles. Note also granules (*K*?) resembling kinetochores which stud the elongate-centrioles in (*d*) and (*e*). In (*a*), the distal portion of the left elongate-centriole (*E*) is in dark compensation, in (*b*) and (*c*) in bright compensation. The basal portion of the same elongate-centriole, which makes a right angle turn to the upper right and runs parallel to the parabasal axostylar lamella, is less clearly displayed but can be seen in reverse contrast (also, see Fig. 2 in Part II). The twisted course taken by the spindle fibers appears prominently in the polarized light images. See legend to Fig. 4 regarding the astral rays and chromosomal spindle fibers. *L*, polarizer and analyzer axes and quadrant containing slow axis of compensator. Time in h:min of day on 74g26. Bar, 30 μm × 690.

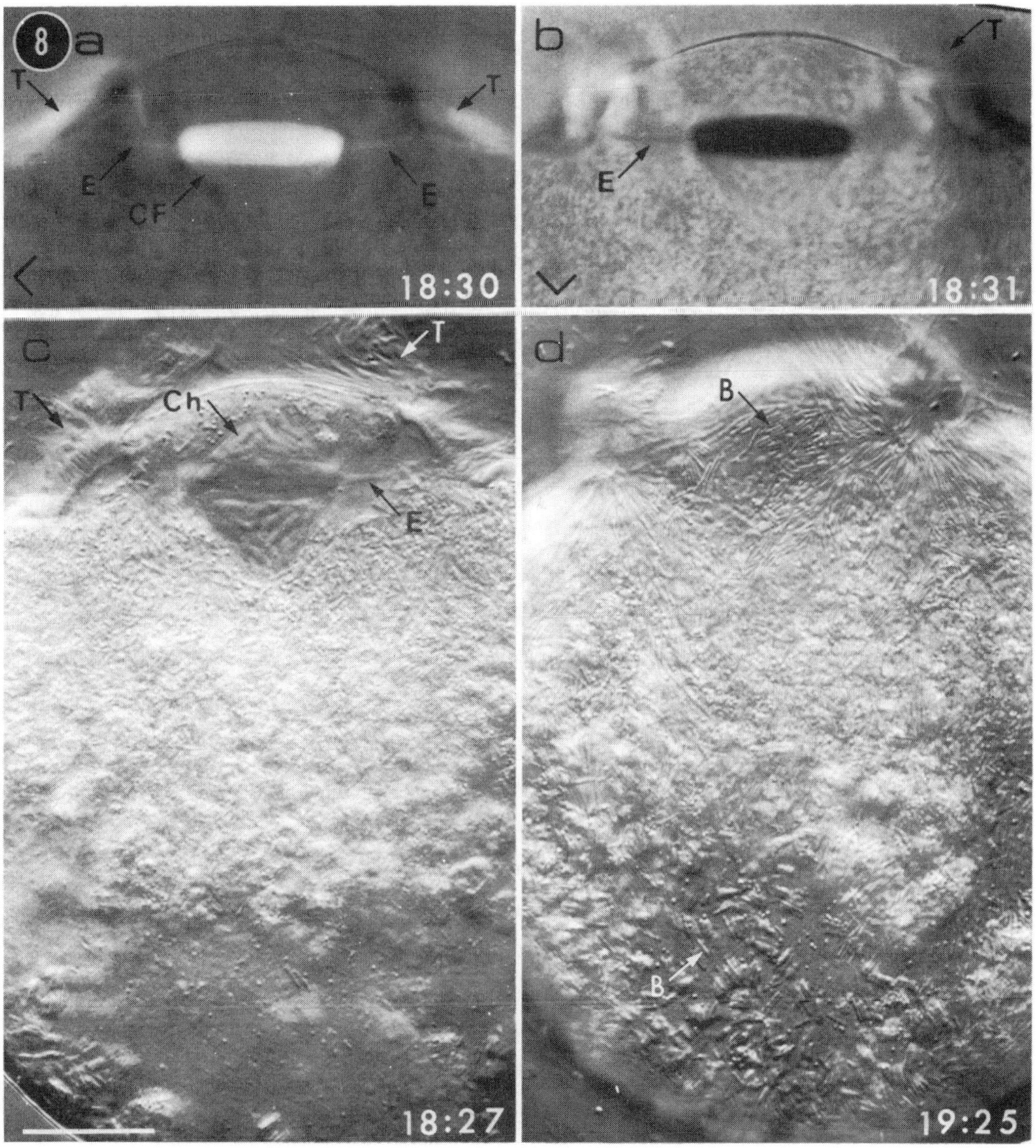

FIGURE 8 Mature central spindle of *Barbulanympha* in anaphase-A. (*a*), (*b*), In polarized light, with compensator orientation reversed. Flagellar tufts (*T*) are quite separated and the distal portions of the elongate-centrioles (*E*) now lie almost in line with the spindle. Short chromosomal spindle fibers (*CF*) run medially just below the spindle. (*c*), (*d*), Differential interference contrast view of the same cell. (*c*), Focused on right hand spindle pole and elongate-centriole. The nucleus has completely enveloped the central spindle so that chromosomes (*Ch*) are now seen anterior to the spindle as well as posteriorly. Chromosomes are in mid-anaphase-A; their tails have not yet separated. (*d*), Focused high on the same cell, showing characteristic, profuse bacterial growth (*B*) on the cell surface. h:min in time of day on 75h8. Continued to Figs. 4, 5, 6, and 7*a* in Part II. Bars, 30 μm. × 590.

(D.K.) believes the microtubule organizing capability not to reside in the elongate-centriole proper but in a material closely ensheathing its fibrous core.

We have indicated that the movement of individual chromosomes and their kinetochores toward the poles is not always direct nor is it monotonic. These kinetochore movements, repetitively toward then away from the poles of the achromatic apparatus, suggest a shuttling process strongly resembling or even corresponding to the prometaphase activity seen in many other cell types. The fluctuation of birefringence seen as "Northern Lightslike" flickering in spermatocytes (28, 41), as well as shuttling activities of chromosomes (1, 18, 35) and perhaps also their prometaphase stretch (26), appear to reflect recurrent growth and contraction of the dynamic chromosomal spindle fibers.

In summary, we have succeeded in devising a long-awaited in vitro culture technique for *Cryptocercus* hindgut microorganisms, especially *Barbulanympha*. We have made use of through-focus serial recordings of living cells by rectified polarized light and differential interference contrast microscopy, supplemented by electron microscopy of thin sections. These advances have permitted us to describe the growth and establishment of the astral rays, central spindle fibers, and chromosomal fibers; the role played by the distal end of the elongate-centriole as an organizing center; the microscopic and fine structure of the anaphase-A central spindle; the engagement and forward hoisting of the nucleus via the kinetochores embedded in the nuclear envelope; and the dynamic features of recurrent transport of chromosomes via the kinetochores toward the spindle poles.

For our collection of *Cryptocercus*, we thank Susan Rogers and Catherine Sale who provided access to their Tumbling Waters property, Clayton, Georgia; also, the United States Department of the Interior, National Park Service for collecting permits specific for the Great Smoky Mountains National Park and the Asheville, Northern Carolina areas of the Blue Ridge Parkway.

The work reported in these two articles was supported in part by grants BMS 7500473 from the National Science Foundation and 5 R01 GM23475 from the National Institutes of Health awarded to Shiny Inoué. This work was also supported by the Office of General Research, University of Georgia, Athens, Georgia.

We thank the Rockefeller University Press for cooperating with us on experiments to optimize half-tone reproduction of polarization micrographs which contain extreme tonal ranges combined with subtle textural details. Based on these experiments, the micrographs in these two articles were skillfully printed by Christopher W. Inoué to optimize image sharpness and tonal ranges. Additionally, in Figs. 3*c*, 4*a*, *b*, 7*a*, *b*, and in Part II, Figs. 2*a*, *e*, *f*, 9*e*, 14*a*, *c*, *e*, and 15*b* and *d*, the very bright central spindle region was "burned in" to bring out image detail without other distortion of the image. Our thanks are also due Dr. Greenfield Sluder for help on making the time-lapse motion picture which displayed chromosome churning in *Barbulanympha* anaphase. We are grateful to the staff of the Biophysical Cytology Program without whose dependent cooperation these reports would not have materialized.

Received for publication 29 July 1977, and in revised form 3 January 1978.

REFERENCES

1. Bajer, A. S., and J. Molè-Bajer. 1972. Spindle dynamics and chromosome movements. *Int. Rev. Cytol.* **3** (Suppl.):1–271.
2. Boveri, T. 1900. Über die Natur der Centrosomen. Zellen-Studien. Heft 4. Gustav Fischer Verlag KG. Jena, East Germany. 1–220.
3. Brinkley, B. R., and E. Stubblefield. 1970. Ultrastructure and interaction of the kinetochore and centriole in mitosis and meiosis. *In* Advances in Cell Biology, Vol. 1. D. M. Prescott, L. Goldstein, and E. McConkey, editors. Appleton-Century-Crofts, New York. 119–185.
4. Cleveland, L. R., S. R. Hall, E. P. Sanders, and J. Collier. 1934. The wood-feeding roach *Cryptocercus*, its protozoa, and the symbiosis between protozoa and roach. *Mem. Am. Acad. Arts Sci.* **17:**185–342 (and 60 plates).
5. Cleveland, L. R. 1938. Origin and development of the achromatic figure. *Biol. Bull.* (*Woods Hole*). **50:**41–55.
6. Cleveland, L. R. 1953. Hormone-induced sexual cycles of flagellates. IX. Haploid gametogenesis and fertilization in *Barbulanympha*. *J. Morphol.* **93:**371–404.
7. Cleveland, L. R. 1953. Studies on chromosomes and nuclear division. IV. Photomicrographs of living cells during meiotic divisions. *Trans. Am. Philos. Soc.* **43:**848–869.
8. Cleveland, L. R. 1954. Hormone-induced sexual cycles of flagellates. X. Autogamy and endomitosis in *Barbulanympha* resulting from interruption of haploid gametogenesis. *J. Morphol.* **95:**189–212.
9. Cleveland, L. R. 1954. Hormone-induced sexual cycles of flagellates. XI. Reorganization in the zygote of *Barbulanympha* without nuclear or cytoplasmic division. *J. Morphol.* **95:**213–235.
10. Cleveland, L. R. 1954. Hormone-induced sexual cycles of flagellates. XII. Meiosis in *Barbulanympha* following fertilization, autogamy and endomitosis. *J. Morphol.* **95:**557–620.
11. Cleveland, L. R. 1955. Hormone-induced sexual cycles of flagellates. XIII. Unusual behavior of gametes and centrioles of *Barbulanympha*. *J. Morphol.* **97:**511–542.
12. Cleveland, L. R. 1956. Effects of temperature and tension on oxygen toxicity for the protozoa of *Cryptocercus*. *J. Protozool.* **3:**74–77.

13. CLEVELAND, L. R. 1957. Correlation between the molting period of *Cryptocercus* and sexuality in its protozoa. *J. Protozool.* **4:**168–175.
14. CLEVELAND, L. R. 1957. Achromatic figure formation by multiple centrioles of *Barbulanympha. J. Protozool.* **4:**241-248.
15. CLEVELAND, L. R. 1958. A factual analysis of chromosomal movement in *Barbulanympha. J. Protozol.* **5:**47-62.
16. CLEVELAND, L. R. 1959. Sex induced with ecdysone. *Proc. Natl. Acad. Sci. U. S. A.* **45:**747-753.
17. CLEVELAND, L. R. 1963. Functions of flagellate and other centrioles in cell reproduction. *In* The Cell in Mitosis. L. Levine, editor. Academic Press, Inc., New York. 3-31.
18. DIETZ, R. 1969. Bau and Funktion des Spindelapparats. *Naturwissenschaften.* **56:**237-248.
19. DVORAK, J. A., and W. F. STOTLER. 1971. A controlled-environment culture system for high resolution light microscopy. *Exp. Cell Res.* **68:**144-148.
20. FUSELER, J. W. 1975. Mitosis in *Tilia americana* endosperm. *J. Cell Biol.* **64:**159-171.
21. HOLLANDE, A., and J. VALENTIN. 1967. Interprétation des structures dites centriolaires chez les Hypermastigines symbiontes des Termites et du *Cryptocercus. C. R. Acad. Sci.* **264:**1868-1871.
22. HOLLANDE, A., and J. VALENTIN. 1967. Relations entre cinétosomes, "atractophores" et complexe fibrillaire axostyloparabasal, chez les Hypermastigines du genre *Barbulanympha. C.R. Acad. Sci.* **264:**3020-3022.
23. HOLLANDE, A., and J. VALENTIN. 1967. Morphologie et infrastructure du genre *Barbulanympha*, hypermastigine symbiontique de *Cryptocercus punctulatus* Scudder. *Protistologica.* **3:**257-267 (and 11 plates).
24. HOLLANDE, A., and J. VALENTIN. 1968. Infrastructure des centromères et déroulement de la pleuromitose chez les Hypermastigines. *C.R. Acad. Sci.* **266:**367-370.
25. HOLLANDE, A., and J. CARRUETTE-VALENTIN. 1971. Les atractophores, l'induction du fuseau et la division cellulaire chez les Hypermastigines. Étude infrastructurale et révision systématique des Trichonymphines et des Spirotrichonymphines. *Protistologica.* **7:**5-100.
26. HUGHES-SCHRADER, S. 1947. The "pre-metaphase stretch" and kinetochore orientation in phasmids. *Chromosoma (Berl.).* **3:**1-21.
27. HUGHES, A. F., and M. M. SWANN. 1948. Anaphase movements in the living cell. A study with phase contrast and polarized light on chick tissue cultures. *J. Exp. Biol.,* **25:**45-70 (and 2 plates.).
28. INOUÉ, S. 1964. Organization and function of the mitotic spindle. *In* Primitive Motile Systems in Cell Biology. R. D. Allen and N. Kamiya, editors. Academic Press, Inc., New York 549-598.
29. INOUÉ, S. 1969. The physics of structural organization in living cells. *In* Biology and the Physical Sciences. Samuel Devons, editor. Columbia University Press, New York. 139-171.
30. INOUÉ, S. 1961. Polarizing microscope: design for maximum sensitivity. *In* The Encyclopedia of Microscopy. G. L. Clarke, editor. Reinhold Publishing Corp., New York. 480-485.
31. INOUÉ, S., and W. L. HYDE. 1957. Studies on depolarization of light at microscope lens surfaces. II. The simultaneous realization of high resolution and high sensitivity with the polarizing microscope. *J. Biophys. Biochem. Cytol.* **3:**831-838.
32. INOUÉ, S., and H. RITTER, JR. 1975. Dynamics of mitotic spindle organization and function. *In* Molecules and Cell Movement. S. Inoué and R. E. Stephens, editors. Raven Press, New York. 3-30.
33. Kubai, D. F. 1973. Unorthodox mitosis in *Trichonympha agilis*: kinetochore differentiation and chromosome movement. *J. Cell Sci.* **13:**511-552.
34. MCKITTRICK, F. A. 1964. Evolutionary studies of cockroaches. Memoir 389. Cornell Univ. Agr. Exp. Sta. Ithaca, N. Y. 1-197.
35. NICKLAS, R. B. 1971. Mitosis. *In* Advances in Cell Biology, Vol. 2. D. M. Prescott, L. Goldstein, and E. McConkey, editors. Appleton-Century-Crofts, New York. 225-297.
36. RITTER, H. 1961. Glutathione-controlled anaerobiosis in *Cryptocercus* and its detection by polarography. *Biol. Bull. (Woods Hole).* **121:**330-346.
37. RITTER, H. 1974. A fluid system for the cultivation, light microscope examination and manipulation of obligate anaerobes. *J. Protozool.* **21:**565-568.
38. RITTER, H., and W. J. HUMPHREYS. 1971. A cell membrane adhering bacterial symbiont of the flagellate protozoan *Barbulanympha*, a mutualist in the hind-gut of the wood-feeding cockroach *Cryptocercus*. Proceedings of the 11th Annual Meeting of the American Society for Cell Biology. **51:**247*a*. (Abstr.).
39. SATO, H. 1975. The mitotic spindle. *In* Aging Gametes. R. J. Blandau, editor. Karger AG., S. Basel, Switzerland. 19-49.
40. SATO, H., G. W. ELLIS, and S. INOUÉ. 1975. Microtubular origin of mitotic spindle form birefringence. *J. Cell Biol.* **67:**501-517.
41. SATO, H., and K. IZUTSU. 1974. Birefringence in Meiosis of Grasshopper Spermatocytes: *Chrysochraon japonicus* and *Trilophidia annulata*. Timelapse motion picture. Available from George W. Colburn Laboratory, Inc., Chicago, Illinois 60606.
42. SCHRADER, F. 1953. Mitosis: The Movements of Chromosomes in Cell Division. Columbia University Press, New York 2nd edition. 1-170.
43. WENT, H. A. 1966. The behavior of centrioles and the structure and formation of the achromatic figure. *Protoplasmatologia.* **6:**1-109.
44. WILSON, E. B. 1925. The Cell in Development and Heredity. Macmillan, Inc., New York 3rd edition. 1-1232.

Reprinted from *Journal of Cell Biology*, Vol. 77(3), pp. 655–684, 1978.

MITOSIS IN *BARBULANYMPHA*

II. Dynamics of a Two-Stage Anaphase, Nuclear Morphogenesis, and Cytokinesis

SHINYA INOUÉ AND HOPE RITTER, JR.

From the Program in Biophysical Cytology, Department of Biology, University of Pennsylvania G7, Philadelphia, Pennsylvania 19104, and the Department of Zoology, University of Georgia, Athens, Georgia 30602

ABSTRACT

Anaphase in *Barbulanympha* proceeds in two discrete steps. In anaphase-A, chromosomal spindle fibers shorten and chromosomes move to the stationary centrosomes. In anaphase-B, the central spindle elongates and ("telophasic") bouquets of chromosomes, with kinetochores still connected by the shortened chromosomal fibers to the centrosomes, are moved far apart.

The length, width, and birefringence of the central spindle remain unchanged throughout anaphase-A. In anaphase-B, the central spindle elongates up to fivefold. During elongation, the peripheral fibers of the central spindle splay, first anteriorly and then laterally. The remaining central spindle progressively becomes thinner and the retardation decreases; however, the coefficient of birefringence stays approximately constant.

The nuclear envelope persists throughout mitosis in *Barbulanympha* and the nucleus undergoes an intricate morphological change. In prophase, the nucleus engulfs the spindle; in early anaphase-A, the nuclear envelope forms a seam anterior to the spindle, the nucleus thus transforms into a complete sleeve surrounding the central spindle. In late anaphase-A, the middle of the seam opens up in a cleft as the lips part; in anaphase-B, the cleft expands posteriorly, progressively exposing the central spindle. Finally, the cleft partitions the nucleus into two.

The nuclear envelope shows an apparent elasticity and two-dimensional fluidity. Localized, transient deformations of the nuclear envelope indicate poleward and counter-poleward forces acting on the kinetochores embedded in the envelope. These forces appear responsible for nuclear morphogenesis as well as anaphase chromosome movement.

At the end of anaphase-B, the two rostrate *Barbulanympha* may swim apart or be poked apart into two daughter cells by another organism cohabiting the host's hindgut.

KEY WORDS *Barbulanympha* · mitotic spindle · living cell · chromosome movement · nuclear envelope · cytokinesis · birefringence

In Part I we described a preparation of *Barbulanympha* which permits extended observation of mitosis in individual living cells in this unusually favorable material. The optical system we assembled for alternate through-focus observation and recording in differential interference contrast and rectified polarized light microscopy is also described. The formation and fine structure of the magnificent central spindle, the forward hoisting of the nucleus to meet the spindle surface, and the churning of chromosomes in the prophase nucleus are reported.

In Part II we report the behavior of chromosomes, kinetochores, nuclear envelope, mitotic spindle fibers, and astral rays during late prophase, the two anaphases, karyokinesis, and cytokinesis in *Barbulanympha.*

RESULTS

Late Prophase

With progress of prophase, the nucleus is continuously hoisted forward. As detailed later, the nucleus is deformed and eventually wrapped around the central spindle, forming a complete sleeve (schematically shown in Fig. 11). Within this tubular nucleus, the prophase chromosomes point polewards with the chromatid arms still unseparated. Their kinetochores remain embedded in the nuclear envelope. In this configuration, the kinetochores slowly move toward and away from the centrosomes individually, continuing the churning of chromosomes described in Part I. Then, without ever entering a regular metaphase configuration, the cell leaves prophase and enters anaphase.

Anaphase-A

At the onset of anaphase, kinetochores are positioned at varying distances from the centrosomes. In differential interference contrast micrographs, the minute, discrete kinetochores sharply delineate the poleward apex of each chromosome (e.g., Fig. 1*d;* also see Figs. 9, 10, and 12). In Fig. 1*a* and *b*, some kinetochores already appear near the centrosome and others are still seen near the equator of the spindle.

With progression of time, arms of the sister chromatids gradually separate. The chromosomes then move, kinetochore foremost, steadily to the spindle poles (Figs. 1*a* to *d*; also 5*a* to *d*; 13*a* to *f*).

As Figs. 3 and 8 show, each kinetochore moves at similar velocities ranging from 0.3 to 0.5 μm/min at 18 ± 1°C. Inasmuch as there is no metaphase as such, and because the kinetochores have varying distances to travel, the chromosomes move poleward over a considerable span of time.

With the polarizing microscope, the nuclear membrane-embedded kinetochores are seen linked to the centrosome by positively birefringent chromosomal spindle fibers (Figs. 2*a, b, e,* 4*a,* and *c*). The individual fibers are difficult to document photographically because the birefringence of the central spindle is so overwhelming (10–15 nm in retardation). The chromosomal fibers also tend to lie parallel to the central spindle fibers or to the large bundle of astral rays enveloping the nucleus. Birefringence due to bundles of chromosomal spindle fibers can be seen in Figs. 2*c* to *e* and 4*a* to *d*. With our microscope equipped with rectified polarization optics, the dark-adapted eye can clearly distinguish the individual chromosomal fibers when the λ/25 Brace-Köhler compensator is rapidly turned back and forth. Furthermore, each fiber is seen to terminate on an individual kinetochore which appears as a positively birefringent, miniature spherulite.

During anaphase-A, the weakly birefringent chromosomal fibers shorten. Concurrently, chromosomes move with kinetochores foremost to the centrosomal surface. Finally, the chromosomal fibers become short birefringent stubs which link the kinetochore and centrosomal surface (Fig. 4*d* to *f*; also shown schematically in Fig. 11).

Eventually, all the kinetochores crown the centrosome. In optical sections given by differential interference contrast, the kinetochores can be seen aligned in an arc, as though a string of pearls (Figs. 5*e, g, h*; and especially 6*b, c* left pole). As the kinetochores become aligned the chromosomes shorten, forming a tight ("telophasic") bouquet around the centrosome.

The arc of kinetochores centers around the distal tip of the elongate-centriole. In polarized light, we see a cone made up of thin, radiating, birefringent strands between the tip of the elongate-centriole and the centrosomal surface crowned by the kinetochores (Fig. 2*a, b, f*; also see Figs. 14, 15, and 16*b*). As discussed in Part I, these strands which show positive birefringence appear to include intracentrosomal portions of the chromosomal spindle fibers (also see text figure 20 in reference 8).

Throughout this somewhat haphazard migration of chromosomes to the centrosome, the bire-

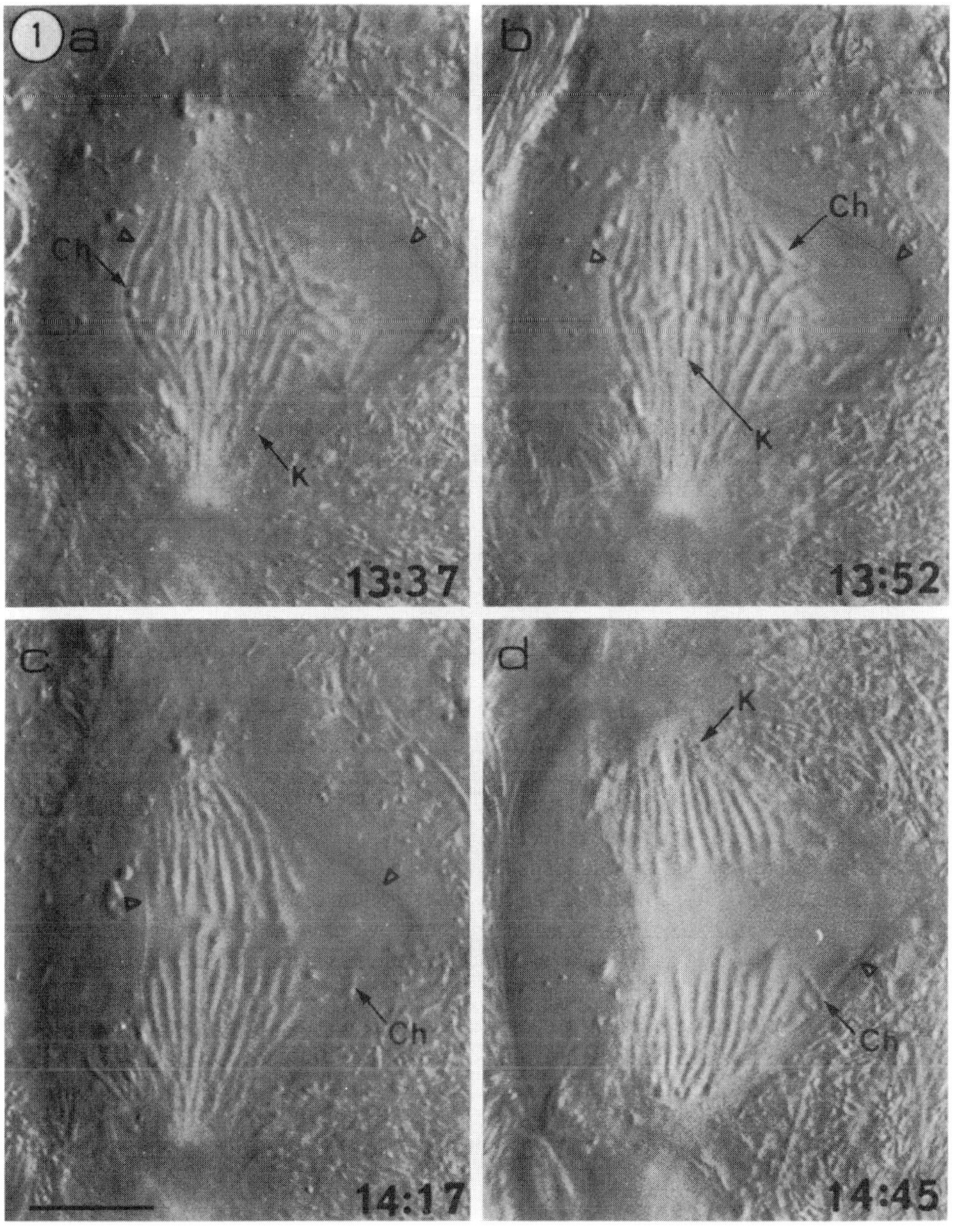

FIGURE 1 Differential interference contrast image of *Barbulanympha* anaphase-A in side view, focused on chromosomes above the central spindle. (*a*), Early anaphase-A. Chromosomes (*Ch*) for the most part are still attached at their distal ends. Kinetochores (*K*, distinct dots at chromosome apices near lower pole) appear at varying distances from the spindle poles. Anterior margin (left side in these photos) of the nuclear envelope is convex outward (triangle). (*b*), Mid-anaphase-A. Arms of several chromosomes are just separating. Their kinetochores are closer to the poles, but others are still lagging. Notice one kinetochore near the axis of the spindle only a fifth of the way from the equator to the lower spindle pole! Anterior margin of the nuclear envelope is less convex than in (*a*). (*c*), Late mid-anaphase-A. Most chromosome arms have separated and their kinetochores are approaching the spindle poles. The anterior margin of the nuclear envelope is now concave (triangle). In the lower right hand part of the nucleus, note the lone pair of chromosomes (*Ch*) that appear as though they had moved away from the lower spindle pole. (*d*), Late anaphase-A. Chromosome arms are contracting. The kinetochores lie near the centrosomal surface but are not yet aligned in a "string of pearls" configuration. The lone pair of chromosomes is moving poleward. h:min in time of day on 74g28. Bar, 20 μm. × 800.

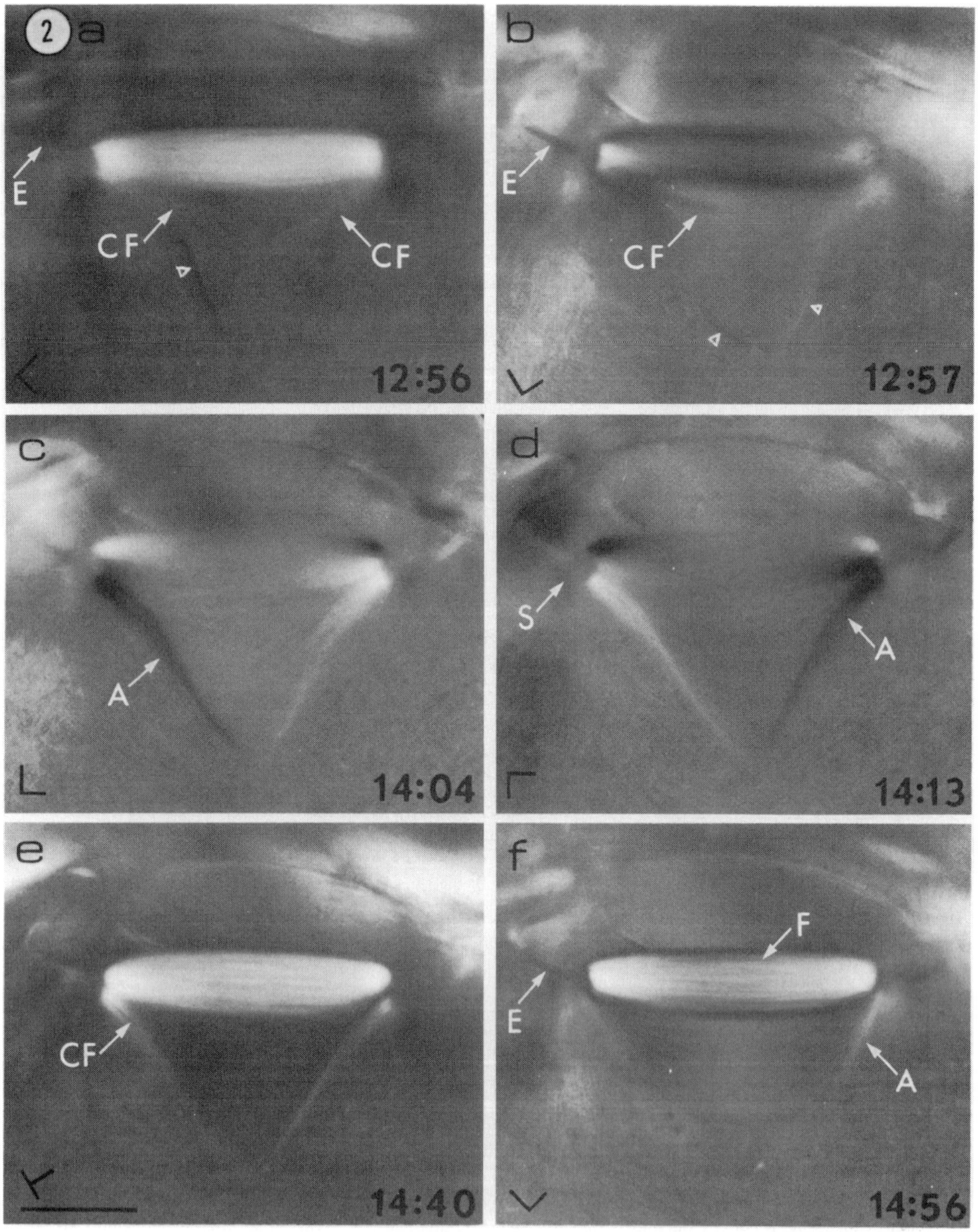

FIGURE 2 Anaphase-A side view of the same cell shown in Fig. 1, observed in polarized light and focused on the central spindle; turned 90° to Fig. 1. Over the 2 h of anaphase-A covered in these photos, the dimensions and birefringence of the central spindle have remained unchanged. Polarizer and compensator axes are oriented variously relative to the spindle axis to enhance different fibers of the mitotic apparatus; central spindle "continuous" fibers (*F* in (*f*), chromosomal spindle fibers (*CF* in (*a*), (*b*), (*e*)) and perinuclear astral rays (*A* in (*c*), (*d*), (*f*)). Compare with Figs. 1 and 3 which indicate location of chromosomes. In (*a*), (*b*), and (*f*), note the cone of weakly positively birefringent strands which link the distal tip of the left hand elongate-centriole (*E*) to the central spindle fibers and astral rays abutting the centrosomes (*S*). The hooked-shape characteristic of the elongate-centriole in anaphase can be seen clearly to the left of the spindle. L, orientation of polarizer and analyzer and quadrant containing slow axis of compensator. Time in h:min on 74g28. Bar, 20 μm. × 835.

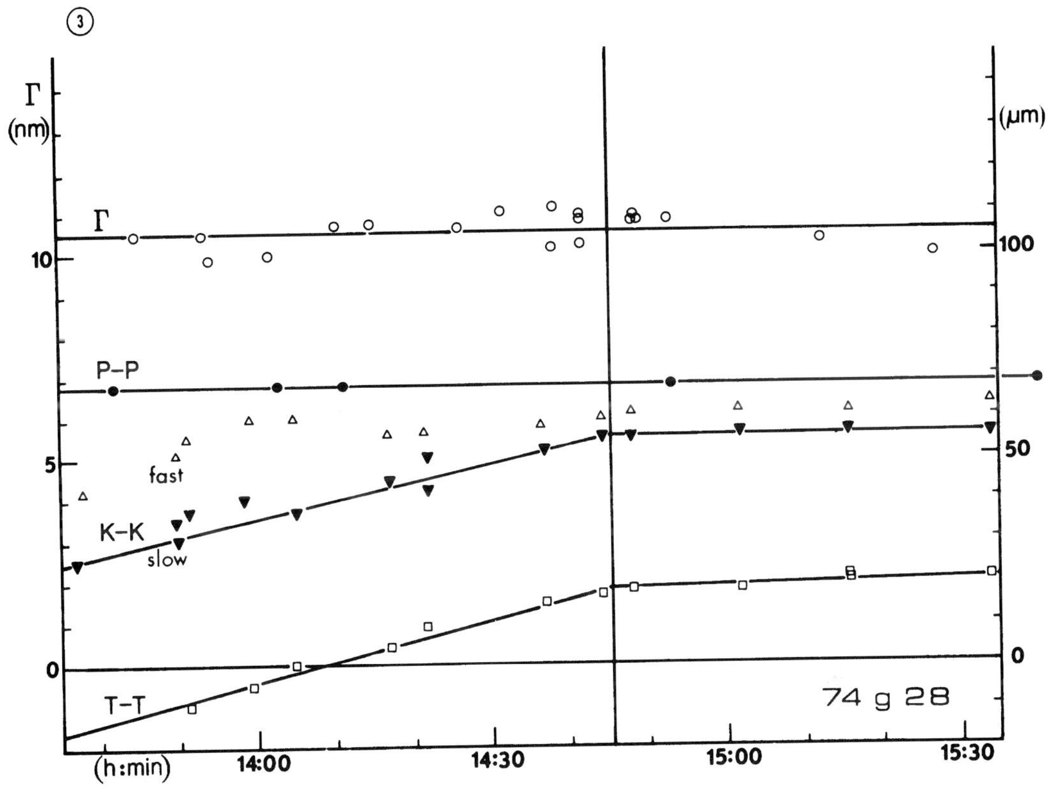

FIGURE 3 Measurements of anaphase-A taken on the same cell 74g28 photographed in Figs. 1 and 2. *P-P*, spindle pole-to-pole length, which clearly remains unchanged throughout anaphase-A. *K-K*, interkinetochore separation for chromosomes moving to poles early (fast), and late (slow). *T-T*, separation of tails of "slow" chromosomes. Γ, retardation of central spindle. h:min in time of day on 74g28.

fringence, shape, and size of the extranuclear central spindle do not change (Fig. 2*a* to *f*; also compare Fig. 8*a* to *c* in Part I and Fig. 5*a* in Part II with Figs. 4*a*, *b*, 5*b*, and *c*, graphed in Fig. 8). The retardation (Figs. 3 and 17), the width (Fig. 17) and pole-to-pole distance of the central spindle (Figs. 3, 8, and 17) remain constant until the chromosomes all reach the spindle poles. While the chromosomal spindle fibers shorten and their birefringence diminishes, their retardation contributes negligibly to the measured retardation of the central spindle.

With the arrival of all of the kinetochores at the centrosomal surface, the first stage of anaphase terminates. This stage thus involves chromosomal segregation and kinetochore poleward migration led by shortening chromosomal spindle fibers. We have named this stage of chromosome segregation, anaphase-A. For *Barbulanympha,* this occurs in the total absence of central spindle elongation.

Some *Barbulanympha* cells, as the one illustrated in Figs. 1–3, did not proceed beyond the end of anaphase-A. Others proceeded to anaphase-B (see below) after varying intervals of time (Figs. 4–6, 8, 9–17). The use of an orange filter in our microscope illuminating system to eliminate even a trace of stray blue light, and other precautions to minimize exposure to intense illumination as described in Part I, Materials and Methods, increased the frequency of cells progressing into anaphase-B. In cells which did not proceed to anaphase-B, the flagellar tufts did not separate farther, but otherwise such cells appeared perfectly healthy. Their flagella beat vigorously and the cells swam about for many hours, continuing to display a normal-looking, birefringent, anaphase-A central spindle.

Anaphase-B

At the end of anaphase-A the chromosomes are well separated, forming ("telophasic") bouquets with their kinetochores aligned at each spindle pole. Up to this stage of anaphase, the length and shape of the birefringent central spindle have not changed since prophase.

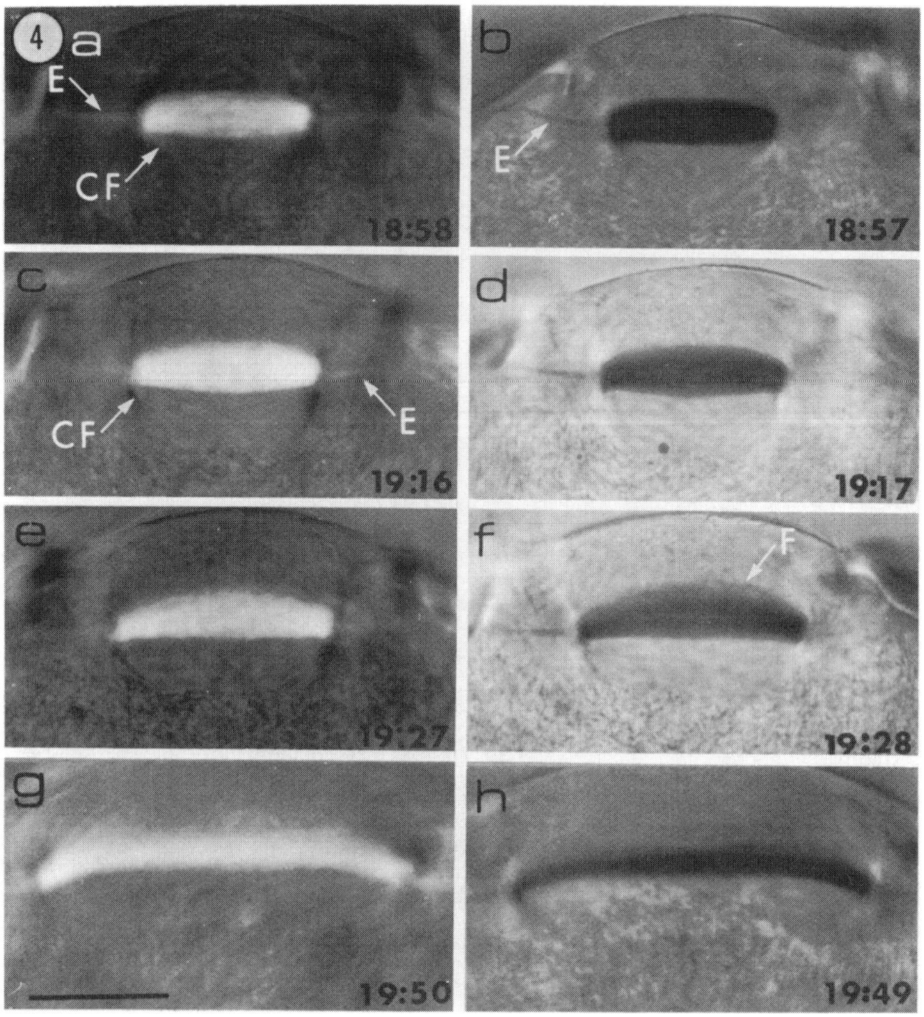

FIGURE 4 Late anaphase-A to mid anaphse-B spindle; side view in polarized light. (*a*), (*b*), Cell in mid-anaphase-A, showing typical anaphase-A spindle. (*c*), (*d*), Anaphase-A is practically completed and the spindle is becoming somewhat banana-shaped. The posterior margin of the spindle is still sharply delineated but the anterior margin (above in these photos) has become fuzzy. (*e*), (*f*), Anaphase-B has already started. The central spindle is longer and the fibers (*F*) have started to splay anteriorly. (*g*), (*h*), Central spindle continues to elongate, becomes thinner and less birefringent. Continued to Fig. 6*a* and *d*. Near the poles, short chromosomal spindle fibers (*CF*) are visible just below the spindle. Spindle axis 45° to polarizer, compensator orientation reversed for each pair of photos. h:min time of day on 75h8. Bar, 30 μm. × 630.

The onset of anaphase-B is signaled in the polarizing microscope by a subtle change in shape and contour of the central spindle. The birefringent fibers at the anterior margin become fuzzier and bowed out, and the spindle's posterior margin remains sharp (Fig. 4*c, d*). Characteristically, this transformation to a banana-shaped spindle is seen at the beginning of each anaphase-B.

The spindle then starts growing, and may extend to as much as five times its original length (Fig. 14 in reference 21). This spectacular growth of the central spindle is shown in the polarized light images in Figs. 4*c* through *h* to 6*a* and *d* (also see Fig. 12 in Reference 21). At 18° ± 1°C, the spindle shown here grew at a steady rate of 1.5 μm/min (Fig. 8), and the spindle in the larger species shown earlier reached an extension rate of 5.9 μm/min (Fig. 14 in reference 21).

As the central spindle elongates, the chromosomes are separated farther. The several-fold separation of the chromosome bouquets is clearly seen in the differential interference contrast im-

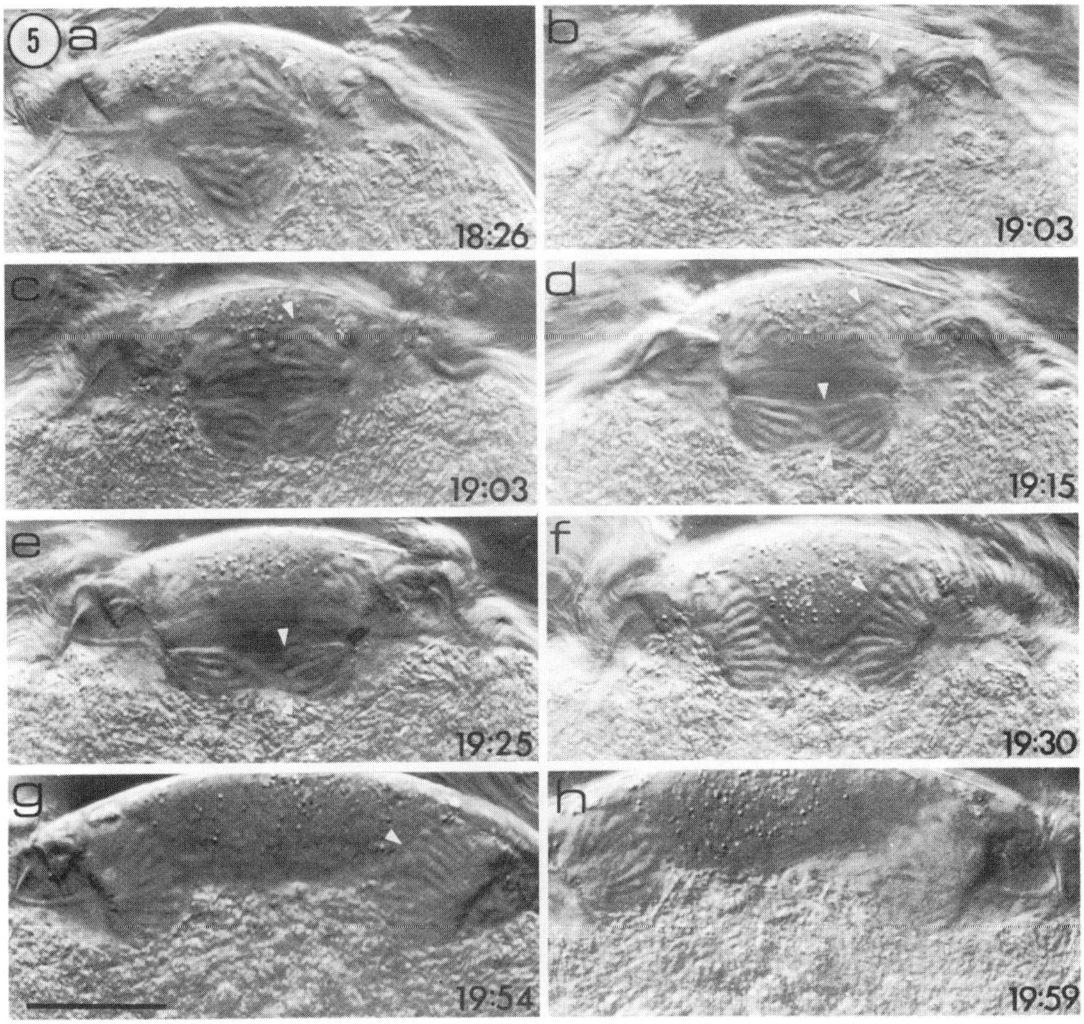

FIGURE 5 Same cell as shown in Fig. 4 but viewed in differential interference contrast. (*a*) to (*d*), Anaphase-A. (*e*) to (*h*), Anaphase-B (continued to Fig. 6*b* and *c*). Note change of nuclear contour (triangles) with progression of mitosis. The nuclear envelope never breaks down in *Barbulanympha* but instead undergoes a complex morphogenesis (see Fig. 11). Time of day in h:min on 75h8. Bar, 30 μm. × 600.

ages (Figs. 5*d* to *h*, 6*b*, *c*; also see Fig. 13 in reference 21). In these figures, the arched row of kinetochores shows clearly (see especially Fig. 6*b*, *c*). Throughout anaphase-B, the kinetochores, which lead the chromosome arms, remain anchored to the centrosomal surface by short stubs of chromosomal spindle fibers (Figs. 4*a* to *h* and 6*a* and *d*).

As the birefringent spindle elongates, its anterior margin becomes fuzzier as though fibers were splaying (Fig. 4*e* to *h*; also Fig. 12 in reference 21). The remaining spindle stem becomes progressively thinner while the retardation also decreases (Figs. 4, 6*a*, *d*, 7*a*; also Figs. 12 and 14 in reference 21).

Decrease in retardation of the central spindle remnant is nearly proportional to the decrease of spindle diameter (Fig. 17; also Fig. 14 in reference 21). Thus, the coefficient of birefringence (= retardation/thickness) remains relatively constant; presumably, fewer microtubules make up the stem of the central spindle while their packing density remains unchanged.

During the early part of anaphase-B, the caudal margin of the spindle remains reasonably straight (Fig. 4*c*, *d*; also Figs. 12, 12:35, and 12:51 in reference 21). Soon, however, the spindle becomes bowed near the poles as the bouquets of chromosomes tilt outwards (Figs. 4*e* to *h*, 5*e* to *h*; also Figs. 12, and 13, 13:01 *et seq*. in reference

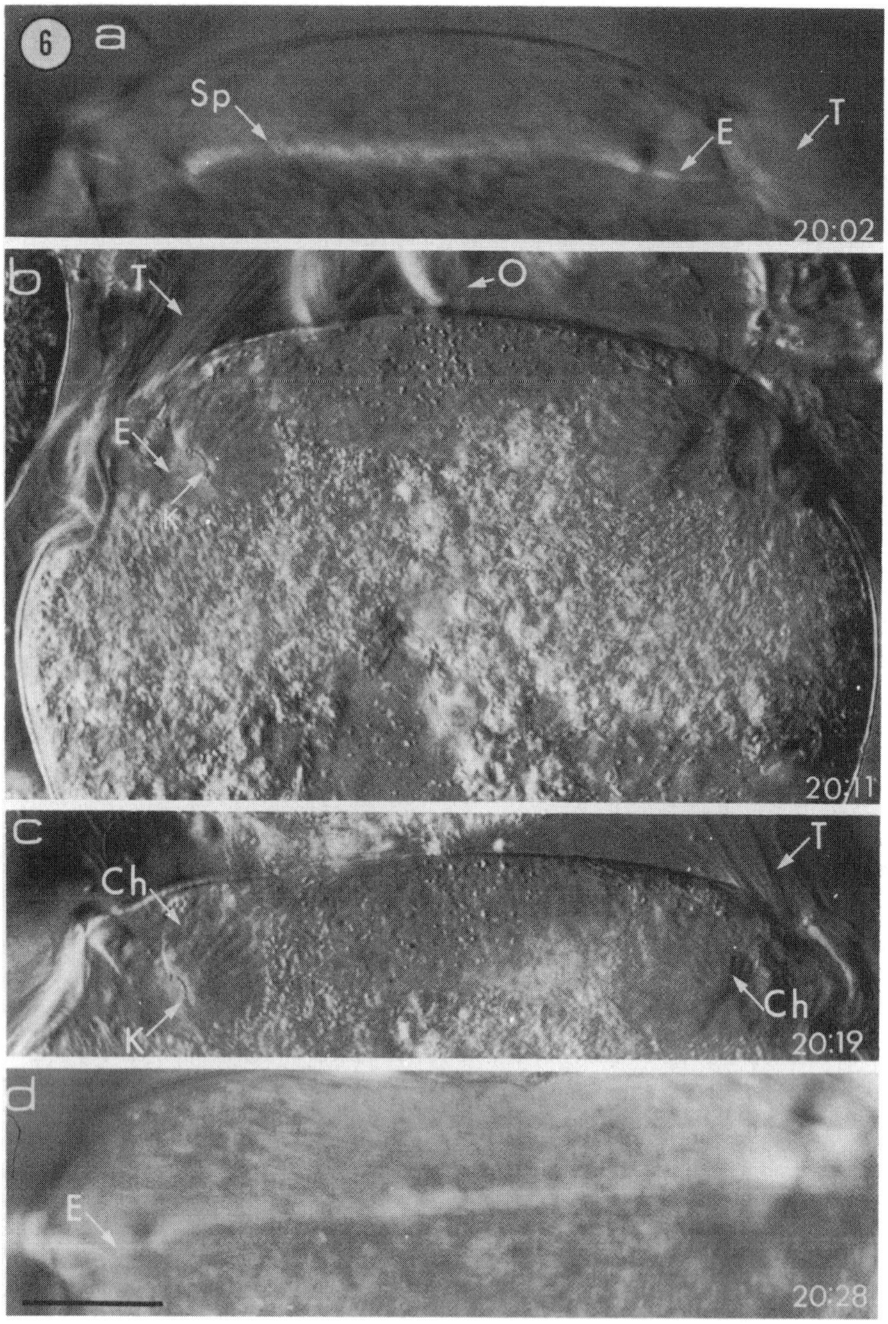

FIGURE 6 Anaphase-B continued from Figs. 4 and 5. (*a*), (*d*), In polarized light showing highly extended central spindle. Elongate-centrioles (*E*) are clear at spindle poles. (*b*), (*c*), Differential interference contrast. Note pearl string configuration of kinetochores (*K*) strikingly displayed by the left hand bouquet of chromosomes (*Ch*). The arc of kinetochores, focused on the elongate-centriole, abuts and outlines the centrosome. The distance between the flagellar tufts (*T*) is progressively increasing from (*a*) to (*d*). Time of day in h:min on 75h8. Bar, 30 μm. × 600.

21). Later, the spindle becomes wavy (Figs. 6*a*, 7*a*) or more or less straight (Fig. 6*d*) depending on transient variations of the interpolar distance. The distance varies in late anaphase-B as *Barbulanympha* swivels its two rostra independently and with increasing frequency.

At the commencement of anaphase-B, two flagellar tufts lie anterio-lateral to the spindle poles.

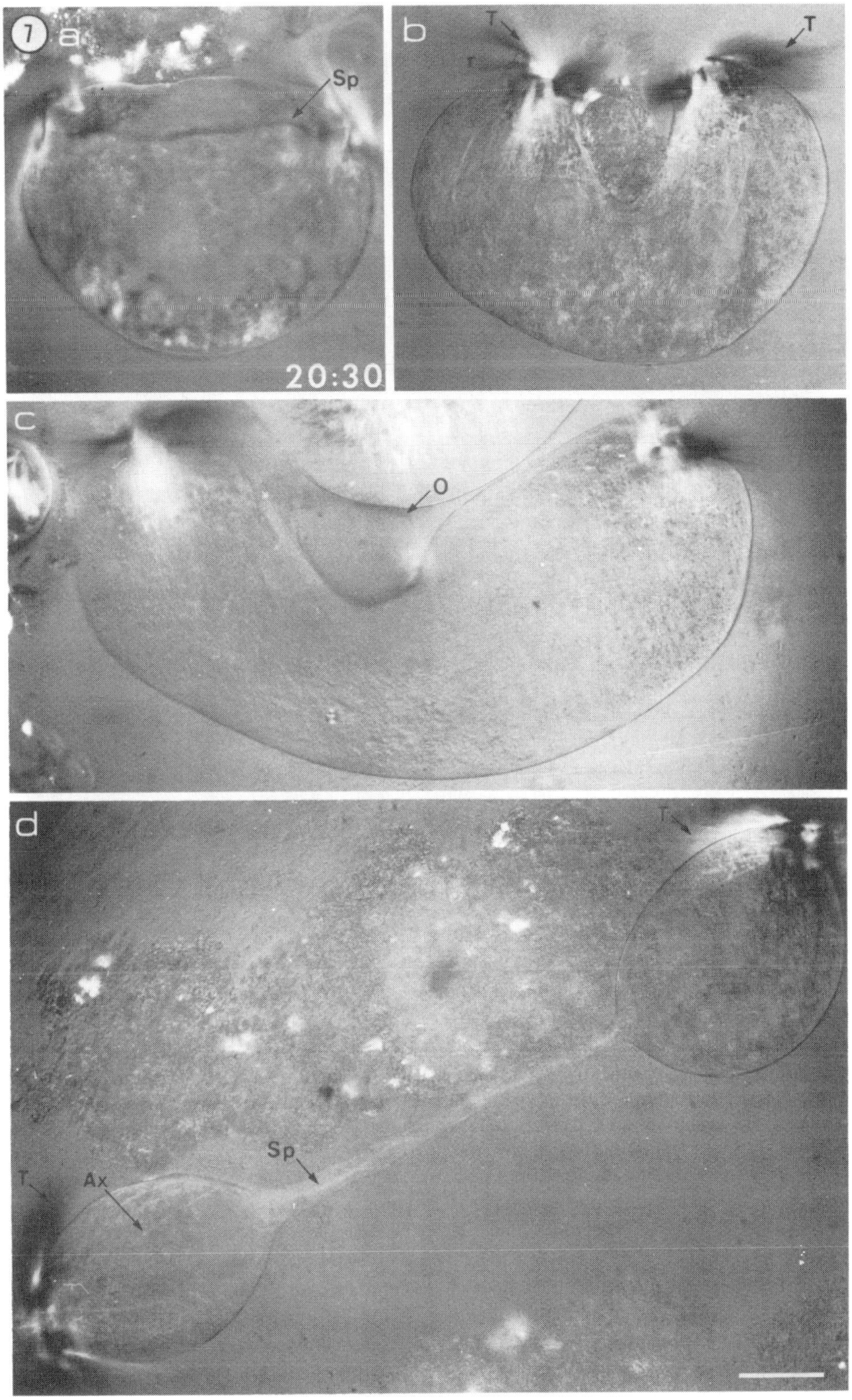

Figure 7 Polarized light views of late anaphase-B to cleavage. (*a*), Lower magnification of the same cell continued from Figs. 4 and 6. Spindle (*Sp*) remnant appears in dark compensation. During cytokinesis, ***Barbulanympha*** are often found pushing against debris as in (*a*), (*b*), or being assisted by another organism (*O*, *Trichonympha* as in (*c*), also Fig. 6*b*), which rams against the side of the ***Barbulanympha*** and seems to push in a cleavage furrow! (*d*), Terminal cleavage stage. A thin spindle remnant is still visible in bright compensation through the length of the long stalk connecting the two daughter cells. The daughter cells are swimming in opposite directions propelled by their flagellar tufts (*T*). In the daughter cells, parabasals and axostyles (*Ax*) have elongated and appear respectively as twisted ribbons and distinct, positively birefringent threads. Bar, 50 μm. × 250.

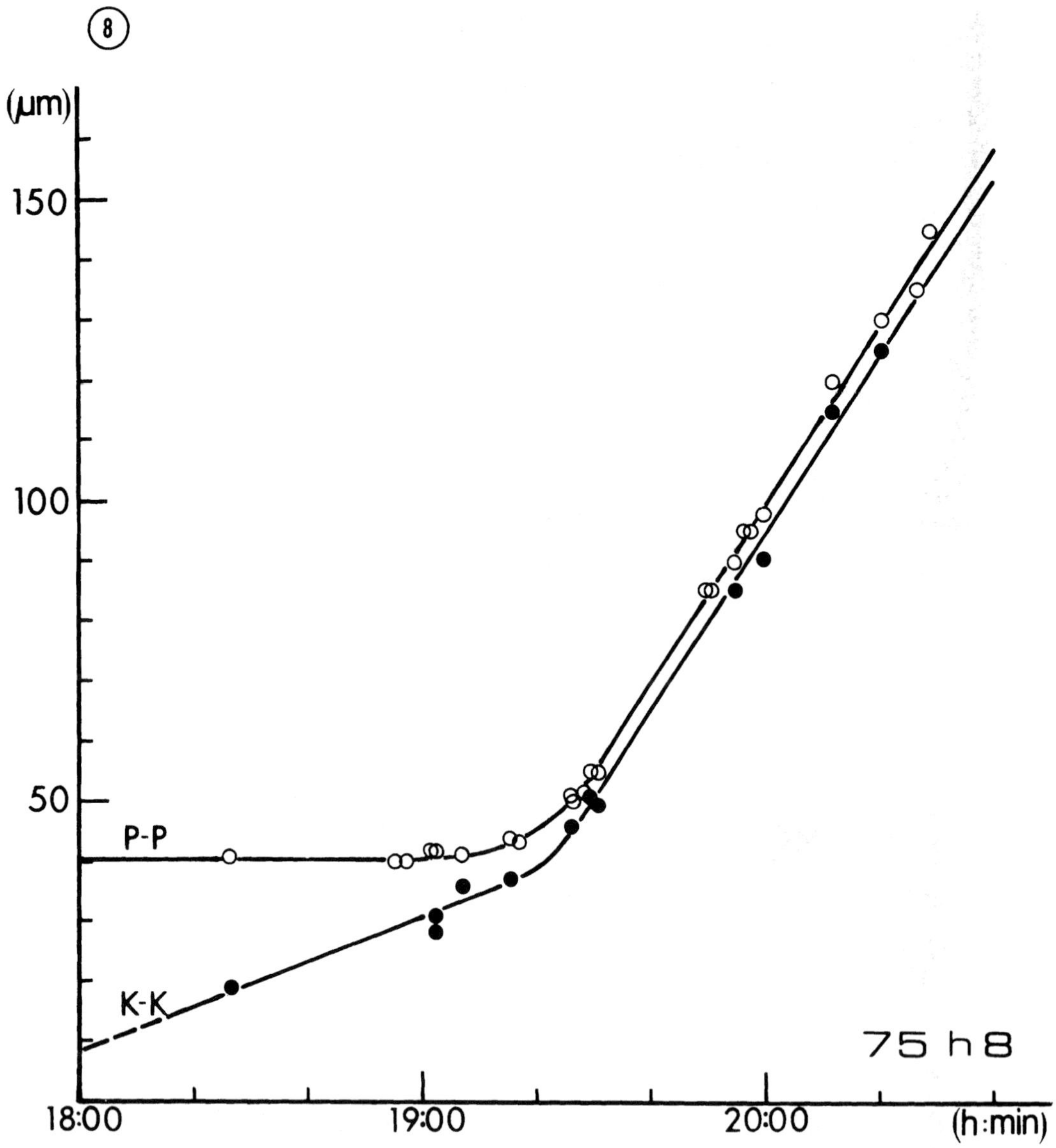

FIGURE 8 Spindle length and kinetochore separation of cell (75h8) shown in Figs. 4–7*a* and Fig. 8 in Part I. *P-P*, spindle pole-to-pole length, which remains stationary in anaphase-A and elongates severalfold in anaphase-B. *K-K*, kinetochore velocity in anaphase-A was 0.4 µm/min, for anaphase-B it was 1.5 µm/min at 18°C.

Each centrosome at the spindle poles is connected by its elongate-centriole to the parabasal axostylar lamella and to the basal structure of the flagellar tufts (Figs. 2*f*, 4*b*; also see Part I and reference 8). As the spindle extends in anaphase-B, the anterior surface of the cell between the flagellar tufts also extends. During spindle elongation the elongate-centrioles sometimes orient in line with the spindle axis as if being pulled into alignment by the diverging flagellar bundles. In other cases, the elongate-centrioles lie at an angle to the spindle axis, suggesting that spindle length and distance between the flagellar bundles each independently increases at nearly the same rate. During anaphase-B, the shape and orientation of the elongate-centrioles change with the temporarily imposed tension as the anterior rostra of the organism actively swivel about.

In the absence of nuclear envelope breakdown, there exists no typical telophase in *Barbulanympha*. The nuclear envelope does not have to be reformed and chromosomes, although contracting, do not decondense as in typical mitosis. Rather, mitosis would appear to terminate undramatically with the end of karyokinesis. In fact, however, the nuclear envelope has undergone a dramatic morphogenetic behavior. This behavior of the nuclear envelope which culminates in the orderly packaging of sister chromosomes will be next described together with changes in the three-dimensional shape of the central spindle.

Nuclear Morphogenesis – Introduction

The exact three-dimensional events of karyokinesis in *Barbulanympha* were clarified through a fortunate position of a cell displaying its spindle in exact anterior view (e.g., Fig. 9*c*). In this position, the cell proceeded from prophase through anaphases-A and -B. Over the entire period, we were able to obtain through-focus time-sequence photographs, alternately in differential interference contrast and rectified polarized light microscopy.[1] While photographically recording the morphological and dimensional changes, we also measured the birefringence retardation of the spindle by adjusting the compensator to visually extinguish the most retarding parts of the central spindle halfway between the two spindle poles. The retardation measurement allows us to determine the concentration of microtubules responsible for the spindle birefringence.

Before describing the anterior view, we shall highlight the shape changes of the nucleus seen in lateral view.

Changing Contour of the Nucleus in Lateral View

As the asters grow in prophase, the nearly spherical nucleus migrates anteriorly (In Part I Figs. 1, 2*a*, *b*, 3*a*, and *b*). Soon the nucleus abuts the posterior margin of the central spindle which has been developing between the centrosomes. The anterior margin of the nuclear envelope flattens and appears as a sharp straight line at the base of the spindle (Part I, Fig. 3*a*).

[1] For this sequence, the differential interference contrast objective lenses were especially carefully oriented to align their Wollastron prisms to the condenser Wollaston. A $\lambda/25$ compensator was introduced, and the polarizers and compensator were oriented so as to achieve sufficiently high extinction to simultaneously display the spindle birefringence and the differential interference contrast image.

With time, the nucleus moves farther forward, wraps around the spindle, and becomes horseshoe-shaped in cross section (also see Figs. 51, 53, 54, etc. in reference 8). At this stage, the nuclear envelope abuts the posterior spindle margin along its whole length. The exterior shape of the nucleus changes from nearly spherical to a hanging drop or inverted bell shape (Part I, Figs. 4, 7).

By the onset of anaphase-A, the nucleus completely engulfs the spindle except at the centrosomes. The nucleus therefore becomes a sleeve surrounding the spindle. The sleeve bulges out in the mid-region so that the anterior as well as posterior margin of the nucleus now appears bell-shaped (Figs. 1*a*, *b*, and 5*a*; also Fig. 8*c* in Part I).

At the end of anaphase-A, chromosomes form a ("telophasic") bouquet around each centrosome. The shortening chromosomal fibers align the kinetochores into a crown capping each centrosome. The poleward margins of the persistent nuclear envelope, in which the kinetochores are embedded, consequently balloon out (Fig. 5*a* to *d*).

The astral rays which run from the centrosomes tangent to the nuclear envelope are shoved together by the ballooning nucleus. The hollow cones formed around the spindle poles by these astral rays increase in angle from $< 90°$ to $180°$, as with an umbrella turning inside out. Concurrently, the mid-region of the anterior nuclear envelope turns from convex outward to convex inward and becomes saddle-shaped (Figs. 1*a* to *d*, 5*a* to *f*).

At the onset of anaphase-B, the inner posterior surface of the nuclear tube stands out in differential interference contrast as a straight line (Fig. 5*b*; also see Figs. 13, 12:42, and 12:52 in reference 21). Here, the nuclear envelope is apposed tightly to the central spindle; in polarized light the posterior margin of the spindle also displays a sharp contour (Fig. 4*a* to *d*; also Figs. 12, 12:35, and 12:51 in reference 21).

In early anaphase-B, the bouquet of chromosomes tilts anteriorly. Concurrently, the nuclear envelope expands more anteriorly than posteriorly (Fig. 5*e*, *f*; also Figs. 13, 12:52 in reference 21). The tilting of the bouquet is accompanied by a localized posteriad bending of the spindle axis (Fig. 4*c* to *h*; Figs. 12, 12:51, 13:04 in reference 21).

With progression of anaphase-B, the posterior surface of the nuclear envelope also invaginates and becomes saddle-shaped (Fig. 5*d* to *f*). Upon casual observation, the nucleus seems to be simply drawn apart into two. In fact, the nuclear envelope undergoes an intricate topological maneuver, as became evident from the anterior view of *Barbulanympha* mitosis which is described below.

Anterior View and Three-Dimensional Interpretation of Nuclear Morphogenesis

Fig. 9 displays prophase views of the *Barbulanympha* cell that gave the best anterior view of mitosis. An old flagellar tuft and rostrum appear at the left end of the central spindle (Fig. 9*c*), and a developing new flagellar tuft, still lying in the cell's interior, is seen at the upper right of the spindle (see lettered tracing, Fig. 10).

In Fig. 9*c*, a differential interference contrast optical section through the spindle axis, many chromosomes fill the bell-shaped area outlined by the nuclear envelope directly lateral to (above and below) the spindle. Both elongate-centrioles show at the spindle poles. To the left, an elongate-centriole is connected to the parabasal axostylar lamella beneath the row of basal granules at the base of an old flagellar tuft. Many axostyles and parabasals surrounding the nucleus trail to the right. All photographs shown in Figs. 9 through 16 are of this same highly flattened individual whose division in anterior view we recorded for 8 h.

With the microscope focused to give a high optical section (i.e. focused anterior to the spindle), this prophase cell displayed a pair of lines running parallel to the spindle axis (Fig. 9*a*). The lines run about one-third the spindle length in the spindle mid-region. At both ends, the parallel lines open into Y's which soon terminate with apices each marked by a kinetochore (see diagram Fig. 10*a*). Through-focus serial micrographs reveal that the parallel lines are apposed lips of the nuclear envelope which have met anteriad to the central spindle after the nucleus had wrapped around the spindle. We have named this structure the nuclear "seam." The seam is shown in some of Cleveland's drawings (e.g. Figs. 6 and 7 in reference 9), but no mention of such a structure is found in his writings.

The nuclear envelope can be traced from the seam to its lateral aspects (toward the top and bottom of each of these photographs) by following optical sections taken at successive levels (e.g., Fig. 12). There, the exterior contour of the nucleus is more or less bell-shaped, but the smooth contour is interrupted in places by protrusions and recesses. The protrusions contain chromosomes, with kinetochores located at the poleward apex of each pointed protrusion (Figs. 9*a* to *c*, 10*a* to *c*; also see Fig. 35 in reference 10). Also, at the pits of the recesses, kinetochores are found embedded in the nuclear envelope (Figs. 9*a* to *c*, 10*a* to *c*; also see Figs. 50 and 73 in reference 8). In living cells, we see these protrusions and recesses slowly appearing and then disappearing over a course of several minutes. These changes in the nuclear envelope occur in association with the churning of chromosomes mentioned in Part I.

With time, the kinetochores gradually move poleward and the nuclear seam elongates (Figs. 9*d*, 10*d*). In this manner, the nucleus becomes a more complete sleeve surrounding the central spindle. Our three-dimensional interpretation of the mitotic apparatus at this stage (less the chromosomes) is shown in Fig. 11, top.

In the individual *Barbulanympha* recorded here, the chromosomes continued to churn and the cell would not proceed to anaphase-A for several hours. So at 4 h and 45 min after initial observation, we cold-shocked the cell hoping to nudge it into anaphase. The first chilling cycle started at 22:13.5 h and lasted 5 min in a 12.5°C refrigerator. No change of spindle birefringence was observed, and the chromosomes continued to churn. The slide was then placed in a 7°C refrigerator at 22:44.4 h for 4 min. Immediately after the slide was removed from the refrigerator, and the condensed moisture was blown off from the slide surface with a jet of dry Freon 12, no difference in central spindle birefringence was observed (cf. Figs. 9*e* and 14*c*). However, this time the chromosomes appeared somewhat thinner, the kinetochores started to recede from the poles, and the nuclear envelope became recessed in several places where kinetochores were inserted (22:56 h to 23:02 h). Some chromosome arms started to separate at 23:00 h and several kinetochores became more or less aligned at 23:05 h. The cell then entered anaphase (see graph, Fig. 17).

Progression of anaphase-A is shown in Figs. 12 and 13*a* to *f*. Anaphase-B is shown in Figs. 13*g*, *h*, 15*c*, *e*, *g*, and 16*a*. The advancing locations of chromosomes and kinetochores are summarized in the graph, Fig. 17.

In general, the chromosomes and kinetochores moved as described for the lateral views. However, presumably due to extreme cell flattening, spindle elongation was limited and the chromosomes did not move rapidly in the latter phases of anaphase-B, especially after 00:19 h.

At the onset of anaphase-A, the nuclear seam spans about three-quarters of the anterior spindle margin between the two centrosomes (Fig. 12*a*, *b*). Because the kinetochores have not yet quite reached the centrosomes, a short gap remains between the centrosomal surface and the apex of the nuclear envelope (follow the rows of kinetochores in the successive optical sections in Fig. 12). As anaphase-A proceeds and kinetochores reach the centrosomes, the seam spans the full length of the spindle (Fig. 13*a*). But just after that, the seam opens up in the middle, as though with parting of lips (Fig. 13*c*). The gap thus formed rapidly opens laterally and towards the spindle poles (Fig. 13*e*, *g*). This gap which appeared in the seam, we call the nuclear "cleft."

In the microscope image, only those regions of the nuclear envelope that lie along the direction of sight show up in optical section. By mentally reconstructing the optical sections, we thus arrive at the three-dimensional shape of the *Barbulanympha* nucleus at an early nuclear cleft stage (Fig. 11, middle).

Proceeding into anaphase-B, the cleft, which outlines the anterior margin of the nuclear envelope, progresses latero-posteriad exposing the lateral surface of the central spindle (Figs. 13*h*, 15*c*, *e*, and *g*). As the cleft passes farther posteriad, the tapered connections, earlier seen in median optical section between the left and right lobes of the nucleus, disappear (cf. Fig. 15*c* with *e*). At this focus, the nucleus takes on the shape of two pairs of insect wings (Fig. 15*g*). The three-dimensional structure of the mitotic apparatus and nuclear

FIGURES 9 and 10 Anterior view of prophase and nuclear seam formation in *Barbulanympha*. (*a*), Midprophase. The nucleus has already moved anteriad (up towards viewer in these photos) and wrapped around the central spindle. (cf. schematic drawing in Fig. 11). In this high focus view, a pair of lines appears parallel to the spindle axis in the mid-region of the nuclear envelope (between arrows in Fig. 10*a*). This is an optical section of the "nuclear seam", which is found anteriad to the spindle where the nucleus has completely wrapped around the spindle and come into contact with itself. Laterally, the nuclear envelope appears shield-shaped. Within the nucleus many chromosomes are visible. In the tracing Fig. 10, those parts of the nuclear envelope visible in Fig. 9 are indicated by broken lines, and kinetochores are indicated by solid dots. (*b*), (*c*), Median focus optical section showing birefringence of central spindle outlined by differential interference contrast image of the nucleus. The lateral aspects of the nuclear envelope appear bell-shaped. Chromosomes are moving apart, but their arms have mostly not yet separated. Their kinetochores remain embedded in the persistent nuclear envelope. Some of the kinetochores appear at protrusions (*Kp*) and in recesses (*Kr*) of the nuclear envelope. (Note: some of the kinetochores can be seen more easily by turning the page by 90°.) In (*c*), one of the two old flagellar tufts appears prominently on the left side of the picture (*F-F*). At its base, the left hand elongate-centriole (*E1*) joins the parabasal axostylar lamella (*L*) and from there runs to the left pole of the central spindle. The centrosome here is surrounded by a large, granular-inclusion-free spherical zone, roughly twice the diameter of the centrosome (cf. Fig. 12*h*). A new flagellar tuft (*T*), still not yet externalized but already beating within the cytoplasm, appears above the right hand elongate-centriole (*E2*). The old flagellar tuft on this side can only be seen at a much higher focus. (*d*), Higher focus view 42 min after (*a*). As the kinetochores are pulled closer to the centrosomes, the seam is extended (between arrows in Fig. 10*d*; compare with 10*a*). (*e*), Median focus in polarization optics. Elongate-centriole (*E*) and margin of the central spindle are compensated (by subtractive action of the compensator) or are extinguished (by being oriented parallel to the polarizer or analyzer axis) and appear dark. The central spindle was completely formed before the stage showed in (*a*), and from that time on has maintained this shape, size, and birefringence. Asters radiating northerly and southerly from the centrosomes (*S*) appear in bright (additive) compensation. *L*, polarizer-analyzer axes and quadrant containing slow axis of compensator. Fig. 9*a* to *d* and other similar figures in Part II were obtained by carefully aligning a Leitz Smith T system differential interference contrast 40× objective and condenser on a high extinction polarizing microscope. We were thus able to display the birefringence of the central spindle simultaneously with the differential interference contrast image. The exceptionally shallow depth of focus of the differential interference contrast image yields striking optical sections of the nuclear envelope, chromosomes, etc, and the polarized light image brings out the spindle fibers. Time in h:min of day on 74g30. Bar, 20 μm. × 720.

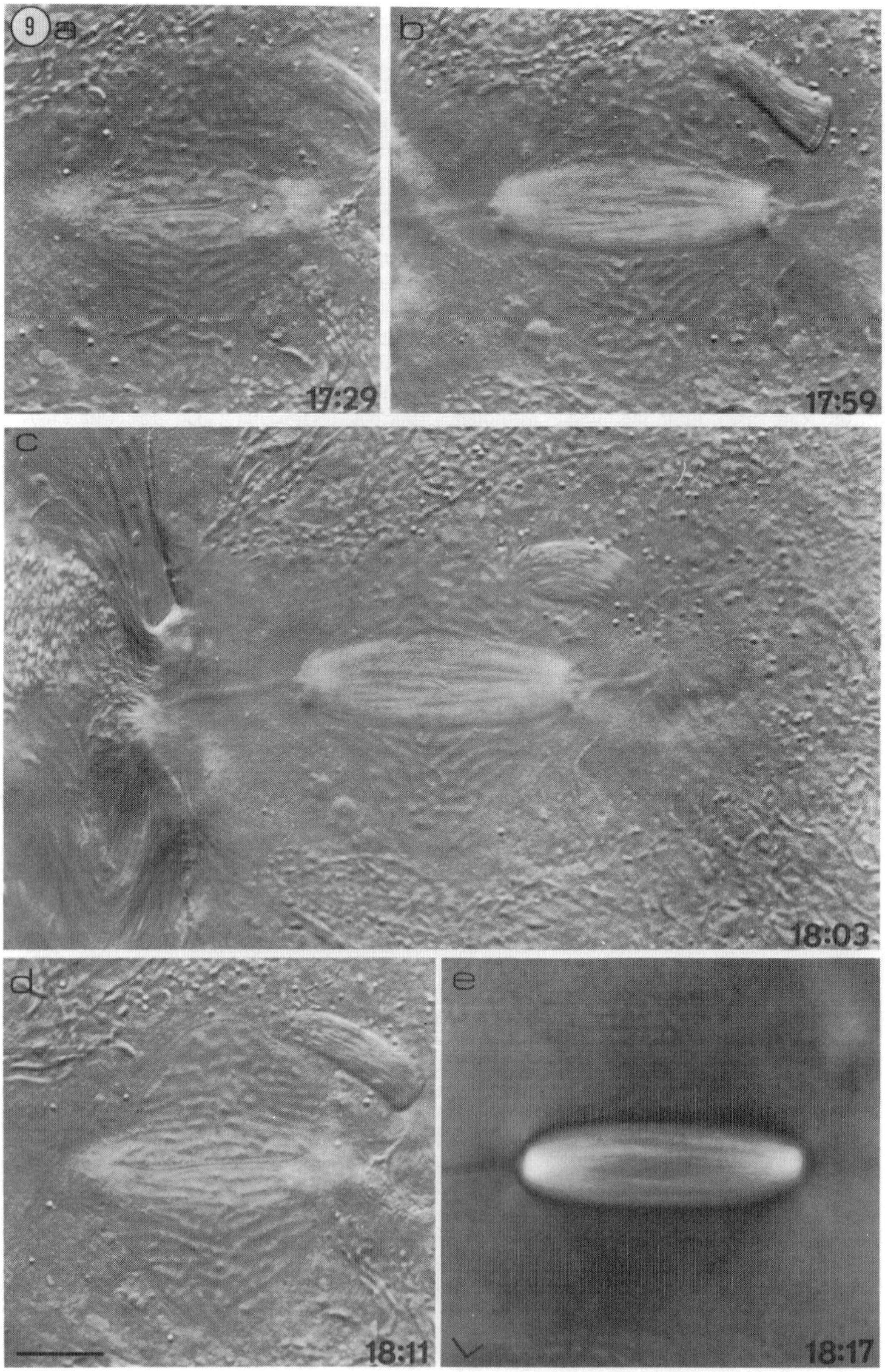
9
a
17:29
b
17:59
c
18:03
d
18:11
e
18:17

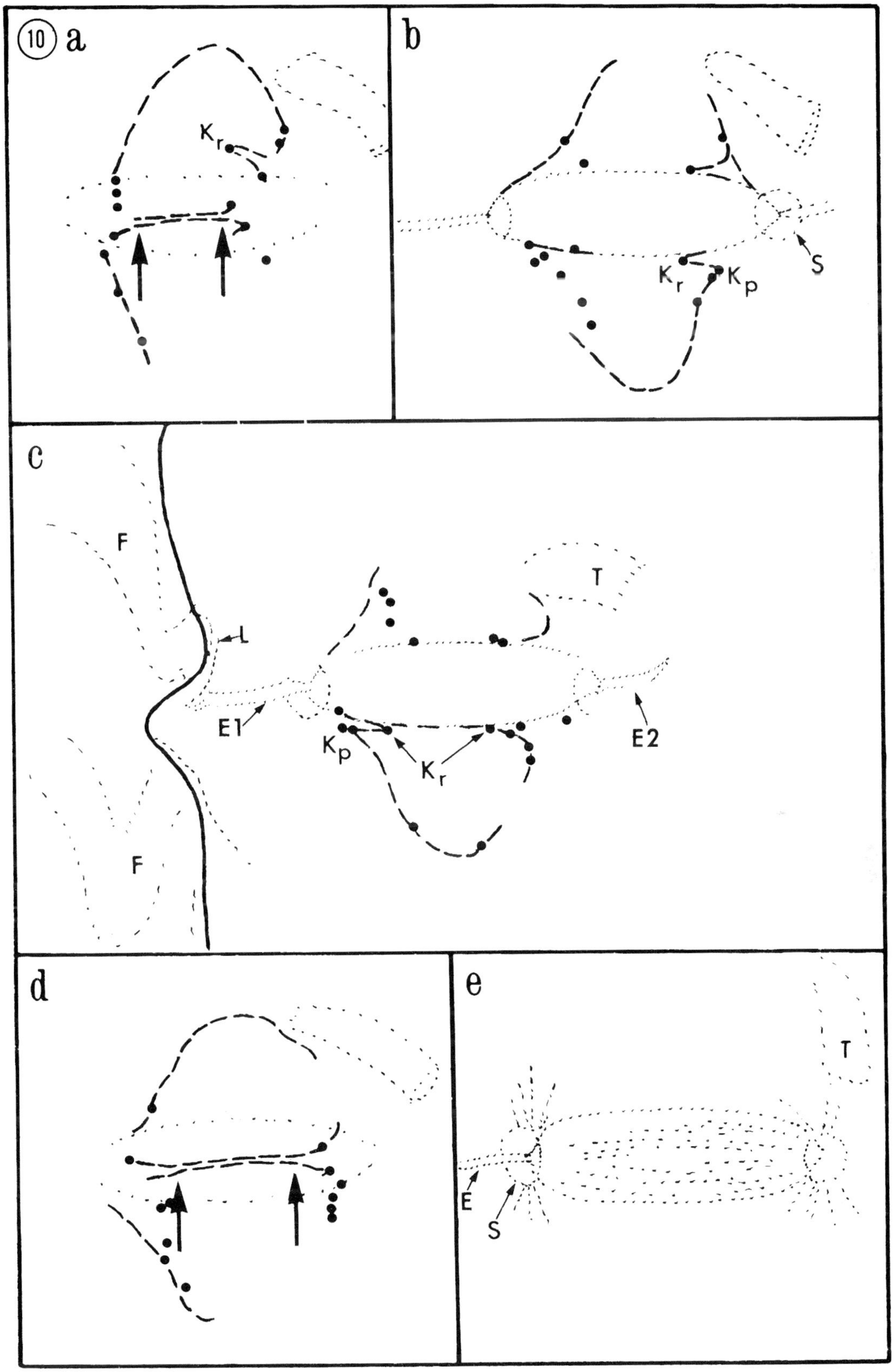
10
a
Kr
b
Kr
Kp
S
c
F
L
T
E1
Kp
Kr
E2
F
d
e
T
E
S

11

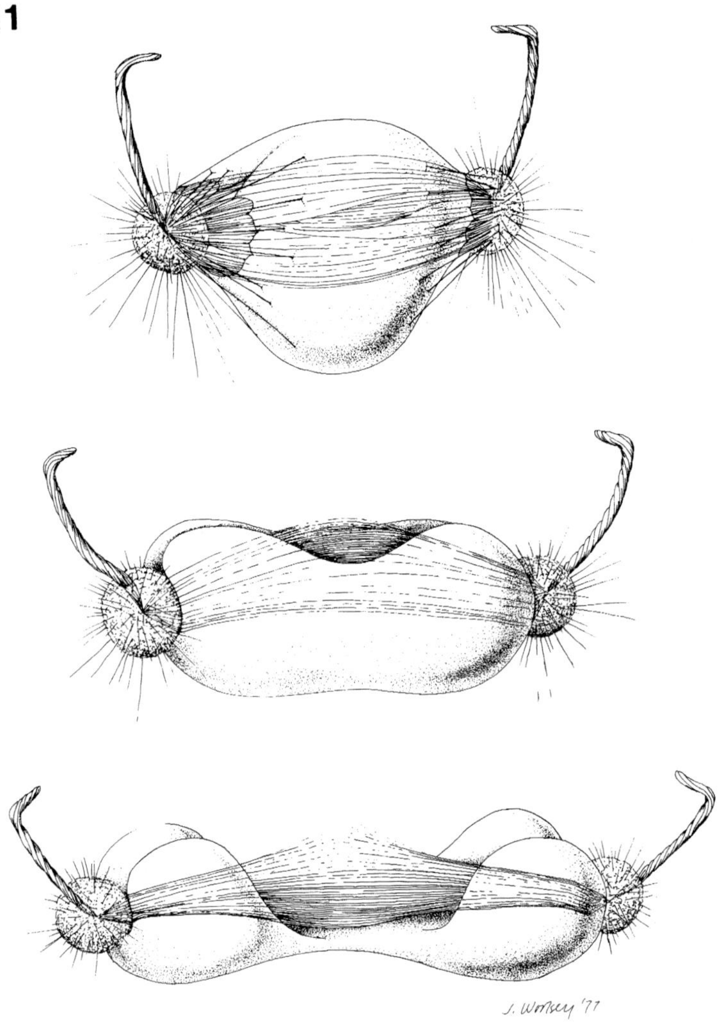

FIGURE 11 Schematic diagram of seam and cleft formation in *Barbulanympha*. *Top*, prophase to early anaphase-A. Those astral rays which have become attached to the kinetochores, embedded in the persistent nuclear envelope, become the chromosomal spindle fibers (chromosomes not shown). These fibers have repetitively shortened, hoisted up and deformed the nucleus, and wrapped it around the central spindle. The nuclear seam is formed where the nucleus contacts itself anterior to (above in these pictures) the central spindle. (For optical sections of the seam, see Figs. 9*a, d,* 12*a, b, c,* and 13*a*.) *Middle*, early to mid-anaphase-B. Chromosomal spindle fibers have shortened during anaphase-A, and the kinetochores as well as the poleward surface of the persistent nuclear envelope now crown the centrosomes. The nuclear seam has extended all the way to the centrosomes and has started to part in the middle. This opening, in the shape of parting lips, has been called the "nuclear cleft." (For optical section, see Fig. 13*c*, *e*, and *g*.) *Bottom*, mid to late anaphase-B. As the central spindle elongates, the kinetochores and nuclear envelope continue to adhere closely to the centrosomes. The nuclear cleft progresses towards the spindle poles and then caudad (towards bottom of picture) around the spindle. A horizontal slice through the middle of the spindle at this stage makes the nuclear envelope appear bilobed at both spindle poles (e.g. Figs. 13*h*, 15*e*) but continuous at a lower level (Fig. 15*c*). The spindle fibers splay where they are no longer constrained by the nuclear envelope, first anteriorly and then laterally (see Figs. 14 and 15). Artwork by John Woolsey.

envelope at this stage is interpreted in Fig. 11, bottom.

Eventually, the cleft progresses posteriorly around the spindle until the nucleus is separated into two parts. Karyokinesis is thus accomplished by an intricate topological maneuver of the persistent nuclear envelope.

At the completion of karyokinesis, each daughter nucleus in median optical section appears to be made up of two parts connected with a strand of kinetochores (Fig. 16*a*). However, from the foregoing description of karyokinesis, it can be inferred that each daughter nucleus is in fact horseshoe-shaped in vertical section. As one focuses the microscope, the narrow space seen in optical section between the "fly wings" opens anteriorly and the nucleus merges into "one piece" posteriorly. This groove in the daughter nucleus contains a ribbon of birefringent fibers attached at their base to the centrosomes (Fig. 16*b*). These are remnants of the central spindle fibers as will be described next.

Shape Change of the Central Spindle in Anaphase—Anterior View in Polarized Light

In anterior view as in lateral view, the width, length, and shape of the central spindle remain unchanged from the time the spindle is completed in prophase until the end of anaphase-A (Fig. 17; see Fig. 3 for graph of lateral view). For the individual *Barbulanympha* recorded here, this spanned a period of over 6.5 h.

The bilaterally symmetric, anaphase-A central spindle is composed of a bundle of fibers that emanate radially from the centrosomes and run parallel to the spindle axis towards the equator (Fig. 14*a* to *d*). The packing density and birefringence of spindle fibers are greatest at the centrosomal surface, and decrease away from the centrosome as the spindle width increases. Toward the lateral margins of the spindle, where the light rays traverse decreasing thicknesses of the spindle, the retardation is proportionately reduced. In Fig. 14*a* and *c*, the compensator slow axis is oriented perpendicular to the spindle long axis so that these less retarding spindle regions are compensated and appear dark (for an illustration of how spindle fiber contrast changes with compensator orientation, and for the numerical relation of spindle birefringence to microtubule packing density, see reference 36). The mid-region of the spindle also shows less birefringence and gives the appearance of possessing a longitudinal gap. The spindle seems to be made up of left and right bundles of spindle fibers with less fibers running through the middle, possibly because of an interfering nuclear sleeve (Fig. 13*b, d, f, h*? In Part I, Fig. 3*a* to *d*. See Fig. 37 and text Fig. 16 in reference 8 for illustration of the nuclear sleeve, a progressively narrowing funnel running anteriorly from the nucleus).

The microtubular origin of the spindle fiber birefringence was described in Part I. The microtubular bundles which make up the positively birefringent fibers of the central spindle are concentrated at the centrosomal surface where they are disposed radially. When the spindle axis is oriented parallel to one of the polarizer axes, the diagonally oriented microtubules contribute to the high contrast near the spindle poles (Fig. 14*b, d*). When the spindle axis is oriented at 45° to the polarizer axes, those same microtubules which now lie parallel to the polarizer axes, are therefore extinguished, and appear dark (Fig. 14*a, c, f*).

Those asters which appear bright in Fig. 14*a* are made up of rays lying in a quadrant at right angles to the spindle axis. Here, their slow axis lies in the same quadrant as the compensator slow axis in additive compensation, hence the asters appear as white fans with greatest brightness adjoining the surface of the centrosome where their microtubules are most concentrated. When the microscope stage is turned 45°, the bundle of astral rays and chromosomal fibers running tangent to the nuclear envelope show in striking bright or dark contrast (Fig. 14*b, d*).

In early to mid-anaphase-A, these astral bundles and chromosomal fibers form a 90° hollow cone surrounding the nucleus at the spindle poles (Fig. 14*b*). Toward the end of anaphase-A, the nuclear envelope becomes ballooned poleward by the chromosomes as they congregate (Fig. 13*b* to *f*) and the astral rays are accordingly displaced (Fig. 14*b* to *h*).

The sleeve-shaped nucleus ensheathes the lateral margins of the central spindle up to early anaphase-B. The spindle in anterior view retains a sharp, boat-shaped contour (Fig. 14*a* to *d*). At this stage, the anterior spindle margin has already started to bulge and become fuzzy as seen in lateral view (Fig. 4*c, d*).

As the nuclear cleft moves posteriorly and exposes the lateral aspect of the spindle (see models in Fig. 11), the birefringent fibers start to splay laterally (Fig. 14*e* to *h*). As seen in Figs. 14*c* to *h* and 15*a, b, d, f,* and *h*, the fibers act as though they were no longer bundled together in the middle of the spindle. The fibers appear to first straighten out (Figs. 14*e* to *h*, 15*a, b*) and then curve outwards as the contour length of each fiber increases (Fig. 15*d, f*).

As the fibers splay, the retardation of the spin-

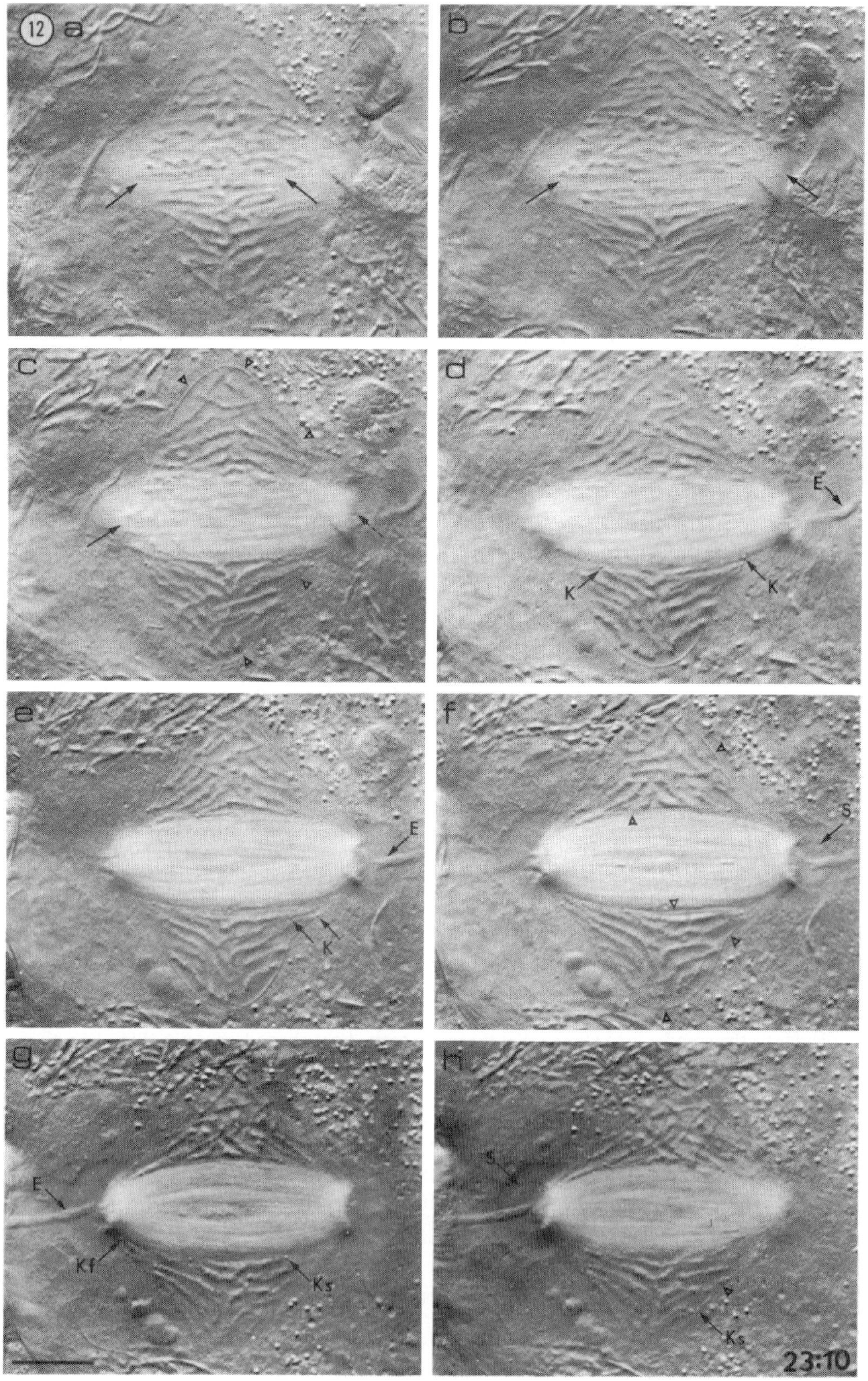

12 a
b
c
d
E
K
K
e
E
K
f
S
g
E
Kf
Ks
h
S
Ks
23:10

dle mid-region decays precipitously (Figs. 14*g*, *h*, 15*b*, *d*, *f*, and *h*). The measured retardation and width of the birefringent core at the mid-region of the spindle decrease proportionately as plotted in Fig. 17.

Eventually the retardation in anterior view becomes very low at the mid-region of the spindle. Only near the poles, where the spindle remnant is sandwiched in the groove of the daughter nucleus (Figs. 15*g*, 16*a*), is appreciable birefringence observed (Figs. 15*h*, 16*b*). As the daughter nuclei are rocked by the undulating flagella, the contrast of the spindle remnants fluctuates. The retardation is greatest when the spindle remnant appears thinnest, undoubtedly because the light ray then has to traverse the maximum distance through the ribbon-shaped spindle remnant.

Within the centrosome, between the centrosomal surface where the birefringent ribbon is anchored and the tip of the elongate-centriole, a set of weakly positively birefringent radial strands persist (Figs. 15*h*, 16*b*). As described for the lateral view, these strands appear to include the intracentrosomal portions of the chromosomal spindle fibers which continue to link the kinetochores to the elongate-centriole tips.

Cytokinesis

In late anaphase-B, the *Barbulanympha* eventually transforms into an organism with two heads (the rostra and flagellar tufts) pointing in opposite directions (Figs. 6, 7). Flagellar activity has increased to a point where, in spite of the compression between slide and cover glass, the two-rostrate organism swims about. It then often comes to rest in a position with the former anterior end pushing against a mass of debris (Figs. 6*c*, 7*a*, *b*, and *d*). Conversely, smaller organisms such as *Trichonympha* may come and poke against the same region (Figs. 6*b*, 7*c*). With assistance from these organisms or after considerable extension of a cytoplasmic bridge, the cell eventually pulls apart into two (Fig. 7*d*). It should be noted that *Barbulanympha* in their natural habitat in the *Cryptocercus* hindgut live in a thick soup of organisms. Only in our culture slides are they so sparsely distributed.

DISCUSSION

In Part II of this report, we have described with respect to spindle function: (*a*) chromosome movement occurring in two sequential steps, anaphase-A (poleward migration of kinetochores) and anaphase-B (kinetochore separation by spindle pole-to-pole elongation); (*b*) the contiguity throughout anaphases-A and -B of the positively birefringent chromosomal spindle fibers with the centriole and individual kinetochores (which exhibit a radially positive spherulitic birefringence); (*c*) protrusion and recessing of the nuclear envelope, with kinetochores positioned in the apex or pit of the deformation; (*d*) the repetitive poleward migration of the chromosomes before and during anaphase-A, led by their kinetochores embedded in the persistent nuclear envelope; (*e*) at the end of anaphase-A, the eventual arrival and alignment of all the kinetochores very close to or up against the centrosomal surface, forcing the nuclear envelope and kinetochores into a crown (string of pearls in optical section) and the chromosomes into a ("telophasic") bouquet configuration; (*f*) the constant length, width, and birefringence of the central spindle throughout anaphase-A; (*g*) in anaphase-B, an elongation of up to fivefold of the central spindle leading far apart the daughter chromosomes already partitioned in anaphase-A;

FIGURE 12 Through-focal series anterior view at onset of anaphase-A. Same cell as in Fig. 9, shown here at 23:10 h after the cell was exposed to cold at 22:44 h (see text). Combined birefringence and differential interference contrast image. In successive optical sections, the elongate-centrioles [*E*, (*c*)-(*f*), right pole; (*f*)-(*h*), left pole], centrosomes [*S*, (*f*), right pole; (*h*), left pole], the nuclear seam [arrows, (*a*)-(*c*)] and other features of the nuclear envelope (triangles) can be traced. Chromosomes and their kinetochores (*K*) show at varying distances from the spindle poles. (Turn left side of page towards you to better visualize kinetochores). Note that the chromosomes have rejoined their arms and have moved farther away from the spindle poles than before the cold treatment (cf. Fig. 9*d*). Similar behavior of chromosomes was observed in anaphase pollen mother cells of Easter lilies when their chromosomal spindle fibers lost their birefringence upon exposure to cold (Fig. 52 in reference 18). The regions of the chromosomal spindle fibers within the centrosomes show exceptionally clearly in (*f*) (right pole) and (*g*) and (*h*) (left pole). 74h8. Bar, 20 μm. × 620.

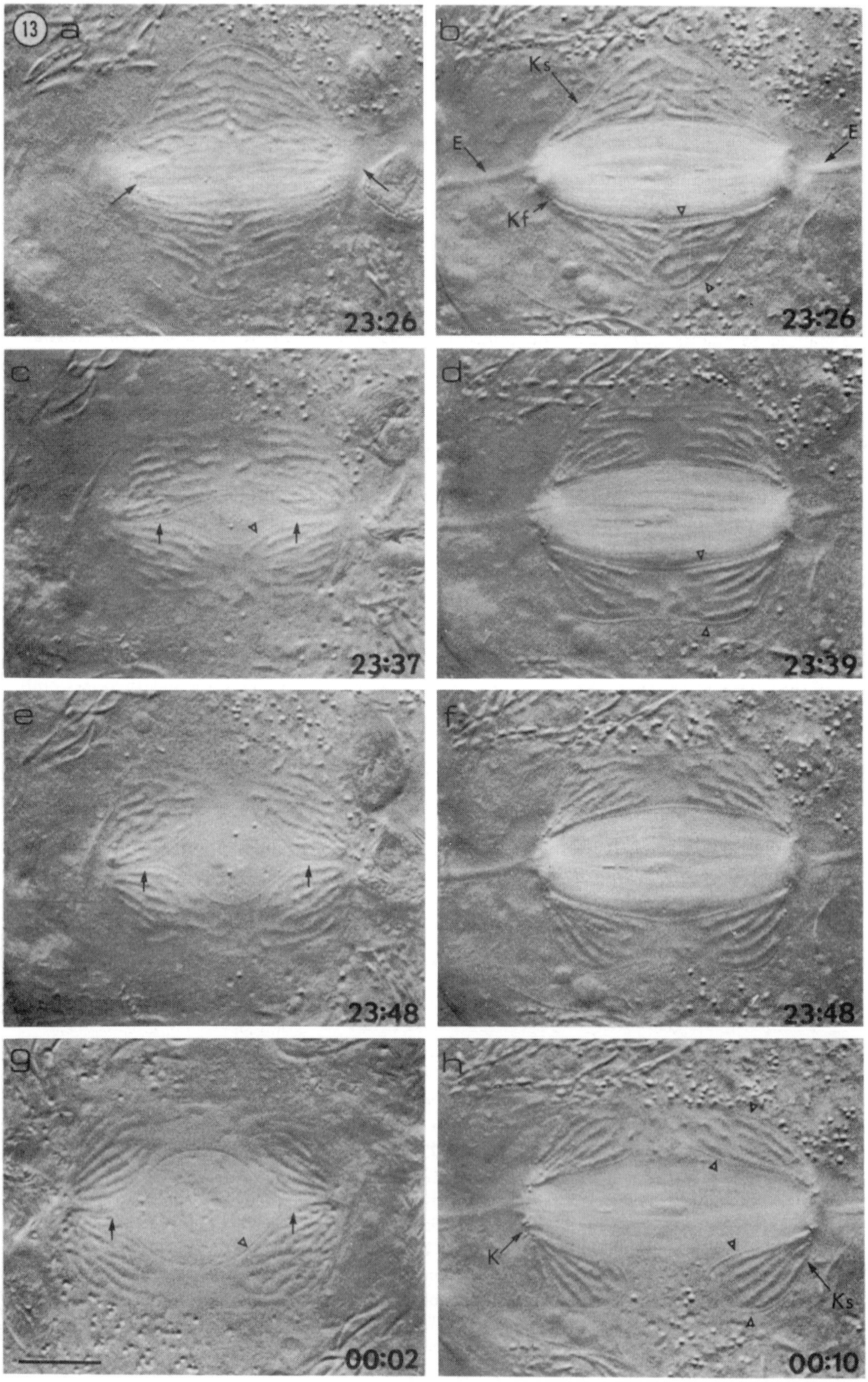
13
a
23:26
b
Ks
E
E
Kf
23:26
c
23:37
d
23:39
e
23:48
f
23:48
g
00:02
h
K
Ks
00:10

(*h*) persistence of the kinetochore grouping and the bouquet configuration of chromosomes to late anaphase-B; and (*i*) the splaying of the central spindle fibers initially anteriorly then progressively laterally which results in the proportionate decrease in width and retardation of the central spindle remnant.

In Part II of this report, we have described with respect to nuclear morphogenesis and cytokinesis: (*a*) the formation and poleward extension (led by apical kinetochores) of the nuclear seam, the juxtaposing surface of the nuclear envelope anterior to the central spindle; (*b*) appearance in late anaphase-A of the nuclear cleft, a parting gap in the middle of the nuclear seam; (*c*) the laterad and posteriad expansion of the nuclear cleft in anaphase-B, culminating in the total partitioning of daughter chromosomes and completion of karyokinesis; and (*d*) the swimming apart of the two daughter cells, whose cytokinesis is often assisted by a clump of debris or by smaller organisms pushing against the prospective furrow region.

The Generality of Anaphase-A and -B

In *Barbulanympha,* we have described a clear-cut separation of anaphases-A and -B. We have demonstrated unambiguously that chromosomes in anaphase-A proceed to the centrosomal surface in the complete absence of central spindle elongation. Shortening of the chromosomal spindle fibers alone is associated with chromosome movement. In anaphase-B, there is no further shortening of the distance between the kinetochores and the tips of the elongate-centrioles. Anaphase-B chromosome separation proceeds hand-in-hand with central spindle elongation. Some cells are blocked at the end of anaphase-A and do not progress into anaphase-B but otherwise appear to be perfectly healthy. Other cells pass directly from anaphase-A into anaphase-B. Therefore, in *Barbulanympha,* anaphase chromosomes are moved apart by two spatially and temporally divisible events, chromosomal spindle fiber shortening and central spindle elongation.

In cells other than *Barbulanympha,* the involvement of these two events, either singly or in overlapping sequence, has long been recognized (e.g., in *Amoeba proteus,* 26; aphids, 34; newt fibroblast, 5; fungus, 2. For reviews, see 29, 35, 37). In polar body division of oocytes, chromosomes commonly separate by anaphase-A alone with no detectable anaphase-B. Moreover, Ris has shown, on grasshopper spermatocytes, that chloral hydrate can block spindle elongation without affecting chromosomal fiber contraction (35). Oppenheim, Hauschka, and McIntosh report a similar observation in tissue-cultured cells treated with colchicine (32). In such material, the two processes of anaphase-B and -A normally take place in overlapping sequence (32, 35). With the addition of the *Barbulanympha* example, we should now seriously consider the likelihood that anaphase-A and -B are driven by two distinct physiological or molecular mechanisms.

FIGURE 13 Nuclear cleft formation and anaphase-A anterior view. Same cell continued from Figs. 9 and 12. Smith T system differential interference contrast 40× objective adjusted to simultaneously display birefringence of central spindle fibers. (*a*), High focus. The nuclear seam now runs much of the distance between the two spindle poles (hold the page sidewise and sight between the tips of the arrowheads). Anaphase-A is progressing and chromosome ends are just separating. Several kinetochores have reached approximately one-half spindle diameter away from the poles. (*b*), Median focus of same stage. Triangles, optical section of nuclear envelope. Both elongate-centrioles are visible (*E*). "Fast" kinetochores (*Kf*) are not quite at centrosomal surface, "slow" kinetochores (*Ks*) lag behind. (*c*), High focus. In the middle of the seam, an opening cleft is seen as though with parting of lips [short arrows in (*c*), (*e*), and (*g*)]. Chromosomal arms have separated. (*d*), Median focus mid-anaphase-A. Many kinetochores have approached the centrosomal surface. Optical section of the nuclear envelope is clear adjacent to spindle surface (upper triangle) and on the outer face of the nucleus, where its shape is now concave (lower triangle). (*e*), High focus. The cleft is considerably wider than in (*c*), but the seam persists near the spindle poles. (*f*), Median focus of late anaphase-A. Chromosome arms are well separated. (*g*), High median focus. The cleft has advanced further. At a slightly lower focus, the inner aspects of the nuclear envelope appear as two arcs which do not reach the poles. (*h*), The cleft has now advanced past the median margin of the spindle. In the lower part of the picture, continuous lines resembling insect wings depict median optical sections of the nuclear envelope (triangles). Many kinetochores (*K*) have clustered at the surface of the centrosome but some "slow" kinetochores (*Ks*) are still lagging. h:min in time of day on 74g30. Bar, 20 μm. × 620.

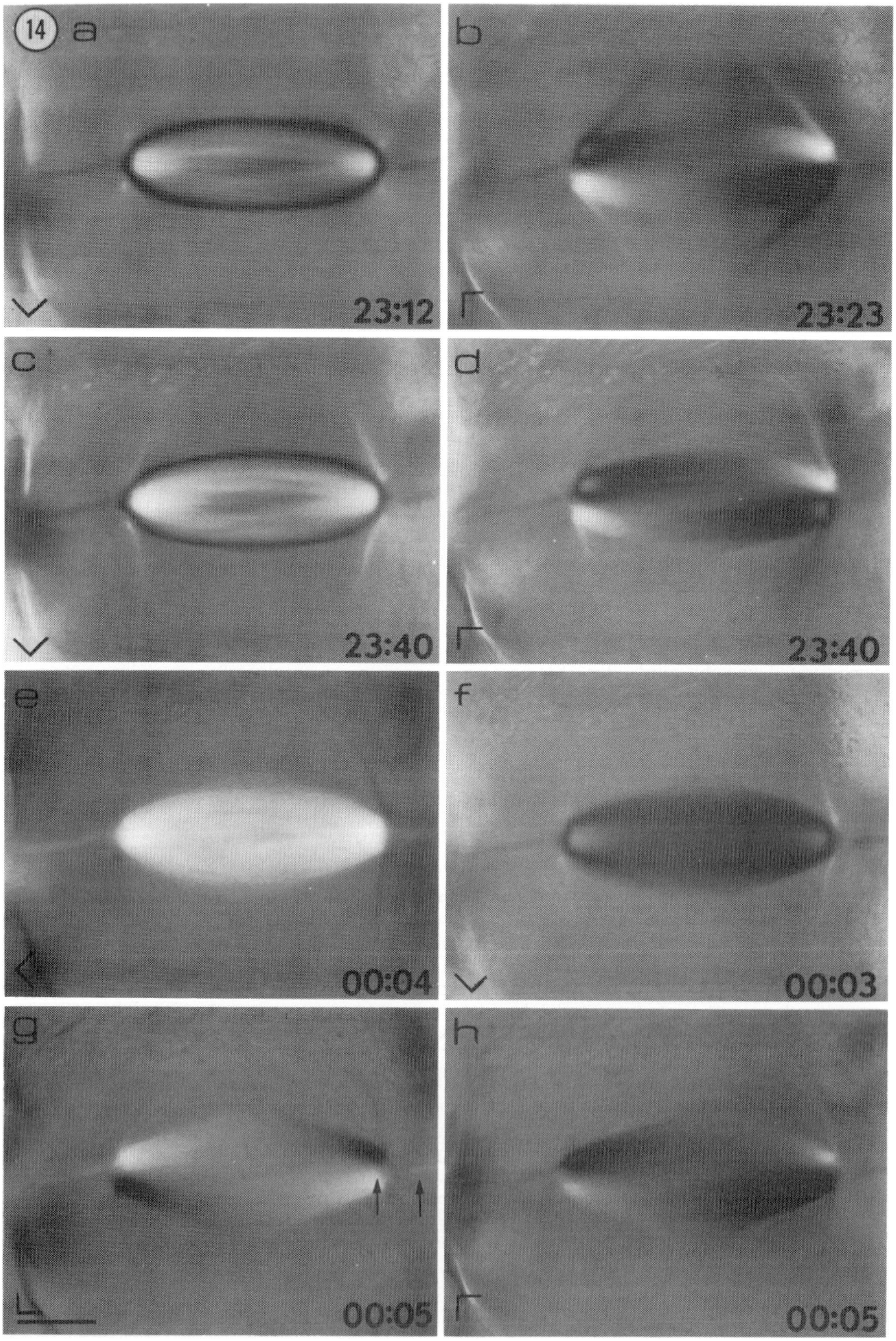
14
a
23:12
b
23:23
c
23:40
d
23:40
e
00:04
f
00:03
g
00:05
h
00:05

Anaphase-A

At the light microscope level, we confirm Cleveland's observation that *Barbulanympha* chromosomal spindle fibers are generated by growth of astral rays. The astral rays attach to each kinetochore embedded in the persistent nuclear envelope, thus establishing a connection between the chromosome, centrosome, and centriole (e.g., 8).

At the fine-structural level, we still do not know how the microtubules, which make up the linear elements of each chromosomal fiber, arise. Growing astral microtubules may directly attach to the kinetochore, or they may interact and align with other microtubules growing from the kinetochore.

Be that as it may, the chromosomal spindle fibers are dynamic. While fibers of the mature central spindle and the spindle pole-to-pole distance remain unaltered, chromosomes are individually observed to both approach and recede from the spindle poles before and during early anaphase-A.

During the several minutes of poleward movement, the poleward pull on a kinetochore is manifested by the point protrusion of the nuclear envelope immediately surrounding that kinetochore. The chromosome arms within the nucleus follow as the particular chromosomal fiber shortens and the kinetochore moves poleward.

When the chromosome moves away from the pole, the kinetochore follows, lying at the pit of a dimple in the nuclear envelope.

As expressed by the cyclic point protrusion and dimpling of the nuclear envelope, the poleward force on each kinetochore varies independently and recurrently after the central spindle has reached its final, stable full length.

Therefore, the chromosomal fibers (astral rays) appear able to (*a*) repeatedly grow and lengthen; (*b*) repeatedly pull the chromosomes poleward; and (*c*) possibly, repeatedly establish new contact with a single kinetochore, thereby establishing new kinetochore-to-pole connections.

The forces exerted on the kinetochores by chromosomal spindle fibers may be produced directly by polymerizing or depolymerizing micro-

FIGURE 14 Anterior view of birefringent central spindle and astral rays in late anaphase-A [(*a*) to (*d*)] and early anaphase-B [(*e*) to (*h*)]. Rectified polarized light image continued from Fig. 9*e*. In (*a*), (*c*), and (*f*) the spindle is oriented at 45° to the polarizer axis. The slow axis of the Brace-Köhler compensator lies at various angles within the quadrant perpendicular to the spindle axis. Therefore, the elongate-centrioles and the (optimally compensated) less retarding regions of the central spindle, as well as those fibers parallel to the P-A axes, appear dark. Astral rays which point northerly or southerly appear bright, especially adjacent to the centrosomes where their microtubule density is high. In (*e*) the spindle axis also lies at 45° to the polarizer but the compensator slow axis is oriented parallel to spindle long axis. The central spindle fibers and the elongate-centrioles now appear bright. Astral rays at right angle to the spindle appear dark. Near each pole, the spindle is brightest where the microtubules converge to the centrosomal surface (cf. Fig. 5 in Part I). A less bright, positively birefringent cone connects this region to the tip of the centriole [also see (*g*), (*h*), and (*f*) in which the compensator orientation is reversed]. The intracentrosomal microtubules described in Part I are believed to contribute to the birefringence of this cone. Extent of centrosome is indicated by two arrows in (*g*). In (*b*), (*d*), (*g*), and (*h*), the spindle is oriented parallel to the polarizer axis so that spindle fibers and astral rays at any single spindle pole may appear bright or dark depending on their orientation and retardation. In (*b*) and (*d*), bright "eyes" appear in the dark NE and SW quadrants near the poles of the central spindle. Here, as along the spindle axis in (*a*), (*c*), and (*f*), the spindle retardation is too great to be (subtractively) compensated; microtubules are packed with a high density (cf. Fig. 5 in Part I) and they are oriented in a direction that introduces maximum alterations of the polarized light beam. In (*a*), the nucleus remains bell-shaped as seen in Fig. 12*c* to *f*. So long as the nucleus is thus shaped, the astral rays, except for those lying tangent to the nuclear envelope, fan out as seen here and in Fig. 9*e*. The rays which lie tangent to the nuclear envelope show clearly in (*b*), but not in (*a*), as they run parallel to the polarizer-analyzer axes in (*a*) and therefore contribute no contrast. As anaphase-A progresses and the shape of the nucleus changes (cf. Figs. 11 and 13), astral rays are pushed poleward as seen in (*b*) through (*h*). Concurrently, their birefringence decreases. (*e*) through (*h*) show the progressive splaying of the lateral spindle fibers as the nuclear cleft progresses caudad and exposes the lateral aspects of the central spindle in anaphase-B (cf. Figs. 13*h*; 11 bottom). The central spindle is growing in length and its birefringence is diminishing, initially anteriorly then progressively laterally and caudally. Nikon rectified 40× objective. h:min in time of day on 74g30. Bar, 20 μm. × 630.

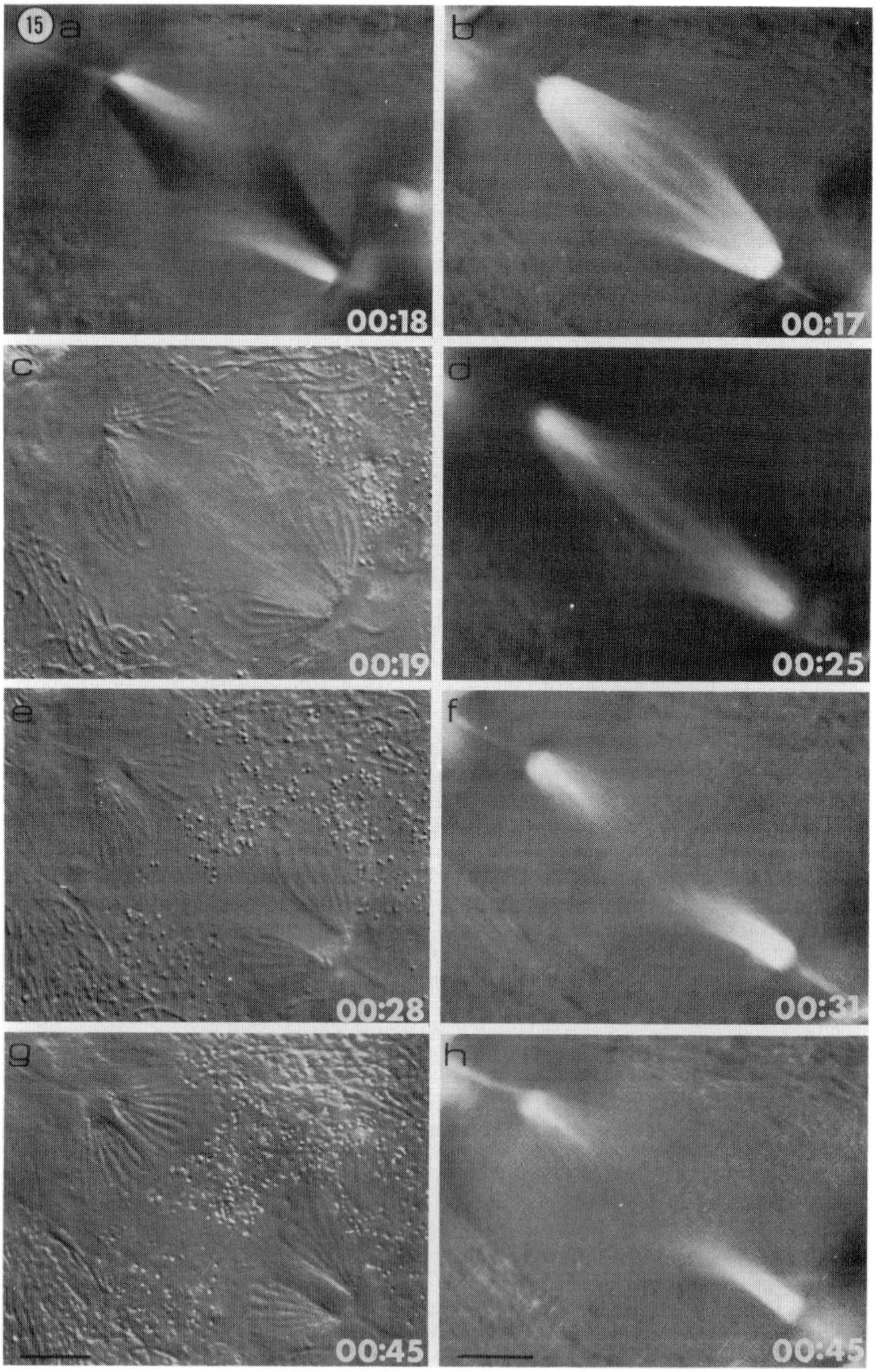

FIGURE 15 Anaphase-B continued from Figs. 13 and 14. The polarized light images, (*a*), (*b*), (*d*), (*f*), and (*h*) show progressive lateral splaying of the central spindle fibers and loss of their birefringence. The central spindle is elongating. Differential interference contrast images (*c*), (*e*), (*g*) show progressive separation of the nuclear envelope as the nuclear cleft traverses caudad (cf. lower diagrams in Fig. 11). In (*c*), at this low median focus, the nuclear envelope still appears connected; in (*e*) and (*g*), they appear separated. At the upper spindle pole in (*e*), note the clusters of kinetochores and their apposition to the centrosome around the tip of the elongate-centrioles. h:min in time of day on 74g30. Bars, 20 μm. × 600.

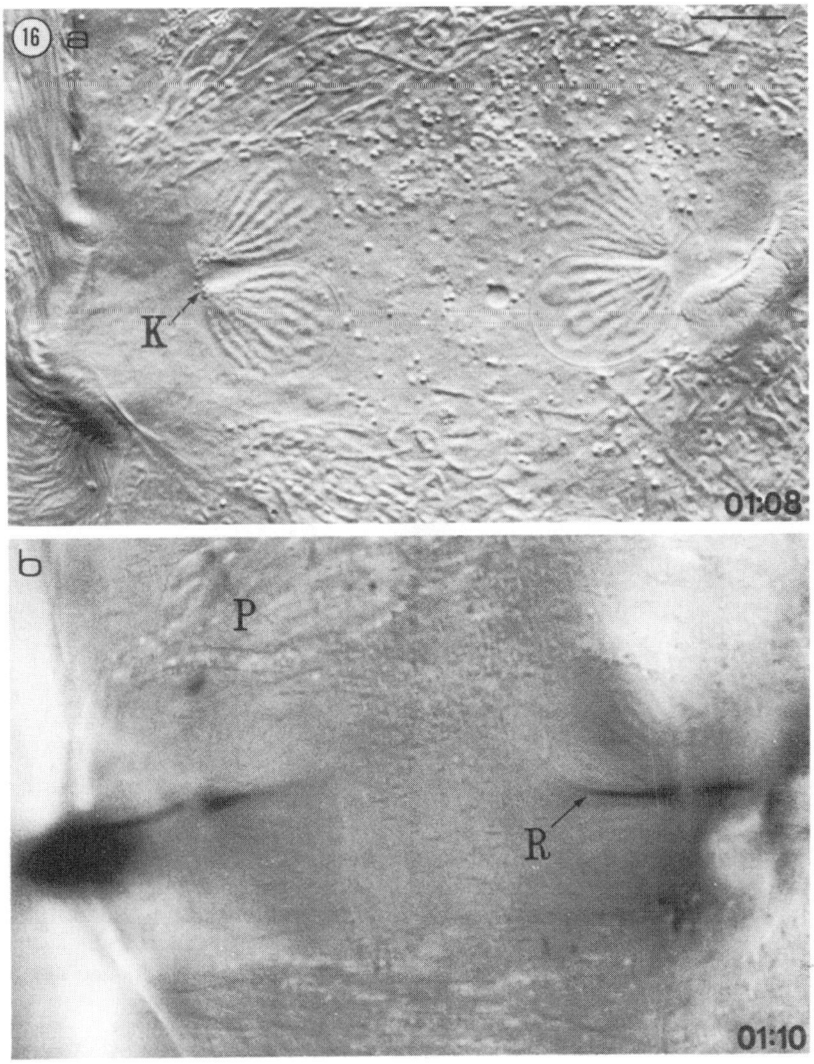

FIGURE 16 Late anaphase-B anterior view continued from Fig. 15. Karyokinesis is now complete (*a*). In differential interference contrast (*a*), the row of kinetochores (*K*) shows prominently at the left spindle pole. (For lateral view of kinetochore rows, see Fig. 6*b, c* left pole.) The shortening chromosome arms, nucleoli, and the wing-shaped optical section of the nuclear envelope are also clear. Parabasals and axostyles have regrown. Only a thin remnant (*R*) of the central spindle persists (*b*). In the polarized light image (*b*) the (negatively birefringent?) helical parabasals (*P*) appear as though rows of bright birefringent dots, and the dark, positively birefringent thin axostyles curve gently. At this stage, the flagella start beating vigorously again and contribute to cytokinesis. The new flagellar bundle (to the right) is externalized and only its basal structures are visible in this picture. h:min in time of day on 74g30. Bar, 20 μm. × 620.

tubules. Or, if there are two sets of microtubules, astral and kinetochoric, they may slide relative to one another. Alternatively, polymerizing microtubules may extend or align another component or components that subsequently draw the chromosomes poleward as the microtubules slowly depolymerize. We have already stressed the occurrence and need for microtubule depolymerization in anaphase-A generally (19–22). Because the dynamic changes take place after the central spindle is established, chromosomal fibers almost certainly could not contract as proposed by Cleveland (p. 27 in reference 12) by release of elastic energy stored in them through prior extension by the growing central spindle.

Polarization of Kinetochores

The polarization of kinetochores to opposite

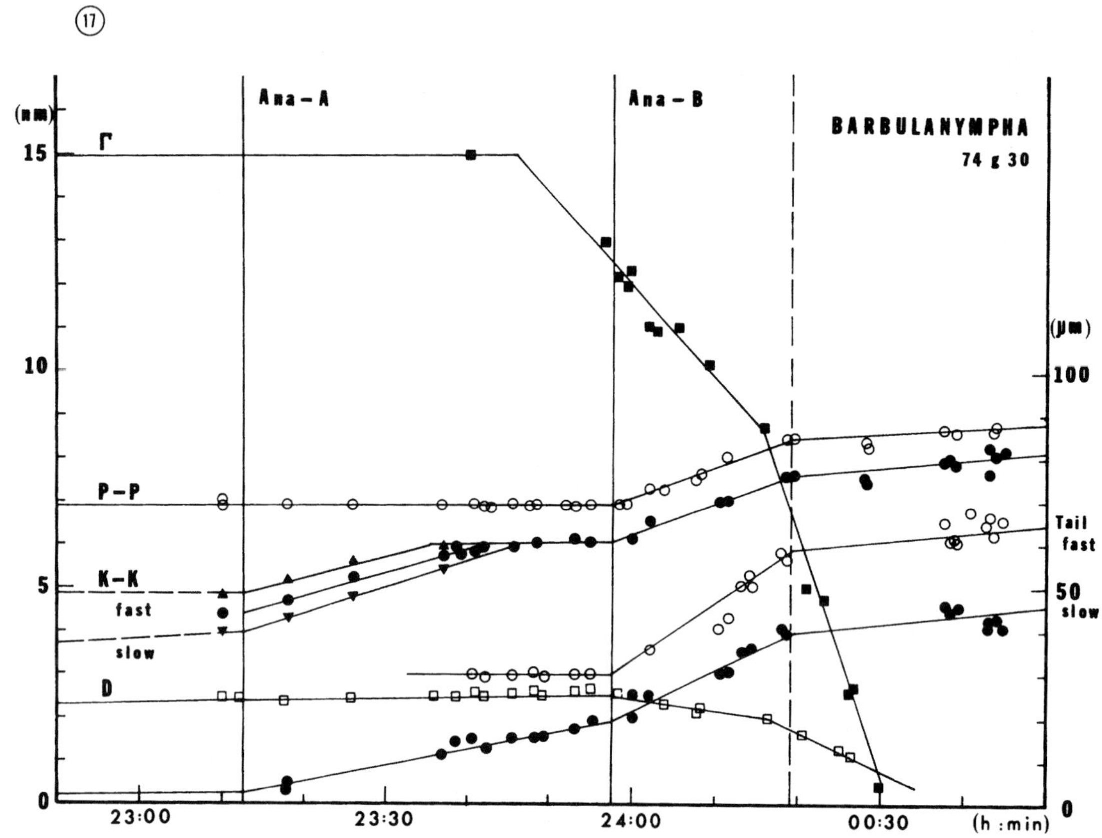

FIGURE 17 Birefringence retardation, dimensions of central spindle, and chromosome positions measured on cells shown in Figs. 9–16. Γ, retardation of spindle mid-region. *P-P*, spindle pole-to-pole length measured between distal tips of centrioles. *D*, width of unsplayed portion of central spindle. *K-K*, distance between kinetochores migrating to opposite poles, for "fast" and "slow" chromosomes. *Tail*, separation of tails of chromosome arms, for "fast" and "slow" chromosomes. Note: presumably because of compression of the cell which made possible recording of this fortuitous anterior view, the extreme elongation of the central spindle commonly found in anaphase-B was not seen. Instead, spindle elongation and chromosome movement slowed down after 00:19 h.

spindle poles is essential for successful completion of mitosis. In *Barbulanympha,* the mechanism for this important event cannot be understood by simple application of Östergren's otherwise elegant metaphase equilibrium hypothesis (31, 33). There simply is no normal metaphase stage in *Barbulanympha* mitosis, nor does Kubai's proposal (24) for kinetochore segregation in the *Trichonympha* nuclear envelope without aide of the astral rays appear applicable. Cleveland (Figs. 11 and 12 in reference 12) shows that where the astral rays fail to reach a nucleus, there is no tendency for the paired chromosomes or indeed their kinetochores to move apart. In the same cell, kinetochores in a second nucleus within reach of the asters do separate.

Cleveland further shows that a monopolar spindle is sufficient for kinetochores to separate from their sisters. After initial separation of kinetochores and chromatids, the chromosomes appear to all polarize more or less away from the single centrosome or from the tip of the elongate-centriole, somewhat reminiscent of movements described for the paternal set of chromosomes in the monopolar division of *Sciara* spermatocytes (30, 37).

Cleveland goes on to state (reference 12, p. 27) that "the anaphase daughter chromosomes—do not move to opposite poles when they have a choice of more than two poles (Fig. 10 of reference 12). When only two poles are present they have no choice; they must move to one or the other---". In a multipolar mitotic figure, the kinetochores may even polarize and separate towards two poles not joined together by a central spindle, and in a direction more or less at right angles to the existing central spindles.

We view Cleveland's work on *Barbulanympha* to strongly suggest that (*a*) kinetochore separation does not take place in the absence of (astral) microtubules; (*b*) initial separation of kinetochores requires only the presence of astral rays and is not dependent on the presence of two centrioles or centrosomes; and (*c*) a bipolar spindle is essential for the orderly, functional segregation of kinetochores and chromosomes. A normal spindle provides the two poles required, and the central spindle fibers presumably function as support which keeps the two poles apart.

Anaphase-B

At the beginning of anaphase-B, the extending central spindle fibers push apart the bouquet of chromosomes formed at the end of anaphase-A. The kinetochores by then firmly crown the centrosomes. Short chromosomal spindle fibers persist throughout anaphase-B and hold the packed crown of kinetochores, the bouquet of chromosome, and nuclear envelope up against or close to the centrosomal surface.

Later in anaphase-B, the central spindle becomes highly attenuated. One may wonder whether such a slender structure could continue to push the daughter nuclei apart. Or might the spindle itself be passively drawn apart by the cortical cytoplasm extending between the two rostra bearing the flagellar tufts (Figs. 6, 7)? At times, the elongate-centrioles and central spindle are pulled into a straight line when the two-rostrate organism swivels its rostra. At other times, the elongate-centrioles and central spindle are each gently curved and no tension on them is apparent. During the later stages of anaphase-B, the central spindle continues to elongate and the rostra and flagellar tufts separate as though the two events were proceeding at a parallel pace but not because they were mechanically coupled together. According to Cleveland (11), the central spindle of related hypermastigote *Macrospironympha* also attenuates and elongates extensively, but in the absence of cortical changes or any other indication of extensile forces acting on the poles of the spindle that are embedded deep in the cytoplasm (also see reference 26). It seems most likely that throughout anaphase-B the central spindle actively elongates and "pushes apart" the chromosome bouquets.

The birefringence retardation of the *Barbulanympha* central spindle decays in anaphase-B, more or less proportionately with the width of the central spindle. The coefficient of birefringence, or the packing densities of microtubules (number per unit spindle cross section, see reference 36), thus remains approximately constant as the spindle remnant becomes longer and thinner. These data for anaphase-B are not incompatible with the proposal that central spindle fibers extend by the sliding of microtubules. However, the morphology and degree of extension (as much as five times) of the central spindle suggest that the extension cannot be brought about by the relative sliding of two sets of nongrowing microtubules each anchored at their poles (28). Instead, the central spindle microtubules must be telescoping or are in fact growing longer. In other words, anaphase-B in *Barbulanympha* appears to be brought about by (*a*) the growth of central spindle microtubules per se, (*b*) microtubular growth coupled with their sliding, or (*c*) telescopic sliding of microtubules.

In thin-section electron micrographs of the anaphase-A spindle of *Barbulanympha* taken by Dr. Kubai, we have demonstrated by a new optical enhancement method the presence of extensive intermicrotubular periodic structures (21). Because the anaphase-A spindles are extremely stable in dimension and birefringence, these intertubular structures may well be structural linkers of the type found between axopodial microtubules in heliozoa (23, 40) and within the rows of hypermastigote axostyles (15, review in 27). We find that *Barbulanympha* central spindles in anaphase-A do not show cold lability (down to 4°C) as is often found in spindle fibers in cells of other organisms (e.g., 18, 19, 22). Generally, the nonkinetochoric microtubules such as the "continuous" central spindle fibers are more sensitive to cold or colchicine and depolymerize faster than kinetochore microtubules (3, 7, 13).

Instead of being stable linkers, some of the intertubular structures in *Barbulanympha* might be arms, such as dynein, involved in force production. They may then generate relative sliding of central spindle microtubules, telescoping in a manner similar to the trypsin-treated flagellar axonemes demonstrated by Summers and the Gibbons (14, 38, 39). Establishment of the nature and functional role of this intermicrotubular material in the gigantic central spindle of *Barbulanympha* may well resolve the question generally of whether central spindle elongation, i.e. anaphase-B, can or does take place by sliding in addition to growth of microtubules.

Morphogenesis of Nuclear Envelope

The morphogenetic movement of the nuclear envelope we recorded may at first appear unique. In fact, however, this may reflect a general pattern of karyokinesis in evolutionarily embryonic eukaryotes. In prokaryotes, the daughter DNA strands are thought to be carried along by parting mesosomes connected to the cell membrane. The mechanism of mesosome partitioning itself is not clear. As reviewed by Kubai (25), mitosis in many dinoflagellates, lower algae, and hypermastigotes is believed to be mediated by kinetochores, or functionally equivalent regions of chromosomes embedded in or attached to the nuclear envelope. As discussed above, polarization of *Barbulanympha* kinetochores is governed by microtubule bundles. If the association of microtubules with kinetochores in the nuclear envelope reflects an early evolutionary event, we may in *Barbulanympha* witness an ancient experiment of nature. Despite the "permanence" of the nuclear envelope, the pattern of mitosis involving the shortening chromosomal fibers and elongating central spindle is essentially that seen in "orthomitosis" (17) of "higher" organisms. The latter may have abandoned the complex maneuver necessitated by the persistence of the nuclear envelope in a multichromosomal organism.

The nuclear seam that we describe is pictured in some of Cleveland's drawings, although without comment. On the other hand, the appearance and opening of the cleft appears never to have been noticed before our report. The later stages of karyokinesis depicted by Cleveland (9) and by Hollande and Valentin (16) are nevertheless totally consistent with the mode of karyokinesis via cleft expansion that we describe.

The deformation of the nuclear envelope during anaphase-A and -B, including the point protrusions, recesses, and the overall shapes (also see reference 10), suggests that the nuclear envelope possesses an apparent elasticity. In fact, the envelope may well be a constrained interfacial membrane. Interfacial membranes such as soap bubbles do give the impression of possessing elasticity. A constrained "hanging drop" or "rising bubble" takes on a bell shape similar to that of the *Barbulanympha* nuclear envelope (reference 1, pages 365–367; also, see reference 6 for an illuminating discussion on the properties of soap bubbles). In addition to the apparent elasticity, the movements of embedded kinetochores suggest a fluidity to the nuclear envelope. Many aspects of the shape change observed in the *Barbulanympha* nuclear envelope as well as its fluidity in two dimensions can be seen or inferred by comparison with patterns exhibited by soap bubbles.

As discussed earlier, there is little doubt that shortening chromosomal fibers acting on individual kinetochores are responsible for the poleward point protrusion of the nuclear envelope and for the associated chromosome movement. On the other hand, the dimples away from the pole may reflect (*a*) a drop of poleward tension and concomitant elastic recoil of the undissociated chromatid pairs, (*b*) an active extension and push by chromosomal spindle fibers, or (*c*) a disequilibrium resulting from a greater tug on the oppositely directed kinetochore.

The sharp protrusions found at each poleward apex of the nuclear seam (Figs. 9*a*, *d*, 12*a*, *b*, and 13*a*) suggest that the seam may also be formed by poleward forces acting on the kinetochores. The recurrent movement of the prophase and early anaphase kinetochores suggests that new astral rays may sequentially attach to the kinetochores. The poleward forces exerted on the kinetochore by the shortening astral rays (chromosomal fibers) would be counteracted by a radial resistive force on the nuclear envelope arising from the presence of the central spindle and of the chromosomes within the nucleus. The balance of poleward and radial forces would eventually wrap the nucleus around the spindle. Where the nucleus abuts itself, a seam would appear.

Likewise, it is not unlikely that the opening of the cleft also results from the interplay between (*a*) poleward forces acting on kinetochores, (*b*) the tendency of the chromosome bouquet to resist collapse of the nuclear envelope, (*c*) radial support provided by the central spindle, and (*d*) the tendency of the elastic nuclear envelope to maintain, with minimum nuclear volume change, a minimum surface area in the presence of these constraints.

Central Spindle Splaying

We observed the splaying of the central spindle which progressed concurrently with, and conformed with the shape of, the opening nuclear cleft and its posteriad extension. The elongating (or sliding) microtubular bundles may be bulging laterally as a constraining girdle is released or, in turn, spindle fiber splaying may contribute to the formation of the cleft. In either case, the integrity of the early anaphase-A spindle, before envelopment by the nucleus and formation of the seam, suggests that there is a cellular mechanism that operates independently to control the lateral affinity of spindle microtubules (also see discussions by Bajer and Molè-Bajer, 4).

Cytokinesis

The astonishing observation that *Trichonympha* assists the cytokinesis of *Barbulanympha* appears not to be accidental. *Trichonympha* have repeatedly been observed to poke their rostra specifically against the mid-region of the late anaphase-B *Barbulanympha* as though attracted to the region. One of us (H.R.) has on several occasions seen a *Trichonympha* literally ram through the two-rostrate *Barbulanympha,* completing the *Barbulanympha's* cytokinesis. For organisms which must have attained obligatory symbiotic relationships so long ago, perhaps we should not be surprised at the occurrence of even such an amazing mutualism.

Concluding Remarks

In conclusion, extended observations of individual *Barbulanympha* cells in mitosis is now possible. Sequential through-focus observations with differential interference contrast and rectified polarized light microscopy have provided insight into aspects of mitosis that had not been revealed in other species in such complete detail. In addition to points of interest for mitotic mechanisms generally, the unique features displayed in this presumably ancient organism may shed light into the evolution of mitosis. Some striking features displayed in *Barbulanympha* suggest intriguing opportunities for further experimental studies of mitosis utilizing this unusual but revealing cell type.

Received for publication 29 July 1977, and in revised form 3 January 1978.

REFERENCES

1. Adam, N. K. 1941. The Physics and Chemistry of Surfaces. Oxford University Press, London. 3rd edition. 1-436.
2. Aist, J. R., and P. H. Williams. 1972. Ultrastructure and time course of mitosis in the fungus *Fusarium oxysporum. J. Cell Biol.* **55:**368-389.
3. Bajer, A. S., and J. Molè-Bajer. 1972. Spindle dynamics and chromosome movements. *Int. Rev. Cytol.* **3** (Suppl.):1-271.
4. Bajer, A. S., and J. Molè-Bajer. 1975. Lateral movements in the spindle and the mechanism of mitosis. *In* Molecules and Cell Movement. S. Inoué and R. E. Stephens, editors. Raven Press, New York. 77-96.
5. Boss, J. 1954. Mitosis in cultures of newt tissues. III. Cleavage and chromosome movements in anaphase. *Exp. Cell Res.* **7:**443-456.
6. Boys, C. V. 1959. Soap-bubbles: Their Colours and the Forces which Mold Them. Dover Publications, Inc., New York. 1-192.
7. Brinkley, B. R., E. Stubblefield, and T. C. Hsu. 1967. The effects of colcemid inhibition and reversal on the fine structure of the mitotic apparatus of Chinese hamster cells *in vitro. J. Ultrastruct. Res.* **19:**1-18.
8. Cleveland, L. R., S. R. Hall, E. P. Sanders, and J. Collier. 1934. The wood-feeding roach *Cryptocercus,* its protozoa, and the symbiosis between protozoa and roach. *Mem. Am. Acad. Arts Sci.* **17:**185-342 (and 60 plates).
9. Cleveland, L. R. 1953. Hormone-induced sexual cycles of flagellates. IX. Haploid gametogenesis and fertilization in *Barbulanympha. J. Morphol.* **93:**371-403.
10. Cleveland, L. R. 1954. Hormone-induced sexual cycles of flagellates. XII. Meiosis in *Barbulanympha* following fertilization, autogamy, and endomitosis. *J. Morphol.* **95:**557-619.
11. Cleveland, L. R. 1956. Hormone-induced sexual cycles of flagellates. XIV. Gametic meiosis and fertilization in *Macrospironympha. Arch. Protistenkd.* **101:**99-169.
12. Cleveland, L. R. 1963. Functions of flagellate and other centrioles in cell reproduction. *In* The Cell in Mitosis. L. Levine, editor. Academic Press, Inc., New York. 3-31.
13. Fuseler, J. W. 1975. Temperature dependence of anaphase chromosome velocity and microtubule depolymerization. *J. Cell Biol.* **67:**789-800.
14. Gibbons, B. H., and I. R. Gibbons. 1972. Flagellar movement and adenosine triphosphatase activity in sea urchin sperm extracted with Triton X-100. *J. Cell Biol.* **54:**75-97.
15. Grimstone, A. V., and L. R. Cleveland. 1965. The fine structure and function of the contractile axostyles of certain flagellates. *J. Cell Biol.* **24:**387-400.
16. Hollande, A., and J. Carruette-Valentin. 1971. Les atractophores, l'induction du fuseau et la division cellulaire chez les Hypermastigines. Étude infrastructurale et révision systématique des Trichonymphines et des Spirotrichonymphines. *Protistologica.* **7:**5-100.
17. Hollande, A., and J. Valentin. 1968. Infrastructure des centromères et déroulement de la pleurom-

itose chez les Hypermastigines. *C.R. Acad. Sci.* **266:**367–370.

18. Inoué, S. 1964. Organization and function of the mitotic spindle. *In* Primitive Motile Systems in Cell Biology. R. D. Allen and N. Kamiya, editors. Academic Press, Inc., New York 549–598.
19. Inoué, S., and H. Sato. 1967. Cell motility by labile association of molecules. The nature of mitotic spindle fibers and their role in chromosome movement. *J. Gen. Physiol.* **50:**259–292.
20. Inoué, S. 1976. Chromosome movement by reversible assembly of microtubules. *In* Cell Motility, Book C. Microtubules and Related Proteins. R. Goldman, T. Pollard, and J. Rosenbaum, editors. Cold Spring Harbor Laboratory, Cold Spring Harbor, N.Y. 1317–1328.
21. Inoué, S., and H. Ritter, Jr. 1975. Dynamics of mitotic spindle organization and function. *In* Molecules and Cell Movement. S. Inoué and R. E. Stephens, editors. Raven Press, New York 3–30.
22. Inoué, S., J. W. Fuseler, E. D. Salmon, and G. W. Ellis. 1975. Functional organization of mitotic microtubules: physical chemistry of the *in vivo* equilibrium system. *Biophys. J.* **15:**725–744.
23. Kitching, J. A. 1964. The axopods of the sun animalcule *Actinophrys sol* (Heliozoa). *In* Primitive Motile Systems in Cell Biology. R. D. Allen and N. Kamiya, editors. Academic Press, Inc., New York. 445–456.
24. Kubai, D. F. 1973. Unorthodox mitosis in *Trichonympha agilis:* kinetochore differentiation and chromosome movement. *J. Cell Sci.* **13:**511–552.
25. Kubai, D. F. 1975. The evolution of the mitotic spindle. *Int. Rev. Cytol.* **43:**167–227.
26. Liesche, W. 1938. Die Kern- und Fortpflanzungsverhältnisse von *Amoeba proteus* (Pall.). *Arch. Protistenkd.* **91:**135–186.
27. McIntosh, J. R. 1974. Bridges between microtubules. *J. Cell Biol.* **61:**166–187.
28. McIntosh, J. R., P. K. Hepler, and D. G. Van Wie. 1969. Model for mitosis. *Nature* (*Lond.*). **224:**658–663.
29. Mazia, D. 1961. Mitosis and the physiology of cell division. *In* The Cell, Vol. III. J. Brachet and A. E. Mirsky, editors. Academic Press, Inc., New York. 77–412.
30. Metz, C. W. 1933. Monocentric mitosis with segregation of chromosomes in *Sciara* and its bearing on the mechanism of mitosis. I. The normal monocentric mitosis. II. Experimental modification of the monocentric mitosis. *Biol. Bull.* (*Woods Hole*). **64:**333–347.
31. Nicklas, R. B. 1971. Mitosis. *In* Advances in Cell Biology, Vol. II. D. M. Prescott, L. Goldstein, and E. McConkey, editors. Appleton-Century-Crofts, New York. 225–297.
32. Oppenheim, D. S., B. T. Hauschka, and J. R. McIntosh. 1973. Anaphase motions in dilute colchicine: evidence of two phases in chromosome segregation. *Exp. Cell Res.* **79:**95–105.
33. Östergren, G. 1949. *Luzula* and the mechanism of chromosome movements. *Hereditas.* **35:**445–468.
34. Ris, H. 1943. A quantitative study of anaphase movement in the aphid *Tamalia. Biol. Bull.* (*Woods Hole*). **85:**164–178.
35. Ris, H. 1949. The anaphase movement of chromosomes in the spermatocytes of the grasshopper. *Biol. Bull.* (*Woods Hole*). **96:**90–106.
36. Sato, H., G. W. Ellis, and S. Inoué. 1975. Microtubular origin of mitotic spindle form birefringence: demonstration of the applicability of Wiener's equation. *J. Cell Biol.* **67:**501–517.
37. Schrader, F. 1953. Mitosis: The Movements of Chromosomes in Cell Division. Columbia University Press, New York. 2nd edition. 1–170.
38. Summers, K. E., and I. R. Gibbons. 1971. Adenosine triphosphate-induced sliding of tubules in trypsin-treated flagella of sea urchin sperm. *Proc. Natl. Acad. Sci. U. S. A.* **68:**3092–3096.
39. Summers, K. E., and I. R. Gibbons. 1973. Effects of trypsin digestion on flagellar structures and their relationship to motility. *J. Cell Biol.* **58:**618–629.
40. Tilney, L. G. 1971. How microtubule patterns are generated: the relative importance of nucleation and bridging of microtubules in the formation of the axoneme of *Raphidiophrys. J. Cell Biol.* **51:**837–854.

Reprinted from *Journal of Cell Biology*, Vol. 83, p. 376a, 1979.

UNEXPECTED INCREASE IN POLEWARD VELOCITIES OF MITOTIC CHROMOSOMES AFTER UV IRRADIATION OF THEIR KINETOCHORE FIBERS

Gerald W Gordon and Shinya Inoué

UV microbeam irradiation produces local areas of reduced birefringence (arbs) in metaphase and anaphase kinetochore (K) fibers of grasshopper primary spermatocytes. Arbs first appear in the irradiated area. Then they move poleward and disappear by moving off the fiber's end in a manner similar to arbs in a crane fly (Forer, *J Cell Biol* **25**: 95, 1965). Arb movement is thought to show microtubule (MT) assembly at the K and disassembly at the pole (Inoué and Sato, *J Gen Physiol* **50**: 259, 1967). Quantitative analysis of K and pole movements reveals subtle behavior which is not apparent visually. A K whose fiber has an arb (arb K) accelerates transiently (ca. 2 min.) causing an unexpected displacement toward the pole it faces. In metaphase, an arb K and its homolog are displaced in tandem toward the arb pole. After the arb disappears (ca. 10 min.), the arb bivalent tends to return to the equator. Non-arb bivalents remain equatorial throughout. For about two minutes after anaphase irradiation, an arb K accelerates poleward relative to its non-arb neighbors. Then it slows down. The non-arb Ks maintain a remarkably constant poleward velocity even if the spindle length is changing. Thus, disruption of some of a fiber's MTs is associated with increased poleward velocity of its K. This suggests that a fiber with a reduced MT count exerts increased force on its K. A simple stochastic model of MT disassembly is proposed. It accounts for the unexpected acceleration of arb Ks. The model assumes non-independent disassembly of MTs within a K fiber and does not specify the origin of mitotic forces. Grant support: NIH IT32-HD07152 to GWG, NIH PHS 9 RO1 GM23475 and NSF BMS 75-00473 to SI.

The following note was added by Shinya Inoué in September of 2006:

This brief abstract is the only published account of the most interesting and thought-provoking thesis research, carried out by Gerald W. Gordon relating to force-generating mechanisms by meiotic microtubules. It should be noted that Gerry's findings with UV microbeam irradiation on grasshopper spermatocytes differ considerably in detail from similar experiments performed on crane fly spermatocytes. The extensive illustrations and detailed descriptions of the experiments and findings which document this research are, unfortunately, only available in Gerry's PhD thesis accepted by the Department of Biology, University of Pennsylvania, in 1980.

Reprinted from *Journal of Cell Biology*, Vol. 80(3), pp. 521–538, 1979.

MOTILITY OF THE MICROTUBULAR AXOSTYLE IN *PYRSONYMPHA*

GEORGE M. LANGFORD and SHINYA INOUÉ

From the Department of Anatomy, College of Medicine, Howard University, Washington, D. C. 20059, and the Program in Biophysical Cytology, Department of Biology, University of Pennsylvania, Philadelphia, Pennsylvania 19104

ABSTRACT

The rhythmic movement of the microtubular axostyle in the termite flagellate, *Pyrsonympha vertens*, was analyzed with polarization and electron microscopy. The protozoan axostyle is birefringent as a result of the semi-crystalline alignment of ~2,000 microtubules. The birefringence of the organelle permits analysis of the beat pattern in vivo. Modifications of the beat pattern were achieved with visible and UV microbeam irradiation.

The beating axostyle is helically twisted and has two principal movements, one resembling ciliary and the other flagellar beating. The anterior portion of the beating axostyle has effective and recovery phases with each beat thereby simulating the flexural motion of a beating cilium. Undulations develop from the flexural flipping motion of the anterior segment and travel along the axostyle like flagellar waves. The shape of the waves differs from that of flagellar waves, however, and are described as sawtooth waves. The propagating sawtooth waves contain a sharp bend, ~3 μm in length, made up of two opposing flexures followed by a straight helical segment ~23 μm long. The average wavelength is ~25 μm, and three to four sawtooth waves travel along the axostyle at one time. The bends are nearly planar and can travel in either direction along the axostyle with equal velocity. At temperatures between 5° and 30°C, one sees a proportionate increase or decrease in wave propagation velocity as the temperature is raised or lowered. Beating stops below 5°C but will resume if the preparation is warmed.

A microbeam of visible light shone on a small segment of the axostyle causes the typical sawtooth waves to transform into short sine-like waves that accumulate in the area irradiated. Waves entering the affected region appear to stimulate waves already accumulated there to move, and waves that emerge take on the normal sawtooth wave pattern. The effective wavelengths of visible light capable of modifying the wave pattern is in the blue region of the spectrum. The axostyle is severed when irradiated with an intense microbeam of UV light. Short segments of axostyle produced by severing it at two places with a UV microbeam can curl upon themselves into shapes resembling lockwashers. We propose that the sawtooth waves in the axostyle of *P. vertens* are generated by interrow cross-bridges which are active in the straight regions.

KEY WORDS motility · axostyle · microtubules · *Pyrsonympha* · sawtooth wave · birefringence

Microtubule-containing motile organelles such as cilia and flagella propagate undulatory bending waves along their length. The mechanism of bend formation is thought to be the result of active sliding between adjacent doublet microtubules (for reviews, see references 15, 32). In order for the sliding movements between tubules to produce regular bending waves, however, it is necessary for the sliding to be properly localized and coordinated. Direct evidence concerning the mechanisms that regulate sliding has been difficult to obtain. In cilia and flagella, it has been postulated that the central tubules together with the central sheath and the radial links that join them to the outer doublets are involved in the coordination process (33, 34). How these components might interact to produce a system for the coordination of bending is not understood.

The microtubule-containing organelle known as the axostyle found in certain zooflagellates (8, 9, 10, 11) also propagates undulatory bending waves (18, 19). This organelle is appealing as a model of microtubule-associated movement because of its structural simplicity and large size. Electron microscopy studies show that the motile axostyle in the wood roach protozoans *Saccinobaculus* (19, 24) and the termite protozoans *Pyrsonympha* (4, 7, 13, 18, 20, 30, 31) contains several thousand singlet microtubules interconnected by cross-bridges. The microtubules are organized into rows, and the microtubules within the rows are connected to each other by regularly occurring linkers or intrarow bridges. In turn, the rows of microtubules are interconnected by less regularly occurring cross-bridges or interrow bridges. The intrarow bridges appear periodic along the tubules with a spacing of 16 nm. The interrow bridges are not strictly periodic and can be oriented at varying angles to the axis of the microtubule (3, 24). The axostyle, therefore, is both organizationally simple and contains a small number of accessory components.

The bending of the axostyle in *Saccinobaculus* appears to be accompanied by sliding of rows of microtubules relative to one another (2, 25). This axostyle contains a protein with enzymatic activity and molecular weight that are similar to those of ciliary and flagellar dynein and a protein with a molecular weight similar to that of flagellar nexin (26). The dyneinlike protein is thought to be located in the interrow bridges (2), and the nexinlike protein in the intrarow bridges. The movement of the axostyle in *Saccinobaculus* is interpreted as the propagation of circular arcs down a helical ribbon (25). The axostyle therefore has been shown to have both structural and functional features in common with cilia and flagella.

The motion and organization of the axostyle in *Pyrsonympha* are different from those of the axostyle in *Saccinobaculus*. In *Pyrsonympha*, the rows of microtubules are shorter and more numerous and the bends on the axostyle are separated by long unbent regions (18, 23, 31). The packing order of the microtubules has been shown to differ in the bent and straight regions of the axostyle (31) but a detailed analysis of this bending motion has not been described. In this paper, we show that the axostyle exhibits both cilia- and flagella-like movements and that the undulatory waves are sawtooth in shape. We also describe perturbations of the axostyle's activity, analyze the effects of temperature on the beat characteristics, and report changes in the wave form after blue and ultraviolet microbeam irradiation.

MATERIALS AND METHODS

The termites used in this study were collected along the Atlantic coast from North Carolina to Cape Cod, but primarily in the vicinity of Philadelphia. They were collected in infected wood and stored in plastic boxes in the laboratory.

To release the anaerobic protozoa from the hindgut, the abdomen of a worker termite was placed in a drop of medium on a glass slide and gently opened with fine dissecting needles. This operation was performed in a CO_2 atmosphere to maintain anaerobiosis. The exposed hindgut was then ruptured, releasing its contents into the drop of medium. The drop of medium was covered by a glass coverslip, each corner of which was supported by a small fragment of coverslip to prevent flattening the cells. The edges of the coverslip were sealed to the slide with valap (vaseline, lanolin, paraffin; 1:1:1). Cells could be maintained for several hours in such a preparation before the axostyle beat was noticeably affected.

Another method of preparation involved squeezing the abdomen of the termite with fine-tip forceps, forcing out some of the hindgut content. The droplet of fluid was mixed into a drop of medium on a glass slide, and the preparation sealed with valap. This type of preparation had the advantage of being free of tissue and debris from the termite abdomen which interfered with observations in polarized light. If preparations were sealed quickly enough, a CO_2 atmosphere was not necessary.

The medium routinely used was a modified Hungate

medium formulated by S. Inoué and T. Punnett, according to the rules of Provasoli (28), and contains the following chemicals: 22.5 mM Na_2HPO_4, 1.0 mM $CaCl_2$, 15 mM NaH_2PO, 0.01 mM $MgSO_4$, 10 mM KCl, 0.01 mM $(NH_4)_2SO_4$, 5 μM $FeSO_4$, 1 μM $CuSO_4$, 25 μM H_3BO_3, 1 μM $ZnSO_4$, 1 μM $MnCl_2$, 1 μM MoO_3. This medium was prepared in Ballantine distilled water (1) and the pH was adjusted to 7.0. Human saliva was sometimes used to sustain the cells. Saliva is slightly hypotonic for these cells, and they tend to gradually swell in it.

Observations were made with either a Zeiss Universal phase contrast microscope or a Leitz Ortholux polarizing microscope equipped with American Optical strain-free rectified objectives and condenser. HN22 polaroid sheets (Polaroid, Inc., Cambridge, Mass.) were used for polarizers and $\sim\lambda/20$ mica plate (Brace-Kohler design) for the compensator. A high pressure quartz mercury arc lamp (HBO/200, Osram), together with heat-absorbing filters (Edmund's Scientific Co., Barrington, N. J.) and monochromatic green multilayer interference filter transmitting the bulk of the mercury 546 nm line (Baird-Atomic, Inc., Cambridge, Mass.), was used.

For studies involving visible light and long wavelength UV microbeam irradiation, a slit several tenths of a millimeter in width was placed at the field stop. The image of this slit projected onto the specimen by the condenser provided the proper size microbeam of ~10 μm. To irradiate, the green interference filter and polarizer, but not the heat-absorbing filter, were removed, thus illuminating the cell with visible white light. The glass lenses, filters and slide effectively eliminated the UV and infrared regions of the spectrum, permitting only light with wavelengths between 320 and 770 nm to pass. For UV studies, the glass condenser was replaced with a quartz condenser, and a quartz coverslip was used instead of a glass slide. To produce the desired effects, the axostyle was irradiated 10–30 s.

Temperature effects were obtained with a temperature-controlled microscope slide through which water of various temperatures was circulated.

Motion pictures of the movement in polarized light were made with an Arriflex movie camera (Arriflex Corp. of America, Westbury, N. Y.). Obtaining sufficient light through the polarizing microscope for proper exposure was a problem. Nevertheless, a framing rate of up to 20/s was obtainable, using Kodak 16-mm Tri-x reversal film processed in Kodak D-19 developer for the recommended time.

Measurements of wave displacement, basal inclination, and velocity were taken from the cine film with a Vanguard motion analyzer. The beat frequency of the axostyle was quantitated with a photomultiplier on the microscope as follows: A photomultiplier tube was attached to the ocular of the polarizing microscope, and a small segment of the axostyle only was illuminated by closing down the field diaphragm. The output of the photomultiplier was amplified by a simple two-step DC amplifier and recorded on a strip chart recorder capable of responding to ~20 cps with a 10-mm pen travel. As the azimuth angle of the illuminated portion of the axostyle changed relative to the polarizer and compensator angles, corresponding to a beat, the brightness of the axostyle changed, and the subsequent intensity variation was recorded on the graph paper. Thereby, each spike on the chart represented a beat.

Cells for electron microscopy were fixed for 1.25 h in a 2% glutaraldehyde solution made up in 0.05 M cacodylate buffer and 1% $CaCl_2$ at pH 7.4. The cells were washed in the cacodylate buffer with $CaCl_2$ plus 10% sucrose for 30 min and postfixed in 1% OsO_4 made up in the same buffer for 45 min. The cells were then dehydrated through a graded series of acetone solutions and, when in 100% acetone, embedded in Araldite. Thin sections were cut on a Sorvall Porter-Blum ultramicrotome II (DuPont Instruments-Sorvall, DuPont Co., Wilmington, Del.), stained with uranyl acetate and lead citrate, and viewed with a Philips EM 200 electron microscope.

RESULTS

Description of the Axostyle

In *Pyrsonympha* (Fig. 1), the axostyle originates from a rodlike structure at the anterior tip of the organism. This curved rod (19) has been shown to contain a single row of cross-bridged microtubules associated with a cross-hatched set of 5- to 10-nm filaments and a pair of centrioles surrounded by dense amorphous material (4, 7, 20, 30). In the phase contrast microscope, the

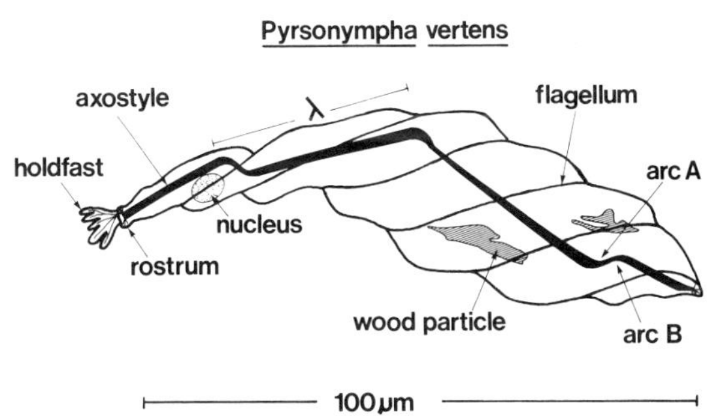

FIGURE 1 A diagram of *P. vertens*. The dominant features of the cell, as seen in the polarizing microscope, are shown. Three waves are depicted on the axostyle. The waves are sawtooth in shape with a sharp bend followed by a long straight region. λ refers to one wavelength. The bends consist of arcs A and B; arc A has a positive radius of curvature, and arc B has negative curvature. The irregularly-shaped objects in the posterior half of the organism are vesicles containing wood particles. Four flagella originate in the rostrum and wrap spirally around the organism. As the organism swims, it rotates about its long axis.

centriolar complex is visible as a 3-μm long, thin rod aligned perpendicular to the long axis of the axostyle (Fig. 2). In the polarizing microscope, this rod is positively birefringent (Figs. 3 and 11). Because it is aligned with its long axis normal (90°) to the length of the axostyle, which is also positively birefringent, the rod appears opposite in contrast to the axostyle. This rodshaped structure is termed the axial ribbon-basal body complex (7) and serves as an important marker for analysis of the axostyle's motion.

The singlet microtubules of the axostyle are aligned parallel to one another along the length of the organelle. In cross-sections of the axostyle, one sees that the microtubules are uniformly distributed in a semicrystalline array (Fig. 4). The microtubules of the axostyle are interconnected by the intrarow and interrow cross-bridges. The axostyle has ~75 rows of microtubules with 25–30 microtubules per row.

In the polarizing microscope, the retardation ([coefficient of birefringence] × [thickness]) of the axostyle is a measure of its thickness in the direction of the light path since the microtubules, which are responsible for the birefringence, have a uniform distribution and, hence, a uniform coefficient of birefringence. Because the axostyle is about three times wider than thick, it appears brighter or darker when it is viewed on its side (its narrow aspect), than when viewed from top or bottom (its wide aspect) (Fig. 3). It is possible to compare maximum and minimum apparent widths with the retardation at the corresponding points on the axostyle. When this is done, one finds that retardation varies inversely with width as predicted.

The axostyle, which measures ~100 μm in length, is helically twisted. The intrinsic twist in the axostyle is shown in the phase contrast micrograph of Fig. 2. One sees, between the two bends in the axostyle of this cell, differences in contrast which correspond to differences in width and thickness of the organelle. These variations in width and thickness result from the fact that the axostyle is helically twisted.

The degree of twist can be ascertained by measuring the distance between a point on the axostyle of maximum and an adjacent one of minimum width, or of minimum and maximum birefringence at a given instant in time. This length constitutes a rotation of 90°. When this measurement is made, the axostyle is found to be intrinsically twisted at a pitch of 133 ± 10 μm, or with an average twist of 2.7°/μm. As a result of the twist on the axostyle, planes defined by, for example, the bend region of two successive waves will be rotated ~68° with respect to one another.

Movements of the Axostyle

Axostyle-associated motility in *P. vertens* is characterized by a rhythmic flipping motion of the anterior region of the organism, and the movement of sharp bends along the helically twisted organelle. From analysis of axostyle movement on cine film (see set of frames in Fig. 3) these activities can be resolved into four basic movements: two primary motions, undulatory and flexual, and two secondary motions, rotational and precessional. The undulatory motion resembles wave propagation along flagella, and the flexual motion simulates the bending of cilia.

THE UNDULATORY AND ROTATIONAL COMPONENTS OF THE MOVEMENT: The undulatory component of the axostyle's motion constitutes the propagation of bending waves at an average frequency of 2/s. The propagated waves most closely resemble sawtooth waves whose wavelength varies between 20 and 35 μm. When actively beating, three or four waves travel along the axostyle at any given time (Fig. 3).

Each sawtooth wave contains a long straight segment which is ~22 μm and a bend which has two circular arcs with a combined arc length of ~3 μm (Fig. 1). The straight region appears to be stiff since it usually shows no tendency to bow. In the bend region, the two arcs are contiguous, one (arc A) having a positive and the other (arc B) a negative radius of curvature (Figs. 1 and 3). Arcs A and B may be functionally identical to the principal and reverse bends of flagellar waves (16). The radius, which averages 1.4 μm, of the two arcs varies independently as the wave propagates. The arcs subtend angles which range between 40° and 60°, and arc A is usually less variable than arc B. This small variation in angle appears to accommodate the cell contour and the ingested wood particles.

As bends propagate, they appear in different orientations as they travel along the twisted axostyle (Fig. 3). The bends are nearly planar and the plane of the bend is perpendicular to the plane which defines the width of the axostyle (Fig. 5) (22). Therefore, when a bend is observed in a region which displays the maximum width of the axostyle, the arcs of the bend are barely detectable (see wave in middle of axostyle in Fig. 3*n*–*p*).

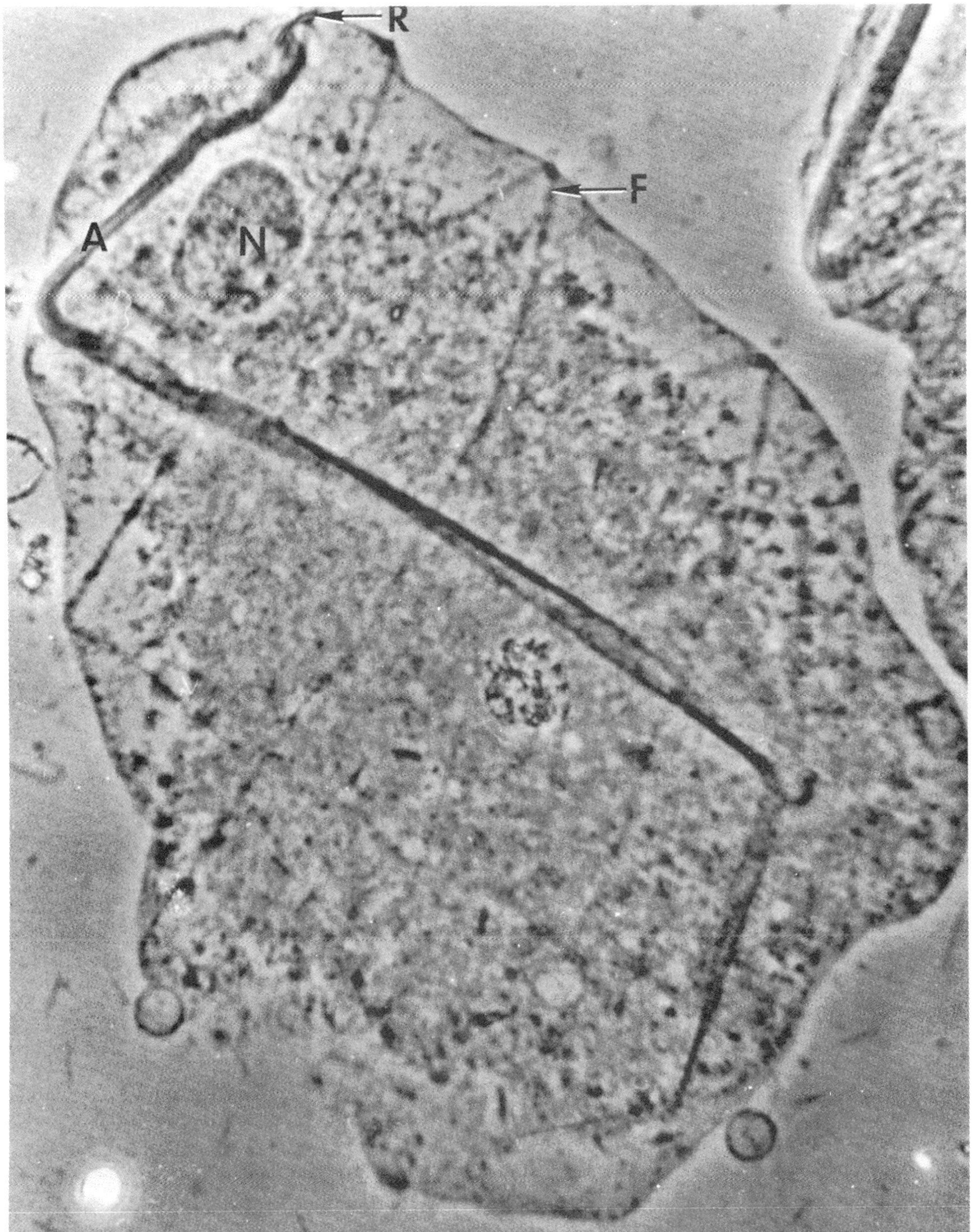

FIGURE 2 Phase contrast micrograph of *P. vertens* showing the helical twist in the axostyle. This cell is flattened to improve the image quality. Flattening prevented the cell from beating, and two inactive bends are present in the axostyle. The twist in the ribbon-shaped axostyle is visible as a variation in width and contrast of the organelle. The axostyle is ~3 × wider than thick. *A*, axostyle; *F*, flagellum; *N*, nucleus; *R*, axial ribbon-basal body complex.

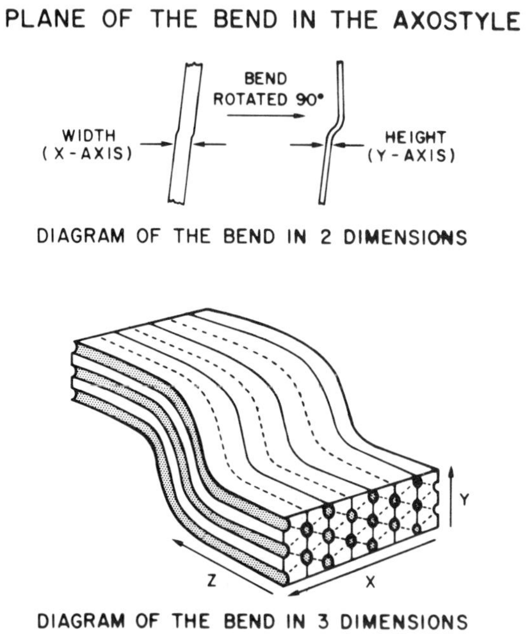

FIGURE 5 Diagram of the plane of the bend in the axostyle of *P. vertens*. In two dimensions, the bend is shown to be almost planar. When the axostyle is viewed with its width to the viewer, the bend is barely visible. When the axostyle is rotated 90°, the two contiguous arcs are seen. In three dimensions, orthogonal axes representing the length, height, and width of the axostyle are shown. A row of microtubules is aligned along the y axis, and the bend is in the y–z plane.

sees a 20% change in width at any given point on the axostyle as a wave approaches and passes that point. The change in width reflects exposure of different aspects of the axostyle which is shaped like a flattened rod with a width-to-height ratio of ~3:1.

In sections through the straight portion of the axostyle, the individual rows of microtubules are all curved, with curvature becoming greater as one progresses from the top of a row (the top of the axostyle is adjacent to the cell surface) to the bottom of it (bound by a row of microtubules set apart from the axostyle proper) (Fig. 10). In sections through the axostyle that are in or near a bend, the rows are not curved but straight (Fig. 4). Apparently, the curvature of the rows, which straightens only in the bend, serves to maintain the stiffness of the axostyle in the unbent regions.

Wave velocity often varies along the length of the axostyle, usually with a maximum velocity approaching 60 μm/s and an average velocity of 40 μm/s at 21° ± 1°C. A typical velocity profile is shown in Fig. 6. The position dependency of velocity is a result of obstructions in the path of the wave. When the wave experiences obstacles such as large cytoplasmic vesicles containing wood chips, neighboring cells, or solid extracellular debris, the velocity decreases. The wave is quite susceptible to impediment for two reasons. Firstly, the wave has a long straight region that is not deformable, and secondly, it follows a helical rather than a straight path. Therefore, when a wave tries to pass an obstruction, it usually cannot do so without a change in wavelength. To achieve this, wave velocity is retarded with a resultant wavelength change. When one wave is impeded, often the effect is transmitted to the oncoming wave. When waves do not experience obstructions, the velocity does not fluctuate along the length of the axostyle but remains constant.

Obstruction to wave passage can be imposed by gently flattening the cell between coverslip and slide. When this is done, an otherwise constant wave velocity can be caused to become variable along the axostyle. A simple reorientation of the cell between coverslip and slide can relieve the obstruction, and the velocity becomes constant again.

THE FLEXURAL AND PRECESSIONAL COMPONENTS OF THE MOVEMENT: The second most obvious movement of the axostyle is the flipping motion of the proximal end. This motion, which occurs with the formation of each new wave, is described as the flexural component of the beating axostyle. This cilialike movement has recovery and effective phases with each beat.

To illustrate this movement, a diagram showing

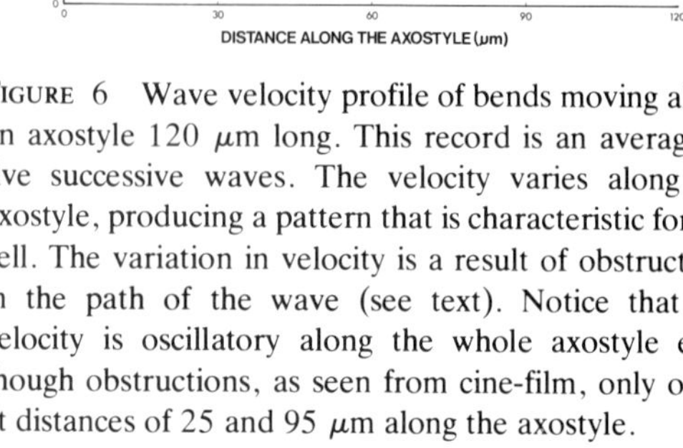

FIGURE 6 Wave velocity profile of bends moving along an axostyle 120 μm long. This record is an average of five successive waves. The velocity varies along the axostyle, producing a pattern that is characteristic for the cell. The variation in velocity is a result of obstructions in the path of the wave (see text). Notice that the velocity is oscillatory along the whole axostyle even though obstructions, as seen from cine-film, only occur at distances of 25 and 95 μm along the axostyle.

a series of profiles of the anterior one-fifth of the axostyle at different time intervals during its beat is given in Fig. 7. Also shown is a graphic record of (*a*) the time change of the angular inclination of the anterior one-fifth of the axostyle with the axial ribbon basal body complex, and (*b*) the position of the bend along the axis of the axostyle. The measurements were made on prints obtained from cine films of the axostyle taken at 18 or 20 frames/s. The detailed timing of the beat of an axostyle is often not constant from on cycle to the next, and so the record given is a typical one, since it is not possible to average out the individual differences between beats.

The recovery phase of the motion involves the initiation and movement of a bend a few micrometers along the axostyle (Fig. 7*a*, times 0 and 0.35 s). During this initial phase of movement the bend contains a single arc only (arc A) which has a

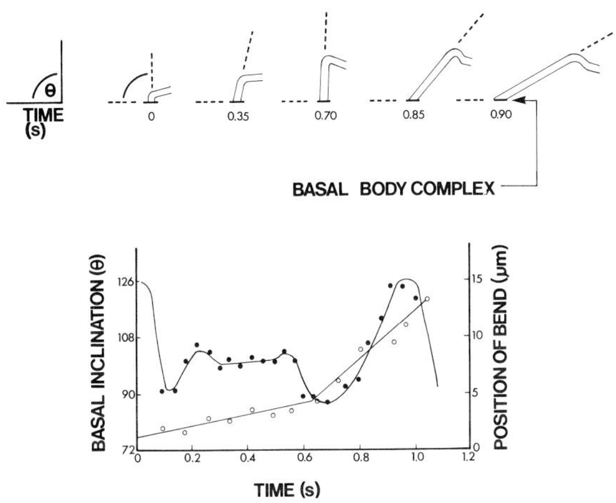

FIGURE 7 A diagram of the cilialike behavior of the anterior segment of the axostyle during the initiation of a wave. (*a*) A diagram of the anterior 1/5 of the axostyle. A thin bar at the base of the axostyle represents the axial ribbon-basal body complex. The angular inclination of the axostyle with the basal body complex is shown at five time intervals. The early phase of the beat (times 0, 0.35, and 0.7 s) is the recovery phase. The effective phase of the beat begins after time 0.7 s. (*b*) A graph of the information depicted in *a* showing the angle Θ, measured in degrees as a function of time (closed circles) and the position of the bend measured in μm, as a function of time (open circles). During the recovery phase, angle Θ increases from 90° to 100° and remains at 100°. Just before the effective phase, angle *θ* returns to 90° and quickly increases to 126°. Bend displacement (open circles) is diphasic, exhibiting a slow rate during the recovery phase and a fast rate during the effective phase.

"positive" radius of curvature. Just before the onset of the effective stroke, another arc (arc B) with a "negative" radius of curvature appears immediately distal to the initial arc which has now travelled a few micrometers from the proximal tip (Fig. 7*a*, time 0.85 s). With the formation of arc B, the axostyle swings through the effective stroke (Fig. 7*a*, times 0.85 and 0.90 s). The end of the effective stroke occurs with the formation of another bend at the base signalling the formation of a new wave.

The angular inclination of the axostyle to the axial ribbon basal body complex is shown in the graph in Fig. 7*b*. The axial ribbon basal body complex is shown in Fig. 7*a* as a thin rod at the very base of the axostyle aligned perpendicular to its long axis. As is seen in the graph, when the bend is first initiated at time 0.1 s the basal inclination is 90°. It quickly rises to 100° and stays at about this inclination for 0.5 s. Then the inclination quickly falls back to 90° which signals the onset of the effective stroke. This is also the time at which arc B is being formed. After returning to 90° the axostyle swings through the effective stroke forming an angle of 126°. This causes the initiation of another bend. The angle of inclination returns to 90° and the sequence is repeated with each new wave.

The movement of the bend along the axostyle during this flexual motion is not constant, but appears to be diphasic. During the recovery phase of the stroke and when the bend contains only one arc, the bend moves slowly (Fig. 7*b*). However, after the second arc is formed and during the effective phase, the bend moves more rapidly.

The fact that, during the recovery phase, the angle of inclination rises and then falls may be apparent rather than real. It may result from the rotation of the axostyle which affects the measured angle.

The fourth component of the axostyle movement is the precessional component which is exhibited ony at the rostrum. The whole anterior tip of the axostyle, together with the axial ribbon-basal body complex, precesses synchronously with the initiation of each wave. The precessing rostrum sweeps out an ellipse that could well be the compression of a circular path that appears elliptical, because of the orientation of the cell.

These four movements describe the vigorous beating of the axostyle. The flexural motion has the most profound effect on the locomotion of the organism. The cell uses this motion to reorient its path, particularly when it encounters other cells and debris.

PERTURBATIONS IN THE MOTION OF THE AXOSTYLE: Quite often when preparations are made of *P. vertens,* one observes deviations from the normal behavior associated with the axostyle. Some of these perturbations are described below.

Rotating Polygon Appearance

Sometimes, upon being released from the termite gut, these cells swell and lose their pear shape, becoming partially spherical. This happens frequently when saliva is used as the medium. When the cell takes on a more or less spherical shape, the axostyle is caused to form what approaches a loop. Waves continue to propagate on the axostyle, however. Waves on the axostyle in this configuration give it the appearance of an open rotating polygon (Fig. 8). The wave shape and properties are not affected. The angles of the arcs in the bend region change to accommodate this orientation. Again, arc A is less affected than arc B. At times, arc B is only slightly evident, to the point of being almost absent.

Local Loss of Stiffness

A frequent anomaly that affects the axostyle is the loss of stiffness over a short region. The region affected is usually the same in all cells, and it is the region in the area of the nucleus. The effect of the stiffness loss is to increase the radius of curvature of the bend region of the wave as it passes the affected region and to allow the straight region of the wave to bow (Fig. 8). After a wave passes the affected region, the wave shape returns to normal. The cause of this stiffness loss is not understood.

Reversal of Propagation Direction

Cells that have been beating in culture media for a few hours begin to show a decrease in both beat frequency and propagation velocity. The frequency becomes irregular and waves sometimes stop on the axostyle. On occasion, these waves reverse their propagation direction. Usually, when this occurs, the waves simply back up without any apparent change in the wave shape. In this case, the wave moves towards the basal body with arc A travelling ahead of arc B. On other occasions, the wave shape is seen to reverse before reversal of propagation direction. Velocity in the reverse direction can be as great as in the forward direction. The propagation direction can change back and forth many times, usually with cessation of activity between shifts.

Wave Initiation other than at the Base

Waves can be initiated at any point along the axostyle including the distal end, in which case the waves travel toward the base. Wave initiation other than at the base usually occurs after cells have been in culture medium for a few hours. Most frequently, waves initiate from preexisting waves that are temporarily stopped on the axo-

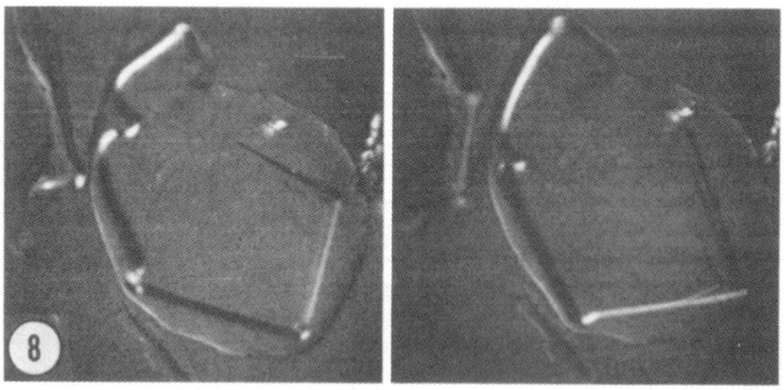

FIGURE 8 Polarized light micrographs of an axostyle. The axostyle has assumed a rotating polygon configuration. Waves continue to propagate along the axostyle in this configuration. They are reminiscent of the movement of the polygons observed by Kamiya (21) in the cytoplasm expressed from the fresh water alga *Nitella.* Note also that the segment between the second and third bends is bowed, appearing to exhibit a local reduction in stiffness. The bends in this region have larger radii of curvature than normal. This stiffness loss is localized to the region including these two bends. Waves past that region have the normal shape. × 500.

style. Many waves can initiate from this site. At other times, waves can be initiated by the axostyle's being deformed, because of contact with neighboring protozoans. In one instance, when a *Trichonympha* bumped into the axostyle of a *P. vertens*, a wave was initiated from that region both towards the centriole and towards the distal end. Sometimes, waves are initiated *de novo*. Wave shape and properties are the same, regardless of the point of origin.

Splitting of the Axostyle into Two or More Parts

Axostyles are frequently seen split into fragments. The splitting occurs along the longitudinal axis. Usually, the first split seen in an axostyle occurs more or less along the center, dividing the axostyle into two halves of equal widths (Fig. 9). These halves often divide several more times. Waves continue to propagate on the fragments. As would be expected, the small fragments are not nearly as stiff as the intact axostyle, and as a result there is a tendency for the straight portion of the wave to bow and for the arcs to take on a larger radius of curvature. Propagation velocity is not affected. Splitting occurs between the microtubular rows (Fig. 10).

Effects of Temperature on the Axostyle Beat

Changes in temperature have an immediate effect on the rate of wave initiation and the propagation velocity. Using a temperature control slide and a photomultiplier to record beat frequency (as described in Materials and Methods), one finds that at temperatures below ~5°C the movement stops altogether, but as soon as the temperature is raised the axostyle starts to beat again, first sluggishly, and then vigorously in the course of a minute or so. Birefringence, however, is not altered by low temperature. In general, at temperatures between 5° and 30°C, one sees a proportionate increase or decrease in wave propagation velocity as temperature is raised or lowered.

After lowering the temperature and then returning it to ambient, propagation velocity usually returns to normal before frequency does. As a result, axostyles recovering from lowered temperature will have only one or two waves travelling on them during the early part of the recovery stage. As the axostyle accommodates to ambient temperature, the number of waves approaches normal. The general wave shape is not affected during this process.

Microbeam Experiments

IRRADIATION WITH VISIBLE LIGHT: Axostyles illuminated with intense visible white light become sluggish and quickly stop beating. By irradiating only a small part of the axostyle with a microbeam of visible light (described in Materials and Methods), a local effect can be produced. When a 10-μm segment of the axostyle is irradiated, the following effect is observed. The waves on the axostyle up to the spot of illumination are not altered. Their shape, wavelength, and propagation are unchanged. At the point of irradiation, however, the shape of the wave changes from a sawtooth to a short wavelength "sinelike" wave,

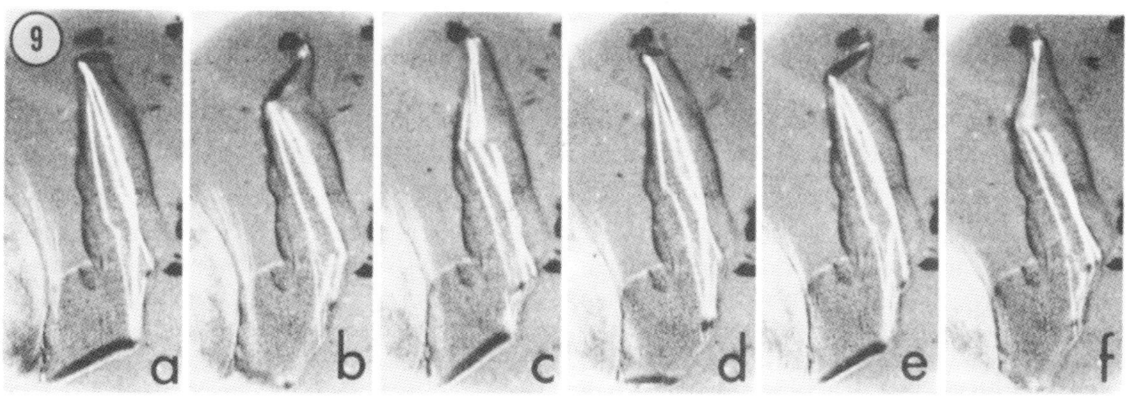

FIGURE 9 Polarized light micrographs of a cell whose axostyle is split along its center into three fragments. The fragments can be seen to beat somewhat out of phase. Sequential frames, 18 frames/s. × 500.

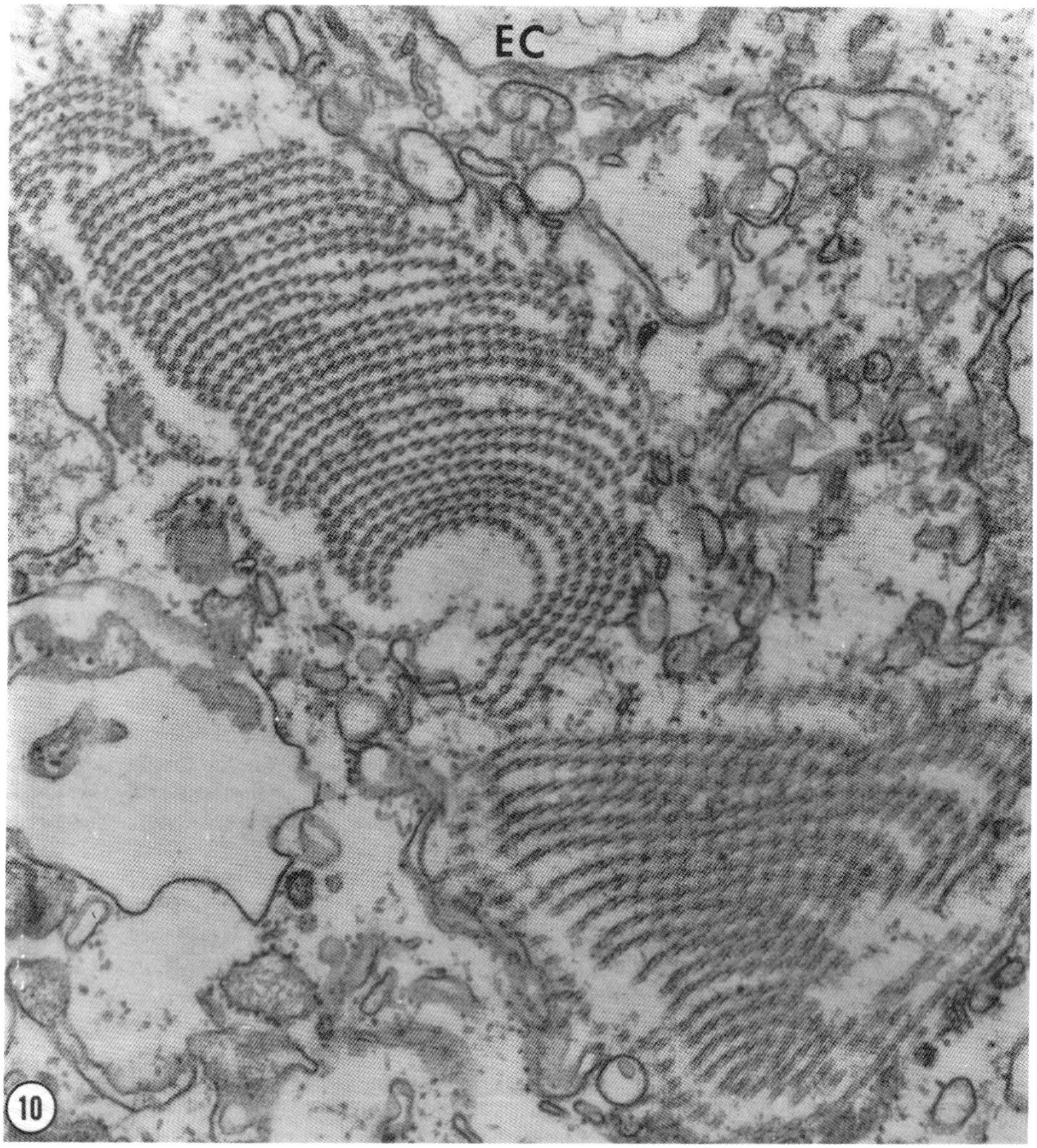

FIGURE 10 Electron micrograph cross-section of an axostyle that is split into two pieces. The splitting occurs between rows and usually not within rows. *EC*, extracellular space. × 54,000.

and this "modified" wave usually does not propagate. As a result, sinelike waves accumulate in the irradiated area (Fig. 11). At times, when enough waves arrive at that area to fill it, some "slip through" in spurts and continue into the distal end of the axostyle (Fig. 11). It appears as though waves entering the affected region push along, or stimulate the propagation of waves accumulated there. The distal end then beats as a sawtooth wave, indicating that only the irradiated region of the axostyle is altered. The usual cessation of propagation of the short wavelength sinelike waves in the irradiated area makes it appear as though the irradiated area has acquired a higher viscosity or stiffness, and the waves have greater difficulty traversing that region. A slight degree of reversibility is sometimes observed in the irradiated area, but generally the effect is not reversible.

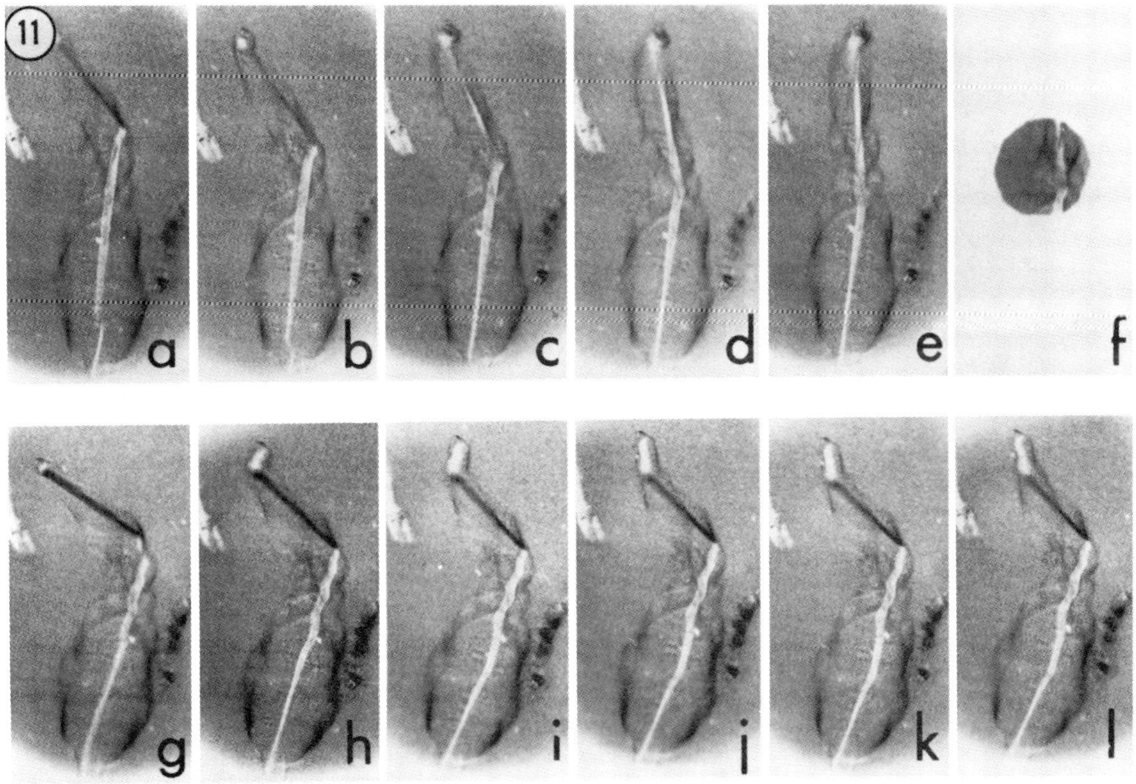

FIGURE 11 Polarized light micrographs of an axostyle irradiated with a microbeam of blue light. The first row of micrographs (*a–e*) shows the normal beating axostyle. Micrograph *f* shows the area of the axostyle irradiated with the blue light microbeam. The second row of micrographs (*g–l*) shows the effect of the irradiation. In the area irradiated, short sine waves accumulate and fill the region. Waves that slip through the microbeamed region assume the normal sawtooth waveshape. There is no reduction of birefringence in the area irradiated. Sequential frames, 18 frames/s. × 500.

Once the propagation stops in the affected region, very seldom does it start up again. When waves can no longer slip through the irradiated region, a wave sometimes spontaneously starts from the distal end of the axostyle and travels up to that region. A rough estimation of the action spectrum for this effect was determined by using either a blue, green, or red glass filter over the light source when irradiating cells. Wave modification was observed only when the axostyle was irradiated with blue light. We therefore describe this modification as a blue light effect.

IRRADIATION WITH ULTRAVIOLET: An axostyle irradiated with a microbeam of UV light responds differently than when irradiated with blue light. UV light severs the axostyle into two pieces (Fig. 12). The birefringence in the microbeamed region disappears, which means that the elements responsible for the birefringence, i.e., the microtubules, must have broken down. The anterior end continues to initiate and propagate waves, the posterior half does not. Short exposures to UV, insufficient to sever the axostyle, sometimes caused waves to stop short of the irradiated region. Low dosages of UV did not, however, duplicate the blue light effect.

A surprising observation was made after some of the axostyles were severed with UV. In several cases, the severed ends began to move together. This movement did not appear to be passive, resulting from forces in the cell cytoplasm or cell surface, but to be movement resulting from activity in the irradiated region itself. This apparently "active" movement could be explained if there were a few microtubules connecting the two severed pieces that could serve as elements for somehow moving the pieces together.

An interesting phenomenon occurs when an axostyle is severed in two places, separated by about one wavelength. The short segment pro-

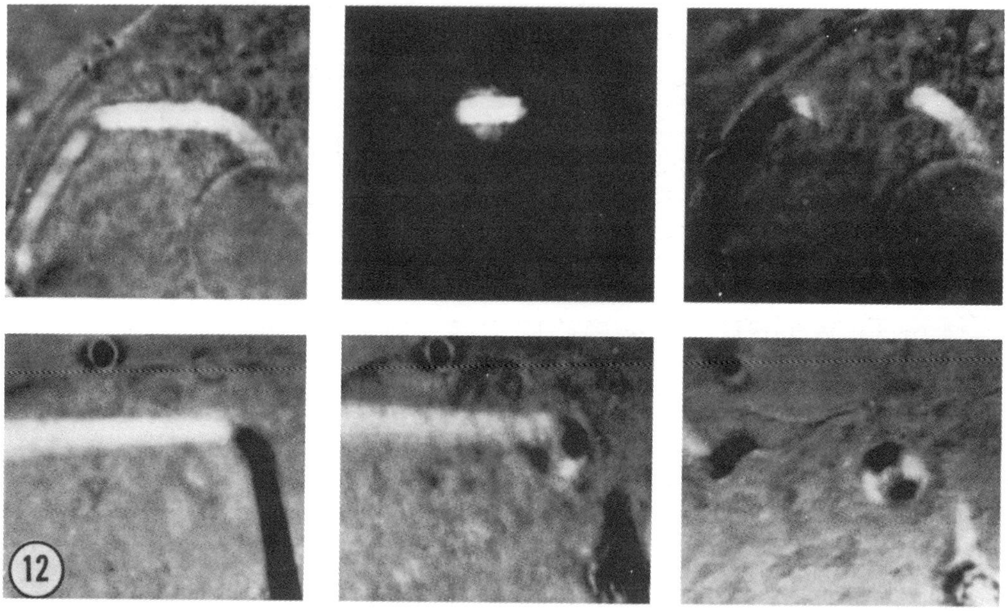

FIGURE 12 Polarized light micrographs of the axostyle irradiated with a UV microbeam. The top row of three micrographs indicates before, during, and after irradiation. Before irradiation, the axostyle had stopped beating. After irradiation, the birefringence in the microbeamed region is gone, and the severed ends curl. The bottom row of three micrographs shows an axostyle that has been irradiated at two points on either side of a bend that had momentarily stopped on the axostyle. The third micrograph of this series shows that the short piece curls upon itself to form a lock-washer with radius similar to normal bend radius. (The ends are displaced in the plane perpendicular to the plane of the photograph.) × 1,200.

duced by the UV microbeam curls upon itself into a lock-washer configuration (Fig. 12). This curling phenomenon was also observed at the two ends, produced when the axostyle is severed in only one place. The ability to curl is significant in that the mechanism may be similar to that involved in wave formation.

DISCUSSION

Cilia- and Flagella-like Movements of the Axostyle

The ribbon-shaped axostyle of *Pyrsonympha* contains a highly ordered bundle of microtubules. The semicrystalline organization of the microtubules of this axostyle has permitted an analysis of its movements in vivo. In cross-section, the axostyle is ~3 × wider than thick and hence the retardation, as measured with the polarized light microscope, varies according to whether the light beam passes through the width or thickness of the axostyle. Therefore it has been possible to deduce the exact three-dimensional waveform of the axostylar motion by coupling polarized light microscopy and cinematography.

We show that the axostyle has both cilia- and flagella-like movements. As with a beating cilium (29), the bending of the anterior one-fifth of the axostyle has a recovery and an effective phase. Much like the principal and reverse bends of flagella (16), the propagating waves contain two contiguous arcs with curvature in opposite directions. These S-shaped bends or arcs are separated by straight regions, the net appearance being that of sawtooth-shaped waves that flow rhythmically along the axostyle.

The bends on the *Pyrsonympha* axostyle were previously described by Smith and Arnott (31) as half sine waves. This term does not adequately describe the entire wave form which contains both a bent and a straight region. Therefore, the more descriptive term, sawtooth waves, was adopted in this paper.

The flexual bending of the anterior part of the axostyle develops the sawtooth wave and causes the anterior tip of the organism to precess. The propagating sawtooth waves cause the axostyle to roll back and forth about its long axis. The precessional and rotational movements occur because the axostyle is intrinsically twisted into a helix of low pitch.

The propagating bends are nearly planar and the twist in the axostyle appears to dissipate in the region of the bend. Three or four sawtooth waves travel along the axostyle at any given time at an average velocity of 40 μm/s. The velocity varies along the axostyle apparently because of differences in the physical conditions that the waves experience.

The undulations in the axostyle in *Pyrsonympha* and *Saccinobaculus* resemble each other but differ in several respects. In both organisms, the axostyle bends to produce a circular arc which propagates down the twisted organelle. Each wave in the axostyle of *Pyrsonympha*, however, contains two arcs with opposite curvature while the wave in the axostyle in *Saccinobaculus* usually contains a single propagating arc (25). The radii of the arcs on the two axostyles differ too. In *Saccinobaculus*, the axostyle bends to a radius of ~8 μm with an arc ranging from 60° to 180°, while in *Pyrsonympha* the axostyle forms two, sharp, contiguous arcs, each with a radius of ~1.4 μm and each with an arc ranging from 40° to 60°. In *Saccinobaculus*, therefore, a larger portion of the axostyle is included in each bend than is true in *Pyrsonympha*.

The Plane of Bending in the Axostyle and the Position of the Dynein Cross-Bridges

The plane of bending in relation to the alignment of the microtubular rows differs in the axostyles of *Saccinobaculus* and *Pyrsonympha*. In *Pyrsonympha*, we show that the plane of the bend is perpendicular to the width of the axostyle. The diagram in Fig. 5 shows orthogonal axes labeled x, y, and z representing the width, height, and length of the axostyle, respectively. In cross-section, microtubular rows appear in succession along the x axis (Fig. 5). The microtubular rows themselves lie in the y–z plane. The bending also takes place in the y–z plane. This means that bending is parallel to the plane of the rows and that microtubules within a row must be displaced relative to one another during bending. This situation does not pertain in the axostyle in *Saccinobaculus*. There, the plane in which bending occurs is perpendicular to the axostyle's width and to the plane that contains the rows of microtubules.Therefore, in the *Saccinobaculus* axostyle, relative sliding of rows is expected.

Since bending is parallel to the rows in the axostyle of *Pyrsonympha*, adjacent microtubules within a row must be displaced during bending. This means that the intrarow cross-bridges must either detach or deform during the bending process. On the basis of electron microscope observations, Smith and Arnott (31) report that in the bend region of the *Pyrsonympha* axostyle the intrarow cross-bridges appear not to completely span the space between adjacent microtubules within a row. This observation may suggest that intrarow cross-bridges detach during bending.

On the other hand, the intrarow cross-bridges may not detach but stretch to accommodate the bend. This would require that the intrarow cross-bridges stretch to about twice their rest length or by ~20 nm. This is the maximum displacement that should develop between adjacent microtubules within a row when the axostyle bends to a radius of curvature in the y–z plane of 1.5 μm and forms an arc which extends over 60°. In fact, such an elastic deformation may even explain the S-shape of the bend. The possibility exists that the B curvature forms to relieve the strain on the intrarow bridges produced by the A curvature. Elastic deformation of the intrarow cross-bridges may be coupled with contraction of the tubules on the inside of the bend. Microtubule contraction has been demonstrated in the axostyle of *Saccinobaculus* (25).

The position of the dyneinlike bridges in the *Pyrsonympha* axostyle has not been firmly established. The difficulty in establishing this fact is a result of the inability to grow these anaerobic protozoa in culture so as to obtain enough cells for biochemical studies. The existing evidence suggests, however, that the dyneinlike enzyme is in the intrarow bridges (2, 26) and that nexin is in the intrarow bridges (26). These data are tenuous at best and are based principally on the appearance of the cross linkers in electron micrographs. Dyneinlike bridges are thought to be more delicate and less regular in appearance than structural bridges.

If the dyneinlike cross bridges generate force within the bend regions of the axostyle, these force-producing cross-bridges should be perpendicular to the width. The one set of bridges, which are aligned at 90° to the width of the axostyle and are delicate in appearance like putative dynein bridges, are the linkers which connect a single row of tubules running along the inner face of the axostyle to the rest of the axostyle. This row of tubules, seen in the cross-sections of the axostyle in Figs. 4 and 10, has strong (intrarow) bridges parallel to the width and weaker, more irregular bridges normal to the width of the axostyle. These delicate bridges attached to this row have the proper orientation to serve as the force-generating cross-bridges for bending. However, force generation may not occur in the bend region as will be discussed below.

The uniform distribution of dyneinlike cross-bridges along the axostyle is supported by observation of the wave motion in the *Pyrsonympha* axostyle. We show that sawtooth waves initiate from any point on the axostyle including the distal end, an observation that corroborates the observation of others (4, 31). In addition, we show that the amplitude of the bend curvature is not dampened as waves propagate, and that waves can speed up in transit after having been slowed by blue light irradiation or by physical constraints in the cytoplasm of the cell or in the medium. These observations suggest that the dyneinlike force-producing cross-bridges (14, 17) which convert chemical energy into mechanical work are uniformly distributed along the axostyle.

We show that the sawtooth waves contain two contiguous arcs with curvature in opposite directions. This could mean that each active cross-bridge can generate force in either direction depending on the direction of curvature. This possibility has generally not been favored (5, 6) because it is inconsistent with our knowledge of the properties of the cross-bridges in striated muscle, in which case the myosin heads are polarized and are believed to only generate force in one direction. However, our data from the axostyle of *Pyrsonympha* strongly suggest that active sliding in either direction is possible.

The propagation of sawtooth waves becomes more difficult to explain on the basis of unidirectional force generation when one considers that waves sometimes reverse propagation direction without wave shape reversal. We observed that, when the direction of movement of some waves reversed, the relationships of the arcs remained unchanged so that the leading arc (arc B) before the reversal maintained its negative curvature as it became the trailing arc after reversal. This reversal of propagation without a concomitant wave shape reversal again suggests that active sliding in either direction is possible.

The microtubules of the axostyle in *Pyrsonympha* originate from an anteriorly placed axial ribbon (4, 7, 20, 31). Unlike axonemal microtubules whose proximal ends are anchored by a basal body, the proximal ends of the axostylar microtubules are simply embedded in an amorphous matrix. The nature of the amorphous material around the tubule ends is unclear, but the lack of a basal-body-like attachment may permit displacement of the tubule ends at the proximal tip. The absence of a basal-body attachment of the axostylar microtubules at the proximal end, in contrast to the situation which pertains in cilia and flagella, may be an important consideration for axostyle movement.

Perturbations of Axostyle Motility

The motions of the axostyle are perturbed by UV and blue light microbeam irradiation and other environmental factors such as changes in temperature. These perturbations are of interest because they help identify constraints on models that are devised to explain the mechanism and coordination of bending. One important modifier of axostyle motility is temperature, and we report here that velocity and wave initiation respond differently to changes in temperature. When the temperature of the medium is lowered to 5°C for a short time and then returned to 21°C, wave velocity returns to normal before frequency does. This is interpreted to mean that the factor(s) which controls the initiation event (an ionic gradient perhaps) is different from the factor which controls the on-off activity of the cross-bridges (bend curvature perhaps). If cross-bridge control is mechanical (6) while wave initiation is chemical, one may expect that the cell membrane or vesicle membranes which regulate the ionic gradient are sensitive to temperature while the mechanical events would not be.

Blue light, another perturbant of axostyle motility, affects both velocity and wave form simultaneously. The waves which accumulate in the irradiated region arise as and are similar in form to the A and B arcs which travel into that region. The birefringence of the irradiated region does not change, suggesting that the microtubules themselves are intact. Our data do not allow us to determine which of the two sets of bridges, if either, is affected. Nevertheless, the fact that wave form and velocity revert to normal beyond a blue light-affected region of the axostyle strongly suggests that the factors which regulate wave form and velocity are local properties of the axostyle.

The blue light effect is likely a result of the

presence of a photosensitizer of either an endogenous or exogenous source in close association with the axostyle. The photosensitizer appears to be intimately associated with the axostyle because the response to irradiation is rapid and irradiation of the cytoplasm near, but not including, the axostyle does not produce a response. Birefringence of the axostyle is not affected by blue light and, therefore, the overall structure of the microtubules is not affected. The blue light response is definitely not a result of a local increase in temperature. Using, as a model, a liquid crystal whose color changes sensitively with temperature (27), we found no temperature increase by the same dosages of irradiation. In addition, when a theoretical calculation is made based on the spot size of the microbeam and the conductive properties of water, we determined that the energy available is insufficient to raise the temperature anywhere near 1°C.

A microbeam of UV light severs the axostyle in two, and the severed ends can coil into circular arcs much like the arcs which propagate on the axostyle. The coiling of microtubules at the severed ends may be the result of the tendency of tubules to coil when not otherwise restrained (12) or the result of active cross-bridges which normally produce sliding.

Perturbations, which are not yet clear to us, cause the axostyle to lose stiffness. The reduction in stiffness results in a larger radius of bending which occurs in a region of the axostyle near the nucleus. This region of the axostyle is where the second of the two arcs of a sawtooth wave (arc B) forms and the flexual motion becomes undulatory. The loss of stiffness may reflect a loss of curvature of the microtubular rows and a loss of the crescent or cupped shape of the axostyle.

The Position of Force Production in the Sawtooth Waves

The position of active force production in the axostyle of *P. vertens* is of special interest. If one accepts the assumption that force is produced by the interrow cross-bridges and that relative sliding of microtubular rows takes place, it is possible that active sliding between microtubular rows may not take place in the S-shaped bends but rather in the non-bent segments of the axostyle. While intuitively the S-shaped bends may appear to be the active regions (in analogy with the bends of sperm flagella, for example), the following observations on the axostyle and a ribbon model of the axostyle suggest that the converse may, in fact, be true.

The rows of microtubules in the axostyle of *P. vertens* are straight (orthogonally arranged) in cross-sections of bends, and the axostyle bends in a plane which includes the straight rows, rather than at right angles to the rows. Conversely, in the unbent segments of the axostyle, the microtubular rows are curved and the cross-section of the axostyle itself is cupped. Therefore, corresponding microtubules in adjacent rows are displaced relative to their arrangement in the bend region. Given these observations, a model which simulates the beating axostyle can be made from a flexible ribbon of paper. A 2-cm wide ribbon of bond paper, ~30 cm long, is considerably stiffer when its cross-section is cupped (and its moment increased) compared to when it is not cupped, but flat. (Consider the rigidity of a thin metal tape measure [25] which becomes flat and more easily bent when an external bending force exceeding a critical value is applied.) The flat ribbon of paper held upright at its base cannot support its own weight and will bow at its base. If at the ribbon's base a curvature along the width of the ribbon is introduced by squeezing the width of the ribbon, the moment of the ribbon increases and the ribbon erects itself progressively from the base upwards. The bend travels upwards as the ribbon is squeezed. When the moment is reduced by squeezing the ribbon less, the ribbon can no longer support itself and the bend travels down. Such a progressive stiffening, which results in the travel of a bend, seems precisely to take place rhythmically at the rostral end of the axostyle where arc A is formed.

Now, if the paper ribbon model is constrained by loosely holding the distal end of the ribbon in a horizontal direction, then a second bend equivalent to arc B appears as arc A is made to travel from the base upwards. If the ribbon is somewhat twisted, a back and forth rotation of it coupled with rhythmically increased stiffness induced at its base, sends a flipping sawtooth wave down the length of the ribbon. This model flips in a lifelike imitation of the beating axostyle in *P. vertens*.

We may thus conjecture that the sawtooth wave in the axostyle of *P. vertens* is generated by interrow cross-bridges which are active in the straight regions. The active cross-bridges curve the microtubular rows and cup the axostyle, thereby stiffening the latter. The axostyle, upon encountering its own twist and the imposed internal and cytoplasmic constraints, assumes the A and B arcs of the S-shaped bend which is pushed ahead of the active stiff (straight) region. We might then interpret the several bends, accumulated in the blue light-irradiated region, as bends which pile up in the absence of active sliding between interrow tubules or as the inability of interrow linkers to dissociate or reassociate.

In conclusion, the axostyle is shown to have many features in common with cilia and flagella but to be simpler in structure and organization and therefore to be a good model of microtubule-associated movement. Since the axostyle in *Pyrsonympha* exhibits both cilia- and flagella-like behavior, this dualism further emphasizes the similarities between these two bending processes. The axostyle has a semicrystalline arrangement of microtubules, and therefore it may be suitable for analysis by x-ray diffraction. Studies which will expand our knowledge of the composition of each of the different cross-bridges within the axostyle will significantly help understand the mechanism which localizes and coordinates bending in the axostyle. A radial spoke-central sheath complex is not required for bend formation, since these structures are not present in the axostyle.

We thank Dr. T. Punnett for his help with the formulation of the protozoan medium, Mrs. Bush for her expert technical assistance with electron microscopy, Ira Sabran for his contributions to the UV microbeam experiments, and Drs. G. Ellis and H. Sato for their continued advice and assistance during the course of this work.

This work was partly supported by National Institutes of Health (NIH) Postdoctoral Fellowship GM 49352 and Marine Biological Laboratory, (MBL) Steps Toward Independence Fellowship to G. Langford; NIH grant CA 10171 and National Science Foundation grant GB 31739 awarded to S. Inoué.

Received for publication 15 September 1978, and in revised form 30 October 1978.

REFERENCES

1. Ballantine, R. 1954. High efficiency still for pure water. *Anal. Chem.* **26:**549–550.
2. Bloodgood, R. A. 1975. Biochemical analysis of axostyle motility. *Cytobios.* **14:**101–120.
3. Bloodgood, R. A., and K. R. Miller. 1974. Freeze-fracture of microtubules and bridges in motile axostyles. *J. Cell Biol.* **62:**660–671.
4. Bloodgood, R. A., K. R. Miller, T. B. Fitzharris, and J. R. McIntosh. 1974. The Ultrastructure of *Pyrsonympha* and its associated microorganisms. *J. Morphol.* **143:**77–106.
5. Brokaw, C. J. 1971. Bend propagation by a sliding filament model for flagella. *J. Exp. Biol.* **55:**289–304.
6. Brokaw, C. J. 1972. Flagellar movement: a sliding filament model. *Science (Wash. D. C.).* **178:**455–462.
7. Brugerolle, G. 1970. Sur l'ultrastructure et la position systematique de *Pyrsonympha vertens* (Zooflagellata Pyrsonymphina). *C. R. Hebd. Seances Acad. Sci. Ser. D. Sci. Nat.* **270:**966–969.
8. Cleveland, L. R. 1923. Correlation between the food and morphology of termites and the presence of intestinal protozoa. *Am. J. Hyg.* **3:** 444.
9. Cleveland, L. R. 1924. The physiological and symbiotic relationships between the intestinal protozoa of termites and their host, with special reference to *Reticulitermes flavipes* Kollar. *Biol. Bull. (Woods Hole).* **46:**178–201.
10. Cleveland, L. R. 1950. Hormone-induced sexual cycles of flagellates. III. Gametogenesis, fertilization, and one-division meiosis in *Saccinobaculus. J. Morphol.* **86:**215–227.
11. Cleveland, L. R., S. R. Hall, E. P. Saunders, and J. Collier. 1934. The wood-feeding roach, *Cryptocercus*, its protozoa, and the symbiosis between protozoa and roach. *Mem. Am. Acad. Arts Sci.* **17:** 185–342.
12. Costello, D. P. 1973. A new theory on the mechanics of ciliary and flagellar motility. II. Theoretical considerations. *Biol. Bull. (Woods Hole).* **145:**292–309.
13. Fitzharris, T. R., R. A. Bloodgood, and J. R. McIntosh. 1972. The effect of fixation on the wave propagation of the protozoan axostyle. *Tissue Cell.* **4:**219–225.
14. Gibbons, I. R. 1963. Studies on the protein components of cilia from *Tetrahymena pyriformis. Proc. Natl. Acad. Sci. U. S. A.* **50:**1002–1010.
15. Gibbons, I. R. 1977. Structure and function of microtubules. *In* International Cell Biology 1976–1977. B. R. Brinkley and K. R. Porter, editors. The Rockefeller University Press, New York. 348–357.
16. Gibbons, B. H., and I. R. Gibbons. 1972. Flagellar movement and adenosine triphosphatase activity in sea urchin sperm extracted with Triton X-100. *J. Cell Biol.* **54:**75–97.
17. Gibbons, I. R., and A. J. Rowe. 1965. Dynein: a protein with adenosine triphosphatase activity from cilia. *Science (Wash. D. C.).* **149:**424–426.
18. Grasse, P. P. 1956. L'ultrastructure de *Pyrsonympha vertens* (zooflagellata Pyrsonymphina): les flagelles et leur coaptation avec le corps, l'axostyle contractile, le paraxostyle, le cytoplasme. *Arch. Biol.* **67:** 595–611.
19. Grimstone, A. V., and L. R. Cleveland. 1965. The fine structure and function of the contractile axostyles of certain flagellates. *J. Cell Biol.* **24:**387–400.
20. Hollande, A., and J. Carruette-Valentin. 1970. La lignee des Pyrsonymphines et les caracteres infrastructuraux communs aux genres *Opisthomitus, Oxymonas, Saccinobaculus, Pyrsonympha* et *Steblomastix. C. R. Hebd. Seances Acad. Sci. Ser. D Sci. Nat.* **270:**1587–1590.
21. Kamiya, N. 1959. Protoplasmic streaming. *Protoplasmatologia.* **3a:** 150–155.
22. Langford, G. M. 1975. The plane of bending in the axostyle of *Pyrsonympha vertens. Biosystems.* **7:**370–371.
23. Langford, G. M., S. Inoue, and I. R. Sabran. 1973. Analysis of axostyle motility in *Pyrsonympha vertens. J. Cell Biol.* **59**(2, Pt. 2): 185a. (Abstr.).
24. McIntosh, J. R., E. S. Ogata, and S. C. Landis. 1973. The axostyle of *Saccinobaculus*. I. Structure of the organism and its microtubule bundle. *J. Cell Biol.* **56:**304–323.
25. McIntosh, J. R. 1973. The axostyle of *Saccinobaculus*. II. Motion of the microtubule bundle and a structural comparison of straight and bent axostyles. *J. Cell Biol.* **56:**324–339.
26. Mooseker, M. S., and L. G. Tilney. 1973. Isolation and reactivation of the axostyle: Evidence for a dynein-like ATPase in the axostyle. *J. Cell Biol.* **56:**13–26.
27. Nicklas, R. B. 1973. Methods for gentle, differential heating of part of a single living cell. *J. Cell Biol.* **59:**595–600.
28. Provasoli, L. 1958. Nutrition and ecology of protozoa and algae. *Ann. Rev. Microbiol.* **12:**279–308.
29. Sleigh, M. 1968. Patterns of ciliary beating. *Symp. Soc. Exp. Biol.* **22:**131–150.
30. Smith, H. E. 1970. Axostyle structure and function in *Pyrsonympha vertens* with description of flagella, scales, and associated microorganisms. Ph.D. Thesis. University of Texas at Austin, Austin, Texas.
31. Smith, H. E., and H. J. Arnott. 1974. Axostyle structure in the termite protozoa *Pyrsonympha vertens. Tissue Cell.* **6:**193–207.
32. Stephens, R. E., and K. T. Edds. 1976. Microtubules: structure, chemistry and function. *Physiol. Rev.* **56:**709–777.
33. Summers, K. E., and I. R. Gibbons. 1971. Adenosine triphosphate-induced sliding of tubules in trypsin-treated flagella of sea-urchin sperm. *Proc. Natl. Acad. Sci. U. S. A.* **68:**3092–3096.
34. Warner, F. D., and P. Satir. 1974. The structural basis of ciliary bend formation: radial spoke positional changes accompanying microtubule sliding. *J. Cell Biol.* **63:**35–63.

The following note was added by Shinya Inoué in September of 2006:

Much of the work described in this article was carried out by Dr. George Langford while he was a postdoctoral fellow in our Biophysical Cytology Program in the Biology Department at the University of Pennsylvania. Using electron microscopy of glutaraldehyde-fixed material as well as polarization microscopy of live *Pyrsonympha*, he expanded extensively on the earlier work that I had carried out while I was a graduate student at Princeton.

In my earlier work (unpublished) using the polarizing microscope that I had built (Article 7 in this *Collected Works)*, I found the saw-tooth-shaped beating pattern of *Pyrsonympha*'s birefringent axostyle and its conversion to a sine wave upon local irradiation with blue light. In 1950, I also had the pleasure of demonstrating these striking features of the axostyle to Professor Fauré-Fremiet who was visiting Princeton from Paris. This world-renowned protozoologist was kind enough to remember and refer to the exciting experience at Princeton in a memoir he wrote some ten years later.

Interestingly, in the current report by George Langford, some details of the beating pattern, and especially the planes of the sharp bends described (Fig. 5), appear somewhat different from the pattern that I recall observing earlier. In the organisms that I examined at Princeton, the axostyle had a ca. 2 to 1 cross section (more or less as can be inferred from Fig. 4 in the present paper). Thus, one would see a narrow but more highly birefringent profile or a wider but less birefringent profile depending on which way the axostyle was tilted around its long axis. Using those criteria, I could tell that the axostyle underwent a 90° twist within a single straight region of a saw-tooth wave, presenting a narrow profile at the beginning (anterior end) of the straight region which gradually became wider until it presented its widest profile just before the next sharp bend. At that point, or as it entered the first flex of the bend, the axostyle underwent a sudden 90° twist so that the sharp bend itself presented a narrow profile. Following the double bend, the straight region of a saw-tooth wave started again with a narrow profile. Thus, at least the second part of the sharp bend must be a natural 90° flex of the type illustrated in Fig. 5.

The sudden twist of the axostyle at, or just before entering, the first part of the sharp bend represents a most unlikely deformation! Interestingly, this whole pattern of deformation of the *Pyrsonympha* axostyle can be modeled by taking a metal tape measure (which is manufactured with a slight concavity on its front surface so that it can be held out without flexing). By holding two parts of such a tape measure several inches apart, with one hand located somewhat above the other, and by pushing the two parts of the tape measure closer, one can generate a double bend of the type seen in Fig. 5. This configuration of the tape measure is quite unstable and requires considerable force to maintain. But when one side of the tape measure is twisted relative to the bend, the instability becomes even greater, and the bends tend to move rapidly away from the twisted side. As the model shows, the cup-shaped cross section of the axostyle may well explain the "saw tooth followed by sharp bend" propagation pattern of the *Pyrsonympha* axostyle. It may likewise explain what is changing in the region irradiated with blue light, as follows.

In contrast to the long straight regions, the sine waves accumulated in the irradiated region showed birefringence and bend patterns that represent no lengthwise twisting of the axostyle. The inability of the irradiated region to generate the twist (probably by losing its cup-shaped cross section) may well explain why the sinusoidal waves accumulate only one at a time as a new wave arrives in the irradiated region. That could also explain how, once the irradiated region is full of sine waves, the final push from a new saw-tooth wave preceding the irradiated region initiates a saw-tooth wave in the non-irradiated, posterior region (where the axostyle cross section is presumably cup-shaped).

In spite of the striking birefringence exhibited by the *Pyrsonympha* axostyle, my attempt at electron microscopy, which I carried out together with Mike Watson (?) while at the University of Rochester, failed to show any fine structure that reflected the axostyle's birefringence. No sign of an ordered structure, or presence of microtubules (which were not yet discovered in those days), was found in our attempt in the mid 1950s using osmium fixation, as was then standard for electron microscopy. In fact, what we found by observing cells ***while*** they were being fixed was that the birefringence of the axostyle disappeared completely in the osmium fixative in less than one minute!

It is ironic that, with the use of glutaraldehyde fixation, the axostyle is now known to be made up of a semi-crystalline array of a large number of microtubules (as shown by Gibbons, McIntosh, Langford, and others in the 1970s). Thus, a detailed analysis of the difference in the state of the side arms (motor proteins?) linking the microtubules in the irradiated and non-irradiated region may well reveal how the blue-light-irradiated region loses its ability to propagate saw-tooth waves and in turn confirm how the saw-tooth wave itself is propagated.

Reprinted from *Journal of Cell Biology*, Vol. 89(2), pp. 346–356, 1981.

Video Image Processing Greatly Enhances Contrast, Quality, and Speed in Polarization-based Microscopy

SHINYA INOUÉ
Marine Biological Laboratory, Woods Hole, Massachusetts 02543, and Department of Biology, University of Pennsylvania, Philadelphia, Pennsylvania 19104

ABSTRACT Video cameras with contrast and black level controls can yield polarized light and differential interference contrast microscope images with unprecedented image quality, resolution, and recording speed. The theoretical basis and practical aspects of video polarization and differential interference contrast microscopy are discussed and several applications in cell biology are illustrated. These include: birefringence of cortical structures and beating cilia in *Stentor*, birefringence of rotating flagella on a single bacterium, growth and morphogenesis of echinoderm skeletal spicules in culture, ciliary and electrical activity in a balancing organ of a nudibranch snail, and acrosomal reaction in activated sperm.

In the last decade, closed circuit television has increasingly been applied to the light microscope for studies of cell structure and behavior. Exceedingly low light level microscope images have been detected and recorded by Reynolds and co-workers (21, 22), who pioneered in the use of electronic image intensifier tubes coupled to vidicon and other video cameras. For example, in a medaka (fish) egg previously injected with aequorin, a photoprotein that luminesces upon exposure to micromolar free Ca^{++}, they were able to record the appearance of a faint luminescent patch at the point of entry of the sperm followed by the migration of a luminescent ring away from that point around the surface of the fertilized egg. Thus they succeeded in visualizing and demonstrating the presence of a wave of intracellular Ca^{++} release that traverses the surface of the egg after fertilization (7). Likewise, Rose and Lowenstein (24) used an image intensifier and video recording to demonstrate the limited diffusion of Ca^{++} in epithelial cells previously microinjected with aequorin (for additional references, see 22, 32, and 33).

Image intensifiers with video output have also been applied to dark-field microscopy, for example, for visualizing the sliding of individual doublet microtubules in a *Chlamydomonas* flagellum or the shape change induced in isolated bacterial flagella (20). Also, in fluorescence microscopy, video cameras using very high sensitivity image tubes, such as the silicon intensified target tube, have been used to record images that are so dim as to be barely perceptible to the observer (18, 35).

For differential interference contrast microscopy, Dvorak et al. (5) devised a video system in 1975, using an image orthicon in a high-gain video camera whose output was processed with an image enhancer. With this system they were able to generate sufficient image contrast and resolution at the low light level needed to obtain time-lapse records of the invasion of human erythrocytes by live malaria parasites.

With the recent availability of moderately priced instrumentation, video cameras, and tape recorders, video systems are finding broader use in light microscopy. Schatten (29) and others have made effective use of moderately priced video equipment attached to phase-contrast and differential interference contrast microscopes. Recordings made in time-lapse and played back immediately allowed repeated observations, and discussion by several individuals, of experiments on extracted cell models and cells in division and early development (also see references 2 and 28). The general advantages of video microscopy are discussed in an informative survey by Butterfield (4).

I now report some striking improvements in the performance and utility of the polarized light and differential interference

THE JOURNAL OF CELL BIOLOGY · VOLUME 89 MAY 1981 346-356

contrast microscopes that we were able to attain by the optimized application of video systems.[1] The new combination provides clear images with outstanding contrast, resolution, diffraction image quality, and sensitivity. Objects with exceedingly weak birefringence and minute optical path differences can be visualized at the maximum resolution of the light microscope, and their rapid changes can be recorded and analyzed in real time, freeze-frame, slow motion, or time-lapse. Some gain is also made in visualizing weak birefringence and small optical path differences through opaque tissues and cells.

These improvements open up new possibilities for polarized light and differential interference contrast microscopy. Subtle, hitherto unobserved, dynamic changes in cellular organization and fine structure can now be analyzed with precision and speed in living cells. The improvements should also prove useful in the materials sciences and other related fields.

In the following I shall open with some general observations about the subject, describe the optical and electronic characteristics required for video high-extinction polarized light microscopy, illustrate the advances achieved, and discuss the limitations and advantages of the system.

The Polarizing Microscope in Cell Biology

In cell biology, the polarizing microscope has been used to detect crystalline inclusions as well as fibers and membranes that are regularly arrayed at a submicroscopic dimension. Examples include myofibrils, mitotic spindle fibers, chloroplasts, and retinal rods, etc. (27). Those structures display a birefringence (and dichroism) whose sign and magnitude reflect the nature of the constituent oriented molecules or fine structure and their concentrations (25, 26). The birefringence can be detected and measured without perturbing the cell, so that dynamic changes that occur at the molecular levels can be monitored directly in living cells, albeit with an image resolution limited to just below a half wavelength of light.

The polarizing microscope complements the electron microscope and X-ray diffractometer, which are used to resolve spacings far smaller than the wavelength of light. With the latter methods, however, the greater resolution is achieved at a cost of either the need to fix and dehydrate the cell or to utilize crystalline domains far greater than the wave length of light, often much greater than the size of a whole living cell.

To take full advantage of the polarizing microscope for studying the molecular and fine structural orientations in living cells, one needs to detect exceedingly weak birefringence (often <0.1 nm in retardation) at the very limit of resolution of the light microscope. Both theoretical and practical limitations tend to interfere with the achievement of these goals. Several biologists have investigated the factors commonly responsible for these limitations and have improved upon the performance of conventional polarizing microscopes (11, 31).

The ultimate limitations were found to originate in the rotation of polarized light at the interfaces of the lenses, slides, and cover slips, etc., lying between the crossed polarizers (1, 9, 36). As seen by examining the extinction pattern at the back aperture of the objective lens, the sense of rotation is reversed in the adjacent quadrants defined by the polarizer axes. The magnitude of rotation varies with azimuth angle and numerical aperture, and at the high numerical apertures essential for high resolution, the rotation can reach several degrees. The rotation thus drastically lowers the extinction factor, E.F. ([Intensity with parallel polarizers] ÷ [Intensity with crossed polarizers]), of a polarizing microscope and increases the amount of stray light. This severely limits the image contrast of weakly birefringent objects under the very condition required to achieve high resolution.

The rotation of polarized light at the air-glass interface was substantially corrected by the introduction of the polarization rectifier (12). The image transfer function of the polarizing microscope was consequently improved, so that the diffraction anomaly and image error that used to accompany the images of weakly retarding objects viewed with an unrectified polarizing microscope were eliminated (14, 19).

Although the polarization rectifier substantially improved the combined resolution, sensitivity, and image quality of the polarizing microscope and expanded its utility in biological research (e.g., references 15, 16, and 23) it was still not possible to use objective lenses with the superior correction that gave the very best resolution and image quality. For example, plan apochromatic objectives that provide superior, low aberration images with objective and condenser numerical apertures (NA) of up to 1.4 utilize fluorite lens elements. These crystalline elements invariably contain heterogeneous crystalline domains that substantially reduce the extinction factor.

Furthermore, even with rectified optics, it has often been impossible to gain enough light to record the images of weakly birefringent motile cellular structures photographically even though they could be seen with the eye.

These practical but serious limitations of the polarizing microscope (and other instruments requiring high extinction conditions) have now been overcome by the use of certain video systems.

Comparison of Visual, Photographic, and Video Imaging

We shall now consider the differences between polarization optical images perceived visually, produced photographically, and aided by video.

For polarized light (and differential interference contrast) microscopes, the image brightness, I, produced by a retardation, R, located between crossed polarizers is given by: $I = I_{\parallel} \times \sin^2(R/2) + I_{\perp}$, where $I_{\parallel}$ is the image brightness with the polarizer and analyzer axes parallel and $I_{\perp}$ the brightness with their axes crossed (Fig. 1). Because the extinction factor $E.F.$ is defined as $I_{\parallel}/I_{\perp}$ (13, 31), $I = I_{\parallel}\{\sin^2(R/2) + 1/E.F.\}$ (Eq. 1). The extinction factor varies with various parameters including the NA of the objective and condenser lenses, and the presence or absence of rectification.

The contrast and details detectable in the final image are affected by a number of factors, including the extinction factor. For direct visual observation, the detected contrast is proportional to the brightness difference, ΔI, of the two image areas to be compared, divided by the average image brightness, I. Thus the eye responds linearly to $\Delta I/I$, or approximately to the log of the image brightness. This is true provided the

[1] These results were first communicated publically at the Tuesday Evening Seminar on August 19, 1980, at the Marine Biological Laboratory, Woods Hole, Mass., and subsequently at the 1980 annual meeting of the American Society for Cell Biology in Cincinnati, Ohio. At these same meetings, Allen et al. reported their parallel development in which image improvements were also made for polarized light and differential interference contrast microscopy by the use of video. Their theoretical framework and choice and adjustments of the polarization optical and video systems are, however, somewhat different (see Discussion).

average brightness of the image is moderately high and the two areas compared (i.e., the image detail) subtend a sufficiently large area of view (3).

Therefore, in a polarizing optical system, the response of the eye to a small increment in retardation, ΔR, is proportional to $\Delta \log I/\Delta R$, where I is defined by Eq. 1. For a given ΔR, the response is greater the higher the extinction factor (Fig. 2).

For a given extinction factor, the sensitivity ($\Delta \log I/\Delta R$, where ΔR approaches zero) varies with the amount of bias or offset retardation (R_b) by which ΔR is displaced from $R = 0$. At $R_b = 0$, the sensitivity becomes zero. As R_b is increased, $\Delta \log I/\Delta R$ increases, reaches a maximum and gradually decreases again. The maximum is reached at a bias retardation that roughly doubles the brightness of the field, $I_\perp$, at maximum extinction (11, 13).

For the detection of weakly birefringent objects, it is therefore essential that a suitable compensator be used to introduce an appropriate amount of bias retardation. As evident from Fig. 2, the bias retardation that maximizes $\Delta \log I/\Delta R$ varies with the extinction factor.

In addition to altering the contrast of a birefringent specimen, the compensator also alters (increases) the average field brightness (Fig. 2). With high-extinction polarization microscopy, one inevitably works at a low field brightness even with a high intensity source. The gain in brightness provided by the compensator improves our ability to detect small retardations by raising the contrast discriminating ability of the eye. Thus we find that, even in a darkened room, the optimum bias retardation for visual observation is somewhat greater than that which provides maximum $\Delta \log I/\Delta R$.

Photographic emulsions respond more or less in the same manner as the eye. Over a moderate range of image brightness, characterized by the Hurter and Driffield (H and D) curve for the particular emulsion, the density is approximately proportional to $\log I$. However, compared to the eye, the photographic emulsion generally fails even faster to detect contrast as we exceed the optimal intensity ranges.

For photographing weakly birefringent cellular structures, one generally chooses emulsions and processing methods that provide moderately fine grain in order to record the needed image detail combined with an adequate grey scale range. The exposure is kept as short as possible to avoid image movement. Given these constraints, when weakly birefringent objects are photographed, the bias retardation is even further increased than the optimum bias retardation used for visual observation. Only then can the intensity range be made compatible with the emulsion (however, see Jones' remarks [17] on correcting the "underloading" of photographic emulsion). Even so, the ex-

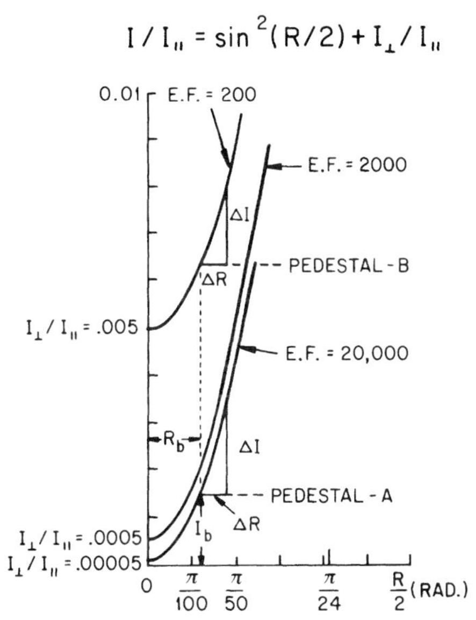

Figure 1

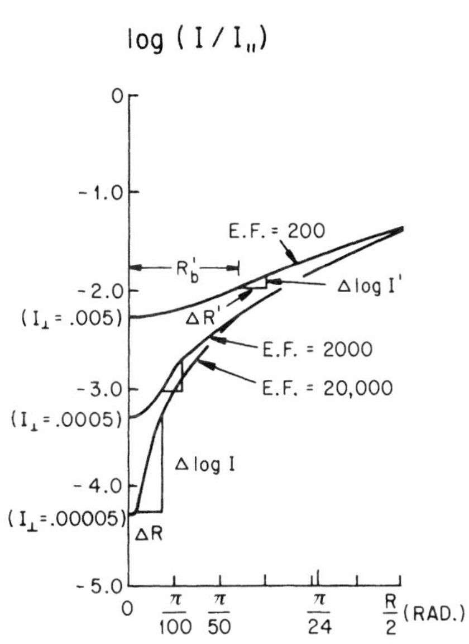

Figure 2

FIGURE 1 Brightness of retarders placed between crossed polarizers. For different extinction factors ($E.F. = I_\parallel/I_\perp$; where $I_\parallel$ is the field brightness with the polarizers oriented parallel to each other, and $I_\perp$ the brightness with zero retardation and the polarizers crossed), the curves are displaced vertically without change in shape. For a given bias retardation, R_b, the increment of brightness, ΔI, introduced by an increment in retardation, ΔR, is not affected by the extinction factor. For a very small ΔR, ΔI converges to 0 at $R_b = 0$. ΔI increases with R_b up to a maximum at $R_b = \lambda/4$, or as shown in the figure, $R_b/2 = \pi/4$. With an idealized photodetector that responds linearly without limit to the light intensity, ΔR would give rise to the same signal independent of E.F. at a given R_b, and the signal would be maximum at $\lambda/4$. The background brightness I_b could be subtracted by setting appropriate pedestal levels. The response of actual vidicon cameras appears to follow curves lying part way between those in Figs. 1 and 2.

FIGURE 2 Log plots of brightness of retarders placed between crossed polarizers. With the intensity plotted on a log scale, the curves for different E.F.s take on different shapes and slopes (cf. Fig. 1). Contrast detectors such as the eye and photographic emulsions respond more or less linearly to the incremental brightness, ΔI, over the background brightness, I_b (see Fig. 1). Thus they respond to the log of brightness. $\Delta \log I/\Delta R$ determines the sensitivity for such detectors. For contrast detectors, $\Delta \log I/\Delta R$ (where $\Delta R \rightarrow 0$) is also 0 at $R_b = 0$. With idealized detectors, the contrast reaches a maximum at $I_b = 2 \times I_\perp$. For systems with different E.F.s, the maximum $\Delta \log I/\Delta R$ as well as the optimum R_b vary considerably.

posure commonly lies in the range of a few to many seconds, and significant image detail must often be recorded in the region of the H and D curve below the linear region, often in the very low negative density range. The contrast is regained up to a point during photographic processing, but this approach is limited by the increase of photographic noise, or graininess, that worsens as the contrast is raised.

With electronic detectors, including video imaging tubes, the voltage output or response is more nearly proportional to the light intensity. Therefore, in contrast to the approximately logarithmic response of the eye and photographic emulsion, we can treat the video camera as an approximately linear device.

With a device whose response is truly linear with the input light intensity, the response should follow Eq. 1, which relates light intensity to specimen or compensator retardation. The curves for different extinction factors would then not change in shape or slope as is the case for detectors with logarithmic responses (Fig. 2). Instead, as Fig. 1 shows, the curves should simply be displaced vertically along the intensity axis. Therefore, with an idealized linear detector, ΔI could in principle remain unaffected by the extinction factor.[2]

In a video system, the electrical signal arising from an increment of brightness, ΔI, is superimposed on an electrical signal that corresponds to the background brightness, I_b. Within a limited range, the background signal can be subtracted electrically by applying an offset—the "pedestal" or "black level" adjustment (Fig. 1). With an appropriate pedestal, the incremental brightness, ΔI, would then stand out in the video image against a dark grey or black background and we would have effectively obtained high extinction.

In addition to the pedestal level, the response of the video system to intensity can be regulated by adjustment of an amplifier gain ("gamma-control") that affects contrast. In a video system equipped with gain and pedestal adjustments, one can regulate the contrast of the image by adjusting the gain and simultaneously cancelling out the signal originating in the stray light of the polarization optical system with the pedestal or black level control.

A suitable video system could thus be used to cure or minimize the extinction defects of a polarization optical system. A dark background and high-contrast video image should be attainable even under some conditions where the extinction factor cannot be made high enough for visual observation or photographic recording of weakly birefringent specimens. The same argument holds for differential interference contrast and other microscopes requiring high extinction. Conversely, with a microscope whose extinction factor is already high, the sensitivity for detecting and measuring small retardations and minute optical path differences could be made even greater by the use of video.

To a considerable extent, these theoretical considerations are materialized with some modern video equipment. With the video camera we used, the regulation of contrast and black level could be achieved automatically by the electronic circuit built into the video camera. Combined with our optical system, the auto-black effectively suppresses the stray background light, and together with the auto-gain control provides exceptionally crisp, sharp images of weakly retarding, minute specimen regions.

[2] In fact the dynamic range for vidicon tubes such as the Newvicon and Chalnicon is quite limited. Their response (sensitivity to light) also varies considerably with face-plate illumination. See the Discussion for further clarification of these points.

MATERIALS AND METHODS

The Microscope

The microscope used in these studies is a specially built, inverted universal polarizing microscope designed jointly by Gordon W. Ellis, Edward Horn, and myself. The optical elements and mechanical components are arranged in a manner similar to that described earlier and generally used with Köhler illumination (10). In brief, a high-intensity 100-W concentrated arc mercury lamp is placed at the top of the optical bench microscope; the projected beam is collimated and filtered to remove heat rays and (for living cells) to provide monochromatic 546-nm green illumination; the beam is polarized with a Glan-Thompson high-extinction, high-transmission calcite prism (Karl Lambrecht Corp., Chicago, Ill.); passes a $\lambda/10$ Brace-Köhler compensator (made to special order; Nikon Instrument Division, Garden City, N. Y.); and is collected to the specimen with a rectified strain-free condenser (e.g., Nikon 8-mm rectified oil-immersion condenser to which a second rectifier was added). The image of the specimen placed on a precision revolving stage is used with selected plan apochromatic objectives, through a Glan-Thompson prism analyzer equipped with stigmatizing lenses, and projected by the ocular to the faceplate of the video camera.

For differential interference contrast, the rectified condenser is replaced with a low-strain-birefringence differential interference contrast condenser with built-in Sénarmont compensator (Leitz-Smith T system). The objective lenses with built-in Wollaston prisms were carefully oriented to maximally darken the field.

The Video System

The video camera used in these studies (model 65 II; Dage-MTI Inc., Michigan City, Ind.) is equipped with a 1-inch Newvicon tube, circuitry with auto-gamma and auto-black controls, and an option for driving the sync signal from an external source (for split-screen comparison or insertion of a second image onto the same field by the use of a special-effects generator).

The output of the video camera feeds through a video processor, a sync stripper, and a video analyzer (nos. 604, 302-2, and 321; Colorado Video, Inc., Boulder, Colo.); a time-date generator (ET 202; Cramer Video, Boston, Mass.); and then to a time-lapse tape recorder and the main monitor (Fig. 3). The image analyzer permits the quantitation of image intensities and the determination of x-y coordinates of image points; the image processor provides a nonlinear gain function that allows manual control of image signal level, black level, and contrast; and the sync stripper allows the concurrent use of these components.

The ¾-inch cassette, time-lapse video tape recorder (TVO 9000; Sony Corp., Long Island City, N. Y.) permits real-time and time-lapse recording, and playback or freeze-frame display of the recording. Although the recorder output is of good quality, the horizontal image resolution is limited by the recorder circuitry to ~320 TV lines (or 160 line pairs). Therefore, in addition to the main monitor (4290; Sanyo Electric, Inc., Compton, Calif.) displaying the signal coming from the recorder, another monitor is placed in the circuit before the video recorder. The second monitor provides an image with a horizontal resolution of >700 lines (350 line pairs) nearly matching the capability of the video camera itself. The monitor chosen (WV 5310; Panasonic Co., Secaucus, N. J.) has an "underscan" capability, so that we can photograph the full field scanned by the video camera rather than that trimmed down by 20% or even more in many monitors. The second monitor also permits comparison of events occurring under the microscope at one time with a scene that had taken place some time earlier, which is displayed from the recorder onto the main monitor. From either monitor, photographs of good image quality could be recorded with a 35-mm camera (Nikon FM equipped with an f/3.5 55-mm Macro lens) with a ¼-s exposure on Kodak Plus-X film, developed according to Kodak prescription with Microdol-X diluted 1:3. A third monitor with a built-in disk recorder (Sony SVM-1010 motion analyzer) allows us to slow down rapid motion or to analyze the recorded events frame-by-frame.

RESULTS

Polarized Light Microscopy

Fig. 4 shows a diatom, embedded in a medium with nearly matched refractive index, viewed in polarized light through a 40-power 0.95 NA plan apochromatic objective. The picture to the left depicts the appearance of the low-contrast image that is seen through the ocular. The picture to the right was taken without readjusting the microscope, except that the picture was taken from the monitor connected to the video camera attached to the microscope. The video camera was equipped with auto-black and auto-gain circuits. The improvements introduced by the video system are clear.

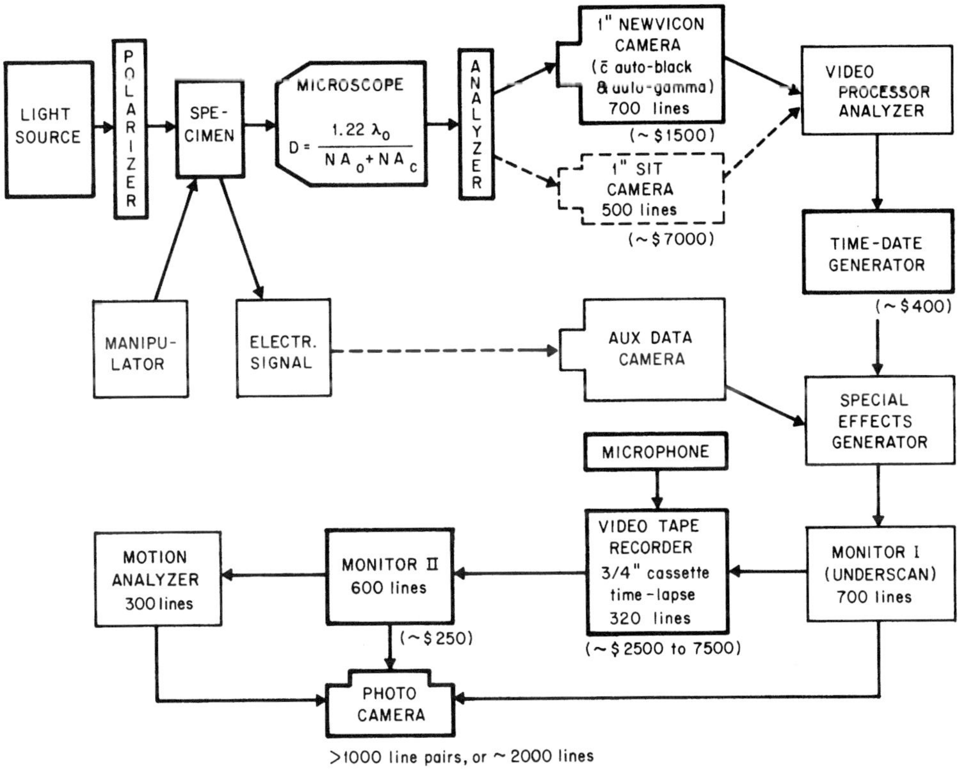

FIGURE 3 Schematics of video-polarizing microscope. Resolution of standard components is expressed as number of lines resolved horizontally over a distance of the vertical video field (one-half this number gives the number of line pairs resolved). The minimum essential components are indicated in bold outlines. (For these components, approximate 1980 U. S. prices are provided as guidelines. Naturally, prices vary substantially depending on manufacturer and features.) We use the SIT (Silicon Intensified Target tube) camera only when the light level is too low for the Newvicon (e.g., fluorescence microscopy); also see footnote 3.

In taking these pictures, the rectifier in the 1.15 NA oil-immersion condenser was doubled so that the rotation of polarized light introduced by the microscope slide and the nonrectified objective lens were also eliminated. The iris of the condenser was adjusted to match the NA of the objective lens, and a $\lambda/10$ Brace-Köhler compensator, located between the crossed polarizers, was oriented to provide a bias retardation of ~15 nm.

The extinction factor of the microscope used with the 40/0.95 plan apochromatic objective was 1.25×10^3 with the system rectified as described above. A higher extinction factor could not be attained at full objective and condenser NAs because of the nonuniform crystalline fluorite elements present in (even this selected) plan apochromatic objective. Nevertheless, the image on the video monitor provides a very high contrast for the birefringent regions of the specimen and in general a high resolution and good correction.

Figs. 5 and 6 show recorded video images taken with the same polarization optical system as for Fig. 4. The cilia and cirri on the *Stentor* were captured by photographing single frozen frames of the recorded video tape. While I used a ¼-s photographic exposure to record the image frozen on the video monitor, the scene on the monitor is formed by a single frame of video scan. Two 1/60-s fields are interlaced to make up a single video frame so that these photographs represent video exposures of 1/30-s each. I do not believe such detailed birefringence of organellar structures in swimming protozoa has even been captured photographically or perhaps even been observed before.

Fig. 7 illustrates the high resolution achieved in video polarized light microscopy with a 100/1.35 oil-immersion plan apochromatic objective lens. The surface ridges of the human oral epithelial cell (about one quarter of the cell is visible) is depicted with unprecedented image detail. The contrast in this image reflects the detailed distribution of birefringence of the surface ridges, whose retardation measured <0.2 nm. Although I have observed many oral epithelial cells as test objects for microscopy, I have never before observed such detailed surface architecture with any mode of microscopy.

Not only do we obtain a fine image, but the contrast of the ridges is precisely reversed without shift in focus when the compensator is turned. In an unrectified system, the rotation of polarized light by the lenses introduces a diffraction image anomaly. The rotation modifies the aperture transfer function of the objective lens; thus fine specimen details may be improperly represented in the image (14, 19). Also, the focus of a weakly birefringent object shifts and necessitates the refocusing of the objective lens when contrast of a weakly birefringent specimen is reversed by turning the compensator. In the present video optical system, the rotation of polarized light and the

[3] After the manuscript was submitted for review, I was introduced to a video camera with built-in image intensifiers, and equipped with automatic controls for contrast, black level, and sensitivity. The TV-2 M camera just developed by Venus Scientific Inc., Farmingdale, N. Y., provided the sensitivity needed to work at full extinction of our rectified polarizing microscope and, for a $\lambda/1,000$ retardation specimen, yielded an image with good contrast and resolution with the compensator set at extinction.

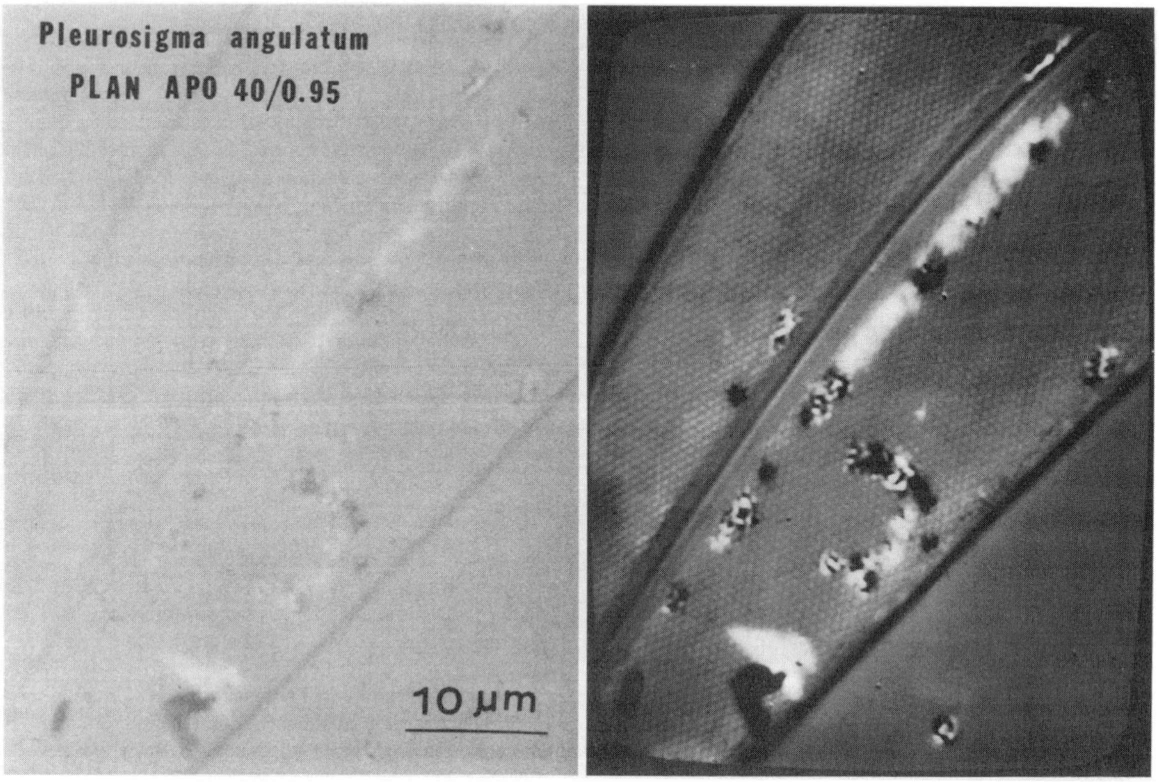

FIGURE 4 Appearance of a diatom in video-polarized light microscopy, using a plan apochromatic objective, as seen through the ocular (left) and on the video monitor (right). The right scene was photographed off the video monitor with the compensator adjusted to provide optimum contrast. Without change in the compensator setting, the left scene was photographed off another monitor whose brightness and contrast were adjusted by an unbiased observer to match the image seen through the ocular. Photographs taken directly through the ocular can be improved upon the left scene by giving a much longer exposure, at a somewhat reduced bias compensation, followed by high-contrast processing.

diffraction image anomaly have been corrected by introducing a stronger rectifier in the condenser without placing a rectifier in the objective lens. The extinction factor of the 100/1.35 plan apochromatic objective lens combined with the doubly rectified condenser of 1.15 NA was 8.5×10^2.

Fig. 8 illustrates another example of high-resolution polarized-light video images obtained with the 100/1.35 plan apochromatic objective. The photographs show a live bacterium attached to the surface of an anaerobic protozoan ("rubberneckia").

The very weak birefringence of the rotating bacterial flagella is clearly seen in alternating contrast trailing below the main body of the bacterium. As the spiral bundle of flagella rotates, the birefringent black and white stripes travel "away" from the bacterial body as in a rotating barber pole. Each black and white region corresponds to a portion of the flagellar spiral tilted to the right or the left. The contrast is produced because the local flagellar axes alternately lie in the opposite and same quadrant as the slow axis of the compensator.

The width of this bacterium as well as the wavelength of the flagellar spiral is only 0.5 μm, while the average amplitude of the wave is ~0.1 μm. Even though the amplitude of the flagellar wave is well below the Abbe limit of resolution, the alternately tilted regions of the wave clearly define the contrast of those flagellar regions in the compensated polarized-light video image. In the past I have occasionally observed birefringent waves of live bacterial flagella with high-resolution rectified optics, but never before these video recordings has it been possible to photographically document or analyze the birefringent waves.

For images of the video monitor magnified to the extent shown here, the scan lines can detract considerably, as in the upper row of pictures in Fig. 8. In the lower row, the distracting scan lines have been nearly eliminated, by rephotographing the upper pictures through a Ronchi grating.[4]

[4] A Ronchi glass grating with 50–100 black lines ruled per inch (Rolyn Optics Co., Arcadia, Calif.) is oriented with the rulings parallel to the video scan lines and placed 4–10 inches above the print or video monitor to be copied. The Ronchi grating is initially positioned so that diffraction by the grating roughly doubles the frequency of the video scan lines. The final position of the grating is set to where the scan lines disappear from the image in the viewfinder of the copy camera. The Ronchi grating can also be placed in the projection beam of the enlarger. The use of the Ronchi grating was kindly suggested to me by Raymond E. Stephens of the Marine Biological Laboratory, who reminded me of a related use that I had made of the grating earlier (6, 8).

Besides using Ronchi gratings, which invariably introduce some image degradation, the scan lines can be made less conspicuous by photographing the video screen just when the scan lines (whose positions periodically fluctuate) are not prominent, and by photographing several sequential or frozen video frames with a ¼- to ⅛-s photographic exposure.

The most dramatic removal of the scan lines, without loss of even very fine video image details, was achieved by a method of optical filtration devised by Gordon W. Ellis of the University of Pennsylvania. This method will be reported elsewhere.

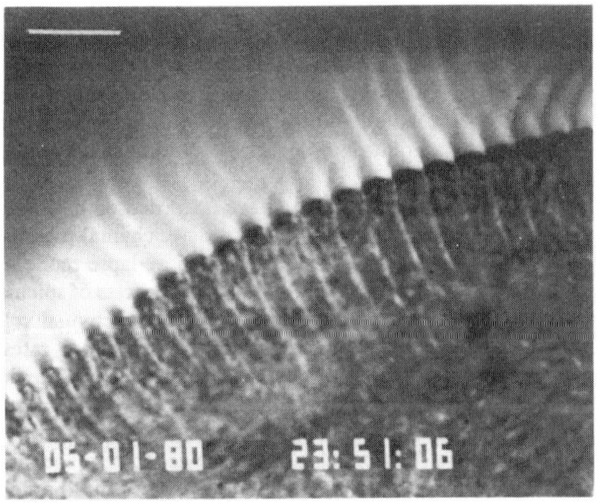

FIGURE 5 Birefringence of ciliary rootlets and beating cilia in *Stentor*. The video-polarized light microscope image was taken with a 40 ×, 0.95 NA nonrectified Nikon plan apochromatic objective lens with correction collar. The objective lens was coupled with a doubly rectified 1.15 NA oil-immersion condenser, illuminated with a narrow-band 546-nm mercury green line. The image was photographed off a 9-inch monitor displaying, in freeze-frame mode, a selected scene from the ¾-inch video cassette tape record. Two 1/60-s interlaced fields make up this single video frame (Inoué and Chen, unpublished photograph). Bar, 10 μm. × 1,300.

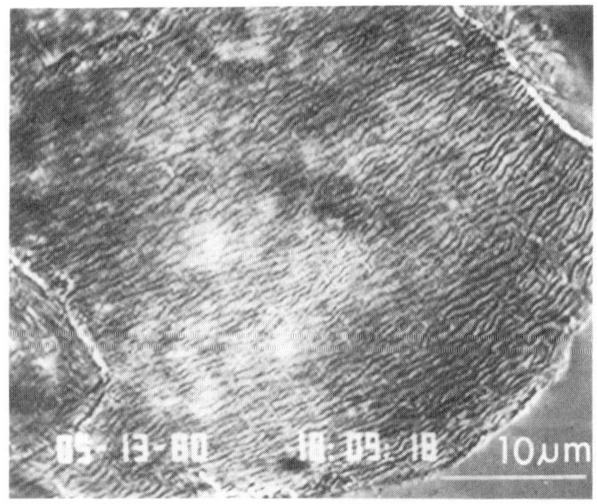

FIGURE 7 High resolution, high extinction video-polarized light microscope image of an oral epithelial cell. The 1-inch Newvicon video camera output was coupled directly to a 9-inch video monitor to minimize the loss of image resolution. The birefringent ridge structure on the cell surface is resolved in great detail. A Nikon 100/1.35 oil-immersion plan apo (nonrectified) objective was combined with a doubly rectified 1.15 NA oil-immersion condenser and illuminated with the mercury green line. Bar, 10 μm. × 2,000.

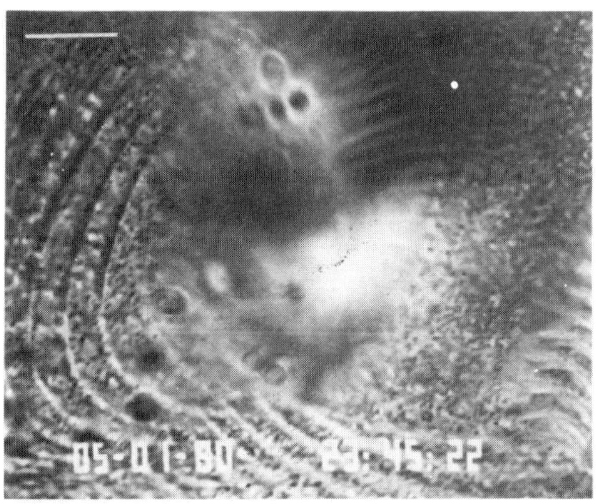

FIGURE 6 Swimming *Stentor* showing birefringence of km fibers, ciliary rootlets, and beating aboral membranelle. Single frame from ¾-inch video cassette record as in Fig. 5 (Inoué and Chen, unpublished photograph). Bar, 10 μm. × 1,300.

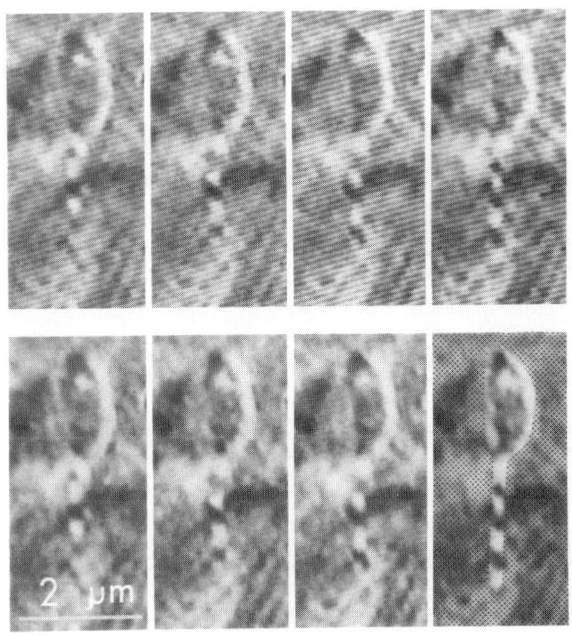

FIGURE 8 Birefringence of rotating flagella on a single bacterium. (In the last frame, a stippled mask outlines the bacterium.) The amplitude of the flagellar wave is <100 nm. Nevertheless, those regions of the flagella tilted to the right and the left show in black and white, in reverse compensation. Time interval between the first two frames, 17 ms; frames 2 and 3, 17 ms; frames 3 and 4, 33 ms. The pictures in the top row were taken directly off the monitor in freeze-frame. Those in the bottom row were obtained by photographing the top row through a Ronchi grating to eliminate the intrusive scan lines[4] (Inoué and Tamm, unpublished photographs). Bar, 2 μm. × 9,500.

Differential Interference Contrast Microscopy

As noted earlier, image contrast in differential interference contrast microscopy is also affected by the extinction factor of the system and the setting of the compensator. We would therefore expect the differential interference contrast images also to be improved by the use of video, which could at the same time provide several other practical advantages characteristic of electronic recording.

Fig. 9 shows the growth of a calcareous skeletal spicule isolated from a prism-stage sea urchin embryo. The three mesenchyme cells that support the in vitro growth and mor-

Rose's and Loewenstein's findings (176) that when calcium ions are microinjected into *Chironomus* salivary gland cells previously loaded with aequorin (a calcium-dependent, light-emitting protein), light is emitted only in that portion of the cell into which the calcium solution is directly applied. In other words, although it is a small, diffusible ion, calcium is sequestered so rapidly in the cytoplasm that the relatively high injected concentration is limited to the region of the cell that receives the microinjection directly.

The cytoplasmic membranes and mitochondria are likely candidates as calcium sequestrants. The calcium-dependent aequorin glow is no longer limited to the site of injection, but is spread out in the presence of respiratory poisons (176). Similarly, Sawada and Rebhun (177) have found that the birefringence of the spindle in some cells is abolished when the cell is exposed to respiratory poisons or uncouplers of oxidative phosphorylation. These agents, as well as caffeine, probably poison the calcium-pumping ATPase, making the cell membranes leaky to calcium ions and inducing the mitochondria and endoplasmic reticulum to dump their accumulated calcium ions. In caffeine-microinjected cells, there is a drop in spindle birefringence that is not sharply delineated, but diffuse, presumably because in contrast with calcium ions, caffeine is not sequestered by the cytoplasm and therefore diffuses normally (165).

Petzelt's calcium-dependent ATPase appears to be membrane bound (178). Membrane-delimited vesicles and cisternae are seen by electron microscopy to be concentrated at the spindle pole and in a sheath surrounding the spindle (163). Thus, as postulated by Harris (164), these membranes may well play an important role in calcium regulation of the cytoplasm, and they may do so in highly localized cell regions. Welsh et al. (179) observed, by fluorescent antibody staining, a higher concentration of calmodulin in the half-spindle. The anaphase location of calmodulin near the spindle poles, and the changes observed in their distribution during late anaphase, suggested to Marcum et al. (167) that calcium is an endogenous regulator of microtubule assembly through the activity of calmodulin (also, 180).

Models for Mitosis

We shall now consider some current models which have been proposed to account for the (anaphase) movement of chromosomes. The models have been reviewed (23, 134, 181) and extensively discussed (e.g., 79–81, 144, 182), except for the recent model that incorporates the treadmilling of tubulin along microtubules (183).

ACTIN AND MYOSIN: One of the oldest models for poleward movement of chromosomes invoked the contraction of a muscle-like fiber which linked the chromosome to the spindle pole (pages 178–184 in [2], pages 70–75 in [6]). From the earliest days, however, this model has been repeatedly questioned. Shortening chromosomal fibers generally do not get thicker; anaphase velocity is so much lower than the contraction velocity of skeletal muscle (of the order of 10 nm/s, in contrast with 100 μm/s for muscle); and the lability of the spindle fibers does not fit the properties of muscle.

For many years there was no reason to resurrect this model, although the counter arguments were not airtight. But recently the model has again gained some support. Actin and myosin, the two major proteins responsible for force production by muscle fibrils (review; e.g., 184, 185), were detected in the half-spindle regions of glycerinated cells. Fluorescent antibodies made against these proteins stained the spindle (186, 187), as did fluorescein-conjugated heavy meromyosin or subfragment-1, which carry the active ATPase sites of the myosin molecule (188–191). (However, the levels of immunofluorescent staining for actin and myosin are not greater in the spindle, according to the latest report from Aubin et al. [192].) In electron micrographs of glycerinated cells, some actin filaments were seen to terminate at or about the kinetochore and to run approximately parallel to microtubules in the half-spindle (e.g., 190, 193).

Whereas these observations on glycerinated cells are suggestive and have attracted considerable attention, the data in themselves do not imply a functional role for actomyosin or an actin system in the poleward movement of chromosomes. Like tubulin, actin is one of the major protein constituents of most cells (each at times amounting to several percent or more of the total cell protein). The spindle region in dividing cells excludes granular organelles such as mitochondria and yolk, so that, on this basis alone, one might expect to find a somewhat higher concentration of nonparticulate cytoplasmic constituents, including actin, in the spindle region of some cells (165, 194; also see Fig. 2 in 195 for a model demonstrating this point).

Two types of tests for the functional role of actomyosin and actin in mitosis have yielded negative results. The microinjection of an antibody against starfish-egg myosin (previously shown to suppress hydrolysis of ATP by egg myosin) prevented many successive cleavage divisions but did not interfere with mitosis in the same starfish eggs (Mabuchi and Okuno [196]). This is consistent with the strong evidence that cleavage is brought about by an actomyosin contractile ring (review, 197). In the eggs injected with anti-egg myosin (Fig. 9), birefringent spindles formed at regular intervals, chromosomes moved toward the spindle poles and spindles elongated normally in anaphase, and nuclear envelopes were reconstituted on schedule, despite the absence of eight or more cleavages (Kiehart [165]).

Likewise, in Cande's detergent-permeabilized tissue culture cells, cleavage was suppressed by the application of heavy meromyosin or subfragment-1, which had been treated with *N*-ethyl maleamide (198). These treated fractions of myosin bind to actin competitively and prevent the interaction of actin with normal myosin (199). In the permeabilized cells, cleavage was arrested, but anaphase movement was not affected by application of the modified myosin fragments. Chromosome movement, especially the part dependent on spindle pole-to-pole elongation was, however, reversibly inhibited by vanadate, a potent inhibitor of ciliary dynein ATPase (200). This ion showed little effect on cleavage in lysed cells, reinforcing the idea that different molecular mechanisms are operating in chromosome movement and cleavage.

In living cells, cleavage is suppressed, or regresses, when cells are bathed in solutions that affect actin gelation. Cytochalasin B and D are reported to weaken actin gels (201, 202), but not to affect mitosis when applied in concentrations adequate to suppress cleavage (e.g., 197). In contrast, colchicine and podophylotoxin, which prevent mitosis and even disassemble mitotic microtubules, do not affect cleavage once the cleavage message has been delivered from the spindle to the potential furrow (26, 203). Actin and myosin have been extracted from the cell cortex, and electron micrographs show a clear band of actin filaments in the cell cortex oriented circumferentially in the cleavage furrow (review, 196). Taken together, these data strongly support the role of an actomyosin system in cytoplas-

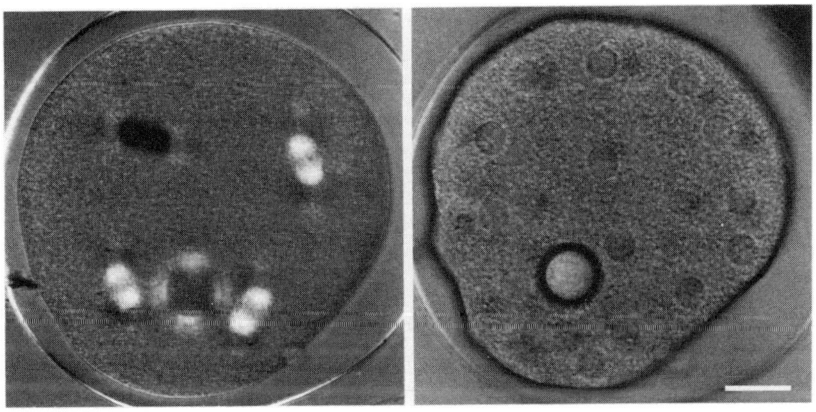

FIGURE 9 Egg of a starfish, *Asterias forbesi*, microinjected with an antibody made by Mabuchi and Okuno (196) against myosin of another species of starfish, *A. amurensis*. Cleavage is suppressed for up to nine divisions, but mitosis is unaffected. One nanogram of IgG, containing the antimyosin, was injected before first cleavage. (*Left*) Third division spindles seen in polarized light. (*Right*) Over 30 nuclei are visible 2.4 hours later. From Inoué and Kiehart (109). Bar, 30 μm.

mic cleavage, but do not favor the involvement of actin or an actomyosin system in mitotic chromosome movement.

ASSEMBLY-DISASSEMBLY OF MICROTUBULES: This model postulates that assembling microtubules, by their extension, push organelles apart, and slowly disassembling microtubules, by their shortening, pull organelles together. I conceived the model through the observation of *Chaetopterus* oocytes exposed to colchicine or cold. As the spindle-fiber material slowly depolymerized in these metaphase-arrested cells, the chromosomes and the inner spindle pole were transported toward the outer spindle pole, which is anchored to the cell cortex (Fig. 3). As the fibers reassembled upon removal of the depolymerizing agent, the chromosomes and inner pole were transported away from the outer spindle pole. Too high a dose of colchicine or overrapid chilling simply caused the spindle fibers to fall apart without appreciable displacement of chromosomes or pole (24, 25). This model at once seemed to explain the labile, yet cohesive, nature of the forces that held together the ephemeral fibrils of the spindle (36), as well as the slowness of chromosome movements.

The subsequent discovery of labile microtubules added credibility to the assembly-disassembly (or dynamic equilibrium) hypothesis (35, 49, 204), but its validity has been repeatedly questioned, presumably in part because the proposal is not intuitively compatible with macroscopic mechanics.[14] Nevertheless, Salmon (74) and Fuseler (126) performed experiments utilizing hydrostatic pressure and cold as microtubule depolymerizing agents, and confirmed that spindle shortening and chromosome movements are induced in living metaphase cells by the depolymerizing agents.[15] They also demonstrated a strict proportionality between the velocity of induced spindle-fiber shortening (and of natural anaphase movement) with the rate of microtubule depolymerization. Although slow depolymerization of microtubules induced a shortening of chromosomal fibers, the induced movement ceased altogether when the rate of microtubule depolymerization became too great, whether the depolymerization was induced by pressure, temperature, or colchicine (review, 96, 205). In Salmon's words (74), "Polymerization of microtubules does produce pushing force and, if controlled microtubule depolymerization does not actually produce pulling forces, at least it governs the velocity of chromosome-to-pole movement."

Whether or not it turns out that shortening microtubules can exert pulling forces in addition to the pushing forces generated by their growth, the dynamic anchorage of microtubules is essential for force transmission through the assembling and disassembling microtubules. In this context, dynein, the ATPase associated with ciliary and flagellar microtubules (206, 207) has lately received much attention. Cytoplasmic dynein has just been isolated and characterized (208), and its role in anaphase movement, at least in pole-to-pole elongation, finds much experimental and observational support, as discussed below.

THE SLIDING MODELS: In 1969 McIntosh et al. (209) proposed that mitotic chromosome movement was brought about by a combination of microtubule sliding (in analogy with muscle contraction and ciliary and flagellar beat) and microtubule assembly and disassembly. Though the details of this model soon needed to be revised (23, 210), it nevertheless struck a favorable chord with many investigators. The labile and dynamic attributes of the spindle fibers were ascribed to demonstrated properties of microtubules, and force production could be attributed to (dynein) cross bridges, whose ability to induce relative sliding of ciliary microtubules was soon to be established (211; review, 212).

While seeming to provide a rational model, some predictions of which were quite readily testable, the model failed to account for certain properties of some spindles. In studying mitosis in yeast and other lower eukaryotes,[16] Roos (213), Peterson and Ris (214), Heath (215) and others came upon connections between chromosome and pole (plaque) that consist of single microtubules! Some of the nonkinetochore microtubules that make up the central spindle appeared to span the whole distance between the spindle-pole structures, even as the spin-

[14] Contractile force production by a disassembling microtubule can be explained by viewing the labile microtubule as a cylindrical micelle, as explained in Inoué and Ritter (96).

[15] Salmon has now induced spindle-shortening and chromosome-to-pole movement in isolated metaphase spindles (of the type shown in Fig. 1), by depolymerizing the labile microtubules with micromolecular concentration of calcium ions.

[16] For mitosis in lower eukaryotes, which show many interesting and potentially instructive variations on mitosis and mitotic organelles, see (4, 20, 55, 215–218) and the excellent review by Kubai (219).

dle elongated. Further, the central spindle microtubules appeared not to be in locations where they could interact with the "kinetochore" microtubules.[17] The single "kinetochore" microtubules appeared to shorten and to bring the "chromosomes" to the spindle pole independent of spindle pole-to-pole elongation.

In serial sections of cultured cells, Brinkley and Cartwright (221) did not find the number distribution of microtubule cross sections predicted by the McIntosh et al. model. On the other hand McIntosh et al. (222, 223) do report finding the distribution of microtubules appropriate for their model. Manton et al. (224) in an elegant study on mitosis and meiosis in a centric diatom, counted microtubule numbers in serial sections that are compatible with the overlapping of central spindle microtubules. In contrast, central spindles in some protozoa are known to elongate much more than twice their initial length, so sliding alone cannot account for spindle elongation; at least some spindle fiber growth is required (e.g., 55).

Despite these reservations, Tippit et al. (225) and McDonald et al. (226, 227) provide a most striking illustration of the overlapping microtubules in a configuration highly suggestive of interactions between oppositely polarized microtubules. In the mid-region of a diatom (*Melosira*) spindle, the cross sections of microtubules are arranged in a regular orthogonal array. And every other microtubule appeared to be connected to opposite spindle poles! In later anaphase, the overlap of the central spindle microtubules progressively decreased, although concomitantly there was growth of some microtubules (also see 228).

Although dynein has not yet been unambiguously demonstrated between opposing microtubules (summary regarding intertubule arms [229]), extensive periodic cross bridges have been seen in the metaphase (extranuclear) central spindle in a hypermastigote protozoan (Fig. 10). These observations, which might suggest the involvement of dynein-mediated sliding, are in fact fortified by two functional tests.

In isolated mitotic apparatuses, Sakai et al. (230) observed chromosome movement, which, while considerably slower than in living cells (but see improved movement reported in [231]), appeared to exhibit general features of in vivo anaphase movements. The movement which required a labile mitotic apparatus was prevented by excess tubulin in the medium and especially by vanadate ions and antibodies formed against dynein (review, 232).

These conclusions were complemented by observations in another type of cell model by Cande and Wolniak (200). In detergent-extracted rat-kangaroo cells in tissue culture, chromosome movement could be stopped by vanadate ions in the +5 oxidation state. After the chromosomes had stopped, they could be restarted by converting the vanadate to the inactive +4 state via the addition of norepinephrine. As shown by Gibbons et al. (233), the +5 vanadate is a potent inhibitor of ciliary and flagellar ATPase (however, see [234] for a lower vanadate sensitivity of cytoplasmic dynein).

Whereas these ongoing experiments require further confirmation,[18] the experiments on isolated spindles and extracted cell models strongly suggest the involvement of dynein in anaphase chromosome movement. It would seem quite likely that a dynein-mediated sliding mechanism is at least in part responsible for the pole-to-pole extension of the anaphase spindle. It is not as clear whether a dynein-mediated sliding is involved in the chromosome-to-pole movement. Perhaps dynein is a dynamic anchor for the kinetochore microtubules, but then we still must ask, how does microtubule disassembly govern the velocity of poleward chromosome movement?

FIGURE 10 Electron micrograph of central spindle in a hypermastigote protozoan *Barbulanympha* sp. Extensive cross bridges are seen in this micrograph, reinforced by superimposing two transparencies of the same image translated once along the microtubule axes. From Inoué and Ritter (96); also see Ritter et al. (55).

TREADMILLING: Recently, an alternative to the McIntosh et al. model, placing a greater emphasis on the role of microtubule assembly-disassembly was introduced by Margolis et al. (183). These authors found in vitro a "treadmilling" of tubulin through microtubules that were in an assembly steady state with soluble tubulin dimers in the presence of an adequate and continuous supply of GTP.[19] They postulate for dividing cells that all mitotic microtubules add microtubule subunits at the equatorial region of the spindle and at the kinetochores, and that the microtubules lose subunits at the spindle poles. They also propose that nonkinetochore microtubules, which overlap at the equator, slide by each other at a rate needed to keep the spindle poles separated. Kinetochore microtubules are thought to form a parallel linkage to the treadmilling interpolar microtubules (183). Depending on the relative rates of tubulin incorporation into the kinetochore and nonkinetochore microtubules and their rates of disassembly, one could increase or decrease spindle length as well as the distance between kinetochores and the spindle poles. Margolis and Wilson (103) report that the in vitro rate of microtubule treadmilling can come close to anaphase chromosome velocity. Because dynein

[17] In yeast, mitosis takes place within the nuclear envelope. The poles of the intranuclear spindle lack centrioles but are organized around spindle pole plaques on the nuclear envelope. The trilaminar kinetochore structure generally seen on chromosomes of higher eukaryotes has not been observed in yeast and chromatin appears to associate directly with microtubules (213). Ris and Witt believe that even in organisms with a trilaminar kinetochore, the kinetochore microtubules are directly linked with the centromeric chromatin (220).

[18] Both the Sakai et al. isolates (231), and the Cande cell models (198) have yet to be refined before the inference of these results is fully accepted. Both models run "downhill" rapidly, and to the best of my knowledge, vanadate and dynein antibody inhibition of mitosis has not yet been observed in living cells.

[19] Weisenberg (personal communication), Bergen and Borisy (235), and Karr and Purich (104) emphasize that treadmilling is not caused simply by assembly at one end of the microtubule and disassembly at the other. Rather, both assembly and disassembly take place at each end, but their rates differ in such a way that the *net* assembly at one end is greater than the *net* disassembly at the other end.

has now been used successfully to "decorate" microtubules and to indicate their polarity (Haimo and Rosenbaum [236]; also [101, 237, and Haimo and Rosenbaum, this volume]), we should soon be able to learn whether the polarities of mitotic microtubules conform to those stipulated in the Margolis and Wilson or McIntosh et al. model.

OTHER PROPOSALS: Before we leave the models for anaphase chromosome movement, we should take special note of the work by the Bajers and co-workers who emphasize the lateral transport seen in spindles and the lateral interaction believed to take place between mitotic microtubules (summary, 131, 134, 238). Since the late 1940s, the Bajers have extensively analyzed chromosome and particle movements and spindle behavior directly in healthy, dividing plant and animal cells and in cells treated with a variety of antimitotic agents. Their analyses on electron micrographs of cells that had been followed and recorded up to the time of fixation with time-lapse cinematography, suggested the importance of the changing lateral association of microtubules seen within individual kinetochore fibers, as well as between kinetochores and nonkinetochores microtubules (134). In general, the presence of intrakinetochore fiber association correlated with cessation of chromosome movement, whereas association between kinetochore and nonkinetochore microtubules was evident whenever chromosomes were moving poleward (but also see [93]).

Little is yet known of how these microtubular organizations are controlled, nor whether there is or is not sliding of microtubules associated with the lateral interactions, but the Bajers alert us to the possible role of microtubule interactions that could play an important role in anaphase chromosome movement (238). In contrast, Thornburg (239) proposed that viscous coupling associated with *intra*microtubular conformational change might propel the microtubules with their attached chromosomes.

Coordination of Cytokinesis with Mitosis

Once the chromosomes are partitioned into two equivalent (or, in meiosis, nonequivalent) groups by mitosis, how are the daughter nuclei placed in the proper cytoplasmic environment? This question is not only important for the successful completion of cell division, but also for determining the future role of the nucleus, because it is the cytoplasm surrounding the nucleus, rather than the unequal division of the nucleus itself, that generally determines how a particular cell is to differentiate (2, page 1059), (also see [5, 240–242]).

In astral mitosis,[20] a close correlation has long been noted between the metaphase spindle axis and the cleavage plane. When a cell was left undisturbed, the cleavage furrow almost always started from the cell surface nearest the spindle and in a plane bisecting the pole-to-pole axis of the spindle (Fig. 11). In centrifuged eggs, the cleavage furrow would appear in a new location dictated by the displaced spindle (e.g., 145). In fact, the correlation was so universal that most postulates for cleavage-furrow induction ascribed a major role to the mitotic spindle (2, 105, 203, 245, 246). Not only was there present a spatial correlation between spindle axis and cleavage plane; in the late 1030s to 1040s, Katsuma Dan and his co-workers showed a striking geometrical relationship between the extending anaphase spindle and the progression of cleavage. In sea urchin and jelly fish eggs, small particulate markers, which were applied directly to the cell surface near the impending cleavage furrow, moved along the exact path predicted from the separation of the astral centers as the spindle elongated. Dan assumed that the cell cortex was connected to the astral centers by interdigitating, inextensible astral rays by which the elongating anaphase spindle drew in the cell cortex, thus forming the cleavage furrow (summary in 247). Whereas this hypothesis could also account for the many unexpected cleavage patterns found in eggs deformed into toroids (248), two sets of experiments negated Dan's hypothesis.

In 1953, Swann and Mitchison (26) applied high doses of colchicine (3 mM in seawater) to metaphase sea urchin eggs and showed that cleavage nevertheless proceeded after destruction of the birefringent asters and spindle, providing the chromosomes had progressed to mid-anaphase (also see 27). Further, in 1956, Hiramoto (249) managed to suck out the entire spindle and asters from a dividing sea urchin egg and showed that a cleavage furrow appeared in the expected location so long as the cell had reached metaphase before spindle extraction (further, detailed analysis in [250]). He thus eliminated the possibility that cleavage was mechanically effected through a noncolchicine-sensitive element of the spindle and aster. These experiments clearly showed that the late metaphase-to-anaphase spindle and asters were unnecessary for cleavage, but it was equally clear that at an earlier stage the mitotic figure did determine the cleavage plane.

If the spindle before metaphase was artificially reoriented, the cleavage furrow appeared perpendicular to and bisecting the spindle in its new position (summary in 105). During a brief critical period, the spindle could even initiate up to ten cleavage furrows in succession, if the spindle were squeezed along the length of a sand-dollar egg previously deformed into a cylinder (251)! Clearly, then, a message[21] must be sent from the spindle to the cell surface where the cortical layer contracts and produces the cleavage furrow[22] (but see 254 and 255).

While the furrow was normally localized where the two asters (whose foci lay at the two poles of a spindle) overlapped, two asters not joined by a spindle could also induce cleavage (203, 246). The message for cleavage induction therefore comes not directly from the spindle itself, but from the spindle poles or astral centers. The speed of the message, and the duration required for the cortex to respond, were determined by displacing the spindle (251). Interestingly, the message travels along the astral rays at about 6 μm/min, approximately the rate at which microtubules grow.

In anastral mitoses,[20] especially in cells of vascular plants, a large number of microtubules appears between the separating chromosomes in late anaphase (138, 256). The ellipsoidal bundle of microtubules, the phragmoplast, has long been thought to arise from central spindle fibers (e.g., Strasburger, 1888, in

[20] Astral mitosis: with asters (Figs. 1 and 11). Typically, but not always nor exclusively, found in animal cells. Anastral mitosis: without asters (Fig. 6). Typically found in, but not limited to, higher plant cells. See Dietz (243) for experimental dissociation of asters from the spindle pole in a living cell, and Aronson (244) for analysis of attractive forces between (astral) centers and nuclei.

[21] The notion of a cleavage-inducing message (substance X which acted through polar relaxation) that traveled along the astral rays was proposed by Swann (18) and Mitchison (252).

[22] Sadly, I must leave out a series of intriguing accounts on the search for the mechanism of cleavage itself. Many interesting experiments were performed and ingenious hypotheses constructed (excellent summaries in [203, 246, 251, 253]). For our present purpose, we proceed by accepting the finding that the "contraction" of a cortical actomysin system is responsible for cell cleavage see (e.g., Schroeder. [197]).

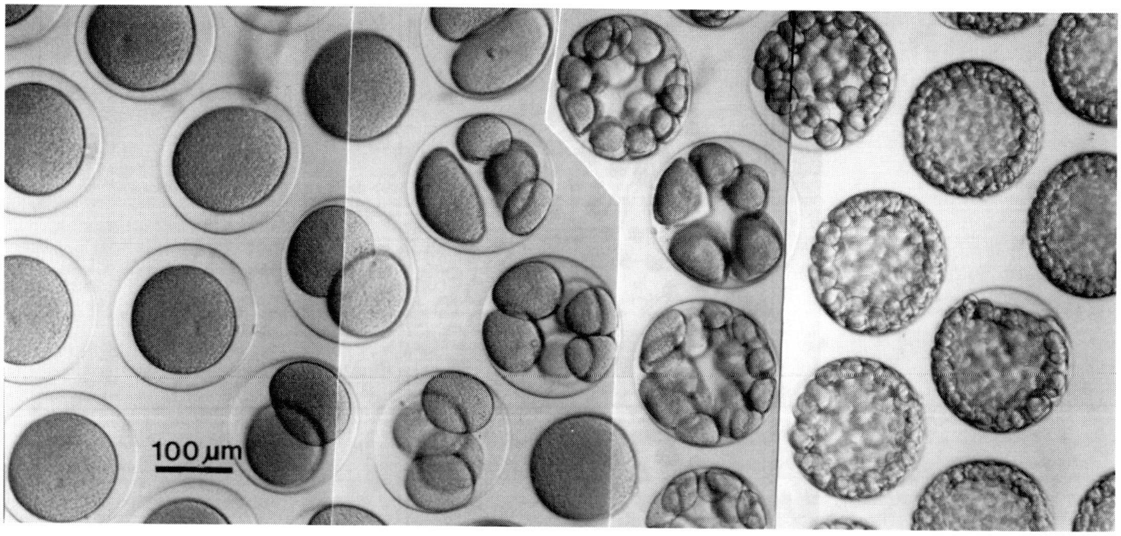

Fig. 4. Larger view of portions of transition zone in micro-drop. June, 1982, data from student project.

sharp that within some single developing embryos, approximately 180 μm in diameter, the outer layer of cells divided twice more often than those facing the middle of the drop (Figs. 2 to 4). In other words, there developed a gradient so sharp that successive cell divisions were arrested within a distance of some 60 μm to 120 μm. Since these star fish eggs normally went through their successive cleavage divisions at ca. 30-minute intervals, the gradient that inhibited cell division appears to have traveled at a rate of some 2 μm/min to 4 μm/min (Figs. 5 and 6).

It was tempting to conclude that one was seeing a wave front of anoxia traveling at that velocity. But it did not seem reasonable that the cells could generate such a sharp gradient, or wave front, of anoxia against oxygen that was diffusing in from the airspace surrounding the drop. That seemed so far fetched that I set aside the 1968 data for a further test at a later date.

In the spring of 1982, I brought the first group of half a dozen undergraduate students from the University of Pennsylvania General Honors Program to Woods Hole to try a miniature course patterned after the MBL Physiology course. In successive years, I was fortunate to have a young faculty member (Doug Lutz, Dan Kiehart, Dave Begg, Tim Otter, in turn) join me to guide the students.

For the first week, we held lecture discussion sessions followed by all-day laboratory exercises to introduce the students primarily to developmental physiology of marine organisms. Then, for the next two weeks, the students tackled individual research projects according to their own choice. Of the first group of students, Carol Brown and Sonia Potrebic decided to follow up on my micro-drop observations with the crowded star fish eggs.

To my complete surprise and delight, these undergraduate students, who had little prior experience in independent research, actually resolved my 14-year-old puzzle. Their experiments left little doubt that we were indeed observing the effect of a sharp gradient of anoxia. By managing to perfuse the air space in the micro-drop with seawater-saturated oxygen gas, they even showed how the anoxia-induced, arrested cell division could be completely prevented or reversed (Figs. 7 and 8).

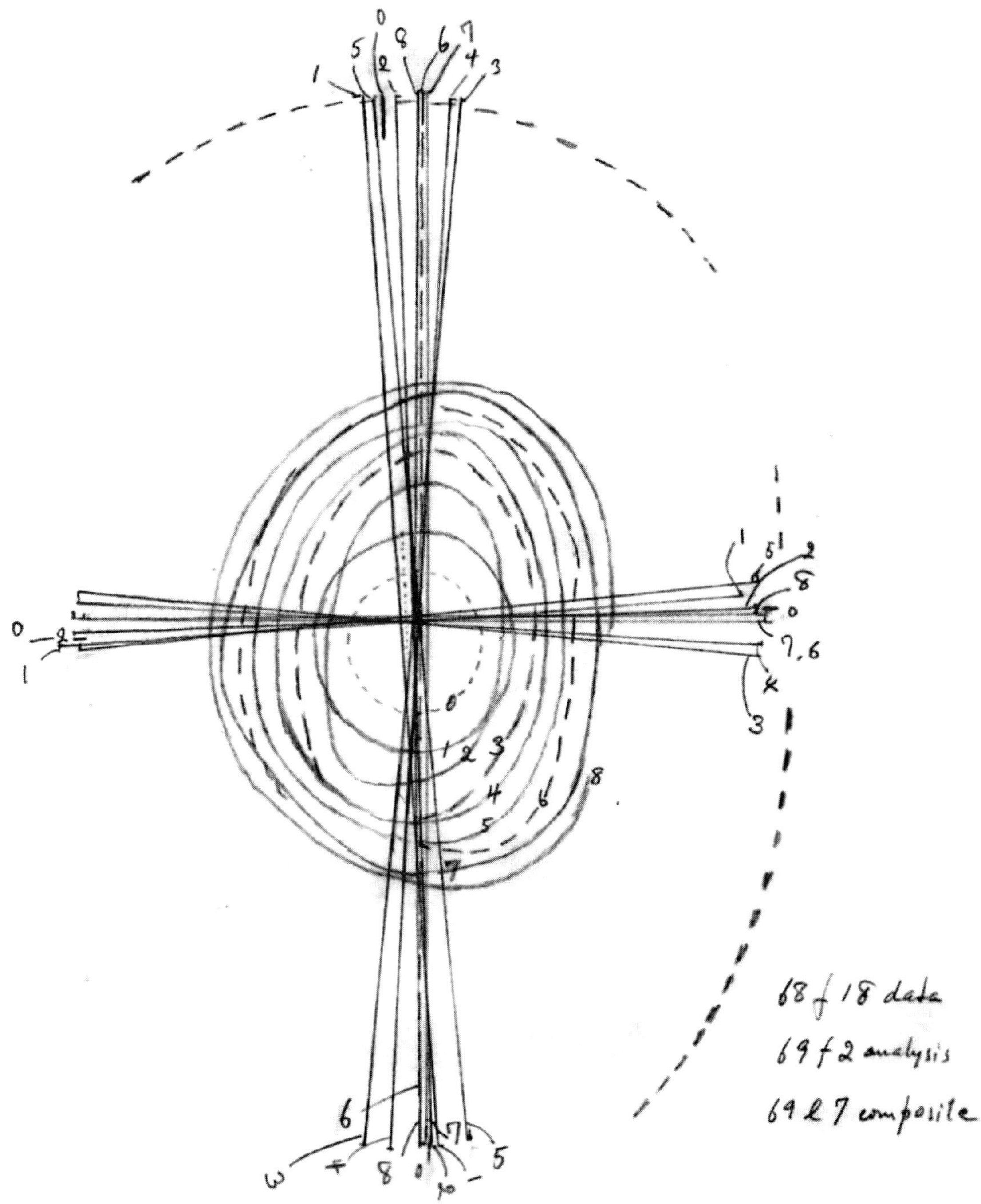

Fig. 5. Composite plot of eggs that have undergone the indicated number of divisions. The shape of the micro-drop had turned from a circle to an oval shape with time. Data from June 18, 1968, plotted December 9, 1969.

The outcome of these students' work, carried out in just two weeks, was so striking that they were reported at both the General Scientific Meetings at the MBL in August, and at the Annual Meeting of the American Society for Cell Biology in Baltimore, Maryland, that December. The abstract of the latter presentation is shown here as Article 36, but we never had a chance to publish an illustrated full paper. Thus, this note is added, together

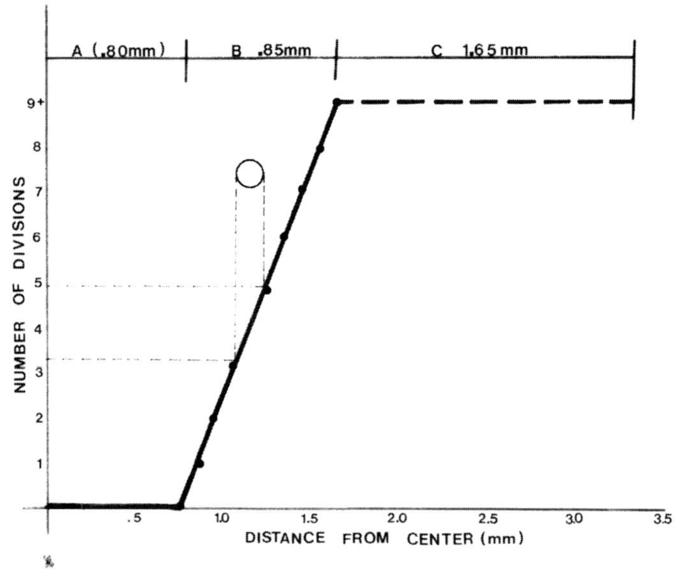

Fig. 6. Averaged composite plot of number of divisions versus distance from the center of the micro-drop. June, 1982, data from student project.

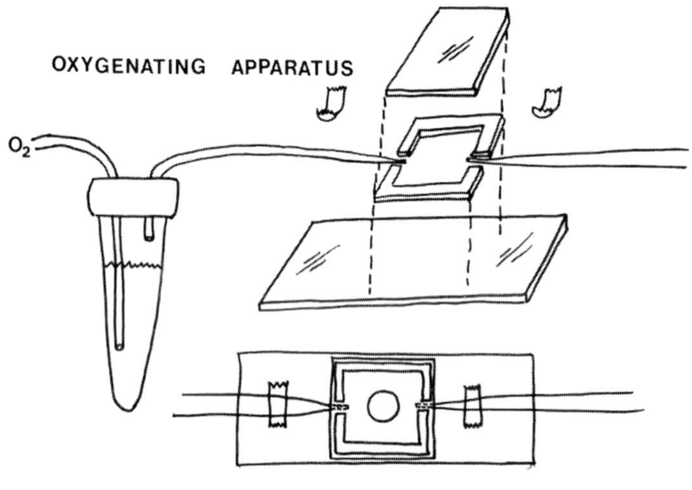

Fig. 7. Modification of the micro-drop chamber to permit oxygen perfusion through the air space in the chamber. A thin polyethylene tube was gently heated and drawn out to make a fine tip that would fit into the space between the two halves of the filter paper ring, then secured to the slide with two small pieces of Scotch tape. Then oxygen gas was bubbled through seawater to equilibrate its vapor pressure to that of the micro-drop. June, 1982, data from student project.

with the methods and data recorded by the students.

More recently, Judah Folkman at the Harvard Medical School has shown how neoangiogenesis (i.e., formation of new blood vessels which supply oxygen and remove waste products beyond what is accomplished by diffusion) is essential for the growth of tumors beyond a few mm in size.

In 1995, John Bonner *et al.* at Princeton University showed how a sharp gradient of differentiation is established at the open end of a

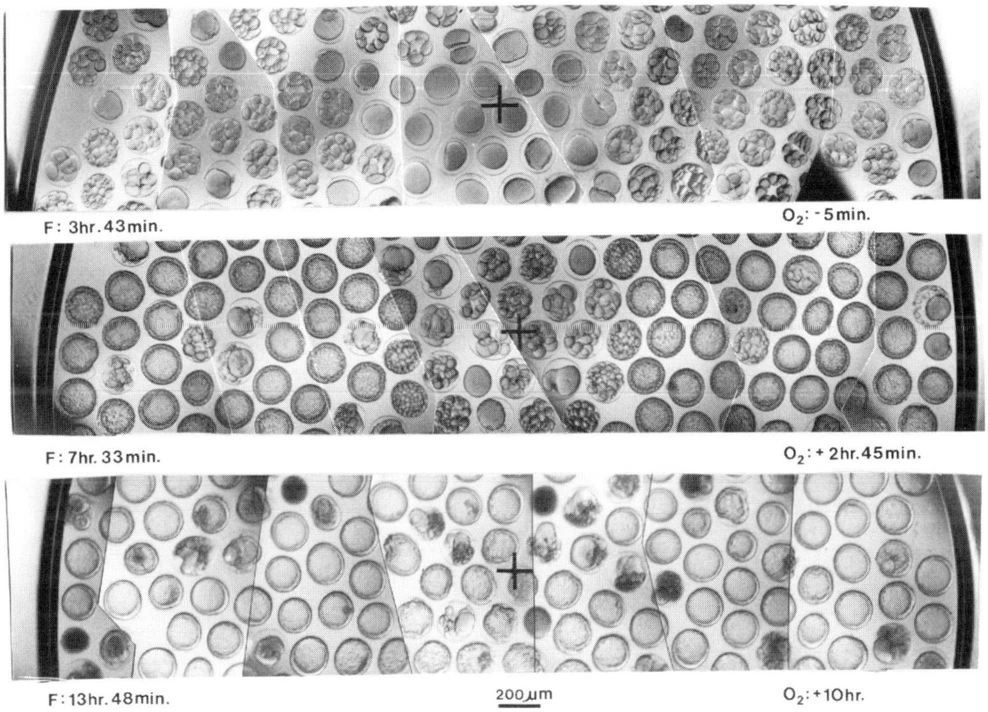

Fig. 8. Reversal of anoxia-generated inhibition of development by oxygen perfusion. F: time after fertilization. O_2: time after start of oxygen perfusion. June, 1982, data from student project.

capillary tube filled with *Dictyostelium* amoebae (*PNAS* **92**: 8249–8253). Again, this sharp gradient of differentiation was no doubt generated by an interplay between oxygen diffusion and cell metabolism.

Both of these observations, together with ours, suggest the important role that a surprisingly sharp gradient of cell-induced anoxia can play in the division of cells and differentiation of tissues. They also point to the need for tissue cells to be no more than one or a few cell layers away from access to their blood, lymph, or supply of bathing medium.

Reprinted from the *Journal of Cell Biology*, Vol. 93(3), pp. 812–819, 1982, with permission from The Rockefeller University Press.

Acrosomal Reaction of *Thyone* Sperm. I. Changes in the Sperm Head Visualized by High Resolution Video Microscopy

SHINYA INOUÉ and LEWIS G. TILNEY
Marine Biological Laboratory, Woods Hole, Massachusetts 02543; and the Department of Biology, University of Pennsylvania, Philadelphia, Pennsylvania 19104

ABSTRACT Structural changes inside the head of *Thyone* sperm undergoing the acrosomal reaction were followed with a high-resolution, differential interference contrast (DIC) video microscope. The beating sperm, adhering by their midpiece to the cover slip of a wedge perfusion chamber, were activated by a calcium ionophore (20 μM A23187) suspended in sea water containing 50 mM excess $CaCl_2$.

Before activation of the sperm, the acrosomal region appears as a 1.1-μm diameter sphere, slightly less dense than the rest of the sperm head. Upon activation, the acrosome pops; the acrosomal region suddenly swells and its refractive index drops. After ~1 s, a crescent-shaped periacrosomal cup appears behind the acrosomal vacuole. In the next several seconds, the cup loses more refractive index and expands forward as the acrosomal process extends. The acrosomal vacuole becomes smaller, but without appreciable drop in refractive index. These observations, coupled with the behavior of the extending acrosomal process reported in the companion paper, and in electron microscopy (EM) and early physiological studies, suggest that the acrosomal process is extended by a combination of the explosive polymerization of actin and the osmotic swelling of the periacrosomal cup material.

In this paper, we also consider the meaning of the enhanced DIC image seen in the high-resolution video microscope, and discuss the reliability of measurements on small linear dimensions made with the DIC microscope.

For successful fertilization, a sperm encountering the egg must first traverse the egg's protective jelly and other coats. Within seconds after the arrival of sperm, a delicate thread is observed in the jelly of some echinoderm eggs, extending from the sperm head to the cytoplasmic surface of the egg. At the base of the thread, the surface of the egg may protrude, or display a "fertilization-" or "entrance-cone."

Some earlier observers believed that the fertilization cone, and the thread that was believed to be spun out of the cone, were the egg's way of lassoing the sperm into the ooplasm (4, 14). Others did not find the thread and also argued that the cone formed only after the sperm came into contact with the egg or its envelope (e.g., reference 27, p. 409).

In the 1950's, however, J. C. Dan (8) and the Colwins (5) showed decisively that the fine thread that had been seen in the egg jelly was produced by the sperm rather than the egg. Even in the absence of the egg, they could induce the thread to grow from the sperm, specifically from its acrosome, a small bi- or tripartite body found at the apex of the sperm (1). The sperm was said to go through an acrosomal reaction, and the fine thread produced at the tip of the sperm was named the acrosomal process. Thus, in the normal process of fertilization, the sperm encountering the egg jelly undergoes an acrosomal reaction and the acrosomal process of the sperm extends and penetrates the jelly and vitelline envelope. After the sperm and egg plasma membrane have united at the tip of the acrosomal process, the nucleus and cytoplasm of the sperm are drawn into the egg.

In the acrosomal reaction, the sperm first reacts by a change in the cell membrane that lies in contact with the "acrosomal granule," or acrosomal vacuole (Fig. 1). In a reaction requiring the presence of Ca^{++}, the membranes of the acrosomal vacuole and the sperm fuse in a ring-shaped annulus at the apex of the

Dedicated to Jean Clark Dan, humanist, friend, pioneer in research on the acrosomal reaction.

THE JOURNAL OF CELL BIOLOGY · VOLUME 93 JUNE 1982 812–819

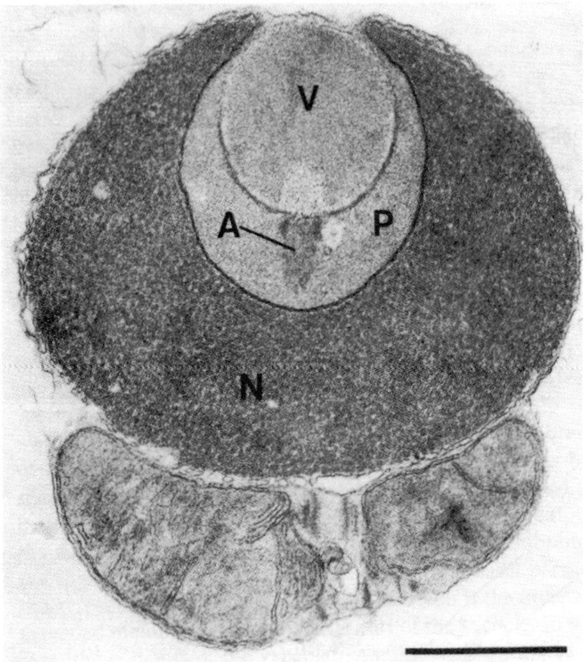

FIGURE 1 Thin section electron micrograph showing the head and midpiece of *Thyone briareus* sperm before acrosomal discharge. The acrosomal vacuole (*V*) is surrounded by the cup-shaped periacrosomal region (*P*) within the pocket of the sperm nucleus (*N*). The plasma membrane at the anterior of the sperm (top) and the vacuolar membrane are closely apposed and are still intact. The apposed membranes in this region fuse at their rim and are cast off when the acrosome pops. The anterior portion of the actomere (*A*), onto which the actin molecules polymerize, is seen within the periacrosomal cup. Electron microscope (EM) procedures used to obtain this micrograph can be found in Tilney and Inoué (25). Bar, 500 nm, × 49,000.

sperm head. The membrane permeability changes and, according to Dan et al. (10, 12), water enters the acrosome. Unless the pH of the medium is too low, the membrane bounded by the annulus is cast off (6, 11). At this time the outer face of the sperm membrane has become contiguous with the inner face of the membrane surrounding the acrosomal vacuole.

Meanwhile, actin that is contained in the cup-shaped cytoplasmic space, lateral to and immediately posterior to the acrosomal vacuole, starts to polymerize onto a rod-shaped organelle, the actomere, located at the base of the cytoplasmic cup (23). The newly forming filaments extend rapidly in parallel to form a thin axial bundle that is believed to provide the force for the extension of the acrosomal process. The axial filaments rapidly extend forward by addition of more actin subunits at the anterior tip of the growing acrosomal process (25, 26).

In this manner, the posterior membrane of the acrosomal vacuole, now contiguous with the plasma membrane of the sperm, is everted and surrounds the growing axial filaments. The new acrosome membrane is in turn covered by the contents of the acrosomal vacuole which is believed to contain the lysins needed to penetrate the jelly coat and vitelline envelope (for review, see references 6, 9), and bindin, a protein thought to attach sperm to the egg (see references 19). The whole acrosomal reaction, including the extension of the process, is completed in a few seconds. Even in the sperm of the sea cucumber, *Thyone*, that produces an acrosomal process 90 μm long, the reaction is completed in 10 s.

The acrosomal reaction can be artificially induced in echinoderm sperm with sea water that contains extracts of the egg jelly, or excess Ca^{++} or ammonia, or whose pH is raised to 9.5 (8). Sometimes the acrosomal process is produced spontaneously when the sperm lands on a solid substrate.

We studied the dynamic changes and kinetics of this rapid reaction of the sperm by recording and analyzing the events with a high-extinction video microscope. With this new instrument, we could obtain excellent quality, high-resolution, DIC images of *Thyone* sperm undergoing the acrosomal reaction. The recorded video image could be analyzed sequentially at time intervals as short as 17 ms (17).

In the first of these two reports, we discuss the methods devised for these studies, as well as describe the morphological and optical changes that take place inside the *Thyone* sperm head. The sperm were observed by video in a newly designed perfusion chamber, and triggered with a Ca^{++} ionophore in the presence of excess Ca^{++}. We conclude that the polymerization of actin and osmotic influx of water may both contribute to the extension of the acrosomal process.

In the second paper, we describe the kinetics of growth and the morphological changes of the growing acrosomal process. On the basis of these data, we analyze the diffusion and polymerization of actin within the acrosomal process as well as the addition of membrane to the process. We then discuss the balance of the various forces that must underly the production and transient maintenance of the extremely slender and long acrosomal process.

MATERIALS AND METHODS

Obtaining Sperm

Thyone briareus were collected by the Marine Resources Department of the Marine Biological Laboratory, Woods Hole, MA. The testes were removed and minced in sea water. The suspension was filtered through cheese cloth and the supernate centrifuged at 5,000 *g* for 5 min to pellet the sperm. The pelleted sperm, stored in the refrigerator, were used either the same day or the next day.

Triggering of the Acrosomal Reaction

The acrosomal reaction of *Thyone* sperm can be triggered by adding an ionophore such as A23187 (CalBiochem, La Jolla, CA) to the sea water. Although the acrosomal processes produced are often shorter than normal (they must exceed 50 μm in length to penetrate the jelly layer which is that thick), all the sperm react. We found that the addition of excess Ca^{++} to the sea water induced longer processes, or processes of normal length. Accordingly, a small amount of the pellet of sperm was suspended in sea water containing an additional 50 mM $CaCl_2$. The dilute suspension of sperm, which was only slightly milky, was then added to the perfusion slide and within a minute high calcium sea water containing A23187 was perfused across the slide. A 1 mg/ml stock of A23187 was made by dissolving A23187 in dimethyl sulfoxide. Immediately before perfusion 10 μl of this stock was added to each ml of high calcium sea water. Thus, the final concentration of A23187 in sea water was ~20 μM.

Perfusion Chamber

The *Thyone* sperm were studied in a simple, metal-free perfusion chamber that we devised. In this chamber, the slide and cover slip form a wedge-shaped rather than a plane-parallel space (Fig. 2). The cells are trapped in the wedge and the flow is forced to spread out to provide a uniform exchange without stagnant zones (18). The chamber provides for thorough and rapid exchange of media, and permits high resolution observations of individual cells on an inverted microscope.

Diluted suspension of sperm is introduced into a clean chamber, and sperm that have attached to the cover slip and which are not drawn out by additional perfusion are used for observations. With slides and cover slips that were cleaned with neutral detergent and rinsed thoroughly according to our standard laboratory procedures (15), healthy *Thyone* and *Arbacia* sperm become "attached" to the glass surface by their midpieces.

Video Observations and Analyses

Sperm in the perfusion chamber were observed with a 40/0.65 dry- or 100/1.30 oil immersion differential interference contrast (DIC objective lens of the Leitz Smith T-system, E. Leitz, Inc., Rockleigh, NJ). The DIC condenser and objective lenses were oriented to provide maximum extinction on a specially built inverted polarizing microscope (17). The specimen was illuminated with monochromatic green light of 546 nm by filtering the output of a 100 watt, concentrated mercury arc burner (III-100, Illumination Industries, Inc., Sunnyvale, CA or Osram HBO-100, Opto-Systems, Inc., Jenkintown, PA) through two layers of glass heat cut filters (Schott KG-1, Fish-Schurman Corp., New Rochelle, NY) and a high (>70%) transmission narrow bandpass multilayer interference filter (Baird Atomic, Bedford, MA). The deSénarmont compensator of the DIC condenser was oriented at between 10° and 20° off extinction to optimize the video image contrast of the acrosomal process or the contents of the sperm.

The microscope image was projected by the ocular onto the 1 inch Newvicon image tube in a video camera equipped with circuits that automatically adjust the gain and the pedestal level (model 65 II, Dage-MTI, Michigan City, IN). The signal from the video camera was looped through a sync stripper and video processor (nos. 302-2 and 604, Colorado Video, Inc., Boulder, CO) for manually optimizing the contrast and pedestal levels. After insertion of the time and date signals (Model ET 202, Cramer Video, Newton, MA), the video signal was stored on a ¾ inch cassette, time-lapse tape recorder (TVO-9000, Sony Corp., Long Island City, NY) recording at normal speed, then displayed on a 9 inch monochrome video monitor (Sanyo 4290, Compton, CA).

For analysis of morphological changes in the sperm head or of the acrosomal process, the video tape was played back at normal speed, or in the 96 h mode for study field-by-field, or re-recorded onto a video motion analyzer (Sony SVM 1010). We studied the 10-s sequence stored on the disk in the motion analyzer, either field-by-field or at various speeds, backwards and forwards.

The lengths of the acrosomal process were determined from frozen images of the video tape record. X- and Y- video markers generated by a video analyzer (Colorado Video No. 321) were positioned onto the tip and base of the acrosomal process, and their coordinates recorded together with the time of the scene. The length of the process was calculated from the coordinates which had been calibrated with images of a stage micrometer.

The exact time of a particular scene was determined by counting the number of video fields elapsed after the seconds digit had last advanced. This method provides the time point with a 16.7 ms precision since each second is represented by exactly 60 fields in standard United States video equipment.

Changes in the dimensions of complex structures were measured from photographed scenes of the monitor, calibrated with scenes of the stage micrometer. The monitor was photographed on Kodak Plus-X or Agfa Super-Pan film with a Nikon FM, single lens reflex camera, equipped with a Micro-Nikkor 55 mm f/3.5 macro lens. Some scenes (e.g., Fig. 3) were photographed through a 50-lines-per-inch Ronchi grating (Rolyn Optical Co., Arcadia, CA) to suppress the intrusive video scan lines (17). The film was developed according to Kodak specifications in Microdol-X diluted 1:1. Further details of the video equipment and the theory of video enhancement for polarized light and DIC microscopy are described elsewhere (17; also see reference 3).

RESULTS

Thyone sperm, suspended in sea water and introduced into the perfusion chamber, attach in large numbers to the clean surfaces of the slide and cover slip. While the attachment is secure enough so that most sperm, once attached, are not dislodged by additional perfusion, the tail and head, nevertheless, appear to remain free. The tail of the attached sperm can beat at its normal frequency, and the tail and head are often seen swinging around the sperm as though pivoted at the attachment. Sperm of *Arbacia* clearly pivot around the midpiece, while *Thyone* sperm appear to pivot either around the midpiece, or the junction of the midpiece and the sperm head. In some preparations the sperm tail also stuck to the slide, but with no apparent effect on the acrosomal reaction.

In an unreacted sperm, the acrosomal region (i.e., the acrosomal vacuole plus the periacrosomal cup) is visible as a small sphere of somewhat lower refractive index than the rest of the sperm head (Fig. 3*A*). The acrosomal region in the living, unreacted sperm measures ~1.1 μm in diameter.

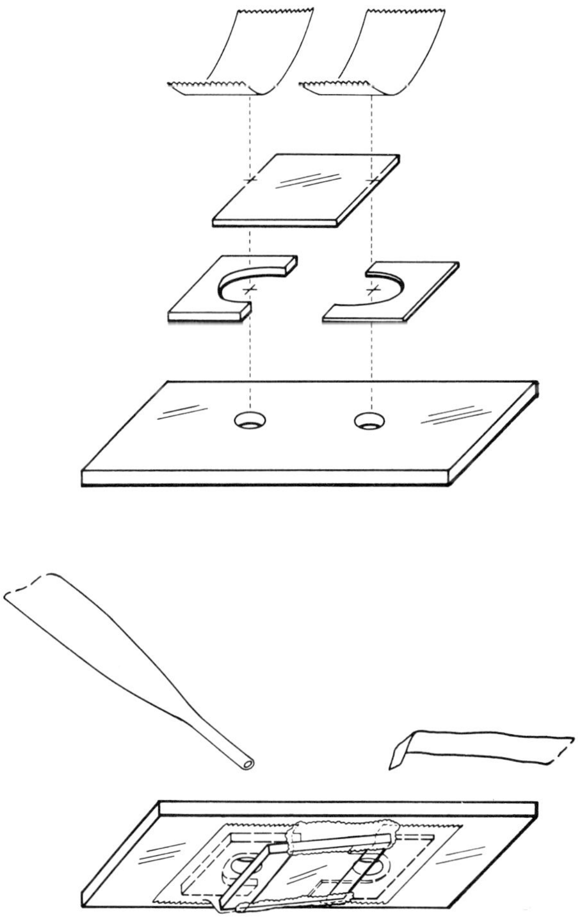

FIGURE 2 Design of the wedge perfusion chamber for use on an inverted microscope. Top: In the process of assembly. Bottom: The inverted, assembled chamber as seen in perspective from below. Two holes, 3–4 mm in diameter, are drilled in stress-free, standard size (1" × 3" × 1.0–1.2 mm thick) microscope slides, with their centers spaced apart by 18 mm, the width of the cover slip. On a clean perforated slide we place 2 U-shaped plastic spacers, with the open ends of the U's clearing the holes in the slide, and facing each other. One is ~12-μm thick (cut out from DuPont Mylar 48-S sheet), and the other is ~100-μm thick (DuPont Mylar 400-S sheet). A clean 18 × 18 mm cover slip is placed on top of the two spacers, spanning the two holes in the slide. The cover slip and spacers are taped onto the slide with two 1" long pieces of ½" wide Scotch tape. The tape seals the spacers, and the entrance and exit sides of the cover slip, leaving the middle 80% of the cover slip clear. The remaining exposed sides of the cover slip and the small recess under the Scotch tape are sealed with molten Valap-B, a 1:1:1 simmered mixture of vaseline, lanolin, and bees wax. The assembled slide is inverted, and a 5–6 mm diameter circle is drawn with a china-marking wax pencil around the hole above the thicker spacer. The wax pencil circle indicates the entrance hole, and also prevents the perfusion solutions from flooding over the slide. The other hole receives the filter paper wick that draws the solution out from the perfusion chamber. The tapered piece of filter paper is replaced every few minutes, generally more frequently when fresh solutions are added to the entrance port. The angle of taper of the filter paper, and the frequency of its replacement, control the rate of perfusion.

Soon after high-calcium ionophore sea water is introduced into the perfusion chamber, the sperm tail starts to beat more vigorously as though stimulated by a signal. Then, the acro-

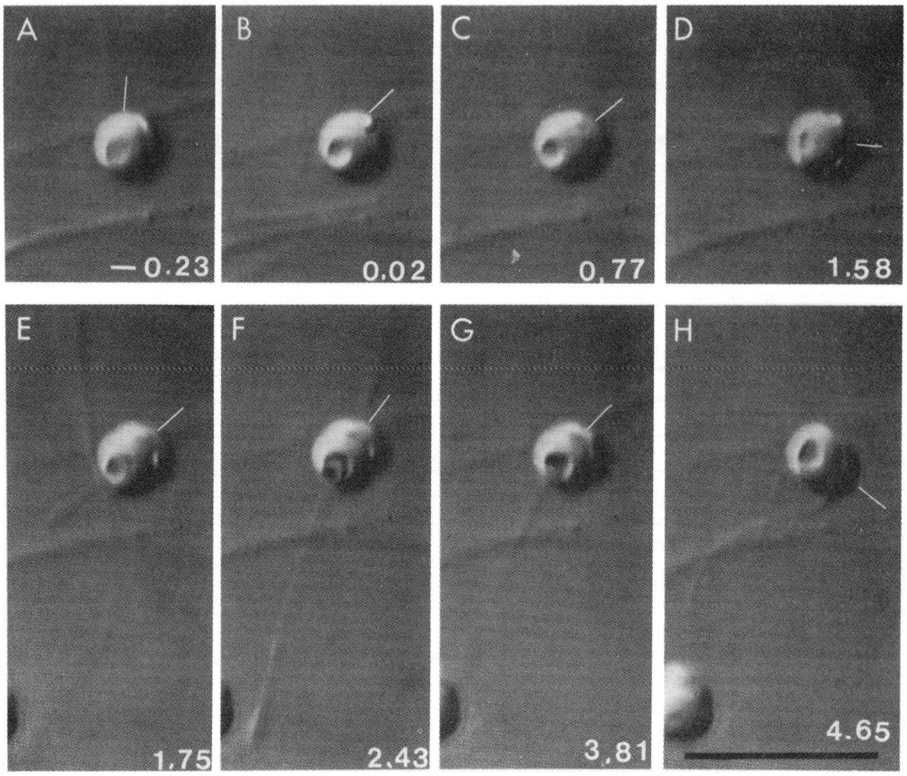

FIGURE 3 High magnification video sequence of changes taking place in the *Thyone* sperm head during the acrosomal reaction. White lines indicate the location and orientation of the sperm tail that would be seen if the tail were to project straight back from the sperm, which it does not. These pictures were taken through a Leitz 100/1.30 Smith T-system DIC objective, with the condenser oil immersed and used at an N.A. of 0.91. The video monitor was photographed through a Ronchi grating to reduce the prominence of the scan lines (for detail of the method, see reference 17). Time shown is in seconds after the acrosome has popped. × 2,800. *A*: 0.23 s before the acrosome pops, the refractive index of the acrosomal region is somewhat less than the remainder of the sperm head. *B*: After the acrosome has popped, the acrosomal region is slightly larger and shows more contrast; its refractive index has dropped. *C*: The acrosomal process has still not started to elongate. *D*: The sperm head has turned and the process is just starting to elongate (to the left). The bundle of actin filaments is visible in the vacuole. The periacrosomal cup is just becoming discernible. *E*: The sperm is turning again. The process is now about as long as the sperm head is wide. Two compartments, the acrosomal vacuole and the periacrosomal cup, are now clearly visible. *F*: The acrosomal process has grown to about four times the diameter of the sperm head. The spear-shaped tip of the acrosomal process has reached the lower left corner of the scene. Some blebs are detectable on the process, which is somewhat out of focus. The refractive index of the periacrosomal cup has dropped dramatically. The acrosomal vacuole is moving forward, but without appreciable loss of refractive index. The tail of the sperm has come into focus at the upper right. *G*: The periacrosomal cup has swelled considerably and appears heart shaped. The remmant of the acrosomal vacuole is now considerably smaller. *H*: The sperm head has turned again and presents a different view of the periacrosomal cup. Within it, the actomere and/or the actin filament bundle is visible. The sperm tail is now whipping down below, and in this scene appears S-shaped, crossing the thin, acrosomal process.

some "pops." That is, all of a sudden, the acrosomal region becomes more prominent, in part due to a drop in its refractive index, and in part due to a slight increase in diameter of ~25% (Fig. 3 *A* and *B*). The sudden drop in refractive index and increase in diameter are completed in three to four video fields, or in 50–70 ms. At the same time that the acrosomal region becomes more prominent, the tip of the acrosome bursts open into a faint inverted funnel, and the acrosome spits forward a fine mist. Sometimes what appears to be a very pale vacuole, ~0.5 μm in diameter, floats down stream from the popped acrosome, but the image of the pale structure is too faint to permit a more definite description.

With the exceptionally high contrast provided by the video DIC image, some of these changes were quite obvious, while we became aware of other features only after repeated viewing of the recorded video sequence.

One to two seconds after the acrosome pops, the acrosomal process comes into sight at the tip of the sperm (Fig. 3 *D*). As detailed in the next paper (25), the process grows rapidly into a slender thread many tens of micrometers long over the next several seconds.

Early during the growth of the acrosomal process, two compartments become visible in the acrosomal region (Fig. 3 *D* and *E*). The anterior part is a sphere with a diameter of ~1.2 μm. It is located at the apex of the sperm head, at the base of the acrosomal process. The posterior part is shaped as a cup that partly surrounds the anterior sphere.

The location and size of the anterior sphere clearly identifies it as the "acrosomal granule," or acrosomal vacuole, seen in electron micrographs (Fig. 1). The cup-shaped compartment with a somewhat cone-shaped posterior contour corresponds, in shape and location, to the periacrosomal cup seated in the anterior depression of the nucleus. In favorable video frames a refractile rodlet, presumably the actomere, is detected in that region.

In the first few seconds after the acrosomal process comes into sight, the cup-shaped periacrosomal compartment expands forward. At the same time, its refractive index drops sharply (Fig. 3 *E*–*G*). The posterior contour of the cup becomes increasingly clear, but is unchanged in shape. Only its anterior contour, i.e., the boundary with the sphere, moves forward so that the cup expands into a heart-shaped chamber.

As the posterior chamber expands forward, the anterior sphere gradually becomes smaller and perhaps somewhat flattened. However, the refractive index of the "sphere" does not appreciably change (Fig. 3 *E*–*G*). Meanwhile, the refractive index of the posterior chamber continues to drop so that the boundary between the two chambers becomes increasingly more pronounced (Fig. 3 *D–H*).

The rate of expansion of the posterior chamber correlated closely with the growth of the acrosomal process. The height of the periacrosomal cup, i.e., the gap between the base of the recess in the nucleus and the posterior margin of the acrosomal vacuole (Fig. 3), is plotted in Fig. 4 against the length of the acrosomal process. As the acrosomal process becomes extended tens of micrometers, the gap expands from ~0.4 to 1.0 μm.

When the size of the gap is plotted against time in seconds after the acrosome has popped, the gap is seen to expand linearly for the first few seconds (Fig. 5). While the acrosomal process and the gap do not become visible until over a full second after the acrosome has popped, the size of the expanding gap interestingly extrapolates to zero just when the acrosome pops.

DISCUSSION

Interpretation of the Image

The application of high-extinction, differential interference contrast video microscopy has allowed us to visualize detailed structures within living sperm with unprecedented clarity. Video recordings of the *Thyone* sperm, triggered by the Ca^{++} ionophore in an appropriate perfusion chamber, has allowed us to analyze rapid, fractional-second events that take place within an individual sperm undergoing the acrosomal reaction.

While these observations serve to illustrate the power of modern video microscopy, and provide the data needed for analyzing the kinetics of acrosomal process growth (25), some of the optical and morphological changes that we found within the *Thyone* sperm were also quite unexpected.

Before considering these biological findings, however, we shall first examine the nature and reliability of the highly magnified video images that are produced by differential interference contrast microscopy (see references 16 and 21 for studies on the nature of DIC images). Much of our observations on the structural changes, and distribution of refractive indexes, in the reacting *Thyone* sperm were made on live and recorded video displays. Just how reliable are the enhanced and recorded video displays, and what can we deduce about the morphological details of the specimen? How reliably can we deduce the distribution of refractive indexes from the differential interference contrast microscope images?

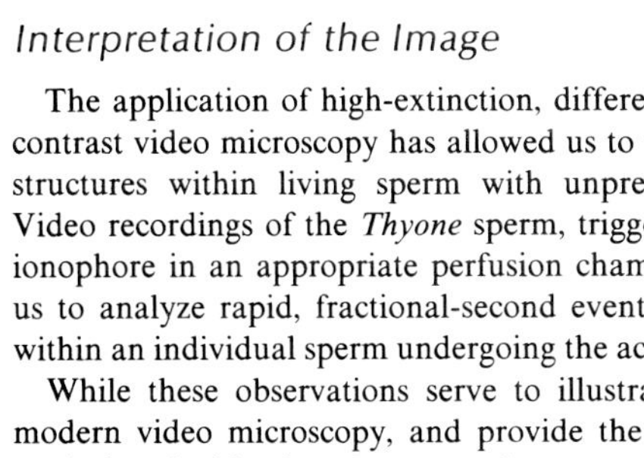

FIGURE 4 Height of the periacrosomal cup (i.e., the gap between the base of the recess in the nucleus and the posterior margin of the acrosomal vacuole) vs. length of the acrosomal process. The data were taken from the sperm illustrated in Fig. 3.

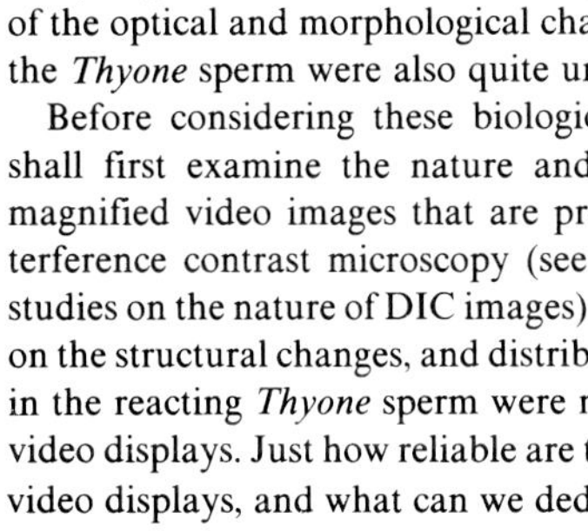

FIGURE 5 Height of periacrosomal cup (gap) plotted against time, in seconds, after popping. Data taken from the sperm shown in Fig. 3.

With video images in general, the distribution of image brightnesses seen by the video camera across a contrasting, or sharply bounded, edge may or may not be maintained in the image displayed on the monitor. The electronic frequency response characteristics of the video camera, recorder, and monitor can all modify the appearance of an edge, depending on the degree of contrast at that edge, the direction of the edge relative to the scan lines, the absolute brightness, and the electrical signal level. Edge sharpening, or compensating, circuits are commonly built into video equipment, especially the monitor, to sharpen up the edges and provide a crisper looking image.

In view of these characteristics of video systems, the interpretation of the image detail on the video monitor needs to be approached with some caution. We have examined the overall system performance of our microscope-optics-and-video-combination by examining the images of sharp edges of thin metal films deposited on microscope slides. Abbe test plates and some stage micrometers (e.g., those made by Bausch and Lomb Optical Company, Rochester, NY) are made in this fashion and provide cover slip thicknesses that match the optical corrections demanded by the objective lenses.

We find that with careful alignment of the light source and the optical axes of the condenser and objective lenses, and by eliminating the flare from various sources in the microscope (especially that arising from between the eyepiece and the video camera image tube), that we can obtain images that are reasonably reliable. For the video train, it was also necessary to keep the video signal at each stage within certain bounds, and to refrain from raising the monitor gain (contrast) too high. With too strong a video signal, horizontal bands of altered intensities appeared and, in the extreme, even distorted the vertical outlines of the image into wavy patterns. With too high a gain for the monitor, an electrical overshoot results in a spurious bright line following (i.e., in going from left to right on a normal monitor) a sharp transition from a dark to bright area in the image.

Once these distortions and spurious contrast arising from the instruments are reduced to an acceptable level, what information can we glean from the differential interference contrast video image?

In differential interference, contrast is provided by the gradient of optical path differences experienced in the specimen by two beams, or wave fronts, sheared laterally by slightly less than the minimum distance resolvable by the particular objective lens (20). Since the Optical Path = (Refractive Index) × (Thickness), the contrast depends on the cross-sectional shape of the edges or boundaries, and the gradients of refractive indexes in the specimen. The contrast in a differential interference microscope image also depends on the setting of the compensator and the direction of the boundaries, the contrast being maximum when the optical shear direction runs perpendicular to the boundary (2, 16, 21).

Once the intensity distribution in the image is provided by differential interference contrast, a psychological factor also enters into the viewing of the image (21). A circular grey disk with a light crescent on top and dark crescent below is interpreted as though it were a white sphere in a dark room illuminated from above. A white sphere illuminated from below, and the corresponding disk with a dark crescent on top, are interpreted as hollow bowls in the absence of additional cues. The differential interference contrast image gives a similar sense of three-dimensional relief.

For interpreting the changes we see in the activated sperm head, we know e.g., that (*a*) the sperm head and tail have refractive indexes greater than that of the surrounding sea water, and (*b*) both are more or less cylindrically symmetric around the long axis of the sperm. Starting with these familiar relationships we can interpret the contrast distribution of other regions of the sperm seen in differential interference contrast with reasonable certainty.

Even before the sperm is exposed to the ionophore, the acrosomal region shows a faint relief, with the shadowing reversed from the sperm head and tail. Therefore, the refractive index of the unreacted acrosome is somewhat less than the index of the rest of the sperm head.

When the acrosome "pops", the shading of that region becomes more pronounced. Since the direction of the shading remains the same as with the unreacted acrosome, the popped acrosome must also have a refractive index below the surrounding parts of the sperm head, but with even a greater difference of refractive index. The refractive index of an organelle above that of the solvent (≃ water) is proportional to the solute concentration (13). Therefore, the contents of the early acrosome must have been diluted or swollen at popping.

Similar arguments hold for the periacrosomal compartment and its relation to the sperm nucleus, as well as to the forward sphere, or acrosomal vacuole.

In our observations, we do not think there is any ambiguity in the deductions we have made regarding the concentrations of material in various parts of the acrosomal region save one. That exception is just at the base of the extending acrosomal process, at the anterior tip of the acrosomal vacuole (Fig. 3 *F*). The region looks as though it were a "hole" that opens anteriorly, although that is probably not the correct interpretation. However, the shading and curvature of this particular minute region are so complex that the structure of the region cannot be interpreted unambiguously.

While we feel comfortable about the qualitative interpretations of the morphological and refractive index changes, we are not so sanguine about the quantitative measurements of the absolute sizes of the compartments in the sperm.

The location of sharp lines, or the distance between sharp boundaries, can generally be determined with a precision considerably greater than the diffraction limited optical resolution of the microscope. On the other hand, the boundaries we deal with in differential interference microscopy are graded and not sharply discontinuous (16). Enhancement with video could make the edges appear sharp and discontinuous, but the level of grey scale that gives rise to the discontinuity can be set arbitrarily. Even without such extreme enhancement, where do we define the edges of a sphere with greater or less refraction than its surroundings?

A simple experiment points up the difficulty and subjective nature of the criteria used for defining the edges. With a pair of precision micrometer calipers we measured the diameter or height of the various acrosomal regions of interest. Repeated measurements, e.g., of photographs shown in Fig. 3, generally agreed to within 2% of each other. That is so until the photographs are turned upside down and new measurements are made. Again, the repeated values agreed to about 2%, but not with the measurements made with the picture turned the other way around. Those two sets of values were easily off by 10 to 15%!

Therefore the quantitative statements made in this paper regarding the absolute dimensions inside the sperm head or acrosomal parts have less precision and are more meaningful in terms of relative changes. The 10 to 15% uncertainty experienced here, however, may not compare unfavorably with the errors encountered in determining absolute dimensions by conventional cytological studies and electron microscopy where errors can be introduced by fixation, dehydration, knife compression, and lens hysteresis.

Popping

In all of the many *Thyone* sperm that responded to the ionophore sea water and produced an acrosomal process, the acrosome popped, consistently and explosively, about 1–2 s before the acrosomal process came into sight. The behavior and duration of excitement of the sperm before popping and process extension varied somewhat. Therefore, it was generally not possible to predict beforehand exactly when a sperm would pop. While we often missed these events at first viewing, reviewing the video tape allowed us to study these early changes.

At first sight, the sudden pop, the bursting open into an inverted funnel, and associated forward spitting of the fine mist, might be taken to mean that the contents of the acrosomal vacuole are suddenly expelled at the start of the acrosomal reaction. Such a view would be consistent with the loss of the membranes at the anterior tips of the sperm and the acrosomal vacuole, as well as the loss of the acrosomal contents inferred from electron microscopy (e.g., reference 7, Fig. 6; reference 24, Fig. 6).

According to our present observations on living sperm, however, the initial drop of refractive index of the whole acrosomal region is rather slight, and can probably be accounted for by a roughly twofold increase in volume (the diameter increases by ~25%). It is not until a second or two later, and then only in the periacrosomal chamber, that a major drop in refractive index is observed. Furthermore, the spherical forward compartment of the acrosome, or the acrosomal vacuole, does not become appreciably less refractive after its initial change.

We therefore interpret the popping to indicate an initial slight swelling of the whole acrosomal region, including the periacrosomal cup. The bulk of the contents of the acrosomal vacuole are not expelled. The mist, the faint inverted funnel, and the pale minute vesicle that are sometimes seen, are possibly the membrane released from the acrosomal vacuole and the apex of the sperm when they fuse and dehiss (to use the Colwins' 1961 expression).

The Periacrosomal Cup

Earlier, Tilney demonstrated that the periacrosomal cup containing the profilamentous actin can be isolated as a compact structure seated in the pocket of the nucleus (22). Our present observation on living sperm shows clearly that the material in that region swells dramatically as the acrosomal process extends. The refractive index of the cup drops precipitously as the periacrosomal cup swells into a heart-shaped cavity. Therefore, in vivo, the periacrosomal material imbibes water rapidly during the acrosomal reaction. Within the expanding compartment, what appears to be the actomere and the base of the polymerizing actin filament bundle that polymerizes onto the actomere, are seen running through the compartment in favorable video frames (Fig. 3 *D*, *H*).

The Acrosomal Vacuole

As the acrosomal process extends and the periacrosomal chamber swells, the acrosomal vacuole, while becoming smaller, persists in front of the swelling periacrosomal chamber. Furthermore, the bulk of the material making up the vacuolar contents is not, as suggested by electron micrographs, expelled at the time of popping. Instead, the material is gradually lost without appreciable decrease in concentration after the initial influx of water at popping.

Interpretation of Structural Changes

Our observations on living sperm by video microscopy, coupled with the EM and physiological studies of earlier workers, lead us to the following schematic (Fig. 6) of the events that take place within the acrosomal region of an activated sperm.

(*a*) Associated with a change in the permeability of the sperm and acrosomal vacuolar membrane, the whole acrosomal region imbibes some water (as suggested by Dan et al., 10, 12) over an initial reaction period of 50 to 70 ms (Fig. 6 *A* and *B*).

(*b*) Early during this initial imbibition of water, the anterior portions of the sperm plasma membrane and vacuolar membrane dehiss (6), and the membrane is cast off explosively within a few milliseconds (Fig. 6 *B*). However, the contents of the acrosomal vacuole remain undischarged.

(*c*) As profilamentous actin is solubilized and released in the periacrosomal compartment, actin filaments polymerize onto the actomere (23) anchored to the base of the pocket in the nucleus (Fig. 6 *B*). The growing actin filament bundle pushes into, and deforms, the posterior surface of the acrosomal vacuole (Fig. 6 *C*).

(*d*) The posterior membrane of the vacuole is now contiguous at its anterior margin with the exterior plasma membrane of the sperm (Figs. 6 *B* and *C*, as emphasized by Colwin et al. [7] Fig. 1). The posterior vacuolar membrane is indented as a cap that covers the tip of the growing actin filament bundle and a thin layer of surrounding cytoplasm (Fig. 6 *C*).

(*e*) The periacrosomal material continues to imbibe water, pushing the acrosomal vacuole forward (Figs. 6 *D* and *E*).

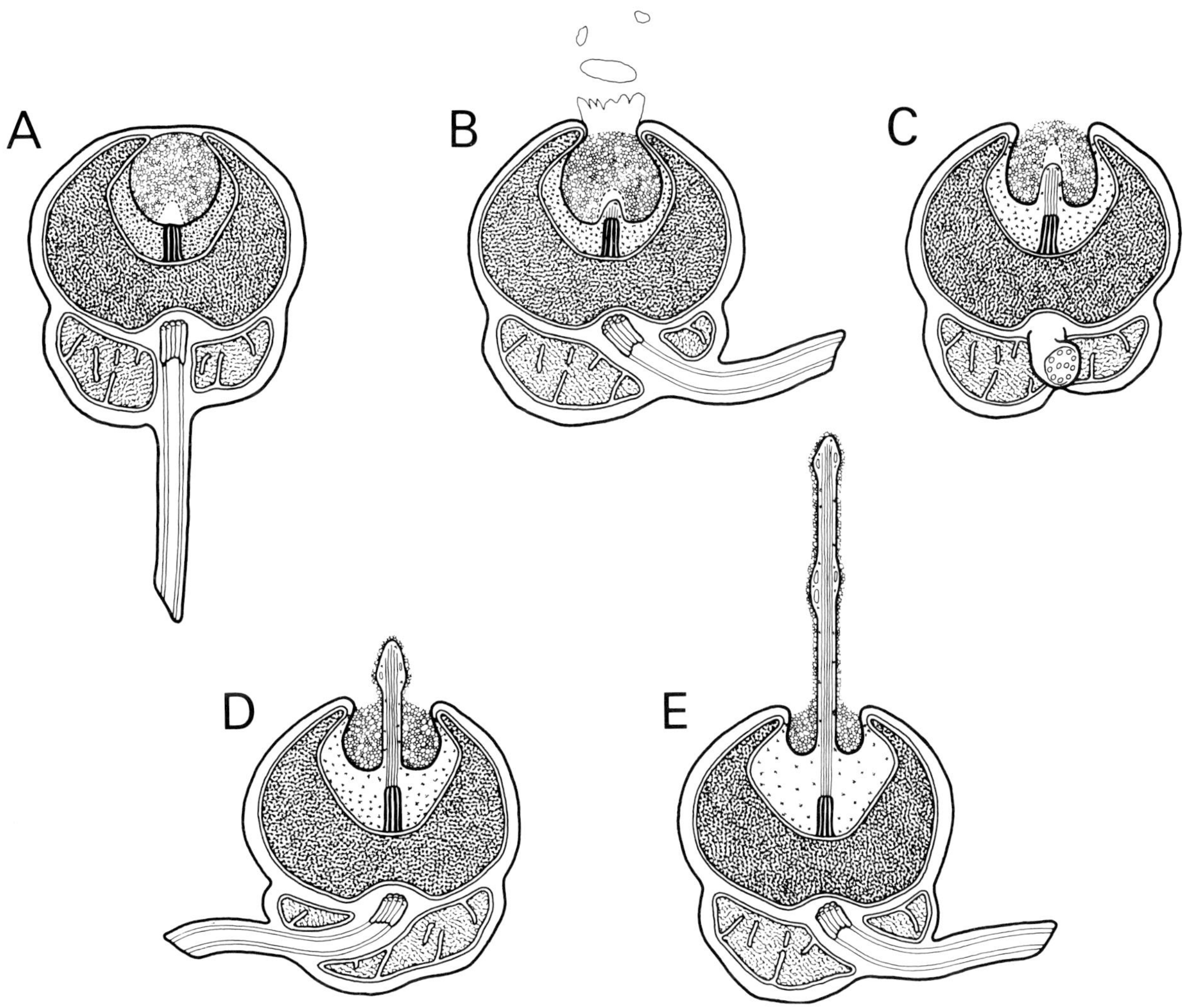

FIGURE 6 Schematics of structural changes in *Thyone* sperm undergoing the acrosomal reaction. The schematics were derived from our observation on living sperm by DIC video microscopy, combined with published EM and physiological accounts of the processes.

(*f*) Simultaneously with the extension of the actin filaments, the influx of water contributes to the growth of the acrosomal process through the donut-shaped acrosomal vacuole (Fig. 6 *D*).

(*g*) As the acrosomal process continues to extend, the external face of its membrane (the inverted membrane from the posterior part of the vacuolar wall) is coated with a thin layer of material arising from inside the acrosomal vacuole (Fig. 6 *E*).

The intermediate stages (Figs. 6 *B* through *E*) have not been seen by electron microscopy of sectioned sperm, even though our Fig. 6 *C*, for example, is superficially similar to Figs. 8–11 in Tilney (23). Those figures show actin filaments within a membrane bound tunnel piercing the acrosomal vacuole. Tilney's electron micrographs, however, show sections of *Thyone* sperm fixed 1.5 min after the addition of ionophore to sea water at pH 7.0. At this low a pH, the apical plasma membrane of the sperm and the vacuolar membrane do not dehiss but remain intact, and prevent the axial filament from extending beyond the apex of the sperm. Under this condition, the tip of the actin filament bundle is surrounded by three membranes: the sperm plasma membrane, the anterior, and basal parts of the acrosomal vacuole; the innermost of which is the inverted, extended membrane from the basal part of the vacuole. In this case, the contents of the vacuole naturally remain trapped in place.

Our observations on living sperm require that the contents of the acrosomal vacuole not be expelled as the acrosome pops. Rather, even after the vacuolar contents become exposed to sea water, they must remain in place for a few seconds until the vacuolar contents are drawn out and coat the growing acrosomal process.

The stages shown in Figs. 6 *B–E* may not have been seen by electron microscopy because: (*a*) these dynamic changes, which do not occur synchronously in different sperm, last for only a few seconds; (*b*) the contents of the acrosomal vacuole are not bounded anteriorly by a membrane. The fixative, or dehydrating, or other reagents may well extract and not preserve the exposed contents of the vacuole. Since no image corresponding to our video observations were found in EM thin sections deliberately prepared after the video observations, we feel that the second alternative is the more likely. We are now making attempts to capture these transient intermediate events by alternative methods of specimen preparation for electron microscopy.

We thank Ed Horn of the University of Pennsylvania for making the perforated slides used in the wedge perfusion chamber. We also thank Christopher Inoué for painstakingly printing the video photographs and Robert and Linda Golder of the MBL Photo Lab for producing the exquisite line drawings.

Supports by grants 5-RO1 GM 23475-16 from the National Institutes of Health (NIH) and PCM 7922136 from the National Science Foundation (to S. Inoué), and HD 14474 from NIH (to L. G. Tilney) are gratefully acknowledged.

REFERENCES

1. Afzelius, B., and A. Murray. 1957. The acrosome reaction of spermatozoa during fertilization or treatment with egg water. *Exp. Cell Res.* 12:325–337.
2. Allen, R. D., G. B. David, and G. Nomarski. 1969. The Zeiss-Nomarski differential interference equipment for transmitted-light microscopy. *Z. Wiss. Mikrosk. Mikrosk. Tech.* 69:193–221.
3. Allen, R. D., N. S. Allen, and J. L. Travis. 1981. Video-enhanced contrast, differential interference contrast (AVEC-DIC) microscopy: a new method capable of analyzing microtubule-related motility in the reticulopodial network of *Allogromia laticollaris. Cell Motil.* 1:291–302.
4. Chambers, R. 1923. The mechanism of the entrance of sperm into the starfish egg. *J. Gen. Physiol.* 5:821–829.
5. Colwin, A. L., and L. H. Colwin. 1955. Sperm entry and the acrosome filament (*Holothuria atria* and *Asterias amurensis*). *J. Morphol.* 97:543–567.
6. Colwin, L. H., and A. L. Colwin. 1961. Changes in the spermatozoan during fertilization in *Hydroides hexagonus* (Annelida) I. Passage of the acrosomal region through the vitelline membrane. *J. Biophys. Biochem. Cytol.* 10:231–254.
7. Colwin, A. L., L. H. Colwin, and R. G. Summers. 1975. The acrosomal region and the beginning of fertilization in the holothurian, *Thyone briareus. In:* The Functional Anatomy of the Spermatozoan. B. A. Afzelius, editor. Pergamon Press, New York. 27–38.
8. Dan, J. C. 1954. Studies on the acrosome II. Acrosome reaction in starfish spermatozoa. *Biol. Bull.* 107:203–218.
9. Dan, J. C. 1967. Acrosome reaction and Lysins. *In:* Fertilization I. C. Metz and A. Monroy, editors. Academic Press, New York. 237–293.
10. Dan, J. C., and Y. Hagiwara. 1967. Studies on the acrosome IX. Course of acrosome reaction in the starfish. *J. Ultrastruct. Res.* 18:562–579.
11. Dan, J. C., Y. Kakizawa, H. Kushida, and K. Fujita. 1972. Acrosomal triggers. *Exp. Cell Res.* 72:60–68.
12. Dan, J. C., Y. Ohori, and H. Kushida. 1964. Studies on the acrosome VII. Formation of the acrosomal process in sea urchin spermatozoa. *J. Ultrastruct. Res.* 11:508–524.
13. Davies, H. G. 1958. The determination of mass and concentration by microscope interferometry. *In:* General Cytochemical Methods. Vol. I. J. F. Danelli, editor. Academic Press, Inc., New York. 55–161.
14. Fol, H. 1879. Recherches sur la fécondation et le commencement de l'hénogénue chez divers animaux. *Mem. Soc. Phys. Hist. Nat. Genève* 26:89–397.
15. Fuseler, J. W. 1975. Mitosis in *Tilia americana* endosperm. *J. Cell Biol.* 64:159–171.
16. Galbraith, W., and G. B. David. 1976. An aid to understanding differential interference contrast microscopy: computer simulation. *J. Microsc.* 108:147–176.
17. Inoué, S. 1981. Video image processing greatly enhances contrast, quality, and speed in polarization-based microscopy. *J. Cell Biol.* 89:346–356.
18. Inoué, S., G. B. Borisy, and D. P. Kiehart. 1974. Growth and lability of *Chaetopterus* oocyte mitotic spindle isolated in the presence of porcine brain tubulin. *J. Cell Biol.* 62:175–184.
19. Moy, G. W., and V. D. Vacquier. 1979. Immunoperoxidase localization of bindin during adhesion of sperm to sea urchin eggs. *Curr. Top. Dev. Biol.* 13:31–44.
20. Nomarski, G. 1955. Microinterféromètre différential à ondes polarisées. *J. Phys. Radium* 16:9–13.
21. Padawer, J. 1968. The Nomarski interference-contrast microscope. An experimental basis for image interpretation. *J. R. Microsc. Soc.* 83:305–349.
22. Tilney, L. G. 1976. The polymerization of actin III. Aggregates of nonfilamentous actin and its associated proteins: a storage form of actin. *J. Cell Biol.* 69:73–89.
23. Tilney, L. G. 1978. The polymerization of actin V. A new organelle, the actomere, that initiates the assembly of actin filaments in *Thyone* sperm. *J. Cell Biol.* 77:551–564.
24. Tilney, L. G. 1979. Actin, motility, and membranes. *In:* Membrane Transduction Mechanisms. R. A. Cone and J. E. Dowling, editors. Raven Press, New York. 163–186.
25. Tilney, L. G., and S. Inoué. 1981. The acrosome reaction of *Thyone* sperm II. The kinetics and possible mechanism of acrosomal process elongation. *J. Cell Biol.* 93:820–827.
26. Tilney, L. G., and N. Kallenbach. 1979. Polymerization of actin VI. The polarity of actin filaments in the acrosomal process and how it might be generated. *J. Cell Biol.* 81:608–623.
27. Wilson, E. B. 1928. The Cell in Development and Heredity. *MacMillan, New York.*

Reprinted from the *Journal of Cell Biology*, Vol. 93(3), pp. 820–827, 1982, with permission from The Rockefeller University Press.

Acrosomal Reaction of *Thyone* Sperm. II. The Kinetics and Possible Mechanism of Acrosomal Process Elongation

LEWIS G. TILNEY and SHINYA INOUÉ
Department of Biology, University of Pennsylvania, Philadelphia, Pennsylvania 19104; and the Marine Biological Laboratory, Woods Hole, Massachusetts 02543

ABSTRACT *Thyone* sperm were induced to undergo the acrosomal reaction with a calcium inophore A23187 in sea water containing 50 mM excess $CaCl_2$, and the extension of the acrosomal process was recorded with high-resolution, differential interference contrast video microscopy at 60 fields/sec. The length of the acrosomal process was measured at 0.25-s intervals on nine sperm. When the data were plotted as $(length)^2$ vs. time, the points fell exactly on a straight line except for the initial and very final stages of elongation. Cytochalasin B alters the rate of elongation of the acrosomal process in a dose-dependent way, inhibiting the elongation completely at high concentrations (20 μg/ml). However, no inhibition was observed unless excess Ca^{++} was added to sea water. The concentration of actin in the periacrosomal cup of the unreacted sperm is as high as 160 mg/ml; we calculate this concentration from the number and lengths of the actin filaments in a fully reacted sperm, and the volume of the periacrosomal cup in the unreacted sperm. These results are consistent with the hypothesis proposed earlier that monomers add to the ends of the actin filaments situated at the tip of the growing acrosomal process (the preferred end for monomer addition), and that the rate of elongation of the process is limited by diffusion of monomers from the sperm head (periacrosomal cup) to the tip of the elongating process.

During the extension of the acrosomal process, a few blebs distributed along its lengths move out with the process. These blebs maintain a constant distance from the tip of the growing process. At maximum length, the straight acrosomal process slackens into a bow, and numerous new blebs appear. A few seconds later, the process suddenly straightens out again and sometimes actually contracts. The behavior of the blebs indicates that membrane is inserted at the base of the growing acrosomal process, and that membrane assembly and water uptake must be coupled to actin assembly during elongation. We discuss how the dynamic balance of forces seems to determine the shape of the growing acrosomal process, and how actin assembly may be controlled during the acrosomal reaction.

One of the most fascinating biological processes yet described is the acrosomal reaction, a reaction the sperm undergoes when it comes in contact with the cellular material surrounding the egg. This reaction, which is particularly dramatic in invertebrate sperm, consists of the opening of the acrosomal vacuole, a reaction which can occur in less than 50 ms (7, 8, 15), followed by the formation of a process which can exceed 90 μm in length, yet forms in <10 s! The rate of elongation of this process is comparable to the rate of contraction of skeletal muscle. What is also appealing about this system is that this reaction occurs in a cell which is one of the simplest in the body and is designed to work just once. Thus the experimenter has some hope of really understanding what takes place, because not only can sperm be obtained in prodigious quantities, but they are so specialized that the reaction can easily be studied without the confusion of simultaneous reactions that occur in other cells.

According to earlier studies, the force needed to extend this process appears to reside in the explosive, yet controlled, assembly of actin that pushes out the acrosomal membrane (16, 19). The assembly of actin is initiated by a rise in the internal pH of the cell, followed by the formation of filaments that appear to be nucleated by a cytoplasmic organelle, the actomere (16, 20). In vitro and in vivo evidence suggests that

THE JOURNAL OF CELL BIOLOGY · VOLUME 93 JUNE 1982 820-827

the monomeric actin adds to the filaments at the tip of the elongating acrosomal process. Since a process 90-μm long can form in 10 s, the question immediately arises as to whether or not the monomer, liberated in the periacrosomal cup, has enough time to diffuse to the tip of the growing process. Using equations derived from Hermans (6), Tilney and Kallenbach (20) calculated that if the viscosity of the cytoplasm were that of water, then such a process could form in 12 s if the concentration of monomer in the sperm were 100 times that at the tip of the process. From Hermans' equations it is clear that if the elongation of the acrosomal process were limited by diffusion, then the length of the process plotted as a function of $(\text{time})^{1/2}$ would be a straight line, a prediction which we can now test by video recording the elongation of the acrosomal process.

Accordingly, we taped the elongation of the acrosomal process using differential interference contrast (DIC) optics and a video recording system that greatly increases the contrast and thus the visibility of this magnificent event (7, 8). When we plotted length vs. $(\text{time})^{1/2}$ from measurements of the video records, we found that indeed this plot gave a straight line. Addition of cytochalasin B markedly affected the rate of this extension and inhibited it altogether at high concentrations, as might be expected, since cytochalasin acts to cap actin filaments, thereby preventing further assembly (10). These results are consistent with the hypothesis that the elongation of the acrosomal process is a diffusion limited event.

In addition to these kinetic analyses, we observed the behavior of blebs that appear along the length of the acrosomal process. Based on these observations, we discuss (*a*) the site of insertion of membranes into the acrosomal process, and (*b*) the relative contributions of membrane addition, actin polymerization, and water influx, to the elongation of the acrosomal process.

MATERIALS AND METHODS

Details about procuring sperm from *Thyone briareus*, perfusion techniques, and the optical and video systems, are described in the preceding paper (8). We also describe there, how accurate measurements of the length of the growing acrosomal process, which is only 50-nm wide in places, can be made by video microscopy.

Induction of the Acrosomal Reaction

Thyone sperm were suspended in sea water containing 50 mM excess $CaCl_2$ (calcium sea water), placed on the perfusion slide, and perfused with calcium sea water containing the ionophore A23187 (Calbiochem, La Jolla, CA). 10 μl of a stock of 1 mg/ml A23187 in dimethyl sulfoxide was added to each 1 ml of calcium sea water.

For the experiments with cytochalasin B, we preincubated the sperm for 1 min in cytochalasin B (Calbiochem, La Jolla, CA) in calcium sea water, then induced the acrosomal reaction with the ionophore solution containing cytochalasin B. A stock of cytochalasin B (1 mg/ml) was prepared by dissolving it in dimethyl sulfoxide. We never added >20 μl of dimethyl sulfoxide to the sperm suspension, an amount that by itself had no effect on the sperm.

Electron Microscopy

Thyone sperm were suspended in calcium sea water and 10 μl of a 1 mg/ml stock of A23187 was added to each 1 ml of sea water. 1.5 min later the sperm were fixed by the addition of sufficient glutaraldehyde (an 8% stock from Electron Microscope Sciences, Fort Washington, PA) to the sea water to make the solution 1% glutaraldehyde. The sperm were fixed at room temperature for 30 min, concentrated by centrifugation, washed briefly in sea water, and post fixed in 1% OsO_4 in 0.1 M phosphate buffer at pH 6.0 for 30 min at 0°C. The fixed sperm were washed three times in cold water and en bloc stained with 0.5% uranyl acetate overnight. They were then rapidly dehydrated in acetone and embedded in Epon 812 (Ernst Fullam, Inc., Burlington, VT). Thin sections were cut with a Sorvall Porter Blum II ultramicrotome (DuPont Co., Newtown, CT), stained with uranyl acetate and lead citrate, and examined in a Philips 200 electron microscope.

RESULTS

Induction of the Acrosomal Reaction

Thyone sperm were selected because they produce the longest (up to 90 μm) processes on record (4, 15) greatly exceeding those of starfish sperm (20 μm in length) or sea urchin sperm (<1 μm in length).

The ideal way to induce the acrosomal reaction would be to use natural stimuli such as eggs or the jelly coats which surround the eggs. Unfortunately, ripe eggs of *Thyone* are difficult to obtain. Thus it is necessary to induce the acrosomal reaction by artificial means such as with an ionophore. Induction of the acrosomal reaction of *Thyone* sperm with the ionophore A23187 in calcium sea water produces many processes 90 μm in length, most exceeding 50 μm, the length needed in nature for fertilization. Thus we believe we are recording a natural event.

Kinetics of Elongation of the Acrosomal Process

As we perfuse calcium sea water containing the ionophore past the sperm the first observable reaction is an increased motility of the flagellum of the sperm. Shortly thereafter, the acrosome pops; there is a rapid and dramatic change in the refractive index of the acrosomal region (see the accompanying paper [8]). ~1–2 s later the acrosomal process is detectable.

We taped the events of acrosomal process elongation in about 50 sperm and examined them qualitatively. Of these, we analyzed nine quantitatively, measuring the length of the acrosomal process every 15 video fields, or every 0.25 s. These sperm were selected because the tip of the process remained in focus so that accurate measurements could be made. An example of the type of image we measured is illustrated in Fig. 1. The montage was assembled from images of single fields of the video monitor photographed every 0.75 s.

In analyzing the kinetics of elongation of the acrosomal process, we found it more convenient to plot our data as $(\text{length})^2$ vs. time, rather than as length vs. $(\text{time})^{1/2}$, which is the immediate form expressing the diffusion limited relationship in Eq. 1 (see Discussion). The data from six sperm are illustrated in Fig. 2. What is amazing, is that, except for the early stages in induction or the very latest stages when the extension is essentially complete, all the values lie on a straight line consistent with the view that the extension of this process is diffusion limited. This is true for all nine sperm. To be sure that $(\text{length})^2$ vs time is really a straight line, we also plotted the data with length as a linear function of time. The points did not fall on a straight line, rather giving a curve with a sigmoid shape.

In six out of the nine sperm, the slopes of the plot, $(\text{length})^2$ vs time, were similar, ranging from 790 μm^2/s to 960 μm^2/s (Table 1) with a mean of 850 μm^2/sec. Four of these are illustrated in Fig. 2. Of the three remaining sperm, two produced only short processes (one is illustrated in Fig. 2—sperm *D*) and one produced a process 33 μm long (Fig. 2, sperm *F*). The slopes of these three sperm were about one half the slopes of the other six (Table 1).

Effect of Cytochalasin B on the Kinetics of Extension of the Acrosomal Process

For studying the effect of cytochalasin B, sperm were suspended in calcium sea water containing cytochalasin B and immediately introduced into the perfusion chamber. 1 min

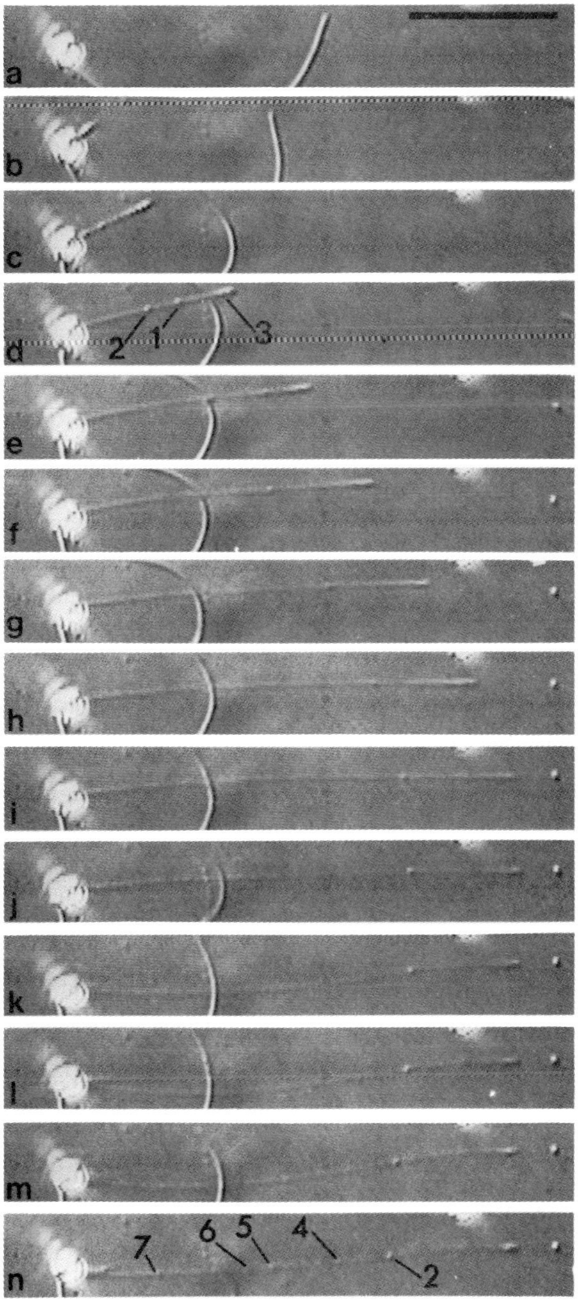

FIGURE 1 Stages in the elongation of the acrosomal reaction. Single fields of sperm C were photographed off the TV monitor every 0.75 s. Frame *a* was taken 1.95 s after the acrosome had popped. In frame *a*, the acrosomal vacuole has opened and just the beginning of a process can be seen. By frame *j*, the process has reached its maximum length. In *k*, the process begins to bow and blebs appear throughout the length. In *d*, the numbers indicate the three blebs whose motion relative to the tip is illustrated in Fig. 5 *a*, and in *n* the numbers indicate blebs whose motion relative to the tip is illustrated in Fig. 5 *b*. The jaggedness of the early, thick process seen in the frames *b*, *c*, and *d* result from the discrete steps of video scan that affect the edges that are oriented diagonally. Much of the jaggedness, therefore, does not represent real structural features. The distinct arc to the right of the sperm head is a portion of the sperm tail which is twitching as the process elongates. Bar, 20 μm. × 1,100.

later calcium sea water containing A23187 and cytochalasin B was perfused. The sperm reacted to the ionophore within a minute. With this cytochalasin B treatment, the maximum length attained by the acrosomal process, and the rate of elongation of the process, were markedly reduced as shown in Fig. 3. The effect of cytochalasin B concentration on the maximum length attained by the process is plotted in Fig. 4.

In these sperm, the (length)^2 vs. time plot no longer fell on a straight line (Fig. 3), in marked contrast to sperm activated in the absence of cytochalasin B (Fig. 2). When length, rather than (length)^2, was plotted against time, some of the data fit more closely to a straight line, but in almost all cases there were breaks in the curve where the rate of elongation was clearly changing. This was particularly clear in sperm undergoing the acrosomal reaction in low concentrations of cytochalasin B. Measurements of the rates of elongation show that, at low concentrations of cytochalasin B (i.e., 2 μg/ml), the maximum rate is similar to that of the slowly growing processes of untreated cells, i.e. 240–300 $\mu m^2/s$, but at higher concentrations of cytochalasin B (i.e., 3.3 or 5 μg/ml), the elongation rate was only 67–88 $\mu m^2/s$. When sperm were preincubated for as long as 3 min in calcium sea water containing 10 μg/ml cytochalasin B, no acrosomal process was formed when we perfused calcium sea water containing A23187 and cytochalasin B. Since this suggests that the cytochalasin B effect is limited by its permeability, we also tested the effect of applying cytochalasin B at the same time as A23187 in calcium sea water. In this case the acrosomal process grew longer and faster than when the sperm were preincubated with the same concentration of cytochalasin B. However, the (length)^2 vs. time curves were irregular. For example, if 10 μg/ml of cytochalasin B were added at the same time as A23187 in calcium sea water, slopes of between 53 $\mu m^2/s$ and 23 $\mu m^2/s$ were recorded. Thus, the effect of cytochalasin is not instantaneous and it appears that time is needed for its penetration.

When excess calcium was not supplied, cytochalasin B had no effect on the acrosomal process elongation. When *Thyone* sperm, preincubated with cytochalasin B in normal sea water, were treated with cytochalasin B and A23187 in normal sea water, the acrosomal process grew to their normal lengths, even though the cells were preincubated in cytochalasin B for periods of up to 10 min. They also grew at the normal rates, even though we used concentrations of cytochalasin B as high as 10 μg/ml. In other words, it is necessary to have Ca^{++} in excess of the concentration present in sea water for cytochalasin B to exert its effect on the elongation of the acrosomal process.

Changes in the Surface Morphology of the Acrosomal Process during Elongation

In the early stages of growth, the acrosomal process extends slowly, is fat and nonuniform in diameter (Fig. 1 *b*). As the process grows, the rate of elongation accelerates. The process becomes slender and more uniform in diameter except at its somewhat bulbous, spear-shaped tip, and in a few locations along the length marked by spherical "blebs" (Fig. 1 *d*). Once in the rapid growth phase, the contour of the slender process is nearly straight, or barely curved.

As the acrosomal process elongates, the blebs move forward with the process. During the earlier phases of process elongation, the distances between the tip of the process and the blebs, and in between adjoining blebs, increase somewhat (Fig. 1 *c*–*e*). Some of blebs also disappear, leaving a smooth uniform

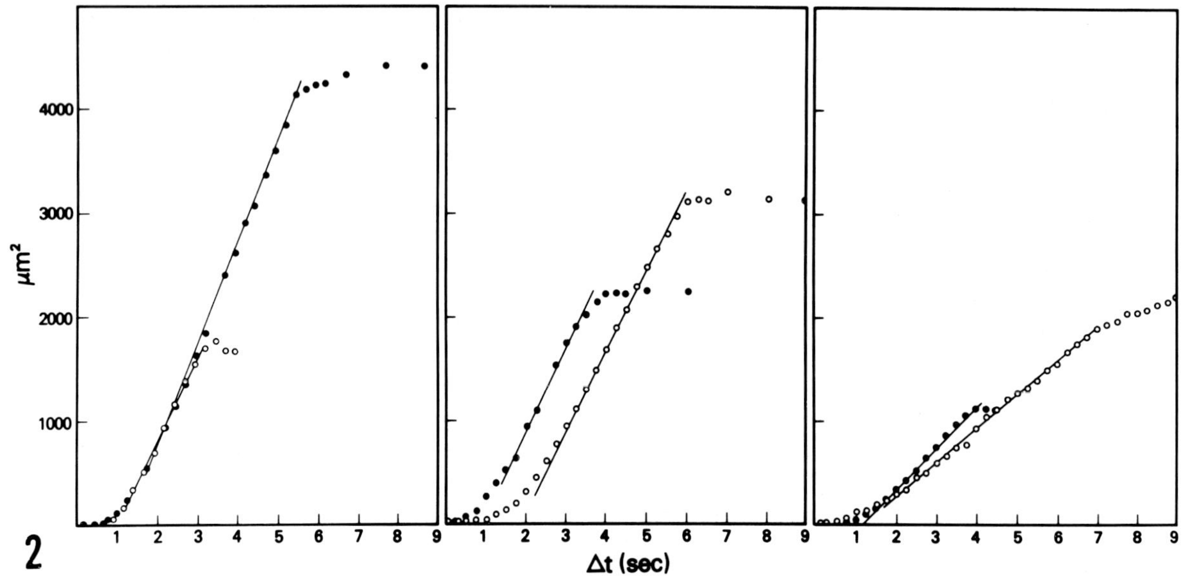

FIGURE 2 Plots of (length of the acrosomal process)² in square micrometers as a function of time in seconds after the process first becomes visible. Of note is that, except for the beginning and very end of the reaction, the points fall exactly on a straight line. In the left hand curve sperm B (closed rcircles) and J (open circles) are illustrated; in the middle, we see sperm D (closed circles) and C (open circles); and in the right hand graph, sperm F (closed circles) and A (open circles) are seen.

TABLE I

Analysis of the Kinetics of Elongation of the Acrosomal Process

Sperm	Rate of elongation	Maximum length
	$\mu m^2/s$	μm
B	960	67
E	950	62.5
D	810	47
J	800	41
H	800	38
C	790	56.5
G	420	28
F	400	33
A	330	47

contour as though the bleb were incorporated into the growing acrosomal process (Fig. 1 *d–f*); these blebs seem to flow forward and disappear at the same time. However, once in the rapid growth phase, the distance between the tip of the process and the persisting blebs remains unchanged (Figs. 1 *e–i*, and 5 *a*).

Eventually, the process stops elongating (Fig. 1 *j*), and about a second later tends to slacken into a bow (Fig. 1 *k*). As the acrosomal process starts to bow, new blebs become visible along the length of the process (Fig. 1 *k*).

Many of these blebs glide backwards towards the sperm cell body. While all of these blebs tend to move backwards, they move at different times and at different rates. Thus, the distances between some blebs increase while others decrease (Fig. 5 *b*). The fully bowed process is covered by many blebs interspersed with slender regions whose thickness appears to be fairly uniform (Fig. 1 *l*).

Finally, ~10 to 12 s after the acrosome had popped, the process abruptly (in ~50 ms) straightens out as though contracting (Fig. 1 *n*); in some cases the process in fact shortens somewhat. All of these changes in the structure of the acrosomal process are observed whether or not the process sticks to the cover glass or is freely waving in the solution.

Concentration of actin in the Sperm Before Activation

Since we can determine (*a*) the volume occupied by the unpolymerized actin before activation, (*b*) the number of actin

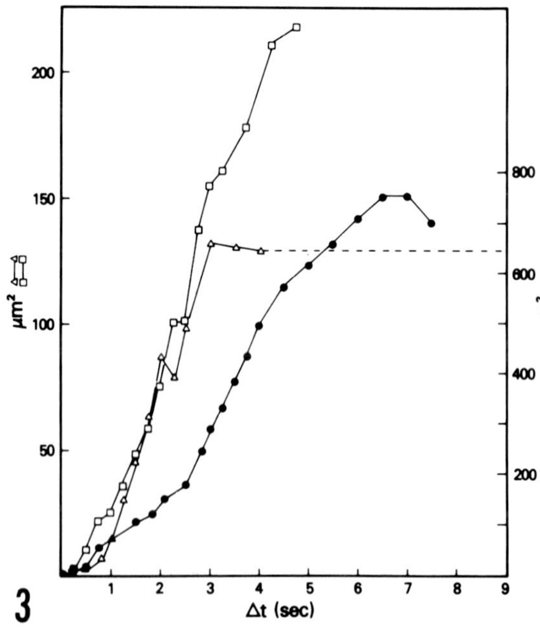

FIGURE 3 The (length of the acrosomal process)² in square micrometers plotted as a function of time (as in Fig. 2) for *Thyone* sperm treated with cytochalasin B. Sperm were treated in calcium sea water containing cytochalasin B for 1.5 min, and then acrosomal reaction was induced with A23187 in the presence of cytochalasin B. 2 μg/ml of cytochalasin B (closed circle), 3.3 μg/ml (squares), 5.0 μg/ml (triangles).

filaments in the acrosomal process, and (*c*) the final length of the acrosomal process, we can calculate the approximate concentration of actin in the periacrosomal cup before activation of the sperm. The unpolymerized actin is contained in the periacrosomal cup, in the form of profilactin, before activation of the sperm (15).

The volume of the periacrosomal cup is calculated as follows. From longitudinal sections through a sperm head of *Thyone* such as is illustrated in Fig. 2 of Tilney (15), and Fig. 1 of the accompanying paper (8), we see that the acrosomal region, i.e., the space occupied by the acrosomal vacuole and the periacrosomal cup, is nearly spherical. By subtracting the volume occupied by the spherical acrosomal vacuole from the total volume of the acrosomal region, we can determine the volume of the periacrosomal cup, or the space occupied by the profilactin. To be more specific, we determined from electron micrographs that the diameter of the vacuole was 0.49 μm, and the diameter of the acrosomal region including the vacuole was 0.76 μm. Therefore, the volume occupied by the profilactin is 1.68×10^{-13} cm^3. This value is actually too small because we know that the tissue shrinks during dehydration and embedding for electron microscopy. The amount of shrinkage depends on a number of factors, but by comparing recent optical and x-ray diffraction patterns of unfixed and dehydrated actin paracrystals (18) with optical diffraction patterns of electron microscope (EM) thin sections, we know that there is a 20 to 40% (linear) shrinkage in the thin sections. Shrinkage can also be calculated by comparing electron micrographs with light micrographs of living sperm. In the living sperm, the unpopped acrosomal region measures 1.1 μm in diameter in the differential interference contrast image (see Fig. 3, and Discussion in the accompanying paper [8]). Therefore, we calculate a shrinkage of 30% ([1.1 − 0.76]/1.1). This is in agreement with the value determined from the diffraction studies. If we take this 30% shrinkage in diameter into account, the volume occupied by the profilactin region in the living sperm should be 5×10^{-13} cm^3.

Since we have been unable to determine the exact amount of actin per sperm chemically or from serial sections, we

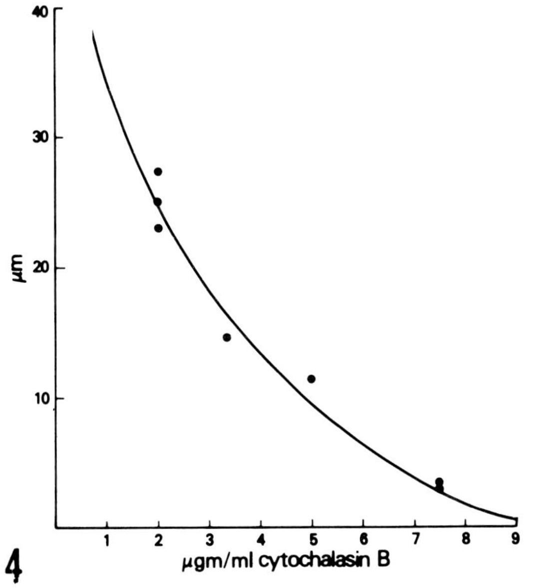

4

FIGURE 4 The final length of the acrosomal process in micrometers as a function of the concentration of cytochalasin B in μg/ml.

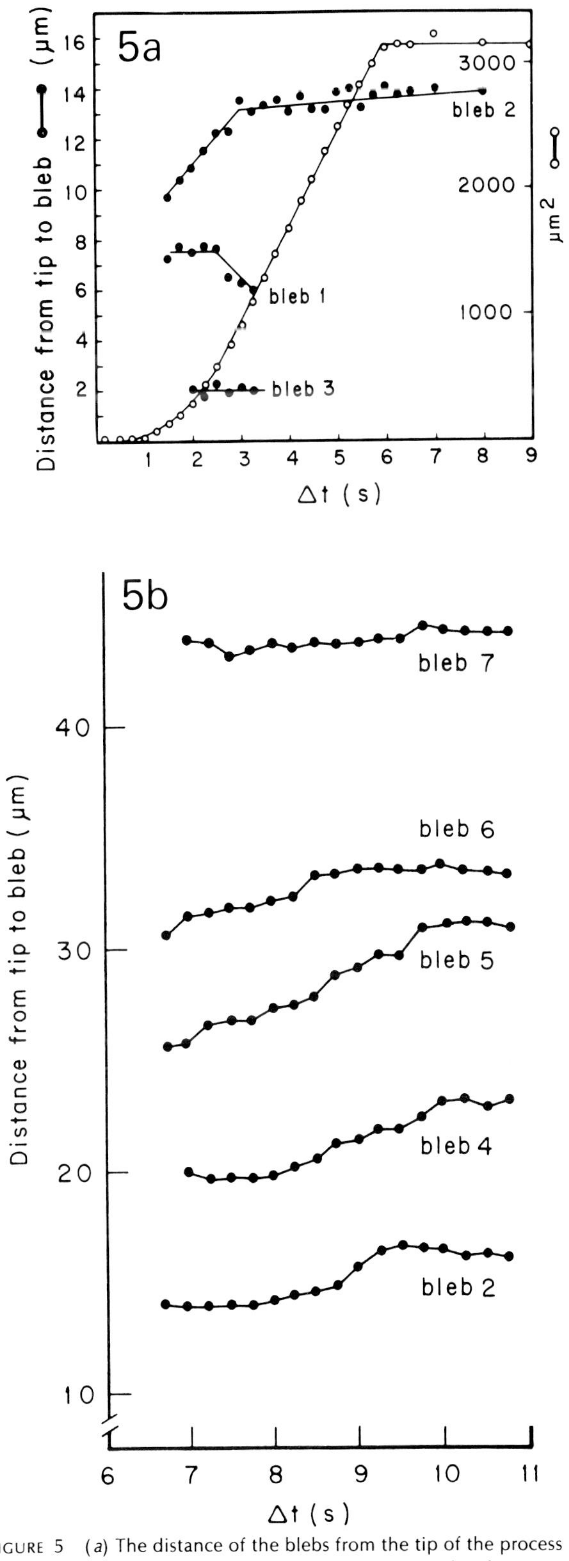

FIGURE 5 (*a*) The distance of the blebs from the tip of the process (μm) plotted as a function of time after the process first becomes visible (closed circles). The blebs considered here are seen in Fig. 1 and are indicated in Fig. 1 *d*. This is sperm C. The (length of the process)^2 as a function of time is illustrated in this figure (opened circles) as well as in Fig. 2. (*b*) Continuation of *a* showing the behavior of blebs after the acrosomal process has maximally elongated. Blebs 4 through 7 first become visible at ~7 s after the acrosomal process first appeared.

bowing of the acrosomal process. It is during this process that the blebs glide a short way back towards the cell body.

As described in Results (Fig. 5 *b*), different blebs on the same acrosomal process glide at various rates, to various extents, and at slightly different times. Thus, some blebs move farther apart and others move closer together until they finally end up in stable positions. This dynamic behavior clearly shows that the blebs are not clumps of cytoplasm that are fixed onto the axial filaments.

About the time the blebs have become stabilized, we have seen a sudden shortening, or straightening, of some acrosomal processes. While the degree of shortening is not great, the speed reminds one of a muscle twitch. We have no idea what mechanism could account for this contraction.

In summary, we propose that the shape of the slender acrosomal process is governed by a dynamic equilibrium between: (*a*) the extending filaments of actin that are polymerizing at the tip of the process; (*b*) the surface tension of the fluid plasma membrane that is being replenished at the base of the extending acrosomal process; (*c*) the drawing out of the acrosomal cytoplasm by the tip of the growing axial bundle; and (*d*) the influx of water. The rate of water influx and consequent swelling of the contents of the periacrosomal cup, described in the accompanying paper (8), presumably govern the rate of supply of the cytoplasm into the acrosomal process.

How Might the Assembly of Actin be Controlled during the Acrosomal Reaction

In an earlier publication, Tilney et al. (21) presented evidence that a change in the internal pH was necessary for actin assembly. The rise in pH presumably releases the actin from its storage form so that it can assemble into filaments. In the present study, we demonstrated that the concentration of diffusable actin at the base of the elongating process is enormous, being at least 160 mg/ml. At these concentrations, the actin would nucleate spontaneously unless some controls are present. Thus monomeric actin would never be able to diffuse to the tip of the process; instead the monomers would be assembled directly into filaments inside the periacrosomal cup. Since that does not happen, the cell must control two events: (*a*) The release of monomeric actin from its storage form, and (*b*) inhibition of spontaneous nucleation of the released actin.

Tilney earlier showed that the cup of profilactin contains, besides actin, four additional proteins whose molecular weights are 16,000, 25,000, 230,000, and 250,000 (16). Preliminary evidence indicates that the 16,000-dalton protein resembles the profilin described by Carlsson et al. (3), Reichstein and Korn (11), and Runge et al. (12). Although not precisely the same molcular weight as profilin in amoebae (see Runge et al.), it lowers the rate of actin assembly in vitro, without affecting the critical concentration for actin assembly. It also occurs in a 1:1 stoichiometry with actin. One hypothesis then is that the 16,000-dalton protein in *Thyone* sperm decreases the rate of nucleation of actin and, at the same time, being small, would not appreciably affect the diffusability of actin. The small size is important, as a larger molecule or molecules could drastically change D in Eq. 1. The other proteins present may singly or collectively form an insoluble complex with actin, sequestering it in the unreacted sperm.

Thus our scheme for filament formation requires a two-step reaction. The first step releases the actin from an insoluble complex. The actin, still bound to an inhibitor of nucleation, then diffuses to the tip of the elongating process. There it assembles onto the extending filaments. As discussed in the previous section, a delicate balance between the extending actin filaments, membrane addition, and water influx, appears to govern the shape of the growing acrosomal process.

Preliminary video taping of the elongation of the acrosomal process of starfish sperm was attempted during the summer of 1979 at the Station Biologique, Roscoff, France. We wish to thank Christian Sardet, Richard Christian, and Jean Panlévé for helping L. G. Tilney obtain promising sequences. These initial results sparked our enthusiasm for further studies. We would also like to express our gratitude to Neville Kallenbach whom we hope survived many exciting discussions in trying to explain what Hermans' formulae really meant, and to Jim Spudich, Tom Pollard, Ed Bonder, and Mary Porter for their help in calculating or showing one of us (L. G. Tilney) how to calculate. We are also extremely grateful to the reviewers whose queries and comments led us to reexamine our data and tighten up the Discussion. We also thank Christopher Inoué for producing the expertly matched photographic prints, and the painstakingly assembled montage.

This work was supported by National Institutes of Health (NIH) grant HD 14474 to L. G. Tilney, and NIH 5 R01 GM 23475-16 and National Science Foundation PCM 7922136 to S. Inoué.

Received for publication 1 October 1981, and in revised form 4 February 1982.

REFERENCES

1. Aebi, W., W. E. Fowler, G. Isenberg, T. D. Pollard, and P. R. Smith. 1981. Crystalline actin sheets: their structure and polymorphism. *J. Cell Biol.* 91:340–351.
2. Boys, D. 1959. Soap-Bubbles: Their Colors and the Forces which Mold Them. Dover Press, New York.
3. Carlsson, L., L. E. Nystrom, I. Sundkvist, F. Markey, and U. Linberg. 1977. Actin polymerization is influenced by profilin, a low molecular weight protein in nonmuscle cells. *J. Mol. Biol.* 115:465–483.
4. Colwin, L. H., and A. L. Colwin. 1956. The acrosome filaments and sperm entry in *Thyone briareus* (Holothuria) and *Asterias*. *Biol. Bull.* 110:243–257.
5. Cooke, R. 1975. The role of bound nucleotide in the polymerization of actin. *Biochemistry.* 14:3250–3256.
6. Hermans, J. J. 1947. Diffusion with discontinuous bonding. *J. Colloid Sci.* 2:387–398.
7. Inoué, S. 1981. Video image processing greatly enhances contrast, quality, and speed in polarization based microscopy. *J. Cell Biol.* 89:346–356.
8. Inoué, S., and L. G. Tilney. 1981. The acrosomal reaction of *Thyone* sperm. I. Changes in the sperm head visualized by high-resolution microscopy. *J. Cell Biol.* 93:812–819.
9. Matthews, B. W. 1968. Solvent content of protein crystals. *J. Mol. Biol.* 33:491–497.
10. Pollard, T. D., and M. S. Mooseker. 1981. Direct measurement of actin polymerization rate constants by electron microscopy of actin filaments nucleated by isolated microvillus cores. *J. Cell Biol.* 88:654–658.
11. Reichstein, E., and E. D. Korn. 1979. *Acanthamoeba* profilin. A protein of low molecular weight from *Acanthamoeba castellanic* that inhibits actin. *J. Biol. Chem.* 254:6174–6179.
12. Runge, M. S., P. C. Tseng, R. C. Williams, Jr., J. A. Cooper, and T. D. Pollard. 1981. Characterization of profilin and profilin-actin interactions. *J. Cell Biol.* 91 (2, Pt. 2):300*a* (Abstr.).
13. Sanger, J. W., and J. M. Sanger. 1976. Polymerization of sperm actin in the presence of cytochalasin B. *J. Exp. Zool.* 193:441–447.
14. Sardet, C., and L. G. Tilney. 1977. Origin of the membrane for the acrosomal process: is actin complexed with membrane precursors? *Cell Biol. Int. Rep.* 1:193–200.
15. Tilney, L. G. 1975. The role of actin in nonmuscle cell motility. *In:* Molecules and Cell Movement. S. Inoué and R. E. Stephens, editors. Raven Press, New York. 339–388.
16. Tilney, L. G. 1979. Actin, motility, and membranes. *In:* Membrane Transduction Mechanisms. R. A. Cone and J. E. Dowling, editors. Raven Press, New York. 163–186.
17. Tilney, L. G., J. G. Clain, and M. S. Tilney. 1979. Membrane events in the acrosomal reaction of *Limulus* sperm. Membrane fusion, filament-membrane particle attachment, and the source and formation of new membrane surface. *J. Cell Biol.* 81:229–253.
18. Tilney, L. G., D. J. DeRosier, and M. J. Mulroy. 1980. The organization of actin filaments in the stereocilia of cochlear hair cells. *J. Cell Biol.* 86:244–259.
19. Tilney, L. G., S. Hatano, H. Ishikawa, and M. S. Mooseker. 1973. The polymerization of actin: its role in the generation of the acrosomal process of certain echinoderm sperm. *J. Cell Biol.* 59:109–126.
20. Tilney, L. G., and N. Kallenbach. 1979. The polymerization of actin VI. The polarity of the actin filaments in the acrosomal process and how it might be determined. *J. Cell Biol.* 81:608–623.
21. Tilney, L. G., D. Kiehart, C. Sardet, and M. Tilney. 1978. Polymerization of actin IV. The role of Ca^{++} and H^+ in the assembly of actin and in membrane fusion in the acrosomal reaction of echinoderm sperm. *J. Cell Biol.* 77:536–550.

Reprinted from the *Journal of Cell Biology*, Vol. 100(4), pp. 1273–1283, with permission from The Rockefeller University Press.

Acrosomal Reaction of the *Thyone* Sperm. III. The Relationship between Actin Assembly and Water Influx during the Extension of the Acrosomal Process

LEWIS G. TILNEY and SHINYA INOUÉ
Department of Biology, University of Pennsylvania, Philadelphia, Pennsylvania 19104; and the Marine Biological Laboratory, Woods Hole, Massachusetts 02543

ABSTRACT In an attempt to investigate the role of water influx in the extension of the acrosomal process of *Thyone* sperm, we induced the acrosomal reaction in sea water whose osmolarity varied from 50 to 150% of that of sea water. (*a*) Video sequences of the elongation of the acrosomal processes were made; plots of the length of the acrosomal process as a function of $(\text{time})^{1/2}$ produced a straight line except at the beginning of elongation and at the end in both hypotonic and hypertonic sea water (up to 1.33 times the osmolarity of sea water), although the rate of elongation was fastest in hypotonic sea water and was progressively slower as the tonicity was raised. (*b*) Close examination of the video sequences revealed that regardless of the tonicity of the sea water, the morphology of the acrosomal processes were similar. (*c*) From thin sections of fixed sperm, the amount of actin polymerization that takes place is roughly coupled to the length of the acrosomal process formed so that sperm with short processes only polymerize a portion of the actin that must be present in those sperm. From these facts we conclude that the influx of water and the release of actin monomers from their storage form in the profilactin (so that these monomers can polymerize) are coupled. The exact role of water influx, why it occurs, and whether it could contribute to the extension of the acrosomal process by a hydrostatic pressure mechanism is discussed.

When a *Thyone* sperm comes in contact with the outer surfaces of an egg, a series of reactions are elicited which culminate in the growth of a process which can form in less than 10 s, yet exceed 90 μm in length. In fact, the rate of elongation of this process, the acrosomal process, is comparable to the rate of contraction of some of the fastest skeletal muscles. Of the many questions that one could ask about this extraordinary motile event perhaps the most basic is, What generates the force required to drive the extension process? In earlier publications, we demonstrated that at all stages during the growth of the acrosomal process there is an explosive, yet controlled assembly of actin and concluded that the assembly per se provides the force for the elongation of the process. The assembly of actin is initiated by a rise in internal pH (20); this is followed by the nucleation of actin monomers on a cytoplasmic organelle, the actomere, which simultaneously nucleates and directs the assembly of filaments (19) to form a directed spear used to penetrate the jelly which surrounds the egg. Thus, it is clear that actin polymerization is involved in the generation of the acrosomal extension, but there remains the mechanical issue of how a polymerization reaction can "push" a membrane protuberance in front of it, particularly as we know that the site of assembly of monomers to the elongating filaments occurs at the membrane protuberance (17, 18).

Beginning 20 years ago, Dan and her co-workers (4, 5) published two detailed fine structural studies on the early events in the acrosomal reaction. They demonstrated in sea urchin sperm that there is roughly a threefold increase in volume in the region which will be occupied by the actin filaments (5) and an even larger increase in volume in starfish sperm (4) and suggested that water influx may play the "major role in extending the process membrane" by an osmotic mechanism. More recently, we recorded and analyzed, through the use of high extinction video microscopy on living sperm, sequences in the initial formation (11) and elongation of the acrosomal process (18); we found that within 50 ms after induction, there is a roughly twofold increase in volume in the acrosomal region and that within the next 1–2 s there is a precipitous drop in the refractive index of the periacrosomal region, the region from which actin monomers are released to polymerize into filaments. We observed, in association with this decrease in dry mass, a pronounced increase in volume of the periacrosomal region. Therefore, in vivo, the periacrosomal material imbibes water and the elongation of the acrosomal process is accompanied by a large increase in volume due to an influx of water into the cell.

Based on these results, Oster et al. (13) suggested that the entire "extension process may be osmotically driven and that actin polymerization merely 'keeps up' with the extension reinforcing it against bending moments." Accordingly, they proposed a simplified mathematical model based upon hydrostatic pressure that was consistent with the kinetic data on the rate of elongation of the acrosomal process as previously measured (18). If hydrostatic pressure plays a role in the elongation of the acrosomal process, then one would predict that changes in the tonicity of the medium outside the sperm should greatly affect the rate and kinetics of elongation of the acrosomal process, parameters that we can accurately measure using the high extinction video microscopy techniques already mentioned.

In an earlier publication in this series (18), we demonstrated that a plot of the length of the acrosomal process as a function of $(\text{time})^{1/2}$ produces a straight line except at the beginning of elongation and at its completion. These observations are consistent with a model in which actin monomers, when released from their bound state adjacent to the nucleus, diffuse to the tip of the elongating acrosomal process where they assemble on the growing filaments, a model that is now confirmed by recent in vitro observations (17).

In this publication, we will show that we can markedly affect the rate of elongation of the acrosomal process by changing the osmolarity of the sea water surrounding the sperm, though plots of the length of the acrosomal process as a function of $(\text{time})^{1/2}$ show a straight line if one ignores the beginning and end of elongation. Thus, the rate of release of monomer from its bound state (so that it can migrate to the tip of the elongating process in order to polymerize) is tightly coupled, not only to the rate of elongation of the acrosomal process, but also to the tonicity of the medium or, in other words, to water entry. This surprising result reinforces the concept that water entry, perhaps hydration of the storage form of unpolymerized actin in the cell, may play a pivotal role in the elongation of the acrosomal process.

Such considerations as these on the relative contributions of osmotic forces and actin assembly on motile processes are, in fact, timely because several reports have appeared implicating that osmotically driven mechanisms may be operational in actin-related events such as blebbing (7), lamellipod extension (6), and the formation of sea urchin microvilli (1).

MATERIALS AND METHODS

Obtaining Sperm: *Thyone briareus* were collected by the Marine Resources Department, Marine Biological Laboratories, Woods Hole, MA. The testes were removed and minced in sea water. The suspension was filtered through cheese cloth and the supernate was centrifuged at 5,000 *g* for 5 min to pellet the sperm. The pelleted sperm, stored in the refrigerator, were used either the same day or the next day.

Induction of the Acrosomal Reaction: *Thyone* sperm were suspended in sea water, introduced into the perfusion slide, and perfused for 1.5 min in sea water which contained mannitol or sucrose, or sea water which had been made hypotonic by the addition of distilled water. We added 50 mM excess $CaCl_2$ to these perfusing solutions because we found earlier that ionophore-induced discharges were longer and more reproducible in excess $CaCl_2$ (18). After 1.5 min, we added to each milliliter of the perfusate 10 μl of a 1-mg/ml stock solution of the ionophore, A23187 (Calbiochem-Behring Corp., San Diego, CA), dissolved in dimethylsulfoxide.

Perfusion Chamber: The design of the chamber is described in detail in a previous article (11).

Video Observations and Analysis: Details of the video techniques and differential interference microscopy used here are described by Inoué (10) and Inoué and Tilney (11). For our kinetic analysis we rerecorded the video sequence on a video motion analyser (Sony SVM1010). We studied the 10-s sequence stored on the disk in the analyser either field by field or at various speeds backwards and forwards. The lengths of the acrosomal processes were determined from frozen images of the video tape record. X and Y video markers were positioned at the tip and the base of the acrosomal process and their coordinates recorded together with the time of the scene (Colorado Video No. 321). The length of the process was calculated from the coordinates which had been calibrated with images of a stage micrometer. The exact time of a particular sequence was determined by counting the number of video fields elapsed after the "seconds" digit had last advanced. This method provides the time point with a 16.7-ms precision since each second is represented by exactly 60 fields in standard United States video equipment.

Electron Microscopy: *Thyone* sperm were suspended in sea water or sea water which contained 1 M sucrose or distilled water and excess $CaCl_2$ for 1.5 min; the acrosomal reaction was then induced with 10 μl of A23187 dissolved in dimethylsulfoxide per milliliter of sperm suspension. After 1.5 min, the sperm were fixed by the addition of sufficient glutaraldehyde (an 8% stock from Electron Microscope Sciences, Fort Washington, PA) to the solution to make the solution 1% glutaraldehyde. The sperm were fixed at room temperature for 30 min, concentrated by centrifugation, washed briefly in sea water, and postfixed in 1% OsO_4 in 0.1 M phosphate buffer at pH 6.2 for 30 min at 0°C. The fixed sperm were washed three times in cold water and en bloc–stained with 0.5% uranyl acetate overnight. They were rapidly dehydrated in acetone and embedded in Araldite 502.

It became apparent that although the above fixation procedure preserved actin filaments satisfactorily, it did not stop the swelling that occurs when the sperm suspension in 50% sea water remains in the ionophore-glutaraldehyde mixture. Therefore, towards the end of this study, we carried out a different protocol. Sperm were handled as before, however, 30 s after the addition of the ionophore to the suspension, we added an equal volume of fluid that contained two parts of what was in the suspension, one part of 4% OsO_4 in sea water, and one part of 4% glutaraldehyde. Thus, the final fixation solution contained 1% OsO_4 and 1% glutaraldehyde in sea water. The fixatives were made up just before use and fixation was carried out for 2 min at 0°C. After 2 min, the sperm were concentrated by centrifugation at 2,000 *g* for 2 min and the fixative was changed to 1% OsO_4 in 0.1 M phosphate buffer at pH 6.2. Fixation in this new medium was continued for 25 additional minutes. The sperm were then washed three times in cold water and en bloc–stained in 0.5% uranyl acetate for 3 h. They were rapidly dehydrated in acetone and embedded in Araldite 502.

RESULTS

The Kinetics of Elongation of the Acrosomal Process

As documented in an earlier publication (18), *Thyone* sperm were induced to undergo the acrosomal reaction with the ionophore, A23187, in high calcium sea water. As this solution is perfused past the sperm, there is an increased motility of the flagellum, the acrosome pops, and 1–2 s later the acrosomal process emerges. We measured the length of the acrosomal process under each experimental condition every 15 video fields or every 0.25 s. We selected those sperm in which the tip of the process and the head of the sperm remained in focus during the entire elongation procedure; this allowed for accurate measurements. In analysing the kinetics of elongation of the acrosomal process (18), we found it most convenient to plot the data as $(\text{length})^2$ vs. time, rather than as length vs. $(\text{time})^{1/2}$. This produces a straight line for all the sperm induced to discharge in sea water (18) except at the beginning of elongation of the acrosomal process and at the very end of extension. Two populations were encountered:

those that grew very fast (850 $\mu m^2/s$) and those that grew at approximately ½ the maximal rate (400 $\mu m^2/s$).

All the biological recordings listed below were made during early July, 1982. Although the sperm were active, could be induced to undergo the acrosomal reaction with A23187, popped normally, and produced processes with linear kinetics when the data were plotted as $(length)^2$ vs. time, the rate of elongation corresponded to that of the population of sperm in our earlier study that elongated at ½ the maximal rate (e.g., 400 $\mu m^2/sec$) (18). The absence of sperm that grew processes at the maximal rate seen in our earlier study (18) was perhaps due to the fact that the animals are maintained in an "over-ripe" condition by reducing the temperature of the sea water to 12°C because in nature gametes are released in early June and possibly due to the fact that the winter of 1982 was exceptionally mild leading to a severe underproduction of gametes. Nevertheless, because the 1982 population was consistent in firing and was normal in all other regards, including the formation of acrosomal processes that exceeded 90 μm in length, and because it corresponded to one population described in our earlier publications, we view the 1982 sample as completely normal.

Two sets of controls were run for the experiments that are outlined in this report. In the first, sperm were incubated in sea water to which excess Ca^{++} had been added and then 1.5 min later induced to undergo the acrosomal reaction by the addition of A23187 to the Ca^{++} sea water (Fig. 1 *a*). Our second control was to perfuse sperm for 1.5 min with a solution that was isotonic with sea water, yet contained one part sea water and one part 1 M sucrose to which excess Ca^{++} had been added. After 1.5 min, A23187 was added to this perfusate (Fig. 1 *b*). The latter is an important control because in the experiments that follow sucrose is added to sea water and it is important to establish that sucrose by itself does not change the kinetics of elongation. As seen in Fig. 1, plots of $(length)^2$ vs. time are straight lines except at the beginning of elongation and at the completion of the reaction for both controls. This deviation at the beginning and end of elongation was also noted in our earlier publication (18). In Table I, the slopes of the plots, $(length)^2$ vs. time, for the controls can be seen to be similar ranging from 188–455 $\mu m^2/s$ with a mean of 343 $\mu m^2/s$ for the sucrose control which compares favorably with the sea water control of 370 $\mu m^2/s$.

TABLE I

Analysis of the Kinetics of Elongation of the Acrosomal Process

Sperm	Experimental regime	Rate of elongation	Average rate of elongation	Maximum length
		$\mu m^2/s$	$\mu m^2/s$	μm
		Control		
A	S.W.*	370	370	43
B	Sucrose/S.W.	348		36.6
C		188	330	29.9
D		455		25.4+
		Hypertonic		
L	1.15 osmol	102		26.3
M		145	216	29.6
N		400		19.2+
O	1.25 osmol	230		27
P		401	228	27.4+
Q		55		10.7
R	1.33 osmol	65	51	13
S		36		9.9
	1.5 osmol	No growth		
		Hypotonic		
E	0.75 osmol	400		30.3
F		374	368	34+
G		420		27+
H		280		24.7
I	0.50 osmol	615	494	51.5
J		470		25
K		398		31.9

* S.W., sea water.

Effect of Hypertonicity on the Kinetics of Extension of the Acrosomal Process

To study the effect of hypertonicity, we suspended sperm in sea water, then introduced them into the perfusion chamber, and perfused them with sea water containing varying concentrations of sucrose for 1.5 min; excess calcium (up to 50 mM) was added to the perfusate. The sperm were then induced to undergo the acrosomal reaction with A23187.

At tonicities of 1.15 osmol and 1.25 osmol plots of $(length)^2$ vs. time produced mostly straight lines except at the beginning of elongation and at its completion (Figs. 2–4), as was seen in the controls. In some cases, e.g., the left curve of Fig. 3, the data might be fitted better with a continuously curved line of increasing slope rather than the straight line that is drawn. The significance of a deviation from a straight line such as seen in Fig. 3 and at the beginning and end of elongation will be discussed in more detail in the Discussion. In most cases, the rate of elongation of the acrosomal process was slower than that of the controls and in some cases was very much reduced. More specifically, as seen in Table I, at a tonicity of 1.15 osmol, the average rate of elongation was 216 $\mu m^2/s$ with a range of 102–400 $\mu m^2/s$. At a tonicity of 1.25 osmol, the range was 55–401 $\mu m^2/s$ with a mean of 228 $\mu m^2/s$ (Figs. 3 and 4).

Interestingly, the overall length of the acrosomal process tends to be significantly shorter than that in the controls with an average length of only about 20–25 μm compared with the control curves in which the average length was ~37 μm; this is nearly double that seen in the samples whose tonicities were 1.15 or 1.25 osmol (see Table I).

At a tonicity of 1.33 osmol there was a dramatic change not only in the kinetics of elongation, but also in the extent. Some sperm produced only short blebs. Others produced short processes such as illustrated in Fig. 5 in which the extent of elongation is from 10 to 13 μm. In no instances did we get processes that exceeded 15 μm in length. Plots of $(length)^2$ vs. time show irregular curves with some linear regions followed by regions that slow down, etc. Measurements of the slopes of the linear segments reveal values of 36–65 $\mu m^2/s$ or a mean of 50 $\mu m^2/s$.

At a tonicity of 1.5 osmol, the acrosomal vacuole popped, but processes did not extend.

Effect of Hypotonicity on the Kinetics of Extension of the Acrosomal Process

To study the effect of hypotonicity, we suspended sperm in sea water, introduced them into the perfusion chamber, and then perfused them with sea water that had been diluted with

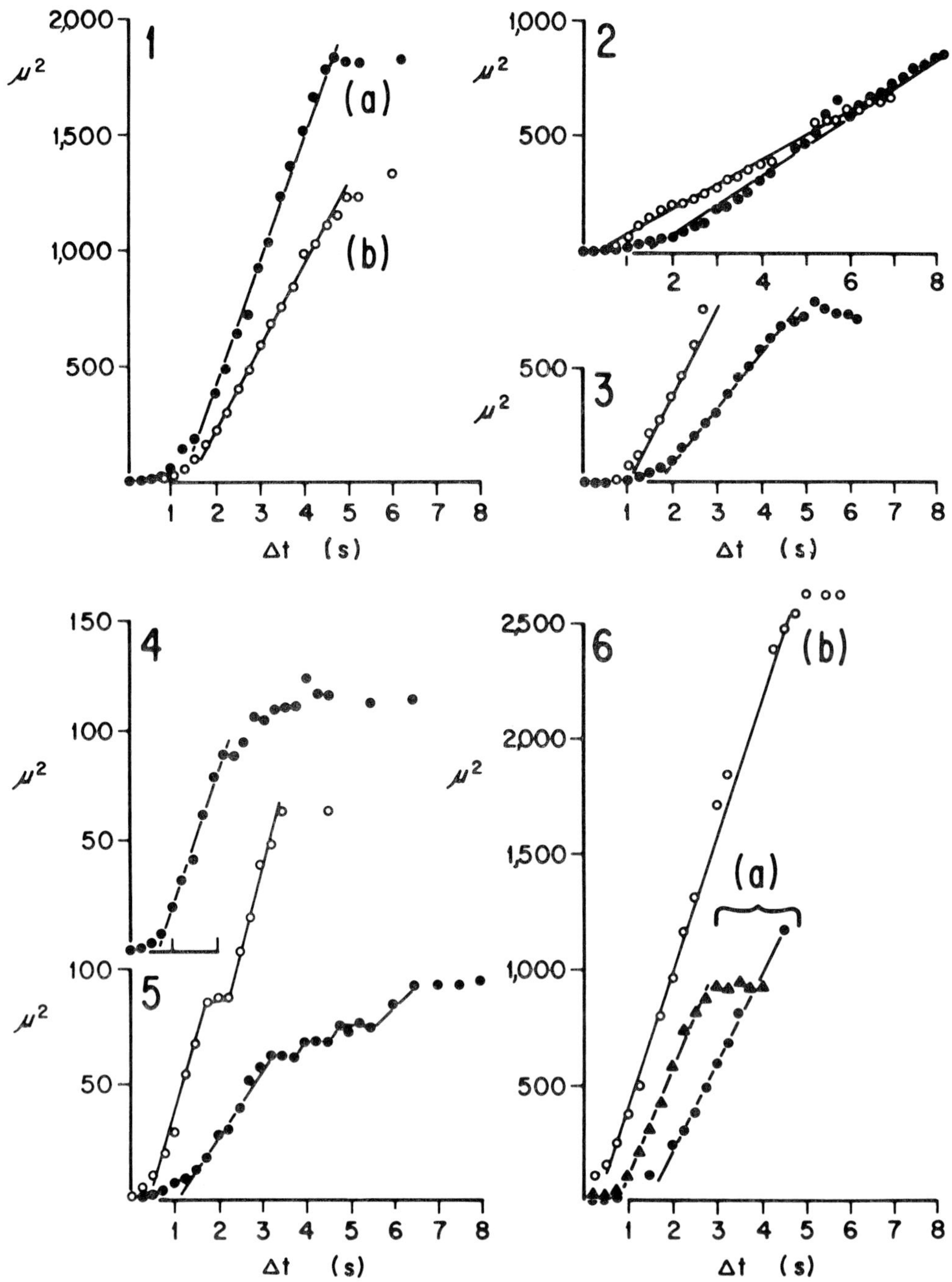
2,000
1,500
1,000
500
μ²
1
(a)
(b)
1 2 3 4 5 6 7 8
Δt (s)
1,000
500
μ²
2
2 4 6 8
500
μ²
3
1 2 3 4 5 6 7 8
Δt (s)
150
100
50
μ²
4
100
50
μ²
5
1 2 3 4 5 6 7 8
Δt (s)
2,500
2,000
1,500
1,000
500
μ²
6
(b)
(a)
1 2 3 4 5 6 7 8
Δt (s)

distilled water. Excess $CaCl_2$ (50 mM) was added after perfusion for 1.5 min. Sperm were then induced to undergo the acrosomal reaction with A23187.

At a tonicity of 0.75 osmol, plots of $(length)^2$ vs. time were linear except at the beginning and end of the reaction with slopes ranging from 280 to 400 $\mu m^2/s$ with a mean of 368 $\mu m^2/s$ (Fig. 6*a* and Table I). The processes generated were 25 μm to well over 35 μm in length and in two instances were still elongating although the overall extent could not be measured because the processes extended out of the field.

At a tonicity of 0.50 osmol, plots of $(length)^2$ vs. time were also linear with slopes ranging from 398 to 615 $\mu m^2/s$ with a mean of 494 (Table I). The lengths of the processes generated were up to 51.5 μm (Fig. 6*b*).

The Relationship between the Rate of Elongation of the Acrosomal Process ($[length]^2$ vs. time) and the Osmolarity of the Medium

A plot of the rate of elongation of the acrosomal process, $(length)^2$ vs. time, as a function of the osmolarity of the medium is depicted in Fig. 7. In this figure, it is obvious that there is considerable variation from sperm to sperm but, nevertheless, what we see is that at low osmolarity, the sperm, in fact, elongate processes at higher rates than in sea water alone, and also that as the osmolarity increases, the rate of elongation decreases dramatically, ceasing altogether at osmolarities of 1,500 mosmol.

Observations on the Surface Morphology of the Acrosomal Process during Elongation at Varying Osmolarities

In an earlier publication (18), we demonstrated that during the early stages of growth the acrosomal process extended very slowly, and was fat and non-uniform in diameter, but as the rate of elongation accelerated, the process became slender and uniform in diameter except at its bulbous tip and at a few locations along the length, marked by "blebs." In this section, we compare these earlier results with the morphology of an acrosomal process extending under hyper- and hypo-osmotic conditions, in particular concentrating on the tips of the acrosomal processes and the blebs.

Of particular interest is that under both hyper- (Fig. 8) and hypo- (Fig. 9) osmotic conditions, the tips of the acrosomal processes are invariably bulbous or tear-shaped as in the controls. They never appear to be needle-like or sharp. An occasional bleb is found along the length of the elongating acrosomal process.

Fine Structural Observations on Sperm Induced to Undergo the Acrosomal Reaction under Hyper- and Hypo-osmotic Conditions

From the preceding sections we have learned that the rate of elongation of the acrosomal process is correlated with the osmolarity of the sea water. At this point we wondered if these differences in rate were related to the amount of actin assembled (perhaps by the osmolarity affecting Z in Hermans' equation [8]) or if all the actin assembles under all conditions and differences in rate are related to some other factor. To distinguish between these two possibilities, sperm were incubated in hyperosmotic (1.33 or 1.25 osmol), iso-osmotic, and hypo-osmotic (50% sea water) conditions, the acrosomal reaction was then induced with A23187, and 1.5 min later the sperm were fixed by the addition of glutaraldehyde to the solution.

Thin sections of sperm fixed under hyperosmotic conditions such as 1.33 osmol showed us that in most instances the membrane that limits the acrosomal vacuole had fused or

FIGURES 1–6 (Fig. 1*a*) Plot of (length of the acrosomal process)^2 in μm^2 as a function of time (in seconds) after the acrosomal process first becomes visible. Of note is that except for the beginning and end of the reaction, the points fall on a straight line. This sperm was induced by the addition of A23187 to sea water. Sperm A. (Fig. 1*b*) Plot of (length of the acrosomal process)^2 in μm^2 as a function of time which serves as a control. In this experiment, one part of sea water was added to one part of a 1-M sucrose solution so that the tonicity remains that of sea water. Note that, as in Fig. 1*a*, all the points except those at the beginning and end of the reaction fall on a straight line. Sperm B. (Fig. 2) Plot of (length of the acrosomal process)^2 in μm^2 as a function of time for two sperm, L (○) and M (●), which were induced to undergo the acrosomal reaction in sea water whose osmolarity was 1.15 that of natural sea water or ~1,150 mosmol. Note that all the points except those at the beginning of the reaction tend to fall on a straight line. (Fig. 3) Plot of (length of the acrosomal process)^2 in μm^2 as a function of time for two sperm, P (○) and O (●), which were induced to undergo the acrosomal reaction in sea water whose osmolarity was 1.25 that of natural sea water or ~1,250 mosmol. Note that all the points except those at the beginning and end of the reaction tend to fall on a straight line. (Fig. 4) Plot of (length of the acrosomal process)^2 in μm^2 as a function of time for sperm Q which was induced to undergo the acrosomal reaction in sea water whose osmolarity was 1.25 that of natural sea water. Notice that the scale on the Y axis of this figure is one-tenth that of Fig. 3; the process formed is very short. Even so, except at the beginning and end of the reaction the points all fall on a straight line. (Fig. 5) Plot of (length of the acrosomal process)^2 in μm^2 as a function of time for two sperm, R (○) and S (●), which were induced to undergo the acrosomal reaction in sea water whose tonicity was 1.33 that of natural sea water or an osmolarity of ~1,330 mosmol. Two points should be noted. First, the scale on the Y axis indicates that the processes that develop are very short, and second, although many of the points fall on a straight line, there are clearly sharp bends where the sperm stop elongating, only to begin to elongate again after 0.75 s. (Fig. 6*a*) Plot of (length of the acrosomal process)^2 in μm^2 as a function of time for two sperm, E (▲) and F (●), which were induced to undergo the acrosomal reaction in sea water whose tonicity was 75% that of natural sea water by being diluted with distilled water. Of interest is that all the points except at the beginning and end of the reaction fall on a straight line. In the case of sperm F the process was still elongating after 4.5 s, but unfortunately, the tip of the acrosomal process disappeared from view by extending out of the video field selected. (Fig. 6*b*) Plot of (length of the acrosomal process)^2 in μm^2 as a function of time for sperm I which was induced to undergo the acrosomal reaction in sea water whose tonicity was 50% that of natural sea water by being diluted with distilled water. Note that the resulting process is long, extends rapidly, and except at the beginning and end of the reaction all the points tend to fall on a straight line.

Reprinted from *Video Microscopy*, pp. 477–510, 1986,
with permission from Springer Science and Business Media.

III

An Introduction to Biological Polarization Microscopy*

III.1. INTRODUCTION

Electron micrographs show fine structural detail and organization in thin sections of fixed cells. Time-lapsed motion pictures show the continually changing shape and distribution of organelles in living cells. How can we combine these observations and follow the changing fine structure in a single living cell as it undergoes physiological activities or experimental alterations?

The relevant dimensions are too small for regular light microscopy since many of the changes of interest take place in the 0.1 to 100 nm range, i.e., the size range of atoms to macromolecular aggregates. From wave optics, the resolution of a light microscope is limited by its NA and the wavelength of light according to equation (5-9, p. 115).† On the other hand, the electron microscope with its very high resolution, practically 1–2 nm for many biological specimens, does not usually permit nondestructive study of cells.

One effective way of closing this gap is to take advantage of anisotropic optical properties, which, albeit at a rather low level, are often exhibited by cellular fine structures. Anisotropic optical properties, such as birefringence and dichroism, measure molecular and fine structural anisotropy as we shall soon see. These properties can be studied *nondestructively* in individual living cells with a sensitive polarizing microscope, so that the time course of fine structural changes can be followed in small parts of a single cell.

From equation (5-9), the smallest distance resolvable with any microscope with a 1.40-NA objective (with a nearly matched NA condenser at $\lambda = 550$ nm) is $\simeq 0.2$ μm, or 200 nm. With a sensitive polarizing microscope, we can determine molecular alignment or changes in fine structure taking place within each resolvable unit area that is smaller than 0.2×2 μm, or less than 1 μm^2.

*Revision of a manuscript originally prepared for a NATO Advanced Summer Institute held at Stresa, Italy, in 1969, but which was never published. Lectures related to this were delivered to the Physiology course and the short course on Analytical and Quantitative Light Microscopy in Biology, Medicine, and the Materials Sciences at the Marine Biological Laboratory, Woods Hole, Massachusetts, and the University of Pennsylvania. The active discussions and willing cooperation of my students and colleagues guided me in developing these lectures and demonstrations. Several of the illustrations in this appendix were initially prepared by Richard Markley, Dr. and Mrs. Hidemi Sato, and Hiroshi Takenaka. They have been refined and finished by Bob and Linda Golder and Ed Horn of the MBL. Their efforts, and support by NIH Grant 5R01-GM-31617 and NSF Grant PCM 8216301, are gratefully acknowledged.

†This does not mean that we cannot form an image of individual organelles, filaments, or particles that are smaller than half the wavelength of light. If they provide enough contrast, and are separated from other comparable structures by a distance greater than the resolution limit, they can be visualized (by their image expanded to the width of their diffraction pattern; Figs. 1-7, 11-3, 11-5, 11-8).

III.2. ANISOTROPY

Anisotropy relates to those properties of matter that have different values when measurements are made in different directions within the same material. It is the opposite of isotropy, where the property is the same regardless of the direction of measurement. A common example of anisotropy is found in a piece of wood. Wood is much stronger parallel to the grain than across. As the humidity changes, wood shrinks and swells across the grain much more than it does along the grain. Also, wood splits more easily along the grain. In these examples, we might have anticipated the anisotropic properties from the grain structure of wood, but there are other anisotropic properties that are not quite so obvious.

Related to the fact that it is more difficult to stretch wood along the grain than across the grain—that is, the coefficient of elasticity is greater parallel to the grain than across—sound waves travel faster parallel to the grain than across the grain. Tyndall pointed out many years ago that, depending on the type of wood, the velocity of sound may be three times as great when the sound is traveling parallel to the grain than across the grain.

We can demonstrate this phenomenon by taking two blocks of wood cut from a single plank (Fig. III-1). The two blocks are made to the same dimensions but one (P) is cut parallel to the grain and the other (S) across. When we tap on the end of a block, a standing wave is formed. The compressive sound wave travels back and forth quickly, and since the ends are open, a longitudinal standing wave is formed with its loops at the ends of the block and a node in the middle. This standing wave takes on a particular pitch or tone. The tone tells us how quickly the wave is traveling in the block. Parallel to the grain we get one tone. Across the grain we get another. In fact, the velocity of sound across the grain is much lower, so that the tone is quite a bit lower (by an octave in one example; see legend to Fig. III-1). This is a dramatic demonstration of acoustic anisotropy, or anisotropic propagation of sound waves. The demonstration tells us how structural anisotropy can relate to wave propagation.

In this case, the grain structure of the wood was obviously visible, and it is not difficult to grasp the reason for the anisotropy; but anisotropy can exist even if we cannot see heterogeneous structures. The anisotropic properties can lie in the material that appears homogeneous at the microscopic level but which is anisotropic at the atomic or molecular levels.

III.3. SOME BASIC FEATURES OF LIGHT WAVES

Before discussing anisotropy further, we will review some basic physical optics needed in the subsequent discussions.

Light is a form of electromagnetic wave. Assume a capacitor that is charged and let it discharge as a spark through two electrodes as shown in Fig. III-2. In the spark, a current flows

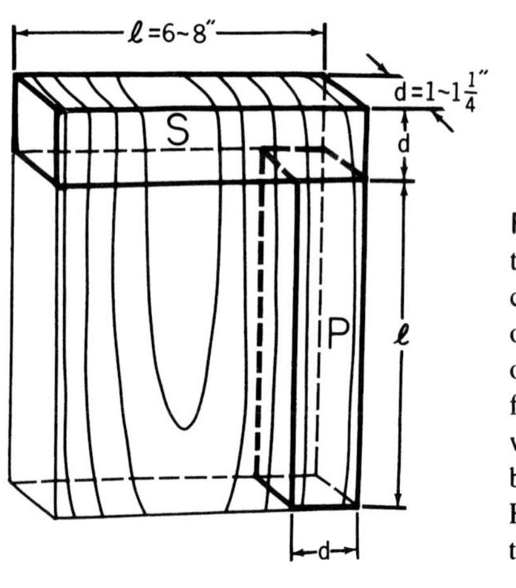

FIGURE III-1. Wooden blocks for demonstrating acoustic anisotropy. Blocks P and S are finished to the same dimensions, but P is cut parallel to the grain while S is cut across grain. The frequency of the standing wave, generated by gently holding onto the middle of the block and tapping (with a steel ball with a flexible handle) from an end, measures the acoustic velocity for the compressive wave that travels through the block. In one example (8-inch-long blocks made of hard oak), P resonated at 4500 Hz and S at 2500 Hz. In this case, the compressive sound wave traveled nearly two times faster parallel to the grain than across.

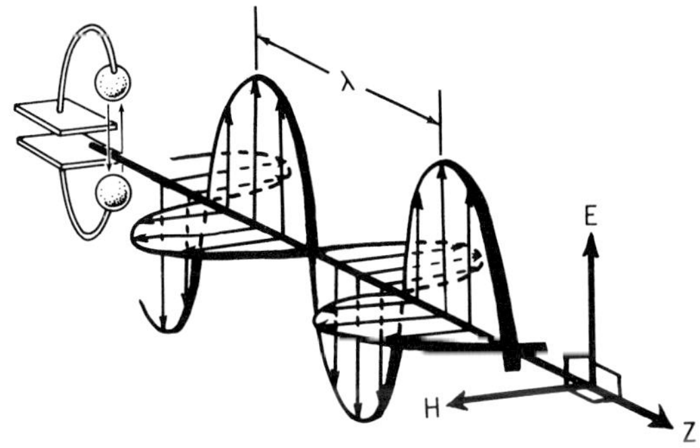

FIGURE III-2. Propagation of electromagnetic wave. The wave is generated by the discharging spark (or oscillating molecular dipole) to the left. The spark current oscillates at a frequency ν characteristic of the circuit. The resultant electromagnetic disturbance is propagated with the electric (E) and magnetic (H) vectors vibrating perpendicular to each other and to the direction of propagation Z. ν is determined by the oscillator, and the wavelength (λ) is given by $\lambda = v/\nu$, where the velocity (v) is given by equation (III-1) or (III-2).*

down for a short time, slows down, but because of the inductance of the circuit, flows back upwards, recharging the capacitor again. The current thus oscillates back and forth at a frequency (ν) that is characteristic of the circuit. The electric current oscillating up and down in the spark gap creates a magnetic field that oscillates in a horizontal plane. A changing magnetic field in turn induces an electric field so that we end up with a series of electrical and magnetic oscillations that propagate as an electromagnetic wave.

As seen in Fig. III-2, the electric field in an electromagnetic wave vibrates with its vectorial force growing stronger and then weaker, pointing in one direction, and then in the other direction, alternating in a sinusoidal fashion. The magnetic field oscillates perpendicular to the electric field, at the same frequency. The electric and the magnetic vectors, which express the amplitude and vibration directions of the two waves, are oriented perpendicular to each other and to the direction of propagation.

From the relationships defining the interaction of the electric and magnetic fields, one can deduce the velocity of the resulting electromagnetic wave. The equations of Maxwell show that the velocity (v) is exactly c (the velocity of light in vacuum = 3×10^{10} cm/sec) divided by the square root of the dielectric constant (ξ) of the medium times the magnetic permeability (μ) of the medium. Thus,

$$v = c/\sqrt{\xi \cdot \mu} \qquad \text{(III-1)}$$

For most materials that occur in living cells, namely for nonconducting material, $\mu = 1$, so that

$$v = c/\sqrt{\xi} \qquad \text{(III-2)}$$

Empirically, we also know that the velocity of light is inversely proportional to the refractive index (n) of the material through which it propagates, i.e.,

$$v = c/n \qquad \text{(III-3)}$$

Equations (III-2) and (III-3) tell us that the refractive index is equal to the square root of the dielectric constant of that material (if the measurements are both made at the same frequency, ν). Thus,

$$n_\nu = \sqrt{\xi_\nu} \qquad \text{(III-4)}$$

This tells us that *optical measurements are, in fact, measurements of electrical properties of the material*. The dielectric properties in turn directly reflect the arrangement of atoms and molecules in a substance.

The direction of interaction between an electromagnetic field and a substance can be consid-

*In the equations, $\xi = \mu = 1$ for a vacuum, and very close to 1 for air.

ered to lie in the direction of the electric vector. That is so whether we consider the electric or the magnetic vectors, since what matters is the effect of the electric or magnetic fields on the electrons in the material medium. (The magnetic field affects those electrons which move in a plane perpendicular to the magnetic field.) From here on, then, we will represent the vibrations of an electromagnetic wave by indicating the direction of the electric field alone. (The magnetic field is, of course, still present perpendicular to the electric field, but we will not show it in our vector diagrams, in order to avoid confusion.)

III.4. OPTICAL ANISOTROPY

The dielectric constant is anisotropic in many substances. Take a cube of such a substance and place it on a Cartesian coordinate, with the sides of the cube parallel to the X, Y, Z axes. The dielectric constant for an electric field measured along the X axis—that is, ξ measured by placing the plates of a capacitor on the faces of the cube parallel to the Y–Z plane (the plane that includes the Y and Z axes, or perpendicular to the X axis)—would have one value, ξ_x. Along the Y axis, it would have another value, ξ_y, and along the Z axis, a third value, ξ_z.

We now take the square roots of these three dielectric constants and make the lengths of the X, Y, and Z axes proportional to these values. Then we draw an ellipsoid whose radii coincide with the X, Y, and Z axes as just defined (Fig. III-3). What we obtain is known as the Fresnel ellipsoid. This ellipsoid describes the dielectric properties measured in all directions in the material. Each radius provides the $\sqrt{\xi}$ value for an electromagnetic wave *whose electric vector lies in the direction of that radius*.

Since $\sqrt{\xi} = n$ from equation (III-4), the Fresnel ellipsoid is, in fact, a refractive index ellipsoid. $\sqrt{\xi_x}$ is the refractive index for waves whose electric fields are *vibrating* in the X-axis direction, $\sqrt{\xi_y}$ for waves with electric fields *vibrating* along the Y axis, and $\sqrt{\xi_z}$ for waves with electric fields *vibrating* along the Z-axis direction.

The value of the refractive index given by the radius of the Fresnel ellipsoid is valid for all waves whose electric vector vibrates in the direction of that radius regardless of the wave's direction of propagation. For example, waves with their electric vector in the X-axis direction may be propagating along the Y- or Z-axis direction. Therefore, given the direction of the electric vector

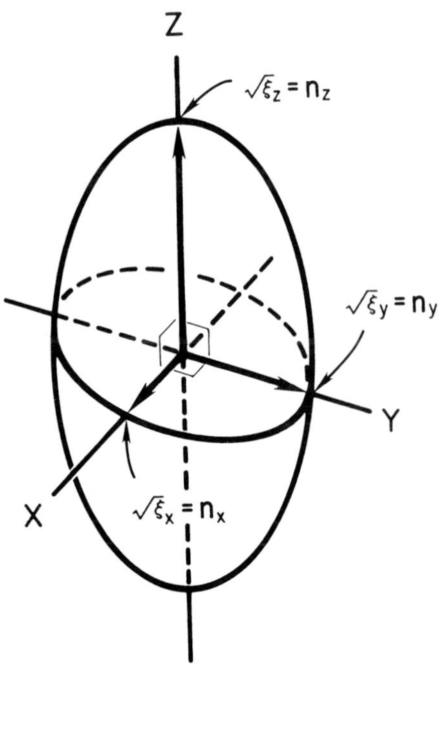

FIGURE III-3. The refractive index, or the Fresnel, ellipsoid. The radius of the Fresnel *ellipsoid* gives the refractive index (n), or the square root of the dielectric constant ($\sqrt{\xi}$), for waves whose electric displacement vectors lie in the direction of the radius of the ellipsoid within an anisotropic medium.

A cross section through the center of the ellipsoid gives the refractive index *ellipse* for waves traveling normal to that section. The major and minor axes of the ellipse denote the refractive indices encountered by the slow and fast waves, which vibrate with their electric displacement vectors along those two axes.

(or more properly, the **electric displacement vector** in the material medium), the refractive index suffered by a wave is defined regardless of its direction of propagation.

III.5. BIREFRINGENCE

How do we use this ellipsoid? If light is traveling along the X-axis direction, for example, then the refractive indices for the waves vibrating in the material are given by an *index ellipse,* which is *the cross section of the three-dimensional ellipsoid cut perpendicular to the direction of propagation* (the X-axis direction in this example). For light traveling in any direction, we obtain an index ellipse by making a major cross section of the ellipsoid perpendicular to that direction of travel. The index ellipse gives us the refractive indices and shows the vibration direction (of the electric vector) of the light waves, because *the waves vibrate with the electric vectors along the major and minor axes of the index ellipse. The refractive indices that the two waves suffer are given by the major and minor radii of the ellipse.* In other words, *for each direction of travel in an anisotropic medium, there are two plane-polarized light waves that vibrate perpendicular to each other and travel at different velocities.* This is, in fact, the phenomenon of birefringence.

To restate what we have just seen, as light travels through an anisotropic material, the wave gets split into two vibrations. The two vibrations (remember, we are just considering the electric vector directions and not the magnetic vectors) are mutually perpendicular to each other and perpendicular to the direction that the wave travels. The wave whose electric vector vibrates along the major axis of the index ellipse is called the slow wave, because the refractive index for this wave is greater than the refractive index for the other wave.* The wave vibrating perpendicular to the slow wave is the fast wave.

Each index ellipse then provides the slow and fast vibration axes for waves traveling perpendicular to the plane of the ellipse. The index ellipse (the cross section of the index ellipsoid) is commonly used to designate the birefringence of a material observed with light propagating perpendicular to that plane (e.g., Figs. III-14, III-17).

We shall now examine another manifestation of birefringence, taking advantage of another demonstration. We place a piece of cardboard with a small hole in front of a light source and on top of the hole, a crystal of calcite. Calcite is a form of calcium carbonate, whose rhombohedral shape is given in Fig. III-4. When we place the calcite crystal on top of the hole, we find that the image seen through the crystal is doubled. As we turn the crystal on the cardboard, we find that one of the images is stationary, while the other one precesses around the first.

This observation is explained by the birefringence of calcite. The birefringence is so strong that we not only get two waves, but even the directions of travel of the two waves become separated (Fig. III-5). One of the waves is traveling straight through—its image remains stationary when the crystal is turned. That is called the ordinary ray, or the o-ray, because it behaves (is refracted) in an ordinary fashion. The other wave, the precessing one, refracts in an extraordinary fashion, and is called the extraordinary ray, or the e-ray.

Not only is the light split into two, but as we argued before, each must be vibrating in a unique direction. The two must also be vibrating with their electric vectors perpendicular to each other. It turns out that the e-ray always vibrates in the plane that joins it and the o-ray. The o-ray always vibrates at right angles to this, as we will now verify.

We orient the crystal so that the e-ray image (which is the one that precesses) appears on top (Fig. III-4). The direction of vibration of the e-ray then should be in the direction joining it and the o-ray image, namely up and down. The o-ray should have its electric vector horizontally. If we use a polarizer that transmits waves with horizontally oscillating electric fields, the top image should disappear; if we use one that transmits electric fields with vertical vibration, the bottom image should disappear.

One handy, inexpensive standard polarizer, or polar, is a pair of Polaroid sunglasses. It shows no apparent anisotropy if we examine an ordinary source of light. No difference is apparent as we

*Equation (III-3) shows the relation of the refractive index to the (phase) velocity of travel.

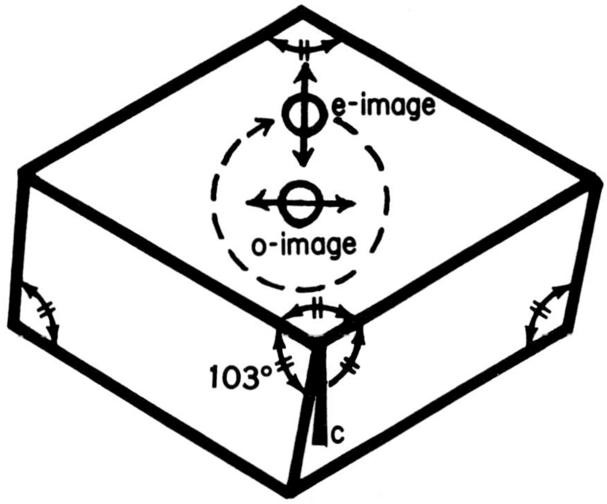

FIGURE III-4. Double image of a single light spot seen through a calcite crystal. As the crystal is turned, the e-ray image precesses around the o-ray image. The e-wave vibrates in the plane that includes the c-axis (the principal section). The o-wave vibrates perpendicular to it. The c-, or optic-axis, of the crystal is indicated by c. In calcite, it is the axis of threefold symmetry. It makes an equal angle with all three of the crystal faces that join at the two corners, where all edges lie at 103° angles with each other. The c-axis lies in the direction of the semi-ionic bond that links the planar CO_3 groups and Ca atoms in the calcite ($CaCO_3$) lattice. (Evans, 1976; Inoué and Okazaki, 1978.)

turn the sunglasses in different directions. But if we look at the surface of water, or at the glare on the road or painted surfaces, the reflected light is cut out as the Polaroid glasses are intended to do. Light that is reflected from the surface of water or any nonconducting material is plane polarized, and especially strongly so at a particular angle of incidence—the Brewster angle. It is polarized so that the electric vector is vibrating parallel to the surface from which it is reflected, and not vibrating perpendicular to the surface. This behavior can be explained from a series of equations of Fresnel, but an easier way is to use the stick model proposed by Robert W. Wood. Consider a stick of wood in place of the electric vector. If it hits the water surface at an angle, the stick goes into the water and is not reflected. If the stick comes in parallel to the surface, it can bounce back. Since in nature we are dealing with horizontal surfaces, it will be a horizontal vibration that is reflected. We want to cut the glare, so Polaroid sunglasses are made to remove the horizontal vibrations and transmit the

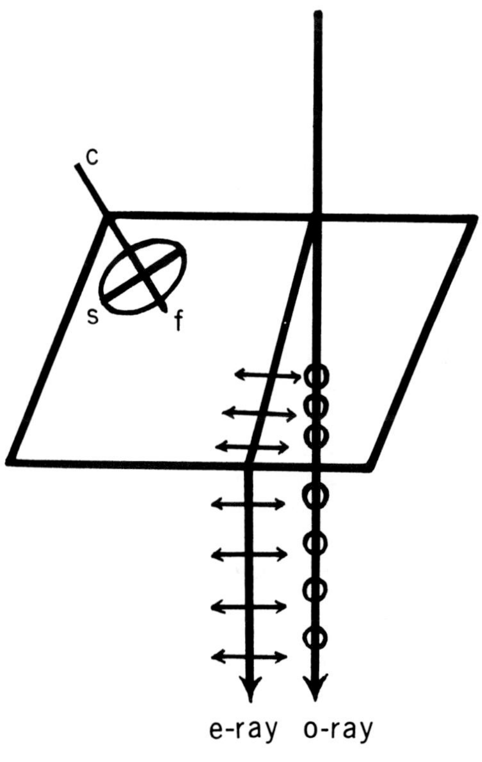

FIGURE III-5. Path of light rays through a calcite crystal. The crystal is shown through a principal section—that is, through a plane including the optic axis, c. The o-ray is refracted as though it were traveling through an isotropic medium; hence, it proceeds without deviation after normal incidence. The e-ray direction is deviated (in the principal section) even for normal incidence, hence the name extraordinary ray. The c-axis is the axis of symmetry of the index ellipsoid. Calcite is a negatively birefringent crystal so that the o-ray is the slow wave. The o-wave always vibrates in a plane perpendicular to the principal section while the e-wave vibrates in the principal section.

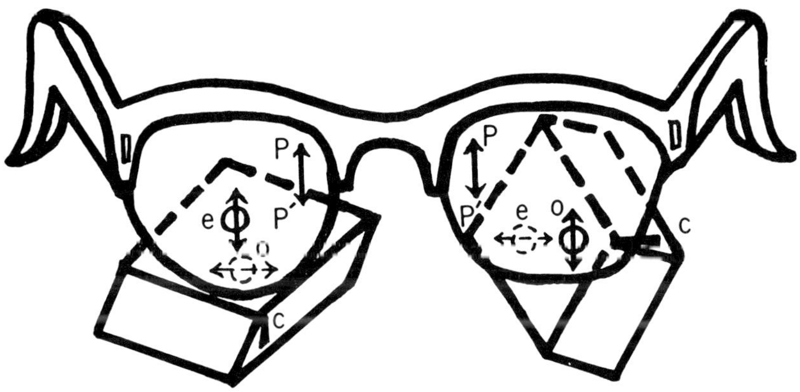

FIGURE III-6. Determining the electric vector directions for the e- and o-ray in calcite with Polaroid sunglasses. Polaroid sunglasses are made to absorb and remove the horizontal electric vectors and transmit the vertical vibrations, PP'. On the left side, the e-ray image from the calcite crystal is transmitted but the o-ray image is absorbed. On the right side, the calcite crystal is turned 90° and the e-ray image is absorbed. At an intermediate angle, both images are partially transmitted following the cosine squared law.

electric vector that is vibrating vertically. In these glasses, then, we have a standard for transmission of the electric vector (Fig. III-6).

Returning to the calcite crystal (Fig. III-4), the top image has its e-vector vertically, so that when we look through our sunglasses, the bottom image should disappear (Fig. III-6, left). We find that is exactly the case. If we turn the sunglasses or the crystal slowly, two images appear for a while, until at 90° the "top" image becomes extinguished (Fig. III-6, right). Thus, with birefringence a light beam is split into two waves, each of the waves vibrates with its electric vector perpendicular to the other, and they travel at different velocities. (Close inspection will reveal that the e-ray image appears farther through the crystal than the o-ray image, indicating that the o-ray image suffered greater refraction, or that $n_e < n_o$ in calcite.)

Calcite crystals can be used as very effective polarizers; in fact, Nicol, Glan–Thompson, Ahrens, and other prisms are made of calcite to isolate and transmit one of the polarized waves.*

For UV microbeam work, it is much less expensive and more effective to use a simple cleaved piece of calcite crystal that has been optically polished on two opposite faces. Placed in front of the microbeam source, the crystal splits the beam into two and gives two equally intense images of the microbeam source (Fig. III-7). One can choose an image with known vibration (or use both) because the e-ray, whose electric wave vibration joins the two images, can be identified by its precession when the crystal is turned around the axis of the microscope. The o-ray vibration is perpendicular to it. One now has two polarized microbeam sources with equal intensities, which can be used for analyzing e.g., the arrangement of DNA bases as discussed later.

Down to which wavelength will calcite transmit? Generally, about 240 nm, but this depends on the calcite. One can get calcite that transmits to somewhat shorter wavelengths, depending on the source and purity of the crystal.

In material such as calcite, the two beams were visibly split into two, but for most material, especially the kind of biological material making up cytoplasmic or nuclear structures in a living cell, the birefringence is so weak that the two beams are not visibly split. Even when they are not visibly split, the extraordinary and the ordinary waves are present, with their individual vibration planes and separate velocities. Before we look into the properties of these two waves and how they can be used to detect birefringence in living cells, let us briefly consider the molecular and atomic meaning of *dichroism*, another form of optical anisotropy.

*It is unfortunate that the wave transmitted by these otherwise superior polars is the e-wave. The velocity of the e-wave, or refractive index of the e-ray, varies with direction of propagation, so that these calcite prisms introduce *astigmatism* unless all the beams travel parallel to each other through the crystal.

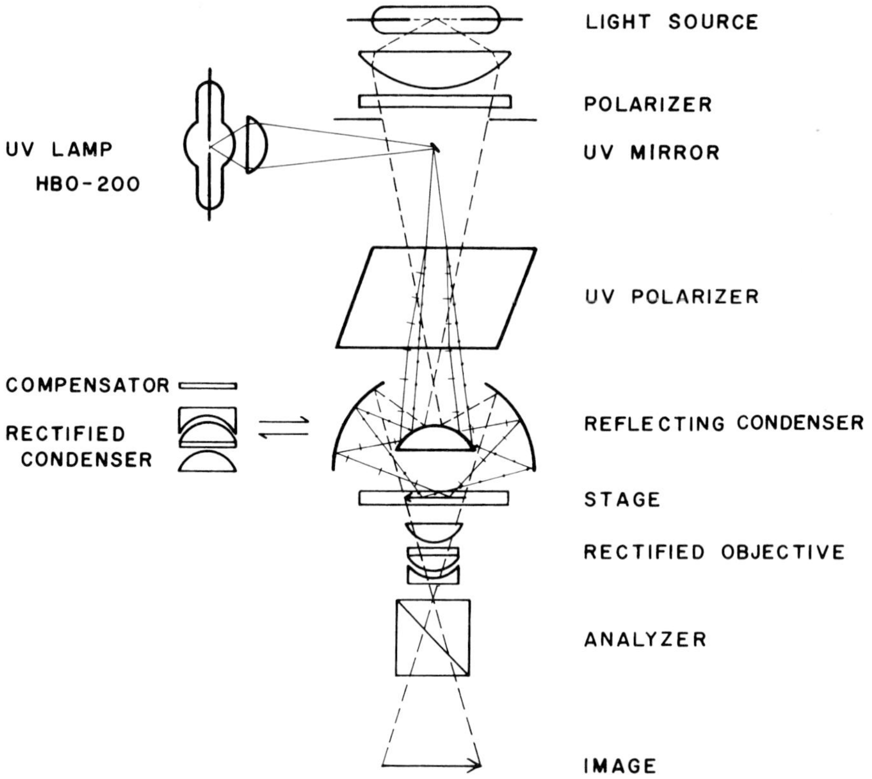

FIGURE III-7. Schematic diagram of polarized UV microbeam apparatus. Compare with schematics of inverted polarizing microscope (Fig. III-21) and photograph of rectified instrument (Fig. III-22). Each ray from the UV source (a small, first-surface mirror mounted on a dovetail slide beneath the polarizer for visible light illumination), as well as from the visible light source diaphragm (which lies between the polarizer and the UV mirror), is split into e- and o-rays by the UV polarizer. The UV polarizer is a cleaved crystal of calcite optically polished on the top and bottom faces. The reflecting condenser focuses the double image of the UV mirror and the visible light source diaphragm onto the specimen plane, each image with its electric vector as defined in Figs. III-4 to III-6. The image of the UV mirror with the desired polarization is superimposed on the specimen to be microbeamed. (From Inoué and Sato, 1966b.)

III.6. DICHROISM

Again, we start with a demonstration. A small source of light, say a flashlight bulb, is placed behind a crystal of tourmaline. Tourmaline is a prismatic crystal with a threefold rotational symmetry, and the type we use for this demonstration is a clear green crystal. Tourmaline crystals are dichroic and transmit in the green (or blue) as long as the axis of symmetry is oriented perpendicular to the light beam (Fig. III-8). But if we turn the crystal and look down the axis of symmetry, no light, or very little light with a brownish tint, is transmitted.

We can understand the wave-optical basis of these observations by returning to an analysis similar to the one used for birefringence. Dichroic crystals are also birefringent, so that light is again split into two waves in the crystal. As shown in Fig. III-9, for each direction of travel, the light wave is split into two waves, whose electric vectors are oriented perpendicular to each other and to the direction of travel. For light traveling along the *Z* axis, we have one wave (*Y*) with its electric vector vibrating in the *Y–Z* plane and another (*X*) with its electric vector vibrating in the *X–Z* plane. Both of these must be absorbed because light does not come through the *Z*-axis direction. This means that the electronic disturbances caused by both the *y* electric vector and the *x* electric vector result in

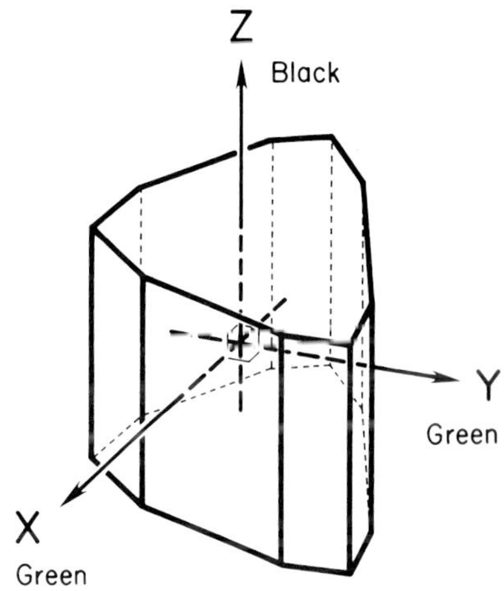

FIGURE III-8. Transmission of light through a dichroic crystal—in this illustration, a clear green crystal of tourmaline. Viewed along the X and Y axes, the crystal is clear green. Viewed along the Z axis, no light is transmitted and the crystal is black or dark brown, even when the crystal is considerably shorter along the Z axis than along the X and Y axes.

absorption of the energy. The chromophores in the crystal are oriented such that they absorb light whose electric fields are vibrating in the X–Y plane.

If we examine the light traveling along the X direction, we again know that light must vibrate in two directions perpendicular to that—the Z direction and the Y direction. We already know that the Y vibration is absorbed, and therefore the green light we see must have its electric vector in the Z direction. Likewise for the Y propagation, the x electric vectors are absorbed and only the z vectors must make up the transmitted green light. We thus deduce that the light coming through the tourmaline crystal is plane polarized with its electric vector parallel to the crystalline (Z) axis of symmetry. This is readily confirmed with our standard Polaroid sunglasses.*

This is the basis of dichroism. As we saw, both dichroism and birefringence can be manifested as macroscopically recognizable optical properties, which vary with the direction of propagation of light, but they are at the same time a reflection of fine-structural, molecular, or atomic anisotropies. For dichroism it is the orientation of chromophores that determines anisotropic absorptions. For birefringence it is the anisotropy of dielectric polarizabilities. The anisotropic polarizability can originate in the fine structure or within the molecules.

III.7. DETERMINATION OF CRYSTALLINE AXES

We are now almost ready to apply these principles to the polarizing microscope to study biological material. Let us place two pieces of plastic sheet Polaroid in front of a diffuse light source and orient them at right angles to each other so that the light is extinguished. In between the "crossed polars" we insert our samples, just as we do with a polarizing microscope. When, for example, a piece of cellophane is placed between crossed polars, we observe light through that area (Fig. III-10). The phenomenon is due to birefringence, as explained below, even though we do not see a double image.

As in Fig. III-11, let the light (with an amplitude of OP) from the polarizer (PP′) vibrate vertically. Upon entering the specimen. OP is vectorially split into two vibrations, OS and OF. These vibration directions are uniquely defined by the crystalline structure of the specimen since

*Polaroid sunglasses and filters are themselves made from a dichroic material. One common form uses a stretched film of polyvinyl alcohol (PVA) impregnated with polyiodide crystals. In the stretched PVA film, the micelles and molecular backbones are oriented parallel to the direction of stretch (Fig. III-17). The needle-shaped, dichroic polyiodide crystals are deposited in the interstices, parallel to the PVA molecules. All of the polyiodide crystals, therefore, become oriented parallel to each other, and one obtains a large sheet of dichroic material (Land, 1951; Land and West, 1946).

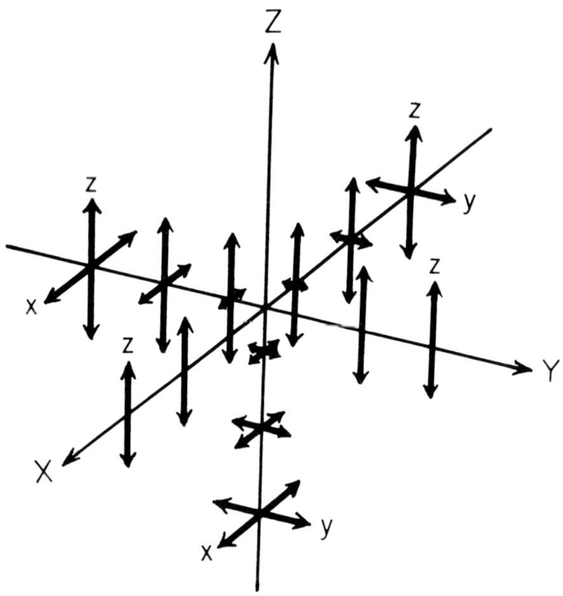

FIGURE III-9. Explanation of dichroic absorption in tourmaline. The x and y electric vectors are absorbed regardless of the direction of propagation. Only the z vectors are not absorbed, so that the transmitted light is plane polarized parallel to the Z axis.

they lie in the directions of the major and minor axes of the index ellipse. In the crystal (cellophane), these two waves travel at different velocities so that when they come out of the crystal, they (O′S′ and O′F′) are out of phase (Δ) relative to each other.* The combination of these two waves, which remain out of phase and continue to oscillate in planes perpendicular to each other in air, yields an elliptically polarized wave (O″P″–O″A″).

Light that is elliptically polarized no longer vibrates only along the PP′ axis, as it has a component that exists along the AA′ axis. That component (O″A″) of the elliptically polarized light can pass through the second polar (the analyzer) and give rise to the light that we observe.

For the birefringent specimen between crossed polars, the elliptically polarized wave is produced by the splitting of the original plane-polarized wave into two vectors. Therefore, we should not get elliptical polarization if, for some reason, one of the two vectors were missing. This happens when the slow or fast specimen axes become oriented parallel to the polarizer axis. At that orientation of the crystal, OP′ cannot be split into two vectors, so OP′ emerges from the crystal unaltered. This wave is then absorbed completely by the analyzer and no light comes through. Therefore, by rotating the specimen between crossed polars and observing the orientation where the specimen turns dark (Fig. III-10), we can determine its crystalline axes. *When the specimen turns dark between crossed polars, the orientations of the major and minor axes of the index ellipse are parallel to the axes of the polars.* (Note that these axes may or may not coincide with the geometrical axes or cleavage planes of the specimen.)

We have now established the directions of the two orthogonal axes in the crystal, but we still do not know which is the fast axis and which is the slow axis. In order to determine this, we

*Δ, expressed in fraction of wavelength (of the monochromatic light wave in air; λ_0), is proportional to the thickness of the crystal (d) and to its "coefficient of birefringence" ($n_e - n_o$). Thus,

$$\Delta = (n_e - n_o)d/\lambda_0 \qquad \text{(III-5)}$$

where n_e is the refractive index for the extraordinary ray, and n_o that of the ordinary ray. The e-ray may either be the slow or fast wave depending on the crystal type. If the e-ray is the slow wave (as it is in quartz and cellophane), the crystal is said to be positively birefringent; if the e-ray is the fast wave (as it is in calcite), the crystal is said to be negatively birefringent. The Fresnel ellipsoid for a positively birefringent material is prolate, while the one for a negatively birefringent material is oblate.

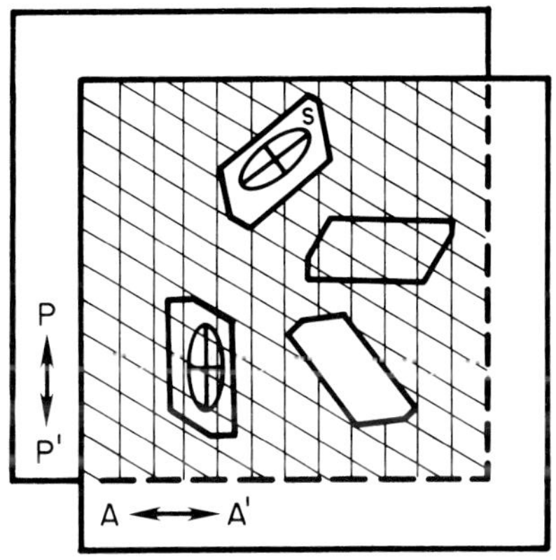

FIGURE III-10. Appearance of a birefringent crystal between crossed polars. Polarizer transmission PP′ and analyzer transmission AA′ are at 90° to each other and extinguish the background light. The crystal is viewed in various orientations. When the optical axes of the crystal are parallel to PP′ or AA′, the crystal is dark. When the axes are at any other orientation, the crystal is brighter.

superimpose a crystalline material whose slow and fast axis directions are already known. Such a crystal is known as a compensator.

Let us assume in our next demonstration that a particular piece of plastic—the compensator—has its slow and fast axes oriented as shown in Fig. III-12. We superimpose this on an unknown sample; both are placed between a pair of crossed polars. If the slow axis of the unknown sample and that of the compensator lie perpendicular to each other, then the two effects cancel each other out. Where the compensator and the sample are superimposed on each other, we observe that the light is extinguished; the specimen appears dark (Fig. III-12, the crystal in the lower right position). What has happened is that the elliptically polarized light produced by the specimen was restored by the compensator to the original plane-polarized light (Fig. III-13). Therefore, no vectorial component is present in the analyzer transmission direction, so that all the light is absorbed by the analyzer.

On the other hand, if the specimen and the compensator slow axes were parallel to each other, the two retardations would add up together, and we would get more light through the analyzer than before (Fig. III-12, the crystal oriented as in the top). By finding the orientation of the compensator

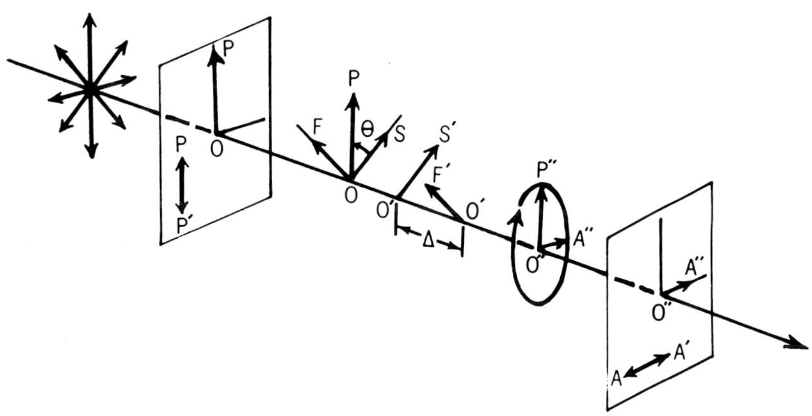

FIGURE III-11. Explanation of the phenomenon in Fig. III-10. So long as the angle θ between the slow axis (OS) of the crystal and OP (the electric vector from the polarizer) is not 0 or 90°, OP is split into vectors OS and OF by the birefringent crystal. Passing through the crystal, they acquire a phase difference, Δ, which remains constant once in air (an isotropic medium) up to the analyzer. Viewed from the analyzer direction, the O′S′ and O′F′ vibrations combine to describe an elliptically polarized wave, O″P″–O″A″. The O″P″ component is absorbed by the analyzer, but the O″A″ component is transmitted, and hence the crystal appears bright on a dark background. Where θ is 0 or 90°, the OF or OS vector is missing, so that O″P″ = OP and O″A″ → 0. Hence, no light passes the analyzer.

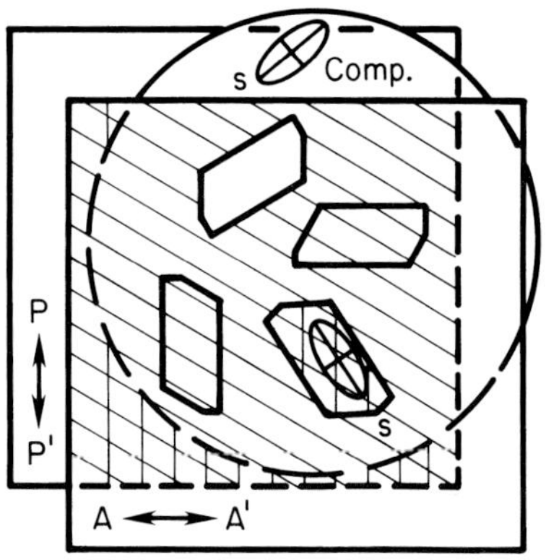

FIGURE III-12. Birefringent crystal between crossed polars, in the presence of a compensator. The compensator (large circle) introduces some light between crossed polars similar to the crystal in Figs. III-10 and III-11. On this gray background, the crystal appears darker (lower right orientation) when its *s* axis is in the opposite quadrant from the *s* axis of the compensator (subtractive orientation; see Fig. III-13). It is completely extinguished when the conditions in equation (III-6) or (III-7) are satisfied. When the crystal and compensator *s* axes are in the same quadrant (top orientation), the crystal appears considerably brighter than the gray background. The contrast of the crystal against the gray background disappears when the axes of the crystal coincide with PP′ and AA′.

that extinguishes or increases light from a birefringent specimen, we can then *determine the slow and fast axis directions* of the specimen.*

At this point I would like to point out that birefringence is not to be confused with *optical rotation*. Optical rotation is a phenomenon that we would encounter, for exmaple, if we observed a solution of sugar instead of our crystalline specimen (or if we observed quartz along its optical axis, the axis of symmetry). A solution of sugar has no directionality, but it can twist polarized light (in a left-handed or right-handed direction depending on its chemical nature). The result is as though the polarizer PP′ were turned; so the light can be *extinguished by turning the analyzer* AA′ by that same amount. With birefringence, you can generally not extinguish the light by turning the analyzer, but with optical rotation you can (for monochromatic light). Optical rotation reflects the three-dimensional asymmetry of the individual molecules, which themselves can be randomly oriented (as in the solution of sugar). The dispersion (variation with different wavelengths of light) is usually very substantial for optical rotation but rather small (parctically negligible) for birefringence.

III.8. MOLECULAR STRUCTURE AND BIREFRINGENCE

The next demonstration will let us correlate molecular and micellar structure with optical and mechanical anisotropy. We take a sheet of polyvinyl alcohol (PVA) produced by casting a hot water sol of the polymer onto a clean piece of glass. The basic structure of PVA is shown in Fig. III-14. In a cast (dried gel) sheet of PVA viewed normal to the surface, the polymers and their micelles would

*Quantitatively, we obtain extinction when

$$\sin\frac{\Delta}{2}\sin 2\theta = -\sin\frac{R_c}{2}\sin 2\theta_c \qquad \text{(III-6)}$$

where Δ and R_c are the retardances of the specimen and the compensator, respectively, expressed in degrees of arc ($1\lambda = 360°$), and θ and θ_c are respectively the angle between the slow axes of the specimen and of the compensator relative to OP, the polarizer transmission direction (Fig. III-13). We use this formula extensively for the photographic photometric analysis of retardance and azimuth angles of DNA microcrystalline domains in a cave cricket sperm (Figs. III-24, III-25; Inoué and Sato, 1966b).

Equation (III-6) can be simplified as follows. The specimen is generally oriented so that $\theta = 45 \pm 5°$. When, in addition, Δ and R_c are both not greater than 20°, or less than approximately $\lambda/20$, then to a very close approximation, equation (III-6) can be written:

$$\Delta = R_c \sin 2\theta_c \qquad \text{(III-7)}$$

This is the principle by which specimen retardance is measured with the *Brace–Koehler compensator* (see also Salmon and Ellis, 1976).

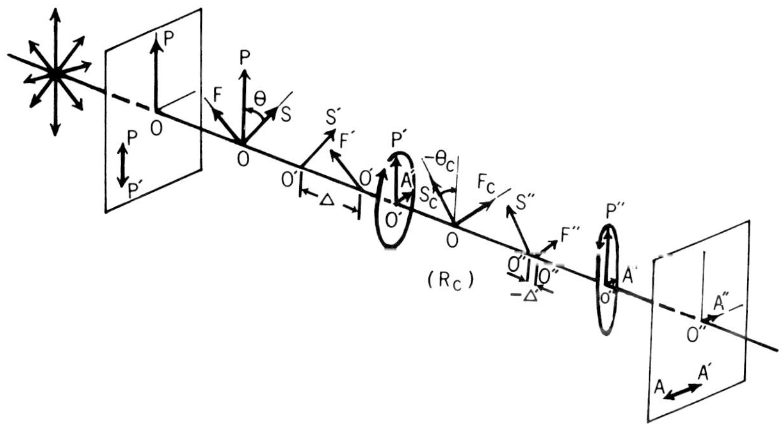

FIGURE III-13. Compensation. The left half of the figure is identical to most of Fig. III-11. However, after the light becomes elliptically polarized O′P′–O′A′) by the specimen, a compensator with a retardance, R_c, is added to the light path, with OS_c, the compensator slow axis, lying in the subtractive orientation, i.e., in the quadrant opposite to OS, the specimen slow axis. R_c is adjusted so as to restore the phase difference, Δ, between O′S′ and O′F′ that was introduced by the specimen. Then, the phase difference, −Δ′, between O″S″ and O″F″ in the waves emerging from the compensator becomes zero and O″S″ + O″F″ = O″P″. The conditions in equation (III-6) or (III-7) are now satisfied—that is, we have achieved compensation. At compensation, O″A″ → 0 and O″P″ = OP, and hence the specimen is extinguished (Fig. III-12, lower right orientation). If −Δ′ is not 0, O″A″ is also not 0, and light whose amplitude is O″A″ passes the analyzer. The sine squared relation in equation (III-9) explains why the crystal is much brighter than the background when OS and OS_c are parallel. [Substitute (Δ + R_c) for Δ in equation (III-9).]

be expected to be arranged helter-skelter as shown in Fig. III-15. Such a structure would be optically isotropic, as we can see by the lack of birefringence—no light passes the cast PVA sheet placed between crossed polars at any orientation. Such a structure should also be mechanically isotropic and difficult to tear in any direction, since covalent molecular backbones traverse in every direction. Indeed, cast sheets of PVA are very tough.

The micelles and molecular chains of the PVA sheet can be aligned by stretching a sheet that has been softened by mild heating with an electric iron. Figure III-16 shows the film before (A) and after (B) stretching. The arrangement of the micelles and molecular backbones after stretch is shown in Fig. III-17. The stretched film is highly birefringent and produces several wavelengths of

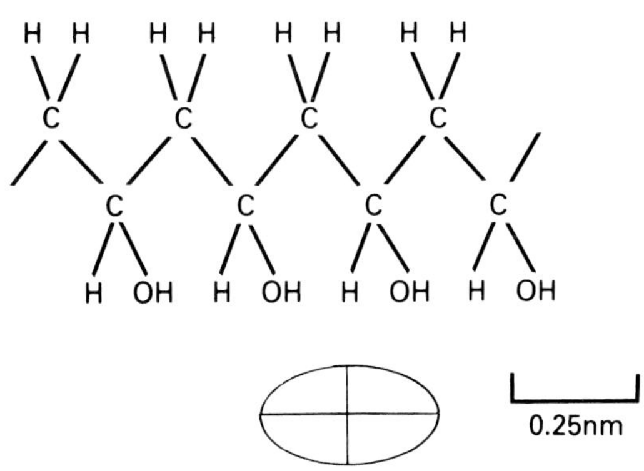

FIGURE III-14. Chemical structure of PVA. PVA is a long, linear, polymeric chain of $-(CH_2-CH\cdot OH)-$. In the absence of external constraints, rotation around the C–C bonds allows the chain to fold into random-shaped strands, except where adjacent chains lie in close proximity parallel to each other and form a minute crystalline domain or a "micelle" (Fig. III-15).

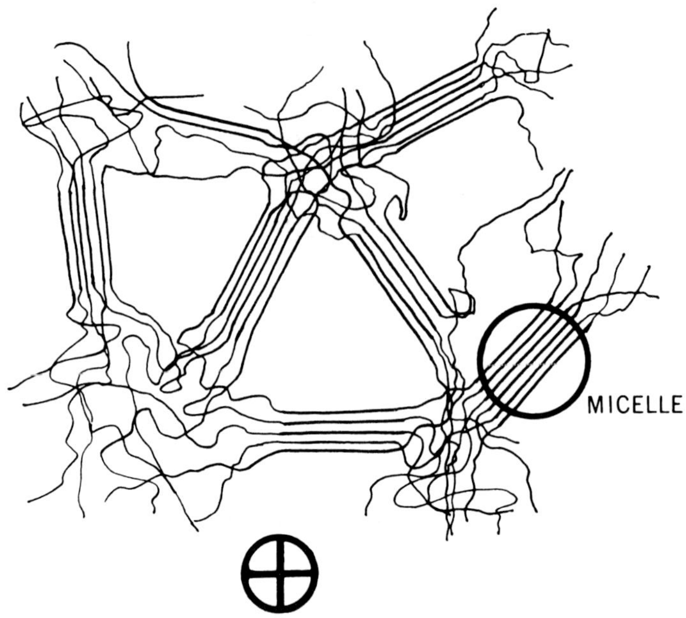

FIGURE III-15. "Fringed micelle" structure of a cast gel of PVA viewed from its face. The circle to the right indicates a micelle where several PVA molecules run parallel and form a minute crystalline region. The micelles, and the polymer chains of PVA in between, are randomly oriented. The structure is isotropic as indicated by the index ellipse at the bottom.

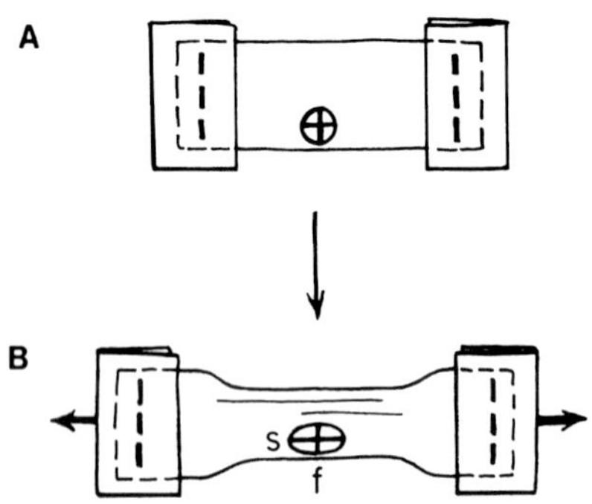

FIGURE III-16. Stretching PVA film. (A) A thin rectangular sheet of isotropic PVA is supported on both ends, taped and stapled onto pieces of cardboard. (B) The sheet is mildly heated, for example with a tacking iron in the middle, and then stretched. The stretched portion becomes optically anisotropic (birefringent) with the slow axis in the stretch direction. It is also mechanically anisotropic and tends to split parallel to the stretch direction.

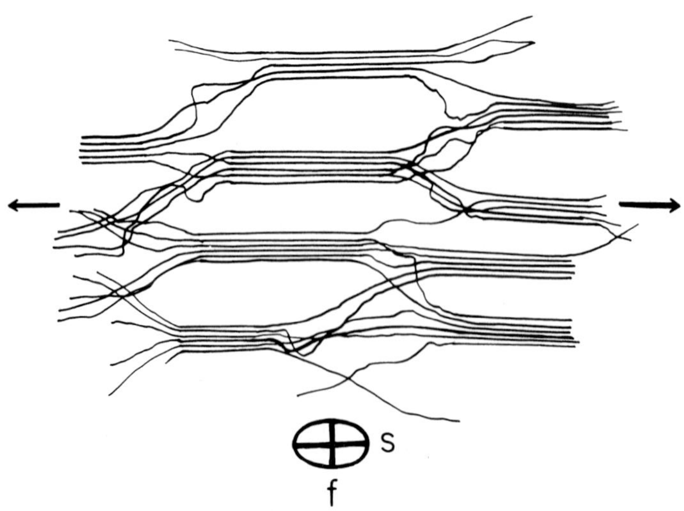

FIGURE III-17. Fringed micelles in stretched PVA. As manifested in the optical and mechanical anisotropy (Fig. III-16), the fringed micelles and the PVA molecules in general have become aligned with their long axes parallel to the stretch direction.

retardation as seen from its color* between crossed polars. In spite of the strong birefringence, light is extinguished when the film is oriented with its stretch direction parallel to the axis of the crossed polars.

Compensation reveals that it is the slow vibration axis that lies parallel to the stretch direction. This agrees with the molecular polarizability expected in PVA. In a polymer, with a covalently linked backbone and with small side groups† as in PVA, the polarizability (hence, the dielectric constant) is considerably greater parallel to the covalent chain backbone than across it.

A striking mechanical anisotropy of the stretched film of PVA also reflects the micellar and molecular arrangements. The film is even tougher than before when one attempts to tear across the stretch direction. However, in response to a rap on the tautly held film, it readily splits into strips parallel to the stretch direction as shown in Fig. III-16B.

The mechanical and optical anisotropies and the directions of their axes reflect the arrangements and anisotropic properties of the underlying molecules. In some biological samples, the molecules may take on a more complex arrangement than the homogeneous distribution seen in this PVA model. Nevertheless, the dielectric and optical properties specific to the particular molecular species provide important clues regarding the biological fine structure.

For example, in a protein molecule with extended chains in the β-form, or with a collagen helix, the slow axis lies parallel to the long axis of the polypeptide chain; the molecular polarizability is considerably greater in that direction than across the chain.

In B-form DNA, the slow axis is perpendicular to the backbone of the Watson–Crick helix. The conjugated purine and pyrimidine bases exhibit a greater UV absorbance and electrical polarizability in their plane than perpendicular to the planes. Because the base planes in B-form DNA are oriented at right angles to the molecular backbone, they show a characteristic UV negative dichroism and strong negative birefringence.

In lipids, the slow birefringence axis of the molecule lies parallel to its backbone. But since they tend to make layered sheets with the molecular axes lying perpendicular to the plane of the sheets, the layered structure also introduces *form birefringence*.‡ In form birefringence of platelets, the slow axis is parallel to the plane of the plates. The axis of symmetry is perpendicular to the plates, and hence platelet-form birefringence always has a negative sign (greater refractive index perpendicular to the axis of symmetry, or $n_e < n_o$). Lipid bilayers (or multilayers in the Schwann sheath of myelinated nerve) show a combination of intrinsic positive and form negative birefringence. Depending on the refractive index of the imbibing medium, the negative-form birefringence of the plates may become so strong as to overcome the intrinsic positive birefringence of the lipid molecules.

*See equation (III-5). In white light that is composed of various wavelengths (λ_0's), the thickness (d) of the crystal that gives rise to a full-wavelength retardation [$(n_e - n_o)d = \mathrm{N}\,\lambda_0$, where N = 1, 2, 3 . . .] varies with λ_0. Where the retardation equals N λ_0, the light exiting the crystal returns to a plane-polarized wave identical to the one before it entered the crystal. Here it is extinguished by the analyzer and that wavelength is missing from the light transmitted through the analyzer. Thus, we get white minus that wavelength, or a series of "interference colors," from strongly birefringent specimens observed between crossed polars in white light.

†As the side groups become larger, or if there are side groups with greater polarizabilities—often associated with light-absorbing *conjugated bonds*—the birefringence of the polymer becomes weaker, and can even reverse in sign. For example, the degree of nitration of nitrocellulose is monitored by observing its birefringence, which changes from positive to zero to negative with increased nitration. The birefringence of DNA is strongly negative as discussed later.

‡Form birefringence, also known as textural birefringence, arises when platelets or rodlets of submicroscopic dimensions are stacked. The platelets or rodlets must be regularly aligned, with spacings that are considerably smaller than the wavelength of light. As shown by Wiener (1912) and by Bragg and Pippard (1953), such bodies generally exhibit anisotropy of electrical polarizability and therefore of refractive index. The anisotropy is stronger, the greater the difference of refractive index between the medium lying between the rodlets or platelets and the rodlets or platelets themselves. The anisotropy disappears or becomes least strong when the refractive index of the medium matches that of the rodlet or platelet. Form birefringence is due to the shape and orientation of molecules or molecular aggregates, and is independent of the birefringence that is intrinsic to the constituent molecules themselves. In contrast to form birefringence the value of *intrinsic birefringence* usually does not change with the refractive index of the imbibing medium.

In contrast to platelet-form birefringence, form birefringence of rodlets is positive. Positive-form birefringence is observed in microtubules, which are thin, elongated structures approximately 24 nm in diameter, made up of rows of globular protein molecules approximately 5 nm in diameter, or in actin, a twisted double cable of globular protein molecules, which themselves possess low molecular asymmetry.

III.9. THE POLARIZING MICROSCOPE

All the principles discussed so far provide the basis for analyzing molecular arrangements and cellular fine structure with the polarizing microscope. A polarizing microscope is nothing more than an ordinary microscope equipped with a polarizer underneath the condenser, an analyzer above the objective, and somewhere in the path, in a convenient place between the polarizer and analyzer, a compensator (Fig. III-18). The compensator can come before or after the specimen. The specimen is supported on a rotatable stage. In principle, that is all there is to a polarizing microscope.

However, if one decides to study cytoplasmic or nuclear structure with a polarizing microscope and tries with one borrowed from a friend in crystallography or mineralogy, chances are that it will be a quite disappointing experience. Other than a few crystalline inclusions, little of interest is seen in the living cell. One reason is that the amount of light that strays through an oridinary polarizing microscope between crossed polars often cannot be reduced enough to see the weak birefringence exhibited by the filaments and membranes of interest to us.

The degree to which we can darken the field is expressed quantitatively as the *extinction factor* (EF), which is defined as

$$\mathrm{EF} = I_p/I_s \tag{III-8}$$

where I_p is the intensity of light that comes through a polarizing device when the polarizer and analyzer transmission directions are parallel, and I_s the minimum intensity that can be obtained when the polarizer and analyzer are crossed. For an ordinary polarizing microscope, the EF is often of the order of 10^3 or even lower with high-NA lenses. *What we need for cell study is an EF of at least 10^4.*

In order to achieve this high an EF, we must attain a number of conditions concurrently. The polarizer and analyzer themselves obviously must be of good quality. They can be of selected sheet Polaroids, as far as the extinction goes (but their transmittance is low, and the rippled surface can deteriorate the image). We cannot use optical elements between the polars that are strained. As seen by stressing a piece of glass or plastic between crossed polars, the strain introduces birefringence. This must not take place because that birefringence can be much higher than the specimen birefringence. So we must use lenses, slides, and coverslips that are free from strain. A little speck of airborne lint (e.g., from our clothing or lens tissue) can be highly birefringent, so we must keep our system meticulously clean. Also, since almost all the light is extinguished by the analyzer, the light source must be very bright (yet harmless to our specimen), and we should work in a darkened area to improve the sensitivity of our eye. Alignment of the optical components is critical, and Koehler illumination must be used to gain minimum stray light coupled with maximum brightness of the field.

The reason for all of this care is that the light that comes from the specimen is only a minute fraction of the original light. If Δ is the retardance of the specimen and θ its azimuth orientation, both expressed in degrees, the luminance (I) due to specimen birefringence (Δ) is given by

$$I = I_p \sin^2 \frac{\Delta}{2} \tag{III-9}$$

where I_p is the luminance of the field with the polarizer and analyzer transmission parallel. I turns out to be of the order of 10^{-4} to $10^{-6} \times I_p$ for the range of specimen retardations of interest to us. So we are trying to see very dim light from the specimen—as though we were trying to see starlight

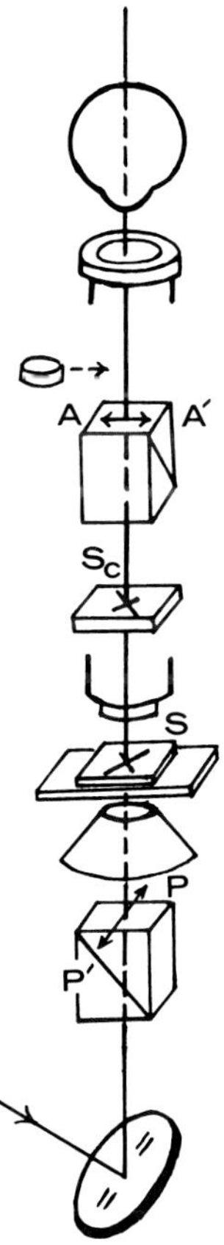

FIGURE III-18. Arrangement of conventional polarizing microscope. The polarizer was commonly oriented with its transmission axis PP′ north–south to the observer. AA′ was oriented east–west.* The Bertrand lens converts the ocular into a telescope and allows one to view an image of the objective lens back aperture. This provides a "conoscopic" image of the specimen, showing the interference pattern of convergent polarized light passing through the specimen at different angles. The Fraunhofer diffraction pattern produced by the specimen is also seen there. With the Bertrand lens out of the light path, the polarizing microscope gives a regular "orthoscopic" image.

during the daytime. The light is there, but there is so much more light around our specimen that we cannot see it. We have to somehow make the background dark enough so we can see the object. We must increase the EF and improve the specimen contrast.

The situation is improved by good use of a compensator. This point can be demonstrated by placing, between crossed polars, a weakly birefringent specimen that we can just barely see. The birefringent region is barely brighter than the background (Fig. III-10). When we introduce the compensator, the background is now gray and the specimen is darker or much brighter than the background (Fig. III-12). In the presence of the compensator, the image brightness and the contrast can be vastly improved as discussed in Chapter 11 (Fig. 11-4B), and images of weakly birefringent objects can now be displayed clearly (Figs. III-19, III-20, III-25).

III.10. RECTIFICATION

There is yet one more improvement required for detecting weakly birefringent specimens at high resolution in polarization microscopy. Following equation (5-9), we require high objective *and* condenser NA to obtain high resolving power. But *each time the NA is increased by 0.2, stray light increases tenfold even when we use strain-free lenses.* With objective and condenser NAs at 1.25, the EF may drop to nearly 10^2. This lowered extinction (increased stray light) at high NA is due to the rotation of the plane of polarization at *every* oblique-incidence interface between the polarizer and the analyzer. It is an inherent physical optical phenomenon that had been considered uncorrectable until we developed the polarization rectifier (Figs. III-21; Inoué and Hyde, 1957; see also Huxley, 1960). With the rectifier the extinction is very good up to high NAs, and we are finally able

*More recently, the orientations of PP′ and AA′ have been reversed in most commercial instruments.

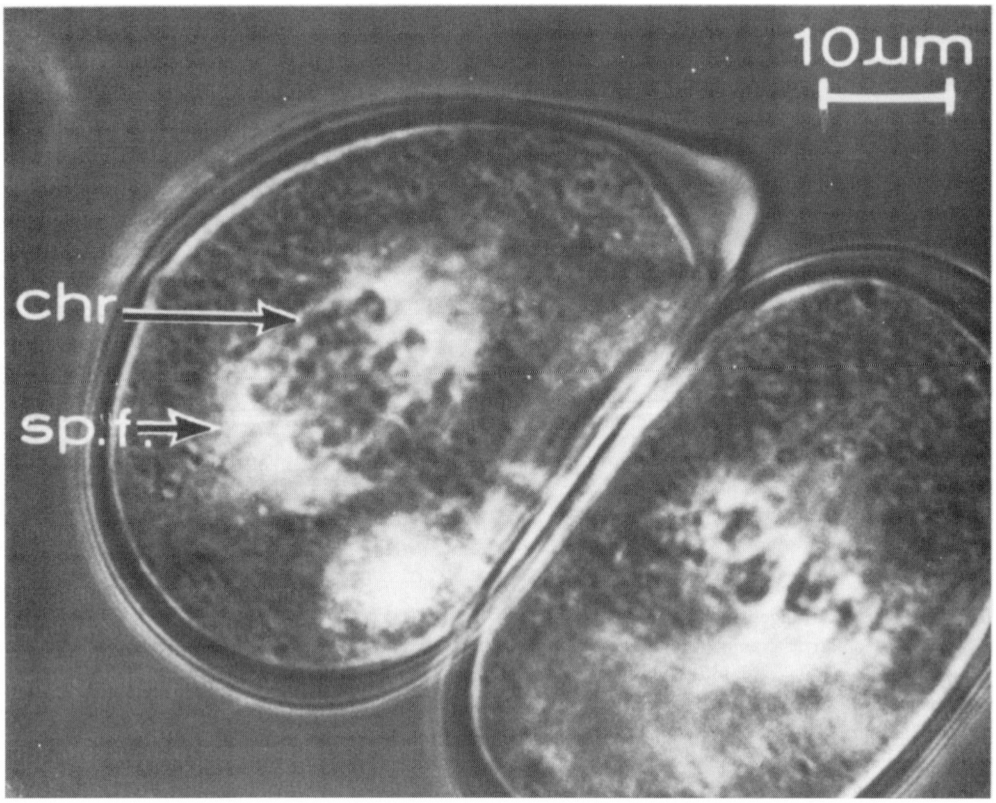

FIGURE III-19. Living pollen mother cell of Easter lily (*Lilium longiflorum*). Spindle fibers (sp.f.) show stronger birefringence adjacent to kinetochores (spindle fiber attachment point) of helical chromosomes (chr) which show little birefringence. The maximum retardance of the spindle is less than λ/100. (From Inoué, 1953.)

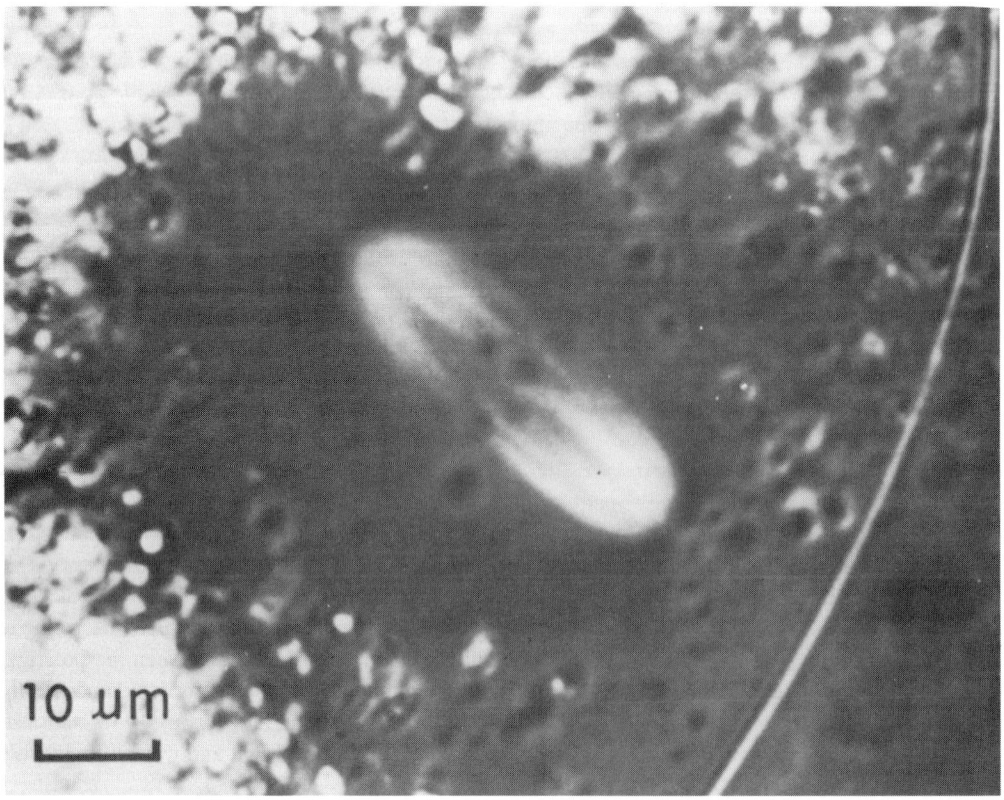

FIGURE III-20. Birefringent spindle fibers in living oocyte of marine worm (*Chaetopterus pergamentaceus*). Chromosomal spindle fibers, white; astral rays, oriented at right angles to spindle axis, dark. The maximum retardance of these spindle fibers is about 3.5 nm. (From Inoué, 1953.)

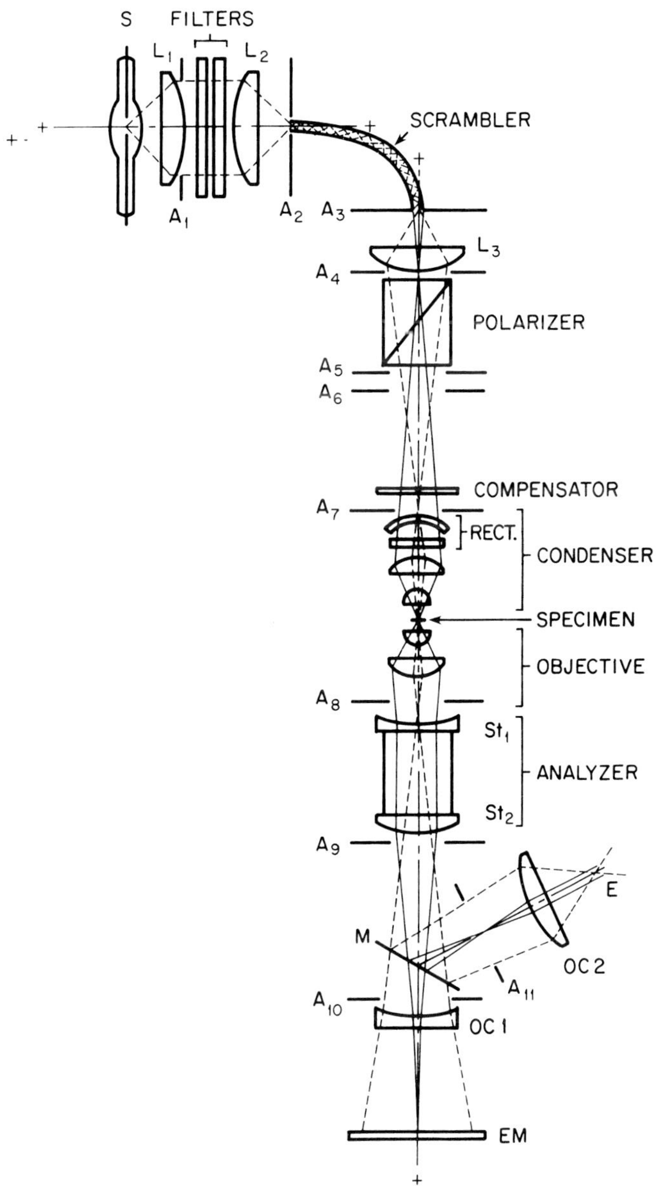

FIGURE III-21. Schematic optical path of transilluminating, universal polarizing microscope designed to provide maximum sensitivity and superior image quality. Inverted system with light source (S) on top and detectors (EM and E) at the bottom. Light from a high-pressure arc lamp is filtered (to remove infrared and provide monochromatic illumination) and focused by L_1 and L_2 onto a fiber-optic light scrambler at A_2. The fiber scrambles the image of concentrated mercury arc (footnote p. 127) and provides a uniform circular patch of light that acts as the effective light source at A_3. This source, projected by the zoom lens (L_3), is made to just fill the condenser aperture diaphragm (A_7). Illuminance of the field can be regulated without affecting the cone angle of illumination (or disturbing the color temperature in white light) by adjusting the iris (A_1). The polarizing Glan–Thompson prism is placed behind stop A_5 away from the condenser to prevent light scattered by the polarizer from entering the condenser. Half-shade and other special plates are placed at level A_6, and compensators above the condenser. Depolarization by rotation of polarized light in the condenser and objective lenses, slide, and coverslips are corrected by the rectifier (RECT.). The image of the field diaphragm (A_4 or A_6) is focused on the specimen plane by the condenser, whose NA can be made equal to that of the objective. Stigmatizing lenses (St_1, St_2), which minimize the astigmatism that is otherwise introduced by the calcite analyzer, are low-reflection coated on their exterior face and cemented directly onto the analyzing Glan–Thompson prism to protect the delicate surfaces of the calcite prism. Stops (A_1–A_{11}), placed at critical points, minimize scattered light from entering the image-forming system. The final image is directed by OC_1 onto a photographic, video, or other sensor (e.g., photomultiplier; EM), or to the eye (E) via mirror (M) and ocular (OC_2). Components A_3 through EM are aligned on a single optical axis to minimize degradation of the image.

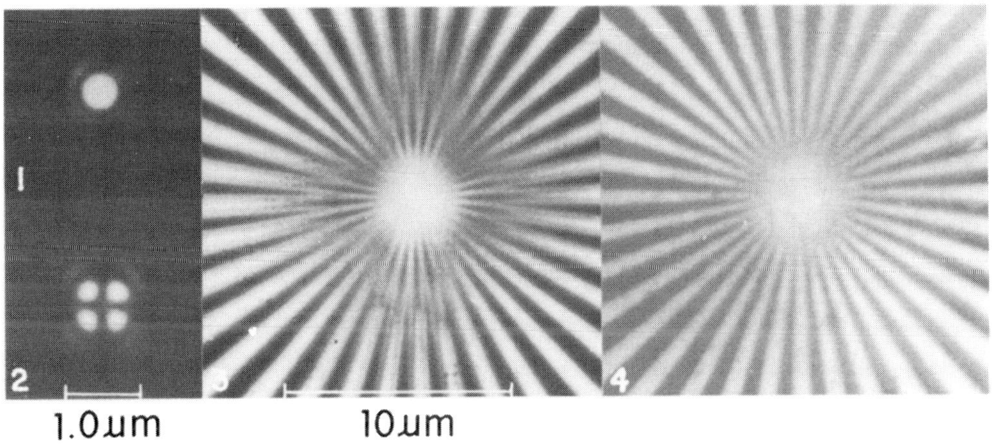

FIGURE III-23. The effect of rectification on image reliability in the polarizing microscope. (1) The Airy disk diffraction pattern of a pinhole viewed through crossed polars with a rectified polarizing microscope of $NA_{obj} = NA_{cond} = 1.25$. The pattern is identical with that obtained with a microscope in nonpolarized light. (2) The same as 1, but with nonrectified lenses between crossed polars. The diffraction pattern is anomalous; each bright point in the specimen is represented in the image by a bright four-leaf clover pattern surrounding a dark cross. (3) A Siemens Test Star pattern viewed between crossed polars in the absence of rectification. Contrast is reversed and spurious toward the center of the image. (4) The same test pattern viewed with the same optics after rectification. The aperture function of the lens is now uniform and the diffraction anomaly has disappeared. (From Inoué and Kubota, 1958.)

FIGURE III-22. Rectified, universal polarizing microscope designed by author, G. W. Ellis, and Ed Horn (see also footnote on p. 145). Optical layout as shown schematically in Fig. III-21. The components, in eight units, are mounted on slides that ride on a 1/10,000th-inch-precision, 3-inch-wide dovetail, (4V + 1H) feet long. The well-aged cast iron (Mehanite) dovetail bench is mounted on a horizontal axis and can be used vertically, horizontally, or at angles in between. Two smaller dovetails, built into the same casting. run precisely parallel to the central dovetail and provide added flexibility (micromanipulators, UV microbeam source, etc. are mounted on these dovetails). Supported on the sturdily built wooden bench, by the horizontal axis near the center of gravity of the massive optical bench, and designed for semikinematic support of the components wherever practical, the microscope is quite immune to vibration.

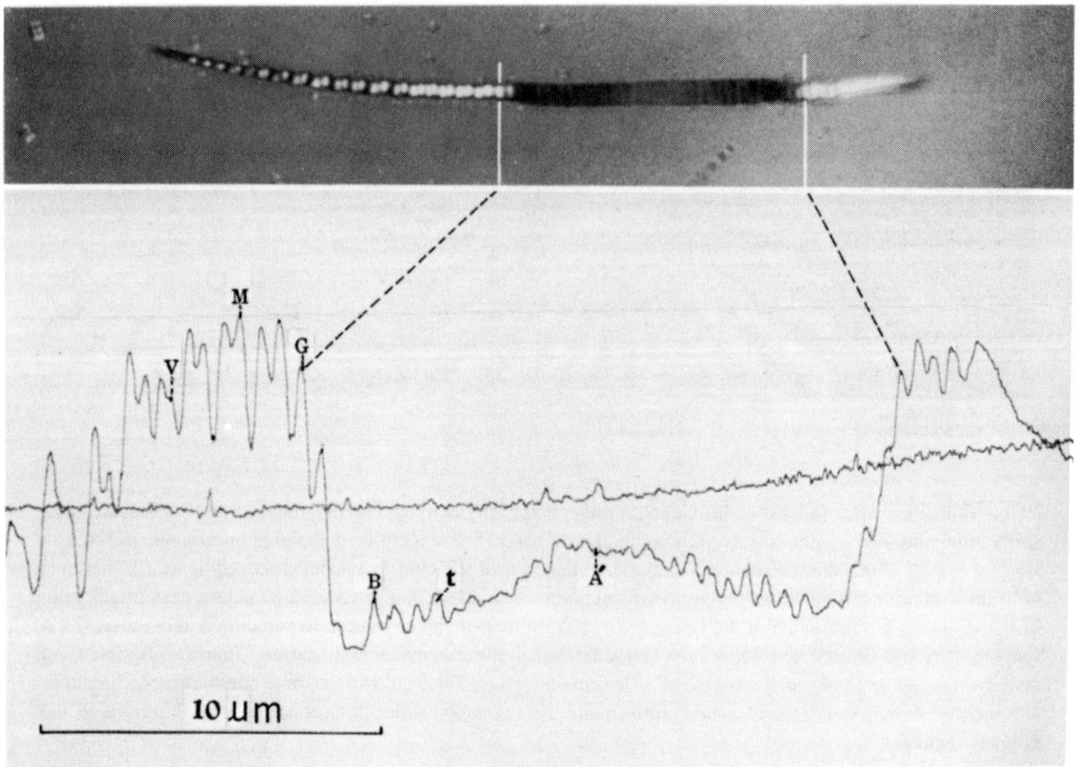

FIGURE III-24. Photograph of living cave cricket sperm in rectified polarizing microscope, with densitometer trace. The broad dark band in the middle of the sperm (where the densitometer trace dips below background) is where the birefringence of the DNA bases were selectively perturbed by irradiation with polarized UV microbeam. The compensator is in the subtractive orientation so that the sperm head is darker than the background except where the specimen retardation is greater than the compensator (especially at the small dumbbell-shaped white patches, which give rise to the M shapes on the densitometer trace). In these regions, the birefringence and optical axes of the "microcrystalline domains" of the sperm DNA are such as to overcome the subtractive effect of the compensator. (From Inoué and Sato, 1966b.)

to use the polarizing microscope to study weakly retarding specimens at the theoretical limit of microscopic resolution. Diffraction image errors, which can be present without rectification, are also corrected with the rectifier (Fig. III-23).*

*As shown in Fig. III-23, rectification corrects for the diffraction error that is introduced by conventional lenses when we observe low-retardation objects between crossed polars. However, the sensitivity for detecting weak retardations at high resolution is so improved by rectification that another hitherto unnoticed optical phenomenon becomes apparent. At the edges of any specimen, including isotropic materials, light is diffracted as though each edge were covered with a double layer of extremely thin, birefringent material; the slow axis on the high index side lying parallel, and on the low index side lying perpendicular, to the edge. We have named this phenomenon *edge birefringence*. It is found at all sharp boundaries (edges) whether the two sides of the boundary are solid, liquid, or gas, so long as there exists a refractive index difference on the two sides of the boundary (e.g., see Fig. 1-7D,F). It is clearly not based on the presence of a membrane at the optical interface, but is a basic diffraction phenomenon taking place at every edge. The electric vectors parallel and perpendicular to the edges must contribute to diffraction in a slightly asymmetric way on both sides of the edge and give rise to edge birefringence. Edge birefringence disappears when the refractive indices on both sides of the boundary are matched. It reverses in sign when the relative magnitude of refractive indices on the two sides of the boundary are reversed.

This behavior of edge birefringence is distinct from form birefringence. Form birefringence becomes zero (reaches a minimum value for rodlets, or maximum absolute value for platelets) when the refractive index of the immersion medium matches that of the rodlets or platelets, but then rises again (parabolically) when the refractive index of the immersion medium exceeds the match point (Ambronn and Frey, 1926; Sato *et al.*, 1975).

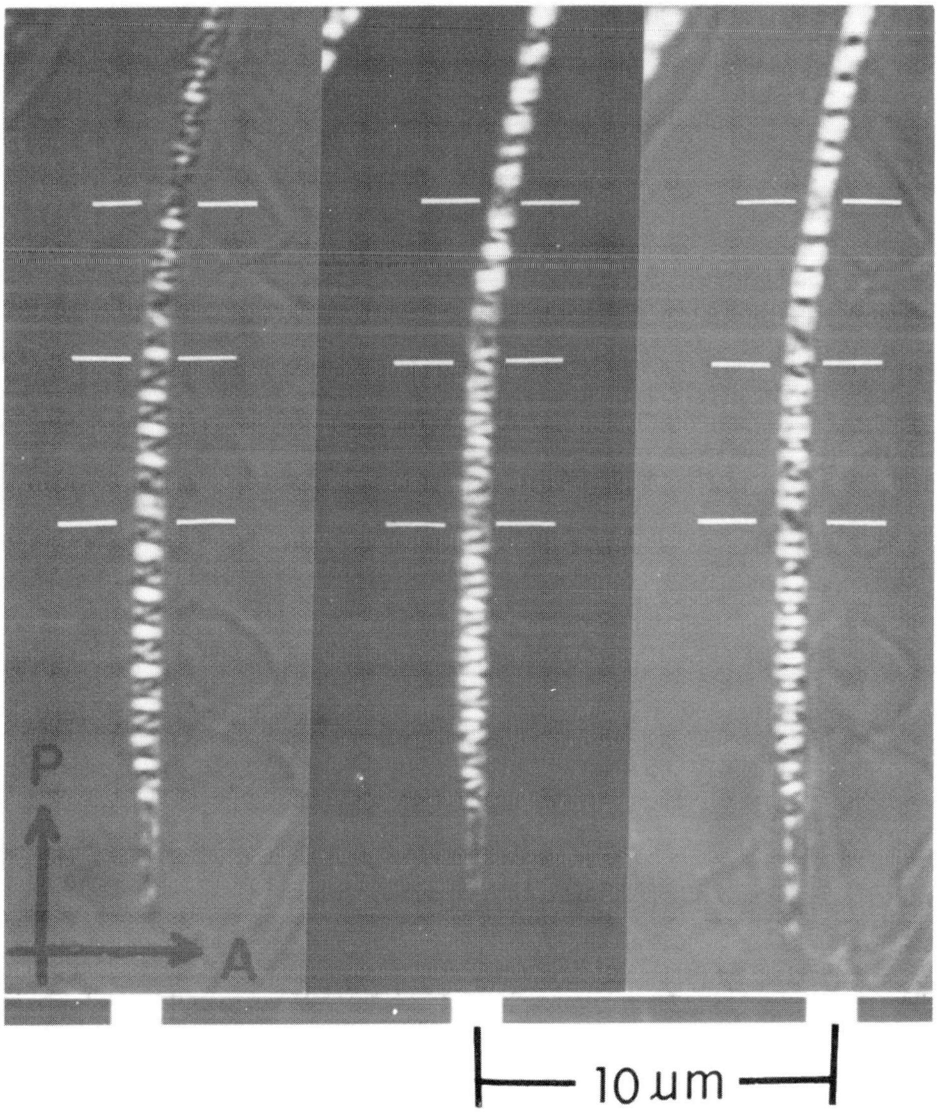

FIGURE III-25. Sperm head of a cave cricket (*Ceuthophilus nigricans*) observed with the rectified polarizing microscope at three different settings of a mica compensator. The detailed distribution of intrinsic birefringence in these chromosomes is shown with great clarity in the specimen immersed in dimethylsulfoxide ($N_D{}^{20}$ = 1.475). Horizontal white bars: positions of helix "breaks" that correspond to the ends of chromosomes. (From Inoué and Sato, 1966b.) Prior to these studies, few chromosomes had been seen in mature sperm of any species.

The improvements achieved with the rectified polarizing microscope have made possible a determination of the complex alignment of DNA molecules in each unit diffraction area of a live insect sperm (Figs. III-24, III-25). The birefringence and azimuth orientation of each DNA micro-domain were measured (both to a precision of 0.1°) from microdensitometer traces (of the type shown in Fig. III-24) by correlating the traces taken at several compensator settings and stage orientations. The changes in birefringence and azimuth angles, following polarized UV microbeam irradiation (which selectively abolishes the birefringence contribution by those bases that absorbed the UV), reveal the helical packing arrangement of the DNA molecules within the microdomains (Inoué and Sato, 1966b).

The rectified polarizing microscope offers many unique opportunities for studying molecular organization, and its changes, in a single living cell undergoing physiological activities and developmental changes.

ANNOTATED BIBLIOGRAPHY*

PHYSICAL OPTICS
- General introductory texts *23,* 119, 223, 264
- Electromagnetic waves 9, *23,* 43, 253
- Polarized light 84, 120, *121, 178,* 182, 208, 216a, 217, 246, 263, 264
- Crystal optics 27, 52, *79,* 80, 144, *178,* 251, 263

MICROSCOPY
- Wave optics and diffraction 66, 104, 129, 147, 152a, 218, *266*
- *Practices of microscopy* 16, *33,* 78, 152a, *216*
- UV microbeam techniques 58, 219, 247

POLARIZING MICROSCOPE
- General 21, 76, *79,* 122, 175, 183, 208, *265*
- Polarizers 75, 122, 136, 137, 138, 142, 216a, 256
- Compensators, general 4, 41, 70, 82, 84, *120,* 121, 127, 175, 182, 183, 193, 216a, *217,* 227, 235, 246
- Compensators, special application, etc. 3, 4, 18, 24, 45, 83, 87, 100, 103, *199, 227*
- Limits and refinements 4, 5, 90, 93, 100, 103, 108, *109,* 227, 265; see also Section 11.3
- Rectification 86, 93, *101,* 104, *129*

MOLECULAR STRUCTURE AND PHYSICAL PROPERTIES
- General 47, 79, 118, 249
- Orbital electrons and optical properties 177, 213
- Polymer structure and physical properties, 71, 128, 130, 146, 181, 245, 255
- Liquid crystals 22, *39,* 74
- Birefringence, intrinsic and general 31, 33, *79, 202,* 248
  - Flow 14, 32, 125, 229, 229a, 240, 250
  - Form *6,* 17, *25,* 32, 161, 195, *199,* 241 (especially see footnote on p. 417 for correction of error), 258, 259
  - Stress/strain 6, 26, 27, 128, 130, *202,* 255
  - Electric 125, 126, 229a
- Dichroism, visible 8, 77a, 138, 144, 166, 169, 185, 206, 247a, 256
  - Infrared and ultraviolet 7, 81, 107, 108, 109, 140, 142, 166, 168, 185, 214, 215, 224
  - Form and flow 69, 81, 268
- Optical Rotation 257

BIOLOGICAL FINE STRUCTURE AND MOLECULAR ORGANIZATION
- General texts and articles 2, 6, 45a, 53, 68, 72, 72a, 79, 98, 111, 152, 164, 165, 171, 181, 184a, 185, *202,* 208, 210, 220, 243a, 248, 261a, 262
- Biocrystals *105,* 162, 163, *201,* 209

MITOSIS
- General review on mitosis 15, 59, *97, 148,* 149, 150, *160,* 176a, *212, 261*
- Spindle birefringence and chromosome movement 19a, 19b, 42, 56, 57, 60, 83, 85, 91, 92 (time-lapse movie), *94,* 95, 99, 100, 106, *110,* 115, 116, 124, 131, 132, 133, 134, 145, 174, 184, 186, 196, 202, 203, 225, 226, 244, 254
- Experimental alterations of spindle birefringence
  - Temperature 30, 89, 94, 95, 96, 113, *114,* 191, 197, *221,* 222
  - Pressure 62, 114, 187, 188, *189,* 190, *192*
  - Mechanical deformation *19a,* 19b, 48, 88, 100, 110, *160,* 160a, 170
  - UV microbeam 29, *56,* 57, 58, 94, 95, 116b, 116c, 138a, 138b, 247, 266a
  - Ca^{2+} 102, *123a*
  - D_2O, etc. 28, 30, 110, 116a, 180, 199a
  - Colchicine, etc. *13,* 83a, *88,* 143, *194,* 217a, 217b, 228, *239*
  - *Antimetabolites* 110, 200
- Isolated spindle 28, 29a, 61, 63, 64, 73, 112, 148a, 179, 180a, 195, 199, 254
- Form birefringence 174, 179, 195, *199*

BIREFRINGENCE AND DICHROISM IN LIVING CELLS
- General review 50, 68, 166, *202, 205,* 207, 209, 210
- Ground cytoplasm and endoplasmic reticulum 1, 2, 65, 123, 157, 158, 159, 171, *202,* 218a, 233, 234, *237,* 238
- Cell membrane and cortex 17, 40, 153, 154, 155, 157, 158, *202,* 207 209, 211, 218a, 228

*Italicized reference numbers indicate material that may be of special interest to the reader.

BIBLIOGRAPHY

1. Allen, R. D. (1972) Pattern of birefringence in the giant amoeba, *Chaos carolinensis*. *Exp. Cell Res.* **72**:34–45.
2. Allen, R. D., and N. Kamiya, eds. (1964) *Primitive Motile Systems in Cell Biology*. Academic Press, New York.
3. Allen, R. D., and L. I. Rebhun (1962) Photoelectric measurement of small fluctuating retardations in weakly birefringent, light-scattering biological objects. I. The revolving tilted compensator method. *Exp. Cell Res.* **29**:583–592.
4. Allen, R. D., J. Brault, and R. D. Moore (1963) A new method of polarizing microscopic analysis. I. Scanning with a birefringence detector system. *J. Cell Biol.* **18**:223–235.
5. Allen, R. D., J. W. Brault, and R. Zeh (1966) Image contrast and phase modulation light methods in polarization and interference microscopy. *Adv. Opt. Electron Microsc.* **1**:77–114.
6. Ambronn, H., and A. Frey, (1926) *Das Polarisationsmikroskop, seine Anwedung in der Kolloidforschung in der Farbarei*. Akademische Verlag, Leipzig.
7. Ambrose, J., A. Elliot, and R. B. Temple (1949) New evidence on the structure of some proteins from measurements with polarized infra-red radiation. *Nature* **163**:859–862.
8. Anderson, S. (1949) Orientation of methylene blue molecules absorbed on solids. *J. Opt. Soc. Am.* **39**:49–56.
9. Andrews, C. L. (1960) *Optics of the Electromagnetic Spectrum*. Prentice–Hall, Englewood Cliffs, N.J.
10. Aronson, J. F. (1961) Sarcomere size in developing muscles of a tarsonemid mite. *J. Biophys. Biochem. Cytol.* **11**:147–156.
11. Aronson, J. F. (1963) Observations on the variations in size of the A regions of arthropod muscle. *J. Cell Biol.* **19**:359–367.
12. Aronson, J. F. (1967) Polarized light observations on striated muscle contraction in a mite. *J. Cell Biol.* **32**:169–179.
13. Aronson, J., and S. Inoué (1970) Reversal by light of the action of N-methyl N-desacetyl colchicine on mitosis. *J. Cell Biol.* **45**:470–477.
14. Asakura, S., M. Kasai, and F. Oosawa (1960) The effect of temperature on the equilibrium state of actin solutions. *J. Poly. Sci.* **44**:35–49.
15. Bajer, A., and J. Molé-Bajer (1972) Spindle dynamics and chromosome movements. *Intntl. Rev. Cytol.* suppl. **3**:1–271.
16. Barer, R., and V. E. Cosslett, eds. (1966) *Advances in Optical and Electron Microscopy*, Volume 1. Academic Press, New York, pp. 77–114.
17. Bear, R. S., and F. O. Schmitt (1936) Optical properties of the axon sheaths of crustacean nerves. *J. Cell Physiol.* **9**:275–288.
18. Bear, R. S., and F. O. Schmitt (1936). The measurement of small retardation with the polarizing microscope. *J. Opt. Soc. Am.* **26**:363–364.
19. Bear, R. S., F. O. Schmitt, and J. Z. Young (1937) The sheath components of the giant nerve fibres of the squid. *Proc. R. Soc. London* **123**:496.

19a. Begg, D. A., and G. W. Ellis (1979) Micromanipulation studies of chromosome movement. I. Chromosome–spindle attachment and the mechanical properties of chromosomal spindle fibers. *J. Cell Biol.* **82:**528–541.
19b. Begg, D. A., and G. W. Ellis (1979) Micromanipulation studies of chromosome movement. II. Birefringent chromosomal fibers and the mechanical attachment of chromosomes to the spindle. *J. Cell Biol.* **82:**542–554.
20. Bendet, I. J., and J. Bearden, Jr. (1972) Birefringence of spermatozoa. II. Form birefringence of bull sperm. *J. Cell Biol.* **55:**501–510.
21. Bennett, H. S. (1950) The microscopical investigation of biological materials with polarized light. In: *Handbook of Microscopical Technique* (C. E. McClung, ed.). Harper & Row (Hoeber), New York, pp. 591–677.
22. Bernal, J. D., and I. Fankuchen (1941) X-ray and crystallographic studies of plant virus preparations. I. Introduction and preparation of specimens. II. Modes of aggregation of the virus particles. *J. Gen. Physiol.* **25:**111–146 and 4 plates.
23. Born, M., and E. Wolf (1980) *Principles of Optics,* 6th ed. Pergamon Press, Elmsford, N.Y.
24. Bozler, E. (1937) The birefringence of muscle and its variation during contraction. *J. Cell. Comp. Physiol.* **10:**165–182.
25. Bragg, W. L., and A. B. Pippard (1953) The form birefringence of macromolecules. *Acta Crystallogr. Sect. B* **6:**865–867.
26. Brewster, D. (1916) On new properties of heat, as exhibited in its propagation along plates of glass. *Philos. Trans.* **106:**46–114 and Plates II–V.
27. Brewster, D. (1818) On the laws of polarization and double refraction in regularly crystallized bodies. *Philos. Trans.* **108:**199–273 and Plates XV–XVI.
28. Bryan, J., and H. Sato (1970) The isolation of the meiosis I spindle from the mature oocyte of *Pisaster ochraceus. Exp. Cell Res.* **59:**371–378.
29. Campbell, R. D., and S. Inoué (1965) Reorganization of spindle components following UV microirradiation. *Biol. Bull.* **129:**401.
29a. Cande, W. Z., and K. L. McDonald (1985) *In vitro* reactivation of anaphase spindle elongation using isolated diatom spindles. *Nature* **316**:168–170.
30. Carolan, R. M., H. Sato, and S. Inoué (1965) A thermodynamic analysis of the effect of D_2O and H_2O on the mitotic spindle. *Biol. Bull.* **129:**402; **131:**385.
31. Cassim, J. Y., and E. W. Taylor (1965) Intrinsic birefringence of poly-γ-benzyl-L-glutamate, a helical polypeptide, and the theory of birefringence. *Biophys. J.* **5:**531–551.
32. Cassim, J. Y., P. S. Tobias, and E. W. Taylor (1968) Birefringence of muscle proteins and the problem of structural birefringence. *Biochim. Biophys. Acta* **168:**463–471.
33. Chamot, E. M., and C. W. Mason (1958) *Handbook of Chemical Microscopy,* 2nd ed. Wiley, New York.
34. Chinn, P., and F. O. Schmitt (1937) On the birefringence of nerve sheaths as studied in cross section. *J. Cell. Comp. Physiol.* **9:**289–296.
35. Cohen, L. B., B. Hille, and R. D. Keynes (1969) Light scattering and birefringence changes during activity in the electric organ of *Electrophorus electricus. J. Physiol. (London)* **203:**489–509.
36. Cohen, L. B., and R. D. Keynes (1969) Optical changes in the voltage-clamped squid axon. *J. Physiol. (London)* **204:**100–101.
37. Cohen, L. B., R. D. Keynes, and B. Hille (1968) Light scattering and birefringence changes during nerve activity. *Nature* **218:**438–441.
38. Colby, R. H. (1971) Intrinsic birefringence of glycerinated myofibrils. *J. Cell Biol.* **51:**763–771.
39. Configurations and Interactions of Macro-Molecules and Liquid Crystals (1958) *Discuss. Faraday Soc.* **25:**1–235.
40. Dan, K., and K. Okazaki (1951) Change in the birefringence of the cortical layer of sea-urchin eggs induced by stretching. *J. Cell. Comp. Physiol.* **38:**427–435.
41. deSénarmont, H. (1840) Sur les modifications que la réflexion spéculaire à la surface des corps métalliques imprime à un rayon de lumière polarisée. *Ann. Chim. Phys.* **73**(Ser. 2)**:**337–362.
42. Dietz, R. (1963) Polarisationsmikroskopische Befunde zur chromosomeninduzierten Spindelbildung bei der Tipulide. *Zool. Anz. Suppl.* **26:**131–138.
43. Ditchburn, R. W. (1963) *Light.* Blackie, London.
44. Dreizel, P. B., and A. Pfleidener (1959) Histochemische und Polarisationsoptische untersuchungen am Amyloid. *Arch. Pathol. Anat. Pathol. Physiol.* **332:**552.
45. Dvorak, J. A., T. R. Clem, and W. F. Stotler (1972) The design and construction of a computer-compatible system to measure and record optical retardation with a polarizing or interference microscope. *J. Microsc.* **96:**109.

45a. Dustin, P. (1978) *Microtubules*. Springer-Verlag, New York.
46. Eberstein, A., and A. Rosenfalck (1963) Birefringence of isolated muscle fibres in twitch and tetanus. *Acta Physiol. scand.* **57**:144–166.
47. Eisenberg, D., and W. Kauzmann (1969) *The Structure and Properties of Water*. Oxford University Press (Clarendon), London.
48. Ellis, G. W., and D. A. Begg (1981) Chromosome micromanipulation studies. In: *Mitosis/Cytokinesis*. (A. M. Zimmerman and A. Forer eds.). Academic Press, New York, pp. 155–179.
49. Engelmann, T. W. (1873) Mikroskopische Untersuchungen über die quergestreifte Muskelsubstanz. I, II. *Pfluegers Arch.* **7**:33–71, 155–188.
50. Engelmann, T. W. (1875) Contractilitat and Doppelbrechung. *Pfluegers Arch.* **11**:432–464.
51. Engelmann, T. W. (1906) Zur Theorie der Contractilitat. *Sitzungsber. K. Preuss. Akad. Wiss.* **39**:694–724.
52. Evans, R. C. (1976) *An Introduction to Crystal Chemistry*. Cambridge University Press, London.
53. Fauré-Fremiet, E. (1970) Microtubules et Mecanmismes Morphopoietiques. *Ann. Biol.* **9**:1–61.
54. Fischer, E. (1936) The submicroscopic structure of muscle and its changes during contraction and stretch. *Cold Spring Harbor Symp. Quant. Biol.* **4**:214–221.
55. Fischer, E. (1938) The birefringence of smooth muscle (*Phascolosoma* and *Thyone*) as related to muscle length, tension and tone. *J. Cell. Comp. Physiol.* **12**:85–101.
56. Forer, A. (1965) .Local reduction in spindle fiber birefringence in living *Nephrotoma suturalis* (Loew) spermatocytes induced by ultra-violet microbeam irradiation. *J. Cell Biol.* **25**(Mitosis Suppl.):95–117.
57. Forer, A. (1966) Characterization of the mitotic traction system, and evidence that birefringent spindle fibers neither produce nor transmit force for chromosome movement. *Chromosoma* **19**:44–98.
58. Forer, A. (1966) A simple conversion of reflecting lenses into phase-contrast condensers for ultraviolet light irradiations. *Exp. Cell Res.* **43**:688–691.
59. Forer, A. (1969) Chromosome movements during cell division. In: *Handbook of Molecular Cytology* (A. Lima-de Faria, ed.). North-Holland, Amsterdam, pp. 553–601.
60. Forer, A. (1976) Actin filaments and birefringent spindle fibers during chromosome movement. In: *Cell Motility,* Volume 3 (R. D. Goldman, T. D. Pollard, and J. L. Rosenbaum, eds.). Cold Spring Harbor Laboratory, Cold Spring Harbor, N.Y., pp. 1273–1293.
61. Forer, A., and R. D. Goldman (1972) The concentrations of dry matter in mitotic apparatuses in vivo and after isolation from sea-urchin zygotes. *J. Cell Sci.* **10**:387–418.
62. Forer, A., and A. M. Zimmerman (1976) Spindle birefringence of isolated mitotic apparatus analysed by pressure treatment. *J. Cell Sci.* **20**:309–327.
63. Forer, A., and A. M. Zimmerman (1976) Spindle birefringence of isolated mitotic apparatus analyzed by treatments with cold, pressure, and diluted isolation medium. *J. Cell Sci.* **20**:329–339.
64. Forer, A., V. I. Kalnins, and A. M. Zimmerman (1976) Spindle birefringence and isolated mitotic apparatus: Further evidence for two birefringent spindle components. *J. Cell Sci.* **22**:115–131.
65. Francis, D. W., and R. D. Allen (1971) Induced birefringence as evidence of endoplasmic viscoelasticity in *Chaos carolinensis*. *J. Mechanochem. Cell Motil.* **1**:1–6.
66. Françon, M. (1961) *Progress in Microscopy*. Row, Peterson, Evanston, Ill.
67. Frey-Wyssling, A. (1943) Doppelbrechung und Dichroismus als Mass der Nukleinsaure-Orientierung in Chromosomen. *Chromosoma* **2**:473–481.
68. Frey-Wyssling, A. (1953) *Submicroscopic Morphology of Protoplasm*. Elsevier, Amsterdam.
69. Frey-Wyssling, A. (1964) Quantitative Bestimmung des Formdichroismus. *Z. Wiss. Mikrosk.* **66**:45–53.
70. Gahm, J. (1964) Quantitative polarisationsoptische Messungen mit Kompensatoren. *Zeiss Mitt. Fortschr. Tech. Opt.* **3**:153–192.
71. Geil, R. H. (1963) *Polymer Single Crystals*. Interscience, New York.
72. Gibbons, I. R. (1967) The organization of cilia and flagella. In: *Molecular Organization and Biological Function* (J. Allen, ed.). Harper & Row, New York, pp. 211–237.
72a. Gibbons, I. R. (1975) The molecular basis of flagellar motility in sea urchin spermatozoa. In: *Molecules and Cell Movement* (S. Inoué and R. E. Stephens, eds.). Raven Press, New York, pp. 207–232.
73. Goldman, R. D., and L. I. Rebhun (1969) The structure and some properties of the isolated mitotic apparatus. *J. Cell Sci.* **4**:179–209.
74. Gray, G. W. (1962) *Molecular Structure and the Properties of Liquid Crystals*. Academic Press, New York.
75. Haase, M. (1961) Optische Eigenschaften neuerer Polarisationsfilter. *Zeiss Mitt. Fortschr. tech. Opt.* **2**:173–181.
76. Hallimond, A. F. (1953) *Manual of the Polarizing Microscope*. Cooke, Troughton & Simms, York.
77. Hárosi, F. (1975) Linear dichroism of rods and cones. In: *Vision in Fishes* (M. A. Ali, ed.). Plenum Press, New York, pp. 55–65.

77a. Hárosi, F. (1981) Microspectrophotometry and optical phenomena: Birefringence, dichroism, and anomalous dispersion. In: *Springer Series in Optical Sciences,* Volume 23 (J. M. Enoch and F. L. Tobey, Jr., eds.). Springer-Verlag, Berlin, pp. 337–399.
78. Hart, R., R. Zeh, and R. D. Allen (1977) Phase-randomized laser illumination for microscopy. *J. Cell Sci.* **23:**335–343.
79. Hartshorne, N. H., and A. Stuart (1960) *Crystals and the Polarising Microscope: A Handbook for Chemists and Others,* 3rd ed. Arnold, London.
80. Hartshorne, N. H., and A. Stuart (1964) *Practical Optical Crystallography.* American Elsevier, New York.
81. Higashi, S., M. Kasai, F. Oosawa, and A. Wada (1963) Ultraviolet dichroism of F-actin oriented by flow. *J. Mol. Biol.* **7:**421–430.
82. Hiramoto, Y., Y. Hamaguchi, Y. Shoji, and S. Shimoda (1981) Quantitative studies on the polarization optical properties of living cells. I. Microphotometric birefringence detection system. *J. Cell Biol.* **89:**115–120.
83. Hiramoto, Y., Y. Hamaguchi, Y. Shoji, T. E. Schroeder, S. Shimoda, and S. Nakamura (1981) Quantitative studies on the polarization optical properties of living cells. II. The role of microtubules in birefringence of the spindle of the sea urchin egg. *J. Cell Biol.* **89:**121–130.
83a. Hiramoto, Y., and Y. Shoji (1982) Location of the motive force for chromosome movement in sand-dollar eggs. In: *Biological Functions of Microtubules and Related Structures* (H. Sakai, H. Mohri, and G. G. Borisy, eds.). Academic Press, New York, pp. 247–249.
84. Hsien Y. H., M. Richartz, and K. L. Yung (1947) A generalized intensity formula for a system of retardation plates. *J. Opt. Soc. Am.* **37:**99–106.
85. Hughes, A. F., and M. M. Swann (1948) Anaphase movements in the living cell: A study with phase contrast and polarized light on chick tissue culture. *J. Exp. Biol.* **25:**45–70.
86. Huxley, A. F. (1960) British patent specification 856,621. Improvements in or relating to polarizing microscopes (applied July 20, 1956).
87. Inoué, S. (1951) A method for measuring small retardations of structures in living cells. *Exp. Cell Res.* **2:**513–517.
88. Inoué, S. (1952) The effect of colchicine on the microscopic and sub-microscopic structure of the mitotic spindle. *Exp. Cell Res. Suppl.* **2:**305–318.
89. Inoué, S. (1952) Effect of temperature on the birefringence of the mitotic spindle. *Biol. Bull.* **103:**316.
90. Inoué, S. (1952) Studies on depolarization of light at microscope lens surfaces. I. The origin of stray light by rotation at the lens surfaces. *Exp. Cell Res.* **3:**199–208.
91. Inoué, S. (1953) Polarization optical studies of the mitotic spindle. I. The demonstration of spindle fibers in living cells. *Chromosoma* **5:**487–500.
92. Inoué, S. (1960) Birefringence in Dividing Cells. Time-lapse motion picture.
93. Inoué, S. (1961) Polarizing microscope: Design for maximum sensitivity. In: *The Encyclopedia of Microscopy* (G. L. Clarke, ed.). Reinhold, New York, pp. 480–485.
94. Inoué, S. (1964) Organization and function of the mitotic spindle. In: *Primitive Motile Systems in Cell Biology* (R. D. Allen and N. Kamiya, eds.). Academic Press, New York, pp. 549–598.
95. Inoué, S. (1969) The physics of structural organization in living cells. In: *Biology and the Physical Sciences* (S. Devons, ed.). Columbia University Press, New York, pp. 139–171.
96. Inoué, S. (1976) Chromosome movement by reversible assembly of microtubules. In: *Cell Motility,* Volume 3 (R. D. Goldman, T. D. Pollard, and J. L. Rosenbaum, eds.). Cold Spring Harbor Laboratory, Cold Spring Harbor, N.Y., pp. 1317–1328.
97. Inoué, S. (1981) Cell division and the mitotic spindle. *J. Cell Biol.* **91**(Part 2)**:**131s–147s.
97a. Inoué, S. (1981) Video image processing greatly enhances contrast, quality, and speed in polarization-based microscopy. *J. Cell Biol.* **89:**346–356.
98. Inoué, S. (1982) The role of self-assembly in the generation of biologic form. In: *Developmental Order: Its Origin and Regulation* (P. B. Green, ed.). Liss, New York, pp. 35–76.
99. Inoué, S., and A. Bajer (1961) Birefringence in endosperm mitosis. *Chromosoma* **12:**48–63.
100. Inoué, S., and K. Dan (1951) Birefringence of the dividing cell. *J. Morphol.* **89:**423–456.
101. Inoué, S., and W. L. Hyde (1957) Studies on depolarization of light at microscope lens surfaces. II. The simultaneous realization of high resolution and high sensitivity with the polarizing microscope. *J. Biophys. Biochem. Cytol.* **3:**831–838.
102. Inoué, S., and D. P. Kiehart (1978) In vivo analysis of mitotic spindle dynamics. In: *Cell Reproduction: In Honor of Daniel Mazia* (E. Dirksen, D. Prescott, and C. F. Fox, eds.). Academic Press, New York, pp. 433–444.
103. Inoué, S., and C. Koester (1959) Optimum half shade angle in polarizing instruments. *J. Opt. Soc. Am.* **49:**556–559.

104. Inoué, S., and H. Kubota (1958) Diffraction anomaly in polarizing microscopes. *Nature* **182:**1725–1726.
105. Inoué, S., and K. Okazaki (1978) Biocrystals. *Sci. Am.* **236**(4)**:**82–92.
106. Inoué, S., and H. Ritter, Jr. (1975) Dynamics of mitotic spindle organization and function. In: Molecules and Cell Movement (S. Inoué and R. E. Stephens, eds.). Raven Press, New York, pp. 3–30.
107. Inoué, S., and H. Sato (1962) Arrangement of DNA in living sperm: A biophysical analysis. *Science* **136:**1122–1124.
108. Inoué, S., and H. Sato (1966) Arrangement of DNA molecules in the sperm nucleus: An optical approach to the analysis of biological fine structure. In: *Biophysical Science Series 3, Progress in Genetics III* (Japanese Biophysical Society ed.). Yoshioka, Kyoto, pp. 151–220.
109. Inoué, S., and H. Sato (1966) Deoxyribonucleic acid arrangements in living sperm. In: *Molecular Architecture in Cell Physiology* (T. Hayashi and A. G. Szent-Gyorgyi, eds.). Prentice–Hall, Englewood Cliffs, N.J., pp. 209–248.
110. Inoué, S., and H. Sato (1967) Cell motility by labile association of molecules: The nature of mitotic spindle fibers and their role in chromosome movement. *J. Gen. Physiol.* **50:**259–292.
111. Inoué, S., and L. G. Tilney (1982) The acrosomal reaction of *Thyone* sperm. I. Changes in the sperm head visualized by high resolution video microscopy. *J. Cell Biol.* **93:**812–819.
112. Inoué, S., G. G. Borisy, and D. P. Kiehart (1974) Growth and lability of *Chaetopterus* oocyte mitotic spindles isolated in the presence of porcine brain tubulin. *J. Cell Biol.* **62:**175–184.
113. Inoué, S., G. W. Ellis, E. D. Salmon, and J. W. Fuseler (1970) Rapid measurement of spindle birefringence during controlled temperature shifts. *J. Cell Biol.* **47:**95a.
114. Inoué, S., J. Fuseler, E. D. Salmon, and G. W. Ellis (1975) Functional organization of mitotic microtubules—Physical chemistry of the in vivo equilibrium system. *Biophys. J.* **15:**725–744.
115. Inoué, S., D. P. Kiehart, I. Mabuchi, and G. W. Ellis (1979) Molecular mechanism of mitotic chromosome movement. In: *Motility in Cell Function, First John M. Marshall Symposium in Cell Biology* (F. A. Pépe, J. W. Sanger, and V. T. Nachmias, eds.). Academic Press, New York, pp. 301–311.
116. Inoué, S., H. Ritter and D. Kubai (1978) Mitosis in *Barbulanympha.* II. Dynamics of a two-stage anaphase, nuclear morphogenesis, and cytokinesis. *J. Cell Biol.* **77:**655–684.
116a. Itoh, T. J., and H. Sato (1984) The effects of deuterium oxide (^2H_2O) on the polymerization of tubulin in vitro. *Biochim. Biophys. Acta* **800**:21–27.
116b. Izutsu, K. (1961) Effects of ultraviolet microbeam irradiation upon division in grasshopper spermatocytes. I. Results of irradiation during prophase and prometaphase I. *Mie Med. J.* **11**(2):199–212.
116c. Izutsu, K. (1961) Effects of ultraviolet microbeam irradiation upon division in grasshopper spermatocytes. II. Results of irradiation during metaphase and anaphase I. *Mie Med. J.* **11**(2):213–232.
117. Jagger, W. S., and P. A. Liebman (1970) Birefringent transients in photoreceptor outer segments. *Biophys. Soc. Abstr. 14th Annu. Meet.* p. 59A.
118. Jelley, E. E. (1945) Microscopy. In: *Physical Methods of Organic Chemistry* (A. Weissberg, ed.). Interscience, New York, pp. 451–524.
119. Jenkins, F. A., and H. White (1957) *Fundamentals of Optics,* 3rd ed. McGraw–Hill, New York.
120. Jerrard, H. G. (1948) Optical compensators for measurement of elliptical polarization. *J. Opt. Soc. Am.* **38:**35–59.
121. Jerrard, H. G. (1954) Transmission of light through birefringent and optically active media: The Poincaré sphere. *J. Opt. Soc. Am.* **44:**634–640.
122. Johansen, A. (1918) *Manual of Petrographic Methods,* 2nd ed. McGraw–Hill, New York.
123. Kautz, J., Q. B. De Marsh, and W. Thornburg (1957) A polarizing and electron microscope study of plasma cells. *Exp. Cell Res.* **13:**596–599.
123a. Kiehart, D. P. (1981) Studies on the in vivo sensitivity of spindle microtubules to calcium ions and evidence for a vesicular calcium-sequestering system. *J. Cell Biol.* **88:**604–617.
124. Kiehart, D. P., I. Mabuchi, and S. Inoué (1982) Evidence that myosin does not contribute or force production in chomosome movement. *J. Cell Biol.* **94:**165–178.
125. Kobayasi, S. (1964) Effect of electric field on F-actin orientated by flow. *Biochim. Biophys. Acta* **88:**541–552.
126. Kobayasi, S., H. Asai, and F. Oosawa (1964) Electric birefringence of actin. *Biochim. Biophys. Acta* **88:**528–540.
127. Köhler, A. (1921) Ein Glimmerplattchen Grau. I. Ordnung zur Untersuchung sehr schwach doppelbrechender Praparate. *Z. Wiss. Mikrosk.* **38:**29–42.
128. Kubo, R. (1947) Statistical theory of rubber-like substances. *J. Colloid Sci.* **2:**527–535.
129. Kubota, H., and S. Inoué (1959) Diffraction images in the polarizing microscope. *J. Opt. Soc. Am.* **49:**191–198.

187. Salmon, E. D. (1975a) Pressure-induced depolymerization of spindle microtubules. I. Spindle birefringence and length changes. *J. Cell Biol.* **65:**603–614.
188. Salmon, E. D. (197b) Pressure-induced depolymerization of brain microtubules in vitro. *Science* **189:**884–886.
189. Salmon, E. D. (1975c) Spindle microtubules: Thermodynamics of in vivo assembly and role in chromosome movement. *Ann. N.Y. Acad. Sci.* **253:**383–406.
190. Salmon, E. D. (1984) Tubulin dynamics in microtubules of the mitotic spindle. In: *Molecular Biology of the Cytoskeleton* (G. Borisy, D. Cleveland, and D. Murphy, eds.). Cold Spring Harbor Laboratory, Cold Spring Harbor, N.Y., pp. 99–109.
191. Salmon, E. D., and D. A. Begg (1980) Functional implications of cold-stable microtubules in kinetochore fibers of insect spermatocytes during anaphase. *J. Cell Biol.* **85:**853–865.
192. Salmon, E. D., and G. W. Ellis (1975) A new miniature hydrostatic pressure chamber for microscopy: Strain-free optical glass windows facilitate phase contrast and polarized light microscopy of living cells. Optional fixture permits simultaneous control of pressure and temperature. *J. Cell Biol.* **65:**587–602.
193. Salmon, E. D., and G. W. Ellis (1976) Compensator transducer increases ease, accuracy and rapidity of measuring changes in specimen birefringence with polarization microscopy. *J. Microsc.* **106:**63–69.
194. Salmon, E. D., M. McKeel, and T. Hays (1984) Rapid rate of tubulin dissociation from microtubules in the mitotic spindle *in vivo* measured by blocking polymerization with colchicine. *J. Cell Biol.* **99:**1066–1075.
195. Sato, H. (1969) Analysis of form birefringence in the mitotic spindle. *Am. Zool.* **9:**592.
196. Sato, H. (1975) The mitotic spindle. In: *Aging Gametes* (R. Blandau, ed.). Karger Basel, pp. 19–49.
197. Sato, H., and J. Bryan (1968) Kinetic analysis of association–dissociation reaction in the mitotic spindle. *J. Cell Biol.* **39:**118a.
198. Sato, H., and S. Inoué (1964) Condensation of the sperm nucleus and alignment of DNA molecules during spermiogenesis in *Loligo pealei*. *Biol. Bull.* **127:**357.
199. Sato, H., G. W. Ellis, and S. Inoué (1975) Microtubular origin of mitotic spindle form birefringence. *J. Cell Biol.* **67:**501–517.

199a. Sato, H., T. Kato, T. C. Takahashi, and T. Ito (1982) Analysis of D_2O effect on *in vivo* and *in vitro* tubulin polymerization and depolymerization. In: Biological Functions of Microtubules and Related Structures (H. Sakai, H. Mohri, and G. G. Borisy, eds.). Academic Press, New York. pp. 211–226.

200. Sawada, N., and L. I. Rebhun (1969) The effect of dinitrophenol and other phosphorylation uncouplers on the birefringence of the mitotic apparatus of marine eggs. *Exp. Cell Res.* **55:**33–38.
201. Schmidt, W. J. (1924) *Die Bausteine des Tierkorpers in polarisiertem Lichte.* Cohen, Bonn.
202. Schmidt, W. J. (1937) *Die Doppelbrechung von Karyoplasma, Zytoplasma und Metaplasma. Protoplasma-Monographien* Volume 11. Borntraeger, Berlin.
203. Schmidt, W. J. (1939) Doppelbrechung der Kernspindel und Zugfasertheorie der Chromosomenbewegung. *Chromosoma* **1:**253–264.
204. Schmidt, W. J. (1941) Einiges uber Optische Anisotropie und Feinbau von Chromatin and Chromosomen. *Chromosoma* **2:**86–100.
205. Schmidt, W. J. (1944) Die Doppelbrechung des Protoplasmas und ihrer Bedeutung fur die Erforschung seines submikroskopischen Baues. *Ergeb. Physiol.* **44:**27.
206. Schmidt, W. J. (1951) Polarisationsoptische analyse der Verknüpfung von Protein-und Lipoidmolekeln, erlautert am Aussenglied der Sehzellen der Wirbeltiere. *Publ. Staz. Zool. Napoli* **23:**158–183.
207. Schmidt, W. J. (1952) Polarisationmikroskopische Beobachtungen an Zellmembrane, Kern, Cytoplasma Tentakel der *Noctiluca miliaris* Suriray. *Zool. Anz.* **148:**115–131.
208. Schmidt, W. J. (1957) Instrumente und Methoden zur mikroskopischen Untersuchung optisch anisotroper Materielen mit Ausschluss der Kristalle. In: *Handbuch der Mikroskopie in der Technik,* Volume I (H. Freund, ed.). Umschau Verlag, Frankfurt am Main, pp. 147–314.
209. Schmidt, W. J. (1964) *Wissenschaftliche Veröffentlichungen.* Justus Leibig-Universität, Giessen/Lahn.
210. Schmitt, F. O. (1955) Cell constitution. In: *Analysis of Development* (B. Willier, P. Weiss, and V. Hamburger, eds.). Saunders, Philadelphia, pp. 39–69.
211. Schmitt, F. O., and N. Geschwind (1957) *The axon surface. Prog. Biophys.* **8:**165–215.
212. Schrader, F. (1953) *Mitosis—The Movements of Chromosomes in Cell Division.* Columbia University Press, New York.
213. Sebera, D. (1964) *Electronic Structure and Chemical Bonding.* Gina (Blaisdell), Boston.
214. Seeds, W. E. (1953) Polarized ultraviolet microspectrography and molecular structure. *Prog. Biophys.* **3:**27–46.
215. Seeds, W. E., and M. H. F. Wilkins (1959) Ultra-violet microspectrographic studies of nucleoproteins and crystals of biological interest. *Biochim. Biophys. Acta* **11:**417–427.
216. Shillaber, C. P. (1944) *Photomicrography in Theory and Practice.* Wiley, New York.

216a. Shurcliff, W. A. (1962) *Polarized Light: Production and Use*. Harvard University Press, Cambridge, Mass.
217. Skinner, C. A. (1925) A universal polarimeter. *J. Opt. Soc. Am.* **10**:491–520.
217a. Sluder, G. (1976) Experimental manipulation of the amount of tubulin available for assembly into the spindle of dividing sea urchin eggs. *J. Cell Biol.* **70**:75–85.
217b. Sluder, G. (1979) Role of spindle microtubules in the control of cell cycle timing. *J. Cell Biol.* **80**:674–691.
218. Smith, L. W. (1960) Diffraction images of disk-shaped particles computed for full Köhler illumination. *J. Opt. Soc. Am.* **50**:369–371.
218a. Soranno, T., and E. Bell (1982) Cytostructural dynamics of spreading and translocating cells. *J. Cell Biol.* **95**:127–136.
219. Stephens, R. E. (1965) Analysis of muscle contraction by ultraviolet microbeam disruption of sarcomere structure. *J. Cell Biol.* **25**:129–139.
220. Stephens, R. E. (1969) Factors influencing the polymerization of outer fiber microtubule protein. *Q. Rev. Biophys.* **1**:377–390.
221. Stephens, R. E. (1972) Studies on the development of the sea urchin *Strongylocentrotus droebachiensis*. II. Regulation of mitotic spindle equilibrium by environmental temperature. *Biol. Bull.* **142**:145–159.
222. Stephens, R. E. (1973) A thermodynamic analysis of mitotic spindle equilibrium at active metaphase. *J. Cell Biol.* **57**:133–147.
223. Strong, J. (1958) *Concepts of Classical Optics*. Freeman, San Francisco.
224. Sutherland, G. B. B. M., and M. Tsuboi (1957) The infra-red spectrum and molecular configuration of sodium deoxyribonucleate. *Proc. R. Soc. London Ser. A.* **239**:446–463.
225. Swann, M. M. (1951) Protoplasmic structure and mitosis. I. The birefringence of the metaphase spindle and asters of the living sea-urchin egg. *J. Exp. Biol.* **28**:417–433.
226. Swann, M. M. (1951) Protoplasmic structure and mitosis. II. The nature and causes of birefringence changes in the sea-urchin egg at anaphase. *J. Exp. Biol.* **28**:434–444.
227. Swann, M. M., and J. M. Mitchison (1950) Refinements in polarized light microscopy. *J. Exp. Biol.* **27**:226–237.
228. Swann, M. M., and J. M. Mitchison (1958) The mechanism of cleavage in animal cells. *Biol. Rev.* **33**:103–135.
229. Takahashi, W. N., and T. E. Rawlins (1933) Rod-shaped particles in tobacco mosaic virus demonstrated by stream double refraction. *Science* **77**:26–27.
229a. Takashima, S. (1968) Optical anisotropy of synthetic polynucleotides. I. Flow birefringence and π-electron polarizability of bases. *Biopolymers* **6**:1437–1452.
230. Tasaki, I., A. Watanabe, R. Sandlin, and L. Carnay (1968) Changes in fluorescence, turbidity and birefringence associated with nerve excitation. *Proc. Natl. Acad. Sci. USA* **61**:883–888.
231. Taylor, D. L. (1975) Birefringence changes in vertebrate striated muscle. *J. Supramol. Struct.* **3**:181–191.
232. Taylor, D. L. (1976) Quantitative studies on the polarization optical properties of striated muscle. I. Birefringence changes of rabbit psoas muscle in the transition from rigor to relaxed state. *J. Cell Biol.* **68**:497–511.
233. Taylor, D. L. (1976) Motile models of amoeboid movement. In: *Cell Motility*, Volume 3. (R. Goldman, T. Pollard, and J. Rosenbaum, eds.). Cold Spring Harbor Laboratory, Cold Spring Harbor, N.Y., pp. 797–821.
234. Taylor, D. L. (1976) The contractile basis of amoeboid movement. IV. The viscoelasticity and contractility of amoeba cytoplasm in vivo. *Exp. Cell Res.* **105**:413–426.
235. Taylor, D. L., and R. M. Zeh (1976) Methods for the measurement of polarization optical properties. I. Birefringence. *J. Microsc.* **108**:251–259.
236. Taylor, D. L., R. D. Allen, and E. P. Benditt (1975) Determination of the polarization optical properties of the amyloid congo red complex by phase modulation microspectrophotometry. *J. Histochem.* **22**:1105–1112.
237. Taylor, D. L., J. S. Condeelis, P. L. Moore, and R. D. Allen (1973) The contractile basis of amoeboid movement. I. The chemical control of motility in isolated cytoplasm. *J. Cell Biol.* **59**:378–394.
238. Taylor, D. L., J. A. Rhodes, and S. A. Hammond (1976) The contractile basis of amoeboid movement. II. Structure and contractility of motile extracts and plasmalemma-ectoplasm ghosts. *J. Cell Biol.* **70**:123–143.
239. Taylor, E. W. (1965) The mechanism of colchicine inhibition of mitosis. *J. Cell Biol.* **25**:145–160.
240. Taylor, E. W., and W. Cramer (1963) Birefringence of protein solutions and biological systems. I. *Biophys. J.* **3**:127–141.
241. Thornburg, W. (1957) The form birefringence of lamellar systems containing three or more components. *J. Biophys. Biochem. Cytol.* **3**:413–419.

242. Thornburg, W., and E. DeRobertis (1956) Polarization and electron microscope study of frog nerve axoplasm. *J. Biophys. Biochem. Cytol.* **2**:475–482.
243. Tilney, L. G., and K. R. Porter (1965) Studies on microtubules in Heliozoa. I. *Protoplasma* **60**:317–344.
243a. Timasheff, S. N., and G. D. Fassman, eds. (1971) *Subunits in Biological Systems.* Marcel Decker, New York.
244. Tippit, D. H., D. Schultz, and J. D. Pickett-Heaps (1978) Analysis of the distribution of spindle microtubules in the diatom *Fragilaria. J. Cell Biol.* **76**:737–763.
245. Tobolsky, A. V. (1967) *Properties and Structures of Polymers.* Wiley, New York.
246. Tuckerman, L. B. (1909) Doubly refracting plates and elliptic analysers. *Univ. Nebraska Stud.* pp. 173–219.
247. Uretz, R. B., and R. P. Perry (1957) Improved ultraviolet microbeam apparatus. *Rev. Sci. Instrum.* **28**:861–866.
247a. Vidal, B. D. C., M. L. S. Mello and E. R. Pimentel (1982) Polarization microscopy and microspectrophotometry of Sirius Red, Picrosirius and Chlorantine Fast Red aggregates and of their complexes with collagen. *Histochem. J.* **14**:857–878.
248. Von Ebner, U. (1882) *Untersuchungen über die Ursache der Anisotropie organischer Substanzen.* Leipzig.
249. Von Hippel, A. R., ed. (1965) *The Molecular Designing of Materials and Devices.* MIT Press, Cambridge, Mass.
250. Von Muralt, A. L., and J. T. Edsall (1930) Flow birefringence of myosin. *Trans. Faraday Soc.* **26**:837–852.
251. Wahlstrom, E. E. (1960) *Optical Crystallography,* 3rd ed. Wiley, New York.
252. Weber, H. H. (1934) Der Feinbau und die mechanischen Eigenschaften der Myosin Fadens. *Pfluegers Arch.* **235**:205–233.
253. Weidner, R. T., and R. L. Sells (1965) *Elementary Modern Physics,* Volume 3. Allyn, Bacon & Cottrell, Boston.
254. Weisenberg, R. C., and A. C. Rosenfeld (1975) In vitro polymerization of microtubules into asters and spindles in homogenates of surf clam eggs. *J. Cell Biol.* **64**:146–158.
255. Weis-Fogh, T. (1961) Molecular interpretation of the elasticity of resilin, a rubber-like protein. *J. Mol. Biol.* **3**:648–667.
256. West, C. D., and R. C. Jones (1951) On the properties of polarization elements as used in optical instruments. I. Fundamental considerations. *J. Opt. Soc. Am.* **41**:976–986.
257. West, S. S. (1970) Optical rotation and the light microscope. In: *Introduction to Quantitative Cytochemistry,* Volume 2 (G. L. Wied and F. G. Bahr, eds.). Academic Press, New York, p. 451.
258. Wiener, O. (1912) Die Theorie Des Mischkörpers für das Feld der stationaren Strömung. I. Die Mittelwertzatze für Kraft, Polarisation und Energie. *Abh. Math. Phys. Kl. Koenigl. Ges. Wiss.* **32**:507–604.
259. Wiener, O. (1926) Formdoppelbrechung bei Absorption. I. Ubersicht. *Kolloidchem. Beih.* **23**:189.
260. Wilkins, M. H. F., and B. Battaglia (1953) Note on the preparation of specimens of oriented sperm heads for X-ray diffraction and infra-red absorption studies and on some pseudo-molecular behaviour of sperm. *Biochim. Biophys. Acta* **11**:412–415.
261. Wilson, E. B. (1925) *The Cell in Development and Heredity,* 3rd ed. Macmillan Co., New York.
261a. Wolman, M. (1970) On the use of polarized light in pathology. In: *Pathology Annual 1970* (S. C. Sommers, ed.). Appleton-Century-Crofts, New York, pp. 381–416.
262. Wolstenholme, G. E. W., and M. O'Connor, eds. (1966) *Principles of Biomolecular Organization.* Churchill, London.
263. Wood, E. A. (1964) *Crystals and Light: An Introduction to Optical Crystallography.* Van Nostrand, Princeton, N.J.
264. Wood, R. E. (1934) *Physical Optics,* 3rd ed. Macmillan Co., New York.
265. Wright, F. E. (1911) *The Methods of Petrographic-Microscopic Research: Their Relative Accuracy and Range of Application.* Carnegie Institution of Washington, Washington, D.C.
266. Zernike, F. (1958) The wave theory of microscopic image formation. In: *Concepts of Classical Optics* (J. Strong, ed.). Freeman, San Francisco, pp. 525–536.
266a. Zirkle, R. E., R. B. Uretz, and R. H. Haynes (1960) Disappearance of spindles and phargmoplasts after microbeam irradiation of cytoplasm. *Ann. N.Y. Acad. Sci.* **90**:435–439.
267. Zirwer, D., E. Buder, W. Schalike, and R. Wetzel (1970) Lineardichroitische Untersuchungen zur DNS-Organisation in Spermienkopfen von *Locusta migratoria* L. *J. Cell Biol.* **45**:431–434.
268. Zocher, H., and F. Jacoby (1972) Flow dichroism. *Kolloid Beih.* **24**:365.

Reprinted from *Methods in Cell Biology*, Vol. 27, pp. 89–110, 1986,
with kind permission from Elsevier.

Chapter 6

Techniques for Observing Living Gametes and Embryos

DOUGLAS A. LUTZ

Department of Anatomy–Histology
University of Toronto
Toronto, Ontario, Canada

SHINYA INOUÉ

Marine Biological Laboratory
Woods Hole, Massachusetts

I. Introduction

In this article, we describe ways of handling relatively small numbers of echinoid gametes and embryos and of microscopically observing their physiology and development. We emphasize ways to keep the embryos healthy for extended periods of observations. Many of the points described should apply equally well to other types of studies.

For microscopic observations, we prefer cells that are optically clear. Thus, we emphasize the use of *Lytechinus variegatus* and *Echinarachnius parma,* whose eggs are among the clearest that we have found on the United States Atlantic coast.

To obtain healthy embryos, it is important to have healthy adults. Therefore, our discussion starts with brief comments on the adult animals.

II. Handling and Care of Animals

We have had good success obtaining *L. variegatus* from two sources: Gulf Specimens Co., Inc., Panacea, FL, and Marine Specimens Unlimited, Big Pine Key, FL. Because of the warmer water temperature in the Florida Keys, the urchins from these areas show less of a population-wide breeding season than those from other areas. The advantage is that shipments can be obtained at least 10 months of the year in which a percentage of the animals are ripe (from late November to early January appears to be the period of the year when this is not true). We have generally tried to order animals just prior to the beginning of the breeding season. By warming our holding tanks (25°C) and supplying generous amounts of food, we ripen the animals in the laboratory. Animals that are shipped at the peak of the breeding season are more easily stressed, and this results in a higher mortality rate and spawning of animals during shipping or shortly after arrival.

Upon receipt of the animals, they are slowly acclimated to the tank's temperature (by floating the plastic bags containing the animals in the tank). Once adjusted they are carefully removed from the bags and placed at the bottom of the tank. They should very soon begin to extend their tube feet and raise their spines. An urchin with drooping spines should be watched closely, as this indicates that they are very stressed and may not survive. Soon after arrival they should be supplied with food, and after about a week the temperature in the tank can be slowly elevated. Once the animals have reached maturity (this can be assessed by electric shock-induced shedding, see Section III), the temperature of the tank should be reduced to a holding level (for *L. variegatus* about 18°C).

To keep our urchins year round, we have been using closed system aquaria equipped with under-gravel filtration. Animals are kept at a density of less than one animal per gallon of seawater. One-half of the volume of the tank is replaced at least once a week when we are using natural seawater. Adequate illumination should be provided for the tanks, especially if one feeds the urchins their natural food, to prevent spoilage.

We are fortunate enough to have running seawater available to us year round, but we have also used aritificial seawater. We have been able to keep animals alive and healthy in our closed system aquaria without changing the seawater for over 1 year using "My Sea" artificial seawater (Jamarin Laboratory, Osaka, Japan). This represents a dramatic improvement over previous results with other artificial or even natural seawaters; using other seawaters, successful maintenance of sea urchins has usually required a replenishment of at least one-half tankful with fresh, tempered seawater every 2 weeks. The specific gravity of the "My Sea" seawater was checked once a week and adjusted with deionized water to make up for evaporation. In addition to "My Sea," Jamarin Laboratory also produces a variety of excellent, culture-quality artificial seawaters and chemicals that we now use routinely in our lab, such as "Jamarin U" and Ca^{2+}-free and Ca^{2+},Mg^{2+}-free mixtures. These are distributed through the Chemical Stockroom of the Marine Biological Laboratory.

We supply our urchins with natural food supplemented with "urchin cakes." The recipe given in Table I is one developed by E. B. Harvey, as modified by B. M. Woodward in our laboratory. It is important to remove excess pieces of food after a day, since they can spoil and contaminate the tanks.

TABLE I

URCHIN CAKE RECIPE[a]

500 ml seawater
500 ml distilled water
30 g agar
Six hard-boiled eggs: Remove shells, dry, and crush them
TetraMin fish flakes (about one-half of a large container)

1. Simmer the agar in the water for 30 minutes
2. Mash eggs through fine screen
3. Add eggs, finely crushed egg shells, and fish flakes to agar. Mix well
4. Pour mixture into several clean Petri dishes
5. Place clean Kimwipes on top of Petri dishes so that they cover the dishes but do not touch the cake mixture, and let cool for several hours
6. Cover dishes and store in the refrigerator
7. For feeding, cut cakes into small pieces and place in urchin tanks

[a] From B. M. Woodward, after E. B. Harvey.

III. Obtaining Gametes

Gametes can be shed through the gonopores, by stimulating ovarian muscles to contract, using one of several methods. Each has associated advantages and disadvantages that will be outlined in the following discussion.

Many echinoderms can be stimulated electrically. This method has the advantage that the number of gametes shed from a given animal can be finely controlled. Only the gametes required for a given experiment are collected, and the same animal may be used repeatedly. Not all species of echinoids respond to electrical stimulation (e.g., *E. parma*), and an alternative method must be employed in these cases.

Figure 1 illustrates, in schematic form, the electrical stimulator we have used. Line voltage is stepped down using a variable transformer. Current is passed through a low wattage light bulb which provides a resistance (to prevent short circuit) as well as visual indicator for the passage of current through the circuit. A volt meter is connected into the circuit to measure the voltage applied to the animal. Two carbon electrodes (used for carbon coating in electron microscopy) are fixed to the ends of the wires, and the connection is insulated so that seawater will only come into contact with the carbon tip. Plastic tubing or tubing connectors make good insulators as well as being convenient handles for the electrodes.

Species vary in the voltage required to stimulate shedding. For *L. variegatus,* we found that 25 to 30 V is required while for *Arbacia punctulata* only 10 V is

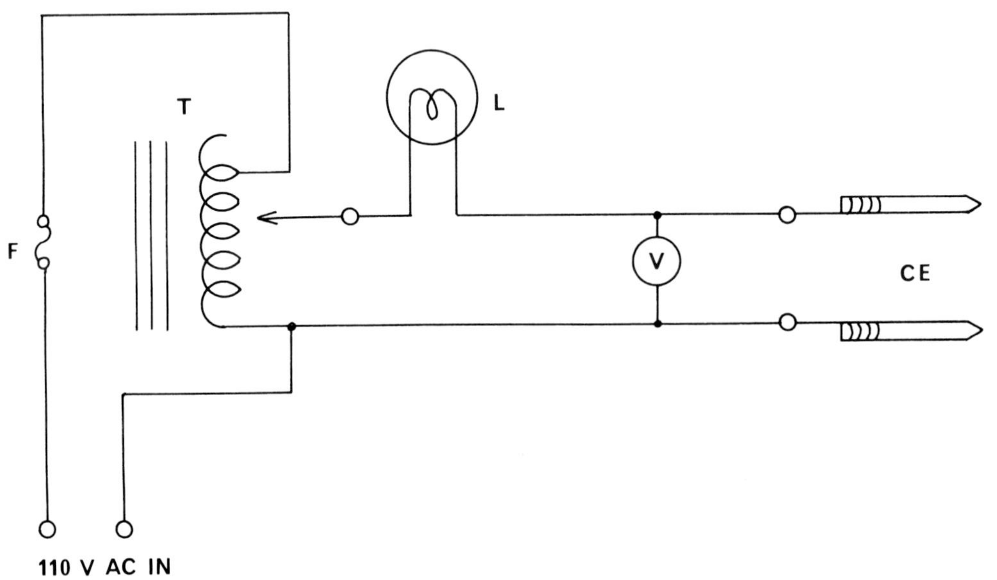

FIG. 1. Circuit diagram for electrical stimulator. F, Fuse; T, variable transformer; L, 120-V 40-W lamp; V, volt meter; CE, carbon electrodes from EM carbon coater.

necessary. If only one species is used in the laboratory, a simplified electrical stimulator may be used consisting of a single output voltage transformer.

For electrical stimulation, the animal is placed in a finger bowl partially filled with seawater, leaving the aboral surface exposed. One electrode is placed in the seawater while the other is lightly touched to the animal's surface 0.5 to 1 cm below a gonopore. In some species the gonopores are quite small, and it is desirable to use a stereomicroscope to observe the placement of the electrode as well as to determine the flow of gametes. We find that the animal is minimally stressed if the stimulation is given in approximately 10-second pulses separated by short pauses rather than in a continuous manner. After several pulses, gametes begin to issue from the gonopore if the animal is gravid. Stimulation is stopped and the gametes are collected from the gonopore region using a mouth-operated capillary tube. Semen is stored "dry" (undiluted) in these tubes while eggs are transferred into a seawater-filled beaker for washing. By using this pulse method of stimulation, problems of overstimulation, resulting in too many gametes being shed, or even complete shedding of the animal, are largely avoided. After collecting a sufficient number of gametes and making sure the animal is no longer shedding, the aboral surface is rinsed with seawater and the animal is returned to its tank. Male and female animals are held in separate tanks to prevent spawning.

As mentioned previously, not all species respond to electrical stimulation. In these cases one can inject 0.1 to 1 ml of a 0.5 *M* KCl or a 10 m*M* acetylcholine solution to induce shedding. A small-gauge needle should be used to inject the stimulant through the peristomal membrane surrounding the mouth into the body cavity. KCl injection usually causes complete shedding of the animal, but a small amount of acetylcholine can be injected instead to release a limited number of gametes. Thus, gametes can be collected from a single animal several times by repeated injections of small volumes of acetylcholine solution.

After injection of either stimulant, the animal is placed aboral side down onto a seawater-filled beaker whose diameter is smaller than the animal. Streams of gametes are then observed to come from the gonopores. Males are removed from the beaker and concentrated semen is collected "dry" from the aboral surface of the animal, as previously described. Eggs are allowed to settle to the bottom of the beaker. A steady stream of individual eggs usually indicates that the animal is ripe and the eggs will be healthy and fertilizable. A discontinuous stream or clumping of eggs as they fall from the gonopore may indicate problems. A small sample of eggs collected from the bottom of the beaker or directly from the gonopore should be checked for germinal vesicle stages or deteriorating eggs. Overstimulation will result in a mixture of "good" and "bad" eggs, and thus injection volumes should be empirically adjusted to provide a minimum dose for shedding.

Eggs should be thoroughly washed to remove debris and any toxic secretions before fertilization is attempted. Washing eggs is accomplished by allowing them to settle to the bottom of a beaker by gravity, decanting the seawater above, and resuspending the eggs in fresh seawater. This process is repeated two to three times. Debris which is heavier than the eggs can be removed by allowing it to settle, then decanting the suspended eggs into a clean beaker. Alternately, one can carefully aspirate the debris from the bottom using a Pasteur pipet. Some eggs withstand washing by centrifugation, but others do not tolerate such rough handling well.

Washed eggs may be stored after shedding. The length of time one can store a batch of eggs and still retain good fertilizability depends on a number of factors. In our experience, storage of eggs in "Jamarin U" artificial seawater increases the storage time by at least a factor of two over other artificial seawaters we have tried. Storing the eggs as a monolayer covered by about 1 cm of seawater appears to be better than maintaining the eggs in more crowded conditions with a thicker covering layer of seawater. Chilling the eggs slightly (5–10°C below tank temperature) also lengthens storage time of warm-water species although lowering the temperature even further, in our experience, decreases their viability and the percentage fertilization.

Semen may be conveniently stored for 1 to 2 days in sealed capillary tubes. Many investigators place sea urchin semen at 4°C for storage, but we found that

TABLE II

DISHWASHING PROCEDURE

1. Never mix other laboratory glassware with glassware for living material. "Living" glassware includes culture dishes, beakers for collection and washing eggs, and glassware used to make up and store seawaters
2. Rinse dishes thoroughly after use in tap water, and leave soaking in tap water. Never allow soaking dishes to be exposed to air: Dishes must be completely covered with water
3. Before starting to wash dishes, be sure you can complete all the steps up to (7). Dishes may be left to soak at this stage
4. Soak dishes in hot detergent solution (Alconox, Micro, or equivalent) for 30 minutes. Do not allow the dishes to stay in the detergent solution longer than 30 minutes. Add only enough detergent to the wash water to make it sudsy: Too much detergent requires additional rinsing
5. Brush each dish thoroughly inside and out
6. Rinse every part of each dish at least 10 times with running tap water. A clean dish begins to "squeak" in your fingers at this point. Check that each dish is visibly clean before proceeding to the next step
7. Transfer tap water-rinsed dishes into a tub filled with tap-distilled water. Dishes may be held at this stage for an extended period if they are completely covered
8. Rinse every part of each dish 10 times in tap-distilled water and immediately rinse every part of the same dish 5 times in glass-distilled water
9. Invert dishes on layers of clean Kimwipes laid on a clean surface, cover with clean Kimwipes, and allow to dry overnight
10. Cover beakers, flasks, graduates, etc., with Parafilm or aluminum foil. Smaller items are most conveniently stored in plastic bags

sperm, like eggs, retain their viability longer if we stored them 5–10°C cooler than the animal's holding temperature.

IV. "Biocleaning" Glassware

We have developed and used the procedure outlined in Table II to clean our culture glass or plastic ware. Depending upon local water conditions, one may be able to reduce the number of rinses at each step. Those listed here are guidelines which have been safely used under a variety of laboratory conditions, and modifications must be evaluated by individual investigators. The soaking steps are carried out in small plastic wash tubs reserved for this purpose.

Biocleaning procedures for cover glasses and slides (Table III) are similar to the one for dishes. Again, the number of rinses at each step may be modified depending upon local water quality. We prefer to store both slides and cover glasses in 80% ethanol, removing them with clean forceps, and drying them with four thicknesses of Kimwipe (to avoid transfer of oils from fingertips to slides) just before use. This lessens the possibility of contamination during storage.

V. Fertilization and Culturing

Methods to fertilize echinoderm eggs have been described in great detail in a number of reviews (Harvey, 1956; Hinegardner, 1975). In the following section,

TABLE III

"BIOCLEANING" SLIDES AND COVER SLIPS

1. Fill 500-ml beaker with hot tap water and enough detergent to make the solution sudsy
2. Drop 30–50 cover glasses or slides *individually* into the beaker. Try to position the slides in a criss-cross arrangement in the beaker
3. Soak for 15–30 minutes
4. Place beaker in ultrasonic washer for 3–5 minutes
5. Rinse many times in tap water to remove the majority of the detergent
6. Using stainless steel forceps, transfer slides or cover glasses individually to another beaker containing tap-distilled water. Repeat this individual transfer rinse 10 times. Sonicate beaker for 3–5 minutes before the last transfer
7. Individually transfer slides or cover glasses into a beaker containing glass-distilled water. Repeat this 10 times. Sonicate beaker for 3–5 minutes before the last transfer
8. Individually transfer slides or cover glasses into bioclean storage jars filled with 80% ethanol for storage
9. Before use, remove slide or cover glass from alcohol using clean forceps and wipe dry with a Kimwipe folded twice (four thicknesses). Do not rub

we briefly outline the procedure we use, adding a number of hints that improve the success of fertilization and the viability of our cultures.

Eggs are placed in a culture dish so that they cover, at most, one-third of the bottom area. Overcrowding results in poorer fertilization success and higher mortality during culture. Concentrated "dry" semen is diluted into seawater before adding to the eggs. First, push out a small amount of sperm from the storage capillary and discard. This semen contains a large number of nonviable sperm due to evaporation. A small drop ($< 20\ \mu l$) is added to 2–3 ml of seawater (dilution $> 1{:}100$) and is gently mixed. A sample of the diluted sperm should be checked under the microscope to make sure that the sperm are actively motile. If a majority of the sperm in the sample is nonmotile, make another dilution from a different batch of sperm until most sperm are swimming.

To fertilize the eggs, add one to three drops of the diluted sperm to the dish and gently mix the culture for approximately 10 seconds. Do not mix by vigorously sucking in and out of a pipet. Avoid adding too much sperm since the eggs become polyspermic. Also, the excess sperm quickly die and encourage bacterial and protozoan growth. Within 1 minute after the addition of sperm, the eggs should be surrounded by a fertilization envelope which fully develops 4–6 minutes after the addition of sperm. Fertilization success may be assessed by scanning the culture dish and evaluating the percentage of eggs in the culture displaying fertilization envelopes. For experiments, we generally use cultures with >95% fertilization success. After insemination, let the eggs settle, aspirate off the supernatant seawater, and add fresh seawater to remove excess sperm.

For a given species, the degree of elevation of the fertilization envelope from the egg surface is a good indicator of egg condition. As the eggs age after shedding, their fertilization envelopes elevate progressively less. This is particularly easy to see in those species whose fertilization envelopes are normally quite high (e.g., *L. variegatus, E. parma*). Another indicator of declining egg condition is the increased concentration of sperm required to inseminate an older batch of eggs.

A. Removal of Fertilization Envelopes

It is often desirable or necessary to remove the fertilization envelope from the eggs. A number of techniques have been published in the literature, and we will list a few in this section. When first starting, we recommend that several techniques be tried before making a decision on which technique to use routinely. Species and investigator differences both play a role in the ease and success of these methods.

While the fertilization envelope expands to its maximum extent in the first minute or two following insemination, it is quite soft and fragile. Shaking the eggs in a test tube during this period of time (approximately the first 4 minutes following fertilization—during the next 4–6 minutes the fertilization envelope becomes progressively tougher and more difficult to remove) breaks open some of the fertilization envelopes. Often, however, the agitation of the eggs cause some to lyse, it is difficult to use with small numbers of eggs, and the percentage removal is quite variable.

Another mechanical method for fertilization envelope removal which is widely used is passing the newly fertilized egg suspension through Nitex, a nylon screen material which comes in a wide variety of mesh sizes (Tetko Inc., Precision Woven Screening Media, Elmsford, NY). Useful mesh size depends on the diameter of the egg and should be about 10 μm smaller than the average egg diameter. Too small a mesh size results in fragmentation of the eggs while too large a mesh size allows eggs with their fertilization envelopes to pass through. Removal of fertilization envelopes is accomplished by gently pouring the egg suspension (about 4 minutes following insemination) through a small piece of Nitex, which is fastened to the bottom of a short plastic tube, into a culture dish. Success with this technique depends on determining the mesh size of the Nytex, the time after fertilization to begin pouring, and the right pouring rate. The filter should be well rinsed with running tap water followed by distilled water after use to keep it free from debris. This method works well with both small and large numbers of eggs: For best results the area of the Nitex should be reduced for small numbers of eggs and increased for larger cultures.

Removal of fertilization envelopes by shearing through Nitex screen can be facilitated by fertilizing the eggs in seawater containing 1 m*M* 3-amino-1,2,4-triazole (Showman and Foerder, 1979). This peroxidase inhibitor blocks fertilization envelope hardening as long as it is present during the first 5 minutes following insemination. The fertilization envelope remains soft and may be removed at any point in development up to hatching. In addition, "stickiness" and clumping of fertilization envelope-free embryos was reduced by letting the embryos establish their initial hyaline layer (approximately 1 hour after insemination) prior to demembranation.

We routinely use an isosmotic solution of glycerol for fertilization envelope removal (Salmon and Segal, 1980). After insemination and once the fertilization envelopes have begun to elevate, the eggs are gently pelleted using a hand centrifuge (about 30 seconds). The seawater supernatant is aspirated off and the eggs are resuspended in 10 to 50 vol of "wash buffer" (1 *M* glycerol, 5 m*M* Tris, pH 8.0–8.2). After the eggs have been in the wash buffer for 30–60 seconds, they are gently pelleted and washed by centrifugation with at least two changes of seawater. In our hands, this method is very effective and reproducible and is not injurious to the subsequent development of the embryos. Another advantage of the method is that it works equally well with eggs producing high or low fertilization envelopes.

B. Handling Small Numbers of Eggs

In order to load our manipulation or observation chambers, we must be able to select, pick up, transfer, and dispense a small number of gametes. In addition, eggs that have had their fertilization envelopes removed have a tendency to stick to glass pipets. We have overcome these problems by making suitably sized and tapered tips of polyethylene tubing (Intramedic polyethylene tubing, Clay Adams, Inc., Parsippany, NJ). To taper the tubing, a small microburner is constructed from an 18-gauge syringe needle placed into the end of a convenient length of tubing with a screw-type tubing clamp used to regulate the gas flow. It takes very little heat to soften the tubing, and unlike glass, the polyethylene tubing has a tendency to contract when heated. The tubing is pulled slowly and steadily to avoid tearing, and this is most easily accomplished by continued mild heating. When the desired taper and end diameter have been reached, remove the tubing from the heat but maintain tension on the tubing until it has cooled (the polyethylene turns from clear to translucent).

Tips pulled from "PE-50" tubing fit onto the ends of most microliter syringe needles. By using a 10-μl syringe, a fine degree of control can be exercised. Because the polyethylene is translucent, eggs can be easily seen in the lumen of the tubing. "PE-160" tubing is useful for making tips that fit on the ends of standard Pasteur pipets. As long as the tips are well rinsed after each use, they can be reused many times before they need to be replaced.

VI. Observation Chambers

A. Microdrop Preparation

Temporary slides can be made by placing a small sample of eggs in seawater on a bioclean slide, covering the drop with a cover slip whose edges have a thin layer of vaseline or silicon high-vacuum grease. The grease seals the preparation to prevent evaporation and also supports the weight of the cover slip. These preparations are useful only for short-term observation since they do not allow for gas exchange and the cells very soon become anoxic. For long-term observations of individual embryos, one of us (S.I.) developed what we call a microdrop preparation. Using this chamber, one can follow the development of individual embryos from just after fertilization through gastrulation to late pluteus stages.

Figure 2 illustrates the construction of this chamber. A bioclean slide and cover slip must be used. A spacer is cut from filter paper slightly smaller than the size of the cover slip, and a hole is punched or cut from the center. The thickness of the spacer used depends upon the diameter of the egg under observation. One

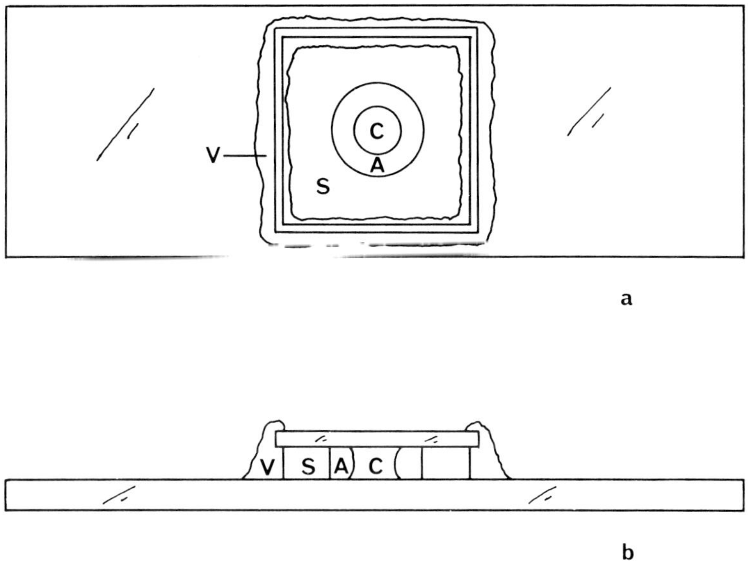

FIG. 2. Microdrop observation chamber, in top (a) and side (b) view. V, Valap sealing wax; S, filter paper spacer; A, air space; C, droplet of culture medium containing cells.

wishes to hold each egg in place without overly compressing it. The spacer must also be made from an absorbent material since its second function is to provide a reservoir of seawater to saturate the gas phase of the chamber. If you measure the thickness of your dry spacer material, remember that the thickness approximately doubles when it is wet.

It is important to construct and seal the chamber as quickly as possible to limit evaporation of the small egg-containing droplet. To this end, we make our preparations in the following sequence. First, place the bioclean slide and cover glass on a support made from applicator sticks. This keeps the slide and cover glass away from the bench top where they will quickly pick up dirt. Next, place the dry filter paper spacer on the center of the slide. Third, draw fresh seawater into a clean Pasteur pipet, and also load a sample of eggs into the polyethylene tubing tip (about 10 eggs in a 20-μl drop) of a microliter syringe. Then, wet the spacer with the fresh seawater, quickly place the drop of eggs in the center of the spacer hole, and gently place the cover slip on top of the spacer. The drop must remain surrounded by air and must not touch the spacer. Finally, the cover slip is held in position while the edges are completely sealed with Tackiwax or valap (a 1:1:1 mixture of vaseline, lanolin, and paraffin or beeswax, simmered on low heat for 2 hours). It is a good idea to check the sealing job using a stereomicroscope and to patch any holes quickly, since even a small pinhole will lead to the preparation drying out.

Instead of placing the drop in the center of the hole on the slide, one can place the drop on the center of the cover slip, then carefully lower the slide with wet spacer on top of the cover slip. The adhesion of the cover slip to the spacer/slide is firm enough to allow the whole preparation to be lifted, inverted, and then sealed. Either method may be used for assembling the preparation. Its success depends upon the speed with which it is assembled, and on the bioclean quality of the components.

As mentioned previously, "Jamarin U" rather than filtered seawater is recommended for suspending the eggs for extended culture. It is the only medium that has allowed *L. variegatus* zygotes to grow into full plutei in the microcrop preparation.

B. Perfusion Chambers

Figure 3 illustrates the construction of a wedge-shaped perfusion chamber designed for use on an upright microscope stand, modeled after the chamber devised by Inoué and Tilney (1982) used on an inverted microscope stand. The wedge-shaped spacer is an improvement over the more familiar parallel spacer in two ways: The cells are trapped in the wedge, and the perfusate is forced to spread out, providing a uniform exchange of media throughout the entire chamber. This chamber permits high-resolution observation of cells while allowing for rapid and complete exchange of media.

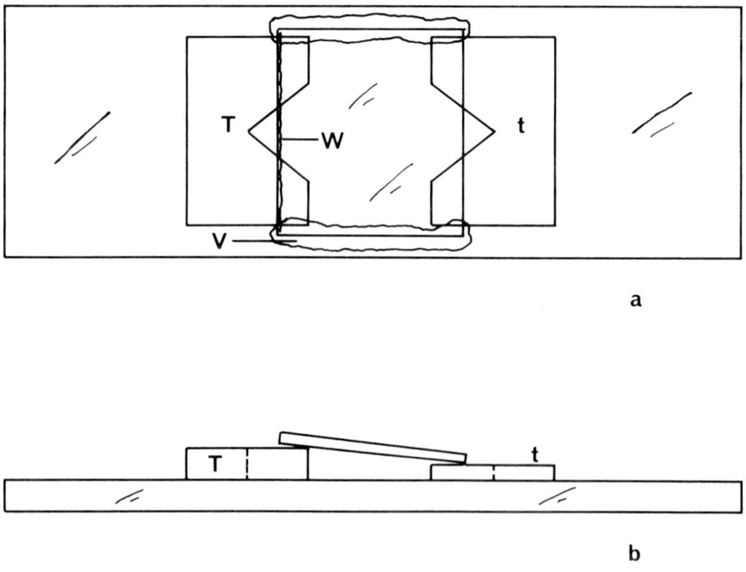

FIG. 3. Wedge perfusion chamber in top (a) and side (b) views. A cover slip is placed on top of a thick (T, entrance side) and a thin (t, exit side) Mylar spacer, forming a wedge-shaped chamber space. The cover slip is sealed to the slide on two sides with valap (V). A drop of perfusate is placed at the opening of the thick side of the chamber and is drawn through the chamber by a filter paper wick placed at the opening of the thin side. A grease pencil line (W) drawn on the entrance edge of the cover slip helps to prevent the perfusate from flooding the cover slip surface.

The spacers are cut from Mylar sheets (sample sheets may be obtained by contacting the Products Division, E. I. DuPont, East Orange, NJ) of different thicknesses, or from Scotch brand Magic tape (3M Co., St. Paul, MN; one tape layer is approximately 60 μm thick). A cover slip is placed on top of the spacers and sealed on two sides with valap. To help prevent drops of perfusate from flooding the cover slip surface, a grease pencil line is drawn on the entrance edge. Tapered filter paper wicks are used to draw solution through the chamber. Flow rate through the chamber can be crudely adjusted by changing the angle of taper of the wicks, and by using different filter paper types (e.g., Whatman #4—medium to fast; Whatman #501—slow to medium).

C. Egg White-Coated Perfusion Chambers

Although the wedge perfusion chamber was ideal for experiments with eggs, in studies of sperm flagellar motility (Schweitzer *et al.*, 1983), we observed that live sperm of some species, as well as detergent-extracted sperm models, quickly became stuck to the glass surfaces along their entire length. To overcome this problem, a coating procedure was developed to lessen the ''stickiness'' of the preparation. Both slides and cover slips were coated with fresh egg white by immersing them for 5–10 minutes. After this time, the slides/cover slips were removed from the egg white, one side was wiped clean, and they were stored in a high-humidity environment. Perfusion chambers were assembled as described previously, with the egg white being wiped away from all areas of the slide except those included in the wedge space. After sealing the chamber with valap, the interior was rinsed with seawater (live sperm) or buffer (sperm models), and the chamber was stored in a moist container in this condition.

Sperm, either models or living, preferentially attach by their midpieces, leaving the flagella free for observation. Although the image is degraded slightly because of the unevenness of the egg white coating, surprisingly high-resolution images can be obtained using this chamber design.

D. Manipulation Chambers

The design and construction of a chamber, designed by G. W. Ellis and used extensively in our laboratory, which allows easy access to the eggs by micropipets or microtools, has already been described in detail by Kiehart (1982). Eggs in this chamber may be observed using high-resolution and high-extinction microscopy modes for extended periods of time without affecting the subsequent development of the embryos. An alternative chamber configuration, and the use of polyanionic coatings of chamber surfaces, is described here.

Rather than the parallel chamber space formed by the coverslip fragment and the support cover slip described by Kiehart (1982), we have used a chamber in which the cover slip fragment and the support overslip form a wedge-profiled space, from front to back (Fig. 4), as described by Hiramoto (1956; also see Kishimoto, this volume). This is accomplished by supporting the fragment only at its front corners by two small tape or Mylar spacers. Using this design, the cells need not be as greatly flattened to hold them in place as in the parallel design.

For micromanipulation work, one of us (D.A.L.) found that neither chamber design held the eggs firmly enough for movement of the microneedle inside the cell. To overcome this problem the surfaces of both the fragment and the support cover slip, in contact with the cells, were coated with either poly-L-lysine (Mazia *et al.*, 1975) or protamine sulfate (Steinhardt *et al.*, 1971) prior to chamber assembly (Table IV). I found species differences in how well these agents worked and recommend that other investigators try each polycation. For example, I found that demembranated *L. variegatus* eggs adhered more tightly to protamine sulfate-coated cover slips than to poly-L-lysine-coated ones. In addition, the embryos grown on the poly-L-lysine-coated cover slips showed tendencies of postgastrulation abnormalities, while eggs grown on protamine sulfate-coated cover slips did not.

VII. Considerations for Microscopic Observation of Development of Single Embryos

A. Separation of Blastomeres

The study of later cleavage stages can be difficult, even in clear eggs, due to the interference produced by out-of-focus blastomeres above or below the cell of

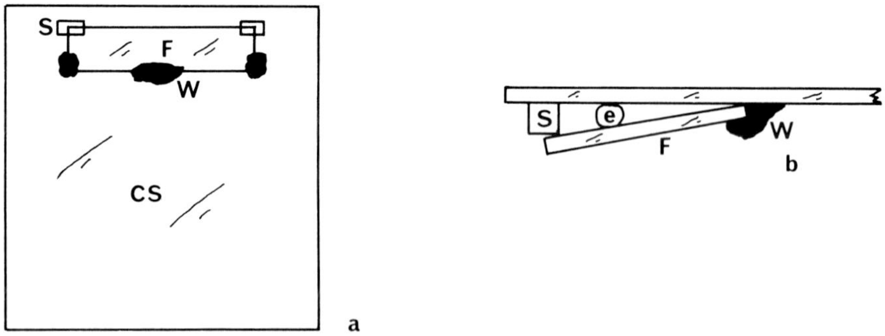

FIG. 4. Wedge manipulation chamber. (a) A cover glass fragment (F) is supported at the front by two small pieces of tape or Mylar (S) while it rests on the main cover slip (CS) at the back. It is positioned 1–2 mm from the front edge of the main cover slip, and is held in place by 3 dabs of tackiwax (W). For additional details of chamber and support slide construction see Kiehart (1982). (b) An egg (e) in the chamber in side view.

TABLE IV

PREPARING COVER SLIPS FOR CELL ADHESION[a]

1. Biocleaned cover slips are immersed in concentrated nitric acid for 30 minutes
2. Rinse well with plenty of glass-distilled water
3. Store in glass-distilled water
4. Remove the cover slip from water and blot off excess water by touching edge of cover slip to filter paper
5. Flood surface of cover slip with several drops of 1 mg/ml poly-L-lysine in glass-distilled water (approximate M_r 220,000; Sigma Chemical Co., St. Louis, MO) or 1% protamine sulfate in glass-distilled water (Gr. A, Calbiochem-Behring, La Jolla, CA). Aliquots of both solutions are kept frozen until use
6. Leave solution on cover slip 2–5 minutes
7. Rinse cover slip with glass-distilled water
8. Blot excess water from cover slip edge with filter paper and air dry

[a] Steps 1–3 may be done ahead of time; steps 4–8 are done *the day of use.*

interest. In a study of the nuclear migration and mitotic apparatus positioning prior to the unequal fourth division in vegetal blastomeres, we used one-half and one-quarter embryos produced by separating blastomeres following first and second division.

Following insemination, fertilization envelopes are removed by one of the previously described methods, and the embryos are cultured in Ca^{2+}-free artificial seawater. Blastomeres are separated from each other after the completion of first (one-half embryos) or second division (one-quarter embryos) by carefully pushing them apart with a fine glass needle. The needle is made by pulling a glass capillary tube in a microburner flame. Once the blastomeres are separated, they are collected and further cultured in calcium-containing artificial seawater. In seawater containing calcium, the hyaline layer reforms between the blastomeres of the subsequent divisions, "glueing" them together.

B. Light Source

For long-term observation of dividing embryos, standard microscope light sources must be appropriately filtered before reaching the specimen. Quartz–halogen, tungsten, mercury, and especially xenon lamps all emit infrared radiations which can heat the specimen and damage optical components (particularly sensitive are polarizing sheet filters). One should insert reliable heat-cut filters between the illuminator and the mirror reflecting the light to the polarizer and condensor (note that differential interference contrast optics also use polarizers).

Short-wavelength UV irradiation can cause cell damage and even death over a short period of time, and the addition of a UV-cut filter may be required if one is using a UV-rich source (e.g., mercury or xenon). Use of an interference filter for a longer wavelength, in combination with the glass lenses in the illuminating optical path, should sufficiently attenuate the UV. However, even wavelengths in the visible blue range can have noticeable detrimental effects on ciliary and flagellar motility as well as on cell division, depending on the species.

The best wavelength to select for illuminating the specimen is 546-nm green or 577/579 yellow light. Not only do they appear to be harmless to cellular activities but they have several other advantages as well. The mercury arc lamp has strong emission lines at these wavelengths. Optical components of the microscope are best corrected in this wavelength range, especially important when using achromatic objective lenses, and our eye has its peak sensitivity in green light (for photopic vision). It is worth the expense to obtain a high quality, 546-nm *narrow band-pass* interference filter for living cell observations. Many microscope manufacturers supply them or one may obtain them from manufacturers of optical components (e.g., Baird-Atomic, Inc., Cambridge, MA; Ealing Optics Corp., South Natick, MA). Upon receipt, it is a good idea to check the transmission spectrum of your filter with a spectrophotometer to make sure that it meets the published specifications. Although transmission of the rest of the visible spectrum would be expected to be quite low, they often transmit a surprising amount of the infrared spectrum, again stressing the need for a high-quality heat-cut filter to be placed directly following the illuminator.

C. Controlling Temperature during Observation

Many species have optimum development temperatures that are below normal room temperatures (e.g., *E. parma,* 16°C; *L. pictus,* 18°C), and therefore some mechanism must be employed to keep the temperature of the embryos under observation near optimum and preferably constant. One simple method is to observe the cultures for no longer than a minute or so at a time, after which they are immediately returned to a low-temperature incubator. However, if long-duration observation is required, another solution must be found.

The entire microscope can be placed in a temperature-regulated room. This is the easiest solution, although investigator comfort begins to be a problem at lower temperatures. The microscope should be placed in the room before the temperature is decreased or it should be insulated so that the temperature change is gradual. A sharp temperature drop can cause stress and strain in the glass and metal components, which can affect their performance. We have also directed the filtered output from an air conditioner, by channeling the cool air through a flexible duct, close to the stage of the microscope. This method has worked quite well in lowering the stage temperature up to 10°C below ambient. Troyer (1974) has described a thermoelectric, temperature-regulated stage providing for either heating or cooling, and we have developed a special slide for rapidly regulating the specimen chamber temperature (Inoué *et al.,* 1975).

D. Microscope Use and Cleaning

A detailed description of the proper use of the light microscope is beyond the scope of this article and we direct you to several other sources for this information. Slayter (1970), Spencer (1982), and Inoué (1986) contain details of image information and practical instructions for microscope alignment and use. Also, the microscope manufacturers, in addition to instructions for use of their own equipment, often have available excellent reference material on different image-forming systems, interpretation of the image, and theoretical aspects of image formation.

Cleanliness is of the utmost importance when using any microscope to its full potential. All optical components, from the illuminator to the oculars or camera, must be kept free of dust, oil, and fingerprints. Dust may be removed by using a soft brush (used only for this purpose), a blower bulb, or a can of compressed air. Commercial cans of compressed air often expel droplets of oil or propellant that are difficult to remove, and they must be used with caution. Oily residues and fingerprints can sometimes be removed by moist breath and a soft piece of lens tissue (hard lens tissues are not recommended as they can scratch the coatings present on lens surfaces). Do not rub the lens surface even with lens tissue. Rather, gently wipe, working from the center of the lens to the edge. Stronger organic solvents, such as ethanol or benzene, may have to be employed for stubborn oil contamination. Never put the solvent directly on the lens surface. Instead, place a drop onto the tip of a rolled piece of lens tissue and gently wipe the surface with the wet tip (see Inoué, 1985, for further details). Slides, when not on the microscope stage, should be supported above the bench top surface. This is easily accomplished by resting the slide on two wooden applicator sticks positioned under each end, and covering the cover slip area with a small Petri dish cover.

E. Types and Selection of Optics

Because of the generation of "phase halos" and the optical noise in the image plane produced by out-of-focus objects, the use of phase-contrast microscopy has slowly declined in recent years in favor of differential interference contrast (DIC) microscopy. However, in certain circumstances it may still be the method of choice. In studies of flagellar motility in living sperm and reactivated detergent models, we found that the phase-contrast image was of higher contrast and more easily recorded than the DIC image. In order to get the most from the phase-contrast microscope it is important to align precisely the phase annulus of the condenser with the phase ring in the objective lens. This is accomplished by observing the objective aperture using a telescope and moving the condenser annulus until it is superimposed over the dark phase ring of the objective. It is especially important to keep the preparation as clean as possible since out-of-focus images of dirt add greatly to the optical noise seen in the image, obscuring specimen detail.

Polarized light microscopy has been used to study the assembly, function, and disassembly of the mitotic spindle during cell division (review Inoué, 1981a) and the morphogenesis of the biocrystalline, spicule skeleton of the pluteus larva (Okazaki and Inoué, 1976). Contrast is produced by the interaction of plane-polarized light with an anisotropic alignment of molecules in the specimen. Using polarized light microscopy, one can quantitatively determine molecular arrangements inside living cells and observe how they change during the cell cycle or during differentiation.

Theoretical and practical aspects of polarized light microscopy are described in the references cited at the beginning of this section and in Bennett (1950). For information on the particular problems associated with biological polarized light microscopy and methods used to overcome them, see Swann and Mitchison (1950), Inoué and Dan (1951), Inoué (1961, 1981b, 1986), and Allen *et al.* (1981a).

A DIC microscope is a type of polarizing interference microscope. The shaded image that one observes is a result of the rate of change of the refractive index experienced by two beams of light separated (sheared) by a distance at or slightly less than the resolving power of the particular objective used. It gives good contrast to fine specimen detail when that structure is oriented perpendicular to the shear direction (the direction that the light appears to be coming from to give rise to the shaded image). In addition, depth of field discrimination is very good, and the specimen may be "optically sectioned."

There are two commercially available DIC systems: the Smith design manufactured by Leitz, and the Nomarski design manufactured by most other microscope companies. The difference between the two designs lies in the physical placement of the Wollaston prisms (the beam splitters). In the Smith system, the Wollastons are located at the condenser and objective aperture planes. This means that prisms are mounted in the objective lenses and these lenses are used exclusively for DIC. The prism design has been slightly modified in the Nomarski system such that the prism does not physically reside within the objective. Thus, the lenses may be used for other types of microscopy.

Fluorescence microscopy techniques are becoming increasingly valuable tools in the study of cell physiology and development. Vital fluorescent dyes have been found which bind or associate with specific locations or structures within the cell (DNA—Hoechst dyes, Latt and Statten, 1976). The fate of a microinjected molecule can be determined in the living cell if the molecule has a fluorophore group attached (Wang *et al.*, 1982; Kreis and Birchmeir, 1982; Salmon *et al.*, 1984). As discussed in the next section, certain fluorescent dyes can be used as cell markers in embryonic cell lineage studies.

Any standard microscope can be easily and inexpensively converted into a fluorescence microscope by inserting an appropriate excitation filter before the specimen, and an appropriate barrier filter after the specimen. While this arrangement is simple and relatively cheap, and high numerical aperature objectives can be used for the efficient collection of the fluorescent light, the separation of the strong excitation light from the weak fluorescent light by the barrier filter is difficult to accomplish, even with the best quality filters.

A dark-field condenser may be used to keep the illuminating excitation light from entering the objective, providing a low background level of light. However, objectives of rather low numerical aperture must be used in conjunction with the dark-field condensers, and the amount of fluorescent light collected by the objective is reduced.

An incident light or epiillumination system helps to solve these problems. In this case both excitation and emitted wavelengths are well separated by a dichroic beam splitter. The objective is also the condenser and high numerical aperture objectives may be used for efficient collection of the fluorescent light. This system is, however, the most expensive.

Selection of objective lenses for fluorescence work is a compromise between magnification (intensity is related to the inverse of the magnification squared) and numerical aperture (intensity is related to numerical aperture to the fourth power for epifluorescence). Many improvements in objective lens design for fluorescence microscopy have been made by the various manufacturers. They have begun to maximize numerical aperture for both low and high magnifications. The use of fluorite and other special optical glass elements has improved the transmission properties of these lenses down to 350 nm. Various immersion lenses have been produced for use with or without coverglasses to limit light loss from reflections at air–glass interfaces.

F. Use of Video for Data Recording

Recent advances in video technology, and the increased availability of high-quality, moderately priced equipment has led many investigators to use closed-circuit television systems in conjunction with a variety of light microscopical techniques. Besides the convenience of these systems for data recording, video imaging has many advantages for the researcher studying cell structure and physiology. In the following section, we will briefly point out some of the advantages of video. Further specific information may be obtained by consulting Allen *et al.* (1981a,b) and Inoué (1981b, 1986).

A dramatic increase in image contrast is obtained using video cameras. This contrast enhancement can be automatically set by the camera, or with many of the cameras now available, the operator has the option of manual control. A standard video camera records one field of information, each one-sixtieth of a

described here a method for rapidly generating stereoscopically paired, image intensity contours. The stereo pair can be generated in about 20 seconds and displayed in monochrome or pseudocolor, with its gray scale or color matching those of the video scenes. As illustrated, the intensity contours give rise to striking stereoscopic images.

ACKNOWLEDGMENT

We thank Dr. Gordon W. Ellis of the Department of Biology, University of Pennsylvania, for the helpful discussions and suggestions he made during the course of this project.

REFERENCES

1. ALLEN, R. D., J. L. TRAVIS, N. S. ALLEN & H. YILMAZ. 1981A. Video-enhanced contrast polarization (AVEC-POL) microscopy: a new method applied to the detection of birefringence in the motile reticulopodial network of *Allogromia laticollaris*. Cell Motil. **1:** 275–289.
2. ALLEN, R. D., N. S. ALLEN & J. L. TRAVIS.1981b. Video-enhanced contrast, differential interference contrast (AVEC-DIC) microscopy: a new method capable of analyzing microtubule-related motility in the reticulopodial network of *Allogromia laticollaris*. Cell Motil. **1:** 291–302.
3. DVORAK, J. A., L. H. MILLER, W. C. WHITEHOUSE & T. SHIROISHI. 1975. Invasion of erythrocytes by malaria merozoites. Science **187:** 748–750.
4. INOUÉ, S. 1981. Video image processing greatly enhances contrast, quality, and speed in polarization-based microscopy. J. Cell Biol. **89:** 346–356.
5. KACHAR, B. 1985. Asymmetric illumination contrast: a new method of image formation for video light microscopy. Science **227:** 766–768.
6. ALLEN, R. D. 1985. New observations on cell architecture and dynamics by video-enhanced contrast optical microscopy. Annu. Rev. Biophys. Biophy. Chem. **14:** 265–290.
7. KAMIYA, R., H. HOTANI & S. ASAKURA. *In* Prokaryotic and Eukaryotic Flagella. W. B. Amos & J. G. Duckett, Eds. 53–76. Cambridge University Press. London.
8. TILNEY, L. G. & S. INOUÉ. 1982. Acrosomal reaction of *Thyone* sperm. II. The kinetics and possible mechanism of acrosomal process elongation. J. Cell Biol. **93:** 820–827.
9. SCHNAPP, B. J., R. D. VALE, M. P. SHEETZ & T. S. REESE. Single microtubules from squid axoplasm support bidirectional movement of organelles. Cell **40:** 455–462.
10. YANAGIDA, T., M. NAKASE, K. NISHIYAMA & F. OOSAWA. 1984. Direct observation of motion of single F-actin filaments in thc presence of myosin. Nature **307:** 58–60.
11. INOUÉ, S. 1986. Video Microscopy. Plenum Press. New York, NY.
12. FRANÇON, M. 1961. Progress in Microscopy. Row, Peterson. Evanston, IL.
13. ELLIS, G. W. 1985. Microscope illuminator with fiber optic source integrator. J. Cell Biol. **101:** 83a.
14. INOUÉ, S., T. D. INOUÉ & G. W. ELLIS. 1985a. Rapid, stereoscopic display of microtubule distribution by a video-processed optical sectioning system. J. Cell Biol. **101(2):** 146a.
15. BAJER, A. S., J. MOLÈ-BAJER & S. INOUÉ. 1985. Three-dimensional distribution of microtubules in *Haemanthus* endosperm cells. J. Cell Biol. **101(2):** 146a.
16. INOUÉ, S., A. S. BAJER, J. MOLÈ-BAJER, M. DEBRABANDER, J. DEMEY, R. NUYDENS, G. W. ELLIS, E. HORN & T. D. INOUÉ. 1985b. Microtubules decorated with 5 nm gold visualised by video-enhanced light microscopy. J. Cell Biol. **101(2):** 146a.
17. ELLIS, G. W., S. INOUÉ & T. D. INOUÉ. 1986. Computer aided light microscopy. *In* Optical Methods in Cell Physiology. P. De Weer & B. M. Salzberg, Eds. Wiley. New York, NY.
18. GAUNT, W. A. & P. N. GAUNT. 1978. Three-Dimensional Reconstruction in Biology. University Park Press. Baltimore, MD.

19. DE MONTEBELLO, R. L. 1969. The RLM synthalyzer technique and instrumentation for optical reconstruction and dissection of structures in three dimensions. Ann. N. Y. Acad. Sci. **157:** 487–496.
20. FUCHS, H., S. M. PIZER, L. -C. TSAI, S. H. BLOOMBERG & E. R. HEINZ. 1982. Adding a true 3-D display to a raster graphics system. IEEE Comput. Graphics App. **2:** 73–78.
21. SHER, L. D. & C. D. BARRY. 1985. The use of an oscillating mirror for three-dimensional displays. *In* New Methodologies in Studies of Protein Configuration. T. T. Wu, Ed. Van Nostrand-Reinhold. Princeton, NJ.
22. AGARD, D. A. 1984. Optical sectioning microscopy: cellular architecture in three dimensions. Annu. Rev. Biophys. Bioeng. **13:** 191–219.
23. FAY, F. S., K. E. FOGARTY & J. M. COGGINS. 1986. Analysis of molecular distribution in single cells using a digital imaging microscope. *In* Optical Methods in Cell Physiology. P. De Weer & B. M. Salzberg, Eds. 51–64. Wiley. New York, NY.

The following note was added by Shinya Inoué in September of 2006:

Figure N-1. This photograph was taken in 1992 when the author was delivering the E.B. Wilson Award lecture at the Annual Meeting of the American Society for Cell Biology in San Francisco.

The audience is wearing Type-II Polaroid stereo-viewing glasses, which transmit left-circular polarized light to the left eye and right-circular polarized light to the right eye. The high-intensity video projector (from StereoGraphics Corporation of San Rafael, CA) is equipped with a liquid-crystal shutter that is synchronized to the incoming video signal and which bestows left- and right-handed circular polarization to the alternate 60-Hertz video field images. In the alternate video fields, the projector also re-expands the left- and right-hand images, which on the video tape had been compressed vertically and recorded half-height on top of each other. (In this way the full-height, left- and right-circular polarized images are projected sequentially in the two fields that make up one video frame; i.e., in 1/30th second.) Thus, the audience wearing Type-II Polaroid eyeware registers the left and right views only with the appropriate eye, which the brain fuses into flicker-free stereoscopic images (see also Article 54 in this *Collected Works*).

By using oppositely rotating circular polarizing filters, one avoids having to orient the eyeware horizontally, so as to match the projected images, as is the case when

Figure N-1.

one uses orthogonal linearly polarizing, Type-I Polaroid eyeware. Also, using the StereoGraphics viewing system, virtually everybody was able to perceive the stereoscopic images, whereas a significant percentage of individuals had trouble gaining stereoscopic views using red/green filters, Type-I Polaroid eyeware, or without filters by convergence/divergence of the eyes.

Various methods for viewing or capturing high resolution stereoscopic images through a microscope equipped with single objectives are described in Section 12.7 in the first edition of the author's *Video Microscopy* (1976). See also Article 2 in this *Collected Works*.

Reprinted from *International Journal of Invertebrate Reproduction and Development*, Vol. 11, pp. 335–354, 1987, with permission from Balaban Publishers.

Studies of Unequal Cleavage in Molluscs II. Asymmetric Nature of the Two Asters

Katsuma Dan* and Shinya Inoué

Marine Biological Laboratory, Woods Hole, MA 02543

and University of Pennsylvania, Philadelphia, PA 19104

** Home Address: 3-19-8 Kami-Yoga, Setagaya-ku, Tokyo 158*

Received November 13, 1986; Accepted December 1, 1986

Summary

By observing living eggs of *Spisula solidissima* with video polarized light microscopy, we confirmed that the metaphase spindle with its two asters (one slightly smaller than the other) forms at the center of the cell and then migrates to one side. The direction of spindle migration is presaged by the tilted apposition plane between the two pronuclei at an earlier stage. Spindle migration is apparently brought about by a centering movement of its larger aster; the smaller aster behaving as though it were pushed against the cortex, its rays being bent flat. After migration, the spindle undergoes a rocking motion pivoted on the centrosome of the central aster, as through "searching" for an anchorage site for the peripheral centrosome. Eventually the peripheral centrosome becomes fixed to the cortex on the side of the future AB-blastomere, about 30°higher than the equator. Thereafter, the cell cleaves asymmetrically producing the smaller AB- and larger CD-cell.

The unequal cleavage is thus achieved in two steps: (1) centering of the larger aster and (2) "searching" oscillation and eventual contact of the centrosome of the smaller aster to a specific site on the cortex.

Unequal cleavage, Astral asymmetry, Spindle anchorage

0168-8170/87 /$03.50

Introduction

Cell differentiation is often accompanied by unequal division which results in the production of one larger and one smaller daughter cell. In atral cleavage, the plane of cell division is determined by the orientation and location of the mitotic (including meiotic) spindle in early anaphase [1;2]. The cleavage furrow arises at telophase, in a plane bisecting the early anaphase spindle axis, and starting with the cell surface closest to that axis. Thus, asymmetric positioning of the early anaphase spindle leads to unequal cleavage, dependent on the location of the displaced spindle.

In bivalve molluscs, such as the surf clam *Spisula solidissima*, the eggs are fertilized before meiosis-I. Shortly after fertilization the first and second meiotic spindles assemble in sequence, leading to the extrusion of the respective polar bodies. In *Spisula*, the cleavages accompanying polar body formation, as well as the first mitotic division and the second and third mitotic divisions giving rise to the macromeres, are all unequal. In contrast, e.g., the division giving rise to the A and B cells in the second division cycle are equal. Such eggs then provide interesting opportunities for studying the relationship between unequal cleavage and spindle and aster behavior.

While the gametes of *Spisula* can be collected readily in large numbers, the eggs are somewhat opaque and interfere with detailed microscopic observation of their spindle and aster behavior in intact eggs. Rebhun [3] describes spindle and aster behavior in *Spisula* eggs vitally stained with methylene blue and toluidine blue, where the stained particles outlined the astral rays and the spindle. He observed the tilted first cleavage spindle appearing between the apposed pronuclei near the center of the egg, followed by a translation of the spindle towards the cell periphery. The spindle then exhibited a rocking motion which lasted until shortly before cleavage.

Dan and coworkers [4,5] isolated *Spisula* mitotic apparati involved in unequal cleavage. In the early stage, the isolated mitotic apparatus (MA) has a pair of asters both of which are similarly radiate in shape except for a slight difference in size between them [5, Fig. 6b–e]. At later stages, the aster closer to the center of the egg tended to be spherical, while the other was smaller and shaped like the umbrella of a shitaké mushroom [5, Fig. 6i–k]. Few astral rays pointed along the spindle axis and the centrosome was "exposed" poleward in the flattened aster. In the isolated MA, the adjoining cell cortex adhered tightly and selectively to the smaller aster [5, Fig. 7c].

In order to study the dynamic behavior of the spindle and asters leading to unequal cleavage in greater detail, we recorded and analyzed their behavior with video polarized light microscopy. The MA in living *Spisula* eggs, which is not readily discernible with ordinary light or phase contrast microscopy, can be detected with polarization optics, owing to the birefringence of the spindle fiber and astral ray microtubules. Interference from light scattering, which tends to obscure the birefringence of the MA in the yolky *Spisula* egg, is substantially reduced by video enhancement.

A preliminary note of this work was published elsewhere [6].

Materials and Methods

Spisula solidissima was acquired through the Marine Resources Department of the Woods Hole Marine Biological Laboratory. Gametes were collected by making incisions in the gonads, the sperm being aspirated dry into a pasteur pipete. The eggs were immediately transferred to sea water, the pH of which was adjusted to 5.3–5.5 by HCl. This medium maintains the fertilizability of the egg for many hours [5].

The theory and operating principles of the video polarized light microscope is described in detail elsewhere [7,8]. For the present work, selected ×20 and ×40 Nikon Plan Apo objectives were used in conjunction with a $\lambda/10$ Brace-Köhler compensator on a custom-built polarizing microscope [8, Figs. III-21, III-22]. The image was projected by a 5 to 8 power projection ocular directly on to the video image tube without an intervening camera lens. Image contrast was improved by the automatic black level and gamma control built into the Dage-MTI (model 65 II) 1-inch Newvicon camera. The video signal was recorded on a 3/4-inch cassette tape recorder (Sony TVO-9000), time-lapsed at 1/10 to 1/40 normal speed.

The behavior of the spindle was analyzed by repeated play back of the tape record, either directly or frame-by-frame after transfer to a motion analyzer (Sony SVM 1010). Spindle position and motion were determined with calibrated x-y fiducial markers placed onto the spindle poles in the video image, [with a Video analyzer (Colorado Video model 321)] or by measurements made on photographs of the video monitor. The images were calibrated with video records of a stage micrometer oriented along the horizontal and vertical video axes. For still records, the still field on the 9-inch monitor (Sanyo model VM-4209) was photographed on Polaroid 3-1/4 × 4 inch film.

Results

Previous findings on fixed eggs mentioned in the Introduction are summarized in Fig. 1. Eggs were fixed in 10% formalin in sea water, sectioned to 8μm and stained with haematoxylin. Figure 1 illustrates: (a,b), formation of the metaphase spindle at the center of the cell; (c), shift of the metaphase spindle to one side; (d), "attachment" of a spindle pole to the cortex, and the onset of chromosome separation. Figure 1d shows the oblique orientation of the spindle with reference to the egg axis. The radiate aster is located near the center of the cell while the flattened aster appears higher up (i.e., closer to the animal pole where the two polar bodies are visible).

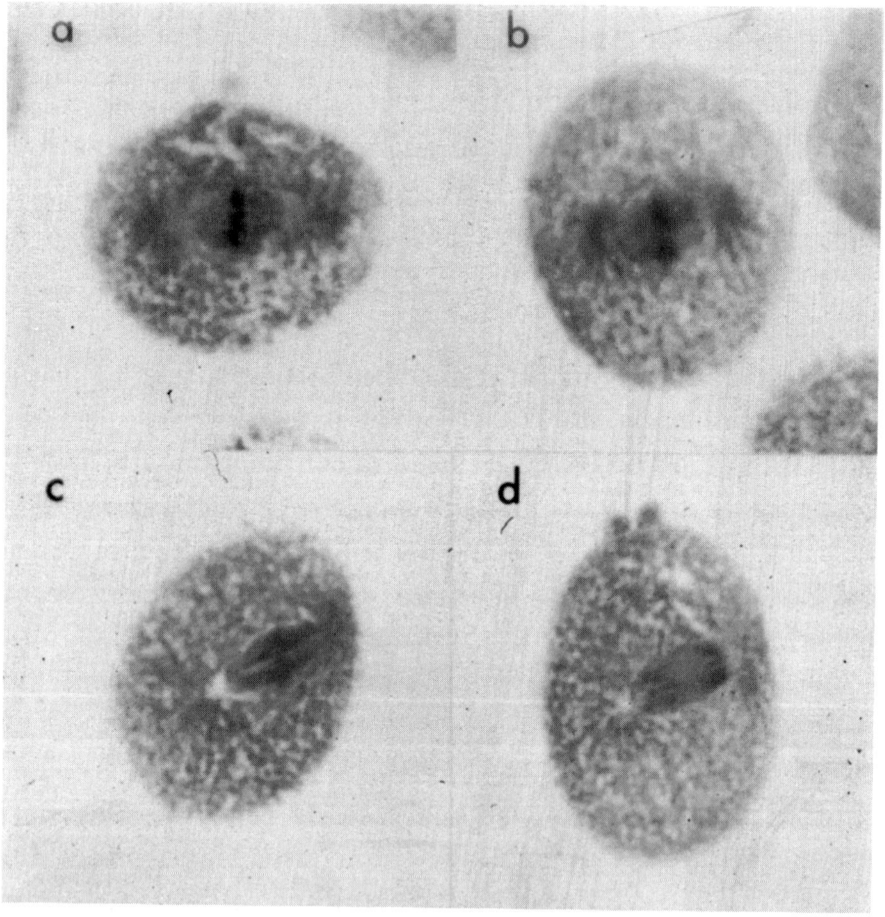

Fig. 1. Sections of eggs of the first cleavage of *Spisula*. Eggs were fixed in 10% formalin in sea water, sectioned to 8μm and stained with haematoxylin. a,b, Metaphase spindles at the center of the egg. c, Migration of the metaphase spindle to one side. d, Attachment of the peripheral pole of the spindle to the cortex and onset of anaphase. Spindle is oblique to the egg axis which is indicated by the location of the polar bodies.

Lillie [9], in his study of *Unio*, and Rebhun [3] in his study of supra-vitally stained *Spisula* zygotes, came to the same conclusion regarding spindle movement in these molluscan eggs. Lillie arranged sections of fixed eggs into temporal series, although he was not aware of the tilt of the spindle. (One of his figures does show an oblique spindle.) Rebhun reports the dynamic behavior of spindle, aster, and centrosome disclosed by the behavior of surrounding particles stained with methylene blue or toluidine blue.

Entering the main part of this report, we show selected frames from the video record to illustrate the behavior of the MA in an unequal cleavage.

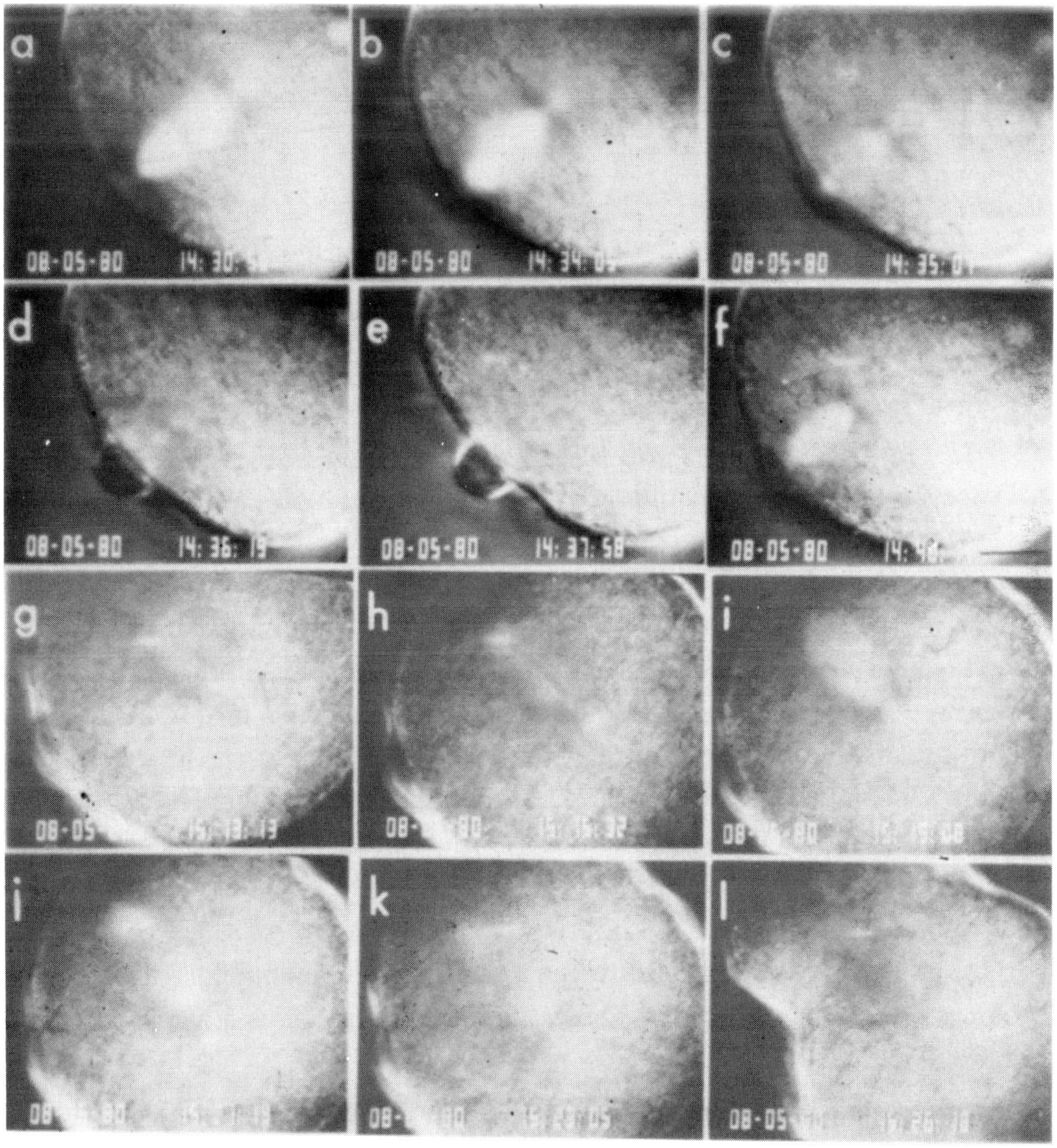

Fig. 2. Video time sequence of *Spisula* zygote division observed with polarized light microscopy. For explanation, see the text.

TABLE I.
Duration of each stage of cleavage in *Spisula solidissima*

Series I

			Duration	Actual Time
1st Cleavage	Metaphase			(15 12'40")
		Spindle at the center	2'20"	
				(15'00")
		Spindle migration	2'00"	
				(17'00")
		Stationary (no rocking)	3'00"	
				(20'00")
	Anaphase		20"	
				(20'20")
	Telophase		2'31"	
				(22'51")
	Furrowing		2'51"	
				(27'42")

( ) = Actual Time

Figure 2: This series (I) starts from the first polar body formation (a–e), through the second polar body formation (f), to the first division (g–l).

Features to be pointed out are: (1) The second polar body spindle (f) is shorter than the first polar body spindle (a), as was noticed in isolated preparations [5, Figs. 2,3]. (2) The dark bar in (g) and (h) is the apposed surface of the pronuclear envelope [3,4,5, Fig. 5] surrounded by the weakly birefringent spindle developing between the two astral centers. For detection of the weak birefringence, the compensator was rotated in quick succession according to standard procedure. This reverses the contrast of birefringence, making birefringent objects to flicker while images of non-birefringent objects remain unchanged. (3) Migration of the metaphase spindle to the periphery (i). (Frames showing the metaphase spindle at the center of the cell have been omitted. Consult Fig. 3a,b.) (4) Slight retreat of the MA from the surface as chromosomes separate (j,k), and finally unequal cleavage (l). The temporal relations of various phases of cleavage in Series I are tabulated in Table I.

In the unequal division of *Spisula*, the furrow invariably passes the animal pole marked by the presence of the polar bodies. Consequently, it is the position of the vegetal side of the furrow that is induced asymmetrically (with respect to the egg axis) by the eccentric and tilted spindle.

Figure 3: (1) Formation of the first cleavage MA at the center of the egg

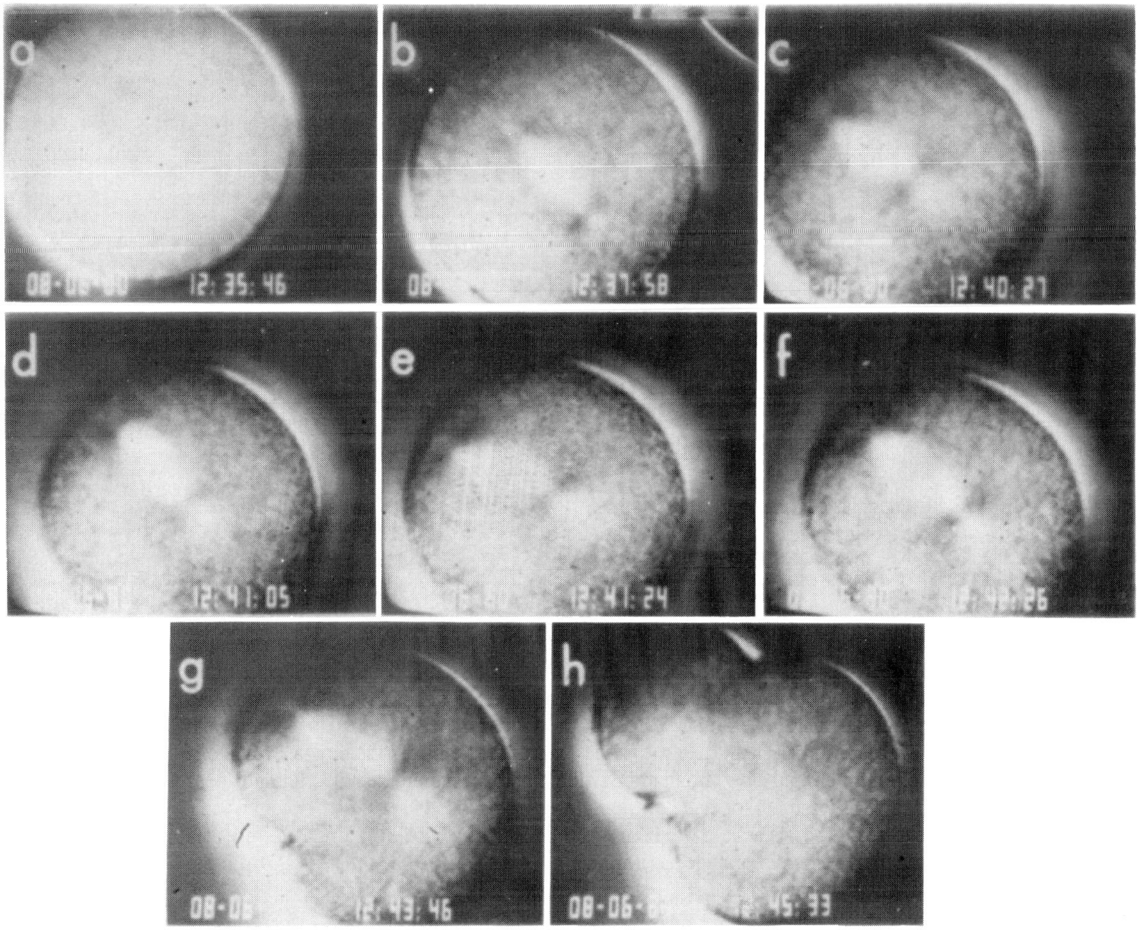

Fig. 3. Video time sequence of *Spisula* zygote division observed with polarized light microscopy. For explanation, see the text.

(a,b). (2) Shifting of the MA to one side (c,d). (3) Rocking motion of the peripheral pole of the spindle (c–e). (4) Cessation of the rocking and onset of anaphase (f). (5) Unequal cleavage (g–h). The remainder of this series (II) is not illustrated since it overlaps with the next series. The data on timing of the events are, however, included in Table II.

Figure 4: This series (III) illustrates (1) considerable rocking of the peripheral pole after the MA has migrated to the periphery (a–e); (2) anaphase while the rocking continues (e–f); (3) the unequal, first cleavage (e–j); (4) late resting stage (k); (5) formation of the MA's of the second cleavage at the center of the cells (1); (6) migration and rocking of the MA in the unequally dividing CD-blastomere but no *net* migration of the MA in the AB-blastomere (m–s); (7) the second cleavage (t–u), the division beginning earlier in the CD-cell than in the AB counterpart; (8) the 4-cell stage, in which the C-

344

Table 2. Duration of each stage in *Spisula solidissima* (continued)

Cleavage	Stage	Series II CD-cell	Series II AB-cell	Series III CD-cell	Series III AB-cell
1st Cleavage	(start)	(12 37'16")			
	Metaphase: Spindle at the center	1'40" — (38'46")			
	Metaphase: Spindle migration	54" — (39'40")			
	Metaphase: Rocking	1'51" — (41'31")		58" (11 50'15") — (51'13")	
	Anaphase	21" — (41'52")		35" — (51'48")	
	Telophase	(41'54") 1'45" — (43'37")		24" — (52'12")	
	Furrowing	3'23" — (47'00")		5'25" — (57'37")	
	Interphase	11'51"		10'52"	
2nd Cleavage	(start)	(58'51")		(12 08'29")	
	Metaphase: Spindle at the center	59" — (59'50")	Stationary at the center (no migration, no rocking)	1'16" — (09'45")	Stationary at the center (no migration, no rocking)
	Metaphase: Spindle migration	40" — (13 01'30")		2'31" — (12'16")	
	Metaphase: Rocking	3'53" — (04'23")	— (06'48")	1'26" — (13'42")	— (15'58")
	Anaphase	27" — (04'50")	38" — (07'26")	51" — (14'33")	13" — (16'11")
	Telophase	2'00" — (06'50")	1'29" — (08'55")	(15'19") 2'14" — (16'47")	1'48" — (17'59")
	Furrowing	3'30" — (10'34")	3'30" — (12'25")	1'52"	1'52" — (19'51")

TABLE II.
Duration of each stage in *Spisula solidissima* (continued)

Whether the eggs of Series I and II represent more normal behavior or whether the egg of Series III is more normal remains to be clarified in the future.

Because of this ambiguity, more accurate data were taken of the first cleavage of Series III by the procedure given at the end of Methods. As shown by the graph (Fig. 5), this spindle undergoes considerable rocking which continues during anaphase but stops before the onset of cleavage.

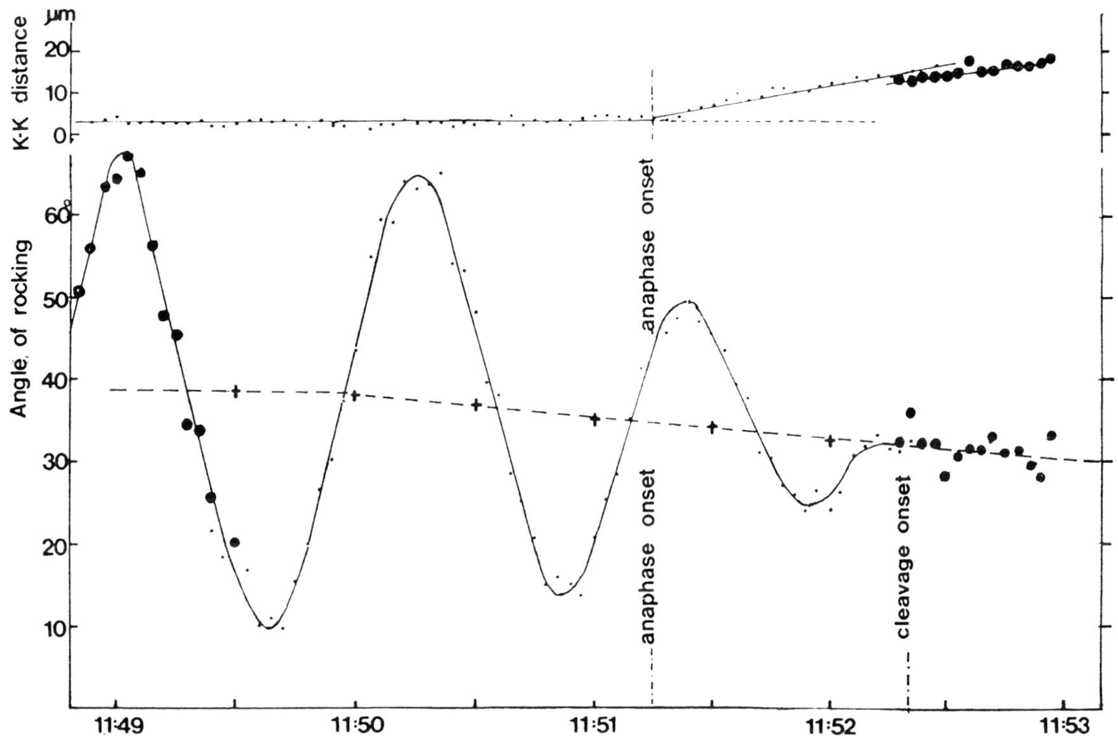

Fig. 5. Rocking angle of the spindle (bottom) and distance between chromosome groups (top) plotted against time in the first cleavage of Series III. •, ●: plotting by different calibration. + : average of the rocking angles.

Considering the entire body of information, it may be safer to think that final attachment of the smaller aster to the cortex (leading to the cessation of rocking) occurs during anaphase. It is particularly so in the light of examples in other animal species such as grasshopper neuroblasts [10] and newt spermatogonia under culture [11].

Returning to the *Spisula* rocking motion, tracings of the positions of the spindle poles and directions of the MA of the first cleavage of Figs. 2 and 3 are respectively superimposed and shown schematically in Fig. 6. The damped oscillations of the peripheral pole is as though the pole were looking for the best place to attach itself (Fig. 5), A further important point is the stability of the location of the central pole, in sharp contrast to the distal pole. The cental pole is held fixed in position while the distal pole is free to move (Fig. 6). This indicates that there is an important difference between the two asters in unequal cleavage.

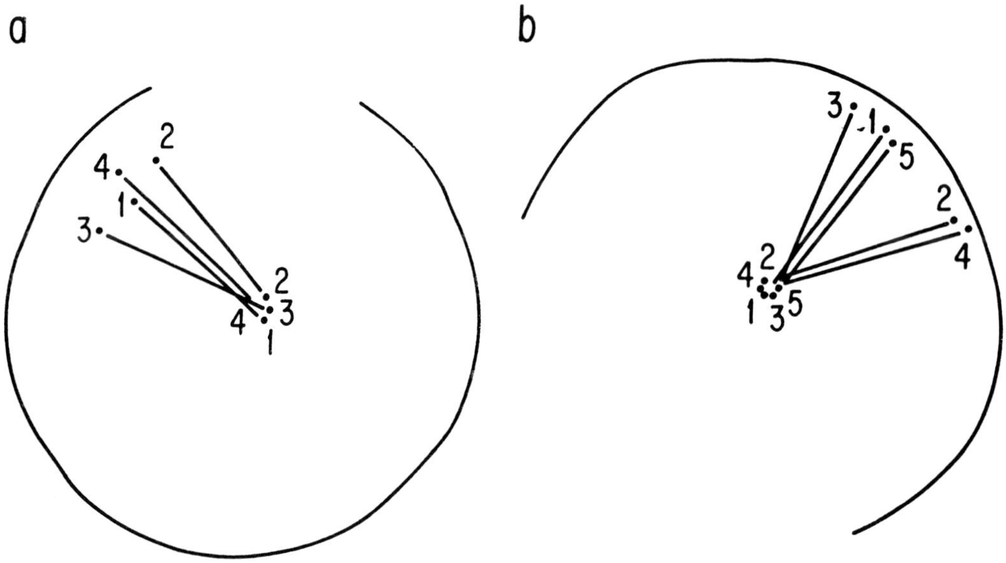

Fig. 6. Superposed tracings of the pole positions and the directions of the spindle during the rocking motion. a, First cleavage of Series II. b, First cleavage of Series III.

In equal cleavage of nematode eggs, oscillation of the first cleavage spindle has been reported [12, p. 385; 13]. In *Diplogaster logicauda* Claus, after the spindle takes a definitive orientation, both poles vibrate in and out of the axis. But by the time the rays reach the cell periphery, the movement stops and a cleavage follows immediately.

Rocking of the whole spindle with chromosomes was recently observed in the secondary spermatogonia of newt in culture [11]. The rocking is symmetrical for both ends, its frequency being 0.9–3.0 rocking/min and the rocking angle being 10–90 degrees at 25°C. The motion stops at mid-anaphase and a division follows. Both Colcemid (1.0μg/ml) and cytochalasin B (4.0μg/ml) stop the motion reversibly.

These findings indicate that in equal division, both asters are concurrently fixed in position before cleavage, while in unequal division, only one of the two asters is initially fixed in position.

A Working Hypothesis

With the above points in mind, we account for the *Spisula* type unequal cleavage in the following manner.

At early metaphase, the two asters are symmetrically situated at the two poles of the spindle located in the center of the cell. They are both radiate in shape, although with a small size difference [4]. This means that the tips of the rays of the larger aster may reach the cell surface while the tips of the rays in the smaller aster may end free (Fig. 7a).

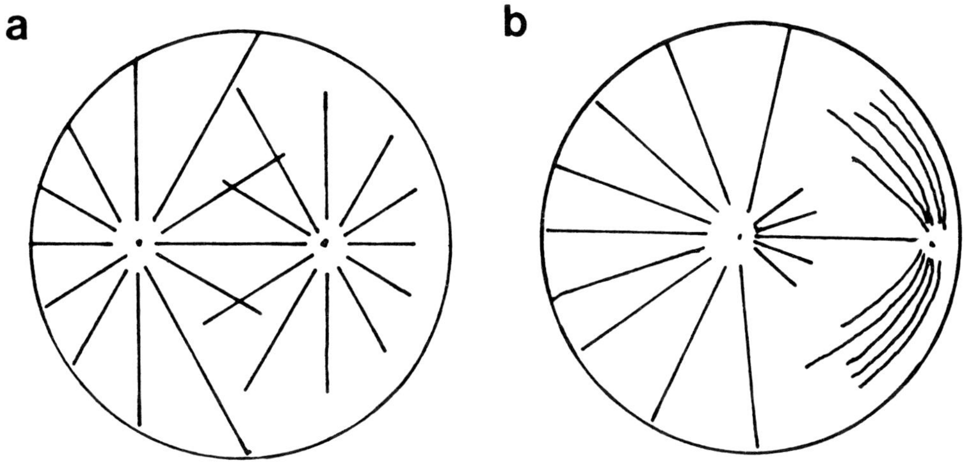

Fig. 7a,b. Models of the spindle behavior in unequal division of *Spisula solidissima*.

At the next stage, the larger aster moves to the center of the cell where its position becomes fixed. The centering of the larger aster could result from the shortening of "contractile" rays*or elongation of stiff rays. (The alternatives will be considered later.) After equilibration of forces among the rays, the final position of the astral center will roughly coincide with the center of the cell as shown in Fig. 7b. The centering larger aster, in turn, forces the spindle to push the free-moving smaller aster to one side, bending its rays and flattening the aster. It is at this stage that the rocking motion of the spindle and gliding of the peripheral centrosome and flattened aster take place.

* For this and the subsequent discussions, it is immaterial whether the "contractile" force is generated by the shortening rays or microtubules themselves, or by a sliding motor at the depolymerizing tip of the microtubules.

The reader may recall that in the fourth unequal division of sea urchin embryos, it is a resting nucleus which is shifted to the vegetal pole in the lower 4 cells of the 8-cell stage [14]. Considering the fact that in this migration in sea urchins, a centrosome is always found to escort and lead the moving nucleus, it can be presumed that a bundle of filamentous structure connecting the centrosome to a specially differentiated part of the vegetal cortex pulls the nucleus to this pre-determined spot. The specially differentiated part of the cortex is discernible in sea urchin embryos by EM [15].

This migration is presumably mediated by microtubules since it is arrested reversibly by Colcemid but not by cytochalasins [16]. In fact, the migration can be arrested by Colcemid until after nuclear envelope breakdown, at which time aster and spindle development can be made to resume by destroying the Colcemid with 366 nm light. Under this condition, microtubules reassemble and both the spindle and asters appear in the center of the vegetal pole cells after which the spindle is transported towards the proper position at the vegetal cortex. Again, the vegetal centrosome leads the migration, but this time, of the spindle rather than the nucleus.

In the metaphase-arrested oocyte in *Chaetopterus*, the meiotic spindle can be detached from its anchorage (to the animal pole cortex) by piercing and pulling at the centrad spindle pole with a micromanipulation needle. After detachment, a new contractile, fibrous link is established between the peripheral spindle pole and the specialized cortical site at the animal pole. The spindle is then drawn at a steady slow pace, precisely back to the specialized site [17]. When the detached spindle is inverted by micromanipulation so that the old centrad pole now becomes the peripheral pole, the new peripheral pole establishes a contractile link with the cortical site. The spindle is then oriented towards and brought back in a straight line to the animal pole, as though a row boat were being towed with a shortening rope to a specific site on the peer.

In the large *Chaetopterus* oocyte, the rays of the centrad aster appear unable to reach the cortex. Disturbing those rays behind the migrating spindle with the micromanipulator needle does not alter the pattern or rate of spindle migration. Whether the spindle has been inverted or not, and whichever centrosome is oriented towards the attachment site, the new link is not established and the spindle remains stranded unless a pole of the detached spindle is brought within approximately 35μm of the attachment site. This must be the limit of astral growth in *Chaetopterus* oocytes under standard conditions.

The fibers that re-establish the link and draw the peripheral centrosome to the specific cortical site in the *Chaetopterus* oocyte appear to be astral rays. They are Colcemid (but not cytochalasin) sensitive, and grow and contract at speeds of the order of 10μm/min. While those rays that link it to the specialized site must be contracting, the rest of the rays in the same aster are bent and lie tangent to the cortex in an umbrella shape.

In the *Spisula* egg, the peripheral astral rays are also bent in an umbrella shape, but the pole is not brought to its attachment site directly. Instead, as the spindle rocks, centered on its inner pole, the flattened peripheral aster and its exposed centrosome appear to glide against the cell cortex, as Rebhun [3] has also noted. In one of Lillie's sketches of a section of *Unio* egg, the centrosome is also shown in direct contact with the cell cortex [9].

As noted earlier, spindle rocking, and oscillation of the peripheral centrosome and aster, appear as a damped sine curve converging to the attachment site (Fig. 5). Perhaps in *Spisula* the unstable orientation of the spindle results from its being pushed from one side by the centering aster. Spindle orientation could eventually be stabilized by some of the rays in the peripheral ater which, while labile, progressively more effectively link the peripheral centrosome to its attachment site on the cortex. Alternatively, we cannot rule out the possibility that the microtubules in the peripheral aster are generating the rocking motion of the spindle by sliding against cortical elements distributed with a gradient peaking at the attachment site.

In general, the asters appear to behave dynamically as thought the rays from each centrosome were filopodia of a radiolarian or heliozoan (but with a difference that the astral rays contain only a single microtubule each). The filopodia grow slowly and then suddenly retract (or try to retract) when they contact prey or some other stimulus. Such a model is not only consistent with the dynamic behavior of mitotic microtubules observed in living cells [e.g., 18], but with very recent data reported on microtubule behavior *in vitro* and *in vivo*. Mitchison and Kirschner [19] showed in an "equilibrium" population of *in vitro* microtubules, grown on an isolated centrosome, that a few microtubules which have reached a critical length appear to suddenly and "catastrophically" fall apart and shorten, while at the same time other microtubules with various shorter lengths slowly add tubulin and grow. Cassimeris et al. [20] showed that microtubules in the interphase cytaster in living human monocytes behave similarly.

These data, taken together with the many cytological observations of spindle and astral behavior which have accumulated over the years, suggest that in an aster, many rays can extend and push, while others can contract and pull. In other words, the behavior of the aster expresses the behavior of its constituent rays and microtubules, which collectively can transport its center and spindle pole by extension of the rays and/or contraction of other rays in the same aster.

In short, we postulate, in the case of *Spisula*, that the forces associated with (1) the centering of the larger aster, and (2) the eventual attachment of the centrosome of the smaller aster to the cortex, are activated in succession to achieve spindle positioning and the consequent asymmetric cleavage. Whether or not a differentiated cortical region like the one in sea urchin embryos is, in fact, present in the *Spisula* eggs is a subject for future study.

Polyspermic Eggs

In a normal, mono-spermic egg of *Spisula* (Fig. 1), even after the attachment of the peripheral pole to the cortex, the radiating rays of the larger (central) aster are more or less straight (Fig. 1a–d). Moreover, the spindle is appropriately tilted toward the side of the future AB-blastomere, as is indicated by a tilting of the contact plane between the two pronuclei (ca. 40°from the egg axis) and of the first cleavage spindle (ca. 60°) [5]. Thus, the direction of the push is invariably toward the AB-side, the rocking motion introducing only a minor adjustment.

In contrast, in polyspermic eggs or in the case of polypolar spindles, one frequently encounters a clear vortex pattern of the rays of the larger aster (Fig. 8a,b). A tentative explanation is as follows: In abnormal polypolar spindles, one of the many poles may not necessarily face exactly the right spot on the cortex; more likely the spot may fall between two adjacent poles (see Fig. 9). If after centering of the large aster and attachment of its rays to the cortex, one pole is attracted to the right spot, the large aster would be twisted. In Fig. 8c, the smaller aster must initially have been directed toward the left side after which it must have been pulled up higher to attach where it is now. As the center rotates clockwise, the lagging rays of the fixed, large aster must be twisted anti-clockwise. In Fig. 8d, the twist of the rays is clockwise.

Painter [21] observed spiral asters in *Paracentrotus lividus* eggs kept in dilute phenyl urethane. But since these spiral asters were found among sister eggs which were already in the 2–4 cell stage, he must have been observing a delayed monaster cycle.

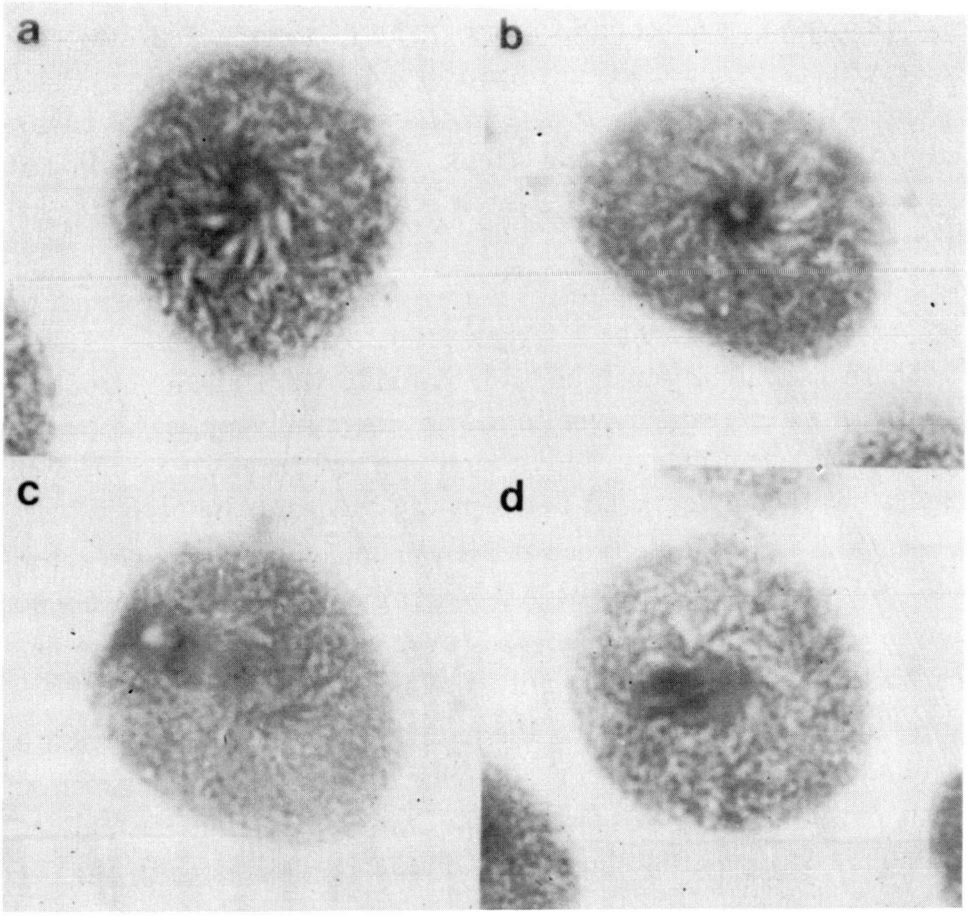

Fig. 8. Sections of polyspermic eggs. a,b, Vortex pattern of the rays. c, Displacement of a mis-directed pole to the right place, rotating the astral center clockwise which, in turn, twists the lagging rays counter-clockwise. d, Clockwise twist of the rays. In the section, although the attachment spot is missed, the direction of the spindle is obvious.

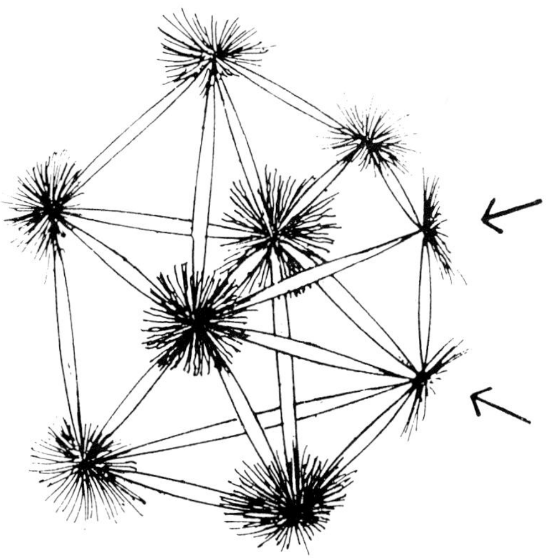

Fig. 9 A sketch of an isolated polyspermic or polypolar spindle. Arrows point to flattened asters.

Harris [22] and her collaborators [23,24] reported spiral fibers in *Strongylocentrotus purpuratus* eggs fixed in 2% glutaraldehyde in 0.45 M Na-acetate (pH 6). But the appearance of the spiral fibers is limited to stages between pronuclear fusion and the streak. In the mitotic stage, only straight rays are found. Therefore, both Painter's and Harris' cases differ from the present case of polyspermy.

On isolating polypolar spindles, it is possible to distinguish which pole has a smaller aster and which has a larger aster. In several cases, polyspermic MA's with two small asters lying side by side were encountered (Fig. 9). The two small asters were never found in diagonally opposite positions or alternating with larger asters. When a polypolar MA shifts to one side, both asters must have been squeezed between the MA and the cortex.

As a result, in taking video records, it was anticipated that in some cases, two small poles might go into the AB-blastomere. But this did not happen. This may be due to the condition that even when two asters are flattened, the attachment can happen with only one aster.

Summarizing, characteristics of unequal cleavage of *Spisula* eggs are based on the following three points: (1) intrinsic size difference between the two asters; (2) fixation of the position of the large aster and free movability of the small aster; (3) eventual anchorage of one smaller aster to the cortex.

Acknowledgements

This is a report of work done during the summer of 1980 when K. Dan was a Lillie Fellow. He expresses his gratitude to the Director of the Woods Hole Marine Biological Laboratory. The work was supported in parts by grants GM-31617 from NIH and DCB-8518672 from NSF.

References

1 Hiramoto, Y., An analysis of the cleavage stimulus by means of micromanipulation of sea urchin eggs, Exp. Cell Res., 68 (1971) 291–298.

2 Rappaport, R., Cytokinesis: cleavage furrow establishment in cylindrical sand dollar eggs, J. Exp. Zool., 217 (1981) 365–375.

3 Rebhun, L.I., Studies of early cleavage in the surf clam *Spisula solidissima*, using methylene blue and toluidine blue as vital stains, Biol. Bull., 117 (1959) 518–545.

4 Dan, K., Ito, S. and Mazia, D., Study of the course of formation of the mitotic apparatus in *Arbacia* and *Mactra* by isolation technique, Biol. Bull., 103 (1952) 292.

5 Dan, K. and Ito, S., Studies of unequal cleavage in molluscs. I. Nuclear behavior and anchorage of spindle pole to the cortex as revealed by isolation technique, Develop. Growth and Differ., 26 (1984) 249–262.
6 Inoué, S. and Dan, K., Mitotic spindle behavior in unequal cleavage of *Spisula solidissima*, Biol. Bull., 159 (1980) 443–444.
7 Inoué, S., Video image processing greatly enhances contrast, quality, and speed in polarization-based microscopy, J. Cell Biol., 84 (1981) 345–356.
8 Inoué, S., Video Microscopy, Plenum Press, New York, 1986.
9 Lillie, F.R., The organization of the egg of *Unio* based on a study of its maturation, fertilization, and cleavage, J. Morph., 17 (1901) 227–292.
10 Carlson, J.G., Anaphase chromosome movement in the unequal dividing grasshopper neuroblast and its relation to anaphase of other cells, Chromosoma, 64 (1977) 191–206.
11 Abe, S. and Nishikawa,A., Periodic rotation of chromosomes during the mitotic division in secondary spermatogonia of newt, *Cynops pyrrhogaster*, Develop. Growth and Differ., 23 (1981) 165–173.
12 Ziegler, H.E., Untersuchungenüber die ersten Entwicklungsvorgänge der Nematoden, Zeitschr. f. wiss. Zool., 60 (1896) 351–410.
13 Tadano, Y., Studies of cleavage in the eggs of nematode, I. Sci. Rept. Tohoku Univ. ser. 4, 19(1951) 100–103.
14 Dan, K., Studies on unequal cleavage in sea urchins. I. Migration of the nuclei to the vegetal pole, Develop. Growth and Differ., 21 (1979) 507–535.
15 Dan, K., Endo, S. and Uemura, I., Studies on unequal cleavage in sea urchins. II.Surface differentiation and the direction of nuclear migration, Develop. Growth and Differ., 25 (1983) 227–237.
16 Lutz, D.A. and Inoué, S., Colcemid but not cytochalasin inhibits asymmetric nuclear positioning prior to unequal cell division, Biol. Bull., 163 (1982) 373.
17 Hamaguchi, Y., Lutz, D.A. and Inoué, S., Cortical differentiation, asymmetric positioning and attachment of the meiotic spindle in *Chaetopterus pergamentaceous* oocytes, J. Cell Biol., 97 (1983) 245a.
18 Inoué, S. and Sato, H., Cell motility by labile association of molecules. The nature of mitotic spindle fibers and their role in chromosome movement, J. Gen. Physiol., 50 (1967) 259–292.
19 Mitchison, T. and Kirschner, M., Microtubule assembly nucleated by isolated centrosomes, Nature, 312 (1984), 232–237.
20 Cassimeris, L., Wadsworth, P. and Salmon, E.D., Dynamic instability and differential stability of cytoplasmic microtubules in human monocytes. In: Microtubules and Microtubule Inhibitors, M. DeBrabander and J. DeMay, Eds., Elsevier/North-Holland, Amsterdam, 1985, pp. 119–125.
21 Painter, T.S., Contribution to the study of cell mechanics. I. Spiral asters. J. Exp. Zool., 20 (1916) 509–527.
22 Harris, P., A spiral cortical fiber system in fertilized sea urchin eggs, Develop. Biol., 68 (1979) 525–532.
23 Harris, P., Osborn, M. and Weber, K., A spiral array of microtubules in fertilized sea urchin egg cortex examined by indirect immunofluorescence and electron microscopy, Exp. Cell Res., 126 (1980a) 227–236.
24 Harris, P., Osborn, M. and Weber, K., Distribution of tubulin-containing structures in the egg of the sea urchin *Strongylocentrotus purpuratus* from fertilization through first cleavage, J. Cell Biol., 84 (1980b) 668–679.

Reprinted from the *Journal of Cell Biology*, Vol. 105(4), p. 226a, 1987, with permission from the Rockefeller University Press.

HIGH-N.A. MICROSCOPE WITH AUTOMATIC CORRECTION FOR SPHERICAL ABERRATION

Shinya Inoué, Robert A Knudson, and Theodore D Inoué

Microscope objective lenses are designed to provide diffraction-limited, aberration-free images but only under narrowly specified optical conditions. High-NA (numerical aperture) dry objective lenses, used with cover slip thickness or working distance departing by only a few µm from those specified, exhibit spherical aberration and suffer from image degradation unless a correction collar is adjusted. Even oil immersion objectives lose spherical corrections when specimens mounted in aqueous media (but not in balsam) are focused more than ca. 10 µm away from the cover slip surface. We have devised a means for electromechanically coupling the motion of the correction collar to the objective lens fine focus; as the microscope is focused, a small computer automatically adjusts a stepper motor attached to the collar. This device substantially extends the focal range over which well corrected images can be formed by high NA objective lenses. Examples of image improvement over wide focal ranges will be demonstrated on several samples including: gold- and silver-stained microtubules and actin filaments in and around mitotic spindles, Golgi-stained neurons, and mitotic figures in living marine zygotes.

Grant support: NIH 2 R37 GM-31617; NSF DCB-8518672.

Reprinted from *Cell Motility and the Cytoskeleton*, Vol. 11, pp. 83–96, 1988.

Micromanipulation Studies of the Asymmetric Positioning of the Maturation Spindle in *Chaetopterus* sp. Oocytes: I. Anchorage of the Spindle to the Cortex and Migration of a Displaced Spindle

Douglas A. Lutz, Yukihisa Hamaguchi, and Shinya Inoué

Department of Biology, University of Pennsylvania, Marine Biological Laboratory, Woods Hole, Massachusetts (D.A.L., S.I.) and Biological Laboratory, Tokyo Institute of Technology (Y.H.)

We investigated the nature of the asymmetric positioning and attachment of *Chaetopterus* oocyte meiotic spindles to the animal pole cortex by micromanipulation. The manipulated spindle's behavior was analyzed in clarified oocyte fragments using video-enhanced polarized light microscopy. As the spindle was drawn towards the cell interior with a microneedle, the cell surface dimpled inwards adjacent to the outer spindle pole. As the spindle was pulled further inwards, the dimple suddenly receded indicating a rupture of a mechanical link between the cell cortex and outer spindle pole. The spindle paused briefly when released from the microneedle; then it spontaneously migrated back to the original attachment site and reassociated with the cell cortex. Positive birefringent astral fibers were seen running between the outer spindle pole and the cortex during the migration. The velocity of the spindle during its migration tended to increase as it came closer to the cortex. Velocities as high as 1.25 μm/sec. were measured. If removed too far from the attachment site cortex (> 35 μm), the spindle remained stationary until pushed closer to the original attachment site. Spindles, inverted by micromanipulation, migrated and reattached to the cortical site by their former inner pole; thus either spindle pole can seek out and migrate to the original attachment site. However, spindle poles pushed against other cortical regions did not attach demonstrating that there is only one unique, localized attachment site for spindle attachment.

Key words: unequal cell division, spindle positioning, spindle anchorage

INTRODUCTION

Early investigators [see reviews Wilson, 1925; Morgan, 1927], relating the cleavage furrow position of a dividing cell and the metaphase spindle axis, observed that the furrow cuts the cell in a plane midway between the two spindle poles. Conklin [1917] emphasized this close relationship when he observed that a spindle displaced from its original position by centrifugation determines a new location for cleavage furrow formation. Hiramoto [1956, 1971] showed, by the micropipet removal of the mitotic apparatus (MA) at various times of the mitotic cycle, that the presence of an early-to-mid metaphase MA is necessary for the determination of the cleavage furrow. Rappaport [review 1971] observed that a cleavage furrow can form between two asters of different spindles and concluded that a message for cleavage induction at the cell surface comes from the spindle pole or centrosome regions, as proposed earlier by Swann and

Received December 8, 1987; accepted May 16, 1988.

Address reprint requests to Dr. Douglas Lutz, Department of Biology, Gilmer Hall, University of Virginia, Charlottesville, VA 22901.

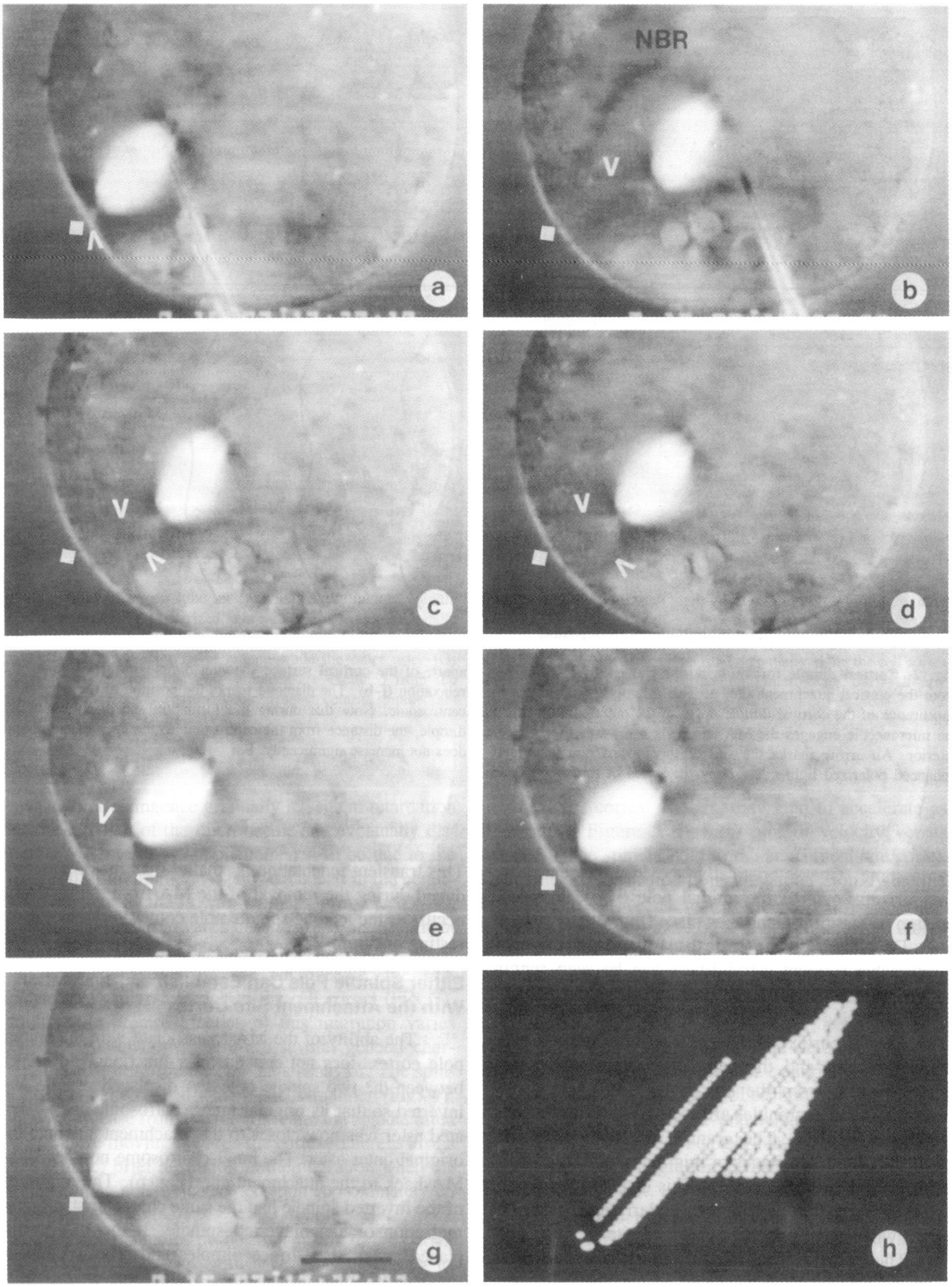

Figure 3

performed on the same spindle, and in each case the centrosome closest to the attachment site on the cortex lead the spindle back and reformed its cortical association.

As *Chaetopterus* fragments age, the inner aster occasionally dissociates giving rise to a "split spindle" with three separate centrosome regions. Any one of these three centrosomes can direct the spindle back to the attachment site and associate with the cortex (data not shown).

The Attachment Site's "Zone of Influence"

If the MA is manipulated too far from its original cortical location, its spontaneous return to the cortex does not occur. This distance, measured from the cortical attachment site to the closest centrosome, has been functionally defined as being >35 μm. Under these circumstances, the spindle remains in one location, exhibiting no directed movement after the needle is removed. If at any time after the original manipulation, the MA is nudged towards the cortical attachment site reducing the distance to <35 μm, the MA now returns to its cortical site spontaneously (Fig. 7). The spontaneous migration occurs with the same characteristics as before (Fig. 8). We have waited for various lengths of time up to 45 min before nudging the spindle towards the attachment site. In each case, the MA did not spontaneously return to the cortex until we moved the leading centrosome to within 35 μm of the attachment site.

The MA Can Only Attach to the Cortex at One, Highly Localized Site

During the course of the manipulation experiments we found that the MA only associates with one, highly

Fig. 3. Detachment and spontaneous return of a displaced meiotic spindle. **a:** As the spindle is moved towards the interior of the cell, a small dimple forms on the cell surface opposite the outer spindle pole. Positive birefringent astral fibers are clearly seen between the outer spindle centrosome and the cortex (arrow). **b:** A region of negative birefringence (NBR) appears in the cytoplasm as the spindle is drawn further into the fragment interior. **c–g:**. The spindle spontaneously returns to the attachment site cortex (white diamond) when the microneedle is removed. Positive birefringent astral fibers are seen in the zone between the advancing outer aster and the attachment site (arrows). Times of micrographs from needle engaging the spindle (min:sec): a, 0:00; b, 0:11; c, 0:38; d, 0:48; e, 1:08; f, 1:28; g, 1:45. Bar = 20 μm. Video-enhanced polarized light microscopy. **h:** This computer generated stick model, drawn from measured locations of the two spindle poles (lines) and the attachment site (dots), illustrates that the migration of a displaced spindle is preceded by its orientation towards the attachment site. After the needle is removed from the cell in b, the long axis of the spindle lies at an angle with respect to a line drawn from the attachment site to the outer centrosome. The outer pole first orients to point towards the attachment site, then the major portion of the migration takes place. The fragment moved slightly during micromanipulation causing the original spindle position to be displaced from the rest of the group.

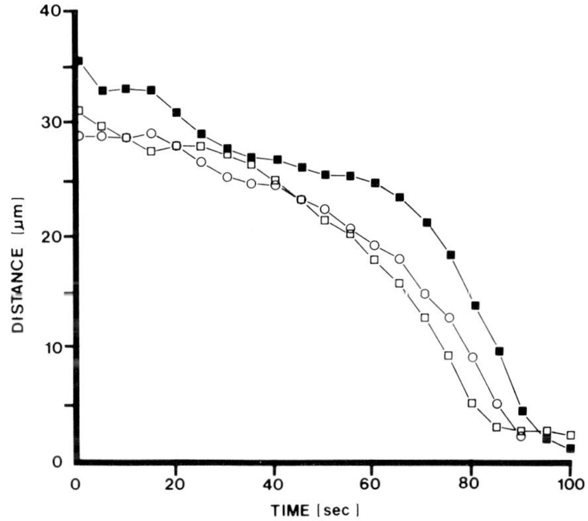

Fig. 4. Distance from the outer centrosome to the attachment site versus time. This graph shows the behavior of the migration during three manipulations of the same spindle. There are three general characteristics common to the different migrations. Following release from the microneedle (time 0), the spindle often pauses before moving towards the cortex. There is a characteristic pause in the migration when the outer centrosome is approximately 25 μm from the cortex. The slope of the curve increases as the spindle gets closer to the attachment site indicating that the migration velocity increases as the distance decreases. The shapes of three curves are quite different although they share the same general characteristics. This indicates that each manipulation is a single event.

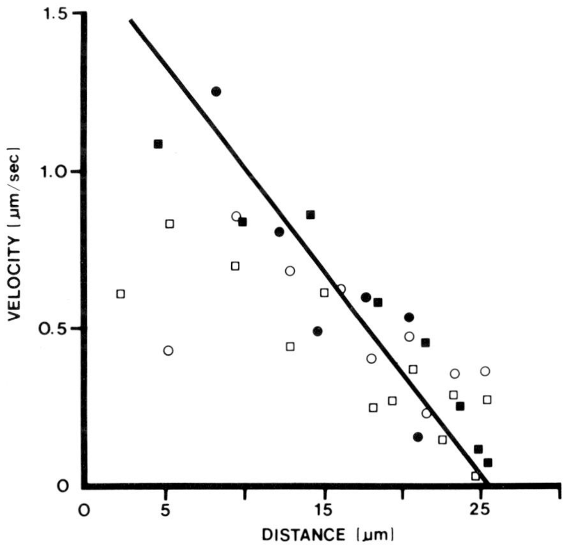

Fig. 5. Velocity of the spindle versus distance from the attachment site cortex. This graph shows the velocities measured from the distance travelled during a 5-sec interval for four separate manipulations. It is clear that the MAs velocity of migration, following the pause at approximately 25 μm, increases as the distance between the leading spindle pole and the cortex decreases. This acceleration continues until the spindle reaches or comes within approximately 5 μm of the starting position. The best fit line was calculated without including two low points ($x = 2.1$ μm, $y = 0.6$ μm/sec; $x = 5.0$ μm, $y = 0.4$ μm/sec).

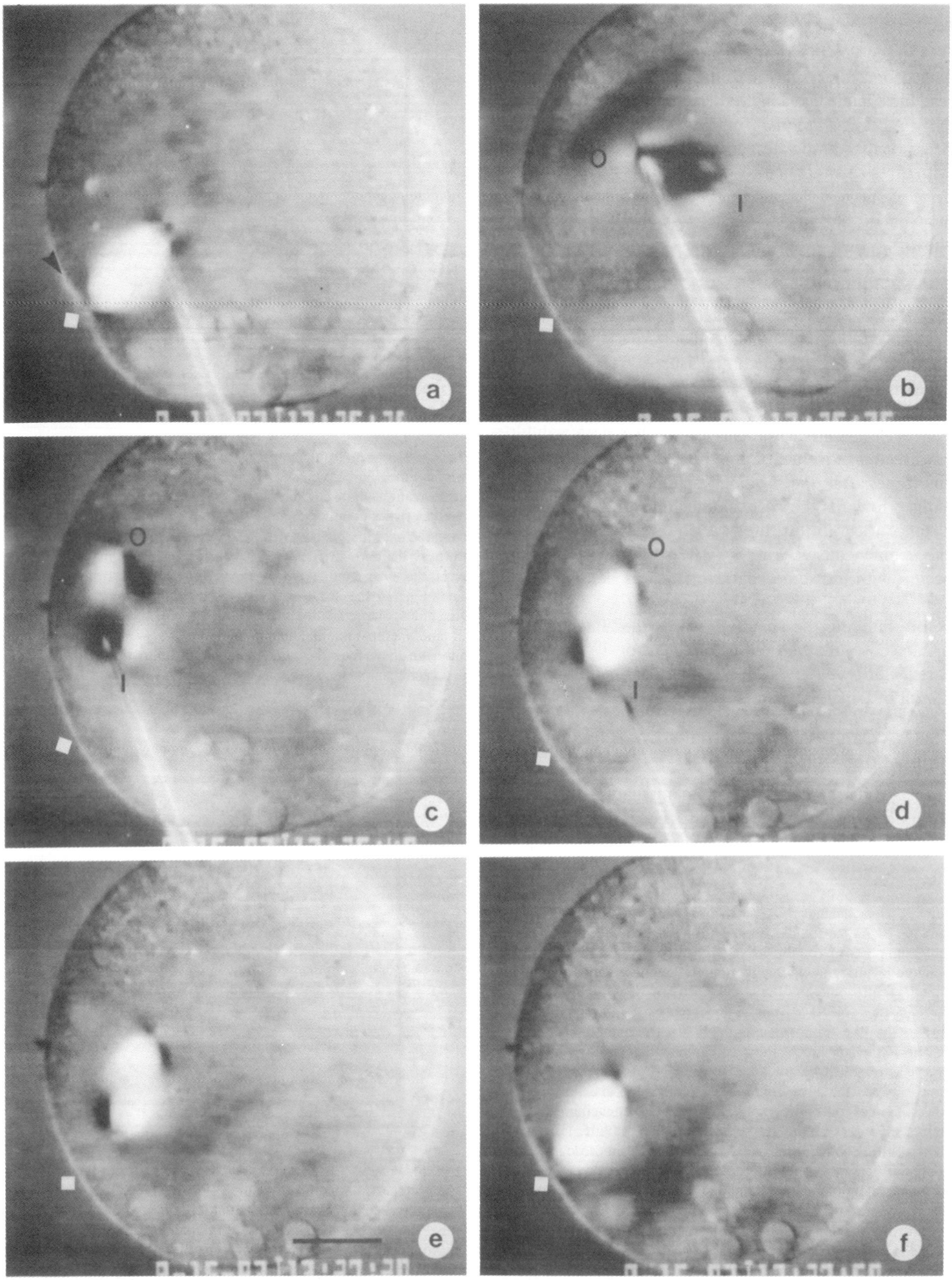

Fig. 6. Either spindle pole can form a link to the attachment site cortex. **a:** Position of the spindle during detachment. Note the small dimple as the spindle is drawn inward (arrow). **b–d:** Using the microneedle the spindle is inverted. The original inner pole (I) is now closer to the attachment site (white diamond) while the original outer pole (O) is in the fragment interior. **e–f:** The original inner pole leads the spindle back to the animal pole cortex and attaches. Times of micrographs from needle engaging the spindle (min:sec): a, 0:02; b, 0:11; c, 0:24; d, 0:36; e, 0:56; f, 1:26. Bar = 20 μm. Video-enhanced polarized light microscopy.

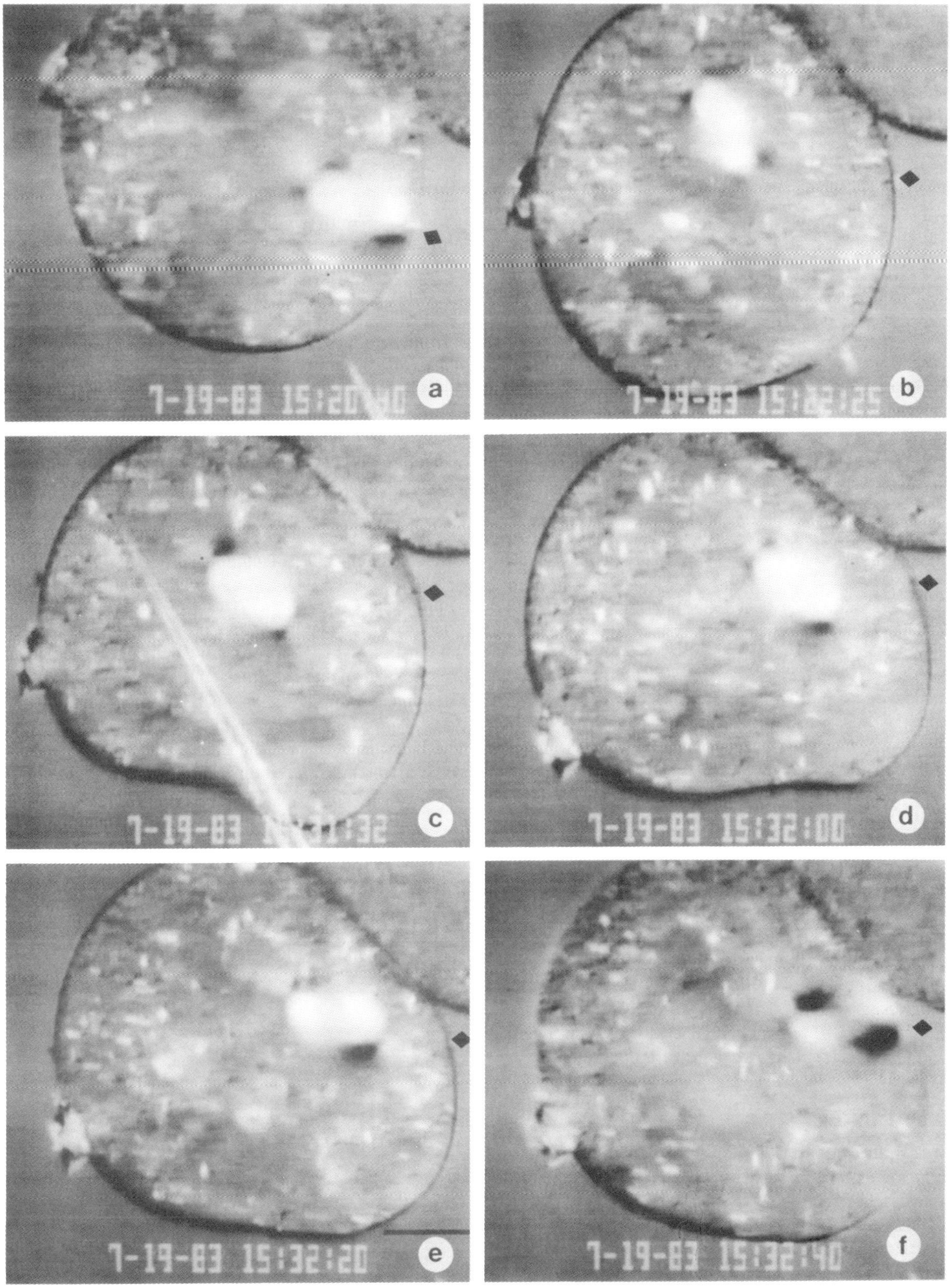

Fig. 7. Spontaneous migration of a displaced spindle does not occur if it is moved too far from the attachment site. **a:** Original spindle position before manipulation. **b:** The spindle is moved 35 μm away from the attachment site. **c:** After 8 min the spindle has not returned to the cortex and remains in the cell interior. The microneedle is inserted to nudge the spindle closer to the original attachment site for a second time. **e–f:** The spindle now migrates and associates with the cortex. Diamonds mark the original attachment site position. The fragment shifted during the manipulation. Times of micrographs from needle engaging the spindle (min:sec): a, −0:20; b, 1:25; c, 10:32; d, 11:00; e, 11:20; f, 11:40. Bar = 20 μm. Video-enhanced polarized light microscopy.

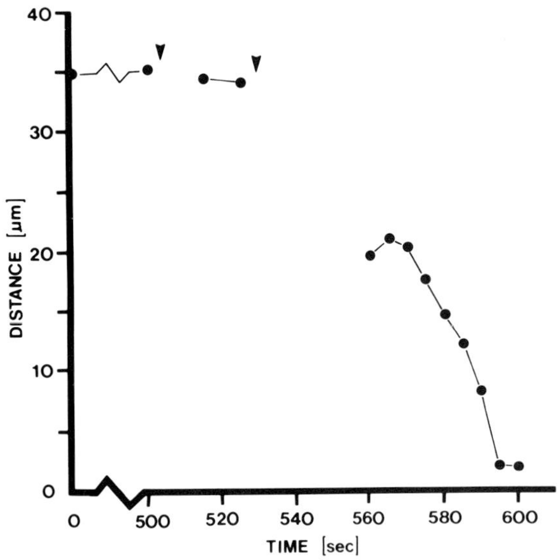

Fig. 8. Distance versus time graph of the manipulation described in Figure 7): The spindle is removed 35 μm away from the attachment site cortex. After 8 min when the MA has not spontaneously returned to the cortex, the MA is nudged closer to the cortex twice (arrows). Following the second nudge the spindle spontaneously returns to the cortex.

localized site on the egg fragment cortex. If a series of manipulations was performed on one fragment, the spindle always returned to exactly the same site on the cortex from which it was removed. To test whether the MA could attach to sites on the cortex other than the original attachment site, the spindle was held against other cortical regions for several minutes with the microneedle (Fig. 9). When the needle was removed, the spindle actually moves away from these other cortical locations. If the MA lies within the "zone of influence," it will orient towards the animal pole, and migrate back exactly to the original attachment site. If the MA lies outside the "zone of influence" as it does in Figure 9, it will move a short distance away from the cortex region it was held against, and will remain in this new location until nudged within the attachment site "zone of influence."

DISCUSSION

Lillie [1906, 1909] observed that when eggs of *Chaetopterus* were centrifuged after the formation and association of the meiotic apparatus to the cortex, occasionally the MA became detached and was found in the clear zone of the stratified egg away from the cell cortex. When these eggs were fertilized, he observed that some made normal polar bodies. A yolk-free channel was seen running from the clear zone through the yolk to the site of polar body formation in sections of these fixed, centrifuged eggs. He suggested that upon fertilization, the detached maturation spindle returned to the cortex and participated in normal polar body formation. The routine preparation of relatively yolk-free fragments of living *Chaetopterus* eggs has enabled us to examine the dynamics of detached spindle behavior in these eggs. The results are discussed below in terms of what they tell us about the nature of the spindle attachment to the cell cortex and about the nature of the attachment site.

The inward dimpling of the cell cortex adjacent to the outer centrosome region upon manipulation and upon the displaced MA's return clearly demonstrates that a mechanical linkage exists between the spindle and the cortex. Early centrifugation experiments by Lillie [1909] and Conklin [1916, 1917] suggested that the maturation spindle was anchored to the cortex in eggs of several species. Dan and coworkers [Dan et al., 1952; Dan, 1960; Dan and Ito, 1984] found pieces of cortex associated with the outer aster, in molluscan eggs undergoing unequal cell division, that remained through their spindle isolation technique. These observations also support that the MA in cells undergoing unequal cell division is firmly attached to the cell cortex. Our observation of cortical dimpling in *Chaetopterus* fragments during each of several successive manipulations of the same spindle further demonstrates that this mechanical linkage is labile and may be broken and reformed several times. The positive birefringence observed between the outer spindle pole and the apex of the dimple indicates that astral fiber microtubules run from the pole to the dimple, and that outer aster fibers are intimately associated with the attachment site cortex.

It is highly unlikely that the spontaneous return of a displaced MA is due to the retraction of a connection maintained between the MA and the attachment site after displacement. First, the extension and relaxation of the cortex dimple during the initial phase of MA manipulation indicates that the mechanical link between the MA and the cortex is broken during manipulation. Second,

Fig. 9. A displaced spindle does not associate with any other region of the cell cortex. **a:** Original spindle position before manipulation. The diamond next to the nub on the cell surface marks the original attachment site. **b–c:** The spindle is held against another region of the cell cortex for 2 min. **d:** After the needle holding the spindle is removed, the MA moves away from the cortex. **e–h:** Following a nudge to move the spindle within the "zone of influence," the MA migrates back to the original attachment site. Note that the MA first orients towards the attachment site before the major portion of the migration takes place. Times of micrographs from needle engaging the spindle (min:sec): a, -0:03; b, 0:15; c, 1:43; d, 4:31; e, 5:02; f, 7:02; g, 8:02; h, 8:49. Bar = 20 μm. Video-enhanced polarized light microscopy.

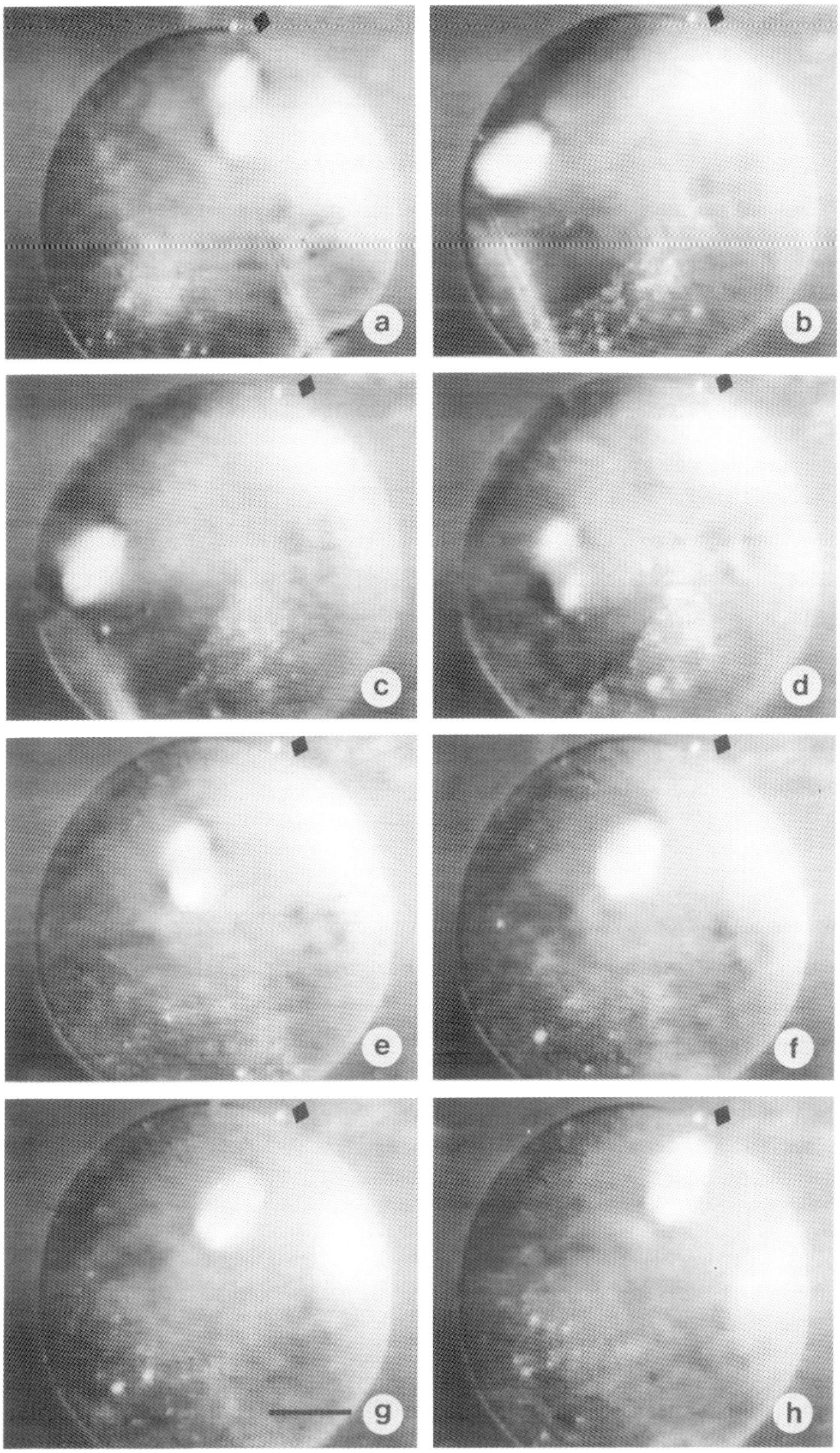

Figure 9

the observation that either spindle pole can lead the spindle's return to the cortical attachment site suggests that a new association takes place between the centrosome closest to the cortex and the attachment site. That a cortical dimple is observed during a successive manipulation clearly demonstrates that a new mechanical linkage between spindle and cortex has been formed.

The spontaneous return of the displaced MA is unique in several ways. First, the rate of migration tends to increase as the spindle gets closer to the cortex. This acceleration indicates that either the force responsible for the movement also increases as the distance between the spindle and the cortex decreases, or that the resistance to spindle movement decreases with decreasing distance. Our experiments thus far cannot clearly distinguish between these two explanations although observations of successive manipulations of the same spindle seem to indicate that the first explanation of an increasing force appears to be the correct interpretation. We observed that the migration kinetics following several manipulations of the same spindle seemed to bear no direct relation to one another. Each manipulation could be considered to be a separate event, and the spindle's past behavior in a previous migration was not correlated to its behavior during the current spontaneous migration. This was true even if the spindle migration followed a similar path back to the cortex in two successive manipulations. One might expect, if the acceleration of the spindle as it approaches the cortex is due to a decrease in resistance to movement, that the resistance to migration would lessen each time the spindle was drawn through the cytoplasm, and each time it migrated back to the cortex. This decrease might be manifested progressively as an increase in maximum migration velocity in each successive manipulation. Such a progressive increase in maximum velocity was not observed.

Second, characteristics of the spontaneous migration differ from those observed for other mitotic apparatus associated motilities. The maximum rate for spindle migration observed in this study is close to an order of magnitude greater than the chromosome-to-pole migration in anaphase A. In anaphase, chromosomes migrate at a constant velocity, whereas the MA in *Chaetopterus* tends to accelerate during a migration following micromanipulation. Salmon [1975] induced the depolymerization of spindle microtubules in *Chaetopterus* by hydrostatic pressure and observed that the inner pole and chromosomes moved to the cortex, at a constant velocity, during a controlled disassembly of the structure. The greater the pressure applied, the faster the movement of the chromosomes to the cortex up to a certain point. Greater pressures past this point resulted in a disassembly of the spindle without generating the force for chromosome-to-cortex movement. Even in this system, where the maximum velocity generated by controlled disassembly was experimentally determined, the maximum rate observed by Salmon is still at least four times slower than the maximum velocity measured during our spindle manipulations.

One spindle associated motility whose rates of movement are in the range and even greater than the rate of the MA migration is the saltatory granule movements observed along the astral rays [Rebhun, 1972; Hamaguchi et al., 1986]. Granule speeds of 0.1–2 μm/sec are common. However, two characteristics of this motility are quite different from the spindle migration. First, the granules tend to move continuously along the astral rays for only short distances. Second, granules are observed to start, stop, and restart their motion, sometimes maintaining the same direction, but at other times reversing their direction of travel.

The migration of the female pronucleus, along astral fibers, to the male pronucleus in fertilized sea urchin eggs occurs initially with increasing velocity [Hamaguchi and Hiramoto, 1980]. In this way pronuclear migration is similar to the MA migration observed in *Chaetopterus* fragments. One difference is that in pronuclear migration, the female pronuclear velocity increase only occurs during the first half of the migration. After that, the female pronuclear velocity decreases until it reaches the male pronucleus. In *Chaetopterus*, the MA migrates with an increasing velocity until it reaches or comes very close to the attachment site cortex.

Our observations of positive birefringent astral fibers running between the outer centrosome and the cortical dimple during the removal of the MA from its cortical site, and between the outer centrosome and the cortex during the spontaneous return of the MA to its attachment site, strongly suggest that the astral fibers are important in both the anchorage of the spindle to the cortex and in the mechanism of the migration of a displaced spindle. The extent of the "zone of influence," functionally determined to be approximately 35 μm, is very similar to the length of the longest astral microtubules measured using anti-tubulin immunofluorescence [Lutz, 1984; Lutz and Inoué, in preparation]. It is tempting to speculate that the reason for a displaced MA to remain stationary within the fragment, if removed too far from the cortical attachment site, is the inability of its astral microtubules to interact with the attachment site cortex because they are too short. If the MA is then moved closer to the attachment site, its astral microtubules are now able to interact with the cortex and the migration ensues.

Czihak [1973] and Kawamura [1977], in studies of unequally dividing cells, have suggested that a partial disassembly of one aster or an asynchronous growth of the asters plays an important role in positioning the MA in its asymmetric location. In both cases the result would be that the aster on the smaller daughter cell side would be smaller than the aster on the larger cell side. This is clearly not the case in the *Chaetopterus* maturation spindle. Astral microtubule lengths of both inner and outer asters are comparable. Either spindle pole and associated aster is able to direct the spindle back to the attachment site with indistinguishable kinetics. The lateral granule movements observed during the return of the MA to the cortex, in the region between the aster and the cortex, suggest that a deformation of the aster along the inner cortical surface accompanies the migration, rather than a reduction in size or disassembly of this aster. Deformation of one of the asters was also suggested by Dan and Ito [1984] and by Dan and Inoué [1987] to be responsible for the asymmetric positioning of the MA during unequal cleavage divisions of *Spisula*. Although either spindle pole could form an attachment with the cortex, the information for the specificity of attachment location is located within the cortex. Attempts to force the spindle to associate with cortical sites other than the attachment site failed. We believe that the behavior we observed in this study reflects the association of the maturation spindle with the animal pole cortex following germinal vesicle breakdown and MA formation. The behavior of the displaced spindle to always return to its original attachment site indicates that it is important that the maturation divisions take place at this position. This in turn may be either a reflection of or an important step in the establishment of embryonic polarity. Whatever force acts to initially position the MA at the animal pole attachment site continues to do so during the meiotic arrested state prior to fertilization.

ACKNOWLEDGMENTS

The authors would like to thank Dr. R. Stephens, Dr. S. Zigmond, Dr. W. Telfer, Dr. E.D. Salmon, and Mr. R. Akins for their helpful comments and suggestions. We also thank M. S. Lutz for her suggestions and for her assistance in the preparation of the manuscript. We gratefully acknowledge the following support: NSERC of Canada Postgraduate fellowship to D.A.L., The Jean and Katsuma Dan Fellowship Program to Y. H., and NIH 2 R37 GM31617-05 and NSF DCB-8518672 to S.I.

REFERENCES

Borisy, G.G., and R.R. Gould (1977): Microtubule-organizing centers of the mitotic spindle. In M. Little, N. Paweletz, C. Petzelt, H. Ponstigl, D. Schroeter, and H.P. Zimmerman (eds.): "Mitosis: Facts and Questions." New York: Springer-Verlag, pp. 78–87.

Conklin, E.G. (1916): Effects of centrifugal force on the polarity of the eggs of *Crepidula*. Proc. Natl. Acad. Sci. U.S.A. 2:87–90.

Conklin, E.G. (1917): Effects of centrifugal force on the structure and development of the eggs of *Crepidula*. J. Exp. Zool. 22:311–419.

Costello, D.P., and C. Henley (1971): Methods for obtaining and handling marine eggs and embryos. 2nd Ed. Woods Hole, MA: Marine Biological Laboratory, 247 p.

Czihak, G. (1973): The role of astral rays in early cleavage of sea urchin eggs. Exp. Cell Res. 83:424–426.

Dan, K. (1960): Cyto-embryology of echinoderms and amphibia. Int Rev. Cytol. 9:321–367.

Dan, K., and S. Inoué (1987): Studies of unequal cleavage in molluscs. II. Asymmetric nature of the two asters. Int. J. Invert. Reprod. Dev. 11:335–354.

Dan, K., and S. Ito. (1984): Studies of unequal cleavage in molluscs: I. Nuclear behavior and anchorage of a spindle pole to cortex as revealed by isolation technique. Dev. Growth Diff. 26:249–262.

Dan, K., and T. Nakajima (1956): On the morphology of the mitotic apparatus isolated from echinoderm eggs. Embryologia 3:187–200.

Dan, K., S. Ito, and D. Mazia (1952): Study of the course of formation of the mitotic apparatus in *Arbacia* and *Mactra* by isolation techniques. Biol. Bull. 103:292.

Haimo, L.T., and J.L. Rosenbaum (1981): Cilia, flagella, and microtubules. J. Cell Biol. 91:125s–130s.

Hamaguchi, M. S. and Y. Hiramoto (1980): Fertilization process in the heart-urchin, *Clypeaster japonicus* observed with a differential interference microscope. Dev. Growth Diff. 22:517–530.

Hamaguchi, M.S., Y. Hamaguchi, and Y. Hiramoto (1986): Microinjected polystyrene beads move along astral rays in sand dollar eggs. Dev. Growth Diff. 28:461–470.

Hamaguchi, Y., D.A. Lutz, and S. Inoué (1983): Cortical differentiation, asymmetric positioning and attachment of the meiotic spindle in *Chaetopterus pergamentaceous* oocytes. J. Cell Biol. 97(5):254a.

Hiramoto, Y. (1956): Cell division without mitotic apparatus in sea urchin eggs. Exp. Cell Res. 11:630–636.

Hiramoto, Y. (1962): Microinjection of the live spermatozoa into sea urchin eggs. Exp. Cell Res. 27:416–426.

Hiramoto, Y. (1971): Analysis of cleavage stimulus by means of micromanipulation of sea urchin eggs. Exp. Cell Res. 68:291–298.

Hiramoto, Y. (1974): A method of microinjection. Exp. Cell Res. 87:403–406.

Inoué, S. (1981b): Video image processing greatly enhances contrast, quality, and speed in polarization based microscopy. J. Cell Biol. 89:346–356.

Inoué, S. (1981a): Cell division and the mitotic spindle. J. Cell Biol. 91:131s–147s.

Inoué, S. (1986): "Video Microscopy." New York: Plenum Press, 584 p.

Kawamura, K. (1977): Microdissection studies on the dividing neuroblast of the grasshopper, with special reference to the mechanism of unequal cytokinesis. Exp. Cell Res. 106:127–137.

Kiehart, D. P. (1982): Microinjection of echinoderm eggs: Apparatus and procedures. In Wilson, L. (ed.): "Methods in Cell Biology," Vol. 25. New York: Academic Press, pp. 13–31.

Kirschner, M. (1978): Microtubule assembly and nucleation. Int. Rev. Cytol. 54:1–71.

Lillie, F. R. (1906): Observations and experiments concerning the elementary phenomena of embryonic development in *Chaetopterus*. J. Exp. Zool. 3:153–268.

96 Lutz et al.

Lillie, F.R. (1909): Polarity and bilaterality of the annelid egg. Experiments with centrifugal force. Biol. Bull. 16:54–79.

Lutz, D.A. (1984): Asymmetric cell center placement prior to unequal cell division. Ph. D. dissertation, Department of Biology, University of Pennsylvania, 148 p.

Lutz, D.A. and Inoué, S. (1986): Techniques for observing living gametes and embryos. In Wilson, L. (ed.): "Methods in Cell Biology," Vol. 27. New York: Academic Press, pp. 89–110.

Mazia, D. and K. Dan (1952): The isolation and biochemical characterization of the mitotic apparatus of dividing cells. Proc. Natl. Acad. Sci. U. S. A. 38:826–838.

Monné, L. (1944): Cytoplasmic structure and cleavage pattern of the sea urchin egg. Ark. Zool. 35A:1–27.

Morgan, T. H. (1927): Experimental embryology. New York: Columbia University Press, 766 p.

Rappaport, R. (1971): Cytokinesis in animal cells. Int. Rev. Cytol. 31:169–213.

Rebhun, L. I. (1972): Polarized intracellular particle transport: Saltatory movements and cytoplasmic streaming. Int. Rev. Cytol. 32:93–137.

Salmon, E.D. (1975): Spindle microtubules: thermodynamics of in vivo assembly and role in chromosome movement. Ann. N. Y. Acad. Sci. 253:383–406.

Swann, M.M., and J.M. Mitchison (1958): The mechanism of cleavage in animal cells. Biol. Rev. Cambridge Phil. Soc. 33:103–135.

Wilson, E.B. (1925): The cell in development and heredity. 3rd Ed. New York: The Macmillan Co., 1232 p.

Reprinted from *Zoological Science*, Vol. 5, pp. 529–538, 1988, with permission from Zoological Society of Japan.

ZOOLOGICAL SCIENCE **5**: 529–538 (1988)

The Living Spindle

SHINYA INOUÉ

Marine Biological Laboratory, Woods Hole, Massachusetts 02543, USA

ABSTRACT—Experience relating to earlier studies on the birefringence of the mitotic spindle and spindle fibers in living cells is first reviewed. This is followed by a description of the dynamic arrangement of micotubules in spindle fibers in living cells studied recently with high resolution video polarized light microscopy.

KD'S LAB

The urchin eggs had been collected, washed gently in a hand centrifuge, and mixed just minutes ago with a dilute sperm suspension. "What's the rate of fertilization?" KD (Dan-san, Dan-sensei, Professor Katsuma Dan) would ask. "75%? Better try another female," he'd say. "You'd just be kidding yourself if you worked with cells that aren't completely healthy." That was the way KD lived his science, and the way he trained the upstarts working in his laboratory.

The second floor laboratory at the Misaki Marine Biological Station, overlooking the waters of Aburatsubo and Moroiso Bay, was where so many of us were exposed to KD's challenges, to his way of biology, and to the enchantment of science. "The goddess of science is terribly jealous," he'd utter when he'd feel that we were not fully engaging our wits in carrying out an experiment. Or, he might even shout, "Try putting yourself in the animals' shoes!" when we'd missed accounting for a parameter that critically affected the organism or embryo.

Most of the time these criticisms seemed born more out of exasperation rather than as tongue lashing from a hard task master. Even those who felt cowed or rebellious at these words were working in KD's lab because that was simply what they wanted to do. They were neither paid assistants nor students who were required to be there. They were a school teacher, a brash graduate student, a junior colleague, or one just fascinated by biology; anyone who cared enough to be there and who dared take on the challenge. That was the makeup of those who frequented KD's and Jean's laboratory in the early post-war days (ca. 1946–1949).

To Jean and KD who had survived the double hardships of raising a young family and keeping up research under the challenging conditions of the bombed-out nation, while coping with the complex adjustments of a couple whose countries were in mortal conflict, those post-war days brought forth incalculable strength, joy, and optimism. Their mood pervaded the Laboratory just returned to the Japanese biologists and students by the occupation forces (with KD's intervention: "The last one to go," now prominently featured at the MBL Library in Woods Hole).

KD's cramped lab, saturated with the scent of marine organisms, bustled with activity oblivious to, and actually challenged by, the absence of all but a few essential tools for research. Some mornings the marine scent would be taken over by the delectable aroma of buttered toast smeared with honey that Jean and KD would share with us. Jean would have rounded up the butter, and the honey had been spun from KD's own hives in Nagai.

In the lab, Kayo was raising miniature adult sand dollars from the four blastomeres she'd cut apart free hand, KD and Jean were busy writing up their papers on the role of spindle elongation in astral cleavage, Endo-san had given up his hardware business to explore fertilization, and I was tinkering with pieces of optics trying to visualize spindles in living sea urchin eggs following Runnström's (in fixed plant cells [1]) and Schmidt's [2]

Accepted April 4, 1988

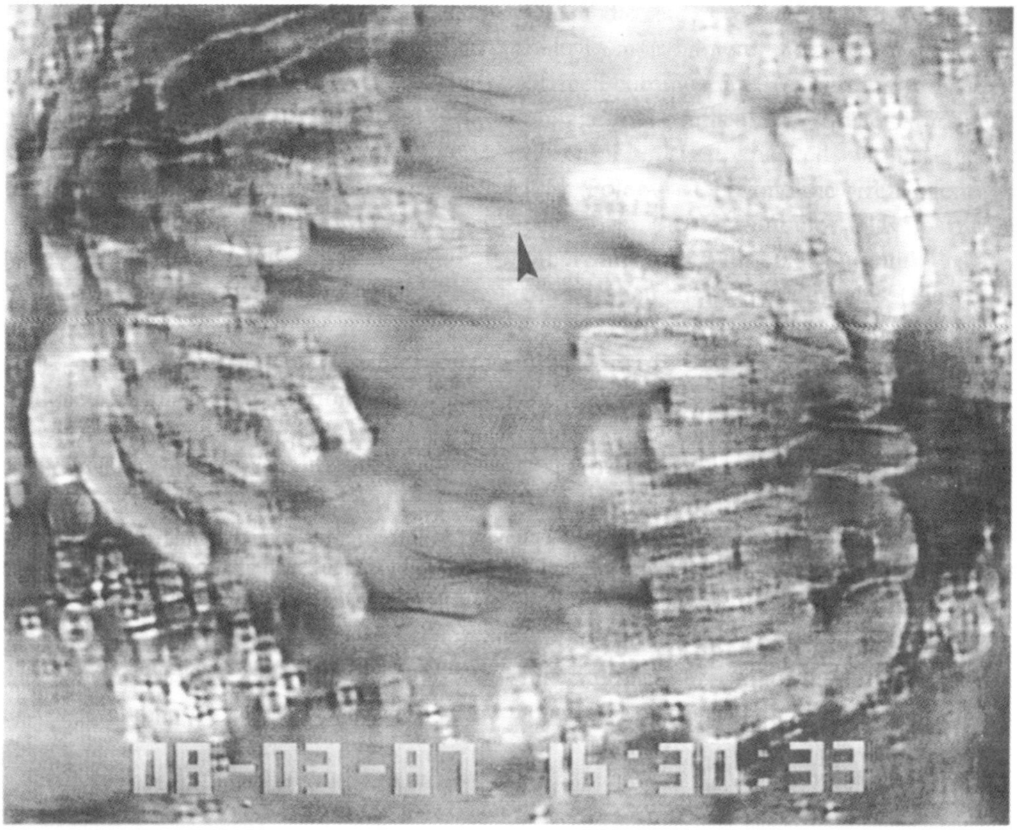

Fig. 8. Telophase of the same cell in Figs. 5 and 7. In such very thin optical section, the density of microtubules is low enough so that individual microtubules are clearly imaged (arrow). ([23]; see [22, 24] on microtubule distribution in *Haemanthus* endosperm, observed in fixed cells stained with immuno-gold antitubulin, also with digitally enhanced high resolution polarized light microscopy).

for only a few seconds, but on the whole they form what could be considered a fringed micellar structure (Figs. 6 and 7).

According to these observations, what was called a chromosomal spindle fiber in the classical literature (e.g., see Schrader [26]), or that we observed with lower resolution in polarized light microscopy [8, 10, 13], turns out to be a mixture of kinetochore and non-kinetochore microtubules. The different microtubules in the chromosomal spindle fiber stochastically associate laterally with each other and presumably form an anastomosing, fringed micellar gel.

We believe that the fringed micelle structure, formed by the short-lived, dynamic lateral association between microtubules, explains the modest mechanical integrity of the whole spindle and of the chromosomal and other spindle fibers. It also accounts for their lability when repetitively teased with a glass microneedle [27,28].

The dynamic fluctuation of microtubules due to transient lateral association (or zipping as the Bajers may prefer to call it), presumably together with the dynamic instability of the microtubules (see below), also accounts for the Northern lights flickering of spindle fiber birefringence that is seen at lower resolution [10].

AND NOW?

Given the dynamic lateral association coupled with the dynamic instability of microtubules (wherein each microtubule extends steadily until it suddenly undergoes rapid shortening as postulated by Tim Mitchison and Mark Kirschner [17] and observed under the light microscope by Horio and Hotani [29] and ourselves—see companion paper by Inoué [30]), it is no wonder that microtubules exhibit complex life-like behavior.

In addition to the dynamic properties of the fibrillar microtubules themselves, also calcium-regulating vesicles, microtubule-associated proteins, and other factors modulate the stability of the microtubules in local parts within a living cell. Together they account for the dynamic, enigmatic behavior of the centrosomes, spindle fibers, and asters, which govern much organizational activity so crucial to the cell's survival and proper function.

In exploring these mysteries, on the one hand, the diversity of pattern expressed by different life forms may provide us with clues, by exaggerating or omitting some elements involved in the complex set of interactions (e.g., see [26, 31–33]).

On the other hand, we need to learn much more about the state of the microtubules and modifying factors, not only as isolated purified elements, but as they function *dynamically* and *locally* in intact living cells.

To this end, direct studies of living cells with the light microscope as a non-invasive, analytical probe should play increasingly important roles, especially when combined with video. Images with high lateral and axial resolution can provide better 3-dimensional insight (Fig. 8), and the video-enhanced contrast can reveal subtler changes in fine structure and molecular organization of living cells with reduced exposure to light [20].

We should also use the microscope more effectively to modify minute targeted regions in living cells, not only with UV or other high energy irradiation, but with wavelengths and energy levels selected to induce subtler chemical modifications [34–36]. Combined with a good choice of reagents that are activated or inhibited by those wavelengths, we can better probe for local factors that govern the intricate, dynamic organization of the living cell and its spindle.

ACKNOWLEDGMENT

This paper is dedicated to Professor Katsuma Dan on his 83rd birthday. I recall with great pleasure the support and challenges provided by KD, which started as early as 1941 when I was still a student at the Musashi Higher School. The recent work described here and preparation of the manuscript were supported by NIH grant R37 GM31617-06 and NSF grant DCB 8518672.

REFERENCES

1 Runnström, J. (1929) Über die Veränderung der Plasmakolloide bei der Entwicklungserregung des Seeigeleies. II. Protoplasma, **5**: 201–310, Fig. & Table 4.

2 Schmidt, W. J. (1937) Die Doppelbrechung von Karyoplasma, Zytoplasma und Metaplasma. Protoplasma-Monographien, Vol. 2. Gebruder Borntraeger, Berlin.

3 Inoué, S. and Dan, K. (1951) Birefringence of the dividing cell. J. Morphol., **89**: 423–455.

4 Swann, M. M. (1951) Protoplasmic structure and mitosis. II. The nature and cause of birefringence changes in the sea-urchin egg at anaphase. J. Exp. Biol., **28**: 434–444.

5 Schmidt, W. J. (1934) Polarisationsoptische Analyse des submikroskopischen Baues von Zellen und Geweben. In "Handbuch der biologischen Arbeitsmethoden." Ed. by E. Abderhalden, Urban und Suhwarzenberg, Berlin and Vienna, Sec. 5, Part 10, pp. 435–665.

6 Dan, K. (1987) "Uni toh kataru." (Dialogue with sea urchin.) Gakkai Shuppan Center, Tokyo.

7 Dan, J. C. (1954) Studies on the acrosome. II. Acrosome reaction in starfish spermatozoa. Biol. Bull., **107**: 203–218.

8 Inoué, S. (1953) Polarization optical studies of the mitotic spindle. I. The demonstration of spindle fibers in living cells. Chromosoma,**5**: 487–500.

9 Lutz, D. A., Hamaguchi, Y. and Inoué, S. (1988) Micromanipulation studies of the asymmetric positioning of the maturation spindle in *Chaetopterus* sp. oocytes. I. Anchorage of the spindle to the cortex and migration of a displaced spindle. Cell Motility and Cytoskeleton, in press.

10 Inoué, S. (1964) Organization and function of the mitotic spindle. In "Primitive Motile Systems in Cell Biology". Ed. by R. D. Allen and N. Kamiya, Academic Press, New York, pp. 549–598.

11 Inoué, S. (1981) Cell division and the mitotic spindle. J. Cell Biol., **91**: 131s–147s.

12 Inoué, S., Fuseler, J., Salmon, E. D. and Ellis, G. W. (1975) Functional organization of mitotic microtubules. Physical chemistry of the in vivo equilibrium system. Biophys. J., **75**: 725–744.

13 Inoué, S. and Sato, H. (1967) Cell motility by labile association of molecules. J. Gen. Physiol., **50**: 259–292.

14 Mazia, D. and Dan, K. (1952) The isolation and biochemical characterization of the mitotic apparatus of dividing cells. Proc. Natl. Acad. Sci., **38**: 826–838.

15 Dan, K. and Ito, S. (1984) Studies of unequal cleavage in molluscs. I. Nuclear behavior and anchorage of spindle pole to the cortex as revealed

by isolation technique. Dev. Growth Differ., **26**: 249–262.

16 Dan, K. and Inoué, S. (1987) Studies of unequal cleavage in molluscs. II. Asymmetric nature of the two asters. Int. J. Invertebr. Reprod. Dev., **11**: 335–353.

17 Mitchison, T. and Kirschner, M. (1984) Dynamic instability of microtubule growth. Nature, **312**: 237–242.

18 Salmon, E. D., Leslie, R. J., Saxton, W. M., Karow, M. L. and McIntosh, J. R. (1984) Spindle microtubule dynamics in sea urchin embryos: Analysis using a fluorescein labelled tubulin and measurements of fluorescence redistribution after photobleaching. J. Cell Biol., **99**: 2165–2174.

19 Ellis, G. W. (1985) Microscope illuminator with fiber optic source integrator. J. Cell Biol., **101**: 83a.

20 Inoué, S. (1986) Video Microscopy. Plenum, New York.

21 Inoué, S. (1988) Imaging of unresolved objects, superresolution, and precision of distance measurement, with video microscopy. In "Fluorescence Microscopy of Living Cells in Culture: Quantitative Fluorescence Microscopy: Imaging and Spectroscopy". Ed. by D. L. Taylor and Y.-L. Wang, Methods in Cell Biology, Vol. 30, Academic Press, New York. In press.

22 Inoué, S. (1987) Video microscopy of living cells and dynamic molecular assemblies. Applied Optics, **26**: 3219–3225.

23 Cassimeris, L., Inoué, S. and Salmon, E. D. (1988) Microtubule dynamics in the chromosomal spindle fiber: Analysis by fluorescence and high resolution polarization microscopy. J. Cell Motil. Cytoskeleton, in press.

24 Inoué, S., Molè-Bajer, J. and Bajer, A. S. (1985) Three-dimensional distribution of microtubules in *Haemanthus* endosperm cell. In "Microtubules and Microtubule Inhibitors". Ed. by M. De Brabander and J. De Mey, Elsevier, Amsterdam, pp. 15–30.

25 Rieder, C. and Bajer, A. S. (1977) Heat-induced reversible hexagonal packing of spindle microtubules. J. Cell Biol., **74**: 717–725.

26 Schrader, F. (1953) Mitosis. The Movements of Chromosomes in Cell Division. Columbia Univ. Press, New York, 2nd ed.

27 Nicklas, R. B., Kubai, D. F. and Hays, T. S. (1982) Spindle microtubules and their mechanical associations after micromanipulation in anaphase. J. Cell Biol., **95**: 91–104.

28 Begg, D. A. and Ellis, G. W. (1979) Micromanipulation studies of chromosome movement. II. Birefringent chromosomal fibers and the mechanical attachment of chromosomes to the spindle. J. Cell Biol., **82**: 542–554.

29 Horio, T. and Hotani, H. (1986) Visualization of the dynamic instability of individual microtubules by dark-field microsocpy. Nature, **321**: 605–607.

30 Inoué, S. (1988) Manipulating single microtubules. Protoplasma: Noburo Kamiya Festschrift, in press.

31 Bělař, K. (1926) Der Formwechsel der Protistenkerne. Ergeb. Fortschr. Zool., **6**: 1–420.

32 Inoué, S. and Ritter, H., Jr. (1978) Mitosis in *Barbulanympha.* II. Dynamics of a two-stage anaphase, nuclear morphogenesis, and cytokinesis. J. Cell Biol., **77**: 655–684.

33 Wilson, E. B. (1928) The Cell in Development and Heredity. Macmillan, New York, 3rd ed.

34 Englemann, Th. W. (1882) Ueber Sauerstoffausscheidung von Pflanzenzellen im Mikrospektrum. Bot. Zeitung, **40**: 419–426.

35 Aronson, J. and Inoué, S. (1970) Reversal by light of the action of N-methyl N-desacethyl colchicine on mitosis. J. Cell Biol., **45**: 470–477.

36 Hiramoto, Y., Hamaguchi, M. S., Nakano, Y. and Shoji, Y. (1984) Colcemid UV-microirradiation method for analyzing the role of microtubules in pronuclear migration and chromosome movement in sand-dollar eggs. Zool. Sci., **1**: 29–34.

Reprinted from *Methods in Cell Biology*, Vol. 30, pp. 85–112,
with permission from Elsevier.

Chapter 3

Imaging of Unresolved Objects, Superresolution, and Precision of Distance Measurement with Video Microscopy

SHINYA INOUÉ

Marine Biological Laboratory
Woods Hole, Massachusetts 02543

The basics of microscope image formation, point and line spread functions, and modulation transfer functions (MTF) are covered in the two previous chapters (see also Castleman, 1979; Hecht, 1987; and Inoué, 1986). In this chapter, we will examine the utility of video microscopy to visualize, resolve, and measure widths and distances to precisions that are conventionally considered to be below the limit of resolution of the light microscope.

I. Visualizing Objects Narrower than the Resolution Limit of the Light Microscope

To start with, we must clearly distinguish between resolving an image and visualizing an object.

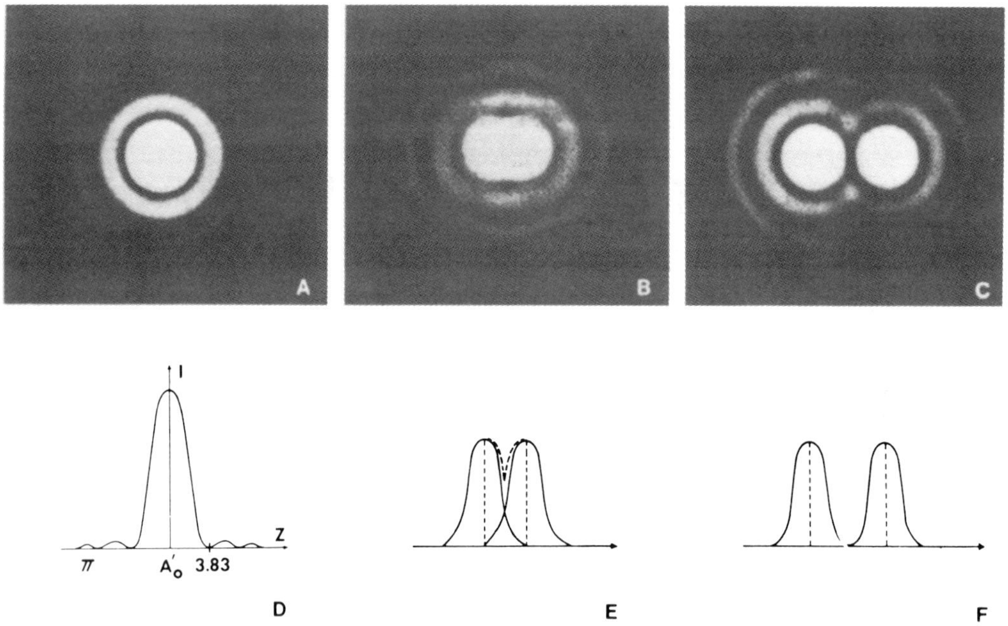

FIG. 1. Airy disk and intensity distribution in the diffraction image. (A) Photograph of Airy disk; (D) intensity distribution. (B, C, E, F) Images of neighboring Airy disks (B, C), and their corresponding intensity distributions (E, F), separated by different distances. In C (and F), where the distance between the disk centers is larger than their radii, the disks (and intensity peaks) are clearly separate, or resolved. In B, the center-to-center distance equals the radius, the disks are just barely distinguishable, and the images are just resolved (the Rayleigh Criterion). [The curves in E and F are schematic only. Had they been drawn exactly as in D, the sum of the two curves in E (dash line) would have dipped in the middle by only 26.5% from their peak.] If the disk centers are closer together than their radii, their intensity distributions merge into a single peak (D), and they are said not to be resolved. (After Françon, 1961.)

In an image-forming system, resolution is generally expressed as a measure of the ability to separate images of two neighboring object points (Born and Wolf, 1980; Sect. 8.b.2; Françon, 1961). In the case of a light microscope, producing diffraction-limited images, the resolution limit is defined as the minimum distance between two self-luminous or incoherently illuminated objects or structures whose diffraction images[1] can visually be distinguished as coming from two points. When the diffraction images of the two points overlap to an extent that they can no longer be distinguished from that of an individual object, the two are said not to be resolved or that the distance is less than the limit of resolution (Fig. 1). The

[1] In this chapters, I use the term diffraction pattern to refer to the pattern produced at the objective lens back aperture by the spatial periodicity of the specimen, and diffraction image to refer to the diffraction limited image (such as the Airy disk) produced in the image plane by the objective lens.

minimum distance (d) between such objects or structures visually resolvable with a light microscope is commonly given by:

$$d = \frac{1.22\lambda_0}{NA_{obj} + NA_{cond}} \quad (1)$$

where λ_0 is the vacuum wavelength of the light used, NA_{obj} is the numerical aperture of the objective lens, and NA_{cond} is the actual working NA of the condenser set by its immersion medium and iris diaphragm setting (Fig. 2). At this separation, the diffraction images of the two

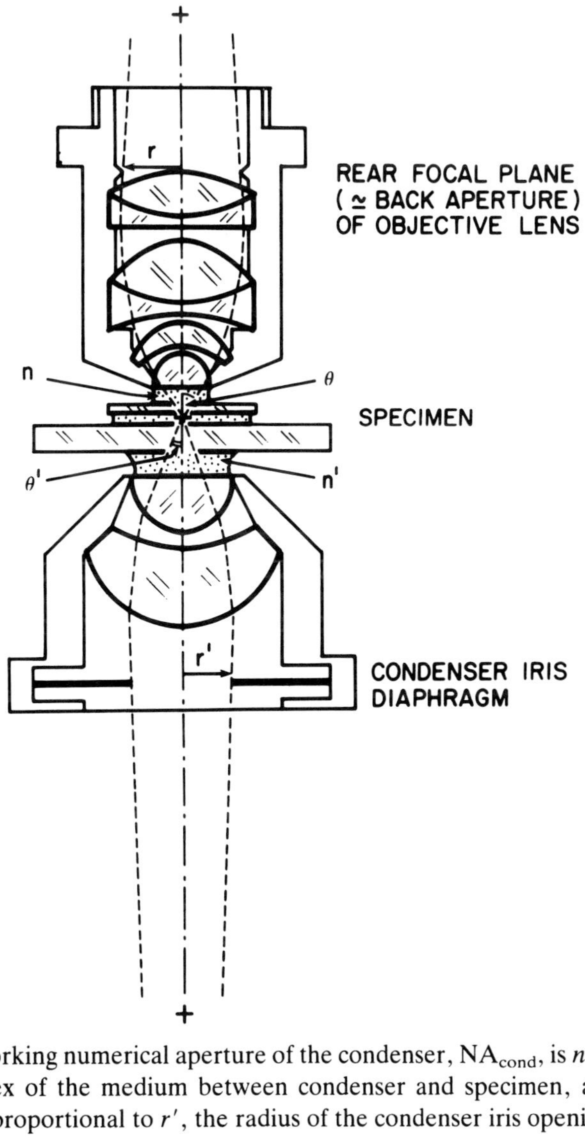

FIG. 2. The working numerical aperture of the condenser, NA_{cond}, is $n' \sin \theta'$, where n' is the refractive index of the medium between condenser and specimen, and θ' is the angle shown. NA_{cond} is proportional to r', the radius of the condenser iris opening. Thus, changing the condenser iris setting changes NA_{cond}. Analogous considerations for the objective lens are also illustrated. (From Inoué, 1986.)

incoherently illuminated objects overlap with a ~26.5% depression that signals the twoness of the object (Rayleigh criterion).[2]

Naturally the Rayleigh criterion assumes that the objective lens is capable of, and is in fact being used with, the correct immersion medium (including the medium bathing the specimen), tube length, coverslip thickness, ocular type, projection distance, and wavelength, etc., that provide a diffraction limited image free from measurable aberrations.[3]

While the lateral resolution in the plane of focus is defined as above [or in some cases using the Sparrow criterion where the central minimum just disappears (Sparrow, 1916)], we can nevertheless visualize isolated objects or structures whose widths are far below the limit of resolution.

Forming an image of an object whose diameter or width is far below the resolution limit of the microscope is not a new event. Dark-field microscopy has been used for decades to visualize the dynamic behavior of smoke and other colloidal particles, bacterial flagella, etc. However, those images could not readily be recorded.

In the past few years the light microscope with a variety of contrast modes has come to be used for visualizing and studying the behavior of unresolved thin objects. The major gain has come from the ability of analog and digital video devices to dramatically boost the image contrast and speed of image acquisition, and to subtract or filter out unwanted optical noise, average out random electronic and photon statistical noise, and to record the image on media that allow immediate playback (e.g., Allen *et al.,* 1981a,b; Inoué, 1981, 1986, 1987a).

Importantly for microscopy, the contrast boosting ability of video allows the use of the best corrected (Plan Apochromatic) objective lenses not only for bright field and fluorescence, but even for polarized light and differential interference contrast (DIC) microscopy. In the past, in these latter contrast modes, such lenses could not be used well because of the poor extinction due to the crystalline elements they contained.

With the contrast boosted electronically, the condenser can also be used at an NA much higher than was previously possible. Thus, the diffraction image becomes narrower and the limit of resolution is improved [Eq. (1)]. At the same time, more light becomes available to the sensor, bringing it

[2] Strictly speaking, we are here talking about periodic or complex objects (rather than just two points) whose separation is *d*. The case of solely distinguishing two points or lines from a single one can be considered a special case, as discussed under the section on superresolution, since in this case we have *a priori* knowledge or assumption about the specimen, i.e., that it is either one or two.

[3] For through-focal photographs of diffraction images free of aberrations and in the presence of various aberrations, see, e.g., Cagnet *et al.* (1962).

into a range where its S/N ratio is more favorable (e.g., see Fig. 7–16 in Inoué, 1986). In addition, as discussed in Section V, the depth of field decreases dramatically with the rise of system NA, so that axial resolution is improved and the influence of out-of-focus objects is reduced.

With these improvements brought about by video, objects (especially filaments) whose width are far below the resolution limit are now routinely visualized by fluorescence, DIC, dark field, and rectified polarized light microscopy. The main reason such thin objects could not be visualized readily without video enhancement was not because their diameters were below the resolution limit of the microscope but because their diffraction images have low contrast. By boosting the contrast (and using conditions that give greater lateral and axial resolution) with video, and in the case of dark-field and fluorescence images with help from the greater light-detecting sensitivity of the intensifier tubes, one sees such dynamic events in real time as: 65-nm-diameter sperm acrosomal process growing (Tilney and Inoué,1982); 25-nm-diameter microtubules assembling and disassembling, gliding or moving particles (Allen *et al.*, 1985; Schnapp *et al.*, 1985; Koonce and Schliwa, 1986); 10-nm-diameter actin filaments sliding (Yanagida *et al.*, 1984; Toyoshima *et al.*, 1987); and lipid bilayers in emulsions extending, deforming, and fusing with each other (Kachar *et al.*, 1984).

II. Diffraction Patterns of Very Narrow Objects

We shall next consider the widths of the images formed by structures whose width approximates or is narrower than the conventional limit of resolution.

For a given optical system, the point or line spread function describes the distribution of intensity in the image of an infinitely narrow point or line (see Young, Chapter 1, this volume). For a well-corrected, diffraction-limited system, the function would essentially be an Airy disk (Fig. 1) for absorbing or self-luminous points (which includes in-focus images in fluorescence and dark-field microscopy). [For DIC images, see Galbraith (1982) and Galbraith and Sanderson (1980); for rectified and nonrectified polarized light images, see Inoué (1986).] What is the width of the diffraction image formed by a single object whose width lies between the infinitely thin and approximately an Airy disk diameter?

Given the proper optical corrections (and appropriate video sampling), these unresolved objects form a diffraction image, expanded to the size of an Airy disk or larger as determined by the microscope optics and size of the object. In detail, the diffraction images vary in peak intensity as well as in width. As the object becomes smaller, the peak intensity decreases

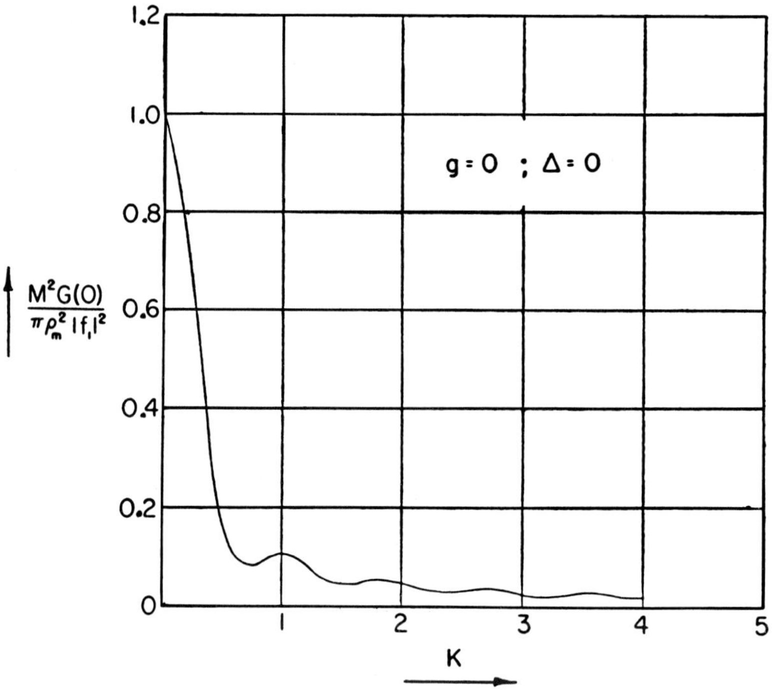

FIG. 3. The central energy density in the images of opaque particles as a function of K the radius of the particles (cf Fig. 6). (From Smith, 1960. See original article for further details.)

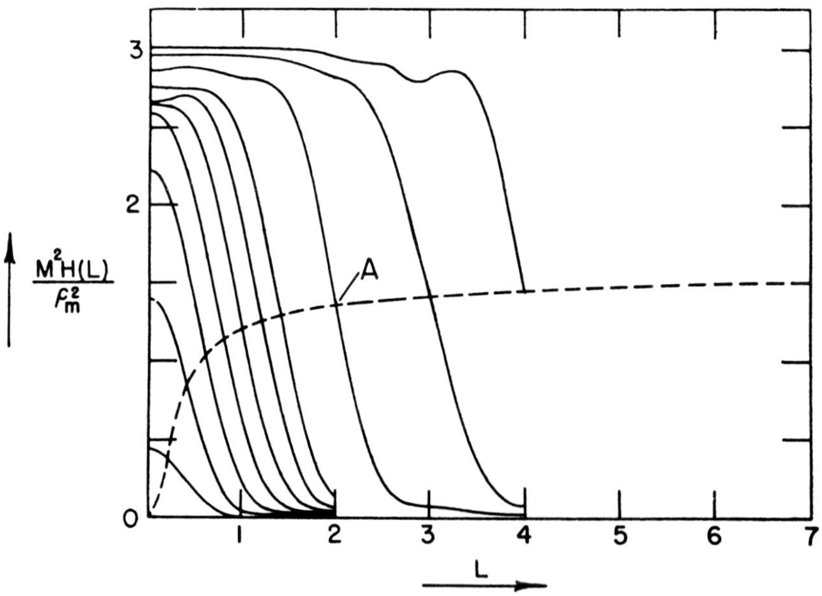

FIG. 4. The solid curves show the distribution of irradiance in the image formed by an Airy-type objective when the source is a self-radiant disk of a specified radius. The abscissa of the intersection of the broken curve with a solid curve indicates the radius in Airy units of the corresponding source disk. For example, the distribution of irradiance in the image of a circular disk 2 Airy units in radius is given by the solid curve passing through the intersection A. (From Smith and Osterberg, 1961. See original article for further details.)

nonmonotonically as shown in Fig. 3. (The curve shown in Fig. 3 is for an absorbing object; cf. Fig. 6.)

The in-focus diffraction image for self-luminous (which includes fluorescence), thin, circular disks of 0.2 to 2.0 Airy units in radius, calculated by Smith and Osterberg (1961), is reproduced in Fig. 4. As shown, the radius of the source disk, for disk diameters of 3 to ~0.5 Airy units, can be approximated by the radius of the diffraction image at half height. For objects with radii below 0.5 Airy units, the half-peak diameter of the diffraction image departs drastically from the diameter of the source.

Figure 5 shows the energy distribution in the diffraction image for a circular hole in an opaque screen, and Fig. 6 the energy distribution for an opaque particle (Smith, 1960). The in-focus diffraction image is again smaller for narrower objects, although the two are not related linearly.

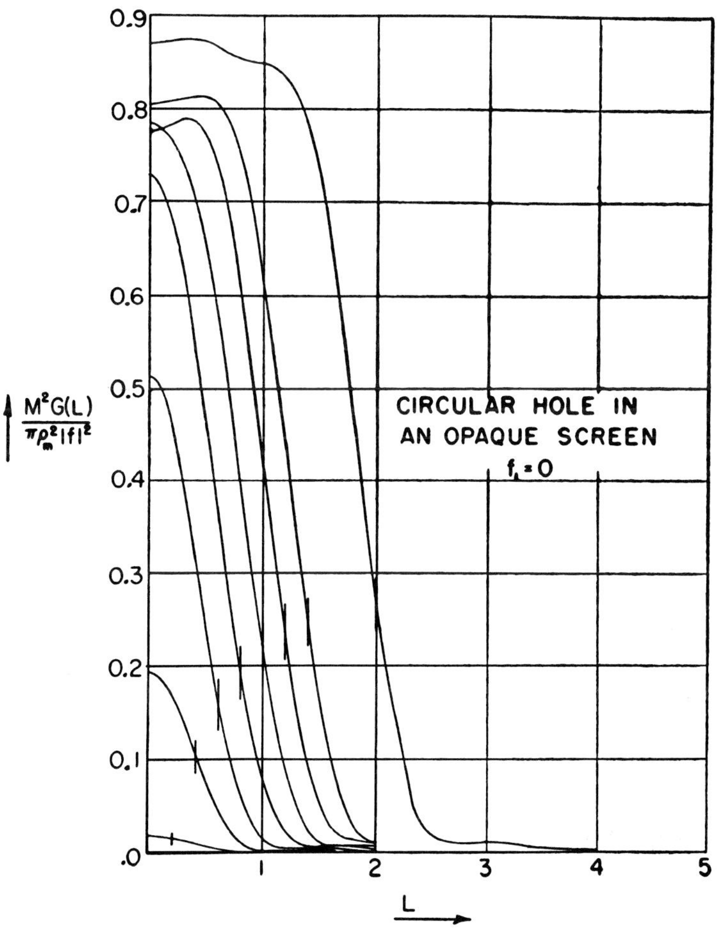

FIG. 5. Diffraction images of circular holes in an opaque screen. The short vertical line on each curve indicates the radius of the corresponding image according to geometrical construction. (From Smith, 1960. For further details see original article.)

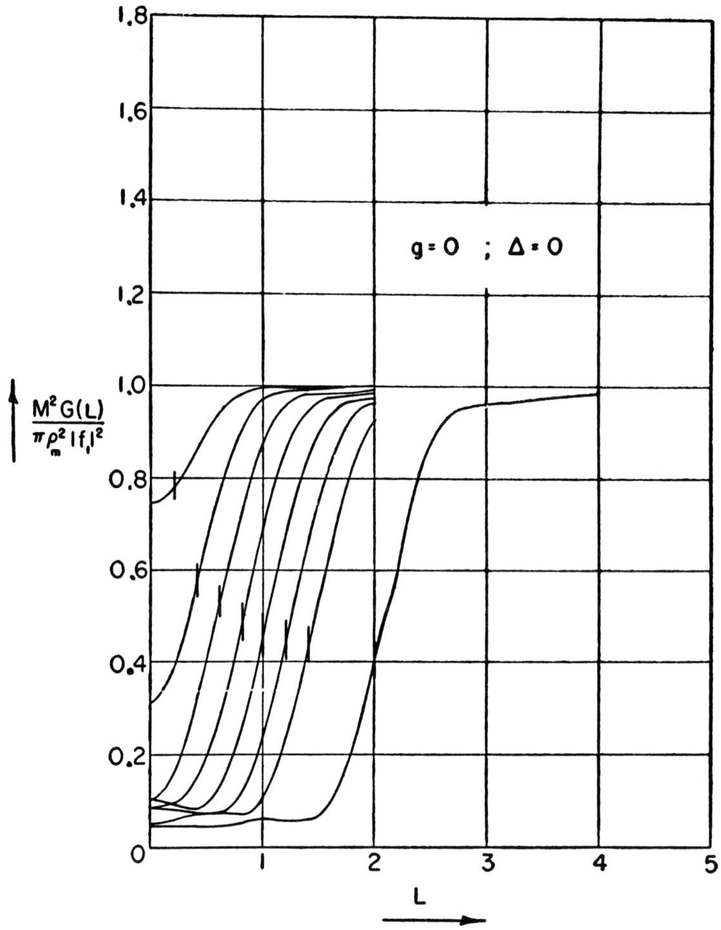

FIG. 6. Diffraction images of opaque particles. The short vertical line on each curve indicates the radius of the corresponding geometrical image. (From Smith, 1960. See original article for further details.)

The width of the diffraction image at half height for self-luminous and small opaque objects thus approximates the geometrical width of the object down to approximately 0.5 Airy disk radius. This relation does not hold for small transmitting apertures nor for particles which only differ in refractive index from their surroundings (Fig. 7).

Another question then arises. In fluorescence, or DIC, or polarized light microscopy, where especially with the aid of video we can visually discern objects as thin as individual microtubules, can we tell, from the diffraction image observed in the microscope, how many individual filaments make up the diffraction image? To paraphrase, when a single microtubule well-separated from others is observed in video-enhanced DIC contrast, its image appears "inflated to" the width of a typical

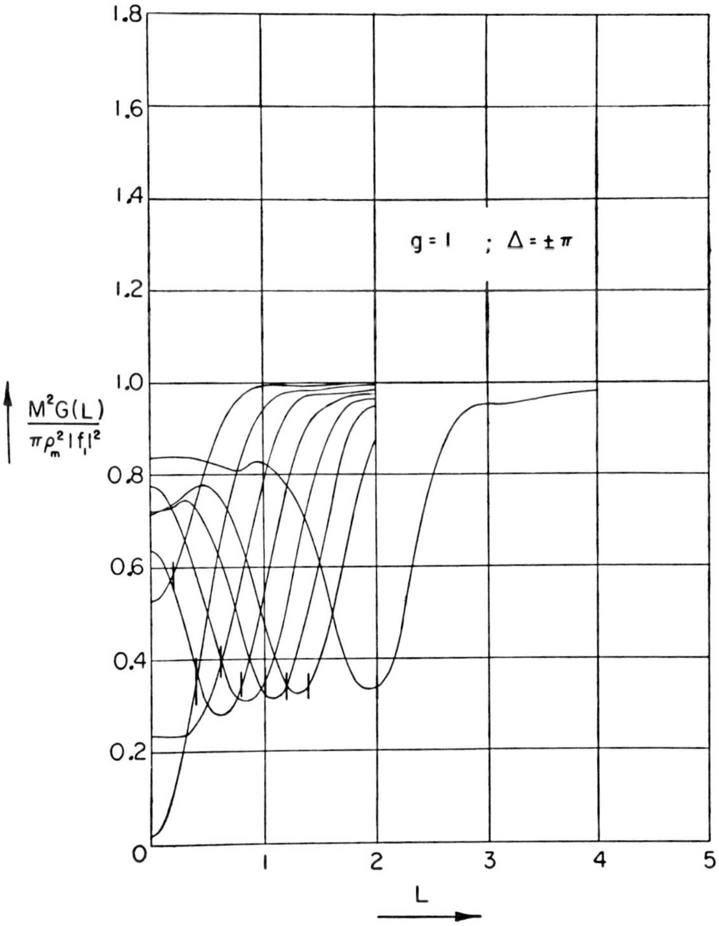

FIG. 7. Diffraction images of particles which differ from the surrounding only by having a different refractive index. The short vertical line on each curve indicates the radius of the corresponding geometrical image. (From Smith, 1960. See original article for further details.)

diffraction image (Fig. 8). What happens to the diffraction image when more microtubules are packed in a width (and depth) less than ~0.5 Airy unit?

One can clearly tell where two microtubules overlap (Fig. 9), and in DIC microscopy, preliminary observation suggests that the area of the diffraction pattern is roughly proportional to the number of microtubules, i.e., the mass of the material contained in the "unresolved" fine thread. Similar relations are expected for fluorescence images and for rectified polarized light microscopy.

In polarization microscopy we have noticed a rather striking difference in the width of images of individual microtubules which were additively

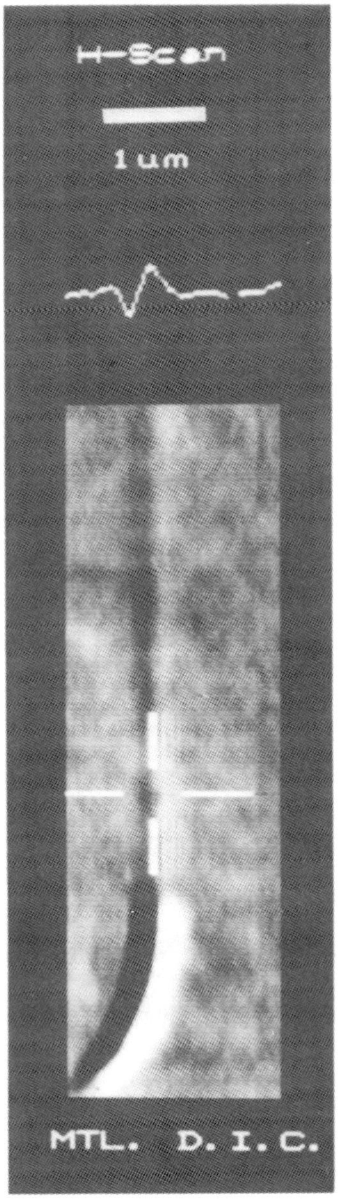

FIG. 8. Intensity distribution in DIC diffraction image of a single MAPs-free microtubule grown on purified axoneme (curved, high-contrast object at bottom). Each point in the line trace (generated by the Image-1/AT processor) represents the intensity across a ten-pixel long, one-pixel wide slit. 100/1.40 Plan Apo objective lens with rectified DIC condenser (Nikon). Magnification, 11,500.

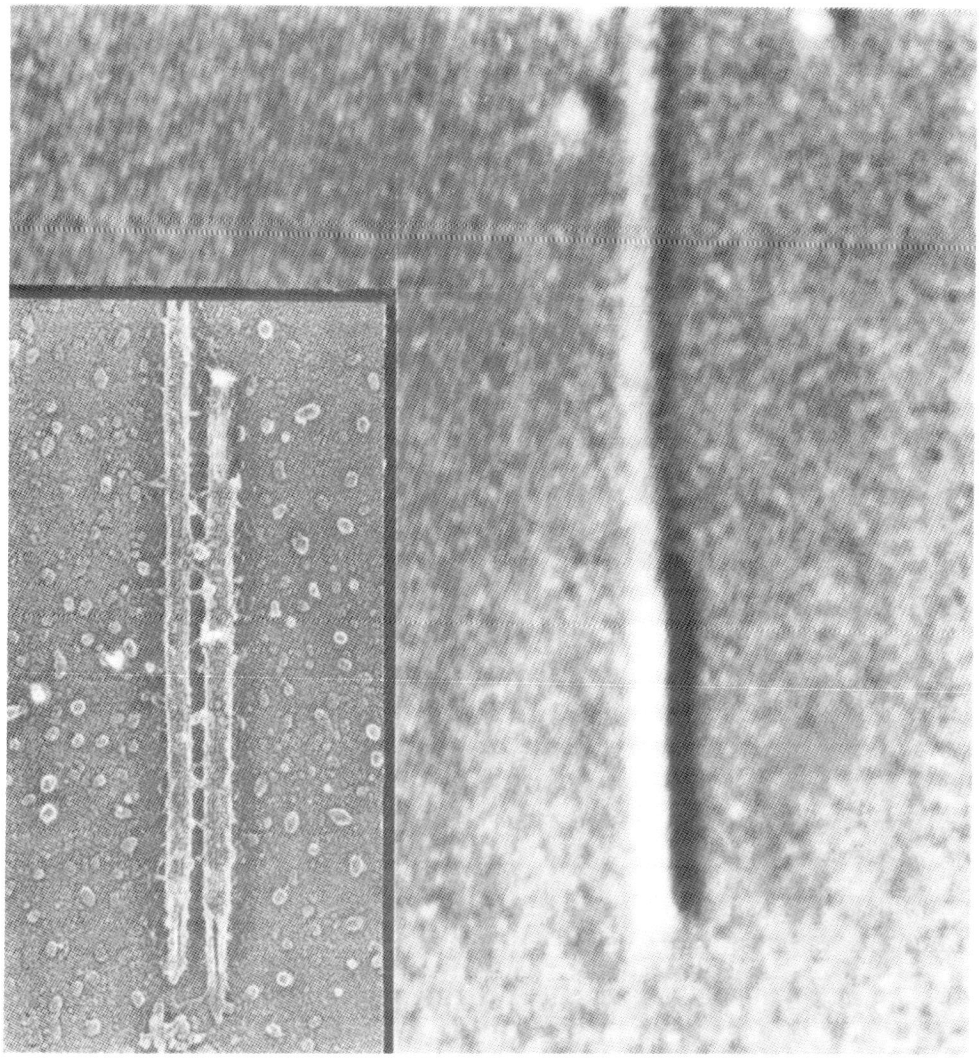

FIG. 9. DIC diffraction image of single and overlapping microtubules. Magnification: approximately 10,000×. Inset shows electron micrograph of overlap region. Magnification: Approximately 100,000×. (Courtesy of B. Schnapp, Marine Biological Laboratory, and National Institutes of Health.)

and subtractively compensated. In the additive compensation the diffraction image of the microtubule (or an axoneme), which is brighter than the gray background, appeared rather broad and with unclearly defined edges (Fig. 10). In the subtractive compensation, the dark image of the microtubule appeared considerably thinner and with rather sharply defined edges. It is not yet clear whether this asymmetric behavior upon compensation reflects our sensory response, an intrinsic physical optical effect, or the

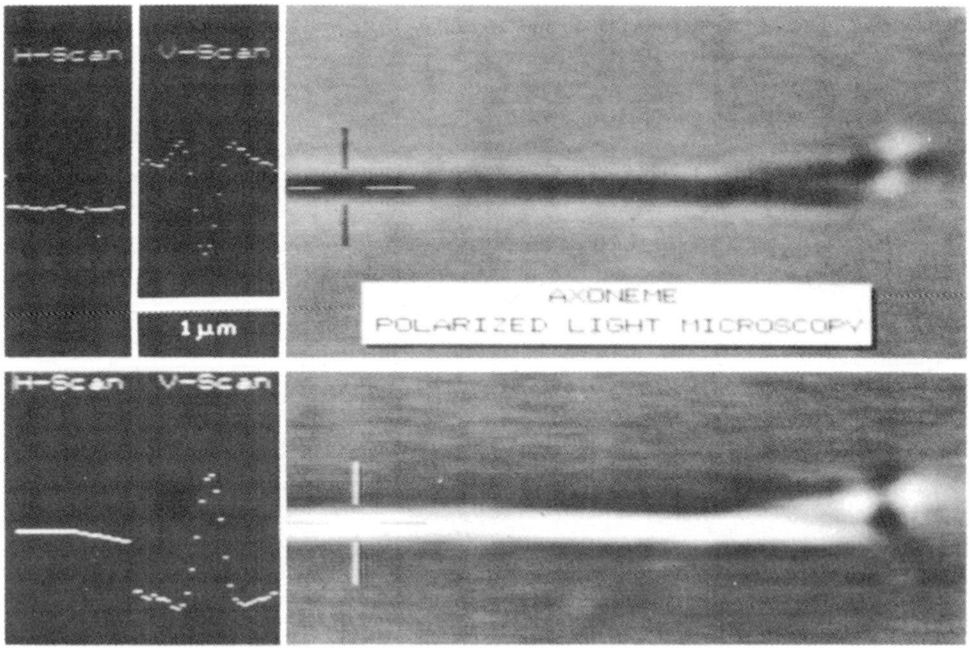

FIG. 10. Rectified polarized light image of axoneme in additive (bottom) and subtractive (top) compensation. Each point in the line trace represents the intensity across a one-pixel wide slit. 100/1.35 NA Plan Apochromatic objective lens with rectified condenser (Nikon). Magnification, 17,000.

property of the instrument.[4] Be that as it may, the birefringence of individual microtubules in the subtractive compensation does yield a narrow-appearing diffraction image that makes them stand out distinctly in video-enhanced, rectified polarized light microscopy.

III. Superresolution

Under some special conditions discussed below, one can obtain "super-resolution" and in fact resolve objects that are closer together than the Rayleigh criterion.

[4]The high NA Plan Apochromatic lenses used to obtain these images do not provide a perfectly uniform aperture function; the rectifier provides an imperfect match so that there exists some residual rotation at the lens surfaces, and the lens also exhibits detectable degrees of birefringence with radial and lateral axes. In addition, we would expect the image to also be affected by "edge birefringence" since the specimen is not embedded in a matched index medium (see Inoué, 1959, 1986 p. 498; Takenaka and Rikukawa, 1974).

One case of superresolution is predicted when the field of view is not infinitely wide (the case of ordinary microscopy), but is confined to a very small area (e.g., Cox *et al.*, 1982; McCutchen, 1967; Fellgett and Linfoot, 1955; Toraldo, 1955). An example of this type of superresolution occurs with confocal scanning microscopy, where the field at any instant is limited to an Airy diffraction pattern of the entrance pinhole formed by the condenser lens. In this case, a superresolution by a factor of two is predicted by Cox *et al.* (1982) and Cox and Sheppard (1986) for fluorescent objects, and a 1.4-fold improvement in resolution is described by Brakenhoff *et al.* (1979) and White *et al.* (1987). The theoretical bases of this type of superresolution is briefly as follows.

From the classical work of Toraldo (1955), expanded on by Harris (1964) and based on information sampling theory, one can show that the overlapping diffraction images produced by two narrow self-luminous objects, lying in a limited microscope field, can be deconvoluted to establish their separations, e.g., down to two-tenths of the Rayleigh criterion distance (Figs. 11 and 12). As Harris explains, the Fourier components of a diffraction pattern can be uniquely defined beyond the physical limit of the objective lens aperture, provided the object lies in a limited narrow field. In other words, given precise enough information about the diffracting waves that do pass the objective lens aperture, the coefficients for the Fourier components which would fall outside of the aperture, i.e., those that are due to spacings which are too fine to be "resolved" by the lens at the specified wavelength, can nevertheless be uniquely approximated theoretically for objects in a small limited field.

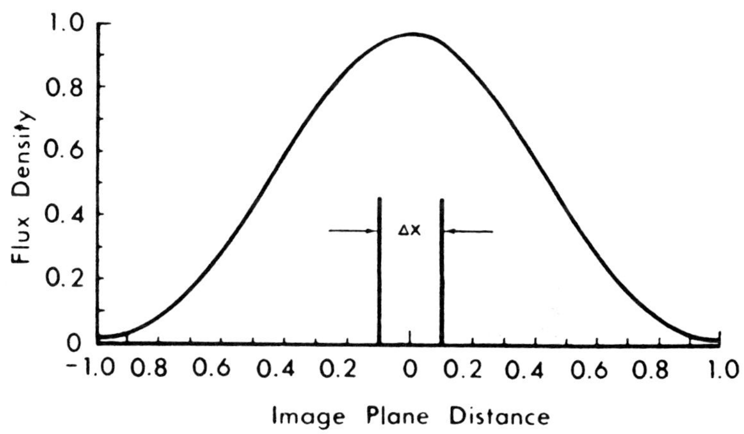

FIG. 11. The composite image formed by two monochromatic mutually incoherent point sources. The vertical lines show the position of the two points. The image plane distance has been normalized to the Rayleigh criterion distance, so that Δx is 0.2 of the Rayleigh criterion separation. (From Harris, 1964.)

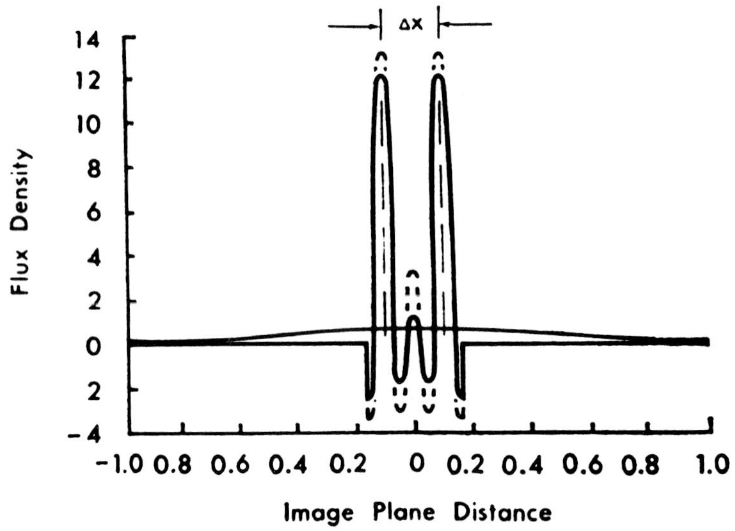

FIG. 12. The restoration of the image of Fig. 11 is shown. The dashed curve is the representation of the two point sources, which is achieved by using the first four terms of a Fourier series over the interval ±0.15. The similar solid curve is the restoration accomplished in the illustrative example. The broad, solid curve is the original image of Fig. 11 replotted to the scale of the image. (From Harris, 1964.)

Thus, the theorem predicts that close study of a diffraction image can reveal structures below the limit of resolution, provided the field is narrowly confined and the noise level is appropriately low [see Cox and Sheppard (1986) for a mathematical treatment of superresolution that includes the noise factor and the two polarization vectors].

Superresolution could also be achieved if the central peak of the diffraction image (point or line spread functions) could be made narrower than the width of the standard Airy disk. This can be achieved to a certain extent by modifying the aperture function of the objective lens. For example, as shown in Fig. 13, in the presence of a central obstruction in the objective lens, more of the diffracted energy appears in the first and higher order fringes, but the zero order fringe becomes narrower (e.g., Born and Wolf, 1980). Thus, while extraneous fringes would obscure the image of a complex structure, the distance between two equally bright points or lines isolated and located near the center of a clean field could be measured with somewhat improved resolution using an objective lens with central obstruction (Boivin and Boivin, 1980).

In some cases, the situation can be further improved by applying a graded series of absorbing rings toward the outer limits of the objective lens aperture. Such "apodizing" treatment, which reduces the energy distributed into the outer diffraction rings (at some cost to the narrowness

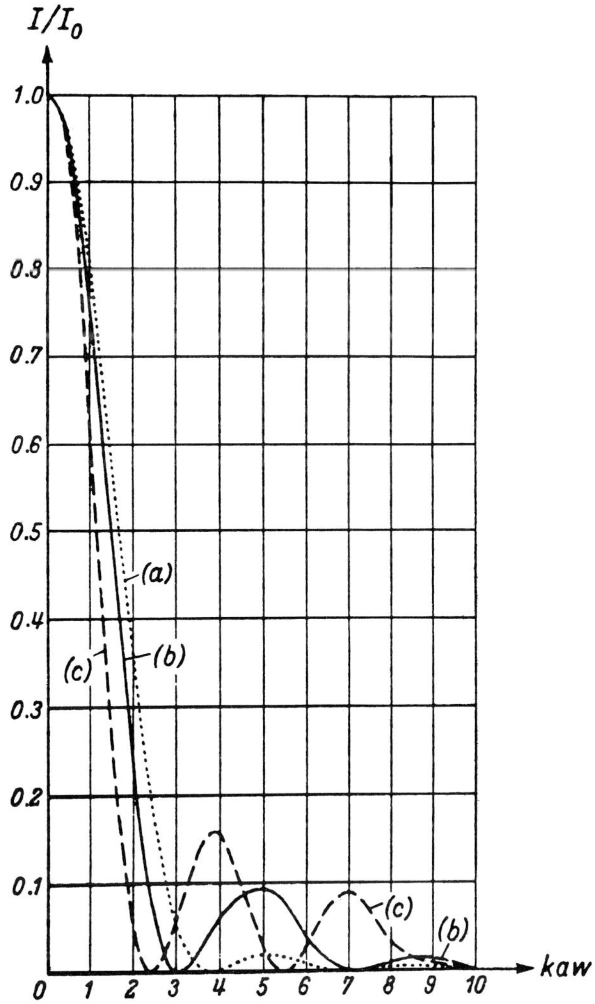

FIG. 13. Illustrating the effect of central obstruction on the resolution. Normalized intensity curves for Fraunhofer patterns (diffraction images) of (a) circular aperture, (b) annular aperture with $\varepsilon = \frac{1}{2}$, and (c) annular aperture with $\varepsilon \to 1$. ε is the ratio of the inner to outer radius of the aperture. (From Born and Wolf, 1980.)

of the central disk diameter), can help to bring out the weak diffraction image of an object lying close to a much brighter object. The dim image of the less bright object would otherwise be masked by the bright first or second order diffraction rings of the brighter object (see, e.g., Hecht and Zajac, 1979).

Another condition giving rise to superresolution occurs when two adjoining structures give rise to diffraction patterns that are reversed in phase. Since I have never formally reported on this condition, which I had formulated and experimentally verified in 1957, I will expand on this subject here.

Consider the double slit diffraction experiment in Fig. 14. Two narrow slits (S_2, S_3), separated by a distance (d), are illuminated by a monochromatic source (of wavelength λ) through slit S_1. The diffraction images of slits S_2 and S_3 are formed by lens (L) whose aperture and λ determine the width of each of the slit images. As d is decreased, the two diffraction images overlap and eventually become unresolvable once their separation is reduced beyond the Rayleigh (or Sparrow) criterion.

We now add the following polarizing components to the experimental setup as shown in Fig. 15. These are a polarizer (P) immediately preceding or following S_1; two birefringent crystals (C_1 and C_2) of equal retardations placed immediately preceding or following S_2 and S_3, respectively, with their slow axes at 90° to each other and lying at azimuths of 45° to the polarizer azimuth PP′ (see lower half of Fig. 15); and a third crystal, or compensator, (C_3) followed by an analyzer (A) with their axes as indi-

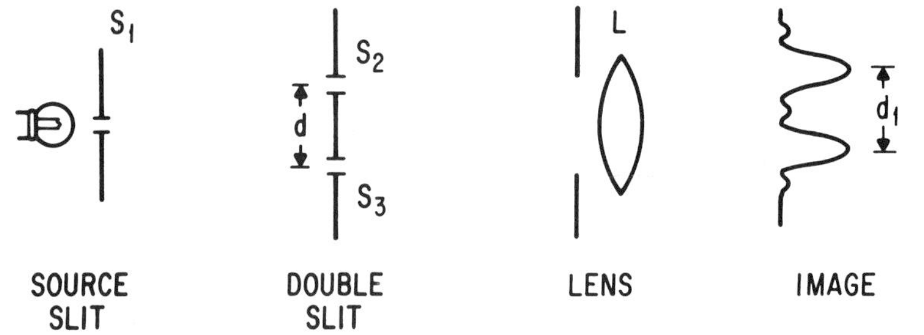

FIG. 14. Diffraction images formed by double slit. See text.

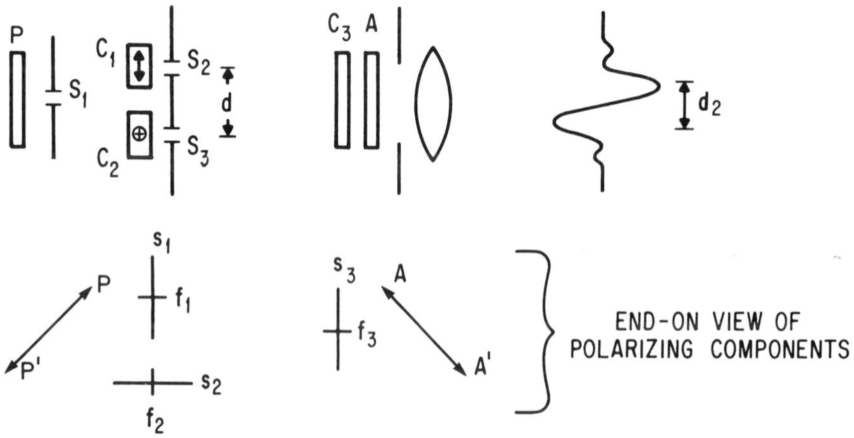

FIG. 15. Double slit experiment coupled with phase reversal of the slit images. Figures in the lower half show the azimuth orientation of the polarizing components used to achieve the phase reversal. See text for further details.

cated. When these polarizing components are added, the phases of the diffraction images of S_2 and S_3 are reversed by 180°. Thus the diffraction images become a bright and dark pair on a gray background. As *d* is reduced below the Rayleigh criterion, they continue to appear as a bright–dark pair but with reduced amplitude, rather than merging into a single unresolved peak. Depending on the orientation of the slow axes S_1 and S_2 relative to S_3, the bright peak appears to one side or the other of the dark peak. Thus the two slits (microscopic objects), separated by distances considerably below the classical resolution limit, produce a white/black or black/white diffraction image on a gray background. In practice, slit separations of one-fifth the Rayleigh criterion produced images that were readily discernible by visual observation in the model experiment.

This experiment shows that the duality and relative positions of two objects can be discerned at separations considerably below the classical resolution limit if the phases of the diffraction patterns are inverted. Likewise, in scanning light microscopy (or other field scanning imaging systems) it should be possible to improve image resolution and precision of position determination by scanning through an aperture that would produce a similar diffraction pattern. Such a pattern can be generated, for example, with half masks placed at the condenser aperture plane that shift the phases of two parts of a coherent illuminating beam by 180° relative to each other.[5]

To summarize this section, the conventional limit of resolution, affected by the objective as well as condenser lens NA, provides a reasonable criterion for defining the smallest distances resolvable in a complex structure. However, when *a priori* knowledge regarding the structure is available, including the fact that it is present in a limited field (as in confocal microscopy) or that the phases of the diffraction images are inverted, the classical limit need not apply. In fact, the diffraction pattern of an isolated object can contain structural information that extends way below the conventional limit of resolution.

IV. Positional Information and Lateral Setting Accuracy

Another problem in light microscopy related to image resolution (but not identical with it) is the precision and accuracy by which we can determine the position of a small object. In addition to defining its spatial

[5] In this case, the beam need only arise from a coherent source and (unlike the case of the experiment discussed above) would not specifically need to be polarized.

coordinates, we can use (relative) position measurements to derive distances, velocities, etc.

The precision (or setting accuracy) for determining the position of an object point, lying in the object plane, reduces to the question "how well can we determine the location of a diffraction image (such as the center of an Airy disk) in the image plane?" In other words, how good a lateral setting accuracy can we achieve? From our earlier discussions it should not be surprising that we can, in fact, determine the center of the Airy disk with extremely high precision.

When two more or less similar points or lines are spaced very close together, the problem of measuring their separation reduces to that of resolution discussed in the previous sections. But if they are well-separated, the centers of corresponding points of the diffraction images can be closely defined (if necessary, by taking symmetric parts, unique points, or the whole envelope of the diffraction images into account, Fig. 16). Thus, one can rather readily measure distances between well-separated points or lines to precisions of the order of one-tenth of the resolution limit of the microscope or better. The measurement is limited by the precision of distance calibration rather than by the size of the diffraction image. Françon (1961) in fact describes a lateral setting accuracy of one-hundredth of the Airy diameter. Likewise, with video enhancement, one can see a small step, of the order of one-tenth of an Airy radius, in the diffraction image of a filament or an edge with an otherwise smooth contour (Inoué and Inoué, unpublished observation).

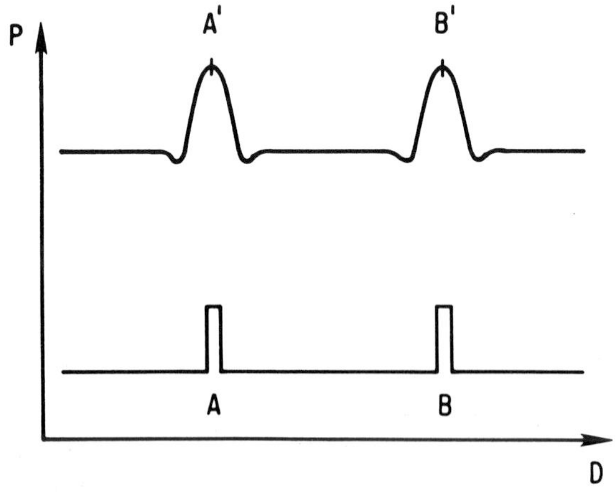

FIG. 16. The distance (D) between two unresolved narrow lines or points (A and B) can be determined with great accuracy (down to one-hundredth or less of the Airy radius) from the separation of the centroids of the diffraction images (A′ and B′) if they are adequately separated.

Denk *et al.* describe a case of very high precision relative position determination using the light microscope. These authors measured the vibrational amplitudes of auditory hair bundles with a sensitivity for the power spectrum density approaching precision of a few picometers/$\sqrt{Hz}$. In their approach, the laser-illuminated phase image of the edge of a hair bundle is projected by the microscope onto a quadrant detector photodiode. Outputs of the quadrants are individually amplified and compared to determine the rapid, few-nanometers displacement of the edge image (W. Denk and W. W. Webb, 1987, and personal communication).

Likewise, Kamimura (1987) describes the measurement of namometer size displacements of 1-μm-diameter polystyrene beads, by photoelectrically determining the displacement of two halves of the microscope image.

Gelles *et al.* (1987) have determined the successive positions of subresolution diameter (kinesin-coated) beads gliding on microtubules down to step sizes of 4 nm. They used video-enhanced DIC microscopy to obtain diffraction images of the gliding bead and determined their centroid by checking the cross-correlation between the bead's image and the previously digitized image of a bead. Thus, by using the whole diffraction image of the unresolved (150-nm diameter) bead, these authors have shown that the light microscope image can directly reveal nanometer-range quantal molecular steps by which particles are transported.

A case which is somewhat more complex than measuring the separation of two points or lines is the measurement of absolute distances between two edges. In general, the edges can be represented by a change in fluorescence, absorbance, reflectance, optical path difference, birefringence, etc., in the microscopic object. These parameters (P) are expressed in the lower parts of Fig. 17.

In contrast to measuring the separation of two points or lines (Figs. 1 and 16), the problem is no longer that of measuring a simple translational distance of identical or similar diffraction images. Instead, we need to interpret the positions of the edges from the diffraction images. Even when the distance between the edges is considerably larger than the Airy radius, we cannot readily determine the exact distance without a detailed knowledge of the relationship between the object's geometrical edge and the shape of the diffraction image. The curves from Smith and Osterberg discussed earlier provide some clues, but further data are needed for different types of objects imaged with alternate contrast modes.

To summarize this section, the position of optically isolated small points or lines can be determined with very high precision—down to distances of the order of a hundredth of the Airy radius. The distance between two separate edges is, however, much more difficult to measure accurately.

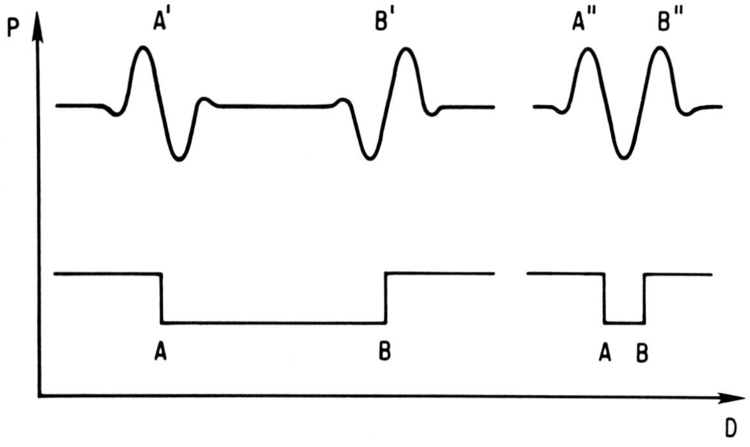

FIG. 17. The distance (D) between two edges is considerably more difficult to measure accurately than between two narrow lines or edges (Fig. 16). Each edge produces an asymmetrical diffraction pattern (A′, B′) from which the location of the true geometrical edges (A, B) must be inferred.

V. Ultrathin Optical Sectioning and Axial Setting Accuracy

Video enhancement has also improved our ability to obtain very thin optical sections, or shallow depth of field, with the light microscope. This means that the axial resolution and axial setting accuracy (both measured along the optical axis of the microscope) are also improved.

While there is no single agreed-upon formula that defines these parameters, Françon (1961; also see Inoué, 1986) expresses the axial setting accuracy (2ξ) as

$$2\xi = \frac{\lambda}{4n \sin^2(u/2)} \tag{2}$$

where λ is the wavelength, n the refractive index of the immersion medium, and u the half angle of the cone of light captured by the objective lens. What is important to note is that the axial setting acurracy, and the related axial resolution and thinness of the depth of field, rises inversely, essentially with the second power of the numerical aperture. From Eq. (2) Françon would predict that, in the ideal case, we should get an axial setting accuracy of 0.29 μm with a 1.4 N A oil immersion lens.

We have in fact obtained optical sections that are 0.1–0.15 μm thick with rectified polarized light, just under 0.2 μm thick in phase contrast microscopy, and approximately 0.2 μm thick with DIC (Fig. 18)! These

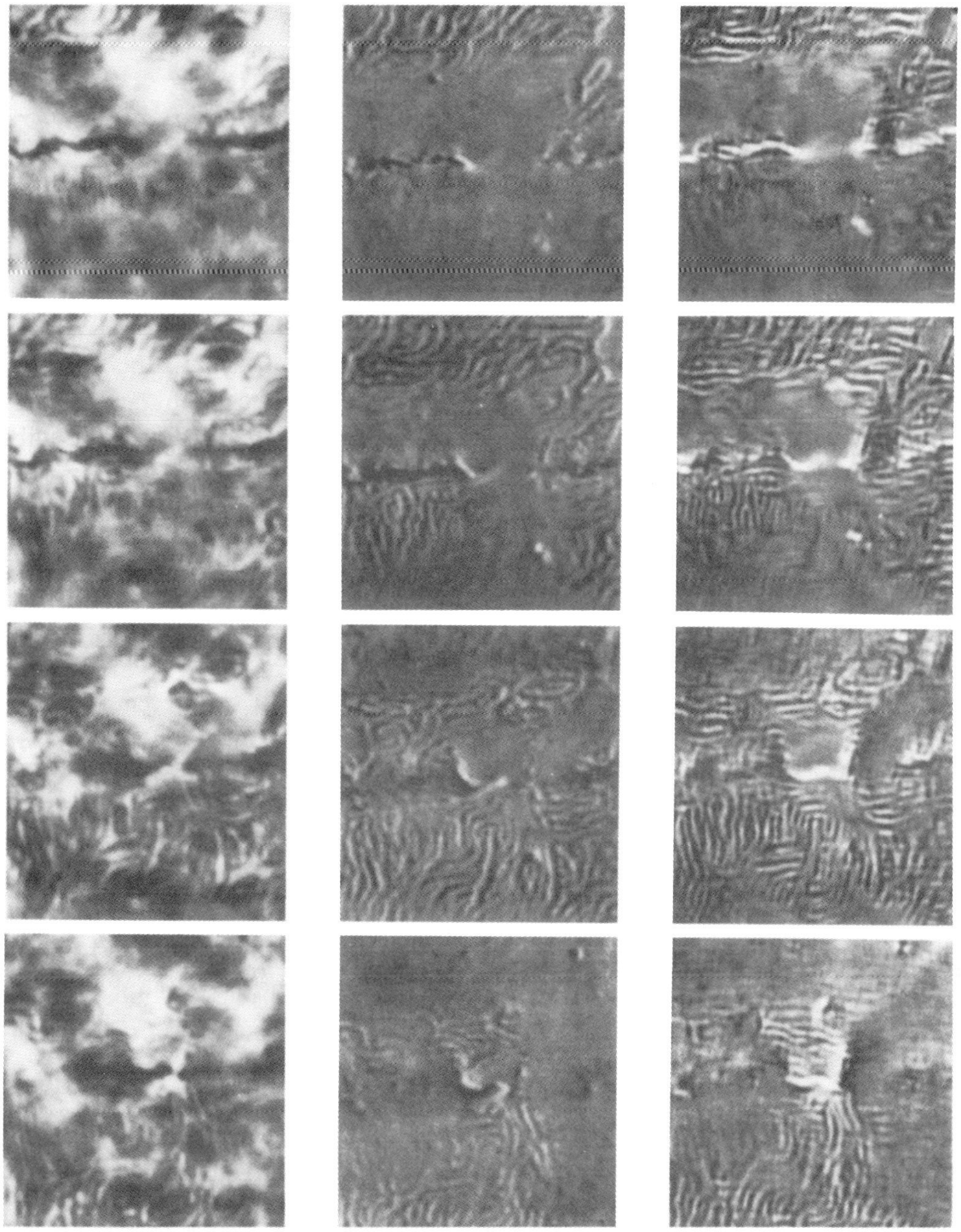

FIG. 18. Shown are 0.22 μm step, through-focal optical sections of buccal epithelial cell surface in phase contrast. The same $\times$100/1.4 NA Plan Apo objective lens (Nikon) was used for all images. Left column: At this stepping distance in phase contrast, the successive optical sections are too far spaced for the image details to be contiguous. In other words, the depth of field is less than 0.2 μm. Middle column: Rectified DIC condenser and Nomarski prism. Details in the consecutive optical sections now just overlap each other. The depth of field is around 0.20–0.25 μm. Right column: Rectified polarized light microscopy. As with phase contrast, the successive optical sections do not overlap, but unlike the phase image, there is much less interference from out-of-focus structures. Scale bar (lower left): 10 μm.

ultrathin sections were achieved by the use of: (1) a new generation of well-corrected, high NA objective lenses (Nikon new series 100/1.4 and 60/1.4, and Zeiss Axiophot 63/1.4, Plan Apochromats); (2) a high NA condenser, fully and uniformly illuminated with high intensity monochromatic (546 nm) light through a light scrambler (Ellis, 1985; Inoué, 1986, Figs. III-21, III-22); (3) optical contrast enhancement and image correction with a condenser rectifier for polarization and DIC optics, together with improved lens coating (Nikon, Inc.); (4) analog enhancement with black level and gain controls on the video camera (DAGE-MTI 65M Newvicon); and (5) further digital enhancement (Universal Imaging Corporation Image-I). Optical sectioning, through-focal sectioning, recording, and analysis were also aided by (6) a solid, optical bench microscope with precision stage (Inoué, 1986); (7) stepper motor drive and timed controller for the fine focus; and (8) a 450-TV-line-resolution laser disk recorder (Panasonic OMDR T2-2021 FBC).

We do not yet have a theory to explain how we obtain such ultrathin optical sections, since the observations exceed the setting accuracy calculated according to Eq. (2) (which one would expect to be considerably smaller than the thickness of the optical section).

One might think that our observation is related to a smaller Airy disk radius and a consequent shrinking of the axial elongation of the three-dimensional diffraction pattern. As discussed earlier (Section III), the radius of the Airy central disk is reduced in a lens with central obstruction. In a high extinction polarizing or DIC system, the back aperture of the objective lens is not completely uniformly extinguished even after rectification; birefringence in the lens elements and ellipticity introduced at the lens surfaces conspire to reduce the extinction at higher NA. Thus to some extent, the aperture function is similar to a lens with a partially absorbing central stop that could reduce the radius of the Airy disk as discussed earlier.

The trouble with the explanation is that the three-dimensional diffraction pattern is supposed to elongate axially, rather than shrink, as the Airy radius is decreased by the introduction of central obstruction (Linfoot and Wolf, 1953). Therefore, the axial resolution and setting accuracy should decrease, and the thickness of the optical sections should in fact increase.

Whatever the theoretical foundation, the ultrathin optical sections now attained with video microscopy allows us to better discriminate objects which are part of a complex three-dimensional structure. For example, the dynamic behavior of individual microtubules amidst a large number of other microtubules can be distinguished in the optical section of living cells viewed with high extinction polarized light microscopy (Cassimeris *et al.*, 1988; Inoué, 1988).

We do not yet have equivalent data for fluorescence microscopy where, as mentioned before, we can consider the light waves arising from adjacent specimen points as being totally incoherent. The point is that, if anything, the resolution obtained on incoherently illuminated objects should be greater than for coherently or partially coherently illuminated objects (Hopkins and Barham, 1950; Born and Wolf, 1980). However, studies using the laser scanning confocal microscope with a 1.4 NA. optical system show that the optical system for Epifluorescence is approximately 0.7 μm thick (White *et al.*, 1987; B. W. Amos, personal communication; see also Brakenhoff *et al.*, 1979).

With fluorescence microscopy, light contribution from out-of-focus objects may affect the in-focus image more severely than with contrast-producing methods based on phase differences. The comparison has, however, not yet been clearly established, especially for diffraction-limited aberration-free optics.

Be that as it may, a number of new developments have made it possible to isolate clean optical sections which are especially helpful for fluorescence microscopy. These include mathematical deconvolution from serial optical sections of fluorescence video images (see Agard *et al.*, Chapter 13, this volume) and confocal microscopy (see Chapter 14 by Brakenhoff *et al.*, this volume).

Whatever means are used to obtain optical sections, serial optical sections provide the basis for three-dimensional and stereoscopic reconstruction. In addition to the chapters mentioned above, articles relating to this topic can be found in Somlyo (1986; see also Inoué *et al.*, 1985).

VI. Other Factors Affecting the Choice of Pixel Dimensions, Calibration, and Choice of Magnification

In contrast to bright-field, dark-field, fluorescence, and phase-contrast microscopy, or conventional photography, in which the apertures should be radially symmetric and resolution is independent of spacing direction, resolution in video is direction dependent. Thus, in conventional raster scan video, the horizontal resolution is generally different from, and independent of, the vertical resolution, which is limited by the number of active video scan lines.

The number of active scan lines, which are those that are visible on an underscanned monitor (i.e., the number of total scan lines per frame, e.g., 525, minus those hidden in the vertical blanking period, e.g., 42) for standard 525/60 video used in the United States and Japan today is around

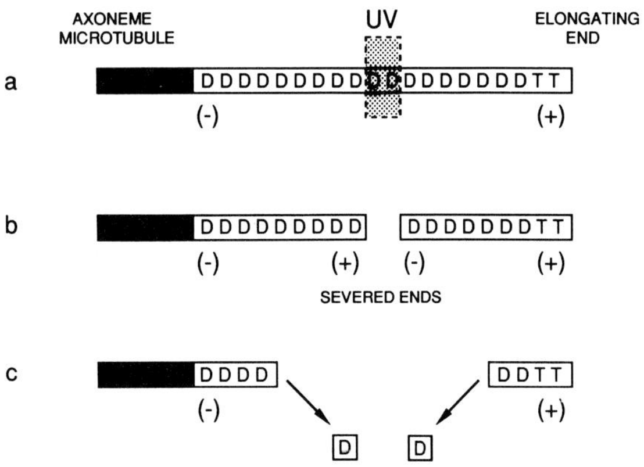

Figure 1. Predicted behavior of severed microtubule ends based on the GTP cap model. Schematic drawing of tubulin subunits in a microtubule growing from one end of an axoneme fragment (*solid area*). According to the GTP cap model (7, 19, 20, 34), the incorporation of GTP-tubulin (*T*) subunits at the elongating end of the microtubule stabilizes a labile core of GDP-tubulin (*D*) (the result of hydrolysis following subunit addition) (*a*). When the microtubule core is cut by UV irradiation (*b*), two severed ends are created, one plus, the other minus. The GTP cap model predicts that both severed ends will rapidly shorten (*c*) because they are no longer stabilized by GTP-tubulin.

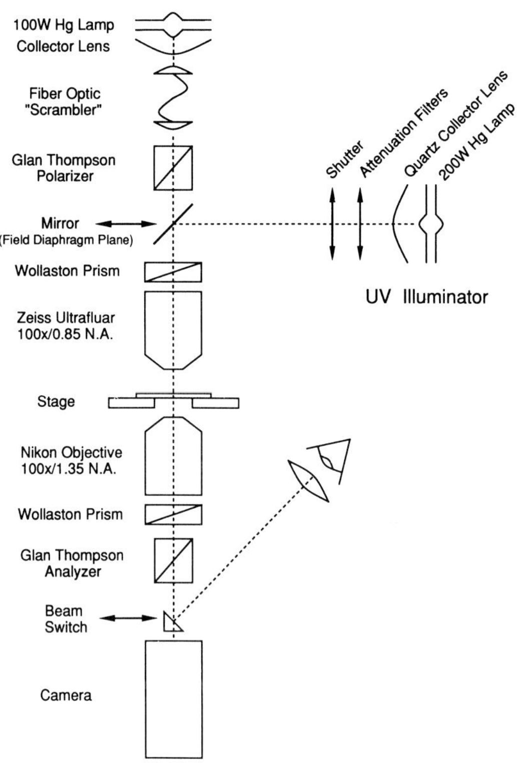

Figure 2. Schematic diagram of the UV microbeam apparatus. A 200-W mercury arc lamp served as the UV source. The microbeam was directed onto the specimen plane from the condenser side by a mirror inserted at the field diaphragm plane. A 100×/0.85 NA Zeiss Ultrafluar (Thornwood, NY) was used as the condenser. A 2-μm-wide image of the UV mirror was projected onto the specimen. The Wollaston DIC prism transmitted sufficient UV (~20%) to sever microtubules with a 3-s exposure. Shutters controlled the UV irradiation. Design details for the differential interference contrast optics, video recording, and digital image processing are described in the text.

GTP cap model is correct, these severed ends should begin rapid shortening immediately after cutting.

To test the GTP cap prediction illustrated in Fig. 1, we have used a UV microbeam to sever individual, elongating microtubules. A UV source and the necessary UV-transmitting optics were incorporated into a custom-designed light microscope (Fig. 2). Purified tubulin was assembled onto the plus and minus ends of sea urchin flagellar axoneme fragments using the methods of Walker et al. (45). The behavior of severed ends was observed and recorded in real time using video-enhanced differential interference contrast (DIC) microscopy and digital image processing. Contrary to the prediction of the GTP cap model, the plus and minus ends behave quite differently.

Materials and Methods

Tubulin and Axoneme Preparation

Porcine brain tubulin was purified by two cycles of assembly and disassembly in a buffer of 100 mM Pipes, 2 mM EGTA, 1 mM $MgSO_4$, 0.5 mM GTP, pH 6.9 (PM buffer) and the resulting pellets were overlaid with a buffer of 100 mM 2[*N*-morpholino]ethanesulfonic acid (MES), 1 mM EGTA, 0.5 mM $MgSO_4$, 3.4 M glycerol, pH 6.6. The tubulin was resuspended and passed over phosphocellulose and further purified by a cycle of assembly in 1 M Na^+-glutamate as described previously (44). The tubulin was then resuspended immediately in PM buffer and frozen in small aliquots. Based on the SDS-PAGE methods described in Walker et al. (45), microtubule-associated proteins constituted ~0.6% of the purified tubulin preparation. Flagellar axoneme fragments were prepared from *Lytechinus pictus* according to the method of Bell et al. (1). Axonemes were osmotically demembranated and mechanically separated from sperm heads by homogenization in a solution of 20% sucrose in distilled water with a hand-held glass homogenizer. Axonemes were resuspended and washed in isolation buffer composed of 0.1 mM NaCl, 4 mM $MgSO_4$, 1 mM EDTA, 7 mM β-mercaptoethanol, and 10 mM Hepes (pH 7.0). Dynein outer arms were removed by incubation in isolation buffer adjusted to 0.6 M NaCl for 30 min at 4°C. Residual sperm heads were removed by sedimentation of the axonemes through an 80% sucrose cushion (16,000 *g*, 10 min). Axonemes were stored at −20°C in a 1:1 solution of isolation buffer/glycerol. Axonemes were washed and resuspended in PM before use.

Microscopy and the UV Microbeam

Preparations were viewed by DIC microscopy, using a custom-built, inverted polarization microscope (22, 24, 25). The image of the UV source, an HBO 200-W mercury arc lamp, was projected onto a 0.2 × 0.7 mm mirror inserted in the field diaphragm plane of the microscope after the visible illuminating beam had passed the polarizing prism (Fig. 2). A Zeiss Ultrafluar 100×/0.85 NA glycerin immersion objective (Thornwood, NY) served as the condenser and was equipped with a Wollaston prism from a Zeiss 63×/1.4 NA objective lens. This combination provided a reasonable match to the Nomarski type prism made for Nikon Plan Apo objectives. The Ultrafluar projected a 2-μm-wide image of the illuminated mirror (the UV slit) onto the specimen plane, superimposed on the visible light DIC image of the specimen. The mirror could be slid in and out of the light path. With the mirror in the light path, the image of the UV source was first focused onto the specimen through UV blocking and neutral density filters. The filters were removed to expose the microtubule to the UV microbeam. A Nikon Plan Apo 100×/1.35 NA oil immersion objective projected the specimen image through a Nikon Nomarski DIC prism and analyzing prism to either the oculars or a video camera.

Image contrast was enhanced by video and digital processing. Video enhancement was provided by a newvicon video camera with high gain and offset (model 65, Dage-MTI Inc., Michigan City, IN), followed by digital enhancement provided by an Image-I/AT processor (Universal Imaging Corp., Media, PA). An exponential average of two frames with background mottle subtraction was used on-line to reduce electronic and optical noise in the image. Processed images were recorded on 3/4 inch U-matic tape (Sony model VO-5800H videocassette recorder) and optical disk (Panasonic model TQ-2021FBC optical memory disk recorder).

UV Irradiation Experiments

Purified tubulin and axonemes were mixed and then diluted with cold PM to final concentrations of 16 μM and 2.7×10^7 ml^{-1}, respectively. Preparations contained 1 mM GTP. The tubulin-axoneme preparation was held at 4°C until needed.

A 5-μl sample of the preparation was added to a biologically clean (32) 22-mm^2 quartz coverslip (thickness No. 1.5) mounted on a stainless steel holder, then covered with a biologically clean 22-mm^2 glass coverslip (thickness No. 1.5) and sealed with valap (1:1:1 mixture of beeswax, lanolin, and petrolatum) to prevent drying and to prevent flow within the chamber. The typical separation between inner chamber surfaces was 10–20 μm. The double coverslip chamber was inverted and glycerol and oil contacted to the Ultrafluar and Plan Apo lenses, respectively.

Microtubules were assembled at 21–22°C. The axoneme fragments adhered to the clean chamber surfaces but the microtubules elongating off of these fragments generally remained in solution. Microtubules that did adhere to the coverslip surface could not be severed by UV irradiation.

Microtubules were typically irradiated 15–40 min after initiation of assembly. Microtubules irradiated at longer times behaved identically (in terms of response to cutting) to microtubules irradiated soon after initiation of assembly.

Exposures of 2–3 s of unfiltered UV irradiation faithfully severed elongating microtubules. Based on previous studies (31, 48), irradiation in the 260–300 nm range severs microtubules in vivo.

Data Analysis

Microtubule elongation and plus end rapid shortening rates were measured from 3/4 inch U-matic videotape recordings played on a Sony VO-5800H. We used a computer-based analysis system to follow microtubule length changes in real time (45). Minus end shortening rates were difficult to measure in real time because of the brief duration of rapid shortening episodes. Minus end shortening rates were therefore calculated as (change in length)/(time of rapid shortening) for each episode.

Plus and minus ends were identified based on rate of elongation (45). Rates in the text are given as mean ± SEM.

Results

Experiments were performed at 21–22°C to prevent thermal damage to the microscope optics. At the tubulin concentration (16 μM) used in this study, a spontaneous catastrophe occurred at elongating plus ends about once every 4 min of elongation. Rescue was infrequent for plus end microtubules, and rapid shortening usually proceeded to the axoneme seed. At the minus end, a spontaneous catastrophe occurred about once every 15 min of elongation. Shortening

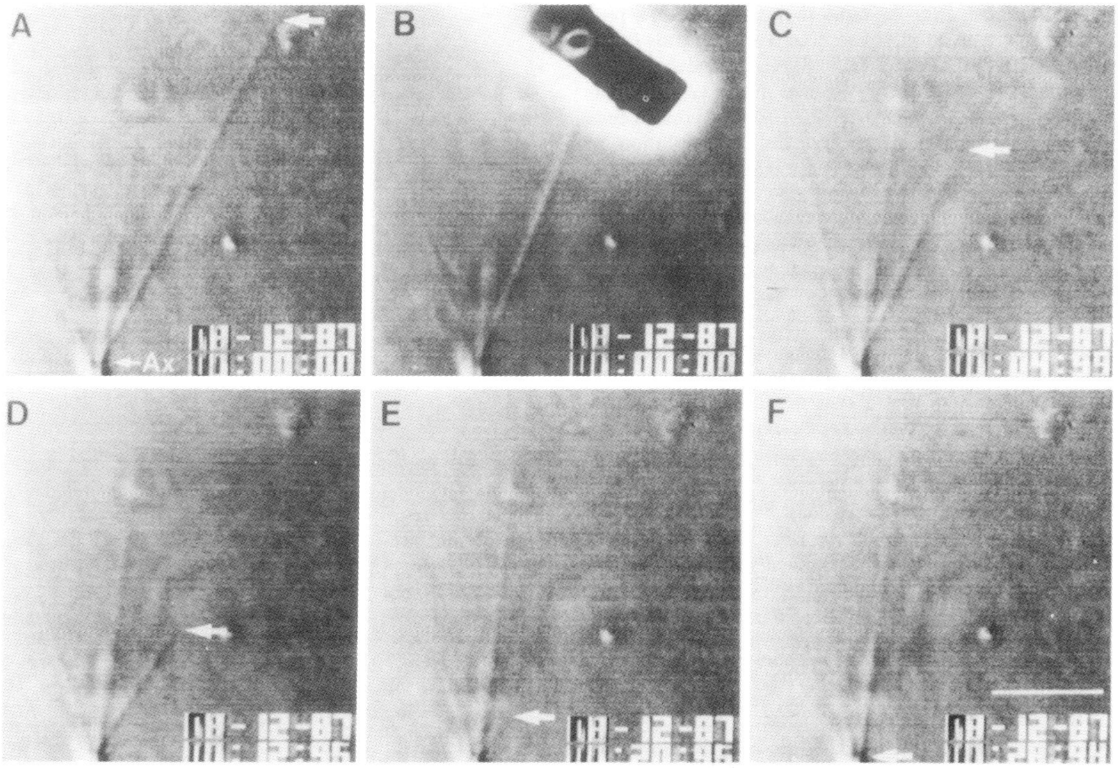

Figure 3. A severed plus end rapidly shortens immediately after UV cutting. The distal tip of an elongating plus end microtubule was cut off by 3 s UV irradiation. (*A*) Field before irradiation, Ax is plus end of axoneme fragment; (*B*) mirror in place just before UV irradiation; (*C*) 4 s after irradiation. An image of the irradiation beam persists on the camera tube. Arrowhead indicates the position of the severed plus end; (*D–F*) the severed plus end continues rapid shortening until it disappears. Time in minutes/seconds/0.01 seconds is given at the bottom right of each video frame. Bar, 5 μm.

minus end microtubules usually underwent rescue, and the average length lost during a shortening phase was 3.2 μm (based on average time of shortening and the mean rate of shortening).

Microtubule Irradiation

Each UV irradiation of an individual microtubule created two severed ends: one plus, the other minus. In practice, we were able to measure the resultant length changes of only the microtubules that remained tethered to the axoneme fragment. The other severed end, now one end of a free microtubule, was impossible to observe because the nascent free microtubule rapidly moved out of the plane of focus. Therefore, the behavior of severed plus end microtubules was observed by cutting microtubules which elongated off of the plus end of an axoneme fragment, whereas the behavior of severed minus end microtubules was observed by cutting microtubules which elongated off of an axoneme fragment's minus end.

All plus end microtubules severed by the UV microbeam immediately began rapid shortening (n = 16) (Figs. 3 and 5 *a*). The onset of rapid shortening was independent of the position of the cut zone relative to either the axoneme seed or the microtubule end. The minimum distance that we were able to remove from an elongating plus end microtubule was 0.8 μm. The rate of rapid shortening of a severed plus end microtubule (20.9 ± 1.5 μm/min [n = 15]) was not significantly different from the rapid shortening rate of a plus end microtubule that had experienced a spontaneous catastrophe (22.2 ± 1.7 μm/min [n = 8]) (independent t-test).

In contrast to the plus end, severed minus end microtubules never rapidly shortened (n = 29). Severed minus ends always remained within 0.2 μm (our limit of resolution) of the cut zone (Figs. 4 and 5, *b* and *c*) and then resumed elongation (Fig. 5, *b* and *c*). The stability of a severed minus end was independent of the amount of polymer removed from the elongating end as shown in Fig. 5 *b*. In this example, an elongating minus end microtubule was cut at successive times at distances progressively closer to the axoneme, but the severed minus ends never rapidly shortened.

The tubulin lattice at severed minus ends did not appear significantly altered because severed minus ends always immediately began elongation at rates (0.40 ± 0.03 μm/min [n = 28]) typical of normal minus end elongation (0.41 ± 0.04 μm/min [n = 14]) (independent t-test). Further, after reelongation, the region of a microtubule at the original cut site was no more stable than elsewhere along the microtubule. As shown in Fig. 5 *c*, severed minus end microtubules were observed to elongate until a spontaneous catastrophe occurred, then to rapidly shorten without interruption through the previous cut site. These observations demonstrate that UV irradiation did not prevent rapid shortening of minus ends by irreversibly cross-linking the microtubule lattice.

Discussion

According to the GTP cap model, a microtubule contains GDP-tubulin all along its length, except for a short region of GTP-tubulin at the elongating end(s) (Fig. 1). Cutting a microtubule at any site along its length (away from the GTP-tubulin cap) will produce severed ends with terminal GDP-tubulin subunits which will immediately dissociate. We have

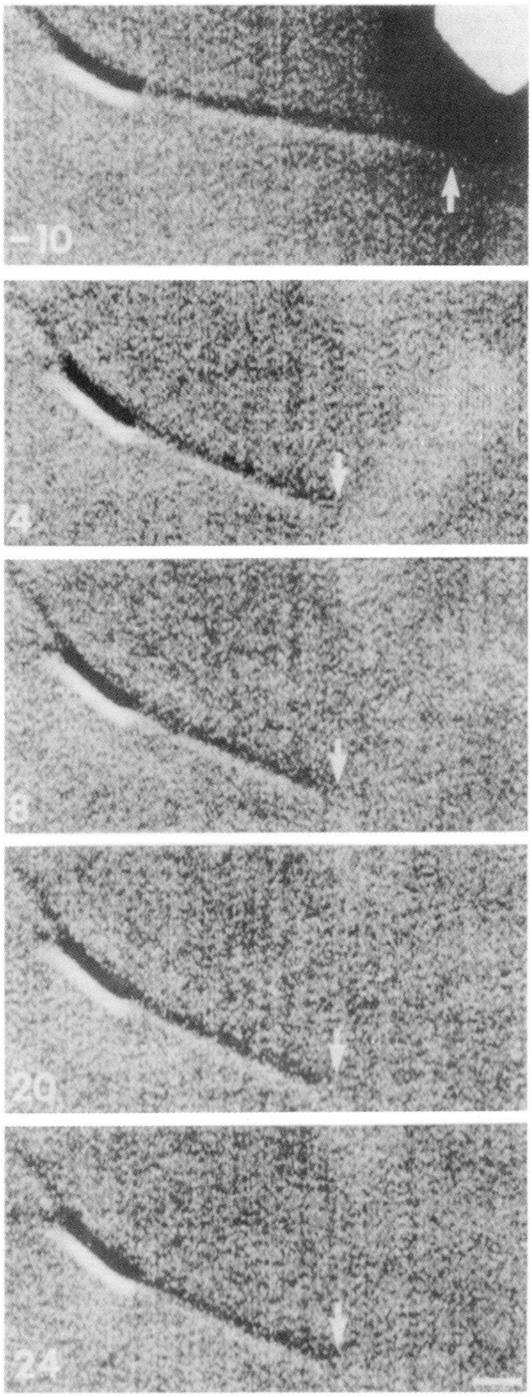

Figure 4. A severed minus end does not rapidly shorten after UV cutting. The distal tip of an elongating minus end microtubule was cut off by 3 s UV irradiation. Time is given in seconds on each video frame with the time of UV irradiation set to zero time. At 4 s, the image of the cutting zone (the UV irradiation area) persists on the video camera tube. The axoneme was slightly reoriented before exposure to the UV microbeam. The arrows indicate the microtubule end. Bar = 1 μm.

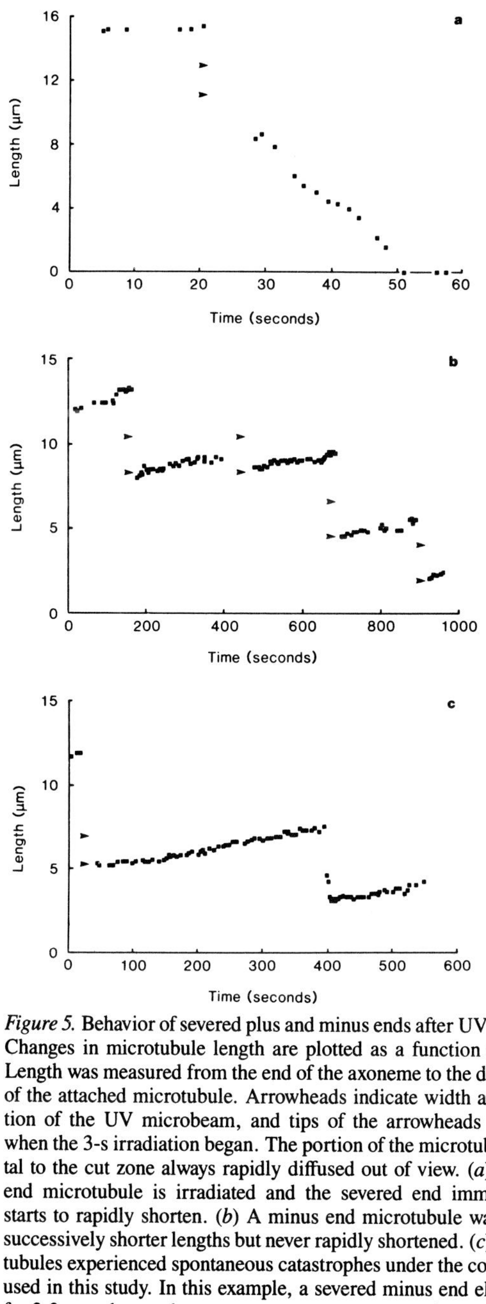

Figure 5. Behavior of severed plus and minus ends after UV cutting. Changes in microtubule length are plotted as a function of time. Length was measured from the end of the axoneme to the distal end of the attached microtubule. Arrowheads indicate width and location of the UV microbeam, and tips of the arrowheads indicate when the 3-s irradiation began. The portion of the microtubule distal to the cut zone always rapidly diffused out of view. (*a*) A plus end microtubule is irradiated and the severed end immediately starts to rapidly shorten. (*b*) A minus end microtubule was cut to successively shorter lengths but never rapidly shortened. (*c*) Microtubules experienced spontaneous catastrophes under the conditions used in this study. In this example, a severed minus end elongated for 2.3 µm, then underwent a spontaneous catastrophe and rapidly shortened for 4.4 µm. Note that the microtubule elongated from and shortened through the original cut site.

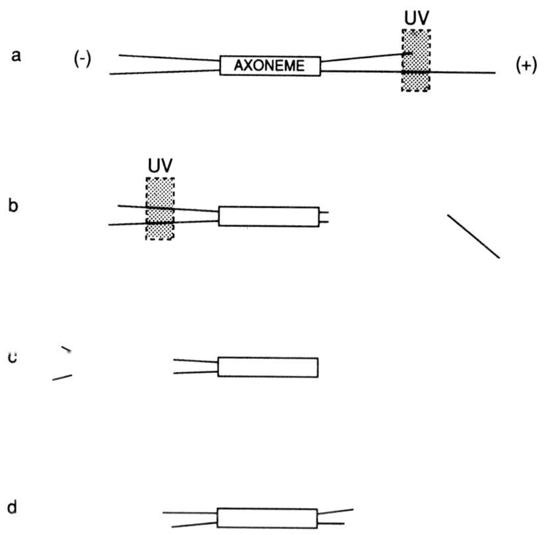

Figure 6. Summary of UV cutting experiments. Severed plus end microtubules (*a*), rapidly shortened (*b*), and frequently disappeared (*c*). Later, new plus end microtubules reelongated off of the axoneme fragment (*d*). In contrast, severed minus end microtubules (*b*), did not rapidly shorten, but immediately resumed elongation at a rate typical of minus end growth (*c* and *d*).

shown that severed plus end microtubules, as predicted by the GTP cap model, immediately begin to disassemble (summarized in Fig. 6, *a–c*). Rapid shortening was extensive and occurred at a constant rate both independent of the length of the microtubule and typical of spontaneous rapid shortening. Our results clearly demonstrate that a short region (<0.8 µm or 1,300 subunits) at the plus end of an elongating microtubule can stabilize the entire polymer.

In contrast to the labile nature of severed plus ends, the stability of severed minus ends was not predicted by the GTP cap model. No shortening of severed minus end microtubules was detected; rather, these microtubules behaved as if capped by GTP-tubulin and immediately resumed elongation at rates typical of unsevered minus end microtubules (summarized in Fig. 6, *b–d*). Because the bulk of a microtubule is GDP-tubulin (17, 29, 30, 36, 46, 47), and because severed plus end microtubules initially appear to have terminal GDP-tubulin subunits (based on the behavior of severed plus ends), severed minus end microtubules must also initially have terminal GDP-tubulin subunits. How did severed minus end microtubules quickly regain a GTP cap? Our results show that UV irradiation did not irreversibly stabilize the microtubule lattice at the severed end. Further, the stability of severed minus ends was clearly not due to the "normal" mechanism of rescue (presumably addition of GTP-tubulin to GDP-tubulin ends), because the average extent of minus end rapid shortening after a spontaneous catastrophe was 3.2 µm, whereas the maximum extent of shortening for each severed minus end was 0.2 µm (our limit of resolution). Alternatively, the lattice structure of plus and minus end microtubules assembled onto an axoneme seed may be different. However, this does not seem likely because preliminary observations of UV severed microtubules self-assembled in the absence of axoneme seeds also show an asymmetric behavior: one end rapidly shortens while the other end is stable (Walker, R. A., and E. D. Salmon, manuscript in preparation).

There are two different explanations, based on the GTP cap model, for the difference in stability of severed plus and minus ends. The first assumes that formation of a severed end immediately produces a GDP-tubulin end capable of rapid shortening. That is, a severed end is initially an end in the rapid shortening phase. This is the basic prediction of the current GTP cap models. If both plus and minus severed ends are initially in the rapid shortening phase, then the extremely rapid rescue of severed minus ends must be produced by a UV-activated mechanism not available at plus ends. According to this hypothesis, UV irradiation somehow promotes rapid reformation of a GTP-tubulin cap at minus, but not plus, severed ends. One possibility is that the exchangeable site for GTP on the tubulin dimer is exposed at the minus end but not the plus end. UV irradiation could rapidly displace the GDP from the exchangeable site, allowing free GTP to bind and stabilize the polymer before significant rapid shortening occurs. Because the exchangeable site is located on the β subunit of the tubulin dimer (18, 27, 35), this mechanism predicts that the β subunit is oriented towards the minus end of a microtubule and the α subunit is oriented towards the plus end. Unfortunately, the orientation of the tubulin dimer within the microtubule lattice is not yet known.

An alternative explanation is that loss of the GTP cap is not sufficient for rapid shortening. A second reaction may be required before GDP-tubulin subunits at the end of a microtubule can dissociate. Thus, there could be an intermediate phase between elongation and rapid shortening. According to this model, catastrophe is a two-step process. The first step, the transition from the elongation phase to the intermediate phase, depends on the presence or absence of a GTP cap. The second step, the transition from the intermediate phase to rapid shortening, may involve a structural transformation at the end of the polymer lattice (28) which, once initiated, produces rapid dissociation of GDP-tubulin subunits as the transformation propagates rearward along the microtubule lattice. Cutting an elongating microtubule can be viewed as creating severed ends in the intermediate phase. The behavior of a severed end will reflect the probabilities of transition from this intermediate phase to either elongation or rapid shortening. At the plus end, conversion from the intermediate phase to the rapid shortening phase is highly favored, since a severed plus end microtubule immediately begins rapid shortening. However, at the minus end, conversion from the intermediate phase to the elongation phase is favored, since a severed minus end microtubule immediately resumes elongation.

The dynamics of microtubule assembly has been studied in living cells using UV microbeam irradiation to sever microtubules (14, 16, 23, 31, 43, 48). It is worth reconsidering the results from these studies with respect to our in vitro findings. We have demonstrated here, for microtubules assembled from pure brain tubulin, that severed minus ends immediately start to elongate while severed plus ends rapidly shorten. A similar behavior can be seen in reports of the effect of UV microbeam irradiation on microtubules in the cytoplasmic microtubule complex in mammalian tissue culture cells (43), the central spindle microtubules of diatoms (31), the chromosomal fibers in the mitotic endosperm of *Haemanthus* (23), and the chromosomal fibers in the first meiotic spermatocytes of the grasshopper, *Trimeratropis maritima* (16). In all these examples, the minus ends of severed microtubules appeared stable while the plus ends rapidly shortened to various extents. The results of UV microbeam irradiation of chromosomal fibers in first meiotic spermatocytes by Forer and co-workers appears to be more complex (14, 48). Recent electron microscopy studies (48) of areas of reduced birefringence (ARBs) produced by UV irradiation of chromosomal fibers indicate that the poleward movement of ARBs may be the consequence of the difference in assembly dynamics of severed plus and minus ends rather than poleward flow of spindle fiber material as originally proposed (14) (later interpreted as treadmilling of tubulin subunits [26, 33]; see Forer [15] and Wilson and Forer [48] for discussion). However, this interesting question remains unresolved.

Overall, the above analysis indicates that because minus ends are relatively stable, microtubule dynamics in living cells is likely to be governed by the dynamics of their plus ends.

We thank Brenda Bourns for her careful preparation of the purified tubulin used in this study and for her assistance in the preparation of this paper. Vicki Petrie and Susan Whitfield assisted with figure preparation. Drs. Tim O'Brien, Bruce Telzer, and Leah Haimo provided many helpful comments. Finally, we are grateful to Lynne Cassimeris and Nancy Pryer for discussions about melted microtubule ends, and to Bob Knudson for assistance in the design and construction of the microbeam apparatus.

This work was supported by National Institutes of Health (NIH) grant GM 24364 and National Science Foundation (NSF) grant DCB-8616621 to E. D. Salmon and NIH grant R37 GN31617-06 and NSF grant DCB-8518672 to S. Inoué.

Received for publication 15 August 1988 and in revised form 2 November 1988.

References

1. Bell, C. W., C. Fraser, W. S. Sale, W.-J. Y. Tang, and I. R. Gibbons. 1982. Preparation and purification of dynein. *Methods Cell Biol.* 24: 373–397.
2. Caplow, M., and R. Reid. 1985. Directed elongation model for microtubule GTP hydrolysis. *Proc. Natl. Acad. Sci. USA.* 82:3267–3276.
3. Caplow, M., and J. Shanks. 1987. GTP requirement for in vitro and in vivo microtubule assembly and stability. *In* The Cytoskeleton in Cell Differentiation and Development. R. B. Maccioni and J. Arechaga, editors. IRL Press, Oxford, UK. 63–73.
4. Caplow, M., J. Shanks, and B. D. Brylawski. 1985. Concerning the location of the GTP hydrolysis site on microtubules. *Can. J. Biochem. Cell Biol.* 63:422–429.
5. Carlier, M.-F., and D. Pantaloni. 1981. Kinetic analysis of guanosine 5′-triphosphate hydrolysis associated with tubulin polymerization. *Biochemistry.* 20:1918–1924.
6. Carlier, M.-F., T. L. Hill, and Y.-D. Chen. 1984. Interference of GTP hydrolysis in the mechanism of microtubule assembly: an experimental study. *Proc. Natl. Acad. Sci. USA.* 81:771–775.
7. Carlier, M.-F., D. Didry, and D. Pantaloni. 1987. Microtubule elongation and guanosine 5′-triphosphate hydrolysis. Role of guanine nucleotides in microtubule dynamics. *Biochemistry.* 26:4428–4437.
8. Carlier, M.-F., D. Didry, R. Melki, M. Chabre, and D. Pantaloni. 1988. Stabilization of microtubules by inorganic phosphate and its structural analogues, the fluoride complexes of aluminum and beryllium. *Biochemistry.* 27:3555–3559.
9. Cassimeris, L. U., P. Wadsworth, and E. D. Salmon. 1986. Dynamics of microtubule depolymerization in monocytes. *J. Cell Biol.* 102:2023–2032.
10. Cassimeris, L. U., R. A. Walker, N. K. Pryer, and E. D. Salmon. 1987. Dynamic instability of microtubules. *Bioessays.* 7:149–154.
11. Cassimeris, L., N. K. Pryer, and E. D. Salmon. 1988. Real-time observations of microtubule dynamic instability in living cells. *J. Cell Biol.* 107:2223– 2231.
12. Chen, Y., and T. L. Hill. 1985. Monte Carlo study of the GTP cap in a five-start helix model of a microtubule. *Proc. Natl. Acad. Sci. USA.* 82: 1131–1135.
13. David-Pfeuty, T., H. P. Erickson, and D. Pantaloni. 1977. Guanosinetriphosphatase activity of tubulin associated with microtubule assembly. *Proc. Natl. Acad. Sci. USA.* 74:5372–5376.
14. Forer, A. 1965. Local reduction of spindle fiber birefringence in living *Nephrotoma surturalis* (Leow) spermatocytes induced by ultraviolet microbeam irradiation. *J. Cell Biol.* 25:95–117.
15. Forer, A. 1985. Does actin produce the force that moves a chromosome to the pole during anaphase? *Can. J. Biochem. Cell Biol.* 63:585–598.
16. Gordon, G. G. 1980. The control of mitotic motility as influenced by ultraviolet microbeam irradiation of kinetochore fibers. Ph.D. dissertation. University of Pennsylvania.

17. Hamel, E., J. K. Batra, A. B. Huang, and C. M. Lin. 1986. Effects of pH on tubulin-nucleotide interactions. *Arch. Biochem. Biophys.* 245:316–330.
18. Hesse, J., H. Maruta, and G. Isenberg. 1985. Monoclonal antibodies localize the exchangeable GTP-binding site in β- and not α-tubulins. *FEBS (Fed. Eur. Biochem. Soc.) Lett.* 179:91–95.
19. Hill, T. L. 1984. Introductory analysis of the GTP-cap phase-change kinetics at the end of a microtubule. *Proc. Natl. Acad. Sci. USA.* 81:6728–6732.
20. Hill, T. L., and Y.-D. Chen. 1984. Phase changes at the end of a microtubule with a GTP cap. *Proc. Natl. Acad. Sci. USA.* 81:5772–5776.
21. Horio, T., and H. Hotani. 1986. Visualization of the dynamic instability of individual microtubules by dark-field microscopy. *Nature (Lond.).* 321:605–607.
22. Inoué, S. 1961. Polarizing microscope: design for maximum sensitivity. *In* Encyclopedia of Microscopy. G. L. Clarke, editor. Reinhold Publishing Corp., New York. 480–485.
23. Inoué, S. 1964. Organization and function of the mitotic spindle. *In* Primitive Motile Systems in Cell Biology. R. Allen and N. Kamiya, editors. Academic Press, New York, 549–598.
24. Inoué, S. 1981. Video image processing greatly enhances contrast, quality, and speed in polarization-based microscopy. *J. Cell Biol.* 89:346–356.
25. Inoué, S. 1986. Video Microscopy. Plenum Publishing Corp., New York. 495.
26. Inoué, S., and H. Ritter. 1975. Dynamics of mitotic spindle organization and function. *In* Molecules and Cell Movement. S. Inoué and R. Stephens, editors. Raven Press, New York. 3–29.
27. Kirchner, K., and E.-M. Mandelkow. 1985. Tubulin domains responsible for assembly of dimers and protofilaments. *EMBO (Eur. Mol. Biol. Organ.) J.* 4:2397–2402.
28. Kirschner, M. W., and T. Mitchison. 1986. Microtubule dynamics. *Nature (Lond.).* 324:621.
29. Kobayashi, T. 1974. Nucleotides bound to brain tubulin and reconstituted microtubules. *J. Biochem. (Tokyo).* 76:201–204.
30. Kobayashi, T. 1975. Dephosphorylation of tubulin-bound guanosine triphosphate during microtubule assembly. *J. Biochem. (Tokyo).* 77:1193–1197.
31. Leslie, R. J., and J. D. Pickett-Heaps. 1984. Spindle microtubule dynamics following ultraviolet-microbeam irradiations of mitotic diatoms. *Cell.* 36:717–727.
32. Lutz, D. A., and S. Inoué. 1986. Techniques for observing living gametes and embryos. *Methods Cell Biol.* 27:89–110.
33. Margolis, R. L., L. Wilson, and B. I. Kiefer. 1978. Mitotic mechanism based on intrinsic microtubule behavior. *Nature (Lond.).* 272:450–452.
34. Mitchison, T., and M. Kirschner. 1984. Dynamic instability of microtubule growth. *Nature (Lond.).* 232:237–242.
35. Nath, J. P., G. F. Eagle, and R. H. Himes. 1985. Direct photoaffinity labeling of tubulin and guanosine 5′-triphosphate. *Biochemistry.* 24:1555–1560.
36. O'Brien, E. T., W. A. Voter, and H. P. Erickson. 1987. GTP hydrolysis during microtubule assembly. *Biochemistry.* 26:4148–4156.
37. Salmon, E. D., and P. Wadsworth. 1986. Fluorescence studies of tubulin and microtubule dynamics in living cells. *In* Applications of Fluorescence in the Biomedical Sciences. D. L. Taylor, A. S. Waggoner, R. F. Murphy, F. Lanni, R. R. Birge, editors. Alan R. Liss, Inc., New York. 377–403.
38. Sammak, P. J., and G. G. Borisy. 1988. Direct observation of microtubule dynamics in living cells. *Nature (Lond.).* 332:724–726.
39. Sammak, P. J., G. J. Gorbsky, and G. G. Borisy. 1987. Microtubule dynamics in vivo: a test of mechanisms of turnover. *J. Cell Biol.* 104:395–405.
40. Schilstra, M. J., S. R. Martin, and P. M. Bayley. 1987. On the relationship between nucleotide hydrolysis and microtubule assembly: studies with a GTP-regenerating system. *Biochem. Biophys. Res. Commun.* 147:588–595.
41. Schulze, E., and M. Kirschner. 1986. Microtubule dynamics in interphase cells. *J. Cell Biol.* 102:1020–1031.
42. Schulze, E., and M. Kirschner. 1988. New features of microtubule behavior observed in vivo. *Nature (Lond.).* 334:356–359.
43. Tao, W., R. J. Walter, and M. W. Berns. 1987. Laser-transected microtubules exhibit individuality of regrowth, however most free new ends of the microtubules are stable. *J. Cell Biol.* 107:1025–1035.
44. Voter, W. A., and H. P. Erickson. 1984. The kinetics of microtubule assembly. *J. Biol. Chem.* 259:10430–10438.
45. Walker, R. A., E. T. O'Brien, N. K. Pryer, M. Soboeiro, W. A. Voter, H. P. Erickson, and E. D. Salmon. 1988. Dynamic instability of individual microtubules analyzed by video light microscopy: rate constants and transition frequencies. *J. Cell Biol.* 107:1437–1448.
46. Weisenberg, R. C., G. G. Borisy, and E. W. Taylor. 1968. The colchicine-binding protein of mammalian brain and its relation to microtubules. *Biochemistry.* 7:4466–4479.
47. Weisenberg, R. C., W. J. Deery, and P. J. Dickinson. 1976. Tubulin nucleotide interactions during the polymerization and depolymerization of microtubules. *Biochemistry.* 15:4248–4254.
48. Wilson, P. J., and A. Forer. 1988. Ultraviolet microbeam irradiation of chromosomal spindle fibers produces an area of reduced birefringence and shears the microtubules, allowing study of the dynamic behavior of new free ends in vivo. *J. Cell Sci.* 91:455–468.

Reprinted from *Development*, Vol. 105, pp. 237–249, 1989, with permission from The Company of Biologists.

Fertilization and ooplasmic movements in the ascidian egg

CHRISTIAN SARDET[1], JOHANNA SPEKSNIJDER[2,*], SHINYA INOUE[2] and LIONEL JAFFE[2]

[1]*Unité de Biologie Cellulaire Marine CNRS/Paris VI, Station Zoologique, Villfranche-sur-mer, 06230 France*
[2]*Marine Biological Laboratory, Woods Hole, MA 02543, USA*

* Present address: Department of Experimental Zoology, University of Utrecht, Padualaan 8, 3584 CH Utrecht, The Netherlands

Summary

Using light microscopy techniques, we have studied the movements that follow fertilization in the denuded egg of the ascidian *Phallusia mammillata*. In particular, our observations show that, as a result of a series of movements described below, the mitochondria-rich subcortical myoplasm is split in two parts during the second phase of ooplasmic segregation. This offers a potential explanation for the origin of larval muscle cells from both posterior and anterior blastomeres.

The first visible event at fertilization is a bulging at the animal pole of the egg, which is immediately followed by a wave of contraction, travelling towards the vegetal pole with a surface velocity of 1·4 μm s^{-1}. This wave accompanies the first phase of ooplasmic segregation of the mitochondria-rich subcortical myoplasm. After this contraction wave has reached the vegetal pole after about 2 min, a transient cytoplasmic lobe remains there until 6 min after fertilization. Several new features of the morphogenetic movements were then observed: between the extrusion of the first and second polar body (at 5 and 24–29 min, respectively), a series of transient animal protrusions form at regular intervals. Each animal protrusion involves a flow of the centrally located cytoplasm in the animal direction. Shortly before the second polar body is extruded, a second transient vegetal lobe ('the vegetal button') forms, which, like the first, resembles a protostome polar lobe. Immediately after the second polar body is extruded, three events occur almost simultaneously: first, the sperm aster moves from the vegetal hemisphere to the equator. Second, the bulk of the vegetally located myoplasm moves with the sperm aster towards the future posterior pole, but interestingly about 20 % remains behind at the anterior side of the embryo. This second phase of myoplasmic movement shows two distinct subphases: a first, oscillatory subphase with an average velocity of about 6 μm min^{-1}, and a second steady subphase with a velocity of about 26 μm min^{-1}. The myoplasm reaches its final position as the male pronucleus with its surrounding aster moves towards the centre of the egg. Third, the female pronucleus moves towards the centre of the egg to meet with the male pronucleus. Like the myoplasm, the migrations of both the sperm aster and the female pronucleus shows two subphases with distinctly different velocities. Finally, the pronuclear membranes dissolve, a small mitotic spindle is formed with very large asters, and at about 60–65 min after fertilization, the egg cleaves.

Key words: ascidian development, fertilization, ooplasmic segregation, contraction waves, myoplasm, video microscopy.

Introduction

The unfertilized ascidian egg is usually arrested in the first meiotic division and the small meiotic spindle lies close to the surface at a point that defines the animal pole. Moreover, a mitochondria-rich layer is excluded from this meiotic region (Reverberi, 1956; Mancuso, 1963). Despite this visible polarity, any half of the unfertilized egg can develop into a normal larva (Reverberi & Ortolani, 1962; Bates & Jeffery, 1987). Thus, in the unfertilized ascidian egg, relatively little developmental pattern exists. However, after fertilization such a pattern is formed with remarkable speed. Well before first cleavage, the future embryonic axes of the ascidian embryo become apparent during a series of cytoplasmic movements known as ooplasmic segregation (Conklin, 1905; Jeffery, 1984). Three to five visible plasms become segregated into different regions of the egg cell, and are subsequently partitioned unequally during the early cleavages. As a result, these plasms become localized in different cells of the 64-cell-stage embryo. The most obvious of these plasms is the mitochondria-rich (and in some species pigment-rich) myoplasm localized in the cortex of the unfertilized egg.

Conklin's classical account describes two distinct phases of myoplasmic segregation: during the first phase, the myoplasm concentrates in the vegetal pole region and, during the second phase, it shifts from the vegetal pole to the equatorial region of the zygote, where it marks the future posterior pole of the embryo (Conklin, 1905). The myoplasm is subsequently partitioned in the blastomeres that give rise to muscle cells. Since the components that specify muscle cell fate are still unknown, we conceive of the myoplasm as an entity that can be roughly identified as the mitochondria-rich mass. It is clear, however, that the mitochondria themselves are not the muscle cell determinants (Conklin, 1931).

In recent years, the mechanisms of ooplasmic segregation have been studied in several laboratories, and it has become clear that the first phase is driven by the contraction of a cortical actin network (Sawada & Osanai, 1981, 1984, 1985; Jeffery & Meier, 1983, 1984; Sawada, 1983; Jeffery, 1984). This cortical contraction is probably triggered by the wave of elevated free calcium that travels through the egg following fertilization (Speksnijder *et al.* 1986, 1988). The second phase of segregation seems to depend upon microtubule integrity (Sawada & Schatten, 1985, 1988).

We have investigated the movements that follow fertilization in the transparent denuded egg of *Phallusia mammillata* using epifluorescence microscopy and videomicroscopy. Besides establishing exact speeds and times for various movements, we have discovered new features that may have significant embryological implications. These include a series of animal protrusions in the period between the formation of the two polar bodies, the formation of a second cytoplasmic lobe reminiscent of a protostome polar lobe, the suggestion that there exists a physical connection between the myoplasm and the sperm aster, a splitting of the myoplasm into two parts which end up at opposite poles of the embryo and a division of the second phase of ooplasmic segregation into two subphases with distinctively different velocities. This work has been reported in part in abstract form (Sardet *et al.* 1986). A separate report on the calcium waves that we observed following fertilization and during meiosis in *Phallusia* eggs is in preparation (see also Speksnijder *et al.* 1986, 1988).

Materials and methods

Experimental animal and gametes

The European ascidian, *Phallusia mammillata*, was collected in France on the Meditteranean coast in Sète (Bassin de Thau), or on the Atlantic coast near Roscoff (Brittany). They were kept in aquaria in Villefranche-sur-mer or Woods Hole at 15–22 °C. Ripe gametes can be obtained from the gonoducts of these animals throughout the year. Concentrated sperm can be kept several days at 5 °C and diluted as needed. The eggs were washed in millipore-filtered seawater and dechorionated by incubation with trypsin (0·1 %) in seawater buffered to pH 8·0 with 10 mM-TAPS (Tris [hydroxymethyl] methylaminopropane sulphonic acid) (TAPS SW) for 2 h at 18 °C with gentle agitation (Zalokar, 1979). The eggs were extensively washed thereafter in TAPS SW and used within 6 h, although they can still be successfully fertilized 24 to 48 h later. To prevent adhesion and subsequent lysis of the dechorionated eggs, all surfaces were treated with gelatin as follows; all glassware and plastic dishes were immersed in a solution of 0·1 % gelatin and 0·1 % formaldehyde, drained, air dried, and subsequently rinsed extensively in running tap water. Provided they are prevented from adhering to each other, most trypsin-dechorionated eggs undergo normal development into swimming tadpoles (Zalokar, 1979; Zalokar & Sardet, 1984).

Light microscopic observation of fertilization

For the observation of fertilization, we perfused dechorionated eggs, which were slightly compressed between a slide and a coverslip, with a dilute suspension of sperm that had been preincubated for 30 min with 'chorionated' eggs, i.e. eggs from which the chorion had not been removed. This greatly improves the synchrony of fertilization of naked eggs. We also observed eggs using the microdrop method described by Lutz & Inoué (1986). Mitochondria were visualized by incubating the eggs for 30 min in the fluorescent carbocyanine dye $DiO(C_2)_3$ (Molecular Probes) at a concentration of 5 $\mu g\,ml^{-1}$, after which they were washed in TAPS SW, and observed under epi-illumination (excitation filter: 450–490 nm, emission filter: 520–560 nm, barrier filter: 510 nm). These vitally stained eggs develop normally after fertilization (Zalokar & Sardet, 1984).

Light microscopic observations were made on a Leitz Orthoplan, a Zeiss IM 35 and a Zeiss Axiophot (differential interference contrast and epifluorescence), or on a specially designed microscope (rectified differential interference contrast, polarization optics; see Inoué, 1986 figs III-21,22). DAGE/MTI Newicon and SIT cameras were used to record images on a time-lapse recorder (GYYR or Sony Umatic TV0 9000) or on an Optical Memory Disc Recorder (OMDR Panasonic 2021F). The observations reported in this paper were made over the last three years on many different eggs. Unless otherwise indicated, all experiments were done at 18–21 °C.

Results

Fertilization of Phallusia mammillata *eggs*

The eggs of ascidians are surrounded by an inner layer of test cells and an acellular chorion with adhering follicle cells. These extracellular structures greatly impair visualization of the events that follow fertilization in the living egg. Therefore, their removal is required for the detailed observation of these fertilization events. Usually, this involves dechorionation with microneedles, which is necessarily limited to a few eggs. An additional problem is that fertilization of naked ascidian eggs is not always reliable. We found, however, that large numbers of chemically dechorionated *Phallusia* eggs could be reliably and synchronously fertilized by sperm that had been previously exposed to chorionated eggs (data not shown). Using this method, we have been able to observe that sperm does not penetrate preferentially in the vegetal pole area as previously thought (Conklin, 1905). By fixing eggs soon after insemination and observing the relative position of female and male pronuclei, we have found that in fact sperm show a tendency to fuse with the egg in the animal hemisphere (Speksnijder *et al.* 1987; J. E. Speksnijder, C. Sardet & L. F. Jaffe, submitted). Unfortunately, there are no obvious surface modifications, such as fertilization cones, associated with sperm entry that allow us to determine readily the point of sperm entry in living eggs, although we were able to get a glimpse of it on one exceptional occasion (arrowhead in Fig. 1A).

Polarity of the Phallusia *egg*

The egg of *Phallusia* shows a distinct animal–vegetal polarity in its organization, which can be readily visualized. The chromosomes, which are in metaphase of the first meiotic division, are located right underneath the egg surface at a point that defines the animal pole. The meiotic spindle is clearly visible in polarized optics (arrow in Fig. 1A). A clear zone surrounding the meiotic spindle can be detected with differential interference contrast optics (not shown). Finally, the subcortical mitochondria-rich mass is located underneath the entire egg surface except for the most animal part, and is thus organized as a basket with its opening at the animal pole. This mass can be visualized by staining with the vital dye $DiO(C_2)_3$ (Fig. 2, see also Zalokar & Sardet, 1984), or by its autofluorescence under u.v. epi-illumination (Deno, 1987).

The contraction wave and first phase of ooplasmic segregation

The first visible sign of fertilization in *Phallusia*, as in other ascidians (Conklin, 1905; Ortolani, 1955; Reverberi, 1971; Sawada & Osanai, 1981; Villa & Patricolo, 1987), is the formation of a bulge at the animal pole (Fig. 1A,B). Simultaneously, a constriction appears near the animal pole, which travels towards the vegetal pole in 2·0 min (S.E.M. = 0·04; $n = 4$). The velocity of this contraction wave ranges from 1·2–1·6 $\mu m\, s^{-1}$, with a mean of 1·4 $\mu m\, s^{-1}$ (S.E.M. = 0·1; $n = 5$). After the contraction wave has reached the vegetal pole area, a transient cytoplasmic lobe remains present from 2 to 6 min after fertilization (Fig. 1F–H).

The contraction wave accompanies the segregation of the subcortical mitochondria-rich myoplasm towards the vegetal pole of the egg. At the end of the contraction wave, this mass is entirely located in the lower third of the vegetal hemisphere (Fig. 2). Upon focusing through the mass in eggs stained with $DiO(C_2)_3$, one can observe that it shows characteristic folds in its most vegetal part, and is located further away from the plasma membrane in this vegetal area than in its most equatorial parts.

It is possible to follow the movement of the mitochondria-rich mass simultaneously with the movement of small Nile blue particles attached to the egg surface (Fig. 2A–D). These two elements, located on the internal and the external sides of the egg plasma membrane, move simultaneously and in a coordinated fashion with an average velocity of 1·4 $\mu m\, s^{-1}$ (S.E.M. = 0·1; $n = 4$) which is a value comparable to the one we calculated for the speed of the contraction wave.

In accordance with previous observations (Zalokar, 1974; Reverberi, 1975; Sawada & Osanai, 1981), we observed that the contraction wave and the segregation of the myoplasm are both blocked by cytochalasins (preincubation for one hour and insemination in 10 $\mu g\, ml^{-1}$ cytochalasin B, or 2 $\mu g\, ml^{-1}$ cytochalasin D). The movement of particles attached to the egg surface is also inhibited by these drugs (data not shown).

Meiosis

At 18 °C, the first polar body forms at about 5 min after fertilization (mean = 5·1; S.E.M. = 0·1; $n = 7$), and the second polar body at 24–29 min (mean = 26; S.E.M. = 1; $n = 7$). During either the entire period, or part of it, we observed from 6 to 15 pulsating movements consisting of alternating protrusions and retractions at the animal pole every 60–90 s (see Fig. 1I–N). These pulsating movements accompany a flow of the central cytoplasm towards the animal pole and back, which is clearly visible in time lapse. Furthermore, shortly after first polar body extrusion, we observed the formation of a small, transient, polar body-like protrusion at the animal pole (7–9 min after fertilization, see arrowhead in Fig. 1J). By this time, a sperm aster starts forming in the vegetal hemisphere (arrowhead in Fig. 1K). Its position relative to the vegetal pole is variable and probably depends on the point of sperm entry (Speksnijder *et al.* 1987); the more animally the sperm enters, the closer to the equator it is likely to end up after the first segregation phase. The aster progressively grows (Fig. 1K–N), and the male pronucleus forms by coalescence of karyomeres at the time of second polar body formation (arrowhead in Fig. 1O). Meanwhile, the finely granular cytoplasm has progressively changed to a coarser texture (Fig. 1I–M). The periodic oscillations at the animal egg surface and the internal cytoplasm cease just before the extrusion of the second polar body (Fig. 1N). In differential interference contrast optics, the mitochondria-rich mass extends 5–10 μm below the surface in the vegetal hemisphere (Fig. 1K–N). By focusing through the mass, one can see that it is folded and seems detached from the surface near the vegetal pole, which confirms our observations in $DiO(C_2)_3$-labelled eggs. In its more equatorial position, the mass tapers off and is more closely associated with the surface. The mitochondria-rich mass can also be readily visualized in polarized optics (arrowhead in Fig. 1W).

Just before second polar body formation, the eggs start to flatten at the vegetal pole (Figs 1O and 3A), and subsequently form a second transient cytoplasmic lobe in this area (arrowhead in Fig. 3B–D). This small polar lobe-like structure lasts 7–8 min, and varies in size depending on the time of the year and the batch of eggs. It does not seem to contain granules or mitochondria (Figs 3B–D and 4B–D). We have called this structure 'the vegetal button'. This structure has not been described before, probably because it is difficult to observe in chorionated eggs. This vegetal lobe is very different from the one described in lectin-activated *Phallusia mammillata* eggs which in fact corresponds to the transient vegetal lobe normally formed 2–6 min after fertilization but is delayed by about 15 min during lectin activation (Zalokar, 1979; Sardet & Speksnijder, unpublished observations).

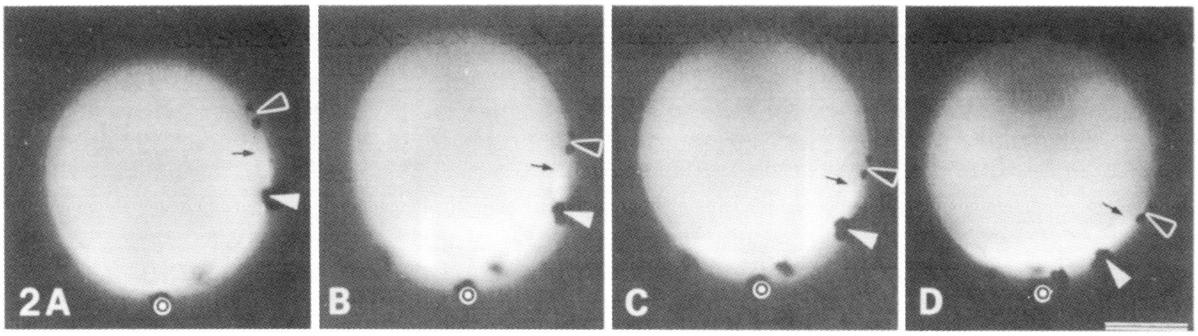

Fig. 2. The contraction wave and the first segregation phase of the mitochondria-rich myoplasm. The mitochondria-rich myoplasm is revealed in epifluorescence after staining with the vital dye $DiO(C_2)_3$). Nile blue particles (arrowheads) are attached to the surface and move toward the vegetal pole (dot marker) with the edge of the subcortical myoplasm (thin arrow). A–D were taken at approximately 20 s intervals, with A being close to the onset of the contraction wave. Bar, 50 μm.

The second phase of ooplasmic segregation

After the second polar body has been extruded, a series of complex events occur, which involve the movement of at least three different entities, i.e. the mitochondria-rich myoplasm, the sperm aster and the female pronucleus. We have observed these various events in many eggs, on the basis of which we describe the movements in a more general way. In addition, we were able to analyse the movements in more detail in two cases, in which the orientation of the egg, which obviously is very critical at this stage, was just right for the observation of the entire process.

Within a minute after the second polar body has been extruded, the vegetally localized myoplasm starts moving towards the equator of the egg. This movement occurs in two distinct subphases: the first one is an oscillatory movement over a period of about 15 min (see Fig. 1N–S), with an average speed of the anterior edge of the myoplasm of 6 μm min^{-1} (6·1 ± 0·3 and 5·7 ± 0·8 μm min^{-1}, respectively, in the two cases analysed; see Table 1). The second subphase of this movement is a smoother and faster motion of the mass towards its final position at the equator (Fig. 1S–U), which occurs over a period of about 5 min with an average speed of its anterior edge of 26 μm min^{-1} (25·9 ± 4·5 and 27·1 ± 9·0 μm min^{-1}, respectively; see Table 1). The direction and the velocity of the final movement of the mass are clearly related to the movement of the male pronucleus and surrounding aster towards the female pronucleus and the centre of the egg (observe the deformation of the egg surface in Fig. 1T–U). Photographic and video records strongly suggest that there is a cable-like structural connection between the myoplasm and the male pronuclear region (see arrowhead in Fig. 1R, and shape of the fluorescent mass in Fig. 4C–F).

Interestingly, we found that not all the mitochondria-rich myoplasm moves towards the equator and future posterior pole of the egg. About 20 % of the mass, which is located at the most anterior side of the egg, gets separated from the remainder of the mass and remains located in the anterior vegetal region as the bulk of the myoplasm moves posteriorly (see arrowhead in Fig. 4C,F). This finding may have major implications for the lineage of the muscle cell line, as will be discussed later.

In contrast to the first segregation phase, which is accompanied by the movement of surface components (Fig. 2), there are no detectable surface movements associated with this second segregation phase of the mitochondria-rich mass. As the mass moves towards the future posterior pole of the embryo, particles attached on the surface remain still (Figs 5 and 6).

The movements of the pronuclei

Several minutes after the formation of the second polar body, the female pronucleus becomes visible at the animal pole of the egg, starts to migrate towards the centre of the egg and then curves to meet with the male pronucleus (Fig. 1P–V; and arrows in Fig. 3C–E). Similar to the movement of the myoplasm, this movement also seems to occur in two subphases, a slow and a fast phase (Table 1). During the first 7 to 8 min, it moves with an average speed of 4 μm min^{-1} (3·8 ± 0·4 and 4·8 ± 1·1 μm min^{-1}, respectively, in the two cases analysed), whereas during the final movement towards the male pronucleus, which takes several minutes, it travels at an average velocity of about 12 μm min^{-1} (9·3 ± 2·2 and 15·1 ± 1·1 μm min^{-1}).

The very large and asymmetric sperm aster, which is still closely associated with the egg cortex in the vegetal hemisphere at the time of the polar body formation, starts moving up together with the mitochondria-rich mass towards the equator of the egg (Fig. 1O–Q). The exact time at which the sperm aster starts moving up seems somewhat variable (being either simultaneous with, or a few minutes later than, the movement of the myoplasm) and may well depend on its position relative to the vegetal pole after the first segregation phase (and thus ultimately on the point of sperm entry; see previous section on Meiosis). The movement of the sperm pronucleus and surrounding aster again shows two distinct subphases (Table 1); a first subphase directed towards the equator with an average velocity of 3 μm min^{-1} (4·0 ± 0·2 and 2·5 ± 0·1 μm min^{-1}, respectively) and a second subphase towards the centre of the egg (Fig. 1R–U) with an average velocity of 11 μm min^{-1} (10·9 ± 1·0 and 10·4 ± 2·5 μm min^{-1}, respectively). The second subphase of both the female and the male pronucleus starts at about the same time, as illustrated in Fig. 7, in which the distance between the two pronuclei is plotted as a function of time. During the first subphase, the pronuclear distance decreases at a rate of about 5 μm min^{-1} (Table 1) and, as both pronuclei speed up and curve towards the centre of the egg, their distance decreases at about 13 μm min^{-1} (Fig. 7 and Table 1). At the final stage of these pronuclear movements, both pronuclei meet in the centre of the egg, after which their membranes break down at approximately 40 min after fertilization (at 23–25 °C) (Fig. 1U–V).

Mitosis and cleavage

The mitotic apparatus is characterized by very large asters that radiate to the cortex and a small mitotic apparatus (Fig. 1W–X). Cytokinesis starts at the animal pole at about 62 min at 18–21 °C (*versus* about 50 min at 23–25 °C). The newly formed blastomeres possess large and prominent asters (Fig. 1X). The mitochondria-rich myoplasm is divided equally between the two daughter cells (Fig. 5D).

Discussion

In the present study, we have investigated the various cytoplasmic and surface movements that accompany fertilization in ascidians. Using microscopy techniques on the extremely clear and naked eggs of *Phallusia mammillata*, we were able to observe several hitherto undescribed aspects of fertilization and ooplasmic segregation in the ascidian egg, which will be discussed in the following sections. A schematic representation of the various events is shown in Fig. 8.

Fertilization of Phallusia mammillata *eggs*

As in other ascidian eggs (Conklin, 1905; Mancuso, 1963; Sawada & Osanai, 1981, 1985), the unfertilized egg of *Phallusia mammillata* is arrested in metaphase of the first meiotic division and displays a distinct animal–vegetal polarity. After staining with the vital dye DiO$(C_2)_3$, a subcortical layer rich in mitochondria can be observed under the entire egg surface, except for the animal pole area where only granules and endoplasmic reticulum are found in the cortex (Reverberi, 1956; Sardet & Gualtieri, in preparation).

The eggs of *Phallusia mammillata* can be obtained throughout the year and can be easily dechorinated in large batches without affecting early development. However, in order to ensure rapid, reliable and synchronous fertilization, we found that it was necessary to preincubate the sperm with eggs that had kept their chorions. We presume that some activation of sperm occurs upon contact with the chorion, which facilitates fusion of the sperm with the egg (Honegger, 1986). Sperm do not seem to attach lastingly to the naked egg surface, and we could not detect any modifications of the egg surface at the site of sperm fusion. By using such preactivated sperm, we have previously shown that, in contrast to what was generally believed, the sperm does not enter at the vegetal pole of the ascidian egg, but rather tends to fuse with the egg in the animal hemisphere (Speksnijder *et al.* 1987). Subsequently, it is dragged towards the vegetal pole during the cortical contraction wave that accompanies the first phase of ooplasmic segregation (see Fig. 8).

The first phase of ooplasmic segregation

In *Phallusia mammillata*, as in other ascidians, whether chorionated or dechorionated, the first manifestation of fertilization is a flow of cytoplasm towards the animal pole and a constriction in the animal hemisphere, resulting in a bulging of the animal pole region (Sawada, 1983; Jeffery, 1984; Sawada & Schatten, 1988). It has been convincingly demonstrated that a network of actin microfilaments is present in the cortex of ascidian eggs (Jeffery & Meier, 1983; Sawada & Osanai, 1985). In *Ciona*, this actin network is organized as a basket with its opening towards the animal pole. The width of the animal actin-free area corresponds with the diameter of the constriction at the earliest stage of the contraction wave (Sawada & Osanai, 1985; Sawada, 1986). As the contraction wave travels towards the vegetal pole, the actin network regresses with the constriction into the vegetal hemisphere. This finding, and the observation that cytochalasins block the contraction wave (Zalokar, 1974; Reverberi, 1975; Sawada & Osanai, 1981; this paper), suggest that it is indeed a contraction of the cortical actin network that provides the force and direction for the contraction wave (Sawada, 1983; Jeffery, 1984). It is possible that, as in *Xenopus* eggs (Christensen *et al.* 1984), an actomyosin contractile mechanism is involved in this cortical contraction, but so far myosin has not been demonstrated in the cortex of ascidian eggs. The wave of elevated free calcium that we have observed at fertilization in both *Ciona* and *Phallusia* eggs (Speksnijder *et al.* 1986, 1988) is a likely candidate for the trigger of the contraction wave.

In many aspects, the wave of contraction in ascidian eggs resembles the mechanisms involved in the locomotion of cells like neutrophils and amoebae. Recently, Bray & White (1988) have proposed an interesting model for 'cortical flow' in animal cells based on a cortically located motor which drives processes such as cell locomotion, growth cone migration and cytokinesis. The contraction wave in ascidian eggs seems to be another example of such a process.

The contraction wave travels over most of the eggs at an average velocity of 1·4 μm s^{-1}. From our results, it is clear that particles attached to the egg surface as well as the subcortical mitochondria-rich mass travel in a coordinated fashion at the same speed during the first phase of ooplasmic segregation. We presume that the sperm that enter in the animal hemisphere also travel at the same speed towards the vegetal pole (Speksnijder *et*

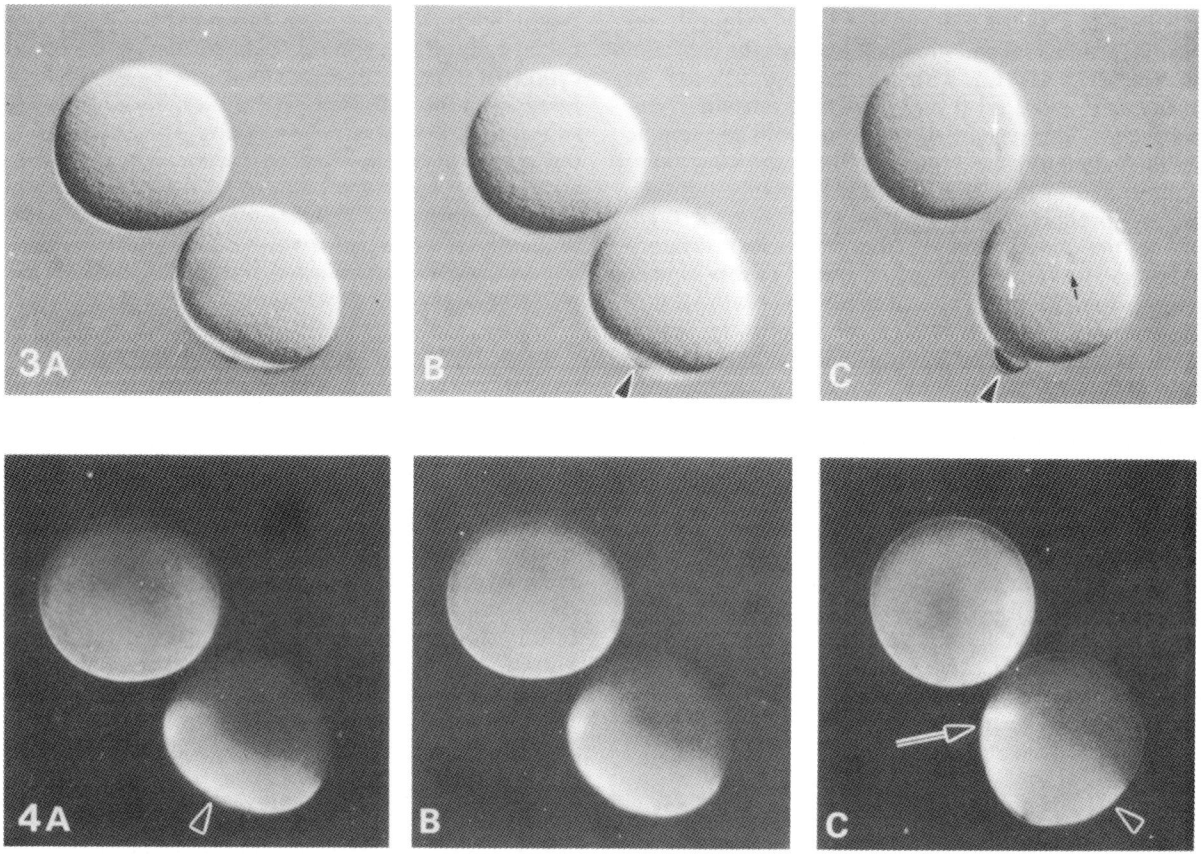

Figs 3 and 4. The second phase of ooplasmic segregation. Two eggs are followed in differential interference contrast (Fig. 3) and epifluorescence after staining of the mitochondria with $DiO(C_2)_3$ (Fig. 4). Times after fertilization of Figs 3 and 4, respectively, are indicated in the legend. (A) The mitochondria-rich myoplasm (arrowhead in Fig. 4A) is situated beneath the surface in the vegetal pole region (16 and 17 min). (B) The second polar body has just been extruded and a cytoplasmic lobe ('the vegetal button') appears at the vegetal pole (arrowhead in Fig. 3B). The bulk of the myoplasm moves with the male pronuclear aster towards the equator (20 and 21 min). (C,D) Male (thin white arrows) and female (thin black arrows) pronuclei are moving closer to each other, the 'vegetal button' reaches its maximum size (arrowhead in Fig. 3C), and then regresses quickly (Fig. 3D). Note that mitochondria are excluded from the button (Fig. 4C). The bulk of the myoplasm has moved with the sperm aster towards the equator (arrow in Fig. 4C). A small fraction remains in the future anterior side of the embryo (arrowhead in Fig. 4C). Note the fluorescent area in the most animal part of the myoplasm which seems to correspond with the structural connection observed in Fig. 1R (25 and 26, and 27 and 28 min). (E,F) The pronuclei have met in the centre of the egg, and their pronuclear membranes have broken down. The myoplasm has split in two parts; the bulk of it is now situated at the future posterior pole and a small part is left behind at the opposite pole (arrowhead in Fig. 4F) (30 and 31 min, 38 and 39 min). Bar, 50 μm.

al. 1987). It must be noted, however, that toward the end of the contraction wave, the subcortical myoplasm lags behind the surface particles (see Fig. 2), confirming the notion that the subcortical components are progressively dragged down during the contraction of the egg cortex (Sawada, 1983).

It is interesting to note that, in the eggs of the Oligochaete *Tubifex*, a similar mechanism for ooplasmic segregation has been described (Shimizu, 1986). As in ascidians, the cortical actin network is believed to generate the motive force as well as determine the polarity of ooplasmic segregation and, in addition, a cosegregation of mitochondria and surface components with the cortical actin network has been demonstrated.

Meiosis

After the contraction wave has reached the vegetal pole of the ascidian egg, a cytoplasmic lobe remains present until shortly after the formation of the first polar body at about 5 min after fertilization. Then, at the time of second polar body formation, a second cytoplasmic lobe is formed at the vegetal pole. Both in their appearance and timing, these cytoplasmic lobes resemble the meiotic polar lobes formed in protostomes such as the gastropod *Ilyanassa obsoleta* (Morgan, 1933; Cather, 1963). The function of these meiotic polar lobes is unknown, but, like other modifications of the vegetal pole during meiosis in molluscan eggs, they may well be part of the general phenomenon of cytoplasmic localiz-

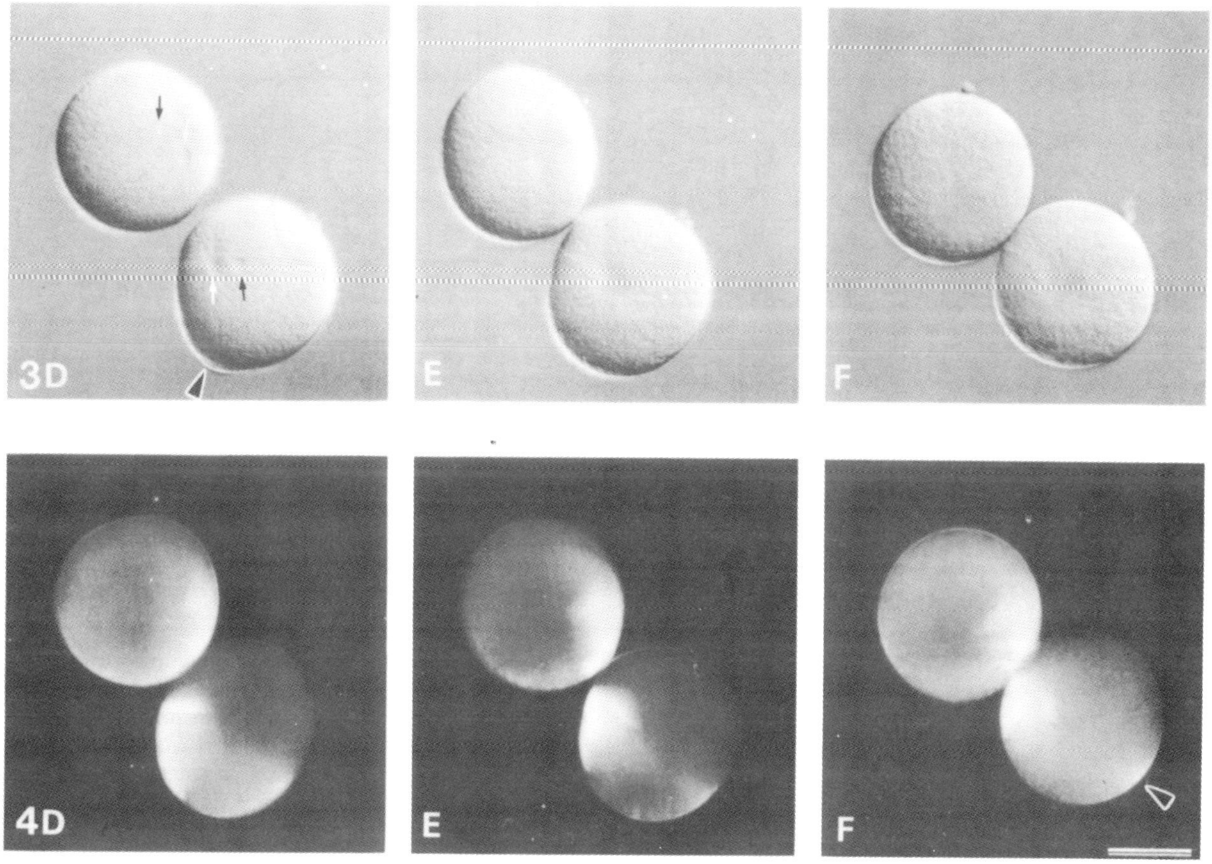

Table 1. *Mean velocity (in* $\mu m\ min^{-1}$*) of the motions of the myoplasm, female pronucleus and sperm aster; and the decrease in distance between the female pronucleus and the centre of the sperm aster (in* $\mu m\ min^{-1}$*) during the second phase of ooplasmic segregation in* Phallusia *eggs*

		Sequence no. 1*	Sequence no. 2*
Myoplasm	slow subphase	6·1 ± 0·3 (11)	5·7 ± 0·8 (5)
	fast subphase	25·9 ± 4·5 (2)	27·1 ± 9·0 (3)
Female pronucleus	slow subphase	3·8 ± 0·4 (7)	4·8 ± 1·1 (2)
	fast subphase	9·3 ± 2·2 (4)	15·1 ± 1·1 (4)
Sperm aster	slow subphase	4·0 ± 0·2 (3)	2·5 ± 0·1 (2)
	fast subphase	10·9 ± 1·0 (4)	10·4 ± 2·5 (3)
Decrease distance between female pronucleus and centre of the sperm aster	slow subphase	3·6 ± 0·6 (7)	5·8 ± 0·3 (2)
	fast subphase	13·2 ± 1·7 (4)	13·2 ± 0·4 (3)

*These data were obtained from two separate sequences – recorded with an Optical Memory Disk Recorder (OMDR) – in which the orientation of the egg was optimal for the analysis of these motions. In each egg, the position of the anterior edge of the myoplasm, the female pronucleus and the centre of the sperm aster was marked at 1- to 3-min intervals, and the average velocity of each of the motions as well as the decrease in distance between the female pronucleus and the centre of the sperm aster over that time interval was calculated. From these values, the mean velocity was determined. Indicated are the mean ± S.E.M. (*n*); *n* is the number of time intervals over which the average velocities were calculated.

ation during meiotic maturation (Longo, 1983). As in many polar lobe-forming species (Dohmen, 1983; Speksnijder & Dohmen, 1983; Speksnijder *et al.* 1985*a*,*b*), ascidian eggs display distinct surface differentiations at their vegetal pole after fertilization, such as a higher density of microvilli and lectin-binding sites (Sawada & Osanai, 1981; Zalokar, 1980; O'Dell *et al.* 1973). In addition, recent experiments by Bates & Jeffery (1987) have demonstrated the transient localization of specific (axial) determinants at the vegetal pole of the ascidian *Styela* during meiosis. The similarity in both appearance and timing of these two types of lobes in two groups of animals (ascidians *versus* molluscs and annelids) that are evolutionarily so far apart, is intriguing.

Another interesting aspect of meiosis is the occurrence of oscillatory cytoplasmic and surface movements during the period between the formation of the first and

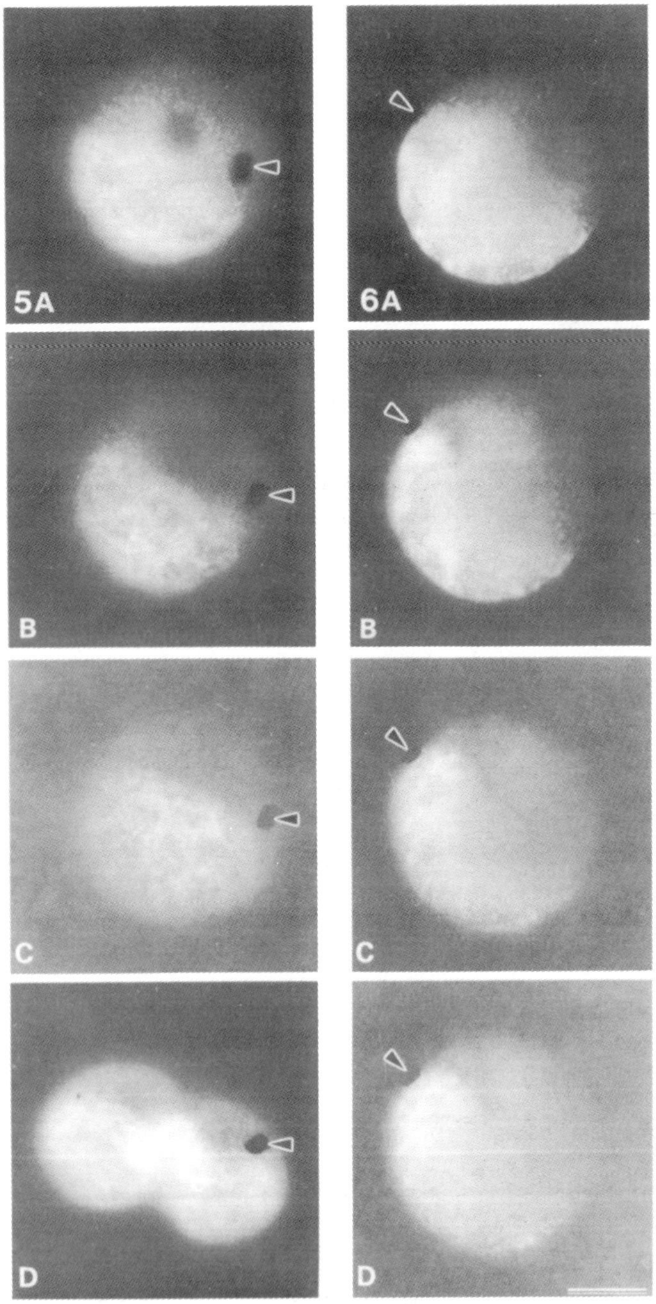

Figs 5 and 6. Coordinated motion of the myoplasm and surface particles during the second phase of ooplasmic segregation. Two complementary views of this process are shown under epifluorescent illumination (Fig. 5: face-on view; Fig. 6: side view). Mitochondria are visualized after accumulation of the vital dye DiO(C_2)$_3$; arrowheads indicate Nile blue particles attached to the egg surface. Times after fertilization for Fig. 5 are 32, 42, 47 and 55 min; and for Fig. 6: 34, 43, 48 and 53 min. Bar, 50 μm.

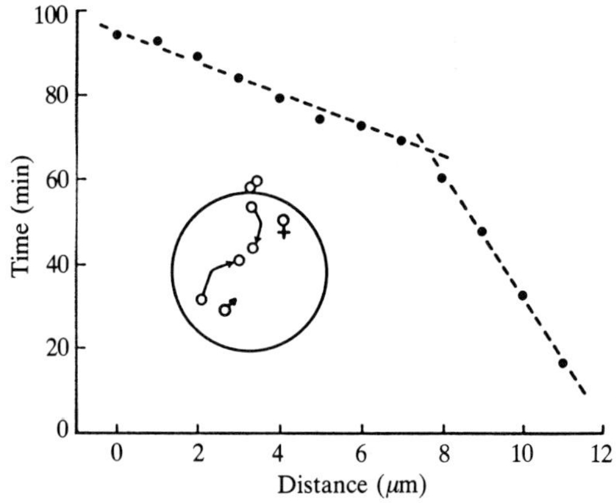

Fig. 7. Distance between the female pronucleus and the centre of the sperm aster as a function of time. The inset shows the path of each of the pronuclei. At $t = 0$, the female pronucleus is first visible. At about 7–8 min, both pronuclei speed up and curve towards the centre of the egg to meet at about 12 min. These data were obtained from the video sequence shown in Fig. 1 (sequence no. 1 in Table 1).

second polar body. Preliminary experiments suggest that these periodic oscillations are coupled to oscillations in the concentration of intracellular free calcium (Speksnijder *et al.* 1986, 1988). At present, we do not know what their function is, but it is noteworthy that periodic oscillations in free calcium concentration have also been observed in mammalian eggs after fertilization and in hepatocytes after hormonal stimulation (Cuthbertson & Cobbold, 1985; Miyazaki *et al.* 1986; Woods *et al.* 1986). However, no associated cytoplasmic or surface motions have yet been reported in mammalian cells.

The second phase of ooplasmic segregation

The mitochondria-rich mass constitutes a distinct cytoplasmic domain which extends up to the growing sperm aster. Our observations suggest they may be physically connected (see Fig. 1R–U and Fig. 3C,4C). After second polar body formation, the bulk of the myoplasm slowly migrates with the sperm aster towards the equator in a succession of oscillatory motions. A faster steady motion then carries the bulk of the myoplasm towards its final position at the future posterior pole of the egg. This motion seems tied to the movement of the male and female pronuclei to the centre of the egg. It is not accompanied by surface motions. It has recently been reported that this second phase of ooplasmic segregation is sensitive to the microtubule inhibitor colcemid (Sawada & Schatten, 1985, 1988). Therefore, it seems likely that the microtubules of the sperm aster are involved in localizing the myoplasm in its posterior position. However, the exact mechanism of this microtubule-controlled movement is not yet clear. Sawada & Schatten (1988) have shown that microtubules of the sperm aster form a three-dimensional array extending through and beyond the myopla m, and that the majority of these microtubules run parallel to the egg surface. If such microtubules were localized in the cortex in a parallel array, they could serve as tracks for shearing forces to drive the subcortical myoplasm, similar to the mechanism suggested for cortical rotation in fertilized *Xenopus* eggs (Elinson & Rowning, 1988). Alternatively, the myoplasmic domain might be pulled

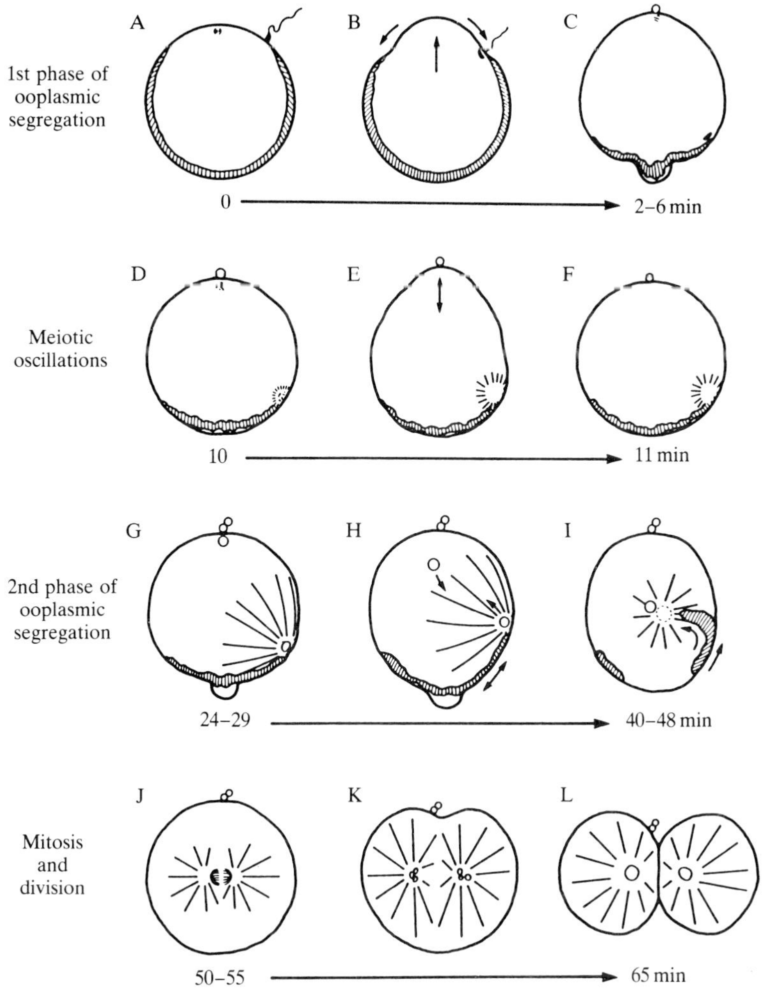

Fig. 8. Fertilization and morphogenetic movements in the ascidian egg. (A–C) The first phase of ooplasmic segregation. (A) The meiotic spindle defines the animal pole. A subcortical layer rich in mitochondria ('the myoplasm') is excluded from the animal pole region. Sperm have a tendency to enter in the animal hemisphere. (B) A wave of cortical contraction propagates from the animal to the vegetal pole regions at an average velocity of $1{\cdot}4\,\mu m\,s^{-1}$, dragging the sperm nucleus and the myoplasm down and pushing the central cytoplasm up towards the animal pole. (C) The first phase of ooplasmic segregation is completed 2 min after fertilization. The first polar body appears at about 5 min and a cytoplasmic lobe remains present at the vegetal pole from 2 to 6 min after fertilization. (D–F) Meiotic oscillations. Between first and second polar body formation (at 5 and 24–29 min, respectively), the animal pole region undergoes periodic cycles of protrusion and retraction movements, simultaneous with movements of the centrally located cytoplasm towards the animal pole and back, with a period of 60–90 s (an example is given in the sequence D,E,F). The sperm aster, located in the vegetal hemisphere, enlarges. (G–I). The second phase of ooplasmic segregation. (G) The second polar body is extruded at 24–29 min after fertilization, a second vegetal cytoplasmic lobe is formed and the female pronucleus migrates towards the centre of the egg to meet the male pronucleus. (H) Oscillating motions move the myoplasm towards the equator. (I) A smooth steady motion carries the bulk of the myoplasm towards the future posterior pole of the embryo. A small part of the myoplasm remains in a subcortical anterior position. Pronuclei meet in the centre of the egg, and their membranes break down. (J–L) Mitosis and division. (J) A mitotic spindle with extremely large asters is formed. (K). Cleavage starts at the animal pole. While karyomeres are fusing, the astral rays can be seen to extend all the way to the egg cortex. (L) Division. Nuclear membrane reforms. The myoplasm is divided equally between the daughter cells (not shown).

by microtubules attached to the male pronucleus or by growing astral microtubules.

An intriguing fact is that the myoplasm – with its much larger mass – seems to move about 2–3 times faster than the sperm aster and the female pronucleus (see Table 1). This suggests an independent motile mechanism such as a shearing force along microtubular tracks. In addition, all three entities (myoplasm, sperm aster and female pronucleus) show two distinct phases in their movement; a first relatively slow subphase (average velocity of 6, 4 and $3\,\mu m\,min^{-1}$, respectively) and a second, faster subphase (average velocity is 26, 12 and $11\,\mu m\,min^{-1}$, respectively). The velocities during the slow subphase correspond to rates of growth and shrinkage of microtubules observed *in vivo* (see e.g. Schulze & Kirshner, 1986; Cassimeris *et al.* 1987). In addition, the velocities measured for the fast subphase of the pronuclear movements correspond with values reported for the movement of beads, organelles and female pronuclei along astral rays in sand dollar and sea urchin eggs (Schatten, 1982; Hamaguchi *et al.* 1986; Wadsworth, 1987). Therefore, all evidence points in the direction of microtubules as being the driving force for the various motions during the second phase of ooplasmic segregation (Sawada & Schatten, 1985, 1988; this paper), but clearly our results suggest a complex mechanism for these microtubule-based motile events in the ascidian egg.

The myoplasm and the origin of larval muscle cells

From a developmental point of view, our most interesting observation is that the mitochondria-rich mass becomes separated in two portions during the second phase of ooplasmic segregation. As the bulk of the mass moves towards the future posterior pole of the embryo, a small portion which is located most anteriorly, remains behind, and thus will be inherited by blastomeres of the A4.1 cell lineage. A similar observation has been made on the mitochondria-rich mass in *Ciona* (Deno, 1987). This finding may have important embryological implications. It was thought until recently that muscle cell determinants (situated in the so-called 'myoplasm' or 'yellow crescent' region) were only localized in the descendants of the posterior B4.1 blastomeres (Reverberi, 1971; Whittaker, 1979; Jeffery *et al.* 1984). However, work by Nishida & Satoh (Nishida & Satoh, 1983, 1985; Nishida, 1987), ourselves (Zalokar & Sardet, 1984), and recent careful reexaminations of traditional cell lineages (Meedel *et al.* 1987), have shown that descendants other than those of the B4.1 blastomere are able to differentiate into muscle cells in the embryo or even in isolation. Our observation that some of the myoplasm (an entity that roughly corresponds to the mitochondria-rich mass) also moves into the vegetal anterior part of the embryo would explain why descendants of the A4.1 blastomeres can differentiate into muscle cells. However, additional muscle cells are apparently also formed from the descendants of the b4.2 (animal posterior) blastomere. As a possible explanation, we suggest that, in cases where the sperm aster forms toward the equator rather than close to the vegetal pole, the myoplasm will end up at, and also slightly above, the equator. As a result, some of the myoplasm will be included in the b4.2 cells after the first equatorial division (i.e. third cleavage). This explanation seems plausible since the position of the sperm aster with respect to the equator will depend on the point of sperm entry which we have shown to be preferentially in the animal hemisphere (Speksnijder *et al.* 1987).

It remains to be defined what structures, organelles or molecules constitute the particular components that determine cell fate as they are segregated into specific areas of the egg and embryo during successive cleavages or by experimental manipulations (see Whittaker, 1980, 1982, 1987; Deno *et al.* 1984; Crowther & Whittaker, 1986; Jeffery *et al.* 1986; Nishikata *et al.* 1987). Our observations of the morphogenetic movements in the ascidian egg open the way for new studies of this unresolved problem.

We acknowledge the support of NATO ('international collaboration in research' program to C. Sardet, L. F. Jaffe & J. E. Speksnijder). We also thank Janna Knudson for photography and Anne Tworkowski for typing. Supporting by NIH grant R-37 GM31617 and NSF grant DCB 8518672 to S.I., and NIH grants to L.F.J.

References

Bates, W. R. & Jeffery, W. R. (1987). Localization of axial determinants in the vegetal pole region of ascidian eggs. *Devl Biol.* **124**, 65–67.

Bray, D. & White, J. G. (1988). Cortical flow in animal cells. *Science* **239**, 883–888.

Cassimeris, L. U., Walker, R. A., Pryer, N. K. & Salmon, E. D. (1987). Dynamic instability of microtubules. *BioEssays* **7**, 149–154.

Cather, J. N. (1963). A time schedule of the meiotic and early mitotic stages of *Ilyanassa*. *Caryologia* **16**, 663–670.

Christensen, K., Sauterer, R. & Merriam, R. W. (1984). Role of soluble myosin in cortical contractions of *Xenopus* eggs. *Nature, Lond.* **310**, 150–151.

Conklin, E. G. (1905). The organization and cell lineage of the ascidian egg. *J. Acad. Natl. Sci. Phila.* **13**, 1–126.

Conklin, E. G. (1931). The development of centrifuged eggs of ascidians. *J. exp. Zool.* **60**, 1–119.

Crowther, R. J. & Whittaker, J. R. (1986). Differentiation without cleavage: multiple cytospecific ultrastructure expressions in individual one-celled ascidian embryos. *Devl Biol.* **117**, 114–126.

Cuthbertson, K. S. R. & Cobbold, P. H. (1985). Phorbol ester and sperm activate oocytes by inducing sustained oscillations in cell Ca^{2+}. *Nature, Lond.* **316**, 541–542.

Deno, T. (1987). Autonomous fluorescence of eggs of the ascidian *Ciona intestinalis*. *J. exp. Zool.* **241**, 71–79.

Deno, T., Nishida, M. & Satoh, N. (1984). Autonomous muscle cell differentiation in partial ascidian embryos according to the newly verified cell lineages. *Devl Biol.* **104**, 322–328.

Dohmen, M. R. (1983). The polar lobe in eggs of molluscs and annelids: structure, composition and function. In *Time, Space and Pattern in Embryonic Development* (ed. W. R. Jeffery & R. A. Raff), pp. 197–220. New York: Alan R. Liss.

Elinson, R. P. & Rowning, B. (1988). A transient array of parallel microtubules in frog eggs: potential tracks for a cytoplasmic rotation that specifies the dorso-ventral axis. *Devl Biol.* **128**, 179–185.

Hamaguchi, M. S., Hamaguchi, Y. & Hiramoto, Y. (1986). Microinjected polystyrene beads move along astral rays in sand dollar eggs. *Develop. Growth Diff.* **28**, 461–470.

Honegger, T. G. (1986). Fertilization in ascidians: studies on the egg envelope, sperm and gamete interactions in *Phallusia mammillata*. *Devl Biol.* **118**, 118–128.

Inoué, S. I. (1986). *Video Microscopy*. New York: Plenum Press.

Jeffery, W. R. (1984). Pattern formation by ooplasmic segregation in ascidian eggs. *Biol. Bull. mar. biol. Lab., Woods Hole* **166**, 277–298.

Jeffery, W. R., Bates, W. R., Beach, R. L. & Tomlinson, C. R. (1986). Is maternal mRNA a determinant of tissue-specific proteins in ascidians embryos? *J. Embryol. exp. Morph.* **97 Supplement**, 1–14.

Jeffery, W. R. & Meier, S. (1983). A yellow crescent cytoskeletal domain in ascidian eggs and its role in early development. *Devl Biol.* **96**, 125–143.

Jeffery, W. R. & Meier, S. (1984). Ooplasmic segregation of the myoplasmic actin network in stratified ascidian eggs. *Wilhelm Roux's Arch. devl Biol.* **193**, 257–262.

JEFFERY, W. R., TOMLINSON, C. R., BRODEUR, R. D. & MEIER, S. (1984). The yellow crescent of ascidian eggs: molecular organization localization and role in early development. In *Molecular Aspects of Early Development* (ed. G. M. Malacinski & W. H. Klein), pp. 1-37. New York: Plenum Publ. Corp.

LONGO, F. J. (1983). Meiotic maturation and fertilization. In *The Mollusca*, vol. 3 (ed. N. H. Verdonk, J. A. M. von den Biggelaar & A. S. Tompa), pp. 49–89. New York: Academic Press.

LUTZ, D. & INOUE, S. (1986). Techniques for observing living gametes and embryos. In *Methods in Cell Biology*, vol. 27. *Echinoderm gametes and embryos*. (ed. T. Schroeder), pp. 89–110.

MANCUSO, V. (1963). Distribution of the components of normal unfertilized eggs of *Ciona intestinalis* examined at the electron microscope. *Acta Embryol. Morph. exp.* **6**, 260–274.

MEEDEL, T. H., CROWTHER, R. J. & WHITTAKER, J. R. (1987). Determinative properties of muscle lineages in ascidian embryos. *Development* **100**, 245–260.

MIYAZAKI, S. I., HASHIMOTO, N., YOSHIMOTO, Y., KISHIMOTO, T., IGUSA, Y. & HIRAMOTO, Y. (1986). Temporal and spatial dynamics of the periodic increase in intracellular free calcium at fertilization of golden hamster eggs. *Devl Biol.* **118**, 259–267.

MORGAN, T. H. (1933). The formation of the antipolar lobe in *Ilyanassa*. *J. exp. Zool.* **64**, 433–467.

NISHIDA, H. (1987). Cell lineage analysis in ascidian embryos by intracellular injection of a tracer enzyme. III. Up to the tissue restricted stage. *Devl Biol.* **121**, 526–541.

NISHIDA, H. & SATOH, N. (1983). Cell lineage analysis in ascidian embryos by intracellular injection of a tracer enzyme. I. Up to the eight cell stage. *Devl Biol.* **99**, 382–394.

NISHIDA, H. & SATOH, N. (1985). Cell lineage analysis in ascidian embryos by intracellular injection of a tracer enzyme. II. The 16- and 32-cell stages. *Devl Biol.* **110**, 440–454.

NISHIKATA, T., MITA-MIYAZAWA, I., DENO, T. & SATOH, N. (1987). Monoclonal antibodies against components of the myoplasm of eggs of the ascidian *Ciona intestinalis* partially block the development of muscle-specific acetylcholinesterase. *Development* **100**, 577–586.

O'DELL, D. S., ORTOLANI, G. & MONROY, A. (1973). Increased binding of radioactive Concanavalin A during maturation of *Ascidia* eggs. *Expl Cell Res.* **83**, 408–411.

ORTOLANI, G. (1955). I movementi corticali dell uovo di ascidie alla fecondazione. *Riv. Biol.* **47**, 169–177.

REVERBERI, G. (1956). The mitochondrial pattern in the development of the ascidian egg. *Experientia* **12**, 55–56.

REVERBERI, G. (ed.) (1971). *Experimental Embryology of Marine and Freshwater Invertebrates.* New York: Elsevier-North Holland.

REVERBERI, G. (1975). On some effects of cytochalasin B on the eggs and tadpoles of the ascidians. *Acta Embryol. exp.* **2**, 137–158.

REVERBERI, G. & ORTOLONI, G. (1962). Twin larvae from halves of the same egg in ascidians. *Devl Biol.* **5**, 84–100.

SARDET, C., INOUE, S., JAFFE, L. F. & SPEKSNIJDER, J. E. (1986). Surface and internal movements in fertilizing *Phallusia* eggs. *Biol. Bull. mar. Biol. Lab., Woods Hole* **171**, 488 (abst.).

SAWADA, T. (1983). How ooplasm segregates bipolarly in ascidian eggs. *Bull. Mar. Biol. Station, Asamushi, Tohoku Univ.* **17**, 123–140.

SAWADA, T. (1986). Cortical actin filaments in unfertilized eggs of *Ciona savignyi*. *Zool. Sci.* **3**, 1038 (abst.).

SAWADA, T. & OSANAI, K. (1981). The cortical contraction related to the ooplasmic segregation in *Ciona intestinalis* eggs. *Wilhelm Roux's Arch. devl Biol.* **190**, 208–214.

SAWADA, T. & OSANAI, K. (1984). Cortical contraction and ooplasmic movement in centrifuged or artificially constricted eggs of *Ciona intestinalis*. *Wilhelm Roux's Arch. devl Biol.* **193**, 127–132.

SAWADA, T. & OSANAI, K. (1985). Distribution of actin filaments in fertilized egg of the ascidian *Ciona intestinalis*. *Devl Biol.* **111**, 260–265.

SAWADA, T. & SCHATTEN, G. (1985). Relation between the microtubule system and cytoplasmic movement in ascidian egg. *Zool. Sci.* **2**, 948 (abst.).

SAWADA, T. & SCHATTEN, G. (1988). Microtubules in ascidian eggs during meiosis and fertilization. *Cell Mot. Cytoskel.* **9**, 219–231.

SCHATTEN, G. (1982). Motility during fertilization. *Int. Rev. Cytol.* **79**, 59–163.

SCHULZE, E. & KIRSCHNER, M. (1986). Microtubule dynamics in interphase cells. *J. Cell Biol.* **102**, 1020–1031.

SHIMIZU, T. (1986). Bipolar segregation of mitochondria, actin network, and surface in the *Tubifex* egg: role of cortical polarity. *Devl Biol.* **116**, 241–251.

SPEKSNIJDER, J. E., CORSON, D. W., JAFFE, L. F. & SARDET, C. (1986). Calcium pulses and waves through ascidian eggs. *Biol. Bull. mar. Biol. Lab., Woods Hole* **171**, 488 (abst.).

SPEKSNIJDER, J. E. & DOHMEN, M. R. (1983). Local surface modulation correlated with ooplasmic segregation in eggs of *Sabellaria alveolata* (Annelida, Polychaeta). *Wilhelm Roux's Arch. devl Biol.* **192**, 248–255.

SPEKSNIJDER, J. E., DOHMEN, M. R., TERTOOLEN, L. G. J. & DE LAAT, S. W. (1985*a*). Regional differences in the lateral mobility of plasma membrane lipids in a molluscan embryo. *Devl Biol.* **110**, 207–216.

SPEKSNIJDER, J. E., MULDER, M. M., DOHMEN, M. R., HAGE, W. J. & BLUEMINK, J. G. (1985*b*). Animal-vegetal polarity in the plasma membrane of a molluscan egg: a quantitative freeze-fracture study. *Devl Biol.* **108**, 38–48.

SPEKSNIJDER, J. E., SARDET, C. & JAFFE, L. F. (1987). Entry of sperm into the animal pole of the egg of the ascidian *Phallusia mammillata*. *Biol. Bull. mar. Biol. Lab., Woods Hole.* **173**, 427 (abst.).

SPEKSNIJDER, J. E., SARDET, C. & JAFFE, L. F. (1988). Calcium waves at fertilization in ascidian eggs. *Proc. 4th Int. Congress Cell Biol.*, Montreal, p. 407.

VILLA, L. & PATRICOLO, E. (1987). A scanning electron microscope study of Ascidia malaca egg (tunicate). Changes in the cell surface morphology at fertilization. *Biol. Bull. mar. Biol. Lab., Woods Hole* **173**, 355–366.

WADSWORTH, P. (1987). Microinjected carboxylated beads move predominantly poleward in sea urchin eggs. *Cell Mot. Cytoskel.* **8**, 293–301.

WHITTAKER, J. R. (1979). Cytoplasmic determinants of tissue differentiation in the ascidian egg. In *Determinants of Spatial Organization* (ed. S. Subtelny & I. R. Konigsberg), pp. 29–51. New York: Academic Press.

WHITTAKER, J. R. (1980). Acetylcholinesterase development in extra cells caused by changing the distribution of myoplasm in ascidian embryos. *J. Embryol. exp. Morph.* **55**, 343–354.

WHITTAKER, J. R. (1982). Muscle lineage cytoplasm can change the developmental expression in epidermal lineage cells of ascidian embryos. *Devl Biol.* **93**, 463–470.

WHITTAKER, J. R. (1987). Cell lineages and determinants of cell fate in development. *Am. Zool.* **27**, 607–622.

WOODS, N. M., CUTHBERTSON, K. S. R. & COBBOLD, P. H. (1986). Repetitive transient rises in cytoplasmic free calcium in hormone stimulated hepatocytes. *Nature, Lond.* **319**, 600–602.

ZALOKAR, M. (1974). Effect of colchicine and cytochalasin B on ooplasmic segregation of ascidian eggs. *Wilhelm Roux's EntwMech. Org.* **175**, 243–248.

ZALOKAR, M. (1979). Effect of cell-surface binding on development of ascidian egg. *Wilhelm Roux's Arch. devl Biol.* **187**, 35–47.

ZALOKAR, M. (1980). Activation of ascidian eggs with lectins. *Devl Biol.* **79**, 232–237.

ZALOKAR, M. & SARDET, C. (1984). Tracing of cell lineage in embryonic development of *Phallusia mammillata* (Ascidia) by vital staining of mitochondria. *Devl Biol.* **102**, 195–205.

(Accepted 2 November 1988)

Note added in proof:

Several important details within the cell were not visible in Figure 1 of the original journal article. Therefore, we generated a new figure for Article 49 by extracting frames from a video sequence of the same cell as was used for the journal. In the new figure, each panel is captured at nearly the same time point as in the journal. The caption stands unchanged.

The time-lapsed video itself (which runs up to nuclear repositioning at the 4-cell stage) is presented as Slide 34 on the DVD disk appended to this volume. It can be accessed by clicking on "View slides and movies" in the first figure that appears when the DVD disk is inserted, then by clicking on the right button of your mouse and choosing "Go to slide" . "34." For further instructions, see pp. xv to xvii in "DVD Contents" of this volume.

Reprinted from *The Biological Bulletin*, Vol. 177(2), p. 318, 1989.

ANALYSIS OF EDGE BIREFRINGENCE OBSERVED NEAR REFRACTIVE INDEX STEPS IN MYOFIBRILS AND KCL CRYSTALS USING HIGH RESOLUTION POLARIZED LIGHT MICROSCOPY AND SPATIAL FOURIER FILTERING

Rudolf Oldenbourg and Shinya Inoué

In conventional polarizing microscopes, vertebrate striated muscles show a pattern of alternating strongly birefringent A-bands and weakly birefringent I-bands; both bands exhibit positive birefringence (slow axis parallel to fiber axis). Each I-band is bisected by a less birefringent Z-line. When observed at high resolution with polarization rectifiers, the Z-band appeared split into three narrow zones, with two outer positively birefringent zones flanking a central negatively birefringent, or isotropic, zone. In addition, each interface between the A- and I-bands seemed to be composed of two narrow zones of different birefringence values. These observations were made upon muscle fibers suspended in standard salt solutions that had a lower refractive index than the filaments in the fibers. When the medium was replaced with one that matched the refractive index of the filaments, the A- and I-bands each appeared uniform and exhibited their characteristic high and low intrinsic birefringence.

The appearance of narrow birefringent zones near refractive index steps in muscle is similar to the patterns observed at the edges of isotropic crystals: the side with the high refractive index medium exhibits what appears to be a thin birefringent layer with the slow axis parallel to the interface; and the low index side exhibits a birefringent layer with the slow axis perpendicular to the interface.

We explored the phenomenon of edge birefringence with inhibition measurements using thin KCl crystals and myofibrils of rabbit psoas muscle. Video images of myofibrils, several sarcomeres long, were recorded and processed using a desktop computer. Spatial filtering was applied by multiplying the Fourier spectrum of the fiber image with a mask retaining only those frequencies that were an integral multiple of the inverse sarcomere length. The back transform of the filtered spectrum gave the fiber image averaged over all sarcomeres.

Supported by grants NIH R37 GM31617 and NSF DCB8518672.

The following note was added by Shinya Inoué in September of 2006:

This abstract records a brief presentation at the General Scientific Meeting of the Marine Biological Laboratory, Woods Hole, MA, in 1989. An extended article covering the subject matter appears in Oldenbourg R, *Biophys J* **60**: 629–641, 1991.

Several important details within the cell were not visible in Figure 1 of the original journal article. Therefore, we generated a new figure for Article 49 by extracting frames from a video sequence of the same cell as was used for the journal. In the new figure, each panel is captured at nearly the same time point as in the journal. The caption stands unchanged.

The time-lapsed video itself (which runs up to nuclear repositioning at the 4-cell stage) is presented as Slide 34 on the DVD disk appended to this volume. It can be accessed by clicking on "View slides and movies" in the first figure that appears when the DVD disk is inserted, then by clicking on the right button of your mouse and choosing "Go to slide" "34." For further instructions, see pp. xv to xvii in "DVD Contents" of this volume.

Reprinted from *Cytokinesis: Mechanisms of Furrow Formation During Cell Division*, Vol. 582, pp. 1–14, 1990, with permission from Blackwell Publishing.

Dynamics of Mitosis and Cleavage[a]

SHINYA INOUÉ

Marine Biological Laboratory
Woods Hole, Massachusetts 02543

As suggested by the organizers of this conference, some of the central questions relating to furrow formation in astral mitosis are: (1) What determines the location and orientation of the mitotic spindle (whose positioning defines the plane of cleavage)? (2) At what stage of mitosis does the spindle define the cleavage plane? (3) How is the contractile ring, which is responsible for cleavage, generated and maintained?

In astral mitosis, the cell generally cleaves in a plane that includes the metaphase plate of the spindle, that is, a plane bisecting the spindle axis, and starting with the cell surface closest to the spindle axis. Thus the location, orientation, and size of the mitotic spindle determine the plane of the cleavage furrow as well as its shape. Cleavage, in turn, divides the ooplasm which is progressively segregated.[1-3]

The cell cleaves in different patterns depending on: whether the spindle is located at the center or away from the center of the cell; how the spindle axis is oriented relative to the cell axis or to the proximal cell surface; and how large the spindle is. Thus, as shown in FIGURE 1, a centrally positioned spindle results in equal cleavage (panels a and b); a spindle displaced towards the animal pole of an egg cell with the spindle axis oriented perpendicular to the animal-vegetal axis (i.e., parallel to the cell surface) yields a heart-shaped cleavage (panels c and d; see also ref. 4); a spindle oriented parallel to the animal-vegetal axis and displaced to the animal pole produces the polar body of the reduction divisions of oocytes, and those displaced to the vegetal pole give rise to the micromeres common at the fourth division of echinoids (panels e and f).

In vegetal sea urchin cells, when the nuclei or spindle fails to migrate to the vegetal pole but remains centrally located (e.g., as a result of treatment with microtubule assembly inhibitors[5]), the cells cleave equally and micromeres do not form.[6] When *Chaetopterus* meiosis I oocytes are treated with pH 9.5 ammonia-containing seawater, the size of the spindle is reduced, perhaps to a quarter of its normal length, and minute polar bodies only 2-3 μ in diameter are formed (panels g and h). (Inoué and Fuseler, unpublished data. We have searched in vain for our July 1969 photographs of the ammonia/D_2O experiments. The foregoing description, therefore, depends somewhat on distant memory, but we believe it is reasonably accurate.)

In this paper I will describe our observations on three types of cells which will hopefully shed light on, or at least raise interesting points that relate to, the foregoing questions regarding cleavage furrow formation in cell division.

[a]This work was supported in part by grants from NIH (R37 GM31617) and NSF (DCB-8518672).

MICROMANIPULATION OF THE *CHAETOPTERUS* MEIOSIS I SPINDLE

When eggs of a marine annelid *Chaetopterus pergamentaceous* are shed into seawater, the germinal vesicle breaks down, and the meiosis I spindle forms and progresses to metaphase. The spindle migrates and attaches to the cortical layer of the oocyte at the animal pole (FIG. 2a, small triangle). There it remains in metaphase for several hours until the egg is activated. Micromanipulation of this spindle reveals some factors that are responsible for spindle positioning.[7]

When we insert a glass microneedle into the inner spindle pole and exert a gentle tug (FIG. 2b), the cell surface dimples until an attachment between the outer spindle pole and the cell surface is broken by further displacement of the spindle towards the cell interior (panel c). Until detachment, a thin, positively birefringent strand can be seen between the outer spindle pole and the base of the dimple. (Photographs of these video observations are presented in ref. 7.) This suggests that the meiotic spindle is attached reasonably firmly, via its outer astral ray, to a small, localized region at the animal pole of the cell cortex (to which the cell membrane in turn is anchored). Once the outer aster of the spindle is detached from the cortex, the cell surface rounds up again (panel c).

When the microneedle is removed, the released spindle hesitates briefly, then migrates back to the same point where it was previously attached (small triangle, panels c, f, and a). When a detached spindle is turned end to end with the microneedle so that the original outer pole now faces away from the animal pole and the spindle is released (panels d and e), the original inner pole swings toward the animal pole. The spindle, inverted from its original orientation, then migrates back to the original attachment site (panels f and a).

The spindle fails to reorient or return to the attachment site in the presence of microtubule assembly inhibitors such as Colcemid or nocodazole, but its return is not prevented by cytochalasins. In other words, there is strong inference that microtubules, but no actin filaments, are involved in the orientation and positioning of the *Chae-*

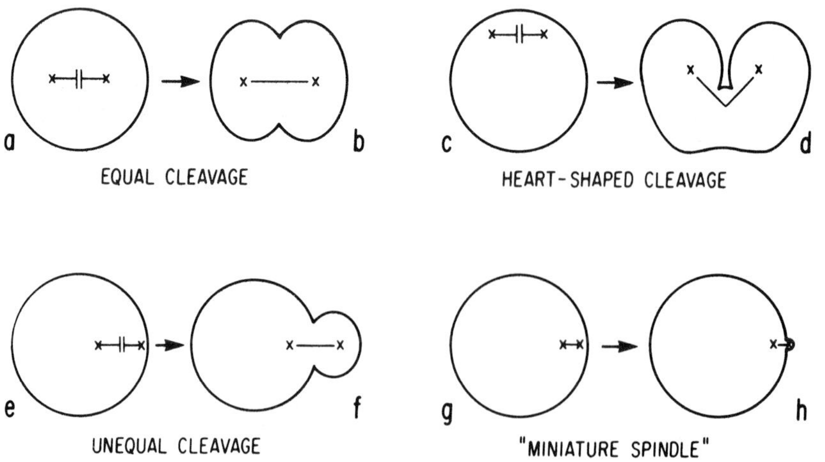

FIGURE 1. Schematics relating the position, orientation, and size of the spindle to the plane of cleavage.

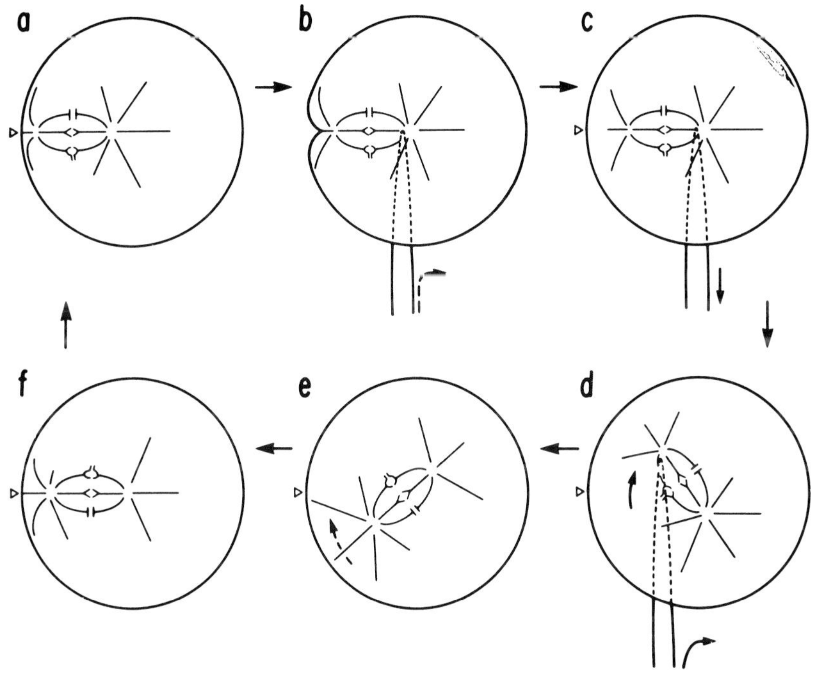

FIGURE 2. Schematics of detachment and spontaneous return of *Chaetopterus* oocyte meiotic spindle to the animal pole.

topterus meiotic spindle (Lutz *et al.,* manuscript in preparation; but see Longo and Chen[8] who conclude otherwise for mouse oocytes).

The experiments also demonstrate that the cell surface (membrane? cortex?) at the animal pole of the oocyte acts as a unique attachment site. In this connection, note the interesting observation by Longo and Chen[8] that cortical and cell surface differentiation (microvillus-free zone localized at the animal pole in mouse oocytes) is reversibly missing when meiotic spindle migration is prevented by application of cytochalasin-B. This poses an interesting chicken-and-egg question—which came first: animal pole differentiation or spindle attachment to the animal pole? Continuous observations made directly on spindle and cortical behavior in living mouse oocytes may well clarify the situation.

ROLE OF MICROTUBULES IN SPINDLE POSITIONING

At this point, I will briefly review some properties of mitotic microtubules and consider their probable involvement in spindle positioning. In brief, polarized light, electron, and fluorescence microscopy show the major linear element of the astral rays and spindle fibers to be microtubules. *In vivo* and *in vitro,* microtubules can be shown to be in a steady-state equilibrium with a pool of tubulin, the assembly of GTP-tubulin generating microtubules, and their disassembly generally yielding GDP-tubulin.

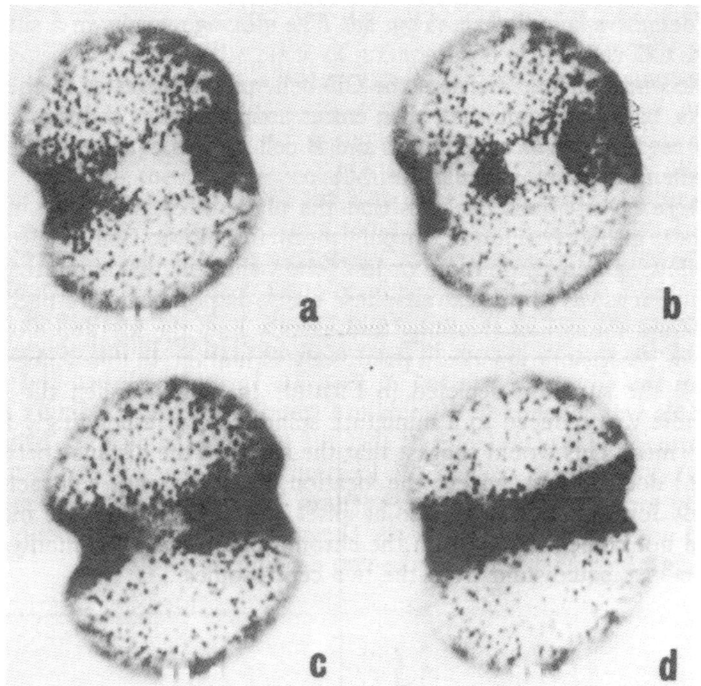

FIGURE 7. Cytochalasin-induced breakage and subsequent repair of cleavage contractile ring revealed by cortical pigment granules in *Arbacia punctulata* egg.

the granules move with velocities as high as 10-15 μm per second, then gradually slow down over the next several seconds (FIG. 8).

Thus the behavior of the granules and the shape of the cell surface under the influence of cytochalasin indicate that about an 8-micrometer-wide belt, in which the granules are embedded, suddenly yields locally and snaps apart "elastically." The yielding belt is undoubtedly the contractile ring. Judging from the shape of the break, the cytochalasin-sensitive actin filaments of the contractile ring (that run circumferentially near the bottom of the cleavage furrow) would appear to be linked together laterally.

The apparent elasticity of the belt must reflect a longitudinal tension residing in the filament array. The tension is most likely counteracting the forces at the surface of the cleaving egg that would tend to return the egg to a sphere when the cleavage contractile ring is broken.[31-33]

While the pigment granules move predominantly along the circumference of the furrow (i.e., parallel to the length of the actin filaments) as the contractile ring yields, the granules do not retrace this movement backwards when the contractile ring is being reformed. Instead, as the egg starts to cleave again after the ring has stopped yielding, the pigment granules migrate into the band-shaped void from various directions (FIGS. 9 and 10). During this process the granules migrate at velocities of 0.2-0.4 μm per second, an order of magnitude slower than that when the contractile ring snapped apart.

These observations suggest that (1) the furrowing force, or the force that prevents the cell from rounding up during cleavage, is indeed localized in a discrete belt in the cleavage furrow, that is, in the contractile ring. (2) The material for the contractile ring recruited from the cortex adjacent to furrow. Thus there is an irreversibility, or temporal directionality, to the establishment of, and force generation in, the contractile ring.

To reconcile these observations, we should perhaps envision a cleavage mechanism in which the shortening of the cleavage contractile ring is intimately coupled with its establishment and slow dismantling. In other words, rather than the contractile ring first becoming established and then shortening to cleave the cell, the processes of establishing and slowly dismantling the contractile ring may of themselves generate the force for cleavage. Unlike muscle and cilia which tend to be poised to generate forces immediately upon call, the cleavage contractile ring, like the mitotic spindle, may well generate forces by mechanisms that are directly coupled with the generation, shortening, and slow dismantling of organized filament arrays.

ACKNOWLEDGMENTS

I would like to thank Dr. Yuichiro Tanaka (Honda R&D Co., Wako-shi, Saitama, Japan) for allowing me to use our unpublished data on *Arbacia* eggs, and Dr. John

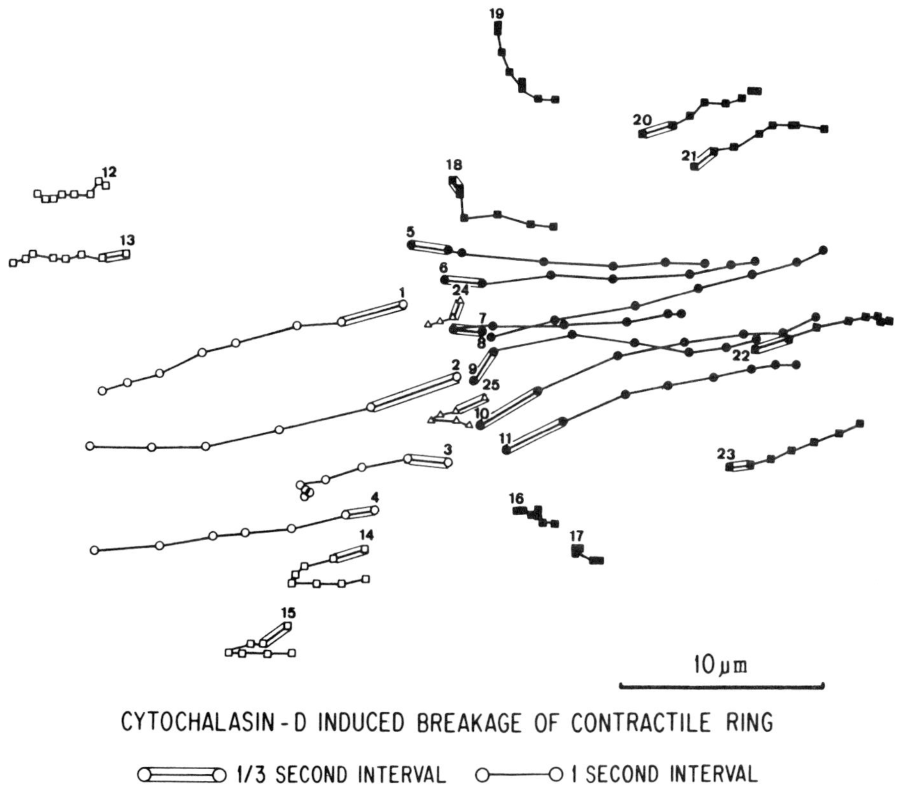

FIGURE 8. Movement of individual pigment granules during cytochalasin-D induced breakage of contractile ring, traced from video records.

Fuseler (Louisiana State University Medical Center, Shreveport, Louisiana) for spending the time to search for and write up detailed recollections of our *Chaetopterus* ammonia spindle experiments. I also thank Bob Golder of the MBL Photo Lab for making the finished drawings.

Note added in proof

At the Mount Desert meeting on cytokinesis, there was some mention of the fact that the cleavage contractile ring is often not detectable until the cell starts to furrow or, in any event, until the cell starts to elongate along the spindle axis at the beginning of cleavage. However, I felt that we did not adequately discuss this important observation. Perhaps this fact, as well as the sharp delineation of the width of the cleavage furrow, can be better understood in the light of Dan's 'astral mechanism."[34] The compression exerted on the furrow cortex by criss-crossing astral rays, radiating from the opposite poles of the spindle extending in anaphase-B, could impose the anisotropic deformation of the cortex localized to the furrow region. That would tend to align the actin filaments, which were previously oriented randomly, along the presumptive contractile ring. Once such an anisotropic alignment is initiated, the contractile ring could become self-propagating (as I had suggested in the M.B.L. Physiology Course lectures).

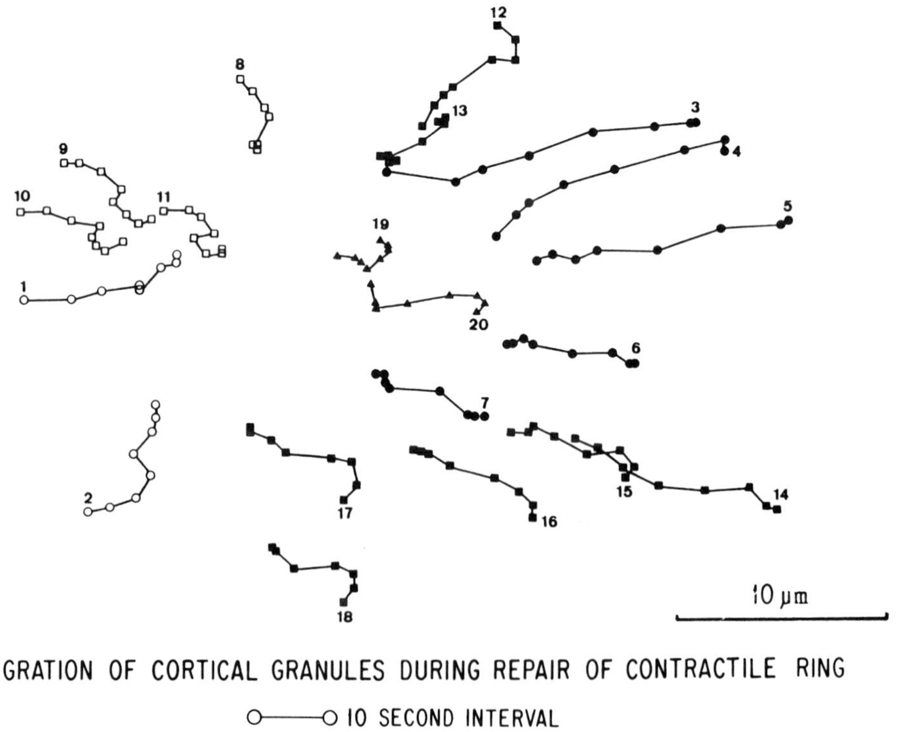

FIGURE 9. Migration of individual pigment granules during repair of contractile ring, traced from video records.

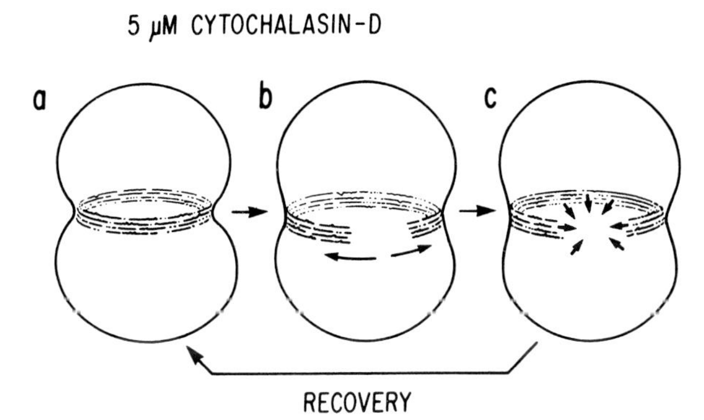

FIGURE 10. Schematics of cytochalasin-induced breakage and subsequent recovery of cleavage contractile ring.

REFERENCES

1. COSTELLO, D. P. 1945. Segregation of ooplasmic constituents. J. Elisha Mitchell Scientific Soc. **61:** 277-289.
2. HUXLEY, J. S. & G. R. DEBEER. 1934. The Elements of Experimental Embryology. Cambridge University Press. Cambridge, England.
3. WILSON, E. B. 1928. The Cell in Development and Heredity, 3rd ed. MacMillan. New York, NY.
4. RAPPAPORT, R. 1982. Cytokinesis: The effect of initial distance between mitotic apparatus and surface on the rate of subsequent cleavage furrow progress. J. Exp. Zool. **221:** 399-403.
5. LUTZ, D. A. & S. INOUÉ. 1989. Microtubule-dependent asymmetric nuclear positioning prior to macromere-micromere formation in the sea urchin embryo. Dev. Biol. (in press).
6. TANAKA, Y. 1976. Effects of the surfactants on the cleavage and further development of the sea urchin embryos. 1. The inhibition of micromere formation at the fourth cleavage. Develop. Growth & Differ. **18:** 113-122.
7. LUTZ, D. A., Y. HAMAGUCHI & S. INOUÉ. 1988. Micromanipulation studies of the asymmetric positioning of the maturation spindle in *Chaetopterus* sp. oocytes: I. Anchorage of the spindle to the cortex and migration of a displaced spindle. Cell Motil. Cytoskeleton **11:** 83-96.
8. LONGO, F. J. & D.-Y. CHEN. 1985. Development of cortical polarity in mouse eggs: Involvement of the meiotic apparatus. Dev. Biol. **107:** 382-394.
9. INOUÉ, S. 1981. Cell division and the mitotic spindle. J. Cell Biol. **91:** 131s-147s.

9a. HYAMS, J. S. & B. R. BRINKLEY. 1989. Mitosis: Molecules and Mechanics. Academic Press. New York, NY.

9b. MCINTOSH, J. R. & M. P. KOONCE. 1989. Mitosis. Science **246:** 622-628.

10. MITCHISON, T. & M. KIRSCHNER. 1984. Dynamic instability of microtubule growth. Nature **312:** 237-242.
11. CASSIMERIS, L., N. K. PRYER & E. D. SALMON. 1988b. Real-time observations of microtubule dynamic instability in living cells. J. Cell Biol. **107:** 2223-2231.
12. KOSHLAND, D. W., T. J. MITCHISON & M. W. KIRSCHNER. 1988. Polewards chromosome movement driven by microtubule depolymerization *in vitro.* Nature **331:** 499-504.
13. GORBSKY, G. J., P. J. SAMMAK & G. G. BORISY. 1987. Chromosomes move poleward in anaphase along stationary microtubules that coordinately disassemble from their kinetochore ends. J. Cell Biol. **104:** 9-18.

14. Bajer, A. S. & J. Molè-Bajer. 1979. Anaphase-stage unknown. *In* Cell Motility: Molecules and Organization. S. Hatano, H. Ishikawa & H. Sato, eds.: 569-591. University of Tokyo Press. Tokyo, Japan.
15. Cassimeris, L., S. Inoué & E. D. Salmon. 1988a. Microtubule dynamics in the chromosomal spindle fiber: Analysis by fluorescence and high-resolution polarization microscopy. Cell Motil. Cytoskeleton **10:** 185-196.
16. McDonald, K., J. D. Pickett-Heaps, J. R. McIntosh & D. H. Tippit. 1977. On the mechanism of anaphase spindle elongation in *Diatoma vulgare.* J. Cell Biol. **74:** 377-388.
17. Wilson, E. B. 1895. An Atlas of the Fertilization and Karyokinesis of the Ovum, by E. B. Wilson, with the Co-operation of E. Leaming. MacMillan. New York, NY.
18. Inoué, S. & K. Dan. 1951. Birefringence of the dividing cell. J. Morphol. **89:** 423-455.
19. Dan, K. 1987. Studies on unequal cleavage in sea urchins. III. Micromere formation under compression. Develop. Growth & Differ. **29:** 503-515.
20. Jennings, H. S. 1896. The early development of *Asplancha herrickii* deGuerne. A contribution to developmental mechanics. Bull. Mus. Comp. Zool. **30:** 1-117 and 10 plates.
21. Costello, D. P. 1961. On the orientation of centrioles in dividing cells, and its significance: A new contribution to spindle mechanics. Biol. Bull. **120:** 285-312.
22. Swann, M. M. & J. M. Mitchison. 1953. Cleavage of sea-urchin eggs in colchicine. J. Exp. Biol. **30:** 506-514.
23. Hiramoto, Y. 1971. Analysis of cleavage stimulus by means of micromanipulation of sea urchin eggs. Exp. Cell Res. **68:** 291-298.
24. Dan, K. & S. Inoué. 1987. Studies of unequal cleavage in molluscs. II. Asymmetric nature of the two asters. J. Invertebr. Repro. & Develop. **11:** 335-354.
25. Inoué, S. 1952. The effect of colchicine on the microscopic and submicroscopic structure of the mitotic spindle. Exp. Cell Res. (Suppl. 2): 305-318.
26. Wassarman, P. M., W. J. Josefowicz & G. E. LeTourneau. 1976. Meiotic maturation of mouse oocytes *in vitro:* Inhibition of maturation at specific stages of nuclear progression. J. Cell Sci. **22:** 531-545.
27. Wolpert, L. 1960. The mechanics and mechanism of cleavage. Int. Rev. Cytol. **10:** 163-216.
28. Dan, K. & S. Ito. 1984. Studies of unequal cleavage in molluscs: I. Nuclear behavior and anchorage of a spindle pole to cortex as revealed by isolation technique. Develop. Growth & Differ. **26:** 249-262.
29. Dan, K. 1954. The cortical movement in *Arbacia punctulata* eggs through cleavage cycles. Embryologia **2:** 115-122.
30. Tanaka, Y. & S. Inoué. 1981. Does cytochalasin-D induce reversible disruption of the cleavage contractile ring? Biol. Bull. **161:** 311.
31. Hiramoto, Y. 1979. Mechanical properties of the dividing sea urchin egg. *In* Cell Motility: Molecules and Organization. S. Hatano, H. Ishikawa & H. Sato, eds.: 653-663. Baltimore University Park. Baltimore, MD.
32. Schroeder, T. E. 1981. Interrelations between the cell surface and the cytoskeleton in cleaving sea urchin eggs. *In* Cytoskeletal Elements and Plasma Membrane Organization. G. Poste & G. L. Nicholson, eds.: 170-216. Elsevier/North-Holland Biomedical Press.
33. Mitchison, J. M. 1953. Microdissection experiments on sea-urchin eggs at cleavage. J. Exp. Biol. **30:** 515-524.
34. Dan, K. 1988. Mechanism of equal cleavage of sea urchin egg: Transposition from astral mechanism to constricting mechanism. Zool. Sci. **5:** 507-517.

IN VIVO VISUALIZATION OF INTERCALARY MICROTUBULE BREAKDOWN BY POLARIZING MICROSCOPY AND VERY HIGH SPEED VIDEO

Febvre J,[1] Febvre-Chevalier C,[1] Knudson RA,[2] Takeshita T,[3] and Inoué S[2]

(Abstract #144 presented at IX International Congress of Protozoology, Berlin, Germany, July 25-31, 1993)*

Actinocoryne contractilis is a stalked contractile Heliozoan consisting of a base, a long stalk (mean length 150 μm), and a head surrounded by numerous slender axopodia. The head contains a single MTOC (microtubule organizing center) that generates patterned arrays of unstable microtubules (MTs). At rest, the MT-rods stiffening the stalk and the axopodia are continuous. On stimulation, the axopodia and stalk shorten exponentially (half time: 4 ms ± 2). This is accompanied by rapid MT-disassembly. Such rapid kinetics and the presence in thin sections of granular bands crossing the axonemes after mild destabilization suggest a model of intercalary destabilization and breakdown of MTs (Febvre-Chevalier and Febvre, *J Cell Biol* **118**: 585–594, 1992). Using very sensitive polarizing microscopy (Inoué, *Video Microscopy*, Plenum Press, 1986) associated with ultra-high speed video (3.10^6 fr.sec.-1) (Takeshita *et al.*, *SPIE* **1539**: 2–10, 1991), modifications of birefringence seem to be visible in untreated cells. These occur during the first milliseconds suggesting that some molecular changes may take place in the body of the MT-arrays during contraction.

---

* Unpublished (1993).

[1] URA 671 CNRS, Observatoire de Villefranche/Mer, 06230, France
[2] MBL, Woods Hole, MA 02543, USA
[3] Hamamatsu Photonic Systems, Bridgewater, NJ 08807, USA

Reprinted from the *Journal of Microscopy*, Vol. 172, pp. 31–39, 1993, with permission from Blackwell Publishing.

Image sharpness and contrast transfer in coherent confocal microscopy

R. OLDENBOURG,*† H. TERADA,‡ R. TIBERIO§ & S. INOUÉ*
**Marine Biological Laboratory, Woods Hole, MA, U.S.A.*
†Martin Fisher School of Physics, Brandeis University, Waltham, MA, U.S.A.
‡Hamamatsu Photonics Kabushiki Kaisha, Hamamatsu City, Japan
§National Nanofabrication Facility, Cornell University, Ithaca, NY, U.S.A.

Key words. Confocal microscope, resolution, test grating, contrast transfer, laser scanning, discrete Fourier analysis.

Summary

Confocal microscopes provide clear, thin optical sections with little disturbance from regions of the specimen that are not in focus. In addition, they appear to provide somewhat greater lateral and axial image resolution than with non-confocal microscope optics. To address the question of resolution and contrast transfer of light microscopes, a new test slide that enables the direct measurement of the contrast transfer characteristics (CTC) of microscope optics at the highest numerical aperature has been developed. With this new test slide, the performance of a confocal scanning laser microscope operating in the confocal reflection mode and the non-confocal transmission mode was examined. The CTC curves show that the confocal instrument maintains exceptionally high contrast (up to twice that with non-confocal optics) as the dimension of the object approaches the diffraction limit of resolution; at these dimensions, image detail is lost with non-confocal microscopes owing to a progressive loss of image contrast. Furthermore, we have calculated theoretical CTC curves by modelling the confocal and non-confocal imaging modes using discrete Fourier analysis. The close agreement between the theoretical and experimental CTC curves supports the earlier prediction that the coherent confocal and the incoherent non-confocal imaging mode have the same limit of resolution (defined here as the inverse of the spatial frequency at which the contrast transfer converges to zero). The apparently greater image resolution of the coherent confocal optics is a consequence of the improved contrast transfer at spacings which are close to the resolution limit.

Introduction

In the confocal scanning laser microscope (CSLM) the specimen is rapidly scanned in a raster fashion (as in a TV screen) by a diffraction-limited point of light that originates from a source pinhole. Light given off by the illuminated point of the specimen is focused by the objective lens onto an exit pinhole which is confocal with the source pinhole and passes only light originating from the intentionally illuminated point. The exit pinhole thus excludes light scattered by other portions of the specimen, including those out of focus, and by other elements in the microscope optical train. As a result, confocal microscopes provide optical sections that are considerably sharper than those taken with non-confocal instruments. The improvement is especially striking in fluorescence and other modes of microscopy where, without confocal optics, light contributed by out-of-focus objects tends to become superimposed and degrade the in-focus image.

Thus, the sharpness of image in a confocal system is generally attributed to its ability to provide exceptionally thin, clean optical sections by the optical arrangement that discriminates against light contributed by objects lying outside of the plane of focus (for details of the optical principles, virtues, critiques and historical background of the development of confocal microscopes see Petráň *et al.*, 1968; Wilson & Sheppard, 1984; Minsky, 1988; Pawley, 1990; Wilson, 1990a). While this factor clearly contributes to the enhanced sharpness of the confocal microscope image, earlier work by Sheppard & Choudhury (1977) and van der Voort *et al.* (1988) suggested that the contrast transfer characteristics of the confocal system may also be better than those of non-confocal systems. This was based on the argument that the exit pinhole reshapes and sharpens the diffraction image of the illuminated spot and, as a consequence, improves the microscope's ability to reproduce fine specimen detail (Sheppard & Choudhury, 1977; Wilson & Sheppard, 1984; van der Voort *et al.*, 1988; Hegedus, 1990; Wilson, 1990b). However, because of the absence of a suitable test target, no direct measure-

function $t(f)$] × [object spectrum $s(f)$]. The amplitude transfer function $t(f)$ is a dimensionless weighing factor defining the fraction of amplitude that is transmitted through the optics as a function of spatial frequency f in object space. The object spectrum $s(f)$ is the Fourier transform of the object amplitude $A(x)$, the amplitude of the light wave in a plane directly adjacent to the object under observation (we use a simplification with x representing distances in the object plane and image plane, i.e. the optical magnification factor is unity).

Equation (1) is typical for a coherent imaging system. In coherent imaging, image intensity is calculated by first summing the amplitudes of the light waves associated with different spatial frequencies at a given image point and then squaring the result to obtain the image intensity at that point.

Scanned images obtained with confocal optics in reflection mode are indeed coherent images. Qualitatively, this can be explained simply by the confocality of the source and the exit pinhole. The illuminating light, when leaving the source pinhole, is necessarily coherent. From the source pinhole the light passes through the optical set-up and illuminates the object with a diffraction-limited spot, the Airy pattern. Part of the illuminating light is diffracted back by the object and passes, as imaging light, through the same optical system as the illuminating light, only in reversed order. The imaging light which passes through the exit pinhole must have the same state of coherence as the illuminating light at the source pinhole, because both pinholes are confocal and therefore at optically equivalent points. Hence, because the illuminating light is coherent at the source pinhole, all the imaging light arriving at the exit pinhole is coherent, i.e. light waves associated with different spatial frequencies have fixed-phase relationships that are constant in time. Therefore, images produced by the confocal reflection microscope are coherent images.

In contrast to the reflection mode, the transmission mode of the scanning laser microscope is a non-confocal, near incoherent imaging mode (see Discussion). In incoherent imaging the intensities (rather than amplitudes) of light waves associated with different spatial frequencies are added directly to obtain the intensity at a given image point, hence:

$$I(x) = \int_{-\infty}^{+\infty} t'(f)s'(f)e^{-2\pi ifx}\,df \qquad (2)$$

with $t'(f)$ denoting the incoherent or intensity transfer function and $s'(f)$ the object intensity spectrum.

The intensity transfer function of the non-confocal transmission mode and the amplitude transfer function of the confocal reflection mode have the same functional form as shown in Fig. 7 (Wilson & Sheppard, 1984).

The object spectra of both imaging modes are closely related. As stated earlier, the amplitude object spectrum

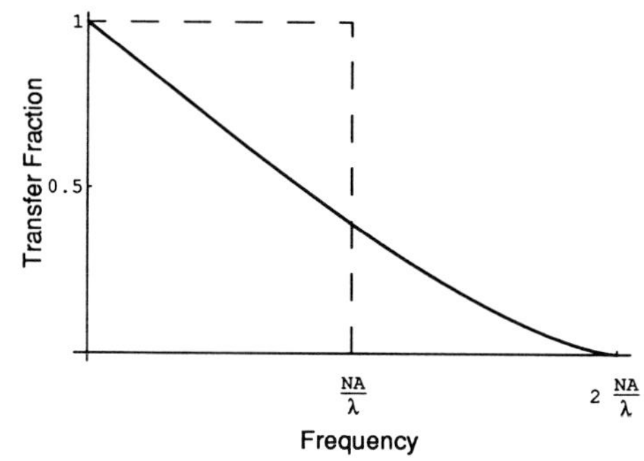

Fig. 7. Optical transfer functions of the coherent confocal and incoherent non-confocal microscope (both continuous curve) and of the coherent non-confocal microscope (dashed curve). The amplitude transfer function of the coherent confocal and the intensity transfer function of the non-confocal transmission scanning microscope have the same functional form under the following imaging conditions: (i) objective lenses in both microscopes have the same numerical aperture (NA) and light of the same wavelength (λ) is used; (ii) in the non-confocal scanning microscope all the light transmitted by the specimen must be sampled by the detector, while the coherent confocal optics must employ an illuminating lens with the same NA as the objective lens; (iii) absence of defocus and lens aberrations. Under these conditions, the diffraction limit of resolution is given by $\lambda/(2\ \mathrm{NA})$, the inverse of the spatial frequency at which the transfer function converges to zero for both types of microscopes. The coherent non-confocal microscope, however, equipped with the same objective lens, has a limiting frequency of NA/λ at which the box-like transfer function drops to zero.

$s(f)$ is the Fourier transform of the object amplitude $A(x)$. In the case of the bar patterns that we observed in reflected light, $A(x)$ is a square wave with a period equal to the grating period and field values changing between zero and a constant value which we arbitrarily set to 1. $A(x)$ is zero in clear spaces and is 1 at locations with metal deposit (Fig. 8A). Because the phase of the reflected field is constant, we considered the metal gratings as pure amplitude objects.

The intensity object spectrum $s'(f)$ is the Fourier transform of the object intensity, which is simply the square of the absolute value of the object amplitude.

To solve Eqs. (1) and (2) for the image intensities we used discrete Fourier analysis, which requires discrete sampling values for the transfer functions and the object spectra. For our calculations, the object amplitude $A(x)$ was sampled with a resolution of 0·01 μm over a length of 40·96 μm, resulting in 4096 sample points (Fig. 8A). The corresponding resolution in Fourier space is 1/(40·96 μm), which defines the sampling frequency for the transfer functions. Figure 7 shows only one-half of the complete transfer function which, for mathematical reasons, extends between negative and positive frequency values and is symmetrical with respect to the origin. In solving for Eq. (1), for example, the sampled transfer function $t(f)$ was first multiplied with

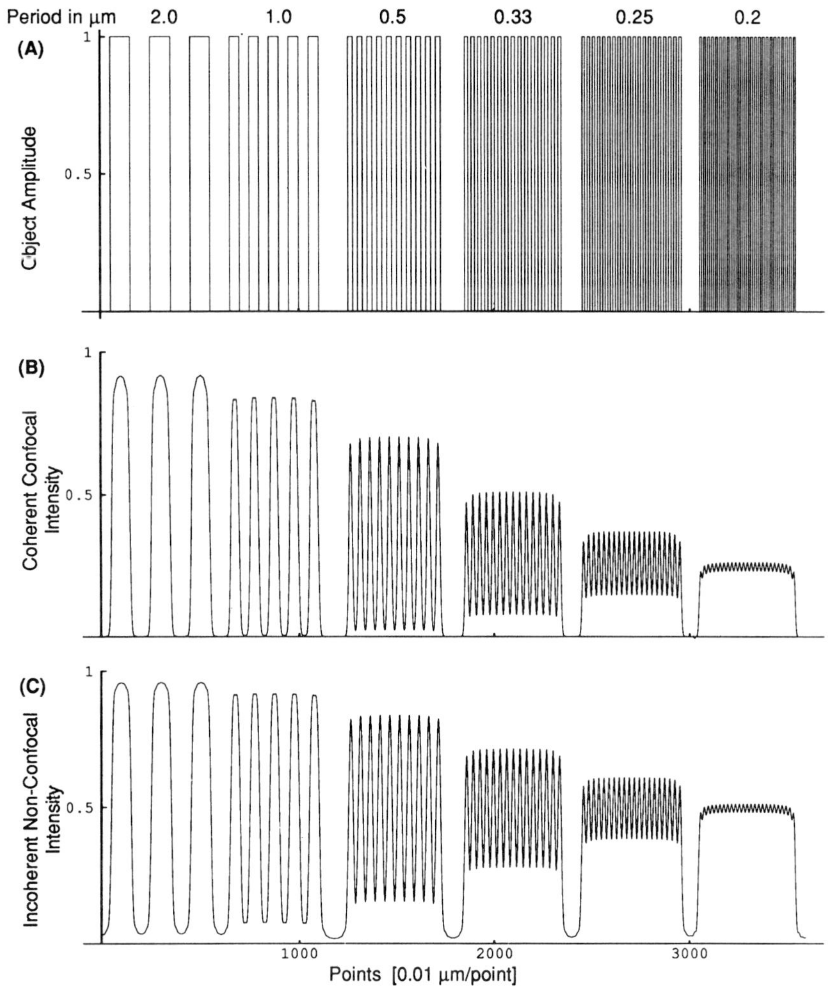

Fig. 8. Object amplitude and calculated image intensities of linear gratings as a function of distance. The object amplitude in (A) represents a series of square gratings (the periods of the gratings are indicated on top). Below, calculated image intensities are shown assuming (B) coherent confocal imaging and (C) incoherent non-confocal imaging. Image intensities in (B) and (C) were calculated with the same object amplitude function (A), but using Eqs. (1) and (2), respectively.

the discrete object spectrum $s(f)$. Then, the product was Fourier transformed by a fast Fourier transform algorithm and the absolute value of the result squared to obtain the image intensities $I(x)$ for the confocal reflection mode. Figure 8 shows several square gratings as object functions and the computed images for the coherent confocal mode and the incoherent non-confocal mode. All calculations were performed on a Macintosh IIx computer (Apple Computer Inc., Cupertino, CA) using Mathematica, a software tool for performing mathematics by computer (Wolfram Research Inc., Champaign, IL).

We calculated the contrast modulations of computed images in the same fashion as for experimental images. For each grating period we obtained I_{min} and I_{max} from the computed images and calculated the quotient $(I_{max} - I_{min})/(I_{max} + I_{min} - 2\,I_g)$, with the background intensity $I_g = 0$. To construct a smooth contrast transfer function, we computed images for many different gratings with closely spaced periods and interpolated between the numerical contrast values obtained for each grating. Results of these computations are shown in Fig. 6, together with the experimental data.

Discussion

It is not always recognized that the coherent confocal and the incoherent non-confocal microscope have the same optical transfer functions. For the former it is the amplitude transfer function and for the latter it is the intensity transfer function. Because both transfer functions are identical, both imaging modes have the same limit of resolution, $\lambda/(2\,\text{NA})$, which is the inverse of the cut-off frequency in Fig. 7. [Note that the diffraction limit of resolution is defined using the cut-off frequency of the transfer function, while the Rayleigh limit of resolution uses the radius of the Airy disc $= 1{\cdot}22\,\lambda/(2\,\text{NA})$.] The limits of resolution are identical, in spite of the narrower intensity point-spread function of a single object point in the coherent confocal microscope compared to the incoherent non-confocal microscope (Wilson & Sheppard, 1984). The image of two closely spaced points in incoherent imaging is simply the sum of the two intensity point-spread functions. In coherent imaging, however, the amplitude point-spread functions are added and the result is squared to obtain the image intensities of two points. Thus, the coherent imaging process is not linear with intensity and the intensity point-spread function cannot be used directly to determine resolution in the coherent imaging mode.

Even though the limit of resolution is the same, the contrast transfer for frequencies below the resolution limit is up to twice as high in coherent confocal as compared to incoherent non-confocal imaging. In Fig. 6(C), the calculated CTC curves for both imaging modes start out with zero contrast at the same limiting period equal to $\lambda/(2\,\mathrm{NA})$. For spacings larger but close to this limit, the slope of the contrast transfer characteristics for the coherent confocal mode are twice the slope for the incoherent non-confocal mode. At larger periods, the contrast transfer saturates and approaches 100% for both imaging modes.

The much improved contrast transfer of the coherent confocal microscope for frequencies close to the resolution limit leads to an improved resolution in real, noisy images, compared to incoherent non-confocal images with the same noise level. If image intensities are subject to random fluctuations with a certain standard deviation, a finite intensity modulation or contrast in the image is required to distinguish the intensity variation induced by the object from image noise.

For the coherent confocal mode, the theoretical curves in Fig. 6(A) and (C) follow the experimental results remarkably well. (Note that the theoretical curves were computed without any adjustable parameters.) The contrast modulations measured with the highest NA objectives, however, seem to lie consistently at higher contrast values than predicted by the theory. We have no good explanation for this result. We speculate that the observed slight asymmetry of the point-spread function of the high-NA lenses may account for the focal spot being somewhat sharper than what is theoretically predicted.

For the transmission mode, experimental contrast transfer values are either above (at low frequencies) or below (at high frequencies) the theoretical curves computed for fully incoherent non-confocal imaging (Fig. 6B, C). This crossover can be explained qualitatively by the observation that the non-confocal transmission mode of the scanning laser microscope is not a truly incoherent, but a partially coherent imaging mode (Sheppard & Choudhury, 1977). At high spatial frequencies, the limit of resolution of partially coherent imaging is expected to be lower than the limit of resolution of fully incoherent imaging. At low frequencies, however, partially coherent imaging has a better contrast transfer than fully incoherent imaging.

Not only the contrast modulation, but also the mean intensity level observed for each grating is strikingly different in the confocal reflection and the non-confocal transmission mode (see Figs. 4 and 5 and their legends). While for the non-confocal transmission mode the mean intensity level is independent of the grating period, for the confocal reflection mode the mean intensity level observed for fine, barely resolved gratings is about half the mean intensity level of the coarse, well-resolved gratings. As shown in the computed intensities of Fig. 8(B), the drop in mean intensity from coarse to fine gratings observed in the confocal reflection mode is a consequence of the coherence of the image-forming waves.

The presence or absence of a drop in mean intensity from coarse to fine gratings can then be used as in indicator to decide between coherent and incoherent imaging. Extending this argument, we may well be able to measure the degree of coherence of the image-forming waves by observing the change in mean intensity level as a function of spatial frequency. When there is no change, the imaging light is fully incoherent. When the mean intensity measured first in a low-frequency, fully modulated grating and then in a high-frequency, barely resolved or unresolved grating decreases by less than 50%, the imaging light could be

Fig. 9. Portion of a diatom shell. The specimen *Amphipleura pellucida* frustule was observed at a wavelength of 515 nm with an HPK prototype confocal scanning laser microscope (Fig. 3) equipped with a Nikon 60×/1·4-NA plan apochromatic objective lens. The optical sections, recorded in confocal reflection mode, were too thin to display an adequate number of frustule pores in a single section. Therefore, three consecutive optical sections, acquired 0·2 μm apart, were superimposed in the final image, each pixel representing the maximum brightness value out of those in the three planes. The prominent striations that occur at 90° to the length (along the central bright line) of the diatom are spaced 0·265 μm apart. The bright dots within each striation arise from the minute pores which are spaced on average 0·205 μm apart along the striations. As can be surmised from Fig. 6(C), the spacing of these dots is generally not discernible at this wavelength with non-confocal optics, but yields clearly resolved, high-contrast images with confocal microscopy.

partially coherent. When this relative change reaches 50%, then the imaging light is coherent.

The ability of the microscope to retain contrast modulation at high spatial frequencies can be improved by the use of special optics such as phase-contrast, differential interference contrast (DIC or "Nomarski"), single-sideband edge enhancement, etc., albeit with a loss of contrast modulation at low spatial frequencies and a potential reduction in the limit of resolution (except in the third case) (figs. 5-24 and 5-25 in Inoué, 1986). Therefore, none of these approaches improves the contrast transfer achieved with confocal microscopy. Figure 9 shows an actual image that illustrates the exceptionally high image contrast for spacings in the object approaching the resolution limit of the optics.

We are currently exploring the contrast transfer characteristics of fluorescence microscopes, where the image is incoherent by the very nature of fluorescent imaging.

Acknowledgments

The experimental studies were carried out at the Marine Biological Laboratory, Woods Hole, MA, during the winter of 1990–1991 on a prototype CSLM developed by Hamamatsu Photonics Kabushiki Kaisha (HPK) of Hamamatsu City, Japan. The test targets were fabricated at the National Nanofabrication Facility (NNF) at the Knight Laboratory at Cornell University, Ithaca, NY. We wish to thank Gordon W. Ellis of the University of Pennsylvania for fruitful discussions concerning the modulation transfer function of microscope objectives, Louis Kerr of the MBL for his help with the scanning electron microscopy, Jane Leighton of STARS for her effective secretarial help, the management and staff of HPK and NNF for their cooperation, and NIH grant R-37 GM31617 and NSF grant DCB-8908169 to S.I. and NSF grant ECS-8619049 to the Cornell NNF for support of several aspects of this project.

References

Hegedus, Z.S. (1990) Pupil filters in confocal imaging. *Confocal Microscopy* (ed. by T. Wilson), pp. 171–183. Academic Press, London.

Inoué, S. (1986) *Video Microscopy*. Plenum Press, New York.

Minsky, M. (1988) Memoir on inventing the confocal scanning microscope. *Scanning*, **10**, 128–138.

Pawley, J.B. (ed.) (1990) *Handbook of Biological Confocal Microscopy*. Plenum Press, New York.

Petráň, M., Hadravsky, M., Egger, D. & Galambos, R. (1968) Tandem-scanning reflected-light microscope. *J. Opt. Soc. Am.* **58**, 661–664.

Sheppard, C.J.R. & Choudhury, A. (1977) Image formation in the scanning microscope. *Opt. Acta* **24**, 1051–1073.

van der Voort, H.T.M., Brakenhoff, G.J. & Janssen, G.C.A.M. (1988) Determination of the 3-dimensional optical properties of a confocal scanning laser microscope. *Optik*, **78**, 48–53.

Wilson, T. (ed.) (1990a) *Confocal Microscopy*. Academic Press, London.

Wilson, T. (1990b) Optical aspects of confocal microscopy. *Confocal Microscopy* (ed. by T. Wilson), pp. 93–141. Academic Press, London.

Wilson, T. & Sheppard, C. (1984) *Theory and Practice of Scanning Optical Microscopy*. Academic Press, London.

Young, I.T. (1989) Image fidelity: characterizing the imaging transfer function. *Methods in Cell Biology* (ed. by D.L. Taylor and Y.-L. Wang), pp. 1–45. Academic Press, New York.

Reprinted from *Three-Dimensional Confocal Microscopy: Volume Investigation of Biological Specimens*, pp. 397–419, 1994, with permission from Elsevier.

CHAPTER 17

Ultrathin Optical Sectioning and Dynamic Volume Investigation with Conventional Light Microscopy

Shinya Inoué

Marine Biological Laboratory
Woods Hole, Massachusetts

I. INTRODUCTION

Together with a rapid surge in confocal microscopy, significant advances have been made in nonconfocal imaging, or conventional light microscopy. Image contrast is improved substantially; better corrected, high-numerical aperture (NA) optics yield images with excellent lateral and axial resolution; dynamic behavior of unresolved objects are captured; and the improved axial resolution gives rise to ultrathin optical sections.

Many of these advances in conventional light microscopy arose by its combination with modern video and digital image processing. In fact, the improvements brought about by electronic imaging and processing are such an integral part of the improved performance of the light microscope today, that in this chapter the expression "conventional light microscopy" is used to include the acquisition of light microscope images using video and, more often than not, processed digitally.

In addition to benefiting from advances in imaging and processing technology, both confocal and conventional light microscopy have benefited immensely by the array of target-specific immunological, genetic, and chemical probes, caged compounds, and so on, that have been developed. Those topics are discussed by others elsewhere in this volume.

In this chapter, we compare the merits of confocal vs nonconfocal imaging with the light microscope, and discuss the conditions needed to obtain ultrathin optical sections and three- and four-dimensional images with the conventional light microscope. We emphasize instrumentation that affects applications in biology, especially analyses of the three-dimensional architecture of tissues, cells, organelles, macromolecular assemblies, and their dynamic changes with time, at high microscope resolution. Some of the techniques are currently in existence; others are natural extensions of ongoing efforts.

We shall start our discussion by reviewing the optical performance of nonconfocal light microscopes that have been improved by video imaging and digital image processing.

II. CONVENTIONAL LIGHT MICROSCOPY

A. Resolution and Contrast

An image must be adequately resolved and endowed with sufficient contrast for the desired details in the specimen to be visible or to be meaningfully recorded. Resolution in a nonconfocal light microscope, properly used to provide a diffraction-limited image, is expressed by Eq. (1):

$$d = 1.22\lambda/(\mathrm{NA}_{\mathrm{obj}} + \mathrm{NA}_{\mathrm{cond}}) \tag{1}$$

where d is the lateral distance between two incoherently illuminated, equally bright points in the specimen that can just be recognized as being separate (i.e., the limit of resolution according to the Rayleigh criterion); λ is the vacuum wavelength of the imaging light; and $\mathrm{NA}_{\mathrm{obj}}$ and $\mathrm{NA}_{\mathrm{cond}}$ are, respectively, the numerical aperture of the objective lens and the NA that is actually provided by the condenser.

As Eq. (1) shows, the actual NA provided by the condenser plays a significant role in determining the limit of lateral resolution that can be attained. The NA of the condenser is affected by adjustment of its focus, immersion, setting of its iris, and the pattern of illumination at its aperture plane. These are often neglected but important points to consider because too small a condenser NA reduces lateral resolution, adds to the depth of field, and introduces high-contrast diffraction rings and refraction effects that confuse the image.

The NA of the condenser has a particularly striking effect on the depth of field, and hence the thickness of the optical section that can be attained, because the axial resolution of the microscope rises with the square of the system NA. Thus, lower NA illumination results in a disproportionate loss in axial resolution and the ability to obtain thin optical sections. Conversely, high-NA, full-cone illumination of a specimen viewed with a plan apochromatic objective lens and illuminated with a matching cone of light from a well-corrected, oil immersion condenser can yield surprisingly thin optical sections even with conventional light microscopy. This important point is discussed further in Section IV.

Despite these major advantages, it has long been common practice to use light microscopes with the condenser at a reduced NA by closing down its iris, a condition that makes the image more easily visible, albeit with the losses described above. When the condenser iris is opened and its NA is raised to approach that of the objective lens, image contrast tends to drop too far for comfortable observation. If the contrast of the image becomes so low as to be invisible or undetectable, a meaningful magnification is not achieved with the microscope in spite of the theoretical improvement in resolution. Visually, we can no longer detect contrast below 2% between moderately bright image regions occupying a sufficiently large visual field and bounded sharply and directly by its background (i.e., without a brighter or darker border line). As the image becomes darker, or the size of the image becomes smaller, or its boundary becomes fuzzier or acquires a boundary line, our ability to detect contrast diminishes drastically. We may require well over 100% contrast for the object or structure to be visible. (Contrast is here defined as the difference of the brightness of the two areas under consideration divided by their average brightness; e.g., see Young, 1989.)

B. Video Imaging

One major advantage of video (and other forms of electronic) imaging compared to photographic recording or direct visual observation is that video allows one to subtract electronically the unwanted background signal and raise image contrast. Whereas photochemical sensors such as the photographic emulsion and the eye respond to the log of the light intensity, photoelectric transducers can respond linearly to light. This linear response gives rise to the opportunity to subtract away or to reduce the (condition 6). [For an illustration of a stable microscope stand with flexible arrangements of the optical components, see Inoué (1986).]

B. Comments on Video and Image Processing Equipment

The following remarks relate to video image capture and digital image processing. Although special emphasis is placed on conditions that pertain to ultrathin optical sectioning with conventional light microscopy, some of the remarks [which include updates from those appearing in Inoué (1986)] may prove useful to those using video imaging for microscopy in general.

9. A clean instrumentation-grade video camera is necessary. The performance of video cameras employing solid state "chip" sensors, and intensifier cameras using microchannel plates and/or electrostatic intensifiers, have been significantly improved over the past few years. The number of picture elements in the chips is steadily increasing; chilled charge-coupled device (CCD) cameras with low noise levels and high intrascene dynamic range are becoming more accessible (see, e.g., Aikens *et al.*, 1989); and for color recording, new professional and consumer-oriented camcorders with improved resolution, sensitivity, and color fidelity continue to appear on the market, some at reasonable prices.

In contrast to these improvements in solid state video cameras, cameras with tube-type sensors have not shown as dramatic an improvement since the mid-1980s. Of the cameras that are still the mainstay for video microscopy, namely those equipped with Newvicon or silicon intensifier target (SIT) tubes (primarily owing to their greater sensitivity at video-rate image capture compared to the CCD cameras), we see improvements in the camera electronics that yield somewhat less noise and easier access to shading correction, and overall improved performance of SIT camera images, in part owing to better camera tubes.

Whether one uses solid state or tube-type video cameras, it is desirable to have a sensor with reasonably uniform sensitivity and a face plate free from dust particles and other optical defects. Dust particles can be removed by careful cleaning (see Inoué, 1986), but the presence of other optical defects such as interference patterns (from the antireflection coating, etc.) may not be apparent without an actual test of the camera on a microscope using the particular monochromatic illumination that is desired. Similarly, chicken wire-like patterns and other image defects introduced by intensifier elements may not become apparent until the camera is used for microscopy. These fixed image defects, unless correctable by reference frame subtraction (or for some applications, division by the reference frame), can prove to be particularly annoying in through-focal imaging of low-contrast scenes.

10. A number of digital image processors and analyzers are available today, or can be assembled by adding software to commercial video processing boards that plug into selected computer buses (see, e.g., Walter and Berns, 1986; Inoué, 1986). Some modestly priced processor/analyzers, such as the Argus-10 Image Processor from Hamamatsu Photonics (Bridgewater, NJ), the DSP processors from Dage-MTI, and the Image-1 from Universal Imaging (West Chester, PA) are particularly suited for online processing and analysis of video microscope images. The Image-1 system also serves as the central control unit for timed movements of the focusing stepper motors, and synchronized acquisition of video images onto a laser disk recorder, and so on, in the high-resolution, through-focal system described in Section V.

11. Quality video recorders are necessary. Optical memory disk recorders (OMDRs) and video cassette recorders (VCRs) have been steadily improving over the past several years. The OMDRs, which allow single-frame or video-rate recording as well as glitch-free playback in frozen frame, or at video rate, or slower or faster, are especially helpful for recording and analyzing through-focal images. The OMDRs, however, may intermittently show some synchronization problems and require the use of (full-frame) time-base correctors for optimum recording and playback. The memory capacity and price per memory of the computer hard disks, especially the magnetooptical disk units, have been improving exponentially and serve as excellent media for error-free storage of digitized images. Unfortunately, the data transfer rate for these hard disks is still too slow by an order of magnitude for video-rate image capture or playback.

Among the VCRs, tape format and recording format continue to change and become superseded by those with gradually improving performance. The ED-Beta (extended definition, not high-definition, Beta format from Sony), which uses standard NTSC signals on 1/2-in. wide tape, provides 500 TV lines of horizontal resolution, exceeding the 450-line resolution of the 3/4-in. U-matic formats used on high-resolution monochrome recorders. The stability and quality of the image in single-field playback mode has also been significantly improved. With the ability of the VCR to record 2hr of video without interruption (against the limited recording time of the OMDR), and the economical price of the tape compared to the comparatively high price of the OMDR disks, VCRs, such as those of the ED-Beta or super VHS format, serve well for continuous video-rate recording of through-focal images taken over extended periods of time.

12 and 13. Monitors, printers, and photographic recording: The quality of the final video image that is transmitted through a chain of video

equipment is a function of the product of the modulation transfer functions (MTFs) of all of the pieces of video equipment. Therefore, the equipment with the lowest MTF becomes the weak link in this chain. Conversely, if the budget permits, it is often advantageous to use equipment whose MTFs are considerably greater than those with the limiting, or average, MTF. The resolution and image quality provided by the monitor and printer, in addition to those of the camera and primary recorder, are cases in point. High-resolution monitors, such as the HR-2000 from Dage-MTI, that allow a wide range of contrast and black level adjustments and that have the capacity to synchronize well to VCR outputs, can provide microscope images, especially of ultrathin optical sections with excellent detail.

V. EXAMPLES OF THREE- AND FOUR-DIMENSIONAL IMAGING AND ANALYSIS WITH CONVENTIONAL LIGHT MICROSCOPY

Confocal microscopy is clearly the method of choice for acquiring serial optical sections of fluorescence or reflection contrast images to be reconstructed for three-dimensional (3D) imaging. However, for DIC, polarization, phase-contrast, and some other contrast-generating modes, conventional light microscopy is currently the method of choice. As described in Section IV, these contrast modes can generate exceptionally thin optical sections, and their full-field images can be captured at rates considerably faster than with confocal imaging.

The method for 3D reconstruction from serial optical sections in conventional light microscopy, except for the deconvolution methods used extensively in fluorescence microscopy (Agard *et al.*, 1989; Carrington *et al.*, 1990), shares much in common with that used in confocal microscopy (Brakenhoff *et al.*, 1989). In high-resolution, stereoscopic and dynamic 3D reconstruction of Golgi-stained neurons (Fig. 17.5), we first acquired 81 through-focal optical sections of Golgi-stained (silver-impregnated) neurons (with a rectified polarization system with analyzer off-crossed by $\sim 15°$) recorded at 0.5-μm steps with an NA 1.4 Plan Apo system combined with a matched NA condenser whose aperture was fully and uniformly illuminated with a 546-nm wavelength source. From the set of serial optical sections, we reconstructed aspect ratio-compensated plane projections for every 4° of rotated view angles (using an Image-I program, as described in Inoué, 1990b). Such calculated projections (whose background is remarkably devoid of shadows owing to the logic minimum algorithm used to select the darkest pixel in each stack) can then be played back as OMDR records and viewed in rapid sequence on a standard monitor to provide a parallax-based sense of rotation. The detailed 3D arrangements of minute structures in the image, such as the direction of extension, and the pattern of bending, of the 0.1-μm thick neuronal spine necks, are better revealed when two calculated projections, 8° apart, are viewed as stereo pairs. They show up even more dramatically when the stereo pairs calculated for sequential view angles (every 4° for a complete rotation of 360° around the vertical axis) were viewed in continuous sequences from an OMDR recording, thus yielding a rotating set of stereo images.

In principle, the stereo pairs can be viewed as red–green anaglyphs through a single, color video monitor or projector, or as side-by-side stereo pairs. In practice, we have found the StereoGraphics (San Rafael, CA) system to be considerably more effective, both for viewing the stereo image on a video monitor and for stereo projection for a large audience. In this system, the stereo pairs are recorded in each frame of standard video as vertically compressed top/bottom stacks. The horizontal resolution of the scene is retained, but the vertical resolution is sacrificed by a factor of two. For viewing, the top and bottom images are each expanded back to video frame height and played back alternately at 120 fields/sec. Synchronized with the field displayed, a "*Z* screen" superimposes a left-circular and right-circular polarizer to the alternating left and right images. Viewed through right-circular and left-circular polarizing glasses (Polaroid type-II stereo viewing glasses, Theatric Support Co., Hollywood, CA), the left and right eye each see the screen image for alternating 1/120-sec intervals.

This system not only yields stereoscopic images of dynamically changing scenes (in monochrome or in color, without as much deterioration of the vertical resolution as might be expected from the vertical compression), but also seems to provide stereoscopy successfully even to that significant fraction of the population having trouble gaining a 3D view by the stereoscopic parallax provided by anaglyphs, stereo-viewing glasses, or standard polarizing projection devices. The success of the StereoGraphics and other systems that effectively employ rapid left/right eye shuttering devices may arise from both of the eyes being forced to view one image at a time; hence the dominant eye does not elicit an inhibitory reaction from the other eye.

As noted in Section IV,A, we have developed a method for repeatedly acquiring high-resolution, through-focal optical sections of live specimens. The image is scanned axially over a specimen depth of 30 μm in just over 1 sec, or at a video-field image-capture rate of once every 0.6 μm. With the through-focal scan repeated once every 5 sec, playback of the video records allows us to follow four-dimensional events during early embryogenesis, such as pronuclear migration and fusion, mitotic chromo-

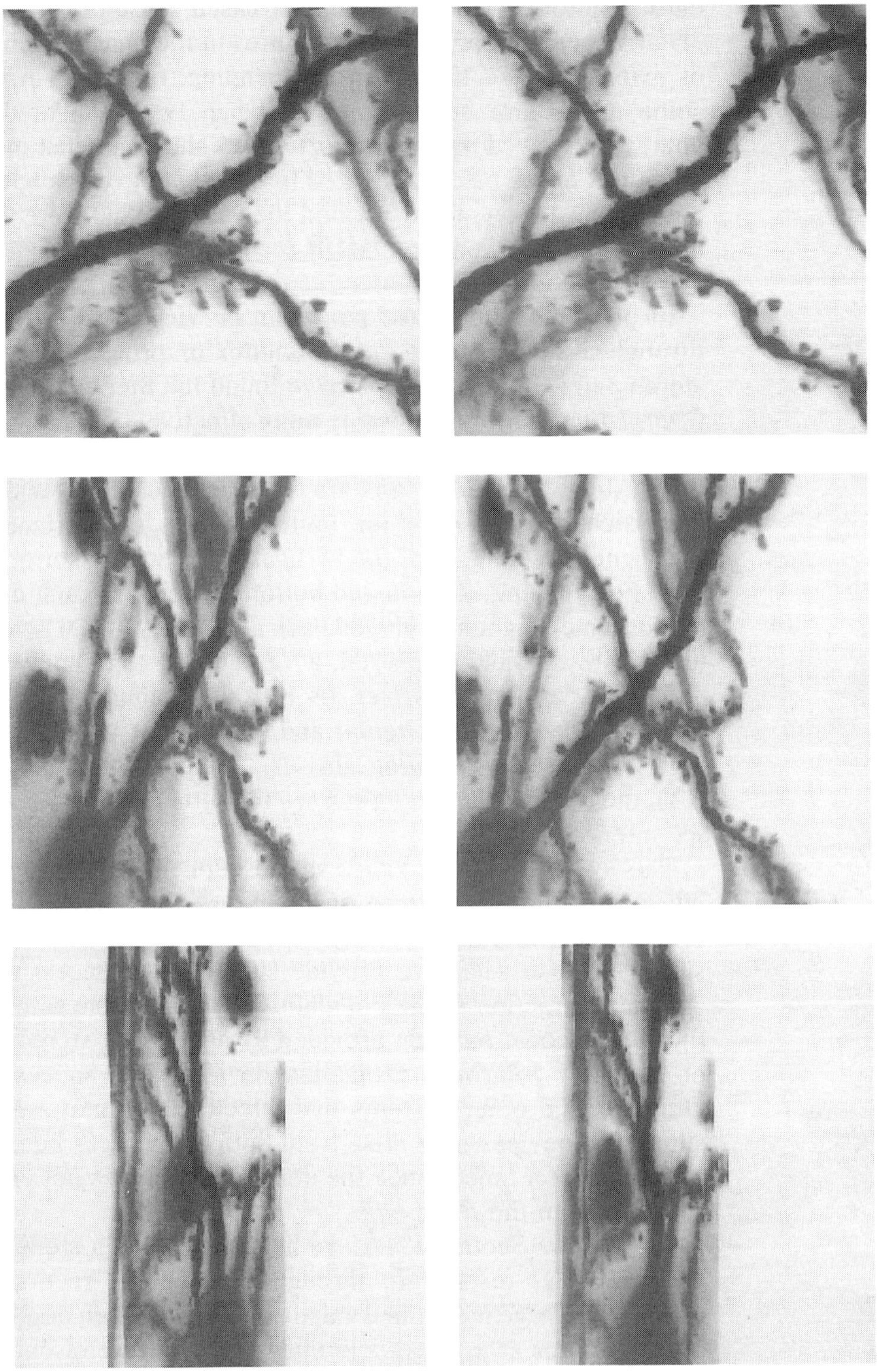

Fig. 17.5. Example of high-resolution stereo-pair images reconstructed from through-focal optical sections acquired without confocal optics. The 81 through-focal optical sections of the Golgi-stained neurons (from the thalamus of embryonic rat brain; microscope slide

some movements, karyomere formation, and restructuring of the nuclear envelope, in marine zygotes (Inoué *et al.*, 1991).

By taking advantage of the capability of the light microscope to provide serial, ultrathin optical sections without damage to the specimen, and the capacity of digital image processors to reconstruct three-dimensional or stereoscopic displays from the optical sections, we are now in a position to follow, with time, changes in the detailed three-dimensional architecture of developing embryos and functional cells and cell parts. Such displays should provide hitherto unforeseen views of the geometric relationships and dynamic functional interactions between tissues, cells, organelles, and molecular assemblies. Thus, we can anticipate exciting new images and insights of biological structure and function to emerge as fruits of four-dimensional light microscopy.

ACKNOWLEDGMENTS

The author thanks Dr. G. W. Ellis of the University of Pennsylvania and Dr. R. Oldenbourg of the Marine Biological Laboratory for the many helpful discussions and for providing original figures. Bob Knudson of the MBL provided much input

---

was a gift from S. Senft, Washington University) were acquired between ca. 15° off-crossed polarizers with a ×100/NA 1.35 planapochromatic objective lens combined with an NA 1.35 rectified aplanatic condenser (Nikon) whose back aperture was fully and uniformly illuminated [using a fiber optic light scrambler (Ellis, 1985); see Inoué, 1986] with 546-nm wavelength light. The digital image processor (Image-1/AT; Universal Imaging Corp.) acquired the video signal from a Newvicon camera (Hamamatsu Photonics), processed the signal on line (16-frame sum minus 16-frame summed background, digital contrast adjust), stored the image to hard disk, then incremented the objective lens and condenser-focusing stepper motors at 0.5 μm (i.e., 0.5 μm/1.5 = 0.33-μm optical path shift per step), repeating the sequence 80 times (see Inoué *et al.*, 1991). Using the same processor and the full stack of stored images for each view angle, plane projections were calculated (with the slice thickness arbitrarily chosen to be 2 pixels) for view angles of from 0 to 180° (as measured from the original viewing direction, i.e., the optical axis of the microscope) at 4° intervals. This choice of slice thickness led to the z compression seen in the 90° view, but even with this degree of compression the Z distance appears exaggerated for most other view angles when the stereo images were viewed on a large projection screen. In addition to calculating the X-axis geometric compression and displacements associated with change of view angle for each of the 525 × 480 ray pencils, the processor used a logic minimum algorithm to select the darkest pixel (which accounts for the unusually clear background of the final images; see Inoué, 1990b). Fine, three-dimensional detail of the neuronal spines that grow off the dendritic branches, including the cup shape of some of the spine heads and the L shape of their 0.1-μm diameter necks, are clearly visible in the projected, rotating stereo images. In the three representative stereo pairs shown here (arranged for cross-eyed viewing), the view angles used to calculate the top pair are 172 and 180°; for the middle pair they are 112 and 120°; for the bottom pair they are 88 and 92° (i.e., at right angles to the original viewing direction). Image height is 28 μm for each panel.

Jan. 16th, 1946

Dear Don:

Hello Don! No matter how many years I havn't written to you, still I have to start in the usual way. How are you? One day toward the end of last year, Lieutenant Coe dropped in, bringing your message. He came all the way from Tokyo in a jeep and must be quite tired, for he didn't stay long. But we gathered enough information about you and we were quite happy.

We got another friend here. Can you guss who it is? Edwin Helwig. But since he has been away from home for three years, the informations we can get through him are slightly stale. ....

In the last four years, we simply had lots of things. But ducking under bombs was not so bad. Rather it was a great excitement. Hide and seek at the expense of your life cann't help being exciting. There was, however, an awful side to it too.

During first three years of the war, our everyday life was not fundamentally affected and we could work regularly. Then, in January, a year ago from now, the Japanese navy took over the station and changed it to a temporary submarine base. As a result, the laboratory moved to a near-by village. Although, we succeeded in getting a wooden building built, we had a hard time to install sea-water pumps and so forth. When the war was over, the American troops occupied the station, for the reason that the Japanese navy had been there. When American officers came there for the first time to take over the place, I went there. It was a funny experience. On one side of a long table, three American officers were sitting and on the opposite side two Japanese navy officers and I were sitting. They served beer and canned asparagus with tomato ketchup. This slightly cute menu made me smile. But, oh boy, the both sides were pretty much excited. I am sure they were really scared of each other. They yelled whatever they wanted to say at the top of their voice but never listened to the other side. And an interpreter translated off and on, paying no attention whether it made sense or not. I was partly absorbed in watching the chaos and partly in the asparagus and was still partly absorbing the beer. Neither side understood the other. But to start with neither side knew what they were going to say. Toward the end, I was loudly laughing which nobody noticed. Somehow the meeting came to end. So I started to work. I stuck to a major and explained to him that this building originally belonged to the University etc., etc. In ten minutes, he began to see the situation. As soon as I saw the sign of dawning in his chaotic mind, I ran back to the building, wrote a psoter asking soldiers to take good care of the place because it is a research institute, pasted it on the wall and took leave from the back door leaving the noisy bunch there. There was no way to know what influence the poster excerted on the American soldiers who came there. But to my greatest surprise, in a copy of over-sea edition of Time issued in the beginning of December, I came across my own words all printed. Moreover, it had a t'tle "Appeal to Gothe".

....

From Katsuma

Reprinted from *Handbook of Biological Confocal Microscopy*, 2nd Edition, pp. 1–17, 1995, with permission from Springer Science and Business Media.

1

Foundations of Confocal Scanned Imaging in Light Microscopy

Shinya Inoué

Seldom has the introduction of a new instrument generated as instant an excitement among biologists as the laser-scanning confocal microscope. With the new microscope, one can slice incredibly clean thin optical sections out of thick fluorescent specimens; view specimens in planes tilted to, and even running parallel to, the line of sight; penetrate deep light-scattering tissues; gain impressive three-dimensional (3D) views at very high resolution; obtain differential interference or phase-contrast images in exact register with confocal fluorescence images; and improve the precision of microphotometry.

While the instruments that engendered such excitement mostly became commercially available first in 1987, the optical and electronic theory and the technology that led to this sudden emergence had been brewing for several decades. The development of this microscope stems from several roots, including light microscopy, confocal imaging, video and scanning microscopy, and coherent or laser-illuminated optics (see historic overview in Table 1). In this chapter, I will first discuss some basic principles relating to lateral and axial resolution as well as depth of field in light microscopy, highlight some history that lays a foundation to the development of laser-scanning confocal microscopy, and end with some general remarks regarding the new microscope.

LIGHT MICROSCOPY

Lateral Resolution*

The foundation of light microscopy was established a century ago by Ernst Abbe (1873, 1884). He demonstrated how the diffraction of light by the specimen, and by the objective lens, determined image resolution; defined the conditions needed to design a lens whose resolution was diffraction-limited (rather than limited by chromatic and spherical aberrations); and established the role of the objective and condenser numerical apertures (NA) on image resolution (Eq. 1). Thus,

$$d_{min} = \frac{\lambda_o}{NA_{obj} + NA_{cond}} \quad (1)$$

where d_{min} is the minimum spacing in a periodic grating that can just be resolved, d_{min} is expressed as lateral distance in the specimen space; λ_o is the wavelength of light in vacuum; and NA_{obj} and NA_{cond} are the numerical apertures of the objective and condenser lenses respectively. The *NA* is the product of the sine of the half-angle (α) of the cone of light either acceptable by the objective lens or emerging from the condenser lens and the refractive indexes (η) of the imbibing medium between the specimen and the objective or condenser lens, respectively.

Equation 1 demonstrates that, in addition to the wavelength and the NA of the objective lens, the condenser NA also affects image resolution in the microscope. For objects that are illuminated fully coherently (a condition that pertains when NA_{cond} approaches 0, namely when the condenser iris is closed down to a pinhole), the minimum resolvable spacing **increases (or the resolution decreases)** by a factor of two compared to the case when the condenser iris is opened so that $NA_{cond} = NA_{obj}$. As the condenser iris is opened up and NA_{cond} becomes larger, the illumination becomes progressively less coherent and resolution increases. [Note, however, that laser beams tend to illuminate objects coherently even when the condenser iris is not closed down (Chapter 5, *this volume*).]

Equation 1 describes the relation between NA and resolution for line grating objects. A complementary method of defining the limit of resolution uses point objects instead of line gratings. The image of an infinitely small luminous object point is itself not infinitely small, but is a circular Airy diffraction image with a central bright disk and progressively weaker concentric dark and bright rings. The radius r_{Airy} of the first dark ring around the central disk of the Airy diffraction image depends on λ and the NA of the objective:

$$r_{Airy} = 0.61 \frac{\lambda_o}{NA_{obj}} \quad (2)$$

where r_{Airy} is expressed as distance in the specimen plane.

When there exist two equally bright points of light separated by a small distance *d* in the specimen plane, their diffraction images lie side by side in the image plane. The images are said to be resolved if d is larger or equal to the radius of the Airy disk. This is the Rayleigh criterion, and it relies on the assumption that the two point sources radiate incoherently. If the two point sources emit light coherently, their amplitude rather than their intensity distri-

* For a more rigorous treatment of the optical principles and applications of light microscopy than is appropriate for this revised chapter and extensive discussions on modern microscope lens design and aberrations, refer to a complementary chapter in *Handbook of Optics* (Inoué and Oldenbourg, 1994).

Shinya Inoué • Marine Biological Laboratory, Woods Hole, Massachusetts 02543.

TABLE 1. Historic Overview[a]

Confocal Microscopy	Microscopy	Video (Microscopy)
	Abbe (1873, 1884)[a]	Nipkow (1884)
	Berek (1927)[d]	
	Zernicke (1935)[a,c]	Zworykin (1934)
	Gabor (1968)[a]	
	Hopkins (1951)	Flory (1951)
	Linfoot and Wolf (1953)	Young/Roberts (1951)
	3D diffraction by annul. apert.[a,d]	Flying spot[c]
	Tolardo di Francia (1955)	Montgomery *et al.* (1956)
	Limited field[b]	Flying spot UV[c]
	Nomarski (1955)[c]	
	Linfoot and Wolf (1956)	
	3-D diffraction pattern[a,d]	
	Ingelstam (1956)	
	Resolution and info. theory[a]	
Minsky Patent (1957)		
Insight[a–d]		
Stage scanning[f]	Kubota and Inoue (1959)[a]	
	Smith and Osterberg (1961)[a]	
	Harris (1964)[a,b]	Freed and Engle (1962)
		Flying spot[c]
	Ellis (1966)	
Petráň *et al.* (1968)	Holomicrography	
Tandem scanning[d]		
Davidovits and Egger (1971)		
Laser illumination		
Lens scan[d]	Hoffman and Gross (1975)	
	Modulation contrast[c]	
Sheppard and Choudhury (1977)		
Theory[a,b,d]		
Sheppard *et al.* (1978)	Ellis (1978)	
Stage scanning	Single sideband edge enhancement microscopy[c]	
Cremer and Cremer (1978)		
Auto-focus stage scanning,[d]		
"4-π-point illumination"		
Brakenhoff *et al.* (1979)		Castleman (1979)
Specimen scan[e](1985)		Digital image processing[a,c,e]
Koester (1980)	Quate (1980)	
Scanning mirror[d]	Acoustic microscopy[a,c]	
		Inoué (1981)[a,c]
		Allen *et al.* (1981a,b)[a,c]
		Fuchs *et al.* (1982)[f]
		Agard and Sedat (1983)[a,c,d]
Boyde (1985a)		
Nipkow type[d,e]		
Cox and Sheppard (1983)		
Digital recording[d,e]		
Åslund *et al.* (1983)		
2-mirror laser scanning[d]		
Hamilton *et al.* (1984)		
Differential phase[c,d]		
Wilson and Sheppard (1984)		Sher and Barry (1985)[f]
Extended depth of field[c–f]		
Carlsson *et al.* (1985)		Inoué (1986)
Laser scan, Stacks of confocal images[d,e]		Overview, how to[a,c,e,f]
Wijnaendts van Resandt *et al.* (1985)	Ellis (1985)	Fay *et al.* (1985)[a,c,d]
xz-view[d,e]	Light scrambler[d]	
Suzuki-Horikawa (1986)	Cox and Sheppard (1986)[b]	Castleman (1987)[a]
Video rate laser scan		
Acousto optical modulator		
No exit pinhole		
Xiao and Kino (1987)		

TABLE 1. (*Continued*)

Nipkow type[d]	
McCarthy and Walker (1988)	
Nipkow type[d]	
Amos *et al.* (1987)[c–e]	Ellis (1988)
	Scanned aperture phase contrast[a,c,d]

[a]Diffraction theory.
[b]Superresolution.
[c]Contrast modes.
[d]Optical sectioning/depth of field.
[e]Stereo.
[f]3-D in objective space.

bution in the image must be considered, and resolution generally decreases. The impact of the quality and NA of the condenser on the resolution of point objects was considered by Hopkins and Barham (1950). Their results are similar to, but not strictly identical with, the case of line-grating objects (see Born and Wolf, 1980).

It is important to realize that these resolution criteria apply only to objective lenses used under conditions in which the image is free from significant aberrations (see Chapters 6, 7, 8, 9, and 11, *this volume*; Chapter 5 of Inoué, 1986; Inoué and Oldenbourg, 1994). This implies several things:

- A well-corrected clean objective lens is used within the wavelengths of light and diameter of field for which the lens was designed (commonly in conjunction with specific oculars and/or tube lenses).
- The refractive index, dispersion, and thickness of the immersion media and coverslip are those specified for the particular objective lens.
- The correct tube length and ancillary optics are used and the optics are axially aligned.
- The full complement of image-forming rays and light waves leaving all points of the objective aperture converge to the image plane without obstruction.
- The condenser aperture is homogeneously and fully illuminated.
- The condenser is properly focused to produce Kohler illumination.

These considerations for resolution assume that the object is viewed in conventional wide-field (WF) microscopy. When the (instantaneous) field of view becomes extremely small, as in confocal microscopy, the resolution can in fact be greater than when the field of view is not so limited. We shall return to this point later.

Axial Resolution

We now turn to the axial (z-axis) resolution, measured along the optical axis of the microscope, i.e., perpendicular to the plane of focus.

To define axial resolution, it is customary to use the 3D diffraction image of a point source that is formed near the focal plane. In the case of lateral resolution, i.e., the resolution in the plane of focus, the Rayleigh criterion makes use of the in-focus diffraction images (the central cross section of the 3D pattern) of two point sources and the minimum distance that they can approach each other laterally, yet still be distinguished as two. Similarly, axial resolution can be defined by the minimum distance that the diffraction images of two points can approach each other along the axis of the microscope, yet still be seen as two. To define this minimum distance, we use again the diffraction image of an infinitely small point object and ask for the location of the first minimum along the axis of the microscope.

The precise distribution of energy in the image-forming light above and below focus, especially for high NA objective lenses, cannot be deduced by geometric ray tracing but must be derived from wave optics. The wave optical studies of Linfoot and Wolf (1956) show that the image of a point source produced by a diffraction-limited optical system (e.g., a well-designed and properly used light microscope) is not only periodic around the point of focus in the focal plane, but is also periodic above and below the focal plane along the axis of the microscope. [Such 3D diffraction images (including those produced in the presence of lens aberrations) are presented photographically by Cagnet *et al.* (1962); also see Chapter 7 *this volume*, Fig. 4). The intensity distribution calculated by Linfoot and Wolf for an aberration-free system is reproduced in Born and Wolf (1980) and also in Inoué (1986, Figs. 5–21). The 3D pattern of a point source formed by a lens possessing an annular aperture was calculated by Linfoot and Wolf (1953).

The distance from the center of the 3D diffraction pattern to the first axial minimum (in object space dimensions) is given by

$$z_{\min} = \frac{2\lambda_o \eta}{(NA_{\text{obj}})^2} \qquad \textbf{(3)}$$

where η is the refractive index of the object medium. $z_{\min}$ corresponds to the distance by which we have to raise the microscope objective in order to focus (onto the small detector pinhole) the first intensity minimum observed along the axis of the 3D diffraction pattern instead of the central maximum.

As with the lateral resolution limit, we can use $z_{\min}$ as a measure of the limit of axial resolution of the microscope optics. Note, however, that $z_{\min}$ shrinks inversely proportionally with the *square* of the NA_{obj}, in contrast to the lateral resolution limit which shrinks with the first power of the NA_{obj}. Thus, the ratio of axial-to-lateral resolution ($z_{\min}/r_{\text{Airy}} = 3.28\eta/NA_{\text{obj}}$) is substantially larger than one and is inversely proportional to the NA of the objective.

instrument, a confocal microscope is especially susceptible to vibration and problems of mechanical distortion. Indeed, once the complex optical, electro-optical, mechanical, and electronic systems have been appropriately designed, the success of one confocal instrument over another may well depend on its immunity to vibration, in addition to the friendliness of its user interface.

CONTRAST TRANSFER AND RESOLUTION IN CONFOCAL VS. NONCONFOCAL MICROSCOPY

In addition to designing and successfully demonstrating the power of stage-scanning confocal microscopy, Wilson and colleagues have extensively analyzed the theoretical foundations of confocal microscope imaging (Wilson and Sheppard, 1984; Wilson, 1990; Chapter 11, *this volume*). Their mathematical treatment leads to the somewhat surprising conclusion that the ultimate limit of resolution (i.e., the cutoff spatial frequency or spacing at which image contrast of periodic objects drops to zero) obtainable with coherent confocal microscopy is identical with that for incoherent nonconfocal microscopy. However, compared to incoherent, nonconfocal optics, image contrast should rise much more sharply with coherent confocal optics as the spatial period is increased. Therefore, the practical resolution attained at threshold contrast (i.e., the minimum contrast required for the spacing to, in fact, be detected) was expected to be significantly greater with confocal optics than with conventional nonconfocal optics.

We have recently confirmed these predictions by direct measurement on test gratings that we fabricated by electron lithography with spacings down to 0.1 μm. Indeed, with confocal optics equipped with a small exit pinhole, the contrast transfer efficiency rose to 80% at twice (and reached 100% at three times) the cutoff spacing. With incoherent nonconfocal imaging, 80% contrast transfer was not attained until four times and did not even reach the 100% transfer rate at eight times the cutoff spacing (Fig. 7 from Oldenbourg *et al.*, 1993)!

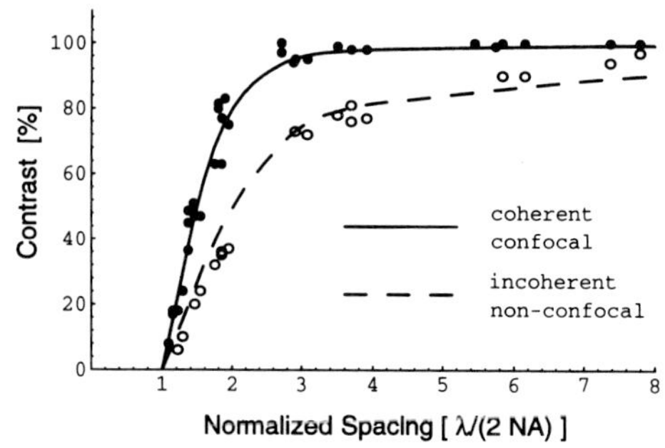

FIGURE 7. Experimental contrast transfer values measured as a function of the spatial period of line gratings using a laser beam-scanning microscope in the confocal reflection mode (solid points) and in the nonconfocal transmission mode (circles). In both imaging modes, Plan Apo objective lenses (Nikon Inc., Melville, NY) with numerical apertures (NAs) ranging from 0.45 to 1.4 and laser wavelengths (λ) of 514.5 or 488 nm were used. Spatial periods are expressed in units of the limiting wavelength, λ/2(NA), to normalize the data taken with different laser λ and lenses of different NA. We call data presented in this fashion the contrast transfer characteristic (CTC).

Continuous and broken lines are theoretical curves displaying calculated CTCs for the coherent confocal and the incoherent, nonconfocal imaging mode. Comparison of the two CTC curves shows that, while the limiting resolutions are identical for both imaging modes, the contrast due to fine detail in the specimen is maintained much better with confocal optics. (The microscope was a prototype built by Hamamatsu Photonics Kabushiki Kaisha, Japan. It was used with the detector pinhole diameter reduced to a small fraction of the Airy disk diameter. (From Oldenbourg *et al.*, 1993.)

SUMMARY

- The limiting resolution of all microscopes depends on the λ of the light used and the NA of the objective and condenser lenses. Dirty or misaligned optics or vibration, or both, can reduce the achieved resolution. Test resolution regularly, and especially pay attention to the iris setting and full illumination of the condenser aperture, to assure optimal performance.
- A small detector pinhole in the confocal microscope is essential if the maximum optical sectioning capability and resolution of the instrument are to be realized concurrently. Correct alignment and use of this control is very important. However, a larger pinhole may be required to get more signal when there is limited light level, motion in the specimen, or fading of fluorescence.
- By opening the confocal exit pinhole (to not much greater than the Airy disk diameter), one loses the advantage of higher resolution in fluorescence microscopy but retains much of the capability for rejecting out-of-focus information while also gaining in the fluorescence signal.
- The ultimate resolution of a CLSM in the reflection mode is the same as that of a conventional light microscope, but the contrast that it produces from features is higher. For fluorescence imaging, the resolution in a confocal microscope can be ~$\sqrt{2}$ greater than with conventional microscopy, but only *if* the confocal detector pinhole is appreciably smaller than the Airy disk produced by a point fluorescent object.
- CLSM is not a cure-all for all biological studies. Its sampling speed is limited, and it does not lend itself to using either interference effects to produce contrast, such as phase or DIC, which have been found to be relatively innocuous to living cells, or polarization contrast that can reveal fine structural dynamics noninvasively.
- Video microscopy can take advantage of the various types of interference and polarizing contrast not easily implemented in the CLSM. Therefore, it is ideal for dynamic, nonfluorescence, high-resolution observations of living specimens and the behavior of macromolecular assemblies.
- Holographic microscopy is a field with intriguing promise that is so far beset by practical difficulties.
- The CLSM is the most efficient way of obtaining clear optical sections of the distribution of fluorescence in 2D planes of a fluorescent 3D specimen.

ACKNOWLEDGMENT

This chapter is an updated version of the chapter in the first edition. Aside from numerous small changes, it includes a completely reworked section on light microscopy and a new section on Contrast transfer and resolution comparing confocal vs. nonconfocal microscopy. For both sections, I am indebted to Dr. Rudolf Oldenbourg of the Marine Biological Laboratory, Woods Hole, Mass., for his most helpful input.

I would also like to thank the following individuals for their valuable comments during the preparation of this article: Drs. Gordon W. Ellis and John Murray, University of Pennsylvania; Dr. Stephen Smith, Stanford University School of Medicine; Dr. Jeff Lichtman, Washington University; Dr. Sture Wahlsten, Sarastro, Inc.; and Dr. James Pawley, University of Wisconsin; and also Dr. H. Kapitza of Carl Zeiss, Oberkochen, and Dr. N. Åslund of the Royal Institute of Technology in Sweden for kindly permitting me to use the figures provided by them.

The preparation of this chapter was supported in part by NIH grant R37 GM 31617–13 and NSF grant MCB-8908169.

REFERENCES

Abbe, E., 1873, Beiträge zur Theorie des Mikroskops und der mikroskopischen Wahrnehmung, *Schultzes Arc. f. Mikr. Anat.* 9:413–468.

Abbe, E., 1884, Note on the proper definition of the amplifying power of a lens or a lens-system, *J. R. Microsc. Soc.* 4:348–351.

Agard, D.A., and Sedat, J.W., 1983, Three dimensional architecture of a polytene nucleus, *Nature* 302:676–681.

Agard, D.A., Hiraoka, Y., Shaw, P., and Sedat, J.W., 1989, Fluorescence microscopy in three dimensions, *Methods Cell Biol.* 30:353–377.

Allen, R.D., 1985, New observations on cell architecture and dynamics by video-enhanced contrast optical microscopy. *Annu. Rev. Biophys. Biophysical Chem.* 14:265–290.

Allen, R.D., Travis, J.L., Allen, N.S., and Yilmaz, H., 1981a, Video-enhanced contrast polarization (AVEC-POL) microscopy: A new method applied to the detection of birefringence in the motile reticulopodial network of *Allogromia laticollaris*, *Cell Motil.* 1:275–289.

Allen, R.D., Allen, N.S., and Travis, J.L., 1981b, Video-enhanced contrast, differential interference contrast (AVEC-DIC) microscopy: A new method capable of analyzing microtubule-related motility in the reticulopodial network of *Allogromia laticollaris*, *Cell Motil.* 1:291–302.

Amos, W.B., White, J.G., and Fordham, M., 1987, Use of confocal imaging in the study of biological structures, *Appl. Opt.* 26:3239–3243.

Åslund, N., Carlsson, K., Liljeborg, A., and Majlof, L., 1983, PHOIBOS, a microscope scanner designed for micro-fluorometric applications, using laser induced fluorescence. In: *Proceedings of the Third Scandanavian Conference on Image Analysis*, *Studentliteratur*, Lund, p. 338.

Åslund, N., Liljeborg, A., Forsgren, P.-O., and Wahlsten, S., 1987, Three dimensional digital microscopy using the PHOIBOS scanner, *Scanning* 9:227–235.

Baxes, G.A., 1984, *Digital Image Processing: A Practical Primer*, Prentice-Hall, Englewood Cliffs, New Jersey.

Berek, M., 1927, Grundlagen der Tiefenwahrnehmung im Mikroskop, *Marburg Sitzungs Ber.* 62:189–223.

Born, M., and Wolf, E., 1980, *Principles of Optics*, 6th ed., Pergamon Press, Oxford, England.

Boyde, A., 1985a, Tandem scanning reflected light microscopy (TSRLM). Part 2: Pre-MICRO 84 applications at UCL, *Proc. R. Microsc. Soc.* 20:131–139.

Boyde, A., 1985b, Stereoscopic images in confocal (tandem scanning) microscopy, *Science* 230:1270–1272.

Boyde, A., 1987, Colour-coded stereo images from the tandem scanning reflected light microscope (TSRLM), *J. Microsc.* 146:137–142.

Brakenhoff, G.J., Blom, P., and Barends, P., 1979, Confocal scanning light microscopy with high aperture immersion lenses, *J. Micros.* 117:219–232.

Brakenhoff, G.J., van der Voort, H.T.M., van Spronsen, E.A., Linnemans, W.A.M., and Nanninga, N., 1985, Three dimensional chromatin distribution in neuroblastoma nuclei shown by confocal scanning laser microscopy, *Nature* 317:748–749.

Brakenhoff, G.J., van der Voort, H.T.M., van Spronsen, E.A., and Nanninga, N., 1986, Three dimensional imaging by confocal scanning fluorescence microscopy. In: *Recent Advances in Electron and Light Optical Imaging in Biology and Medicine*, Vol. 483 (A. Somlyo, ed.), Ann. N.Y. Acad. Sci., New York, pp. 405–414.

Brakenhoff, G.J., van Spronsen, E.A., van der Voort, H.T.M., and Nanninga, N., 1989, Three dimensional confocal fluorescence microscopy, *Methods Cell Biol.* 30:379–398.

Bright, G.R., Fisher, G.W., Rogowska, J., and Taylor, D.L., 1989, Fluorescence ratio imaging microscopy, *Methods Cell Biol.* 30:157–192.

Cagnet, M., Françon, M., and Thrierr, J.C., 1962, *Atlas of Optical Phenomena*, Springer-Verlag, Berlin.

Carlsson, K., Danielsson, P., Lenz, R., Liljeborg, A., Majlof, L., and Åslund, N., 1985, Three-dimensional microscopy using a confocal laser scanning microscope, *Opt. Lett.* 10:53–55.

Castleman, K.R., 1979, *Digital Image Processing*, Prentice-Hall, Englewood Cliffs, New Jersey.

Castleman, K.R., 1987, Spatial and photometric resolution and calibration requirements for cell image analysis instruments, *Appl. Opt.* 26:3338–3342.

Castleman, K.R., 1993, Resolution and sampling requirements for digital image processing, analysis, and display. In: *Electronic Light Microscopy* (D. Shotton, ed.), Wiley-Liss, New York, pp. 71–94.

Cox, I.J., and Sheppard, C.J.R., 1983, Scanning optical microscope incorporating a digital framestore and microcomputer, *Appl. Opt.* 22:1474–1478.

Cox, I.J., and Sheppard, C.J.R., 1986, Information capacity and resolution in an optical system, *J. Opt. Soc. Am.* 3:1152–1158.

Cox, G., and Sheppard, C., 1993, Effects of image deconvolution on optical sectioning in conventional and confocal microscopes, *Bioimaging* 1:82–95.

Cremer, C., and Cremer, T. 1978, Considerations on a laser-scanning-microscope with high resolution and depth of field, *Microsc. Acta* 81:31–44.

Davidovits, P., and Egger, M.D., 1971, Scanning laser microscope for biological investigations, *Appl. Opt.* 10:1615–1619.

Davidovits, P., and Egger, M.D., 1972, U.S. Patent #3,643,015, Scanning Optical Microscope.

Egger, M.D., 1989, The development of confocal microscopy, *Trends Neurosci.* 12:11.

Egger, M.D., and Petráň, M., 1967, New reflected-light microscope for viewing unstained brain and ganglion cells, *Science* 157:305–307.

Ellis, G.W., 1966, Holomicrography: Transformation of image during reconstruction a posteriori, *Science* 154:1195–1196.

Ellis, G.W. 1978, Advances in visualization of mitosis *in vivo*. In: *Cell Reproduction, in Honor of Daniel Mazia* (E. Dirksen, D. Prescott, and C.F. Fox, eds.), Academic Press, San Diego, pp. 465–476.

Ellis, G.W., 1979, A fiber-optic phase-randomizer for microscope illumination by laser, *J. Cell Biol.* 83:303a.

Ellis, G.W., 1985, Microscope illuminator with fiber optic source integrator, *J. Cell Biol.* 101:83a.

Ellis, G.W., 1988, Scanned aperture light microscopy. In: *Proceedings of the Forty-Sixth Annual Meeting of EMSA*, San Francisco Press, San Francisco, pp. 48–49.

Fay, F.S., Fogarty, K.E., and Coggins, J.M., 1985, Analysis of molecular distribution in single cells using a digital imaging microscope. In: *Optical Methods in Cell Physiology* (P. De Weer and B.M. Salzberg, eds.), John Wiley & Sons, New York.

Flory, L.E., 1951, The television microscope, *Cold Spring Harbor Symp. Quant. Biol.* 16:505–509.

Freed, J.J. and Engle, J.L., 1962, Development of the vibrating-mirror flying spot microscope for ultraviolet spectrophotometry, *Ann. N.Y. Acad. Sci.* 97:412–448.

Fuchs, H., Pizer, S.M., Heinz, E.R., Bloomberg, S.H., Tsai, L-C., and Strickland, D.C., 1982, Design and image editing with a space-filling 3D display based on a standard raster graphics system, *Proc. Soc. Photo. Opt. Instrum. Eng.* 367:117–127.

Gabor, D., 1948, A new microscope principle, *Nature* 161:777–778.

Goldstein, S., 1989. In: *Digitized Video Microscopy* (B. Herman and K. Jacobson, eds.), Alan R. Liss, New York.

Gonzales, R.C., and Wintz, P., 1987, *Digital Image Processing*, 2nd ed., Addison-Wesley, Reading, Massachusetts.

Hamilton, D.K., and Wilson, T., 1984, Two dimensional phase imaging in the scanning optical microscope, *Appl. Opt.* 23:348–352.

Hansen, E.W., 1986, Appendex II. In: *Video Microscopy* (S. Inoué, ed.), Plenum Press, New York, pp. 467–475.

Hard, R., Zeh, R., and Allen, R.D., 1977, Phase-randomized laser illumination for microscopy, *J. Cell Sci.* 23:335–343.

Harris, J.L., 1964, Diffraction and resolving power, *J. Opt. Soc. Am.* 54:931–936.

Hecht, E., 1987, *Optics*, 2nd ed., Addison-Wesley, Reading, Massachusetts.

Hell, S., Reiner, G., Cremer, C., and Stelzer, E.H.K., 1993, Aberrations in confocal fluorescence microscopy induced by mismatches in refractive index, *J. Microsc.* 169:391–405.

Hoffman, R., and Gross, L., 1975, Modulation contrast microscopy, *Appl. Opt.* 14:1169–1176.

Hopkins, H.H., 1951, The concept of partial coherence in optics, *Proc. R. Soc. Lond.* 208A:263.

Hopkins, H.H., and Barham, P.M., 1950, The influence of the condenser on microscopic resolution, *Proc. R. Soc. Lond.* 63B:737–744.

Ingelstam, E., 1956, Different forms of optical information and some interrelations between them. In: *Problems in Contemporary Optics*, Istituto Nazionale di Ottica, Arcetri-Firenze, pp. 128–143.

Inoué, S., 1981, Video image processing greatly enhances contrast, quality, and speed in polarization-based microscopy, *J. Cell Biol.* 89:346–356.

Inoué, S., 1986, *Video Microscopy*, Plenum Press, New York.

Inoué, S., 1988, Progress in video microscopy, *Cell Motil. Cytoskel.* 10:13–17.

Inoué, S., 1989a, Imaging of unresolved objects, supperresolution, and precision of distance measurement, with video microscopy, *Methods Cell Biol.* 30:85–112.

Inoué, S., 1989b, Video enhancement and image processing in light microscopy. Part I: Video microscopy. Part II: Digital image processing, *American Laboratory* (April 1989), pp. 52–70.

Inoué, S., 1989c, Whither video microscopy? Towards 4D imaging at the highest resolution of the light microscope. In: *Digitized Video Microscopy* (B. Herman and K. Jacobson, eds.), Alan R. Liss, New York.

Inoué, S., 1994, Ultra-thin optical sectioning and dynamic volume investigation with conventional light microscopy. In: *Three-Dimensional Confocal Microscopy* (J.K. Stevens, L.R. Mills, and J. Trogadis, eds.), Academic Press, San Diego, pp. 397–419.

Inoué, S., and Inoué, T.D., 1986, Computer-aided stereoscopic video reconstruction and serial display from high-resolution light-microscope optical sections. In: *Recent Advances in Electron and Light Optical Imaging in Biology and Medicine* (A. Somlyo, ed.), Vol. 483, Ann. N.Y. Acad. Sci., New York, pp. 392–404.

Inoué, S., and Oldenbourg, R., 1994, Optical instruments: Microscopes. In: *Handbook of Optics* (M. Bass, ed.), 2nd ed., Vol. 2, McGraw-Hill, New York, Chapter 17.

Koester, C.J., 1980, Scanning mirror microscope with optical sectioning characteristics: Applications in ophthalmology, *Appl. Opt.* 19:1749–1757.

Kubota, H., and Inoué, S., 1959, Diffraction images in the polarizing microscope, *J. Opt. Soc. Am.* 49:191–198.

Leith, E.N., and Upatnieks, J., 1963, Wavefront reconstruction with continuous-tone objects, *J. Opt. Soc. Am.* 53:1377–1381.

Leith, E.N., and Upatnieks, J., 1964, Wavefront reconstruction with diffused illumination and 3D objects, *J. Opt. Soc. Am.* 54:1295–1301.

Lewin, R., 1985, New horizons for light microscopy, *Science* 230:1258–1262.

Linfoot, E.H., and Wolf, E., 1953, Diffraction images in systems with an annular aperture, *Proc. Phys. Soc.* B 66:145–149.

Linfoot, E.H., and Wolf, E., 1956, Phase distribution near focus in an aberration-free diffraction image, *Proc. Phys. Soc.* B 69:823–832.

McCarthy, J.J., and Walker, J.S., 1988, Scanning confocal optical microscopy, *EMSA Bull.* 18:75–79.

Minsky, M., 1957, U.S. Patent #3013467, Microscopy Apparatus.

Minsky, M., 1988, Memoir on inventing the confocal scanning microscope, *Scanning* 10:128–138.

Montgomery, P.O., Roberts, F., and Bonner, W., 1956, The flying-spot monochromatic ultra-violet television microscope, *Nature* 177:1172.

Nipkow, P., 1884, German Patent #30,105.

Nomarski, G., 1955, Microinterférométre différentiel à ondes polarisées, *J. Phys. Radium* 16:S9–S13.

Oldenbourg, R., Terada, H., Tiberio, R., and Inoué, S., 1993, Image sharpness and contrast transfer in coherent confocal microscopy. *J. Microsc.* 172:31–39.

Petráň, M., Hadravsky, M., Egger, D., and Galambos, R., 1968, Tandem-scanning reflected-light microscope, *J. Opt. Soc. Am.* 58:661–664.

Quate, C.F., 1980, Microwaves, acoustic and scanning microscopy. In: *Scanned Image Microscopy* (E.A. Ash, ed.), Academic Press, San Diego, pp. 23–55.

Schotten, D., ed., 1993, *Electronic light microscopy: The Principles and Practice of Video-enhanced Contrast, Digital Intensified Fluorescence, and Confocal Scanning Light Microscopy*, John Wiley & Sons, New York.

Sharnoff, M., Brehm, L., and Henry, R., 1986, Dynamic structures through microdifferential holography, *Biophys. J.* 49:281–291.

Sheppard, C.J.R., and Choudhury, A., 1977, Image formation in the scanning microscope, *Optica* 24:1051.

Sheppard, C.J.R., Gannaway, J.N., Walsh, D., and Wilson, T., 1978, *Scanning Optical Microscope for the Inspection of Electronic Devices*, Microcircuit Engineering Conference, Cambridge.

Sher, L.D., and Barry, C.D., 1985, The use of an oscillating mirror for 3D displays. In: *New Methodologies in Studies of Protein Configuration* (T.T. Wu, ed.), Van Nostrand-Reinhold, Princeton, New Jersey.

Shimizu, Y., and Takenaka, H., 1994, Microscope objective design. In: *Advances in Optical and Electron Microscopy* (C. Sheppard and T. Mulvey, eds.), Academic Press, San Diego, Vol. 14, pp. 249–334.

Smith, L.W., and Osterberg, H., 1961, Diffraction images of circular self-radiant disks, *J. Opt. Soc. Am.* 51:412–414.

Stevens, J.K., Mills, L.R., and Trogadis, J., 1994, *Three-Dimensional Confocal Microscopy*, Academic Press, San Diego.

Streibl, N., 1985, Three dimensional imaging by a microscope, *J. Opt. Soc. Am.* A 2:121–127.

Suzuki, T., and Hirokawa, Y., 1986, Development of a real-time scanning lasermicroscope for biological use, *Appl. Opt.* 25:4115–4121.

Tanasugarn, L., McNeil, P., Reynolds, G.T., and Taylor, D.L., 1984, Microspectrofluorometry by digital image processing: Measurement of cytoplasmic pH, *J. Cell Biol.* 89:717–724.

Tolardo di Francia, G., 1955, Resolving power and information, *J. Opt. Soc. Am.* 45:497–501.

Tsien, R.Y., 1989, Fluorescent indicators of ion concentration, *Methods Cell Biol.* 30:127–156.

White, J.G., Amos, W.B., and Fordham, M., 1987, An evaluation of confocal versus conventional imaging of biological structures by fluorescence light microscopy, *J. Cell Biol.* 105:41–48.

Wijnaendts van Resandt, R.W., Marsman, H.J.B., Kaplan, R., Davoust, J., Stelzer, E.H.K., and Strickler, R., 1985, Optical fluorescence microscopy in three dimensions: Microtomoscopy, *J. Microsc.* 138:29–34.

Wilke, V., Gödecke, U., and Seidel, P., 1983, Laser-scan-mikroskop, *Laser Optoelecktron.* 15:93–101.

Wilson, T., 1985, Scanning optical microscopy, *Scanning* 7:79–87.

Wilson, T., 1990, *Confocal Microscopy*, Academic Press, London.

Wilson, T., and Sheppard, C., 1984, *Theory and Practice of Scanning Optical Microscopy*, Academic Press, London.

Wilson, T., Gannaway, J.N., and Johnson, P., 1980, A scanning optical microscope for the inspection of semiconductor materials and devices, *J. Microsc.* 118:390–314.

Xiao, G.Q., and Kino, G.S., 1987, A real-time confocal scanning optical microscope. In: *Proc. SPIE*, Vol. 809, *Scanning Imaging Technology* (T. Wilson and L. Balk, eds.), pp. 107–113.

Young, J.Z., and Roberts, F., 1951, A flying-spot microscope, *Nature* 167:231.

Zernicke, V.F., 1935, Das Phasenkontrastverfahren bei der mikroskopischen Beobachtung, *Z. Tech. Phys.* 16:454–457.

Zworykin, V.K., 1934, The iconoscope—a modern version of the electric eye, *Proc. IRE* 22:16–32.

Reprinted from *Molecular Biology of the Cell*, Vol. 6, pp. 1619–1640, 1995,
with permission from ASCB.

Molecular Biology of the Cell
Vol. 6, 1619–1640, December 1995

Essay

Force Generation by Microtubule Assembly/Disassembly in Mitosis and Related Movements

Shinya Inoué*† and Edward D. Salmon‡

*Marine Biological Laboratory, Woods Hole, Massachusetts 02543; and ‡Department of Biology, University of North Carolina, Chapel Hill, North Carolina 27599

Submitted June 2, 1995; Accepted August 30, 1995
Monitoring Editor: Thomas D. Pollard

In this article, we review the dynamic nature of the filaments (microtubules) that make up the labile fibers of the mitotic spindle and asters, we discuss the roles that assembly and disassembly of microtubules play in mitosis, and we consider how such assembling and disassembling polymer filaments can generate forces that are utilized by the living cell in mitosis and related movements.

1. EARLY HISTORY: THE DYNAMIC EQUILIBRIUM MODEL

The orderly segregation of chromosomes at every cell division, and the placement of the resulting daughter nuclei into appropriate cytoplasmic environments, are essential for the normal development of an organism, the generation of functional tissues and gametes, and indeed for the continuity of life itself. Such bipolar segregation of chromosomes in mitosis, and the movement of centrosomes that position the daughter nuclei into appropriately partitioned regions of the cytoplasm in animal cells are accomplished by a bipolar mitotic spindle and its associated astral rays.

Although the significance of the spindle and astral rays for mitosis and for the coordination of mitosis with cell cleavage were well recognized by the early cytologists, much controversy abounded in the first half of this century regarding the actual nature of the mitotic spindle fibers (Wilson, 1928; Schrader, 1953). Although visible in cells exposed to acidic or protein-precipitating fixatives, the fibers were not visible by bright field or phase contrast microscopy in most living cells and were absent in cells observed by light or electron microscopy after fixation with what were considered to be better structure-preserving reagents. There were indications that the fibers would disappear reversibly in cells treated with low temperature or with ethyl ether, and that they were so labile that chromosomes having to traverse laterally to accomplish their anaphase separation could even cut right through the fibers (Östergren, 1949). Yet without the bipolar fibrous organization, what would move or guide the chromosomes to the spindle poles?

The reality in living cells of spindle fibers, and the fibrils that as a bundle made up the fibers, were established by Inoué by observing a variety of living, animal and plant cells in division with a sensitive polarized light microscope (Inoué, 1951–1953, 1964). The birefringence of the fibers measured and photographed through the sensitive polarizing microscope depicted the distribution, concentration, and appearance and disappearance of oriented fibrils in the spindle fibers in living cells. In addition to the dynamic formation, fluctuation, and disappearance of these fibers and fibrils during the normal course of cell division, the birefringence revealed the labile nature of the fibrous spindle elements in cells exposed to cold or to a mitosis-inhibiting alkaloid, colchicine (Inoué, 1952, 1964). From these findings, Inoué postulated that the spindle fibers and fibrils were made up of a loosely coupled, linear chain of protein molecules, which were in a temperature-sensitive dynamic equilibrium with their pool of subunits that make up the chain. The assembly was entropy driven, and the equilibrium toward polymer formation was favored at elevated temperatures and less favored at a lower temperature (Inoué, 1951, 1959).

Furthermore, by making the fibers and fibrils depolymerize slowly by chilling a living cell, exposing them to elevated hydrostatic pressure, or by applying moderate doses of colchicine solutions to metaphase

† Corresponding author.

cells, Inoué and his coworkers demonstrated that the depolymerizing spindle fibers and fibrils could generate enough force to transport chromosomes and the whole spindles through the cytoplasm to an anchorage site at cell surface (Figure 1). Conversely, the assembly of subunits from the pool into the growing protein fibrils could generate a force that would push chromosomes, and the whole spindle, away from the anchorage site (Inoué, 1952; Inoué and Sato, 1967; Inoué *et al.*, 1975; Salmon *et al.*, 1976).

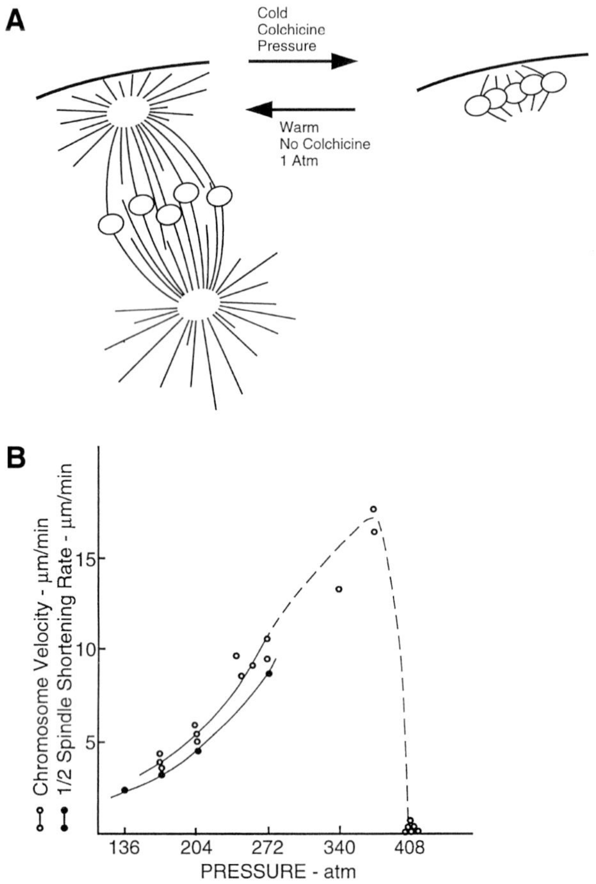

Figure 1. Chromosome movement induced by microtubule assembly/disassembly in meiotic metaphase-arrested *Chaetopterous* spindles. (A) Movements of chromosomes toward the cell surface induced by shortening of kinetochore fibers and astral rays accompanying loss of birefringence (number of microtubules). Microtubule disassembly is induced reversibly by cooling, or by application of colchicine or hydrostatic pressure. The chromosomes move back away from the attached pole upon spindle reassembly (Inoué, 1952; Inoué *et al.*, 1975; Salmon, 1976). (B) Velocities of chromosomes as they move toward the cell surface during hydrostatic pressure-induced depolymerization of kinetochore microtubules in the metaphase spindle. The greater the hydrostatic pressure applied, the faster the microtubules depolymerize and the faster the chromosomes and the spindle pole are transported toward the anchorage site on the cell surface. Above about 370 atm, microtubule depolymerization was too rapid to generate any pulling force (Salmon, 1976).

By the mid 1960's, microtubules, which earlier had not been seen in the spindle region, were clearly visualized by electron microscopy using improved, osmium- and glutaraldehyde-containing fixatives (Harris, 1962; reviewed by Porter, 1966). The distribution and behavior of the microtubules closely paralleled those of the birefringent fibrils that had been described earlier within the spindle fibers. In 1967, Inoué and Sato (1967) were able to explicitly propose that depolymerizing microtubules could perform mechanical work by pulling, and conversely, polymerizing and growing microtubules could exert a pushing force. The former mechanism could draw the chromosomes via their spindle fiber attachment site, the kinetochore, to the spindle pole. The latter could, for example, extend the distance between the two poles of the spindle.

In 1972, Weisenberg (1972) developed a method for isolating microtubules whose properties finally resembled the behavior of the spindle fibrils in living cells. Unlike all earlier methods for isolating microtubules or the spindle apparatuses, which yielded only stabilized structures that were unresponsive to cold or to colchicine, microtubules isolated according to Weisenberg's method yielded microtubules that disassembled into their subunit tubulin dimers upon chilling. Furthermore, they reassembled into their tubular polymer state (the microtubules, Figure 2A) at room temperature in the presence of magnesium ions, GTP, and neutral organic buffer, provided the calcium ion concentration was kept below about 100 nM. The reassembled microtubules would depolymerize again when exposed to low temperature or high hydrostatic pressure (Borisy *et al.*, 1975; Olmsted and Borisy, 1975; Salmon, 1975; Weisenberg, 1972). The reassembly was inhibited by colchicine, and accelerated by D_20 that increases the birefringence of the spindle in living cells. The appearance and disappearance, as well as the number concentration of microtubules seen with the electron microscope after fixation with birefringence-maintaining fixative were found to parallel the (form) birefringence of the spindle fibers measured in living cells (Sato *et al.*, 1975).

Finally then, the main linear fibrils that make up the spindle fibers could be equated with microtubules, which exhibited assembly/disassembly properties similar to the labile spindle fibers in dynamic equilibrium with their subunits (tubulin) in living cells.

Since the late 1960's, following the discovery of dynein (Gibbons and Rowe, 1965), the focus on force generation for mitosis in most laboratories shifted to translocator motor proteins that generate sliding forces between or along microtubules (McIntosh *et al.*, 1969; McIntosh and Koonce, 1989; McIntosh and Pfarr, 1991; Sawin and Endow, 1993). Since the discovery of kinesin in 1985, at least 52 kinesin-related proteins and

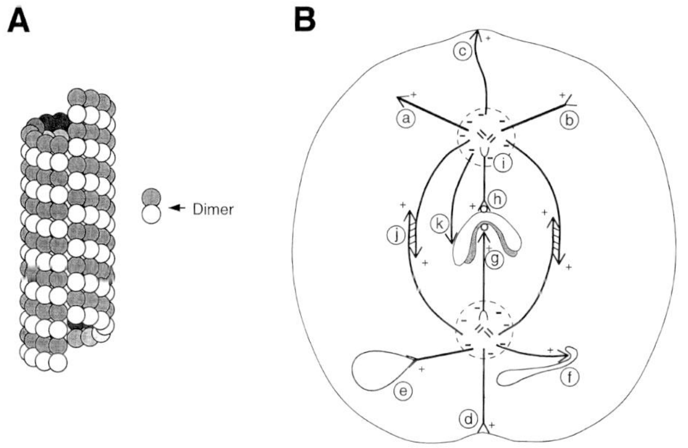

Figure 2. Microtubule structure, polarity, dynamic instability, and probable sites of force generation associated with microtubule assembly/disassembly. (A) Sketch of the structure of a 13-protofilament microtubule showing the head-to-tail alignment of the α and β tubulin dimers along the protofilaments. A single 4×8-nm tubulin dimer is shown on the right. (Modified from Mandelkow *et al.*, 1986; Song and Mandelkow, 1995). (B) Schematic summary of how microtubule assembly/disassembly is likely to be involved with chromosome movement and centrosome positioning in living cells. Some microtubules with free ends grow (arrowheads) at a steady pace, incorporating tubulin-GTP subunits at their free plus ends (a). At other moments, an elongating microtubule suddenly switches to shortening and as it disassembles, also at its plus end (forked), it releases tubulin-GDP subunits (b). Some attached to the cell cortex (c) continue to grow, push against the cortex and centrosome, and deform the cell surface as well as the microtubule itself. Others attach to the cell cortex, start to shorten at the attachment site (d), and exert a pulling force on the cell surface and the centrosome. Microtubules whose plus ends attach to vesicles and other organelles deform the organelles so that they point toward the centrosome (by the organelles being pulled against obstructing structures) as the plus end depolymerizes and pulls the organelles centripetally (e). Growth of plus ends at the organelle attachment site pushes the organelle away from the pole (f). The plus ends of microtubules attached to the chromosome at the kinetochore can continue to grow at that attachment site (g) and contribute to an increase in kinetochore to pole distance, or as they disassemble at the kinetochore (h) or at the pole (i), they can exert a tension between the kinetochore and the center. Plus end growth of overlapping microtubules from opposite spindle poles helps to push the poles apart (j). Arms of chromosomes may also be pushed away from the center by growing plus ends (k). Modified from Inoué (1990).

a cytoplasmic form of dynein have been found in living cells (Goldstein, 1993; Sawin and Endow, 1993; Vallee, 1993; Goodson *et al.*, 1994). Several motor proteins have been found, by immuno-chemical staining, to co-localize with specific regions of chromosomes, kinetochores, centrosomes, and spindle fiber microtubules or membrane vesicles (McIntosh and Pfarr, 1991; Goldstein, 1993; Sawin and Endow, 1993; reviewed in Fuller, 1995). In addition, genetic mutants of motor proteins, studied in particular in yeast and *Drosophlia*, have been shown to suppress or modify mitosis, indicating that they are important for spindle assembly, separation of the spindle poles, and chromosome movement (reviewed in: Goldstein, 1993; Sawin and Endow, 1993; Hoyt, 1994; Snyder, 1994; Fuller, 1995).

Despite this justified shift in emphasis toward seeking the roles of motor proteins in chromosome movements for the past two and a half decades, interest in the role played by assembly and disassembly of microtubules in force generation for mitosis and related movements (Figure 2B) has been rekindled in recent years. It is probable that multiple mechanisms and force producers have evolved to ensure accurate segregation of chromosomes. Both microtubule assembly/disassembly and microtubule motors could contribute together or in different ways in the production and regulation of forces, both in mitosis and other microtubule-dependent movements. For example, there is now evidence that both active and inactive microtubule motor proteins can couple cargo to depolymerizing microtubule ends (see Sections 4, 6, and 7).

In the following sections, we review the observations and experiments that have led to the renewed interest in the assembly/disassembly mechanism for force production, which had earlier been considered by many to be counter intuitive and thus untenable. In this discussion, we focus on how the acts of assembly and disassembly in and of themselves could contribute to generating forces required for mitosis and related movements. In this light, we discuss possible force-generating mechanisms at the kinetochore and how these mechanisms may also regulate the velocity and onset of anaphase chromosome movement. We begin with the discovery of microtubule "dynamic instability," because it both accounts for the dynamic nature of the spindle and it is fundamental to force generation by assembly/disassembly.

2. MICROTUBULE STRUCTURE, POLARITY, AND DYNAMIC INSTABILITY

Microtubules are linear, polarized, polymers of tubulin dimers (Figure 2A). Each dimer is a 100-kDa complex of closely related α and β tubulin polypeptides. The 24-nm diameter cylindrical wall of the microtubule is made up of protofilaments of tubulin dimers. Cytoplasmic and spindle microtubules typically have 13 protofilaments. One end of a microtubule has a crown of α tubulin while the other end has a crown of β tubulin (Mandelkow *et al.*, 1986; Song and Mandelkow, 1995).

The "head-to-tail" arrangement of the dimers along the protofilaments gives the microtubules an intrinsic structural polarity. In the early 1980's, several structural methods were developed to determine microtubule polarity (Heidemann and McIntosh, 1980; Telzer and Haimo, 1981; McIntosh and Euteneuer, 1984). In ciliary axonemes, all microtubules have the same polarity; the ends distal to the basal body are called "plus" while those proximal are called "minus." Similar methods were used to show that during mitosis the centrosome at each spindle pole nucleates microtubules with their plus ends pointing away from the pole (Figure 2B). A subset of these, called kinetochore microtubules, become attached by their plus ends to the kinetochore regions of the chromosomes (Figure 2B, g and h). It is not yet clear if plus ends have crowns of α or β tubulin; there is evidence favoring both possibilities (Oakley, 1992; Mitchison, 1993; Song and Mandelkow, 1995). Most of the kinesin-related proteins have been shown to be plus end–directed motors (Walker and Sheetz, 1993; Goodson *et al.*, 1994), but members of one group, the ncd family, are minus end–directed like the dyneins (McDonald *et al.*, 1990; Walker *et al.*, 1990; Vallee, 1993; Endow *et al.*, 1994).

A big surprise in the early 1980's was evidence that polymerized tubulin throughout the mitotic spindle exchanges rapidly with the pool of unassembled tubulin dimers. In the laboratories of McIntosh and Salmon (Salmon *et al.*, 1984; Saxton *et al.*, 1984; Wadsworth and Salmon, 1986), fluorescently labeled tubulins, added to the cellular tubulin pool, were used as a tracer, in combination with measurements of fluorescence redistribution after photobleaching, to measure the rate of microtubule turnover at steady state. These studies showed that the half-life of tubulin turnover in the mitotic spindle of sea urchins and mammalian cells is fast, occurring within 20–60 s. Subsequent studies (Mitchison *et al.*, 1986) showed that this fast turnover is the property of microtubules whose plus ends are free; kinetochore microtubules are differentially stable (Brinkley and Cartwright, 1975; Salmon *et al.*, 1976; reviewed in Salmon, 1989) with half-lives of several minutes (Mitchison *et al.*, 1986; Mitchison, 1989; Cassimeris *et al.*, 1990). Previously, the assembly/disassembly of microtubules in vitro had been shown to occur only at the ends of microtubules (Margolis and Wilson, 1978; Bergen and Borisy, 1980; Gelfand and Bershadsky, 1991). How then could tubulin subunits appear to be exchanging rapidly throughout the array of spindle microtubules?

The origin of this fast turnover turned out to be a unique property of microtubule assembly called dynamic instability, which was discovered by Mitchison and Kirschner (1984a,b). In an in vitro study, they examined how microtubule lengths become redistributed in a population self-assembled from pure tubulin. The key observation was that in a population of microtubules whose average length was declining, a subset of microtubules continued to grow longer. On this basis, they proposed that microtubule ends, in a population at steady-state assembly, have the ability to alternate between persistent phases of growth and shortening. Subsequently, the dynamic instability of individual microtubule ends was directly seen in vitro, in real-time, using dark-field microscopy by Horio and Hotani (1986), and by video-enhanced DIC (differential interference, or Normarski, contrast) microscopy by Walker *et al.* (1988) (Figure 3, B–D). The growth and shortening (shrinking) phases of dynamic instability are persistent because thousands of tubulin subunits add during a single growth phase or dissociate during a single shortening phase. The abrupt transition between growth and shortening is termed catastrophe, while the switch from shortening back to growth is termed rescue (Walker *et al.*, 1988) (Figure 3A).

In living cells, both fluorescence microscopy and video-enhanced DIC microscopy have been used to measure the kinetics of dynamic instability for individual microtubule plus ends (Gelfand and Bershadsky, 1991; reviewed in Caplow, 1992; Erickson and O'Brien, 1992; Wadsworth, 1993). In interphase and mitotic vertebrate cells, growth velocity of free ends is about 7–14 μm/min, and shortening velocity is about 20 μm/min (Figure 4, A and B, left) (Cassimeris *et al.*, 1988b; Hayden *et al.*, 1990; Spurck *et al.*, 1990; Shelden and Wadsworth, 1993). In interphase, microtubules become very long and have lifetimes of many minutes. In comparison, spindle microtubules are short and non-kinetochore microtubules have lifetimes of 1 min or less. This change in microtubule length and dynamics between interphase and mitosis depends mainly on changes in the transition frequencies of dynamic instability, catastrophe, and rescue (Gliksman *et al.*, 1993).

Interestingly, the frequencies of catastrophe and rescue have been shown to depend somehow on the kinase activity of the cell cycle regulator, mitosis-promoting factor (MPF; the active form of $p34^{CDC2}$-cyclin B complexes). In interphase, MPF kinase activity is low. Catastrophes occur, but rescue is more frequent, and microtubules achieve long lengths and rarely shorten all the way back to the spindle poles (Figure 3C) (Cassimeris *et al.*, 1988b; Belmont *et al.*, 1990; Gliksman *et al.*, 1992; Simon *et al.*, 1992; Verde *et al.*, 1992). Upon activation of MPF activity in mitosis or inhibition of phosphatase activity by okadaic acid in interphase, the frequency of catastrophe increases and rescue becomes rare (Figure 3D) (Belmont *et al.*, 1990; Gliksman *et al.*, 1992; Verde *et al.*, 1992). In the spindle, when plus ends switch from growth to shortening, they appear to shorten all the way back to the spindle poles, where nucleation and regrowth occurs (Figure 4, A and B, left) (Mitchison *et al.*, 1986; Wadsworth and Salmon, 1986). Phosphorylation of microtubule-associated proteins (MAPs) by MPF appears to be one mechanism that induces the loss of rescue in mitosis (Ookata *et al.*, 1995).

For the thermodynamic analysis of the spindle dynamic equilibrium, Inoué assumed that tubulin subunits exchanged all along the length of spindle fibers (Inoué, 1959–1964). Although there is evidence that tubulin can very slowly dissociate from the walls of

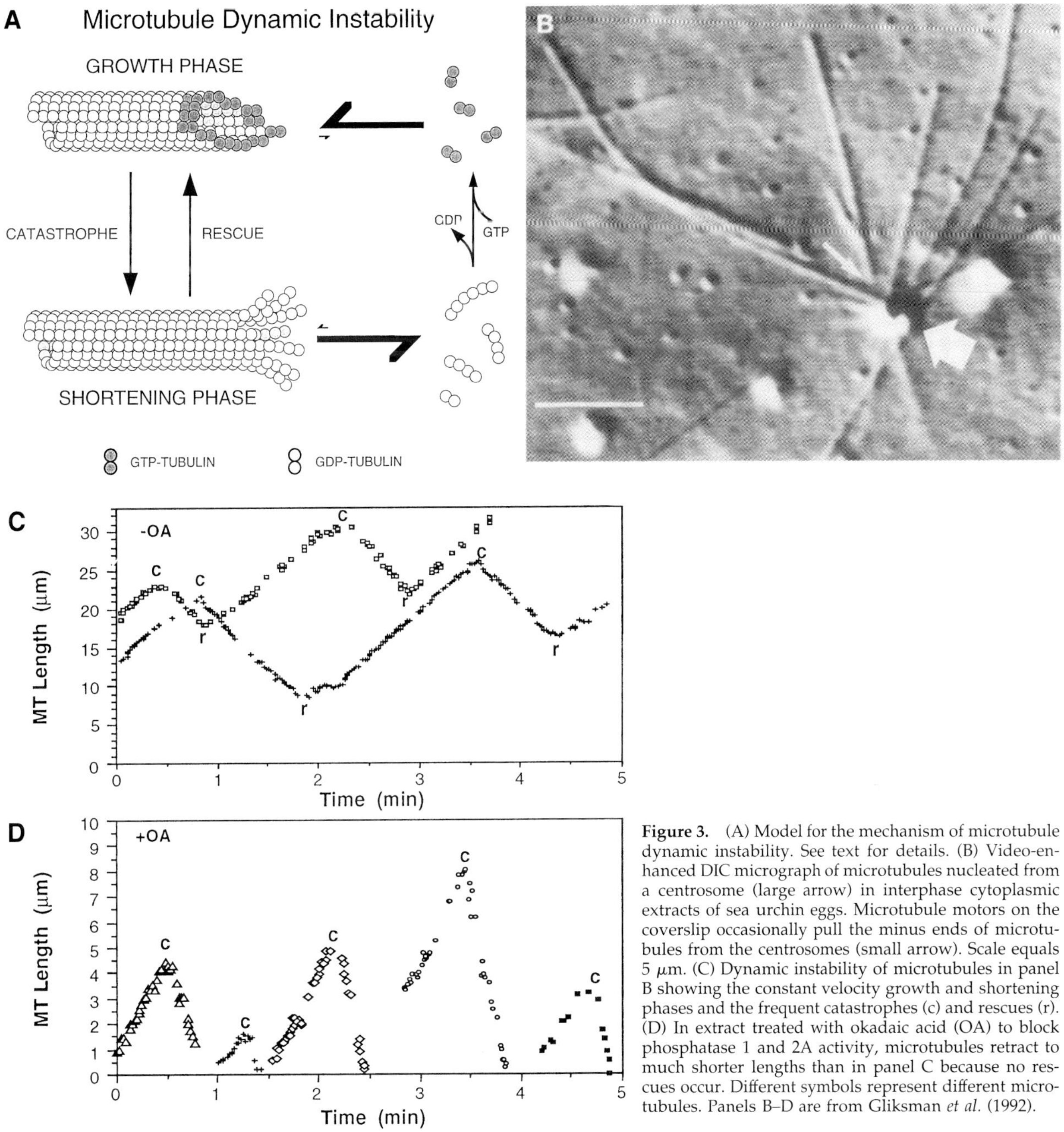

Figure 3. (A) Model for the mechanism of microtubule dynamic instability. See text for details. (B) Video-enhanced DIC micrograph of microtubules nucleated from a centrosome (large arrow) in interphase cytoplasmic extracts of sea urchin eggs. Microtubule motors on the coverslip occasionally pull the minus ends of microtubules from the centrosomes (small arrow). Scale equals 5 μm. (C) Dynamic instability of microtubules in panel B showing the constant velocity growth and shortening phases and the frequent catastrophes (c) and rescues (r). (D) In extract treated with okadaic acid (OA) to block phosphatase 1 and 2A activity, microtubules retract to much shorter lengths than in panel C because no rescues occur. Different symbols represent different microtubules. Panels B–D are from Gliksman *et al.* (1992).

microtubules (Dye *et al.*, 1992), there is no evidence to date that rapid and extensive tubulin exchange with microtubules occurs at sites other than at their ends. The dynamic instability of microtubule plus ends does make the spindle dynamic and it does explain the tubulin subunit turnover throughout the spindle fibers.

Microtubule dynamic instability fundamentally depends on the hydrolysis of GTP bound to tubulin as initially proposed by Mitchison and Kirschner (1984a,b). Figure 3A summarizes the current view of how GTP hydrolysis changes the conformation of tubulin and generates dynamic instability (Stewart *et al.*, 1990; Carlier, 1991; Walker *et al.*, 1991; Caplow, 1992;

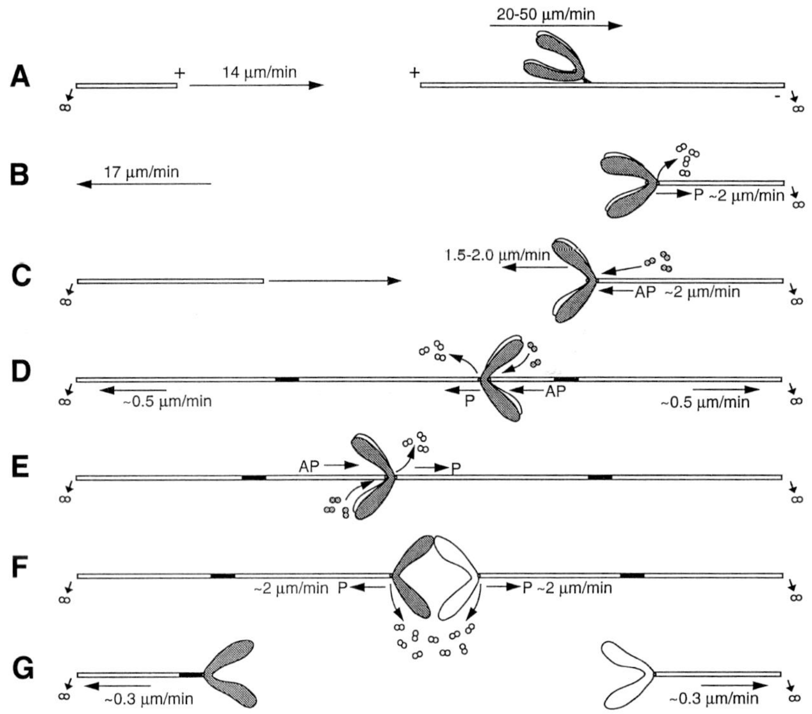

Figure 4. Schematic of kinetochore motility and microtubule assembly/disassembly in vertebrate mitotic cells. The kinetochore of a chromosome may initially become attached to the side of a polar microtubule and slide rapidly along its wall (A). Once the tips of microtubules attach to a kinetochore, the chromosome is slowly pulled poleward (B) or pushed away from the pole (C) alternating every few minutes. Once the sister kinetochore becomes attached to microtubules from the opposite pole (D), it persists in poleward movement (P) while the initially attached kinetochore persists in away from the pole movement (AP) toward the spindle equator. Near the equator, sister kinetochores continue to alternate between poleward and away from the pole movements (E) until chromosome separation at the onset of anaphase (F). Then both sister kinetochores persist in poleward movement (G), although occasional episodes of away from the pole oscillation occur. Fluorescent marks (dark bars) made on kinetochore fibers in prometaphase, metaphase, and anaphase reveal only slow poleward flux of kinetochore microtubules (D–G). Small curved arrows show the microtubule shortening at the kinetochore coupled to poleward movement, or microtubule growth at the kinetochore coupled to away from the pole movement, or slower microtubule shortening at the polar minus ends coupled to the poleward flux. Velocities indicated in the figure include contributions from flux.

Erickson and O'Brien, 1992; Caplow *et al.*, 1994; Drechsel and Kirschner, 1994).

The phase of a microtubule (growth or shortening) is thought to be determined by the presence of GTP or GDP on the tubulin subunits at the end of the polymer. There is one exchangeable site for GTP on β tubulin (E site) that must be occupied by GTP for dimer polymerization. The rate of polymerization depends on the concentration of GTP-bound tubulin (tubulin-GTP) and the binding constant for growing ends. As tubulin-GTP dimers are incorporated into the microtubule end, GTP is hydrolyzed to GDP and phosphate is released. This produces a core lattice of tubulin-GDP dimers capped at the end by newly associating tubulin-GTP dimers.

A catastrophe is thought to occur when the terminal tubulin-GTP cap is lost through hydrolysis or dissociation. Loss of the cap substantially inhibits tubulin-GTP association and allows rapid dissociation of the labile core of tubulin-GDP dimers. Rescue, from the shortening phase back to the growing phase of dynamic instability, is thought to occur when the end becomes re-stabilized by a new cap of tubulin-GTP.

As predicted by the model in Figure 3A, hydrolysis of GTP is not required for microtubule assembly, but it is required for dynamic instability. Microtubules assembled from tubulin bound to a slowly hydrolyzable analogue of GTP (GMPCPP) grow at rates typical of tubulin-GTP, but they do not exhibit catastrophes and shortening phases (Hyman *et al.*, 1992b; Caplow *et al.*, 1994). Tubulin-GMPCPP microtubules also shorten in the absence of tubulin at a rate 1000-fold or more slower than do tubulin-GDP microtubules (Caplow *et al.*, 1994).

The tubulin-GTP cap appears to be confined to the growing tip. When microtubules are assembled from tubulin-GTP, cutting off the growing tip at plus ends with a microbeam induces immediate shortening of the newly exposed plus cut end (Walker *et al.*, 1989). Alternatively, when microtubules are diluted to block tubulin-GTP association, within seconds they loose their terminal stabilizing caps and convert to the shortening phase (Voter and Erickson, 1991; Walker *et al.*, 1991). A brief pulse of tubulin-GMPCPP before dilution is sufficient to cap and stabilize the otherwise labile core of tubulin-GDP (Drechsel and Kirschner, 1994). So far, biochemical methods have been unable to detect any unhydrolyzed GTP (E-site) in microtubules (Stewart *et al.*, 1990; Erickson and O'Brien, 1992). These kinetic and biochemical assays indicate that the tubulin-GTP stabilizing cap may be only one or a few dimers thick at a growing end in solution (Stewart *et al.*, 1990; Walker *et al.*, 1991; Erickson and O'Brien, 1992; Drechsel and Kirschner, 1994).

The hydrolysis of GTP bound to tubulin induces a conformational change in the tubulin dimer as indicated by the different structures of growing and shortening ends that can be seen in electron micrographs (Erickson, 1974; Erickson and O'Brien, 1992; Kirschner *et al.*, 1974, 1975; Simon and Salmon, 1990; Simon *et al.*, 1991; Mandelkow *et al.*, 1991; Mandelkow and Mandelkow, 1992). Growing ends in solution have straight or slightly curved protofilaments that laterally contact each other (Figure 3A, Growth). In contrast, shortening ends often have protofilaments that have lost their lateral contacts, curl radially outward like "ram's-horns" and break off as curved oligimers (Figure 3A, Shortening).

Growing ends in solution also exhibit sheets of protofilaments that have not yet closed completely into a cylinder (Simon and Salmon, 1990; Simon *et al.*, 1991; Mandelkow *et al.*, 1991; Chrétien *et al.*, 1995). As seen in Figures 2A and 3A, there is a "seam" along the length of the microtubule between protofilaments where α and β tubulins are laterally associated; for all other adjacent protofilaments, α is adjacent to α and β is adjacent to β (Mandelkow *et al.*, 1986; Song and Mandelkow, 1995). This seam may be where cylindrical closure occurs; weaker lateral interactions there may also explain why cooling can cause cylindrical microtubules to open up into 13 protofilament sheets (Simon and Salmon, 1990). Nevertheless, the straight conformation of the tubulin-GTP dimer appears important for the tight lateral association of tubulin-GTP dimers at growing ends. For example, the growing ends of microtubules assembled from tubulin-GMPCPP have straight protofilaments and tubulin dissociation is extremely slow (~0.1 dimers/s) (Hyman *et al.*, 1992b, 1995; Caplow *et al.*, 1994)

In contrast, the curvature of the protofilaments at shortening ends, where tubulin dissociation can occur at ~1000 dimers/s, is typical of the 28 degree curvature of individual tubulin-GDP protofilaments in solution (Howard and Timasheff, 1986; Melki *et al.*, 1989). Hyman *et al.* (1995) have recently shown that the lattice structure of microtubules made up of tubulin-GDP and tubulin-GMPCPP are very similar. This result indicates that the strain energy in the tubulin dimer produced by hydrolysis of GTP is stored in the microtubule lattice and only becomes released when the tubulin-GDP subunits are exposed at microtubule ends.

We will later consider the potential magnitude of force and the work that can be produced by the growth and shortening phases of dynamic instability, but first we review the evidence showing that kinetochore motility in mitosis is tightly coupled to microtubule assembly dynamics.

3. KINETOCHORE AND CHROMOSOME MOVEMENT ARE COUPLED TO MICROTUBULE ASSEMBLY/DISASSEMBLY

It is now well documented that microtubule assembly is required to establish the bipolar spindle in prometaphase, and to increase the separation of the spindle poles during anaphase (Inoué, 1981; Salmon, 1989; McIntosh, 1994). Assembly of spindle polar microtubule arrays in prometaphase is also involved with pushing chromosome arms away from the spindle poles (Rieder *et al.*, 1986; Rieder and Salmon, 1994; Leslie, 1992; Theurkauf and Hawley, 1992; Ault and Rieder, 1994; Cassimeris *et al.*, 1994). These "polar ejection forces" or "polar winds" on the arms (Figure 2B, j) are thought to contribute to the congression of chromosomes to the metaphase plate in animal cells (Salmon, 1988, 1989; Rieder and Salmon, 1994). In addition to microtubule assembly, motor proteins bound to the chromosome arms may contribute to the polar ejection force (Theurkauf and Hawley, 1992; Murphy and Karpen, 1995). In this connection, several kinesin-related motor proteins have very recently been found to be bound to chromosome arms: nod in *Drosophila* (Afshar *et al.*, 1995); chromokinesin in chicken tissue cells (Wang and Adler, 1995), and Xklp1 in *Xenopus* embryos (Vernos *et al.*, 1995). It will be interesting to know the polarity of their movement along microtubules and how much they function in generating the polar ejection forces.

The assembly dynamics of kinetochore microtubules are essential for fully understanding the movements of chromosomes relative to their spindle poles (Figure 4, B–G; see also Figure 9). Ultrastructural studies of vertebrate cells by Rieder (1981) and McDonald *et al.* (1992) have shown that many if not most kinetochore microtubules extend all the way between the spindle pole and the kinetochore where they penetrate the outer layer of the trilaminate kinetochore. Vertebrate kinetochores on average capture about 20–25 microtubules (Rieder, 1982). As chromosomes move toward the spindle equator during congression, the microtubules attached to sister kinetochores increase and decrease in length, until they achieve approximately similar lengths when the chromosomes arrive near the equator (Figure 4E). Upon sister chromosome separation at the onset of anaphase, the kinetochore microtubules shorten as their chromosomes move poleward, led by their kinetochores and centromere region (Figure 4, F and G).

In addition to the major shifts in the lengths of kinetochore microtubules during congression and anaphase, high resolution video microscopy by Skibbens *et al.* (1993) and Cassimeris *et al.* (1994) has shown that sister kinetochores in vertebrate tissue cells oscillate back and forth over 1- to 3-μm distances relative to their spindle poles throughout mitosis following their attachment to microtubule plus ends (Figure 4, B–G). An attached kinetochore exhibits poleward movement at a constant velocity of 1- to 3-μm/min, then abruptly switches to constant velocity away from the pole movement. The direction of movement of a kinetochore can be independent of the movement of the sister kinetochore or the chromosome arms. This switching of the kinetochore between poleward and away from the pole motility states has been termed "kinetochore directional instability" (Skibbens *et al.*, 1993).

During kinetochore directional instability, we find that the centromere region of the chromosome with its apical kinetochore protrudes poleward during poleward movement of the chromosome. The centromere region shortens or dimples inward as the chromosome moves away from the spindle pole (Cassimeris *et al.*, 1994; Skibbens *et al.*, 1993, 1995). These and related observations by Bajer (1982) and others (reviewed in Rieder and Salmon, 1994) strongly suggest that localized pulling and pushing forces applied to the kinetochore are primarily responsible for movement of the chromosome toward and away from the spindle pole once microtubules have become attached to the kinetochore; the polar ejection forces only contribute to push arms away from the pole.

Although the fact that localized forces act on the chromosome's kinetochore and are associated with shortening and growth of microtubules appears to be well established, this still leaves open the question of where the forces for mitotic chromosome movement are generated. An important clue to this question, as well as to the nature of the force-generating mechanism, is provided by studies that reveal the site(s) of kinetochore microtubule assembly and disassembly, as discussed next.

The sites of kinetochore microtubule growth and shortening have been determined by Borisy, Mitchison, Salmon, Wadsworth, and their coworkers by examining the sites at which labeled tubulin is incorporated into kinetochore microtubules (Mitchison *et al.*, 1986; Mitchison, 1988; Wadsworth *et al.*, 1989; Wise *et al.*, 1991; Sheldon and Wadsworth, 1992). Alternatively, the positions (and motility, if any) of fluorescent tubulin markers, placed along a limited stretch of the kinetochore microtubules were monitored (Wadsworth and Salmon, 1986; Gorbsky *et al.*, 1987, 1988; Cassimeris *et al.*, 1988a; Mitchison, 1989; Cassimeris and Salmon, 1991; Centonze and Borisy, 1991; Mitchison and Salmon, 1992).

The general picture that has emerged for vertebrate tissue cells is diagrammed in Figure 4, B–G. All movement of kinetochores away from the pole is coupled to assembly of kinetochore microtubules at the kinetochore. Most kinetochore poleward movement, either during congression or anaphase-A, is coupled to disassembly of kinetochore microtubules, again at the kinetochore. A minor fraction of kinetochore poleward movement is coupled to a slow (0.3–0.5 μm/min in newt cells) poleward flux (Figure 4, D–G) of kinetochore microtubules as they disassemble at their minus ends at the spindle poles.

In early anaphase, available data indicate that the bulk of kinetochore poleward movement occurs at 1- to 3-μm/min velocity along relatively stationary kinetochore microtubules (Gorbsky *et al.*, 1987, 1988; Mitchison and Salmon, 1992) (Figure 4, F). In late anaphase, the microtubules are no longer disassembled at the kinetochore (in newt cells), but the chromosomes continue to move poleward at the slow steady rates exhibited by the poleward flux of the kinetochore microtubules (0.3 μm/min in newt cells, Figure 4, G) (Mitchison and Salmon, 1992).

Thus, two mechanisms have evolved for growth and shortening of kinetochore microtubules: those associated with plus end assembly/disassembly at the kinetochore, and those associated with minus end disassembly near the poles. In vertebrate tissue cells, assembly/disassembly at the kinetochore before late anaphase dominates kinetochore microtubule assembly dynamics. In other cell types, minus end disassembly at the spindle poles may play a much more significant role in assembly dynamics and pulling chromosomes poleward (Forer, 1965, 1966; Mitchison and Salmon, 1992; Sawin and Mitchison, 1994; Wilson *et al.*, 1994).

The direction of kinetochore movement is sensitive to the conditions of microtubule assembly at their plus ends (Figure 5). If in prometaphase or metaphase, microtubule assembly is blocked with drugs such as colchicine or nocodazole, kinetochores switch to poleward movement (Inoué, 1952; Cassimeris and Salmon, 1991). The kinetochore fibers shorten much as occurs in anaphase at about 2 μm/min, primarily disassembling at the kinetochore (Figure 5A). When a bolus of labeled tubulin is microinjected into metaphase cells, the label is incorporated into kinetochore microtubules only at the kinetochore (Mitchison *et al.*, 1986; Mitchison, 1988; Wise *et al.*, 1991) (Figure 5B).

In anaphase, Sheldon and Wadsworth (1992) found that microinjection of a large bolus of labeled tubulin causes all the kinetochores to reverse direction and move away from the pole. The labeled tubulin is incorporated at the kinetochore (Figure 5C). Eventually, the kinetochores switch back to anaphase poleward movement as kinetochore microtubules revert to disassembling at the kinetochores when the free tubulin concentration drops.

Taken together, these studies, and others (Rieder and Salmon, 1994), have shown that the kinetochore: 1) is the major site for assembly/disassembly

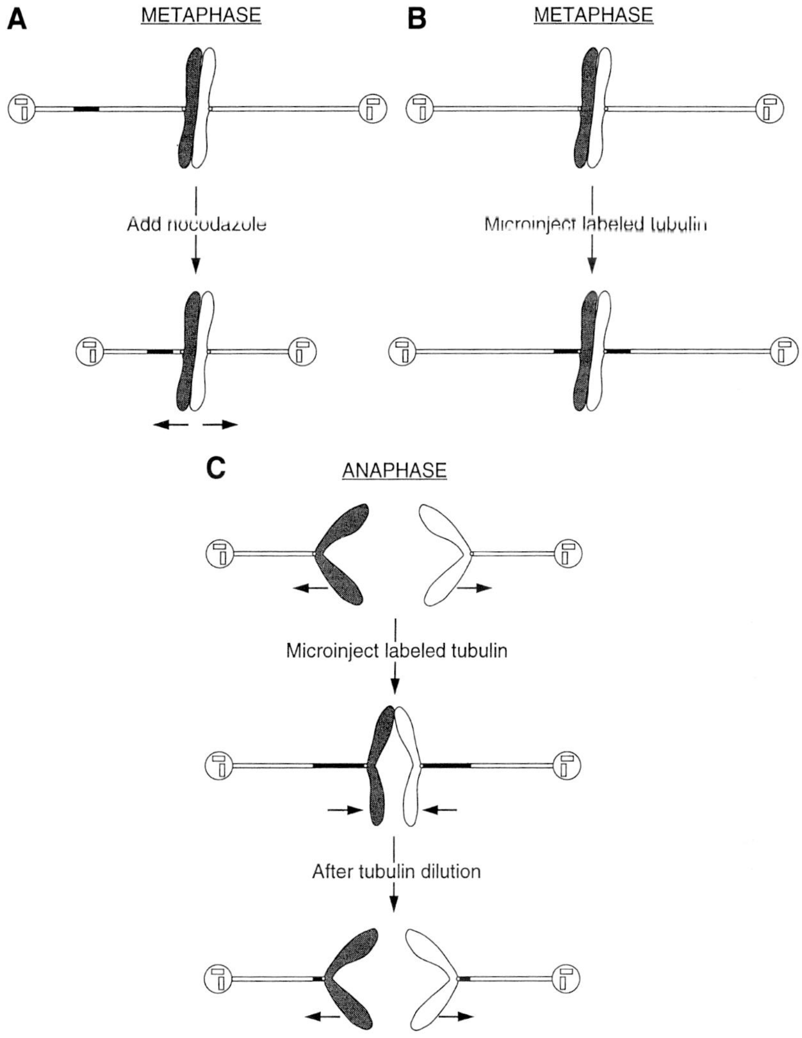

Figure 5. The direction of kinetochore movement is sensitive to manipulation of microtubule assembly. (A) When metaphase cells are treated with nocodazole, kinetochores persist in poleward movement toward fluorescent marks (dark bars) on their kinetochore microtubules. (B) Microinjection of a bolus of labeled tubulin into metaphase spindles results in incorporation of labeled tubulin into kinetochore fibers at the kinetochores. (C) Microinjection of a bolus of labeled tubulin into a cell that has already entered anaphase results in incorporation of labeled tubulin at the kinetochores and movement of the kinetochores *away* from the poles. Eventually, as the tubulin concentration drops, the kinetochores switch back to poleward movement as microtubules disassemble at the kinetochore.

of kinetochore microtubules; 2) can abruptly switch between poleward and away from the pole motility at slow constant velocities; and 3) can switch direction of motility in response to conditions that affect microtubule assembly.

Thus, assembly/disassembly of microtubules at the kinetochore is intimately coupled to force generation for chromosome congression to the spindle equator as well as for anaphase-A segregation of chromosomes to the spindle poles (at least in vertebrate tissue cells). (The qualifier is added here because in some lower invertebrate and plant cells kinetochore microtubule poleward flux (treadmilling) and their disassembly at the minus ends may play a much more important role).

4. KINETOCHORE MOTILITY AND RELATED FORCES MAY BE GENERATED BY MOTOR PROTEINS, BY MICROTUBULE ASSEMBLY/DISASSEMBLY, OR BY BOTH

Both microtubule-dependent motor proteins as well as microtubule assembly/disassembly have been proposed to contribute to force generation for kinetochore poleward and away from the pole movements and the polar ejection forces that act on the chromosome arms. The roles in force generation, attachment, or regulation played by each and how they vary in different cell types are major unresolved issues. The inventory of microtubule motor proteins at the kinetochore has only begun and little information is available about their specific functions. Both cytoplasmic dynein

(Pfarr *et al.*, 1990; Steuer *et al.*, 1990) and the kinesin-related proteins CENPE (Yen *et al.*, 1991, 1992) and MCAK (Wordeman and Mitchison, 1995) have been localized by antibody staining to mammalian kinetochores. In the yeast, *Saccharomyces cerevisiae*, both genetic studies and biochemical assays indicate that the kinesin-related protein kar3p is a minus end–directed motor that may function in generating pole-directed force in mitosis (Saunders and Hoyt, 1992; Endow *et al.*, 1994; Middleton and Carbon, 1994). In in vitro motility assays, kar3p has been shown to bind to centromere DNA and move it along microtubules (Hyman *et al.*, 1992a). Both biochemical and genetic approaches are likely to identify other unknown microtubule-dependent motor proteins at the kinetochore (Earnshaw, 1994) that could function in kinetochore motility either in force generation or attachment to kinetochore microtubules.

One model proposed by several authors for motility generated at the kinetochore is shown in Figure 6 (Skibbens *et al.*, 1993; Murray and Mitchison, 1994). During kinetochore poleward movement, a minus end–directed motor is active and movement is coupled to plus end disassembly of the kinetochore microtubules. During movement of the kinetochore away from the pole, a plus end motor is active and movement is coupled to plus end assembly of kinetochore microtubules.

In support of kinetochore force production by activity of motor proteins, Rieder *et al.* (1990) and Merdes and DeMey (1990) have reported that kinetochores slide along microtubules toward their minus ends (i.e., toward the spindle poles, Figure 4A, right) in early

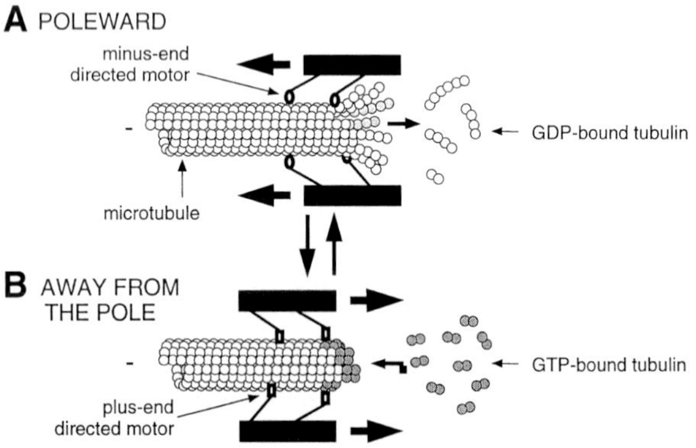

Figure 6. A model for a kinetochore motor (cf. Figure 8). An individual microtubule-binding site (cylinder section shown as two dark bars) of a kinetochore undergoing poleward movement (A) or away from the pole movement (B). During poleward movement, the microtubule ends depolymerize, releasing tubulin-GDP subunits while motors directed toward the "minus" (centrosomal) end of the microtubule are activated. During movement away from the pole, microtubules polymerize tubulin-GTP and plus end–directed motors are activated. Modified from Murray and Mitchison (1994).

prometaphase before the kinetochores become attached to microtubule plus ends. This poleward translocation occurs at a fast rate of about 30–60 μm/min and is thought to depend on cytoplasmic dynein located at the kinetochore. Hyman and Mitchison (1991) used in vitro motility assays to show that kinetochores, on isolated metaphase chromosomes from mammalian cells, can attach to the walls of microtubules and translocate also at fast velocities in a minus direction in the presence of ATP.

Pre-treatment of isolated chromosomes with ATPγS promotes phosphorylation of kinetochores. In this case, ATP-dependent microtubule translocation occurs in a plus direction at slow velocities, 1–4 μm/min (Hyman and Mitchison, 1991). This experiment shows that slow plus end–directed motor activity is nacently present in the kinetochores and that phosphorylation can switch between minus and plus end–directed motor activity.

In living cells, the majority of kinetochore movement occurs at the plus ends of kinetochore microtubules, with the movement tightly coupled to the assembly dynamics of the kinetochore microtubule at their plus ends (Section 3). Thus a major function of the minus and plus end–directed "motor proteins" at the kinetochore could be to maintain attachment to shortening and growing plus ends of the microtubules rather than to produce sliding forces (Skibbens *et al.*, 1993; Desai and Mitchison, 1995; Lombillo *et al.*, 1995a). The pulling and pushing forces could in turn be produced by disassembly and assembly reactions. As discussed above (Figure 5), blocking microtubule assembly induces persistent movement of the kinetochore toward the spindle pole (or shortening of the kinetochore to pole distance); promoting assembly by raising the local tubulin concentration induces persistent motion of the kinetochore away from the pole. (Whether there are accompanying changes in kinetochore phosphorylation is unknown.) These experimental results clearly show that the direction of kinetochore motility can be controlled by changing the assembly state of the plus ends of the kinetochore microtubules: growth induces movement away from the pole, shortening induces poleward movement.

Taken together, these observations strongly suggest both the presence of "motor proteins" at the kinetochore and an intimate coupling of microtubule assembly/disassembly to kinetochore movement. We shall now examine further in vitro evidence that microtubule assembly of itself can generate pushing forces, that disassembly can generate pulling forces, and that active and inactive motor proteins can couple cargo to depolymerizing microtubule ends.

5. IN VITRO EVIDENCE FOR PUSHING BY MICROTUBULE ASSEMBLY

There is growing evidence that the polymerization of microtubules as well as actin filaments can push. Liposomes encapsulating either tubulin dimers (Miyamoto and Hotani, 1988; Hotani and Miyamoto, 1990; Fygenson, 1995) or G-actin (Cortese *et al.*, 1989; Janmey *et al.*, 1992; Miyata and Hotani, 1992) have been shown to become extended when either of these subunit proteins are induced to polymerize inside the liposome. Fygenson (1995), in particular, found conditions where microtubule dynamic instability persists within pure lipid vesicles. The membrane is pushed outward into narrow protrusions during the growth phase (Figure 7A), which retract during the shortening phases of dynamic instability.

More recently, Waterman-Storer *et al.* (1995) have found that membranes (mainly endoplasmic reticulum) from undiluted extracts of *Xenopus* eggs become attached to growing plus ends of microtubules. Growth at the attachment site extends these membranes into narrow tubes at velocities typical of free plus end growth rates, ~20 μm/min (Figure 7B). Addition of inhibitors of microtubule motor translocation (5 mM AMPPNP, 250 μM orthovanadate, or ATP depletion), which block sliding movement along microtubules (Allan and Vale, 1991, 1994), has no effect on the velocity of membrane extension by microtubule growth. These results show the following: 1) membranes of endoplasmic reticulum have attachment sites for plus microtubule ends, presumably on their outer (cytosol) surfaces, and 2) these membrane microtubule tip attachment complexes are able to maintain attachment to ends that are growing at about 540 dimers/s.

6. IN VITRO EVIDENCE FOR PULLING BY MICROTUBULE DISASSEMBLY

Koshland *et al.* (1988) were the first to provide in vitro evidence that microtubule plus ends can remain attached to kinetochores on isolated chromosomes as the microtubule plus ends shorten in the presence or absence of ATP (Figure 7C). Coue *et al.*

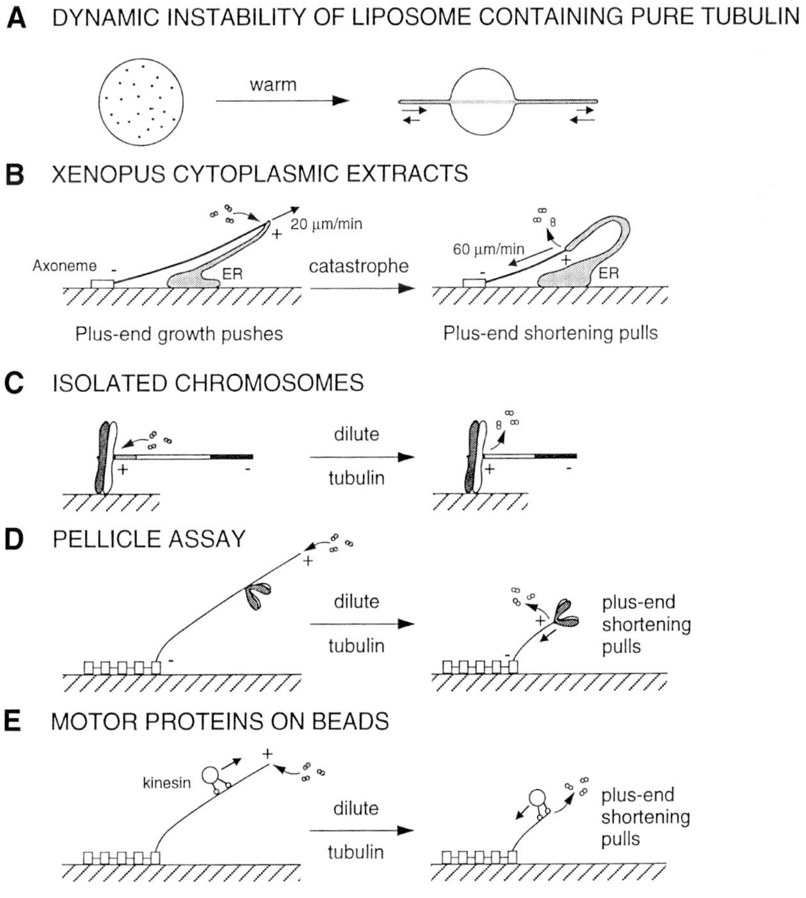

Figure 7. In vitro evidence for pushing and pulling forces generated by microtubule assembly/disassembly. (A) Microtubule undergoing dynamic instability within liposomes reversibly pushes the membrane into narrow protrusions. (B) In *Xenopus* egg cytoplasmic extracts, growing microtubule plus ends become attached to membranes and push them into long tubes. When catastrophe takes place, the depolymerizing end then pulls the membrane tubes toward their minus ends, which are anchored to the coverslip. (C) Microtubules attached by their plus ends to isolated chromosomes shorten at their plus end attachment sites as indicated by fluorescent marks on the microtubules. (D) Chromosomes attached at their centromere (kinetochore?) region to the walls of microtubules grown from protozoan pellicles are pulled toward the pellicle as the microtubule plus end, attached to the chromosome, is depolymerized by dilution of the tubulin. (E) Plastic beads coated with kinesin initially translocate toward microtubule plus ends. But upon tubulin dilution in a high salt buffer, the beads become pulled in a minus end direction toward the pellicle when the beads reach, and become attached, to the microtubule plus ends.

(1991) have investigated this issue further by using a novel in vitro motility assay (Figure 7D). In their assay, microtubules are grown from pure tubulin onto the basal bodies of isolated protozoan pellicles attached to the inner surface of a coverslip. This produces a high density of long microtubules extending many microns and all with their plus ends pointing away from the stationary pellicle. Isolated chromosomes are added by perfusion until they bind to the microtubule wall. Then the microtubules are made to disassemble by perfusion with buffer lacking tubulin. When the shortening end of a microtubule reached the centromere region of the attached chromosome, the chromosome began moving toward the pellicle attached to the end of the microtubule as the end continued to disassemble at velocities similar to free ends (i.e., at 30–40 μm/min). Chromosomes were pulled to the pellicle, with or without ATP, against drag forces estimated at ~10 pN (Coue *et al.*, 1991).

In another series of experiments, Lombillo *et al.* (1995a) attached several different types of microtubule motor proteins to 1-μm diameter plastic beads (Figure 7E). Beads with active axonemal dynein, cytoplasmic dynein, or kinesin translocate along microtubules in standard motility buffers with ATP, but they did not remain attached once they reached the microtubule's shortening end.

Under conditions that weakened motor attachment to the microtubule lattice, however, the beads behaved differently. Examples include axonemal dynein inhibited with orthovanadate, and an inactive hybrid molecular complex with a kinesin motor domain attached to the tail of a kinesin-related protein, ncd. These beads did not show their normal unidirectional gliding activity, but both remained attached to shortening ends of microtubules, in the presence or absence of ATP.

Finally, increasing salt concentration to reduce binding strength to the microtubule lattice also allowed native kinesin to remain attached to a shortening end. Remarkably, under these conditions and in the presence of ATP, the bead with native kinesin moved toward the microtubule plus end until the shortening end reached the bead. Then, the bead was dragged, attached to the shortening end, in a minus direction (Figure 7E)! These results showed that even after active translocation by the plus end–directed motor has ceased, the motor was capable of remaining attached to the shortening plus end of the microtubule, resulting in a minus end–directed motility (Lombillo *et al.*, 1995a).

Particles and membranes in cytoplasmic egg extracts have also been observed to be pulled toward the microtubule minus ends by their attachment to depolymerizing plus ends (Gliksman and Salmon, 1993; Waterman-Storer *et al.*, 1995). For example, in *Xenopus* cytoplasmic extracts (see Section 5), Waterman-Storer *et al.* (1995) also found that when a growing microtubule end switched spontaneously to the shortening phase of dynamic instability, membrane tubes remained attached to the microtubule end (Figure 7B). The membrane tube was pulled through the extract at velocities similar to the velocity of free end shortening, 50–60 μm/min, whether or not inhibitors (5 mM AMPPNP, 250 μM orthovanadate, or ATP depletion) of microtubule-dependent motor activity were present. Thus, as in the kinetochore or bead assays above, in the absence of active motor translocation, the *Xenopus* membranes can be pulled through cytosol attached to shortening plus ends of microtubules (which are losing dimers rapidly, at about 1600 dimers/s in the *Xenopus* cytoplasm).

7. THE NATURE OF MICROTUBULE TIP ATTACHMENT COMPLEXES

The above experiments demonstrate that microtubule growing ends can push, while shortening ends can pull without becoming detached from their load. They also show that both active and inactive motor proteins can couple objects to shortening ends, allowing rapid dimer dissociation while remaining attached to the end. Whether motor proteins, either active or inactive, can couple cargo to growing ends has yet to be demonstrated.

The actual molecules within tip attachment complexes that couple membranes or kinetochores to growing and shortening ends have not been definitely identified. Nevertheless, Lombillo *et al.* (1995b) have shown that antibodies to CENPE, a kinetochore-associated, kinesin-related protein, disrupt attachment of kinetochores to shortening ends in their in vitro motility assays. The result implies that this kinetochore protein is involved in attachment. Other possible candidates include cytoplasmic dynein, MCAK, nod, XKLP1, kar3p, chromokinesin, or other as yet unknown kinesin-related proteins associated with the kinetochore.

The attachment molecules need not be active motor proteins. Microtubules have been shown to exhibit Brownian motion along their long axis, i.e., without any lateral movement when weakly bound to inactive axonemal dynein (orthovanadate inhibited) or to nonmotile kinesin-related proteins attached to glass surfaces (Vale *et al.*, 1989; Chandra *et al.*, 1993; Stewart *et al.*, 1993; Lombillo *et al.*, 1995a). These observations indicate that a sufficient concentration of nonmotor, microtubule-binding proteins might also be able to couple cargo to shortening ends. An interesting candidate is the protein CLIP 170, which has been shown by Kreis and coworkers to localize to the plus ends of microtubules in HeLa cells (Scheel and Kreis, 1991; Pierre *et al.*, 1992).

Also unknown are the sites on the tubulin dimer at which the TAC is attached. "Binding machines" that follow the ends of growing and shortening microtubules presumably have two features: 1) the activation energy for binding and dissociation must be low enough to allow the rapid cycles of binding and release to follow a moving microtubule end without disrupting the assembly dynamics; and 2) the binding machine must have multiple binding sites or rapid enough binding so that dissociation of a single site does not lead to detachment from a shortening end. We don't know how few these multiple sites could be. As diagrammed in Figure 8A, most models (see next section) envision some form of lateral attachment that permits translocation of the tip attachment complex over the lattice coupled to growth and shortening. However, there is no evidence to rule out the possibility that effective attachment occurs at the distal tip of the tubulin dimers at the microtubule end (Figure 8B).

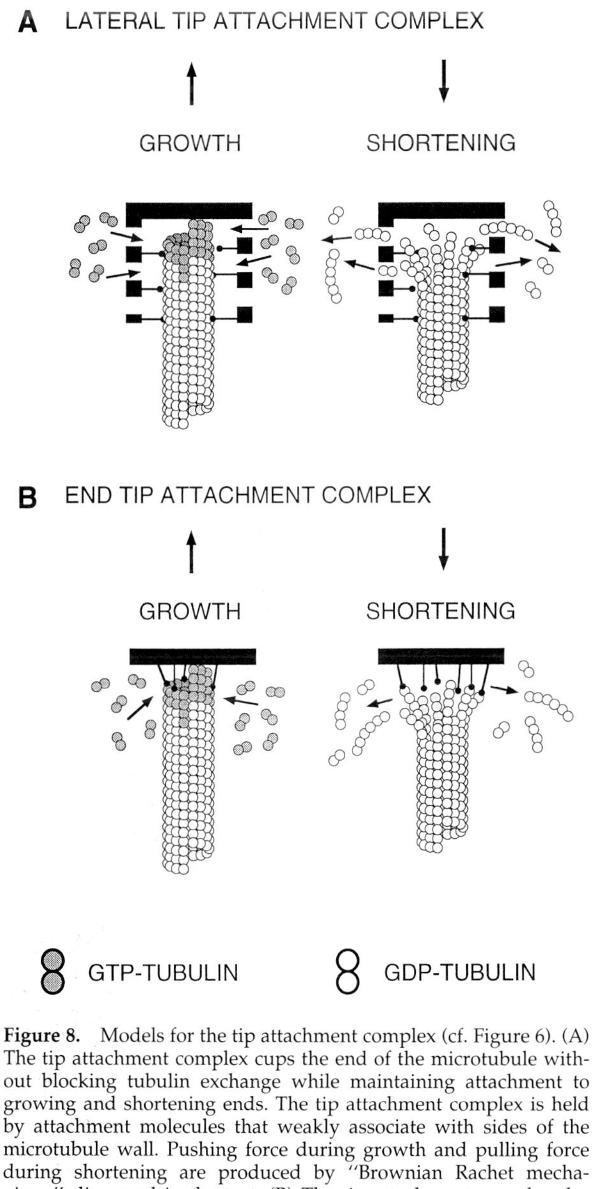

Figure 8. Models for the tip attachment complex (cf. Figure 6). (A) The tip attachment complex cups the end of the microtubule without blocking tubulin exchange while maintaining attachment to growing and shortening ends. The tip attachment complex is held by attachment molecules that weakly associate with sides of the microtubule wall. Pushing force during growth and pulling force during shortening are produced by "Brownian Rachet mechanisms" discussed in the text. (B) The tip attachment complex dynamically binds to tubulin dimers at the microtubule tip in both the growing and shortening phases of dynamic instability.

8. MODELS FOR THE MICROTUBULE ASSEMBLY/DISASSEMBLY ENGINE

A major question is the mechanism by which the energetics of assembly/disassembly is transduced into work at the attachment site. Motion is probably generated by some type of molecular "Brownian Rachet" for either growth or shortening in which the highly irreversible addition of tubulin-GTP to growing ends and the highly irreversible dissociation of tubulin-GDP from shortening ends rectifies thermal motion of the tip attachment complex. There are several theoretical considerations about how such Brownian Rachet mechanisms can work to produce pushing during growth (Peskin *et al.*, 1993) and pulling during shortening (Hill, 1985; Koshland *et al.*, 1988; Mitchison, 1988; Desai and Mitchison, 1995). They are, so far, all based on some form of lateral tip attachment complex.

The growth mechanism is easier to visualize for a tip attachment complex which "cups" the growing end (Figure 8A, Growth). Within this geometry, the tight binding of tubulin at the end biases the Brownian motion of the tip attachment complex in the direction of growth.

Thermodynamically, the maximum pushing force (the stall force, F) available from the free energy of growth is related to the kinetics of tubulin association/dissociation (Hill and Kirschner, 1982; Peskin *et al.*, 1993):

$$F = (k_B T/d)\ln(k_{on}/k_{off}) \quad (1)$$

where k_BT is Boltzman's constant times absolute temperature (4.1 pN-nm at room temperature), d is the average displacement of the microtubule end against a load (0.61 nm on average for a 13-protofilament microtubule, for association or dissociation of the 8 nm long dimer), k_{on} is the dimer association rate (the product of the association rate constant and the tubulin concentration), and k_{off} is the dimer dissociation rate during growth. In the growth phase, Eqn. 1 predicts pushing stall forces of F = 0, 16, 31, and 46 pN for ratios of k_{on}/k_{off} = 1, 10, 100, and 1000, respectively. The binding of GTP to unassembled tubulin subunits makes GTP-tubulin association with growing ends highly favored. Stall forces of at least 16 pN, and more likely 32 pN, are possible in cells, because MAPs promote k_{on} and suppress k_{off} (Dreschel *et al.*, 1992; Pryer *et al.*, 1992; Vasquez *et al.*, 1994).

The shortening mechanism is harder to visualize. One concept as originally proposed by Hill (1985) is that cargo movement is driven by the energetics involved with maximizing the number of weak attachment sites. This movement to achieve maximum overlap of the sleeve with the microtubule wall biases Brownian motion of the tip attachment complex toward the intact portion of the microtubule as it shortens (Figure 8A, Shortening).

The outward bending of the tubulin-GDP dimers at the shortening ends may also be important for maintaining attachment and biasing Brownian motion of the TAC in the direction of shortening (Figure 8A, Shortening). This mechanism was termed the "conformational wave" model by Koshland *et al.* (1988) and Mitchison (1988). High resolution microscopy of growing and shortening ends attached to cargo, like kinetochores or ER membranes, may reveal how much the curvature of the protofilaments contributes to force production. In this regard, we are still in the dark ages of kinetochore structure in comparison to the high resolution view available for actin-myosin interactions in skeletal muscle (Rayment, 1993; Rayment *et al.*, 1993).

Thermodynamically, Eqn 1 can be used to predict the potential force that can be generated by depolymerizing ends where the k_{on}/k_{off} ratio is now the value during the shortening phase of dynamic instability. For free plus ends of microtubules assembled in vivo or in vitro, the velocity of microtubule shortening is rapid (often > 500 dimers/s) and in vitro, velocity has been shown to be independent of tubulin concentration (Walker *et al.*, 1988, 1991). During shortening, k_{on} appears near zero and k_{on}/k_{off} appears possibly 0.001 or less. From Eqn. 1, this ratio yields a potential pulling force (minus sign) of 32 pN or more.

Hydrolysis of the GTP bound to tubulin within the microtubule is responsible for the huge change in the k_{on}/k_{off} ratio between growth and shortening phases (Section 2). As a consequence, the potential force available from the shortening phase can be estimated from the energy of GTP hydrolysis stored in the microtubule lattice (Koshland *et al.*, 1988). Caplow *et al.* (1994) report that for the assembly of pure tubulin in solution, the free energy of GTP hydrolysis is −5.18 kcal/mol. Within the microtubule lattice he estimates an energy release of only −0.9 kcal/mol. The difference between these two values, about 4 kcal/mol or up to 26 pN-nm/dimer (1kcal/mol = 6.6 pN-nm/molecule), should be the strain energy stored in microtubule lattice from the hydrolysis of GTP. This energy is available to do work, such as producing the curvature in the tubulin-GDP protofilaments seen at shortening ends. It predicts a stall force of about 43 pN if each 8-nm dimer biases cargo movement in the direction of shortening on the average by 0.61 nm (8 nm/13 protofilaments).

The above analysis provides a rough estimate of the potential forces available from either growth or shortening phases of dynamic instability, on the order of 40 pN or less per microtubule end. For reference, the stall force for a single kinesin molecule is about 5 pN (Kuo and Sheetz, 1993; Hunt *et al.*, 1994; Svoboda and Block, 1994). Nicklas (1988), using a compliant microneedle, has directly measured the force that stalls anaphase poleward kinetochore movement and kinetochore fiber shortening in meiotic grasshopper spermatocytes. This force is estimated to be about 10 pN per kinetochore microtubule, the same order of magnitude, and less than the force available from the energy of GTP hydrolysis stored in the microtubule lattice. Direct measurements of the forces that stall microtubule growth and shortening in cell-free systems have yet to be achieved.

9. THE KINETOCHORE AS A GOVERNOR

The kinetochore in living cells appears to govern several different activities. These include the rates of microtubule growth and shortening at the kinetochore, the switching between kinetochore microtubule growth and shortening, and the checkpoint that controls anaphase onset.

A major puzzle is why kinetochore motility in cells is much slower (~2 μm/min) than the growth and shortening velocities of free plus ends of microtubules in cells or plus ends attached to membrane tip attachment complexes in the *Xenopus* extracts (10–50 μm/min). The rate that kinetochores move in vitro when coupled to shortening plus ends (20–60 μm/min) is also much faster than kinetochore motility in vivo. In other words, once attached to kinetochores *in the cell*, microtubules grow and shorten at a rate that is an order of magnitude slower!

This difference is clearly not due to drag forces or polar ejection forces on the chromosome arms. The velocities of kinetochore poleward and away from the pole movement is the same for kinetochores on tiny centromere fragments severed by a microbeam from the bulk of the chromosome arms (Figure 9) (Skibbens *et al.*, 1993, 1995). So, what mechanisms limit the plus end growth and shortening rates of kinetochore microtubules inside living cells to a constant velocity of about 2 μm/min?

Could the velocity be regulated by nonmotor MAPs? Neuronal MAP2 and Tau do slow the rate of microtubule shortening, but these MAPs also promote faster growth rates (Dreschel *et al.*, 1992; Pryer *et al.*, 1992). The *Xenopus* XMAP protein makes both growth and shortening velocities faster than occurs for micro-

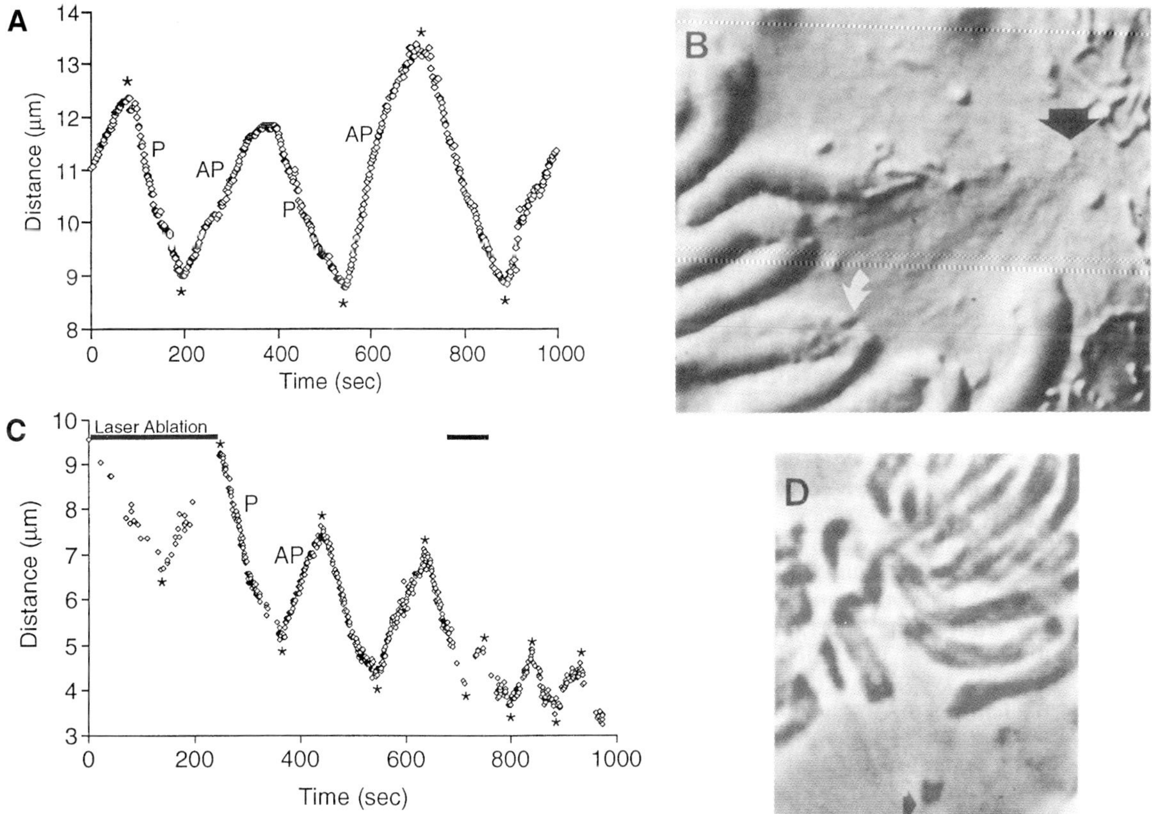

Figure 9. The velocities and abrupt switching of kinetochore motility between poleward and away from the pole movement is independent of the size of the chromosome. (A) Kinetochore directional instability for a mono-oriented chromosome (connected to only one spindle pole as in Figure 4, B and C) in a mitotic newt cell shown in panel B. (C) Kinetochore directional instability for a kinetochore on a small chromosome fragment (short black arrow in panel D, which was severed from the bulk of the chromosome arms and its sister kinetochore by laser microsurgery in the cell shown in panel D. Horizontal lines mark periods of laser surgery. The second irradiation severed the tenuous connection remaining between the centromere region and the chromosome arms. (B and D) Smaller black or white arrows show the positions of the kinetochores while the large black or clear arrows show the position of the spindle poles. Note that in the kinetic plots (A) and (C), the transitions (*) between poleward and away from the pole phases of kinetochore motility are quite abrupt. Modified from Skibbens *et al.* (1993, 1995).

tubule assembly with pure tubulin alone (Vasquez *et al.*, 1994). So MAPs may not be the answer. Something about the structure of the attachment site could limit diffusion, restricting tubulin access to or escape from a deeply embedded microtubule tip. In addition, an obvious possibility is that the growth and shortening rates of the microtubules at the kinetochores are regulated by slow (~2 μm/min) motors that translocate along the microtubule lattice at the plus end attachment sites (models in Figure 6 and Figure 8A).

It is fairly easy to see how slow advancement of a plus end capping assembly (either model in Figure 8) could slow microtubule growth. It is harder to envision how slow minus-end directed movement of the same complex during depolymerization can slow tubulin dissociation without detachment of the complex from the end. In the lateral attachment model in Figure 8A, the tip attachment complex could slow depolymerization without detachment if only subunits distal to the attachment sites are capable of dissociation. In the distal tip attachment model in Figure 8B, the binding of the attachment molecules could stabilize lateral interactions between adjacent tubulin dimers, slowing dissociation.

By whatever mechanism that operates at the kinetochore, the cell seems to have chosen to use the kinetochore as a governor, to slow down and synchronize the shortening and growth of the microtubules that collectively must bring the full set of daughter chromosomes to the spindle pole.

Another puzzle is what does the kinetochore sense within the spindle to control its direction of movement during prometaphase congression and anaphase segregation of the chromosomes? There is evidence that motor activity can destabilize the ends of microtubules (Endow *et al.*, 1994; Lombillo *et al.*, 1995a). Also, combinations of plus and minus motors have been shown to generate directional instability in microtubule gliding in in vitro motility assays (Vale *et al.*, 1993). Thus, an interesting possibility is that a combination of plus and minus motors at kinetochores may function not only as a mechanism for attachment to dynamic plus ends (Skibbens *et al.*, 1993; Lombillo *et al.*, 1995a), but also to synchronously switch kinetochore microtubules between growth and shortening phases of dynamic instability.

There is a long history of experimental evidence indicating that tension at the kinetochore controls the direction of motility during chromosome congression and segregation; low tension promotes switching to poleward movement or shortening of kinetochore microtubules and high tension promotes switching to movement away from the pole or growth of kinetochore microtubules (Inoué, 1952; Rieder and Salmon, 1994; Skibbens *et al.*, 1995). But the molecular pathways between changes in kinetochore tension and changes in the direction of kinetochore motility are still unknown. Micromanipulation studies by Nicklas and Ward (1994) have also shown that tension stabilizes kinetochore microtubule attachment.

There is evidence that the onset of anaphase depends on microtubule dynamics at the kinetochore. In many cell types, anaphase onset is delayed until all chromosomes have become attached to opposite poles by kinetochore microtubules (Rieder *et al.*, 1994). Anaphase onset is delayed by disruption of microtubule attachment to kinetochores by drugs like nocodozole and taxol, which inhibit or stabilize microtubule assembly respectively (Jordan *et al.*, 1992; Wendell *et al.*, 1993; Rieder *et al.*, 1994), by detachment produced by chromosome micromanipulation (Nicklas *et al.*, 1995), and by antibodies to centromere proteins like CENPC (Tomkiel *et al.*, 1994). Genetic analysis in yeast systems (Murray, 1994, 1995) has identified genes (MAD1-MAD3 and BUB1-BUB3) whose products are required for the anaphase onset checkpoint that detects defective spindle assembly. It will be interesting to know how these proteins interact with the kinetochore.

Very recently, intriguing light has been shed on the role that mechanical tension, acting on the kinetochores and their microtubules, appears to serve as the "checkpoint" for the cell's entry into anaphase as proposed by Hartwell and Weinert (1989) (see also McIntosh, 1991). When chromosomes become attached to opposite spindle poles by kinetochore microtubules, a net tension develops across the centromere, even for the oscillating kinetochores in vertebrate cells (Skibbens *et al.*, 1994). Through ingenious experiments using mantid spermatocytes, Li and Nicklas (1995) demonstrated that a lack of tension on the kinetochore of one of the two X chromosomes, which had failed to pair with the Y chromosome, prevents the mitotic cell from completing metaphase, thus being arrested without being able to enter anaphase. Such a cell can, however, be made to proceed into anaphase (after a short delay) if the missing tension is applied to the stray X chromosome by stretching the chromosome with a microneedle. Thus the lack of tension on a kinetochore can somehow signal that an error of meiosis has been detected and eventually lead to degeneration of the cell.

A phosphorylated kinetochore epitope recognized by 3F3 antibody appears involved with this tension-dependent signal pathway. Gorbsky and Ricketts (1993) initially discovered that in PtK1 tissue culture cells, unattached kinetochores stain brightly with the 3F3 antibody, but attached kinetochores on chromosomes near the metaphase plate and in anaphase stain dimly. Nicklas *et al.* (1995) have reported a similar staining pattern for kinetochores in grasshopper meiosis-I spermatocytes and gone on to provide evidence by micromanipulation experiments that it is tension, and not simply the attachment of kinetochore microtubules, which reversibly turns staining from bright to dim! In this regard, it is worth calling attention to "Teinochemical Systems" where mechanical stretching of an amphoteric gel strand changes its chemical affinity, for example, to charged ions (Kuhn *et al.*, 1960).

10. CONCLUDING REMARKS

It is clear from the above survey that depolymerizing, or disassembling, microtubules can generate a pulling, or "contractile" force, and that polymerizing, or assembling, microtubules can generate a pushing, or extensive, force. These forces are of sufficient magnitude and with appropriate directionality to bipartition chromosomes in mitosis as well as position centrosomes and other organelles that are involved in the establishment of cell polarity and differentiation, albeit through the transient establishment of highly dynamic architectures in which the component subunits are rapidly turning over within the functional edifice in the living cells.

Several questions that need answering are as follows. What are the maximum forces that can be generated by individual microtubule assembly/disassembly, and how does the force depend on velocity of elongation or shortening? Is the energy stored by hydrolysis of GTP during assembly of the microtubule essential for the contractile force production? Or, does the thermodynamics of the assembly system provide adequate energy? What fraction of force generation in mitosis and related motility is based on assembly/disassembly of microtubules as opposed to the sliding of microtubules or organelles powered by motor proteins such as dynein or kinesin?

What is the nature of the forces that hold the microtubule "tip" to the kinetochore, organelle or membrane, while still permitting the subunits of microtubules to enter or escape from this extending or contracting site? Is there a labile glue or wetting agent that intervenes between the microtubule tip and the attachment site? Are modified motor proteins involved in these dynamic attachments? Or, can nonmotor tubulin or microtubule-binding proteins attach at the contacting site? An interesting suggestion made by Andrew Murray (personal communication) is that "motors initially arose as nonhydrolytic components of rachet mechanisms and only later in evolution learned how to use ATP hydrolysis to walk along lateral surfaces."

What of the minus ends of the microtubules at the spindle pole? What is the nature of their anchorage, or ability not to be drawn in toward the kinetochore? As at the kinetochore, the minus end attachment site must also be dynamic because kinetochore microtubules flux toward the spindle poles in vertebrate tissue cells (Figure 4) and nonkinetochore microtubule flux has been measured in embryonic *Xenopus* spindles (Sawin and Mitchison, 1994). We are only beginning to understand the nature of minus end complexes and their role in force generation for chromosome movement and spindle morphogenesis.

What controls catastrophe and rescue at the kinetochore to generate synchronous behavior of the many microtubules attached to a single kinetochore during poleward and away from the pole motility? How does tension affect catastrophe and rescue? Do catastrophe and rescue take place stocastically, or are there local regulators? What are the molecular mechanisms at the kinetochore that sense tension to control the direction of kinetochore motility and anaphase onset? How does the dynamic instability of microtubules participate in moving or positioning organelles, including the centrosomes, other than chromosomes?

Many of these questions will no doubt be answered by further biochemical studies and amino acid sequence analyses, as well as through ingeniously contrived experiments on cell-free systems directly being observed, perhaps with the aid of electronically enhanced light microscopy.

At the same time we hope that the applicability of the models and paradigms emerging from in vitro experiments will be tested directly in living cells. Microtubule and organellar behavior should be closely observed at high resolution in several cell types undergoing mitosis or meiosis that can be proven to be normal, as well as in cells that are experimentally manipulated in highly localized regions, such as might be achieved by activating appropriate caged compounds with a microbeam of light, and/or with reagents with a high degree of target specificity, including altered genes. In addition to experimenting with cell types that are widely used by investigators, we should remember that nature sometimes reveals her most well kept secrets through exaggerated displays found only in exotic cell types.

After all, through extensive trial and error, nature has chosen an intricately interacting, dynamic system to achieve mitosis, and to safeguard the propagation and unfolding of life despite its myriad forms. While we are becoming privy to some of nature's surprising ways today, we need, in addition to dissecting the molecules further, to listen ever more carefully to the living cell, and be prepared to be taught further unexpected paradigms, which will undoubtedly be essential for clearer understanding of the physico-chemical and biological basis of cellular organization, life, and disease.

ACKNOWLEDGMENTS

We are grateful to Andrew Murray, Bruce Nicklas, Tim Mitchison, and Conly Rieder for their comments and criticisms of early versions of this essay. Supported by National Institutes of Health grants GM-24364 to E.D.S. and GM-316617 to S.I.

REFERENCES

Afshar, K., Barton, N.R., Hawley, R.S., and Goldstein, L.S.B. (1995). DNA binding and meiotic chromosomal localization of the *Drosophila* Nod kinesin-like protein. Cell *81*, 129–138.

Allan, V.J., and Vale, R.D. (1991). Cell cycle control of microtubule-based transport and tubule formation in vitro. J. Cell Biol. *113*, 347–359.

Allan, V.J., and Vale, R.D. (1994). Movement of membrane tubules along microtubules in vitro: evidence for specialized sites of motor attachment. J. Cell Sci. *107*, 1885–1897.

Ault, J.G., and Rieder, C.L. (1994). Centrosome and kinetochore movement during mitosis. Curr. Opin. Cell Biol. *6*, 41–49.

Bajer, A.S. (1982). Functional autonomy of monopolar spindle and evidence for oscillatory movement in mitosis. J. Cell Biol. *93*, 33–48.

Belmont, L.D., Hyman, A.A., Sawin, K.E., and Mitchison, T.J. (1990). Real-time visualization of cell cycle-dependent changes in microtubule dynamics in cytoplasmic extracts. Cell *62*, 579–589.

Bergen, L.G., and Borisy, G.G. (1980). Head-to-tail microtubule polymerization in vitro: electron microscopic analysis of seeded assembly. J. Cell Biol. *84*, 141–150.

Borisy, G.G., Marcum, J.M., Olmsted, J.B., Murphy, D.B., and Johnson, K.A. (1975). Microtubule assembly in vitro? Ann. NY Acad. Sci. *253*, 107–132.

Brinkley, B.R., and Cartwright, J. (1975). Cold-labile and cold-stable microtubules in the mitotic spindle of mammalian cells. Ann. NY Acad. Sci. *253*, 428–439.

Caplow, M. (1992). Microtubule dynamics. Curr. Opin. Cell Biol. *4*, 58–65.

Caplow, M., Ruhlen, R.L., and Shanks, J. (1994). The free energy for hydrolysis of a microtubule-bound nucleotide triphosphate is near zero: all of the free energy for hydrolysis is stored in the microtubule lattice. J. Cell Biol. *127*, 779–788.

Carlier, M.F. (1991). Nucleotide hydrolysis in cytoskeletal assembly. Curr. Opin. Cell Biol. *3*, 12–17.

Cassimeris, L., Inoué, S., and Salmon, E.D. (1988a). Microtubule dynamics in the chromosomal spindle fiber: analysis by fluorescence and high-resolution polarization microscopy. Cell Motil. Cytoskeleton *10*, 185–196.

Cassimeris, L., Pryer, N.K., and Salmon, E.D. (1988b). Real-time observations of microtubule dynamic instability in living cells. J. Cell Biol. *107*, 2223–2231.

Cassimeris, L., Rieder, C.L., Rupp, G., and Salmon, E.D. (1990). Stability of microtubule attachment to metaphase kinetochores in PtK_1 cells. J. Cell Sci. *96*, 9–15.

Cassimeris, L., Rieder, C.L., and Salmon, E.D. (1994). Microtubule assembly and kinetochore directional instability in vertebrate monopolar spindles: implications for the mechanism of chromosome congression. J. Cell Sci. *107*, 285–297.

Cassimeris, L., and Salmon, E.D. (1991). Kinetochore microtubules shorten by loss of subunits at the kinetochores of prometaphase chromosomes. J. Cell Sci. *98*, 151–158.

Centonze, V.E., and Borisy, G.G. (1991). Pole-to-chromosome movements induced at metaphase: sites of microtubule disassembly. J. Cell Sci. *100*, 205–211.

Chandra, R., Endow, S.A., Skeen, V., and Salmon, E.D. (1993). An N-terminal truncation of the ncd motor protein supports diffusional movement of microtubules in motility assays. J. Cell Sci. *104*, 899–906.

Chrétien, D., Fuller, S.D., and Karsenti, E. (1995). Structure of growing microtubule ends: two dimensional sheets close into tubule at variable rates. J. Cell Biol. *129*, 1311–1328.

Cortese, J.D., Schwab, B., Friden, C., and Elson, E.L. (1989). Actin polymerization induces a shape change in actin-containing vesicles. Proc. Natl. Acad. Sci. USA *86*, 5773–5777.

Coue, M., Lombillo, V.A., and McIntosh, J.R. (1991). Microtubule depolymerization promotes particle and chromosome movement in vitro. J. Cell Biol. *112*, 1165–1175.

Desai, A., and Mitchison, T.J. (1995). A new role for motor proteins as couplers to depolymerizing microtubules. J. Cell Biol. *128*, 1–4.

Dreschel, D.N., Hyman, A.A., Cobb, M.H., and Kirschner, M.W. (1992). Modulation of the dynamic instability of tubulin assembly by the microtubule-associated protein tau. Mol. Biol. Cell *3*, 1141–1154.

Drechsel, D.N., and Kirschner, M.W. (1994). The minimum GTP cap required to stabilize microtubules. Curr. Biol. *4*, 1053–1061.

Dye, R.B., Flicker, P.F., Lien, D.Y., and Williams, R.C., Jr. (1992). End-stabilized microtubules observed in vitro: stability, subunit interchange, and breakage. Cell Motil. Cytoskeleton *21*, 171–186.

Earnshaw, W.C. (1994). Structure and molecular biology of the kinetochore. In: Microtubules, ed. J. Hyams and C. Loyyd, New York: Wiley-Liss, 393–412.

Endow, S.A., Kang, S.J., Satterwhite, L.L., Rose, M.D., Skeen, V.P., and Salmon, E.D. (1994). Yeast kar3 is a minus-end microtubule motor protein that destabilizes microtubules preferentially at the minus ends. EMBO J. *13*, 2708–2713.

Erickson, H. (1974). Assembly of microtubules from preformed, ring-shaped protofilaments and 6-S tubulin. J. Supramol. Struct. *2*, 393–411.

Erickson, H.P., and O'Brien, E.T. (1992). Microtubule dynamic instability and GTP hydrolysis. Annu. Rev. Biophys. Biomol. Struct. *21*, 145–166.

Forer, A. (1965). Local reduction of spindle fiber birefringence in living *nephrotoma suturalis* spermatocytes induced by ultraviolet irradiation. J. Cell Biol. *25*, 95–117.

Forer, A. (1966). Characterization of the mitotic traction system and evidence that birefringent spindle fibers neither produce nor transmit force for chromosome movement. Chromosoma *19*, 44–98.

Fuller, M.T. (1995). Riding the polar winds: chromosomes motor down east. Cell *81*, 5–8.

Fygenson, D.K. (1995). Microtubules: the rhythm of assembly and the evolution of form. Ph.D. Thesis, Princeton, NJ: Princeton University.

Gelfand, V.I., and Bershadsky, A.D. (1991). Microtubule dynamics: mechanism, regulation, and function. Annu. Rev. Cell Biol. *7*, 93–116.

Gibbons, I.R., and Rowe, A.J. (1965). Dynein: a protein with adenosine triphosphate activity from cilia. Science *149*, 424–426.

Gliksman, N.R., Parsons, S.F., and Salmon, E.D. (1992). Okadaic acid induces interphase to mitotic-like microtubule dynamic instability by inactivating rescue. J. Cell Biol. *119*, 1271–1276.

Gliksman, N.R., and Salmon, E.D. (1993). Microtubule-associated motility in cytoplasmic extracts of sea urchin eggs. Cell Motil. Cytoskeleton *24*, 167–178.

Gliksman, N.R., Skibbens, R.V., and Salmon, E.D. (1993). How the transition frequencies of microtubule dynamic instability (nucleation, catastrophe, and rescue) regulate microtubule dynamics in interphase and mitosis: analysis using a Monte Carlo computer simulation. Mol. Biol. Cell *4*, 1035–1050.

Goldstein, L.S. (1993). With apologies to Scheherazade: tails of 1001 kinesin motors. Annu. Rev. Gen. *27*, 319–351.

Goodson, H.V., Kang, S.J., and Endow, S.A. (1994). Molecular phylogeny of the kinesin family of microtubule motor proteins. J. Cell Sci. *107*, 1875–1884.

Gorbsky, G.H., Sammak, P.J., and Borisy, G.G. (1987). Chromosomes move poleward in anaphase along stationary microtubules that coordinately disassemble from their kinetochores ends. J. Cell Biol. *104*, 9–18.

Gorbsky, G.J., Sammak, P.J., and Borisy, G.G. (1988). Microtubule dynamics and chromosome motion visualized in living anaphase cells. J. Cell Biol. *106*, 1185–1192.

Gorbsky, G.J., and Ricketts, W.A. (1993). Differential expression of a phosphoepitope at the kinetochores of moving chromosomes. J. Cell Biol. *122*, 1311–1321.

Harris, P. (1962). Some structural and functional aspects of the mitotic apparatus in sea urchin embryos. J. Cell Biol. *14*, 475–485.

Hartwell, L.H., and Weinert, T.A. (1989). Checkpoints: controls that ensure the order of cell cycle events. Science *246*, 629–634.

Hayden, J.H., Bowser, S.S., and Rieder, C.L. (1990). Kinetochores capture astral microtubules during chromosome attachment to the mitotic spindle: direct visualization in live newt lung cells. J. Cell Biol. *111*, 1039–1045.

Heidemann, S.R., and McIntosh, J.R. (1980). Visualization of the structural polarity of microtubules. Nature *286*, 517–519.

Hill, T., and Kirschner, M. (1982). Bioenergetics and kinetics of microtubule and actin filament assembly and disassembly. Int. Rev. Cytol. *78*, 1–125.

Hill, T.L. (1985). Theoretical problems related to the attachment of microtubules to kinetochores. Proc. Natl. Acad. Sci. USA *82*, 4404–4408.

Horio, T., and Hotani, H. (1986). Visualization of the dynamic instability of individual microtubules by darkfield microscopy. Nature *321*, 605–607.

Hotani, H., and Miyamoto, H. (1990). Dynamic features of microtubules as visualized by dark-field microscopy. Adv. Biophys. *26*, 135–156.

Howard, W.D., and Timasheff, S.N. (1986). GDP state of tubulin: stabilization of double rings. Biochemisty *25*, 8292–8300.

Hoyt, M.A. (1994). Cellular roles of kinesin and related proteins. Curr. Opin. Cell Biol. *6*, 63–68.

Hunt, A.J., Gittes, F., and Howard, J. (1994). The force exerted by a single kinesin molecule against a viscous load. Biophys. J. *67*, 766–781.

Hyman, A.A., Chrétien, D., Arnal, I., and Wade, R.H. (1995). Structural changes accompanying GTP hydrolysis in microtubules: information from a slowly hydrolyzable analogue guanylyl-(a,b)-methylene-diphosphate. J. Cell Biol. *128*, 117–125.

Hyman, A.A., Middleton, K., Centola, M., Mitchison, T.J., and Carbon, J. (1992a). Microtubule-motor activity of a yeast centromere-binding protein complex. Nature *359*, 533–536.

Hyman, A.A., and Mitchison, T.J. (1991). Two different microtubule-based motor activities with opposite polarities in kinetochores. Nature *351*, 206–211.

Hyman, A.A., Salser, S., Drechsel, D., Unwin, N., and Mitchison, T.J. (1992b). Role of GTP hydrolysis in microtubule dynamics: information from a slowly hydrolyzable analogue, GMPCPP. Mol. Biol. Cell *3*, 1155–1167.

Inoué, S. (1951). Ph.D Thesis. Studies of the structure of the mitotic spindle in living cells with an improved polarization microscope. Princeton, NJ: Princeton University.

Inoué, S. (1952). The effect of colchicine on the microscopic and sub-microscopic structure of the mitotic spindle. Exp. Cell Res. Suppl. *2*, 305–312.

Inoué, S. (1953). Polarization optical studies of the mitotic spindle. I. The demonstration of the spindle fibers in living cells. Chromosoma *5*, 199–208.

Inoué, S. (1959). Motility of cilia and the mechanism of mitosis. Rev. Mod. Phys. *31*, 402–408.

Inoué, S. (1960). On the physical properties of the mitotic spindle. Ann. NY Acad. Sci. *90*, 529–530.

Inoué, S. (1964). Organization and function of the mitotic spindle. In: Primitive Motile Systems in Cell Biology, ed. R.D. Allen and N. Kamiya, New York: Academic Press, 549–598.

Inoué, S. (1981). Cell division and the mitotic spindle. J. Cell Biol. *91*, 131–147.

Inoué, S. (1990). Dynamics of mitosis and cleavage. Ann. NY Acad. Sci. *582*, 1–14.

Inoué, S., Fuseler, J., Salmon, E.D., and Ellis, G.W. (1975). Functional organization of mitotic microtubules: physical chemistry of the in vivo equilibrium system. Biophys. J. *15*, 725–744.

Inoué, S., and Sato, H. (1967). Cell motility by labile association of molecules: the nature of mitotic spindle fibers and their role in chromosome movement. J. Gen. Physiol. *50*, 259–292.

Janmey, P., Cunningham, C., Oster, G., and Stossel, T. (1992). Cytoskeletal networks and osmotic pressure in relation to cell structure and motility. In: Swelling Mechanics: From Clays to Living Cells and Tissues, ed. T. Karalis, Heidelberg: Springer Verlag.

Jordan, M.A., Thrower, D., and Wilson, L. (1992). Effects of vinblastine, podophyllotoxin and nocodazole on mitotic spindles. J. Cell Sci. *102*, 401–416.

Kirschner, M., Honig, L., and Williams, R. (1975). Quantitative electron microscopy of microtubule assembly in vitro. J. Mol. Biol. *99*, 263–276.

Kirschner, M., Williams, R., Weingarten, M., and Gerhart, J. (1974). Microtubules from mammalian brain: some properties of their depolymerization products and a proposed mechanism of assembly and disassembly. Proc. Natl. Acad. Sci. USA *71*, 1159–1163.

Koshland, D.E., Mitchison, T.J., and Kirschner, M.W. (1988). Polewards chromosome movement driven by microtubule depolymerization in vitro. Nature *331*, 499–504.

Kuhn, W., Ramel, A., Walters, D.H., Ebner, G., and Kuhn, H.J. (1960). The production of mechanical energy from different forms of chemical energy with homogeneous and cross-striated high polymer systems. Fortschr. Hochpolym.-Forsch. *1*, 540–592.

Kuo, S.C., and Sheetz, M.P. (1993). Force of single molecules measured with optical tweezers. Science *260*, 232–234.

Leslie, R.J. (1992). Chromosomes attain a metaphase postion on half-spindles in the absence of an opposing spindle pole. J. Cell Sci. *103*, 125–130.

Li, X., and Nicklas, R.B. (1995). Mitotic forces control a cell-cycle checkpoint. Nature *373*, 630–632.

Lombillo, V.A., Stewart, R.J., and McIntosh, J.R. (1995a). Kinesin supports minus end-directed, depolymerization-driven motility of microspheres coupled to shortening microtubules. Nature *373*, 161–164.

Lombillo, V.A., Nislow, C., Yen, T.J., Gelfand, V.I., and McIntosh, J.R. (1995b). Antibodies to the kinesin motor domain and CENP-E inhibit microtubule depolymerization-dependent motion of chromosomes in vitro. J. Cell Biol. *128*, 107–115.

Mandelkow, E.-M., and Mandeldow, E. (1992). Microtubule oscillations. Cell Motil. Cytoskeleton *22*, 235–244.

Mandelkow, E.-M., Mandelkow, E., and Milligan, R.A. (1991). Microtubule dynamics and microtubule caps: a time-resolved cryo-electron microscopy study. J. Cell Biol. *114*, 977–991.

Mandelkow, E.M., Schultheiss, R., Rapp, R., Muller, M., and Mandelkow, E. (1986). On the surface lattice of microtubules: helix starts, protofilament number, seam, and handedness. J. Cell Biol. *102*, 1067–1073.

Margolis, R.L., and Wilson, L. (1978). Opposite end assembly and disassembly of microtubules at steady state in vitro. Cell *13*, 1–8.

McDonald, H.B., Stewart, R.J., and Goldstein, L.S.B. (1990). The kinesin-like *ncd* protein of *Drosophila* is a minus end-directed microtubule motor. Cell *63*, 1159–1165.

McDonald, K.L., O'Toole, E.T., Mastronarde, D.N., and McIntosh, J.R. (1992). Kinetochore microtubules in PTK cells. J. Cell Biol. *118*, 369–383.

McIntosh, J.R. (1991). Structural and mechanical control of mitotic progression. Cold Spring Harbor Symp. Quant. Biol. *56*, 613–619.

McIntosh, J.R. (1994). The roles of microtubules in chromosome movement. In: Microtubules, ed. J. Hyams and C. Loyyd, New York: Wiley-Liss, 413–434.

McIntosh, J.R., and Euteneuer, U. (1984). Tubulin hooks as probes of microtubule polarity: an analysis of the method and an evaluation of data on microtubule polarity in the mitotic spindle. J. Cell Biol. *98*, 525–533.

McIntosh, J.R., Hepler, P.K., and Van Wie, D.G. (1969). Model for mitosis. Nature *224*, 659–663.

McIntosh, J.R., and Koonce, M.P. (1989). Mitosis. Science *246*, 622–628.

Wadsworth, P., Shelden, E., Rupp, G., and Rieder, C.L. (1989). Biotin-tubulin incorporates into kinetochore fiber microtubules during early but not late anaphase. J. Cell Biol. *109*, 2257–2265.

Walker, R.A., Inoué, S., and Salmon, E.D. (1989). Asymmetric behavior of severed microtubule ends after ultraviolet-microbeam irradiation of individual microtubules in vitro. J. Cell Biol. *108*, 931–937.

Walker, R.A., O'Brien, E.T., Pryer, N.K., Soboeiro, M.F., Voter, W.A., Erickson, H.P., and Salmon, E.D. (1988). Dynamic instability of individual microtubules analyzed by video light microscopy: rate constants and transition frequencies. J. Cell Biol. *107*, 1437–1448.

Walker, R.A., Pryer, N.K., and Salmon, E.D. (1991). Dilution of individual microtubules observed in real time in vitro: evidence that cap size is small and independent of elongation rate. J. Cell Biol. *114*, 73–81.

Walker, R.A., Salmon, E.D., and Endow, S.A. (1990). The *Drosophila claret* segregation protein is a minus end-directed motor molecule. Nature *347*, 780–782.

Walker, R.W., and Sheetz, M.P. (1993). Cytoplasmic microtubule-associated motors. Annu. Rev. Biochem. *62*, 429–451.

Wang, S.-Z., and Adler, R. (1995). Chromokinesin: a DNA-binding, kinesin-like nuclear protein. J. Cell Biol. *128*, 761–768.

Waterman-Storer, C.M., Gregory, J., Parsons, S.F., and Salmon, E.D. (1995). Membrane/microtubule tip attachment complexes (TACs) allow the assembly dynamics of plus ends to push and pull membranes into tubulovesicular networks in interphase *Xenopus* egg extracts. J. Cell Biol. *130*, 1161–1169.

Weisenberg, R.C. (1972). Microtubule formation in vitro in solutions containing low calcium concentrations. Science *177*, 1104–1105.

Wendell, K.L., Wilson, L., and Jordan, M.A. (1993). Mitotic block in HeLa cells by vinblastine: ultrastructural changes in kinetochore-microtubule attachment and in centrosomes. J. Cell Sci. *104*, 261–274.

Wilson, E.B. (1928). The Cell in Development and Heredity, 3rd ed., MacMillan, New York.

Wilson, P.J., Forer, A., and Leggiadro, C. (1994). Evidence that kinetochore microtubules in crane-fly spermatocytes disassemble during anaphase primarily at the poleward end. J. Cell Sci. *107*, 3015–3027.

Wise, D., Cassimeris, L., Rieder, C.L., Wadsworth, P., and Salmon, E.D. (1991). Chromosome fiber dynamics and congression oscillations in metaphase PtK_2 cells at 23 degrees celsius. Cell Motil. Cytoskeleton *18*, 1–12.

Wordeman, L., and Mitchison, T.J. (1995). Identification and partial characterization of mitotic centromere-associated kinesin, a kinesin-related protein that associates with centromeres during mitosis. J. Cell Biol. *128*, 95–105.

Yen, T.J., Compton, D.A., Wise, D., Zinowski, R.P., Brinkley, B.R., Earnshaw, W.C., and Cleveland, D.W. (1991). CENP-E, a novel human centromere-associated protein required for progression from metaphase to anaphase. EMBO J. *10*, 1245–1254.

Yen, T.J., Li, G., Schaar, B.T., Szilak, I., and Cleveland, D.W. (1992). CENP-E is a putative kinetochore motor that accumulates just before mitosis. Nature *359*, 536–539.

Reprinted from *Nanofabrication and Biosystems*, pp. 123–138, 1996, with permission from Cambridge University Press.

8

Standard test targets for high-resolution light microscopy

RUDOLF OLDENBOURG

Marine Biological Laboratory, Woods Hole, MA 02543 and
Martin Fisher School of Physics, Brandeis University, Waltham, MA 02254

SHINYA INOUÉ

Marine Biological Laboratory, Woods Hole, MA 02543

RICHARD TIBERIO

National Nanofabrication Facility, Cornell University, Ithaca, NY 14853

ANDREAS STEMMER

Marine Biological Laboratory, Woods Hole, MA 02543

GUANG MEI

Marine Biological Laboratory, Woods Hole, MA 02543

MICHAEL SKVARLA

National Nanofabrication Facility, Cornell University, Ithaca, NY 14853

8.1 Introduction

The light microscope, aided by analog and digital image enhancement, is now used to visualize objects, and to measure events, at dimensions (including the third dimension along the microscope axis) near and considerably below the Abbe limit of resolution. Therefore, it is increasingly important that we assess experimentally, and understand quantitatively, images formed by high-resolution microscope optics from simple, well-characterized test objects. This is necessary both to avoid misinterpreting images and to gain further insight into the specimen fine structure. To address this problem, we are developing and fabricating test slides for the following two purposes: (1) to improve our understanding of the optical transfer functions and reliability of the image generated in three dimensions by wide-field and confocal microscope optics in various contrast modes, and (2) to provide a standard for assessing, and to help improve, the quality of microscope optics, electronic imaging equipment, and digital image processors.

Traditional test targets include fluorescent microspheres, pinholes, or jagged edges of lines ruled in metalized slides, silica shells of diatoms, and so forth. Fluorescent microspheres and pinholes act as point sources of light in the specimen plane, whose three-dimensional images – also called the point-spread functions (that are seen as one focuses above and below the in-focus image of the point source) – serve as sensitive indicators of optical aberration. Oblique-illumination images of the jagged edges in thin metal films in the Abbe test plate (which incorporates a slightly wedged coverglass whose graded thickness is calibrated) are also used for detecting aberrations. The silica shells of diatoms are decorated with rows of perforation (the frustule pores) whose spatial periods are, to a rather close approximation, characteristic of the diatom species. These rows of pores, and the pores within a row (which have a somewhat smaller spacing), are used as test targets for determining the resolution and image quality provided by a microscope objective lens.

The point-spread function is also used to calculate a thin optical section, or the "true" intensity distribution in the image attributable to a single specimen plane within a three-dimensional object (Agard, 1984; Carrington et al., 1990). In principle, the contrast transfer characteristics of the optical system, which includes the coverglass, can be calculated from the point-spread function measured on pinholes or the line-spread function measured on sharp edges in metalized slides. However, the calculations are not straightforward, and they involve assumptions that may not be readily quantifiable, especially in the presence of aberrations. We therefore fabricated simple test targets for contrast transfer measurements and image calibrations of object spacings down to, and somewhat beyond, the resolution limit of high NA (numerical aperture), highly corrected microscope objective lenses.

The test targets that we have fabricated contain line gratings with accurate periods down to 100 nm. The gratings allow, for the first time, direct measurement of contrast transfer characteristics of high-resolution microscope objectives with numerical apertures of up to 1.4. We have fabricated test targets for different contrast modes, including transmission and reflection microscopy (metallic targets), phase contrast, polarizing and differential interference microscopy (phase targets made of quartz), and fluorescence microscopy (targets made of fluorescently doped resist). Recently, we made gratings that can be mounted at a tilt angle to the focus plane of the microscope optics. With the tilted gratings we are able to measure three-dimensional imaging characteristics.

All test targets described here were developed jointly at the Marine Biological Laboratory and the National Nanofabrication Facility at Cornell University; therefore, we have named these targets MBL/NNF test targets.

We will first describe the fabrication procedures of the test targets and then report on our contrast transfer measurements imaging these targets with different light microscope optics. At the end, we show an image of a square grid used to document geometric image distortions, for example, in scanning probe (force) microscopy.

8.2 Fabrication of test targets

All test targets were fabricated at the National Nanofabrication Facility at Cornell University, Ithaca, N.Y. We used electron lithography techniques to make test patterns, such as bar gratings with accurate periods from 2.00 μm down to 0.10 μm, which is just beyond the resolution limit of light microscope optics of even the highest numerical aperture. Fabricated test patterns were checked for accuracy and consistency using a scanning electron microscope (Figure 8.1). As seen in Figure 8.1, targets contain bar gratings as well as single object features, such as double lines, single lines, double dots, and single dots. An additional, very useful test object that is incorporated is the Siemens star, which represents bar gratings of continuously varying periodicity and orientation (Figure 8.2). Furthermore, a square grating of 1-μm-spaced horizontal and vertical lines is included to assess geometric image distortions.

Each test target was fabricated on the surface of a microscope coverglass, selected for 0.17 ±0.005-mm thickness (the thickness assumed for lens design calculations to minimize aberrations). The coverglass is a stable substrate for the exceedingly thin

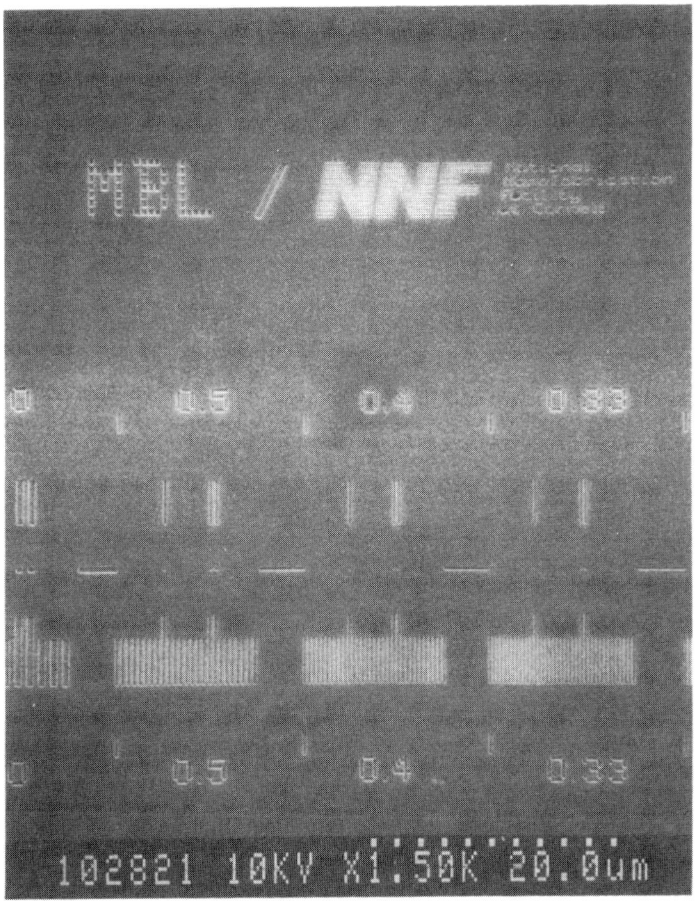

Figure 8.1. Scanning electron micrograph of portion of the MBL/NNF test target. Bar gratings, lines, and dots were etched into the 50-nm-thick aluminum using electron lithography techniques. Numbers above and below the edged features indicate bar grating periods in micrometers. Single and double lines, associated with each bar grating, have the same dimensions as the grating lines. Dot diameters and spacings are equal to the line widths and spacings of the associated line patterns.

test target itself and, at the same time, it is a required optical component for use with many highly corrected microscope objectives, (Inoué and Oldenbourg, 1994).

The targets mentioned so far extend in the focus plane of the microscope, that is, the entire target will be in focus for a given focus position. The targets are only 20- to 100-nm thick, much thinner than the z-resolution of an optical microscope. Therefore, these targets are ideal for measuring two-dimensional transfer characteristics of microscope optics. To determine three-dimensional imaging performance, we have fabricated oblique test targets that are tilted at angles of 45° or 90° to the focus plane. With the tilted targets we obtained experimental contrast transfer values using well-defined test objects that produced out-of-focus information at any given focus level. Next, we will briefly describe the fabrication procedures of the different types of test targets.

8.2.1 Metallic test target

Metallic test targets consist of an aluminum film, 50-nm thick, on a microscope coverglass. Etched into the aluminum film were test patterns created by nanolithographic processes: After evaporation of a uniform aluminum film onto the clean coverglass (glass plate, typically 18 x 18 mm^2, 0.17-mm thick), a thin polymer resist film (polymethyl methacrylate, PMMA, typically 100-nm thick) was spin coated on the aluminum. The resist layer was exposed with the test patterns using electron lithography. The exposed resist was developed, forming a mask for the subsequent reactive ion etching process. During reactive ion etching, the exposed parts of the aluminum were etched away, leaving most of the film intact. Hence, in a standard transmission light microscope the test patterns appear bright against a dark background (Figure 8.2). A method that we reported earlier produced dark test patterns against a white background (Oldenbourg et al., 1993).

8.2.2 Phase target

The fabrication of phase targets is very similar to the fabrication of metallic test targets. Instead of the metal film, a 90-nm-thick layer of SiO_2 was deposited directly on the clean coverglass surface. A thin PMMA resist layer was spin coated and then topped by a sputtered film of Au/Pd, which served as a conducting layer during the electron lithography exposure. After the lithographic exposure, the sputtered film was dissolved away, the resist was developed, and the SiO_2 layer was etched using the reactive ion etching process. It is worth mentioning that pure SiO_2 etches very well, whereas regular glass etches very slowly using standard reactive ion etching methods. Thus, the glass provides an accurate edge stop. As an alternative to SiO_2, other transparent dielectrics can be deposited onto the glass surface to obtain higher refractive index layers. A particularly simple target to make is the developed polymer resist mask itself which can be used as a

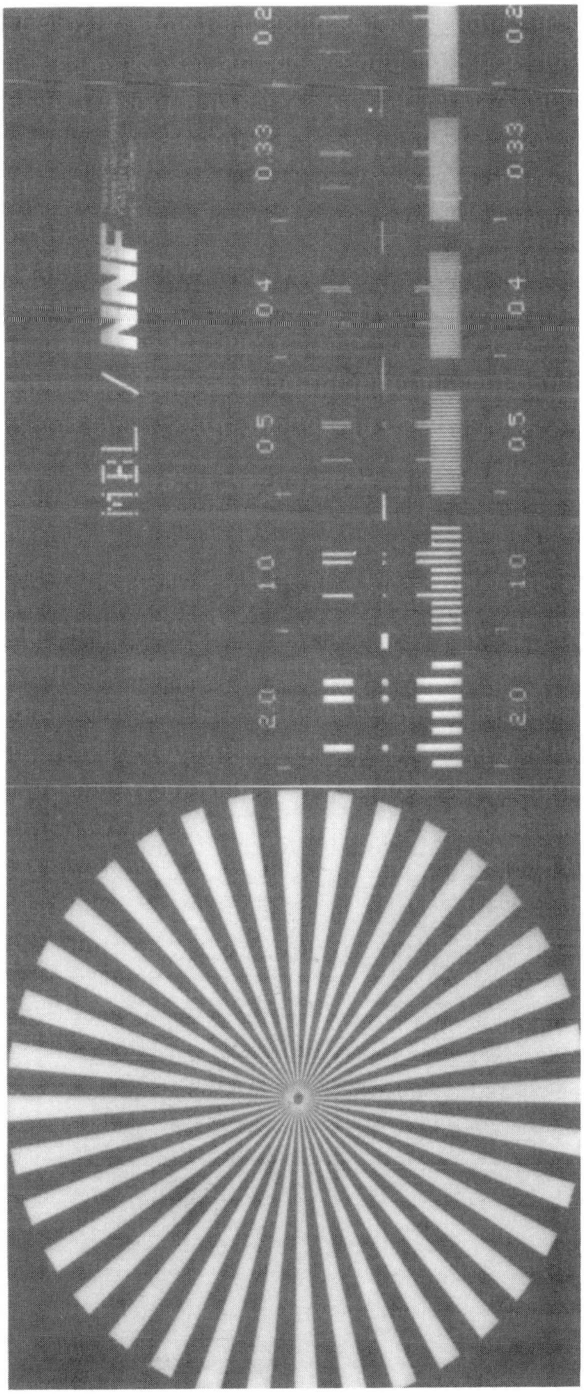

Figure 8.2. Siemens star, and line and dot patterns of the MBL/NNF test target imaged with a wide-field light microscope using transmitted light and a 60x/1.4 NA Plan Apo oil immersion objective lens (Nikon Inc.). The dark background is due to the low transmissivity of the 50-nm-thick aluminum film. Bright features were edged into the film (see text; also see legend to Figure 8.1). The Siemens star consists of 36 wedge pairs, with an outer diameter of 75 μm. The period near the outer edge is 6.5 μm, decreasing continuously toward the center. The smallest period is 0.1 μm and is near the inner black disk (1.2-μm diameter).

dielectric layer containing the test patterns (see Figure 8.3). However, the polymer resist mask is not stable against contact with solvents, including water, which tend to lift the mask off the substrate.

8.2.3 Fluorescent target

Of all the test targets we fabricated, the fluorescent targets were the easiest to make, and the easiest to loose, due to the bleaching of fluorescent dopants. We made fluorescent targets simply by adding a fluorescent, either rhodamine or fluorescein-like dopant (Molecular Probes, Eugene, Ore.), to the polymer resist solution, and then bleaching the test patterns into a 100-nm-thick fluorescent resist layer using electron beam irradiation. The fluorescent resist layer was coated on the coverglass by the standard procedure of dispensing some dissolved resist (PMMA in chlorobenzene) with fluorescent dopant on the coverglass, which was spun at 1,500 rpm for 1 min and then baked at 170°C for 1 h. To make the specimen suitable for electron lithography, a thin conducting layer of gold/palladium was sputtered onto the resist surface. None of these procedures noticeably affected the fluorescence efficiency of the dopant. The exposure of the test patterns by electron lithography bleached the coarse and fine gratings, Siemens star, and other parts of the pattern into the hardened, uniformly fluorescent layer (see Figure 8.4).

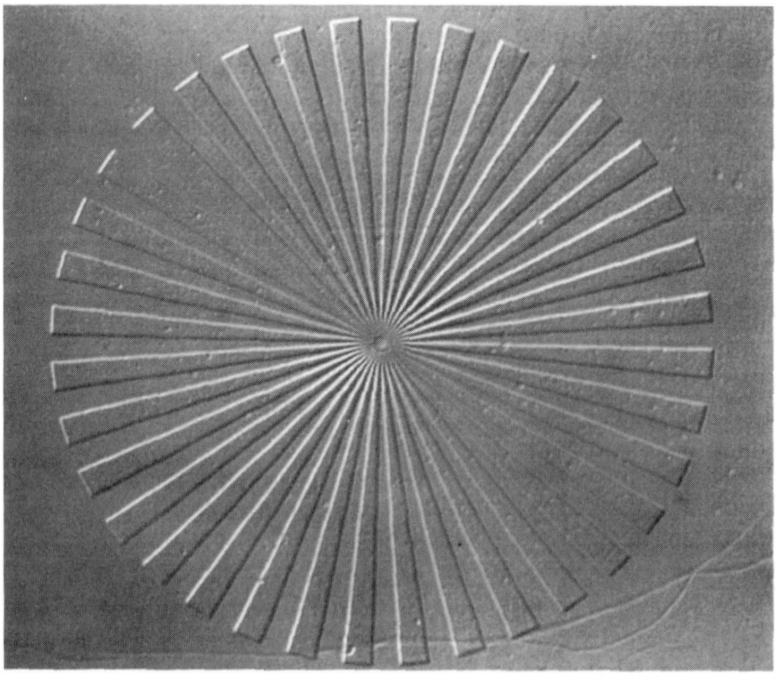

Figure 8.3. Siemens star fabricated as phase target (resist mask on coverglass) and imaged with a differential interference contrast microscope using a 60x/1.4 NA Plan Apo oil immersion objective lens.

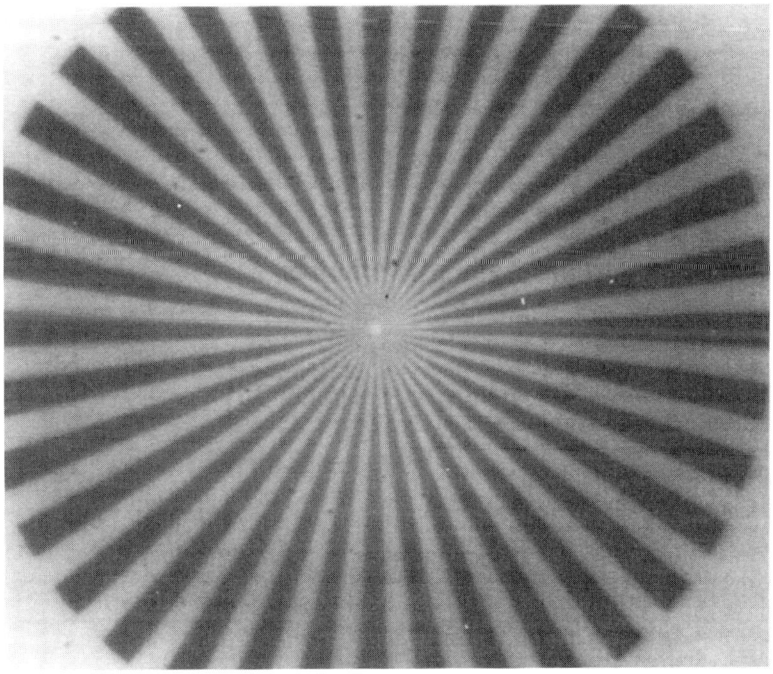

Figure 8.4. Fluorescent target with rhodamine-like dopant in resist layer imaged with epi-illumination microscope using 60x/1.4 NA Plan Apo objective lens and silicon intensified video camera (SIT). Wedges of the star pattern were bleached by electron lithography. Geometric distortions in the image are due to the intensifier stage of the video camera.

Although the fabrication is relatively simple and straightforward, the solid, fluorescent layer as a whole is progressively bleached when imaged in a regular fluorescence microscope. By keeping the light exposure to a minimum, however, we could preserve our fluorescent targets for several measurement cycles.

8.2.4 Tilted test target

Tilted test targets were fabricated on specially prepared edges of microscope coverglass. Several coverglass panels were glued together to form a glass block, which was firmly secured and glued to an aluminum chuck used for grinding and polishing one edge surface of the block (Figure 8.5a). The grinding surface was oriented either at 45° or 90° to the individual glass panels in the block. After the grinding, lapping, and cleaning of the edge surface, a 50-nm-thick aluminum layer was evaporated onto it. Bar gratings were sputtered from the aluminum film using a focused ion beam. The focused ion beam removes the aluminum directly on the sample surface during the exposure, at a resolution of about 100 nm, eliminating many steps of the electron lithography process described earlier. Furthermore, patterns that are written with high ion-beam intensity can be imaged immediately after writing using a low beam intensity. (Compared to electron lithography, however, ion-beam "milling"

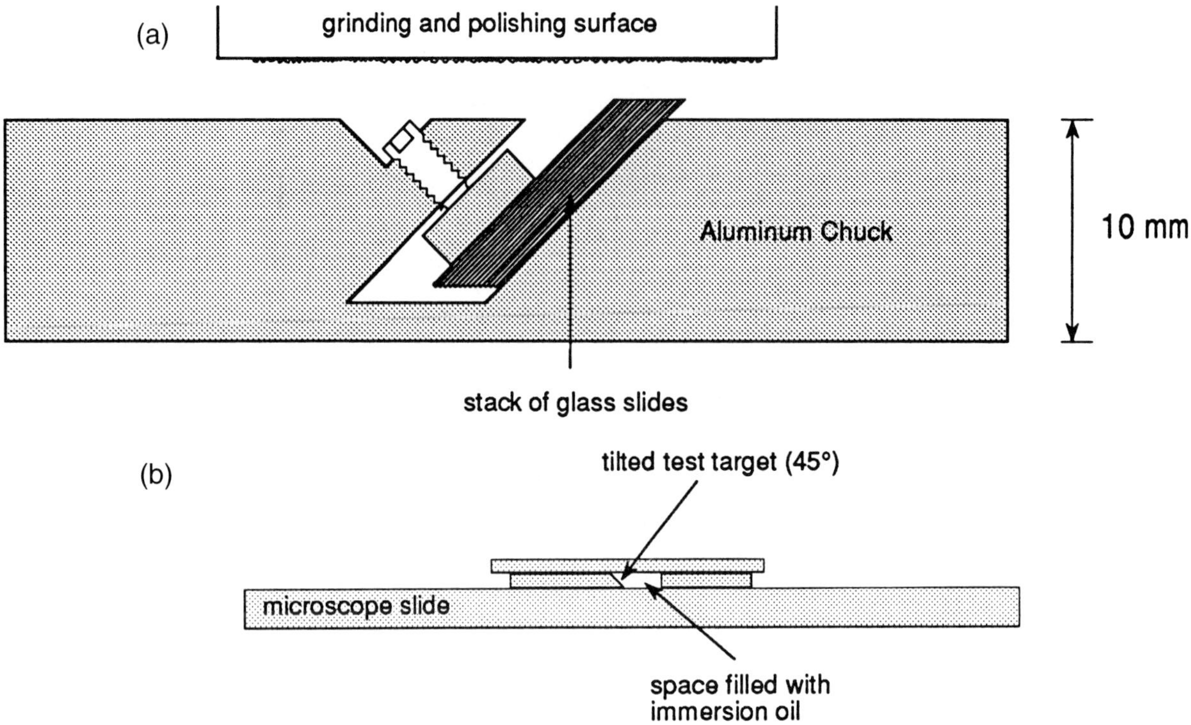

Figure 8.5. (a) Cross section showing block of coverglass panels secured in aluminum chuck for grinding and polishing. The edge surface of the coverglass stack is tilted at 45° to the grinding surface. (b) Schematic of tilted test target mounted on microscope slide and with top coverglass for observation by light microscope.

is still in an experimental stage, has lower resolution by at least a factor of two, and takes longer machine time; therefore, it is more costly in the exposure process.) After patterns were sputtered into the aluminum film, the glass block with aluminum chuck was soaked in acetone overnight to dissolve the glue ("Krazy Glue," cyanoacrylate ester) and recover the individual coverglass pieces with aluminized edges and test patterns. Individual coverglass pieces were mounted as shown in Figure 8.5b for contrast transfer measurements with the light microscope.

8.3 Applications of test targets

8.3.1 Contrast transfer of coherent confocal vs. incoherent nonconfocal optics

Figure 8.6 shows an intensity scan taken through a portion of the MBL/NNF test target as seen in Figure 8.2. The gratings with large periods are well resolved, but the ones with smaller periods are barely resolved. As the intensity scan shows, the signal from the well-resolved pattern is well modulated (the intensity varies between a large maximum, I_{max}, to a small minimum, I_{min}). As the spacing approaches the limit of resolution, the contrast transfer value $(I_{max} - I_{min})/(I_{max} + I_{min} - 2I_g)$

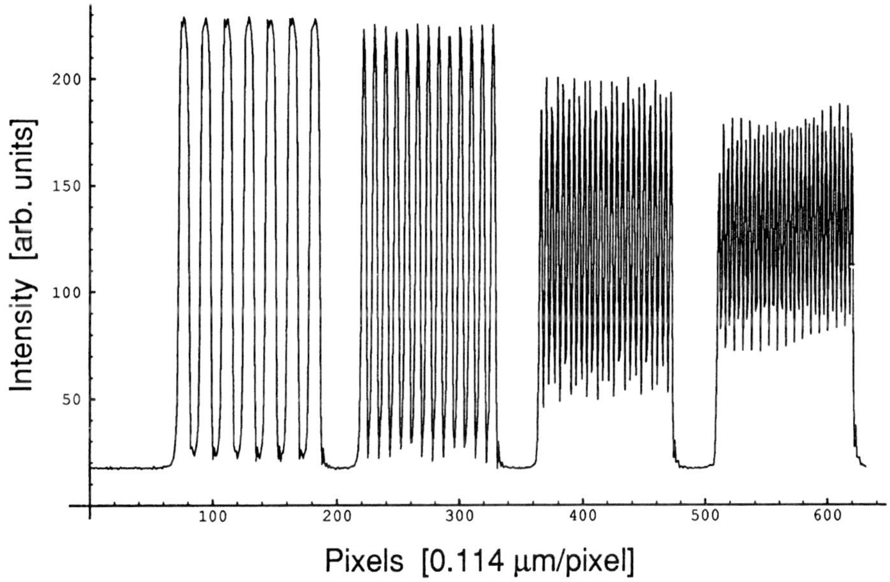

Figure 8.6. Sample of intensity scan through gratings of the MBL/NNF test target with grating periods (from left to right) of 2.0, 1.0, 0.5, and 0.4 μm. The target was imaged with a 40x/1.3 NA Fluor objective lens (Nikon Inc.) using light of 546-nm wavelength. Intensities were averaged over the dimension parallel to the grating lines.

decreases. At the background level, I_g, there is only weak transmission because of the low transmissivity of the aluminum film.

In a recent study (Oldenbourg et al., 1993) we have measured contrast transfer values using both confocal and nonconfocal microscope optics to investigate why images recorded in the confocal reflection mode seemed to be crisper and to resolve finer detail than images of the same specimens taken with nonconfocal optics. This is in contrary to the theoretical prediction that the confocal reflection mode has the same limit of resolution as the incoherent nonconfocal imaging mode (Wilson and Sheppard, 1984). Our results are summarized in Figure 8.7, which compares the ability of confocal versus nonconfocal optics to retain image contrast as a function of specimen spacing. These "contrast transfer characteristics" curves were measured for several plan apochromatic objective lenses on a single confocal microscope (prototype instrument developed by Hamamatsu Photonics K.K. of Hamamatsu City, Japan). The confocal transfer characteristic was measured in the reflection contrast mode with the exit pinhole closed down to a fraction of the Airy disk image diameter. The nonconfocal characteristic was measured in the transmission mode, which lacks an exit pinhole. As Figure 8.7 shows, our measurements confirm the resolution limit to be equal for both imaging modes. However, the contrast transfer for fine specimen detail that is close to the resolution limit is up to twice as efficient in the confocal compared to the nonconfocal imaging mode. This explains the improved contrast and resolution of the image in reflection confocal microscopy.

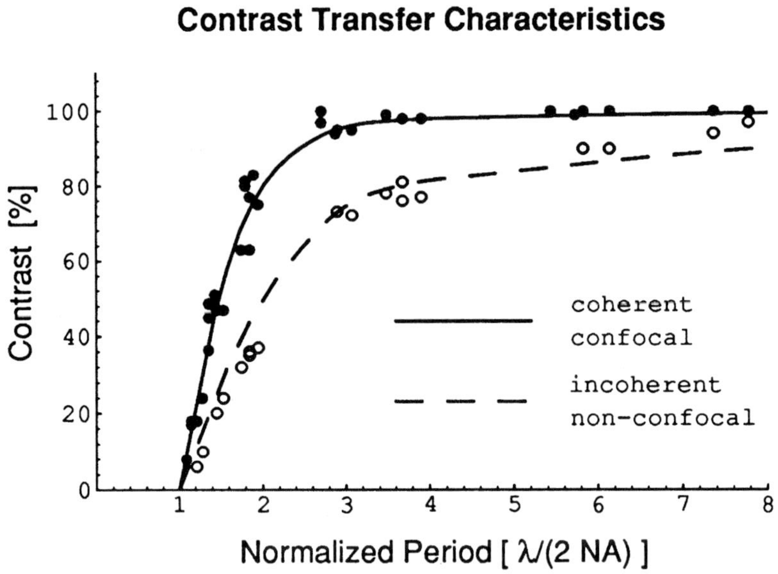

Figure 8.7. Contrast transfer characteristics measured in coherent confocal (•) and incoherent nonconfocal (○) imaging. The gratings were imaged with Plan Apo objective lenses (Nikon Inc., Melville, N.Y.) with numerical apertures (NAs) ranging from 0.45 to 1.4; laser wavelengths (λ) of 514.5 nm or 488 nm were used. Grating periods are expressed in units of the limiting wavelength, λ/2(NA), to normalize the data taken with different laser λ and lenses of different NA. (From Oldenbourg et al., 1993)

8.3.2 Contrast transfer of oil-immersion versus water-immersion objective

Objective lenses with the highest numerical aperture (NA > 1.3) are designed for use with homogeneous immersion; that is, the oil-immersion medium contacting the front element of the objective lens and the coverglass, the coverglass itself, and the medium imbibing the specimen are all expected to possess a refractive index (~1.515) equal to that of the front element of the objective lens. With the refractive indexes of all these layers equaling each other, the rays pass from the specimen to the hemispherical rear surface of the front element in the objective without being refracted or lost by total internal reflection. Thus, the front element of the objective lens is designed to capture the highest NA rays and to satisfy the sine condition that leads to correction for spherical aberration and coma (Inoué and Oldenbourg, 1994)

If one wishes to view living cells or functional cell-free extracts, the specimen must generally be imbibed in an aqueous medium. If an oil-immersion lens is used to view such a specimen, the lower refractive index (1.33–1.35) of the imbibing medium voids the assumption used in designing a homogeneous immersion lens. This gives rise to three undesirable optical effects: (1) The highest NA rays from the specimen are lost by being totally internally reflected at the water–glass interface; (2) the incidence angle of the rays that do enter the objective lens is restricted by refraction at the water–glass interface; and (3) the sine condition no longer holds for the objective lens front element. This third effect introduces spherical aberration and coma unless the region of interest in the specimen is positioned directly against the coverglass. The aberration becomes worse the further one focuses into the aqueous medium. The consequence is loss of resolution and image contrast, as clearly shown in the lower panel of Figure 8.8.

One should be able to avoid these losses by using a well-corrected, high-NA water-immersion lens, that is, an objective lens that is designed to use water

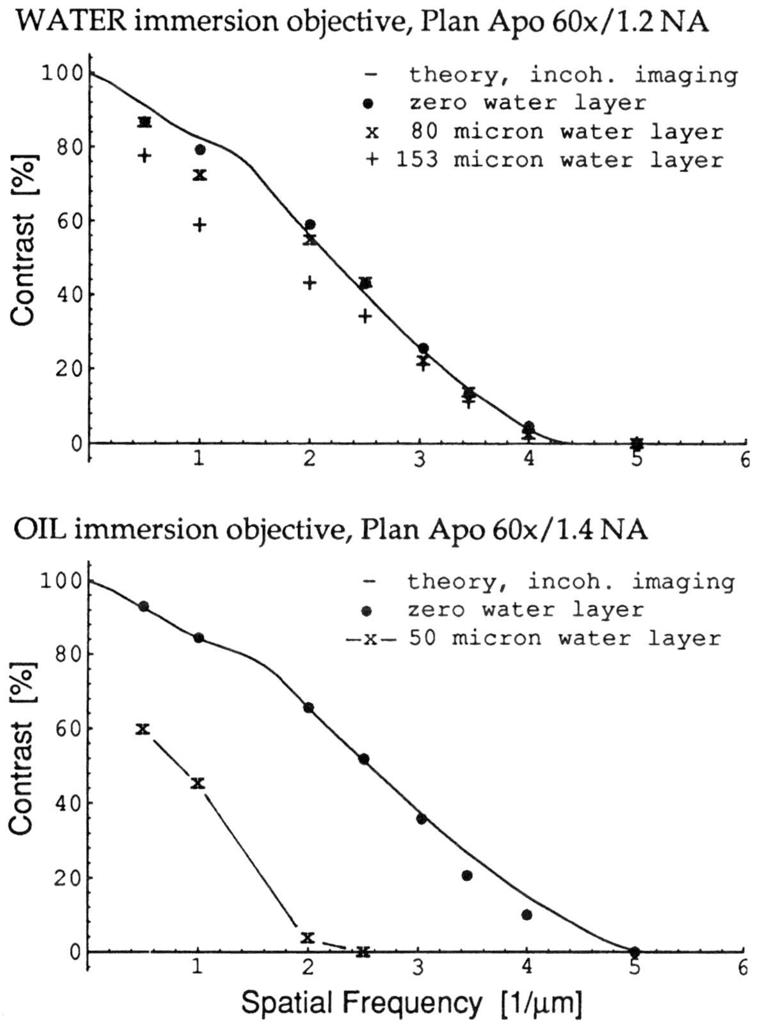

Figure 8.8. Contrast transfer values of (**upper panel**) 60x/1.2 NA water-immersion and (**lower panel**) 60x/1.4 NA oil-immersion objectives (both Plan Apochromats), measured with and without water layer between coverglass and test gratings. The graphs demonstrate that the water-immersion objective performs at the theoretical limit of contrast transfer even with water layers as thick as 153 µm, whereas the contrast transfer of the oil-immersion objective is dramatically reduced and the resolution limit is cut by half when the test grating is imaged through a 50-µm-thick water layer. The continuous lines are theoretically calculated contrast transfer functions of aberration-free objective lenses of corresponding NA at a wavelength of 546 nm (Oldenbourg and Inoué, 1994, in Brenner, 1994).

instead of immersion oil to fill the space between the objective front element and the coverglass. The upper panel in Figure 8.8 shows the contrast transfer property measured on a prototype of a recently developed Nikon Plan Apo 60x/1.2 NA water-immersion objective lens equipped with a correction collar (to accommodate slight variations in medium refractive index or coverglass thickness). Clearly the resolution and contrast transfer capability remain high even when the specimen needs to be focused through a thick layer of water between it and the coverglass. (For a sample DIC [differential interference contrast] image of a diatom frustule located under a 220-µm water layer beneath the coverglass, see Inoué and Stemmer [1994] in Brenner [1994].)

Although correction for the aberration introduced by the aqueous layer is important for observing details in living cells and tissue by conventional modes of microscopy, it is especially significant in confocal microscopy. In the latter mode, confocal efficiency is affected by the imaging quality not only of the imaging rays, but also of the illuminating rays that form the scanning image of the source pinhole. In the presence of aberration, the specimen is not only imaged with lower contrast and resolution, but scanned by a fuzzy volume of light instead of by a tight diffraction-limited image of the source pinhole. Thus with confocal microscopy in the presence of aberration, there is a dual source of loss in resolution and contrast as well as of loss in luminous efficiency.

8.3.3 Contrast transfer of bright-field microscope using tilted test targets

Recently, we recorded bright-field images of test targets tilted at 45° to the focus plane (Figure 8.9). We used a Nikon Microphot-SA microscope and a 40x Fluor oil-immersion objective lens with aperture iris (Nikon). The bar gratings with periods of 5.0, 2.0, 1.0, and 0.5 µm were illuminated with 546-nm light and images were recorded with a charge coupled device video camera (Dage-MTI CCD C72). Computer-assisted image analysis (NIH-Image, a public domain image analysis program for Macintosh computers, available by anonymous ftp from zippy.nimh.nih.gov/pub/nih-image) provided measurements of contrast transfer values at various grating periods (Figure 8.10).

The left panel in Figure 8.9 shows the image of a grating of the tilted test target recorded at a single focus level. Only one row of pixels near the center shows a small portion of the test grating in focus; the other rows represent out-of-focus images of other portions. The right panel of Figure 8.9 shows an extended-focus image of the same grating. The extended-focus image was assembled using the in-focus rows of images taken at different focus levels. Extended-focus images were used to measure contrast transfer values by the same procedure as discussed earlier. As can be seen in Figure 8.10, contrast transfer values for gratings tilted at 45° to the focus plane are generally lower than for horizontal gratings of the same peri-

Figure 8.9. Tilted test target (45°) imaged at a single focus level (left panel) and with extended focus (right panel). Grating lines run parallel to the tilt axis. Grating period, measured in the plane of the grating, is 2.0 μm. The image was recorded with a 40x oil-immersion objective with both objective and condenser NA adjusted to 1.0 (wavelength 546 nm). In the left panel the focus level is near the center part of the grating. Due to the shallow depth of field of the high-NA objective, grating parts with increasing distance from the central, in-focus part are progressively more out of focus. The extended-focus image is a composite of image parts taken at single focus levels.

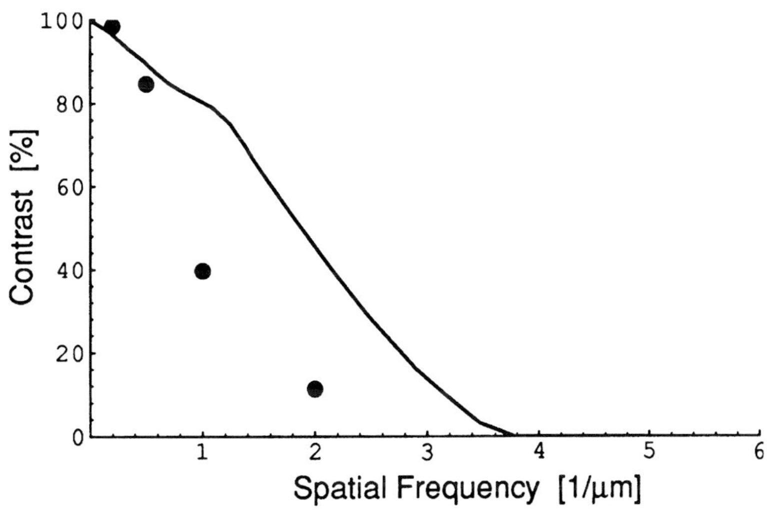

Figure 8.10. Contrast transfer values measured with tilted (45°) test gratings using extended-focus images such as the one shown in the right panel of Figure 8.9. The horizontal axis indicates the spatial frequency of the bar gratings as measured in the plane of the gratings. The continuous curve represents the theoretical contrast transfer function for *horizontal* bar gratings, which are located entirely in the focus plane, assuming imaging conditions that are identical to the ones used for the tilted targets (see legend to Figure 8.9).

Inoué, S. and Oldenbourg, R. (1994) Microscopes. In *Handbook of Optics*, 2nd Edition, ed. M. Bass, pp 17.1–17.45. New York: McGraw-Hill, Inc.

Oldenbourg, R., Terada, H., Tiberio, R., and Inoué, S. (1993) Image sharpness and contrast transfer in coherent confocal microscopy. *J. Microsc.* 172:31–39.

Wilson, T. and Sheppard, C. (1984) *Theory and Practice of Scanning Optical Microscopy.* London: Academic Press.

Reprinted from *Cell Motility and the Cytoskeleton*, Vol. 36, pp. 339–354, 1997, with permission from John Wiley & Sons.

Video Supplement

Cell Motility and the Cytoskeleton 36:339–354 (1997)

Amoeboid Movement Anchored by Eupodia, New Actin-Rich Knobby Feet in *Dictyostelium*

Yoshio Fukui[1]* and Shinya Inoué[2]

[1]*Cell and Molecular Biology, Northwestern University Medical School, Chicago, Illinois*
[2]*Marine Biological Laboratory, Woods Hole, Massachusetts*

To date, protrusion of pseudopodia has been considered to be primarily responsible for translocation of free-living amoebae and leukocytes of higher organisms. Although there is little question that the pseudopodium plays an important role, little attention has been given to the cortical structures that are responsible for cell-substratum anchorage in amoeboid movement. Here, we report on a new knobby foot-like structure in amoebae of a cellular slime mold, *Dictyostelium discoideum*. These feet, each about 1 μm in diameter, appear transiently in multiple units at the base of certain pseudopodia where the amoeba contacts a partially deformable substrate. The feet were discovered, and their spatial and temporal behavior relative to pseudopodial anchorage and invasive locomotion were observed, by examining *Dictyostelium* amoebae using a DIC video microscope providing an 0.3 μm depth of field. Key evidence for the anchoring role of the knobby feet was obtained by investigating amoebae, flattened in a specially devised observation chamber, and attracted by chemotaxis towards 3',5' cyclic-adenosine monophosphate (cAMP). The cAMP was released by highly localized, pulsed UV-microbeam irradiation of caged cAMP. We show by indirect immunofluorescence that the knobby feet contain a high concentration of filamentous (F-) actin, myoB (a member of *Dictyostelium* myosin-I family), and α-actinin (an actin-binding protein). Interestingly, myoB exhibits a circular disposition around each foot. Neither myosin-II (conventional myosin) nor the 269 kD protein, which has been recently identified as a talin homologue of *Dictyostelium* [Kreitmeier et al., 1995: *J. Cell Biol.* 129:179–188], are concentrated at the feet. We propose that the knobby feet provide anchorage to the substratum needed by lamellipodia to exert projectile forces for invading narrow spaces or otherwise for a flattened amoeba to secure itself to the deformable substratum. Some forms of adhesion plaques in higher organisms such as "podosomes" or "invadopodia" may perform functions similar to the knobby feet, but appear to differ in life time, cytoskeletal organization and composition. We have named the knobby foot "eupodium." Cell Motil. Cytoskeleton 36:339–354, 1997. © 1997 Wiley-Liss, Inc.

Key words: caged cAMP; chemotaxis; pseudopodia; ultrathin optical section; UV microbeam

INTRODUCTION

The term, pseudopodium refers to motile protrusive structures such as lobopodia (broad, thick protrusions found in amoeboid cells), filopodia (slender, finger-like projections found in a wide variety cells), axopodia (radial array of thin, long protrusions of *Heliozoa*), and myxopodia or rhizopodia (anastomosing cytoplasmic networks of *Myxomycetes* and *Rhizopoda*) [Hartmann, 1953]. Thin, veil-like protrusions formed by tissue cul-

Video tape documenting these observations has been accepted to be incorporated into a video supplement.

Contract grant sponsor: NIH; Contract grant numbers: ROI-GM39548 and R37-GM31617.

*Correspondence to: Dr. Yoshio Fukui, CMB, Northwestern Medical School, Ward 8-132, 303 East Chicago Avenue, Chicago, IL 60611-3008. E-mail: yoshifk@casbah.acns.nwu.edu

Received 17 July 1996; accepted 11 November 1996.

ture cells exhibiting ruffling movement (lamellipodia) are also classified as pseudopodium [Abercrombie et al., 1971]. All these are transient, actin-supported structures except axopodia and myxopodia, which are microtubule-supported semi-permanent (though rapidly contractile) structures.

Since Dejardin [1835], *Amoeba proteus* has served as a paradigm for studying the mechanism of amoeboid movement. The classic literature focused primarily on cytoplasmic streaming and gel-sol transformation and did not address the significance of amoeba-substrate attachment [Hyman, 1917; Mast, 1923; Pantin, 1923; review, Fukui and Yumura, 1986]. Nevertheless, it should be noted that Mast [1926], and earlier Dellinger [1906] in a most interesting paper, described an intriguing observation from the side of the amoeba in which they suggested the role played by a downward branch of the anterior pseudopodium towards the substratum in the crawling translocation of the amoeba. These workers interpreted the local contact of the anterior extension to the substratum as a means for supporting the buoyant weight, and providing a pivot for a forward tumbling by the giant amoebae that they were studying.

More than 20 years ago, Taylor et al. [1973] demonstrated that amoeboid streaming occurs in demembranated *A. proteus*. In that study, they implicated an important role of attachment of plasma membrane, along with glycocalix, to the substratum for amoeboid locomotion. More recently, Grebecki [1984, 1986] indicated the occurrence of a direct connection between the adhesion sites and the cytoskeleton in *A. proteus* based on detailed video microscopic analysis [review: Grebecki, 1994]. However, amazingly little is known about the structure, dynamics, and components of the attachment sites in such amoeboid cells as *A. proteus, Dictyostelium discoideum,* and leukocytes.

In tissue cultured cells, adhesion plaques were first identified as focal points formed along stress fibers in slowly moving cells such as fibroblasts and epithelial cells [Abercrombie et al., 1971]. They are 2–10 μm long, 0.25–0.5 μm wide structures which are believed to play crucial role in anchoring the stress fibers to the substratum [review; Burridge et al., 1988]. Adhesion plaques are only formed by cells cultured on a solid substratum but not by fibroblasts grown in collagen gels where the cells do not form discernible stress fibers [review; Burridge et al., 1988]. It is known that there is a sophisticated mechanical linkage and signal transduction network in the adhesion plaques involving proteins such as F-actin, α-actinin, vimentin, talin, α, β-integrins, and fibronectin (or vitronectin). The role of talin has been an issue of controversy because of its weak interaction with vinculin, which connects talin to F-actin [review Turner and Burridge, 1991].

Podosomes are dot-shaped sites of cell-substrate adhesion originally found in fibroblasts transformed by Rous sarcoma virus (RSV) [Tarone et al., 1985; Nermut et al., 1991], and more recently also called invadopodia [Mueller et al., 1992]. They are similar to adhesion plaques in composition but differ in several aspects including speed of formation (i.e., they appear in the initial 60 min of attachment while normal adhesion plaques require at least 180 min), and structure (i.e., podosomes are formed independent of stress fibers) [review Burridge et al., 1988]. Podosome-like structures are also formed by normal blood cells when they are penetrating the narrow space between endothelial cells [Wolosewick, 1984], or by osteoclasts during calcium resorption [Lakkakorpi and Väänänen, 1991]. In short, podosomes appear to be invasive protrusions formed from the ventral surface toward the substrate, and are considered to be a phenotypic variant of adhesion plaques [Tarone et al., 1985].

In *D. discoideum*, filopodia, lamellipodia, and lobopodia have been the only reported cortical structures so far believed to be involved in amoeboid locomotion. We report here that *Dictyostelium* amoeba also form foot-like, knobby cortical protrusions, in a reversible and dynamic manner, when it appears to require a firm anchorage to the substratum. To demonstrate that these cortical protrusions provide anchorage for invasive pseudopodial extension, we devised a new observation chamber and a method for attracting amoebae into thin spaces by highly localized, controlled release of cAMP using UV-microbeam uncaging of "caged" cAMP.

In the following we also show that the knobby cortical protrusions contain a high concentration of F-actin, α-actinin, and myosin-I, but not talin. In several aspects, this structure is different from adhesion plaques and related structures. Based on its structure and function, we name this structure *"eupodium"* (Gr., *eu* = true, primitive, *podos* = foot) as contrasted to *"pseudopodium."* A preliminary account of this structure appears in Fukui and Inoué (Biol. Bull., Oct., 1995) and Inoué and Fukui (Mol. Biol. Cell, 6, 261a, 1995).

MATERIALS AND METHODS

Cells and Cultures

Wild type amoebae (strain A x 3) of *D. discoideum* were cultured in a plastic flask with a 25 cm^2 bottom area (Falcon 3013; Becton Dickinson Labware, NY) containing 4.5 ml of HL5 medium made of proteose peptone, yeast extract, D-glucose, and Na/K-PO_4 (pH 6.5) [Cocucci and Sussman, 1970]. At late exponential phase, the medium was replaced with Bonner's salt solution containing 10 mM NaCl, 10 mM KCl, and 3 mM $CaCl_2$ [Bonner, 1947]. The cells were incubated for 2–3 h at 22°C, and stationary stage amoebae were scraped off from the bottom of the flask with a silicone rubber policeman. Aggregation stage amoebae were prepared by incubating a confluent culture in Bonner's solution for 18 h at 18°C before scraping off. The scraping does not adversely affect the viability, motility, or chemotaxis of the amoebae.

Observation Chamber

2.25% agarose-M (LKB Produkter, Bromma, Sweden) was dissolved in Bonner's salt solution and allowed to gel between two pieces of clean microscope slides supported with 50 μm-thick spacers. A small piece of the agarose film (50 μm-thick, 5 x 5 mm^2) was cut out and placed in the middle of a 22 x 22 mm^2 glass coverslip (no. 1½, 170 μm-thick; Corning Glass Works, Corning, NY) before applying the cells. A 1 μl aliquot of cell suspension containing about 200 amoebae was spread over the surface of the agarose film and excess fluid was sucked off with a tiny wedge of filter paper. For experiments involving UV-microbeam uncaging of caged cAMP (c^2AMP; see below), the agarose was pre-saturated with 100 mM c^2AMP. The sample overlaid with the cells was turned over and placed on a 22 x 40 mm^2 glass coverslip (no. 1½ , 170 μm-thick) supported by a ring-shaped 50 μm-thick spacer made of Mylar 200D plastic film (Du Pont, Wilmington, PA) and sealed with VALAB (a 1:1:1 mixture of vaselline, lanolin, and bee's wax). The air space between the agarose and the opening in the Mylar spacer provided adequate oxygen supply for the amoebae to continue to divide for more than a day. This chamber design provided good viability of the flattened amoebae as well as the optical conditions needed for high resolution DIC and UV imaging. The sample was then attached to an aluminum holder with dental tacky wax (Fig. 1b).

Video Microscopy

For visual-light observation and recording, the specimen was imaged through a universal inverted polarizing microscope [Inoué, 1986] illuminated through a quartz fiber optic light scrambler with 546-nm monochromatic light. The microscope was equipped with a rectified, 1.4 NA oil immersion DIC condenser (Nikon Inc., Melville, NY) and a 60x or 100x, 1.4 NA plan apo oil immersion objective lens (Nikon, Inc.) with DIC optics (Fig. 1a). The images were captured with a compact CCD camera (see below). Under this condition, the x-y and z-axis resolutions were 0.2 μm and 0.3 μm respectively [Inoué, 1988], and the temporal resolution was 16.5 milliseconds per video field.

UV Microbeam Irradiation

An auxiliary UV illuminator was attached to the universal polarizing microscope for controlled release of cAMP from the caged compound (Fig. 1a). The amoebae were placed in an observation chamber (Fig. 1b) on the 5 x 5 mm^2 agarose film saturated with 100 μM cAMP [Nerbonne et al., 1984] dissolved with 5 mM Hepes (pH 6.7) in Bonner's salt solution. The microscope system was modified from Inoué [Fig. III-21, 1986] in a manner similar to that described in Walker et al. [1989]. The modification includes the use of a mercury-Xenon arc lamp (model HPK L2422-01; Hamamatsu Photonics Systems, Bridgewater, NJ) as the visible light source, and replacement of the condenser with a glycerol-immersion, UV-transmitting objective lens (Ultrafluar 100x, NA 1.25; Carl Zeiss, Inc.) equipped with a DIC prism. A high-pressure HBO 100W mercury arc with quartz collector lens attached to a side rail of the microscope was used as the auxiliary UV source. The lamp output was filtered through a 366-nm band-pass filter (a combination of a Corning or Edmund Scientific [Barrington, NJ] heat cut filter that removes wavelength below 350 nm and a UG2 fluorescence excitation filter that removes wavelength above 380 nm). The UV beam, controlled with an electronic shutter (UniBlitz model D122; Vincent Associates, Rochester, NY), was reflected from a first-surface micromirror (0.15 x 0.35 mm^2 or 0.15 x 0.25 mm^2) placed in front of the field stop (Fig. 1a). The micromirror was positioned about 175 mm (instead of the standard projection distance of 150 mm) from the shoulder of the Ultrafluar lens used in place of a condenser. With this setting, the 60x or 100x plan apo objective provided a good UV image (e.g., Fig. 5b) of the micromirror through the 170 μm-thick glass coverslip and 50 μm-thick agarose (see above). The CCD camera had adequate sensitivity to capture the 546 nm green DIC image together with the image of the intense ultraviolet microbeam even after attenuation. To prevent excess UV light from entering the CCD camera, a "UV cut-off" filter (LL400-F-N963; Corion Corp., Holliston, MA) was inserted between the zoom ocular and the camera. The microbeam was delivered as a series of 3 or 30 msec flashes repeated every 650 msec.

The cells exposed to the intense 546 nm green light had to be kept at a room temperature no warmer than 21°C. At that temperature, we continuously observed healthy cells undergoing amoeboid movement for many hours, whether or not the amoebae were exposed to the c^2AMP-uncaging UV flashes. At room temperature of 23°C, the amoebae rounded up and stopped pseudopodia formation in several minutes under 546-nm green monochromatic illumination alone.

Imaging Technique

Images of living cells were recorded onto video tape using a Sony Extended Definition (ED)-Beta VCR and a Sony optical memory disk recorder (Sony Corp., USA, Paramus, NJ) in real time through a Hamamatsu Photonics Systems' model C-2400 CCD video camera. The image was processed with an Image-1/AT Digital Imaging Processor (Universal Imaging Corp., West

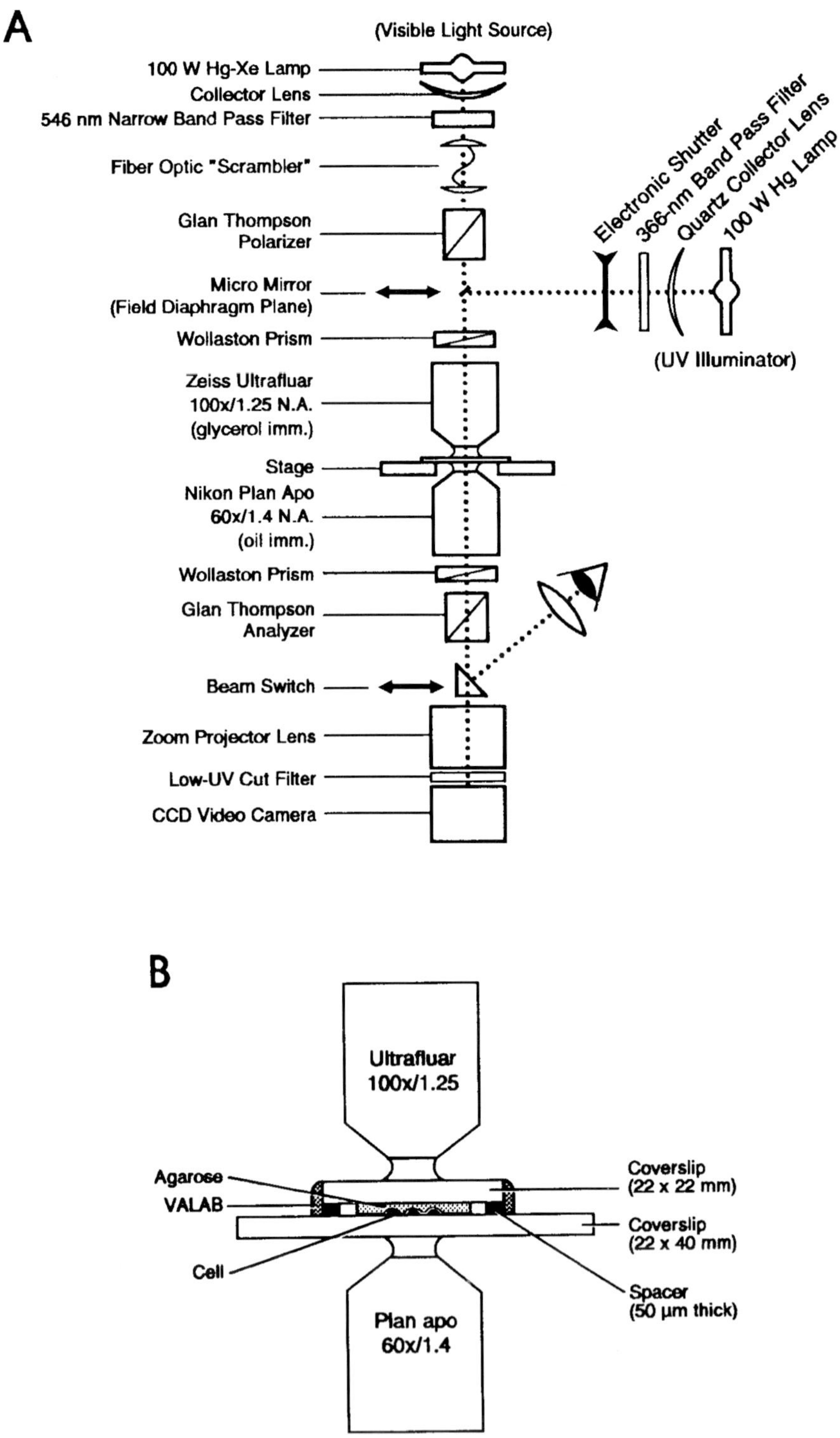

Fig. 1. Diagrams showing the microscope system. **a:** A schematic drawing of the universal polarizing microscope with an attached auxiliary UV illuminator. The UV, reflected from a minute first-surface mirror, is focused by a Zeiss Ultrafluar lens to produce a microspot that releases cAMP from c^2AMP. The basic system is similar to that described in Walker et al. [1989]. For detail, see Material and Methods. **b:** Diagram of the observation chamber. Note that the vertical and horizontal scales are not proportional. The amoebae are placed between a glass coverslip and a small (5 x 5 mm^2) piece of 50 µm-thick agarose sheet which is saturated with 100 µM c^2AMP in 5 mM Hepes buffer (pH 6.7). A Mylar spacer with a circular opening (12 mm in diameter) prevents the sample from being excessively flattened and provides air space around the agar. In addtion to providing excellent optical conditions for microscopy, the amoebae sealed in this chamber survive for several days. The lower coverslip is glued on an aluminum holder (not shown) and placed on the stage of the inverted microscope.

Chester, PA), edited using Photoshop (Adobe Systems, Inc., Mountain View, CA) and printed using a dye sublimation printer (model Phaser-II SDX; Tektronix, Wilsonville, OR).

Immunofluorescence Microscopy

For immunofluorescence, the samples were fixed with methanol containing 1% formaldehyde for 5 min at -15°C, and washed with phosphate-buffered saline (PBS: 138 mM NaCl, 2.5 mM KCL, 8 mM Na_2HPO_4, 1.5 mM KH_2PO_4, pH 7.2) three times for 5 min each. Primary antibody was applied to the samples and incubated for 30 min at 36°C. The samples were then washed with PBS three times for 5 min each. Secondary antibody was applied to the samples and incubated for 30 min at 36°C. The stained samples were then washed with PBS three times for 5 min each, followed by a brief rinse with distilled water. For double staining with rhodamine-phalloidin (rh-ph), rh-ph was mixed with secondary antibody at a final concentration of 3 μM. The samples were mounted with the mounting medium made of a 1:2:4 mixture of polyvinyl alcohol, glycerol and PBS containing 1% (w/v) of an anti-oxidant DABCO (diazabicyclo[2.2.2.]octane) (Aldrich Chemical Co., Milwaukee, WI). For fluorescence microscopy, a Zeiss Axioskop-50 or Photomicroscope-III (Carl Zeiss, Inc., Thornwood, NY) equipped with a 63x, NA 1.4 plan apo objective was used. The image was recorded through a cooled CCD camera (model PXL; Photometrics, Tucson, AZ) equipped with Kodak KAF1400 chips, or onto Kodak Tmax-400 monochrome negative films.

Chemicals and Probes

Caged cAMP (c^2AMP; adenosine-3', 5',-cyclic monophosphate, P^1-[2-nitrophenyl]-ethyl ester) with an absorption maximum at 350 nm was obtained from Calbiochem (La Jolla, CA). A mouse monoclonal anti-*Dictyostelium* myosin [DM-2; Yumura and Fukui, 1985] and a mouse monoclonal anti-*Dictyostelium* actin (HB-80; ATCC) were produced in our laboratory. A rabbit polyclonal anti-*Dictyostelium* myosin-I primarily recognizing myoB was generated by Dr. Thomas Lynch in Dr. Edward Kohn's laboratory at NIH [Fukui et al., 1989]. The mouse monoclonal anti-*Dictyostelium* α-actinin [Schleicher et al., 1988; Witke et al., 1992] and mouse monoclonal anti-*Dictyostelium* talin [Kreitmeier et al., 1995] were generously provided by Dr. Michael Schleicher (Ludwig-Maximilians University, Germany), and Dr. Günther Gerisch (Max-Plank-Institute, Germany), respectively. The rhodamine-labeled phalloidin (rh-ph) was obtained from Molecular Probes, Inc. (Eugene, OR).

RESULTS

Knobby Feet in the Stationary Stage Amoebae

The amoebae which were prepared between the agarose sheet and coverslip were flattened to different degrees. For example, stationary stage amoebae (between the vegetative and aggregation stages) shown in Figure 2, had an overall thickness of only about 3 μm. At mid focus, they clearly displayed several types of organelles (mitochondria and different sizes of vesicles and granules) actively streaming to and from the jaxtanuclear centrosome, located half way between the two nucleoli. The depth of field of the high (condenser and objective) NA DIC system we used was sufficiently shallow (less or equal to 0.3 μm) that these organelles became completely blurred or disappeared from view when the focus was "lowered" to the surface of the amoebae contacting the surface of the agarose layer (Fig. 3d). Instead, at that focus, we discovered the presence of small, knobby cortical structures that pressed into the agarose layer Fig. 3e).

The stationary stage amoebae actively protruded and retracted large pseudopodia. Occasionally, the amoebae formed more than one pseudopodium, but only one of them continued to develop while the others retracted prematurely. Full development of the pseudopodium took about 30 seconds. When a flattened amoeba protruded a pseudopodium, it produced knobby cortical structures at the cell-agarose interface (Fig. 2; small black arrows). The initial appearance of a single or two knobs was followed by the formation of multiple knobs (Fig. 3d,e). The size of the knobby feet was 0.95 ± 0.21 μm (n = 50) in cross section. Occasionally they appeared to be larger possibly due to close aposition (Fig. 2a,c,f,g; arrowheads). The knobby feet remained absolutely stationary relative to the substratum and never displayed lateral movement.

Interestingly, the knobby cortical structures usually exhibited an arc-like arrangement at the base of pseudopodium with the opening of the arc pointing towards the direction of the forming pseudopodium (Fig. 2a,c,d,e,g; white arrows, Fig. 3d,e). Within 30 seconds, the pseudopodium reached its maturity and started to retract. Coincidentally, the number and the size of the knobby cortical structures decreased, and they completely disappeared when a new pseudopodium developed from different parts of the cortex (Fig. 2b,f). The flattened shape of the pseudopodium (lamellipodium), the timing of formation and disappearance of the knobby cortical structures, and their stationary location relative to the substratum suggested that they play a role in anchoring the base of the forming lamellipodium to the substratum to help the pseudopodium penetrate into tight spaces. The relative positions of eupodia to pseudopodium is schematically

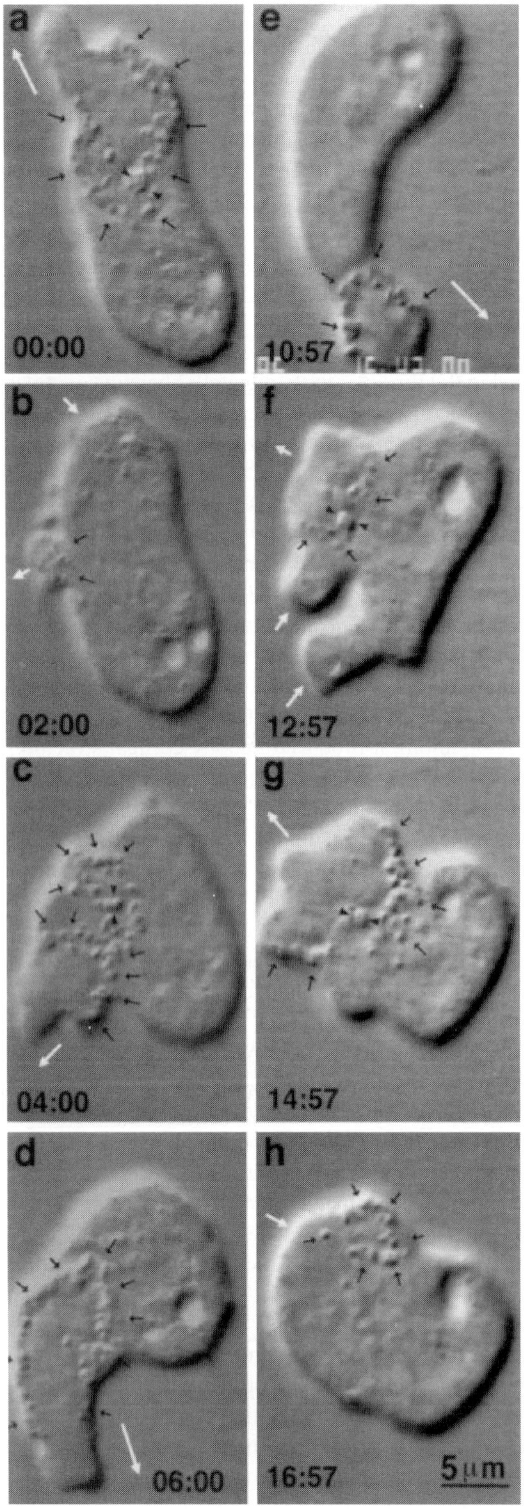

Fig. 2. Time lapse sequence showing the dynamics of eupodia formation in a stationary stage amoeba flattened to ca. 3 μm thickness between the coverslip and agarose sheet. Visualized at the amoeba-agarose interface using the very shallow depth of field (ca. 0.3 μm) achieved in our DIC system that provides an exceptionally high objective and condenser NA (= 1.4). **a–d:** Sequence during the first 6 min printed once every 2 min. **e–h:** Sequence of the same amoeba starting at ca. 5 min after panel d. Note that new eupodia, or knobby cortical feet (small black arrows) are formed aligned in an open arc at the base of each newly forming pseudopodium. Each foot is about 1 μm in diameter and does not change its position relative to the substratum. Some feet appear larger perhaps due to close aposition (arrowheads; a, c, f, g). The feet are resorbed as the amoeba retracts the pseudopodium. White arrows indicate the direction of forming or retracting pseudopodia. Time in min:seconds. Scale bar = 5 μm.

shown in the diagram viewed from the side of the sample (Fig. 4a).

We defined the surface of the amoeba contacting the agarose as ventral because the amoeba always remained attached on this side even when the depth of the medium was as high as 20 μm and almost the entire body was exposed to the fluid. In such cases, the amoebae stayed on the agarose surface and exhibited dynamic three-dimensional (3-D) shape changes by stretching multiple pseudopodia (lobopodia and filopodia) towards the fluid phase between the agarose and glass surfaces. When the pseudopodia contacted the glass surface, the amoebae quickly retracted the pseudopodia. Usually, such amoebae had a "euroid" (a globular, cytoplasmic appendage attached to the rear cortex of the amoeba) and several "retraction fibers" (long, straight, spiny structures attached to the euroid or the rear cortex).

Induced Feet in Aggregating Amoebae

We then used aggregation stage amoebae and induced cAMP-mediated chemotaxis by pulsed UV-microbeam irradiation of caged cAMP (c^2AMP) that was applied to the agar. This new technique allowed us to explore the spatial and temporal relationships between the dynamics of pseudopodial extension and formation of the knobby cortical structures, or feet.

When the microbeam, 1.5 x 3.5 μm^2 at the focal plane, was placed about 5-10 μm to the side and front of an aggregation-competent amoeba (Fig. 5a), the amoeba responded by 1) extending a lobopodium towards the pulsed UV spot, 2) completely turning towards, then 3) covering the focused UV microbeam (Fig. 5b), the source of cAMP. This behavioral response was consistent with the cAMP-mediated chemotaxis observed by application of cAMP through a micropipette, particularly the two-step responses originally observed by Swanson and Taylor [1982]. The chemotactic effects of UV flashes were specific to released cAMP since, unlike aggregation stage amoebae, stationary stage amoebae (which do not

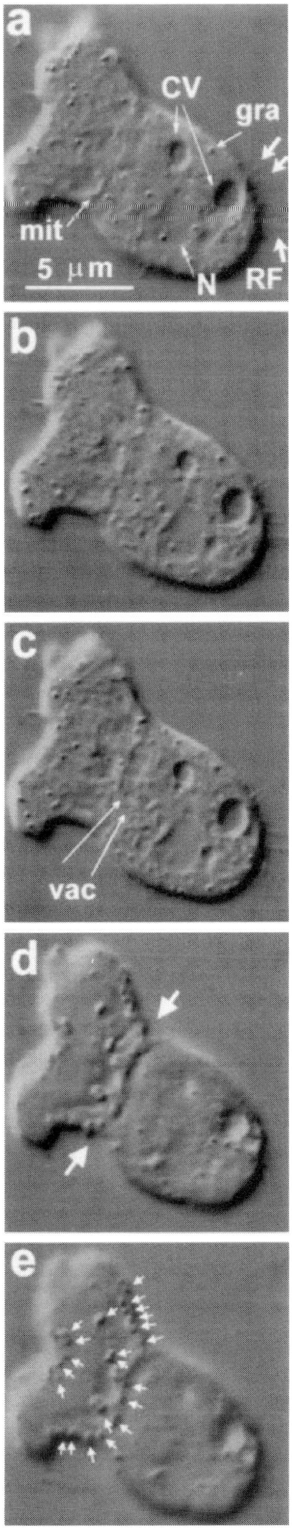

Fig. 3. A through-focus series of DIC images showing that the knobby feet project into the agarose matrix. The panels represent frozen frames selected from a video rate recording of optical sections of a total of 3.5 μm z-axis movement of the microscope stage. The height of the section from the dorsal surface is 0.5 μm (**a**), 1.5 μm (**b**), 2.5 μm (**c**), 3.0 μm (**d**), and 3.3 μm (**e**). In the middle cytoplasm (b,c), numerous organelles show saltatory (gra, vac) or random (CV, N) modes of movement. At the ventral cortex contacting the surface of agarose (d), an array of the knobby feet ("eupodia") exhibits a continental fold-like appearance (large arrows). Focusing 0.3 μm below from the ventral surface (e), the unitary appearance of eupodia (small arrows) becomes obvious. Also note a mottled appearance of the agarose matrix surrounding the eupodia (e). CV: contractile vacuole; gra: granule; mit: mitochondrion; N: nucleus; RF: retraction fiber; vac: vacuole. Scale bar = 5 μm.

have cAMP receptors and therefore are insensitive to cAMP) did not show any response. Furthermore, amoebae in neither stage responded to UV microbeam flashes applied in the absence of c^2AMP.

In the monopodial, aggregation stage amoebae that were relatively unflattened and which continued to migrate (cell #1 and #2), we did not observe the formation of knobby feet (Fig. 5a,b). Our observation on the stationary stage amoebae (Fig. 2), nevertheless, suggested that knobby feet are essential when firm anchorage of the cell to the substratum is required.

The above suggestion was reinforced by observing the amoeba that was located at the center of an induced aggregation, immediately beneath the UV-microbeam (Fig. 5c,d). When the microbeam was precisely focused at the amoeba-agarose interface, the centrally-located amoeba stopped migrating, became frozen in location and extended into a disk shape (cell #1 in Fig. 5c). At this point, the amoeba, which has become flatter than the migrating monopodial amoebae, form knobby feet at the lower middle surface just beneath the microbeam (arrows in cell #1 in Fig. 5d; diagram in Fig. 4b). When the microbeam was turned off, the feet disappeared within a matter of a minute and the amoeba became motile again (Fig. 5d,e). The induction of the feet was reversible; when the UV flashes were restarted, they reappeared very rapidly under the beam center. Irradiation of the UV microbeam with 30 msec exposure and 650 msec interval for at least a half hour did not cause any visible damage to the amoebae.

As shown in Figure 5, amoebae other than the central one just described, also responded to cAMP and formed an "aggregate" by surrounding the first amoeba now affixed beneath the microbeam. Continuous, long term observations showed that the surrounding amoebae, which remained more elongated and did not form knobby feet, exhibited a swirling movement around the central amoeba. As long as the UV flashes continued, the aggregate continued to increase in size until several scores of amoebae swirled in a stream around the central affixed amoeba.

Dynamics of the Knobby Feet Formed in Thin Lamellipodia

A key evidence for the role played by the knobby feet in amoeba-substratum anchorage was achieved by observing amoebae that projected extremely thin lamellipodia into a tight space (Fig. 6; schematic side view in Fig. 4c). Amoebae crawling in a naturally formed channel in the agarose were induced to project lamellipodia into a thin space between the coverslip and agarose adjacent to the channel. When the microbeam source of cAMP was placed at about 5–10 μm from the channel, the amoeba responded by 1) ceasing its migration (Fig. 6a), 2) forming several filopodia and lamellipodia at time 0:10, and 3) extending an extremely thin lamellipodium towards the image of the UV beam (Fig. 6b; schematic side view in Fig. 4c-1). Coincidentally, knobby feet were formed on the penetrating lamellipodium (arrows in Fig. 6b,c; schematic side view in Fig. 4c-2 to c-3). The knobby feet quickly disappeared when the UV microbeam was turned off and as the lamellipodium started to retract (Fig. 6d,e; schematic side view in Fig. 4c-4). Again, the position of the feet was fixed relative to the substratum and showed no sign of lateral movement. When the c^2AMP uncaging UV microbeam was activated again, the retracting lamellipodium re-expanded towards the cAMP source as it formed new knobby feet. Thus, there is a close correlation between feet formation and invasive extension of the lamellipodium.

Cytoskeletal Organization of the Feet

In bright field microscopy of fixed amoebae, the knobby feet, or eupodia, appears as dark bodies indicating their high refractive index (Fig. 7a,d,g). Immunofluorescence staining demonstrates that the feet are highly enriched in the major cytoskeletal component actin (Fig. 7b). Staining with rh-ph showed that most actin in the feet is filamentous (F-actin) (Fig. 7e,h). Double-staining for actin and myosin-I demonstrated that myosin-I (myoB) is also localized in the feet. Myosin-I exhibits a ring-like structure, indicating that it is concentrated at the periphery of the feet (Fig. 7b,c). In contrast, α-actinin shows a pattern of homogeneous staining within the feet suggesting a more uniform association with F-actin filaments located there (Fig. 7e,f). Certain variations in the occurrence between F-actin and myosin-I as well as F-actin and α-actinin were noted. Particularly interesting was that some of the knobby feet only showed high accumulation of F-actin, but not of myosin-I (arrowhead, Fig. 7a–c).

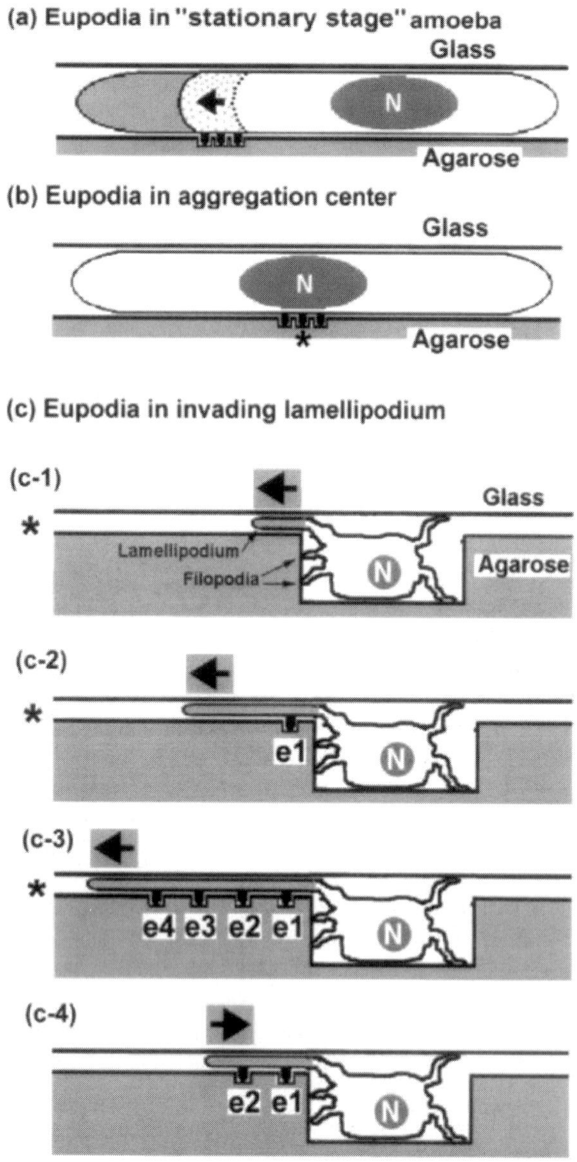

Fig. 4. Schematic diagrams of flattened amoebae with eupodia in side view. **a:** Naturally forming eupodia at the base of an extending pseudopodium in randomly moving "stationary stage" amoeba. The schematic corresponds to the image shown in Figure 2e. **b:** Eupodia induced beneath the source of cAMP at the ventral surface of the aggregation stage amoeba located in the center of the swirling aggregate. The schematic corresponds to Figure 4d. **c:** Eupodia formed on a very thin lamellipodium invading into a narrow space. The amoeba stops in the channel and extends its lamellipodium towards the source of cAMP released by the UV-microbeam. **c-1**: The lamellipodium begins to invade into the thin space. **c-2**: The first eupodium is formed. **c-3**: As the lamellipodium continues to develop, more eupodia are formed. Some of them disappear while others are formed; but they do not show any lateral movement relative to the substratum. **c-4**: The lamellipodium is retracting as the UV flashes have been turned off and the amoeba retreats to the channel. The eupodia disappear as the lamellipodium retracts. The image sequence i shown in Figure 5. The figures in (c) are drawn smaller than those in (a) and (b). For simplicity, the drawing of the cell body in c-1–c-4 does not show its dynamic shape changes. Arrows a,c = direction of expansion of lamellipodium. N: nucleus. *: c^2AMP uncaging UV flashes. (e-1–e-4): eupodia.

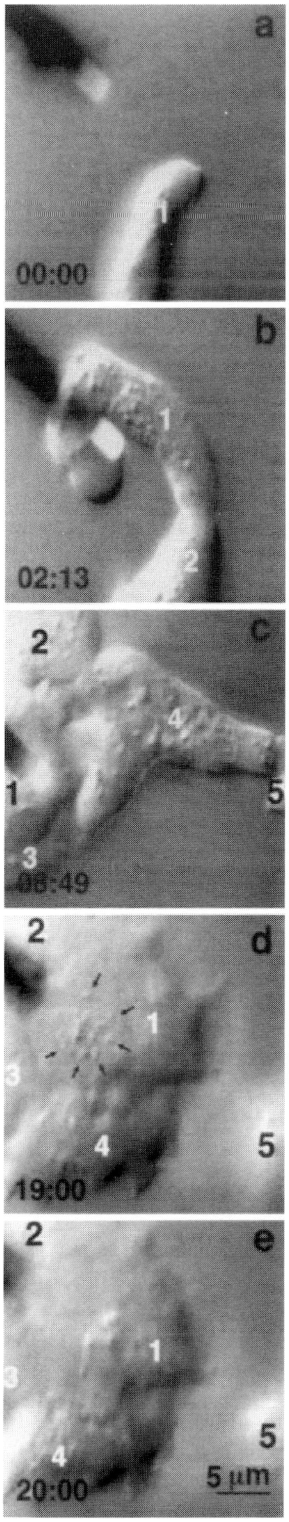

The localization of a talin homologue of *Dictyostelium* was determined by using two mouse monoclonal antibodies: 1) mab 169-477-5 raised against purified *Dictyostelium* talin [Kreitmeier et al., 1995], and 2) mab 227-341-3 raised against a fusion protein containing N-terminus of *Dictyostelium* talin [Jens Niewöhner, Max-Planck-Institute, unpublished]. Both antibodies showed primarily diffuse cytoplasmic staining with a higher accumulation at the cortex (Fig. 7i). Double staining with rh-ph demonstrates that *Dictyostelium* talin is not concentrated at the eupodia (Fig. 7h,i).

These results reveal that the knobby feet are rich in F-actin associated with actin-binding proteins such as myosin-I and α-actinin. In contrast, a *Dictyostelium* homologue of talin is not concentrated in the eupodia. We have also examined the localization of conventional myosin (myosin-II) and find that it is not concentrated in the eupodia (data not shown).

At high resolution, fluorescence micrographs demonstrate that F-actin is disposed in a radial array after emerging from the eupodia (Fig. 8). It appears that most of these radially arrayed F-actin filaments run parallel to the ventral cortex. Some of the filaments appear to interact with others from neighboring eupodia at the periphery of the radial array (curved arrow in a; double arrows in b). Actin filaments also interact with the axial bundle of F-actin cables emerging from the tip of the lamellipodium (arrowheads, a,b) suggesting the presence

Fig. 5. Chemotaxis towards uncaged cAMP, and eupodia formation in an amoeba located at the center of a forming aggregate. When activated, the electronic shutter was opened for 30 msec every 650 msec exposing the preparation to intermittent 366-nm UV microbeam, approximately 1.5 x 3.5 µm in size. The microbeam released cAMP from the agarose containing 100 µM c^2AMP and acted as the source of a transient gradient of cAMP which should dissipate rapidly. The microbeam (white rectangle) was located somewhat above the center of the field in panels a and b. The dark rod is the shadow of the metal arm supporting the micromirror whose actual size was about 150 x 350 µm (see Fig. 1). **a:** The image of the micromirror was placed about 10 µm to the side of the anterior lobopodium of cell #1 and pulsed irradiation started. **b:** The amoeba turned to the beam where the uncaged cAMP was released and started a spiral motion around the beam center. A second cell (#2) followed the first cell (#1). **c:** The same amoeba (#1) stopped migrating, changed shape from elongated to pancake shaped, and stayed just beneath the UV beam. Note that other amoebae (#2–5) were also attracted to the cAMP source (or cAMP released from congregating amoebae) and started to form a small aggregate encircling the first amoeba. **d:** In cell #1, a group of eupodia was formed beneath the UV microbeam (which had been turned off and removed from the field just 3 sec before this frame was recorded). Small black arrows point to representative eupodia. **e:** 63 sec after the UV microbeam was turned off. The eupodia are disappearing. In (a–c) focused near the mid-plane of the amoeba, organelles are visible. No eupodia are present at this stage. d,e: Focused at the agar-amoeba interface, show the eupodia that are present only in cell #1 during, and for a short while after, cAMP is released by the UV microbeam. Time in min:sec. Scale bar = 5 µm.

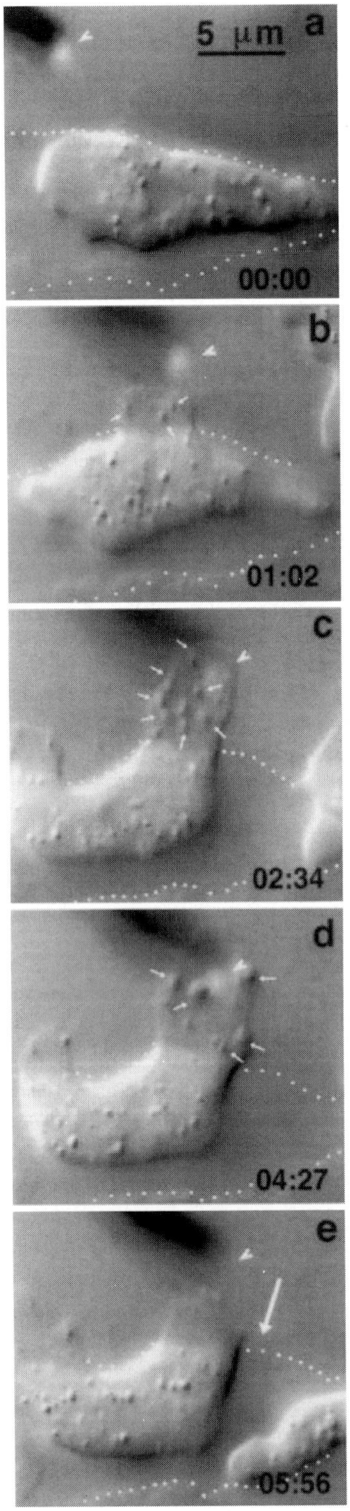

of mechanical linkage between eupodia and the lamellipodium through F-actin cables. Interestingly, several long, very thin, F-actin-containing fibers were also observed, connecting the eupodium and the leading edge of the pseudopodium, as represented by the one indicated by the triple small arrowheads in Figure 8b.

DISCUSSION

Is Eupodium a True Foot?

As demonstrated in this study, the knobby feet are stationary relative to the substratum and never show lateral movement. This structure usually appears as multiple units arranged approximately in an arc of a circle opened in the direction of a forming lamellipodium invading a narrow space. They disappear as the lamellipodium retracts. The knobby feet also formed at the ventral surface of an aggregation stage amoeba, located and anchored in the center of an aggregate just above the cAMP uncaging UV microbeam. Other amoebae lacking knobby feet continued to actively swirl around this central amoeba. The knobby feet were resorbed and the central amoeba began to migrate again very soon after the UV microbeam was turned off.

Thus the knobby feet are formed only when a firm cell-substratum anchorage appears to be required. These characteristics suggest an obvious function of eupodia, i.e., transient anchorage to deformable substratum. Therefore, we named this structure eupodium. In natural habitat, we believe the eupodia could anchor pseudopodia that extend into tight spaces when the amoebae crawls in the soil and amidst rotten leaves. The absence of eupodia in rapidly moving aggregation stage amoebae suggest that amoebae in that stage use locomotory mechanisms

Fig. 6. Dynamics of eupodia formed on a thin lamellipodium invading an exceptionally narrow space between agarose and coverslip. **a:** An aggregation stage amoeba exhibiting a monopodial lobopodium and migrating to the left of the field along a narrow channel in the agarose. A 1.5 x 2.5 μm^2 UV microbeam (white arrowhead) was placed to the side of the channel and 3 msec pulses started 18 sec before this frame. **b:** A thin lamellipodium projects toward the source of the uncaged cAMP. The small white arrows point to the first three eupodia developing. **c:** The lamellipodium has reached the microbeam as prominent eupodia are formed (white arrows). **d:** The thin lamellipodium exhibited dynamic shape change during the 2 min period between c and d, and eupodia were formed at different positions (small arrows). **e:** The UV microbeam was turned off within 1 second after (d). The lamellipodium is quickly retracting and coincidentally all the eupodia are disappearing. All of the spherical structures other than eupodia are organelles in the endoplasm as determined by through focus observation. In contrast to the eupodia whose locations do not change, the organelles actively move around in the endoplasm; only those that are within 0.3 μm of the focal plane are imaged. Arrowhead: UV microbeam releasing cAMP. Dotted line: edge of the channel. Small arrows: eupodia. Large arrow: retracting lamellipodium. Time in min:sec. Scale bar = 5 μm.

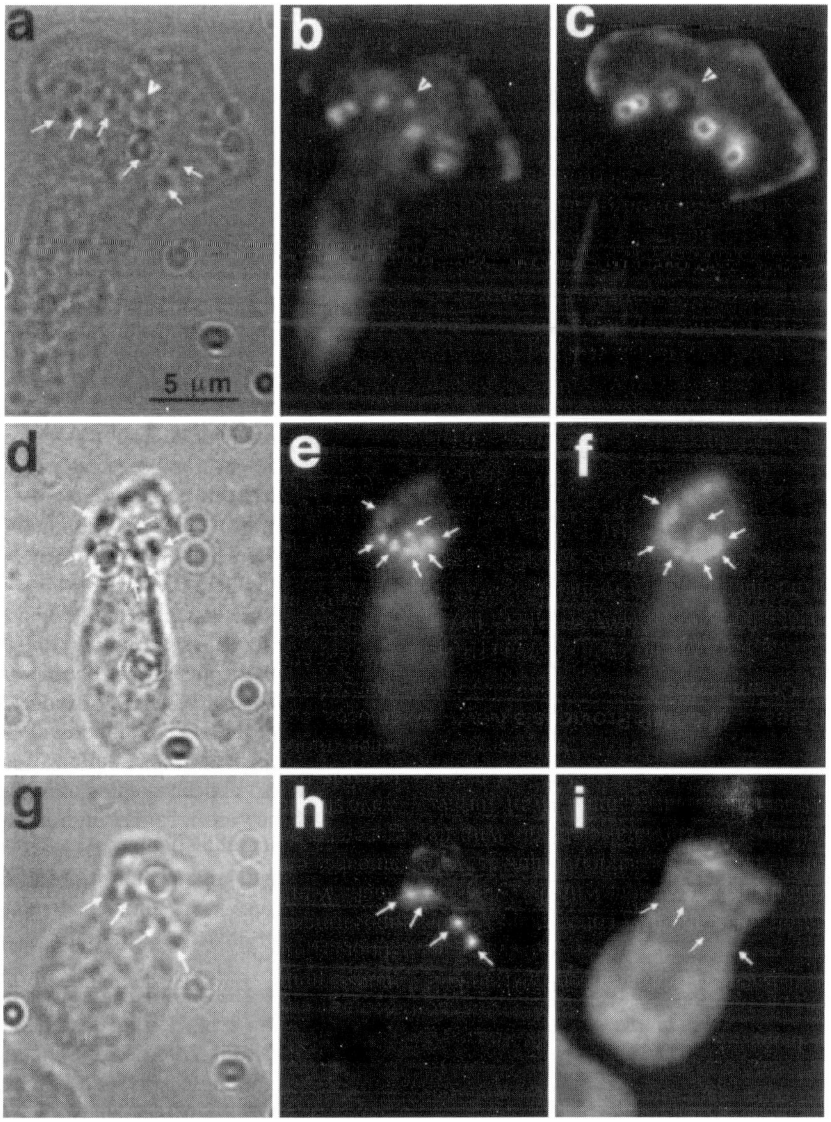

Fig. 7. Immunofluorescence localization of actin and related cytoskeletal components at the eupodia of the stationary stage amoebae. a–c: Double staining for actin and myosin-I. **a**: Bright field. **b**: Actin as stained with a mouse monoclonal anti-*Dictyostelium* actin. **c**: Myosin-I (myoB) as stained with a polyclonal anti-*Dictyostelium* myosin-I. Myosin-I shows a ring-like distribution indicating that it is located beneath the membrane of the eupodium. One eupodium (arrowhead) is rich in F-actin (b), but not in myosin-I (c), indicating that these cytoskeletal components may be integrated into, or leaving, the eupodium at different times. d–f: Double staining for F-actin and α-actinin. **d**: Bright field. **e**: F-actin as stained with rh-ph. **f**: Staining with a mouse monoclonal anti-*Dictyostelium* α-actinin demonstrating that it is highly concentrated throughout the eupodium. g–i: Double staining for F-actin and talin. **g**: Bright field. **h**: F-actin as stained with rh-ph. **i**: Staining with a mouse monoclonal anti-*Dictyostelium* talin showing that *Dictyostelium* talin is not concentrated at the eupodia. Arrows: eupodia. Scale bar = 5 μm.

similar to leukocytes moving in a non-invasive mode, i.e., by simple pseudopodial projection and retraction [Kolega et al., 1982].

Is Eupodium a Dictyostelium Adhesion Plaque?

The eupodium is a structure distinctive from adhesion plaques of higher organisms based on the following criteria. 1) While adhesion plaques are always attached to stress fibers, *Dictyostelium* does not form stress fibers. 2) While it takes at least 180 min for adhesion plaques to establish, eupodia is formed in less than 2 min. 3) While adhesion plaques are only formed on solid substratum (but not in collagen gel) [Burridge et al., 1988], eupodia are formed on partially deformable substratum. 4) While

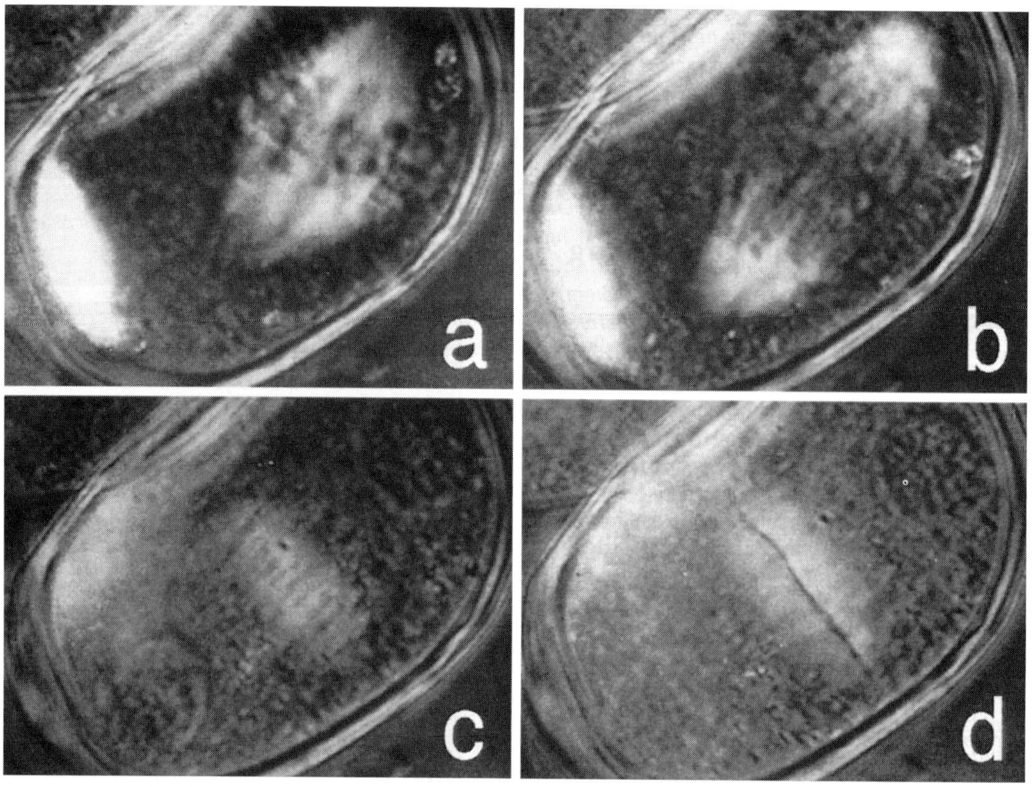

Figure 2. Mitosis and cell plate formation in centrifuged pollen mother cell of the Easter lily, *Lilium longiflorum*, observed with polarization microscopy. Reproduced from *The Journal of Cell Biology*, vol 130, 687–700, 1995, by copyright permission of The Rockefeller University Press.

where molecules are regularly aligned, i.e., birefringent regions, become highlighted (Figures 2, 5, and 6).

As with the Bajer's endosperm cell series, this series on the pollen mother cell mitosis was initially time-lapse recorded on 16-mm ciné film. The elapsed time

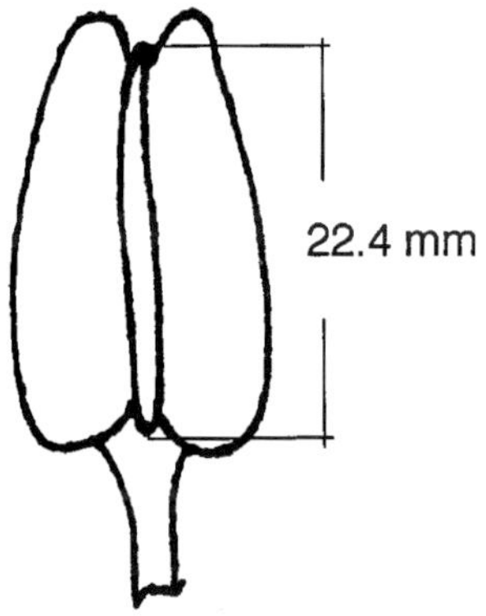

Figure 3. Length of flower bud of *Lilium longiflorum* in which pollen mother cells undergo their first division (after Erickson, 1948).

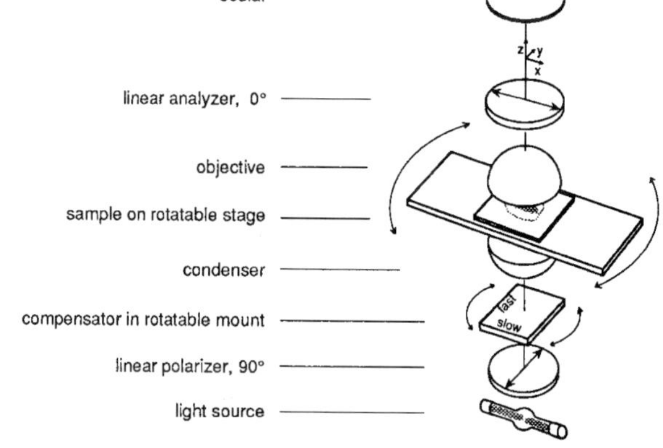

Figure 4. Schematic of a polarizing microscope with crossed polarizers and a compensator. Reproduced from *The Journal of Cell Biology*, vol 139, 985–994, 1997, by copyright permission from The Rockefeller University Press.

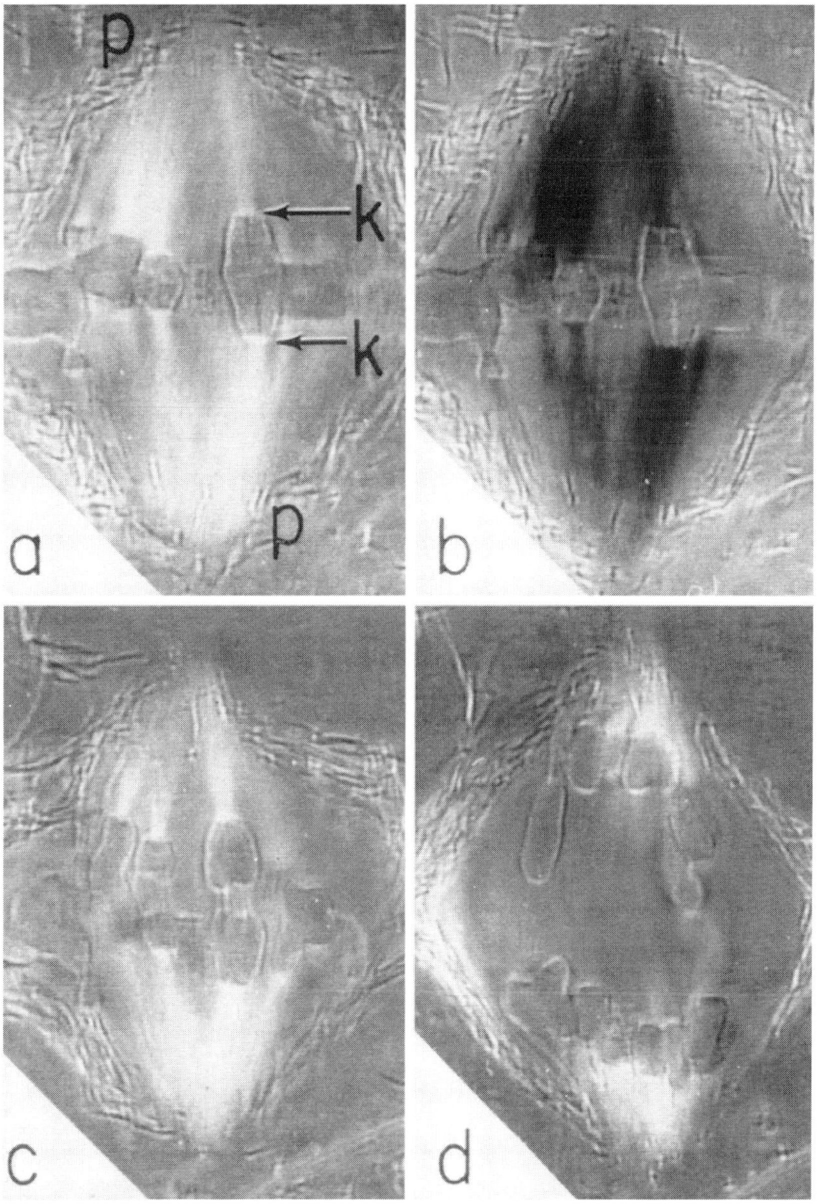

Figure 5. Primary spermatocyte of *Pardalophora apiculata* observed with a rectified polarizing microscope (from Nicklas, 1971). Reproduced from *Advances in Cell Biology*, vol 2, 225–298, copyright 1971 Appleton-Century-Crofts.

from breakdown of the nuclear envelope to formation of the cell plate was ~2 h. The ciné records were transferred to video some 40 years later.

The polarizing microscope view of the pollen mother cells distinctly shows the spindle fibers that were not visible with phase-contrast microscopy (for polarizing microscope images of *Haemanthus* endosperm cells, see Inoué and Bajer, 1961; Inoué, 1964). Phase-contrast microscopy clearly shows the chromosome and nucleoli because of their higher refractive index, but not the spindle fibers that lead the chromosomes apart to the spindle poles or the phragmoplast fibers that bring the vacuoles to the cell plate. The refractive index of these fibers is too close to that of the surrounding cytoplasm. They nevertheless show clearly in a well-tuned polarizing microscope, because

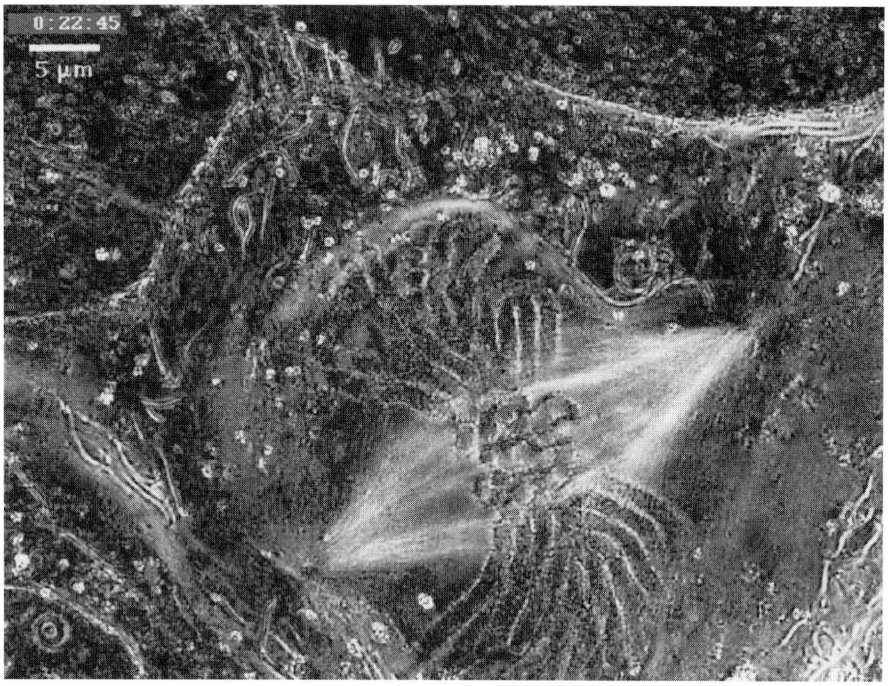

Figure 6. Mitosis in tissue-cultured lung cell of a newt, *Taricha granulosa*, recorded with the new Pol-Scope. Reproduced from *The Journal of Cell Biology*, vol 139, 985–994, 1997, by copyright permission of The Rockefeller University Press.

the fibers are birefringent, being made up of a bundle of regularly aligned molecular filaments. This sequence, taken by Inoué in 1950, demonstrated, for the very first time, the reality of spindle fibers and fibrils in living cells (Inoué, 1953, 1964) as well as the highly dynamic, labile nature of the molecular filaments (later identified as microtubules).

The microtubules disassembled reversibly when cells were exposed to cold, to high hydrostatic pressure, or to antimitotic drugs such as colchicine (reviewed in Inoué, 1964, 1981). During slow depolymerization of microtubules by these agents, metaphase-arrested chromosomes were pulled to a spindle pole anchored to the cell surface. After removal of the depolymerizing agent, growing spindle fibers pushed the chromosomes toward the metaphase plate. Thus arose the notion that chromosome movement toward the metaphase plate was associated with (and powered by) assembly and growth of microtubules, whereas movement of the chromosomes toward the spindle poles was associated with (and powered by) disassembly and shortening of the microtubules attached to the kinetochore of each sister chromosome (recent evidence and discussions summarized in Inoué and Salmon, 1995).

These polarized light microscopy studies on the birefringence of dividing cells demonstrated the assembly properties of microtubules and their dynamic function in living cells long before microtubules themselves were discovered or their assembly properties were characterized in vitro (reviewed in Inoué, 1981).

The microtubules that make up the spindle fibers and their shortening in anaphase can be seen more distinctly in the sequence of high-resolution images of a grasshopper spermatocyte (*Pardalophora apiculata*; Figure 5) taken by Nicklas (1971) with a rectified polarizing microscope. Rectification provides a higher-resolution image, restoring the needed extinction and correcting for the image error found in conventional polarizing microscopes when high–numerical aperture lenses are used (Inoué and Hyde, 1957).

The final video sequence shows a dividing newt (*Taricha granulosa*) lung epithelial cell (Figure 6) recorded by R. Oldenbourg, P.T. Tran, and E.D. Salmon with the new Pol-Scope. With the Pol-Scope, each image is generated by an image-processing computer from four video images taken in rapid succession at different settings of two electronically driven liquid crystal compensators. In the images thus displayed by the computer, the brightness of each pixel is strictly proportional to the birefringence of the specimen point and independent of orientation of the birefringence axis (Oldenbourg, 1996). Thus, in addition to providing displays with exquisitely high resolution and definition, Oldenbourg's new Pol-Scope provides highly sensitive, dynamic image information on molecular alignment that is strictly quantitative.

In addition to being of historic interest, considered use of polarized light microscopy should continue to reveal much regarding the behavior of molecular and fine structural dynamics, noninvasively, in dividing, developing, and otherwise actively functioning living cells (Oldenbourg, 1998).

ACKNOWLEDGMENTS

Several of the video sequences accompanying this article were reproduced, with permission, from Video Supplement 2 of the journal *Cell Motility and the Cytoskeleton*. In the journal see "Cellular Motile Processes: Molecules and Mechanisms" (1990. Cell Motil. Cytoskeleton *17*, 356–372) and the accompanying VHS videotape edited by Jean M. Sanger and Joseph W. Sanger for extensive additional material.

REFERENCES

Bajer, A., and Molè-Bajer, J. (1956). Ciné-micrographic studies on mitosis in endosperm. II. Chromosome, cytoplasm and Brownian movements. Chromosoma *7*, 558–607.

Bajer, A., and Molè-Bajer, J. (1986). Reorganization of microtubules in endosperm cells and cell fragments of the higher plant *Haemanthus* in vivo. J. Cell Biol. *102*, 263–281.

Erickson, R.O. (1948). Cytological and growth correlations in the flower bud and anther of *Lilium longiflorum*. Am. J. Bot. *35*, 729–739.

Inoué, S. (1953). Polarization optical studies of the mitotic spindle. I. The demonstration of spindle fibers in living cells. Chromosoma *5*, 487–500.

Inoué, S. (1964). Organization and function of the mitotic spindle. In: Primitive Motile Systems in Cell Biology, ed. R.D. Allen and N. Kamiya, New York: Academic Press, 549–598.

Inoué, S. (1981). Cell division and the mitotic spindle. J. Cell Biol. (special issue, Discovery in Cell Biology) *91*, 131s–147s.

Inoué, S., and Bajer, A. (1961). Birefringence in endosperm mitosis. Chromosoma *12*, 48–63.

Inoué, S., and Hyde, W.L. (1957). Studies on depolarization of light at microscope lens surfaces. II. The simultaneous realization of high resolution and high sensitivity with the polarizing microscope. J. Biophys. Biochem. Cytol. *3*, 831–838.

Inoué, S., and Salmon, E.D. (1995). Force generation by microtubule assembly/disassembly in mitosis and related movements. Mol. Biol. Cell *6*, 1619–1640.

Nicklas, R.B. (1971). In: Advances in Cell Biology, vol. 2, ed. D.M. Prescott, L. Goldstein, and E.H. McConkey, New York: Appelton-Century-Crofts, 225–298.

Oldenbourg, R. (1996). A new view on polarization microscopy. Nature *381*, 811–812.

Oldenbourg, R. (1998). Polarized light microscopy of spindles. In: Methods in Cell Biology: Structure, Composition and Function of the Mitotic/Meiotic Spindle, ed. C. Rieder, New York: Academic Press (*in press*).

Reprinted with permission from *The FASEB Journal*, Vol. 13, p. S179, 1999, special supplement in honor of July 24–25, 1998, symposium held at the Marine Biological Laboratory, Woods Hole, MA.

INTRODUCTION

A Half-Century of Advances in Microscopy

From the first known use of lenses for observing living organisms to the first electron micrographs of the fine structure of cells, to our powerful, contemporary approaches that reveal the dynamic interactions of individual molecules, microscopy has played—and is playing—an essential role in biology. This is particularly true today, where powerful contemporary techniques enable us to reveal dynamic interactions at the level of individual molecules.

It is now just over 50 years since the late Keith R. Porter published the first electron micrographs of the fine structure of cells. Over the past half-century, microscopy has been a seminal tool in providing new insights into the relationship amongst structure and function of cells. In contemporary studies, the use of microscopy has extended from routine examinations to new methods of observation and manipulation of cells and molecules once thought to be in the realm of fancy.

This meeting provided an overview of the past and current uses of microscopy. It also served as an education to future avenues of study of cell and molecular structure and function into the next century. The scope of microscopy has extended enormously since Dr. Porter's early efforts.

The international roster of speakers (pictured above) provides us with but a vignette of the many exciting fields and wonderful developments and the many pioneers at the MBL and elsewhere around the world in this dynamic and continually evolving discipline. We extend our thanks to the many individuals who provided us with helpful suggestions, guidance and assistance in organizing and conducting this meeting.

The symposium examined our use of microscopy in its many forms, sophisticated methods for manipulation of cells and their components, and video and computational imaging methods to glean deeper insights into the biology of the cell.

We welcome you to this supplement to *The FASEB Journal*, which records the papers of the symposium.

Shinya Inoué
George D. Pappas
Robert B. Silver
Co-organizers

0892-6638/99/0013-S179/$02.25 © FASEB

Reprinted with permission from *The FASEB Journal*, Vol. 13, pp. S185–S190, 1999.

Windows to dynamic fine structures, then and now

SHINYA INOUÉ[1]
Marine Biological Laboratory, Woods Hole, Massachusetts 02543, USA

ABSTRACT **How can we learn about dynamic fine structures that are far too small to be resolved with the light microscope without destroying the active living cell? Examples spanning the last half century show how polarized light microscopy can–and should–continue to provide an attractive window for such studies. Long before microtubules were found with electron microscopy, or their assembly properties were biochemically characterized in isolated cell-free systems, the dynamic fine structure of the mitotic spindle and assembly properties of its microtubules were revealed in living cells by polarized light microscopy. More recently, the polarizing microscope was improved, by invention of the new Pol-Scope, so that quantitative measurements of birefringence retardation and axes could be made rapidly for all image pixels independent of their birefringence axis orientation. In addition, the centrifuge polarizing microscope, just developed, allows us to follow the dynamic ordering of fine structures in living cells as they become stratified or restructured by centrifugal acceleration of up to ten thousand times gravity. The significance of these technological advances is discussed—Inoué, S. Windows to dynamic fine structures, then and now. *FASEB J.* 13 (Suppl.), S185–S190 (1999)**

PARTLY BECAUSE OF my personal interest and partly because I believe it to be an important approach for learning about the physico-chemical basis of life, my efforts over the past half century have stressed how we could nondestructively glimpse the dynamic aspects of fine structures in living cells, i.e., those structures and organizations that are too small to be resolved (even though perhaps detectable) with the light microscope (LM) and yet whose rapid changes in space and time, or dynamic individualistic behavior, call for close monitoring of their optical behavior using the LM.

In the late 1940s and early 1950s, amid the discovery of a series of respiratory enzymes and other wonderful proteins, many physiologists and biochemists considered a living cell to be a sackful of enzymes. Despite the careful earlier work by E. B. Wilson, Karl Bělăr, and other cytologists, the presence of a structural framework in the cell was seriously debated, and I recall the real heat with which Dr. L. Victor Heilbrunn greeted the observations of Eric Kao (who was working with Dr. Robert Chambers). Kao showed that microinjected oil drops took on an elongated, nonspherical shape, implying the presence of microscopically invisible fibrils in neurons and muscle cells (1). Said Heilbrunn: "I determined the viscosity of the living cytoplasm using the standard method by measuring the sedimentation rate of particles in the living cell, including neurons, and it is virtually the same as water. There cannot be any invisible filaments." As a brash graduate student, I myself was harshly reprimanded for asking: "But Dr. Heilbrunn, how do you explain the shape of those oil drops?"

Around the same time, I was able to demonstrate the reality of spindle fibers and fibrils and their dynamic nature in active living cells by using polarized light microscopy, whose sensitivity and resolution I had been improving (2). Because with polarized light microscopy contrast is generated noninvasively by the anisotropic alignment of molecules, I argued that we were observing the actual formation and changes in the fine structure inside living cells. Neither those fine structural filaments, nor their bundle, spindle fibers, had been visible in living cells earlier with the microscope unless the cells were fixed and stained, or the cells had been made acidic.

The notion that spindle fibers and fibrils seen after fixation and staining are fixation artifacts was so strong that, after seeing my time-lapse movies showing the birefringence of those structures and their changes in actively dividing pollen mother cells of Easter lilies, Dr. Ethyl Brown Harvey asked: "Were those cells alive?"

A film clip from those days shows mitosis in living endosperm cells of *Hemanthus* captured by Andrew and Wishia Bajer with phase contrast microscopy. The chromosomes, which show clearly because of their higher refractive index, move mysteriously to the spindle poles in anaphase (video)[2]. A second

---

[1] Correspondence: Shinya Inoué, Marine Biological Laboratory, 7 MBL Street, Woods Hole, MA 02543-1015.

[2] Both film clips, together with mitosis in a newt lung cell observed with Oldenbourg's new Pol-Scope, can be seen online at http://www.molbiolcell.org (3). Uncompressed versions of the first two scenes, together with many others showing experimental modifications of the spindle microtubule assembly, and chromosome movement, can be seen in Video Supplement II of *Cell Motility, and the Cytoskeleton* (4). (The *FASEB Journal* cannot guarantee the availability of references to the Web.)

0892-6638/99/0013-S185/$02.25 © FASEB

film clip shows mitosis and cell plate formation in Lily pollen mother cells, which I recorded with a sensitive polarizing microscope (video)[2]. In conjunction with these observations, I have done experiments using antimitotic agents such as colchicine and cold, and demonstrated that what we were seeing as birefringent spindle fibers were a bundle of protein filaments that could be reversibly depolymerized by the anti-mitotic agents.

Thus, in the living cell, we had a pool of material that could be transiently polymerized around orienting centers as needed by the cell, only to be taken apart again (5–7). I was able to say polymerization states, using the LM despite its limited resolving power, because I was using the polarizing microscope to deduce the reversible dynamic changes of fine structure by measuring the birefringence and changes in dimensions of the aligned fibers in the spindle (8, 9).

That was in 1950–1951. But for 15 years after that, Dr. Keith Porter, whose many contributions we celebrate today, kept telling me that what I was observing in the spindle was not protein filaments but membranes. And my explanation that the sign of birefringence would be wrong for membranes fell on deaf ears. I pointed out that if these were membranes, the slow axis would be perpendicular to what is observed for the spindle fibers. Our friendly dispute lasted until the spindle fiber region that had appeared empty in the electron microscope (EM), except for a few membrane fragments, was found to be full of microtubules once Sabatani came out with an improved fixative using glutaraldehyde. Then in the mid-1960s, Dr. Porter gave the famous Ciba Foundation Symposium talk (10), where he demonstrated the wide presence of microtubules, including in spindle fibers. And to show that the electron micrographs were not fixation artifacts, he came to borrow my slides that showed their birefringence in living cells (11)!

So microtubules are in spindle fibers, but, people argued, they couldn't possibly polymerize and depolymerize the way I said they would because protein fibers just don't do such things. So it took another several years before Dick Weisenberg figured out how to obtain *LABILE* microtubules from cells (by sequestering the calcium ions that would otherwise disassemble, for example, the spindle microtubules) (12). This was immediately followed by Joanna Olmsted and Gary Borisy's classical studies that showed how the isolated microtubules could be made to reversibly disassemble and assemble by cold and other agents (13). So now it was acceptable because these changes could be made to happen outside the cell, rather than be based on observations inside the cell.

In the past few years, model experiments performed by some ingenious investigators have also confirmed that depolymerizing microtubules can remain attached, and by the very act of depolymerizing and shortening pull organelles, as I had postulated on the basis of polarization optical studies of metaphase-arrested spindles treated with colchicine (14, 15). Of course, I am happy to learn that the conclusions that we had drawn from polarization microscopy of living cells could finally be accepted (although after decades) by being verified also in isolated, cell-free systems.

Skipping to the present and future, we have seen ample, remarkable examples of how microscopy is shedding ever more light on dynamic cell fine structure, including at the level of individual molecules. And we may ask: Does polarized light microscopy still offer attractive windows for studying dynamic fine structures in living cells? My belief is a resounding yes. While, unfortunately, very few biologists still have a good intuitive grasp of the interaction of polarized light with molecular and atomic order, and perhaps even less in terms of how in practice one takes advantage of such interactions, my personal conviction is that we are still at the foothills of a grand mountain of treasures that are awaiting to be explored using polarized light microscopy, by those who are prepared. Fortunately, more individuals are better prepared in physics and chemistry today, so I hope these treasurers will be explored.

Naturally, the specificity and sensitivity of fluorescent markers have great appeal, as does the much higher image resolution given by electron microscopy. Fluorescence microscopy is now yielding spectacular images signaling the activity of single protein motors and of unexpected molecular fluxes in membranes and mitotic microtubules. Combined with video enhancement and ingenious analyses, differential interference is shedding light even on transcription rates along single strands of DNA. And the resolution by cryo-EM now competes with X-ray analyses of dynamic atomic arrangements, using much smaller, 2-dimensional samples than the 3-dimensional crystals that are needed for X-ray diffraction analyses.

Following are some of the images obtained with the new Pol-Scope developed by Rudolf Oldenbourg in our Architectural Dynamics in Living Cells Program at the MBL. With the new Pol-Scope, unlike the images formed with classical polarization microscopy, image contrast is strictly proportional to the local specimen birefringence and independent of slow axis orientation (**Fig. 1**). This shows the distribution of microtubules in an isolated sea-urchin spindle. It typifies the sensitivity and striking image quality of the new system.

The videotape shows the dynamic behavior of the microtubule bundles and organelles in a dividing

Figure 1. Birefringence of isolated sea-urchin mitotic spindle imaged with Oldenbourg's new Pol-Scope (16).

newt lung epithelial cell taken by Ted Salmon, Phong Tran, and Rudolf Oldenbourg. The degree of resolution here is truly remarkable, and the measurement of birefringence changes along the length of each fiber should tell us much about their local concentration kinetics.

The idea behind this new Pol-Scope is that, instead of using an ordinary compensator, one uses 2 liquid-crystal plates to which voltage is applied to generate a half-wave plate and a quarter-wave plate plus-minus delta phase differences. That effectively changes the crossed polarizer images into those with a regular Brace-Koehler compensator turned clockwise, then counterclockwise, then parallel to the initial polarizer axes, and finally in crossed circularly polarized light. A computer grabs and calculates from all of these 4 images (which together are acquired in 0.25 s without involving any mechanical movements) and displays the retardance image (16). The images seen in the video were grabbed as a time-lapsed series every few seconds. Unlike polarized light images in the past, all the microtubules in the plane of focus show their birefringence regardless of their orientation, and the intensity that you observe in the image is strictly proportional to the retardance at the image point. So you can count, for example, the number of microtubules or actin filaments that are making up a particular part of a fiber by measuring the image brightness. Also, one can obtain an image that maps the slow axis orientation and one can follow the fine structures change dynamically.

Another example of dynamic fine structures that Kaoru Katoh and Rudolf have captured with the new Pol-Scope is the growth cone of *Aplysia* neurons (in collaboration with Peter Smith's Biocurrents Research Center group at the MBL). In a low-power video, we saw the active movements and changes in these filaments. At higher power, the video shows the generation, treadmilling, and dynamic fusion of actin filament bundles (17). As mentioned above, one can count the number of the unresolvable actin filaments that make up each part of the filament bundle and follow their changes.

Another video sequence was taken with a centrifuge polarizing microscope (CPM), an instrument that I had dreamed of being able to use since I first observed the spindle fibers and fibrils in *Chaetopterus* and *Lilium* nearly half a century ago. Because if we could apply centrifugal force, we could stratify and visualize the structure that had been obscured by the light-scattering cellular contents. We could also follow how molecules and fine structures become aligned or stretched by centrifugal forces inside the living cells and perhaps even discover new dynamic fine structures. By collaborating with Hamamatsu Photonics (Hamamatsu City, Japan) and Olympus Optical (Hachioji, Japan), we were finally able to develop a prototype CPM that allows us to gain contrast from fine structures that are stratified and ordered in living cells under up to 10,000 times gravity.

Looking at cells in a centrifuge microscope as such is not at all new (18). In fact, here at the MBL in the 1930s and 1940s, Ethyl Brown Harvey, using a centrifuge microscope that her husband E. Newton Harvey and Bill Loomis had designed, was able to observe *Arbacia* eggs under centrifugation (**Fig. 2**). The eggs were made to float on a gradient between seawater and isotonic sucrose solution. She was able to show the stratification, from the light side, of oil drop layer, nucleus, clear cytoplasm, mitochondrial layer, yolk, pigment, and so on. And as the centrifugation continued, she showed that the eggs turn into a dumbbell shape and eventually are pinched into a white half and a red half, then into 4 quarters. The real surprise came when she removed from the

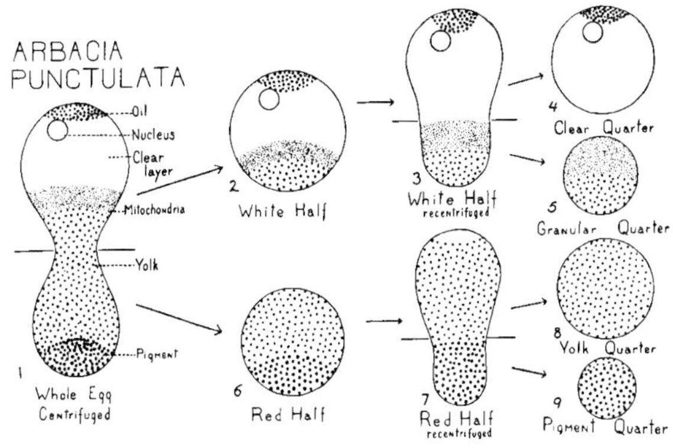

Figure 2. Schematic showing stratification and fragmentation of *Arbacia* eggs observed by E. B. Harvey in E. B. Harvey and A. L. Loomis' centrifuge microscope (19).

17. Katoh, K., Langford, G., Hammar, K., Smith, P. J. S., and Oldenbourg, R. (1997) Actin bundles in neuronal growth cone observed with the Pol-Scope. *Biol. Bull.* **193,** 219–220
18. Hiramoto, Y., and Kamitsubo, E. (1995) Centrifuge microscope as a tool in the study of cell motility. *Int. Rev. Cytol.* **157,** 99–128
19. Harvey, E. B. (1941) Vital staining of the centrifuged *Arbacia punctulata* egg. *Biol. Bull.* **114,** 118
20. Harvey, E. B. (1940) A comparison of the development of nucleate and non-nucleate eggs of *Arbacia punctulata. Biol. Bull.* **166,** 187
21. Inoué, S., Knudson, R. A., Suzuki, K., Okada, N., Takahashi, H., Iida, M., and Yamanaka, K. (1998) Centrifuge polarizing microscope. *Microsc. Microanal.* **4,** 36–37
22. Edgerton, H. (1987) In *Stopping Time: The Photographs of Harold Edgerton* (Kayafas, G., ed.), p. 126, Harry N. Abrams, New York
23. Inoué, S., and Spring, K. R. (1997) *Video Microscopy–The Fundamentals,* 2nd Ed, Plenum, New York
24. Goda, M., Inoué, S., and Knudson, R. A. (1998) Oocyte maturation in *Chaetopterus pergamentaceous* observed with centrifuge polarizing microscope. *Biol. Bull.* **195,** 212–214

Reprinted from *PNAS*, Vol. 97, pp. 10020–10025, 2000, with permission from National Academy of Sciences, USA.

How well can an amoeba climb?

Yoshio Fukui*†‡, Taro Q. P. Uyeda§, Chikako Kitayama§, and Shinya Inoué†

*Cell and Molecular Biology, Northwestern University Medical School, Chicago, IL 60611-3008; §Biomolecular Research Group, National Institute for Advanced Interdisciplinary Research, Tsukuba, Ibaraki 305-8562, Japan; and †Marine Biological Laboratory, Woods Hole, MA 02543-1005

Contributed by Shinya Inoué, June 22, 2000

We report here our efforts to measure the crawling force generated by cells undergoing amoeboid locomotion. In a centrifuge microscope, acceleration was increased until amoebae of *Dictyostelium discoideum* were "stalled" or no longer able to "climb up." The "apparent weight" of the amoebae at stalling rpm in myosin mutants depended on the presence of myosin II (but not myosins IA and IB) and paralleled the cortical strength of the cells. Surprisingly, however, the cell stalled not only in low-density media as expected but also in media with densities greater than the cell density where the buoyant force should push the amoeba upward. We find that the leading pseudopod is bent under centrifugal force in all stalled amoebae, suggesting that this pseudopod is very dense indeed. This finding also suggests that directional cell locomotion against resistive forces requires a turgid forward-pointing pseudopod, most likely sustained by cortical actomyosin II.

Generation of mechanical forces is essential for cell locomotion, division, embryonic development, and morphogenesis (1–5). Although the forces involved in some of these biological activities have been measured as mechanical properties in *local* regions of living cells (6–9), few measurements have been made of the maximum ability of an *entire* cell to propel itself. An example includes the maximum propulsive force of 7×10^3 pN generated by a swimming ciliated protozoan, *Paramecium caudatum*, measured by using a centrifuge microscope (10). Little is known, in particular, of the propulsive forces that can be generated by any cell undergoing amoeboid movement.

In the present paper, we report the maximum "apparent weight," or centrifugal force against which wild-type and myosin mutants of *Dictyostelium discoideum* amoebae were able to crawl "upward." The small mass of the amoebae required the use of a recently developed centrifuge polarizing microscope capable of generating fields of greater than $11{,}465 \times g$ (Earth's gravitational acceleration), with image resolution of better than 1 μm in differential interference or Nomarsky contrast microscopy (11).

As described below, mutant amoebae stall or cease to be able to crawl up against the imposed apparent weight at characteristic centrifugal accelerations, so they are at least able to overcome that much external force. Those lacking the muscle type myosin, myosin II, stall at very much lower centrifugal acceleration.

However, we will show that the mechanism of stalling, or inability of the amoeba to maintain directional locomotion against the centrifugal field, in fact depends on the very high local density of its leading pseudopod rather than the apparent weight felt by the whole amoeba. Even in media whose density is greater than that of the whole amoeba, amoebae lacking myosin II are unable to sustain the forward protrusion of the high-density pseudopod that is apparently needed to sustain directional amoeboid locomotion against the external field.

Materials and Methods

Cells and Cell Culture. *D. discoideum* wild type NC4 (12), axenic strain Ax3 (13, 14), myosin II heavy chain knockout mutant HS1 (*mhcA*$^-$) (15), and a triple myosin knockout mutant A5 (*mhcA*$^-$/*myoIA*$^-$/*myoIB*$^-$) were cultured as described previously (12–15). Before observation, the growth phase cells were washed from medium by centrifugation ($100 \times g$, 1.5 min) and incubated overnight at 18C° in a standard buffer (10 mM NaCl/10 mM KCl/3 mM $CaCl_2$/2.5 mM Pipes, pH 6.8).

Generation of Myosin Knockout Mutants. Generation of a mutant cell line (A5) that lacks heavy chain genes for myosin II, myosin IA, and myosin IB will be described in detail elsewhere (C.K. and T.Q.P.U., unpublished work). Briefly, a plasmid carrying *Dictyostelium* myosin II heavy chain gene (15) was modified such that a fragment corresponding to the carboxyl quarter of the motor domain and the amino half of the tail were replaced with a blasticidin resistance cassette. The DNA fragment of the disrupted gene was excised and electroporated into a cell line lacking myosins IA and IB (16). Blasticidin resistant colonies were isolated, and the double-crossover gene disruption was confirmed by Southern hybridization. Absence of myosin II heavy chain was further confirmed by Western blot and phenotypic assays.

Centrifuge Polarizing Microscope. The centrifuge polarizing microscope was designed by S.I. and developed in collaboration with Hamamatsu Photonics (Hamamatsu City, Japan) and Olympus Optical Company (Tokyo) (11). Cells were suspended in standard buffer and allowed to settle on the strain-free glass cover of the centrifuge observation chamber. The image of the specimen spinning in the rotor (at a radius of 7.5 cm) at up to 11,700 rpm is frozen stroboscopically by brief (6-ns) laser flashes that illuminate the specimen as it transits between the stationary condenser and objective lenses. The 532-nm wavelength image, formed by a $\times 40/0.55$ numerical aperture objective lens, was captured at video rate by a Hamamatsu interference-fringe-free CCD camera. The original image was recorded into Sony ED-Beta tapes as Y/C signals at video rate through a digital signal converter (Sony model DSC-1024G).

Calibration of Medium Density. The density of stock and diluted Percoll solution (17) (Amersham Pharmacia Biotech) was calculated from the equations below:

$$\rho_{100} = (V_a \times \rho_a) + (V_b \times \rho_b)/(V_a + V_b)$$
$$\rho_i = (V_{100} \times \rho_{100}) + (V_b \times \rho_b)/(V_{100} + V_b)$$

Where

V_a = volume of 23% (v/v) Percoll (from bottle).
V_b = volume of $\times 20$ standard buffer (appropriately diluted to provide final $\times 1$ standard buffer).
V_{100} = volume of stock Percoll (21.85%).

ρ_a = density of 23% (v/v) Percoll = 1.130 gram/cm^3.
ρ_b = density of $\times 20$ standard buffer = 1.005 gram/cm^3.
ρ_{100} = density of stock Percoll = 1.124 gram/cm^3.
ρ_i = density of diluted Percoll.

‡To whom reprint requests should be addressed at: Cell and Molecular Biology, Ward 7-342, Northwestern University Medical School, 303 East Chicago Avenue, Chicago, IL 60611-3008. E-mail: y-fukui@northwestern.edu.

The publication costs of this article were defrayed in part by page charge payment. This article must therefore be hereby marked "*advertisement*" in accordance with 18 U.S.C. §1734 solely to indicate this fact.

Table 1. Apparent weight and stalling rpm for wild-type and myosin knockout mutants in *Dictyostelium*

Cell lines*	NC4	Ax3	HS1	A5
Calculation of reduced mass of amoebae				
Volume† ($\times 10^{-10}$ cm^3)	3.78 ± 1.44	4.87 ± 1.80	4.26 ± 1.34	4.87 ± 1.69
Density‡ (gram/cm^3)	1.066	1.065	1.063	1.063
ΔMass§ ($\times 10^{-11}$ gram)	2.31 ± 0.70	2.92 ± 0.70	2.47 ± 0.78	2.82 ± 0.98
Apparent weight at stalling rpm¶				
Maximum rotation, rpm	>11,700	6,400	3,500	3,400
Stalling acceleration (×g)	>11,465	3,431	1,025	968
Apparent weight at Stall‖ (× 10^3 pN)	>2.59 ± 0.78	0.99 ± 0.24	0.25 ± 0.08	0.27 ± 0.09

The apparent weight was calculated by multiplying the amoeba's reduced mass and the stalling acceleration. The stalling acceleration was calculated from the rpm beyond which the amoebae were unable to crawl centripetally and the distance (7.5 cm) between the center of the rotor to the center of the observation chamber.
*NC4: wild type (12), Ax3: axenic strain (13, 14), HS1: myosin II knockout (*mhcA*⁻) (15), A5; triple knockout (*mhcA*⁻/*myoIA*⁻/*myoIB*⁻).
†Calculated from measurements of radii of 200 cells each; each cell was assumed to be a 5-μm-high disk.
‡Apparent cell density (δ) was determined as isopycnic density (17).
§ΔMass; reduced mass = (cell density − medium density) × cell volume. Medium density of the standard buffer (ρ_0) was 1.005 gram/cm^3 as measured with an Ostwald's pycnometer (21).
‖(ΔMass) × (stalling acceleration); standard deviation, each based on measurements of radii of 200 cells.

Calibration of Apparent Weight of Amoeba. The forces on the amoebae were calibrated by Newton's equation of motion as below:

$$\text{Force (F)} = \Delta m \times a \times 10^7 (\text{pN}).$$

Where Δm (reduced mass) = [cell density (δ) − medium density (ρ)] × cell volume (gram/cm^3).

$$\text{Acceleration } (a) = \{(2\pi \times R/60)^2 \times r\} \ (\text{cm/sec}^2).$$

Where R = rpm and r = distance between the center of the centrifuge rotor and center of the specimen chamber (in centimeters).

Cell density was measured by linear density gradient centrifugation. A 0.5-ml aliquot of cell suspension (2×10^7 cells/ml in the standard buffer) was loaded on a preformed Percoll gradient. The gradient was made by centrifugation of 23% Percoll (ρ = 1.130 gram/cm^3) (17) in silicone-coated Ultra-Clear centrifuge tube (Beckman Coulter) at 20,000 × *g* for 30 min in an angle-head rotor (JA-20) on a J2–21 M centrifuge (Beckman Coulter). The apparent densities were calibrated by using Density Marker Beads (Amersham Pharmacia Biotech).

Calculation of Equivalent Actin Concentration. The actin concentration (Ai) with densities equivalent to Percoll solutions was calculated by the following equation:

$$\text{Ai} = (\rho_i - \rho_0) \times 1.2/50 \text{ mM}$$

Where ρ_i = density of diluted Percoll at 0, 10, 25, 50, 75, and 100% fraction.

$$\rho_0 = \text{density of actin buffer}$$

Because 50 mg/cm^3 monomeric actin should give rise to 1.2 mM concentration (18), the actin concentrations equivalent to 0–100% Percoll fractions would be 0.12, 0.17, 0.62, 1.4, 2.1, and 2.9 mM (see Table 2 and Fig. 3).

Fluorescence Microscopy. The cells were fixed in methanol containing 1% formalin at −15°C for 5 min and stained with 3 μM tetramethylrhodamine isothiocyanate-phalloidin (rh-ph) (#R-415; Molecular Probes) for 30 min at 36°C. The sample was observed under an epifluorescence microscope (Axioskop-50; Zeiss) equipped with an oil-immersion ×63 plan apo objective (numerical aperture 1.4). The image was acquired with a cooled CCD (PXL; Photometrics, Tucson, AZ) equipped with a Kodak KAF 1400 chip. The fluorescence and phase-contrast images were acquired at an 8-bit depth of gray scale by using an integrated image acquisition and processing system (MetaMorph; Universal Imaging Corporation, West Chester, PA). The spatial resolution of the system is calculated to be better than 250 nm (19, 20).

Results

Behavior of Cells Under High Centrifugal Fields. In the centrifuge polarizing microscope, the glass windows of the specimen chamber lie in the plane of rotation of the rotor. After amoebae that had settled on a glass window started to migrate randomly, we gradually increased the rotational speed (in rpm) of the centrifuge until no amoebae were crawling upward, i.e., against the centrifugal field. At that "stalling" rpm, we found that the amoebae could crawl only sideways or downward. In those amoebae that started to stretch their pseudopods upward and attempted to crawl in that direction, the extended pseudopods would not remain straight up but became bent. The amoebae thereupon started crawling either sideways or downward. Others lost hold of the substrate and fell to the bottom of the centrifuge chamber. The stalling rpm was surprisingly uniform for amoebae of each strain; at only 100 rpm above the stalling rpm, essentially no amoeba was able to crawl upward.

Thus we measure the stalling rpm as the rpm of the centrifuge in which the cell's geometric center (centroid) fails to move centripetally. Because the cells are attached to the glass substrate aligned parallel to the centrifugal field, we calculated the forces on the amoeba at stalling rpm as (reduced mass of the amoeba) × (the centrifugal acceleration/*g*), where g = 980 cm/sec^2 (see *Materials and Methods* for details). Except in those cases noted, we found that all cell lines were able to remain adherent to the glass surface up to maximum rotation of the centrifuge polarizing microscope.

Apparent Weight of Amoeba at Stalling rpm. Table 1 shows the reduced mass of the amoebae and centrifugal forces (= apparent weight) at which different strains of amoebae were stalled. We found that wild-type amoebae (NC4) (12) could still migrate centripetally under maximum centrifugation. Under this condition, an NC4 amoeba must generate more than 2.59×10^3 pN of

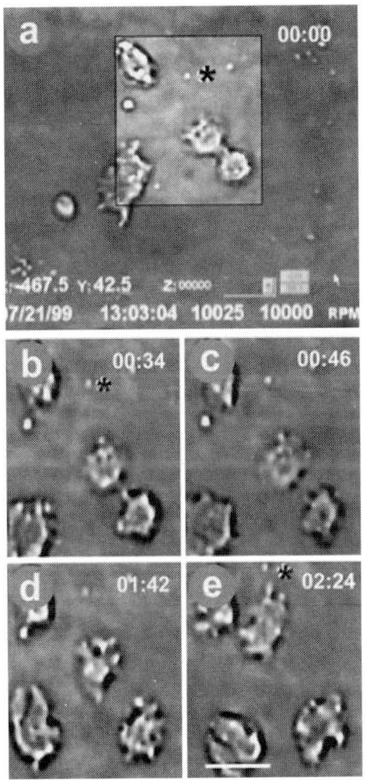

Fig. 1. Behavior of cells under high centrifugal fields. A representative centrifuge microscope image in differential interference or Nomarski contrast showing a wild type NC4 cell undergoing cytokinesis at 10,000 rpm (8,376 × *g*). (*a*) The full video screen; (*b–e*) a region of interest extracted sequence acquired at elapsed times indicated (*Upper Right* corner, *a–e*). After dividing, the upper daughter cell migrated centripetally. A movie showing the full sequence can be seen in the supplementary material (www.pnas.org). *, stable marker on the substrate. Elapsed time, minute/second. Bar = 5 μm.

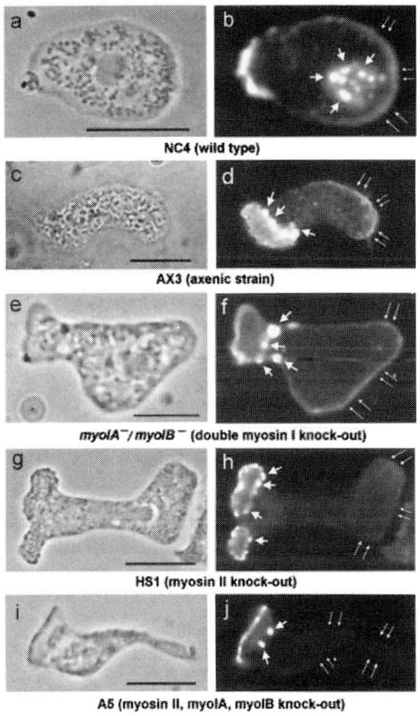

Fig. 2. Distribution of filamentous (F-) actin in migrating *Dictyostelium* amoebae. (*Left*) Phase-contrast images; (*Right*) rhodamine–phalloidin fluorescence. (*a* and *b*) Wild type (NC4) (12); (*c* and *d*) axenic strain (Ax3) (13, 14); (*e* and *f*) double myosin I knockout (*myoIA*$^-$/*myoIB*) (16); (*g* and *h*) myosin II knockout (HS1) (15); (*i* and *j*) triple myosin knockout (A5) strains. Note that the trailing cortex of NC4, Ax3, and *myoIA*$^-$/*myoIB*$^-$ contains a substantial F-actin layer (double thin arrows, *b*, *d*, and *f*). In contrast, the cortical F-actin in myosin II knockout cells (HS1 and A5) is weak (*h* and *j*). The globular fluorescence dots (thick short arrows) represent the cell-substrate anchoring structures "eupodia" (19, 20), which are similar to "podosomes" or "invadopodia" in invasive mammalian cells (30, 31). In contrast to the trailing cortex, actin in the eupodia and leading pseudopod is well established in all five strains. Bar = 5 μm.

climbing force to overcome the downward pull by the centrifugal force. Even so, the wild-type amoebae could divide by apparently normal cytokinesis (Fig. 1) and could even show aggregation behavior (22–24) (video data available as supplementary material; see www.pnas.org).

In contrast to NC4, the axenic strain amoebae (Ax3) (13, 14) stalled at 3,431 × *g*. At higher centrifugal forces, Ax3 cells moved only laterally or centrifugally. The apparent weight of Ax3 at stalling rpm was 0.99×10^3 pN, or less than 40% of NC4. These measurements support the proposition that axenic strains in fact are not wild type but are motility mutants of NC4 (25, 26).

As with Ax3 and NC4, the myosin II knockout mutant (HS1) (15) could remain attached to the substrate up to maximum speed of the centrifuge. This mutant, lacking myosin II, could migrate only laterally or centrifugally and not centripetally at speeds higher than 3,500 rpm (at 1,025 × *g* Table 1 *Lower*). The apparent weight for HS1 at stalling rpm was 0.25×10^3 pN, or 25% of the parental Ax3.

To examine possible contribution from two "mini"-myosin isoforms (myosins IA and IB), we measured the apparent weight at stalling rpm of a triple myosin knockout mutant (A5) from which myosin IA and IB as well as myosin II were removed. A5 cells were found capable of migrating against the centrifugal field up to 968 × *g*. Under this rpm, the apparent weight of A5 cells (0.27×10^3 pN) against which the cells can generate upward migration forces is similar to that of HS1, indicating that myosin IA and IB do not add extra migration forces to the myosin II knock-out mutant.

The above results indicate that the ability of *Dictyostelium* amoebae to generate forces that counter their apparent weight are myosin II dependent to a major extent. In this context, we note that the ratio of apparent weight at stalling rpm in the myosin II knockout mutant to the parental strain Ax3 (i.e., 25%) more or less parallels published ratios of cortical tensions measured by poking with a microneedle (68%; ref. 6) or sucking with a microcapillary (50%; 7, 30%; ref. 8). Because myosin II is reported to be localized in the posterior (trailing) cortex of *Dictyostelium* amoebae (4, 27–29), we next examined the organization of cortical F-actin (the major constituent of the cortex) in those mutants whose stalling we observed.

Weak F-Actin Cortex in Myosin II Knockout Mutants. By fluorescence microscopy, we find that the cortical F-actin in the layer surrounding the trailing cell body (double thin arrows) is well established in wild-type NC4 (Fig. 2 *a* and *b*), the axenic strain Ax3 (*c* and *d*), and the double myosin I knockout mutant (*myoIA*$^-$/*myoIB*$^-$) (*e* and *f*). However, it is poorly organized in the myosin II knockout mutants (HS1, *g* and *h*; A5, *i* and *j*). In contrast, in the leading pseudopod and the eupodia, F-actin is highly concentrated in all five strains. These results, together with decreased migration forces in HS1 and A5, suggest that the strength of the cortical actomyosin II in the posterior cortex plays an important role in generating directional migration forces in *Dictyostelium*.

Table 2. Medium density vs. stalling acceleration measured in myosin II knockout mutant (HS1)

Fraction of stock percoll, %*	0	10	25	50	75	100
Medium density (gram/cm³)†	1.005	1.012	1.031	1.062	1.093	1.124
Stalling revolution, rpm	3,500	4,100	4,900	6,800	7,800	8,600
Stalling acceleration, × g	1,025	1,408	2,011	3,873	5,096	6,195
Buoyant force‡ (× 10³ pN)	−0.25	−0.31	−0.27	−0.02	0.65	1.61

*Percents (v/v) of stock Percoll solution. The stock Percoll was 21.85% of Percoll containing the standard buffer (see *Materials and Methods*).
†Density was calibrated by using Density Marker Beads (17).
‡Calculated using Δmass in various Percoll concentrations. In high-density media, the cells are exposed to buoyant forces in excess of their apparent weight, yet their upward migration is still progressively limited by the centrifugal field (see also Fig. 3).

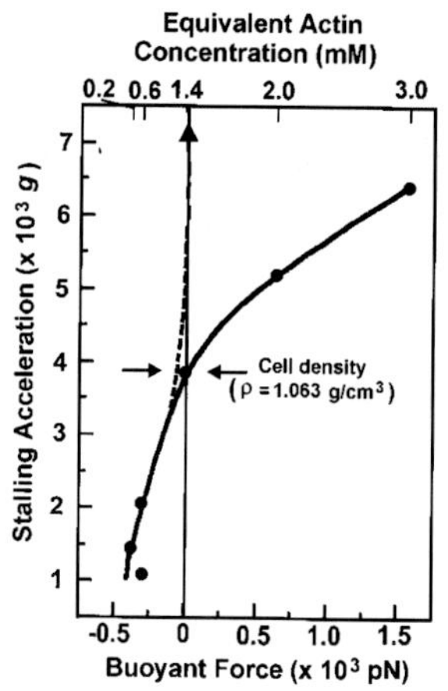

Fig. 3. Relationship between stalling acceleration and buoyant force experienced by amoebae in media with different densities. The stalling acceleration was measured on HS1. After the cells adhered to the glass surface in the centrifuge observation chamber, the medium was replaced with 10, 25, 50, 75, and 100% stock Percoll. Absolute concentrations of Percoll and their densities (ρ) are shown in Table 2. The buoyant force changes from negative to positive when the medium density approaches the cell density (δ = 1.063 grams/cm³). The graph shows that the cells stall even in media whose density is higher than the average cell density. Dashed line indicates the anticipated result extrapolated from stalling at lower media density, assuming that the apparent weight of the cells determines the stalling acceleration. The actin concentrations (*Upper*) represent calculated actin concentrations whose densities are equivalent to the Percoll solutions (see *Materials and Methods*).

Stalling in High-Density Medium. If the apparent weights accurately reflect the ability of cells to migrate directionally against an external force, the cells should be able to resist higher centrifugal forces when they are surrounded by media with elevated density and become more buoyant. We find that this assumption is in fact correct. The density of the medium was changed by using various concentrations of Percoll (see *Materials and Methods*). Observations were made on the myosin II knockout mutant (HS1cells) because they stall under moderate centrifugal fields (Table 1 *Lower*).¶

As predicted, the HS1 cells kept migrating centripetally at much higher rpm in Percoll solutions than in standard buffer solution (Table 2). Surprisingly, however, the cells would stall even when exposed to media whose density was higher than the cell's density (Fig. 3). This result was unexpected because in those Percoll solutions, the cells are actually being pushed upward by their buoyancy. In fact, amoebae that became detached in those solutions floated up instead of sinking. In 50% Percoll, the cells should have *no* apparent weight and should be receiving no centrifugal forces even at the maximum rotation because the reduced mass is nearly zero. Beyond 50%, the cells are pushed upward harder when the rotational speed increases (because of increasing buoyant forces). Nevertheless, in 50% Percoll medium, the cells stalled at 3,873 × g (Table 2). At higher Percoll density, the stalling acceleration continued to rise monotonically, continuing the tendency seen for Percoll densities less than the average density of the amoeba (Table 2, Fig. 3). These results clearly show that the apparent weight at stalling rpm is *not* a simple measure of the maximum directional migration force that the amoebae can generate but must reflect some other event taking place in or on the amoeba.

Bending Pseudopods Under High Centrifugal Fields. As noted, we found that when the cell fails to migrate "upward" against the centrifugal field, a pseudopod protruding upward from the leading cell body is bent down. In fact, this behavior of the pseudopods was observed in all strains when the cells fail to migrate centripetally, *including* in those cells that were exposed to media with elevated density (Fig. 4*A*). Thus it is not the apparent weight, or the centrifugal force applied to the whole amoeba, that determines the stalling rpm. Rather, it is the centrifugal field under which the leading pseudopod is bent that stalls the "upward" directional migration of the amoeba.

These observations show that the density of the pseudopod itself must be very high indeed (>3 mM equivalent of actin; see Fig. 3). It also suggests that forward protrusion of the pseudopod is a prerequisite for directional amoeboid locomotion.

Discussion and Conclusion

Because those mutant cells lacking myosin II stall at low rpm and also show a weak trailing cortex, we postulate that beyond the stalling rpm their fragile cortex is unable to support the needed upward thrust of the leading pseudopod. We argue that the rigidity of the pseudopod will very likely depend on the circumferential strength of its tubular cortex plus a sufficiently high hydrostatic pressure provided by the actomyosin II-containing contracting cortex surrounding the trailing cell body. If the cortical contracting force is inadequate, the turgor pressure in the cell would be too weak to support the leading pseudopod that is needed to propel the amoeba against a load that has to be overcome (Fig. 4*B*).

In summary we conclude that myosin II, but not myosins IA and IB, plays a crucial role in the generation of cortical contraction forces. Because there are more than 10 myosin I (minimyosin) genes identified in *Dictyostelium* (32), we are unable to decipher the direct or "compensatory" roles (33) of the minimyosins as a whole. Nevertheless, there is no question that the cells lacking myosin II have considerable less ability to generate forward migrating forces.

We believe that the forward protrusion of the leading pseudopod is not simply a phenomenon observed in *Dictyostelium* and other amoebae, but that it is an essential feature for the directional migration of cells undergoing amoeboid locomotion in general (1, 4). Once the direction of propagation is defined by some cue [e.g., cyclic adenosine-3′,5′-monophosphate gradient

¶Under the brief periods of centrifugation required for our observations, we found no signs of Percoll stratification or differences in rpm for the amoebae stalling at various heights within the centrifuge chamber.

CELL BIOLOGY

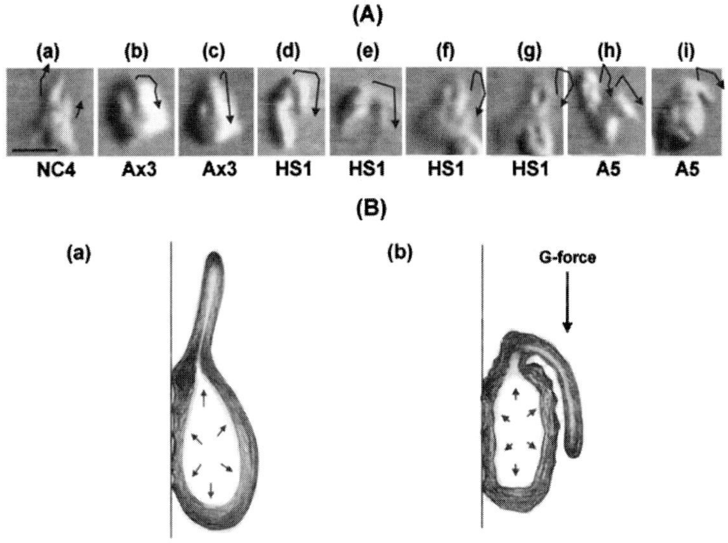

Fig. 4. Behavior of pseudopods and proposed stalling mechanism. (*A*) Selected video frames showing behavior of pseudopods. (*a*) Control. When an amoeba migrates centripetally, the leading pseudopod(s) continues to protrude upward. (*b–i*) At stalling rpm, where the amoebae can no longer migrate against the centrifugal force, the pseudopod(s) bends down. NC4, original wild type; Ax3, parental axenic strain; HS1, myosin II knockout (*mhcA*$^-$); A5, triple myosin knockout (*mhcA*$^-$/*myoIA*$^-$/*myoIB*$^-$). Arrows, approximate shape of pseudopods. Bar = 5 μm. (*B*) Diagram illustrating the suggested stalling mechanism. (*a*) The cortical tension represents a sum of the tangential stress rendered by the cortical structure and the hydrostatic pressure in the cell. Contractile force generated by the cortical actomyosin II, which is bound to the semipermeable plasma membrane, provides a turgor pressure to support a protruding pseudopod. Because the pseudopod is not attached to the substrate and must be supported by the cell body, the cortical F-actin framework must support the protruding pseudopod. (*b*) Lacking myosin II, the cortical F-actin layer cannot produce sufficient hydrostatic pressure to support the forward protrusion of the pseudopod. When the weak cortex cannot provide enough turgor to support the F-actin-rich dense leading pseudopod, it bends down under the strong G-forces generated by the centrifuge. If the bent pseudopod manages to attach to the substrate, the cell can migrate sideways or downward but not upward. Otherwise the cell detaches and falls down or, in media whose density exceeds that of the cell, floats up.

for *Dictyostelium* amoebae (22–24)] and a pseudopod starts forming in that direction (34, 35), we suggest that the contractile force generated by the trailing cell cortex must provide adequate support for the pseudopod to penetrate into that direction without collapsing against the external force, whether gravitational or a barrier presented, for example against leukocytes, parasitic protista, or migrating embryonic cells by a tissue layer.

Although we have learned that the mechanism by which a cell stalls in a centrifugal field is more sophisticated than initially assumed, we believe that the stall forces measured, as apparent weight with a centrifuge microscope, nevertheless provide valid measures for the relative crawling force that an amoeboid cell can generate for *directional* migration. The relative crawling force as well as the stalling acceleration depends on the presence of myosin II (but not myosins IA and IB) which, in association with actin filaments, provides for a robust trailing cortical layer. We postulate that the directional locomotion of an amoeboid cell requires the contractile cortical framework to provide the turgidity needed for the leading pseudopod to direct the locomotion in that direction.

This paper is dedicated to a memory of Prof. Noburo Kamiya (1913–1999), a pioneer in the research of biological forces, who, among other pioneering explorations, devised the ingenious method of the hydrostatic forces needed for cytoplasmic streaming in syncytial slime molds. We thank Hamamatsu Photonics and the Olympus Optical Company for support of this research and Dr. G. Albrecht-Buehler of Northwestern University Medical School, Chicago, Illinois and Dr. I. Mabuchi of the Tokyo University for discussions about biological forces and actin density. We thank Dr. T. D. Pollard of the Salk Institute for careful reading and astute comments on an earlier version of the manuscript and Dr. R. Oldenbourg of the Marine Biological Laboratory (MBL) for discussions on the forces produced by the apparent weight of an object in a centrifuge. We also thank Dr. L. G. Tilney of the University of Pennsylvania for helpful comments on the manuscript. Thanks are also due to R. A. Knudson, J. MacNeil, and D. Baraby of MBL for their technical and secretarial support.

1. Stossel, T. P. (1982) *Philos. Trans. Soc. London B* **299,** 275–289.
2. Pollard, T. D. & Cooper, J. A. (1986) *Annu. Rev. Biochem.* **55,** 987–1035.
3. Spudich, J. A. (1989) *Cell Regul.* **1,** 1–11.
4. Harris, A. K. (1994) *Int. Rev. Cytol.* **150,** 35–68.
5. Borisy, G. G. & Svitkina, T. M. (2000) *Curr. Opin. Cell Biol.* **12,** 104–112.
6. Pasternak, C., Spudich, J. A. & Elson, E. L. (1989) *Nature (London)* **341,** 549–551.
7. Egelhoff, T. T., Naismith, T. V. & Brozovich, F. V. (1996) *J. Muscle Res. Cell Motil.* **17,** 269–274.
8. Dai, J. H., Ting-Beall, P., Hochmuth, R. M. Sheetz, M. P. & Titus, M. A. (1999) *Biophys. J.* **77,** 1168–1176.
9. Oliver, T., Dembo, M. & Jacobson, K. (1999) *J. Cell Biol.* **145,** 589–604.
10. Kuroda, K. & Kamiya, N. (1989) *Exp. Cell Res.* **184,** 268–272.
11. Inoué. S., Knudson, R. A., Suzuki, K., Okada, N., Takahashi, H., Iida, M. & Yamanaka, K. (1998) *Microsc. Microanal.* **4,** 36–37.
12. Raper, K. B. (1935) *J. Agr. Res.* **50,** 135–147.
13. Free, S. J. & Loomis, W. F. (1974) *Biochimie* **56,** 1525–1528.
14. Franke, J. & Kessin, R. (1977) *Proc. Natl. Acad. Sci. USA* **74,** 2157–2161.
15. Egelhoff, T. T., Manstein, D. J. & Spudich, J. A. (1990) *Dev. Biol.* **137,** 359–367.
16. Novak, K. D., Peterson, M. D., Reedy, M. C. & Titus, M. A. (1995) *J. Cell Biol.* **131,** 1205–1221.
17. In *Percoll: Methodology and Applications* (1995) (Pharmacia LKB Biotechnology, Uppsala, Sweden) pp. 5–12.
18. Tilney, L. G. & Inoué, S. (1985) *J. Cell Biol.* **100,** 1273–1283.
19. Fukui, Y. & Inoué, S. (1997) *Cell Motil. Cytoskeleton* **36,** 339–354.
20. Fukui, Y., de Hostos, E. L., Yumura, S., Kitanishi-Yumura, T. & Inoué, S. (1999) *Exp. Cell Res.* **159,** 141–157.
21. Fukui, Y. (1976) *Dev. Growth Differ.* **18,** 145–155.
22. Bonner, J. T. (1947) *J. Exp. Zool.* **106,** 1–26.
23. Gerisch, G. (1982) *Annu. Rev. Physiol.* **44,** 535–552.

24. Devreotes, P. N. (1989) *Science* **245,** 1054–1058.
25. Williams, K. L., Kessin, R. H. & Newell, P. C. (1974) *Nature (London)* **247,** 142–143.
26. Kayman, S. C. & Clarke, M. (1983) *J. Cell Biol.* **97,** 1001–1010.
27. Yumura, S., Mori, H. & Fukui, Y. (1984) *J. Cell Biol.* **99,** 894–899.
28. Yumura, S. & Fukui, Y. (1985) *Nature (London)* **314,** 194–196.
29. Fukui, Y., Lynch, T. J., Brzeska, H. & Korn, E. D. (1989) *Nature (London)* **341,** 328–331.
30. Tarone, G., Cirillo, D., Giancotti, F. G., Comoglio, P. M. & Marchisio, C. (1985) *Exp. Cell Res.* **151,** 141–157.
31. Chen, W.-T. (1989) *J. Exp. Zool.* **251,** 167–185.
32. Titus, M., A., Kuspa, A. & Loomis, W. F. (1994) *Proc. Natl. Acad. Sci. USA* **91,** 9446–9450.
33. André, E., Brink, M. & Gerisch, G. (1989) *J. Cell Biol.* **108,** 985–995.
34. Condeelis, J. (1993) *Annu. Rev. Cell Biol.* **9,** 441–444.
35. Fukui, Y. (1993) *Int. Rev. Cytol.* **144,** 85–127.

Reprinted from *The Biological Bulletin*, Vol. 199, pp. 213–214, 2000.

Fertilization-induced Changes in the Fine Structure of Stratified *Arbacia* Eggs. II. Observations with Electron Microscopy

Mario H. Burgos[1], *Makoto Goda*[2], *and Shinya Inoué*
(Marine Biological Laboratory, Woods Hole, Massachusetts)

Unfertilized *Arbacia* eggs are stratified by centrifugation; the centripetal pole is occupied by an oil cap, which crowns a large clear zone containing the nucleus (1). When such eggs are observed with the centrifuge polarizing microscope (CPM), a curtain of negatively birefringent material, draping down from the oil cap, is introduced to the upper part of the clear zone (2). When stratified eggs are fertilized or activated by the Ca^{2+} ionophore A23187, this birefringence disappears within a few seconds—even before the fertilization envelope starts to elevate. Its sign, and the fluorescent staining by brefeldin A, suggest that the negative birefringence is due to a stack of membranes, stratified and aligned by centrifugation, and oriented more or less parallel to the direction of the centrifugal force.

To evaluate this proposal further, we investigated the birefringent region of the egg by electron microscopy. We used 2% glutaraldehyde in phosphate-buffered saline made up into 700-m*M* sucrose to prevent swelling of the *Arbacia* egg. Eggs placed in fixative without sucrose swelled up to about eight times the volume of the unfixed egg, lost their microvilli, and (reversibly) lost their negative birefringence.

Thin sections of stratified non-activated eggs, fixed with sucrose-glutaraldehyde, retained their negative birefringence and revealed that the birefringent region is occupied by stacks of smooth and rough endoplasmic reticulum (ER; Fig. 1A). The ER surrounded the nucleus and was aligned more or less parallel to the axis of centrifugation. A small number of Golgi membrane stacks were found amidst the ER, but with random orientation. At the lower region of the ER, we found stacks of annulate lamellae (3,4). These are most likely the refractile rod- and plate-like structures that are seen in centrifuged eggs by light microscopy, especially clearly in DIC. They tended, at first, to lie parallel to the axis of centrifugation, but changed their orientation as time elapsed after the centrifuge was stopped.

In centrifugally stratified eggs fixed about 5 min after fertilization—well after the negative birefringence had disappeared, but before it re-appeared—the distribution of the Golgi and annulate-lamellar material was basically unchanged. However, the ER was no longer in large sheets oriented along the centrifugal axis; rather, the sheets had fragmented into smaller vesicles (Fig. 1B), as was anticipated from their loss of bire-

[1] IHEM. UNC-CONICET, Argentina
[2] Kyoto University, Japan

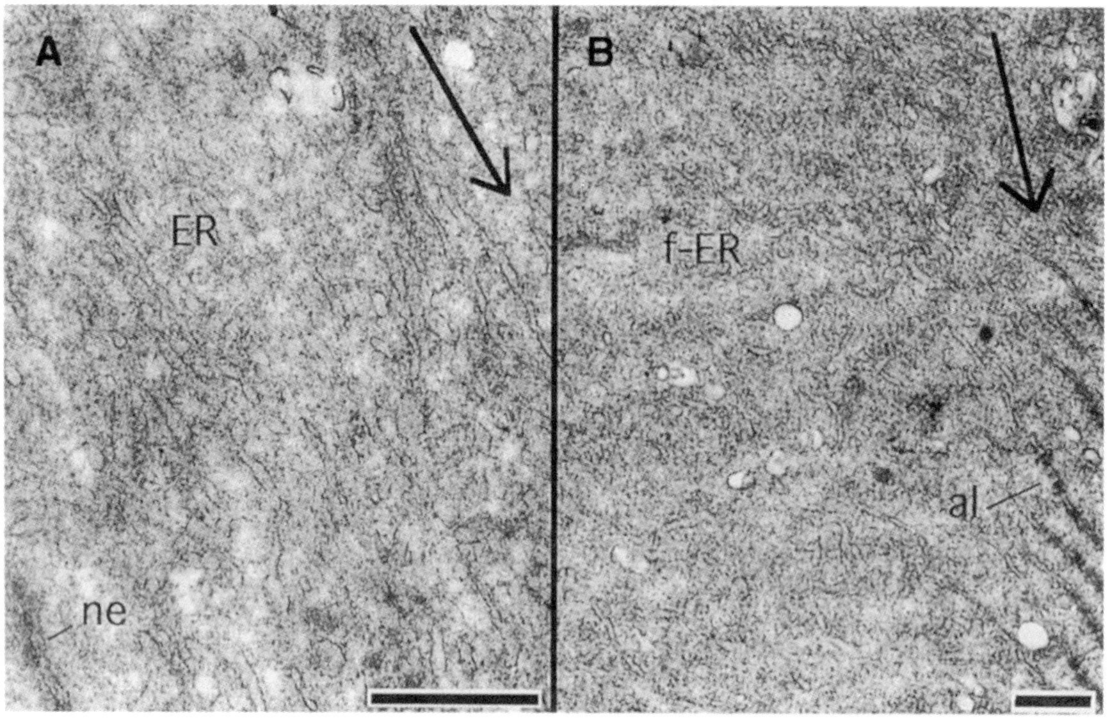

Figure 1. *Ultrastructure of upper clear zone in centrifugally stratified* Arbacia *egg. A. Unfertilized; B. fixed about 5 min after fertilization. After centrifugation on a Percoll-seawater density gradient for about 30 min (A), and after an additional 5 min centrifugation following fertilization (B), the eggs were fixed in 2% glutaraldehyde phosphate buffer made up in 700-mM sucrose solution. After confirming, with 1-μm sections of the unfertilized eggs, that the negative birefringence in the upper part of the clear zone remained intact, the cells were post fixed with osmium, dehydrated, embedded in Epon, and sectioned for electron microscopy. In A, the upper region of the clear zone contains a dense stack of ER, oriented more or less in the direction of the centrifugal force (long arrow). In B, only small pieces of ER remain, consistent with the (transient) loss of the negative birefringence. al: annulate lamellae. ER: endoplasmic reticulum. f-ER: fragmented ER. ne: nuclear envelope. Scale bars 0.5 μm.*

fringence. Because the birefringence of activated live eggs does return in the upper half of the clear zone after about 10 min, albeit with less ordered alignment of the birefringence axes, the EM of cells fixed at that stage would be expected to again show stacks of large ER membrane sheets, but with the stacks oriented along less uniform axes.

These observations suggest that the birefringence observed in live eggs with the CPM is a good indicator of membrane anisotropy, distribution, and especially their dynamic changes. In addition, centrifugally fragmented mini-cells could well prove to be a useful source for several isolated membrane components of the cells.

We thank Hamamatsu Photonics KK, Olympus Optical Company, Kyoto University, and the Marine Biological Laboratory for support of this project. We also thank Louis Kerr and Christina Stamper of the MBL Central Microscope Facility for their cooperative help with electron microscopy. M.B. was supported by an MBL Chairman of the Board Fellowship.

Literature Cited

1. **Harvey, E. B. 1941.** *Biol. Bull.* **81:** 114–118.
2. **Goda, M., M. H. Burgos, and S. Inoué. 2000.** *Biol. Bull.* **199:** 212–213.
3. **Afzelius, B. 1955.** *Exp. Cell Res.* **8:** 147–158.
4. **Swift, H. 1956.** *J. Biophys. Biochem. Cytol. (Suppl.)* **2:** 415–418 and 4 plates.

Reprinted from the *Journal of Microscopy*, Vol. 201(3), pp. 341–356, 2001, with permission from Blackwell Publishing.

Centrifuge polarizing microscope. I. Rationale, design and instrument performance

S. INOUÉ*, R. A. KNUDSON*, M. GODA*§, K. SUZUKI*†, C. NAGANO†, N. OKADA†, H. TAKAHASHI†, K. ICHIE‡, M. IIDA‡ & K. YAMANAKA‡

*Marine Biological Laboratory, Woods Hole, MA 02543, U.S.A.

†Olympus Optical Co., Hachioji-shi, Tokyo 192, Japan

‡Hamamatsu Photonics KK, Hamamatsu City 430, Japan

§Kyoto University, Kyoto 606-8502, Japan

Key words. Birefringence, centrifuge microscope, fluorescence, freeze motion, pulsed laser, strain-free chamber.

Summary

We first describe early uses of the centrifuge for deciphering physical properties and molecular organization within living cells, as well as the development and use of centrifuge microscopes for such studies. The rationale for developing a centrifuge microscope that allows high-extinction polarized light microscopy to observe dynamic fine structures in living cells is next discussed. We then describe a centrifuge polarizing microscope (CPM) that we developed for observing fine structural changes in living cells which are being exposed to up to ≈11 500 times earth's gravitational field (g).

With the specimen housed in a rotor supported on an air spindle motor, and imaged through an external microscope illuminated by a precisely synchronized flash of less than 10 ns duration from a Nd:YAG laser, the image of the spinning object remains steady up to the maximum speed of 11 700 rev min^{-1}, or up to $\approx 11\,500 \times g$. The image is captured, at up to 25 frames s^{-1}, by an interference-fringe-free CCD camera that is synchronized to the centrifuge rotor. At all speeds (in 100 rev min^{-1} increments), the image is resolved to better than 1 µm, while birefringence of the specimen, housed in a specially designed specimen chamber that suffers low-stress birefringence and prevents leakage of the physiological solutions, is detected with a retardance sensitivity of better than 1 nm. Differential interference contrast and fluorescence images (532 nm excitation) of the spinning specimen can also be generated with the CPM.

The second part of this study (Inoué *et al.*, *J. Microsc.* **201** (2001) 357–367, describes several biological applications of the CPM that we have explored. Individual live cells, such as oocytes and blood cells, are supported on a sucrose or Percoll density gradient while other cells, such as cultured fibroblasts and *Dictyostelium* amoebae, are observed crawling on glass surfaces. Observations of these cells exposed to the high G fields (centripetal acceleration/g) in the CPM are yielding many new results that lead to intriguing questions regarding the organization and function of fine structures in living cells and related quasi-fluid systems.

Correspondence to: S. Inoué. Tel.: +1 508 540 5382; fax: +1 508 540 6902; e-mail: jmacneil@mbl.edu

1. Introduction

1.1. Early history: why centrifuge cells?

As early as a century ago, living cells were exposed to centrifugal fields in order to explore their structure, physiology, and effects of centrifugation on development. Thus, Lillie (1909) and Shimamura (1940), among others, proposed physical and fine structural models for the nature of mitotic spindle 'fibres' and astral 'rays' based on their observations regarding the changes in these structures that were noted in cells fixed after centrifugation. While invisible in living cells, the spindle fibres were believed to be involved in the movement of chromosomes to the spindle poles, and the astral rays to the anchoring, positioning and movement of the spindle poles themselves, which in turn determined the plane of cell division.

Lillie proposed from his detailed studies of the metaphase-arrested oocytes of a marine annelid worm, *Chaetopterus pergamentaceous*, fixed within 2–3 min after centrifugation at G fields of a few hundred to a few thousand $\times g$, that these ephemeral fibres and rays are 'composed of orientated particles' rather than 'organic radii attached to the poles of the spindle'. In other words, they represented 'lines of forces' not unlike iron filings lined up along magnetic fields rather than physically integral fibres. In Lillie's words 'the general conclusion that the poles of the spindles are centres of force appears to me to be inevitable; no system of antagonistic fibrillae could behave in such a way' (i.e. they could not appear, re-arrange, or disappear in such a short time if they were true fibres).

By contrast, Shimamura deduced in pollen mother cells (of an Easter lily, *Lilium japonicum*) that were isolated from whole flower buds that he centrifuged at 5200 × *g*, that the spindle fibres are true cohesive physical entities; they resisted extension by the heavy, centrifugally displaced chromosomes that are attached by their kinetochores to the fibres. He also noted that the fibres could trap oil droplets that were being displaced centripetally inside the cell. Interestingly, he noted that these plant spindles deformed in such a way as to suggest that their poles were anchored to the cell surface, a surprising observation because plant spindles are generally thought to lack astral rays and hence lack anchorage of their spindle poles to the cell surface.

In searching for factors in living cells that determine embryonic polarity and differentiation, Morgan and colleagues surveyed whether, and how, stratification of visible particles and organelles in living cells exposed to centrifugal fields affected their subsequent development. Morgan (1927), in an extensive summary, concludes that, with a few exceptions, stratification of an egg cell exposed to (moderate) centrifugal fields does not affect the pattern of development, and can even yield embryos with pigment granules and other cell inclusions displaced to unusual regions of the embryo as a consequence of centrifugation at the single-cell stage.

In the 1930s to 1950s, Heilbrunn and his co-workers extensively used hand-driven centrifuges (that can be accelerated and decelerated very rapidly) to measure sedimentation velocities of cell inclusions (Heilbrunn, 1943). From those measurements they concluded that the 'viscosity' of the ground cytoplasm in sea urchin and other marine eggs was generally not much greater than that of pure water. (During this same period, ultracentrifuges, including those incorporating schlieren optics, were developed and used extensively to separate molecules in density gradients and to determine particle weights, e.g. Svedberg & Pedersen, 1940; Schachman, 1959.)

In 1960, Zalokar placed hyphae of *Neurospora* that were arranged to grow radially (like spokes of a wheel) in an air turbine centrifuge to stratify the organelles and regions of the cytoplasm with different densities (Zalokar, 1960a, b). After several minutes of centrifugation at some 100 000 × *g*, the cell contents stratified into no less than eight layers within the tube-shaped hyphal wall. Their compositions were established cytochemically. By applying a pulse of radioactive element and observing the sequential labelling of radioactivity into the different layers, Zalokar concluded that DNA did not directly serve as a template to code for the sequence of amino acids in a protein molecule. Instead, the nucleotide sequence in the DNA was first transferred to RNA, which, in turn, coded for the amino acid sequence to make the protein. (It is interesting to note that this work precedes by a few years the discovery of the role of transfer and messenger RNA.)

1.2. Centrifuge microscopes

While the observations mentioned above were made on cells and embryos after they were removed from a centrifuge, in the meanwhile instruments were developed that allowed the observation of a specimen directly, as it is spinning in a centrifuge, in other words, through a centrifuge microscope. In 1932, E. Newton Harvey provided a thoughtful survey of alternative designs, potential capabilities and limitations of centrifuge microscopes that should provide reasonably stable microscope images of specimens that are being exposed to centrifugal fields.

A type of centrifuge microscope described in Harvey's survey, in which the microscope objective lens is made to spin together with the specimen (Fig. 1; Harvey & Loomis, 1930; Harvey, 1932), was used by E. N. Harvey (1931) to determine the tension at the surface of cells. He took advantage of the fact that the light oil droplets and heavy yolk granules, which had stratified to the centripetal and centrifugal poles, respectively, within a cell, would stretch the cell suspended in a density gradient. As the volume of the cell remains constant, a sea urchin egg cell, for example, would eventually be pulled into a dumbbell shape. Harvey calculated the tension at the surface of the cell from the known diameter of the neck of the dumbbell and the forces exerted by the lighter and heavier 'balls' of the dumbbell. The tension at the surface thus established was much lower than had been previously expected, amounting to only ≈ 0.19 dyne cm^{-1} for eggs of *Arbacia punctulata*.

In the 1930s and 1940s, Ethel Brown Harvey made extensive use of the Harvey–Loomis centrifuge microscope to study the stratification and formation of heavier and lighter cell fragments. Observing unfertilized sea urchin (*Arbacia*) eggs in seawater layered above isotonic sucrose solution, she found that the stratified eggs would eventually pull apart at the neck of the dumbbells into half and then quarter cell fragments. The lightest quarter contained an oil cap, the nucleus and clear cytoplasm, with the next layer containing mitochondria and other cytoplasmic granules. The third fragment primarily contained the yolk granules and the heaviest quarter pigment granules and some yolk (Harvey, 1936).

Surprisingly, all four fragments were not only viable but would elevate fertilization envelopes and undergo cleavage when fertilized. The fertilized halves, as well as a small fraction of the clear (topmost) quarter, developed into small but normal pluteus larvae (Harvey, 1946). More surprisingly, even the heavier half fragment, devoid of the egg

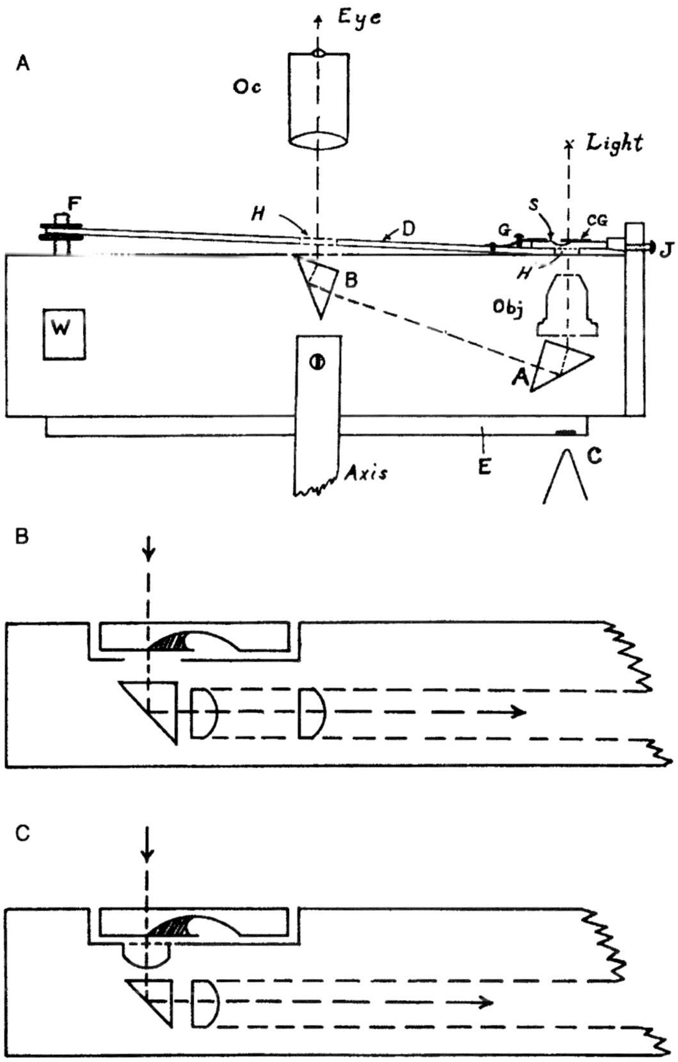

Fig. 1. Harvey–Loomis centrifuge microscope (from Harvey, 1932). In this type of centrifuge microscope, the whole microscope revolves with the specimen in the centrifuge head. Earlier versions used synchronized flash bulbs, which were found unnecessary and later replaced with incandescent tungsten bulbs. (A) Cut out through the rotor. (B) and (C) Schematic of specimen chamber (made out of pieces of depression slides) viewed through low power (B) and higher power (C) objective lenses.

nucleus, could be activated by artificial parthenogenesis (i.e. in the absence of sperm and therefore the nucleus, centrioles and other organelles normally carried into the egg by the sperm). Such 'parthenogenetic merogones' with no paternal or maternal chromatin still developed astral rays and amphiasters and underwent a few cleavage divisions (Harvey, 1940).

In contrast to the Harvey–Loomis electric-motor-driven centrifuge microscope, in which the whole microscope optics spun with the rotor, Pickels (1936) designed a system for observing the specimen through an external microscope. The design was used by Beams & King (1937) in an air-driven microscope centrifuge for observing specimens exposed to very high *g*-forces.

In this system, a first-surface mirror is built into the rotor so that its reflecting surface intersects the rotational axis of the rotor in the plane containing the specimen (Fig. 2 from Beams, 1937). A horizontal microscope views the image of the specimen reflected from the 45° mirror, as though the specimen were aligned with the rotational axis of the turbine, thus keeping the image of the specimen more or less stationary. Riding on a jet of compressed air, and stabilized by the Bernouilli force generated by the air escaping from the tapered narrow space between the rotor and the stator, the air turbine can be spun at hundreds of thousands of revolutions per minute, thus generating very high centrifugal fields.

The air-turbine-driven ultracentrifuge was used to observe the effect of high *G* fields on early development in *Fucus* eggs (a marine alga) and of *Ascaris* (a nematode worm) (Beams & King, 1937). They found in *Ascaris* that the fertilized egg stratified by centrifugation could form

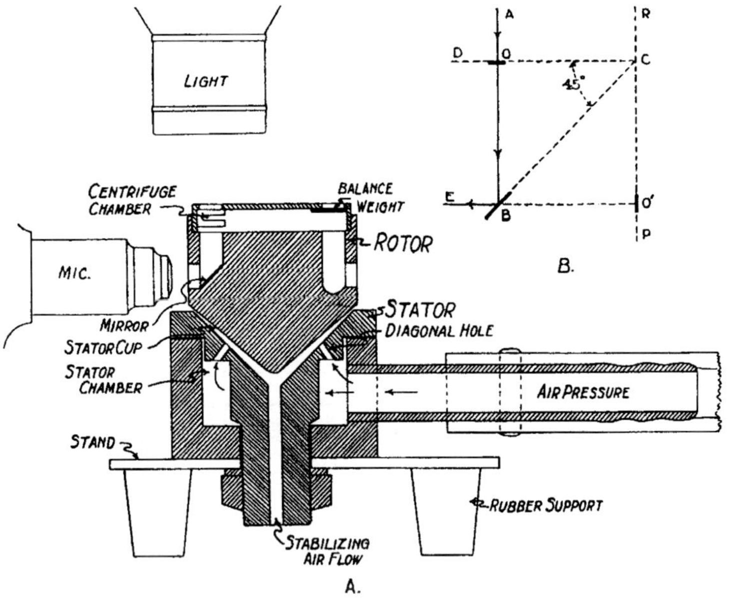

Fig. 2. Beam vs. centrifuge microscope (from Beams, 1937). (A) The specimen, placed in a high-speed air-driven centrifuge, is viewed with an external, horizontal microscope. (B) The 45° mirror on the rotor reflects the image (0′) of the specimen (0) as though it originated on the axis of rotation and hence without extensive blur, even without the use of short stroboscopic illumination.

normal appearing mitotic figures and undergo one to a few cycles of mitoses, but that the ensuing cleavages were absent at 150 000 × *g*, although not at 5000 × *g*. Fertilized *Fucus* eggs appeared to develop normally after centrifuging at 150 000 × *g* for 30 min (Beams, 1937).

More recently, Kuroda & Kamiya (1989) modified the Harvey–Loomis-type centrifuge microscope by building a video camera into the rotor illuminated by a stationary light source (Fig. 3). Using this system, they measured the forces involved in cytoplasmic streaming (chloroplasts movement on actin cables) in plant cells as well as swimming forces of *Paramecium* species.

Kamitsubo and co-workers (Kamitsubo *et al.*, 1989; Kamitsubo & Kikuyama, 1994) and Hiramoto (Hiramoto & Kamitsubo, 1995) designed centrifuge microscopes in which the specimen was observed in Nomarsky optics with a video camera through a stationary microscope illuminated with a xenon flash bulb (flash time = 180 ns) synchronized to the rotation of the spinning rotor. While the relatively long duration of the xenon flash bulb smeared the image of the specimen by several micrometres along the direction of specimen flight, Kamitsubo *et al.* were nevertheless able to measure the shear stress and yield value for the endoplasm of *Nitella*, a large algal cell. In particular, they were able to record the changes in the flow patterns of organelles and cell inclusions along cortical actin fibres, before and after electrical-stimulus-induced cessation of streaming in cells spinning in the centrifuge. Since their review, Hiramoto has proposed an improvement that includes a rocking glass plate placed in front of the objective lens. The plane parallel glass plate rocks (on an axis parallel to the radius of the centrifuge) in synchrony with the rotation of the specimen to deviate the image beam and thus compensate for the blurring of the spinning specimen image (Japanese Patent Application SHO 63-250615 by Nikon Corp.).

Oiwa *et al.* (1990) used this centrifuge microscope (maximum centrifugal acceleration = 1900 × *g*) to measure the force–velocity relation of myosin-coated polystyrene beads on actin cables in similar algal cells. They found that beads suffering 'negative loads' (those flowing in the same direction as the centrifugal force) surprisingly moved slower than those in a normal gravitational field.

Hall *et al.* (1993, 1996) used a similar centrifuge microscope but equipped with a 10 ns stroboscopic illuminator (Xenon Corp., Model 437 Nanopulser) to measure the G-force vs. velocity and pull-off force for demembranated sperm gliding on a sparse lawn of the motor protein, kinesin. They provide detailed analyses of the various forces that are likely to be acting on the model sperm attached to the substrate by a few kinesin molecules in the centrifuge. The diameter of rotation of the 'rotating microscope stage' was $2r = 10.5$ cm and the 'typical rotational speeds' ranged from 0 to 4000 rev min^{-1}, yielding a maximum field of just over 1000 × *g*. While the model 437 Nanopulser is specified as providing a 10 ns (air discharge) pulse duration, it is also described as displaying a 10 ns jitter. At 4000 rev min^{-1} and $r = 5.25$ cm, however, the image blur due to the flash discharge should still be under 0.5 μm.

Hiramoto & Kamitsubo (1995) review the many types of centrifuge microscope developed between Harvey and Loomis' early model (1930) and some recent models using

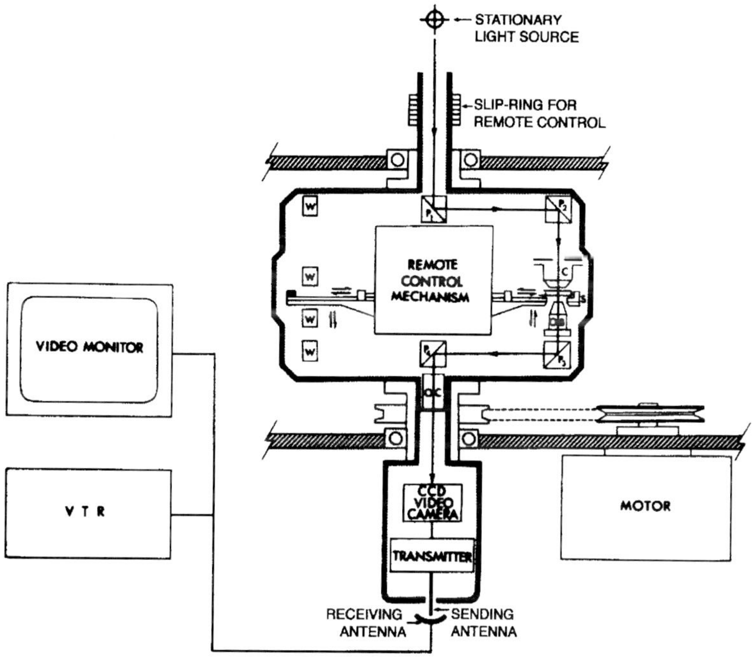

Fig. 3. Centrifuge microscope with built-in video camera (from Kuroda & Kamiya, 1989). In these devices, a video camera, in addition to the microscope, spins together with the specimen. As the specimen is viewed continuously as the rotor turns, the light source need not be extremely bright.

stroboscopic illumination, together with their own applications in studies related to cell motility.

1.3. Advantage of polarized light microscopy

In the centrifuge microscopes described so far, the spinning specimens were observed in bright-field microscopy and more recently in differential interference, or Nomarski, contrast (DIC). However, there is a major advantage in observing the specimen being centrifuged with polarized light microscopy that provides relatively high extinction. While technically challenging, that would allow the detection and measurement of weakly birefringent structures that are present, or became developed, in centrifugally stratified living cells and other quasi-fluid specimens.

Birefringence signals the orderly alignment of submicroscopic structures that are too fine to be resolved with the light microscope and allows one to determine the axes, degree of alignment, and optical character of the molecules involved. In other words, one can measure changes in submicroscopic alignment of molecules and polymerization states of molecular filaments, membranes, etc., whose diameters or thicknesses are far too small to be resolved with the light microscope (Inoué, 1986; Appendix III; Ambronn & Frey, 1926; Schmidt, 1937). Furthermore, with high-extinction polarized light microscopy, one can non-destructively visualize birefringent structures that may otherwise be difficult or impossible to see because of the very small difference in their refractive index relative to their surround (Schmidt, 1937; Inoué, 1953). More recently, video and digital contrast enhancement have dramatically improved our ability to detect weak birefringence (Allen *et al.*, 1981a, b; Inoué, 1981a, 1986).

In fact, high-extinction polarization microscopy of live dividing cells was used to: provide definitive proof of the reality of spindle fibres; demonstrate the dynamic nature of their constituent molecular filaments (microtubules); reveal the thermodynamic shift in assembly states of the molecular filaments; and to propose that by the very act of depolymerizing, microtubules can generate contractile forces adequate to move chromosomes (review, Inoué, 1964, 1981b; Inoué & Sato, 1967; Inoué & Salmon, 1995; Inoué & Oldenbourg, 1998). These studies have helped shed considerable light on the enigmatic cell organization responsible for mitotic chromosome movement alluded to earlier (Lillie, 1909; Shimamura, 1940; see also Wilson, 1928; Schrader, 1953).

Nevertheless, and despite the striking recent advances in biochemical studies, together with resolution of the detailed structures of several major molecules involved (see, e.g. Silver, 1999), there is much to learn regarding the dynamic submicroscopic organization of the living cell. For example, what is the nature of the linkage that directs a spindle pole to move and attach to unique sites on the cell surface, thus defining the ensuing cleavage plane and subsequent developmental fate of the resulting blastomeres (see, e.g. Dan & Inoué, 1987)? How firmly, and through what organizational forces or fine structural elements, are centrosomes

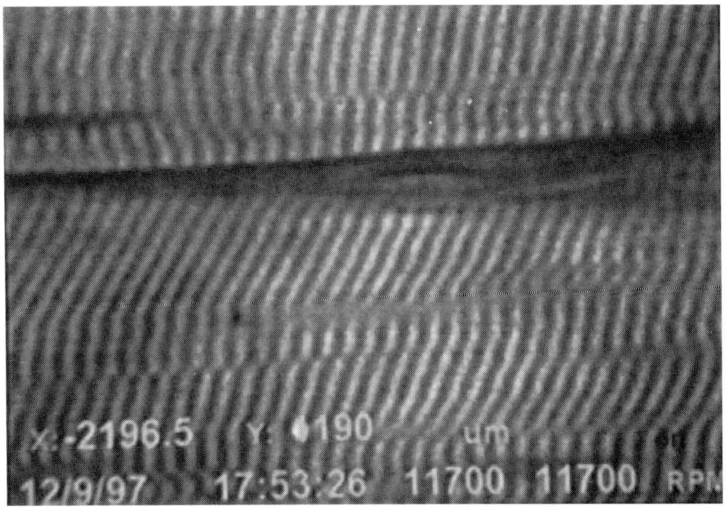

Fig. 10. 1 μm thick section of frog muscle. The 1 nm retardance of the 1.5 μm long birefringent A band (bright cross-striations) is clearly seen at 11 700 rev min^{-1}.

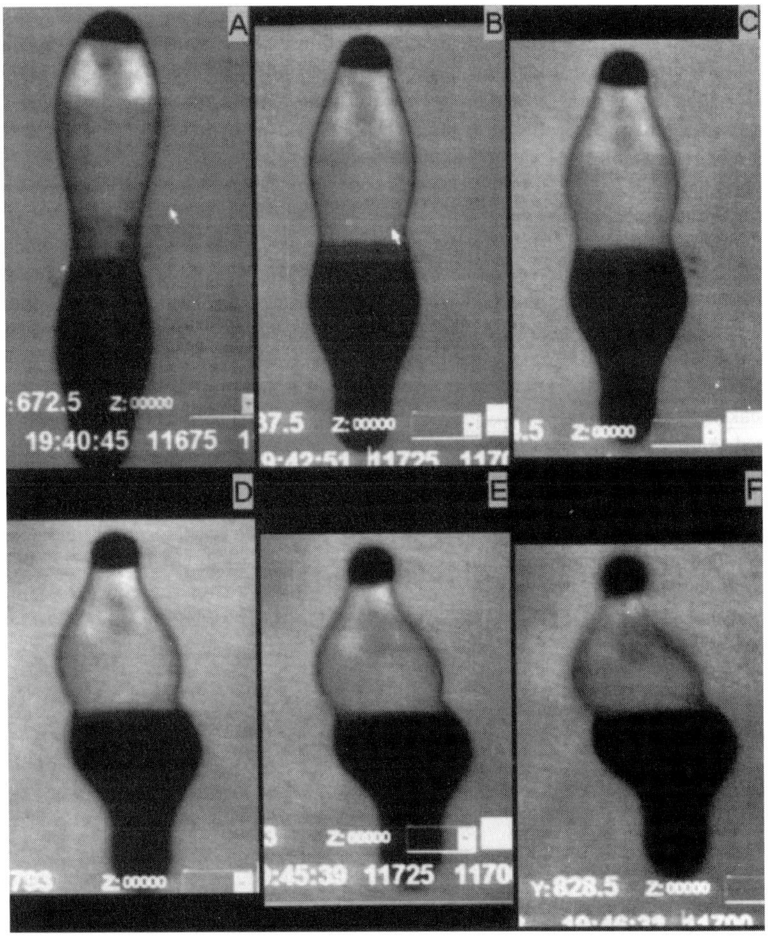

Fig. 11. Squat *Arbacia* eggs on sedimented Percoll layer. After ca. 30–60 min of centrifugation at 11 700 rev min^{-1} (11 488 × *g*), stratified sea urchin and hamster eggs progressively became squat as seen here, from their earlier shape of being elongated up and down. It turns out that the colloidal silicone particles in the Percoll seawater (or physiological saline) solution had sedimented and formed a dense layer on which the cells were forced down. A lateral flow sometimes displaced the eggs, or bent the 'heads' of the squat eggs, to the left or right. Time elapsed from panel A to F: ≈ 6 min (Inoué, Goda and Liu, 1998; unpublished observations.)

is incredibly stable, some changes take place while the speed is increased or decreased. (A) During a rapid, continuous shift of the spin speed, the video image is designed to be momentarily frozen in order to avoid the superposition of unsynchronized images. (B) There can be some shift of the specimen chamber as its cushion is compressed during rising speed and relaxed during falling speed. (C) The variation in window birefringence may need to be compensated as it varies with rotor speed.

With respect to item B, as described earlier, the specimen chamber is now supported via a thin (≈0.2 mm) layer of G.E. RTV 662 silicon rubber on a firm support of acetyl plastic that sits directly on the aluminium rotor and cover plate (Figs 6 and 7). With this new design, the displacement of the chamber is limited to about 10 μm for every 1000 rev min^{-1} change of the rotor speed. Although this is not a large displacement, this drift has to be taken into account when the absolute coordinate of the specimen needs to be known. Markers firmly attached to the inner surface of the cover window can aid in determining and correcting for the drift. Item C was extensively discussed earlier.

During observation of biological samples, such as marine egg cells floating in an isopycnic layer, we often noticed a moderately slow (a few seconds per cycle) rotation of small particles that were suspended in the medium. For some reason, the particles were distributed in layers to the side, and at the level, of the top and bottom of the egg cells being observed. The particles rotated with a few micrometres' radius around a horizontal axis (i.e. an axis at right angles to the g-vector), more or less in line with the top and bottom of the cells under observation.

As the shape of the specimen chamber is not an ideal pie-segment shape of the rotor, there is the possibility that the microcirculation of the small particles is indicating some convection in the density gradient medium (see, e.g. Svedberg & Pedersen, 1940; especially Section III-7 in Schachman, 1959). Alternatively, could the microcirculation result from Coriolis-related forces in the stratified (or stratifying) medium, or possibly be induced by small temperature gradients? The latter seems unlikely, as the overall rotor temperature remained within a 1°C of room temperature even at maximum speed.

Another surprise took place after living egg cells and mouse oocytes were centrifuged for tens of minutes in Percoll gradients at 10 000 × *g*. Instead of continuing to elongate along the centrifugal axis, the cells eventually became foreshortened and squat (Fig. 11). They also tended to be bent or to drift sideways. Unexpectedly, it turns out that the cells had come to lie on a highly concentrated layer of Percoll instead of in an isopycnic gradient, as was the case earlier during the centrifugation (Fig. 11A). With the short height of the specimen chamber (4–8 mm), after nearly an hour of centrifugation at 10 000 × *g*, the silicone sol in the Percoll (mixed with physiological saline or seawater) had settled out at the bottom of the chamber as a discrete layer with high density. The sharp gradient and high refractive index exhibited by this high density layer could be seen visually and by the bending of a collimated beam of light.

Beyond these limiting factors, the image resolution and stability, sensitivity for detecting small birefringence retardations, and contrast-generating ability of the CPM all match or exceed the specifications targeted in the original design.

In Part II of this study (Inoué *et al.*, 2001), we shall describe several sample applications of the CPM that we have explored over the last 2 years.

References

Allen, R.D., Allen, N.S. & Travis, J.L. (1981b) Video-enhanced contrast, differential interference contrast (AVEC-DIC) microscopy: a new method capable of analyzing microtubule-related motility in the reticulopodial network of *Allogromia laticollaris*. *Cell Motil.* **1**, 291–302.

Allen, R.D., Travis, J.L., Allen, N.S. & Yilmaz, H. (1981a) Video-enhanced contrast polarization (AVEC-POL) microscopy: a new method applied to the detection of birefringence in the motile reticulopodial network of *Allogromia laticollaris*. *Cell Motil.* **1**, 275–289.

Ambronn, H. & Frey, A. (1926) *Das Polarisationsmikroskop, Seine Anwedung in der Kolloidforschung in der Farbarei*. Akademische-Verlag, Leipzig.

Aronson, J.F. (1971) Demonstration of a Colcemid-sensitive attractive force acting between the nucleus and a center. *J. Cell Biol.* **51**, 579–583.

Beams, H.W. (1937) The air turbine centrifuge, together with some results upon ultracentrifuging the eggs of *Fucus serratus*. *J. Mar. Biol. Assoc. UK*, **21**, 571–588, and Plate II.

Beams, H.W. & King, R.L. (1937) The suppression of cleavage in *Ascaris* eggs. *Biol. Bull.* **73**, 99–111.

Dan, K. & Inoué, S. (1987) Studies of unequal cleavage in molluscs. II. Asymmetric nature of the two asters. *Int J. Invertebrate Reprod. Development*, **11**, 335–354.

Hall, K., Cole, D.G., Yeh, Y. & Baskin, R.J. (1996) Kinesin force generation measured using a centrifuge microscope sperm-gliding motility assay. *Biophys. J.* **71**, 3467–3476.

Hall, K., Cole, D.G., Yeh, Y., Scholey, J.M. & Baskin, R.J. (1993) Force-velocity relationships in kinesin-driven motility. *Nature*, **364**, 457–459.

Harvey, E.B. (1936) Parthenogenetic merogony or cleavage without nuclei in *Arbacia punctulata*. *Biol. Bull.* **71**, 101–121.

Harvey, E.B. (1940) A comparison of the development of nucleate and non-nucleate eggs of *Arbacia punctulata*. *Biol. Bull.* **79**, 166–187.

Harvey, E.B. (1946) Structure and development of the clear quarter of the *Arbacia punctulata* egg. *J. Exp. Zool.* **102**, 253–276.

Harvey, E.N. (1931) The tension at the surface of marine eggs, especially those of the sea urchin *Arbacia*. *Biol. Bull.* **61**, 273–279.

Harvey, E.N. (1932) The microscope-centrifuge and some of its applications. *J. Franklin Inst.* **214** (1279), 1–24.

Harvey, E.N. & Loomis, A.L. (1930) Scientific apparatus and laboratory methods. *Science*, **72**, 42–44.

Heilbrunn, L.V. (1943) *An Outline of General Physiology*. W.B. Saunders, Philadelphia.

Hiramoto, Y. & Kamitsubo, E. (1995) Centrifuge microscope as a tool in the study of cell motility. *Intl. Rev. Cytology*, **157**, 99–128.

Inoué, S. (1952) Studies on depolarization of light at microscope lens surfaces. I. The origin of stray light by rotation at the lens surfaces. *Exp. Cell Res.* **3**, 199–208.

Inoué, S. (1953) Polarization optical studies of the mitotic spindle. I. The demonstration of spindle fibers in living cells. *Chromosoma*, **5**, 487–500.

Inoué, S. (1964) Organization and function of the mitotic spindle. *Primitive Motile Systems in Cell Biology* (ed. by R. D. Allen and N. Kamiya), pp. 549–598. Academic Press, New York.

Inoué, S. (1981a) Video image processing greatly enhances contrast, quality, and speed in polarization based microscopy. *J. Cell Biol.* **89**, 346–356.

Inoué, S. (1981b) Cell division and the mitotic spindle. *J. Cell Biol. Special Issue, 'Discovery in Cell Biology'*, **91** (2), 131s–147s.

Inoué, S. (1986) *Video Microscopy*. Plenum Press, New York.

Inoué, S. (1999) Windows to dynamic fine structures, then and now. *FASEB J.* **13**, S185–S190.

Inoué, S. & Hyde, W.L. (1957) Studies on depolarization of light at microscope lens surfaces. II. The simultaneous realization of high resolution and high sensitivity with the polarizing microscope. *J. Biophys. Biochem. Cytol.* **3**, 831–838.

Inoué, S. & Oldenbourg, R. (1995) Optical instruments. Microscopes. *Handbook of Optics*, Vol. 2. 2nd edn (ed. by Optical Society of America), pp. 17.1–17.52. McGraw-Hill, Inc, New York.

Inoué, S. & Oldenbourg, R. (1998) Microtubule dynamics in mitotic spindle displayed by polarized light microscopy. *Mol. Biol. Cell*, **9**, 1603–1607.

Inoué, S. & Salmon, E.D. (1995) Force generation by microtubule assembly/disassembly in mitosis and related movements. *Mol. Biol. Cell*, **6**, 1619–1640.

Inoué, S. & Sato, H. (1967) Cell motility by labile association of molecules. The nature of mitotic spindle fibers and their role in chromosome movement. *J. Gen. Physiol.* **50**, 259–292.

Inoué, S. & Spring, K.R. (1997) *Video Microscopy – the Fundamentals*. 2nd edn. Plenum Press, New York.

Inoué, S., Goda, M. & Knudson, R.A. (2001) Centrifuge polarizing microscope. II. Sample biological applications. *J. Microsc.* **201**, 357–367.

Inoué, S., Knudson, R.A., Suzuji, K., Okada, N., Takahashi, H., Iida, M. & Yamanaka, K. (1998) Centrifuge polarizing microscope. *Microsc. Microanal.* **4** (Suppl. 2), 36–37.

Jenkins, F.A. & White, H. (1957) *Fundamentals of Optics*. 3rd edn. McGraw-Hill, New York.

Kamitsubo, E. & Kikuyama, M. (1994) Bidirectional flow of endoplasm in *Physarum* plasmodium under centrifugal acceleration. *Protoplasma*, **182**, 53–58.

Kamitsubo, E., Ohashi, Y. & Kikuyama, M. (1989) Cytoplasmic streaming in internodal cells of *Nitella* under centrifugal acceleration: a study done with a newly constructed centrifuge microscope. *Protoplasma*, **152**, 148–155.

Kuroda, K. & Kamiya, N. (1989) Propulsive force of *Paramecium* as revealed by the video centrifuge microscope. *Exptl. Cell Res.* **184**, 268–272.

Lillie, F.R. (1909) Karyokinetic figures of centrifuged eggs; an experimental test of the center of force hypothesis. *Biol. Bull.* **17**, 101–119.

Monné, L. (1944) Cytoplasmic structure and cleavage pattern of the sea urchin egg. *Ark. f. Zoologi*, **35A**, 1–27.

Morgan, T.H. (1927) *Experimental Embryology*. Columbia University Press, New York.

Oiwa, K., Chaen, S., Kamitsubo, E., Shimmen, T. & Sugi, H. (1990) Steady-state force-velocity relation in the ATP-dependent sliding movement of myosin-coated beads on actin cables *in vitro* studied with a centrifuge microscope. *Proc. Natl. Acad. Sci. USA*, **87**, 7893–7897.

Oldenbourg, R., Inoué, S., Tiberio, R., Stemmer, A., Mei, G. & Skvarla, M. (1996) Standard test targets for high-resolution light microscopy. *Nanofabrication and Biosystems: Integrating Materials Science, Engineering, and Biology* (ed. by H. C. Hoch, L. W. Jelinski and H. G. Craighead), pp. 123–138. Cambridge University Press, New York.

Pfeiffer, H.H. (1938) Double refraction measurements and structural changes in mitotic spindles disturbed by centrifugal force. *Biodynamica*, **2** (35), 1–8.

Pfeiffer, H.H. (1942) Über die Abhängigkeit der Doppelrechung von der Orientierung protoplasmatischer Leptonen durch Zentrifugieren. *Koll. Z.* **100**, 254–263.

Pickels, E.G. (1936) Optical designs for observing objects in centrifugal fields of force. *Science*, **83**, 471–472.

Schachman, H.K. (1959) *Ultracentrifugation in Biochemistry*. Academic Press, New York.

Schmidt, W.J. (1937) *Die Doppelbrechung von Karyoplasma, Zytoplasma und Metaplasma*. Protoplasma-Monographien, 11. Borntraeger, Berlin.

Schrader, F. (1953) *Mitosis—the Movements of Chromosomes in Cell Division*. Columbia University Press, New York.

Shimamura, T. (1940) Studies on the effect of the centrifugal force upon nuclear division. *Cytologia*, **11**, 186–216.

Silver, R.E., ed. (1999) A half-century of advances in microscopy. *FASEB J.* **13** (Special Suppl.), 179–283.

Svedberg, T. & Pedersen, K.O. (1940) *The Ultracentrifuge*. Clarendon Press, Oxford.

Wilson, E.B. (1928) *The Cell in Development and Heredity*. 3rd edn. Macmillan, New York.

Zalokar, M. (1960a) Cytochemistry of centrifuged hyphae of *Neurospora*. *Exptl. Cell Res.* **19**, 114–132.

Zalokar, M. (1960b) Sites of protein and ribonucleic acid synthesis in the cell. *Exptl. Cell Res.* **19**, 559–576.

Reprinted from the *Journal of Microscopy*, Vol. 201(3), pp. 357–367, 2001, with permission from Blackwell Publishing.

Centrifuge polarizing microscope. II. Sample biological applications

S. INOUÉ*, M. GODA*† & R. A. KNUDSON*
*Marine Biological Laboratory, Woods Hole, MA 02543, U.S.A.
†Kyoto University, Kyoto 606 8502, Japan

Key words. Amoeboid movement, fertilization, fibroblast, living cell fine structure, meiosis, mitosis, nematic virus solution, oocyte, organelle stratification, sickle cell, wound healing.

Summary

The rationale, design and general performance of the CPM (centrifuge polarizing microscope) were described in Part I of this study (Inoué *et al. J. Microsc.* **201** (2001) 341–356. In this second part, we describe observations on several biological samples that we have explored over the past two years using the CPM.

As described in the first part of the study, although the CPM was basically designed as a high-extinction centrifuge polarizing microscope, it also allows observations of the specimen exposed to high centrifugal fields up to 10 500 × *g* (earth's gravitational acceleration) in fluorescence (532-nm excitation) and in DIC (differential interference or Nomarski contrast).

1. Stratification of *Arbacia* eggs

Figure 1 shows an egg of a sea urchin *Arbacia punctulata*, observed in the centrifuge polarizing microscope (CPM). As earlier shown by E. B. Harvey (1936), the centrifuged, unfertilized egg supported on an isopycnic gradient stratifies into several discrete layers. These contain from top to bottom: oil cap, clear zone containing the nucleus, mitochondria, yolk and pigment granules. The stratified egg elongates by being pulled up (centripetally) by the lighter oil cap and clear cytoplasm, and by being pulled down (centrifugally) by the heavy yolk and pigment granules. Upon further centrifugation, the egg is pinched apart into lighter and heavier halves and then quarters (see also Section 3, Fig. 6).

Observations in 532 nm excitation fluorescence of *Arbacia* eggs stained with MitoTracker Red (Molecular Probes, Eugene, OR) identifies the mitochondrial layer as a discrete zone at the bottom of the clear zone above the dense yolk layer (Fig. 1B), confirming E.B. Harvey's earlier observation (Harvey, 1956; Chapter 17).

In the CPM, some features that were not seen in Harvey's classical studies (made with a non-polarizing centrifuge microscope) become obvious. In the upper region of the clear zone, a curtain of birefringent material appears draping down from the oil cap and surrounding the nucleus (Figs 1A and C). In the curtain of birefringent material, the axis of greater refractive index appears horizontally, i.e. at right angles to the centrifugal axis, or the 'folded' texture of the curtain. (The birefringence axis with greater refractive index is established, e.g. by comparing the contrast of specimen regions in the presence of a compensator with regions of the tangentially positively birefringent fertilization envelope that elevates after the egg is fertilized, see Fig. 4). It is well established that the slow axis of the birefringent fertilization envelope parallels its tangent (it exhibits a positive birefringence with reference to the tangent; e.g. Inoué & Dan, 1951). Therefore, all one needs to do is to introduce (or turn) a Brace–Köhler compensator and to observe which region of the specimen (e.g. the curtain material) shows the same contrast as regions of the fertilization envelope. For specimen regions that show the same contrast (brighter or darker than the background) as the fertilization envelope, their axes of greater refractive index lie parallel to each other. Where the contrast is reversed, the axes are reversed (Inoué, 1986; Appendix 3). The material, which gives rise to the texture in this curtained material (which must have a low specific density) therefore exhibits a negative birefringence relative to the texture, suggesting that the texture represents a plaited series of cytoplasmic membranes.

Observation of *Arbacia* eggs stained with brefeldin A (Sigma Chemical Co., St. Louis, MO) in the CPM in the fluorescence mode, in fact, shows a concentration of membranes in the negatively birefringent layer, but not in

Correspondence to: S. Inoué. Tel.: +1 508 540 5382; fax: +1 508 540 6902; e-mail: jmacneil@mbl.edu

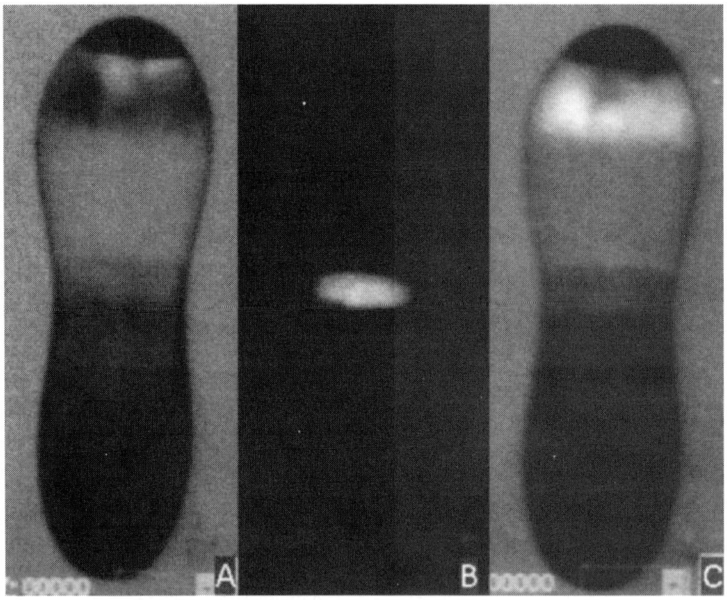

Fig. 1. Centrifuged sea urchin egg viewed in CPM. After some 10–15 min of centrifugation of unfertilized eggs at 5200 rev min^{-1} ($\approx$2300 $\times$ g) on a sucrose density gradient, the unfertilized *Arbacia* egg is stratified and takes on a dumbbell shape. Just below the oil cap on top, a negatively birefringent curtain of low-density material (membranes) emerges in the upper region of the clear zone. The birefringent curtain appears dark in (A) and bright in (C), surrounding the non-birefringent nucleus located immediately below the oil cap. In (A), the slow axis of the compensator is orientated vertically (i.e. N–S), and in (C), horizontally (i.e. E–W) Thus, the slow axis of the curtain material is orientated horizontally, or it displays a negative birefringence with respect to its pleated (vertical) direction. After prolonged centrifugation, some positive birefringent material also appears just below the oil cap. In all figures, the crossed polarizer-analyser axes are orientated at ±45°, and the centrifugal force points down towards the bottom of the page. (B) shows the same stratified egg in the spinning CPM in the fluorescence mode. The MitoTracker fluorescence clearly identifies the layer into which the mitochondria are stratified.

the lower half of the clear zone (Figs 2 and 3A). This lipid staining dye shows the oil cap and yolk layer with high intensity and the mitochondria and upper layer of the clear zone with somewhat less intensity.

Fluorescence is conspicuously absent from the lower part of the clear zone, from the heavy granule layer below the yolk, and from the nucleus (Figs 2 and 3).

2. Early changes in *Arbacia* eggs upon fertilization

A surprising series of changes are seen in the egg when we observe the process of fertilization in the CPM. Within 10–20 s after fertilization, the negatively birefringent material disappears from sight, in some cases progressing as a wave (Figs 4A–D). In other words, fertilization induces the 'membrane material', previously aligned by the centrifugal force, apparently to rapidly fragment and lose its regular orientation. At the same time, all of a sudden the egg starts to float up, i.e. into regions of the medium with less density (Figs 4D–F). Clearly the egg has become 'lighter', or its overall density has become reduced. As the cell continues to float upward slowly, the fertilization envelope is elevated and gains a stronger positive birefringence over the next 3–4 min.

During the summer of 2000, we found by electron microscopy that it was the endoplasmic reticulum that became aligned by the centrifugation and then almost instantaneously became fragmented and disorganized upon fertilization (Burgos *et al.*, 2000). We also found with the CPM that it was primarily the egg jelly that is responsible for the cell's sudden 'levitation' following fertilization (Goda *et al.*, 2000). Nevertheless, contributions by exocytosis and swelling of the gel material beneath the fertilization envelope and reduction of the whole cell's density by some early response to fertilization may also contribute to the egg's levitation. Such swelling of the whole egg could be induced by changes in selective permeability of the cell membrane or by release of osmotically active molecules within the egg. In fact, many somewhat irregularly shaped marine eggs become fully spherical when fertilized whether or not they elevate fertilization envelopes.

3. Stratification and fragmentation of *Chaetopterus* oocytes

Figure 5 shows oocytes of a marine annelid worm *Chaetopterus pergamentaceous* shed into normal seawater

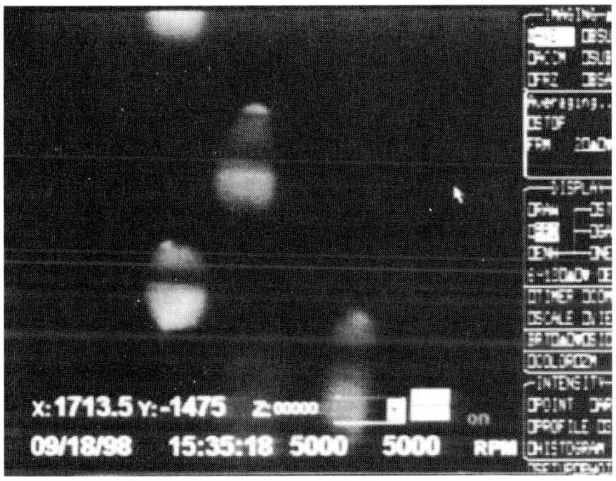

Fig. 2. Brefeldin A fluorescence showing membranes in centrifuged sea urchin eggs. The overview of the monitor display is taken with a 10×/0.30 NA objective lens.

and centrifuged after formation of their meiotic spindle. In such an oocyte, centrifuged on a cushion of isopycnic sucrose or Percoll seawater, the lighter and heavier cell contents, which are stratified to the centripetal and centrifugal poles of the cells, respectively, stretch the cell into an extended dumbbell. As seen in Fig. 5, the positively birefringent meiotic spindle (vertically elongated white structure in panel A, black in panel B), generally located in the lighter clear zone within the cell, becomes stretched with pointed poles and elevated birefringence (compare with Fig. 7H). Clearly the microtubule bundles in the spindle are under tension through forces applied to the spindle poles that are anchored either to the cell cortex or other centrifugally stratified components in the cell.

Again, as with the *Arbacia* eggs, birefringent material becomes aligned at the centripetal pole, but in several layers with complex distributions and signs of birefringence. These layers are more clearly stratified in the egg fragment shown in Fig. 6, showing a quarter egg fragment just as it has lost the oil cap (panel A) until 4 s later when it is again fragmenting, this time into 1/8 egg fragments (panels E, F). As seen here, when the cell is pinched apart, its membrane is sealed and the egg fragments round up (panels A-D) and yield intact cell fragments containing ever more stratified cell components. We look forward to collecting the separated, fractionated cell components and to determining the exact make up of the lighter, centrifugally aligned, birefringent components (perhaps as: smooth ER, Golgi, rough ER, annulate lamellae?).

4. Ca^{2+} activation of *Chaetopterus* oocytes

When *Chaetopterus* oocytes are shed into Ca^{2+}-free seawater, instead of into normal seawater, the giant oocyte nucleus (germinal vesicle) persists without breakdown, and the meiotic spindle is not formed. Figure 7A shows the upper regions of such a non-activated oocyte that has been stratified in the CPM. Immediately below the oil cap on top, one sees the irregularly shaped, very large, germinal vesicle, within whose bottom lies the heavy nucleolus. In the clear zone below the germinal vesicle and surrounding the nucleolus, a zone of negatively birefringent (membrane) material (dark in these figures) has become aligned by the centrifugation.

Shortly after Ca^{2+} ions are added to the sample chamber,

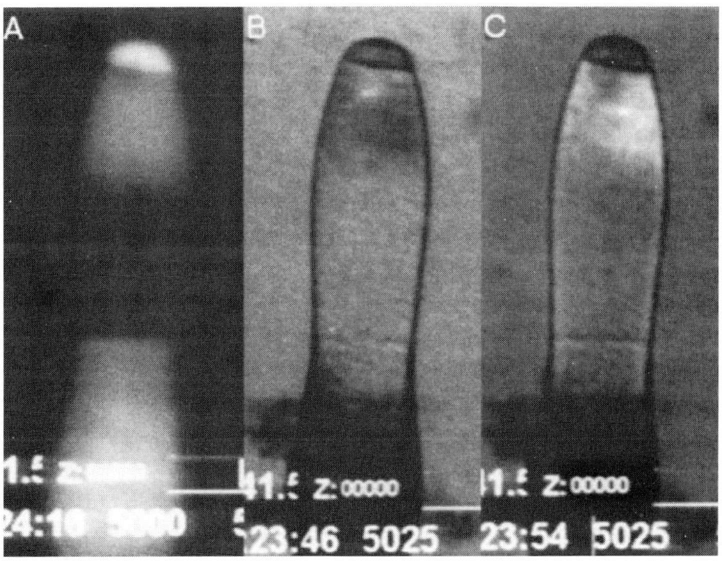

Fig. 3. Brefeldin A fluorescence showing membranes in centrifuged sea urchin eggs taken with a 20×/0.40 objective lens. (A) shows the brefeldin fluorescence. (B) and (C) in polarized light, with the compensator slow axis orientated vertically in (B) and horizontally in (C). The nucleus (which lacks birefringence except for its envelope) shows especially clearly just below the oil cap in (C).

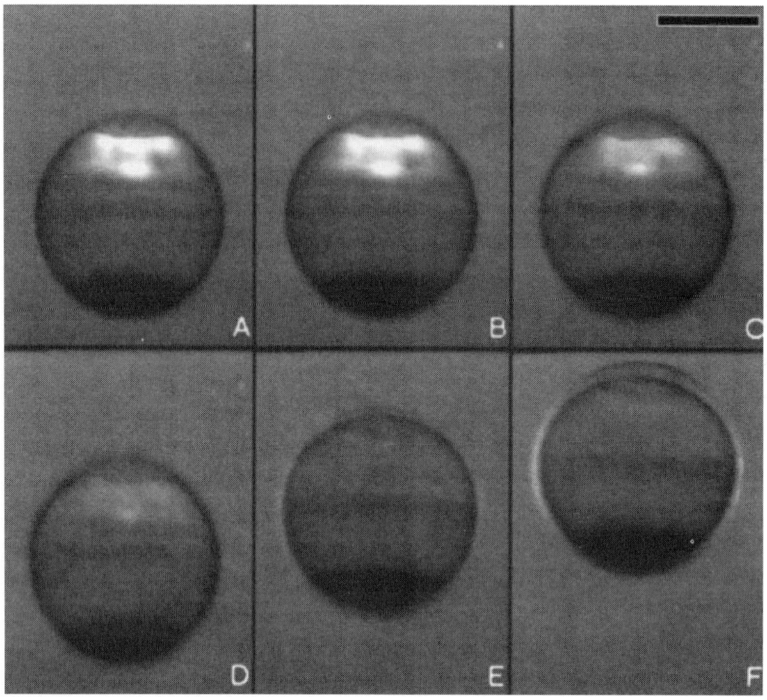

Fig. 4. Loss of negative birefringence (bright region below oil cap), levitation, and elevation of fertilization envelope upon activation of *Arbacia* egg. For the sequence shown here, the CPM was stopped after 50 min of centrifugation at ≈ 5600 rev min^{-1}, and a drop of sperm suspension was added into the top opening of the specimen chamber. (The eggs are still stratified but have rounded up during the few minutes the CPM was stopped.) The sperm started to appear in the field 11.5 min after the rotation was raised to 800 rev min^{-1}. (A) 7 s after the sperm appeared (dark spot at one o'clock on the egg surface). (B) 51 s after (A). (C) 9 s after (B); the negative birefringence in the upper half of the clear zone starts to disappear. (D) 4 s after (C); the negative birefringence is almost gone. (E) 34 s after (D); the egg is floating up as the positively birefringent fertilization envelope starts to appear. (F) 27 s after (D); the egg continues to become lighter as the fertilization envelope rises higher. Characteristically, the egg lies at the bottom of the perivitelline space. Note: the compensator orientation has been reversed between the earlier panels and panels E, F. Bar = 40 μm.

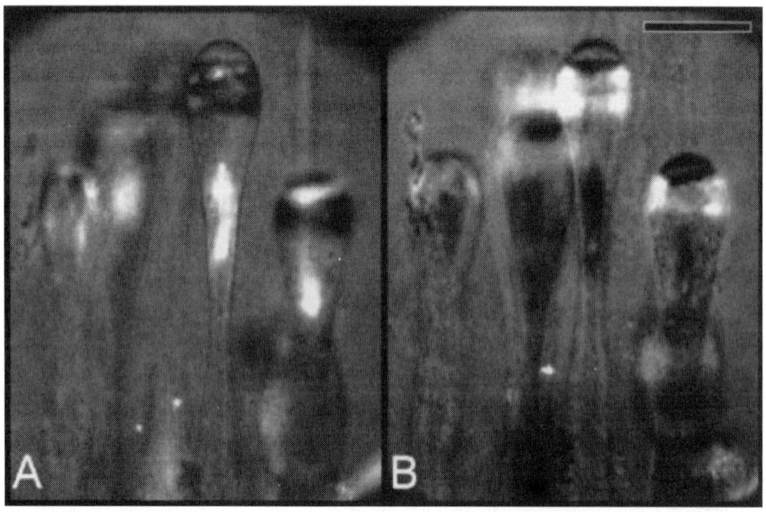

Fig. 5. Birefringence in *Chaetopterus* oocyte centrifuged on a sucrose-density gradient after spindle formation. A complex series of birefringent material with low density has accumulated at the centripetal pole of each cell. Immediately below, in the clear zone (grey), the meiosis I spindle has become stretched, showing pointed poles and a strong positive birefringence (cf. Fig. 7H). For (A), the compensator slow axis runs vertically; for (B), horizontally. Bar = 40 μm.

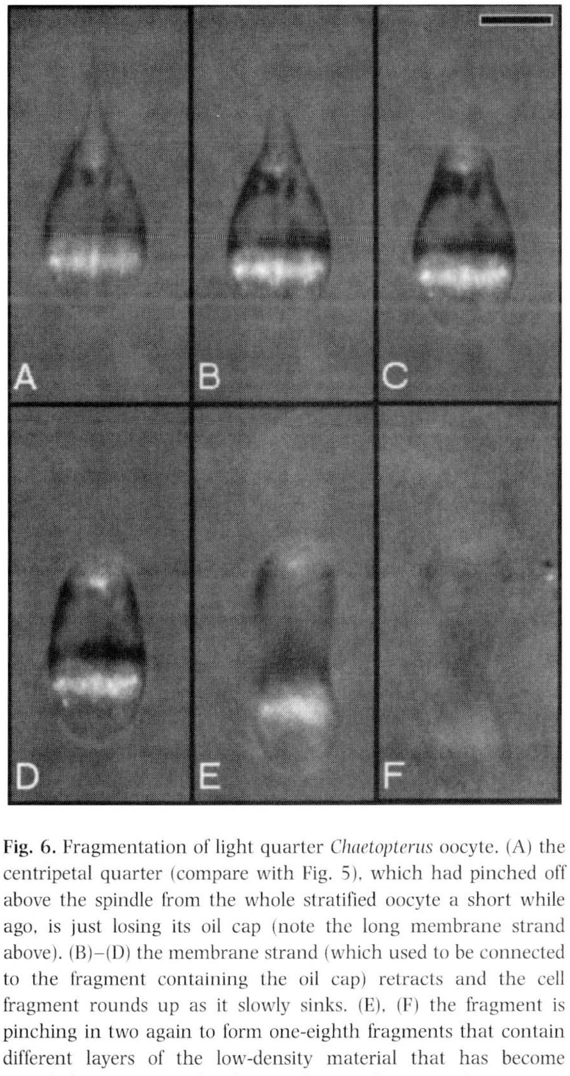

Fig. 6. Fragmentation of light quarter *Chaetopterus* oocyte. (A) the centripetal quarter (compare with Fig. 5), which had pinched off above the spindle from the whole stratified oocyte a short while ago, is just losing its oil cap (note the long membrane strand above). (B)–(D) the membrane strand (which used to be connected to the fragment containing the oil cap) retracts and the cell fragment rounds up as it slowly sinks. (E), (F) the fragment is pinching in two again to form one-eighth fragments that contain different layers of the low-density material that has become stratified into separate birefringent layers. Slow axis of compensator runs north to south. Time elapsed from (A) to (F) is 4 s. Bar = 20 μm.

the germinal vesicle breaks down, the nucleolus drops precipitously, and yolk and other particles (which had been trapped between the germinal vesicle and the oil cap) start 'raining down' (Figs 7B–E). Simultaneously, the negative birefringence of the layer below the nucleus starts to fade, and in its place a small positively birefringent aster starts to appear (panels E, F).

In the next few minutes, the asters grow and separate into two, between which develops the positively birefringent meiotic spindle (Figs 7G and H). Thus, we see the disintegration of the nuclear envelope, disappearance of the negatively birefringent structures, as well as the assembly of microtubules that form the asters and the spindle fibres that push the spindle poles apart (Goda *et al.*, 1998).

Interestingly, the spindle and asters formed within these oocytes, which were activated while being centrifuged, appear not to be under tension as were those that had first been formed before centrifugation. Unlike the latter, they show a normal pattern of birefringence and their poles are rounded rather than pointed (cf. Figs 5 and 7H). It could well be that, once formed, the linkage of the astral rays to the cell cortex is quite firm (despite the fact that the individual astral microtubules are likely to be growing and shortening very actively). By contrast, spindle morphogenesis itself is clearly not affected by the high centrifugal environment or by the resulting stratification or deformation of the cell.

5. Stratification of mouse oocytes (in collaboration with Drs Lin Liu and David Keefe of Marine Biological Laboratory, Woods Hole, and Women and Infants Hospital of Rhode Island)

Figure 8 shows mouse oocytes from which the tough zona pellucida had been removed before centrifugation. These oocytes were stratified in the CPM at 11 400 × g for 30 min on a Percoll gradient. As in the marine eggs and oocytes, one sees, from top to bottom, an oil cap, clear zone, and granular zone. However, unlike in the marine eggs, in the mouse oocytes (which lack yolk) the mitochondria appear at the core of the heavy granular zone (Fig. 7B) rather than as a horizontal layer in the lower part of the clear zone (cf. Fig. 1). Furthermore, a high-density, strong, positively birefringent 'cup' surrounds the mitochondrial core. We saw no sign of the negatively birefringent, labile curtain of material below the oil cap that appeared prominently in the centrifugally stratified *Arbacia* eggs and *Chaetopterus* oocytes.

Lin Liu *et al.* (unpublished results) report an interesting application of the stratification patterns of mitochondria noted above. By centrifuging mouse oocytes on a density gradient of Percoll, they obtained heavy fragments containing mitochondria and lighter ones without. When stimulated by addition of Sr^{2+}, only those fragments containing mitochondria showed rhythmic changes in cytosolic Ca^{2+} concentrations, the other fragments only a general rise in cytosolic Ca^{2+}. This is the first demonstration in mammalian oocytes (or in any cell where the mitochondria-containing cell region had been fragmented) that mitochondria were shown to engage in rhythmic Ca^{2+} pumping activities.

The nucleus in the mouse oocytes tended to appear in the granular zone (data not shown) rather than high up in the clear zone. As interphase nuclei in most types of cell, except in mature spermatids and sperm, tend to be surprisingly light, as well as showing an optical density less than the surrounding cytoplasm (reverse 'shadow cast'

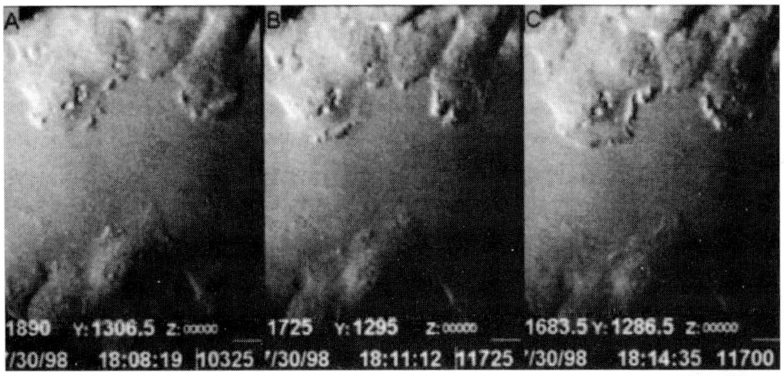

Fig. 9. Fibroblasts in CPM undergoing wound healing. The cells above the wound show exceptionally active membrane ruffling and pinocytosis on their poles facing the wound (their centrifugal faces). The poles of the cells facing the wound from below (i.e. their centripetal faces) are nearly quiescent. DIC, 20×/0.4 NA.

in toadfish erythrocytes that are becoming sickled under mild anoxia (Hárosi *et al.*, 1998). As the haemoglobin crystallizes, the mean density of the cells increases (the cells sink to a higher density layer). We thus observe a net loss of water by the cell, associated with crystallization of the haemoglobin molecules. Does this reaction and the converse, the levitating sea urchin egg upon fertilization (Section 2), both express density changes brought about by transition in the state of water molecules associated with crystallization or by the swelling of molecules or fine structure?

9. Birefringence in fd virus solution (in collaboration with Drs Seth Fraden and Zvonimir Dogic, Brandeis University)

Figure 12 shows spontaneously developing 'comets' found in a concentrated solution (6 mg mL^{-1}) of a thread-shaped DNA virus. The birefringent comets signal the transient alignment of nematic-phase fd viruses as some high-density material appears to fall at a very rapid rate through the viral solution. While we have not been able to ascertain the nature of this high-density material, the comet generally arises from a layer in the upper region of the virus solution (approximately at the level the sedimentation front would be expected to occupy; Dogic *et al.*, 2000).

While striking in the CPM video records, reminding one of repeated flashes of lightning in a heavy thunderstorm, there remain many mysteries about these intriguing 'comets' that suddenly appear and then fade away as they descend through the viral solution.

Why do they repeatedly, and semi-periodically, appear in more or less the same location, and with similar

Table 1. Stalling of upward crawling by myosin mutants of *Dictyostelium discoideum* amoebae measured in the CPM. Mutants lacking myosin II are stalled at a much lower speeds than those that are not missing the gene. Note that in media whose densities have been increased by addition of Percoll, and in which amoebae are pushed upward by buoyant forces (+ signs in sixth column), the amoebae nevertheless stall at given (though higher) spin speeds (after Fukui *et al.*, 2000).

Cell line	Reduced mass of cell ($\times 10^{-11}$ g)	Density of medium	Stalling speed (rev min^{-1})	Apparent weight at stall ($\times 10^3$ pN)	Buoyant force ($\times 10^3$ pN)*
NC4 (Bacteria-grown wild type)	2.31 ± 0.70	1.005	> 11 700	> 2.59 ± 0.78	
Ax3 (Parental axenic strain of A5 and HS-1)	2.92 ± 0.70	1.005	6400	0.99 ± 0.24	
A5 (missing myosin II, myosin IA, & myosin IB)	2.82 ± 0.98	1.005	3400	0.27 ± 0.09	
HI-S (missing myosin II)	2.47 ± 0.78	1.005	3500	0.25 ± 0.08	− 0.25
		1.012	4100		− 0.31
		1.031	4900		− 0.27
		1.062	6800		− 0.02
		1.093	7800		+ 0.65
		1.124	8600		+ 1.61

*Media densities greater than that of amoebae.

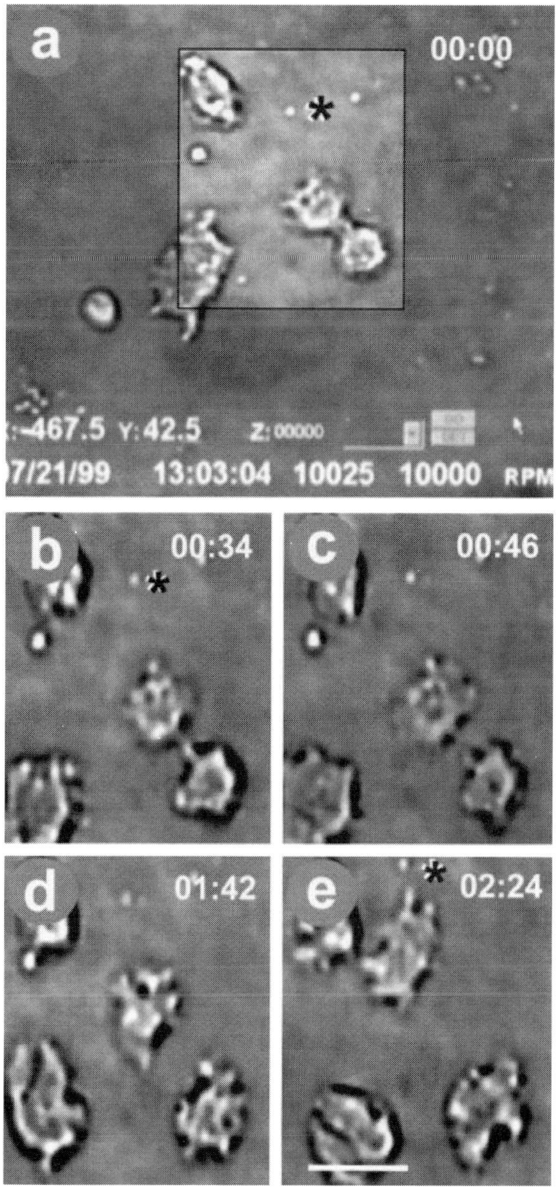

Fig. 10. Amoebae of *Dictyostelium* dividing in the CPM. These DIC images show time sequences of 'wild-type' *Dictyostelium* amoebae crawling on the surface of the glass cover plate of the specimen chamber in the CPM. The centrifugal field lies in the plane of the images shown, and at 10 000 rev min^{-1} the cells are exposed to a strong downward pull (of ≈ 2000 pN). Nevertheless, these cells can crawl about and even undergo cell division as shown here. By contrast, mutant amoebae lacking myosin II are stalled, or unable to crawl upward against the *G* field at characteristically lower speeds (Table 1). 40×/0.55 NA objective, 532 nm, 6 ns illumination. *: fixed reference mark on cover plate. Time in minutes and seconds elapsed after panel (a): on upper right of each panel. Scale bar 5 μm (from Fukui *et al.*, 2000.)

birefringence patterns? How can they achieve such a high initial speed of descent of 5 mm s^{-1}, up to over 1 cm s^{-1}, in an aqueous medium? What material could suddenly gain enough density and mass to fall through and orientate a viral solution at such a high speed, even though under high centrifugal acceleration (11 500 × g)? Why are some flashes clearly comet-shaped (Fig. 12) while others show bright flashes of more or less sickle-shaped transient birefringence? Why, during early centrifugation, do comets arise at the air–water interface, just as descending miniature droplets of water merge with the solution at the interface at the top of the viral solution? Why, later on, do the comets progressively arise from deeper in the solution and are in no way correlated with the entry of droplets into the air–water interface? Why is the birefringence of some of the comets so low (≈1 nm), while others (especially the 'sickles') are so high, far exceeding the (20 nm) retardance of the compensator?

While many questions remain regarding these enigmatic comets, there is no question that the positively birefringent 'comet tails' show transient alignment of nematic viral threads, orientated along the 'lines of flow' in the comet tail (i.e. slow axis of birefringence lies along the length of the tail or sickle). The bright balls at the front of the comet (Fig. 12) represent a region of very high birefringence with reversed sign relative to the tail, most likely reflecting an alignment of the nematic viral threads by compression at right angles to the 'tail' at the rapidly travelling front.

Conclusion

In Part I of this study (Inoué *et al.*, 2001) we described the rationale, design, and performance of a new type of centrifuge capable of imaging, in polarization microscopy at video rate, weakly birefringent objects exposed to high centrifugal fields. (U.S. Patent #5982 535 'Centrifuge Microscope Capable of Realizing Polarized Light Observation' was awarded 9 November 1999. Also a related Patent #5930 033 'Slit Scan Centrifuge Microscope' was awarded on 27 July 1999.) The objects, exposed to up to 11 500 times gravitational acceleration, can also be examined in fluorescence (532 nm excitation), DIC, and brightfield microscopy. In addition to discussions on special considerations needed to use very brief, synchronized laser pulses as light sources for the CPM, we detailed the development of a unique specimen chamber that exhibits low strain birefringence and otherwise functions well under high centrifugal fields.

We have described several examples of observations on biological and related samples. These included: isolated egg and blood cells stratified and activated in isopycnic physiological media; polarized behaviour of cultured cells growing on glass coverslips; different ability of amoeboid cells to crawl up against high *G* fields; and spontaneous dynamic orientation, possibly due to sedimentation of

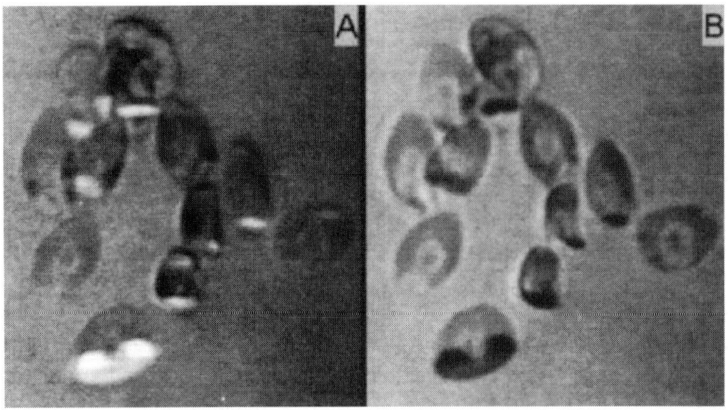

Fig. 11. Haemoglobin crystals in sickled toadfish erythrocytes. The anoxic erythrocytes that have formed the birefringent (and dichroic) haemoglobin crystals sink lower to a somewhat higher density region in the density gradient. The cells without the haemoglobin crystals remain higher up outside of the field of view seen here. Compensator slow axis is orientated horizontally in (A) and vertically in (B). (Inoué *et al.* 1998, unpublished observations.)

concentrated nematic viral particles, in a solution of thread-shaped virus particles.

These explorations of the application of the CPM demonstrate the imaging capability and utility of the new microscope. The CPM images uncover fine-structural events that take place dynamically at domains far below the resolution limit of the light microscope in centrifugally stratified living cells and related quasi-fluid systems. Several surprising observations have been made and lead to intriguing questions regarding the organization and dynamic changes of fine structures, as well as their physiological and functional roles in living cells.

Acknowledgements

We are grateful to the Olympus Optical Co., Ltd, and Hamamatsu Photonics KK for generous support of this project, and especially to Directors Hideo Hiruma and Hitoshi Iida of Hamamatsu for initially approving and continuously encouraging the three-way collaborative development. We also wish to acknowledge the efforts of Mr M. Yamagishi at Olympus for his continued support and encouragement of the project. Chroma Technology provided the efficient band pass filters that made for ready conversion of the CPM to a transmitted light fluorescence system. We thank the following

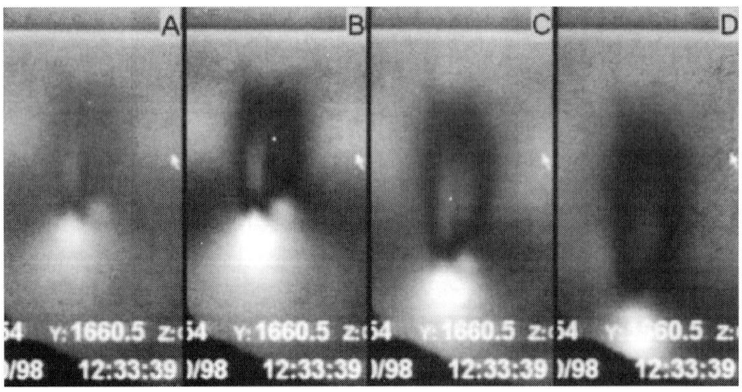

Fig. 12. Birefringent comets observed with the CPM in concentrated fd virus solution. With continuing centrifugation, the comets spontaneously arose increasingly further down from the surface of the virus solution (horizontal line below letters) where the concentration of virus particles is expected to rise sharply at the sedimentation front. A dense material appears to align the elongated virus particles into a comet-shaped birefringent region as it rapidly sinks toward the bottom of the centrifuge chamber. The slow axis of birefringence in the comet 'tail' is orientated vertically, while the slow axis of the $\lambda/30$ compensator is orientated horizontally in these panels, thus making the bulk of the positively birefringent tail appear darker than the background. The thicker core region of the tail, in which the nematic-phase virus particles are also orientated parallel to the length of the tail, appears somewhat brighter since its retardance exceeds that of the compensator. In what appears to be bright 'heads' of the comets, the slow axis runs horizontally, showing that the nematic-phase virus particles in this localized zone are orientated at right angles to those in the tail. Perhaps the nematic virus particles in this zone are responding to a shock wave in front of the rapidly descending front? Observed with 4× objective and identical condenser. Frame intervals: ≈ 40 ms. The frame height is ≈ 800 μm (Fraden *et al.* 1998, unpublished observations.)

collaborators for exploring the capabilities of the CPM with us, and for allowing us to present data from these early studies: Dr Mario H. Burgos of IHEM, CONISET, Mendoza, Argentina (Section 2), Drs Lin Liu and David Keefe of MBL and the Women and Infants Hospital of Rhode Island (Section 5), Dr Greg Gundersen of Columbia University (Section 6), Drs Yoshio Fukui of Northwestern University Medical School (Section 7), Dr Ferenc I. Hárosi of MBL and New College of the University of South Florida (Section 8), and Drs Seth Fraden and Zvonimir Dogic of Brandeis University (Section 9). We are particularly grateful to the efficient, hard-working staff of our MBL Program, especially Jane MacNeil and Diane Baraby, whose superb help has been essential for the progress of this project.

References

Burgos, M.H., Goda, M. & Inoué, S. (2000) Fertilization-induced changes in the fine structure of stratified *Arbacia* eggs. II. Observation with electron microscopy. *Biol. Bull.* **199**, 213–214.

Dogic, Z., Philipse, A.P., Fraden, S. & Dhont, J.K.G. (2001) Concentration dependent sedimentation of colloidal rods. *J. Chem. Phys.* **113**, 8368–8380.

Fukui, Y., Uyeda, T.Q.P., Kitayama, C. & Inoué, S. (2000) How well can an amoeba climb? *Proc. Natl. Acad. Sci. USA*, **97**, 10020–10025.

Goda, M., Burgos, M.H. & Inoué, S. (2000) Fertilization-induced changes in the fine structure of stratified *Arbacia* eggs. I. Observation on live cells with the centrifuge polarizing microscope. *Biol. Bull.* **199**, 212–213.

Goda, M., Inoué, S. & Knudson, R.A. (1998) Oocyte maturation in *Chaetopterus pergamentaceous* observed with centrifuge polarizing microscope. *Biol. Bull.* **195**, 212–214.

Hárosi, F.I., Hunt Von Herbing, I. & Van Keuren, J.R. (1998) Sickling of anoxic red blood cells in fish. *Biol. Bull.* **195**, 5–11.

Harvey, E.B. (1936) Parthenogentic mergony or cleavage without nucleus in *Arbacia punctulata*. *Biol. Bull.* **71**, 101–121.

Harvey, E.B. (1956) *The American Arbacia and Other Sea Urchins.* Princeton University Press, Princeton, NJ.

Inoué, S. (1986) *Video Microscopy.* Plenum Press, New York.

Inoué, S. & Dan, K. (1951) Birefringence of the dividing cell. *J. Morph.* **89**, 423–456.

Inoué, S., Goda, M., Gundersen, G. & Knudson, R.A. (1998) Oocyte maturation and tissue cell adhesion observed with centrifuge polarizing microscope. *Mol. Biol. Cell*, **9** (Suppl.), 105a.

Inoué, S., Knudson, R.A., Goda, M., Suzuki, K., Nagano, C., Okada, N., Takahashi, H., Ichie, K., Iida, M. & Yamanaka, K. (2001) Centrifuge polarizing microscope. I. Rationale, design and intrument performance. *J. Microsc.* **201**, 341–356.

Reprinted from *The Biological Bulletin*, Vol. 201, p. 234, 2001, with permission from Blackwell Publishing.

Reference: *Biol. Bull.* **201:** 234. (October 2001)

Centrifuge Polarizing Microscope with Dual Specimen Chambers and Injection Ports

Robert A. Knudson (Marine Biological Laboratory, Woods Hole, Massachusetts), Shinya Inoué, and Makoto Goda[1]

We reported earlier on a centrifuge polarizing microscope (CPM) that was designed for observing the weak birefringence of organelles and fine structures in living cells as they became stratified and oriented under centrifugal fields of up to 10,500 times Earth's gravitational field. In this earlier model (1), one chamber A contained the specimen under observation, while the contents of the opposed second chamber B, which acted solely to balance the rotor, could not be viewed. We have now improved the CPM so that either chamber can be viewed and selected at the flick of a lever, within the duration of a video frame. In the CPM, an electronic timing circuit synchronizes the firing of the light source laser precisely to the transit of the specimen under the microscope (freezing the image to less than 0.5-μm specimen motion at up to 11,700 rpm, regardless of the speed of the 16-cm diameter rotor). The timing circuit, in turn, is triggered by the signal from a photodiode that picks up the light originating from a stationary diode laser, and reflected by a small mirror (M1) mounted on the spinning rotor near its axis. The complexity of the electronic timing circuit led us to keep the electronic circuit undisturbed and instead to devise an optical system for switching between the display of the two chambers. To this end, we installed a second timing mirror (M2) on the rotor, exactly opposite the one for chamber A, but tilted up by a few degrees, rather than oriented parallel to the rotor axis as is M1. In front of the photodiode we also placed a mounted pair of small mirrors on a "beam switcher" that could either be flipped up out of the way so that the photodiode would capture the light reflected from M1, or flipped down into position so that light reflected from the tilted mirror M2 would be reflected by the mounted pair of beam-switcher mirrors and enter the photodiode. Thus, depending on the position of the beam switcher, the timing light would enter the photodiode, reflected either from mirror M1 or M2. The timing circuit would then trigger the light source laser at precisely (to within a few nanoseconds) the time point required to display a stable image of the specimen in chambers A or B. The response time of the electronic timing circuit and laser firing device turned out to be so short that no video frames were lost in flipping the beam switcher and capturing the images from either of the two chambers.

Figure 1, left panel, shows the recorded image of sea urchin eggs stratified in a density gradient in chamber A, while the right panel shows density-standard beads (Nycomed Amersham, Oslo, Norway) that reveal the gradient of the identically prepared seawater/Percoll mix in chamber B. As the figure shows, the density of the unfertilized *Arbacia* eggs is approximately 1.060. Immediately after fertilization, the negative birefringence disappears from the membranes stacked in the upper half of the clear zone of the stratified eggs (Fig. 1 and Ref. 2). Concurrently, the (de-jellied) egg becomes lighter over the next minute, presumably by influx of water, and starts to float upward in the gradient until its density is somewhat less than 1.040.

In addition to being able to instantly switch between images from chambers A and B, we have devised a method for introducing reagents, sperm, etc., into either chamber while the specimen and control are rotating in the CPM. A plastic unit, notched out to pass the timing light for mirrors M1 and M2, was placed over the support post for these mirrors. The unit incorporated an "injection port" drilled along the rotational axis of the rotor, which in turn could be connected through a selection valve, without stopping the rotor, to either of the two thin pieces of plastic tubing leading to chambers A or B. Thus, we can now observe fine structural and density changes—*e.g.*, in marine eggs upon activation—without having to stop the rotor for a period to remove the specimen chamber and introduce the activating agents. Video records resulting from both of these improvements were presented.

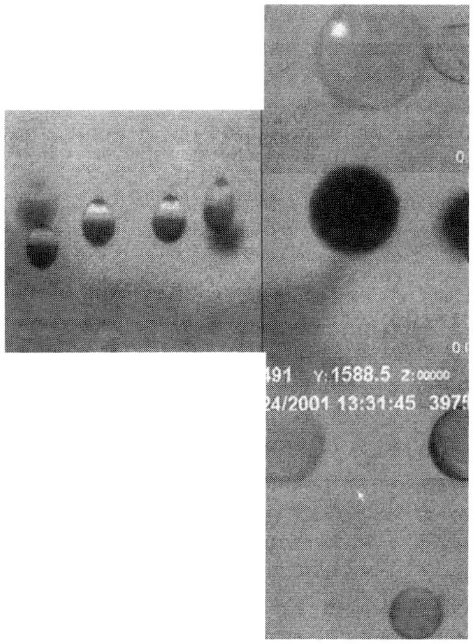

Figure 1. *Left: Unfertilized* Arbacia *eggs stratified in chamber A in a seawater/Percoll density gradient. Right: Four sets of beads of standard densities stratified concurrently in the gradient; observed in chamber B. Nominal densities of the beads are, top to bottom: 1.04, 1.06, 1.09, and 1.10.*

We thank Hamamatsu Photonics K.K. and Olympus Optical Co. for generous support of this project.

Literature Cited

1. **Inoué, S., R. Knudson, M. Goda, K. Suzuki, C. Nagano, N. Okada, H. Takahashi, K. Ichie, M. Iida, and K. Yamanaka. 2001.** *J. Microscopy* **201:** 341–356.
2. **Inoué, S., M. Goda, and R. A. Knudson. 2001.** *J. Microscopy* **201:** 357–367.

[1] AIST, Tokyo, Japan.

The following note was added by Shinya Inoué in September of 2006:

See Article 75, Fig. 1, for photograph of the dual-timing mirror arrangement.

Reprinted from *Optical Engineering*, Vol. 41(5), pp. 943–954, with permission from SPIE.

Polarization aberrations caused by differential transmission and phase shift in high-numerical-aperture lenses: theory, measurement, and rectification

Michael Shribak, MEMBER SPIE
Shinya Inoué
Rudolf Oldenbourg, MEMBER SPIE
Marine Biological Laboratory
Woods Hole, Massachusetts 02543
E-mail: mshribak@mbl.edu

Abstract. We present a theoretical and experimental study of radially symmetric aberration caused by the differential transmission and phase shift of *p*- and *s*-polarized components of an axial beam passing through spherical lenses and plane parallel plates. We give a general description of the aberrations for an axial beam. The extinction is calculated as a function of the numerical aperture for uncoated lenses and for plane parallel plates. In our theoretical analysis, the polarization of output rays is described as a function of the input ray parameters, the shape factor, and refractive index of the lenses used. For rays that are inclined to the optical axis, optimal lens shape factors that minimize the rays' polarization aberrations are found. Techniques for measurement of radially symmetric birefringence in a lens system are described. Finally, we discuss strategies for polarization rectification and introduce new designs including meniscus rectifiers and a liquid crystal rectifier that can actively compensate a wide variety of polarization aberrations. Good correlations between theory and experimental results for microscope optical systems with coated and uncoated optical elements are found. Our results enable us to suppress depolarization and remove anomalous diffraction in a modern microscope equipped with high-numerical-aperture lenses.
 [DOI: 10.1117/1.1467669]

Subject terms: polarization; microscopy; aberration; rectifier; compensator; birefringence; diattenuation.

Paper PL-010 received Sep. 7, 2001; revised manuscript received Dec. 13, 2001; accepted for publication Dec. 31, 2001.

1 Introduction

The extinction factor of a wide-field polarizing microscope rapidly drops as the numerical aperture of the objective and condenser lens is raised, even with a high-quality polarizing system and the use of carefully selected lenses that are free from strain birefringence and birefringent inclusions.[1–3] The loss of extinction originates from the differential transmission (diattenuation) and phase shift (retardance) between the *p*- and *s*-polarization components of rays that pass through steep optical interfaces. Differential transmission and phase shift lead to spatial polarization changes in the exit pupil plane called "polarization aberrations."[4] The polarization aberrations cause undesirable polarization components (depolarization[5]) that reduce the extinction in the image plane. The polarization aberrations result in four bright quadrants separated by a dark cross, known as the Maltese cross, that is seen conoscopically in the exit pupil of a high-numerical-aperture (NA) strain-free lens system between crossed linear polarizers in the absence of a specimen. If crossed circular polarizers are used instead, a dark central disk is surrounded by a bright ring in the exit pupil. Polarization aberrations can also give rise to anomalous diffraction, caused by a four-leaf clover pattern that replaces the Airy disk when imaging weakly birefringent objects between crossed linear polarizers.[6,7] In general, polarization aberrations modify the point spread function and the optical transfer function of optical systems.[8] Various possible theoretical pictures of polarization state distributions in the exit pupil of lens systems based on a paraxial approximation of the effect of polarization aberrations can be found in published papers.[4,9] Other polarization optical devices, where lenses are used between polarizers, such as ellipsometers, polarimetric sensors, and optical disk systems with polarization reading of information,[10–12] must also take into account the depolarization caused by the factors considered here. In a single-point scanning confocal imaging system, however, polarization aberrations that occur in different quadrants of the exit pupil cancel each other and the extinction remains high even when high-NA lenses are used.[13]

2 General Description of Polarization Aberrations of an Axial Beam Passing Through Lens Surfaces

An axial beam can be considered as being composed of individual rays that travel through lens surfaces each under a different plane and angle of incidence. Because of the rotational symmetry of a lens the differential amplitude transmission and phase shift between the *p*- and *s*-polarization components of rays are radially symmetric.

Let us consider an axial beam in a lens optical system. We choose polar coordinates at the center of an entrance pupil that has unit radius. The polarization change of a ray with the radial coordinate ρ $(0\leqslant\rho\leqslant 1)$ and azimuth θ $(0\leqslant\theta\leqslant 2\pi)$ is determined by the Jones matrix $\mathbf{M}(\rho,\theta)$:

$$\mathbf{M}(\rho,\theta)=\begin{pmatrix}\cos\theta & -\sin\theta\\ \sin\theta & \cos\theta\end{pmatrix}\cdot\begin{pmatrix}T_p(\rho)\exp[i\Delta_p(\rho)] & 0\\ 0 & T_s(\rho)\exp[i\Delta_s(\rho)]\end{pmatrix}\cdot\begin{pmatrix}\cos\theta & \sin\theta\\ -\sin\theta & \cos\theta\end{pmatrix}$$
$$=\frac{1}{2}T_s e^{i\Delta_s}\begin{pmatrix}\mu e^{i\Delta}+1+(\mu e^{i\Delta}-1)\cos 2\theta & (\mu e^{i\Delta}-1)\sin 2\theta\\ (\mu e^{i\Delta}-1)\sin 2\theta & \mu e^{i\Delta}+1-(\mu e^{i\Delta}-1)\cos 2\theta\end{pmatrix}. \tag{1}$$

Here T_p and T_s are amplitude transmission coefficients of the p- and s-polarization components, and Δ_s and Δ_p are phases of the components after traversing the optical system. Parameters μ and Δ describe the differential transmission and radial retardance, respectively, where $\mu=T_p/T_s$ and $\Delta=\Delta_p-\Delta_s$. A small angle ξ is also used for the description of the differential transmission $\xi=\tan^{-1}(\mu-1/\mu+1)\approx(1/2)(\mu-1)$. Typically, in the center of the pupil when $\rho=0$, there is no differential transmission or phase shift, and the parameter μ equals 1 and the turn angle ξ is 0 deg. Toward the periphery of the pupil when ρ increases, the values of parameters μ and ξ also increase. Differential transmission is usually stronger in a microscope with dry, high-NA condenser and objective lenses when compared to oil-immersion systems. Radial retardance originates from the oblique passage of a ray through optical multilayer coatings and from radially symmetric stress in lenses.

If the optical system, including the slide and cover slip that lie between the condenser and objective lenses, has N surfaces, then

$$T_p=\prod_{k=1}^{N} t_{pk},\quad T_s=\prod_{k=1}^{N} t_{sk},\quad \mu=\prod_{k=1}^{N}\hat{\mu}_k,$$
$$\xi\approx\sum_{k=1}^{N}\hat{\xi}_k,\quad \Delta=\sum_{k=1}^{N}\delta_k, \tag{2}$$

where t_{pk} and t_{sk} are the amplitude transmission coefficients, and δ_k is the phase shift for the optical surface with number k. Hence, $\hat{\mu}_k=t_{pk}/t_{sk}$ and $\hat{\xi}_k=(1/2)(\hat{\mu}_k-1)$.

Eigenvectors of the matrix of Eq. (1) are linearly polarized. The polarization direction of the p component is along the azimuth angle θ of the rays. Coefficients of the differential transmission and radial retardance only depend on the radius ρ. The differential transmission and phase shift lead to a polarization component in the beam that is orthogonal to its initial polarization.

The orthogonal polarization component reduces the extinction of an optical system. The intensity ratio of the initial polarization I'_x over the orthogonal polarization I'_y of the output beam is called the system extinction factor κ of the complete optical system, such as a microscope. This extinction is measured in the image plane. The extinction factor of the complete system can be derived from the extinction of each individual ray $\eta(\rho,\theta)=[E'_x(E'_x)^*]/[E'_y(E'_y)^*]$, which can be measured in the exit pupil plane:

$$\kappa=\frac{I'_x}{I'_y}=\frac{\int_0^{2\pi}\int_0^1 I(\rho,\theta)(1-\eta^{-1})\rho d\rho d\theta}{\int_0^{2\pi}\int_0^1 I(\rho,\theta)\eta^{-1}\rho d\rho d\theta}\approx\frac{\int_0^{2\pi}\int_0^1 I(\rho,\theta)\rho d\rho d\theta}{\int_0^{2\pi}\int_0^1 I(\rho,\theta)(\rho/\eta)d\rho d\theta}, \tag{3}$$

where $I(\rho,\theta)$ is the intensity distribution of the initial beam in the entrance pupil, and E'_x and E'_y are amplitudes of orthogonal polarization components of the ray after the optical system. Note that there are also other sources of depolarization in optical systems such as the scattering of light by dust, contaminations, and mountings, and the birefringence of glass. These should be removed carefully.

In the strict sense the preceding analysis treats only "on-axis" beams. However, we are here concerned with high-NA microscope systems with large magnification that image only a small field of view. All rays that pass through this small field of view can be considered "on-axis" beams. For example, the objective CFN DIC Plan Achromat 40× with NA=0.7 (Nikon, part 85031) has a focal length of 4 mm and a field of view radius of 0.06 mm, approximately. In this case, the difference angle between a ray from a peripheral point in the object and a ray from its center, which cross in the same pupil point, is less then 0.9 deg in the pupil center and less than 0.6 deg for the pupil periphery. Hence, peripheral object points will show almost the same extinction as the central point and our results are expected to apply for the entire field of view of a high-NA microscope.

Let us consider several particular cases. When the initial beam is linearly polarized $\mathbf{E}=\binom{1}{0}$ then the field amplitude of the output rays can be written as

$$\mathbf{E}'=\frac{1}{2}T_s e^{i\Delta_s}\begin{bmatrix}\mu e^{i\Delta}+1+(\mu e^{i\Delta}-1)\cos 2\theta\\ (\mu e^{i\Delta}-1)\sin 2\theta\end{bmatrix}. \tag{4}$$

This equation is derived by multiplication of the matrix of Eq. (1) and the vector $\mathbf{E}$. The extinction η of a ray is determined by

$$\eta=\frac{(\mu+1)^2+(\mu-1)^2\cos^2 2\theta-4\mu\sin^2(\Delta/2)\sin^2 2\theta}{[(\mu-1)^2+4\mu\sin^2(\Delta/2)]\sin^2 2\theta}\approx\frac{1}{[\tan^2\xi+\sin^2(\Delta/2)]\sin^2 2\theta}. \quad (5)$$

In the last term, the parameter ξ is used instead of parameter μ. That enables us to simplify the formula and its analysis.

As follows from Eq. (5), when the optical system is put between two crossed linear polarizers, the image of the exit pupil has a cross shape. The cross branches are parallel to the polarizers.

If there is differential transmittance only but no retardance, then

$$\mathbf{E}'=\frac{T_p+T_s}{2}\begin{pmatrix}1+\tan\xi\cos 2\theta\\ \tan\xi\sin 2\theta\end{pmatrix}. \quad (6)$$

Thus all output rays will be linearly polarized with a small rotation of the polarization plane α compared to the initial polarization plane:

$$\alpha(\rho,\theta)=\tan^{-1}\left(\frac{E'_x}{E'_y}\right)=\tan^{-1}\left[\frac{\tan\xi(\rho)\sin 2\theta}{1+\tan\xi(\rho)\cos 2\theta}\right]\approx\xi(\rho)\sin 2\theta. \quad (7)$$

Here we took into account that components E'_x and E'_y are real.

The maximal rotation of the polarization plane is near the diagonal directions ($\theta=\pm 45$ deg) and equals about ξ. The direction of rotation is positive in the first and third quadrants and negative in the second and fourth quadrants. The exact value of the maximal rotation, $\alpha_{\max}$, equals $\tan^{-1}[\tan\xi/(1-\tan^2\xi)^{1/2}]$ and is observed at $\theta_{\max}=\pm[45\text{ deg}+(1/2)\alpha_{\max}]$. Hence, the distribution of the polarization plane rotation and the intensity distribution in the Maltese cross are not quite symmetrical to the diagonal axes.

In the case where there is only radial retardance, the field vector of an output ray is

$$\mathbf{E}'=e^{i\Delta_s}\begin{pmatrix}\cos\frac{\Delta}{2}+i\sin\frac{\Delta}{2}\cos 2\theta\\ i\sin\frac{\Delta}{2}\sin 2\theta\end{pmatrix}. \quad (8)$$

The output rays are elliptically polarized with azimuth γ and axes ratio σ:

$$\gamma=\frac{1}{2}\tan^{-1}\left[\frac{\tan^2(\Delta/2)\sin 4\theta}{1+\tan^2(\Delta/2)\cos 4\theta}\right]\approx 0$$

and

$$\sigma=\frac{\sin\Delta\sin 2\theta}{1+(1-\sin^2\Delta\sin^2 2\theta)^{1/2}}\approx(1/2)\sin\Delta\sin 2\theta. \quad (9)$$

Hence, the long axis of the vibration ellipse is almost parallel to the initial polarization plane. The elipticity is maximal in the diagonal directions and is zero for $\theta=0$ and 90 deg. The sign of the ellipses are opposite in neighboring quadrants.

Usually the entrance pupil of a microscope is illuminated with a uniform beam $I(\rho,\theta)=\text{const}$, and $\eta^{-1}(\rho)\ll 1$. So from Eq. (4), we obtain

$$\kappa=\frac{\pi}{\int_0^{2\pi}\int_0^1\rho\{\tan^2\xi(\rho)+\sin^2[\Delta(\rho)/2]\}\sin^2 2\theta d\rho d\theta}=\frac{1}{\int_0^1\rho\{\tan^2\xi(\rho)+\sin^2[\Delta(\rho)/2]\}d\rho}. \quad (10)$$

When the initial beam has left circular polarization $\mathbf{E}=(1/\sqrt{2})\binom{1}{i}$, then the output polarization $\mathbf{E}'$ is most conveniently described as a superposition of left and right circular polarization states $\mathbf{E}'_l$ and $\mathbf{E}'_r$:

$$\mathbf{E}'=\mathbf{E}'_l+\mathbf{E}'_r=\frac{1}{2\sqrt{2}}T_s e^{i\Delta_s}(\mu e^{i\Delta}+1)\begin{pmatrix}1\\ i\end{pmatrix}+\frac{1}{2\sqrt{2}}T_s e^{i\Delta_s}(\mu e^{i\Delta}-1)e^{i2\theta}\begin{pmatrix}1\\ -i\end{pmatrix}. \quad (11)$$

The extinction η of a ray is determined by

$$\eta=\frac{E'_l(E'_l)^*}{E'_r(E'_r)^*}=\frac{(\mu+1)^2-4\mu\sin^2(\Delta/2)}{(\mu-1)^2+4\mu\sin^2(\Delta/2)}\approx\frac{1}{\tan^2\xi+\sin^2(\Delta/2)}. \quad (12)$$

Hence, in the case of crossed circular polarizers, the distribution of light in the exit pupil corresponds to a dark central area surrounded by a bright ring. Extinction factor κ of the complete optical system is determined by the ratio

$$\kappa=\frac{\pi}{\int_0^{2\pi}\int_0^1\rho\{\tan^2\xi(\rho)+\sin^2[\Delta(\rho)/2]\}d\rho d\theta}=\frac{1}{2\int_0^1\rho\{\tan^2\xi(\rho)+\sin^2[\Delta(\rho)/2]\}d\rho}. \quad (13)$$

When comparing Eqs. (10) and (13) we can see that the extinction factor for systems with crossed linear polarizers is twice as great than for crossed circular polarizers under otherwise the same conditions. Moreover, the extinction of a system with crossed circular polarizers can further deteriorate by the use of imperfect quarter-wave plates for the circular polarizers.

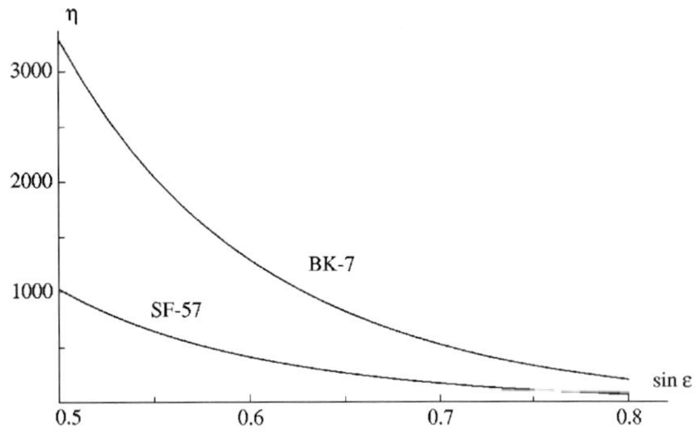

Fig. 1 Ray extinction η versus the sine of the incidence angle ϵ for single dry plane parallel plates made from Schott glasses BK-7 ($n=1.51$) and SF-57 ($n=1.84$).

3 Radially Symmetric Polarization Aberrations of Lenses and Plane Parallel Plates

3.1 *Plane Parallel Plates*

Glass plane parallel plates are often used between the condenser and objective lens of a microscope, for example, as the "slide" glass, cover slip, etc. Between the two, the specimen may be homogeneously immersed or mounted in media of various refractive indices. Usually the glass plates do not have any surface coatings. Let us consider the polarization aberrations contributed by a plate like this. Incidence and refractive angles are noted as ε and ε', respectively. The refractive index of the plate is n. The focused beam has an NA. We note that the multiple beam interference inside the plate can ordinarily be neglected. We assume that there is no phase shift between the p- and s-polarization components (Sec. 1.2 in Ref. 14). As follows from the Fresnel formulas,[14] the amplitude transmission ratio after passing the two surfaces of a dry plane parallel plate is

$$\mu=1+\tan^2(\varepsilon-\varepsilon'). \tag{14}$$

Taking into account that $\sin\varepsilon_{\max}=\text{NA}$ and $\rho=1$ for the marginal ray, we obtain

$$\sin\varepsilon=\rho\text{NA} \quad\text{and}\quad \sin\varepsilon'=\frac{\rho}{n}\text{NA}. \tag{15}$$

So the angle ξ can be found by

$$\tan\xi=\frac{n^2-[\sin^2\varepsilon+\cos\varepsilon(n^2-\sin^2\varepsilon)^{1/2}]^2}{n^2+[\sin^2\varepsilon+\cos\varepsilon(n^2-\sin^2\varepsilon)^{1/2}]^2}. \tag{16}$$

Figure 1 shows the dependence of the ray extinction η on the sine of the incidence angle ε for plates made from Schott glasses BK-7 ($n=1.51$) and SF-57 ($n=1.84$). The extinction applies for a circularly polarized ray or a linearly polarized ray with an azimuth of 45 deg. As we can see, the plate with higher refractive index glass introduces more polarization aberration. Theoretical studies of the rotation of polarized light by plane parallel plates, based on similar principles and with experimental verification, were published by Wright[15] in 1923.

3.2 *Uncoated Lenses*

Let us consider a thin positive lens without coating.[12] Curvature radii of the first and second surfaces are R_1 and R_2. The focal length of the lens is f. Thus, the lens shape factor K is

$$K=(n-1)\frac{f}{R_1}. \tag{17}$$

Incidence and refractive angles on the first and the second surfaces are noted as ε_1, ε_1', ε_2, and ε_2', respectively.

The amplitude transmission ratio after passing the two lens surfaces is

$$\mu=\frac{1}{\cos(\varepsilon_1-\varepsilon_1')\cos(\varepsilon_2-\varepsilon_2')}. \tag{18}$$

As can be shown, a lens with equal differences $\varepsilon_1-\varepsilon_1'$ and $\varepsilon_2-\varepsilon_2'$ induces a minimal angle ξ and correspondingly minimal polarization aberrations. A similar method for minimizing depolarization in lenses was proposed by Wright[15] in 1923.

In the case where a collimated axial beam falls on the lens

$$(\varepsilon_1-\varepsilon_1')-(\varepsilon_2'-\varepsilon_2)=\Omega, \tag{19}$$

where Ω is the angle by which a ray is tilted to the lens axis after passing through the lens. Thus, for an optimal lens the incidence and refractive angles are $(\varepsilon_1-\varepsilon_1')=(\varepsilon_2'-\varepsilon_2)=\Omega/2$ and $\eta=[1+\cos^2(\Omega/2)]^2/\sin^4(\Omega/2)$. For example, if $\sin\Omega=0.3$, then $\eta\approx 7700$.

A plane convex lens with the plane as the first surface has the shape factor $K=0$ and differences $\varepsilon_1-\varepsilon_1'=0$ and $\varepsilon_2'-\varepsilon_2=\Omega$. Therefore its extinction is $\eta=1/[\tan^4(\Omega/2)]$. If $\sin\Omega=0.3$, then $\eta\approx 1900$, that is, four times less than for the optimal lens.

Figure 2 gives the extinction as a function of the shape factor for lenses made from Schott glasses BK-7 ($n=1.51$) and SF-57 ($n=1.84$). This extinction corresponds to a circularly polarized ray or a linearly polarized ray with an azimuth of 45 deg. The initial ray is parallel to the lens axis. The ray angle Ω in the image field corresponds to $\text{NA}=0.3$. A drawing of lenses that conforms to the shape factor K is shown below the graph.

As we can see the uncoated lenses from different refractive glasses have the same maximal extinction of about 7700, which drops to about 1900 when the first surface is plane. Hence, the beam depolarization depends on the lens shape factor. Lenses possessing shape factors between 0.5 and 1.0 have maximal extinction. Hence, lenses that produce about equal bending of the ray at both surfaces have the best extinction. It is interesting to note that those lenses also produce minimal spherical aberration. In contrast, menisci produce maximal polarization aberrations.

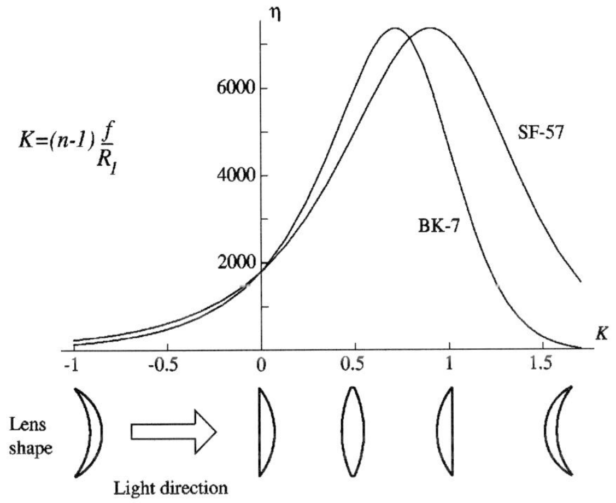

Fig. 2 Ray extinction η of a ray with NA=0.3 versus the shape factor K for single lenses made from Schott glasses BK-7 (n =1.51) and SF-57 (n=1.84). At the bottom of the figure, drawings of lenses are shown that correspond to the shape factors (light passes the lenses from left to right).

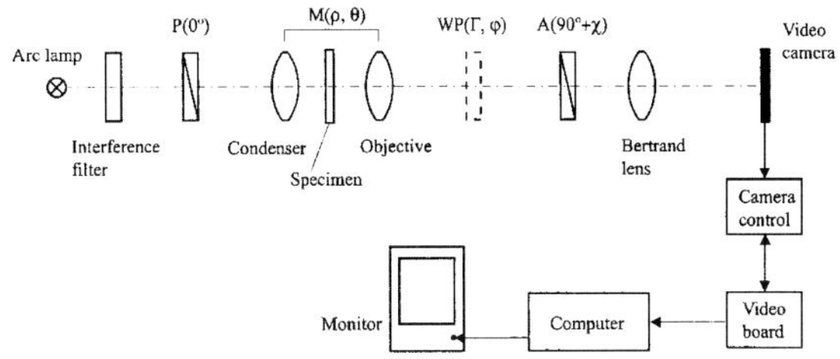

Fig. 3 Optical setup for measuring differential transmissions and phase shifts: $M(\rho,\theta)$, microscope optical system under investigation; P(0 deg), polarizer with an azimuth of 0 deg; $A(90\ \mathrm{deg}+\chi)$, analyzer with an azimuth of 90 deg+χ; WP(Γ,φ); compensation wave plate with small retardance Γ and an azimuth of φ.

4 Measurement of Radially Symmetric Polarization Aberrations

4.1 *Rotated Polarizer Technique*

The first scheme for measuring the differential transmission and phase shift between the p- and s-polarization components is presented in Fig. 3. Here the optical system with radial polarization aberrations is placed between a linear polarizer and analyzer. One of them can be turned. A Bertrand lens creates an image of the exit pupil on a video camera. If the polarizer and analyzer are crossed, the image shows a cross shape (Maltese cross) [see Fig. 4(a)]. When the analyzer is turned by an angle χ from the crossed position, the intensity distribution $I(\rho,\theta)$ in the image is

$$I(\rho,\theta)=\left\{[\sin(2\theta-\chi)\tan\xi(\rho)-\sin\chi]^2 + \sin^2(2\theta-\chi)\sin^2\frac{\Delta(\rho)}{2}\right\}I', \tag{20}$$

where it is assumed that values ξ and Δ are small, and I' is the initial intensity distribution after the polarizer. Hence, the cross transforms into two arcs [see Fig. 4(b)]. The distance between the arcs depends on the analyzer azimuth and the values of differential transmission. It can be shown that if radial retardance is nonzero and overwhelms the differential transmission, then rotating the analyzer will not produce dark arcs but will cause the cross to fade away.

For a point A at radius ρ along the diagonal with azimuth coordinate θ=45 deg the intensity is

$$I_A=\left\{[\tan\xi(\rho)-\sin\chi]^2+\sin^2\frac{\Delta(\rho)}{2}\right\}I'. \tag{21}$$

In this equation, we assume that χ is small and neglect terms with χ^4 and more.

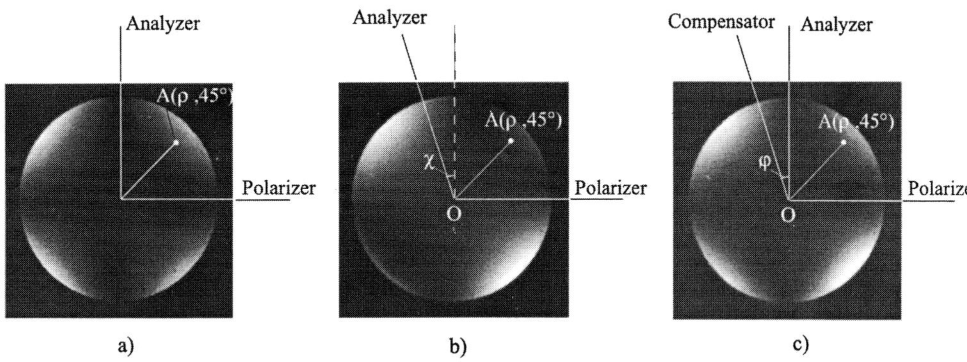

Fig. 4 Images of the back aperture of a microscope objective between linear polarizer and analyzer during measurements of differential transmission and phase shift between the p- and s-polarization components: (a) polarizer and analyzer crossed, (b) analyzer rotated by a small angle, and (c) polarizer and analyzer crossed and compensation wave plate rotated by a small angle.

The intensity at a point A along the diagonal is a minimum $\tilde{I}_A$ when $\chi=\xi(\rho)$:

$$\tilde{I}_A=I' \sin^2\frac{\Delta(\rho)}{2}. \quad (22)$$

The polarization turn angle ξ and retardance Δ are equal to zero at the center point $O(\rho=0)$. The center point intensity I_O is

$$I_O=I' \sin^2 \chi=I' \tan^2 \xi(\rho). \quad (23)$$

Thus, we can find the angle $\xi(\rho)$ by rotating one of the polarizers to find the minimum intensity at a point A with radius ρ and azimuth $\theta=45°$. At minimum intensity we have

$$\xi(\rho)=\chi. \quad (24)$$

If the minimum intensity is not zero, the lens also possesses radial retardance. To determine the radial retardance we measure the minimum intensity $\tilde{I}_A$ at the point A and the intensity $\tilde{I}_O$ in the center of the pattern. Then the radial retardance $\Delta(\rho)$ is

$$|\Delta(\rho)|=2 \sin^{-1}\left[\left(\frac{\tilde{I}_A}{\tilde{I}_O}\right)^{1/2} \tan \chi\right] \approx 2\chi\left(\frac{\tilde{I}_A}{\tilde{I}_O}\right)^{1/2}=2\xi(\rho)\left(\frac{\tilde{I}_A}{\tilde{I}_O}\right)^{1/2}. \quad (25)$$

This technique does not enable us to distinguish the slow and fast axes of the lens retardance. Also, the measurement requires a flat intensity distribution in the entrance pupil plane or Eq. (25) should be corrected.

4.2 *Rotated Compensator Technique*

It is also possible to employ a rotating compensator made of a wave plate with retardance Γ below 20 deg to find the radial differential transmittance and retardance of a lens. The compensator is shown in Fig. 3 by dashed lines. The polarizer and analyzer are crossed. In case of small values ξ and Δ, the intensity distribution $I(\rho,\theta)$ in the image of the exit pupil is described by

$$I(\rho,\varphi)=\left\{\sin^2 2\theta \tan^2 \xi(\rho) + \left[\sin\frac{\Gamma}{2}\sin 2\varphi+\sin\frac{\Delta(\rho)}{2}\sin 2\theta\right]^2\right\} I', \quad (26)$$

where φ is the azimuth of the slow axis of the compensator, and I' is the initial distribution of intensity after the polarizer.

When $\varphi=0$ deg [see Eq. (26)], the intensity is $I(\rho,\theta)=0$ for points with $\theta=0$ deg and $\theta=90$ deg. This results in a cross at the back aperture similar in appearance to the cross generated by differential transmission alone. However, if the compensator is turned by an angle $\varphi\neq 0$, then the cross transforms into two arcs. The distance between the arcs depends on the compensator azimuth and the value of radial retardance. In the presence of differential transmission that overwhelms the radial retardance, the cross fades away when the compensator is rotated.

For the diagonal point A with azimuth coordinate $\theta=45$ deg the intensity is

$$I_A=\left\{\tan^2 \xi(\rho)+\left[\sin\frac{\Gamma}{2}\sin 2\varphi+\sin\frac{\Delta(\rho)}{2}\right]^2\right\} I'. \quad (27)$$

This intensity has a minimum $\tilde{I}_A$ when $\sin(\Gamma/2)\sin 2\varphi=-\sin[\Delta(\rho)/2]$. Hence,

$$\Delta(\rho)=-\Gamma \sin 2\varphi, \text{ and}$$

$$\tilde{I}_A=I' \tan^2 \xi(\rho). \quad (28)$$

Simultaneously the center point intensity I_O is

$$I_O=I' \sin^2\frac{\Gamma}{2}\sin^2 2\varphi=I' \sin^2\frac{\Delta(\rho)}{2}. \quad (29)$$

Thus, we can find the radial retardance $\Delta(\rho)$ by searching for the minimum intensity in a point A with radius ρ and $\theta=45$ deg and then determine the angle $\xi(\rho)$ by measuring the intensities $\tilde{I}_A$ and the center point intensity $\tilde{I}_O$ when the minimum has been found:

$$\xi(\rho)=\tan^{-1}\left[\left(\frac{\tilde{I}_A}{\tilde{I}_O}\right)^{1/2} \sin\frac{\Gamma}{2}\sin 2\varphi\right] \approx \frac{1}{2}\Gamma \sin 2\varphi\left(\frac{\tilde{I}_A}{\tilde{I}_O}\right)^{1/2}=-\frac{\Delta(\rho)}{2}\left(\frac{\tilde{I}_A}{\tilde{I}_O}\right)^{1/2}. \quad (30)$$

This second technique with a rotating compensator enables us to determine the sign of the lens retardance, but it precludes us from finding the sign of the angle $\xi(\rho)$. On the other hand, the first technique with the rotating analyzer enables us to determine the sign of the angle $\xi(\rho)$ but does not give us the possibility to find the sign of the lens retardance.

The described experimental procedures measure polarization aberrations of the complete microscope optical system. Other techniques could be employed to measure the 2-D distribution of retardance and differential transmission, for example, by Mueller matrix imaging polarimetry,[16] polarized light microscopy using the universal liquid crystal compensator,[17] or phase shifting techniques.[18,19] If it is necessary to find aberrations introduced by a single optical component such as the objective or condenser lens separately, the polarization aberrations of the other components in the system should be known beforehand. Otherwise, one can either use a pair of identical lenses as condenser and objective[20] or use a return path technique[21] to measure the differential transmittance and radial retardance of either a single objective or a condenser.

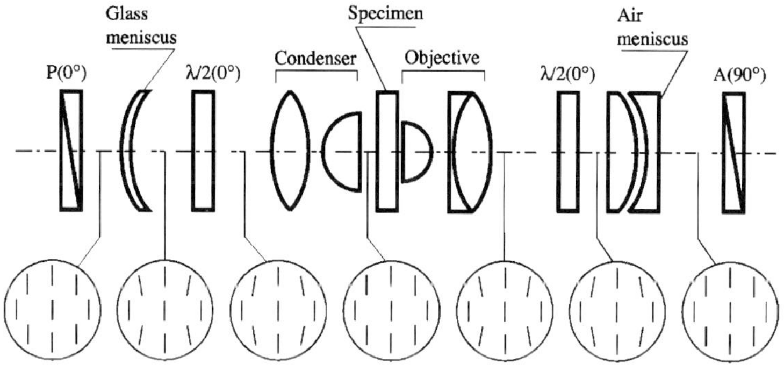

Fig. 5 Meniscus rectifier for linearly polarized beam (Ref. 20).

In Sec. 6 we present measurements of polarization aberrations of high-NA microscope lenses using the experimental procedures just described. However, first we discuss ways to reduce or almost eliminate polarization aberrations in high-NA imaging systems.

5 Rectification of Depolarization in a High-NA Microscope

To reduce the beam depolarization in a high-NA microscope a polarization rectifier can be used.[20] Here two kinds of rectifiers are described. The first contains a zero-power meniscus. The second kind uses a sectored liquid crystal compensator. We can put one rectifier in the illumination path of the microscope before the condenser lens, or in the imaging path after the objective lens or two rectifiers in both places. In the last case different condenser and objective lenses can be combined in the microscope and in addition, a more uniform polarization pattern is produced in the specimen plane.

The first scheme of polarization rectifier[20,22] with a meniscus is shown in Fig. 5. It can be used to decrease the depolarization of a linearly polarized beam. The rectifier consists of a glass or air meniscus with zero optical power and a half-wave plate. The meniscus creates the same distribution of the polarization rotation angle and ellipticity as the compensated optical element. The principal axis of the half-wave plate is parallel to the polarizer P. The half-wave plate flips the rotated polarization with respect to the plate's principal axis and then the condenser or the objective or both compensate this rotation. Thus, in the exit pupil, the beam has the correct linear polarization distribution and will be extinct by the analyzer A. In the optical schematic of Fig. 5, the different distributions of beam polarizations between elements of the microscope are illustrated. The differential transmission of lenses rotates the polarization plane of rays in the direction of the plane of incidence (the radial direction), as shown in the diagrams. The rotation is zero if the electric vector is parallel or perpendicular to the plane of incidence and maximum for $\theta \approx 45$ deg.

The first scheme of meniscus rectifier is described by the matrix equation

$$\mathbf{E}' = \mathbf{M}_{\mathrm{micr}} \cdot \mathbf{M}_{\lambda/2}(0\ \ \mathrm{deg}) \cdot \mathbf{M}_{\mathrm{rect}} \cdot \mathbf{E}, \tag{31}$$

where $\mathbf{E}'$ and $\mathbf{E} = \binom{1}{0}$ are the field vectors of the output and input beams, $\mathbf{M}_{\mathrm{micr}}$ is the Jones matrix of the microscope optical system, $\mathbf{M}_{\lambda/2}(0\ \mathrm{deg})$ is the matrix of the half-wave plate with an azimuth of 0 deg and $\mathbf{M}_{\mathrm{rect}}$ is the matrix of the meniscus rectifier. Taking into account that the matrices of the optical system and rectifier are the same [see Eq. (1)] and $\mathrm{M}_{\lambda/2}(0\ \mathrm{deg}) = \left(\begin{smallmatrix} 1 & 0 \\ 0 & -1 \end{smallmatrix}\right)$ we obtain

$$\mathbf{E}' = \frac{1}{4} T_s^2 \times \begin{bmatrix} (\mu e^{i\Delta}+1)^2 + 2(\mu^2 e^{i2\Delta}-1)\cos 2\theta + (\mu e^{i\Delta}-1)^2 \cos 4\theta \\ (\mu e^{i\Delta}-1)^2 \sin 4\theta \end{bmatrix}. \tag{32}$$

So this rectifier does not remove the orthogonal component completely. The extinction ratio η of a ray is

$$\eta \approx \frac{1}{[\tan^2 \xi + \sin^2(\Delta/2)]^2 \sin^2 4\theta}. \tag{33}$$

If $\tan \xi = \sin(\Delta/2) = 0.05$, then the extinction ratio is improved 100 times compared to the case without the rectifier [see Eq. (5)]. Figure 6 shows pupil images recorded with a microscope equipped with such meniscus rectifiers.

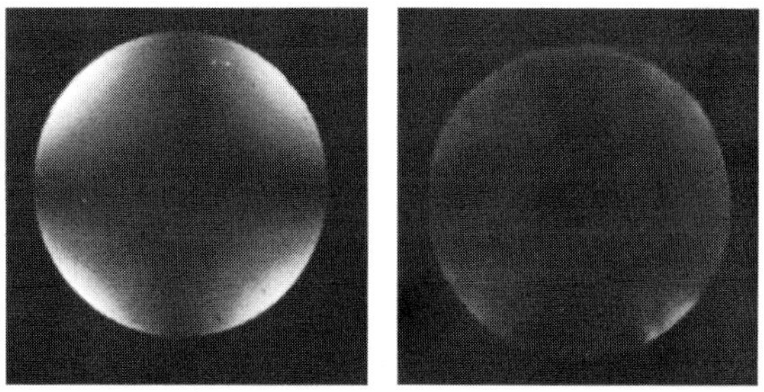

a) b)

Fig. 6 Images of the back aperture of a pair 1.25-NA oil immersion objectives Spencer 97× (American Optical Corporation) with single-layer MgF_2 antireflection coatings between crossed polarizers (a) without rectifiers and (b) equipped with meniscus rectifiers. Exposure time in the second case is increased 10 times.

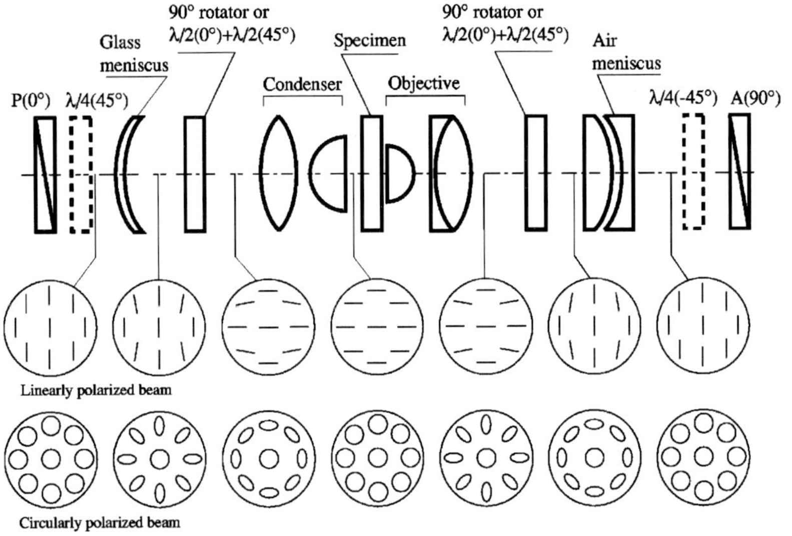

Fig. 7 Optical system with universal meniscus rectifier.

For circularly polarized light, the first rectifier scheme was improved by the addition of a second half-wave plate that is turned by 45 deg with respect to the first half-wave plate.[23] This combination of half-wave plates can be replaced by a 90-deg polarization rotator made from an optically active crystal, which is cut perpendicular to its optic axis. For example, a z-cut quartz plate with thickness 3.8 mm can be used as a 90-deg rotator for $\lambda=546$ nm. In addition, optical coatings that are applied to the meniscus and introduce the same radial retardance as the microscope optical system can compensate the influence of the microscope radial retardance. We call the meniscus with a 90-deg rotator a universal meniscus rectifier (see Fig. 7). Shortly we show that the universal meniscus rectifier enables rectification of a beam with any initial polarization state. Figure 7 shows the different distributions of beam polarizations for a linearly and circularly polarized beam. In the latter case using a circularly polarized beam, the differential transmission of lenses squeezes the polarization ellipse of rays in the tangential direction, as shown in the diagrams.

The universal meniscus rectifier is described by the following matrix equation:

$$\mathbf{E}'=\mathbf{M}_{\mathrm{micr}}\cdot\mathbf{M}_{\mathrm{rot}}(90\ \mathrm{deg})\cdot\mathbf{M}_{\mathrm{rect}}\cdot\mathbf{E}, \tag{34}$$

where $\mathbf{E}'$ and $\mathbf{E}$ are the field vectors of the output and input rays, $\mathbf{M}_{\mathrm{micr}}$ is the Jones matrix of the microscope optical system, $\mathbf{M}_{\mathrm{rot}}(90\ \mathrm{deg})$ is the matrix of the 90-deg rotator, and $\mathbf{M}_{\mathrm{rect}}$ is the matrix of the meniscus rectifier. Here the matrices of the optical system and rectifier are equal and include the differential transmittance and radial retardance terms [see Eq. (1)] and $\mathbf{M}_{\mathrm{rot}}(90\ \mathrm{deg})=\begin{pmatrix}0&1\\-1&0\end{pmatrix}$. After forming the matrix product of the matrices, we find the Jones matrix of the microscope with rectifier:

$$\mathbf{M}_{\mathrm{micr}}\cdot\mathbf{M}_{\mathrm{rot}}(90\ \mathrm{deg})\cdot\mathbf{M}_{\mathrm{rect}} = T_pT_s\exp[i(\Delta_p+\Delta_s)]\begin{pmatrix}0&1\\-1&0\end{pmatrix}. \tag{35}$$

As follows from the obtained matrix, the microscope will preserve the polarization structure of the entrance pupil. This enables us to completely remove the depolarization caused by differential transmission and radial retardance for any initial polarization state. The electric field vectors of the output rays will be turned by 90 deg compared to the initial polarization.

An alternative design for a rectifier uses sectored liquid crystal compensators for correcting a beam with arbitrary polarization distribution (see Fig. 8). It consists of two sectored liquid crystal cells LCA and LCB. Both cells are divided into one central circular sector and eight side sectors as shown in the lower diagrams of Fig. 8. The slow axes of the sectors in a cell have the same orientation. The slow axis of the first cell LCA is turned by 45 deg with respect to the polarizer P.

The second cell LCB has the slow axis oriented parallel to the polarizer P. The retardance magnitude of each sector can be set independently of all the other sectors. Hence, each sector in the polarizer-LCA-LCB assembly functions as an universal compensator, as described earlier by Oldenbourg and Mei.[17] Diagrams under the optical schematic in Fig. 8 illustrate the polarization states of an initially linearly and circularly polarized beam when the radial differential transmittance is the dominant source of depolarization. When the optical system has additional radial retardance, the sectored compensator can also improve the extinction, but the detailed description of its effect becomes more complicated.

In the case of linear polarization, the compensator works in the following manner. The linearly polarized beam

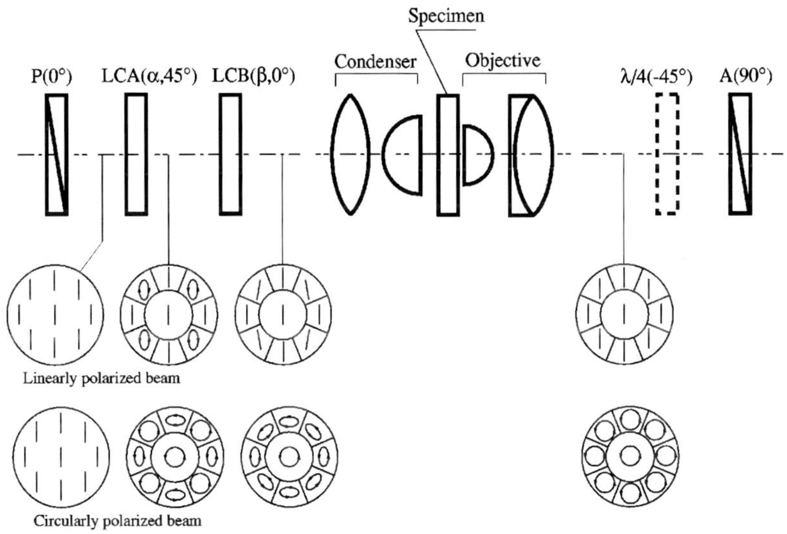

Fig. 8 Optical system with liquid crystal sectored rectifier.

passes through the first cell LCA and then the second cell LCB. The retardance values of the central sector and the horizontal and vertical side sectors are adjusted so as to not change the polarization state of the beam. These sectors have zero retardance for cell LCA and any retardance for cell LCB. The diagonal sectors rotate the beam polarization as it is shown in the diagram (middle row, Fig. 8). The magnitude of this rotation is chosen so as to obtain maximal extinction for each diagonal sector and is about 0.9ξ. Here ξ is the amount of rotation of the polarization plane of the marginal ray at the edge of the aperture and an azimuth of 45 deg. Thus retardance values of the diagonal sectors are $\pm 1.8\xi$ for cell LCA and 90 deg for cell LCB. As one can see, for linearly polarized light, the second cell can be replaced by a quarter-wave plate.

For rectification of circularly polarized light, it is necessary to have a circularly polarized beam after the central sectors and elliptically polarized side beams (bottom diagram, Fig. 8). The axes ratio of the polarization ellipses is $\tan[45\text{ deg}-(\xi/2)]$. This ratio is used to maximize the extinction for each side sector. In cell LCA the central and diagonal sectors have 90-deg retardance, the horizontal sectors have $90\text{ deg}-\xi$ retardance and the vertical sectors have $90\text{ deg}+\xi$ retardance. The central sector and horizontal and vertical side sectors of cell LCB do not change the polarization state of the beam and have zero retardance. The diagonal sectors of cell LCB have retardance 90 deg $\pm\xi$. Hence, the vibration ellipses of side beams have the small axis in the radial direction. The radial differential transmittance of the optical system squeezes the ellipses in the tangential direction and almost circularly polarized side beams fall on the circular polarizer that consists of a quarter-wave plate $\lambda/4(-45\text{ deg})$ and a linear polarizer A(90 deg). Figure 9 shows images of the sectored liquid crystal rectifier built into the front focal plane of the condenser lens of a microscope.

6 Experimental Results

Using techniques described in Sec. 5 we measured the distribution of differential transmission and phase shift in several microscope objective and condenser lenses. Our experimental setup was based on a Nikon Microphot SA microscope equipped with a Brace-Köhler compensator with 23-deg retardance at a wavelength of 546 nm and a bandpass filter (made by Chroma Technology Corp. in Brattleboro, Vt.) with a central wavelength of 546 nm and 12-nm FWHM. For image capture, a Dage-MTI C300 video camera was used. We measured aberrations in the lens combination of an oil immersion objective CFN DIC Plan Apochromat 60×/1.4 NA (Nikon, part 85033) and Nikon Achromat-Aplanat condenser lens with an aperture of 1.4, and the combination of a dry objective CFN DIC Plan Achromat 40×/0.7 NA (Nikon, part 85031) and the same

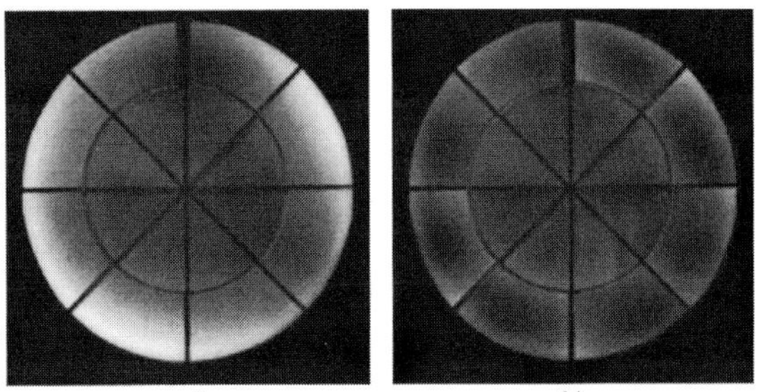

a) b)

Fig. 9 Images of the liquid crystal sectored rectifier that is placed in the front focal plane of the 0.9-NA condenser lens in a Zeiss microscope Axiovert 200M equipped with a dry antireflective coated objective Plan-NEOFLUAR 40×/0, 85Po1 and left circular analyzer: (a) sectors that are part of one cell have same retardance values and (b) each sector has individually adjusted retardance values for maximum extinction.

Table 1 Calculated (η_{calc}) and measured (η_{exp}) extinction, maximum turn angle (ξ_{max}), and retardance (Δ_{max}) measured in the pupil plane of microscope lenses made by American Optical Corporation (United States) and Nikon (Japan). NA is the numerical aperture of the objective and condenser lenses. The ratio η_{exp}/η_{calc} is also given.

Objective	NA	ξ_{max} (deg)	Δ_{max} (deg)	η_{calc}	η_{exp}	η_{exp}/η_{calc}
AO Spencer 97×, SFC	1.25 (oil)	5.5	2.0	600	600	1.0
AO Spencer 97×, SFR (with rectifier)	1.25 (oil)	0.5	2.5	8800	5100	0.58
Nikon Plane 40×	0.7	8.0	2.5	670	600	0.90
Nikon PlanApo 60×	1.40 (oil)	5.5	6.0	450	430	0.96

condenser used without oil immersion. Both objectives and condenser were designated for differential interference contrast (DIC) microscopy and are of good polarization quality. The Nikon objectives and condenser have multilayer antireflection coatings. In both cases, a test sample with diatoms (made by Carolina Optical Supply Co.) was placed between objective and condenser lenses. Also we studied two pairs of oil immersion, strain-free objectives, Spencer 97× with NA=1.25 (made by Optical Corporation of America around 1957). One objective pair included meniscus rectifiers for linearly polarized light. The Spencer objectives are coated by a single-layer antireflection coating of MgF_2. These objectives were first analyzed by Inoué and Hyde[20] in 1957, who first indicated that antireflection coatings and immersion liquids can reduce polarization aberrations in microscopes. Thus, it is possible to compare our results with those obtained over 40 years ago. For image processing and data analysis, we used NIH Image (public domain software, available at http://rsb.info.nih.gov/nih-image/) and Mathematica (Wolfram Research, Champaign, Illinois).

For each combination of objective and condenser lenses, we measured in the exit pupil the intensity along a line through the pupil center and oriented at 45 deg (see Fig. 4). Several such intensity profiles that were measured for different polarizer and compensator settings were combined to yield the turn angle $\xi(\rho)$, which describes differential transmission, and retardance $\Delta(\rho)$, which describes differential phase shift, each as a function of radial position in the pupil. Based on the pupil function measurements we calculated a theoretically predicted extinction factor in the image plane, using Eq. (10). Note that the formulas were obtained assuming a uniform intensity distribution in the pupil plane. But the formulas can easily be corrected for other kinds of radially symmetric intensity distributions, according to Eq. (3), for instance, a Gaussian distribution.

As mentioned in Sec. 2, there are other sources of depolarization in optical system. Therefore, we determined the extinction factor experimentally by measuring intensities of light in the image plane with parallel and crossed polarizer and analyzer. The ratio between the theoretically calculated extinction factor and the experimental one shows the contribution of radially symmetric polarization aberration at total extinction. A more detailed description of the calculation procedure and additional experimental results will be published elsewhere.

Table 1 shows results for microscope lenses made by American Optical Corporation (United States) and Nikon (Japan). It can be seen that high-NA microscope objectives with antireflection coatings have large values of differential transmission and phase shift. The radially symmetric polarization aberrations contribute the dominant part in the total extinction in systems without polarization rectification. Differential transmission is significantly more for a dry system than for an immersion system. The meniscus rectifier reduced the radial differential transmission of the optical system but not its retardance, possibly because the meniscus has no dielectric coating applied. To decrease the radial retardance it is necessary to cover the meniscus by an antireflection coating with the same retardance as the compensated objective. The rectifier enables us to improve the extinction factor by about a factor of 10. Early measurements by Inoué and Hyde[20] using the same objective lenses and a restricted field size showed that the rectifier improved the extinction by a factor of more than 20 (extinction 13,000 compared to 600). The reduced extinction measured now might be caused by additional defects in the objective that have since then developed.

Our findings indicate that the measured value of extinction in the image plane does not depend on the distance from the measured point to the optical axes of the microscope. In other words, the measured extinction is uniform over the field of view. This observation confirms that the theory that is strictly correct for a point on the optical axis also applies to the available image plane.

Examples of the measured turn angle ξ and retardance Δ distribution as functions of the normalized pupil coordinate ρ are shown in Fig. 10. Here measurement results for two pairs of oil immersion, strain-free objectives Spencer 97× with NA=1.25 are presented. Both objectives are coated by a single-layer MgF_2 antireflection coating. As we can see, both turn angle and retardance dependence have almost radial symmetry and rise from a 0 value in the center of the pupil to a maximum measured at the pupil edge. We observed with all objectives that the radial symmetry of the turn angle distribution is more pronounced than the radial symmetry of the retardance distribution. This difference is probably due to a higher susceptibility of the retardance field to inner stress in lenses and other depolarization factors. Furthermore, the radial retardance distribution of the oil immersion objective Nikon PlanApo-60× shows a change of the retardance sign with radial coordinate. The

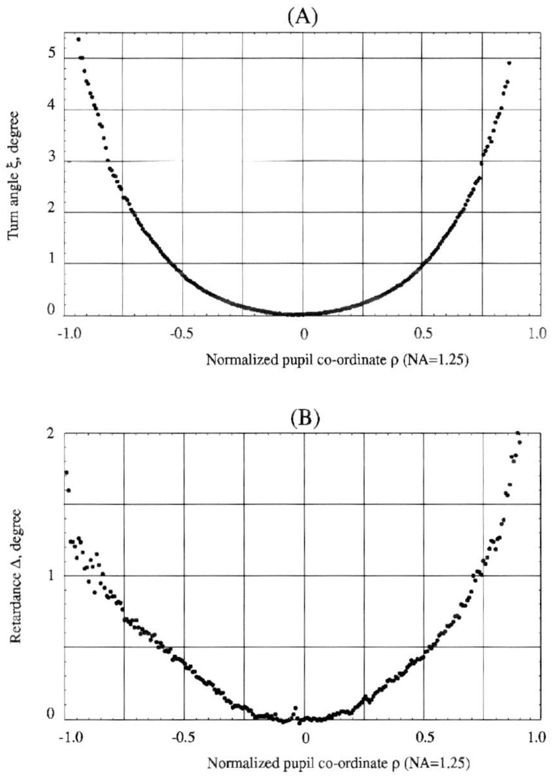

Fig. 10 Turn angle ξ (a) and retardance Δ (b) as a function of normalized pupil coordinate ρ of a pair of 1.25-NA oil immersion objectives Spencer 97× (American Optical Corporation) with single-layer MgF_2 antireflection coatings without rectifiers.

retardance is 0 for the axial point and for points that correspond to rays with NAs of about 0.7. It seems possible that in the case of complex multilayer antireflection coatings, the retardance sign of optical surface members δ_k in the summary retardance Δ [see Eqs. (2)] can be positive or negative dependent on the angle of incidence of rays that pass through the dielectric layers. Hence, the summary retardance can be compensated at some radii. Perhaps the members $\hat{\xi}_k$ in summary turn angle ξ have only one sign because $T_p > T_s$ and cannot compensate each other.

7 Conclusions

The radially symmetric polarization aberration introduced by lenses and plane parallel plates reduces the extinction and can lead to anomalous diffraction and modifies the point spread function. The extinction factor for systems with crossed linear polarizers is twice as great as that for crossed circular polarizers under otherwise the same conditions.

The extinction as a function of the NA for uncoated lenses and for plane parallel plates was calculated. Moreover, the connection between the shape factor of a lens and the extinction was determined. It was shown that a plate made from higher refractive glass introduces more depolarization than if it is made from lower refractive glass. The maximal extinction value of an uncoated lens with optimal shape, however, does not depend on the refractive index. Lenses that possess shape factors between 0.5 and 1.0 give maximal extinction. Menisci produce strong depolarization.

Techniques for measuring differential transmission and radial retardance using a rotatable analyzer or compensator were described. The results of the measurement can be used to choose the optimal design of polarization rectifiers.

The function of polarization rectifiers with a meniscus and a sectored liquid crystal compensator were analyzed. The meniscus rectifier consists of a glass or air meniscus with zero-order optical power and a half-wave plate and was employed to decrease the depolarization of a linearly polarized beam as well as attendant anomalous diffraction.

The introduction of a 90-deg rotator instead of the half-wave plate made it possible to remove polarization aberrations for any initial polarization state. Furthermore, a sectored liquid crystal compensator can create a corrected beam with arbitrary polarization aberration.

Finally, experimental results on polarization aberrations measured in a rectified and a nonrectified microscope system were presented. Good agreement between experiment and theory was found.

Acknowledgment

This research was funded by the National Institutes of Health Grant No. GM49210 awarded to R.O.

References

1. F. E. Wright, *The Methods of Petrographic-Microscopic Research, Their Relative Accuracy and Range of Application*, Carnegie Institution of Washington, Washington, DC (1911).
2. S. Inoué and R. Oldenbourg, "Microscopes," Chap. 17 in *Handbook of Optics*, Vol. II, 2nd ed., M. Bass. Ed., pp. 17.1–17.52, McGraw-Hill, New York (1995).
3. S. Inoué and K. Spring, *Video Microscopy: The Fundamentals*, 2nd ed., pp. 76–86, Plenum Press, New York (1997).
4. R. A. Chipman and L. J. Chipman, "Polarization aberration diagrams," *Opt. Eng.* **28**(2), 100–106 (1989).
5. R. A. Chipman, "Polarimetry," Chap. 22 in *Handbook of Optics*, Vol. II, 2nd ed., M. Bass, Ed., pp. 22.1–22.37, McGraw-Hill, New York (1995).
6. S. Inoué and H. Kubota, "Diffraction anomaly in polarizing microscopes," *Nature (London)* **182**, 1725–1726 (1958).
7. H. Kubota and S. Inoué, "Diffraction images in the polarizing microscope," *J. Opt. Soc. Am.* **49**, 191–198 (1959).
8. J. P. McGuire and R. A. Chipman, "Diffraction image formation in optical systems with polarization aberration. I: Formulation and example," *J. Opt. Soc. Am. A* **7**(9), 1614–1626 (1990).
9. J. P. McGuire and R. A. Chipman, "Polarization aberration. 1. Rotationally symmetric optical systems," *Appl. Opt.* **33**(22), 5080–5100 (1994).
10. J. B. Uri, "Polarizations and interference in optics V. Lenses, imaging properties of lenses," *Optik (Stuttgart)* **49**, 375–378 (1978).
11. P. I. Lamekin and K. G. Predko, "Change of the polarization structure of axial polarized light beams by lens systems," *Opt. Spectrosc.* **60**, 137–142 (1986).
12. Y. M. Klimkov, M. I. Shribak, "Influence of lens shape on polarization modification of axial beam," *Izv. Vyssh. Uchebn. Zaved. USSR. Geodez. Aerophot.* **5**, 128–139 (1990).
13. T. Wilson and R. Juskaitis, "On the extinction coefficient in confocal polarization microscopy," *J. Microsc.* **179**(3), 238–240 (1995).
14. M. Born and E. Wolf, *Principles of Optics*, 6th ed., pp. 38–47, Pergamon Press, Oxford (1987).
15. F. E. Wright, "The formation of interference figures a study of the phenomena exhibited by transparent inactive crystal plates in coherent polarized light," *J. Opt. Soc. Am.* **7**, 778–817 (1923).
16. J. L. Pezzanitti and R. A. Chipman, "Mueller matrix imaging polarimetry," *Opt. Eng.* **34**(6), 1558–1568 (1995).
17. R. Oldenbourg and G. Mei, "New polarized light microscope with precision universal compensator," *J. Microsc.* **180**(2), 140–147 (1995).
18. Y. Otani, T. Shimada, T. Yoshizawa, and N. Umeda, "Two-dimensional birefringence measurement using the phase shifting technique," *Opt. Eng.* **33**(5), 1604–1609 (1994).
19. Y. Otani, T. Shimada, and T. Yoshizawa, "The local-sampling phase

To further explore the underlying events, one of us (M.S.) analyzed the quantitative relationships between the events that should be taking place in the dipoles undergoing polarized absorption and emission, making the first-order assumption that the major dipole axes are oriented parallel to the crystal long axis.

As detailed in supporting information III(i) (on the PNAS web site), with the polarizer present, but in the absence of an analyzer (Fig. 2*A*), the equation becomes:

$$I/I_0 = 0.5(\alpha_{ps} + \alpha_{ss}) + 0.5((\alpha_{pp} + \alpha_{sp}) - (\alpha_{ps} + \alpha_{ss}))\cos^2\phi, \quad [1]$$

where the αs are proportionately coefficients for the intensity vectors of the fluorescence oriented parallel (α_{pp}, α_{ps}) and perpendicular (α_{sp}, α_{ss}) to the crystal long axis, relative to the excitation by blue light of unit intensity oriented parallel (α_{pp}, α_{sp}) or perpendicular (α_{ps}, α_{ss}) to the crystal long axis. ϕ is the angle between the crystal long axis (major dipole axis) and the transmission axes of the polarizer or analyzer as appropriate.

When the analyzer is present but the polarizer is absent [Fig. 2*B*, open circles; supporting information III(ii) on the PNAS web site], the equation becomes:

$$I/I_0 = 0.5(\alpha_{sp} + \alpha_{ss}) + 0.5((\alpha_{pp} + \alpha_{ps}) - (\alpha_{sp} + \alpha_{ss}))\cos^2\phi, \quad [2]$$

again, proportional to $\cos^2\phi$ plus an offset as in Eq. **1**.

For parallel polars, the fluorescence ratio becomes a function that is the sum of a $\cos^4(\phi)$ and a $\cos^2(\phi)$ term plus a small offset, as expressed by the following equation:

$$I/I_0 = 0.5(\alpha_{ss} + (\alpha_{ss} + \alpha_{sp} - 2\alpha_{ss})\cos^2\phi + (\alpha_{pp} + \alpha_{ss} - \alpha_{ps} - \alpha_{sp}))\cos^4\phi. \quad [3]$$

The curve for Eq. **3** [supporting information III(iii) on the PNAS web site], using the four coefficients calculated from the measured fluorescence polarization ratios, is plotted in Fig. 2*C* together with the observed points (open circles). The calculated curve indeed matches the observations very closely.

When the polarizer and analyzer are crossed with each other [Fig. 2*D* (open circles); supporting information III(iv) on the PNAS web site], the equation becomes:

$$I/I_0 = 0.5(\alpha_{ps} + (\alpha_{pp} + \alpha_{ss} - 2\alpha_{ps})\cos^2\phi - (\alpha_{pp} + \alpha_{ss} - \alpha_{ps} - \alpha_{sp}))\cos^4\phi. \quad [4]$$

As predicted by Eq. **4** (where α_{ps} is very nearly equal to α_{sp}), we find that the fluorescence intensity varies only slightly as the crystal orientation is changed between crossed polars. Furthermore, as predicted from the equation, the fluorescence intensity rises and falls four times during a 360° rotation of the crystal axis rather than twice, as was the case for the previous three conditions of measurement. The fluorescence intensity is maximum when the crystal long axis is oriented at approximately 45, 135, 225, and 315° rather than twice at 0 and 180°. Finally, between crossed polars, the fluorescence intensity minimum does not become as low as the minimum for the case of parallel polars (compare Fig. 2 *D* with *C*).

Solving these equations, we arrive at the four relative coefficients as: $\alpha_{pp} = 29.6 \pm 6.7$, $\alpha_{ps} = 4.3 \pm 0.5$, $\alpha_{sp} = 3.8 \pm 0.8$, and $\alpha_{ss} = 1$, with α_{ss} chosen as 1 to normalize these coefficients (see supporting information III on the PNAS web site).

Other Optical Properties of the Crystals. When observed with a polarizing microscope, illuminated through a 546 ± 25-nm interference filter that removes the fluorescence excitation, the GFP crystals showed a weak birefringence, with the refractive index very slightly larger across the crystal. The coefficient of birefringence was -1.6×10^{-3} (nanometer retardance per micrometer thickness, measured at 546-nm illumination). The weak negative birefringence suggests that the β cans of GFP (in which the β sheets run diagonally at an angle somewhat smaller than 45° to the major axis of the β can) are oriented with a large tilt angle relative to the length of the rod-shaped GFP crystal.

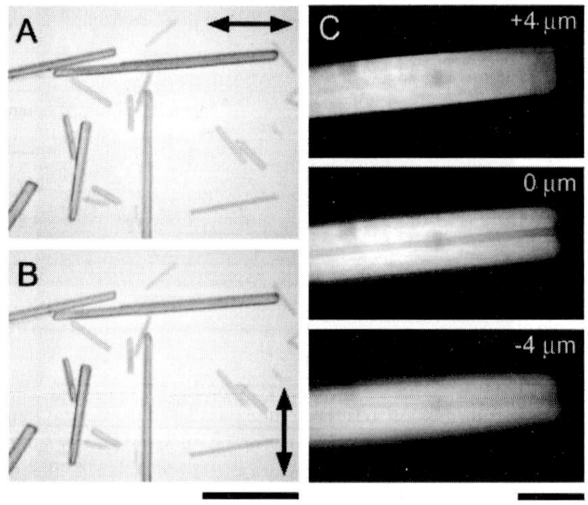

Fig. 3. (*A* and *B*) Dichroism of GFP crystals observed with quartz halogen illuminator. Double-headed arrows: polarizer transmission axis. (Bar = 30 μm.) (*C*) Confocal epifluorescence optical sections taken 4 μm above, at, and 4 μm below the mid plane of GFP crystal with large hollow core. The asymmetry of the image above and below the midplane and the strong flare reflects the high refractive index of the crystal wall material. (Bar = 10 μm.)

Observed in white light, the crystals also display a discernable visible light dichroism (Fig. 3 *A* and *B*). The crystals appeared in a yellowish straw color when the crystal long axis lies parallel to the polarizer or analyzer transmission axis and very pale blue in the perpendicular orientation. This dichroism undoubtedly reflects the orientation of the GFP chromophores whose major blue-absorbing dipoles lie oriented along the length of the crystal.

We noticed that many of the crystals in exact focus showed a weaker fluorescence along the axis of the crystal. At high magnification of a confocal microscope, several of the crystals showed a distinct hollow core (Fig. 3*C*). Despite this hollow cylindrical morphology, x-ray diffraction patterns of our crystals showed an unambiguous three-dimensional lattice (see supporting information II on the PNAS web site).

BIOPHYSICS

Discussion

Utility of the Polarizing Microscope. As demonstrated, a transilluminating polarizing microscope (to which appropriate excitation and barrier filters are added) provides a major advantage for measuring anisotropy of fluorescence. The distribution of specimen fluorescence, in turn, can be rapidly recorded and measured with the aide of a wide dynamic range modern charge-coupled device camera.

With a transilluminating polarizing microscope, one can: (*i*) rapidly switch between different modes of polarized excitation and emission; (*ii*) readily measure specimen fluorescence as a function of stage angles, i.e., at various specimen orientations; and (*iii*) avoid using beam splitters, birefringent elements, and other optical components that could inadvertently alter the polarization states of the illuminating or imaging beam.

From measurements of the relative fluorescence intensities made under different combinations of polarizer, analyzer, and crystal axis orientations, one can derive the coefficients for

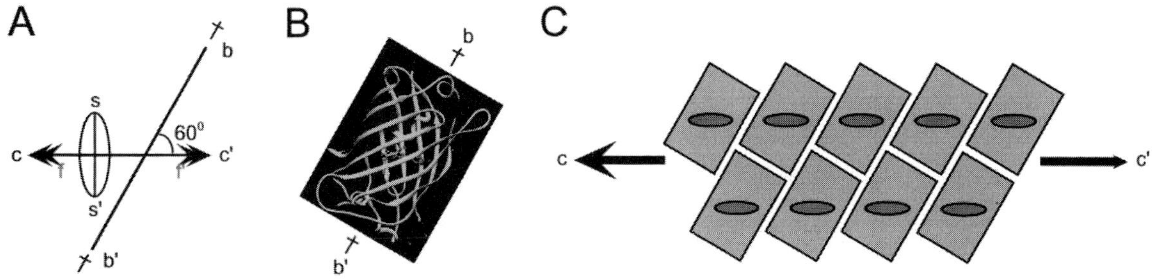

Fig. 4. Schematic of crystal structure. (*A*) Orientations relative to the long crystal axis (*c*–*c*′) of: major dipole for fluorophores (*f*–*f*′), slow birefringence axis (*s*–*s*′), and β can major axis (*b*–*b*′). (*B*) Published structure of GFP molecule based on x-ray crystallographic analyses (6, 7). (*C*) Schematic array of molecules fitting conditions shown in *A* and *B*. While maintaining the parallel alignment of the chromophore ellipsoids' major axes, this schematic could be modified with the backbone of the β cans rotated around the long crystal axis to occupy two or more discrete orientations.

fluorescence excitation and emission parallel and perpendicular to the crystal axis. When the crystal contains well-aligned chromophores, those values, in turn, should reflect the fluorescence quantum efficiencies of the major and minor transition dipoles. Additionally, one could, e.g., establish the degree of coherence between the absorbed excitation and emitted fluorescence and the degree of ellipticity of the fluorescence emitted.

Compared with using a single polarizer or analyzer, the use of *parallel* polars gives rise to a major increase in fluorescence intensity ratios as a function of fluorophore orientation. This new approach should dramatically improve the signal-to-noise ratio and sensitivity for following dynamic changes in orientation, or energy transfer, of individual fluorophores, e.g., when angular orientation changes for portions of single molecules are to be measured by using the polarized emission of single fluorophores (e.g., see refs. 13–15).

Significance of Fluorescence Polarization Ratios. Between parallel polars, the fluorescence of the native GFP crystal became maximum when the crystal long axis was oriented parallel to the transmission axes of the polarizer and analyzer and minimum at right angles to this direction (Fig. 2*C*). The ratio of maximum-to-minimum fluorescence was as high as 30:1 (Table 1). When the fluorescence is maximum, the excitation absorbing and fluorescence emitting dipoles must both be oriented very nearly parallel to the polarizer and the analyzer transmission directions. That we observe a fluorescence polarization ratio that approximates the product of the polarized absorption and emission coefficients suggests not only that the chromophores are well aligned parallel to the crystal axis, but also that there is little dissipative loss of energy between fluorescence excitation and emission.

Between crossed polars, a fluorescence minimum is observed when the crystal axis is oriented perpendicular to the polarizer or analyzer axis (open circles in Fig. 2*D*). These minimum values are, however, several times larger than the minimum observed when the polarizer and analyzer transmission axes are oriented parallel to each other (see Fig. 2*C*). In fact, the very weak fluorescence emitted by a crystal whose long axis is oriented perpendicular to the transmission axes of polars that are oriented parallel to each other increases in intensity several-fold when the analyzer is turned 90°. This somewhat counterintuitive rise in fluorescence is explained by the fact that the component of polarized emission that parallels the crystal axis is now fully transmitted by the analyzer.

The relative coefficients (α) that we derive are phenomenological values, i.e., they relate to the polarization components parallel and perpendicular to the long axis of the crystal. They cannot be directly ascribed to the anisotropies of the absorbing or emitting chromophores of GFP. Nevertheless, our observations suggest that the transition moment of the chromophore in GFP is not so much planar (as the arrangement of the tripeptide alone may suggest), but that it is nearly linear. They may be highly elongated prolate ellipsoids, oriented parallel to the length of the tubular crystal (Fig. 4*C*).

In our analysis, we assumed that the major dipole axes of the GFP chromophores were all uniformly oriented parallel to the crystal long axis. If, instead, we assume that the dipoles, each of which we now assume to possess infinite polarization ratios, were to be oriented with some scatter angle relative to the crystal long axis, what would be the maximum angle they could deviate from the crystal long axis and still show the polarization ratios similar to what we observed? Model calculations (assuming two sets of incoherently illuminated dipoles that are oriented in a plane normal to the light path at various degrees from the crystal axis) show that the maximum angle by which such dipoles could deviate is approximately 22°. These numbers set a limit to the tilt angle that the chromophores could exhibit relative to the crystal long axis, assuming that each dipole exhibits an infinite polarization ratio for fluorescence excitation and emission.

Proposed Arrangement of the GFP Molecules Within the Crystal and Jellyfish Photophore. We envision that the major absorbing and emitting dipoles of the fluorescent chromophores in our water-dialyzed crystals of native GFP are regularly aligned parallel to the crystal long axis (Fig. 4 *A f*–*f*′, *C*). Taking into account reported x-ray crystallographic data showing that the plane of the conjugated bonds of the fluorescent chromophores are oriented approximately 60° to the long axis of the β can (Fig. 4*B*), we come up with a schematic model for the arrangements of the GFP molecules in the crystals as shown in Fig. 4*C* and as expanded in the figure legend.

Within the light-emitting cells of the jellyfish *Aequorea*, GFP and aequorin molecules are tightly packed in small membrane-bound photophores (3). The intact photophores emit only green light characteristic of GFP fluorescence but no blue light from aequorin.

We speculate that within the photophores, sheets of GFP molecules, with chromophores oriented in a regular array as depicted in Fig. 4*C*, alternate with sheets of aequorin molecules so that the two chromophore groups are in close proximity and oriented parallel to each other. Blue energy from the aequorin layer, triggered by Ca^{2+} ions, would then nonradiatively excite the green fluorescence of GFP. Fingers or sheets of cell membrane, layered next to the aequorin layers, could efficiently control the Ca^{2+} environment of aequorin. We hope to test this model by checking the fluorescence anisotropy and ultrastructure of the intact *Aequorea* photophores.

The thin rod-shaped crystals that we used for our optical analyses were formed in ≈2 days by dialysis of highly concentrated GFP against deionized water. The same material, diluted by 3-fold to more slowly produce somewhat larger rod-shaped crystals (of up to 15 μm in width), was used to obtain the x-ray diffraction pattern (see supporting information II on the PNAS web site). By growing crystals with even better formed faces, we may be able to measure the fluorescence polarization ratios for light traveling along different axes of the crystal. The scatter of the data we report in Table 1 may then be explained as representing different ratios of the polarized fluorescence originating from chromophores viewed from different angles.

We thank Dr. Toshiya Senda of the Biological Information Research Center, National Institute of Advanced Industrial Science and Technology, Tokyo, for the x-ray diffraction analysis; Dr. Raymond E. Stephens of the Marine Biological Laboratory (MBL), Woods Hole, MA, for the protein analysis; and Drs. Kensal van Holde of Oregon State University, Carolyn Cohen of Brandeis University, Edward D. Salmon of the University of North Carolina, and Rudolf Oldenbourg of MBL, Woods Hole, for careful reading and comments on draft versions of the manuscript. We also thank Rudi Rottenfusser and Louis Kerr for help with the Zeiss microscope at the MBL Central Microscopy Facility, Bob Knudson for fabricating microscope parts as needed, and Jane MacNeil for typing and editing the paper. Loans or gifts of equipment were provided by Chroma Technology, Dage-MTI, Hamamatsu Photonics, Leica, Nikon, Olympus, Universal Imaging, Yokogawa Electric, and Carl Zeiss. We are also grateful for support provided by Dr. Yoshinori Fujiyoshi of Kyoto University; JEOL; a New Energy and Industrial Technology Development Organization Fellowship (to M.G.); a National Institutes of Health (NIH) Postdoctoral Fellowship (to P.T.T.), and Rudolf Oldenbourg's NIH grant (to M.S.).

1. Johnson, F. H., Shimomura, O., Saga, Y., Gershman, L. C., Reynolds, G. T. & Waters, J. R. (1962) *J. Cell. Comp. Physiol.* **60,** 85–103.
2. Morin, J. G. & Hastings, J. W. (1971) *J. Cell. Physiol.* **77,** 313–318.
3. Morise, H., Shimomura, O., Johnson, F. H. & Winant, J. (1974) *Biochemistry* **13,** 2656–2662.
4. Chalfie, M. & Kain, S., eds. (1998) *Green Fluorescent Protein: Properties, Applications, and Protocols* (Wiley-Liss, New York).
5. Sullivan, K. F. & Kay, S. A., eds. (1999) *Green Fluorescent Proteins, Methods in Cell Biology* (Academic, San Diego), Vol. 58.
6. Ormö, M., Cubitt, A. B., Kallio, K., Gross, L. A., Tsien, R. Y. & Remington, S. J. (1996) *Science* **273,** 1392–1395.
7. Yang, F., Moss, L. G. & Phillips, G. N., Jr. (1996) *Nat. Biotechnol.* **14,** 1246–1251.
8. Ward, W. W. (1998) in *Green Fluorescent Protein: Properties, Applications, and Protocols*, eds. Chalfie, M. & Kain, S. (Wiley-Liss, New York), pp. 45–75.
9. Inoué, S. & Spring, K. R. (1997) *Video Microscopy-The Fundamentals* (Plenum, New York), 2nd Ed., pp. 158–161.
10. Shribak, M., Inoué, S. & Oldenbourg, R. (2002) *Opt. Eng.* **41,** in press.
11. Seeds, W. E. (1953) *Progr. Biophys. Biophys. Chem.* **3,** 27–46.
12. Inoué, S. & Sato, H. (1966) in *Molecular Architecture in Cell Physiology*, eds. Haysahi, T. & Szent-Györgyi, A. G. (Prentice-Hall, New York), pp. 209–248.
13. Corrie, J. E. T., Brandmeier, B. D., Ferguson, R. E., Trentham, D. R., Kendrick-Jones, J., Hopkins, S. C., van der Heide, U. A., Goldman, Y. E., Sabido-David, C., Dale, R. E., *et al.* (1999) *Nature (London)* **400,** 425–430.
14. Noji, H., Yasuda, R., Yoshida, M. & Kinosita, K., Jr. (1997) *Nature (London)* **386,** 299–302.
15. Warshaw, D. M., Hayes, E., Gaffney, D., Lauzon, A., Wu, J., Kennedy, G., Trybus, K., Lowey, S. & Berger, C. (1998) *Proc. Natl. Acad. Sci. USA* **95,** 8034–8039.

BIOPHYSICS

Reprinted from *Current Protocols in Cell Biology*, Vol. 4(9), pp. 1–27, 2002, with permission from John Wiley & Sons.

Polarization Microscopy

UNIT 4.9

INTRODUCTION

Polarization microscopy, also called polarized light microscopy, allows one to nondestructively follow dynamic, anisotropic organization of living cells and tissues at the microscopic as well as submicroscopic levels.

Since, in polarization microscopy, light waves are used to probe the specimen, spatial resolution of the image is limited to ~0.2 µm, even using objective and condenser lenses with numerical apertures approaching 1.4. Nevertheless, the brightness of each resolved area in the image measures the polarization optical property, most commonly birefringence, of the corresponding minute area in the specimen. The birefringence directly reflects the anisotropic organization of the microscopically unresolvable, fine structure—namely, the submicroscopic, molecular, and atomic lattice arrangements of the specimen within each diffraction-limited image area.

Thus, e.g., the assembly and disassembly dynamics of spindle microtubules and the developmental changes in sperm chromatin could be revealed, and the arrangement of the atomic lattice in developing biocrystalline spicules, as well as the packing arrangement of DNA molecules in sperm chromosomes, could be unveiled with polarization microscopy at levels even finer than those conventionally seen with electron microscopy.

Examples in this unit were chosen mostly from those particularly familiar to the author and which have been used successfully by many generations of students and co-workers. Beyond the biological material mentioned below, or those that are commonly used for studying living cell behavior and fine structure of their organelles, a treasure trove of suggestions for other living cells appropriate for study can be found throughout an article by Belar (1928).

The first part of this unit touches on some general procedures, and outlines the background optical theory and relevant literature sources for polarization microscopy. The unit then examines several basic examples for using the polarizing microscope, emphasizing the study of living cells or products, starting with those that can be carried out with relatively simple equipment and proceeding to examples that require more advanced instruments. For instructions on setting up and operating a polarization microscope see Murphy (2001), Oldenbourg (1999), and manufacturer's instructions.

For more comprehensive discussions and an annotated list of references on polarized light and birefringence phenomena, see Inoué (1986, especially Appendix 3). For additional information, see Bennett (1950), Jenkins and White (1957), Hartshorne and Stuart (1960), and Hecht (1998).

Types and Transmission Axis of Polars

Polarizers, or polars, range from plastic-sheet "Polaroids" to calcite polarizers of many designs. Although very expensive and fragile, the calcite polarizers provide the theoretically maximum transmission (i.e., nearly 50% of the incoming nonpolarized light) over a wide wavelength range. Their extinction factor (EF; see in Example 5 below) also tends to be an order of magnitude better than the sheet Polaroids. On the other hand, light transmitted through the calcite polars has to be collimated into parallel beams in order to avoid astigmatism. They are used in expensive polarization optical systems requiring the highest EF and minimum loss of light. Some new sheet polarizers such as "Polacolor" and those available from Spindler and Hoyer are reported to have both high transmission for white light and high EFs.

Sheet Polaroids generally contain a stretched sheet of polyvinyl alcohol, a long-chain polymer whose backbone can be aligned by heating and stretching the polymer sheet that has been cast from an aqueous solution. Commonly they are stained with polyiodide micelles, whose conjugated backbones lie parallel to the aligned polyvinyl alcohol chains. It is the conjugated backbone of the polyiodide that gives rise to the *dichroism* of these Polaroid sheets (Inoué, 1986). They can be made into large sheets and can easily be cut into various shapes with a pair of scissors. This is the material used for Polaroid sunglasses. As polars in microscopes, their soft and somewhat wavy surfaces are protected by mounting them with index-matched cement between antireflection-coated glass cover plates.

The transmission axis (E-vector direction of the transmitted polarized light) for Polaroid sunglasses is oriented vertically in order to reduce the glare from horizontal surfaces, such as the road, water, or hood of a car. Light

Contributed by Shinya Inoué

Current Protocols in Cell Biology (2002) 4.9.1-4.9.27

reflected from such (nonmetallic) horizontal surfaces contains a high proportion of polarized light whose E- (electric-) wave oscillates horizontally and hence is cut out by the Polaroid sunglasses. Thus, the Polaroid sunglasses become convenient, quick references for establishing the E-vector transmission direction of other polars.

Birefringence

Light traveling inside a birefringent material is split into two waves that travel at different speeds. The slower wave suffers greater refraction than the faster wave since the refractive index experienced by each wave is inversely proportional to its velocity of propagation.

The two waves vibrate in planes that lie perpendicular to each other, as dictated by the material's optical axes, and to the direction of propagation of the light ray. For a ray traveling in a particular direction through a birefringent material, the electric (E-) vector direction encountered by the slow wave is called the slow axis, and that of the fast wave the fast axis.

Optical Axes and Optic Axis

Broadly speaking, crystals can be classified into six groups depending on their axes of symmetry. All crystals, except those in the cubic group (e.g., NaCl, KCl), are birefringent, that is, they transmit light whose E-vectors, oscillating in different planes, travel at different velocities (Hartshorne and Stuart, 1960; Inoué, 1986). However, each crystal type has a unique single axis, or two axes, along which light is transmitted at equal speeds regardless of the direction of the E-vector. In other words, light propagating along such an axis does not suffer birefringence. Such an axis is called the optic axis of the crystal or the birefringent material.

In addition to the optic axis, a crystal generally has three principal optical axes (one of which may correspond to the optic axis). Between crossed polars, when the optical axes of a crystal come to lie parallel to the polarizer or analyzer, the entering polarized light is not split into two components, thus encountering no birefringence, and so the crystal appears dark. This extinction is seen intermittently in a swimming pluteus larva of a sea urchin, when the long axis of one of the biocrystalline, skeletal spicules comes to lie along the polarizer or analyzer transmission direction (see Example 1; Fig. 4.9.1).

In the younger, gastrula-stage sea urchin embryo, the triradiate precursor of the spicule may appear not to be birefringent unless viewed from its side. That is because the optic axis of these biocrystals, made of calcite, lies perpendicular to the plane of the triradiate spicule.

With the background between crossed polars made gray by turning a compensator from its extinction orientation, regions of the birefringent specimen whose axes coincide with the transmission axes of the polars appear with similar brightness to the background, i.e., neither brighter nor darker than the background. Four such regions are commonly seen, e.g., on the fertilization envelope, asters at the spindle poles, and starch grains (see Examples 3 and 4 and Fig. 4.9.4).

Sign of Birefringence and Coefficient of Birefringence

Depending on the crystal type, light may travel faster or slower along the optic axis compared to other axes. The former type of crystal is known to be positively birefringent (e.g., quartz, collagen fiber), and the latter negatively birefringent (e.g., calcite, DNA thread; Inoué, 1986). In addition to its sign, each crystal type exhibits a unique amount of birefringence, the coefficient of birefringence, which is the *retardance*-per-unit thickness of crystal (see discussion of Retardance, below). In fact, the signs of birefringence and refractive indices for light traveling along the axes of pure crystals are so unique that they are used in crystallography for identifying their composition and lattice type.

Traditionally, in biological polarization microscopy, the expressions positive and negative birefringence are used relative to some obvious frame of reference (see Examples 3 and 4) rather than to the crystal axis of the object. For example, the fertilization envelope is said to be positively birefringent relative to its tangent, meaning the E-wave oscillating parallel to the tangent of the fertilization envelope (the obvious outline seen outside the equator of the egg; Fig. 4.9.4) is retarded (refracted) more than the E-wave oscillating perpendicular to the envelope. Since the fertilization envelope is a hollow sphere, the axis of symmetry (and, hence, optic axis for any unit area of the envelope) should be its radius rather than its tangent. So, according to the crystallographic definition, the envelope has a negative birefringence. On the other hand, if one were to make a thread of the material that makes up the envelope, the axis of symmetry would be along the length of the thread, so that the traditional nomenclature commonly used in biology does have its utility.

Retardance

Light waves traveling through a birefringent specimen are split into two orthogonally polarized components, one being refracted more, or traveling slower, than the other (Bennett, 1950; Inoué, 1986). The phase difference between these two waves as they emerge from the specimen is known as their retardance, or birefringence retardation. The retardance (Δ) is, in fact, the difference in refractive indices ($n_e - n_o$) experienced by the "extraordinary" and "ordinary" waves multiplied by the thickness (d) of the specimen. Thus $\Delta = (n_e - n_o) \times d$.

The retardance can be expressed in fractions of wavelengths or in nanometer units; e.g., $\lambda/4$, or 136.5 nm (= 546 nm/4), for a quarter-wave plate for the mercury green line. A mitotic spindle may show a retardance of 2.7 nm, or only 1/200 wavelength for the mercury green line.

Intrinsic, Form, and "Accidental" Birefringence

Birefringence of a specimen may arise from anisotropy (difference in physical property exhibited for different directions in the same material) of structural organization at various submicroscopic levels that are all too fine to be resolved with the light microscope. If the ultrastructure of the specimen is anisotropic at a level usually detectable with the electron microscope, but not with the light microscope (e.g., stacks of cell membranes, such as in the Schwann sheaths of nerve or the outer segments of the retina, parallel-oriented filament systems such as the F-actin thin filaments in muscle, microtubules in flagellar axonemes, and spindle fibers), they exhibit an optical anisotropy known as form birefringence. Form birefringence arises from the parallel array of filaments or membranes whose refractive index differs from that of the imbibing medium and therefore, disappears when the refractive indexes of the two match. For filaments, the sign of form birefringence is positive, whereas for membranes it is negative.

When the birefringence arises from anisotropy within the molecules themselves, it is not altered by the refractive index of the imbibing medium, and the material is said to possess intrinsic birefringence. The regularly aligned polypeptide chain in collagen fibers and in the myosin filaments of muscle A-bands, the regularly aligned DNA protein in mature sperm head, and the biocrystalline calcite in the skeletal spicules of sea urchin embryos are some examples of material that exhibit intrinsic birefringence.

While intrinsic birefringence tends to be considerably stronger than form birefringence, the two may contribute more or less equally in some biological structures, either with the same or opposite signs of birefringence. For example, the positive intrinsic birefringence of lipid molecules, or of membrane proteins, may overcome the negative form birefringence introduced by the lipid bilayers when the refractive index of the imbibing medium approaches that of the bilayers (Ambronn and Frey, 1926; Schmidt, 1935, 1937; Frey-Wyssling, 1953).

In addition to the naturally observed intrinsic and form birefringence, one may encounter birefringence induced by flow or by external forces such as stretch or compression or by the application of electrical, magnetic, or other fields. The induced birefringence, in turn, may reflect the intrinsic or form birefringence of the aligned molecules or submicroscopic particles, or may reflect distortion of the electron orbits within the substance itself. In this last regard, strain birefringence can be induced even in naturally isotropic substances such as glass, isotropic crystals, or well annealed plastics. In fact, the strain birefringence in the objective and condenser lenses can be high enough to interfere with observation of small specimen retardances, and special strain-free objectives (designated with P or Pol) are sometimes supplied to minimize such effects.

Edge Birefringence

The edges of objects, viewed with polarization microscopy using high-NA objective lenses, may display what appears to be a double birefringent layer. While in some cases these may represent the birefringence due to membrane fine structure or its molecular organization, in other cases they can be caused merely by the presence of a sharp refractive index gradient at the edges of the object. The object can be purely isotropic; the only condition required is that there exist a sharp gradient of refractive index at the boundary between the object and its surrounding. Such birefringence is called "edge birefringence."

As detailed by Oldenbourg (1991), with edge birefringence the inner, thin "birefringent layer" shows a slow axis (axis with greater refractive index) parallel to the boundary and an outer thin "birefringent layer" with its slow axis perpendicular to the boundary when the refractive index of the object is greater than its surroundings. When the refractive index of the object becomes equal to its surroundings, the edge birefringence disappears. When the refractive index of the object becomes less than its surroundings, the directions of the slow axes are reversed.

As with edge birefringence, form birefringence also disappears when the refractive index of the rodlets or platelets matches that of the imbibing medium. But the sign of form birefringence does not change below and above the matched index. The slow axis direction is unchanged (relative to the specimen axis) above and below index match point. In contrast, with edge birefringence the signs of birefringence (in the two layers on both sides of the optical boundary) are reversed in media whose refractive index is greater than the index match point compared to the situation in media below the match point. In contrast to both edge and form birefringence, the retardance due to intrinsic birefringence is unaffected by the refractive index of the imbibing medium.

It is important to distinguish edge birefringence from form or intrinsic birefringence since edge birefringence reflects an optical effect that has nothing to do with the molecular or fine structural anisotropy that we are probing.

COLOR IN POLARIZED LIGHT

Interference Colors

Many crystals that are normally colorless, including the biocrystalline spicules of sea urchin plutei, show in vivid color when observed in white light between crossed polars (Fig. 4.9.2). The color reflects the birefringence retardation (or retardance, i.e., the phase difference) that the two components of the orthogonally polarized light components suffered in passing through the crystal. For example, if the retardance were 550 nm, or a full wavelength for green light, the two orthogonal components would add back together to form the initial state (i.e., of plane polarized light as before entering the crystal) upon exiting the crystal. Thus, green light is extinguished by the analyzer. Wavelengths other than 550 nm are elliptically polarized and are partially transmitted by the analyzer. Thus, the complementary color of green, i.e., red, is seen through the analyzer (Inoué, 1986).

Similarly, for an object giving a retardance of 490 nm, full wavelength for blue light, blue is extinguished and its complementary color yellow is seen through the analyzer. For a retardance of 568 nm, full wavelength for yellow light, yellow is extinguished and its complementary color blue is seen through the analyzer. For a retardance of 600 nm, full wavelength for red light, red is extinguished and its complementary color green is seen through the analyzer. These interference colors thus measure the retardance of the specimen.

For a cylindrical object like the pluteus skeletal spicule, light would travel the furthest distance across its middle and suffer the most retardation, so that the axis of the spicule would appear with the highest-order interference color. Thus, the axis may appear green, surrounded by blue, then red, then yellow, then gray, etc., as light travels through decreasing thickness towards the outer margin of the cylindrical biocrystal and is thus retarded, proportionately, less and less.

Dichroism

In addition to optical anisotropy that is manifested as birefringence (difference of refractive index for light waves whose E-waves are oscillating in different planes), one may also encounter dichroism (difference of absorbance, or of absorption spectrum, for light waves whose E-waves are oscillating in different planes). Whereas a birefringent material simply changes the phase of polarized light, a dichroic material transmits light to different degrees, depending on the wavelengths and the angle between its axes and the plane of the entering polarized light.

Polaroid films serve as effective polars by taking advantage of the aligned polyiodide chains that show a strong dichroism for most wavelengths in the visible spectrum. The conjugated electrons that make up the backbone of the polyiodide chains strongly absorb the (E-) component of the visible light waves that are polarized parallel to that direction, whereas they absorb little of the component that is polarized at right angles to the backbone.

The dichroism of Polaroid sheets that extends over much of the visible wavelength range is, however, an exception. Most dichroic materials (or pleochroic materials, i.e., many-colored instead of showing two different colors, depending on the vibration direction of the polarization of light) exhibit two different absorption curves depending on the orientation of their chromophores relative to the plane (of the E-wave) of polarized light.

DNA in mature sperm is a biological material that shows prominent dichroism in ultraviolet (UV) light. The conjugated bonds in the purine and pyrimidine bases of B-form DNA, regularly oriented at right angles to the DNA backbone as in steps of a spiral staircase, give rise to a UV dichroism with a dichroic ratio of 4, spanning the wavelength range of ~250 to 380 nm. This dichroism is present because the conjugated bonds in the bases (that resonate in the UV range) lie at right angles to the B-form DNA backbone. So the UV polarized at right angles to the backbone is absorbed four times more effectively than UV that is polarized parallel to the DNA backbone (Wilkins, 1951; Inoué and Sato, 1966). This gives rise to the UV dichroism that has a negative sign (i.e., greater absorbance perpendicular to the backbone compared to parallel to the backbone).

The same conjugated bonds give rise to the much greater polarizability of the electrons, and, hence, to higher refractive indexes for visible light waves whose E-fields lie in the plane of the bases compared to fields along the backbone of DNA. That gives rise to the strong negative birefringence of B-form DNA in the visible wavelength range (Inoué and Sato, 1966; Inoué, 1986). The author took advantage of this dichroism and birefringence, which are both abolished upon polarized UV microbeam irradiation (possibly by dimerization of the thymidine bases), to analyze the detailed packing arrangement of DNA and to uniquely visualize chromosomes, taking advantage of a needle-shaped head of an insect sperm (Fig. 4.9.7; Inoué and Sato, 1966).

Chromophores with such high dichroic ratios can be recognized by a change in light transmittance or absorbance, or of color, as the orientation of a single polar (a polarizer in the absence of an analyzer, or an analyzer in the absence of a polarizer) is changed.

Cell inclusions with regularly oriented natural chromophores—e.g., retinal rods and cones (Schmidt, 1935), sickled blood cells (Perutz and Mitchison, 1950; Hárosi, 1981), and chloroplasts (Breton et al., 1973; Hsu and Lee, 1987)—may also show some degree of dichroism. Cell inclusions, or products with regularly oriented fine structure that have been stained with dyes, may likewise be dichroic.

When the dichroic ratio is low, as in these latter examples, it may require a sensitive spectrophotometer to detect a change in color, or wavelength dependence of light transmittance, as the specimen orientation is changed relative to the orientation of the polarized light (Hárosi and MacNichol, 1974). In such cases, the dichroism may still be detected visually if one uses both a polarizer and an analyzer and orients one of the two a few degrees from the crossed position. The complementary dichroic colors can then be seen visually by turning the polar or compensator in opposite directions from the extinction orientation.

Using the dichroic fluorescence of tetramethylrhodamine that was used to sparsely decorate actin filaments, Kinosita and coworkers were able to demonstrate, and measure, the speed of rotation of individual actin filaments around their own axis as the filaments are propelled by myosin molecules coated on the surface of a microscope coverslip. Through video contrast enhancement of the fluorescence microscope images taken in polarized light, they were able to show the cyclic brightness change in the dichroic fluorescent dye as the single actin filament, <10 nm in diameter, glided along the lawn of myosin heads (Kinosita, 1999).

The sensitivity for measuring the orientation of a dichroic fluorescent molecule, such as green fluorescent protein, can be increased several fold by using parallel polars instead of a single polarizer or analyzer (Inoué and Goda, 2001).

Colloidal Gold, Pinholes

Spindle microtubules in *Haemanthus* endosperm cells decorated with 20 nm-diameter colloidal gold appear as dark black lines when observed in a compensated field between crossed polars in monochromic light. These same microtubules show in striking colors ranging from red to purple, green, orange, and gold when they are observed in white light with the polars off-crossed by a few degrees and the compensator orientated to different directions (Inoué and Spring, 1997, Color Plate III). These colors are attributed to anisotropy of Tyndall light scattering (or di-Tyndallism) by the linear array of colloidal gold particles that are arranged quite densely (not contacting each other, yet acting as an effective conductor, or antenna, for the high-frequency electromagnetic light waves) along the lengths of each microtubule.

Examined at high magnification, pinholes in thin metal film, evaporated on coverslips that are used to test the performance of microscope objective lenses, can also each appear with striking saturated colors when examined in polarized white light. Light from the pinholes becomes alternately extinguished as the microscope stage is rotated. In other words, each pinhole is acting as a polarizer that appears to have a narrow transmission wavelength. No doubt, each pinhole is not completely circular but must be elongated, since they are generally produced by some random dust particle or impediment to sputtering or high-vacuum deposition of the metal film. Thus, each "pinhole" acts as an anisotropic nonconductor in the thin metal film.

In fact, a very narrow crack in a thin metal film preferentially transmits light that is polarized with its E-vector oriented perpendicular to the narrow crack and whose wavelength reflects the local width of the submicroscopic crack (S. Inoué, unpub. observ.). The situation is somewhat the converse of the aligned polyiodide chains or of the colloidal gold-decorated microtubules, in that instead of a submicroscopic thread-shaped conductor, one has a thread-shaped non-conductor surrounded by a conducting film.

Anomalous Birefringence

Stained or colored biological tissue and cell components often exhibit "anomalous birefringence," i.e., the color observed in white light between crossed polars actually changes with adjustment of the compensator. In other words, the birefringence, or the retardance observed, is a function of the wavelength of light. Such a color change, or "anomalous birefringence," is observed even when the retardance of the object is a small fraction of the wavelength of light (Perutz and Mitchison, 1950).

Colorless objects, on the other hand, generally show birefringence that changes only slightly with variation of wavelengths. What color is observed for a colorless object is primarily due to interference colors when the retardance approaches a wavelength or more, as described earlier (Fig. 4.9.2).

On the other hand, with strongly absorbing materials such as a dye or stain, the refractive index drops sharply on the shorter wavelength side and rises precipitously on the longer wavelength side of the absorption peak ("anomalous dispersion"; Jenkins and White, 1957; Hecht, 1998). Since the wavelength-specific absorption of light in a dye or stain is commonly brought about by conjugated ring structures, the birefringence associated with such an anisotropic structure would also be expected to show a strong dispersive character (wavelength dependence). The anisotropic dispersive character would then give rise to anomalous birefringence. Objects exhibiting anomalous birefringence would also be expected to show some dichroism as well.

These several wavelength-dependent polarization optical effects (as with polarization of fluorescence, a topic not covered here) are of considerable potential interest as potent tools for analyzing dynamic changes within, or interactions among, biological molecules and fine structures. Further studies on the physical optical bases of these phenomena should lead to their wider application in cell biology.

OPTICAL ROTATION

Optical rotation (also known as circular birefringence) is a polarization optical effect that may (rarely under the microscope) induce an object to become bright between crossed polars. Unlike (linear) birefringence, the brightness due to optical rotation does not depend on the regular alignment of the molecules but on their internal asymmetry and thus is observed even in solutions of the particular molecules. Optical rotation may also be seen in certain crystalline substances, such as quartz and other material viewed along the helical axes of the molecules that make up the crystal.

An object exhibiting optical rotation does not undergo extinction between crossed polars as the microscope stage revolves. Instead, the object can be made darker by turning the polarizer or analyzer away from their crossed orientation. Except in pure monochromatic light, as a polar is turned, these "optically active" objects or solutions acquire a brownish or bluish hue instead of undergoing complete extinction. The color reflects a high degree of wavelength dependence that is characteristic of optical rotation. Such wavelength dependence of optical rotation is known as optical rotatory dispersion. The infrared optical rotatory dispersion, e.g., of myosin and related proteins in solution, has been used to determine the degree of helix content of protein molecules (Szent-Györgyi et al., 1960).

VIDEO MICROSCOPY

As in many other modes of microscopy, the capabilities of polarization microscopy have been substantially enhanced by the application of video and electronic image processing, analysis, and device control. As described in Allen et al. (1981) and Inoué (1981b) and summarized, e.g., in Shotten (1993), and Inoué and Spring (1997), video microscopy can effectively enhance image contrast, subtract background, reduce noise, and allow image capture with better corrected higher-NA objective lenses and even with relatively short exposures.

The new LC PolScope introduced by Oldenbourg, provides striking, high-contrast images of weakly birefringent cellular regions and does so independently of specimen axis orientation. The recent system introduced by N. Allen and co-workers selectively enhances (or suppresses) contrast in differential interference contrast (DIC) microscopy and should also be applicable for bringing out birefringent regions of the specimen.

Oldenbourg's LC PolScope

Oldenbourg's LC PolScope (which can be adapted to most research-grade microscopes outfitted for polarization microscopy) provides striking, high-contrast images of weakly birefringent objects. Furthermore, the system has the unique attribute of providing maps of birefringence distribution that is independent of orientation of the specimen axis within the focal plane.

The new system uses circularly, instead of linearly polarized light and two electronically controlled liquid crystal compensators instead of a mechanically driven conventional compensator (Oldenbourg and Mei, 1995; Oldenbourg, 1996). Four video images are acquired into computer memory in rapid succession. In acquiring each of these four images, the computer provides appropriate voltages that adjust the retardances of the two compensators. The LC PolScope uses no mechanically moving compensators and records the images in perfect register. From the set of four images, the retardance map is immediately calculated by the computer and displayed on the monitor.

The calculated image generated with the LC PolScope is, in fact, a map of birefringence distribution whose intensity is directly proportional to the retardance of each specimen point. In contrast, with images obtained by standard polarization microscopy, the intensity of each image point is proportional to the square of the retardance of the specimen point. The LC PolScope can also display a map of slow-axis orientations for pixels for each image point.

The LC PolScope has been used to display the dynamic distribution of microtubules in newt lung epithelial cells undergoing mitosis (see on-line video accompanying Inoué and Oldenbourg, 1998) and the highly dynamic behavior of actin filaments and network in cultured growth cones of *Aplysia* neurons (Katoh et al., 1999) with unprecedented clarity.

The LC PolScope system is available from CRI (Cambridge Research and Instrumentation).

Nina Allen's Compensator

In another recent development, Holzwarth et al. (1997, 2000) also use a voltage-driven liquid-crystal compensator for DIC observations with video microscopy. They use crossed linear polars rather than circular polarizers as in Oldenbourg's system. Appropriate voltages, applied by a computer to the compensator, generate an additive and a subtractive shift to the phase difference introduced by the Wollaston or Nomarsky prism. Each of these images is stored in computer memory and subtracted from each other. The difference image displays the DIC effect alone, while subtracting away stationary background noise and image features that did not originate from interference between the two sheared rays.

Although designed for use in DIC and still retaining the orientation-dependent contrast characteristic of images observed between crossed linear polarizers, Allen's new compensator applied to polarization microscopy should also effectively bring forth contrast in weakly birefringent regions of the specimen. Conversely, the modified system could selectively suppress contrast due to birefringence by adding (rather than subtracting) the two images stored in computer memory and selectively bring forth image features not based on optical anisotropy.

Centrifuge Polarizing Microscope

Living cells centrifuged on an isopycnic gradient at several thousand $\times\, g$ (earth's gravitational field) tend to stratify with their lighter components packed at the centripetal pole and heavier inclusions packed at the centrifugal pole of the cell (Harvey, 1940; Zalokar, 1960). In the process, cytoskeletal elements, membranes, and organelles tend to become aligned or deformed. To study the fine structure and function of stratified cell components and to learn about the mechanical linkages of the organelles to the membranes and cytoskeletal elements, the author of this unit has developed a centrifuge polarizing microscope (CPM; Inoué et al., 1998, 2001a).

The CPM allows video-rate polarization microscope observations of the spinning specimen, with image resolution of 1 μm, in living cells suspended in an isopycnic solution or crawling on a glass substrate. Birefringence retardations can be detected and measured down to 1 nm in cells that are exposed to up to 10,500 $\times\, g$.

With the CPM, the author has found striking changes in the conformation of endoplasmic reticulum immediately following fertilization of sea urchin eggs, has followed the process of Ca^{2+}-induced nuclear-envelope breakdown and meiotic spindle formation in a marine annelid egg, measured the climbing forces generated by myosin mutants of slime mold amoebae, and observed the enigmatic formation and rapid precipitation of dense "comets" in a concentrated solution of a thread-shaped virus (Inoué et al., 2001b).

APPLICATIONS OF POLARIZATION MICROSCOPY

Example 1: Examining Biocrystals

Hard tissue generated in living organisms often contains, or is made up of, biocrystals. Biocrystals are indistinguishable (in terms of their atomic lattice and, hence, intrinsic optical properties) from inorganic crystals, yet their morphology is primarily determined by their biological function and seldom reflects the habits (e.g., cleavage faces) of the natural crystal.

A common form of biocrystal, such as the skeletal spicules found in embryonic sea urchins (or the plates of their adult forms), is based on calcite, a highly birefringent, negative uniaxial material that is the main ingredient of chalk.

The birefringence, growth, and morphogenesis of calcareous biocrystals, such as the skeletal spicules generated by the mesenchyme cells in the gastrula- and pluteus-stage sea urchin larvae, can be studied with a simple polarizing microscope that can be easily assembled.

All one needs is a polarizer, inserted in the illumination path before the condenser of the microscope, and an analyzer (crossed relative to the polarizer) placed after a low (10× or 20×) power objective lens. In fact, both of these "polars" may already be present on the microscope. If the microscope is equipped for DIC, simply slide out the Nomarsky prism and set the condenser turret to the bright-field position. Close down the condenser iris so that the condenser NA becomes just below half that of the objective NA. Make sure the polars are carefully crossed so that, with the lamp brightness turned up, the blank field becomes completely (or very) dark in Köhler illumination (see *UNIT 4.1*).

When live gastrula or pluteus larvae are observed between crossed polars, the pair of triradiate or skeletal spicules should alternately twinkle like stars as the ciliated embryo swims about (see Fig. 4.9.1). The thicker parts of the skeletal spicule may show brilliant interference colors when observed in white light (see Fig. 4.9.2).

One can even make these observations using a dissecting microscope of the type that allows transillumination of the specimen. Simply attach a polarizer below the specimen-support glass plate and an analyzer below the objective lenses. The polars can be sheet Polaroids, perhaps those cut to appropriate shape from plastic Polaroid sunglasses. With the analyzer transmission axis oriented horizontally (i.e., turned 90° with respect to the way Polaroid sunglasses would normally be oriented), the polarizer mounted below the glass specimen support plate can be oriented vertically (north-south) by rotating the support plate to achieve extinction. Note that it is important to observe the specimen in a glass container, not in a plastic Petri dish, since most plastic lab ware exhibits very strong birefringence that can overwhelm the specimen birefringence.

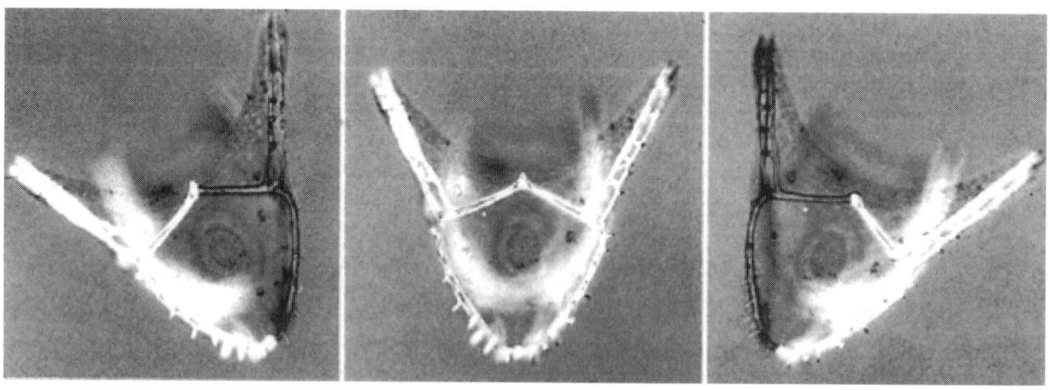

Figure 4.9.1 Pair of skeletal spicules in sea urchin pluteus larva. As the ciliated pluteus swims about in the field of the polarizing microscope, the skeletal spicules alternately shine and become extinguished. Despite its complex shape, the whole of each spicule is extinguished at once when its (calcite) crystal axes come to lie parallel to the axes of the crossed polars. Such extinction pattern is characteristic of biocrystals (from Okazaki et al., 1980).

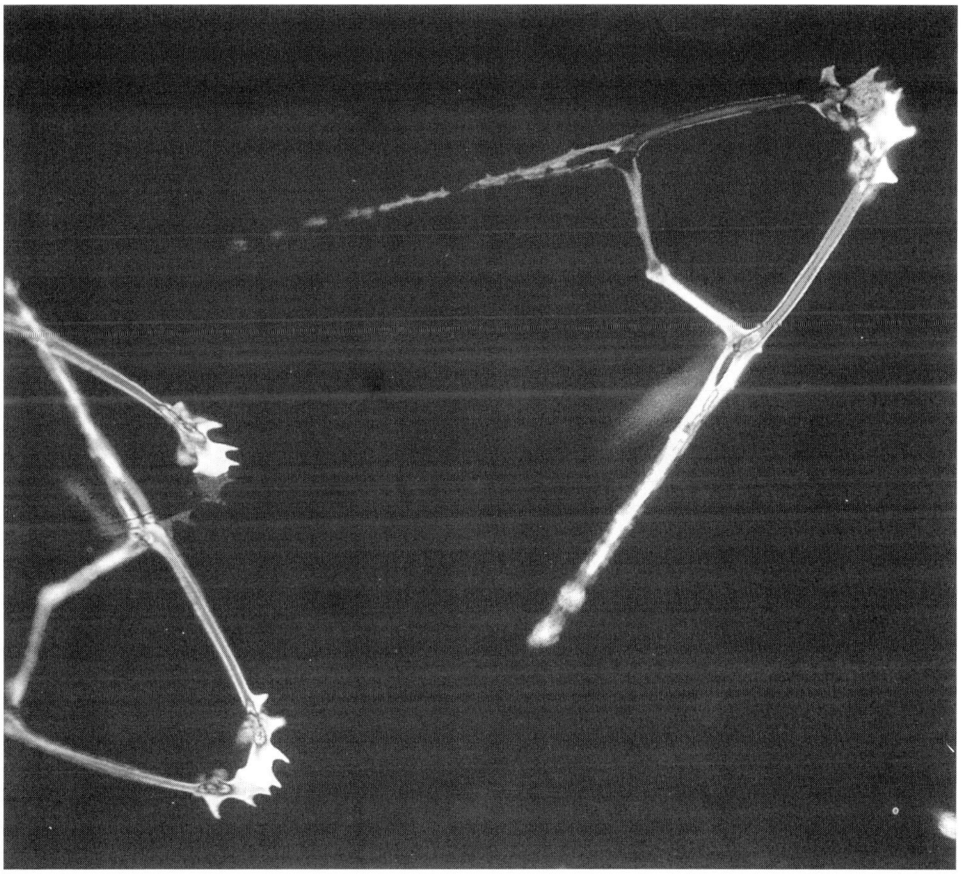

Figure 4.9.2 Skeletal spicules isolated from sea urchin pluteus larvae observed between crossed polars in white light. The calcite spicule is so strongly birefringent (coefficient of birefringence = –0.172) that a region just over 3-μm thick already introduces a 580-nm retardation. In that region, yellow light exiting the spicule is extinguished by the analyzer so the spicule takes on a bluish hue. Figure courtesy of Jan Hinsch, Leica.

While appearing to be a rudimentary exercise, consideration of the polarization optics and close observations of the image of the specimen can, in fact, reveal the directions and homogeneity of optical axes of the biocrystals, yield their sign and coefficient of birefringence, relate the growth of the biocrystal to the activity of the mesenchyme cells, and produce other useful information (Okazaki and Inoué, 1976).

In addition to the calcareous skeletal spicules that consist of a single biocrystal in echinoderm embryos, most bone and teeth contain an array of biocrystals (Schmidt, 1924).

Example 2: Examining Collagen, Muscle, and Histological Preparations

Many transparent organisms and tissue sections exhibit birefringent regions that can be seen with simple polarizing microscopes. The wavy collagen fibers and cross-striated muscle in animal tissues (see Fig. 4.9.3), and cellulose fibrils in the walls of plant cells (e.g., see Green, 1963), stand out prominently sometimes amidst bright crystalline inclusions. Likewise, the keratin layers in hair (which can vary with animal species) and synthetic fibers show strong birefringence.

Some of the crystalline inclusions seen in the background were formed in life, while other randomly scattered crystals could have developed in the medium in which a fixed specimen is embedded. Others, which at first glance may appear to be simple crystals (with simple optical axes as in biocrystals), upon closer inspection may turn out to have complex (more or less radially symmetric) extinction patterns. These may be starch grains with concentric lamellar structures that show characteristic extinction

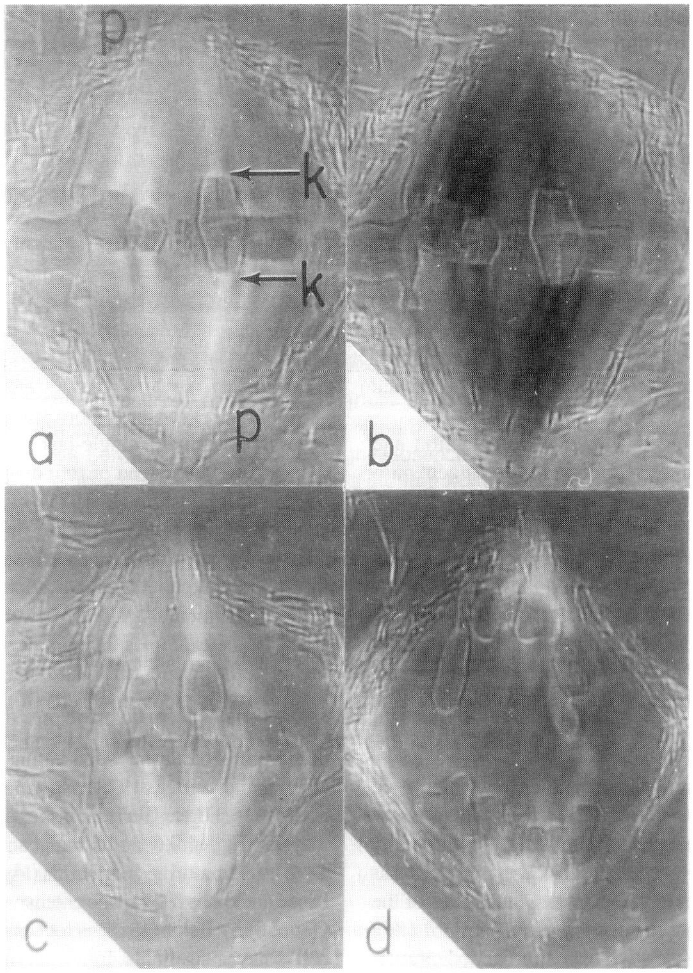

Figure 4.9.6 Meiosis-1 spindle of grasshopper spermatocyte observed with rectified optics. The positive birefringence of the bundle of microtubules that make up the chromosomal spindle fibers shows prominently. The kinetochores (k) also appear to show the same sign of birefringence as the microtubules, but it has not yet been established whether this reflects fine-structural anisotropy of the kinetochore, or primarily the edge birefringence of the chromosome at the kinetochore. The mitochondria also show a longitudinally positive birefringence with overlapping edge birefringence. The compensator slow axis lies parallel to the spindle fibers in panels (**A**), (**C**), and (**D**), so that the positively birefringent microtubule bundles appear brighter than the background. (**C**) The compensator orientation is reversed so that the microtubule bundles appear in reverse contrast. (From Nicklas, 1971.) For online version of time-lapse series of this and other cells in mitosis, see *http://www.molbiolcell.org* and Inoué and Oldenbourg (1998).

Sharp, stainless-steel dissecting needles;
A sharp hypodermic needle mounted on a wooden handle;
Pasteur pipets with rubber bulbs;
Small balls of absorbent cotton;
6- to 8-mm-wide strips of filter paper;
Mineral oil or other innocuous oils such as Halocarbon oil 27;
Valap (see *UNIT 13.1*);
70% ethanol;
Dissecting microscope with trans-illumination.

Preparation instructions for several of these items, along with some parameters critical to performing the experiments described here, are included in the discussion of Ensuring that Cells are "Happy" While They Are Being Observed Under the Microscope, below.

Let the male grasshopper hold onto a small wad of absorbent cotton or a small piece of tissue, so as to absorb the brown fluid expelled

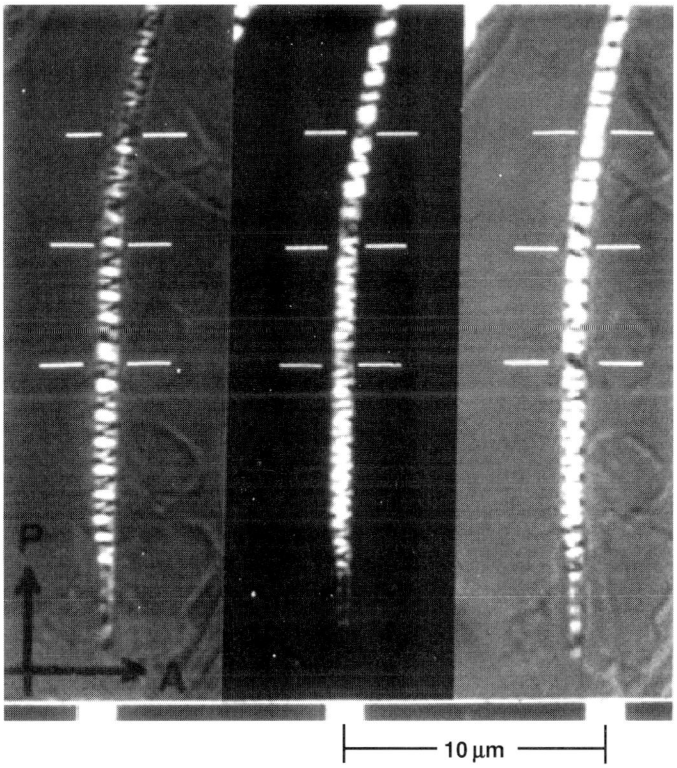

Figure 4.9.7 Live sperm head of cave cricket viewed with a high-resolution polarizing microscope. The sperm head immersed in DMSO (dimethyl sulfoxide) is viewed at three compensator settings using rectified optics (×97/1.25 NA). P, A: transmission axes of the polarizer and analyzer. The compensator slow axis is oriented perpendicular to the sperm axis in the middle panel, turned ~5° clockwise in the left panel and ~4° counterclockwise in the right panel. The helical regions of the DNA, wound in a coil of coil within the chromosomes, appear bright or dark depending on their slow axis orientations relative to that of the compensator. Bars indicate junctions of chromosomes that are packed in tandem in the needle-shaped sperm head. This is the first (and virtually only) mode of microscopy by which the packing arrangement of DNA and the chromosomes have been clearly imaged in live sperm of any species. From Inoué and Sato (1966).

from its mouth. Lift and clip off the wings at their bases. Then, holding onto the grasshopper with his hind legs folded under the abdomen, insert one blade of the iridectomy scissors pointing up and forward at around the third or fourth from the last abdominal segment as shown in Figure 4.9.8. Make an incision that covers four or five segments along the back of the abdomen. Then gently squeeze the lower part of the abdomen until the pair of testes (Fig. 4.9.9; usually colored bright yellow, orange, or red) pop out from the incision. Cut the ducts and place the testes in the saline (grasshopper Ringer's) solution in a watch glass.

Under the dissecting microscope, use a watchmaker's forceps and sharp needle to peel off the colored layer of fat and the Malpigian tubules that surround the testes. They can be toxic to the isolated spermatocytes if not removed from the final preparation. Freed from the covering material, the whitish lobes of the testes should now look like a bundle of bananas. With a piece of absorbent cotton moistened with 70% ethanol, wipe clean the iridectomy scissors, forceps, and needle used, and then transfer the lobe bundles of the testes to a clean Syracuse watch glass containing fresh saline solution.

Again, under the dissecting microscope, use sharp dissecting needles to separate a testis into small bundles containing three to four intact lobes. Cut apart the bundles where they bunch together, i.e., at the proximal end of the lobes where they share a common sperm duct.

Using the Pasteur pipet, transfer one small bundle of lobes together with a small drop of saline onto a clean slide. Place a clean coverslip on top and remove excess solution with a small piece of filter paper or tissue paper until the lobes are just starting to be flattened.

Observe the preparation under the polarizing microscope using a low-power (10×) objective lens. In each intact lobe, there should be located, from the duct end towards the distal end, more or less in the following order, bundles of mature sperm heads, immature sperm heads, and sperm tails, then follicles containing spermatids, secondary spermatocytes, primary spermatocytes, and, finally, spermatogonial cells (Fig. 4.9.10). The birefringence of the heads of the mature and immature sperm, as well as of some of the spermatid, and perhaps of the bundle of sperm tail, should be quite prominent.

By using a Brace-Köhler compensator, establish the sign of birefringence (direction of

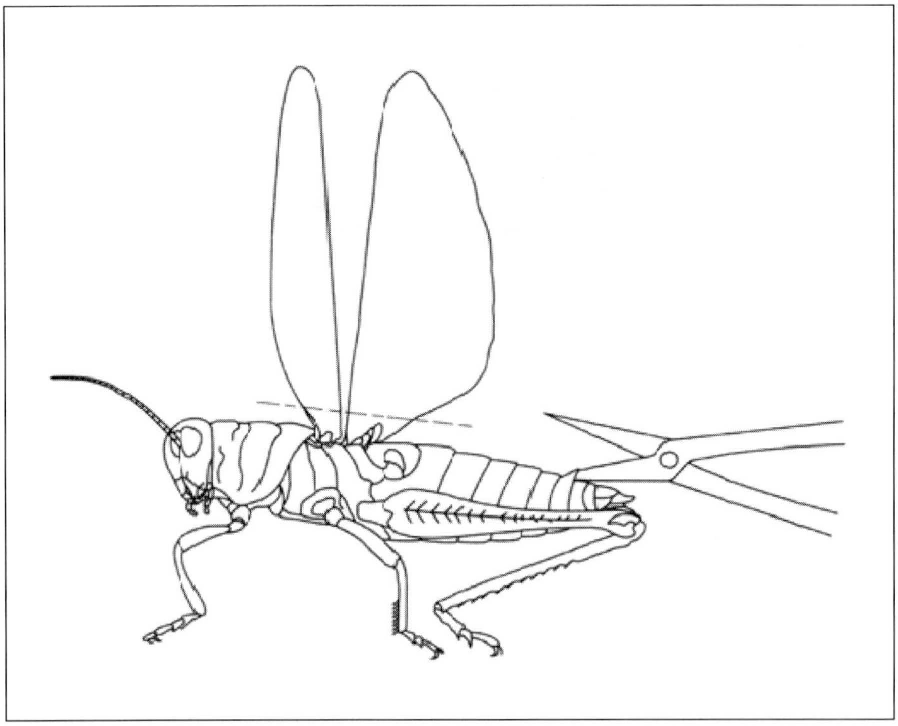

Figure 4.9.8 Dissecting a male grasshopper to expose testes (also see Fig. 4.9.9 and instructions in text).

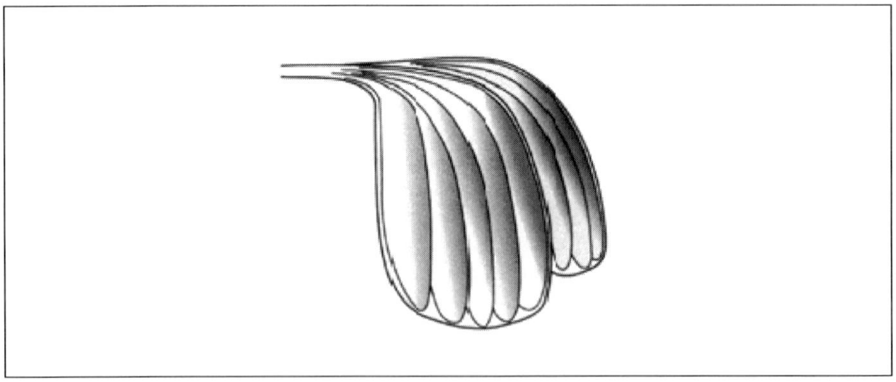

Figure 4.9.9 The pair of testes extracted from a grasshopper. The bundle of lobes in each testis, made like a bunch of bananas and held together where they share a common sperm duct, are covered by a brightly colored sheath. The distal ends of the lobes splay apart when the testicular sheath is peeled off.

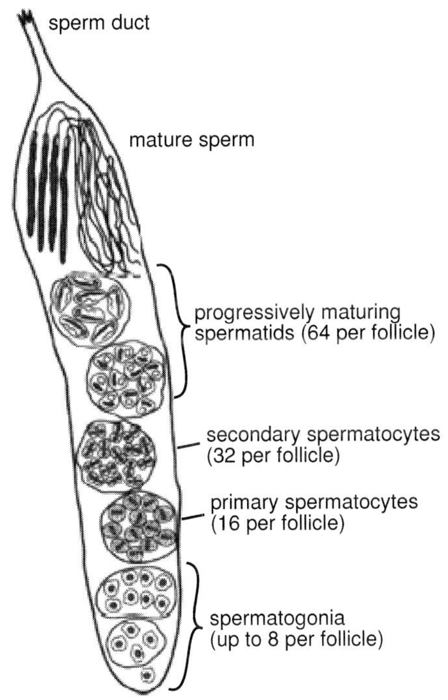

Figure 4.9.10 Arrangement of follicles and cells (not to scale) in the lobe of a grasshopper testis. This highly schematic diagram illustrates one of the many lobes, which are arranged like a bunch of bananas, in a grasshopper testis. The cells, developing synchronously within each follicle in the lobe, are progressively more mature from the distal end of the lobe to the proximal end where the lobes are joined by the sperm ducts. For a photograph of section of fixed lobes, see Figure 1 in Belar (1929).

the slow axes of these structures relative to their long axes). Note that the contrast due to sperm tail birefringence, which is mostly due to the form birefringence of their axonemal microtubules, is reversed from that of the sperm head, which is almost exclusively due to the intrinsic birefringence of the DNA molecules aligned parallel to the length of the sperm head.

Next, make a fresh preparation for observing the spindle fibers and chromosomes and individual sperm with greater resolution. This exercise may require some practice, since a combination of care and speed of preparation is important. Also, do not be frustrated if, at first try, primary spermatocytes at just the right stage of meiotic mitosis are not found. There are a number of follicles at different stages of spermiogenesis in each testicular lobe, and the divisions of the 16 primary spermatocytes in each follicle are synchronized, but you may not come across a lobe that contains a follicle at the right stage. Usually, it is not necessary to try more than two or three lobes.

Using a pair of watchmaker's forceps, transfer a bundle of three to four testicular lobes onto a bio-cleaned microscope coverslip. The coverslip should be resting on a clean microscope slide or some other appropriate holder that lifts it off from the lab bench or other surfaces, which may be a source of lint or other birefringent dust. In making the transfer, be sure to hold onto the duct ends of the lobes and not the other, free end. During the transfer, touch the lobes gently to the clean, dry rim of the watch glass to remove much of the saline solution.

Using a hypodermic needle with a very sharp tip as a knife, cut the lobes (now resting on the coverslip) near their free ends. Then, holding onto the duct end of the lobes with a watchmaker's forceps, draw the lobes in a spiral a few millimeters in radius, making sure that their free, cut ends trail on the surface of the coverslip. As you draw the spiral, cells should flow out together with the testicular fluid, from the cut ends of the lobes onto the surface of the coverslip. Those near the cut end flow out first, then those closer to the ducts that you are holding onto. Pick up and discard the residue of the lobes. The result is a spiral-shaped thin film of cells, more or less arranged in order from least differentiated cells to mature sperm (see Fig. 4.9.10).

Quickly place a drop of oil (Halocarbon 27 or light mineral oil) on the spiral, invert the coverslip, and cover the preparation by contacting onto a clean slide. Do not use any spacers. Remove the excess oil by contacting a strip of filter paper to the rim of the coverslip, then seal the preparation with Valap.

The slide is now ready for observation of the spindle and chromosomes in meiotic divisions and different stages of sperm maturation. The observations can be carried out on the same live cells for many hours. Not only are the cells arranged and clustered, more or less in the order of stages of spermiogenesis, but they should also lie in reasonably flat monolayers making them suitable for observations with high-NA lenses using polarization microscopy, phase contrast, or DIC.

Note, however, that the testicular lobes, from the time they are removed from the saline medium, and especially the very thin spread of cells in the spiral, could almost instantly lose water by evaporation before being covered by the Halocarbon oil. Thus, some practice and speed of operation becomes important for the last phases of the preparation described above.

In successful preparations observed in high-extinction polarization microscopy, one finds meiosis-I spindles (which are considerably larger than those in meiosis-II) with a moderate degree of positive birefringence, somewhat rounded spindle poles (in contrast to being very pointed), and chromosomal spindle fibers that appear somewhat fluffy (rather than too sharp as though drawn as a line). Such cells can be expected to complete meiosis-I as well as meiosis-II in several hours (see Inoué, 1964).

The process of sperm maturation, on the other hand, is very slow, so that one is not likely to see the actual transition of single spermatids from one stage to the next. The successive stages of sperm maturation, including the rise of negative birefringence in the spermatid nucleus associated with their dehydration and shape changes, should, however, be neatly displayed in chronological order on the slide, so that it should not be difficult to visualize the sequence of events.

Example 7: Examining Meiotic Spindles and Chromosomes

Note that in grasshopper spermatocytes, it can take several hours between the time when an apparently mature spindle is formed and the beginning of anaphase. During that interval, known as prometaphase, chromosomes oscillate back and forth to both sides of the metaphase plate (Deitz, 1969). Concurrently, the associated spindle fibers change the strength of their birefringence. With time-lapse cinematography or video, one can also observe considerable fluctuation of the spindle birefringence (the so-called "Northern Lights Phenomenon"; Inoué, 1964), reflecting the dynamic growth and shortening of their constituent microtubules (Mitchison and Kirschner, 1984; Walker et al., 1988).

The sex chromosomes, in particular, may be found to travel all the way form one spindle pole to another. The oscillatory behavior of these prometaphase chromosomes, and the associated stretching and contraction of their kinetochores to which the chromosomal spindle fibers are attached, may be observed more clearly by switching from polarization to DIC or phase-contrast microscopy (Salmon and Segal, 1980; Nicklas et al., 1995).

Once full metaphase is established, and all of the chromosomes have lined up on the metaphase plate, a signal is released and the chromosomes synchronously enter anaphase (Nicklas et al., 1995). Note the change in distribution and strength of birefringence of the spindle fibers and astral rays as anaphase progresses. What can one infer about the behavior of the microtubules from these observations?

Note also, as anaphase progresses, that another bundle of thread-shaped, birefringent elements lines up along the long axis and surrounding the spindle, to be pinched apart into two as the cell cleaves. Note the sign of birefringence of these mitochondrial threads.

In each daughter cell that results from the division of the spermatocytes, the mitochondria come together to form a dense, round body, the Nebenkern, which has a diameter similar to the reconstituted daughter nucleus and is often mistaken for the latter. The Nebenkern shows a higher refractive index than their surrounding cytoplasm, whereas the daughter nuclei themselves tend to have a lower refractive index than the surrounding cytoplasm. Using DIC optics, verify these refractive index differences by observing the shadow-cast appearances at the edges of these structures.

Example 8: Examining The Sperm and Maturing Spermatids

The needle-shaped, mature sperm heads in grasshoppers show a very strong birefringence whose character is reversed from that of the sperm tail and the acrosome at the tip of the head. Establish the sign of birefringence of these three regions.

Also note that, unlike the nuclei in spermatocytes (and many other cells), the refractive index of the sperm head is very high, reflecting the loss of water and condensation of its chromatin during spermatid maturation.

Choose a series of spermatids starting with those that are nearly mature to very young ones. As maturation progresses, the sperm nucleus becomes more and more elongated, while its birefringence increases. The coefficient of birefringence of the mature sperm is, in fact, extremely high, amounting to -2×10^{-2}, approximately that of a thread of pure DNA (Schmidt, 1941; Wilkins, 1951; Inoué and Sato, 1966).

Note that before the spermatid starts to lose water and become elongated, the nucleus is almost nonbirefringent. Most cell nuclei show very little birefringence (with the exception of those in sperm and in certain Dinoflagellates; Schmidt, 1932, 1937; Cachon et al., 1989), despite the fact that they contain a large quantity of DNA.

In maturing spermatids, note also that the axis of birefringence of the sperm head is tipped this way and that and is not totally parallel to the elongated axis of the head, as in the fully mature sperm. In some insects, such as the cave cricket, the axes of negative birefringence, even of the mature sperm head, can still be resolved and seen not to be uniform. Instead, they are disposed more or less in a zigzag fashion (Fig. 4.9.7), reflecting the packing arrangement of their DNA backbone and chromosomes. The detailed packing arrangement of the DNA bases, including at levels far smaller than the resolution limit of the polarizing microscope, was determined by irradiation of the live sperm head with a microbeam of polarized UV and by quantitatively measuring changes in the birefringence and their axes for each resolvable area (Inoué and Sato, 1966).

Ensuring that Cells Are "Happy" While They Are Being Observed Under the Microscope

To observe the behavior or fine structure in living cells under a microscope, there are several factors that demand special attention. This is especially true when observations are to be made for extended periods of time and under conditions that require brilliant illumination, as in polarization microscopy, to visualize weakly birefringent objects. On the other hand, birefringence can be an especially effective monitor for the physiological state, and "happiness," of the cell.

Biological material

Start with healthy organisms and cells. Clearly, one cannot expect to gain healthy cells from diseased or necrotic plants or animals or gain normal fertilization and development when the gametes are not fully ripe.

Culture medium

Be alert to toxic contaminants, especially those that can affect the cells at low concentrations. Heavy metals such as copper, zinc, and silver, which can be present in trace amounts even in conventional reagent-grade sodium chloride, can be sources of trouble since NaCl is used in high concentrations in artificial seawater and physiological saline. Likewise, for making culture media, distilled water made in glass or quartz stills (with adequate vapor, i.e., water droplet, suppression) is preferred over that from metal stills. Deionizers remove trace metals, but one should be careful that the water is not contaminated by mold or bacteria (which can grow in the columns or tubing) and their products including surfactants (which show up as long-lasting foam after the water is shaken or bubbled).

Slide and coverslips

Cells prepared between a slide and coverslip are exposed to a large area of the glass surface relative to the volume of the medium bathing the cell. In addition, the cells are likely to be contacting the glass surface directly. Thus, slight contaminants on the glass surface can disproportionately affect the cells being observed. Eliminate surfactants and heavy metals that may affect the well-being of the cells as well as birefringent contaminants that may interfere with observations by polarization microscopy (see below).

Oxygen supply

In addition to changes in pH of the medium, anoxia (and hyperoxia in the case of some obligatory anaerobes; Ritter et al., 1978) can interfere with proper health and division of cells grown in the confined space between the slide and coverslip. Tissue culture cells are commonly observed through special flow-through microscope chambers where the culture medium is appropriately oxygenated and its pH and temperature are monitored or controlled. Alternatively, developing marine embryos and dividing protists can be observed continuously for many hours using a micro-chamber containing an equilibrated gas phase formed between a slide and coverslip (see Inoué and Spring, 1997). Microscope preparation of cells covered with a layer of nontoxic oil (even under the coverslip) can survive for many hours since many oils, especially silicone and fluorocarbon oils, dissolve gases, including oxygen, very effectively. Halocarbon oils, such as Kel F-10, FC-47, and Voltalef that previously were used to cover cultured cell smears, e.g., of insect spermatocytes, are no longer available. Instead, the Halocarbon oils 27 or 200 (Halocarbon Products; see *SUPPLIERS APPENDIX*), used for immersing developing *Drosophila* embryos, would appear to be reasonable, currently available alternatives. The manufacturer recommends that fresh oils be bubbled with air before use for a day or two to replenish the oxygen that was evacuated during manufacturing.

Temperature

While slides containing cultured cells of mammalian origin generally require heating, most invertebrate cells may need to be kept cool, i.e., no warmer than moderate room temperature. For observing such cells, be especially careful with the illumination used for polarization microscopy. A source with high brightness is needed in order to see or record the weak birefringence in the cell. But most high-brightness sources (not just quartz halogen or other tungsten filament lamps, but also mercury and xenon arc lamps) produce a huge quantity of infrared. Be sure to use a good quality heat-cut filter, and, if practical, a high-transmission (>65%) narrow-band-pass (±15 nm) green interference filter. That is because the preparation can be heated up with the infrared and far red portions of the spectrum, and cells can react adversely to the blue end of the spectrum (Langford and Inoué, 1979).

In addition, use Köhler illumination and keep the illuminated area to the smallest practical size by adjusting the opening of the field diaphragm. Even using a laser beam, it is not easy to heat up a portion of the specimen with a diffraction-limited small spot of light because water carries away the heat very effectively. On the other hand, the light from an unfiltered incandescent or arc source can rapidly heat up the specimen under the microscope if the illumination covers a large area of the specimen.

The birefringence of the mitotic spindle (as well as its shape) can be a sensitive thermometer that reflects the temperature of the specimen, which is otherwise quite difficult to monitor closely. At lower temperatures, the spindle birefringence drops, while at higher (physiologically compatible) temperatures the birefringence rises (showing enhanced polymerization of tubulin) following a remarkably reproducible curve (Inoué et al., 1975; Nicklas, 1979).

Drying

With cells mounted between slide and coverslip in an aqueous medium be especially alert to drying and increased tonicity of the medium. Valap (see *UNIT 13.1*) is an effective material that can be applied to seal cells mounted in aqueous solutions between the slide and coverslip. The distribution of birefringence in the spindle can indicate the tonicity of the medium. In hypertonic media the chromosomal spindle fibers tend to appear as sharp pencil-drawn lines that converge sharply to the spindle poles (a sign of poor health, although partially reversible), while in less concentrated media, they appear fluffier.

Bio-clean slides and coverslips

In preparing cells for long-term observations under a microscope, one needs to be especially careful to prepare media with the appropriate pH and degree of oxygenation, free of heavy metals. It is especially important to use glassware that is "biologically clean." Since the cells being observed under a microscope are bathed in a very small volume of medium and their surfaces contact a disproportionately large area of the glass surface (and the concentration of any contaminant can become exceptionally high), the slide and coverslips in particular must be bio-clean.

Bio-clean glassware should be free from soap and detergent in addition to even minute amounts of heavy metal, fixatives, and other toxic contaminants or residues that may have been left or adsorbed onto the glass surface. Glassware (including slides and coverslips) and plasticware that have been acquired new from suppliers (except specially packaged pipet tips and capped centrifuge tubes, etc.) should not be assumed to be bio-clean. Table 4.9.1 shows the sequence of procedures used in the author's laboratory for preparing bio-clean glassware and slides and coverslips (Lutz and Inoué, 1986).

Glassware once used with fixatives should be separated by marking with clear, indelible F

Table 4.9.1 Procedures for Bio-Cleaning Glassware (after Lutz and Inoué, 1986)

Step	Procedure
1	Fill 500-ml beaker with hot tap water and enough detergent to make the solution sudsy.
2	Drop 30 to 50 slides or coverslips *individually* into the beaker. Try to position the slides in a criss-cross arrangement in the beaker.
3	Soak for 15 to 30 min.
4	Place beaker in ultrasonic washer for 3 to 5 min.
5	Rinse many times in tap water to remove the majority of the detergent.
6	Using stainless-steel forceps, transfer slides or coverslips individually into another beaker containing tap-distilled water. Repeat this individual transfer/rinse ten times. Sonicate beaker 3 to 5 min before the last transfer.
7	Individually transfer slides or coverslips into a beaker containing glass-distilled water. Repeat this ten times. Sonicate beaker for 3 to 5 min before the last transfer.
8	Individually transfer slides or coverslips into bioclean storage jars filled with 80% ethanol for storage.
9[a]	Before use, remove slide or coverslip from alcohol using clean forceps and wipe dry with a Kimwipe folded twice (four thicknesses). Do not rub.

[a]If possible, use a centrifuge ("Spin Dry" available from Technical Video) to avoid the possibility of lint associated with Step 9.

marks, and not mixed with bio-clean glassware, since it may be very difficult to get rid of some ingredients of fixatives.

In addition to preparing bio-clean slides and coverslips, keep their surfaces, as well as optical surfaces of the microscope that lie between the polarizer and analyzer, free from birefringent dirt and dust particles. In particular, be careful to prevent pieces of lint (including cellulose fibers, which often have birefringence retardations close to a wavelength or more) from falling on to, or becoming attached to, the surfaces of slides and coverslips. Such fibers can drastically lower the EF of the whole polarization optics and completely interfere with observation of weak birefringence of intracellular structures. Since clean glass surfaces tend to attract lint particles electrostatically, avoid laying your slide or coverslip directly on a bench top, or on the surface of the glass plate that supports the specimen on a dissecting microscope. Instead, place the slide and coverslip on a spacer, such as a fresh applicator stick bent into a V shape, or prop it up by supporting one end by a couple of millimeters off the surface. Note that Kleenex and lens tissue used to wipe the slide or coverslip can also be a source of cellulose fibers. To avoid wiping, use a centrifuge to spin off the alcohol used for the final rinse of the slide and coverslip (see Figures 3 to 11 in Inoué and Spring, 1997).

Grasshopper Ringer's solution

Table 4.9.2 provides the composition of saline solutions that can be used for preparing the grasshopper testes and for observing meiosis in their spermatocytes for many hours (Nicklas et al., 1979). Note that the final tonicity of the medium (plus testicular fluid) may vary somewhat, depending in part on the atmospheric humidity at the time the contents of the testicular lobe are streaked onto the coverslip. Thus, Belar advises either breathing onto the preparation with one's mouth wide open or blowing with lips nearly closed, depending on whether the preparation should be made more or less dilute before sealing the preparation. As noted before, the morphology and birefringence distribution in the spindles are, in fact, sensitive indicators of the cell's health and whether the preparation is or is not appropriate for long-term observation of the division events.

Valap

Valap is a low-melting-point, nontoxic, waxy material used for sealing coverslips. Valap is made by simmering a 1:1:1 mixture of vaseline, lanolin, and paraffin (using indirect heat to prevent overheating or catching on fire) for several hours (also see *UNIT 13.1*). Use of beeswax instead of paraffin, or Valab, yields a somewhat more adhesive sealant. A major advantage of these sealants is that they can be applied directly to seal the edges of coverslips covering wet specimens.

Valap and Valab can be applied either from a melt using a small brush or with a warm spatula or heating wire made by passing current (e.g., from a low-voltage-microscope lamp transformer) through an ~5-inch length of U-shaped nichrome wire fastened to an insulated holder such as an electric plug fixture (Fig. 4.9.11). Be careful not to pass too much current through the heating wire (if it smokes, it is too hot) since the sealant can spit, and small droplets of this highly birefringent aggregate that land on the coverslip can ruin the extinction of the polarization optical system.

After applying the sealant, examine all four sides of the coverslip with a dissecting microscope or magnifying glass to make sure that there are no imperfect seals or pinholes, since water evaporates rapidly from even a very small hole in the seal.

Effect of fixation

Polarization microscopy, carried out on living cells that continue to divide, develop, or otherwise undergo normal physiological or developmental activities, provides a good

Table 4.9.2 Composition of Insect Ringer Solution[a]

Stock solution[b]
500 mM PIPES
27 mM KCl
67.5 mM $CaCl_2$
2.5 mM $MgCl_2$
Adjust pH to 6.8 to 6.9 with 10 N NaOH[c]

[a]This is the solution currently recommended by Nicklas et al. (1979) and supercedes the earlier formula of Niklas (Nicklas and Staehly, 1967) as well as those of Belar (1929, footnotes on pp. 364 and 387). In his 1929 artcile, Belar describes and illustrates with photographs of live cells the appearance of pseudopodia, chromosomes, and mitochondria of spermatocytes exposed to media with different degrees of tonicity. He also warns that the preparation be kept from being heated by (the infrared rays from) the light source. See Inoué, 1952b, for effect of temperature on the birefringence of spindle fibers (polymerization of spindle microtubules), which serves as a sensitive intracellular thermometer.

[b]Prepare fresh in glass-distilled water each month and store in refrigerator.

[c]On day of use, dilute 5 ml of stock solution to between 27 and 38 ml (start with 32 ml) with glass-distilled water. The final optimum concentration depends on the extent of evaporation during culture preparation and the physiological state of the animals; the required dilution usually does not change over a period of a week or two.

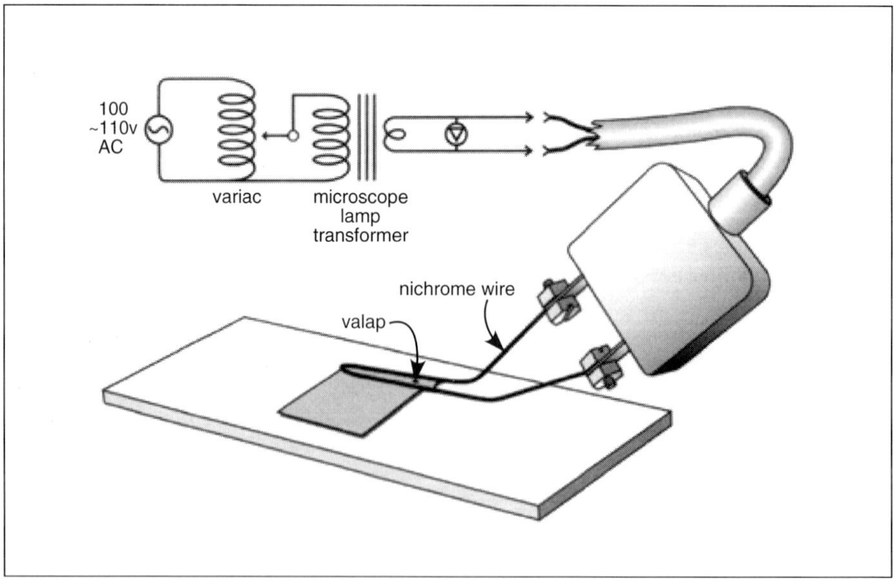

Figure 4.9.11 Electric heater for sealing coverslips with Valap. A 5-in.-long piece of ~20-G nichrome wire is bent as shown and attached, via a convenient handle (e.g., a two-pronged AC plug), to a variable voltage transformer (e.g., one used to power a low voltage microscope lamp). The melted Valap, that has formed a bridge by capillarity between the arms of the nichrome wire loop, is transferred to and painted along the edges of the cover slip.

baseline for understanding the submicroscopic organization and its changes in these cells. That baseline is an important yardstick to heed in considering whether the fine structure of the cell has remained intact when one fixes a cell, or tissue, in preparation for electron microscopy, cytochemistry, or other technique.

When the same cell loses, gains, or changes its birefringence appreciably upon fixation, one needs to wonder whether the fixation itself may not have abolished, created, or altered the fine structure of the cell component. Note that the refractive index of the medium in which the fine structure is immersed also affects the form birefringence of the structure.

For example, prior to introduction of glutaraldehyde as a fixative, spindle and other labile microtubules tended to disappear with standard osmium fixation (Inoué, 1993). Even with glutaraldehyde fixation, there is the possibility that not all of the labile microtubules are preserved (Sato et al., 1975). Given the preparative techniques used for electron microscopy today, we may still be unaware of the loss in fine structural organization encountered by some of the cell's labile filament or membrane systems (e.g., see Burgos et al., 2000).

Acknowledgments

The author is grateful to: Dr. Bruce Nicklas of Duke University for providing updated information on grasshopper Ringer's solutions and on Halocarbon oils, Jan Hinsch of Leica for providing the original for Fig. 4.9.2 and for help with locating references, Drs. Rudolf Oldenbourg and Michael Shribak of the Marine Biological Laboratory for comments on a draft version of the article, Drs. Edward D. Salmon, Nina Allen, and Sid Shaw for help in locating references, the Woods Hole Oceanographic Institution Graphics Services Group for expert help in preparing the figures, and Jane MacNeil of the Marine Biological Laboratory for her tireless help in preparing this article.

LITERATURE CITED

Allen, R.D., Travis, J.L., Allen, N.S., and Yilmaz, H. 1981. Video-enhanced contrast polarization (AVEC-POL) microscopy: A new method applied to the detection of birefringence in the motile reticulopodial network of *Allogromia laticollaris*. *Cell Motil.* 1:275-289.

Ambronn, H. and Frey, A. 1926. Das Polarisationsmikroskop, seine Anwedung in der Kolloidforschung in der Färbarei. Akademische Verlag, Leipzig, Germany.

Belar, K. 1928. Die Technik der Deskriptiven Cytologie. *In* Methodik der Wissenschaftlichen Biologie, Vol. 1 (T. Péterfi, ed.), pp. 638-735. Julius Springer, Berlin.

Belar, K. 1929. Beiträge zur Kausalanalyse der Mitose. II. Untersuchungen an den Spermatocyten von *Chorthippus (Stenobothrus) lineatus* Panz. *Arch. Entwicklungsmech. Organismen* 118:359-484, plus 8 plates.

Bennett, H.S. 1950. The microscopical investigation of biological materials with polarized light. *In* Handbook of Microscopical Technique (C.E. McClung, ed.) pp. 591-677. Harper & Row (Hoeber), New York.

Breton, J., Michel-Villaz, M., and Paillotin, G. 1973. Orientation and structural proteins in the photosynthetic membrane of spinach chloroplasts: A linear dichroism study. *Biochim. Biophys. Acta.* 314:42-56.

Burgos, M.H., Goda, M., and Inoué, S. 2000. Fertilization-induced changes in the fine structure of stratified *Arbacia* eggs. II. Observations with electron microscopy. *Biol. Bull.* 199:213-214.

Cachon, J., Sato, H., Cachon, M., and Sato, Y. 1989. Analysis by polarizing microscopy of chromosomal structure among dinoflagellates and its phylogenetic involvement. *Biol. Cell* 65:51-60.

Costello, D.P., Davidson, M.E., Eggers, A., Fox, M.H., and Henley, C. 1957. Methods for Obtaining and Handling Marine Eggs and Embryos. Lancaster Press, Lancaster, Pa.

Deitz, R. 1969. Bau und Funktion des Spindelapparats. *Naturwissenschaften* 56:237-248.

Forer, A. 1965. Local reduction of spindle fiber birefringence in living *Nephrotoma suturalis* (Loew) spermatocytes induced by ultraviolet microbeam irradiation. *J. Cell Biol.* 25:95-117.

Frey-Wyssling, A. 1953. Fine structure of protoplasmic derivatives. *In* Submicroscopic Morphology of Protoplasm, Chapter 3, pp. 279-370. Elsevier/North-Holland, Amsterdam.

Green, P.B. 1963. On mechanisms of elongation. *In* Cytodifferential and Macromolecular Synthesis (M. Locke, ed.) pp. 203-234. Academic Press, New York.

Hárosi, F.I. 1981. Microspectrophotometry and optical phenomena: Birefringence, dichroism, and anomalous dispersion. *In* Springer Series in Optical Sciences, Vol. 23: Vertebrate Photoreceptor Optics (J.M. Enoch and F.L. Tobey, Jr., eds.) pp. 337-399. Springer-Verlag, Berlin.

Hárosi, F.I. and MacNichol, Jr., E.F. 1974. Dichroic microspectrophotometer: A computer-assisted, rapid, wavelength-scanning photometer for measuring linear dichroism in single cells. *J. Opt. Soc. Am.* 64:903-913.

Hartshorne, N.H. and Stuart, A. 1960. Crystals and the Polarising Microscope: A Handbook for Chemists and Others, 3rd Ed. Arnold, London.

Harvey, E.B. 1940. A comparison of the development of nucleate and non-nucleate eggs of *Arbacia punctulata*. *Biol. Bull.* 79:166-187.

Reprinted from *Methods in Cell Biology*, Vol. 70, pp. 87–127, 2002,
with permission from Elsevier.

CHAPTER 2

Direct-View High-Speed Confocal Scanner: The CSU-10

Shinya Inoué* and Ted Inoué

Marine Biological Laboratory
Woods Hole, Massachusetts

Universal Imaging Corporation
West Chester, Pennsylvania

*This author is a consultant to Yokogawa Electric Corporation.

0091-679X/02 $35.00

I. Introduction

Over the past decade, confocal microscopes have dramatically improved our ability to examine the structural and functional detail of biological tissues and cells. The same characteristics that made these advances possible, namely the ability of confocal microscopes to provide exceptionally clean serial optical sections that are free of out-of-focus flare, have also made it possible to directly generate tilted sections, as well as to generate striking three-dimensional (3-D) images.

Alternative methods, such as the application of computationally intensive deconvolution algorithms to serial sections (obtained with nonconfocal, or wide-field, fluorescence microscopes) can also yield clean optical sections and reconstructed 3-D images (e.g., Agard, 1984; Holmes *et al.,* 1995).

More recently, two- (and multi-) photon optics have been used to further improve optical sectioning, especially for deep, living tissues. With only the fluorophores lying in the focal plane being exposed to the coincident action of the double- (multiple-) wavelength excitation wave, fluorescence is strictly limited to that plane. Meanwhile, the longer wavelength excitation wave dramatically reduces light scattering in general and photobleaching of the fluorophore that lie outside of the focal plane.

Although many of these confocal and deconvolution approaches yield exquisite optical sections and their composites, it may take from many seconds to considerably longer for each two-dimensional image covering a reasonable area to become available.

The CSU-10 described in this chapter is a disk-scanning, direct-view confocal scanning unit.[1] As detailed in the next section, the CSU-10 incorporates, in addition to the main Nipkow disk, a second disk with some 20,000 microlenses that are each aligned with a corresponding pinhole on the main Nipkow disk, thus substantially improving the transmission of the confocal illuminating beam. Thus, one can view confocal fluorescence images in real time, i.e., at video rate or faster, through the eyepiece or captured through a video camera.

The compact unit, attached to an upright or inverted microscope, transforms a research microscope into an exceptionally easy-to-use and effective, direct-view epifluorescence confocal microscope (Fig. 1). Thus, optical sections of fluorescent specimens show with high resolution, and in true color, directly through the ocular as one focuses through the specimen. Sparsely distributed, weakly fluorescent objects are readily brought to view.

In addition to viewing the image through the ocular in real time, the confocal image generated by the CSU-10 can be directly captured by a photographic camera or a video or CCD camera attached to the C-mount on top of the unit.

Since images of the microscope field are scanned at 360 frames/s by the multiple arrays of pinholes on its Nipkow disk, the CSU-10 not only provides very clean, full-frame video-rate images but, by use of high-speed intensified cameras, can even capture frames in as short an interval as 3 ms or less.

In this paper prepared in 1999, we describe the basic design of the CSU-10; some sample applications including video-rate and higher speed confocal imaging; the low

[1]See Chapter 1 for an overview of Nipkow-disk-type confocal microscopes and other video-rate, direct-view confocal systems. See also Amos and White (1995), Juškaitis *et al.* (1996), Lanni and Wilson (1999), and Yuste *et al.* (1999).

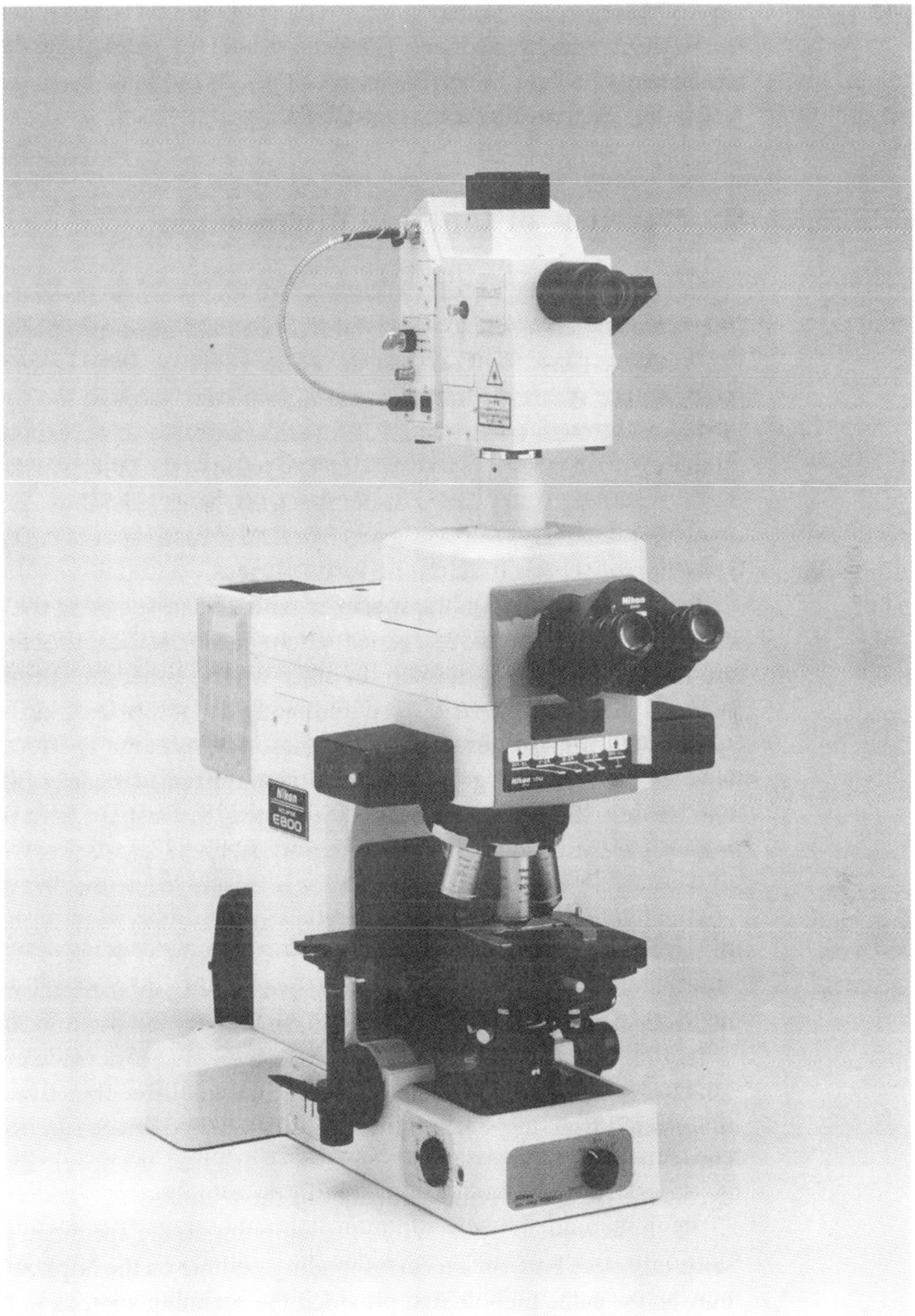

Fig. 1 The CSU-10 confocal scanning unit attached to the C-mount video port on top of an upright microscope (Nikon E-800). The scanner can also be attached to the video port of an inverted microscope. Illumination is conducted through a polarization-maintaining optical fiber from the laser source (not shown). (Figure courtesy of Yokogawa Electric Corporation.)

rate of fluorescence bleaching observed using the CSU-10; rapid digital processing for further haze removal and image sharpening and for generation of dynamic stereo images; mechanical and optical performance of the CSU-10; and potential future applications. Addendum A provides some recent updates.

II. Overview of Confocal Microscopes

In a point-scanning confocal microscope, the specimen is illuminated through a well-corrected objective lens by an intense, reduced image of a point source that is located in the image plane (or its conjugate) of the objective lens. Light emitted by the focused point on the specimen traces the light path back through the objective lens and (after deviation through a dichromatic mirror) progresses to an exit pinhole. The small exit pinhole, which is also placed in the image plane of the objective lens, effectively excludes light emanating from planes in the specimen other than those in focus. Thus, confocal imaging selectively collects signals from the focused spot in the specimen with dramatic reduction of signals from out-of-focus planes.

To achieve an image of the specimen with a point-scanning confocal system, either the specimen itself is moved in a raster pattern or the confocal spot and exit pinhole together are made to scan the specimen. In other words, either the specimen is precisely raster scanned in the x–y plane, or the illuminating and return imaging beam are made to raster scan a stationary specimen by tilting these beams at the aperture plane of the objective lens (commonly with galvanometer-driven mirrors) (see, e.g., this volume, Chapter 1, and Pawley, 1995). The very rapidly changing intensity of light passing the exit pinhole is detected by a photomultiplier tube and captured in a digital frame buffer. The output of the frame buffer, the confocal image, is displayed on a computer monitor.

For a number of reasons, it generally requires a few seconds to generate a low-noise, full-screen confocal fluorescent image with a point-scanning confocal system (see, e.g., Amos and White, 1995). One way of overcoming this speed limitation is to use many points to scan the specimen in parallel, such as by the use of a Nipkow disk.

In Nipkow-disk-type confocal microscopes, the disk with multiple sets of spirally arranged pinholes is placed in the image plane of the objective lens. The pinholes are illuminated from the rear, and their highly reduced images are focused by the objective lens onto the specimen. As the Nipkow disk spins, the specimen is thus raster scanned by successive sets of reduced images of the pinholes.

The light emitted by each illuminated point on the specimen is focused back, by the same objective lens, onto a corresponding pinhole on the Nipkow disk. This exit pinhole may be the same pinhole that provided the scanning spot, as in the Kino-type confocal system (Kino, 1995), or may be one located on the diametrically opposite side of the Nipkow disk as in the Petráň-type confocal microscope (Petráň *et al.,* 1968; see this volume, Chapter 1, Figs. 8 and 9).

With any Nipkow-type confocal system, one needs to maintain a moderately large separation between adjacent pinholes relative to their diameters in order to minimize crosstalk (i.e., leakage) of the return beam through neighboring pinholes. On the other hand, with the pinholes separated by, say, 10 times their diameter, only 1% of the incoming beam is transmitted by the Nipkow disk since the pinholes occupy only that fraction of the disk area. Nipkow disk systems, therefore, generally tend to suffer from low levels of transmission of the beam that illuminates the specimen.

In addition, a major fraction of the illuminating beam can be backscattered by the Nipkow disk and contribute to unwanted background light in the Kino-type arrangement. The Kino system thus includes several design features to minimize this source of unwanted light (see, e.g., Kino, 1995). The Petráň, tandem-scanning-type does not suffer from backscatter of the illuminating beam but does require very high precision of pinhole placement on the Nipkow disk as well as stability of rotational axis since the entrance and exit pinholes are located on opposite sides of the axis of rotation of the disk.

III. Design of the CSU-10

In order to overcome these difficulties encountered with the past Nipkow disk systems, the Yokogawa CSU-10 uses the same pinholes for the entrance and exit beams but is equipped with a second, coaxially aligned Nipkow disk that contains some 20,000 microlenses. Each microlens is precisely aligned with its corresponding pinhole on the main Nipkow disk onto which the illuminating beam is focused. Thus, instead of the 1% or so found with conventional Nipkow disk systems, some 40% to 60% of the light impinging on the disk containing the microlenses becomes transmitted through the pinholes to illuminate the specimen (Fig. 2).

In the CSU-10, each fluorescent beam emitted by the specimen is focused by the objective lens onto the same pinhole that acted as the entrance pinhole. Light passing the pinhole is then reflected by a dichromatic filter placed between the two Nipkow disks, behind the pinhole-containing disk, but before the one with the microlenses (Figs. 2 and 3). Thus, much of the backscattered light that could contaminate the image-forming beam is eliminated before the illumination beam reaches the main Nipkow disk. The image-forming beam reflected from the dichromatic mirror is projected by a collector lens onto the camera faceplate or into the eyepiece.

In the CSU-10, the microlenses and pinholes and arranged into 12 sets of a unique geometrical pattern (constant-pitch helical; see Section VIII). This unique pattern provides image frames that are homogeneously illuminated and traced by uniformly spaced scan lines. The image is thus free from nonuniform distribution of light intensity across the field or from scan-line inhomogeneity (scanning stripes) that can detract from Nipkow disk confocal systems using other pinhole patterns (Ichihara *et al.,* 1996).

With the (standard) CSU-10, the two Nipkow disks revolve together on the axis of an electric motor spinning at 1800 rpm, or 30 rps. Since there are 12 sets of helically arranged pinholes on the Nipkow scanners of the CSU-10, the scanned image is averaged 12 times per video frame (NTSC video repeats 30 frames per second). At any instant,

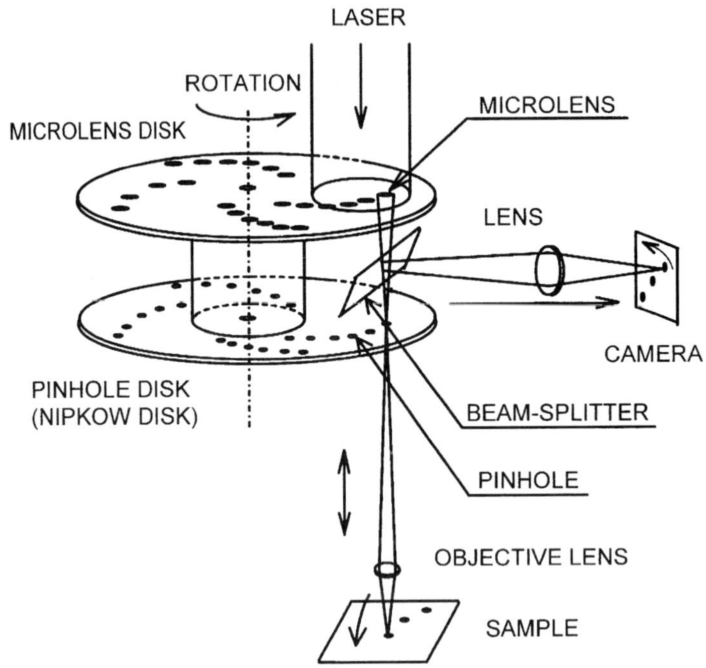

Fig. 2 Schematic of optics in the CSU-10. The expanded and collimated laser beam illuminates the upper Nipkow disk containing some 20,000 microlenses. Each microlens focuses the laser beam onto its corresponding pinhole, thus significantly raising the fraction of the illuminating beam that is transmitted by the main Nipkow disk containing the pinhole array. From the pinholes, the beams progress down to fill the aperture of the objective lens. The objective lens generates a reduced image of the pinholes into the specimen focal plane. Fluorescence given off by the illuminated regions in the specimen is captured by the objective lens and focused back onto the Nipkow disk containing the pinhole array. Each pinhole now acts as the confocal exit pinhole and eliminates fluorescence from out-of-focus regions, thus selectively transmitting fluorescence that originated from the specimen region illuminated by that particular pinhole. (However, for specimens with fluorescence distributed over large depths, some out-of-focus fluorescence can leak through adjacent pinholes in multiple pinhole systems such as the CSU-10.) The rays transmitted by the exit pinholes are deflected by the dichromatic beam splitter, located between the two Nipkow disks, and proceed to the image plane. (Figure courtesy of Yokogawa Electric Corporation.)

some 1200 pinholes scan the field in succession (Fig. 4). This produces a very clean, 12-frame-averaged image for every video frame[2] or as directly viewed through the eyepiece. Image noise due to low number of photons can, if necessary, be further reduced by on-chip averaging on a low-noise CCD camera, or by digital image averaging as described in Section VII.

[2]Whether with the standard speed CSU-10 or those equipped with high-speed motors for faster-than-video-rate recording, the video signal and pinhole locations are synchronized through the "Video Sync" BNC connector located (above the DC power input plug) on the CSU-10. Lack of sync between the video camera V-sweep (or shutter intervals, if present, in CCD cameras) and the pinhole locations can introduce uneven spacing and/or illumination of scan lines that show up as "streaks" or scanning stripes in the video record.

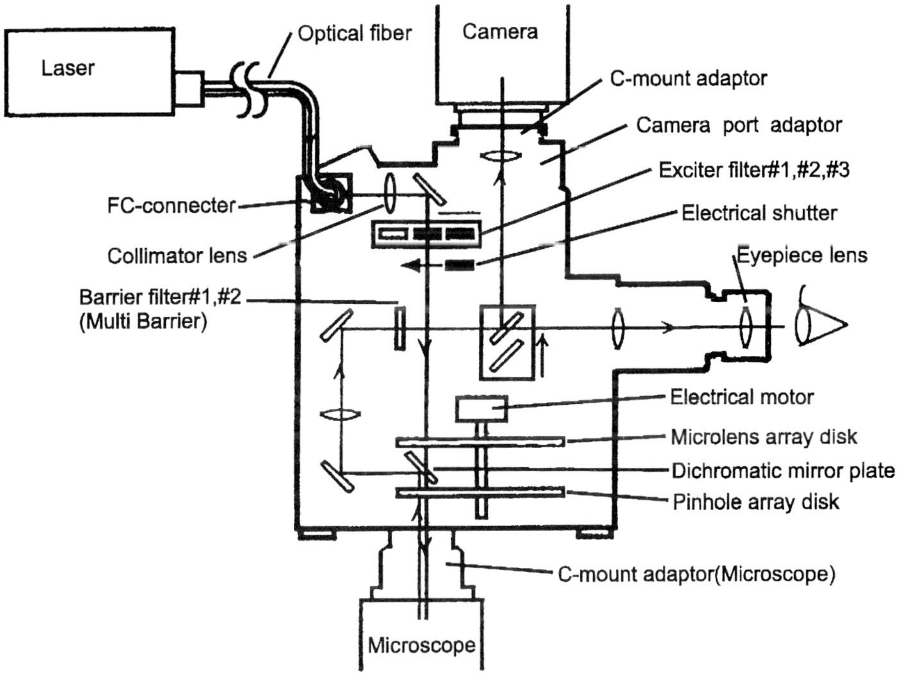

Fig. 3 Light path in the CSU-10. The 488-nm (and 568-nm) light emitted by the argon laser (or argon-krypton gas laser) is conducted through a polarization-maintaining, shielded optical fiber to the FC connector on the CSU-10. The collimated expanded beam, after passing the selected barrier and/or neutral density filters and electrically controlled shutter, impinges on the Nipkow disk containing the microlenses. Each microlens focuses the beam onto its corresponding pinhole and thence onto the specimen and back to the same pinhole, which now acts as the exit pinhole as explained in Fig. 2. The fluorescent light transmitted by the exit pinholes is deviated by the beam-splitting dichromatic filter located between the two spinning disks. Thereafter it passes through the barrier filter and is focused into the image plane in the eyepiece or camera faceplate. In the standard CSU-10, the modified Nipkow disks containing the pinholes and the microlenses spin at 1800 rpm on the shaft of a DC motor. The whole compact scanning unit is coupled to the microscope, as well as the video, CCD, or photographic camera, through female and male C-mount connectors. (Figure courtesy of Yokogawa Electric Corporation.)

IV. Sample Biological Applications

Figure 5 (right column) shows a series of optical sections of the autofluorescence of a dandelion pollen grain viewed through the CSU-10. The images were obtained with 488-nm excitation through a 40/0.8 Fluor objective lens and captured with a monochrome CCD camera. The left column shows optical sections recorded through the same microscope but by switching to standard epi-fluorescence.

Figure 6 shows the autofluorescence of another species of pollen grain in natural color, again excited with a 488-nm laser beam but captured with a chilled three-chip color CCD camera attached to the CSU-10. (See color plate.)

Figure 7 illustrates a high-resolution optical section taken through a rat heart muscle cell showing the distribution of ryanodine receptors in the sarcoplasmic reticulum.

As these figures illustrate, the CSU-10 provides uniformly lit, well-resolved, optical sections of 3-D fluorescent specimens. Without resorting to the use of a

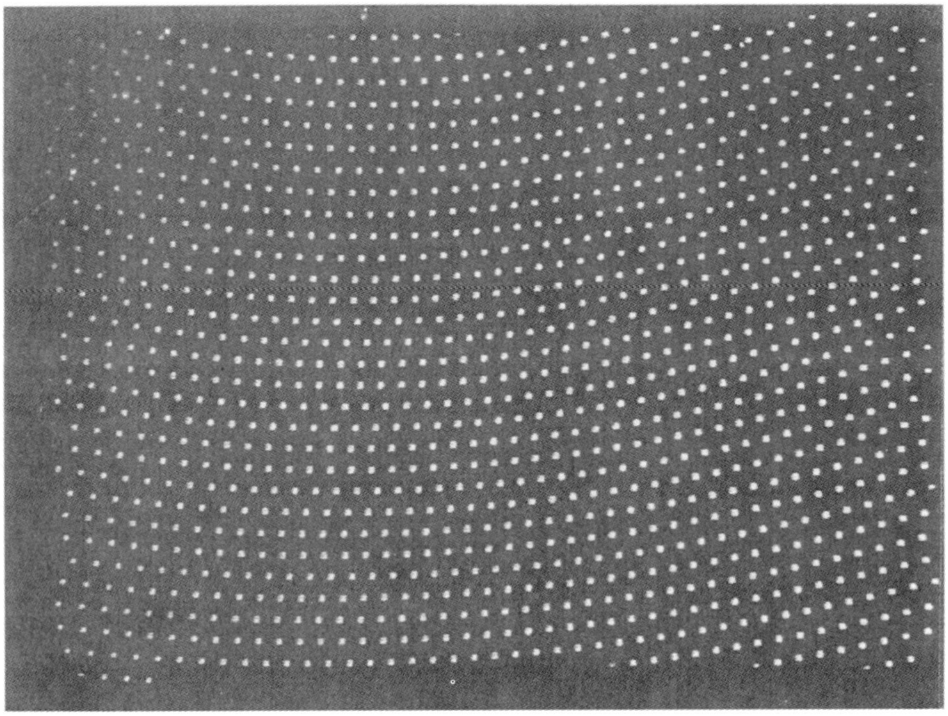

Fig. 4 Pinholes in a single visual field of the CSU-10. The 10-mm wide by 7-mm high visual field of the pinhole arrays on the main Nipkow disk was recorded with a SIT camera while the disks were not spinning. As seen, some 1200 pinholes cover the field at any one instant. When the disks spin, the spirally arrayed pinholes successively scan the specimen. Since the radical scan pitch between the adjacent spirals is arranged in a pattern that offsets the adjoining pinholes by a fraction of the pinhole diameter, the specimen is fully raster-scanned by partially overlapping images of the pinholes.

vibration-isolation bench, the image is free from vibration-induced waviness that sometimes plagues a point scanner system. The image is also free from scan line "streaks" (nonuniform spacing of scan lines) that are common in other disk scanning systems (see footnote 2).

Through the eyepiece of the CSU-10, these detailed images can be seen in natural fluorescence color, one optical section after another, in real time, as one focuses the microscope fine adjustment.

The direct images seen through the CSU-10 eyepiece and the video records impressively represent the specimens' fluorescence distribution in sequential optical sections. Nevertheless, with thick fluorescent specimens, such as those illustrated, they are somewhat contaminated by haze from out-of-focus fluorescence that is quite absent in confocal images generated by point scanners and especially multiphoton systems. In multipinhole systems such as the CSU-10, some out-of-focus light inevitably leaks through adjacent exit pinholes. The haze can be removed and the image sharpened considerably further by fairly simple and rapid digital processing of the direct or stored video signal as described in Section VII.

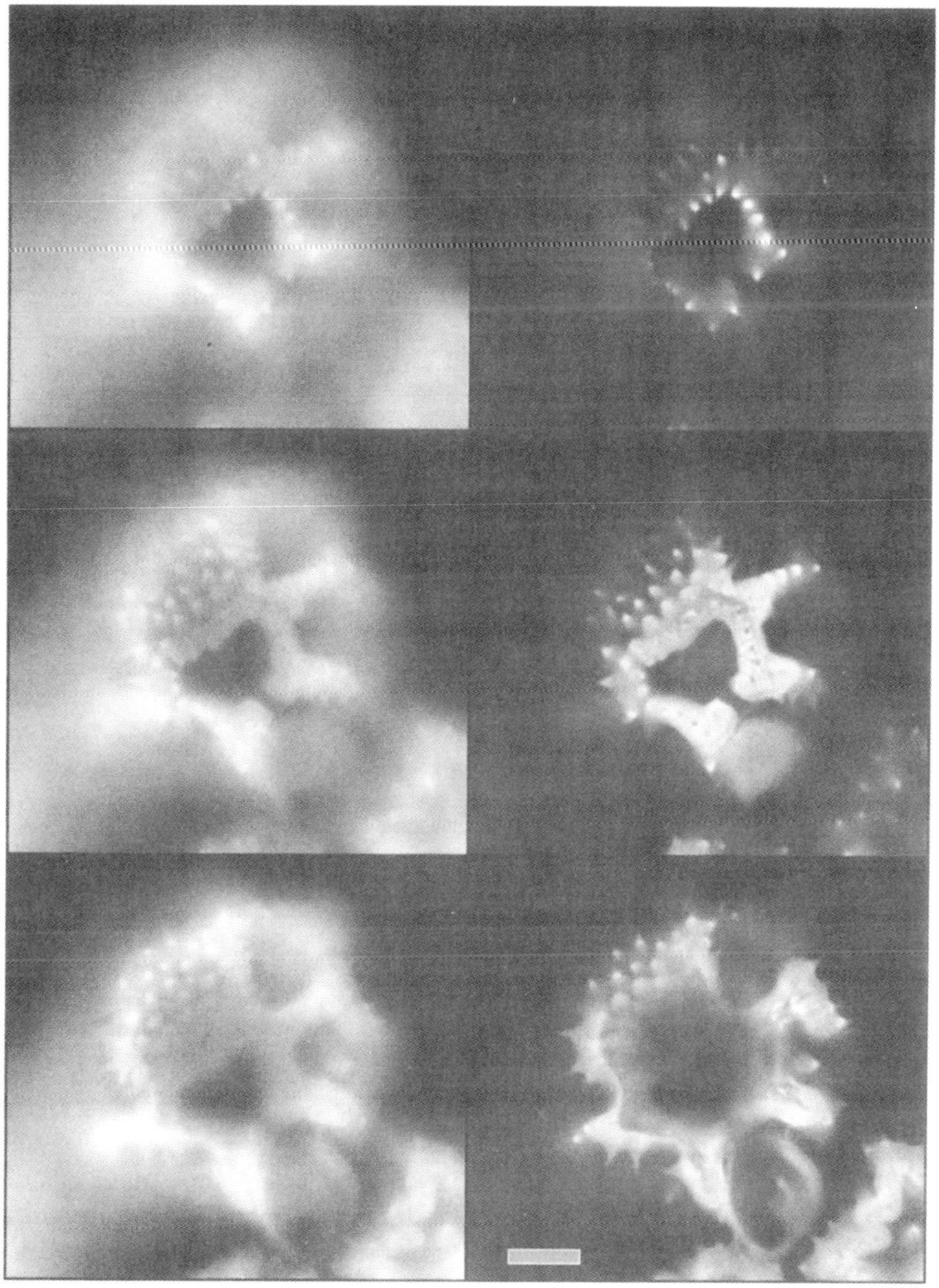

Fig. 5 Autofluorescence of dandelion pollen grain. Specimen slide of dandelion pollen grains boiled in mineral acid courtesy of Dr. Brad Amos, MRC, Cambridge, UK. Right column: Selected frames from serial optical sections spaced 0.5 μm apart were taken with a chilled interline CCD camera (Hamamatsu C5985) through the 488-nm illuminated CSU-10 mounted on a custom-made microscope (Fig. 3-13 in Inoué and Spring, 1997). An Olympus FLA epifluorescence unit was inserted between the calcite analyzer and the trinocular tube, and the specimen was imaged with a Nikon 40/0.85 Fluor objective lens equipped with a correction collar. Left column: Images taken at the same focal levels in normal epifluorescence through the same lens. See Figures 14 to 17 for improvement of the same video signals from the CSU-10 after digital image processing. Scale bar = 10 μm.

Fig. 6 Autofluorescence of pollen grain of mallow. This multicolor autofluorescence was captured in ca. 0.6 s with a chilled, 3-CCD color video camera (Dage-MTI Model 330, Michigan City, IN) mounted on the CSU-10 and illuminated with 488-nm laser light. The video signal, after conversion from RGB to standard (Y/C) format through a scan converter (VIDI/O Box, Truevision Inc., Indianapolis, IN), was captured onto a Sony ED-Beta VCR. For this illustration, a single frame of the VCR output was captured and printed with a Sony Mavigraph video printer. The same full-color image is seen through the eyepiece of the CSU-10 in real time. (See Color Plate.)

Turning to dynamic images of active living cells, Fig. 8 illustrates the process of macropincytosis in a GFP-coronin-labeled slime mold (*Dictyostelium discoideum*) ameba. Time-lapse video sequences of optical sections recorded through the CSU-10 were postprocessed as described in the figure legend. The three panels clearly depict the sequence by which the coronin-rich cortex of the ameba engulfs the fluid medium and then forms the pinocytotic vesicle (Fukui *et al.,* 1999a). Other observations of dynamic changes, seen by optical sections using the CSU-10 in this small ameba expressing GFP-coronin and GFP-actin, are illustrated in the paper just cited for cytokinesis, and in Fukui *et al.* (1999b) for the dynamics of eupodia (true feet) formation and resorption in a crawling ameba.

Figure 9 illustrates mitochondria in a ciliate protozoan *Tetrahymena,* swimming in the confocal field of the CSU-10. The images in the whole panel were taken in 1 s.

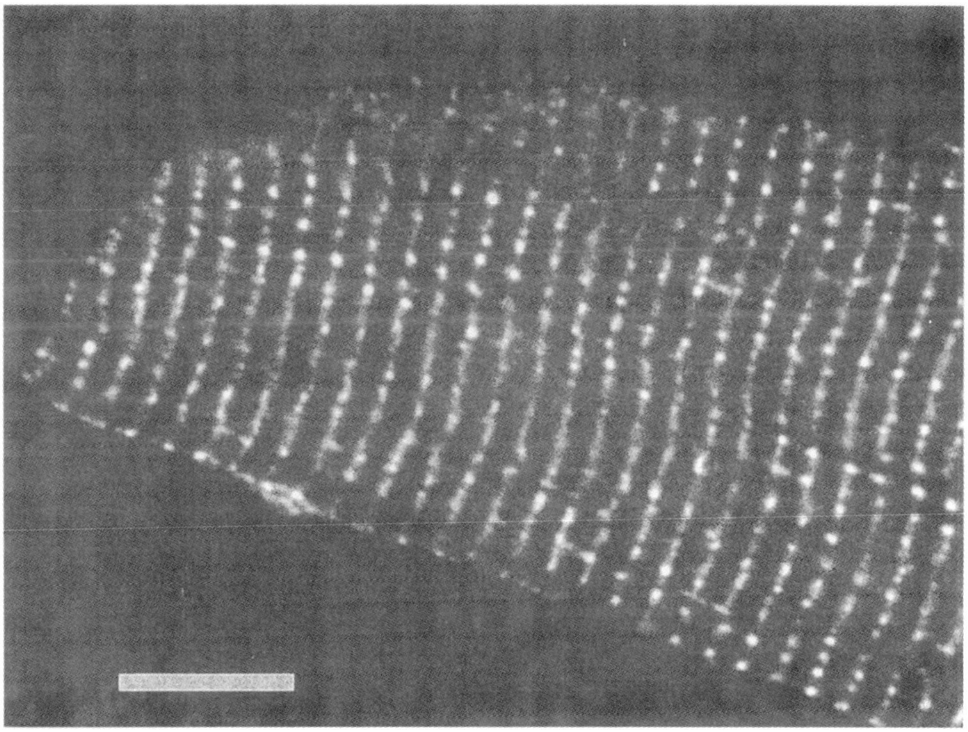

Fig. 7 Ryanodine receptors in heart muscle cell. This optical section of a rat myocyte shows the distribution of type 2 ryanodine receptors. The receptor proteins were stained with a polyclonal antibody against ryanodine 2 and visualized with an FITC-labeled secondary antibody. (The tetrametric receptor proteins form channels in the membranes of the sarcoplasmic reticulum and control the rapid release of Ca^{2+} to the myofibrils.) Scale bar = 10 μm. The confocal optical section of the rat ventricular myocyte was captured by Drs. Peter Lipp, M. Laine, and M. D. Batman of the Babraham Institute, UK, using a modified CSU-10 ("UltraView" by PerkinElmer Wallac; see footnote 7) on a Wallac chilled-CCD camera equipped with a Sony 1300 × 1000 pixel interline CCD chip. Data courtesy of Dr. Baggi Somasundaram (bsomasundaram@wallacus.egginc.com).

As the panels illustrate, one can obtain very sharp, full-frame video images of selected organelles and their dynamics, in high spatial and temporal resolution as the protist swims about (motion slowed down with Protoslo). We believe such detailed fluorescence images of swimming protozoa have never before been observed or recorded.

Figure 10 illustrates the dynamic growth, shortening, motility, and binding to the cell wall of microtubules in the fission yeast of *Schizosaccharomycetes pombe*. The confocal time-lapse frames of these small, living cells clearly show, in high spatial and temporal resolution, the distribution and changes of the microtubules (expressing GFP-tubulin) in optical section. In the optical sections, the intensity of fluorescence reflects the number of microtubules (one to a few) in the microtubule bundles, which coupled with their dynamic behavior reveals the sites of microtubule organizing centers.

Figure 11 illustrates, in natural color, the rapidly changing fluorescence in a clam sperm stained with Eosin-B. (See color plate.) Live cells are commonly thought to be unstained by Eosin-B, and entrance of this dye is considered an indicator of cell death. We nevertheless find that membrane regions of the swimming sperm are stained and exhibit

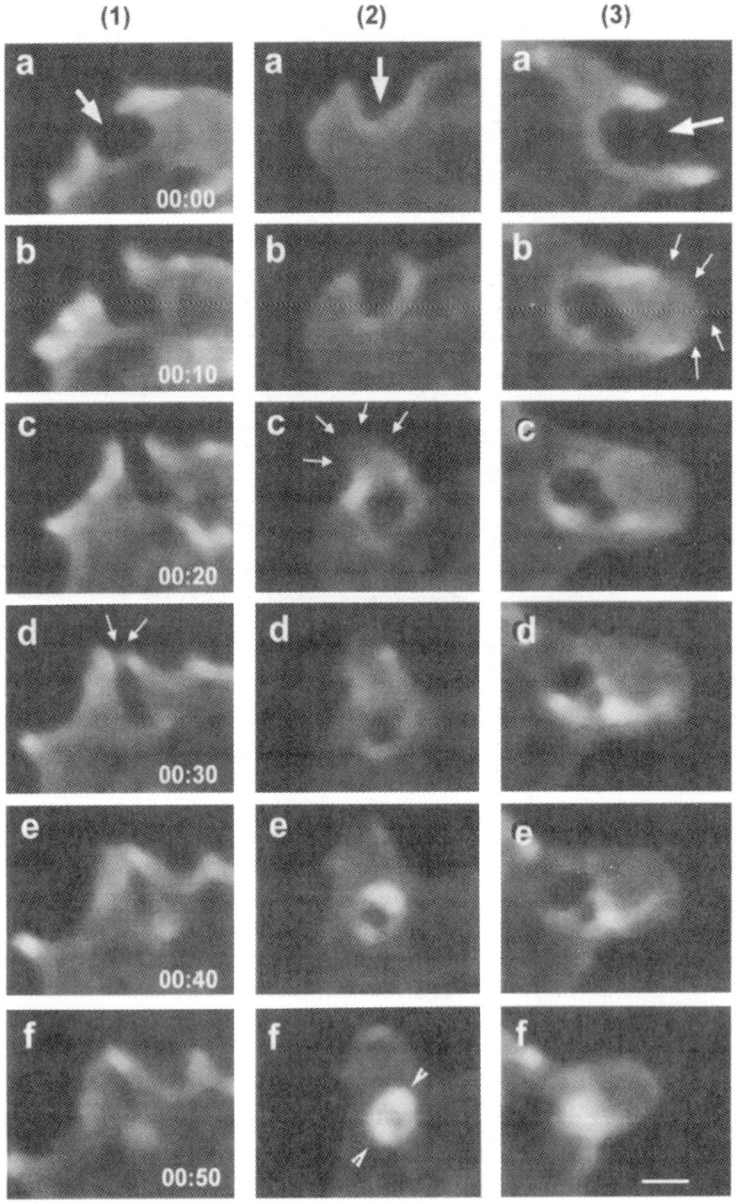

Fig. 8 Three representative sequences of macropinocytosis in amoeba of the slime mold *Dictyostelium*. Fluorescence in these time-lapsed (10-s interval) CSU-10 confocal images is due to GFP-coronin, which accumulates around the pinocytotic crowns and remains surrounding the internalized pinocytotic vesicles as shown. Scale bar = 1 μm. *Technical details:* Vegetative-stage *Dictyostelium discoideum* expressing GFP-coronin fusion protein were overlaid by an agarose sheet and observed in a microchamber using 1.4 NA Plan Apo objective lenses on a Nikon E-800 microscope. The shutter in the CSU-10 was opened for 0.75 s while the signal was integrated on chip in a chilled CCD camera (Hamamatsu C-5985) for 0.5 s once every 10 s. Regions of interest were selected in the image stacks captured on hard disk (in the same image-processing and system-control computer operated by the UIC MetaMorph program that controlled the shutter operation), low-pass filtered, and unsharp masked, and a median filter was applied. (From Fukui, Y., Engler, S., Inoué, S., and de Hostos, E. L. Architectural dynamics and gene replacement of coronin suggest its role in cytokinesis. *Cell Motil. Cytoskel.* **42,** 204–217. Reprinted with permission of Wiley-Liss, Inc., a subsidiary of John Wiley & Sons, Inc.)

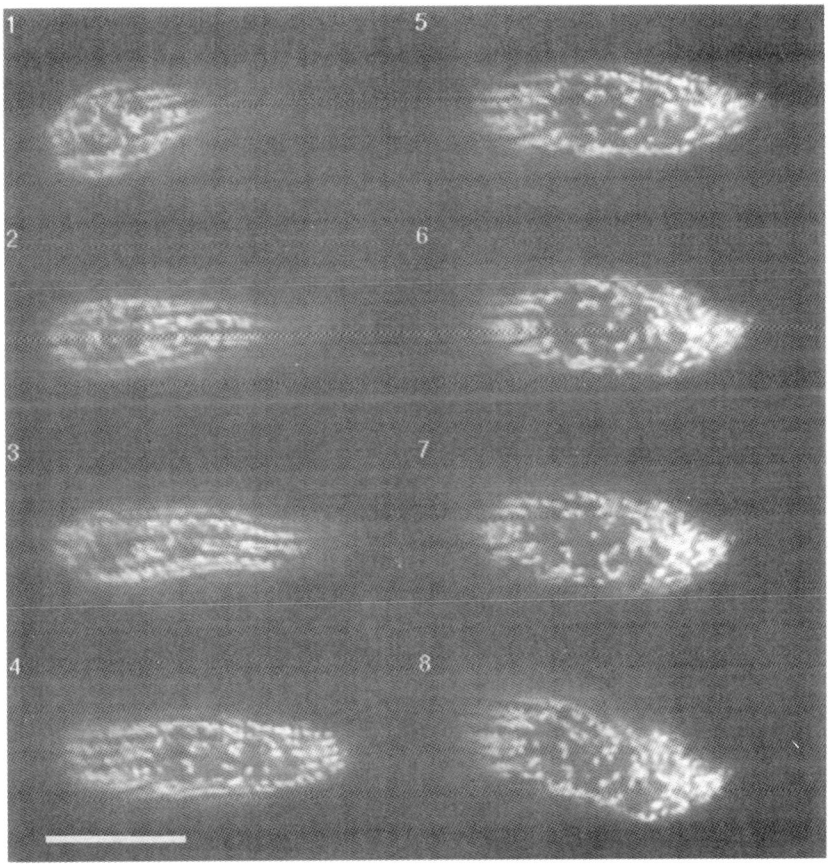

Fig. 9 Mitochondria in a swimming ciliate protozoan. The MitoTracker-G-stained *Tetrahymena pyriformis* generated these sequential series of optical sections as it tumbled through the confocal field viewed with the CSU-10. This whole sequence covers a 1-s period. In panels 1–3, the head (to the right) of the *Tetrahymena* is bent below the confocal plane. In panels 6–8, the head is lifted further up. Where the confocal optical section lies just below the cell surface, mitochondria (that sandwich the linearly arranged basal bodies of cilia, not seen here) show in paired rows. Where a region further inside the cell comes to lie in the optical section, one sees randomly oriented mitochondria. Scale bar = 20 μm[3]. *Technical details:* The mitochondria were selectively stained with the cell-permanent fluorescent dye MitoTracker-G (Molecular Probes, Eugene, OR). Optical sections of the swimming *Tetrahymena,* captured through a Leica 100/1.3 Plan Fluotar oil-immersion objective lens, were recorded directly to a VCR through a low-light-level silicon intensifier target (SIT) camera at video rate. Frame intervals ca 0.1 s except 0.26 s between panels 1 and 2. The *Tetrahymena,* whose swimming motion was slowed down to between 10 and 20 μm/s with Protoslo (Carolina Science, Burlington, NC), was viewed through a Leica 100/1.3 Plan Fluotar oil-immersion objective lens. The CSU-10 confocal unit mounted on top of the Leica DMRX microscope was illuminated with a 488-nm laser, and the image captured with a SIT camera (Dage-MTI 66, Michigan City, IN) was recorded at video rate onto an ED-Beta VCR (Sony EDW 30F, San Jose, CA). Several seconds of the video played back from the VCR was passed through a time-base corrector (For-A FA-400, Newton, MA) and captured into RAM as a "stack" through an InstruTech DVP-32 Digital Video Processor (Port Washington, NY) using the "Adjust Analog Contrast" and "Acquire Stream" functions in MetaMorph (Version 4.0, UIC, West Chester, PA). Using MetaMorph, a sequence of contiguous frames was selected from the stack held in the RAM, from which eight partial frames (region of interest) were chosen to generate the montage. (S. Inoué and B. Matsumoto, 1999, original images.)

[3] The confocal fluorescence images seen through the CSU-10 can easily be superimposed in real time with the brightfield or DIC images of the specimen. All one has to do is to turn on the transilluminating light source to an appropriate brightness. Then the fluorescing mitochondria, for example, that come to lie in the confocal plane shine in succession within the (nonconfocal) brightfield image of the swimming *Tetrahymena* that show the ciliary rows and other cell structures.

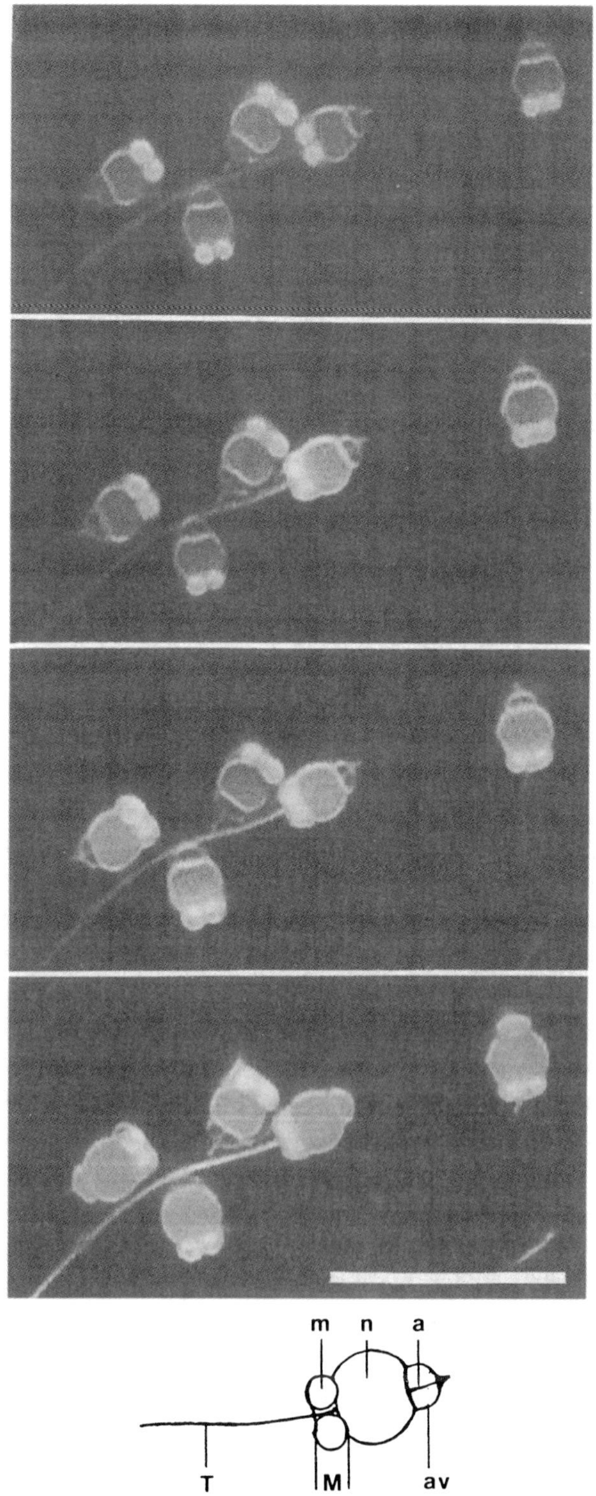
m
n
a
T
M
av

Experts do not seem to agree on whether fluorescence bleaching is linearly dependent on the dose rates of the excitation beam seen by the specimen, especially at the very high doses [brief illumination for near saturation of fluorescence that each specimen part experiences in a point-scanning confocal system (see, e.g., the several articles that discuss bleaching in Pawley, 1989)]. At lower dose rates, it appears that bleaching is, in fact, linearly dependent on dose rates, i.e., the degree of bleaching is determined by (energy absorbed per unit of time) × (duration of exposure) and is independent of whether a higher dose of energy is given over a brief interval of time or a lower dose is given over a longer period of time (Fred Lanni, Carnegie Mellon University, personal communication).

In the CSU-10, the power of the excitation beam that reaches the specimen is quite low. In fact, we find that it is several times lower, using the same objective lens, than in the wide-field fluorescence mode illuminated by an HBO-100 short-arc mercury lamp.

As to the efficiency of transmission of the imaging system, we measured the throughput of the imaging beam through the CSU-10 to be 70% of that compared to the same microscope with the CSU-10 removed.[5]

Coupled with the high throughput of the return beam, the higher quantum efficiency (perhaps by a factor of 2 to 4) and lower noise level of quality chilled-CCD cameras compared to photomultipliers (as used in current point-scanning confocal systems) may explain the greater photon-capture efficiency and better *S/N* ratio for systems such as the

[5] We measured the peak pixel brightness in the image of the pinholes (with the disk stopped) to be 70% of that compared to the CSU-10 removed from the microscope (Inoué and Knudson, unpublished data). These measurements were made (using monochromatic green and red trans-illumination respectively for the 488- and 532-nm excitation filter cubes) by checking the integration time required to generate the same pixel gray values in the middle of the stationary pinhole image compared to the pixel gray values in the same region of the blank, trans-illuminated field (i.e., with the CSU-10 removed but with the microscope optics and illumination otherwise unchanged). The image magnification was confirmed to be unchanged when the CSU-10 was removed from the microscope (by imaging the same object onto the CCD camera). This increased throughput should not be confused with the improved throughput for the entrance pinhole that is brought about by the use of microlenses. The imaging beam from the specimen is diverted to the camera (and eyepiece) before it reaches the microlens-containing disk (see Figs. 2 and 3).

Fig. 11 Selected frames from CSU-10 confocal video sequences showing striking photodynamic changes in Eosin-B-stained *Spisula* (clam) sperm exposed to the 488-nm (fluorescence exciting) laser beam. Top to bottom panels: 13, 14, 15, and 31s after exposure to the laser beam (0.01 mW per μm^2 at the specimen). Except for the fourth sperm from the left, much of the tail is invisible owing to the shallow depth of field of the confocal optics. Scale bar = 10 μm. Bottom: Schematic diagram of *Spisula* sperm. a; Acrosome: av; acrosomal vesicle; M; midpiece; m; mitochondrion; n; nucleus; T; sperm tail (part only shown). *Technical details:* Sperm were suspended in seawater containing 0.1% Eosin B and observed through a Nikon E-800 upright microscope equipped with a 100 × /1.4 NA Plan Apo oil-immersion lens and a CSU-10. The weak fluorescence of the sperm parts excited by the 488-nm laser beam was integrated on chip for 20 frames (0.66 s) on a chilled, 3-chip color camera (Dage-MTI Model 330, Michigan City, IN) attached to the CSU-10 and observed continuously on an RGB color monitor. The RGB output through the monitor was converted by a scan converter (VIDI/O Box, Truevision Inc., Indianapolis, IN) to Y/C format, color balance and background levels adjusted with a video processor (Elite Video BVP-4 Plus, www.elitevideo.com), and recorded to an ED-Beta VCR. (From Inoué *et al.*, 1997, with permission) (See Color Plate.)

CSU-10 compared to point-scanning systems. Those factors would also allow for use of lower levels of excitation light (and hence lower rates of fluorescence bleaching)[6] to obtain comparable low noise images with the CSU-10 using a chilled-CCD camera compared to a point scanner that captures the image signal with a photomultiplier tube.

With wide-field fluorescence using chilled-CCD cameras, comparably low levels of illumination could be used to reduce bleaching as with multipinhole scanning systems such as the CSU-10. However, background fluorescence, e.g., from the dyes used in the culture medium when observing GFP-microtubules in the yeast cells, obscures the desired images of the weakly fluorescent microtubules (Paul Maddox, University of North Carolina, personal communication).

Although we do not have a precise measure as to how much each of these factors contributes to the low fluorescence bleaching, there is no question, empirically, that the CSU-10 can provide excellent quality, optically sectioned, time-lapse images of sensitive, living specimens over extended periods of time (See also Addendum B).

VI. Very-High-Speed Full-Frame Confocal Imaging with the CSU-10

With point-scanning confocal systems one can scan along the horizontal axis at very high speeds. Thus, single-to-a-few sequential horizontal scans can be used to record rapid physiological transients and to generate kymographs (slit images versus time plot) depicting those transient events (see, e.g., Tsien and Bacskai, 1995). However, as mentioned earlier, a few to several seconds are generally required to display or record each low-noise full frame with a point-scanning confocal system. In contrast,with real-time, disk-scanning confocal systems, one can record sequential full-frame images at intervals of 33 ms, i.e., at video rate as already illustrated. Using a different motor to spin the disks in the CSU-10 at higher speeds, one can even generate full-frame signals at millisecond intervals (See also Addendum A).

Thus, Genka *et al.* (1999) modified the CSU-10 to drive the Nipkow disk motor at 5000 instead of 1800 rpm. At this higher speed, each full-frame optical section can be obtained in 1 ms. Combined with a Gen II image intensifier (ILS-3, Gen II, Imco, Essex, UK) and a high-speed CCD camera with controller (model 1000HR, Eastman Kodak, San Diego, CA), the 512 × 384 pixel 8-bit images were stored in the controller memory at 1–4 ms/frame. Such data were used to measure the 3-D diffusion pattern of

[6] When somewhat greater noise levels can be tolerated, the light-capturing efficiency of intensifier video tubes, such as SIT, or Gen-III or Gen IV intensifiers coupled with CCD cameras, also allows the use of low fluorescence excitation combined with high rates of image capture (as illustrated, e.g., in Figs. 9 and 12).

[7] In addition to the striking calcium waves in optical section seen in these excerpted frames, many sparks and more waves can be seen by playing back the full sequence of the 290 frames available. Contact Dr. Baggi Somasundaram at Perkin Elmer Wallac, or see http:\\www.wallac.com, to view this and many other impressive dynamic scenes taken with the Wallac UltraView system (mitosis in Drosophila embryo expressing histone-GFP, endoplasmic reticulum and organelles in onion cells, flow of cortical actin in fibroblasts, etc.). Data courtesy of Dr. Baggi Somasundaram.

Fluo-3-bound calcium ions surrounding the nucleus of contracting myocytes (see also Ishida *et al.*, 1999).

Figure 12 shows vivid pseudocolor images of Ca^{2+} transients in myocytes taken through a standard-speed CSU-10 equipped with an image intensifier. (See color plate.) The images in Ishida *et al.* (1999) and Genka *et al.* (1999), recorded with considerably greater temporal and spatial resolution than in this illustration, display hitherto unknown changes in Ca^{2+} localization and dynamic distribution that take place within the myocytes.

Yamaguchi *et al.* (1997) describe a system for recording confocal images of fluorescein-tagged blood cells at 250 and 500 frames/s flowing in pulmonary capillaries (which were also fluorescently labeled for selected adhesion molecules; see also Aoki *et al.*, 1997). They use the CSU-10 equipped with a 5000-rpm motor, coupled with an image intensifier CCD camera (EktaPro Intensified Imager VSC, Kodak, San Diego, CA), connected with a high-speed video analysis system (EktaPro 1000 Processor, Kodak) to record the high-speed images.

Before continuing our discussion on the performance of the CSU-10 in Section VIII, in the next section we shall examine the advantage, means, and rationale for applying digital image processing to multibeam confocal and related image data. With appropriate digital processing, one can very effectively and nearly instantaneously improve the quality of optical sections by enhancing details in the in-focus image while suppressing the background fluorescence. Other forms of image noise can also be reduced or removed. Such approaches are especially useful in cell biology, for bringing out subtle or fine image details, for preparing a stack of images for 3-D projection, and for processing large quantities of image data, e.g., when many time points are involved and/or when each time point is associated with a substantial stack of serial optical sections.

VII. Haze Removal, Image Sharpening, and Dynamic Stereo Image Generation by Digital Signal Processing

Once digital computers became widely available for image processing, wide-field microscope images were effectively enhanced using simple, neighborhood-based digital convolution operations such as sharpening and unsharp masking. While such simple convolution operations typically may not be as effective at removing haze from images as are iterative, constrained deconvolution algorithms, they can nevertheless significantly improve images to the extent necessary to obtain excellent 3-D reconstructions and stereo pairs. Moreover, their speed makes these operations well suited to 4-D microscopy (3-D imaging over time) where large data sets are common.

The CSU-10, by virtue of its ability to collect images at high speed with low rates of fluorescence bleaching and its ease of use, makes it possible to collect massive quantities of 3-D image data. This allows it to be used quite effectively in 4-D microscopy. As noted earlier, the CSU-10 is a multiple-pinhole, high-light-throughput scanning confocal

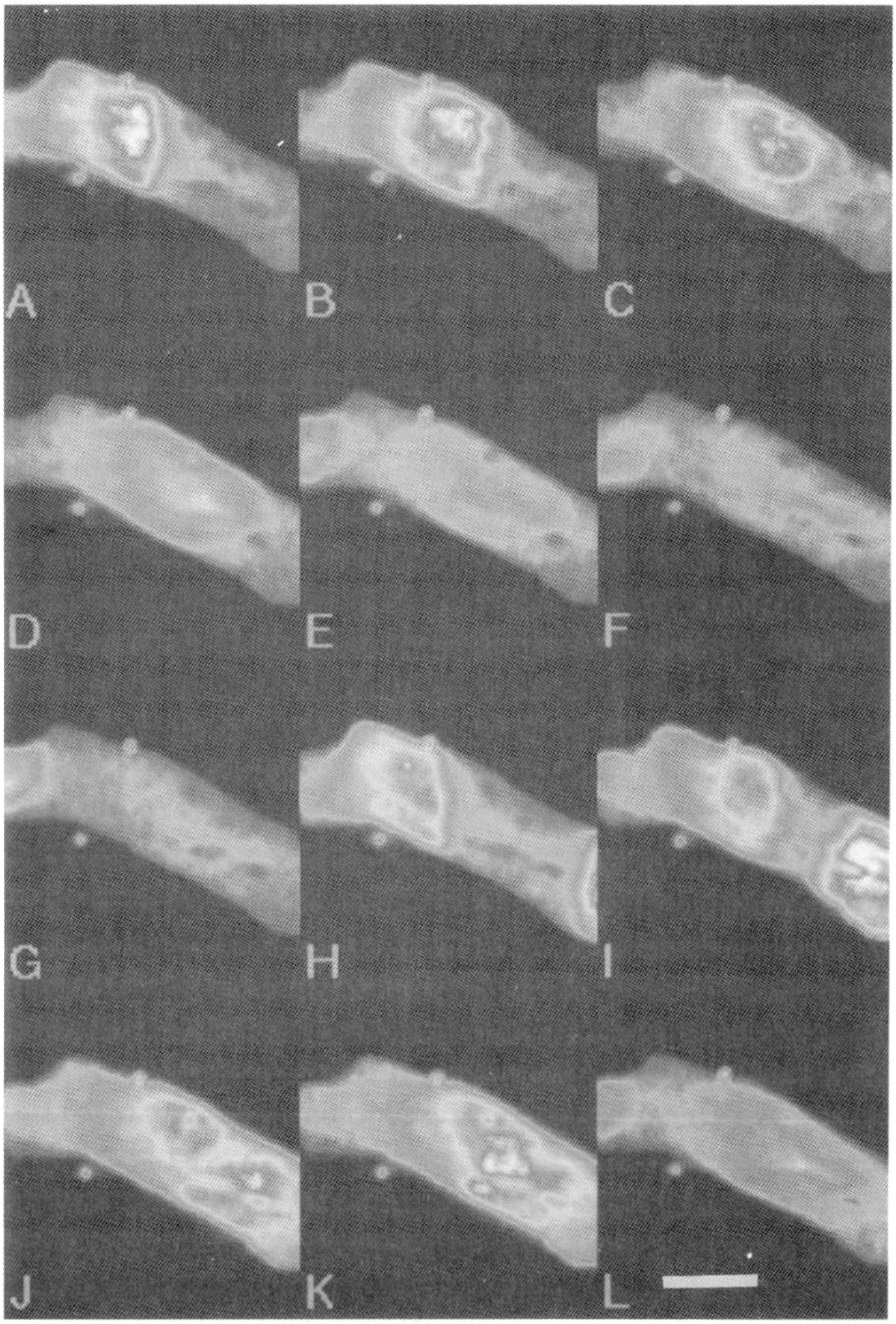

Fig. 12 Mouse cardiac myocyte. Waves and sparks of elevated cytosolic Ca^{2+} are presented in pseudocolor in this heart muscle cell injected with Fluo-3. (Ca^{2+} concentrations rise in order of purple, light-blue, dark-blue, green, yellow, orange, red, and white.) Frame intervals each 200 ms between panels A–F. Then after 1.4s, panel G-L show the approach and annihilation of two calcium waves. Frame intervals: G–H = 360 ms, H–I = 160 ms, I–J = 120 ms, J–K = 40 ms. Scale bar = 5 μm.[7] *Technical details*. The image from the CSU-10 (modified by addition of motorized filter wheel as a part of the "UltraView" system distributed by PerkinElmer Wallac), viewed through an Olympus 60/1.4 oil-immersion objective lens, was intensified with a Videoscope Intensifier. The images, each exposed for 25 ms, were captured at 25 frames per second onto a PerkinElmer FKI-300 cooled-CCD camera equipped with a Kodak Olympics interline chip. (See Color Plate.)

system. The fixed size of its pinhole apertures is necessarily a compromise between light-gathering capacity and confocal rejection. Because of these design choices, images from the CSU-10 can benefit from digital image processing to further suppress residual background fluorescence and to sharpen image detail.

Ideally, one might use image deconvolution operations for precise image restoration. However, the speed of such operations, typically on the order of hours per image stack,[8] is too slow for routine processing of 4-D data sets. Fortunately, when used properly, image processing operations such as sharpening and unsharp masking are adequate, rapidly generating significantly improved results.

A. Neighborhood-Based Processing

Many of the most commonly used image processing programs, such as Universal Imaging's MetaMorph and Adobe's PhotoShop, use *neighborhood*-based convolution algorithms for operations such as image sharpening. Here, neighborhood refers to the fact that these algorithms modify every pixel in an image by replacing it with the weighted sum of the pixel and those surrounding it. The matrix containing these weighing factors is called the *convolution kernel* or *kernel* for short. Typically, the kernel covers a 3 × 3-pixel area, although other sizes, such as 5 × 5 pixels, are also common.

The results of convolution operations depend on the values used in the kernel. For example, the convolution blurs the image if all kernel elements have the same value, e.g., 1. To convolve with this kernel, the brightness of the central pixel and of the surrounding 8 pixels in a 3 × 3-pixel area are each multiplied by 1 and summed together. The result is divided by 9. Then the central pixel is replaced by this average brightness of the 3 × 3-pixel area, and thus the image is blurred. Such operation is also referred to as a low-pass filter or neighborhood average.

The sharpening kernel, on the other hand, subtracts the average brightness of the neighboring pixels from the weighted brightness of the central pixel, resulting in the enhancement of brightness gradients in the image. A typical sharpening kernel, called the Laplacian sharpening kernel (Fig. 13, panel 1), subtracts the sum of the 8 neighboring pixels from the central pixel multiplied by 9. The result is an image containing the same values as the original in areas of uniform brightness, and enhanced details where there are brightness gradients. A variant of this operation only subtracts the sum of the 4 nearest pixels from the central pixel multiplied by 5 (Fig. 13, panel 2) (Castleman, 1979; Inoué and Spring, 1997, Section 12.7; Russ, 1995).

Neighborhood-based computations are well suited for processing stacks of images because current microprocessors, such as the Intel Pentium, AMD Athlon, and Motorola PowerPC, perform the computations required for neighborhood operations very rapidly. The system in the authors' laboratory uses the MetaMorph software under the Windows NT operating system. The computer has 384 megabytes of RAM and an Intel Pentium II running at 400 MHz. This system sharpens 100 images, each 640 × 480 pixels, using

[8] With the advent of the new Gigaflop MacIntosh and other microprocessors, we should see this number drop by a factor of 2 to 5 during the next few years. However, even with this performance boost, deconvolution will be too slow for routine processing of 4-D data.

Panel 1

-1	-1	-1
-1	9	-1
-1	-1	-1

Panel 2

0	-1	0
-1	5	-1
0	-1	0

Fig. 13 The Laplacian sharpening kernel. See text.

a 3 × 3 sharpening kernel, in approximately 5 s, i.e., each frame nearly at video rate. Newer computers using the Intel Pentium 4 running at over two GHz are several times faster, reducing the computation time to approximately one second, or several times faster than video rate!

A typical 4-D imaging experiment might include 200 time points each with 30 images collected through the depth of the sample collected. If the reference system takes 3 s to process each image stack, the entire experiment would be processed in about 10 min, much shorter than the time in which a typical deconvolution system processes a *single* image stack! It is this efficiency that makes neighborhood operations so useful.

B. Sharpening and Unsharp Masking

Whereas sharpening often substantially improves images from the CSU-10, *unsharp masking* provides superior haze removal with only a modest increase in computational complexity.

In unsharp masking the image is enhanced by masking it with an unsharp (blurred) version of the same image. The result of this masking is a brightness reduction in areas that lack fine details. Conversely, unsharp masking increases the contrast of areas that do contain fine details. The resulting image contains enhanced fine details, much like a sharpened image, but also has a reduced background level, very useful for fluorescence applications containing background haze (compare Fig. 14 with Fig. 5 right).

The MetaMorph software provides controls for fine-tuning the unsharp masking process, a feature not typically found in other sharpening methods. One option sets the amount of blurring to use, thereby selecting the size of the features to enhance. This is very useful as it permits control of the degree of sharpening of image details without significantly increasing image noise.

A second option sets the amount of the blurred image to be removed from the original. This parameter adjusts the magnitude of the enhancement that is applied to the image or, to think of it in another way, this sets how much haze is removed. For example, if 100%

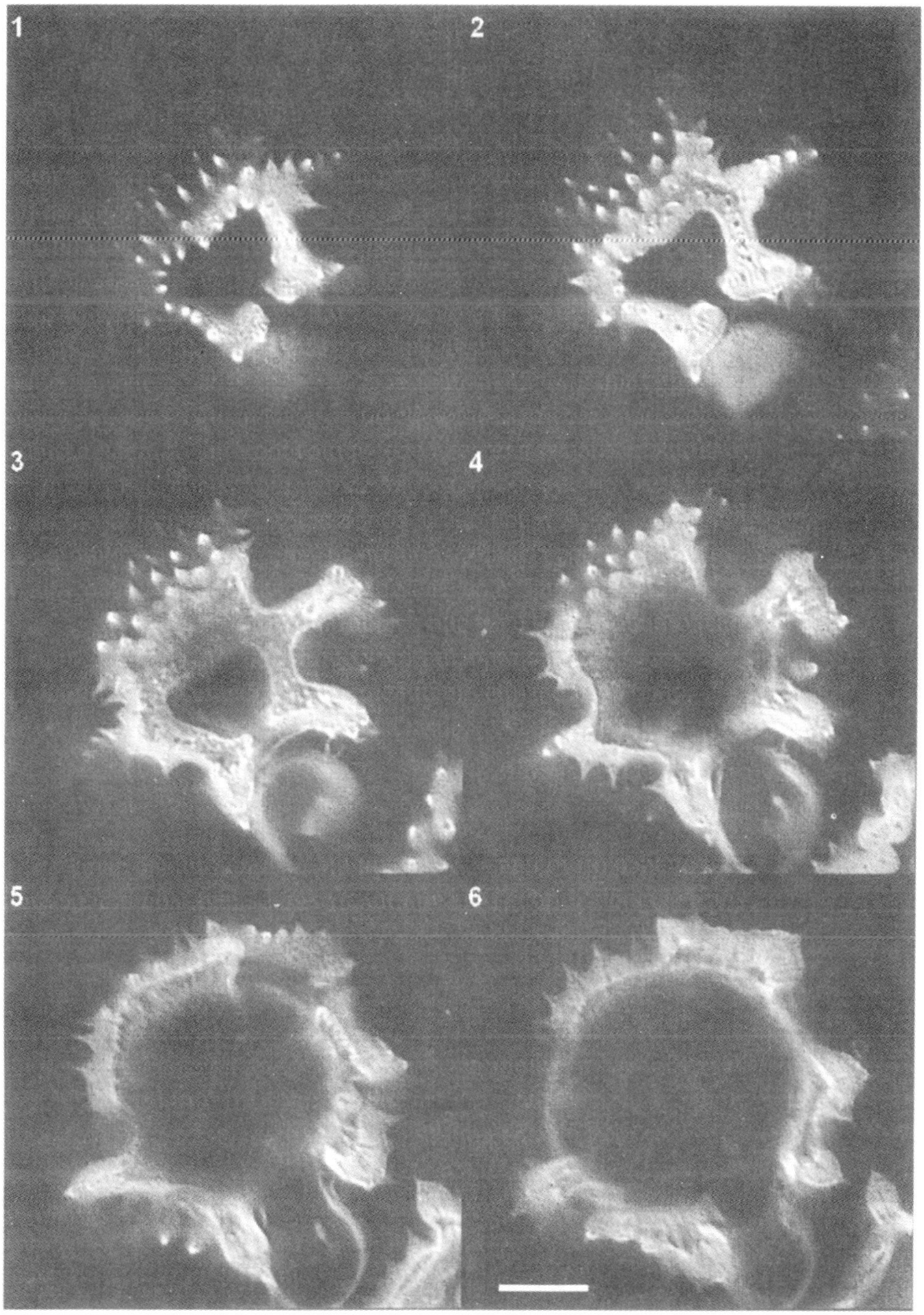

Fig. 14 Unsharp masking of CSU-10 images. Selected images (ca 2 μm apart) of the serial optical sections taken through the CSU-10 were digitally unsharp masked (using a 2 × 2 kernel and 0.90 scaling factor after first applying a 2 × 3 median filter to prevent the appearance of white noise spots). The images in panels 2 and 4 resulted from processing of the two bottom right images in Fig. 5. Scale bar = 10 μm.

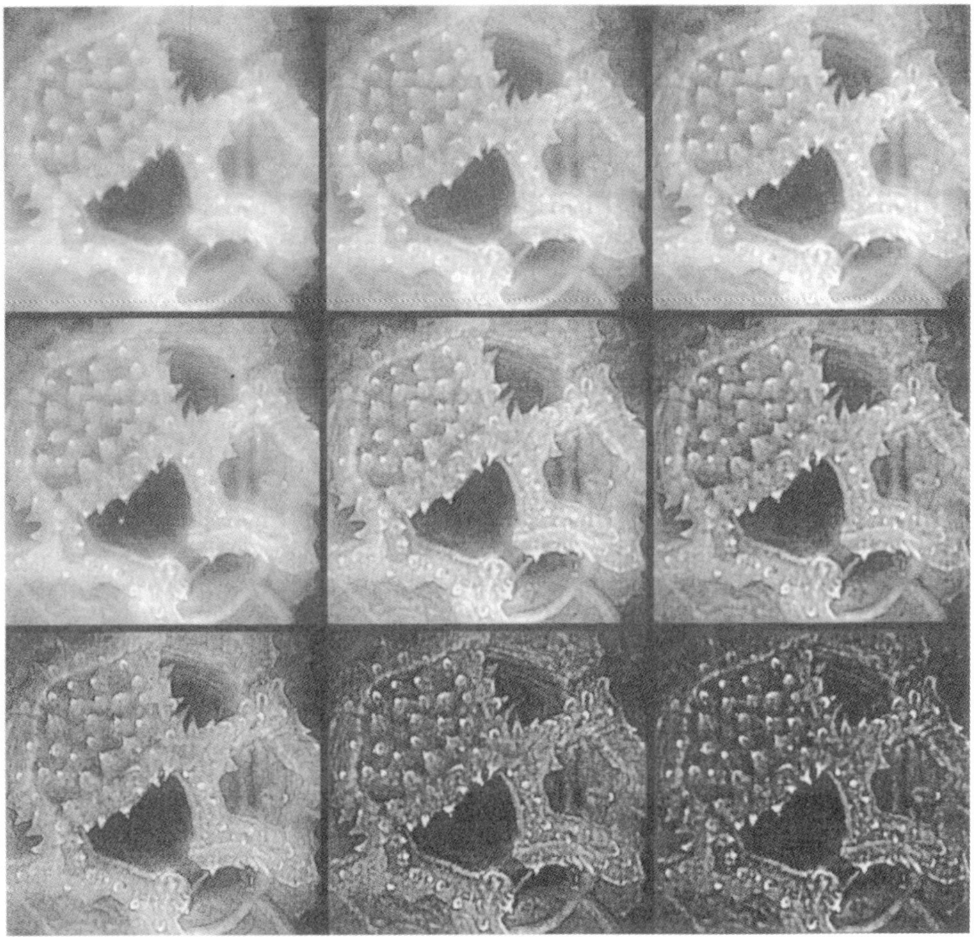

Fig. 15 Influence of kernel size and scaling factor on unsharp masking. Central portion of max-projected images [sum of all optical sections taking the brightest pixels projected through the stack of images (see Section VII.F.)] of the pollen grain shown in Fig. 14. The kernels used to produce the unsharp masks for the three columns were, from left to right, 3 × 3, 7 × 7, and 11 × 11. The scaling factors used to determine the fraction of contrast given the masks were, from top to bottom rows, 50%, 80%, and 90%. The MetaMorph software automatically enhances contrast to fill the gray value ranges.

of the blurred image is subtracted, then only fine details will remain in the resultant image; all regions of uniform brightness, such as background haze, are reduced to black. On the other hand, if only 50% is subtracted, then the resultant image will look much like the original with only a slight enhancement.

Figure 15 shows the results of applying a variety of blur sizes and subtraction percentages. The upper left panel has been processed the least (and is virtually indistinguishable from the original image) whereas the lower right panel represents the highest amount of processing, large blur and large subtraction percentage. For this particular sample, a small amount of blur and a moderate subtraction percentage results in the most esthetically pleasing image.

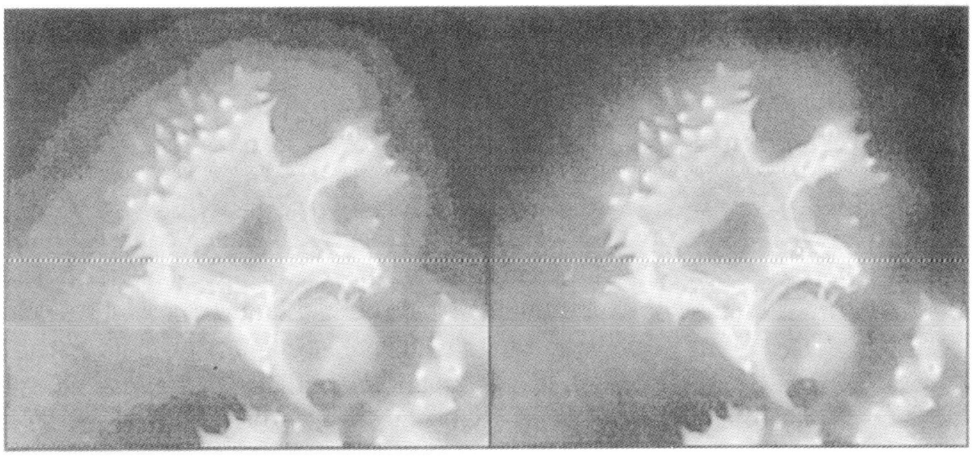

Fig. 16 Digital processing of 8-versus 16-bit images. The left panel shows quantizing error (discrete steps of gray values, especially noticeable in the background) by digital contrast enhancement of an 8-bit image (see text). Right panel: Even though the original image was taken with a CCD camera capable of capturing no more than 8 bits of image gray values, the quantizing error does not appear when the same image is processed after first converting to a 16-bit image.

C. Arithmetic Precision

Because unsharp masking is a subtractive process, the dynamic range of the processed image is necessarily less than that of the original. For example, if 95% of the haze is removed from an original pixel value of 100, then the processed pixel will have a pixel value of roughly 5. With an 8-bit image containing 255 gray values, the best resultant image will have integer gray values ranging only between 0 and 12 (0 to 5% of 255).

Although these values may be scaled through contrast enhancement to fill the full black-to-white dynamic range of the video system, there will never be more than 12 distinct gray values. In most cases this results in unsatisfactory quantization effects (Fig. 16, left panel). Fortunately, we can minimize this undesirable effect by scaling the original image.

If the original image values were first multiplied by 100, then the range of gray values would span the range of 0 to 25,500. Even though there are still only 255 unique gray values, the convolution operation is more accurate when starting with larger values since the computed values retain more precision. In other words, scaling allows integer-based operations to compute results that would have contained fractional values, which are lost in integer computations.

After scaling, even with a 95% haze removal factor, the resultant image could contain pixel values ranging from 0 to 1275, more than enough for excellent quality visualization (Fig. 16, right panel). Most importantly, a scaled image after processing contains a greater number of distinct gray values than an unscaled image after processing, resulting in an image free from quantization artifacts.

It should be noted that such improvements by scaling can be obtained for most convolution operations and many arithmetic operations (image averaging, division, etc.), since such operations typically involve steps that would generate fractional values that are rounded off to integer values in the resultant image data.

D. Significance of Unsharp Masking

Unsharp masking selectively increases the contrast of fine image details, including edges (without the ringing that accompanies conventional high-pass filtering or convolution), while reducing the contrast of low spatial frequency components of the image, suppressing haze and background fluorescence, etc.

Some argue that such sharpening generates an image that is artificial and even artifactual. Indeed, images convolved with any kernel could be viewed with such criticism, especially when one is concerned with exact distribution of intensity in the image. However, one should recognize that the generation of an image through *any* optical instrument, such as a microscope, or the recording or transmission of image data through analog video recorders or image processors, involves a series of hidden convolutions that can differentially attenuate or accentuate the spatial and temporal frequencies present. Not only microscopes using any contrast-generating mode, but our own visual system, including the optics in our eye, the retina, and other parts of our visual neural system, processes an image in such a way as to extract or emphasize certain significant features. Extreme examples would be the loss of contrast of high-frequency components (by the drop in modulation function) as one approaches the resolution limit in any diffraction-limited optical instrument including microscopes; selective enhancement of the higher frequency components in darkfield, phase contrast, DIC, etc.; loss and compensation of losses for certain frequency signals in video format signal transmission and especially recording; and edge and motion detection and image feature filtering, or accentuation, in our own visual system (see, e.g., Inoué and Spring, 1997, Chapters 2, 4, and 11).

In interpreting the significance of any image or image feature, we naturally need to be cognizant of these series of filtering or convolving and deconvolving steps. Yet, as the listed examples show, selection or emphasis of particular image features at the cost of suppressing other features is not only often an advantage, but also a necessity for generating data or information that is meaningful from a particular perspective. The question is, from what perspective or for seeing what image attributes?

Unsharp masking, in part, compensates for the high-frequency image components that were reduced or lost in the optical or electronic transmission and recording process. Thus, e.g., it can significantly restore image details whose contrast was reduced and that *appeared to be* lost in video recording. On the other hand, as with any operation that enhances the high spatial frequency components, it could bring forth undesirable high-frequency image noise. Unlike convolving with simple high-pass kernels, however, there is less tendency to introduce ringing or extraneous high-frequency boundary lines, etc., as described earlier.

If excessive noise, such as random bright spots, is seen after unsharp masking, the noise can often be reduced or eliminated by first appropriately filtering the original image, e.g., by applying a median filter (an operation that replaces each pixel with the median value of the local neighborhood's pixel values). Similarly, periodic noise, such as video scan lines and motion artifacts, is eliminated by Fourier filtering to blur the image before unsharp masking, e.g., by applying a "blur" fast Fourier transform (FFT).

Unsharp masking effectively brings out fine details in the image and suppresses haze, background, and contributions from any large image regions with uniform brightness. Thus, it is most effective for aiding in the visualization or measurement of locations or distances between finer image features. It is clearly not intended for photometric measurements.

E. Image Processing in 3-D Projection and Stereo Image Generation

Unsharp masking is especially effective when building up 3-D or 4-D (3-D with time) image projections, for example, for stereoscopic presentation of complex objects as seen in Fig. 17.

In contrast to unsharp masking and other convolution operations that are applied to images taken at single focal planes, deconvolution operations use images in a whole stack of focal planes to compute each theoretical, unadulterated, in-focus image. Thus, they eliminate, or minimize, haze and other undesirable image features that were present in the original microscope image that arose by superimposition of light waves that originated from out-of-focus planes. Although it is computationally extremely intensive, involving many rounds of iterative operations using the full stack of images obtained at many foci, deconvolution can produce nearly perfect optical sections free from out-of-focus contributions. It is desirable for obtaining very clean optical sections, for photometry, and for 3-D reconstruction of relatively uncomplicated or isolated systems with high image contrast and low noise in the initial image stack.

For generating time lapse, and especially 4-D (time-lapsed 3-D) confocal images, or for preparation for morphometric measurements of specimens with complex 3-D structures, unsharp masking and *nearest-neighbor haze-removal* or *inverse filtering* deconvolution algorithms provide major advantages. Although limited by the caveats already described, data stacks for many time points can be processed with simpler computers and in a much shorter time. Indeed, today one can carry out such operations as fast as the images can be acquired using the faster, commonly available personal computers.

F. From Image Stacks to Projections

Presenting 3-D image stacks is a significant challenge for microscopists. As illustrated in Fig. 14, one can display a few representative images from the stack. Although this results in the clearest display of fine details in each focal plane, it can be difficult to interpret the 3-D structure of the specimen from serial optical sections.

A second, common approach is the *maximum projection* (e.g., Shotton, 1993, Chapter 2). This method compresses a stack into a single image based on the maximum brightness for each pixel through all the planes in the stack. For example, the resulting images illustrated in each panel in Fig. 17 contain information on the most prominent features through the sample. Maximum projection also gives some degree of transparency, as bright objects within the sample have precedence over darker objects in front of them.

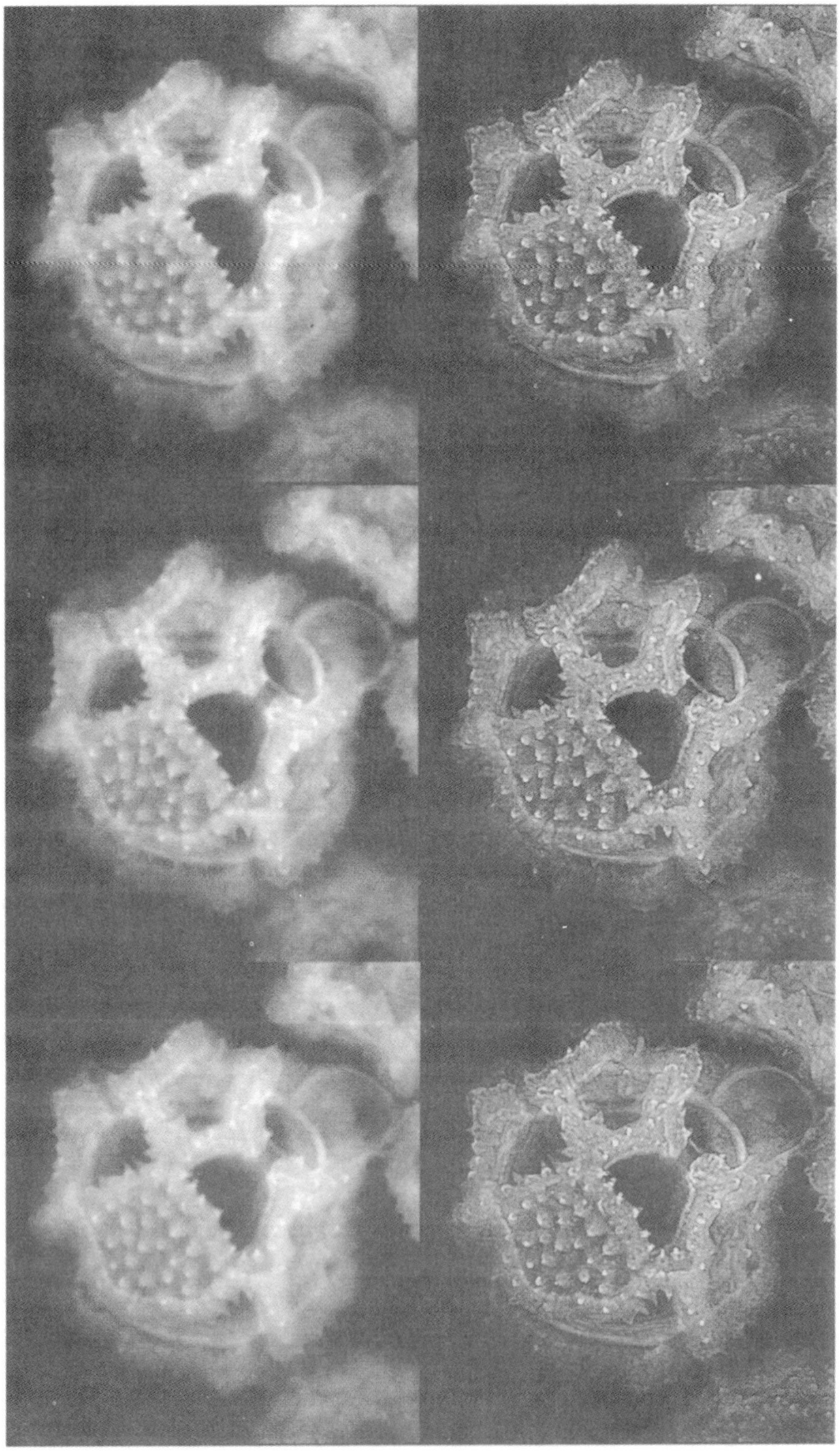

However, as with all two-dimensional representations of real objects, the image appears flat, containing few cues as to the true 3-D structure of the object being observed.

In order to obtain more *depth cues* to assist in the interpretation of the image, the maximum projection may be computed at two or more angles. When viewed as an animation, the projected stack appears to rotate, resulting in a compelling 3-D display. Alternatively, pairs of projections may be displayed as stereo pairs, allowing truly remarkable views of the specimen (Fig. 17).

Rotation of stereo pairs adds even more visual cues. These dynamic stereo images take advantage of our visual system's ability to interpret 3-D structures through relative motion in addition to stereopsis resulting in a clear 3-D view.

G. Viewing of 3-D and 4-D Images

Just as microscopists migrated from static photography to motion pictures, today video microscopists are moving from single image stacks to series of stacks collected over time, popularly called *4-D microscopy.*

However, outside the laboratory, while 2-D movies are the norm, 3-D movies are rare, still only seen at places like Disney World. Why is this?

Importantly, there have been few breakthroughs in 3-D image visualization, even for static 3-D images. Although some promising technologies exist for creating holograms or other *objective space 3-D images* (see, e.g., Inoué, 1986, Section 11.5), most practical methods depend on special glasses that allow the light from only one perspective to reach each eye.

Fig. 17 Stereo pairs of dandelion pollen grain. Fig. 17 is presented as a novel "three-panel stereogram" in which the two end panels are identical. With such an arrangement, one can view the stereo images with the unaided eyes by "cross-eyed" or "parallel-eyed" viewing, or with the aid of stereo viewing glasses. Either way, upon fusion, one sees a set of four panels with the two inner panels being an ortho- and pseudo-stereoscopic (distance cues reversed) view. For cross-eyed viewing, the panels here generate the pseudo-stereo image as the left and the ortho-stereo image as the right of the middle two panels. (In the pseudo-stereo view, the pollen grain appears concave, with the external ridges and spines pointing away, and the pollen grain appears in the ortho-view mound-shaped with the appendages pointing towards and to the side of the viewer.) Observed with stereopticons or wall-eyed viewing, the locations of the ortho- and pseudo-views are reversed. The three-panel stereogram makes the stereoscopic images directly available to those able to fuse the images with their unaided eye (whether cross- or wall-eyed), as well as for those needing to use viewing glasses. In addition to the convenience of access, side-by-side presentation of ortho- and pseudo-stereograms helps one perceive the true three-dimensional structure of the specimen (with the closer and distant features, respectively, pointing towards the observer) without being misled by shadowing, perspectives, and other cues that can otherwise confuse our senses. Technical details. Each panel in Fig. 17 was generated by projecting (at inclinations of ± 5 degrees) the brightest pixel in the line of sight (max-projection) through a stack of 31 optical sections covering the proximal half of the pollen grain. Top pair without, and bottom pair with, unsharp masking (some individual optical sections shown in Fig. 5 and 14). Before unsharp masking, the original stack was converted to 16 bit and Fast Fourier filtered with the "low blur" filter in order to reduce final image noise (see Section VII.D.). Stereo images of the lower panels, generated after unsharp masking, exquisitely display details of the thin, fin-like ridges and the spines with thin pores. Those of the unprocessed upper panels better show some of the larger features, such as the balloon-shaped vellum lying outside of the lower right pore.

For many years, *anaglyphs* (red–green or red–blue stereo pairs) have been used for the presentation of stereo images. These work by printing or otherwise presenting each perspective image in a unique color, usually red and green. The viewer then wears glasses containing green and red filters so that each eye sees only a single image.

Although anaglyphs are convenient for certain types of presentation, they do not work well with analog video equipment. Because anaglyphs depend on displaying pure colors that match colored filters in glasses, they are generally not compatible with video taping and other analog recording systems. Nor do they work well with image compression techniques that can also result in image color shift. These systems encode color information for recording and transmission in such a way that the colors shift, resulting in bleed-through that destroys the stereo effect. However, with the prevalence of purely digital storage and display, anaglyphs are again becoming a feasible stereo display technology.

Far better, however, are 3-D viewing systems based on polarizing glasses (see, e.g., Inoué and Spring, 1997, Section 10.8). There exist at least two methods for displaying and viewing stereo images using polarizing systems.

The first method, used for years in movie theaters, uses two projectors, one for each viewing perspective. The two projectors contain polarizing filters that are oriented so as to polarize the light in exactly opposite orientation from one another. For example, one might contain a horizontal polarizer while the other would contain a vertical polarizer.

The viewer wears glasses with corresponding polarizers so that each eye sees only light from the desired projector. The crossed polarizer blocks the other image. A slight variant of this technique uses left and right circularly polarized light rather than linearly polarized light. This version does not require exact alignment of the axes of the viewing glasses relative to those of the projector.

A second method that produces even better results uses special glasses that have electronically controllable liquid crystal shutters. In this method, the images for the left and right eye are displayed alternately (at twice the normal frame rate to avoid image flicker) and synchronized with the shutters in the glasses. When the left eye's image is displayed, the right eye's shutter is closed and vice versa (Lipton, 1991).

This last method of stereo image presentation has been the most effective of all that we have evaluated, as the shutters force each eye to see the appropriate image. Even people who have difficulty seeing stereoscopic displays using other methods typically see 3-D images with ease when using shuttered stereo goggles.

H. Storage Requirements

In the late 1980s, when we started to acquire stacks of 3-D images digitally, computer memory and hard disks were quite expensive. A typical image stack consisting of 50 images uses 12 megabytes per 3-D data set. At that time, hard drive capacity was only large enough to store approximately 10 of these images, making it impractical to store time-lapsed 3-D images on a hard drive. It was, therefore, often necessary to save image stacks using analog video media such as videotapes and laser disks.

Today in 2002, the tiny Zip disk holds as much data as the previous state-of-the-art hard drives, and current hard drive capacities have swollen to beyond 250 gigabytes, over a 1000-fold increase in less than 12 years. Image sizes have also increased, in many cases to approximately 2 megabytes per image and 100 megabytes per stack. Even with the vastly increased storage capacities, the largest disks still only store approximately 2500 3-D time points. Although this is an impressive amount of storage, one can imagine the difficulties that might be encountered when working with such large data sets.

A primary consideration is data backup. If a single image stack consumes 100 megabytes of storage space, even CD-R disks, with a capacity of 650 megabytes, are inadequate. Newly released devices, such as the writeable DVD technologies, with capacities of approximately 5 gigabytes are considerably better, though still too small for anything but relatively short experiments.

VIII. Mechanical and Optical Performance of the CSU-10

In order to generate a compact scanner that provides an excellent quality image, the designers of the CSU-10 chose a Nipkow disk diameter of 55 mm, on which some 20.000 pinholes were placed in a novel arrangement (Fig. 18; Ichihara *et al.,* 1996). They avoided placing the pinholes in a "fixed-angle helical pattern," which produces reduced illumination towards the periphery of the spinning disk, or a "tetragonal pattern," which while providing uniform illumination, generates an uneven scanning pattern, i.e., scanning stripes or "streaks." Instead they placed the pinholes in a "constant-pitch helical pattern" that gives both uniform illumination and even scanning that is free from streaks (assuming that the Nipkow disk rotation and video camera vertical scanning rates are synchronized; see footnote 2).

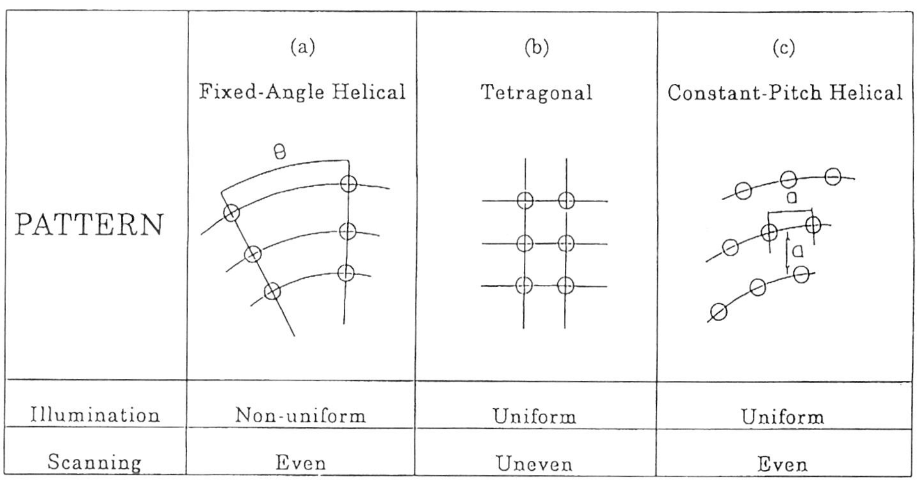

Fig. 18 Influence of pinhole pattern in the Nipkow disk on the uniformity of field illumination and evenness of scanning pattern and patented arrangement of pinholes (constant-pitch helical) used in the CSU-10. (From Ichihara *et al.,* 1996, with permission.)

Indeed, as shown in the illustrations in this paper, the CSU-10 provides a uniformly illuminated confocal field free of scanning stripes.

The diameter of each pinhole in the primary Nipkow disk is 50 μm, and they are spaced 250 μm apart, yielding a pinhole-fill factor of 4% [$(50/250)^2$]. The disk containing the microlenses is mechanically fixed to the second Nipkow disk so that each microlens is well aligned and focused onto the corresponding pinhole. Since the same pinhole acts both as the entrance and exit pinhole, a slight run-out of the axis of disk rotation would have minimum impact on the confocality of the system.

Nevertheless, any mechanical vibration introduced by the disks or the electric motor driving the disks could result in significant image deterioration, especially with the scanner perched on top of a microscope using a high-power objective lens. In fact, we were impressed to find no such sign of image vibration, or deterioration, on any of the three brands of research microscopes that we tested. Indeed, the lateral image resolution through the CSU-10 seemed to be fully determined by the NA of the objective lens used and the wavelength involved (for expected lateral resolution, see, e.g., Inoué and Oldenbourg, 1995).

The depth of field provided by the CSU-10, measured on fluorescent microspheres and a first surface mirror, has been reported to lie between 1.0 and 1.3 μm using $\times 100$, 1.3 to 1.4 NA objective lenses and 488-nm excitation (Yokogawa Research Center, personal communication). Such measurements made on extremely thin test specimen, and with confocal pinholes somewhat larger than the Airy disk produced by the objective lens, should yield a "depth of field" (i.e., the depth, projected into specimen space, where the intensity of the diffraction pattern falls to a half of its peak value) that is more characteristic of the objective lens performance than the confocal system performance. Rather than the depth of field defined by very thin objects, for multiple pinhole systems such as in the CSU-10, our concern is more in terms of the ability, or efficiency, of the system to eliminate or reduce the contribution from out-of-focus regions of the specimen.

The authors so far have no numerical value to ascribe to such efficiency for the CSU-10. Instead, the reader needs to judge the efficacy of the scanner from the illustrations presented here, through copies of dynamic images available on CD disks (see footnote 7), and through direct examination of their own specimens through a microscope equipped with a scanner unit. It may well be that for specimens or scenes in which the specimen is not moving or changing rapidly, or subject to bleaching, a laser point-scanning microscope would be more suited for obtaining better discrimination of out-of-focus regions.

Nevertheless, as illustrated, images obtained with the CSU-10 can rapidly be post processed to reduce out-of-focus contributions, to bring out greater image detail, and/or to generate impressive rotating, depth-stacked, or stereoscopic images.

IX. Potentials of the CSU-10

As stressed in this chapter, a major advantage of real-time, direct-view confocal microscopy is the ability to very easily observe, display, or record moving specimens or scenes that are changing under our eyes.

Another advantage, which may not be as obvious, is the ability to readily find sparsely distributed, weakly fluorescent objects. Such objects are quite difficult to find with point-scanning systems since point scanners yield only one low-noise, full-frame optical section every few seconds.

In contrast, for example, with the CSU-10 equipped with a video-rate intensifier CCD camera, one can focus through the specimen and rapidly find sparsely distributed sources of very weak fluorescence. Such images can be somewhat noisy as characteristic of highly intensified low-light-level video images. Nevertheless, we were able to find images of very sparsely distributed, 20-nm-diameter fluorescent spheres (which are most difficult to find with a point scanner) without any trouble on the video monitor as we focused through the specimen (Smith, S., Stanford University, and Inoué, S., 1998, unpublished observation).

Such ability to rapidly locate and follow weak fluorescence becomes especially important for observing the behavior of a few fluorophores. They may be sparsely distributed in space, and their fluorescence emission may last only very briefly.

Indeed, Tadakuma *et al.* (2001) confirms the possibility of visualizing fluorescence from single molecules of tetramethyl rhodamine in cultured cells of *Xenopus* excited with 514.5-nm argon laser emission in the CSU-10 (see also Funatsu, 1995). Funatsu *et al.* found the intensity of fluorescence (which lasted for several seconds before suddenly disappearing) to be comparable to that established earlier for single fluorophores using total internal reflection (evanescent wave) fluorescence microscopy (Funatsu *et al.,* 1995; Vale *et al.,* 1996). Thus, the CSU-10 can be expected even to serve as a tool for studying, in a living cell, the behavior or activity of single molecules, provided they are not undergoing rapid Brownian motion.

Waterman-Storer and Salmon (1998) have devised a method for visualizing specific regions on a single microtubule where the tubulin molecules were assembling or disassembling. This was accomplished by fluorescent speckle microscopy (FSM) in which the polymer, labeled with a very low concentration of X-rhodamine-conjugated tubulin, was observed with a low-noise CCD camera. The nonuniform statistical distribution of the sparse label then gave rise to a punctate, or speckly, appearance to the individual microtubules when the fluorescence was observed with a high-NA objective lens and recorded digitally with high resolution. The speckle pattern, originating from the small, random numbers of fluorophores that were incorporated per unit length of the polymer (i.e., per length corresponding to the resolution limit of the objective lens), remained sufficiently stable during the observation to serve as unique markers for the particular stretch of polymer. Thus, one can identify which region of the single microtubule had disassembled or is being newly assembled and which remains stable as shown in Fig. 19.[9]

[9] FSM reveals, among other attributes, three events related to the motility and exchange of subunits in microtubules: a) the microtubules may move without exchange of subunits (the whole speckle pattern moves without adding or losing speckles); b) free ends of microtubules may grow or shorten by addition or subtraction of subunits (as in Fig. 19, the existing speckle pattern remains stationary, new speckles are added at growing ends, and old speckles disappear from shortening ends); or c) the subunits may treadmill through the microtubules (as in Fig. 20, the same pattern flows through the microtubule).

A

B

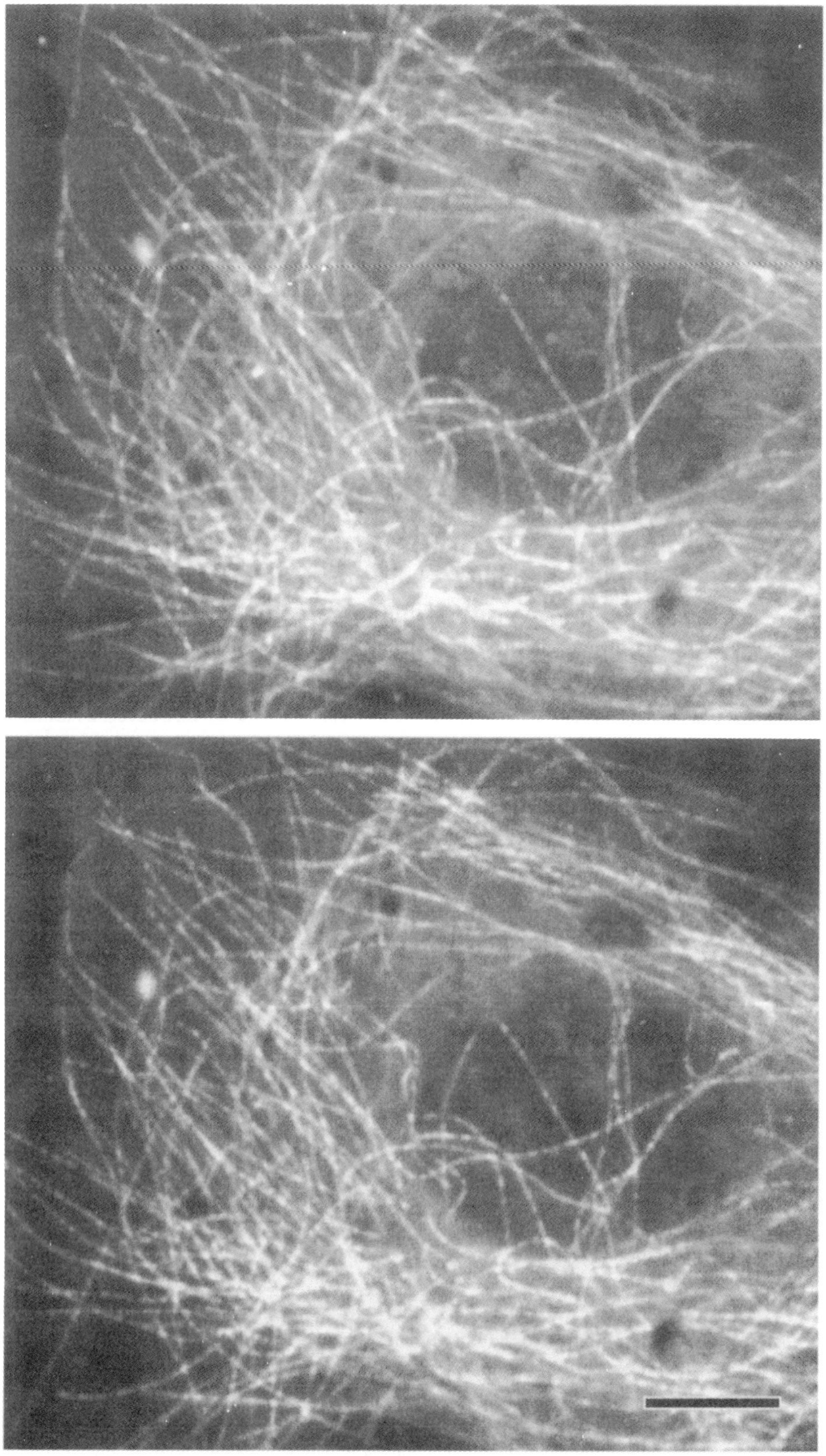

Using FSM, Maddox and co-workers (1999) recorded the poleward flux of tubulin taking place in spindles formed in *Xenopus* egg extracts (see also Waterman-Storer *et al.,* 1998). In order to visualize the flux of speckle pattern through these relatively thick (ca 5 μm) spindles with wide-field fluorescence microscopy (Fig. 20C), it was necessary to lable the microtubules with very low fractions (0.05%) of fluorescent tubulin (Fig. 20D). With such a low fraction of tubulin contributing to the fluorescent image, one could not readily discern the detailed structure of the spindle or discern the individual microtubules.

On the other hand, by observing a more heavily (1%) labeled *Xenopus* spindle (Fig. 20A) through the CSU-10, one could clearly record the poleward tubulin flux in the spindle fibers (Fig. 20B). In dynamic playback of the recorded time-lapsed scenes, such flux, reflecting the distribution of very few molecules of labeled tubulin, could even be observed in single microtubules growing from the spindle pole outside of the central spindle region (Fig. 20A, arrows).

As discussed in Section VI, the capability of the CSU-10 for capturing full-frame optical-sectioned images at considerably greater than video rate (1 to 4 ms/frame), albeit with the use of special high-speed cameras and recording devices, opens up further possibilities for monitoring rapid phenomena hitherto unobservable.

Very recently the CSU-10 reportedly has also been used for fluorescence resonance energy transfer experiments between modified GFP molecules (M. Shimizu, Yokogawa Electric Corporation, personal communication).

In addition to such advanced applications, the real-time, direct-view capability of the CSU-10, the ability to directly view confocal fluorescence images in true color, its adaptability to virtually any upright or inverted research microscope, its compact size and mechanical stability, the high quality of images obtained with brief exposures, its low rate of fluorescence bleaching, and its exceptional ease of use should open up many ready applications in cellular, developmental, and neurobiology, medical screening, etc.

Fig. 19 Confocal fluorescence speckle microscopy of microtubules (MTs) in tissue cultured cell. Panels A and B show two selected frames from a time-lapsed sequence of an interphase PTk2 cell injected with a low concentration of porcine tubulin labeled with X-rhodamine. The cells in panels A and B are the 32nd and 151st of the 152 frames that were recorded once every 4 s. Scale bar = 10 μm. As seen in the two panels, the speckle patterns on the *persistent* interphase MTs (seen in the middle of the cell) are stationary and remain remarkably constant over this long period of time. In contrast, the *free ends* of many of the MTs (pointing toward the cell periphery to the left of the panels) grow and shorten actively. In panel B, several new or longer individual MTs can be seen, while many others present in panel A have shortened or disappeared. These changes are dramatically displayed when the full sequence is played back dynamically. *Technical details:* Using the 568-nm illumination from a Kr/Ar laser, portions of the cell lying next to the coverslip were observed through the CSU-10 with an Olympus 100 × /1.3 NA objective lens. The confocal image (each exposed for 1700 ms) was captured with a Hamamatsu chilled-CCD camera (Orca II) in the 14-bit mode and stored as a stack file in MetaMorph. The somewhat-less sharp signal for panel A was enhanced to match panel B by first applying a median filter (horizontal = vertical size = 3, subsample ratio = 2) to prevent the rise of background noise, and then by applying an unsharp masking operation (low-pass kernel size = 3, scaling factor = 0.5, with auto result scale). (For the meaning of these operations, see Section VII.B of this article.) Data courtesy of Dr. Clare Waterman-Storer, Scripps Research Institute.

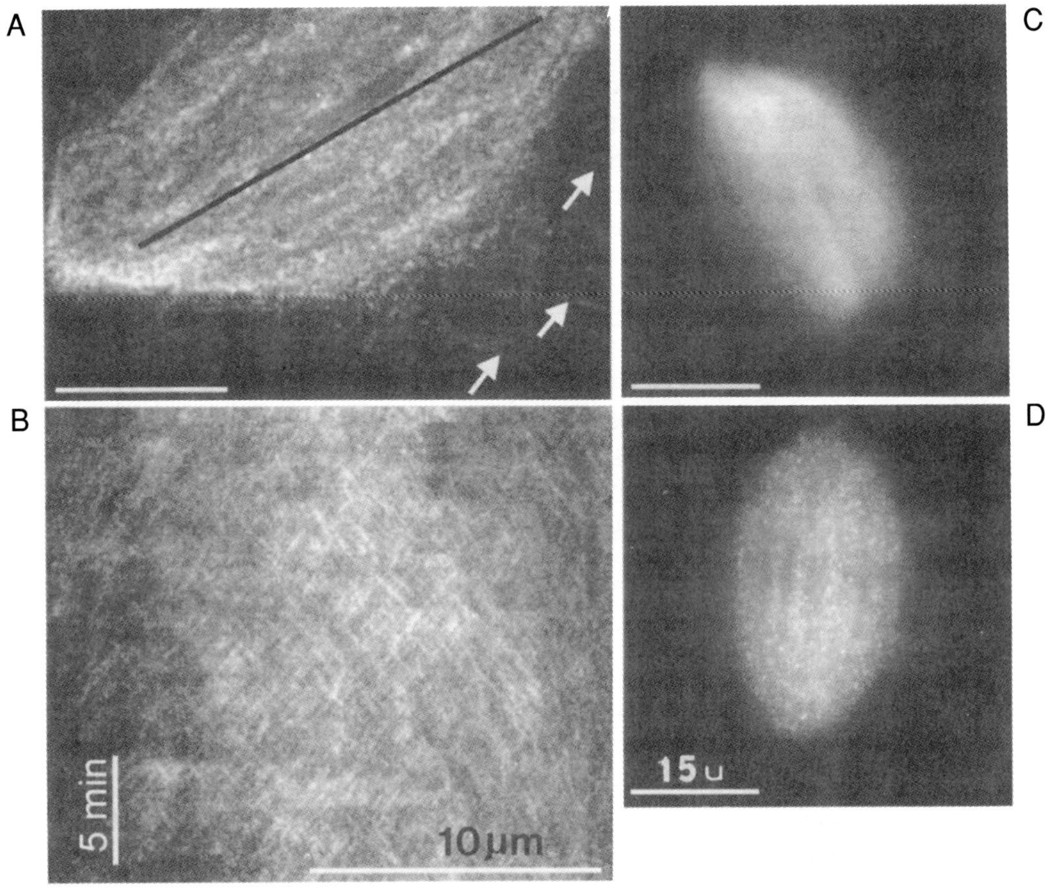

Fig. 20 Fluorescence speckle microscopy of tubulin in spindle formed in *Xenopus* egg extract. (A) Optical section recorded on a low-noise, chilled-CCD camera (Orca I, Hamamatsu Photonics) through CSU-10. Speckles are seen along spindle fibrils as well as in microtubules fanning out (arrows) from the spindle. Approximately 1% of tubulin was labeled with rhodamine-X. (B) Kimograph along the dark line in panel A. The speckles (revealing tubulin flux in the microtubules) move poleward at ca 2.0 μm/min. (C,D) Wide-field epifluorescence of similar spindle. Fluorescence speckles are visible in D in a spindle incorporating only ca 0.05% fluorescent tubulin, but not in C with ca 2% labeled tubulin. Scale bars for A, B = 10 μm, for C, D = 15 μm. (From Maddox *et al.,* 1999, with permission.)

The capability for video-rate and faster recording and display of dynamic events in optical section or as through-focused images, digital processing to rapidly obtain exceptionally sharp and clean optical sections or projections at any tilt angle, and striking depth-stacked or stereoscopic views that are generated using commercially available digital processors open up various possibilities for additional experiments and observations.

X. Addendum A (April, 2002)

Although three years have elapsed since this article was prepared, we believe that much of its content stands as originally presented. Nevertheless, we have appended the following brief descriptions to bring the article up to date.

A. Yokogawa has now introduced the CSU-21, which does not completely supercede but adds functions not present in the CSU-10. In brief, instead of a fixed speed of 1,800 rpm, the CSU-21 incorporates a variable speed motor that allows choice of Nipkow disk rotational speed of up to 5,000 rpm under computer control. Thus up to 1,000 full frame images can be acquired every second with the CSU-21. In addition, the fluorescence excitation filters, emission filters and dichromatic mirrors (3 each), neutral density filters, and illumination beam shutter can all be controlled either using the panel switches located below the eyepiece on the CSU-21 or remotely via computer. Remarkably, these new functions have been introduced without altering the external dimensions or the number of pinholes and microlenses (20,000 each) on the spinning disks. As before, the unit can be attached to any research-grade upright or inverted microscope, and images in full color can be viewed through the eyepiece in real time, or monitored or recorded through appropriate cameras mounted through C-mount threads.

B. The speed and capacity of computers and storage devices continue to improve by leaps and bounds. For example, newer computers using the Intel Pentium 4 running at over two GHz are several times faster than our reference computer, a 400 MHz Pentium II machine, reducing the computation time for a 3 × 3 neighborhood processing operation on 100 images from five to approximately one second, or several times faster than video rate (Section VII.A.). We have updated some numbers in Section VII.H. to reflect these changes.

C. Meanwhile, CCD cameras with considerably improved sensitivity, noise level, effective wavelength range, and general capabilities have become available (see, e.g., Maddox *et al.,* 2002, and articles in Pawley Handbook of Confocal Microscopy, 3rd edition, 2002, in preparation).

D. In an article on a Biorad confocal microscope incorporating SELS (signal-enhancing lens system), Reichelt and Amos compare the performance of their system with the CSU-10 (Microscopy and Microanalysis, Nov. 2002, pp. 9–11). As with the CSU-10, their SELS system operates with high photon capture efficiency, and hence can be used for observations of living cells with low rates of fluorescence bleaching and photon damage. The SELS point-scanning confocal system has the advantage that it can be switched to standard point scanning confocal mode with its greater depth discrimination. However, in either mode with the point scanner, extended scan time is required to capture images with any appreciable frame size. In contrast, the CSU provides full-frame capture at video- and higher-rate-imaging, an essential feature whenever the specimen is changing or moving moderately rapidly. While the authors of the SELS system argue that the SELS provides a much better z-discrimination compared to the CSU, their comparison is based on normalized z-axis intensity profiles obtained on a large thin film of a fluorescence dye (Nile Red). In fact, for cell biology applications where the imaging of fine specimen detail (rather than the location of an "infinite-sized fluorescent layer") is important, we believe that the CSU provides a considerably better depth discrimination than the SELS. For example, for yeast cells (*S. cervisiae*) the authors of the SELS article report that "the depth of focus became so great that most of the depth of each cell was included." In contrast, with the CSU-10 the focal depth was some one fifth of the thickness of the yeast cell, *S. pombe,* whose diameter is only ca. 3 μm (Fig. 10 of our article.) With both the SELS and CSU systems, depth discrimination and specimen detail are improved further by post processing, such as with unsharp masking (Fig. 14) or by applying deconvolution algorithms (Maddox *et al.,* 2002). The CSU is clearly the system of choice whenever full frame images of rapidly changing or moving specimens are to be recorded or viewed (e.g., Figs. 9 to 12), or when serial optical sections need to be acquired in rapid succession (e.g., Maddox *et al.,* 2002).

E. Selected additional references: (i) Members of the Salmon laboratory at the University of North Carolina provide details on the practical use of the CSU-10 for achieving very high resolution images in GFP imaging of a variety of living cells undergoing mitosis, including methods for visualizing the dynamic flow of tubulin molecules within the microtubules (Maddox *et al.,* 2002; see also Cimini *et al.,* 2001; Grego *et al.,* 2001; Waterman-Storer and Salmon, 1998). Maddox *et al.* (2002) also provide important data on the performance of the CSU-10 used in conjunction with selected hardware and software. (ii) Nakano and coworkers have made extensive use of the CSU-10 for analyzing endocytic and exocytic pathways of selected molecules in plant cells, including video-rate observations of the movement of Cop1 molecules between Golgi and ER in yeast cells (Saito *et al.,* 2002; Sato *et al.,* 2001; Ueda *et al.,* 2001; Yahara *et al.,* 2001). (iii) Tadakuma *et al.,* (2001) from the Funatsu group present data showing the real-time imaging capability of the CSU-10 for observing the behavior of fluorescence-labeled single protein molecules.

XI. Addendum B (personal communication from Dr. Kenneth R. Spring, June, 2002)

The extent of bleaching of fluorescence with the CSU-10 during illumination of a test specimen with 488-nm laser light was determined by K. R. Spring. A 1-μm-thick film of eosin-stained polymer was used to determine the fluorescence-bleaching rate caused by continuous illumination through a 40 × /0.75 objective lens at an intensity at the specimen of 45 microwatts/cm^2. The resulting fluorescence intensity decreased exponentially, as expected from photo bleaching with a half time of 445 seconds. When the experiment was repeated under identical conditions with another video-rate confocal attachment, the Noran Odyssey, the half time was not significantly different, 420 seconds.

Since the rate of bleaching is that expected for the illumination intensity, it is worthwhile to consider why users of the CSU-10 have reported less bleaching. We have identified two factors that contribute to the improved S/N of the CSU-10 that others and we have noted. First, because the excitation light is spread over 1200 pinholes in the field of view, each point in the image is irradiated with 1/1200 of the intensity of that in a video-rate point-scanning confocal system. Second, because each point is illuminated 12 times in a video frame, the collected image has already undergone some averaging.

When a point in the specimen is illuminated using the CSU-10, the fluorophores are probably all in their ground state because of the comparatively long time between each period of illumination (about 3 msec), the relatively long illumination period (about 30 μsec), and the relatively low intensity of the exciting light. In video-rate point-scanning of 512 points per horizontal line, each point is excited only once during each video frame for about 80 nsec with an intensity that is 1200 times higher than in the CSU-10. Under these circumstances, most of the fluorophores in the illuminated volume will be excited and cycle as rapidly as possible-between the ground and excited states (a fluorescein molecule could undergo only about 15–20 cycles of absorption and emission during the 8-nsec interval). The total fluorescence emission that a single fluorophore can produce with saturating excitation light in the 80-nsec interval is far less than that emanating from the same molecule illuminated for 12–30-μsec intervals with three orders of magnitude lower excitation light intensity.

The S/N achieved in the CSU-10 is improved to such a great extent that much lower excitation levels can be employed to produce visually acceptable images. The reduced bleaching that is observed with the CSU-10 most probably stems form the need for less excitation light on the specimen rather than form anything special about the manner in which that light is delivered to the specimen through the objective lens.

Acknowledgments

The authors thank Dr. Ken Spring of NHLBI, NIH, for contributing Addendum B, Dr. Brad Amos of Cambridge University for the gift of the dandelion pollen grain slide, Dr. John Murray of the University of Pennsylvania and Rieko Arimoto of Washington University for their efforts to prepare isolated sea urchin spindles, Dr. Brian Matsumoto of the University of California at Santa Barbara for participating in recording the swimming *Tetrahymena,* Dr. Phong Tran of Columbia University for providing records of yeast cells, Dr. Clare Waterman-Storer of Scripps Research Institute and Paul Maddox and co-workers of the University of North Carolina for providing the fluorescence speckle images, Dr. Yoshio Fukui of Northwestern University for providing the panels of *Dictyostelium,* Dr. Baggi Somasundaram of PerkinElmer Wallac for providing extensive image data, and Dr. Hideyuki Ishida of Tokai University and Dr. Akihiko Nakano of Riken Institute for providing information their recent work. We are grateful to members of the Yokogawa Electric Corporation for use of the CSU-10 confocal scanning unit, as well as for providing several original illustrations, information on the CS4-21 (see Addendum A), and list of publications; to members of Leica, Nikon Inc., and Carl Zeiss Inc. for use of their microscopes; and to members of Dage-MTI and Hamamatsu Photonics Systems for making available their chilled-CCD cameras. The authors also thank Dr. Edward D. Salmon and Paul Maddox of the University of North Carolina, Dr. Fred Lanni of the Carnegie Mellon University, Dr. Rudolf Oldenbourg of the Marine Biological Laboratory, Dr. Clare Waterman-Storer of Scripps Research Institute, and Dr. Phong Tran of Columbia University for thoughtful discussion on problems related to fluorescence fading and image-capture efficiency of the CSU-10. Finally we thank David Szent-Gyorgyi of Universal Imaging Corporation for providing extensive help in acquiring high-quality printouts from the MetaMorph imaging system, and Jane MacNeil and Bob Knudson of the MBL for invaluable help throughout the preparation of the many revisions of this article.

References

Agard, D. A. (1984). Optical sectioning microscopy: Cellular architecture in three dimensions. *Annu. Rev. Biophys. Bioeng.* **13,** 191–219.

Amos, W. B., and White, J. G. (1995). Direct view confocal imaging systems using a slit aperture. *In* "Handbook of Biological Confocal Microscopy" (J. B. Pawley, Ed.), pp. 403–415. Plenum Press, New York.

Aoki, T., Suzuki, Y., Nishino, K., Suzuki, K., Miyata, A., ligou, Y., Serizawa, H., Tsumura, H., Ishimura, Y., Suematsu, M., and Yamaguchi, K. (1997). Role of CD18-ICAM-1 in the entrapment of stimulated leukocytes in alveolar capillaries of perfused rat lungs. *Am. J. Physiol.* **273,** H2361–H2371.

Castleman, K. R. (1979). "Digital Image Processing." Prentice-Hall, Englewood Cliffs, NJ.

Cimini, D., Howell, B., Maddox, P., Khodjakov, A., Degrassi, F., and Salmon, E. D. (2001). Merotelic kinetochore orientation is a major mechanism of aneuploidy in mitotic mammalian tissue cells. *J. Cell Biol.* **153**(3), 517–527.

Denk, W., Piston, D. W., and Webb, W. W. (1995). Two-photon molecular excitation in laser-scanning microscopy. *In* "Handbook of Biological Confocal Microscopy" (J. B. Pawley, ed.), pp. 445–458. Plenum Press, New York.

Fukui, Y., Engler, S., Inoué, S., and de Hostos, E. L. (1999a). Architectural dynamics and gene replacement of coronin suggest its role in cytokinesis *Cell Motil. Cytoskel.* **42,** 204–217.

Fukui, Y., de Hostos, E., Yumura, S., Kitanishi-Yumura, T., and Inoué, S. (1999b). Architectural dynamics of F-actin in eupodia suggests their role in invasive locomotion in *Dictyostelium. Exp. Cell Res.* **249,** 33–45.

Funatsu, T., Harada, Y., Tokunaga, M., Saito, K., and Yanagida, T. (1995). Imaging of single fluorescent molecules and individual ATP turnovers by single myosin molecules in aqueous solution. *Nature* **374,** 555–559.

Genka, C., Ishida, H., Ichinori, K., Hirota, Y., Tanaami, T., and Nakazawa, H. (1999). Visualization of biphasic Ca^{2+} diffusion from cytosol to nucleus in contracting adult rat cardiac myocytes with an ultra-fast confocal imaging system. *Cell Calcium* **25,** 199–208.

Grego, S., Cantillana, V., and Salmon, E. D. (2001). Microtubule treadmilling *in vitro* investigated by fluorescence speckle and confocal microscopy. *Biophys. J.* **81,** 66–78.

Holmes, T. J., Bhattacharyya, S., Cooper, J. A., Hanzel, D., Krishnamurthi, V., Lin, W.-C., Roysam, B., Szarowski, D. H., and Turner, J. N. (1995). Light microscopic images reconstructed by maximum likelihood deconvolution. *In* "Handbook of Biological Confocal Microscopy" (J. B. Pawley, ed.), pp. 389–402. Plenum Press, New York.

Ichihara, A, Tanaami, T., Isozaki, K., Sugiyama, Y., Kosugi, Y., Mikuriya, K, Abe, M., and Uemura, I. (1996). High-speed confocal fluorescence microscopy using a Nipkow scanner with microlenses for 3-D imaging of single fluorescent molecule in real time. *Bioimages* **4,** 57–62.

Inoué, S. (1986). "Video Microscopy." Plenum Press, New York.

Inoué, S., and Oldenbourg, R. (1995). Optical instruments. Microscopes. *In* Optical Society of America, ed., "Handbook of Optics," 2nd ed., Vol. 2. McGraw-Hill, Inc., pp. 17.1–17.52.

Inoué, S., and Spring, K. R. (1997). "Video Microscopy," 2nd ed. Plenum Press, New York.

Inoué, S., Tran, P., and Burgos, M. (1997). Photodynamic effect of 488-nm light on Eosin-B-stained *Spisula* sperm. *Biol. Bull.* **193,** 225–226.

Ishida, H., Genka, C., Hirota, Y., Nakazawa, H., and Barry, W. H. (1999). Formation of planar and spiral Ca^{2+} waves in isolated cardiac myocytes. *Biophys. J.* **77,** 2114–2122.

Juškaitis, R., Wilson, T., Neil, M. A. A., and Kozubek, M. (1996). Efficient real-time confocal microscopy with white light sources. *Nature* **383,** 804–806. See also News and Views by Dixon, T., on pp. 760–761.

Kino, G. S. (1995). Intermediate optics in Nipkow disk microscopes. *In* "Handbook of Biological Confocal Microscopy" (J. B. Pawley, Ed.), pp. 155–165. Plenum Press, New York.

Lanni, F., and Wilson, T. (1999). Grating image systems for optical sectioning fluorescence microscopy of cells, tissues, and small organisms. *In* "Imaging Neurons—A Laboratory Manual" (R. Yuste, F. Lanni, and A. Konnerth, Eds.), pp. 8.1–8.9. Cold Spring Harbor Laboratory Press, Cold Spring Harbor, NY.

Lipton, L. (1991). "The Crystal Eyes Handbook." StereoGraphics Corp., San Rafael, CA.

Maddox, P., Desai, A., Salmon, E. D., Mitchison, T. J., Oogema, K., Kapoor, T., Matsumoto, B., and Inoué, S. (1999). Dynamic confocal imaging of mitochondria in swimming *Tetrahymena* and of microtubule poleward flux in *Xenopus* extract spindles. *Biol. Bull.* **197,** 263–265.

Maddox, P. S., Moree, B., Canman, J., and Salmon, E. D. (2002). A spinning disk confocal microscopy system for rapid high resolution, multimode, fluorescence speckle microscopy and GFP imaging in living cells. *Methods in Enzymology,* in press.

Pawley, J. B., Ed. (1995). "Handbook of Biological Confocal Microscopy." Plenum Press, New York.

Petráň, M., Hadravsky, M., Egger, M. D., and Galambos, R. (1968). Tandem scanning reflected light microscope. *J. Opt. Soc. Am.* **58,** 661–664.

Russ, J. C. (1995). "The Image Processing Handbook," 2nd ed. CRC Press, Boca Raton, FL.

Saito, C., Ueda, T., Abe, H., Wada, Y., Kuroiwa, T., Hisada, A., Furuya, M., and Nakano, A. (2002). A complex and mobile structure forms a distinct subregion within the continuous vacuolar membrane in young cotyledons of *Arabidopsis. Plant J.* **29,** 245–255.

Sato, K., Sato, M., and Nakano, A. (2001). Rer1p, a retrieval receptor for ER membrane proteins, is dynamically localized to Golgi by coatomer. *J. Cell Biol.* **152,** 935–944.

Shotton, D. Ed. (1993). "Electronic Light Microscopy." Wiley-Liss, New York.

Squirrell, J. M., Wokosin, D. L., White, J. G., and Bavister, B. D. (1999). Long-term two-photon fluorescence imaging of mammalian embryos without compromising viability *Nat. Biotechnol.* **17,** 763–767.

Tadakuma, H., Yamaguchi, J., Ishihama, Y., and Funatsu, T. (2001). Imaging of single fluorescent molecules using video-rate confocal microscopy. *Biochem. Biophys. Res. Comm.* **287,** 323–327.

Tran, P. T., Maddox, P., Chang, F., and Inoué, S. (1999). Dynamic confocal imaging of interphase and mitotic microtubules in fission yeast, *S. pombe. Biol. Bull.* **197,** 262–263.

Tsien, R., and Bacskai, B. J. (1995). Video-rate confocal microscopy. *In* "Handbook of Biological Confocal Microscopy" (J. B. Pawley, Ed.), pp. 459–478. Plenum Press, New York.

Ueda, T., Yamaguchi, M., Uchimiya, H., and Nakano, A. (2001). Ara6, a plant-unique novel-type Rab GTPase, functions in the endocytic pathway of *Arabidopsis thaliana. EMBO J.* **20,** 4730–4741.

Vale, R. D., Funatsu, T., Pierce, D. W., Romberg, L., Harada, Y., and Yanagida, T. (1996). Direct observation of single kinesin molecules moving along microtubules. *Nature* **380,** 451–453.

Waterman-Storer, C. M., and Salmon, E. D. (1998). How microtubules get fluorescent speckles. *Biophys. J.* **75,** 2059–2069.

Waterman-Storer, C. M., Desai, A., Bulinski, J. C., and Salmon, E. D. (1998). Fluorescent speckle microscopy, a method to visualize the dynamics of protein assemblies in living cells. *Curr. Biol.* **8** 22, 1227–1230.

Yahara, N., Ueda, T., Sato, K., and Nakano, A. (2001). Multiple roles of Arf1 GTPase in the yeast exocytic and endocytic pathways. *Mol. Biol. Cell* **12,** 221–238.

Yamaguchi, K., Nishio, K., Sato, N., Tsumura, H., Ichihara, A., Kudo, H., Aoki, T., Naoki, K., Suzuki, K., Miyata, A., Suzuki, Y., and Morooka, S. (1997). Leukocyte kinetics in the pulmonary microcirculation: observations using real-time confocal luminescence microscopy coupled with high-seed video analysis. *Lab. Invest.* **76**(6), 809–822.

Yuste, R., Lanni, F., and Konnerth, A. Eds. (1999). "Imaging Neurons—A Laboratory Manual." Cold Spring Harbor Laboratory Press, Cold Spring Harbor, NY.

Ch. 2, Fig. 6. Autofluorescence of pollen grain of Mallow. This multicolor autofluorescence was captured in ca. 0.6 sec with a chilled, 3-CCD color video camera (Dage-MTI Model 330, Michigan City, IN) mounted on the CSU-10 and illuminated with 488-nm laser light. The video signal, after conversion from RGB to standard (Y/C) format through a scan converter (Truevision Vid I/O, Pinnacle Systems, Mt. View, CA), was captured onto a Sony ED-Beta VCR. For this illustration, a single frame of the VCR output was captured and printed with a Sony Mavigraph video printer. The same full-color image is seen through the eyepiece of the CSU-10 in real time.

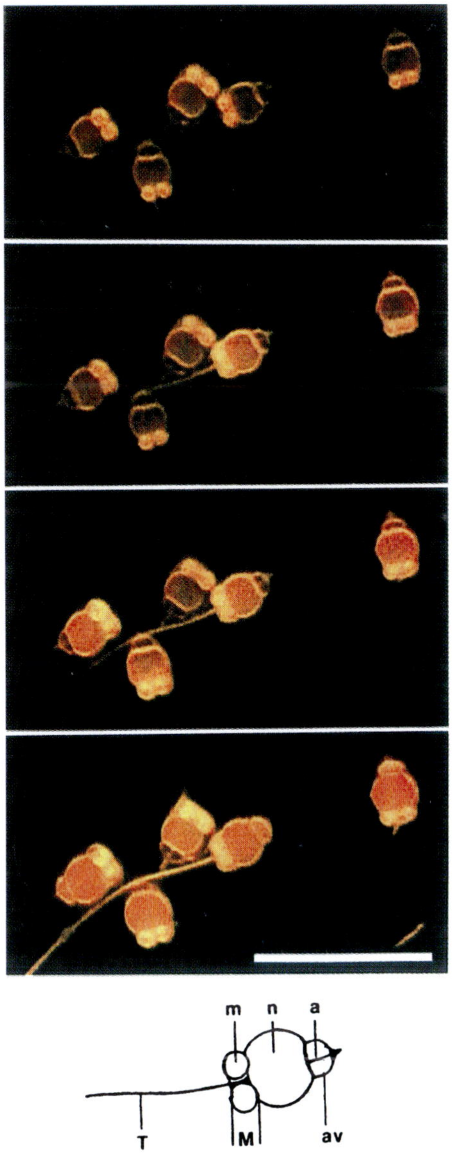

Ch. 2, Fig. 11. Selected frames from CSU-10 confocal video sequences showing striking photodynamic changes in Eosin-B-stained *Spisula* (clam) sperm exposed to the 488-nm (fluorescence exciting) laser beam. Top to bottom panels: 13, 14, 15, 31 s after exposure to the laser beam (0.01 mW per μm^2 at the specimen). Except for the fourth sperm from the left, much of the tail is invisible owing to the shallow depth of field of the confocal optics. Scale bar = 10 μm. Bottom: Schematic diagram of *Spisula* sperm. A, Acrosome: av, acrosomal vesicle; M, midpiece; m, mitochondrion; n, nucleus; T, sperm tail (part only shown). *Technical details.* Sperm were suspended in seawater containing 0.1% Eosin B and observed through a Nikon E-800 upright microscope equipped with a 100×/1.4 NA Plan Apo oil-immersion lens and a CSU-10. The weak fluorescence of the sperm parts excited by the 488-nm laser beam was integrated on chip for 20 frames (0.66 s) on a chilled, 3-chip color camera (Dage-MTI Model 330, Michigan City, IN) attached to the CSU-10 and observed continuously on an RGB color monitor. The RGB output through the monitor was converted by a scan converter (Truevision Vid I/O) to Y/C format, color balance and background levels adjusted with a video processor (Elite Video BVP-4 Plus, www.elitevideo.com), and recorded to an ED-Beta VCR. (From Inoué *et al.*, 1997.)

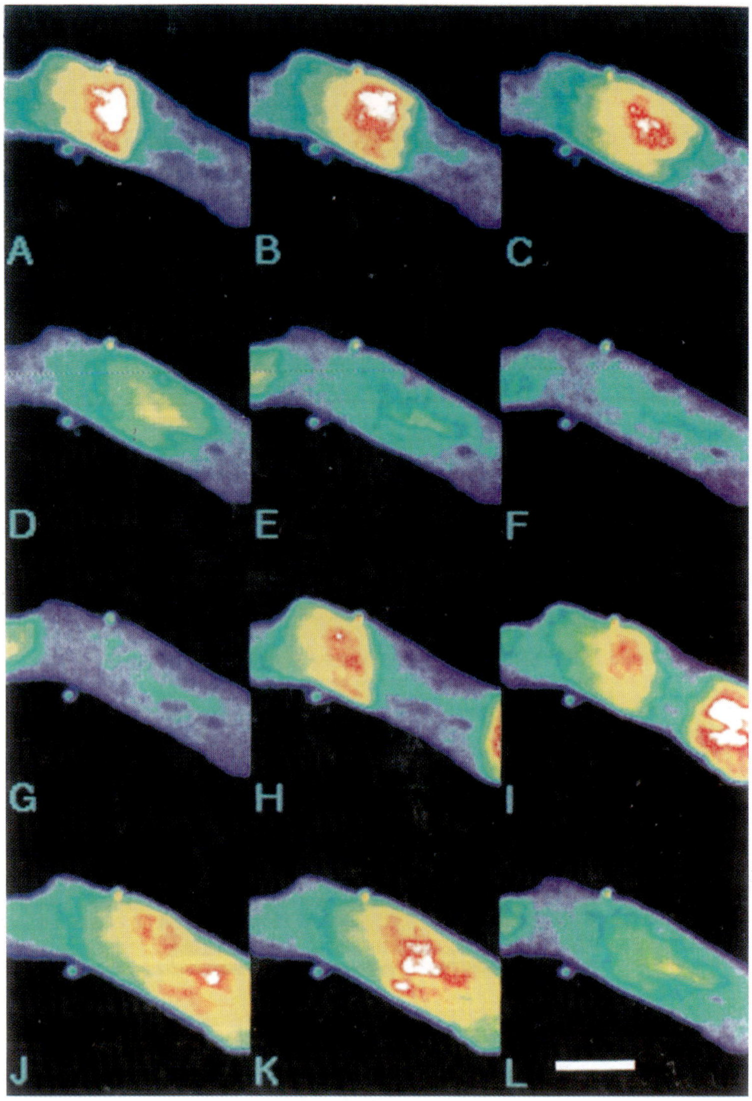

Ch. 2, Fig. 12. Mouse cardiac myocyte. Waves and sparks of elevated cytosolic Ca^{2+} are presented in pseudocolor in this heart muscle and cell injected with Fluo 3. (Ca^{2+} concentrations rise in order of purple, light-blue, dark-blue, green, yellow, orange, red, and white.) Frame intervals each 200 ms between panels A-F. Then after 1.4 s, panel G-L show the approach and annihilation of two calcium waves. Frame intervals: G–H = 360 ms, H–I = 160 ms, I–J = 120 ms, J–K = 40 ms, K–L = 40 ms. Scale bar = 5 μm.[7] *Technical details*. The image from the CSU-10 (modified by addition of motorized filter wheel as a part of the "UltraView" system distributed by PerkinElmer Wallac), viewed through an Olympus 60/1.4 oil-immersion objective lens, was intensified with a Videoscope Intensifier. The images, each exposed for 25 ms, were captured at 25 frames per second onto a PerkinElmer FKI-300 cooled-CCD camera equipped with a Kodak Olympics interline chip.

●第十九回●

国際生物学賞

2003 International Prize for Biology
— Record —

CONTENTS

Committee on the INTERNATIONAL PRIZE FOR BIOLOGY
December 2003

Presentation Ceremony of the 2003 International Prize for Biology

授賞式
Presentation Ceremony

賜品を手にする
井上博士と夫人

Seated next to Mrs. Inoué,
Dr. Inoué, holding Imperial Gift

Address by His Majesty the Emperor

I should like to express my heartfelt congratulations to Dr. Shinya Inoué on being awarded the 19th International Prize for Biology, which is dedicated this year to the field of "Cell Biology."

In an attempt to better understand how cells divide, Dr. Inoué succeeded in making a series of epochal innovations in the development of light microscopy. These advances rendered it possible to directly observe dynamic changes in the supramolecular structure of living cells during cell division. This contributed immensely to advancing research in such fields as cell division, cytoskeleton, and cell motility. Among Dr. Inoué's most remarkable achievements is his use of polarization microscopy to demonstrate that spindles are a real structure within the cell; and that, as a self-assembly system, spindle microtubules are in dynamic equilibrium between assembly and disassembly—a discovery which transformed the prevailing static image of cells to a dynamic one.

Using microscopes that he developed and enhanced himself, Dr. Inoué has achieved a great many research milestones. I have heard, however, that he started his long involvement with microscopes under the severe conditions of Japan's early postwar period. He made use of a discarded rifle pedestal and an old tea can to construct his microscope. When I think of the pains he took back in those days of his youth when Japan was destitute—how he used substitute materials to build a microscope to make his observations—I am deeply impressed by the research efforts that Dr. Inoué continues to exert today, still developing new and ever-improved microscope technologies.

The products of Dr. Inoué's research are widely utilized by researchers around the world, and contribute immensely to the advancement of biological sciences. I sincerely hope that he will enjoy good health and will continue his research activities to achieve improved results.

Reprinted from *Applied Optics*, Vol. 45(3), pp. 460–469, 2006, with permission from OSA.

Orientation-independent differential interference contrast microscopy

Michael Shribak and Shinya Inoué

We describe a new technique for differential interference contrast (DIC) microscopy, which digitally generates phase gradient images independently of gradient orientation. To prove the principle we investigated specimens recorded at different orientations on a microscope equipped with a precision rotating stage and using regular DIC optics. The digitally generated images successfully displayed and measured phase gradients, independently of gradient orientation. One could also generate images showing distribution of optical path differences or enhanced, regular DIC images with any shear direction. Using special DIC prisms, one can switch the bias and shear directions rapidly without mechanically rotating the specimen or the prisms and orientation-independent DIC images are obtained in a fraction of a second. © 2006 Optical Society of America

OCIS codes: 180.0180, 100.2980, 170.1530, 170.3010, 170.3890, 170.3880.

1. Introduction

Differential interference contrast light microscopy (DIC) is widely used to observe structure and motion in unstained living cells and isolated organelles. DIC microscopy uses an interference optical system in which the reference beam is sheared laterally by a small amount, generally by somewhat less than the diameter of the Airy disk. The technique produces a monochromatic shadow-cast image that displays the gradients of optical paths in the specimen. Those regions of the specimen where the optical paths increase along a reference direction appear brighter (or darker), while regions where the path differences decrease appear in reverse contrast. As the gradient of the optical path grows steeper, image contrast is accordingly increased. Another important feature of DIC imaging is that it produces effective optical sectioning.[1] This is particularly obvious when high-numerical-aperture (NA) objectives are used together with high-NA condenser illumination.

The DIC technique was invented by Smith[2] in 1947. He placed between a pair of polarizers one Wollaston prism at the front focal plane of the condenser and another Wollaston prism in the back focal plane of the objective lens. The first Wollaston prism splits the input beams angularly into two orthogonally polarized beams. The condenser makes the beam axes parallel, with a small shear. Then the objective lens joins them in the back focal plane where the second Wollaston prism introduces an angular deviation into the beam and makes them parallel. Thereafter both of the two orthogonally polarized beams are recombined into one beam. This optical configuration creates a polarizing shearing interferometer, by which one visualizes phase gradients of the specimen under investigation.

In conventional medium- to high-NA objective lenses, the back focal plane is located inside the lens system and therefore is not available for insertion of a Wollaston prism. If the Wollaston prism is placed far from the back focal plane, the prism only makes the rays parallel, but those beams are spatially displaced and hence are not recombined. Therefore the Smith DIC scheme requires a special design for the microscope objective lenses such that the Wollaston prism can be incorporated within.

Nomarski, who proposed a modified Wollaston prism in 1952, provides an alternative solution.[3,4] The Nomarski prism simultaneously introduces a spatial displacement and an angular deviation of the orthogonally polarized beams. The prism therefore can be placed outside the objective lens. By using crystal wedges with appropriately oriented axes, the Nomarski prism recombines the two beams (separated by the condenser Wollaston) as though a regu-

The authors are with the Marine Biological Laboratory, Woods Hole, Massachusetts. M. Shribak's e-mail address is mshribak@mbl.edu.

Received 25 February 2005; revised 30 August 2005; accepted 1 September 2005.
0003-6935/06/030460-11$15.00/0

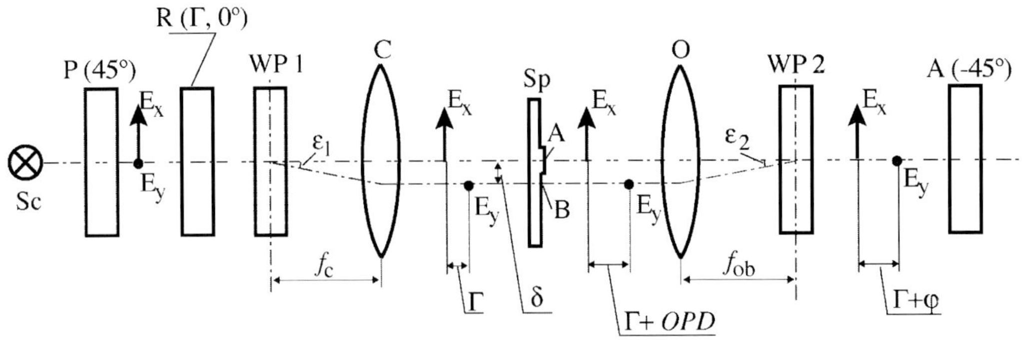

Fig. 1. Conventional DIC microscope setup: Sc, monochromatic light source; P(45°), polarizer at 45° azimuth; R(Γ, 0°), retarder at 0° azimuth and phase shift Γ; WP1, Wollaston prism 1 with splitting angle ε_1; C, condenser lens with focal distance f_c; Sp, object under investigation with ***OPD*** = φ between points A and B; O, objective lens with focal distance f_{ob}; WP2, Wollaston prism 2 with splitting angle ε_2; A(−45°), analyzer at −45° azimuth; E_x, E_y polarization components; δ, amount of shear.

lar Wollaston prism were located in the proper plane in the objective lens. The Nomarski DIC scheme can be used in conjunction with regular high-NA microscope objectives, which exhibit low stress birefringence.

Video-enhanced DIC permits enhancement of contrast while it removes unwanted background signal (such as fixed image noise caused by dust particles, nonuniform illumination, or other imperfections in the optical system) by subtraction of a reference image with no specimen.[5] Also, Inoué[6] and Allen *et al.*[7] applied similar enhancement techniques to improve contrast in polarization microscopy.

Holzwarth *et al.* developed video-enhanced DIC microscopy with polarization modulation.[8,9] Switching the polarization of the incident light in alternate video frames with a computer-controlled liquid-crystal variable retarder increased the contrast signal of unstained biological specimens by a factor of 2 relative to the standard video-enhanced DIC. The modulator switches image highlights into shadows and *vice versa.* Subtracting alternate frames creates a difference DIC image in which contrast is doubled while image defects and noise are canceled.

These regular DIC techniques show the two-dimensional distribution of optical path gradients as observed for a particular shear direction. The contrast of DIC images is thus not symmetrical; it varies proportionally with the cosine of the angle between the azimuth of the specimen gradient and the direction of the wavefront shear. It is therefore prudent to examine unknown objects at several azimuths.[4,10]

Preza used two, four or eight conventional DIC images at different shear directions but with the same bias in an iterative phase-estimation method to calculate the specimen's phase function.[11] She did not use images with inverse biases because she supposed that they can be useful in reduction of noise only. The proposed theory was simulated by mathematical models that employ a large amount of shear (near the Airy radius) and large bias, $\sim\pi/2$. A thin specimen with 0.7 μm thickness was investigated with a 0.5 NA objective lens.

To overcome the limitations of existing systems we devised an orientation-independent DIC system, which requires no mechanical movement of the DIC prisms or of the specimen.[12] To explore the utility of such a system by using immediately available components, we recorded several images with the specimen oriented in different directions, followed by digital alignment and processing of the images. Through computation we generated two separate images, which show, the magnitude distribution of the optical path gradients and their azimuths. As described in Section 3 below, the two images can be viewed separately or combined into one color shadow-cast picture, e.g., with the brightness representing magnitude and color showing azimuth, respectively. The images obtained are independent of specimen orientation and free from gray background. Furthermore, we show how it is possible to generate quantitative phase images (showing distribution of optical paths) as well as enhanced regular DIC images with any shear direction from the orientation-independent DIC data.

2. Principles of Orientation-Independent DIC Microscopy

A. Theoretical Model of DIC Image

A schematic of ray paths in a DIC microscope is shown in Fig. 1. The polarizer creates a linearly polarized beam that contains two components, E_x and E_y, with equal intensities, and the retarder introduces a variable phase shift, Γ, between the components. Then the first Wollaston prism splits the beam and the second Wollaston prism combines the split beam, as described in Section 1.

A DIC image can be modeled as the superposition of one image over an identical copy that is displaced by a differential amount δ and phase shifted by bias Γ. For example, point A in the first image will be superimposed with point B in the copy of the image (Fig. 1). For simplicity consider a pure (nonbirefringent) phase specimen, which is described in Cartesian coordinates *XOY* in the object plane. If the optical path distribution in the first image is $\phi(x, y)$, then the second displaced image has the optical path distribution $\phi(x - \delta_x, y - \delta_y)$. Here δ_x and δ_y are projections of shear vectors on the corresponding coordinate axes. The optical path difference (OPD) between the two images can then be written as

$$\mathrm{OPD}(x, y) = \phi(x, y) - \phi(x - \delta_x, y - \delta_y). \quad (1)$$

Taking into account that the shear distance is a small value (approximately half of the optical resolution limit), we can write the optical path distribution in the displaced copy $\phi(x - \delta_x, y - \delta_y)$ in the following linear approximation:

$$\phi(x - \delta_x, y - \delta_y) = \phi(x, y) - (\nabla\phi(x, y), \boldsymbol{\delta}), \quad (2)$$

where the last term on the right-hand side is the scalar product of two vectors.

$$\nabla\phi(x, y) \equiv \mathrm{grad}[\phi(x, y)] = \left[\frac{\partial\phi(x, y)}{x}, \frac{\partial\phi(x, y)}{y}\right]$$

is a gradient of the optical path and $\boldsymbol{\delta} = (\delta_x, \delta_y)$ is a shear vector.

Thus a phase difference distribution between two overlapping images $\varphi(x, y)$ is described by the formula

$$\begin{aligned}\varphi(x, y) &= \frac{2\pi}{\lambda}(\Gamma + (\nabla\phi(x, y), \boldsymbol{\delta})) \\ &= \frac{2\pi}{\lambda}(\Gamma + \delta\gamma(x, y)\cos\psi(x, y)),\end{aligned} \quad (3)$$

where λ is the wavelength, $\psi(x, y)$ is the angle between the shear direction and the optical path gradient, and $\gamma(x, y) = |\nabla\phi(x, y)|$ is the gradient's magnitude.

The intensity distribution in the image, which is the result of interference of the two beams, is

$$I(x, y) = \frac{1}{2}\tilde{I}\left[1 - \cos\left(\frac{2\pi}{\lambda}\{\Gamma + \delta\gamma(x, y) \times \cos[\theta(x, y) - \sigma]\}\right)\right] + I_{\min}, \quad (4)$$

where $\tilde{I}$ is the beam's intensity before it enters the first DIC prism, $\theta(x, y)$ is the gradient azimuth, and σ is the shear azimuth [here we used $\psi(x, y) = \theta(x, y) - \sigma$].

From Eq. (4), if the shear direction coincides with the optical path gradient ($\psi = 0°$ or $\psi = 180°$) the picture contrast is maximal. When the shear direction is oriented perpendicular to the gradients ($\psi = 90°$ or $\psi = 270°$), the intensity contrast becomes zero.

Equation (4) does not take into consideration the refraction of the beam in a thick nonhomogeneous specimen or changes of intensities and phase owing to birefringence. Therefore a quantitative interpretation of the results will work better with a thin, low-birefringence specimen.

B. Six-Frame Processing Algorithm

To find the two-dimensional distribution of magnitude and azimuth of optical path gradients we can apply a technique similar to the one that we developed for orientation-independent polarization microscopy.[13] For instance, we take the next four raw images with 90° step differences in the shear orientation and with the same bias Γ, and two raw images with no bias:

$\sigma = 0°$,

$$I_1 = \frac{1}{2}\tilde{I}\left\{1 - \cos\left[\frac{2\pi}{\lambda}(\Gamma + \delta\gamma\cos\theta)\right]\right\} + I_{\min};$$

$\sigma = 90°$,

$$I_2 = \frac{1}{2}\tilde{I}\left\{1 - \cos\left[\frac{2\pi}{\lambda}(\Gamma + \delta\gamma\sin\theta)\right]\right\} + I_{\min};$$

$\sigma = 180°$,

$$I_3 = \frac{1}{2}\tilde{I}\left\{1 - \cos\left[\frac{2\pi}{\lambda}(\Gamma - \delta\gamma\cos\theta)\right]\right\} + I_{\min};$$

$\sigma = 270°$,

$$I_4 = \frac{1}{2}\tilde{I}\left\{1 - \cos\left[\frac{2\pi}{\lambda}(\Gamma - \delta\gamma\sin\theta)\right]\right\} + I_{\min};$$

$\sigma = 0°$ or $\sigma = 180°$,

$$I_0' = \frac{1}{2}\tilde{I}\left[1 - \cos\left(\frac{2\pi}{\lambda}\delta\gamma\cos\theta\right)\right] + I_{\min};$$

$\sigma = 90°$ or $\sigma = 270°$,

$$I_0'' = \frac{1}{2}\tilde{I}\left[1 - \cos\left(\frac{2\pi}{\lambda}\delta\gamma\sin\theta\right)\right] + I_{\min}. \quad (5)$$

Then compute the following two terms:

$$A \equiv \frac{I_1 - I_3}{I_1 + I_3 - 2I_0'}\tan\left(\frac{\pi\Gamma}{\lambda}\right) = \tan\left(\frac{2\pi}{\lambda}\delta\gamma\cos\theta\right),$$

$$B \equiv \frac{I_2 - I_4}{I_2 + I_4 - 2I_0''}\tan\left(\frac{\pi\Gamma}{\lambda}\right) = \tan\left(\frac{2\pi}{\lambda}\delta\gamma\sin\theta\right). \quad (6)$$

Using these terms, we can find the exact solution for gradient magnitude and azimuth distributions $\gamma(x, y)$and $\theta(x, y)$, respectively:

$$\gamma(x, y) = \frac{\lambda}{2\pi\delta}[(\arctan A)^2 + (\arctan B)^2]^{1/2},$$

$$\theta(x, y) = \arctan\left(\frac{\arctan B}{\arctan A}\right). \quad (7)$$

1. *Four-Frame Processing Algorithm*

In some cases, fewer than six images are collected to support calculation of orientation-independent values and images. Some approximations may be required for completion of the calculations. For example, if the product of shear distance and gradient magnitude is small $[(2\pi/\lambda)\delta\gamma \ll 1]$, Eq. (4) can be simplified to

$$I(x, y) = \frac{1}{2}\tilde{I}\left\{1 - \cos\left(\frac{2\pi}{\lambda}\Gamma\right) + \frac{2\pi}{\lambda}\delta\gamma \sin\left(\frac{2\pi}{\lambda}\Gamma\right) \times \cos[\theta(x, y) - \sigma]\right\} + I_{\min}. \quad (8)$$

Then four images can be collected and modeled as

$\sigma = 0°$,

$$I_1 = \frac{1}{2}\tilde{I}\left[1 - \cos\left(\frac{2\pi}{\lambda}\Gamma\right) + \frac{2\pi}{\lambda}\delta\gamma \sin\left(\frac{2\pi}{\lambda}\Gamma\right)\cos\theta\right] + I_{\min};$$

$\sigma = 90°$,

$$I_2 = \frac{1}{2}\tilde{I}\left[1 - \cos\left(\frac{2\pi}{\lambda}\Gamma\right) + \frac{2\pi}{\lambda}\delta\gamma \sin\left(\frac{2\pi}{\lambda}\Gamma\right)\sin\theta\right] + I_{\min};$$

$\sigma = 180°$,

$$I_3 = \frac{1}{2}\tilde{I}\left[1 - \cos\left(\frac{2\pi}{\lambda}\Gamma\right) - \frac{2\pi}{\lambda}\delta\gamma \sin\left(\frac{2\pi}{\lambda}\Gamma\right)\cos\theta\right] + I_{\min};$$

$\sigma = 270°$,

$$I_4 = \frac{1}{2}\tilde{I}\left[1 - \cos\left(\frac{2\pi}{\lambda}\Gamma\right) - \frac{2\pi}{\lambda}\delta\gamma \sin\left(\frac{2\pi}{\lambda}\Gamma\right)\sin\theta\right] + I_{\min}; \quad (9)$$

Thus the next two equations permit calculation of the two-dimensional distribution of the gradient magnitude and azimuth of optical paths in the specimen; these can be imaged separately and independently of the shear direction:

$$\gamma(x, y) = \frac{\lambda}{2\pi\delta}\tan\left(\frac{\pi\Gamma}{\lambda}\right)\left[\left(\frac{I_1 - I_3}{I_1 + I_3}\right)^2 + \left(\frac{I_2 - I_4}{I_2 + I_4}\right)^2\right]^{1/2},$$

$$\theta(x, y) = \arctan\left(\frac{I_2 - I_4}{I_1 - I_3}\right). \quad (10)$$

Here we have assumed that the microscope has moderately high extinction ($\tilde{I}/I_{\min} > 200$) and thus have neglected $I_{\min}$ in the divisors of the first of Eqs. (10).

Note that the two algorithms considered above employ ratios between intensities of light that has interacted with the specimen. Therefore they suppress contributions of absorption by the specimen, nonuniformity of illumination, etc., which can otherwise deteriorate a DIC image.

C. Two-Frame Processing Algorithm

Other numbers of specimen images, for instance two or three can also be used for calculating the magnitude and azimuth of the gradient. For example, two specimen images and three background images can be collected and modeled according to Eq. (8):

$\sigma = 0°$,

$$I_1 = \tilde{I}\sin\left(\frac{\pi}{\lambda}\Gamma\right)\left[\sin\left(\frac{\pi}{\lambda}\Gamma\right) + \frac{2\pi}{\lambda}\delta\gamma\cos\left(\frac{\pi}{\lambda}\Gamma\right)\cos\theta\right] + I_{\min};$$

$\sigma = 90°$,

$$I_2 = \tilde{I}\sin\left(\frac{\pi}{\lambda}\Gamma\right)\left[\sin\left(\frac{\pi}{\lambda}\Gamma\right) + \frac{2\pi}{\lambda}\delta\gamma\cos\left(\frac{\pi}{\lambda}\Gamma\right)\sin\theta\right] + I_{\min}; \quad (11)$$

and background images without the specimen:

$$\sigma = 0°, \qquad I_{bg1} = \tilde{I}\sin^2\left(\frac{\pi}{\lambda}\Gamma\right) + I_{\min},$$

$$\sigma = 90°, \qquad I_{bg2} = \tilde{I}\sin^2\left(\frac{\pi}{\lambda}\Gamma\right) + I_{\min}; \quad (12)$$

for $\Gamma = 0$ and an arbitrary σ, $I_{bg0} = I_{\min}$.

Values K and L are calculated as follows:

$$K \equiv \frac{I_1 - I_{bg1}}{I_{bg1} - I_{bg0}}\tan\left(\frac{\pi}{\lambda}\Gamma\right) = \frac{\tilde{I}\frac{2\pi}{\lambda}\delta\gamma\sin[(\pi/\lambda)\Gamma]\cos[(\pi/\lambda)\Gamma]\cos\theta}{\tilde{I}\sin^2[(\pi/\lambda)\Gamma]} \times \tan[(\pi/\lambda)\Gamma] = \frac{2\pi}{\lambda}\delta\gamma\cos\theta,$$

$$L \equiv \frac{I_2 - I_{bg2}}{I_{bg2} - I_{bg0}}\tan\left(\frac{\pi}{\lambda}\Gamma\right) = \frac{\tilde{I}\frac{2\pi}{\lambda}\delta\gamma\sin[(\pi/\lambda)\Gamma]\cos[(\pi/\lambda)\Gamma]\sin\theta}{\tilde{I}\sin^2[(\pi/\lambda)\Gamma]} \times \tan[(\pi/\lambda)\Gamma] = \frac{2\pi}{\lambda}\delta\gamma\sin\theta. \quad (13)$$

Then the gradient magnitudes $\gamma(x, y)$ and azimuth distributions $\theta(x, y)$ can be obtained from the following formulas:

$$\gamma(x, y) = \frac{\lambda}{2\pi\delta}\sqrt{A^2 + B^2},$$

$$\theta(x, y) = \arctan\left(\frac{B}{A}\right). \quad (14)$$

Further approximations can be employed, for example, to simplify the calculations or to reduce the number of collected background images. For example, if I_{bg0} is small it can be omitted from the above equations.

Using just two specimen images for obtaining orientation-independent DIC data increases the speed of measurement. However, the sensitivity becomes lower. Moreover, the algorithm does not use ratios between intensities of light that has interacted with the specimen. Therefore this algorithm can be used only when the specimen's absorption is negligible and illumination is highly stable and uniform.

D. Using Orientation-Independent DIC Data to Obtain Enhanced Regular DIC and Phase Images

After computing the optical path gradient distribution we can restore enhanced regular DIC images $I_{enh}(x, y)$ with any selected shear direction σ:

$$I_{enh}(x, y) = 1 - \cos\left(\frac{2\pi}{\lambda}\{\Gamma + \delta\gamma(x, y)\cos[\theta(x, y) - \sigma]\}\right). \quad (15)$$

The enhanced image, $I_{enh}(x, y)$, provides a calculated image for any desired shear direction σ without the requirement to collect an image directly for that shear direction. Moreover, the enhanced image will have less noise than a regular DIC image. In addition, it suppresses deterioration of the image caused by specimen absorption and illumination nonuniformity.

We can obtain the distribution of optical phase $\Phi(x, y)$ in the specimen by computing the following line integral[14]:

$$\Phi(x, y) = \frac{2\pi}{\lambda}\int_C [\gamma(x', y')\cos\theta(x', y')dx' + \gamma(x', y')\sin\theta(x', y')dy']. \quad (16)$$

As the simplest contour for integration we can choose rectangular ones along the coordinate axes:

$$\Phi(x, y) = \frac{2\pi}{\lambda}\left[\int_0^x \gamma(x', 0)\cos\theta(x', 0)\,dx' + \int_0^y \gamma(x, y')\sin\theta(x, y')dy'\right]. \quad (17)$$

Here we assign zero phase to the beginning point of the coordinate.

Where calculations are performed in terms of pixels, Eq. (17) can have the following form:

$$\Phi_{mn} = \frac{2\pi}{\lambda}\left(\sum_{k=1}^{n} \gamma_{0k}\cos\theta_{0k} + \sum_{p=1}^{m} \gamma_{pn}\sin\theta_{pn}\right), \quad (18)$$

where m and n are row and column numbers of the pixels and Φ_{mn} is the value of the optical phase for pixel m, n.

Other techniques for phase computation can also be used, for instance, iterative computation,[11] noniterative Fourier phase integration,[15] or nonlinear optimization with hierarchical representation.[16]

Biggs[17] and colleagues have developed an iterative deconvolution approach to the computation of a phase image that follows the same principles as the deconvolution techniques normally used to remove out-of-focus haze.[18] In the case of DIC imagery, if the image background is removed such that the data have a zero mean, then the point-spread function can be modeled as a positive and a negative Dirac delta function with small separation between them.[11] If the specimen is not a purely phase object and contains absorbing structures, then four DIC images with 90° differences in orientation should be collected. The two sets of opposite orientation can be used to separate the phase-difference information from the amplitude modulation, which reduces potential artifacts in the reconstruction. This preprocessing is also valuable when small bias retardations are used to maximize image contrast, which also results in a highly nonlinear intensity response to phase differences.

The Biggs optimization is based on minimizing the least-squares error between the observed phase differences and the DIC image recalculated from the estimated phase map. For a two-image reconstruction, the errors from each are combined to form a single metric. We solve the single phase image iteratively by minimizing the error metric, using a gradient descent optimization with acceleration,[19] with typically 100 iterations required. A conjugate gradient or other optimization approach could also be employed. Depending on the specimen used, a penalty term can also be introduced to minimize negative phase values with respect to the background if that information is known *a priori*. The reconstruction process thus far has not taken into account any effect of diffraction or other optical effects that limit the observable resolution in the image. Once the phase map is calculated, the image is further deconvolved by use of an iterative blind deconvolution algorithm that is able to determine the amount of blur directly from the image provided and use this PSF to improve the resolution of the image. Once the point-spread function is known, only five iterations are typically required. The algorithm is commercially available in the AutoDeblur software from AutoQuant Imaging (Troy, New York).

3. Experimental Verification

A. Description of the Experiment

For verification of theoretical analysis, we carried out several experiments, using artificial and biological specimens. We employed the universal (polarized) light microscope (ULM, or ShinyaScope) that has been developed by one of us over five decades and that is one of the most sophisticated and versatile light microscope platforms available today.[1] The microscope is optimized for polarization imaging and includes a precision stage that permits rotation of the specimen with fractional micrometers of eccentricity or focal shift. The ULM is equipped with a high-brightness fiber illuminator. A 100 W concentrated mercury arc lamp coupled with a green 546 nm bandpath-filter was employed during the experiments. For the test published here, we used a Nikon 20× objective with a NA of 0.75 and a Nikon Universal Achromatic-Aplanat condenser with its NA of 1.4 stopped down to match the objective's NA. Images were collected with a Hamamatsu C5985 chilled CCD camera or a Hamamatsu C4742-95 high-resolution digital CCD camera.[20] The resolution of the microscope, which is determined by the ratio $1.22\lambda/(2\ \mathrm{NA})$, was 0.44 μm, and the depth of field according to the ratio $\lambda n/\mathrm{NA}^2$ was 1.5 μm, where n is the refractive index of the objective lens immersion medium.[1]

We acquired four regular DIC images of the specimen with the same bias at 90° orientation difference, and two images without bias, as described in Section 2. In addition, we took corresponding background DIC images without the specimen in the field of view. The regular DIC images were aligned by use of ImageJ software.[21] The background images were used to compensate for nonuniformity in the intensity distribution of the illumination beam, because otherwise each part of the specimen could be exposed to different light intensities during rotation of the specimen.

Using Eqs. (7), (15), and (18), we computed images of gradient magnitude and azimuth distributions of optical paths, optical phase distribution, and enhanced regular DIC image with any chosen shear direction. For the computation we used Mathematica software.[22]

B. Orientation-Independent DIC Images of Microscopic Glass Rods Embedded in an Immersion Medium

First we explored a model specimen, which is optically similar to transparent filaments in living organisms. We took short segments of glass rods, which are used as spacers in liquid-crystal cells, and embedded them in Fisher Permount mounting medium (Fisher Scientific Company, http://www.fishersci.com). The refractive indices of the glass rods and the Permount at wavelength 546 nm were measured with a Jamin–Lebedeff microscope (Zeiss, Germany), and were 1.554 and 1.524, respectively. A drop of the suspension was placed between a microscope slide and a 0.17 mm thick coverslip, and the preparation was placed on the precision rotating microscope stage.

One of the images taken by regular DIC optics is shown in Fig. 2. Also, the figure contains line scans of intensities in two sections; one of them (AA') is scanned along the shear direction and the other (BB') is perpendicular the shear direction. Both scans are normalized to the maximal intensity in the image.

Gray-scale images of gradient magnitude computed from Eqs. (7) are given in Fig. 3. Here brightness is linearly proportional to the magnitude. The absolute magnitude value was not determined, because the exact amount of the shear was unknown. The gradient magnitude values measured along lines A-A′ and B-B′ are shown by filled circles (Fig. 3, right). Unlike in Fig. 2, the two linear scans of magnitude now exhibit virtually no difference. The gradient magnitude image clearly shows the rod boundaries independently of orientation.

For preliminary analysis of the experimental data we employed a simple model for a DIC image of glass microspheres as described by Resnick.[23] For analysis of DIC images he used a model that comprised 2 μm diameter microspheres with a 100×/1.4 objective lens. The spherical model can also be used for a (one-dimensional) glass rod, which is oriented parallel to the object plane. This model does not take into account refraction of rays, microscope resolution, out-

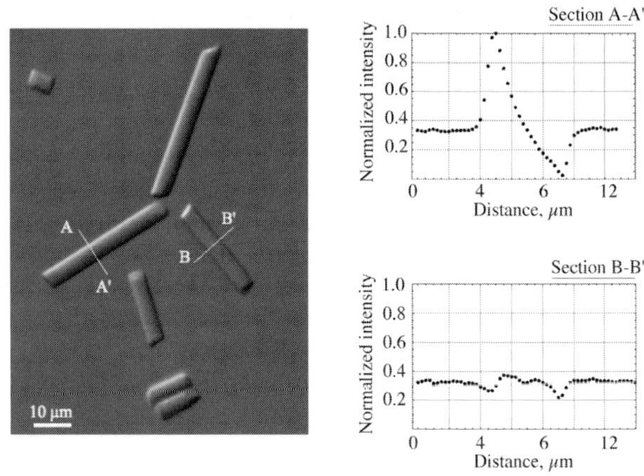

Fig. 2. Regular DIC image of 4 μm diameter glass rods immersed in Permount. The DIC shear direction is parallel to AA' (left). Linear scans of normalized intensity in sections A-A′ and B-B′ are shown at the right.

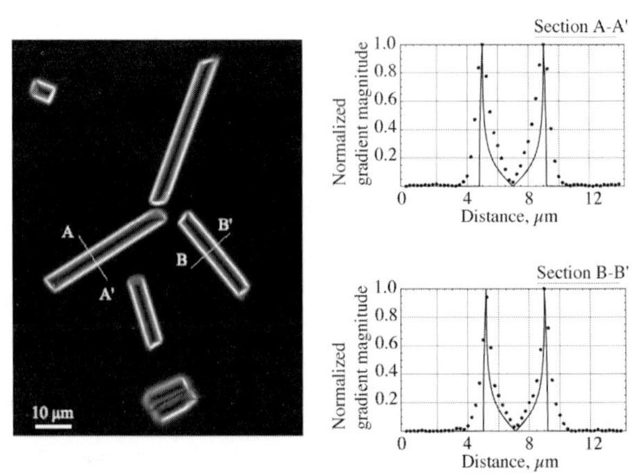

Fig. 3. Computed gradient magnitude image of the glass rods (left). Linear plots of normalized gradient magnitude in sections A-A′ and B-B′ are shown at the right. The experimental data are indicated by filled circles. The solid curves show the theoretical results calculated with simplified assumptions (see text).

of-focus contributions in the thick specimen, changes in polarization, etc.

Assume that the center of the Cartesian coordinate corresponds to the center of the rod, the Y axis is parallel to the rod axis, and the Z axis is directed along the microscope axis. The relative phase retardation Φ (in radians) experienced by a ray traveling through the rod along the microscope axis is as follows[23]:

$$\Phi(x) = \frac{4\pi}{\lambda}(n_r - n_m)\sqrt{r^2 - x^2} = \Phi_{\max}\sqrt{1 - (x/r)^2}, \quad (19)$$

where $\Phi_{\max} = (4\pi/\lambda)(n_r - n_m)r$, r is the diameter of the rod, and n_r and n_m are the refractive indices of the rod and the surrounding media, respectively.

When two rays are spatially displaced along the X axis by a small distance $\delta \ll r$, the magnitude of gradient between them can be determined by the following formula:

$$\gamma(x) = \frac{\Phi(x+\delta) - \Phi(x)}{\delta} = \frac{2}{\delta}(n_r - n_m) \times \begin{cases} 0 & x < -r \text{ or } x \geq r + \delta \\ \sqrt{r^2 - x^2} & -r \leq x < -r + \delta \\ \left|\sqrt{r^2 - x^2} - \sqrt{r^2 - (x-\delta)^2}\right| & -r + \delta \leq x < r \\ \sqrt{r^2 - (x-\delta)^2} & r \leq x < r + \delta \end{cases}. \quad (20)$$

The maxima of function $\gamma(x)$ occur at $x = -r + \delta$ and $x = r$. For $\delta \ll r$,

$$\gamma_{\max} = 2\sqrt{2}(n_r - n_m)\sqrt{r/\delta}.$$

Then the normalized gradient magnitude can be written as

$$\gamma(x)/\gamma_{\max} = \frac{1}{\sqrt{2r\delta}} \times \begin{cases} 0 & x < -r \text{ or } x \geq r + \delta \\ \sqrt{r^2 - x^2} & -r \leq x < -r + \delta \\ \left|\sqrt{r^2 - x^2} - \sqrt{r^2 - (x-\delta)^2}\right| & -r + \delta \leq x < r \\ \sqrt{r^2 - (x-\delta)^2} & r \leq x < r + \delta \end{cases}. \quad (21)$$

Theoretical plots of the normalized gradient magnitude are shown by the solid curves in Fig. 3 (right). We used $r = 2.1$ μm for section A-A′ and $r = 2.0$ μm for section B-B′. The shear distance introduced by the microscope's DIC prisms was unknown and was estimated to be 0.2 μm for these examples. The experimental results show a better correspondence for higher values of the gradient magnitude. However, for lower values the experimental data are wider than the theoretical curve by ~0.5 μm. This fact indicates that the theoretical model needs to be improved by taking into account the Airy disk's radius (0.44 μm at a N.A. of 0.75), the distance (0.31 μm) between the beam and the contact points of the peripheral rays with the rod, etc.

Both the gradient magnitude and the azimuth images can be combined into one color-shaded picture, e.g., with the brightness representing magnitude and the color showing azimuth (Fig. 4). As can be seen from the shaded color, a gradient vector on the rod boundaries is always oriented toward the rods. That means that the rod's refractive index is higher than the refractive index of the Permount.

Using the simple numerical contour integration described in Eq. (17), we computed the two-dimensional phase distributions $\Phi(x, y)$ of rays transmitted through the glass rods (Fig. 5, left). The reconstructed phase image contains some streaks, which can be caused by misalignment of the raw DIC images. The normalized phase values $[\Phi(x, y)/\Phi_{\max}]$ along lines A-A′ and B-B′ are shown by filled circles (Fig. 5, right).

C. Orientation-Independent DIC Imaged Microstructure of the MBL–NNF Test Target

We investigated another specimen: a test target developed at the Marine Biological Laboratory, Woods Hole, Mass., in collaboration with the National Nanofabrication Facility at Cornell University.[24] The specimen has a 90 nm thick SiO_2 film deposited onto a number 1 1/2 glass coverslip. The film was treated to remove SiO_2 from regions exposed to the electron beam by microlithography. Then the coverslip was cemented with the target directly lying against the inside surface of a 1 mm thick regular microscope slide. The target includes a 75 μm diam-

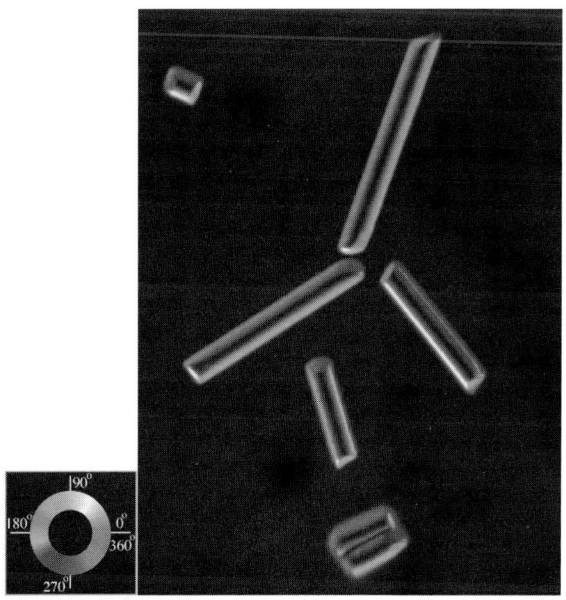

Fig. 4. Computed gradient image of the same group of glass rods with the azimuth in pseudocolor. The color wheel at the lower left corner depicts the gradient azimuth.

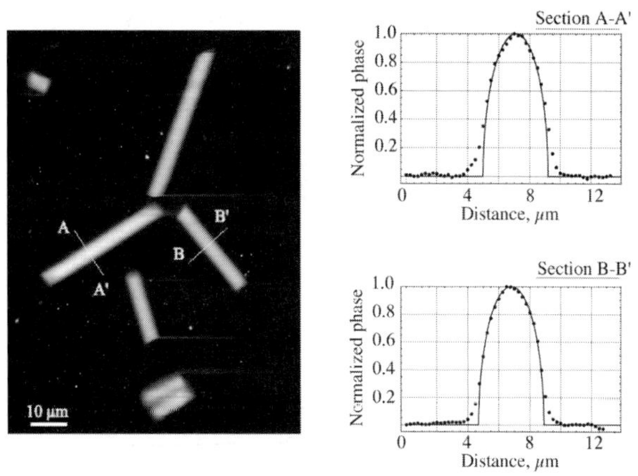

Fig. 5. Computed phase of the glass rods (left). Linear scans of normalized phase in sections A-A′ and B-B′ are shown at the right. The experimental data are indicated by filled circles. The solid curves show the theoretical results.

eter Siemens test star that consists of 36 wedge pairs. The period near the outer edge is 6.5 μm, decreasing linearly toward the center. The optical path difference between the etched structure and the film was measured with a Jamine–Lebedeff microscope. It equals 60 nm.

In Fig. 6 we show a regular DIC image (left), the computed gray-scale gradient magnitude image (center), and the phase of the MBL-NNF test target (right). In the rightmost figure the dark star image and the light background show that the refractive index in the star wedges is lower than in the surrounding silicon dioxide film. The lower portions of the figure show line scans of intensity in the regular DIC image, the gradient magnitude image, and the phase image along the same section. The plots are normalized to the maximum intensity values.

D. Orientation-Independent DIC Images of Fixed Bovine Pulmonary Artery Endothelial Cell

The analytical technique has been used with several biological specimens. One of them was a prepared microscope slide (F-14780) containing bovine pulmonary artery endothelial cells stained with a combination of fluorescent dyes (Molecular Probes, Inc., Eugene, Ore., http://www.probes.com).

Figure 7 demonstrates a regular DIC image of a cell, the computed gradient magnitude image, and the computed phase image. The orientation-independent DIC images clearly reveal the refractive boundaries and detailed structures of the cell. The phase image shows the dry mass distribution of the cell regions under investigation.

In Fig. 8 we present enhanced regular DIC images of the same bovine epithelial cell with various shear directions at every 45°, which are computed from the gradient data. The images show the changes in contrast and gray shading of different parts of the cell as the computed shear direction is altered. In these digitally enhanced DIC images, the empty areas have 31% less optical noise than the original DIC images.

These results demonstrate that the proposed DIC technique can successfully image and measure phase gradients as well as the distribution of phase retardances (OPDs) of transparent specimens, indepen-

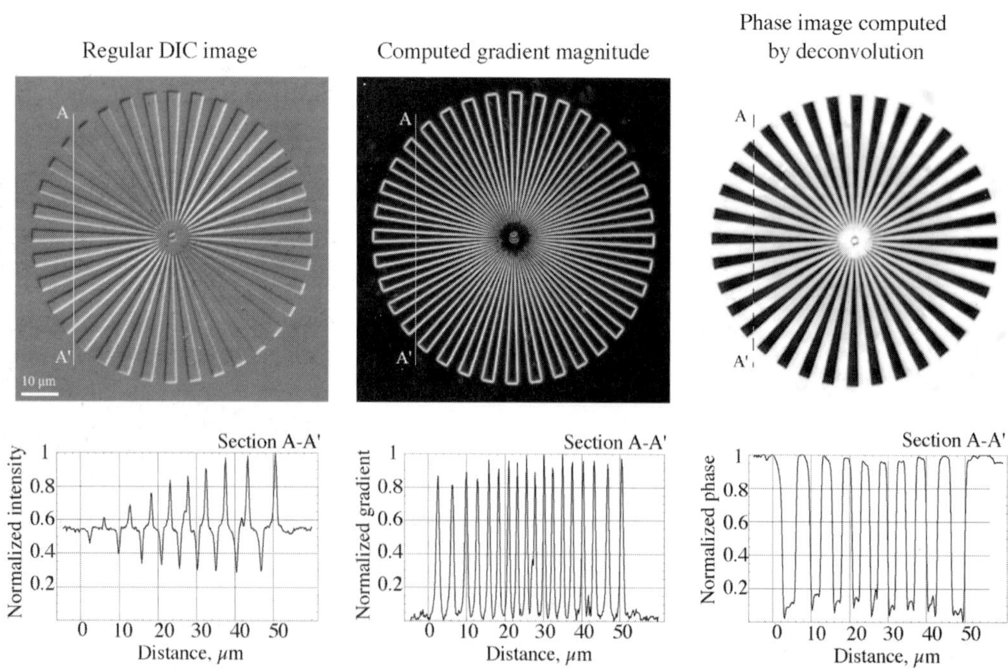

Fig. 6. Regular and computed orientation-independent DIC images of Siemens star and line scans along section AA'. The phase image of the Siemens star was calculated by D. Biggs (AutoQuant Imaging, Inc., Troy, N.Y.).

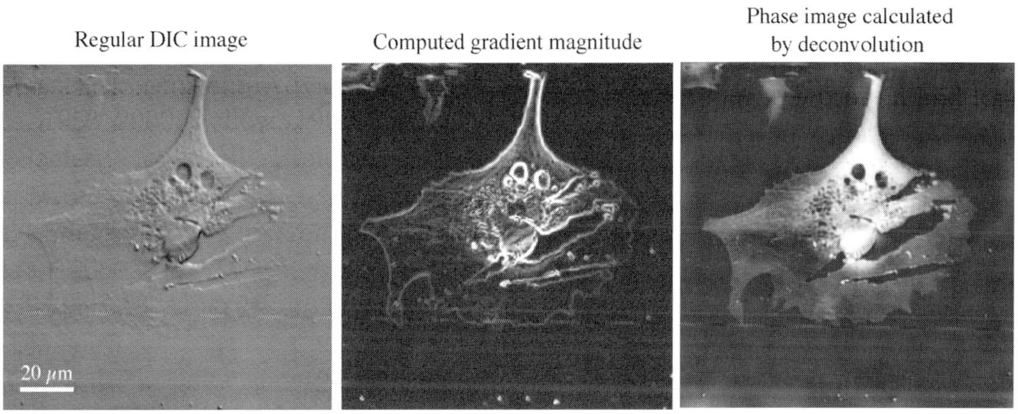

Fig. 7. Regular and computed orientation-independent DIC images of *bovine* pulmonary artery endothelial cell. The phase image of the *bovine* cell was calculated by D. Biggs (AutoQuant Imaging, Inc., Troy, N.Y.).

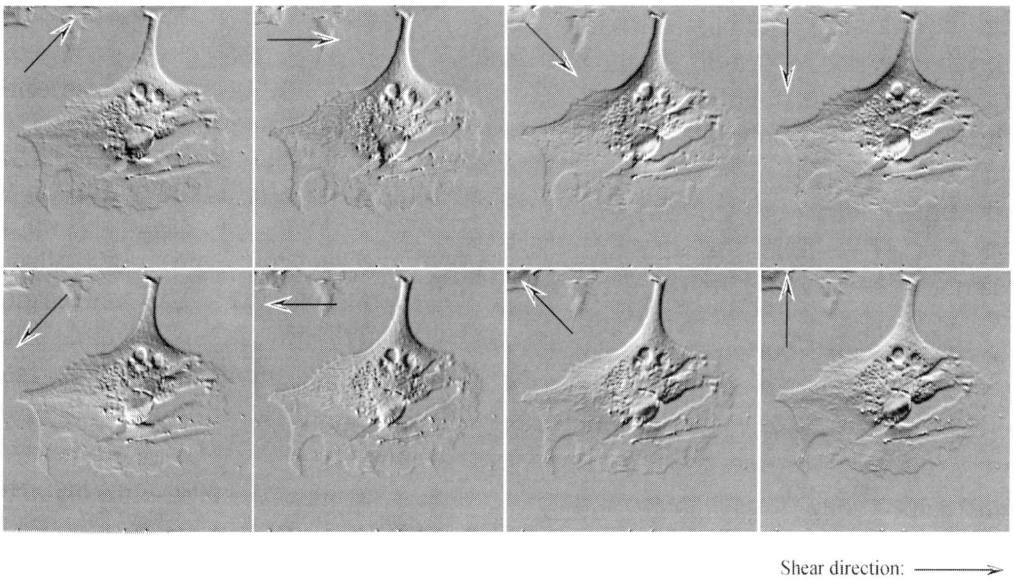

Fig. 8. DIC images of *bovine* pulmonary artery endothelial cell with various shear directions computed from orientation-independent DIC data.

dently of the directions of the gradient and minimally influenced by specimen absorption.

4. Conclusions

1. We have developed the theoretical basis for orientation-independent differential interference contrast microscopy. The new approach facilitates precise analyses of organelle morphology motility, and shape changes as well dry mass distribution, e.g., in unstained living cells.

2. Using regular DIC optics and a microscope equipped with a precision rotating stage, we confirmed the theoretical validity of the proposed technique.

3. Since the manuscript of this paper was submitted for publication, U.S. patent application 2005-0152030, "Orientation-independent differential interference contrast microscopy technique and device," was published on 14 July 2005.[12] The patent describes designs for achieving orientation-independent DIC images by using novel arrangements of optical and electro-optical components, without the need for any mechanical reorientation or adjustments of the DIC prisms or of the specimen. We estimate that using the new device and the algorithms described in the current paper will allow orientation-independent DIC images to be digitally generated in a fraction of a second.

The authors thank David Biggs and Mike Meichle of AutoQuant Imaging, Inc., for helpful discussions and for providing deconvoluted phase images. We are also grateful to Rudolf Oldenbourg of the Marine Biological Laboratory and to James LaFountain of the University of Buffalo for their encouragement and support.

References and Notes

1. S. Inoue and K. R. Spring, *Videomicroscopy: The Fundamentals,* 2nd ed. (Plenum, 1997).
2. F. H. Smith, "Interference microscope," U.S. patent 2,601,175 (17 June 1952).
3. G. Nomarski, "Interferential polarizing device for study of phase object," U.S. patent 2,924,142 (9 February 1960).
4. R. D. Allen, G. B. David, and G. Nomarski. "The Zeiss–Nomarski differential equipment for transmitted light microscopy," Z. Wiss. Mickrosk. **69**(4), 193–221 (1969).
5. R. D. Allen, N. S. Allen, and J. L. Travis, "Video-enhanced contrast differential interference contrast (AVEC-DIC) microscopy: a new method capable of analyzing microtubule-related motility in the reticulopodial network of *Allogromia laticollaris,*" Cell Motil. **1,** 291–302 (1981).
6. S. Inoué, "Video image processing greatly enhances contrast, quality, and speed in polarization-based microscopy," J. Cell Biol. **89,** 246–356 (1981).
7. R. D. Allen, J. L. Travis, N. S. Allen, and H. Y. Imaz, "Video-enhanced contrast polarization (AVEC-POL) microscopy: a new method applied to detection of birefringence in motile reticulopodial network of *Allogromia laticollaris,*" Cell Motil. **1,** 275–289 (1981).
8. G. Holzwarth, S. C. Webb, D. J. Kubinski, and N. S. Allen, "Improving DIC microcopy with polarization modulation," J. Microsc. (Oxford) **188,** 249–254 (1997).
9. G. Holzwarth, D. B. Hill, and E. B. McLaughlin, "Polarization-modulated differential-interference contrast microscopy with a variable retarder," Appl. Opt. **39,** 6288–6294 (2000).
10. M. Pluta, ed., *Specialized Methods*, Vol. 2 of Advanced Light Microscopy (Elsevier, 1989),
11. C. Preza, "Rotational-diversity phase estimation from differential-interference-contrast microscopy images," J. Opt. Soc. Am. A **17,** 415–424 (2000).
12. M. Shribak "Orientation-independent differential interference contrast microscopy technique and device," U.S. patent application 2005-0152030 (14 July 2005).
13. M. Shribak and R. Oldenbourg, "Technique for fast and sensitive measurements of two-dimensional birefringence distribution," Appl. Opt. **42,** 3009–3017 (2003).
14. D. C. Ghiglia and M. D. Pritt, *Two-Dimensional Phase Unwrapping* (Wiley, 1998).
15. M. R. Arnison, K. G. Larkin, C. J. R. Sheppard, N. I. Smith, and C. J. Cogswell, "Linear phase imaging using differential interference contrast microscopy," J. Microsc. (Oxford) **214,** 7–12 (2004).
16. F. Kagalwala and T. Kanade, "Reconstructing specimens using DIC microscope images," IEEE Trans. Syst. Man Cybern. B **33,** 728–737 (2003).
17. D. Biggs, AutoQuant Imaging, Inc., Troy, N.Y. 12180 (personal communication, 2005).
18. T. J. Holmes, S. Bhattacharyya, J. A. Cooper, D. Hanzel, V. Krishnamurthi, W. Lin, B. Roysam, D. H. Szarowski, and J. N. Turner, "Light microscopic images reconstructed by maximum likelihood deconvolution," in *Handbook of Biological Confocal Microscopy,* J. B. Pawley, ed. (Plenum, 1995), pp. 389–402.
19. D. S. C. Biggs and M. Andrews, "Acceleration of iterative image restoration algorithms," Appl. Opt. **36,** 1766–1775 (1997).
20. Descriptions of Hamamatsu C5985 chilled CCD camera and Hamamatsu C4742-95 high-resolution digital CCD camera are available at http://usa.hamamatsu.com.
21. The ImageJ software is available at http://rsb.info.nih.gov/ij/.
22. The Mathematica software is available at http://www.wolfram.com.
23. A. Resnick, "Differential interference contrast microscopy as a polarimetric instrument,," Appl. Opt. **41,** 38–45 (2002).
24. R. Oldenbourg, S. Inoue, R. Tiberio, A. Stemmer, G. Mei, and M. Skvarla, "Standard test targets for high resolution light microscopy," in *Nanofabrication and Biosystems*, H. C. Hoch, L. W. Jelinski, and H. G. Craighead, eds. (Cambridge U. Press, 1996), pp. 123–138.

Reprinted from *PNAS*, Vol. 103(8), pp. 2971–2976, 2006,
with permission from National Academy of Sciences, USA.

Directly observed reversible shape changes and hemoglobin stratification during centrifugation of human and *Amphiuma* red blood cells

Joseph F. Hoffman*† and Shinya Inoué‡

*Department of Cellular and Molecular Physiology, Yale University, New Haven, CT 06520; and ‡Marine Biological Laboratory, Woods Hole, MA 02543

Contributed by Joseph F. Hoffman, December 23, 2005

This paper describes changes that occur in human and *Amphiuma* red blood cells observed during centrifugation with a special microscope. Dilute suspensions of cells were layered, in a centrifuge chamber, above an osmotically matched dense solution, containing Nycodenz, Ficoll, or Percoll (Pharmacia) that formed a density gradient that allowed the cells to slowly settle to an equilibrium position. Biconcave human red blood cells moved downward at low forces with minimum wobble. The cells oriented vertically when the force field was increased and Hb sedimented as the lower part of each cell became bulged and assumed a "bag-like" shape. The upper centripetal portion of the cell became thinner and remained biconcave. These changes occurred rapidly and were completely reversible upon lowering the centrifugal force. Bag-shaped cells, upon touching red cells in rouleau, immediately reverted to biconcave disks as they flipped onto a stack. *Amphiuma* red cells displayed a different type of reversible stratification and deformation at high force fields. Here the cells became stretched, with the nucleus now moving centrifugally, the Hb moving centripetally, and the bottom of the cells becoming thinner and clear. Nevertheless, the distribution of the marginal bands at the cells' rim was unchanged. We conclude that centrifugation, *per se*, while changing a red cell's shape and the distribution of its intracellular constituents, does so in a completely reversible manner. Centrifugation of red cells harboring altered or missing structural elements could provide information on shape determinants that are still unexplained.

centrifuge polarization microscope | Hb sedimentation | shape deformations

This paper is concerned with red blood cells and changes in their shape and intracellular characteristics during centrifugation. These studies were carried out by use of a newly developed centrifuge polarizing microscope (CPM) (see Fig. 1), that provides, at high resolution, direct observation of the red blood cells, at speeds of up to 11,000 rpm (1, 2). It is known that centrifugation, *per se*, has essentially no effect on shape or volume when comparisons are made of human red cells before and after centrifugation. But it is not clear how these cellular parameters are affected during centrifugation with concomitant exposure to shear forces. In the present work, we describe the effects of centrifugation on single red blood cells of humans and *Amphiuma* with additional observations on human red cells in rouleau.

Human red blood cells are normally biconcave disks but, when exposed to shear stress, are known to become deformed, reversibly, as is evident when studied in a rheoscope (3), an ectocytometer (4), or during normal capillary flow. On the other hand, red blood cells in the spleen of rats (5, 6) and in clotted human blood (7), centrifuged at 400,000 × *g* for 30 min before fixation and staining, showed elongation and, interestingly, stratification of Hb in the centrifugal ends. These results in rat red cells could reflect properties of the paracrystalline state of Hb in those cells (8). In neither rat nor human red cells were the shape transformations or the Hb stratifications reversible, results that differ from our observations reported below. Another approach to the deformability of human red cells during centrifugation is to compare the relative packing of "hardened" (acetaldehyde prefixed) red cells with normal red cells. The hardened human red cell pellets were found to contain ≈40% trapped medium, whereas the normal cells trapped <1% (9). This finding was interpreted to indicate that normal cells, being more deformable, packed to a greater degree. A novel variation of this scheme involves centrifuging, horizontally, a dilute suspension of biconcave human red blood cells for 5 sec (2–15,000 × *g*) through a Teflon capillary tube into a 2% acetaldehyde fixation solution (10, 11). The idea was that any deformation of the cells that took place while in the capillary would be preserved on contact with the fixation solution. Typically, the centrifuged cells displayed rounded ends with stretched and somewhat triangular twisted tails with possible unequal distribution of Hb. It is not clear in these experiments what the relationship was between the relative shapes of the cells and their time of fixation. Partial fixation and partial dilutions of the local fixation solution might have introduced sufficient time to alter whatever their cell shapes were while sedimenting in the capillary tube. In addition, it is not known what the orientation of the cells was with respect to the centrifugal field at the time they were fixed. All of abovementioned uncertainties of interpreting red cell shape resulting from exposure to a centrifugal field are removed in the present studies, because the cells were directly observed during their centrifugation. Also, the various shapes we saw were completely different from those described in any of the previously cited studies.

It has been known since the time of Malphighi, Swammardam, and Leuwenhock (see ref. 12) that human blood contained "red globules," which, with improved optics, segued into lenticular-shaped cells (13, 14). That human red blood cells were, in fact, biconcave disks was definitively established in 1827 once the resolution of the microscope was improved by corrections for chromatic aberration (15). Rouleau formation of human red blood cells, in which cells are stacked one upon another like a roll of coins, was first seen in 1776 (16). Our results extend the catalog of classical shapes that human and *Amphiuma* red cells display in demonstrating the changes in form with attendant Hb stratifications that occur dynamically and reversibly during centrifugation.

Results and Discussion

This paper is primarily concerned with the alterations that red blood cells undergo as they are spun at various speeds in a centrifuge microscope. The cells are suspended in buffered saline solutions and centrifuged into an osmotically matched solution of a naturally formed density gradient composed of

Conflict of interest statement: No conflicts declared.

Abbreviation: CPM, centrifuge polarizing microscope.

†To whom correspondence should be addressed. E-mail: joseph.hoffman@yale.edu.

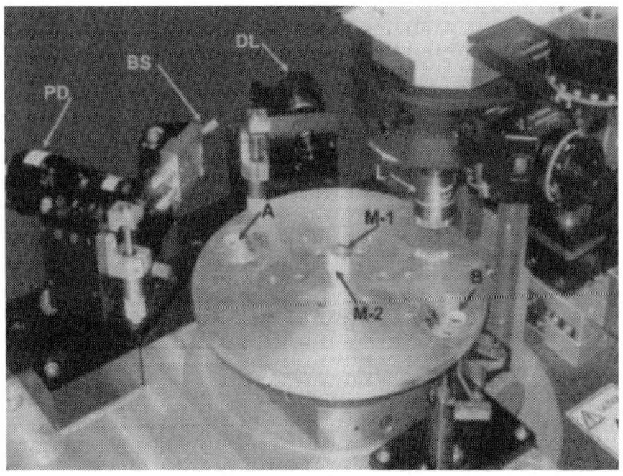

Fig. 1. CPM with dual-specimen chamber. To capture sequential sharp images of the specimen (spinning at peripheral velocities as high as 100 m/s at 10,000 rpm), the light source [neodymium (Nd)-yttrium/aluminum garnet (Nd-YAG)] laser is made to fire a very brief pulse, exactly when the specimen (in chamber A or B) comes under the objective lens (L). To provide exact registration of the sequential images with an image resolution of 0.5 μm, a photodiode (PD) triggers the NdYAG laser to fire a 6-ns pulse through a delay circuit that compensates for the rotor rpm. The PD, in turn, is triggered by light from the diode laser (DL) reflected off a mirror (M-1 or M-2 located on the rotor post) that signals the rotor orientation (1). To distinguish the trigger signals for chambers A and B, mirror M-2 is tilted to reflect the beam from the DL higher than M-1. With the beam switcher (BS) in its higher position (as shown here), light reflected from M-1 passes below the BS and reaches the PD, triggering the laser to fire when chamber A is in position. With the BS in its lower position, the beam from M-1 is blocked, while the beam from M-2 is reflected to enter the PD by the two mirrors on the BS. That provides the trigger for viewing chamber B. Thus, one can instantly select to see the specimen in chamber A or B simply by flipping the BS (2). The inside dimension of the specimen chamber is 8 mm long (column height), 3 mm wide, and 0.5 mm thick. Approximately 10 μl of solution (see *Materials and Methods*) was pipetted into a chamber for experimental observation. The upper window of the chamber is glass, and the lower is annealed acrylic, to cancel out the stress birefringence induced in the windows by the centrifugal force (1).

Nycadenz, Percoll, or Ficoll (Pharmacia). When centrifuged in the CPM at low speeds (<1,000 rpm), human red blood cells are seen to orient either vertically or horizontally in the force field and to slowly move downward, seeking a position in the stationary density gradient (Fig. 2). The cells wobble somewhat as they descend but maintain their initial orientations without tumbling. The slight wobble of the red cells is presumably in response to the shear forces that the cells are subjected to as they move centrifugally. As the speed is increased (>2,000–11,000 rpm), the cells become stationary upon reaching an equilibrium position in the pynotic/density layer and thus are free of any external fluid shear forces. In this situation, most cells orient vertically (Figs. 2 *C–E*, 3*B*, and 4 *A–H*). There are two major changes in the cells' appearance as they are centrifuged at higher speeds. In one, the appearance of the cells changes because the Hb inside the cell sediments; in the other, there is a concomitant bulging of the lower portion of the cell, which *pari passu* assumes a "bag-like" shape. This occurs with the upper part of the red cell remaining biconcave (dimpled) albeit thinner, as well as brighter, due to the movement of Hb out of the upper space into the bag. This result indicates that the cell's biconcave shape is independent of the presence of Hb. The sedimentation of Hb during centrifugation results from the force field overcoming the diffusibility of Hb.

The stratification and shape changes are completely reversible, as shown in Figs. 2–5. As noted in the legend of Fig. 3*B*, human red cells attached to a coverslip during centrifugation show similar changes in shape and Hb sedimentation, as well as reversibility (Fig. 3*C*), as the cells in free suspension. One reason that these "sliding" cells are of interest is that comparable changes may obtain when cells are centrifuged in an angle-head rotor. It should be noted that all centrifugations took place at room temperature (20–23°C), and that there is uncertainty regarding the degree of mixing of the cell suspension with its dense solution, due to uncontrolled convection and/or pipetting. Nevertheless, the cells in every case found an isopycnic boundary/position that varied somewhat across the lower layer of the centrifuge chamber. It also needs to be stated that all of the attendant changes in cell shape and Hb distribution seen in Nycodenz solutions were also seen when the dense solution contained either Ficoll (data not shown) or Percoll (Fig. 4).

In contrast to human, *Amphiuma* red cells undergo cellular elongation with Hb moving upward (centripetally) rather than downward, with a concomitant bulging of the upper part of the cell, as shown in Fig. 5. The nucleus is seen to move downward into a relatively clear cytoplasm. Because of the ability to analyze the birefringence of these cells during their centrifugation, the structural orientation of the marginal bands (17) can clearly be visualized surrounding the internal periphery of the cells, as noted in the legend. Here again, with reduction of the centrifugal force field, the changes in cell shape, nucleus, and Hb distribution were reversible.

A question of interest is to what extent does the centrifugation (force field) *per se* have an effect on human red blood cells, knowing that the steady-state volume of the cells is not affected by centrifugation? It should be understood that, at the highest speeds used (11,700 rpm), the hydrostatic pressure head at the bottom of the centrifuge chamber is <5 atmosphere (atm) (1 atm = 101.3 kPa), with, of course, there being no net pressure gradient across the cell. Human red blood cells are known to be essentially incompressible [3.4×10^{-11} cm^2/dyne (1 dyne = 10 μN; ref. 18)]. Centrifugation of red blood cells between 20,000 × *g* and 190,000 × *g* decreases the hematocrit of packing by 1.7% and between 40,000 × *g* and 190,000 × *g* by 0.7% (19). These decreases presumably represent changes in the volume of serum lodged between the cells, although it is not known that the hematocrit value, i.e., the packed cell volume, is the same during centrifugation as it is afterward when the measurements are made. It is also known (20) that Hb exists inside the cell in random orientation that persists for small perturbations of cell volume, such as those that took place (±10%) in the conditions underlying the results presented in Figs. 2 and 3. The red cells under these conditions are also known to act as perfect osmometers (21). Whether these comments apply to *Amphiuma* red cells is currently not known. Of note, the fragmentation and stratification reported for the red cells of *Necturus* and frogs when centrifuged at forces between 20,000 and 400,000 × *g* (22) were not seen in the present work.

The rouleau formation seen in Fig. 4 took place when the dense solution contained Percoll [polyvinylpyrrolidone (PVP)]. Rouleau is seen in normal human blood, and the physiologic promoter for its formation is the presence of fibrinogen. Many types of uncharged polymers, such as PVP, Dextrans, and polyethylenglycols, are also known to stimulate rouleau formation. Nevertheless, there is no agreement on the mechanism(s) that underlies rouleau formation. The distance between the stacked cells has been shown to vary directly with the size of the inducing agent (23), but it is not known to what extent the glycocaylyx, present on the surface of human red blood cells, is involved (23, 24). Neurominidase treatment of human red cells causes accelerating rouleau formation by reducing their surface charge and electrophoretic mobility.[§] On the other hand, treatment of cells with 4,4′-diisothio-cyanatostilbene-2,2′-disulfonic

[§]Jan, K.-M. & Chien, S. (1972) *Fed. Proc.* **31**, 341.

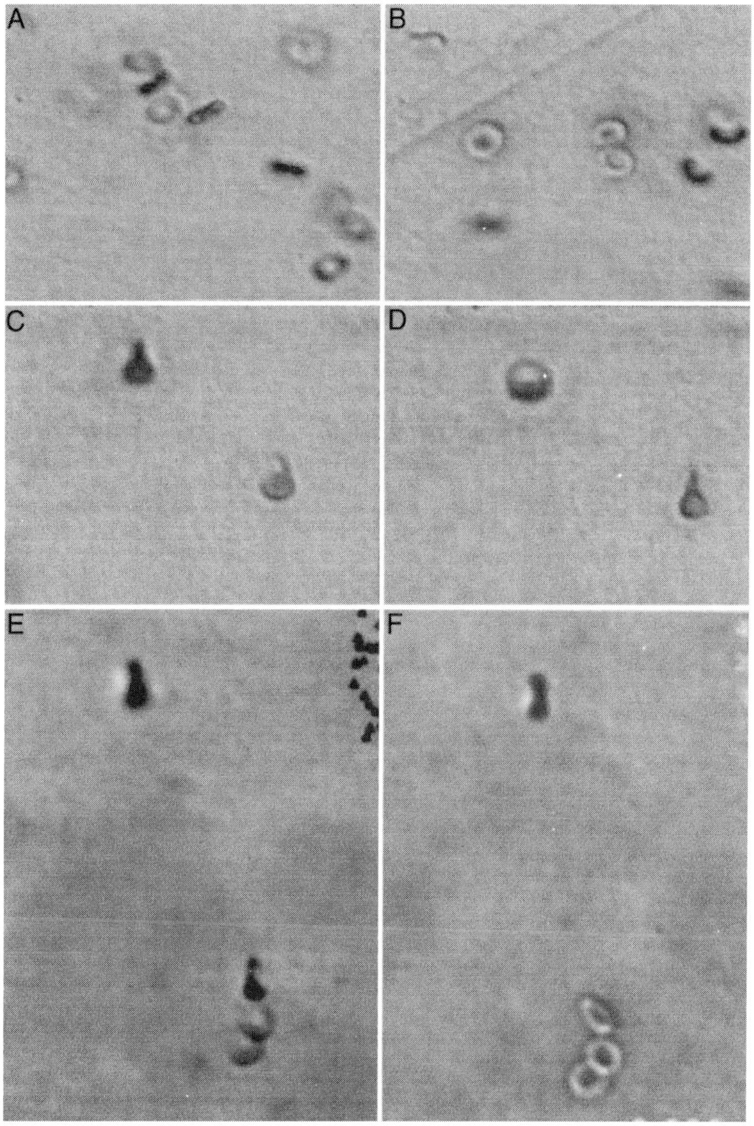

Fig. 2. Hb sedimentation and shape changes of human red bloods during centrifugation and their reversibility. A dilute suspension of human red blood cells in buffered hypotonic saline (solution A) was laid onto the top of a dense solution of Nycodenz in the centrifuge chamber (see *Materials and Methods*). During the initial centrifugation at 700 rpm (*A*), in which the centrifugal force is down (as in Figs. 2–5), the depicted biconcave disks were seen to sediment in either a vertical or horizontal orientation (parallel or perpendicular with respect to the force field). The cells in either orientation tended to wobble slowly but not tumble. When the speed was increased to 2,000 rpm (*B*), the cells remained in their same orientations, but some of the cells became cup-shaped, with the bottom of the cup pointing centrifugally. At higher speeds of rotation (8,700 rpm), the cells oriented vertically (*C* and *D*), and it is evident that the Hb molecules sediment toward the bottom of each cell, so that the cell acquires a prominent bulge. At the same time, the upper part of the cell became or stayed a thin half disk that still is biconcave. Hence the whole cell takes on a "bag" shape. *D* shows, 17 sec later, that the same two cells seen in *C* have turned and are now seen in different perspectives. The cells displayed are free-floating and not in contact with either cover plate. *E* (9,700 rpm) shows four bag-shaped cells in different views. *F*, 100 sec later, shows that the same cells seen in *E* return to their normal biconcave disk shape after the speed is markedly reduced (400 rpm). Depending on the focal level, the thinner biconcave region of the cell may appear brighter or darker than the thicker (light-absorbing and refractive) regions. These results show that the alterations in shape and Hb distribution induced by the centrifugal force are completely reversible.

acid (DIDS), which is known to inhibit anion transport by Band 3, inhibits rouleau formation, presumably by adding net charge or repulsive force to the interface (26). In contrast, rouleau formation with dextrans is known to be biphasic, independent of the size of the dextran used (27, 28), with the interesting result that the dextran polymers change their structure (i.e., coil to network) as a function of concentration and the peak of the biphasic response correlates with the transition in the dextran's molecular structure (28). Evidently, there is no specificity in agents that induce rouleau formation, so long as minimum requirements are met (e.g., ref. 29). On the other hand, the mechanisms(s) and the detailed chemistry responsible for the interaction and the stacking phenomenon have yet to be defined. It is also not clear what forces underlie the shape changes of the

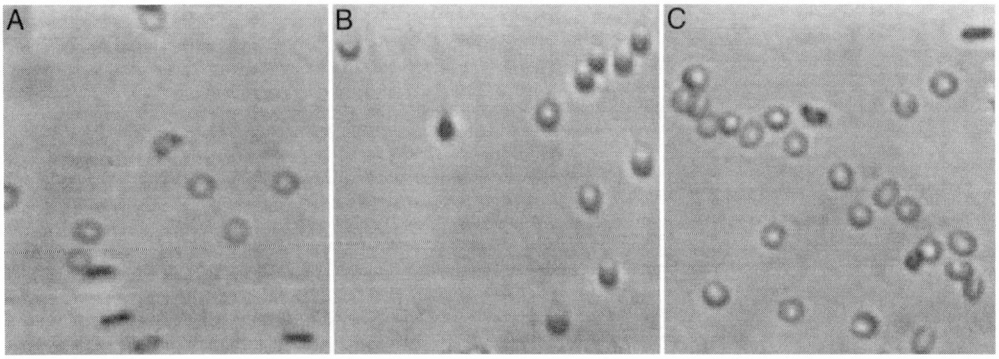

Fig. 3. Displacement and deformation of human red blood cells attached to the bottom plastic window of the chamber. The human red cells were suspended in a hypertonic saline/Nycodenz medium (see Solution B in *Materials and Methods*) in which many of the cells were allowed to settle and attach on the lower (plastic) window of a chamber before rotation of the centrifuge. At low centrifugal force (2,000 rpm), the biconcave cells seen *en face* (*A*) are attached to the window and display uniform Hb distribution. The unattached cells sediment horizontally as seen in Fig. 2. As the speed of rotation was increased (9,700 rpm), the centrifugal force sediments the Hb and stretched the attached cells (*B*), which become narrower and more extremely bag-shaped as they glide down the surface of the window. *C* shows that, as the speed was decreased to 400 rpm, the cells that remained attached to the surface regained their symmetric biconcave shape.

red cells as they attach and flip onto the stack, as depicted in Fig. 4, but it would appear that the rouleau forces that drive the maximal surface contacts between cells in a stack are stronger than the centrifugal forces responsible for the distortion of the cell's shape.

The experimental results reported here have been mainly concerned with the deformability of human and *Amphiuma* red blood cells that occur during their centrifugation. Although we were surprised that the centripetal region of the bag-shaped human red cells still remained biconcave, the results do not

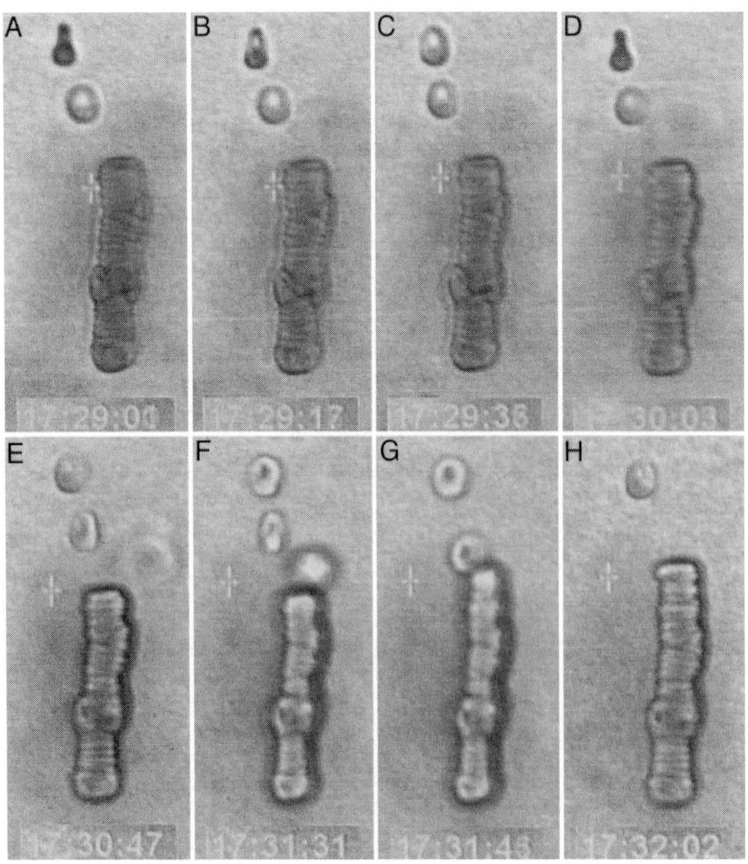

Fig. 4. Rouleau formation. Shown is a timed sequence of two bag cells adding to the top of a stack of human red cells in rouleau formation. The speed is 4,800 rpm and the time of day, shown in the bottom, is in h:min:sec. The cells were suspended in an isotonic saline/Percoll medium (see Solution C in *Materials and Methods*). It should be noted that this rouleau was only seven cells tall 10.5 min before that seen in *A*; in the meantime, cells were added to both its top and bottom. In *A–G*, two bag-shaped cells above the rouleau can be seen in different views as they turn. (Depending on the focal level, the contrast of the thicker and thinner parts can be reversed.) In *E–G*, a third cell, which is out of focus, approaches from the right and joins the stack at the top. In *G* and *H*, the lower of the two original bag-shaped cells became incorporated into the rouleau. As soon as the bag-shaped cell touches the stack, it loses its asymmetric shape and unequal Hb distribution and resumes its original biconcave disk shape as it flips onto the top. (Taking *A* as zero time, *B–H* represent, respectively, 17, 36, 63, 107, 151, 165, and 182 sec later.)

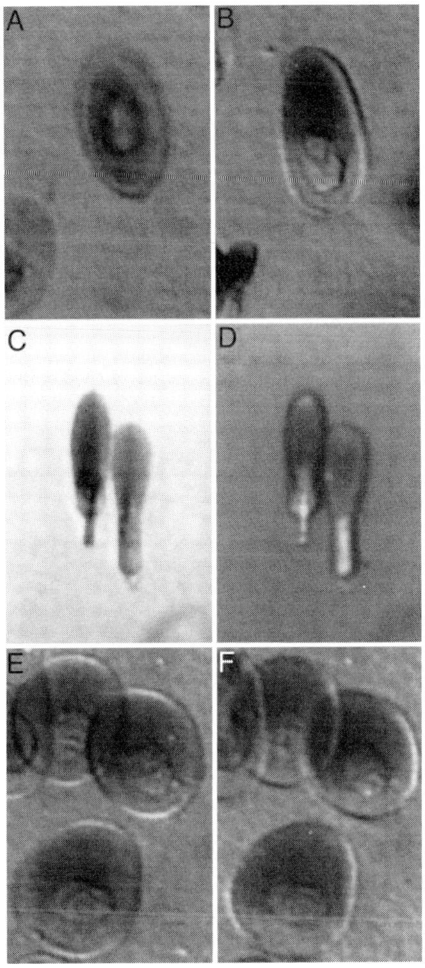

Fig. 5. The effect of centrifugation on *Amphiuma* red blood cells. *Amphiuma* red cells were suspended in an isotonic amphibian Ringer/Ficoll solution, as described in *Materials and Methods*. *A* shows two cells, sedimenting at low speed (1,000 rpm), in which the nucleus is located centrally in each cell similar to their appearance at 0 rpm. *B* shows that under increased centrifugal force (11,700 rpm), the red cells elongate as the Hb stratifies upward, leaving a thin clear lip below, while the nucleus moves down. The descending nucleus tends to be deformed as though still partially anchored to the middle of the cell. The microtubules in the marginal band remain at the periphery of the cell, despite the extreme deformation of the cell. *C* and *D* show side views of two other cells also being centrifuged at 11,700 rpm. *E* and *F* show *Amphiuma* red cells ≈18 min after reduction of the speed to 200 rpm. Here the cells display their reversibility, having returned to become roundish disks with their Hb being almost equally distributed and their nuclei having moved toward the middle of the cells. Typically, as shown, the cells are oriented vertically during their centrifugation. In *B*, *D*, and *F*, the birefringent marginal band appears bright vertically and dark horizontally, whereas in *C* and *E*, they appear in reverse contrast. The contrast is due to the interaction of the positive birefringence of the microtubules (that make up the marginal band) with the Brace–Koehler compensator, whose orientation is reversed between *C* and *E*.

address the interesting and important property of cell shape, particularly the nature of the biconcave disk that characterizes these cells (see ref. 30). The centrifuge microscope could be a useful device to further investigate the basis for the biconcave shape of human red cells by testing the effects of various agents (e.g., DIDS, Diamide, and wheat and germ agglutinin) on cell deformation during centrifugation. The study of human (and mouse) red cells in which abnormalities in their cytoskeletal and surface structures have been characterized (e.g., knockouts and missing or altered cytoskeletal and/or membrane constituents) could provide insight into the molecular basis for the disk shape and cellular elements involved. It is hoped that studies such as these will soon be forthcoming.

Materials and Methods

Red Cell Preparations. Normal human blood was taken by finger puncture, collecting either the third or fourth drop into 3 ml of a heparinized isotonic solution (310 mosM) that contained 150 mM NaCl, 20 mM Tris·Cl, and 0.2% BSA (pH 7.4) at 23°C. The cells were then sequentially diluted ≈50,000-fold into either the same solution or a solution in which the concentration of NaCl was changed to either 125 mM (hypotonic, 272 mosM) or 175 mM (hypertonic, 353 mosM). Three to five microliters of these diluted cell suspensions, containing 100–300 red blood cells, was then pipetted into a centrifuge chamber (see below) and carefully layered on top of 5 μl of a dense solution. The dense solutions were as follows: Solution A, hypotonic: 17 mM NaCl/0.8 mM KCl/0.3 mM $CaCl_2$/3.2 mM 4-morpholinepropanesulfonic acid (pH 7.3 at 23°C)/1.40 g of Nycodenz (Sigma) per 5 ml of solution (275 mosM, density 1.126 g/ml at 23°C). Solution B, hypertonic: the same as Solution A, except that the concentration of Nycodenz was 1.80 g per 5 ml of solution (351 mosM, density 1.162 g/ml). Solution C, isotonic: 130 mM NaCl/15 mM Tris·Cl (pH 7.3 at 23°C)/90% Percoll (wt/vol; Sigma)/302 mosM; density, 1.130 g/ml. The osmolality of the dense solutions matched those saline solutions in which the cells were suspended.

Red blood cells of the three-toed salamander, *Amphiuma tridactylum*, were obtained by cardiac puncture as described (31). These are giant nucleated elliptical red blood cells, 150 times the volume of human red blood cells (87 μm^3, 6–8 μm diameter) whose average dimensions are 61 μm in length, 36 μm in width, and the thickness of the cell 15 μm across the nucleus compared with 8 μm for the cytoplasm. The cells were treated essentially as described above for human red cells but were washed (by spontaneous settling) in a modified isotonic (201 mosM) amphibian Ringer solution whose composition was 95 mM NaCl/2.5 mM KCl/1.8 mM $CaCl_2$/0.2% BSA/10 4-morpholinepropanesulfonic acid (pH 7.0 at 23°C). After suitable dilution (two steps) of the cells in this solution, 3–5 μl of the final suspension containing ≈200–300 cells was pipetted onto the top of a dense solution, as described before. The isotonic dense solution contained Ficoll70 (70,000 M_r, Sigma) in a concentration of 25 g Ficoll in 40 ml of a 1/10 dilution of the wash medium (200 mosM; density, 1.1502 g/ml, pH 7.0, at 23°C).

It should be understood that the composition of the dense solutions was primarily dictated by the requirements of osmolarity and density to provide a buoyant medium and density gradient that optimized viewing and minimized cell volume changes.

CPM. The CPM, developed jointly by Olympus (Melville, NY), Hamamatsu Phototonics ((Hamamatsu City, Japan), and the Marine Biological Laboratory, allows video microscopic observation and recording of intracellular constituent stratification and molecular alignment in living cells spinning at between 100 and 11,700 rpm. Through a special timing circuit, the specimen is momentarily illuminated by a 532-nm wave-length 6-ns laser pulse exactly as it traverses below the objective lens. With a 40 × 0.55-numerical aperture-long working-distance objective lens, the image is resolved to somewhat better than 1 μm (1). By manually switching a timing-trigger control device (added since the original design), one can observe the specimen placed in either one of the two chambers in sequence with no delay (ref. 2; Fig. 1).

PHYSIOLOGY

Contrast of the specimen placed in the chambers, whose windows are designed to suffer only low degrees of stress birefringence, can be generated in bright field, in polarization optics with a sensitivity exceeding 1 nm in retardation, in differential interference contrast (Nomarski contrast), or in fluorescence excited by the 532-nm neodymium-yttrium/aluminum garnet laser illumination (32). Human red cells were generally observed with the compensator rotated sufficiently away from extinction, in other words, essentially as bright-field images, whereas the *Amphiuma* cells were observed either in bright field (Fig. 5*A*) or with appropriate compensation to display the birefringence of the microtubule bundles that make up the marginal band (Fig. 5 *B–F*).

The human and *Amphiuma* red blood cells, suspended in solutions as described above, were observed in the CPM rotating at between 200 and 11,700 rpm ($3.4 \times g$ to $11,500 \times g$, where g = Earth's gravitational acceleration). Data were analyzed by repeated playback of the video sequences at various speeds and by listening to the real-time audio comments made on the videotape during observation.

It is important to mention that, in interpreting the shape of the red blood cells from their images in conventional bright-field microscopy, individual unstained cells do not appear red at all. This is because they are so thin that refraction, rather than absorption by Hb, contributes primarily to the contrast of the image. Even in the green light used in our CPM, the thicker portions of the biconcave cell appear dark only at exact focus. As the focus is shifted, the contrast reverses by refraction effects as in the "Becke line" phenomena (see ref. 25).

We thank Dr. Makoto Goda and Bob Knudson for input to and improvements of the CPM and Hamamatsu Photonics and Olympus for major contributions to the development of the CPM.

1. Inoué, S., Knudson, R.A., Goda, M., Suzuki, K., Nagano, C., Okada, N., Takahashi, H., Ichie, K., Iida, M. & Yamanaka, K. (2000) *J. Microsc.* **201,** 341–356.
2. Knudson, R.A., Inoué, S. & Goda, M. (2001) *Biol. Bull.* **201,** 234.
3. Fischer, T. M., Stöhr-Liesen, M. & Schmid-Schönbein, H. (1978) *Science* **202,** 894–896.
4. Bessis, M. & Mohandas, N. (1975) *Blood Cells* **1,** 307–313.
5. Beams, H. W. & Kessel, R. G. (1966) *Anat. Rec.* **155,** 541–550.
6. Beams, H. W. (1947) *Proc. Soc. Exp. Biol. Med.* **66,** 373–375.
7. Skalak, R. & Branemark, P.-I. (1969) *Science* **164,** 717–719.
8. Ponder, E. (1948) *Hemolysis and Related Phenomena* (Grune and Stratton, New York).
9. Chien, S., Dellenback, R. J., Usami, S., Seaman, G. V. F. & Gregersen, M. I. (1968) *Proc. Soc. Exp. Biol. Med.* **127,** 982–985.
10. Corry, W. D. & Meiselman, H. G. (1978) *Biophys. J.* **21,** 19–34.
11. Corry, W. D. & Meiselman, H. J. (1978) *Blood* **51,** 693–701.
12. Bessis, M. & Delpech, G. (1981) *Blood Cells* **7,** 447–480.
13. Sénac, M. (1749) in *Traité de la Structure du Coeur* (Briasson, Paris), pp. 654–670.
14. Gulliver, G. (1846) *The Works of William Henson* (Sydesham Society, London), Vol. 2, pp. 209–291.
15. Hodgkin, T. & Lister, J. J. (1827) *Philos. Mag.* **2,** 130–138.
16. Della Torre, G. M. (1776) in *Nuove Osservazioni Microscopiche* (C. R. Somasco, Napoli, Italy), pp. 1–135.
17. Cohen, W. D. (1978) *J. Cell Biol.* **78,** 260–273.
18. Weiser, M. A. H. (1981) Ph.D. thesis (Yale Univ., New Haven, CT).
19. Parpart, A. K. & Ballentine, R. (1943) *Science* **98,** 545–547.
20. Bateman, J. B., Hsu, S. S., Knudsen, J. P. & Yudowitch, K. L. (1953) *Arch. Biochem. Biophys.* **45,** 411–422.
21. Freedman, J. C. & Hoffman, J. F. (1979) *J. Gen. Physiol.* **74,** 157–185.
22. Beams, H. W. & King, R. L. (1945) *J. Morphol.* **77,** 63–69.
23. Chien, S. & Jan, K.-M. (1973) *Microvasc. Res.* **5,** 155–166.
24. Neu, B. & Meiselman, H. J. (2002) *Biophys. J.* **83,** 2482–2490.
25. Chamot, E. M. & Mason, C. W. (1958) in *Handbook of Chemical Microscopy* (Wiley, New York), 3rd Ed., pp. 314–315.
26. Norris, S. S., Allen, D. D., Neff, T. P. & Wilkinson, S. L. (1996) *Transfusion* **36,** 109–112.
27. Evans, E. A. & Parsegian, V. A. (1983) in *Annals of the New York Academy of Sciences*, eds. Copley, A. L. & Seaman, G. V. F. (New York Academy of Sciences, New York), Vol. 416, pp. 13–33.
28. Barshtein, G., Tamir, I. & Yedgar, S. (1998) *Eur. Biophys. J.* **27,** 177–181.
29. Armstrong, J. K., Wenby, R. B., Meiselman, H. J. & Fisher, T. C. (2004) *Biophys. J.* **87,** 4259–4270.
30. Hoffman, J. F. (2001) *Blood Cells Mol. Dis.* **27,** 57–61.
31. Hoffman, J. F. & Geibel, J. P. (2005) *Proc. Natl. Acad. Sci. USA* **102,** 921–926.
32. Inoué, S., Goda, M. & Knudson, R. A. (2001) *J. Microsc.* **201,** 357–367.

APPENDIX I

DEVELOPMENT OF THE "SHINYA SCOPES"

Shinya Scope-1

As described in Articles 5 and 56 of this *Collected Works*, and in the video taped lecture on the appended DVD disk, my mentor Katsuma Dan (nicknamed Katy) asked me to visualize the mitotic spindles in dividing sea urchin eggs using a polarizing microscope, as had been shown by W.J. Schmidt in his 1937 monograph. Our first attempt at Katy's home (behind the air-raid blackout curtains necessary during WW-II), using a petrographic microscope that Katy had borrowed from his friend in the Geology Department, proved inconclusive. So upon returning to Katy's laboratory at the Misaki Marine Biological Station in 1947 and 1948, I decided to build my own polarizing microscope.

For the project, Professor Koana of the Physics Department, Tokyo University, loaned me a calcite polarizing prism and, in addition, had special strain-free slides and cover slips fabricated for me in his department's optics shop. Katy let me use one of his Zeiss microscopes, and I myself used a "Dichrome" polarizing filter (that I had used in the projects described in Articles 2 and 3) to make the analyzer. By pasting onto the sub-stage filter holder a sheet of mica, which I pealed to 5 μm thickness using an ivory knife made from an old piano key, I managed to make a Brace-Koehler compensator. A surplus electrical parts store provided the AH-4 mercury arc lamp, which I placed in a tin tea can and served as a bright light source.

But resources, in general, were scarce in those early post-war years. So I scrounged around the woods of the Misaki Station which, thanks to Katy's effort, had been returned to Tokyo University by the Occupation Forces (after it had been taken over by the Japanese Navy as a miniature submarine base for the last year of the Second World War; see Article 56). There, I found a cast-away machine gun base which looked suitable as a firm base on which to assemble the various components of the microscope that I was building. For a short while until then, I had placed the components on stacks of books to align their axes — a rather precarious and imprecise arrangement.

After learning from some mistakes I made about background birefringence (Article 56), I did manage to assemble a scope that worked quite well and which gave us the published results reported in Article 5. It is also the scope with which I demonstrated the twinkling birefringence of spicules in swimming sea urchin plutei (Articles 29 and 73) to Emperor Showa and his royal family, who were visiting the Misaki Station for the first time after the Second World War.

Following my departure to Princeton in the fall of 1948, the scope was used by Kayo Okazaki and Katy Dan, who gave it the nickname "Shinya Scope." This Misaki scope was followed by several new versions that I designed, so I shall refer to this one as Shinya Scope-1.

Also, as reported in Article 3, I had already assembled another polarizing microscope,

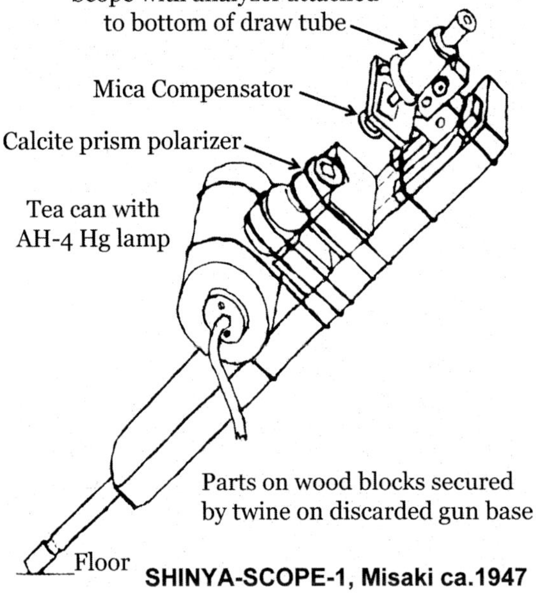

Fig. 1. Sketch of the Shinya Scope-1.

although one that required less sensitivity since it was for studying muscle fiber birefringence. Since that version preceded the one at Misaki, perhaps I should label the earlier one as Shinya Scope-0. The design and use of this 0-th version are detailed in Article 3.

Figure 1 is a sketch of the Shinya Scope-1 which I drew from memory in around 1989. As labeled in the drawing, I secured different components, placed on notched wooden blocks, by twine to the gun base. Also, as indicated, the whole microscope was often used as though one were playing a cello by placing the tip of the base on the floor.

Shinya Scope-2

After Shinya Scope-1, the next version was the one I built while a graduate student at Princeton University. This version is illustrated and described in Article 7 and its performance is analyzed in Article 8. This was the first version of the Shinya Scopes to use inverted optics (in fact, apparently the very first inverted scope made for biological applications), with the light source and condenser above, rather than below, the specimen for reasons described in Article 7. Also, it was the first version on which I actually measured the extinction factor (EF = the ratio of the intensity of light coming through parallel versus crossed polarizers). By measuring the EF at various numerical apertures of the lenses, I could reason what, in principle, would be needed to achieve high sensitivity for detecting very small amounts of birefringence, coupled simultaneously with excellent image resolution.

For Shinya Scope-2, as described in Article 7, Bausch and Lomb fabricated and provided special optical components, while various members of the Princeton faculty provided other components for the microscope. Russell Mycock of the Princeton University Biology shop fabricated the sliders and coupling devices which let me adjust the location of the optical components on the upright stand.

By using this microscope, coupled with a variety of biological specimen, I was able to demonstrate the universal presence of spindle fibers and their labile fibrils (which later turned out to be microtubules), as well as other basic features of mitosis and cell division as illustrated in Articles 10, 11, 12, 13, and 22.

Fig. 2. Overview of Shinya Scope-3.

Shinya Scope-3

As shown in Figs. 2 and 3, this version of the Shinya Scope is built on a massive vertical optical bench, which came from a Ford spring tester. As in the earlier versions, all of the optical components lie on a single straight axis. The location of the components (except for the movie camera at the bottom) is each adjustable on a slider along the 3-foot-long, 3-inch-wide optical bench. In contrast to Shinya Scope-2, which I designed at Princeton using pieces made available to me by faculty members of the Biology Department, the design for version three (including the use of the spring tester base and precision sliders) was suggested to me by an engineer at the University of Washington Medical School shop where the instrument was built. This microscope was used extensively for some of the work reported in Articles 14, 15, 16, 17, 18, and 22.

In Fig. 2, starting at the top, the light source unit includes a custom-designed housing for the AH-6 high pressure mercury arc lamp, and a pair of 16 mm movie projector lenses (placed face to face sandwiching the 2-inch-square monochromatic and heat-cut filter holder) which projects a 1:1 image of the arc source onto a 1-mm-diameter pinhole. The pinhole trims the elongated image of the mercury arc to a small circle and acts as the virtual light source. The image of the virtual source is projected by a 16 mm camera lens through the polarizer and compensator to the condenser

Fig. 3. Close up of central section of Shinya Scope-3.

lens (actually a strain-free objective lens). The polarizer is mounted on a precision goniometer which allows measurement of the polarizer orientation angle to 1/100 degree.

Figure 3 shows a close-up view of the components including the compensator condenser unit, revolving stage, and the body tube unit. As shown here, each of these three units, mounted on individual dovetail sliders, is equipped with its coarse and fine focus controls. The objective lens used as the condenser, as well as the objective lens itself, are each mounted on a centrable quick-change nose piece holder. The upper part of the condenser mount holds a Brace-Koehler compensator, which is designed so that the limits of its back and forth movement can be preset. The body tube contains the analyzing prism (equipped with stigmatizing lenses), whose orientation can be preset, and a first-surface removable mirror, which reflects all the light to the eye piece or allows it to travel without obstruction to the projection eye piece at the bottom of the tube.

The film cameras, which fit into the opening of the spring tester base, are supported on yolks supported with four adjustment screws, which together position each camera kinematically onto two mutually perpendicular V-grooves and a flat that we added to the spring tester base. The 16 mm motion picture camera illustrated in Fig. 2 uses a Kodak Ciné Special film cartridge combined with a special focusing device that I designed. In this focusing device, a sectored first-surface mirror (placed at 45° to the light path) is driven by an external, variable speed motor to rotate synchronously with the film-driving mechanism in the film cartridge. Thus, while the film in the cartridge is being advanced, 100% of the light from the projection ocular is reflected onto the focusing, continually-revolving, ground-glass screen viewed through the eye piece. While the film is stationary, all the light passes the empty portion of the sectored mirror and exposes the 16 mm film without losing any intensity. Full illumination of the focusing eye piece and film was necessary in order to both see and to capture the dim images of the weakly birefringent cell organelles, which required extended exposure. For still images, the motion picture camera was replaced by a 35 mm, single-lens reflex camera (from which the lens was removed).

For Shinya Scope-3, the precision goniometer holding the polarizer was kindly made available to me by the Chairman of the Anatomy Department Dr. H. Stanley Bennett (author of an extended chapter on biological polarization microscopy in McClung's Handbook; see reference in Article 40) who designed the goniometer. The coarse and fine focusing blocks and graduated revolving stage were made available by Ernst Leitz, while Bausch and Lomb provided the specially selected, strain-free coated objectives and the quick-change centrable nose pieces.

Shinya Scope-4

The next version of the microscope was developed in corporation with the American Optical Company and the Institute of Optics at the University of Rochester where I was a faculty member in the Biology Department.

As shown in Figs. 4 and 5, Shinya Scope-4 still uses a casting patterned after the three-foot-long optical bench from a Ford spring tester but without its base. Instead, the main casting is supported on a sturdy, custom-made wooden bench via an axle that penetrates near the casting's center of gravity. In addition to minimizing the influence of external vibration, this design allows the optical axis of the whole microscope to be tilted to any desired angle, including the horizontal (Fig. 4). In addition, two solid, stainless-steel side rails are mounted on the sides of the casting, parallel to the main dovetail. The side rails are used to accommodate special lamps for UV microbeam irradiation

Fig. 4. Shinya Scope-4 in horizontal orientation.

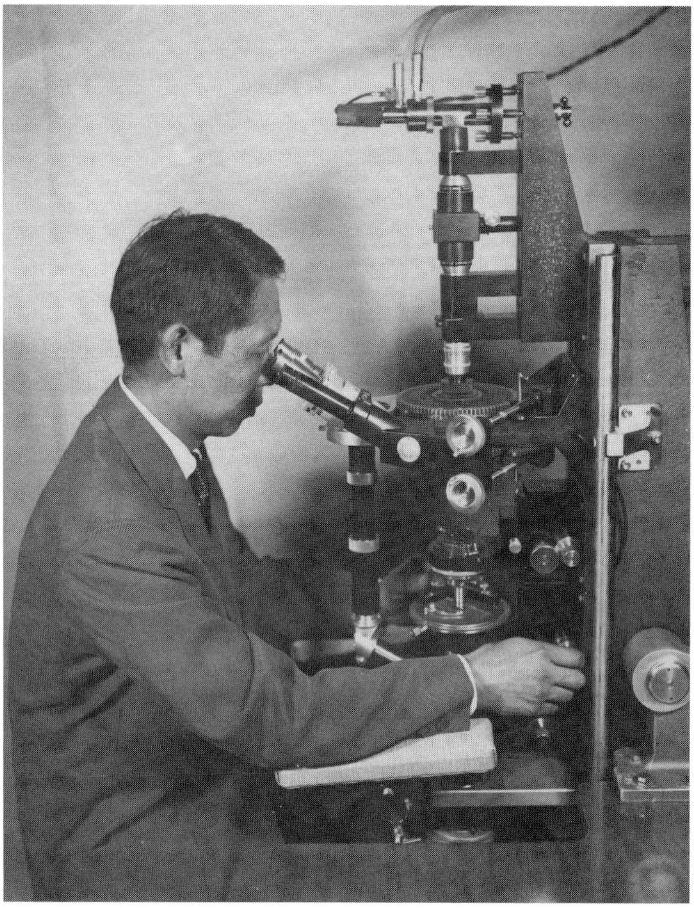

Fig. 5. Close up of Shinya Scope-4, in use by the author.

and other auxiliary components. The sturdy wooden bench with leveling feet also supports two arm rests (Fig. 5), which were made from modified obstetric knee supports.

As in the earlier versions, the optical axis of the whole system remains straight from the light source to the cameras in order to avoid affecting the polarized light beam by the presence of reflecting prism, mirrors, beam splitters, etc.

An improved stainless-steel housing, which I designed for the water-cooled AH-6 high pressure mercury arc lamp, was fabricated by the Instrument Shop of the Institute of Optics, while Stan Bennet's goniometer used in version three was replaced with a higher precision polarizer-comparator mount designed by Warner Engel of the American Optical Research Center in Southbridge, Massachusetts. As designed by Warner, this unit houses a couple of glass circles mounted on high precision tapered bearings, which respectively support the polarizer and comparator units. The glass circles, made for Wilde theodolites, are engraved for each degree with the angle number plus a fiducial line, precisely positioned to 1/1,000 degree. The orientation of the theodolite circle is read through an ocular equipped with a 100 division reticule whose full length covers 1° on the circle. Thus, by interpolation, one can determine the orientation of the polarizer or comparator to 1/1,000 degree.

The comparator unit, placed below the polarizer, is equipped with two 1-inch-wide, rotatable, horizontal dovetails. The dovetails accept sliders supporting a quartz wedge

(Article 3), comparator crystal (Article 9), pinholes (Articles 15 and 17), slits (Article 18), or micro-mirrors for reflecting the UV microbeam (Articles 22, 24, 26, 32, 33, and 48). The images of these components are projected, superimposed with the specimen, by the condenser lens. The polarizer and analyzer were Glan-Thompson calcite prisms with integral anti-reflection coated cover plates or stigmatizing lenses, especially fabricated by Karl Lambrecht of Chicago, Illinois, to provide exceptionally high Extinction Factors (EF > 1×10^6). As illustrated in Fig. 37 by Rinne and Berek (reference in Article 40), the stigmatizing lenses render the main imaging beams parallel through the calcite analyzer, thereby preventing the pronounced astigmatism which would be introduced by the anisotropic calcite prism.

Starting with this version of the Shinya Scope, the objective and condenser lenses incorporated rectified optics, designed by me together with W. Lewis Hyde of the American Optical Research Center. The rectifiers allow the generation of very high Extinction Factors up to large Numerical Apertures. Thus, the polarizing microscope could finally be used with little stray light even when using objective and condenser lenses that provide high resolution. In other words, one could finally achieve high sensitivity for detecting very weak birefringence concurrently with high image resolution (Articles 8, 14, 18, and 40). In addition, rectification abolished the image anomaly that accompanied the use of non-rectified lenses, so that the polarization optical images of weakly birefringent, minute objects could now be observed without the artifacts that are introduced in classical polarization optical systems (Articles 15, 17, 18, and 69).

In addition to studies of fine structure in living cells (Articles 16, 19, 24, 25, and 26), this version of the Shinya Scope was used extensively for basic studies of image formation in polarization optics (Articles 14, 15, 17, and 18), and for polarization optical properties of a variety of thin films used for anti-reflection coating. For these purposes, we wanted to be able to quickly switch between viewing the specimen and reading the polarizer and comparator angles. To make this possible, the specimen image was transferred through relay lenses to the eye pieces, which were raised next to the ocular for viewing the reticule (the black eye tube next to the eyepieces; Figs. 4 and 5).

By the time Shinya Scope-4 was designed, Arriflex 16 mm movie cameras, with built-in 45° sector mirrors similar to the one I thought up for Shinya Scope-3 had become available commercially. So for time-lapsed ciné work, we opted to use the Arriflex camera coupled with a home-built, geared-down motor drive. Also, Polaroid cameras with high-speed, 4 × 5-inch negatives and prints became available (Fig. 4), which we used alternately with fine-grain 35 mm film. The Arriflex, Polaroid, and 35 mm camera and photomultiplier were all mounted interchangeably on a dovetail slider near the base of the main dovetail.

Shinya Scope-5 (and -6)

Shinya Scope-5 was developed, with major input from Gordon Ellis and Bob Knudson, while we were in the Cytology Department at the Dartmouth Medical School and later with further help from Ed Horn in the Biology Department at the University of Pennsylvania. These developments, as well as the biological applications of the new instruments, were supported generously by Dartmouth Medical School and the University of Pennsylvania, which each provided our group with laboratories renovated according to our specifications. In addition, our project received generous support from the NIH and NSF as well as from the American Optical Company, Nikon, Leica, and Karl Lambrecht of Crystal Optics Company in Chicago, Illinois.

In place of the 3-foot-long Ford spring tester bench, for the new scope we had the main casting, now four-feet long, made of carefully stress relieved Meehanite steel by Pohlman Foundry of Rochester, New York. Unlike the Ford bench, which was widest at its base (Figs. 2 and 4), the new bench designed by Gordon Ellis was widest in the middle and tapered down symmetrically on both ends. For this design, we added two 1.5-inch-wide dovetail side rails with their front faces tilted at 30° to the front face of the main three-inch-wide dovetail, all built into the full length of the main casting. After molten salt immersion, the dovetails were all machined and finished by hand scraping by Omega Consolidated Corporation, also of Rochester, New York, to a tolerance of 1/10,000 inch, and straight and parallel to each other within 2.5 μm.

At the top of the main casting, we added a foot-long flat casting with its three horizontal dovetails precisely aligned at 90° to those on the main vertical dovetails. As in version four, the main casting is supported by an axle near its center of gravity on a wooden bench, again with the option of tilting and clamping the whole microscope at any desired angle. The extended length of the main dovetail, the horizontal extension above, and the built-in precision side rails gave considerably greater flexibility to the choice and placement of various components for this new version.

The basic construction of Shinya Scope-5, including the new wooden bench, is the same as seen for Shinya Scope-6 in Fig. III-22 in Article 40 (for schematics, see Fig. III-21 in Article 40). While many new features were added on Shinya Scope-6, developed later at the Marine Biological Laboratory in Woods Hole, Massachusetts, several of the features are shared with Shinya Scope-5. The following describes these features of Shinya Scope-5.

Starting from the top, the light source is now a compact arc, high intensity concentrated mercury arc lamp (Osram 100-W), whose regular housing is further enclosed in an air-circulating light-tight box to eliminate the stray light leaking to the observer or to a light-sensitive detector. Also mounted on the horizontal portion of the optical bench are: a beam switcher that allows use of a quartz halogen bulb as an alternate light source, a filter holder, relay lenses, and finally a first-surface mirror. The mirror directs the beam downwards along the optical axis of the components mounted on the 3-inch vertical dovetail. The image of the light source, focused onto a pinhole (mounted on the auxiliary vertical dovetail) and trimmed into a uniformly bright small circle of light, is projected by a zoom lens through the polarizer, compensator, etc., onto the back aperture of the condenser lens. The zoom lens is used at its longer focal length when one desires to illuminate a large area of the specimen but through a low aperture of the condenser, namely with lower power objectives. Conversely, the zoom lens is used at its shorter focal length when the condenser aperture needs to be illuminated maximally, but only a small field of the specimen needs to be illuminated, a condition for gaining high resolution when using high N.A. objective lenses.

For Shinya Scope-5, the polarizer-comparator unit was transferred from Shinya Scope-4. On the other hand, the compensator-condenser mounts for the new scope were modified to increase their precision and versatility. In addition to being able to accept Nikon phase contrast or rectified condensers, the new unit could alternately accept a four position revolver with individually centrable objective lens holders or kinematically-designed quick-change centrable nose pieces, both made by Leitz.

The graduated revolving stage was also replaced, since no commercially available units held the specimen centration and height to the desired tolerance. The new unit was designed by Gordon Ellis and Ed Horn and made in-house incorporating a pair of high-precision ball bearings made to order by Split Ball Bearing Company. The ball-bearing raceways, which support the stainless-steel graduated stage, are clamped through two sets of 16 supporting screws. Careful tensioning of these screws adjusted the raceways so that the bearings ran true to a tolerance of a micron or better.

Importantly, we also eliminated the coarse and fine focusing block that supported the precision revolving stage, since there were no adjustment blocks that could support the revolving stage with adequate stability. Instead, we clamped the revolving stage, supported securely onto a dovetail slider, directly onto the main dovetail. The center of rotation of the revolving stage thus defined the central axis for the full set of optics on the microscope.

In the 5th and later versions of the Shinya Scope, the objective lenses were mounted on revolving nose pieces which allowed each objective lens to be individually rotated and centered. A unit, consisting only of the revolving nose piece and analyzer (housed in a graduated, 360° rotatable cylinder), was mounted through its own coarse and fine focusing block onto the main dovetail. The imaging beam exiting this unit was coupled through infinity-focused optics to the remainder of the body tube, which was fastened directly by its own slider to the main dovetail (i.e., with no coarse and fine focusing block). By mounting the main, heavier portion of the body tube on a separate slider, the objective lens/analyzer unit became considerably lighter so that we could focus the objective lens much faster and without encountering a large inertial mass.

In the body tube, a three-way mirror on a small, horizontal slider allowed the full beam to travel either to the camera or to the eye piece, or to be split between the two. Another switcher on the body tube allowed insertion and focusing of the Bertrand lens. Unlike conventional Bertrand lenses, the arrangement in this body tube allowed one to focus not only the back aperture of the objective and condenser lenses, but also the full length of the optical train up past the field stop placed above the polarizer (or below as desired). Thus, one could inspect the whole optical train for proper alignment of the components, as well as for detecting any source of dust particles or imperfections which lowered the Extinction Factor or otherwise affected the image quality. The lower part on the body tube supported a zoom lens whose exiting parallel rays were focused by the camera, supported further down on its own slider.

Biological applications of Shinya Scope-5 are highlighted in Articles 24 to 35. As seen by these articles, in addition to close collaborations with Hidemi Sato and Gordon Ellis, who were fellow faculty members of our Biophysical Cytology Program at the University of Pennsylvania, this version of the scope was used extensively with students and visitors to the program. In fact, we finished building two of the Shinya Scope-5s in Philadelphia, where we also acquired a Shinya Scope-4 which was generously released by the University of Rochester. The former were used by Gordon Ellis and me, and the latter by Hidemi Sato and his students.

While the features described so far are shared in both Versions 5 and 6, the latter, developed after introduction of video imaging at Woods Hole in the early 1980s, has several novel features as described next.

Shinya Scope-6

As described in the historical review in my monograph "Video Microscopy" (1986 and 1997 by Plenum Press, see also Articles 34, 42, 47, 54, 57, 63, and 72 in this *Collected Works*), the combination of video with the light microscope in the late 1970s to early 1980s opened up a completely new world. By using an appropriate video camera and monitor, coupled with a simple image processor, one could now see things that previously could barely be seen or not be seen at all through the eye piece. The contrast of extremely small, or weakly retarding, objects could be enhanced and made visible after eliminating background optical noise by subtracting the background image; very low fluorescence could be detected by averaging shot noise; and dynamic scenes could be recorded readily at various speeds together with audio narration. The recorded data could be stored in massive quantities, printed out, or communicated rapidly through electronic media. (For biological and related applications, see also Articles 37–40, 43, 45, 46, 48, 49–53, 58–62, 64, and 66–71 in this *Collected Works.*)

In short, microscopy had entered the new age of electronic imaging and processing, which quickly advanced to digital imaging, processing, recording, and analysis. Furthermore, the same digital computers that performed these functions could now also be used to control various mechanical and optical functions of the microscope. At the same time, laser sources, optical modulators, monitors and projectors employing liquid crystals became widely available, and solid state and integrated circuits with exceptionally high speed and massive capacity continued to be developed.

Shinya Scope-6 modified the mechanical and optical designs of the 5th version in order to take advantage of the new functions made possible by these electro-optical devices and digital computers. Referring to Fig. III-32 in Article 40 once more, the mirror, which had directed the horizontal illumination beam downward to the vertical main axis in the Shinya Scope-5, was replaced with a 1-mm-diameter single optical fiber, the "light scrambler," developed by Gordon Ellis. By passing the illuminating beam through the three-dimensionally bent optical fiber, the inhomogeneously distributed intensity of the concentrated mercury arc and other light sources was converted into a uniform, 1-mm-diameter, circular spot by the time the beam reached the fiber exit. This uniformly bright spot was used as the virtual light source, whose magnified image was projected by a zoom lens onto the condenser aperture. By using the light scrambler, the field *and* aperture of the microscope are both uniformly illuminated. Thus, the light scrambler coupled with video imaging and digital processing brought forth images of even smaller objects and lower brightness than were previously possible.

The zoom ocular with powered controller (seen above the video camera lens in Fig. III-22 in Article 40) allowed us to choose between image magnification and the amount of light available to the camera, or between image resolution and exposure time. This flexibility was expanded further by the emergence of a variety of video and CCD cameras. In succession, intensified, cooled, color, high-quantum-yield, and other cameras capable of on-chip integration, real time background subtraction etc., became available.

In Shinya Scope-6, various cameras were supported stably near the bottom of the main dovetail where ample space is available. Here, they could be readily interchanged to meet the particular requirements of the biological and optical observations or experiments. Exceptional cooperation by Dage-MTI, Hamamatsu Photonics, Venus Scientific, Carl Zeiss, Colorado Video, and Universal Imaging, among others companies made both the biological experiments and technological developments possible.

In addition to the companies just listed, contributions by Leica, Nikon, Olympus, Kodak, Chroma, and many other optical and related companies, together with input from the academic and commercial faculty and students of the MBL microscopy courses, in fact spurred the development of the new field of video microscopy. And, as with the other contributors to the courses, ours was a two way interaction, in which we both taught and learned at the courses. We learned much of what was developing in the industry and elsewhere while we shared our views of the capabilities and needs experienced through the development and use of our own scopes. The friendships that developed as a result of the courses have also helped immensely in our and others' development and research efforts.

Shinya Scope-7

Shinya Scope-7 further modifies the 6th version by adding various motorized controls to the microscope, as well as by making their computer controls "transparent." That is, instead of relying on a keyboard and computer display to control focusing, to adjust the compensator angle, select filters etc., one turns knobs or presses preprogrammed buttons on a control panel as though one were using analog devices. A computer with versatile function, operating behind those control knobs and buttons, sends commands to stepper motors, relays etc., so that one is no longer distracted by interacting with the computer through the keyboard or by watching the monitor display.

To make these functions practical, and the behavior of the condenser and objective lens focus more reliable, we switched the coarse and fine focusing blocks from our long-time stand-by Leitz Ortholux blocks to those made by Nikon for their industrial microscopes (Fig. 7). To each of the spindles of their fine focus control, we added an appropriately geared-down stepper motor. We also added stepper-motor drives to the polarizer, the compensator, and two filter wheels. Two shutters were electrically driven, and the switcher for the three alternate light sources was driven electro-pneumatically. The behavior of these drives was all controlled through the control box (Figs. 6 and 7, in front of the monitor) by the "invisible" computer output.

With the control box, one cannot only adjust the focus, compensator settings etc., but also the sensitivity for each of the focusing or orientation control knobs. Likewise, their zero positions can be set over wide ranges. The condenser and objective lenses can be moved independently at any predetermined rates or they can be coupled to move together. The two can be made to move at the same rate (for low-power observations or for slides observed under completely homogeneous immersion). Alternately, they can be made to move at different rates, calculated to maintain precise confocality between the condenser and objective foci, depending on the refractive indices of the immersion media and medium imbibing the specimen. This feature is particularly useful, since it allows perfect background subtraction by maintaining a constant background image while the specimen is focused to different levels.

These developments were made recently at the MBL with input from Ted Inoué (who designed the control system and needed software) and Robert A. Knudson (of Technical Video, Ltd. who assembled the "Super Micro Controller" and supplies the Fiberoptic Light Scrambler). We also received generous support from Professors Yoshinori Fujiyoshi of Kyoto University and Kazuhiko Kinoshita now of Waseda University. The instrument is currently in use by members of the Architectural Dynamics in Living Cells Program, including Rudolf Oldenbourg for the development of a high-resolution, polarization optical sectioning system, and Michael Shribak for the development of the orientation-independent DIC system.

The following piece may not quite fit into the series of equipment developed as Shinya Scopes. Nevertheless, I have included a brief description here, since it is based on a polarizing microscope for detecting weak birefringence, albeit in living cells and other specimen specifically exposed to high centrifugal fields.

Centrifuge Polarizing Microscope (CPM)

In the late 1990s, I felt the time may have finally arrived when one could develop a polarizing microscope for examining changes in the fine structure of living cells while they were

Fig. 6. Author with Shinya Scope-7.

being exposed to high centrifugal fields. Centrifuge microscopes of various designs had been, in fact, developed and used since the 1930s. However, they were made for observing the stratification of components within live marine eggs and plant cells using conventional microscopy and could not be used for polarized light microscopy.

To develop a CPM, many stringent requirements had to be met. In order to detect the weak birefringence which was expected to reflect fine structural changes in the centrifuged living cells, one needed to eliminate or minimize any microscope optics which themselves would become birefringent by being exposed to the centrifugal force. This required the whole polarization optical system, including the condenser and objective lenses, to remain outside of the rotor, which meant that the rotating specimen had to be illuminated by an exceptionally brief and very bright stroboscopic flash, synchronized to the arrival of the specimen precisely in the microscope field. In addition, the windows of the rotating specimen chamber had to remain essentially isotropic despite the deformation induced by the high centrifugal field.

In order to achieve an image resolution of one micrometer or better with the specimen rotating at 11,000 RPM (exposing the specimen to 10,000 times gravitational acceleration), the exposure had to be no greater than a few nanoseconds. This required the use of a pulsed laser controlled through a special timing device that sensed the rotor speed and orientation. The need to use such a precisely-timed laser source, and other requirements for maintaining image stability and contrast of

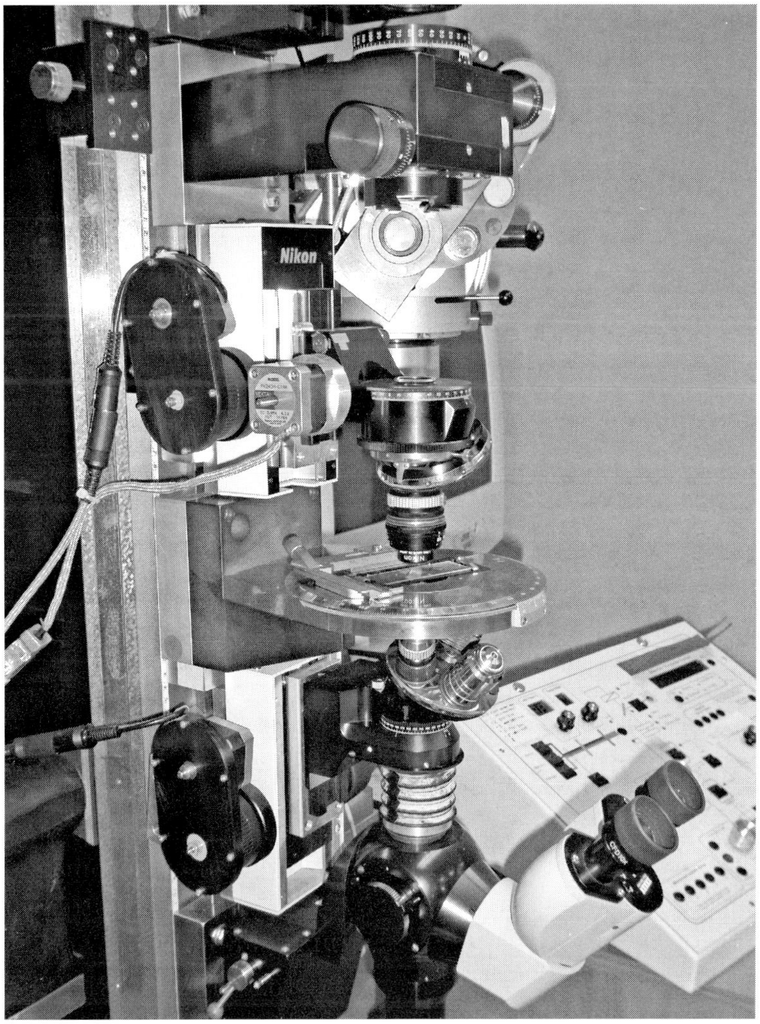

Fig. 7. Close up view of Shinya Scope-7.

the rotating specimen observed in high extinction polarization microscopy, demanded a substantial number of electronic and optical innovations.

Fortunately, Hamamatsu Photonics and Olympus Optical Company provided very generous engineering support and input for this project. We on our part at the Architectural Dynamics in Living Cells Program at the MBL carried out a number of preliminary experiments and added refinements to the instrument fabricated by Olympus and Hamamatsu. The final instrument resulting from our three-way collaboration and its diverse biological applications are described in Articles 64, 66, 67, 68, and 75. Some results are also illustrated by video in the appended DVD.

APPENDIX II

CURRICULUM VITAE

Shinya Inoué

Personal: Born January 5, 1921, London, England
Married, Sylvia McCandless, five children
United States citizen, naturalized 1989

Address: Marine Biological Laboratory
7 MBL Street, Woods Hole, MA 02543, USA
(508) 289–7382

Education: Princeton University, MA (Biology), 1950, PhD (Biology), 1951
Tokyo University, Rigakushi (Zoology), 1944

Academic Positions and Instructorships:

Marine Biological Laboratory, Woods Hole, Architectural Dynamics in Living Cells Program Director, 1992–present; Distinguished Scientist, 1986–present; Senior Scientist, 1980–1986; Short Course on Analytical and Quantitative Light Microscopy in Biology, Medicine, and Materials Sciences, Instructor-in-Chief, 1979–1987; Instructor, 1987–present; Short Course on Optical Microscopy and Imaging in the Biomedical Sciences, Instructor, 1978–present; Physiology Course Instructor, 1961–1966; Lecturer, intermittently 1967–present

University of Pennsylvania, Professor of Biology, Director, Program in Biophysical Cytology, 1966–1982

NATO Summer School, Instructor, Cannes, 1967; Stressa, 1970; Szeged, 1975

Dartmouth Medical School, John LaPorte Given Professor of Cytology, 1965–1966; Professor and Chairman, Department of Cytology, 1959–1966; Chairman, Department of Anatomy, 1959–1963

University of Rochester, Department of Biology, Research Associate, Assistant Professor and Associate Professor, 1954–1959; Instructor in Optics, 1957–1959

Tokyo Metropolitan University, Department of Biology, Assistant Professor, 1953–1954

University of Washington, Department of Anatomy, Instructor, 1951–1953

Honors and Awards:

Lecture and Demonstrations to Emperor Showa, 1947, 1953, 1975

Princeton University, William Grieg Lapham Fellow, 1948–1950; Charlotte Elizabeth Proctor Fellow, 1950–1951

Marine Biological Laboratory, Lalor Fellow, Summer 1952
American Cancer Society, Scholar in Cancer Research, 1955–1958
John Simon Guggenheim Fellow, 1971–1972
American Academy of Arts and Sciences Fellow, 1971–present
Biophysical Society National Lecturer, February 11, 1986
National Institutes of Health, NIGMS, MERIT Award, 1982–1996
Royal Microscopical Society, Honorary Fellowship for Distinguished Service to Science, March 15, 1988
Brandeis University, Lewis S. Rosenstiel Award for Distinguished Work in Basic Medical Research, April 27, 1988
State of New York, Brown-Hazen Award for Outstanding Contributions to the Basic Life Sciences, October 18, 1988
American Society for Cell Biology, EB Wilson Award, November 16, 1992
National Academy of Sciences, Elected, April, 1993
Microscopy Society of America, Distinguished Scientist Award, August 14, 1995
New York Microscopical Society, Ernst Abbe Award, November 18, 1997
Focus on Microscopy Symposium, Kaohsiung, Taiwan, Honorary Chair, April 10, 2002
Japan Society for the Promotion of Science, International Prize for Biology, December 1, 2003

Consultancies and Affiliations:

American Optical Company, Consultant, 1954–1960
National Science Foundation, Molecular Biology Panel, 1962–1965
Marine Biological Laboratory, Corporation Member, 1962–present
Office of Science and Technology, Wooldridge Study Committee on NIH, 1964
National Institutes of Health, Biophysical Sciences Training Committee, 1965–1970
Marine Biological Laboratory, Trustee, 1970–1977, 1981–1985, 1992–1996
Universal Imaging Corporation, President, 1984–1987, Board Member, 1987–2002
Hamamatsu Photonics Kabushiki Kaisha, Consultant, 1988–2002
Nikon Corporation, Tokyo, Consultant, 1994–present
Olympus Optical Company, Ltd, Tokyo, Consultant, 1994–2001
Yokogawa Electric Corporation, Tokyo, Consultant, 1997–present
AutoQuant Imaging, Inc, Consultant, 2000–present

Patents Received:

"Stereoscopic Apparatus," awarded 1944 to S. Inoué, Japanese Patent #166,528
"Polarizing Optical Systems" awarded May 17, 1960, to W.L. Hyde and S. Inoué, U.S. Patent #2,936,673
"Slit Scan Centrifuge Microscope," awarded July 27, 1999, to S. Inoué and K. Suzuki, U.S. Patent #5930033
"Centrifuge Microscope Capable of Realizing Polarized Light Observation," awarded November 9, 1999, to S. Inoué, U.S. Patent #5982535

Editorial Board Memberships:

Biological Bulletin, 1964–1967, Associate Editor, 1998–present
Cell Biology International Reports, 1976–1985
Cell Motility and Cytoskeleton, 1983–1990
Development, Growth and Differentiation, 1994–1999
Journal of Cell Biology, 1972–1974
Journal of Morphology, 1975–1980
Journal of Ultrastructure Research, 1967–1987

Society Memberships:

American Association for the Advancement of Science, 1953–2000; Fellow, 1962–2000
American Society for Cell Biology, 1968–present; Council Member, 1970–1973
Biophysical Society, 1957–present; Council Member, 1968–1971
Microscopy Society of America, 1995–present
Sigma Xi, 1951–present
Society of General Physiologists, 1960–present; Council Member, 1962–1965; President, 1969–1970
Optical Society of America, 1995–present

Research Interests:

Mechanisms of mitosis, cell division and their controls. Molecular and structural organization of cytoplasm and nucleoplasm in living cells. Development of new biophysical approaches and instruments for fine structure analysis directly in living cells. Physical optics. Video microscopy.

Doctoral Students Sponsored:

John Aronson, 1961, PhD, University of Rochester. Sarcomere size in developing muscles of a tarsonemid mite.
Arthur H. Forer, 1964, PhD, Dartmouth Medical School. Evidence of two spindle fiber components: A study of chromosome movement in living crane fly (*Nephrotoma suturalis*) spermatocytes using polarization microscopy & an ultraviolet microbeam.
Raymond E. Stephens, 1965, PhD, Dartmouth College. Characterization of the major mitotic apparatus protein and its subunits.
Louise Marie Gordon, 1971, MA, University of Pennsylvania. Reversal by light of the action of colchicine on the organization of the mitotic spindle.
Jeffrey C. Freedman, 1973, PhD, University of Pennsylvania. Control of solute distribution by erythrocytes during *in vitro* incubation.
John W. Fuseler, 1973, PhD, University of Pennsylvania. The effect of temperature on chromosome movement and the assembly-disassembly process of birefringent spindle fibers in actively dividing plant and animal cells.
Edward D. Salmon, 1973, PhD, University of Pennsylvania. The effects of hydrostatic pressure on the structure and function of the mitotic spindle: An *in vivo* analysis with a newly developed microscope pressure chamber.
Daniel Kiehart, 1978, PhD, University of Pennsylvania. Microinjection studies on spindle assembly and chromosome movement *in vivo*.
Gerald W. Gordon, 1980, PhD, University of Pennsylvania. The control of chromosome motion: UV microbeam irradiation of kinetochore fibers.
Andrew Eisen, 1982, PhD, University of Pennsylvania. Local, intracellular metabolic changes at fertilization in single marine eggs.
Douglas A. Lutz, 1984, PhD, University of Pennsylvania. Asymmetric cell centre placement prior to unequal cell division.

APPENDIX III

LIST OF PRIMARY PUBLICATIONS

Shinya Inoué

[Bracketed numbers refer to articles in this *Collected Works*]

Inoué S, Technical news. A device for altering the distance between slide and cover slip at will, *Botany and Zoology Theoretical and Applied* **11** (8): 669, 1943. [**A 1**]

Inoué S, Japanese Patent #166,528: Stereoscopic apparatus, 1944. [**A 2**]

Inoué S, Studies of the Nereis egg jelly with the polarization microscope, *Biol Bull* **97** (2): 258–259, 1949.

Inoué S, Dan K, Birefringence of the dividing cell, *J Morphol* **89** (3): 423–456, 1951. [**A 5**]

Inoué S, A method for measuring small retardations of structures in living cells, *Exp Cell Res* **2** (3): 513–517, 1951. [**A 9**]

Inoué S, Studies on depolarization of light at microscope lens surfaces. I. The origin of stray light by rotation at the lens surfaces, *Exp Cell Res* **3** (1): 199–208, 1952. [**A 8**]

Inoué S, The effect of colchicine on the microscopic and submicroscopic structure of the mitotic spindle, *Exp Cell Res Suppl* **2**: 305–318, 1952. [**A 11**]

Inoué S, Effect of temperature on the birefringence of the mitotic spindle, *Biol Bull* **103** (2): 316, 1952. [**A 13**]

Inoué S, Polarization optical studies of the mitotic spindle. I. The demonstration of spindle fibers in living cells, *Chromosoma* **5**: 487–500, 1953. [**A 10**]

Inoué S, Hyde WL, Studies on depolarization of light at microscope lens surfaces. II. The simultaneous realization of high resolution and high sensitivity with the polarizing microscope, *J Biophys Biochem Cytol* **3** (6): 831–838, 1957. [**A 14**]

Inoué S, Kubota H, Diffraction anomaly in polarizing microscopes, *Nature* **182**: 1725–1726, 1958. [**A 15**]

Inoué S, Motility of cilia and the mechanism of mitosis, *Rev Mod Phys* **31** (2): 402–408, 1959, and in Oncley JL (ed.), *Biophysical Science — A Study Program*, John Wiley & Sons, NY, pp. 402–408, 1959. [**A 16**]

Inoué S, Koester CJ, Optimum half-shade angle in polarizing instruments, *J Opt Soc Amer* **49**: 556–559, 1959.

Kubota H, Inoué S, Diffraction images in the polarizing microscope, *J Opt Soc Amer* **49** (2): 191–198, 1959. [**A 17**]

Hyde WL, Inoué S, U.S. Patent #2,936,673: Polarizing optical systems, 1960.

Inoué S, Birefringence in Dividing Cells, Time-lapse motion picture, Geo W Colburn Laboratory, Inc, Chicago, IL, 1960.

Inoué S, Polarizing microscope: Design for maximum sensitivity in Clark GL (ed.), *The Encyclopedia of Microscopy*, Reinhold, NY, pp. 480–485, 1961. [**A 18**]

Inoué S, Bajer A, Birefringence in endosperm mitosis, *Chromosoma* **12**: 48–63, 1961. [**A 19**]

Inoué S, Sato H, Arrangement of DNA in living sperm: A biophysical analysis, *Science* **136**: 1122–1124, 1962.

Inoué S, Sato H, Tucker RW, Heavy water enhancement of mitotic spindle birefringence, *Biol Bull* **125** (2): 380–381, 1963. [**A 20**]

Reynolds GT, Allen RD, Eckert R, Hastings JW, Inoué S, Application of an image intensifier tube to the microscopic observation of bioluminescent cells and visualization of weak radioactive source distributions, *Biol Bull* **125** (2), 389–390, 1963.

Tucker RW, Inoué S, Rapid exchange of D_2O and H_2O in sea urchin eggs, *Biol Bull* **125** (2): 395, 1963. [**A 21**]

Inoué S, Organization and function of the mitotic spindle in Allen RD, Kamiya N (eds.), *Primitive Motile Systems in Cell Biology*, Academic Press, New York, NY, pp. 549–598, 1964. [**A 22**]

Inoué S, Studies of cell division with an improved polarizing microscope in Von Hippel (ed.), *The Molecular Designing of Materials and Devices*, MIT Press, Cambridge, MA, pp. 211–214, 1965.

Inoué S, Sato H, Ascher M, Counteraction of colcemid and heavy water on the organization of the mitotic spindle, *Biol Bull* **129** (2): 409–410, 1965. [**A 23**]

Inoué S, Sato H, Arrangement of DNA molecules in the sperm nucleus: An optical approach to the analysis of biological fine structure in Japanese Biophysical Society (edn.), *Biophysical Science Series 3, Progress in Genetics III*, Yoshioka, Kyoto, pp. 151–220, 1966 (*in Japanese*).

Inoué S, Sato H, Deoxyribonucleic acid arrangement in living sperm in Hayashi T, Szent-Györgyi A (eds.), *Molecular Architecture in Cell Physiology*, Prentice Hall, Englewood Cliffs, NJ, pp. 209–248, 1966. [**A 24**]

Stephens RE, Inoué S, Clark JI, Sulfhydryl balance in mitosis: The effect of mercaptoethanol on spindle birefringence, *Biol Bull* **131** (2): 409, 1966.

Inoué S, Sato H, Cell motility by labile association of molecules. The nature of mitotic spindle fibers and their role in chromosome movement, *J Gen Physiol* **50** (6): 259–292, 1967. [**A 25**]

Inoué S, The physics of structural organization in living cells, in Devons S (ed.), *Biology and the Physical Sciences*, Columbia University Press, New York, NY, pp. 139–171, 1969.

Aronson J, Inoué S, Reversal by light of the action of N-methyl N-desacetyl colchicine on mitosis, *J Cell Biol* **45** (2): 470–477, 1970. [**A 26**]

Inoué S, Ellis GW, Salmon ED, Fuseler JW, Rapid measurement of spindle birefringence during controlled temperature shifts, *J Cell Biol* **47**: 95a–96a, 1970.

Aronson JF, Inoué S, Observations on the potential for increasing spindle birefringence during mitosis in *Lytechinus variegatus*, *J Gen Physiol* **58**: 713, 1971.

Langford G, Inoué S, Sabran I, Analysis of axostyle motility in *Pyrsonympha vertens*, *J Cell Biol* **59**: 185a, 1973.

Inoué S, Borisy GG, Kiehart DP, Growth and lability of *Chaetopterus* oocyte mitotic spindles isolated in the presence of porcine brain tubulin, *J Cell Biol* **62**: 175–184, 1974.

Inoué S, Ritter H, Dynamics of mitotic spindle organization and function in Inoué S, Stephens RE (eds.), *Molecules and Cell Movement*, Raven Press, New York, NY, pp. 3–30, 1975.

Inoué S, Fuseler J, Salmon ED, Ellis GW, Functional organization of mitotic microtubules: Physical chemistry of the *in vivo* equilibrium system, *Biophys J* **15**: 725–744, 1975. [**A 27**]

Sato H, Ellis GW, Inoué S, Microtubular origin of mitotic spindle form birefringence: Demonstration of the applicability of

Wiener's equation, *J Cell Biol* **67** (3): 501–517, 1975. [**A 28**]

Okazaki K, Inoué S, Crystal property of the larval sea urchin spicule, *Dev Growth Differ* **18** (4): 413–434, 1976. [**A 29**]

Inoué S, Chromosome movement by reversible assembly of microtubules in Goldman RD, Pollard T, Rosenbaum J (eds.), *Cell Motility*, Vol. 3. Cold Spring Harbor Conference on Cell Proliferation, pp. 1317–1328, 1976.

Inoué S, Okazaki K, Biocrystals, *Sci Am* **236** (4): 82–92, 1977.

Ritter H, Inoué S, Kubai D, Mitosis in *Barbulanympha*. I. Spindle structure, formation and kinetochore engagement, *J Cell Biol* **77** (3): 638–654, 1978. [**A 30**]

Inoué S, Ritter H, Mitosis in *Barbulanympha*. II. Dynamics of a two-stage anaphase, nuclear morphogenesis, and cytokinesis, *J Cell Biol* **77** (3): 655–684, 1978. [**A 31**]

Inoué S, Kiehart DP, *In vivo* analysis of mitotic spindle dynamics in Dirksen, Prescott, Fox (eds.), *Cell Reproduction, In Honor of Daniel Mazia*, Academic Press, New York, NY, pp. 433–444, 1978.

Inoué S, Kiehart DP, Mabuchi I, Ellis GW, Molecular mechanism of mitotic chromosome movement in Pepe, Sanger, Nachmias (eds.), *Motility in Cell Function*, First John M. Marshall Symposium in Cell Biology, Academic Press, New York, NY, pp. 301–311, 1979.

Gordon GW, Inoué S, Unexpected increase in poleward velocities of mitotic chromosomes after UV irradiation of their kinetochore fibers, *J Cell Biol* **83**: 376a, 1979. [**A 32**]

Langford GM, Inoué S, Motility of the microtubular axostyle in *Pyrsonympha*, *J Cell Biol* **80** (3): 521–538, 1979. [**A 33**]

Okazaki K, McDonald K, Inoué S, Sea urchin larval spicule observed with the scanning electron microscope in Omori and Watanable (eds.), *The Mechanisms of Biomineralization in Animals and Plants*, Tokai University Press, Tokyo, pp. 159–168, 1980.

Inoué S, Video image processing greatly enhances contrast, quality, and speed in polarization based microscopy, *J Cell Biol* **89** (2): 346–356, 1981. [**A 34**]

Inoué S, Cell division and the mitotic spindle, *J Cell Biol* special issue, "Discovery in Cell Biology," **91** (3): 131s–147s, 1981. [**A 35**]

Inoué S, Cohen D, Ellis GW, High resolution, stereo video microscopy. *Biol Bull* **161**: 306, 1981.

Inoué S, Potrebic B, Brown CR, Lutz DA, Gradient of cleavage inhibition induced by limited diffusion of oxygen, *J Cell Biol* **95** (2): 136a, 1982. [**A 36**]

Inoué S, Tilney LG, The acrosomal reaction of *Thyone* sperm. I. Changes in the sperm head visualized by high resolution video microscopy, *J Cell Biol* **93** (3): 812–819, 1982. [**A 37**]

Tilney LG, Inoué S, The acrosomal reaction of *Thyone* sperm. II. The kinetics and possible mechanism of acrosomal process elongation, *J Cell Biol* **93** (3): 820–827, 1982. [**A 38**]

Inoué S, The role of self-assembly in the generation of biological form in Subtelny S, Green PB (eds.), *Developmental Order: Its Origin and Regulation*, 40-th Symposium of the Society for Developmental Biology, Alan R. Liss, New York, NY, pp. 35–76, 1982.

Kiehart DP, Mabuchi I, Inoué S, Evidence that myosin does not contribute to force production in chromosome movement, *J Cell Biol* **94**: 165–178, 1982.

Inoué S, Simultaneous video display of fluorescence and polarized light, or DIC, microscope images in real time, *J Cell Biol* **95** (2): 461a, 1982.

Lutz DA, Inoué S, Colcemid but not cytochalasin inhibits asymmetric nuclear positioning prior to unequal cell division. *Biol Bull* **163** (2): 373–374, 1982.

Tanaka Y, Inoué S, Reversible relaxation of cleavage contractile ring by cytochalasins, *J Cell Biol* **95** (2): 289a, 1982.

Woodruff RI, Lutz DA, Inoué S, Lucifer yellow CH as a non-intrusive, *in vivo* fluorescent probe for physiological studies during early development, *Biol Bull* **163**: 379, 1982.

Inoué S, Inoué TD, Ellis GW, Visualizing extremely low contrast images by digital enhancement of selected portions of the image gray scale, *Biol Bull* **165**: 492, 1983.

Hamaguchi Y, Lutz DA, Inoué S, Cortical differentiation, asymmetric positioning and attachment of the meiotic spindle in *Chaetopterus pergamentaceous* oocytes, *J Cell Biol* **97**: 254a, 1983.

Otter T, Lutz DA, Eichen J, Inoué S, Contractile vacuole activity and associated birefringence demonstrated in living *Paramecium caudatum* by video-enhanced DIC and polarized light microscopy, *J Cell Biol* **97**: 248a, 1983.

Akins RE, Lutz DA, Inoué S, Microtubule assembly and pronuclear movement analyzed by microbeam inactivation of colcemid, *Biol Bull* **165**: 501, 1984.

Johnson CH, Inoué S, Flint A, Hastings JW, Compartmentalization of algal bioluminescence: Autofluorescence of bioluminescent particles in the dinoflagellate *Gonyaulax* as studied with image-intensified video microscopy and flow cytometry, *J Cell Biol* **100**: 1435–1446, 1984.

Tilney LG, Inoué S, Acrosomal reaction of the *Thyone* sperm. III. The relationship between actin assembly and water influx during the extension of the acrosomal process, *J Cell Biol* **100**: 1273–1283, 1985. [**A 39**]

Inoué S, Bajer AS, Molè-Bajer J, De Brabander M, De Mey J, Nuydens R, Ellis GW, Horn E, Inoué TD, Microtubules decorated with 5 nm gold visualized by video-enhanced light microscopy, *J Cell Biol* **101**: 146a, 1985.

Inoué S, Inoué TD, Ellis GW, Rapid, stereoscopic display of microtubule distribution by a video-processed optical sectioning system, *J Cell Biol* **101** (2): 146a, 1985.

Inoué S, Molè-Bajer J, Bajer AS, Three-dimensional distribution of microtubules in *Haemanthus* endosperm cells in De Brabander and De Mey (eds.), *Microtubules and Microtubule Inhibitors*, Elsevier, pp. 269–276, 1985.

Inoué S, *Video Microscopy*. Plenum Press, New York, 1986. [**Appendix III as A 40**]

Ellis GW, Inoué S, Inoué TD, Computer aided light microscopy in De Weer P, Salzberg B (eds.), *Optical Methods in Cell Physiology*, John Wiley and Sons, New York, NY, pp. 15–30, 1986.

Lutz DA, Inoué S, Techniques for observing living gametes and embryos in Schroeder T, (ed.), *Echinoderm Gametes and Embryos. Methods Cell Biol.*, Vol. 27, Academic Press, New York, NY, pp. 89–110, 1986. [**A 41**]

Inoué S, Inoué TD, Computer-aided stereoscopic video reconstruction and serial display from high-resolution light-microscope optical sections in Somlyo (ed.), *Recent Advances in Electron and Light Optical Imaging in Biology and Medicine. Ann NY Acad Sci*, Vol. 483, pp. 392–404, 1986. [**A 42**]

Inoué S, Video microscopy of living cells and dynamic molecular assemblies, *Appl Opt* **26** (16): 3219–3225, 1987.

Dan K, Inoué S, Studies of unequal cleavage in molluscs. II. Asymmetric nature of the two asters, *Int J Invertebr Repr Dev* **11**: 335–354, 1987. [**A 43**]

Inoué S, Ultrathin optical-sectioning-tomography achieved with the light microscope, *Biol Bull* **173**: 419, 1987.

Tilney LG, Inoué S, Flagellar gyration and midpiece rotation during extension of the acrosomal process of *Thyone* sperm: How and why this occurs, *J Cell Biol* **104**: 407–415, 1987.

Inoué S, Knudson RA, Inoué TD, High-N.A. microscope with automatic correction for spherical aberration, *J Cell Biol* **105**: 226a, 1987. [**A 44**]

Molè-Bajer J, Bajer AS, Inoué S, Three-dimensional localization and redistribution of F-actin in higher plant mitosis and cell plate formation. *Cell Motil Cytoskeleton* **10**: 217–228, 1988.

Cassimeris L, Inoué S, Salmon ED, Microtubule dynamics in the chromosomal spindle fiber: Analysis by fluorescence and high-resolution polarization microscopy, *Cell Motil Cytoskeleton* **10**: 185–196, 1988.

Lutz DA, Hamaguchi Y, Inoué S, Micromanipulation studies of the asymmetric positioning of the maturation spindle in *Chaetopterus* sp. oocytes: I. Anchorage of the spindle to the cortex and migration of a displaced spindle, *Cell Motil Cytoskeleton* **11**: 83–96, 1988. [**A 45**]

Inoué S, The living spindle, *K Dan Festschrift: Advances in Cell Division Research, Zool Sci* **5**: 529–538, 1988. [**A 46**]

Inoué S, Progress in video microscopy. *Cell Motil Cytoskeleton* **10**: 13–17, 1988.

Inoué S, Manipulating single microtubules. *Protoplasma Suppl* **2**: 57–62, 1988.

Inoué S, Inoué T, Stereoscopy of high resolution video microscope images reconstructed from serial optical sections, *Biol Bull* **175**: 316, 1988.

Inoué S, Imaging of unresolved objects, superresolution, and precision of distance measurement, with video microscopy in Taylor and Wang (eds.), *Fluorescence Microscopy of Living Cells in Culture: Quantitative Fluorescence Microscopy: Imaging and Spectroscopy. Methods Cell Biol*, Vol. 30, Academic Press, New York, NY, pp. 85–112, 1989. [**A 47**]

Walker RA, Inoué S, Salmon ED, Asymmetric behavior of severed microtubule ends after ultraviolet-microbeam irradiation of individual microtubules *in vitro*, *J Cell Biol* **108**: 931–937, 1989. [**A 48**]

Sardet C, Speksnijder J, Inoué S, Jaffe L, Fertilization and ooplasmic movements in the ascidian egg, *Development* **105**: 237–249, 1989. [**A 49**]

Oldenbourg R, Inoué S, Analysis of edge birefringence observed near refractive index steps in myofibrils and KCL crystals using high resolution polarized light microscopy and spatial Fourier filtering, *Biol Bull* **177** (2): 318, 1989. [**A 50**]

Inoué S, Inoué T, Knudson RA, Oldenbourg R, Very high resolution and dynamic stereo images of neurons, *Biol Bull* **177**: 322, 1989.

Silver RE, Shimomura O, Inoué S, Detection of intracellular free calcium with aequorin and single photon video light microscopy, *J Cell Biol* **109**: 90a, 1989.

Inoué S, Foundations of confocal scanned imaging in light microscopy, in Pawley J (ed.), *Handbook of Biological Confocal Microscopy*, Chapter 1, Plenum Press, New York, NY, pp. 1–14, 1990.

Inoué S, Whither video microscopy? Towards 4-D imaging at the highest resolution of the light microscope, in Herman B, Jacobson K (eds.), *Optical Microscopy for Biology*, Wiley-Liss, New York, NY, pp. 497–511, 1990.

Inoué S, Dynamics of mitosis and cleavage, in Conrad GW, Schroeder TE (eds.), *Cytokinesis: Mechanisms of Furrow Formation During Cell Division, Ann NY Acad Sci*, Vol. 582, pp. 1–14, 1990. [**A 51**]

Kiehart DP, Ketchum A, Young P, Lutz D, Alfenito MR, Chang XJ, Awobuluyi M, Pesacreta TC, Inoué S, Stewart CT, Chen TL, Contractile proteins in *Drosophila* development, in Conrad GW, Schroeder TE (eds.), *Cytokinesis: Mechanisms of Furrow Formation During Cell Division, Ann NY Acad Sci*, Vol. 582, pp. 233–251, 1990.

Sanger JM, Sanger JW, video eds., Video Supplement 2. Experiments on the mitotic spindle viewed in polarized light (Inoué S, Bajer AS, Fuseler J, Salmon ED). Labile attachment of meiotic spindle to localized site on cell cortex (Hamaguchi Y, Lutz DA, Inoué S). Reversible inhibition of pronuclear migration in a sea urchin, *Lytechinus variegatus* (Aronson JF, Inoué S). Birefringence of flagella and axostyle in hypermastigotes (Inoué S, Chen V), *Cell Motil Cytoskeleton* **17**: 356–372, 1990.

Fukui Y, Inoué S, Cell division in *Dictyostelium* with special emphasis on actomyosin organization in cytokinesis, *Cell Motil Cytoskeleton* **18**: 41–54, 1991.

Febvre J, Febvre-Chevalier C, Knudson R, Takeshita T, Inoué S, *In vivo* visualization of intercalary microtubule breakdown by polarizing microscopy and very high speed video. Abstract of invited paper presented at IX Intl. Congress of Protozoology, Berlin, Germany, July 25–31, 1993. [**A 52**]

Inoué S, Porter and the fine architecture of dividing cells, in Barlow RB, Jr., Dowling JE, Weissmann G (eds.), *The Biological Century: Friday Evening Talks at the Marine Biological Laboratory*, Harvard University Press, pp. 100–115, 1993.

Oldenbourg R, Terada H, Tiberio R, Inoué S, Image sharpness and contrast transfer in coherent confocal microscopy, *J Microsc* **172** (1): 31–39, 1993. [**A 53**]

Inoué S, *Video Microscopia*. Inoué's 1986 *Video Microscopy* translated into Spanish by Mario H. Burgos, Inca Editorial, Mendoza, Argentina, 1993.

Inoué S, Ultra-thin optical sectioning and dynamic volume investigation with conventional light microscopy, in Stevens JK, Mills LR, Trogadis JE (eds.), *Three-Dimensional Confocal Microscopy: Volume Investigation of Biological Specimens*, Academic Press, pp. 397–419, 1994. [**A 54**]

Inoué S, Recollections Kayo Okazaki, *Dev Growth Differ* **36** (4): 342, 1994. [**A 55**]

Inoué S, A Tribute to Katsuma Dan, *Biol Bull* **187** (2): 125–131, 1994. [**A 56**]

Inoué S, Inoué TD, Through-focal and time-lapse stereoscopic imaging of dividing cells and developing embryos in DIC and polarization microscopy, *Biol Bull* **187**: 232–233, 1994.

Inoué S, Foundations of confocal scanned imaging in light microscopy, in Pawley J, (ed.), *Handbook of Biological Confocal Microscopy*, 2nd edn., Chapter 1, Plenum Press, New York, pp. 1–17, 1995. [**A 57**]

Inoué S, Oldenbourg R, Optical instruments. Microscopes, in Optical Society of America (ed.), *Handbook of Optics*, 2nd edn., Vol. 2, McGraw-Hill, Inc., pp. 17.1–17.52, 1995.

Inoué S, Salmon ED, Force generation by microtubule assembly/disassembly in mitosis and related movements, *Mol Biol Cell* **6**: 1619–1640, 1995. [**A 58**]

Inoué S, Inoué T, Digital unsharp masking reveals fine detail in images obtained with new spinning-disk confocal microscope, *Biol Bull* **191**: 269–270, 1996.

Inoué S, Mitotic organization and force generation by assembly/disassembly of microtubules, *Cell Struct Funct* **21**: 375–379, 1996.

Oldenbourg R, Inoué S, Tiberio R, Stemmer A, Mei G, Skvarla M, Standard test targets for high-resolution light microscopy, in Hoch H, Jelinski L, Craighead H (eds.), *Nanofabrication and Biosystems: Integrating Materials Science, Engineering, and Biology*, Cambridge University Press, New York, NY, pp. 123–138, 1996. [**A 59**]

Inoué S, To Professor Katsuma Dan (Obituary), *Dev Growth Differ* **38**: 450–451, 1996.

Inoué S, The role of microtubule assembly dynamics in mitotic force generation and functional organization of living cells, *J Struct Biol* **118**: 87–93, 1997.

Fukui Y, Inoué S, Amoeboid movement anchored by eupodia, new actin-rich knobby feet in *Dictyostelium*, *Cell Motil Cytoskeleton* **36** (4): 339–354, 1997. [**A 60**]

Inoué S, Spring KR, *Video Microscopy — The Fundamentals*, 2nd edn., Plenum Press, New York, NY, 1997.

Inoué S, Tran P, Burgos MH, Photodynamic effect of 488-nm light on Eosin-B-stained *Spisula* sperm, *Biol Bull* **193** (2): 225–226, 1997. [**A 61**]

Tran PT, Salmon ED, Inoué S, UV cutting of MAPs-bound microtubules, *Biol Bull* **193**: 218–219, 1997.

Tran PT, DePina A, Inoué S, Oldenbourg R, Segal SJ, Burgos MH, Effect of gossypol on *Spisula* sperm observed with real-time confocal microscopy, polarized light microscopy, and video microscopy, *Biol Bull* **193**: 227–228, 1997.

Inoué S, Oldenbourg R, Microtubule dynamics in mitotic spindle displayed by polarized light microscopy, *Mol Biol Cell* **9** (7): 1603–1607 (video essay), 1998. [**A 62**]

Inoué S, Goda M, Gundersen G, Knudson RA, Oocyte maturation and tissue cell adhesion observed with centrifuge polarizing microscope, *Mol Biol Cell* **9**: 105a, 1998.

Goda M, Inoué S, Knudson RA, Oocyte maturation in *Chaetopterus pergamentaceous* observed with centrifuge polarizing microscope, *Biol Bull* **195**: 212–214, 1998.

Inoué S, Knudson RA, Suzuki K, Okada N, Takahashi H, Iida M, Yamanaka K, Centrifuge polarizing microscope, *Microsc Microanal* **4** (Suppl 2): 36–37, 1998 (abstract).

Fukui Y, Engler S, Inoué S, De Hostos EL, Architectural dynamics and gene replacement of coronin suggest its role in cytokinesis, *Cell Motil Cytoskeleton* **42**: 204–217, 1999.

Fukui Y, De Hostos EL, Yumura S, Kitanishi-Yumura T, Inoué S, Architectural dynamics of F-actin in eupodia suggests their role in invasive locomotion in *Dictyostelium*, *Exp Cell Res* **249**: 33–45, 1999.

Fukui Y, Uyeda TQP, Kitayama C, Inoué S, Migration forces in *Dictyostelium* measured by centrifuge DIC microscopy, *Biol Bull* **197**: 260–262, 1999.

Maddox P, Desai A, Salmon ED, Mitchison TJ, Oogema K, Kapoor T, Matsumoto B, Inoué S, Dynamic confocal imaging of mitochondria in swimming *Tetrahymena* and of microtubule poleward flux in *Xenopus* extract spindles, *Biol Bull* **197**: 263–265, 1999.

Inoué S, Windows to dynamic fine structures, then and now, *FASEB J* **13** (9002): S185–S190, 1999. [**A 63**]

Suzuki K, Inoué S, U.S. Patent #5930033: Slit scan centrifuge microscope, 1999.

Inoué S, U.S. Patent #5982535: Centrifuge microscope capable of realizing polarized light observation, 1999.

Fukui Y, Uyeda TQP, Kitayama C, Inoué S, How well can an amoeba climb? *Proc Natl Acad Sci USA* **97** (18): 10020–10025, 2000. [**A 64**]

Goda M, Burgos MH, Inoué S, Fertilization-induced changes in the fine structure of stratified *Arbacia* eggs. I. Observations on live cells with the centrifuge polarizing microscope, *Biol Bull* **199** (2): 212–213, 2000.

Burgos MH, Goda M, Inoué S, Fertilization-induced changes in the fine structure of stratified *Arbacia* eggs. II. Observations with electron microscopy, *Biol Bull* **199** (2): 213–214, 2000. [**A 65**]

Inoué S, Knudson RA, Goda M, Suzuki K, Nagano C, Okada N, Takahashi H, Ichie K, Iida M, Yamanaka K, Centrifuge polarizing microscope. I. Rationale, design, and instrument performance, *J Microsc* **201** (3): 341–356, 2001. [**A 66**]

Inoué S, Goda M, Knudson RA, Centrifuge polarizing microscope. II. Sample biological applications, *J Microsc* **201** (3): 357–367, 2001. [**A 67**]

Tran PT, Marsh L, Doye V, Inoué S, Chang F, A mechanism for nuclear positioning in fission yeast based on microtubule pushing, *J Cell Biol* **153**: 397–411, 2001.

Inoué S, Goda M, Fluorescence polarization of GFP crystals, *Biol Bull* **201** (2): 231–233, 2001.

Knudson RA, Inoué S, Goda M, Centrifuge polarizing microscope with dual specimen chambers and injection ports, *Biol Bull* **201** (2): 234, 2001. [**A 68**]

Shribak M, Inoué S, Oldenbourg R, Rectifiers for suppressing depolarization caused by differential transmission and phase shift in high NA lenses, in Goldstein D, Chenault D, Egan W, Duggin M (eds.), *Polarization Analysis, Measurement, and Remote Sensing IV, Proceedings of Society of Photo-Optical Instrumentation Engineers* **4481**: 163–174, 2002.

Shribak M, Inoué S, Oldenbourg R, Polarization aberrations caused by differential transmission and phase shift in high NA lenses: theory, measurement and rectification, *Opt Eng* **41** (5): 943–954, 2002. [**A 69**]

Inoué S, Shimomura O, Goda M, Shribak M, Tran PT, Fluorescence polarization of green fluorescent protein (GFP), *Proc Natl Acad Sci USA* **99** (7): 4272–4277, 2002. [**A 70**]

Inoué S, Polarization microscopy, in Lippincott-Schwartz J (ed.), *Current Protocols in Cell Biology*, Vol. 1, Suppl. 13, John Wiley & Sons, pp. 4.9.1–4.9.27, 2002. [**A 71**]

Inoué S, Inoué T, Direct-view high-speed confocal scanner — the CSU-10, in Matsumoto B (ed.), *Cell Biological Applications of Confocal Microscopy*, 2nd edn., Academic Press, pp. 87–127, 2002. [**A 72**]

Inoué S, Exploring living cells and molecular dynamics with polarized light microscopy, in Török P, Kao K (eds.), *Optical Imaging and Microscopy*, Vol. 87, Springer-Verlag, Berlin Heidelberg, pp. 3–20, 2003.

Shribak M, Inoué S, Orientation-independent differential interference contrast microscopy, *App Opt* **45** (3): 460–469, 2006. [**A 74**]

Hoffman JF, Inoué S, Directly observed reversible shape changes and hemoglobin stratification during centrifugation of human and *Amphiuma* red blood cells, *Proc Natl Acad Sci USA* **103** (8): 2971–2976, 2006. [**A 75**]